16th International Conference on Modelling of Casting, Welding and Advanced Solidification Processes (MCWASP 2023)

IOP Conference Series: Materials Science and Engineering
Volume 1281

Banff, Canada
18-23 June 2023

Editors:

André Phillion
Hani Henein
Steve Cockcroft
Daan Maijer

ISBN: 978-1-7138-7433-1
ISSN: 1757-8981

Printed from e-media with permission by:

Curran Associates, Inc.
57 Morehouse Lane
Red Hook, NY 12571

CURRAN ASSOCIATES INC. proceedings.com

Some format issues inherent in the e-media version may also appear in this print version.

Institute of Physics
Dirac House, Temple Back
Bristol BS1 6BE UK

Phone: 44 1 17 929 7481
Fax: 44 1 17 920 0979

techtracking@iop.org

Additional copies of this publication are available from:

Curran Associates, Inc.
57 Morehouse Lane
Red Hook, NY 12571 USA
Phone: 845-758-0400
Fax: 845-758-2633
Email: curran@proceedings.com
Web: www.proceedings.com

TABLE OF CONTENTS

ADDITIVE MANUFACTURING AND WELDING

CONTINUOUS CASTING, SHAPE CASTING

DEFECTS

HEAT TRANSFER AND FLUID FLOW

MICROSTRUCTURE FORMATION

ALLOY SEGREGATION

IMAGING OF SOLIDIFICATION

MATERIALS PROPERTIES

THERMOMECHANICS

IOP Conf. Series: Materials Science and Engineering 1281 (2023) 011001 doi:10.1088/1757-899X/1281/1/011001

16th International Conference on the
Modelling of Casting, Welding, and Advanced Solidification Processes

MCWASP XVI

Modelling of Casting, Welding and Advanced Solidification Processes

June 18–23, 2023
Banff, Alberta, Canada

Editors

André Phillion, Hani Henein, Steve Cockcroft, Daan Maijer

Conference Organized by the Metallurgy & Materials Society of the
Canadian Institute of Mining, Metallurgy and Petroleum

MET SOC
CIM ICM
Metallurgy & Materials Society
Société de la métallurgie et des matériaux

IOP Conf. Series: Materials Science and Engineering 1281 (2023) 011001 doi:10.1088/1757-899X/1281/1/011001

Preface

The 16th International Conference on the Modelling of Casting, Welding, and Advanced Solidification Processes (MCWASP XVI) was held from June 18 to 23, 2023, in Banff, Canada, at the Banff Centre for Arts and Creativity. Founded in 1933, the Centre in Treaty 7 Territory within Banff National Park—Canada's first National Park—is a learning organization built upon an extraordinary legacy of excellence in artistic and creative development. The "all-inclusive" nature of the conference and the remote setting meant that participants dined, attended oral and poster presentations, and participated in social activities as a group, fostering outstanding opportunities for networking.

Given that the MCWASP community had not met in person since 2015 in Japan (the 2020 edition of MCWASP was virtual owing to COVID-19), the 2023 conference provided the opportunity to renew old friendships and make new ones as well as discuss the science of solidification and related processes—all within the backdrop of the beautiful Canadian Rocky Mountains.

The technical program comprised more than 70 oral and poster presentations. In addition to content related to modelling of casting, welding, and advanced solidification processes, keynotes were invited to talk about related subjects (artificial intelligence/machine learning, and permeability modelling in shale rock) as well as the rich diversity of fossils, especially dinosaurs, found in Alberta.

The oral technical program was organized with as a single session (i.e., no concurrent presentations). It featured all aspects of solidification modelling, including solidification process technologies (continuous and semi-continuous casting, shape casting, additive manufacturing, and welding), coupled multi-physics simulations, defect formation, fluid flow, micro- and macro-structure formation, numerical methods, and related experimentation, especially in-situ observation of solidification.

The four-day technical program was spread over five days to give participants the opportunity to explore the stunning Canadian Rocky Mountains.

In these proceedings, the papers are organized by major theme. The dominant topics are Additive Manufacturing and Welding and Microstructure Formation, followed by Continuous Casting – Shape Casting, Heat Transfer and Fluid Flow, Alloy Segregation, Defects, Imaging of Solidification, Thermomechanics, and Materials Properties. In these themes, the authors report advances in numerical modelling techniques, new scientific and process developments in solidification, and related in-situ experimentation.

Although significant progress has been made over these past 16 MCWASP conferences covering 43 years, it is clear that the complexity of advanced solidification phenomena as related to conventional and emerging manufacturing technologies still attracts a great deal of scientific and industrial interest to support technological innovation.

André Phillion
Banff, Canada, June 2023
MCWASP XVI 2023

IOP Conf. Series: Materials Science and Engineering 1281 (2023) 011001 doi:10.1088/1757-899X/1281/1/011001

Peer Reviewers

The editors are deeply indebted to the 31 peer reviewers, who reviewed the 78 manuscripts submitted to the conference. We thank them for their dedicated time and service.

Michael Auinger
Christoph Beckermann
Michel Bellet
Steve Cockcroft
Hervé Combeau
Janin Eiken
Charles-André Gandin
Jianzheng Guo
Jesper Hattel
Hani Henein
Alain Jacot
Andrew Kao
Laurens Katgerman
Peter Lee
Enzo Lotti
Andreas Ludwig
Daan Maijer
Patricio Mendez
Charles Monroe
Laurentiu Nastac
Jakob Olofsson
Koulis Pericleous
André Phillion
Begoña Santillana
Marc Schneider
Ingo Steinbach
Brian Thomas
Damien Tourrett
Menghuai Wu
Hideyuki Yasuda
Lang Yuan

IOP Conf. Series: Materials Science and Engineering 1281 (2023) 011001 doi:10.1088/1757-899X/1281/1/011001

Sponsors

The organizing committee of MCWASP XVI would like to thank the following sponsors, who supported this conference*

FLOW-3D® WELD

MAGMA

* as of March 2023

IOP Conf. Series: Materials Science and Engineering 1281 (2023) 011001 doi:10.1088/1757-899X/1281/1/011001

MCWASP XVI Organizers

André Phillion, McMaster University
Hani Henein, University of Alberta
Steve Cockcroft, University of British Columbia
Daan Maijer, University of British Columbia

International Scientific Committee

Christoph Beckermann, University of Iowa – USA
Michel Billet, MINES ParisTech, CEMEF – France
Steve Cockcroft, University of British Columbia – Canada
Hervé Combeau, École des Mines de Nancy – France
Charles-André Gandin, MINES, ParisTech, CEMEF – France
Jianzheng Guo, ESI US R&D – USA
Jesper Hattel, Technical University of Denmark – Denmark
Hani Henein, University of Alberta – Canada
Alain Jacot, ESI Group-Calcom – Switzerland
Mark Jolly, Cranfield University – UK
Laurens Katgerman, Delft University – Netherlands
Matthew Krane, Purdue University – USA
Peter Lee, University College London – UK
Andreas Ludwig, University of Leoben – Austria
Daan Maijer, University of British Columbia – Canada
Charles Monroe, University of Alabama – USA
Laurentiu Nastac, University of Alabama – USA
Jakob Olofsson, Jönköping University – Sweden
Koulis Pericleous, University of Greenwich – UK
André Phillion, McMaster University – Canada
Marc Schneider, MAGMA GmbH – Germany
Ingo Steinbach, Bochum University – Germany
Ingvar Svensson, Jönköping University –Sweden
Brian Thomas, Colorado School of Mines – USA
Hideyuki Yasuda, Kyoto University – Japan

IOP Conf. Series: Materials Science and Engineering 1281 (2023) 011002 doi:10.1088/1757-899X/1281/1/011002

Peer Review Statement

All papers published in this volume have been reviewed through processes administered by the Editors. Reviews were conducted by expert referees to the professional and scientific standards expected of a proceedings journal published by IOP Publishing.

- **Type of peer review:** Single Anonymous
- **Conference submission management system:** Morressier
- **Number of submissions received:** 79
- **Number of submissions sent for review:** 78
- **Number of submissions accepted:** 71
- **Acceptance Rate (Submissions Accepted / Submissions Received × 100):** 89.9
- **Average number of reviews per paper:** 1
- **Total number of reviewers involved:** 31
- **Contact person for queries:**
  Name: Janice Burke
  Email: jburke@cim.org
  Affiliation: Metallurgy & Materials Society of the Canadian Institute of Mining, Metallurgy and Petroleum

IOP Conf. Series: Materials Science and Engineering 1281 (2023) 012001 doi:10.1088/1757-899X/1281/1/012001

The inherent strain method for additive manufacturing: critical analysis and new inherent strain rate method

M Bellet, J Keumo Tematio, Y Zhang

Mines Paris, PSL University, CEMEF – Centre de Mise en Forme des Matériaux, CNRS UMR 7635, Sophia Antipolis, France. Email: michel.bellet@minesparis.psl.eu

Abstract. To reduce computation time, the inherent strain (IS) method is popular. It consists in adding layers, at room temperature, with an "inherent strain" representing the plastic deformation undergone during deposition. An IS method is developed, including a direct determination of the IS tensor based on the transient thermo-elastic-viscoplastic (TEVP) simulation. Validation is achieved: taking full-field inherent strains, the IS method allows retrieving TEVP predictions. However, application to a full part, for which a set of inherent strains must be conserved all along the construction, leads to poor results, far from TEVP. As an alternative, a new "inherent strain rate" (ISR) method is proposed, consisting in linearizing TEVP resolutions. Combined with an on-line learning technique, this strategy leads to results identical to TEVP reference, with still a significant speed-up. This makes the proposed ISR method very promising.

1. Introduction

The paper focuses on thermomechanical simulation of additive manufacturing (AM), especially DED (Directed Energy Deposition). Numerical tools are strategic to optimize manufacturing routes. However, for large parts, and despite some shortcomings (fluid flow ignored in melt pool, material deposited by batch: fractions of layer, groups of several layers), simulation time remains quite long. To maintain computation time in an acceptable range, the inherent strain (IS) method, initially developed for welding simulation by Ueda *et al.* [1], was adapted to AM, first by Keller & Ploshikhin [2]. Since then, different variants of the IS method have been developed (Alvarez *et al.* [3], Liang *et al.* [4]). They differ essentially by the mode of determination of the inherent strains, which has a high impact on results. Because of that, the approach developed here consists in determining the IS in such a way that the IS method replicates the transient thermomechanical simulation. For that, we take as reference a finite element simulation developed and validated separately. This transient thermomechanically coupled simulation, based on an elastic-viscoplastic constitutive model, is not reported here, as it is not essential to the purpose of the present study. It can be found in [5], and it will be referred as STD-TEVP. Throughout the rest of the paper, we will not compare numerical predictions to experimental measurements, but to results obtained by this STD-TEVP simulation.

The paper is structured as follows. Section 2 consists of a short description of the IS method, and of the different methods to determine the inherent strains. In Section 3, the new technique to determine the IS will be presented. It consists of a direct determination based on the STD-TEVP simulation of the first layers deposition of the studied part. However, it will be shown that this method – although validated on these first layers – leads to poor results for the whole part. Therefore, in Section 4, an alternative method will be presented: the inherent strain rate method, which is based on a linearization of the reference STD-TEVP simulation. This new method leads to excellent results, in perfect agreement with the reference ones.

IOP Conf. Series: Materials Science and Engineering 1281 (2023) 012001 doi:10.1088/1757-899X/1281/1/012001

2. Inherent strain method

2.1. Main features, in AM context

In the IS method material addition is conducted layer-by-layer, isothermally, at room temperature, and without any time consideration. This is a huge simplification compared to the STD-TEVP approach in which the progress of the build process is simulated. In the IS method, after each layer addition, the static equilibrium of the growing structure is solved at ambient temperature, by performing a finite element (FE) resolution of the weak form of the vector equation:

$$\nabla \cdot \boldsymbol{\sigma} + \rho \mathbf{g} = 0 \tag{1}$$

where $\boldsymbol{\sigma}$ is the stress tensor, ρ the density, and $\mathbf{g}$ the gravity vector. Owing to the hypotheses made on temperature and time, and to the small deformation context of AM, the behavior law is of elastic-plastic type, with isotropic hardening (for extension to kinematic hardening, refer to Keumo Tematio [5]):

$$\boldsymbol{\varepsilon} = \boldsymbol{\varepsilon}^{el} + \boldsymbol{\varepsilon}^{pl} + \boldsymbol{\varepsilon}^{*} \qquad \boldsymbol{\varepsilon}^{el} = \mathbf{S}^{el}\boldsymbol{\sigma} \qquad \begin{cases} \bar{\sigma} < \sigma_Y + R(\bar{\varepsilon}) \quad \text{and} \quad \mathrm{d}\boldsymbol{\varepsilon}^{pl} = 0 \\ \text{or} \quad \bar{\sigma} = \sigma_Y + R(\bar{\varepsilon}) \quad \text{and} \quad \mathrm{d}\boldsymbol{\varepsilon}^{pl} = \dfrac{3\mathrm{d}\bar{\varepsilon}}{2\bar{\sigma}}\mathbf{s} \end{cases} \tag{2}$$

where $\boldsymbol{\varepsilon}$ is the strain tensor, $\boldsymbol{\varepsilon}^{el}$ and $\boldsymbol{\varepsilon}^{pl}$ its elastic and plastic parts, respectively. $\mathbf{S}^{el}$ is the elastic compliance tensor, $\mathrm{d}\bar{\varepsilon}$ is the generalized strain increment, and $\bar{\sigma}$ is the generalized (von Mises) stress:

$$\mathrm{d}\bar{\varepsilon} = \sqrt{(2/3)\mathrm{d}\boldsymbol{\varepsilon}^{pl}:\mathrm{d}\boldsymbol{\varepsilon}^{pl}} \qquad \bar{\sigma} = \sqrt{(3/2)\mathbf{s}:\mathbf{s}} \tag{3}$$

where σ_Y is the initial plastic yield stress, and R is the strain hardening function. In Eq. (2) $\boldsymbol{\varepsilon}^*$ denotes the inherent strain tensor, which represents the permanent strains undergone by the deposited material. It is assigned to each newly deposited layer and is kept as such until the end of the simulation. For a part made of N layers, the IS resolution simply consists of a series of N FE resolutions of Eq. (1).

2.2. Determination of inherent strains

The key point of the IS method lies in the identification of the inherent strain tensor $\boldsymbol{\varepsilon}^*$. In the literature, essentially two ways can be found to determine $\boldsymbol{\varepsilon}^*$.

2.2.1. Comparison experiment/simulation on test specimen. A small-scale specimen is supposed to be representative of the process, *e.g.* a straight wall of a few superimposed tracks for DED. $\boldsymbol{\varepsilon}^*$ can be estimated by minimizing the difference between measurements and numerical simulation by IS method:

$$\boldsymbol{\varepsilon}^* = \text{argmin} \left\| \mathbf{u}^{sim}(\boldsymbol{\varepsilon}^*) - \mathbf{u}^{exp} \right\| \tag{4}$$

where $\mathbf{u}^{exp}$ denotes a set of distortion measurements, while $\mathbf{u}^{sim}$ denotes the corresponding predictions by IS simulation. This method was used by Setien *et al.* [6], Siewert *et al.* [7], and many others.

2.2.2. Identification by simulation only of a test specimen. A small-scale specimen is considered, for which material deposition is simulated by a transient thermo-mechanical simulation, formerly validated. This allows determining some rules to estimate $\boldsymbol{\varepsilon}^*$. For instance, Liang *et al.* [4] suggested:

$$\boldsymbol{\varepsilon}^* = \boldsymbol{\varepsilon}_i^{el} + \boldsymbol{\varepsilon}_i^{pl} - \boldsymbol{\varepsilon}_s^{el} \tag{5}$$

where $\boldsymbol{\varepsilon}_i^{el}$ and $\boldsymbol{\varepsilon}_i^{pl}$ respectively stand for the calculated elastic and plastic strain tensors in a so-called intermediate state, while $\boldsymbol{\varepsilon}_s^{el}$ stands for the calculated elastic strain tensor after final cooling. The intermediate state is "the state when the heat source just passes by and the local (compressive) mechanical strain reaches the largest amplitude". This method, known as the "modified inherent strain method", has been used to simulate DED (Duan *et al.* [8]) and L-PBF processes [4]. In this framework, the three tensors at the right-hand side of Eq. (5) are directly deduced from the transient thermo-mechanical simulation. Note that contrary to the previous method – where a unique IS tensor $\boldsymbol{\varepsilon}^*$ is determined – here the IS tensor is determined everywhere, in all layers of the test specimen. This brings up the question of selecting one or two layers and averaging $\boldsymbol{\varepsilon}^*$ therein to come down to a unique tensor.

To conclude on inherent strains determination, approaches based on direct measurements appear difficult and time consuming because they require the definition and the realization of complex strain or stress measurements on test specimens. On the contrary, approaches based on thermomechanical simulation of test specimens appear more efficient, provided that the exploitation of such simulations in view of estimating the inherent strains is simple, direct, and achievable in an automatic process. In line with this last remark, a new IS method is proposed. The basic idea is to identify inherent strains such that the IS method could provide the same results as STD-TEVP on the selected test specimen. In the following Section, it will be explained how this can be achieved, by considering the inherent strain tensor as a layer-dependent and space-dependent field, the determination of which being done without any costly inverse finite element procedure, but directly and locally, based on STD-TEVP results.

3. A new inherent strain method

In this Section, prior to application to a full part, we first focus on how to determine inherent strains, based on the simulation of a small-scale test specimen with STD-TEVP formulation.

3.1. Determination of inherent strains from STD-TEVP simulations of small-scale test specimen

In the context of the DED process, the test specimen consists of a few (N) superimposed layers. Hence, N different states are defined, see Figure 1. State n is obtained after i) deposition of layers 1 to n, including inter-layer dwell times, and ii) final cooling to room temperature. In this Section, we will explain how the inherent strains can be directly and locally (in each finite element) deduced from the reference STD-TEVP simulation.

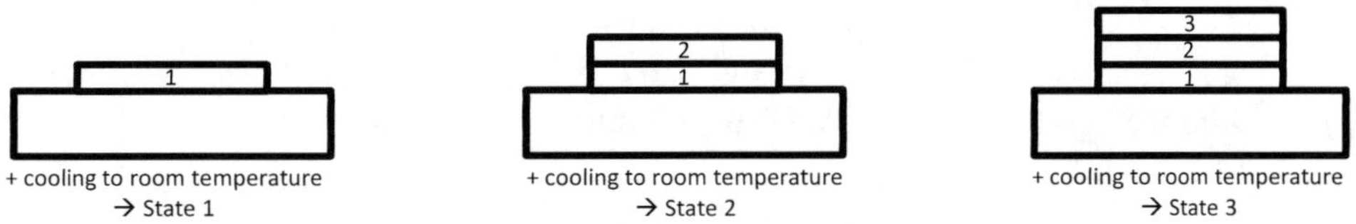

Figure 1. Schema of a small-scale 3-layer specimen, and of the three states considered.

Each of the deposition processes leading to states $1, 2, \cdots, N$ is simulated by means of STD-TEVP simulation. This allows calculating, for each state n, the residual displacement and stress fields: $\mathbf{u}_{(n)}, \boldsymbol{\sigma}_{(n)}$ obtained after deposition of n layers and final cooling. In the context of small deformations, the differences in displacement and strain between state n and state $n-1$ are:

$$\Delta\mathbf{u}_{(n)}^{STD} = \mathbf{u}_{(n)}^{STD} - \mathbf{u}_{(n-1)}^{STD} \qquad \Delta\boldsymbol{\varepsilon}_{(n)}^{STD} = (1/2)\left(\nabla(\Delta\mathbf{u}_{(n)}^{STD}) + \left(\nabla(\Delta\mathbf{u}_{(n)}^{STD})\right)^{\mathrm{T}}\right) \tag{6}$$

Besides, when simulating the different states with the IS approach, we can write, from Eq. (2):

$$\Delta\boldsymbol{\varepsilon}_{(n)}^{IS} = \Delta\boldsymbol{\varepsilon}_{(n)}^{IS,el} + \Delta\boldsymbol{\varepsilon}_{(n)}^{IS,pl} + \Delta\boldsymbol{\varepsilon}_{(n)}^{*} \tag{7}$$

where the increment of inherent strain only affects the newly deposited layer: $\Delta\boldsymbol{\varepsilon}_{(n)}^{*} = 0$ in all layers from 1 to $n-1$. The principle leading to the identification of the inherent strains is then the following: find for each layer n the tensor $\Delta\boldsymbol{\varepsilon}_{(n)}^{*}$ such that:

$$\Delta\boldsymbol{\varepsilon}_{(n)}^{IS} = \Delta\boldsymbol{\varepsilon}_{(n)}^{STD} \quad \text{and} \quad \Delta\boldsymbol{\sigma}_{(n)}^{IS} = \Delta\boldsymbol{\sigma}_{(n)}^{STD} \tag{8}$$

Taking the trace of Eq. (7), $\Delta\boldsymbol{\varepsilon}_{(n)}^{*}$ should be such that:

$$\mathrm{tr}\left(\Delta\boldsymbol{\varepsilon}_{(n)}^{STD}\right) = \mathrm{tr}\left(\Delta\boldsymbol{\varepsilon}_{(n)}^{IS,el}\right) + \mathrm{tr}\left(\Delta\boldsymbol{\varepsilon}_{(n)}^{*}\right) \tag{9}$$

in which $\mathrm{tr}\left(\Delta\boldsymbol{\varepsilon}_{(n)}^{IS,el}\right)$ is expressed by the Hooke's law. We obtain:

$$\mathrm{tr}\left(\Delta\boldsymbol{\varepsilon}_{(n)}^{STD}\right) = -\Delta p_{(n)}^{IS}/\chi + \mathrm{tr}\left(\Delta\boldsymbol{\varepsilon}_{(n)}^{*}\right) \tag{10}$$

where $\chi = E/(3(1-2\nu))$ is the elastic compressibility module, here taken at room temperature, and $\Delta p_{(n)}^{IS}$ is the pressure difference between states n and $n-1$. Because the stress tensors should be the same for the two approaches (Eq. (8)b), the pressure difference is known from the STD-TEVP simulations: $\Delta p^{IS}_{(n)} = \Delta p^{STD}_{(n)} = -(1/3)\mathrm{tr}(\Delta\boldsymbol{\sigma}^{STD}_{(n)})$. This allows a direct determination of the spherical part of $\Delta\boldsymbol{\varepsilon}^*_{(n)}$, from the results of STD-TEVP simulations of states n and $n-1$:

$$\mathrm{tr}(\Delta\boldsymbol{\varepsilon}^*_{(n)}) = \mathrm{tr}(\Delta\boldsymbol{\varepsilon}^{STD}_{(n)}) + (p^{STD}_{(n)} - p^{STD}_{(n-1)})/\chi \tag{11}$$

Taking now the deviatoric part of Eq. (7), we get:

$$\Delta\mathbf{e}^{STD}_{(n)} = \Delta\mathbf{e}^{IS,el}_{(n)} + \Delta\boldsymbol{\varepsilon}^{IS,pl}_{(n)} + \Delta\mathbf{e}^*_{(n)} \tag{12}$$

where $\Delta\mathbf{e}^{STD}_{(n)}$, $\Delta\mathbf{e}^{IS,el}_{(n)}$, and $\Delta\mathbf{e}^*_{(n)}$ are the deviatoric parts of $\Delta\boldsymbol{\varepsilon}^{STD}_{(n)}$, $\Delta\boldsymbol{\varepsilon}^{IS,el}_{(n)}$, and $\Delta\boldsymbol{\varepsilon}^*_{(n)}$, respectively: $\Delta\mathbf{e}^{STD}_{(n)} = \Delta\boldsymbol{\varepsilon}^{STD}_{(n)} - (1/3)\mathrm{tr}(\Delta\boldsymbol{\varepsilon}^{STD}_{(n)})\mathbf{I}$, with $\mathbf{I}$ the identity tensor. The tensor $\Delta\mathbf{e}^{IS,el}_{(n)}$ is in turn expressed by the Hooke's law to obtain:

$$\Delta\mathbf{s}^{IS}_{(n)} = 2\mu\left[\Delta\mathbf{e}^{STD}_{(n)} - \Delta\boldsymbol{\varepsilon}^{IS,pl}_{(n)} - \Delta\mathbf{e}^*_{(n)}\right] \tag{13}$$

where μ is the elastic shear modulus, and $\mathbf{s}$ denotes the deviatoric part of the stress tensor: $\mathbf{s} = \boldsymbol{\sigma} + p\mathbf{I}$. Consider now the incremental plastic law between the two states $n-1$ and n. Because the stress and strain tensors should be the same for the two approaches (Eq. (8)), and using the constitutive equations (Eqs. (2)-(3)), this leads to:

$$[1 + 2\mu\lambda(\Delta\bar{\varepsilon}^{IS}_{(n)})]\mathbf{s}^{STD}_{(n)} = \mathbf{s}^{STD}_{(n-1)} + 2\mu[\Delta\mathbf{e}^{STD}_{(n)} - \Delta\mathbf{e}^*_{(n)}] \quad \text{with} \quad \bar{\sigma}^{IS}_{(n)} \leq \sigma_Y + R(\bar{\varepsilon}^{IS}_{(n)}) \tag{14}$$

It can be shown (details are not reported here, see [5]) that Eq. (14) is a non-linear equation to be solved for $\Delta\mathbf{e}^*_{(n)}$, by means, for instance, of a Newton-Raphson procedure. This being done, considering Eq. (11), the tensor $\Delta\boldsymbol{\varepsilon}^*_{(n)} = \Delta\mathbf{e}^*_{(n)} + (1/3)\mathrm{tr}(\Delta\boldsymbol{\varepsilon}^*_{(n)})\mathbf{I}$ is fully determined. Note that this direct determination of $\Delta\boldsymbol{\varepsilon}^*_{(n)}$ (based on the tensors $\boldsymbol{\varepsilon}_{(n)}$ and $\boldsymbol{\sigma}_{(n)}$ previously calculated by the standard TEVP simulation) can be achieved locally, *i.e.* in each finite element of the small-scale specimen. It does not require any additional global finite element resolution on the n-layer structure. This means that its resolution cost is negligible.

3.2. First validation on a small straight wall made of 8-layers

To validate the proposed inherent strain method, reference is made to the thermomechanical study of Biegler *et al.* of a straight wall made of stainless steel 316L by DED, using a back-and-forth single track deposition strategy [9]. The configuration composed of the first 8 layers is adopted for this first validation. The geometrical, material, and process parameters are documented in Table 1.

Table 1. Simulation data: material and process parameters, geometrical data.

	Properties	Values or reference
Material properties (stainless steel 316L)	Thermophysics	Kim [10]
	Thermomechanics	Muransky *et al.* [11]
Heat exchange	Initial temperature, T_0 [°C]	25
	Ambient temperature, T_{air} [°C]	25
	Convection coefficient, h_{conv} [W m^{-2} K^{-1}]	35
	Emissivity, ε	0.6
	Stefan–Boltzmann constant, σ [W m^{-2} K^{-4}]	5.670374×10^{-8}
Process parameters (Biegler et al. [9])	Nominal laser power, P_L [W]	400
	Scan speed, v_L [mm s^{-1}]	10
	Reflection coefficient, R	0.7
	Spot diameter, ϕ_L [mm]	0.6
	Inter-layer dwell time, t_{dwell} [s]	31
Part geometry (Biegler *et al.* [9])	Length, L [mm]	65
	Layer width, l_D [mm]	1.4
	Layer thickness, h_D [mm]	0.62
	Number of layers, N	30
	Total height, H [mm]	18.6
	Substrate, $L_S \times l_S \times h_S$ [mm^3]	100 x 30 x 8

The average values of the 6 components of $\Delta\boldsymbol{\varepsilon}^*_{(n)}$ in each of the 8 layers are given in Table 2. Except for the first layer, they are globally similar. Shear components are quasi null (<10^{-5}), except for xz, the sign of which changes due to the back-and-forth deposition. Regarding absolute values of diagonal terms, xx components are the largest, while yy and zz components are about twice as small.

Table 2. Average values of $\Delta\boldsymbol{\varepsilon}^*_{(n)}$ components in each of the 8 layers.

Layer n	Average values in layer					
	$\Delta\varepsilon^*_{(n),xx}$	$\Delta\varepsilon^*_{(n),yy}$	$\Delta\varepsilon^*_{(n),zz}$	$\Delta\varepsilon^*_{(n),xy}$	$\Delta\varepsilon^*_{(n),xz}$	$\Delta\varepsilon^*_{(n),yz}$
1	−0.0126	−0.0007	0.0127	0.	−0.0081	0.
2	−0.0144	0.0046	0.0093	0.	0.0079	0.
3	−0.0152	0.0058	0.0090	0.	−0.0079	0.
4	−0.0153	0.0063	0.0085	0.	0.0081	0.
5	−0.0151	0.0067	0.0079	0.	−0.0081	0.
6	−0.0148	0.0067	0.0074	0.	0.0089	0.
7	−0.0144	0.0067	0.0072	0.	−0.0089	0.
8	−0.0141	0.0068	0.0068	0.	0.0092	0.

To further evaluate the spatial distribution of $\Delta\boldsymbol{\varepsilon}^*_{(n)}$ the map of xx component is shown in Figure 2. In fact, in this representation, the 8 fields $\Delta\boldsymbol{\varepsilon}^*_{(n)}$ have been somewhat superimposed, depending on which layer each element belongs to. It is worth noting the heterogeneity in the first layers, but also near the two ends of the wall, where inherent strains clearly vary with height.

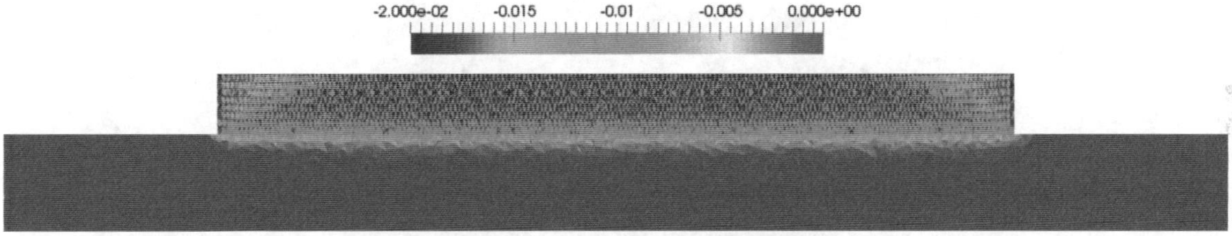

Figure 2. Map of $\Delta\varepsilon^*_{xx}$.throughout the 8-layer wall and its substrate.

To validate the proposed IS method, the full tensor fields $\Delta\boldsymbol{\varepsilon}^*_{(n)}$ (n = 1 to 8) extracted as explained above, are applied for each layer addition, according to Eq. (7). Results are compared with the reference ones obtained by STD-TEVP. As shown in Figure 3, for distortion and stress along the deposition direction, the results obtained by the proposed IS method can hardly be distinguished from those obtained by STD-TEVP. This expresses that the proposed IS method quasi perfectly replicates the STD-TEVP simulation. However, this only validates i) the determination process of $\Delta\boldsymbol{\varepsilon}^*_{(n)}$, and ii) its use in the IS method. The strategy to apply the IS method to a large part is presented in the next Section.

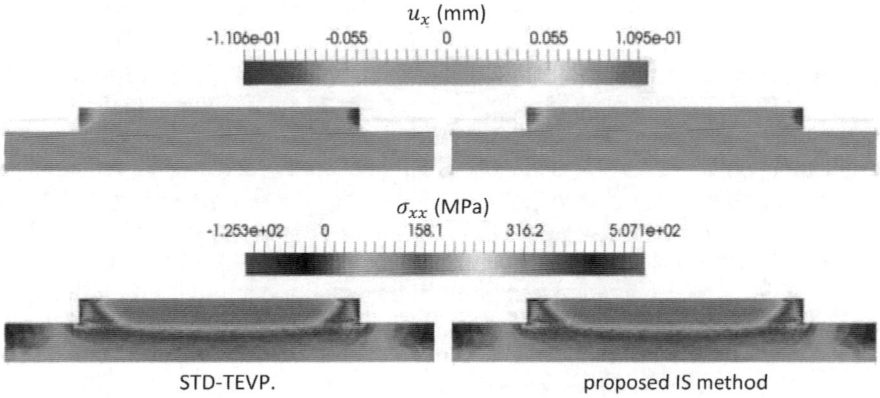

Figure 3. Validation of the proposed IS method on an 8-layer wall. Comparison of IS results (right) with reference STD-TEVP (left), for axial distortion and stress.

3.3. Application to the 30-layer straight wall: learning and transport strategy

Regarding now application of the IS method, especially for parts made of many layers, STD-TEVP simulations should be minimized. This needs to calculate approximate inherent strain fields in a simplified way, instead of calculating the exact $\Delta\boldsymbol{\varepsilon}^*_{(n)}$ for each layer addition as previously done. To

IOP Conf. Series: Materials Science and Engineering 1281 (2023) 012001 doi:10.1088/1757-899X/1281/1/012001

achieve this goal, it is chosen to determine the inherent strains on the first N_l layers – named learning layers – of the studied part. Following the identification of $\Delta\boldsymbol{\varepsilon}^*_{(n)}$, for $n = 1$ to N_l, it is then necessary to define how the inherent strains should be applied to model the addition of the rest of the layers ($n = N_l + 1$ to N). The method that is selected consists in transporting vertically the profile of inherent strains obtained from the learning layers. Taking care about the sign change for shear components due to back-and-forth deposition, we get, at location $\mathbf{x}$ in layer $n > N_l$:

$$\begin{cases} \text{In layer } n \text{ such that } n - N_l \text{ is an even number: } \Delta\boldsymbol{\varepsilon}^*_{(n)}(\mathbf{x}) = \Delta\boldsymbol{\varepsilon}^*_{(N_l)}(\mathbf{x} - (n - N_l)h_D\mathbf{e}_z) \\ \text{In layer } n \text{ such that } n - N_l \text{ is an odd number: } \Delta\boldsymbol{\varepsilon}^*_{(n)}(\mathbf{x}) = \Delta\boldsymbol{\varepsilon}^*_{(N_l-1)}(\mathbf{x} - (n - N_l + 1)h_D\mathbf{e}_z) \end{cases} \quad (15)$$

where h_D is the layer thickness. Figure 4 illustrates the results of the proposed IS method, compared with STD-TEVP simulation, in the case of the 30-layer wall, using 4 learning layers. It can be seen that significant differences now appear. This result expresses the fact that the inherent strains identified in the learning layers do not reflect the inherent strains encountered in the upper layers. In fact, this result could have been anticipated, just looking at Figure 2. Therefore, the conclusion is that the IS method, definitely cannot replicate STD-TEVP.

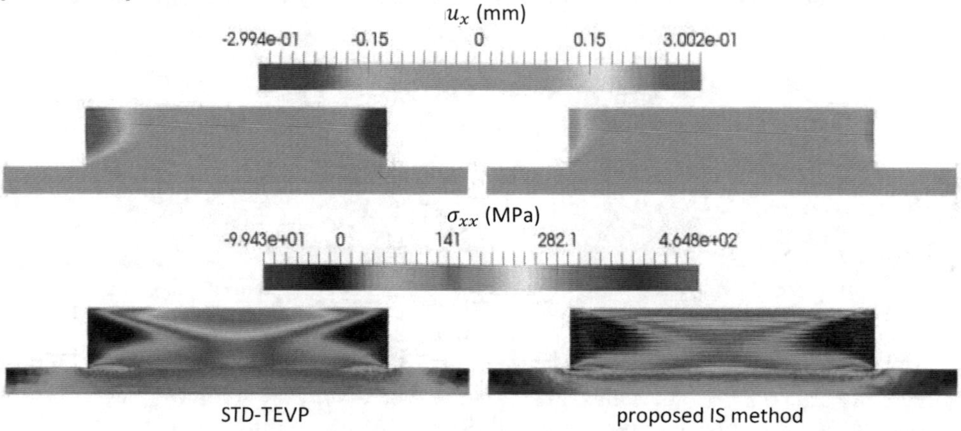

Figure 4. Application of the proposed IS method to the 30-layer wall, using 4 learning layers. Comparison with reference STD-TEVP for axial distortion and stress.

Regarding computer time, the IS method is 19 times faster than STD-TEVP. As a conclusion, the proposed new IS method was validated – by comparison with STD-TEVP – on a small-scale model, when using full-field inherent strains ($\Delta\boldsymbol{\varepsilon}^*_{(n)}$) in the different layers. However, the application of the IS method to the full 30-layer wall, using a few learning layers and transport of inherent strains, led to rather poor results, offering only a qualitative agreement with the reference STD-TEVP simulation. This result is disapointing, because the proposed IS method was especially designed to provide results as close as possible to the reference STD-TEVP. This was proved correct at small scale, but not at a larger scale. The reason for that is the heterogeneity of inherent strains on the learning layers, and, as a consequence, the impossibility to generate a field of inherent strains that would be appropriate on the whole construction. This is, in our opinion, a clear failure acknowledgement of the IS method. Finally, it appears relevant to develop an alternative approach of model reduction. In the following Section, we propose a new method, inspired by the IS method, but which is operated in a quite different way.

4. New method: the inherent strain rate method

The inherent strain rate method, ISR in short, consists of a linearization of the transient thermomechanical elastic-viscoplastic simulation (STD-TEVP). This linearization relies on the approximation of the generalized viscoplastic strain rate. This approximation is named "inherent strain rate" and is an input to the simulations (like the inherent strains in IS simulations). As such, this inherent strain rate (scalar) also requires an identification. But contrary to the IS method, identification is performed on-line, when simulating the studied part. The ISR approach is described hereafter.

4.1. Linearization of the thermo-elastic-viscoplastic constitutive equations

The constitutive equations for thermo-elastic-viscoplasticity are as follows:

$$\dot{\boldsymbol{\varepsilon}} = \dot{\boldsymbol{\varepsilon}}^{el} + \dot{\boldsymbol{\varepsilon}}^{vp} + \dot{\boldsymbol{\varepsilon}}^{th} \tag{16}$$

$$\dot{\boldsymbol{\varepsilon}}^{el} = \left(\mathbf{D}^{el}\right)^{-1}\dot{\boldsymbol{\sigma}} = \frac{1+\upsilon}{E}\dot{\boldsymbol{\sigma}} - \frac{\upsilon}{E}\mathrm{tr}(\dot{\boldsymbol{\sigma}})\mathbf{I} \tag{17}$$

$$\dot{\boldsymbol{\varepsilon}}^{th} = \alpha\dot{T}\mathbf{I} \tag{18}$$

$$\dot{\boldsymbol{\varepsilon}}^{vp} = \frac{3}{2\bar{\sigma}}\left[\frac{\bar{\sigma} - (\sigma_Y + R(\bar{\varepsilon}))}{k}\right]_+^{1/m}\mathbf{s} \tag{19}$$

where the strain rate tensor $\dot{\boldsymbol{\varepsilon}}$ is split into an elastic part, $\dot{\boldsymbol{\varepsilon}}^{el}$, a viscoplastic part, $\dot{\boldsymbol{\varepsilon}}^{vp}$, and a thermal part, $\dot{\boldsymbol{\varepsilon}}^{th}$. α is the one-dimensional thermal dilation coefficient, $\dot{\boldsymbol{\sigma}}$ the time derivative of the stress tensor, k the viscoplastic consistency, m the strain rate sensitivity. The function $[x]_+$ is equal to 0 when x is negative and to x otherwise. From Eq. (19) , when $\dot{\boldsymbol{\varepsilon}}^{vp} \neq \mathbf{0}$, and calculating $\dot{\bar{\varepsilon}} = \sqrt{(2/3)\dot{\boldsymbol{\varepsilon}}^{vp}:\dot{\boldsymbol{\varepsilon}}^{vp}}$, the following scalar relation can be easily obtained:

$$\bar{\sigma} = \sigma_Y + R(\bar{\varepsilon}) + k\dot{\bar{\varepsilon}}^m \tag{20}$$

Consequently, Eq. (19) can be straightforwardly rewritten:

$$\dot{\boldsymbol{\varepsilon}}^{vp} = \frac{3\dot{\bar{\varepsilon}}}{2\bar{\sigma}}\mathbf{s} \tag{21}$$

Non-linearity comes from the viscoplastic behavior law (Eq. (19) or (20)). The proposed linearization process consists in replacing the unknown equivalent strain rate $\dot{\bar{\varepsilon}}$ by an a priori estimate, hereafter denoted $\dot{\bar{\varepsilon}}^*$: the inherent strain rate. Proceeding in this way, we have at each time increment t:

$$\bar{\varepsilon}(t) = \bar{\varepsilon}(t-\Delta t) + \dot{\bar{\varepsilon}}^*\Delta t \quad \text{and} \quad \bar{\sigma}(t) = \sigma_Y + R(\bar{\varepsilon}(t)) + k\dot{\bar{\varepsilon}}^{*m} \tag{22}$$

Therefore, the viscoplastic constitutive equation is linearized

$$\dot{\boldsymbol{\varepsilon}}^{vp} = \frac{3\dot{\bar{\varepsilon}}^*}{2\bar{\sigma}(\dot{\bar{\varepsilon}}^*)}\mathbf{s} \tag{23}$$

Hence, the finite element resolution of the equilibrium equation $\nabla \cdot \boldsymbol{\sigma} = 0$, combined with constitutive equations (16)-(18) and (23) now requires a unique resolution, instead of an iterative solution process. The determination of the inherent strain rate $\dot{\bar{\varepsilon}}^*$ and its application are detailed in the next sub-Sections.

4.2. Determination of inherent strain rates and ISR solution scheme

The inherent strain rate $\dot{\bar{\varepsilon}}^*$ can be deduced from STD-TEVP calculation. When simulating with STD-TEVP, it is essential to remark that the most significant strain rates are found in the neighborhood of the material supply zone. This is evidenced in the left part of Figure 5.

From that, it is quite easy to elaborate simple rules (not detailed here) to define the shape of a simple domain – in red color in the right part of the figure – in which a uniform inherent strain rate $\dot{\bar{\varepsilon}}^*$ will be defined as the volume average of the most significant strain rates.

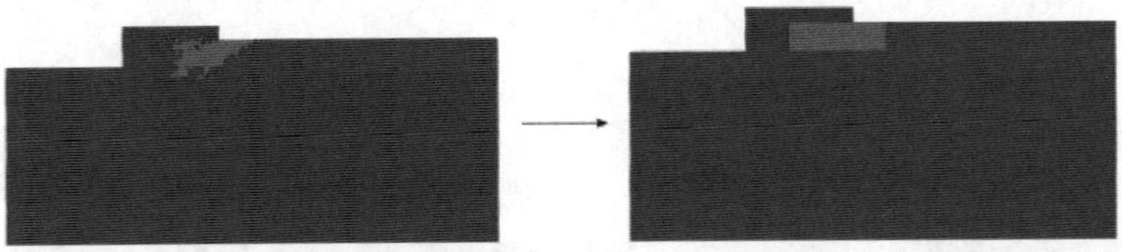

Figure 5. Left: during STD-TEVP simulation of the deposition of the third layer, the red zone encompasses elements showing the top 10% highest strain rates. Right: the red zone indicates the domain in which a uniform inherent strain rate $\dot{\bar{\varepsilon}}^*$ is to be applied.

The strategy to simulate the construction process consists then in a succession of linearized resolutions of the elastic-viscoplastic deformation of the structure, interrupted by some full (non-linear) STD-TEVP resolutions to update the value of the inherent strain rate, and its application domain. In practice, the use of the two solvers is organized as shown in Figure 6. The STD-TEVP resolution is used at each layer for the very first time steps at deposition start, and for a very few time increments at the transition between deposition end and dwell-time. All other time steps are solved with the ISR method. Only the first two layers are treated differently, STD-TEVP being used all along layer deposition.

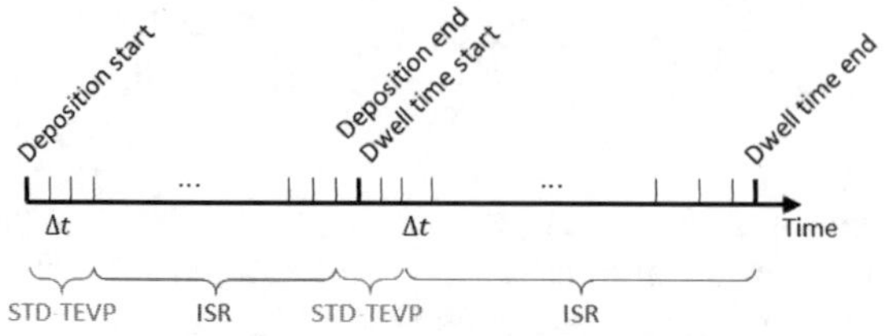

Figure 6. Organisation of the ISR solution method to simulate the deposition of a layer, and the subsequent inter-layer dwell time. STD-TEVP stands for the use of the non-linear resolution, while ISR stands for the use of the linearized elastic-viscoplastic resolution.

4.3. Application of ISR method to the 30-layer wall

Apart from the larger number of layers, geometrical data, process parameters and material properties are identical to those given in Section 3.2. Figure 7 shows the comparison, in terms of predicted distortion and stress fields, between ISR and STD-TEVP. The difference between displacement or stress components is not perceptible (relative difference less than 1%).

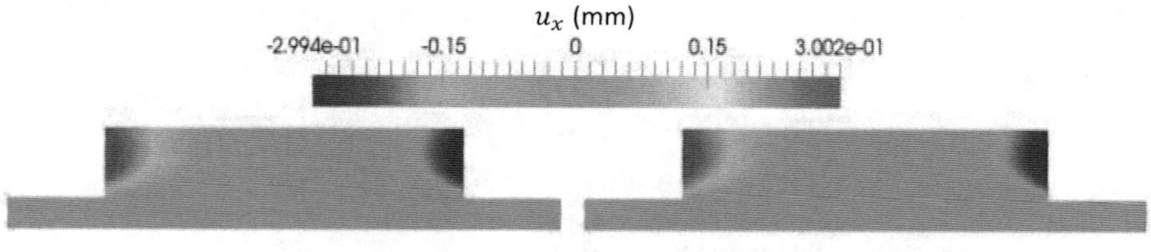

Figure 7. Application of ISR method to the 30-layer wall. Distortion and stress fields (x and z components) after complete cooling.

Regarding computation time, the ISR method shows a speed-up ratio of 5.2 with respect to STD-TEVP on the studied configuration. This is certainly lower than for the IS method, but the accuracy of the results is invaluably better. Also, it can be anticipated that this speed-up ratio could be higher on more complex simulation cases and could be improved by minimizing the number of STD-TEVP resolutions.

5. Conclusion

The lack of fidelity of the IS method with respect to a reference STD-TEVP numerical simulation has been evidenced. This fundamental weakness is due to the fact that the learning process made on a small-scale test part (possibly on a first few layers of the studied part) fails to be representative of the inherent strains undergone by the deposited material during deposition of the upper layers of the construction. Conversely, the new ISR method – based on a linearization process of realistic elastic-viscoplastic constitutive equations, and equipped with an efficient on-line learning process – perfectly replicates the reference calculations. Although less computer efficient, in its current state of development, than the IS method, the ISR method is very promising.

6. References

[1] Ueda Y, Fukuda K, Nakacho K, Endo S 1975 *J. Soc. Naval Architects of Japan* **1975** 499
[2] Keller N, Ploshikhin V 2014 *Proc. 25th Annual International Solid Freeform Fabrication Symposium, U. Texas, Austin,* 1229
[3] Alvarez P, Ecenarro J, Setien I, San Sebastian M, Echeverria A, Eciolaza L 2016 *Int. J. Eng. Research and Science* **2** 39
[4] Liang X, Cheng L, Chen Q, Yang Q, To AC 2018 Additive Manufacturing **23** 471
[5] Keumo Tematio J 2022 *PhD Thesis*, PSL University, Mines Paris
[6] Setien I, Chiumenti M, van der Veen S, San Sebastian M, Garciandía F, Echeverría A 2019 *Comput. and Math. with Appl.* **78** 2282
[7] Siewert M, Neugebauer F, Epp J, Ploshikhin V 2019 *Comput. and Math. with Appl.* **78** 2407
[8] Duan C, Cao X, Luo X 2021 *IOP Conf. Series: Earth and Environmental Science* **714** 032032
[9] Biegler M, Graf B, Rethmeier M 2018 *Additive Manufacturing* **20** 101
[10] Kim CS 1979 *Technical Report Argonne National Laboratory*, Ill., USA
[11] Muransky O, Smith MC, Bendeich PJ, Edwards L 2011 Comp. Mater. Sci. **50** 2203

IOP Conf. Series: Materials Science and Engineering 1281 (2023) 012002 doi:10.1088/1757-899X/1281/1/012002

A 3D Multiscale Model for Prediction of the Microstructure Evolution in Ti6Al4V Components Produced by Laser Powder Bed Fusion

Laurentiu Nastac

The University of Alabama, P. O. Box 870202, Tuscaloosa, AL, 35487, USA

lnastac@eng.ua.edu

Abstract. A transient 3D multiscale model has been created to simulate the evolution of solidification microstructures rapidly and effectively in Ti6Al4V parts produced through Laser Powder Bed Fusion (LPBF). The microstructure simulation tool has been enhanced to account for rapid solidification conditions in Ti6Al4V alloys during processing of multi-layer multi-track LPBF parts. The simulation tool can evaluate the impact of part geometry, lase power input, laser speed, and laser beam shape on the formation of the microstructure in LPBF-processed Ti6Al4V alloy. The multiscale model considers several factors, including preferential crystallographic growth direction, isomorphism, epitaxy, melt pool motion, and temperature gradients to generate the observed texture and morphology of the microstructure in Ti6Al4V components. The model can evaluate various microstructure characteristics, such as grain size and texture. Consequently, it could aid in controlling the formation of solidification microstructures in Additive Manufacturing (AM) processed parts. The simulation tool has been previously validated by using IN625 laser remelting experiments made by National Institute of Standards and Technology. The 3D simulation tool can also be utilized to predict the microstructure formation in Ti6Al4V components produced by the Electron Beam AM processes.

1. Introduction

This paper presents a multiscale model that efficiently simulates the evolution of Ti6Al4V-solidification microstructures in components produced using the Laser Powder Bed Fusion (LPBF) equipment. Recent advancements and difficulties related to controlling the Ti6Al4V microstructures of parts produced using various Additive Manufacturing (AM) techniques were examined in [1].

The present simulation tool was developed by linking a transient macro-model with a microstructure model [2-7]. By connecting a transient macro-model with a microstructure model [2-7], the current simulation tool was created. The macro-model calculates the temperature distribution within the melt pool, which varies with time. As the Richardson number (Ri) has been calculated to be less than 0.1 [8], gravitational convection is deemed insignificant in the melt pool. Nevertheless, it is crucial to consider convective fluid flow, recoil pressure, and Marangoni convection in the melt pool flow development [9]. Notwithstanding these restrictions, the thermal macro-model is capable of predicting the observed tendencies in the melt pool dimensions, and it can be adjusted through calibration [10] to achieve a high level of agreement with the experimental data.

The National Institute of Standards and Technology (NIST) conducted laser remelting benchmark experiments on IN625 alloy using LBPF equipment [11]. These experiments were utilized to verify the

simulation tool's accuracy, as demonstrated in [6], and subsequently employed to anticipate the effect of crucial process parameters on the microstructure morphology on IN718 alloy [7].
The simulation tool is capable of assessing the impact of the deposition of multiple layers and tracks within a layer, as well as the subsequent layers, on the formation of the microstructure of the entire component.
The simulation tool considers the combined influence of preferential crystallographic orientation, isomorphism, epitaxy, melt pool motion, and temperature gradients to generate the observed microstructure morphology. As the simulation tool encompasses the entire geometry of the part, there are no further constraints in accommodating rapid solidification kinetics, remelting, epitaxy, and/or isomorphism conditions, as might be encountered in other proposed models for predicting microstructure evolution during alloy solidification in AM [12-14].
This 3D multiscale simulation tool is capable of predicting experimental observations and evaluating the impact of scanning strategy and melt transients on microstructure evolution. Furthermore, this model can be customized to predict the formation of the solidification microstructure for Ti alloys and superalloys using Direct Energy Deposition (DED) and Electron Beam PBF processes.

2. Multi-Scale Model Description

Figure 1 depicts a schematic diagram with the connection between the macro-model and the micro-model [2-7].

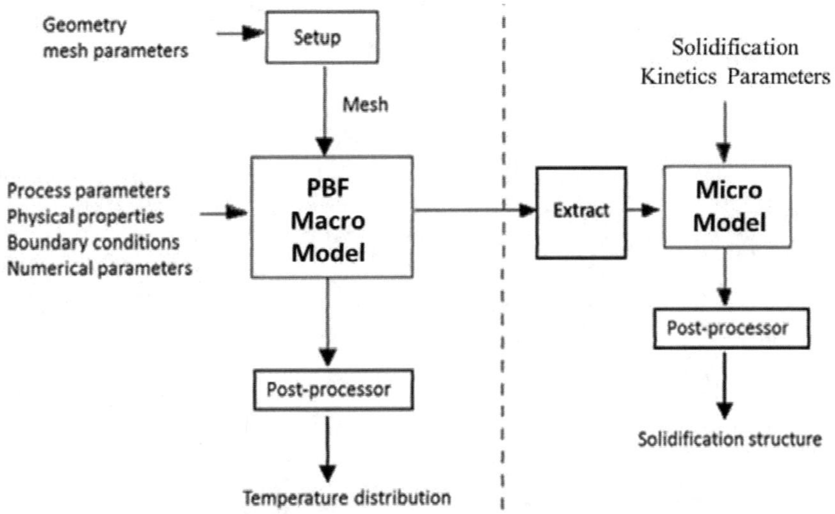

Figure 1. Schematic diagram with the connection between the macro-model and the micro-model.

The "Extract" computational tool illustrated in Fig. 1 is utilized to link the macro-model and the micro-model. The macro-model calculates the time-dependent temperature field during the LPBF melting and solidification process of the entire part geometry. Subsequently, the Extract tool calculates all the necessary data for the micro-model, including temperature gradients, cooling rates and local solidification times. Once the necessary data is extracted by the "Extract" tool, it is passed on to the micro-model, which then calculates the evolution of the solidification microstructure in the LBPF part. The 3D fully transient macro-model is capable of conducting heat transfer simulations for the entire deposited material and factors in the moving energy source that follows the prescribed scanning path.
The current micro-model employs a 3D stochastic methodology, which distinguishes from the Cellular Automata (CA) method [2]. The advantage of this approach is that it considers the temperature history results that are computed by the heat transfer model at the macro-scale level, which enables it to predict to the microstructure in the entire part geometry quickly and efficiently. The 3D stochastic micro-model

accounts for solidification (nucleation and growth) kinetics, growth anisotropy, and grain selection mechanisms [2-5]. The nucleation kinetics parameters are calculated using the cooling rates just below the liquidus temperatures, while the growth kinetics parameters are calculated using the average cooling rates and average time-dependent temperature gradients in the mushy zone. During processing of AM parts, equiaxed and columnar grain morphologies solidified under a variable *G/V* ratio can be created, where *G* and *V* are the local temperature gradients and solid-liquid (S/L) interface velocities in the mushy zone, respectively. The creation of the microstructure morphology and the columnar-to-equiaxed transition are controlled by the nucleation and growth competition of the equiaxed and columnar grains in the mushy region. The stochastic mesoscopic models for solidification of equiaxed and columnar grains solidified processed under a variable *G/V* ratio are described in [2-5]. In this research study, the micro-model described earlier has been modified for rapid solidification conditions and multi-component alloys, such as Ti6Al4V or IN718. The technique used Ti6Al4V micro-modeling under normal solidification conditions was presented in [5]. Under normal cooling rates that are usually below 100 K/s, the radius of the dendrite tip reduces as the solidification velocity increases, as shown in Figure 2. But in the rapid solidification regime where the cooling rate is higher, the dendrite tip radius increases, leading to a decrease in branching. As a result, equiaxed dendrite morphology becomes globular or cellular. To accommodate rapid solidification conditions in the stochastic model (characterized by solidification velocities that usually exceed 0.1 m/s), the partition coefficient (k^*) and the liquidus slope (m_L^*) of each alloying element can be corrected using the equations proposed by Aziz [15] and Baker and Kahn [16]:

$$k^*(V) = \frac{k_e + \delta_i V/D_i}{1 + \delta_i V/D_i} \qquad m_L^*(V) = \frac{m_L}{1-k_e}\left[1 - k^*\left(1 - \ln\frac{k^*}{k_e}\right)\right] \qquad [1]$$

where k_e is the equilibrium partition coefficient, V is the S/L interface velocity, D_i is the interfacial diffusivity, δ_i is the atomic boundary layer thickness, and m_L is the equilibrium liquidus slope. Segregation experiments can be done to determine the unknown δ_i/D_i ratio. The typical critical S/L interface velocity for solute trapping for metals is 5 m/s (using $D_i = 2.5\text{e-}09$ m^2/s and $\delta_i = 0.5\text{e-}9$ m) [15, 17].
The present model considers isomorphism and epitaxy phenomena that occur during solidification of the partially remelted regions between tracks and layers of the Ti6Al4V alloy. Unique grain IDs, which also indicate the grain's crystallographic preferential orientation, were assigned using computer generated random numbers ranging from 0 to 65535.

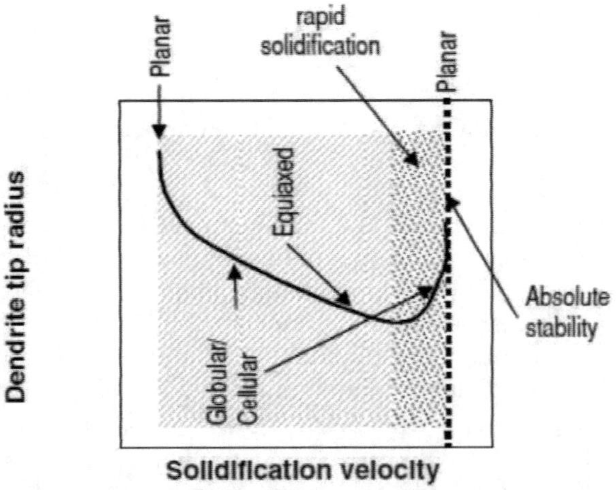

Figure 2. A schematic relationship between solidification velocity and dendrite tip radius [17]. Reprinted from [17] with permission from Springer Nature.

IOP Conf. Series: Materials Science and Engineering 1281 (2023) 012002 doi:10.1088/1757-899X/1281/1/012002

2.1 Model Validation

The model validation was effectively accomplished for the dimensions of the melt pool, as well as for the morphology and texture of the grains. The successful validation was demonstrated in [6], where NIST's IN625 laser remelting benchmark experiments [11] were employed.

3. Results and Discussion

The simulation tool discussed in Section 2 was utilized to simulate the formation of grain microstructure in LPBF-processed Ti6Al4V parts. The thermo-physical and kinetics properties of Ti6Al4V used in the current simulations are presented in Table 1, where T is the temperature in K that ranges from 298 K to 1930 K, ρ is the density, c_p is the specific heat, K is the thermal conductivity, L is the latent heat of solidification, D_L is the liquid diffusivity, Γ is the Gibbs–Thomson coefficient, N_0 is the grain density, m_L is the liquidus slope, k_e is the equilibrium partition coefficient, and the subscripts S and L denote solidus and liquidus, respectively. The equilibrium liquid slope and partition coefficient values of Ti6Al4V-0.2 wt. %Fe are taken from [5]. The gas infiltrated powder bed/layer mixture consists of two phases, Ti6Al4V powder and Ar gas. Specific heat and density of this mixture are calculated by using the volume-weighted average method using a packing density factor that increases linearly with the fraction of solid (f_S) from 0.55 [20] at $f_S = 0$ to 1 for $f_S = 1$. The specific heat and density of the Ar phase as a function of temperature are taken from [21]. The thermal conductivity (K_{pb}) of the Ar gas infiltrated Ti6Al4V powder bed/layer is calculated using the data from [20, 22] as: $K_{pb} = -0.0264 + 8.7\times10^{-4}\ T$. K_{pb} is assumed to increased linearly with temperature from 0.3 W/m/K at 373 K to 1.65 W/m/K at 1930 K. For temperatures higher than 1930 K, K_{pb} increases linearly with f_S from 1.65 W/m/K at $f_S = 0$ to K_L at $f_S = 1$.

Table 1. Thermo-physical and kinetics properties of Ti6Al4V [5, 18-20]

Property	Value	Unit
ρ_S	$4420.0 - 15.4\times10^{-2}$ (T-298.15)	kg/m^3
ρ_L	5342.0 -0.7307 T	kg/m^3
c_{pS}	$411.5 + 0.2\ T + 5\times10^{-7}\ T^2$	J/kg/K
c_{pL}	830.0	J/kg/K
K_S	$0.797+18.2\times10^{-3}\ T-2\times10^{-6}\ T^2$	W/m/K
K_L	34.0	W/m/K
L	3.5×10^5	J/kg
m_L	−0.322	K/wt. %
Γ	2.0×10^{-7}	K/m
D_L	3.0×10^{-9}	m^2/s
k_e	0.444	
T_L	1939.9	K
T_S	1930.0	K
N_0	1.0×10^{10}	m^{-2}

The process and material parameters used in the current simulations are shown in Table 2 where P is the laser power, V_s is the laser scanning speed, Δt is the time step used in the simulations, A is the absorption coefficient, σ is the laser beam radius, and T_i is the initial temperature of the substrate and powder material. The source term equation associated with the laser power source is shown in Eq. (1) in [6]. The length, width, and height of the simulated domain are 610 mm, 308 mm, and 244 mm, respectively. 5 layers and 5 tracks per layer were considered and the laser scanning strategy was bi-directional between successive layers. The mesh size is 2 μm for the macro-model and 1 μm for the micro-model. The entire simulated domain has been meshed using about 5.7 million control volume elements at the macro-scale level and about 45.8 million cells at the micro-scale level. The calculated partition coefficient (k^*) and liquidus slope

(m_L^*) of Ti6Al4V-0.2 wt. %Fe as a function of V (based on Eq. (1)) is presented in Figure 3. Deviations from equilibrium values can be seen for both k^* and m_L^* for S/L interface velocities higher than about 0.1 m/s, which are typically reached during the LPBF process. For the current simulated 3D geometry, the heat transfer boundary conditions during the LPBF process comprise heat losses due to conduction, convection, and radiation at the top, edges, and bottom of the geometry. Additionally, heat input is generated from the laser scanning source at the top of the geometry.

Table 2. Process and material parameters used in the current LPBF simulations

Parameter	Value	Unit
P	200	W
V_s	0.8	m/s
σ	50	µm
A	0.3	-
Δt	2.0×10^{-5}	s
T_i	298.15	K

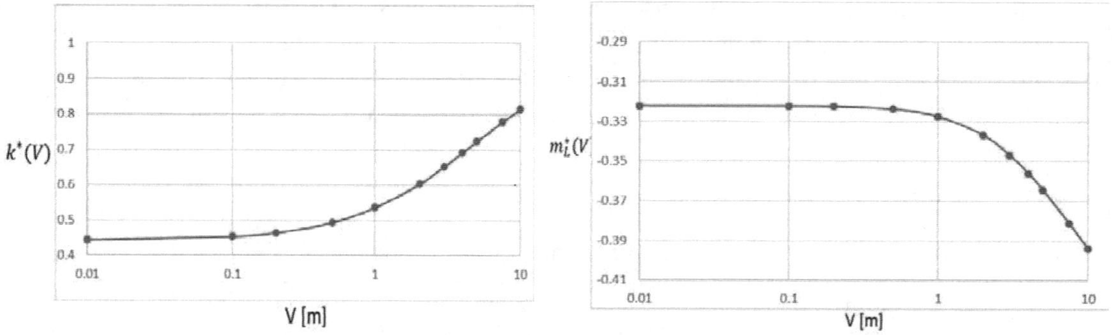

Figure 3. Calculated partition coefficient and liquidus slope of Ti6Al4V-0.2%Fe alloy as a function of S/L interface velocity.

Figures 4 and 5 exhibit the predicted solidification microstructures in ZY and ZX planes, respectively, for a 5-layers 5-tracks case. The simulation results from the ZY and ZX planes were obtained from the middle of the simulated 3D geometry. In both figures, the color-coded legends represent the preferential crystallographic orientation angle of the columnar grains, indicates as 65535 color indices (*CI*). The orientation angles can then be calculated from these legends using the color indices as:

$$\theta = \frac{\pi}{2}\frac{CI}{65535} - \frac{\pi}{4} \tag{2}$$

Thus, when *CI/65535* ratio varies from *0* to *1*, θ ranges from $-\pi/4$ to $\pi/4$.

The most favorable oriented grains with respect to the normal temperature gradients will typically grow to the topmost surface while the least favorable ones will typically disappear. The grains that have the most favorable orientation are identified by their corresponding colors/ids in the middle of the legends presented in Figs. 4 and 5. For the present simulations, the laser power and scanning velocity were set at relatively higher levels, resulting in cooling rates ranging from $2\text{x}10^4$ to $3\text{x}10^5$ K/s and temperature gradients ranging from $3\text{x}10^5$ to $2\text{x}10^7$ K/m. The microstructures produced by the simulations are displayed in Figs. 4 and 5 and are examined in the following paragraph.

In Fig. 4(h), the average size of the columnar grain is approximately 30 microns, with a minimum size of around 10 microns and a maximum size of about 100 microns. Fig. 5 (h) shows that the number of grains decreases from approximately 50 at the bottom of the geometry to about 30 in the middle, and 15 grains at the top. In this 5-layer, 5-track simulation, only up to 30% of the grains nucleated at the bottom survived Based on the results presented in Fig. 5, it can be inferred that the grain selection mechanism in multi-layer multi-track LPBF-processed Ti6Al4V-0.2 wt. %Fe parts is highly effective.

4. Computational Details

The microstructure simulation code was developed using both CUDA C++ for GPU computing and C++ for CPU computing. To improve the efficiency of CPU time and reduce the disk space usage, a sparse format was implemented for input and output files used for the communication between the macro- and the micro-simulation codes: Further computational details are presented below.

- The CUDA 3D-Micro simulator is typically at least one order of magnitude faster than the CPU 3D-Micro simulator based on simulations performed with the Dense Memory Cluster (DMC) available at the Alabama Supercomputer Authority Center (NVidia GPU and CPU Skylake SP processors).
- The CPU 3D-Micro simulator is quite fast, typically taking about 1 h for a 5-layer 5-track simulation based on a 1-μm mesh size. To mesh the entire simulation domain, about 45.8 million micro-cells were used and the required RAM to run the simulation was about 8 GB.
- The CPU 3D-Micro simulator can run maximum 3 trillion cells and require a maximum RAM of 500 GB, which is equivalent of a simulated domain of about 3 mm^3 using a 2-μm mesh size.

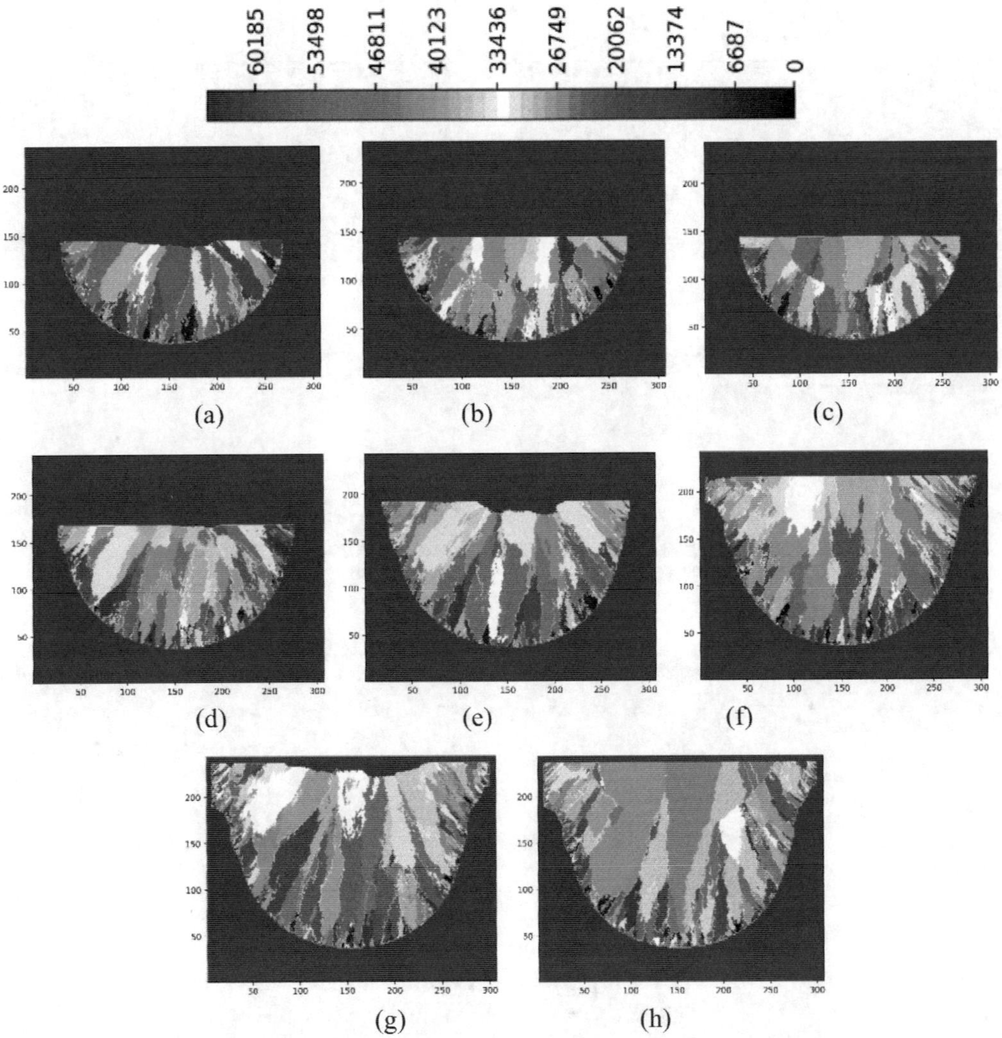

Figure 4 Simulated LPBF Ti6Al4V microstructures (5 layers, 5 tracks) – ZY plane: (a) layer 1 track 1; (b) layer 1 track 3; (c) layer 1 track 5; (d) layer 2 track 2; (e) layer 3 track 2; (f) layer 4 track 4; (g) layer 5 track 3; (h) layer 5 track 5.

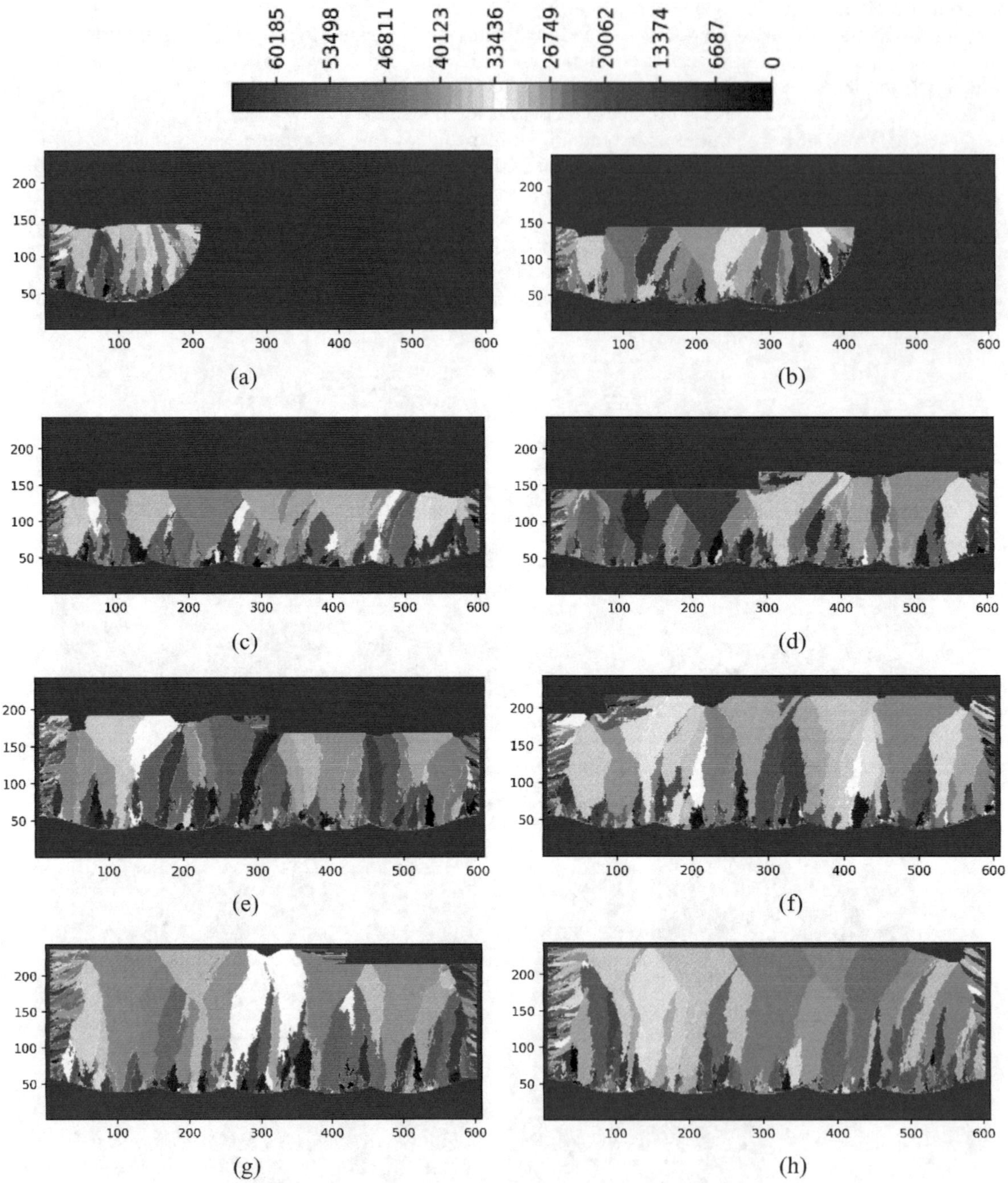

Figure 5. Simulated LPBF Ti6Al4V microstructures (5 layers, 5 tracks) – ZX plane: (a) layer 1 track 1; (b) layer 1 track 3; (c) layer 1 track 5; (d) layer 2 track 2; (e) layer 3 track 2; (f) layer 4 track 4; (g) layer 5 track 3; (h) layer 5 track 5.

5. Concluding Remarks

A 3D transient multiscale simulation tool was created by combining a 3D macro-simulator with a 3D microstructure-simulator, allowing for the successful prediction of microstructure evolution in Ti6Al4V parts produced by LPBF. The simulation tool proved effective in evaluating how process and material conditions impact the microstructure, including grain size and orientation, in Ti6Al4V components.

Using this simulation tool along with previous production experience can provide better understanding of how LPBF processing factors (such as laser power and scanning path) and material characteristics (nucleation potential and chemical composition, including wt. %Fe) affect the formation of Ti6Al4V microstructures Consequently, it has potential to significantly decrease the number of experimental trials required for the development of a new AM practice.

The 3D simulation tool presented in this study can be utilized for exploring the underlaying mechanisms behind the formation of microstructures during rapid solidification of metallic alloys in AM processes. It enables the investigation of the complex interaction between factors such as preferential crystallographic growth direction, epitaxy, isomorphism, melt pool motion and temperature gradients, providing a deeper understanding of the process.

6. Acknowledgement

The author would like to acknowledge the Alabama Supercomputer Authority Center (https://www.asc.edu/) for granting permission to use the DMC to carry out the microstructure simulations.

7. References

[1] Liu S and Shin Y C 2019, *Materials and Design* **164** 107552
[2] Nastac L 2004, *Modeling and Simulation of Microstructure Evolution in Solidifying Alloys* (Springer, New York) ISBN 978-1-4020-7831-6 6
[3] Nastac L 2012, *Proceedings of the CFD Modeling and Simulation in Materials Processing* (Eds. Nastac L et al., The Minerals, Metals & Materials Society, Wiley, A John Wiley and Sons, Inc., Publications, ISBN 978-1-118-29615-8) 123-130
[4] Sundarraj S, Nastac L, Pang Y and Yu K O, 1998 *Proceedings of the Modelling of Casting, Welding and Advanced Solidification Processes VIII,* (The Minerals, Metals & Materials Society) 297-304
[5] Nastac L, Chou J S, and Pang Y 1999 *Proceedings of the International Symposium on Liquid Metals Processing and Casting* (Vacuum Metallurgy Conference) 207-233
[6] Nastac L 2021 *Metals*, **11**(12) 1995 https://doi.org/10.3390/met11121995
[7] Nastac L 2022 *Materials* **15**(24) 8885 https://doi.org/10.3390/ma15248885
[8] Turner J S 1973 *Buoyancy Effects in Fluids* (Cambridge University Press) ISBN 978-0-521-08623-3
[9] Khairallah S A, Anderson A T, Rubenchik A and King W E 2016 *Acta Materialia* **18** 36-45
[10] Goldak J, Chakravarti A and Bibby M 1984 *Metallurgical Transactions B* **15** 299–305
[11] 2018 AM-Bench Test Descriptions for AMB2018-02 NIST (accessed on 19 December 2022) https://www.nist.gov/ambench/amb2018-02-description
[12] Rolchigo M, Stump B, Belak J and Plotkowski A 2020 *Modelling Simul. Mater. Sci. Eng.* **28** 065003
[13] Koepf J A, Gotterbarm M R, Markl M and Koerner C 2018 *Acta Materialia* **152** 119-126
[14] Rodgers T M, Madison J D and Tikare V 2017 *Computational Materials Science*, **135** 78–89
[15] Aziz M J 1982 *J. Appl. Phys* **53** 1158-1168
[16] Baker J C Cahn J W 1971 *Solidification* (ASM Metals Park, OH, USA) p.23
[17] Stefanescu D M 2009 *Science and Engineering of Casting Solidification* (Springer, NY)
[18] Valencia J J and Quested P N 2008 *ASM Handbook: Casting*, ASM, **15** 468-481.
[19] Mills K C 2002 *Recommended Values of Thermo-Physical Properties for Selected Commercial Alloys* (Woodhead Publishing, Cambridge, UK)
[20] Zhang S, Lane B, Whiting J and Chou K 2019 *J. of Manufacturing Processes* **47** 382-392
[21] Thermo-physical properties of Ar: https://webbook.nist.gov/ (accessed on 19 December 2022)
[22] Wei L C, Ehrlich L E, Palm M J P, Montgomery C, Beuth J and Malen J A 2018 *Additive Manufacturing* **21** 201–208

IOP Conf. Series: Materials Science and Engineering 1281 (2023) 012003 doi:10.1088/1757-899X/1281/1/012003

Simulation of the solute transport and microstructure evolution during the selective laser melting process

Neng Ren[1], Jun Li[1,*], Chinnapat Panwisawas[2], Mingxu Xia[1], Hongbiao Dong[3], Jianguo Li[1]

1. Shanghai Key Laboratory of Advanced High-temperature Materials and Precision Forming, School of Material Science and Engineering, Shanghai Jiao Tong University, Shanghai 200240 P.R. China

2. School of Engineering and Materials Science, Queen Mary University of London, London E1 4NS, United Kingdom

3. School of Engineering, University of Leicester, Leicester LE1 7RH, United Kingdom

E-mail: li.jun@sjtu.edu.cn (Jun Li)

Abstract. Selective laser melting is of great expectation to be used in additive manufacturing of aerospace components with complex geometry. However, there are still defects in the built parts, such as solutal segregation and unexpected microstructure, which contribute to cracks and lead to failure. At present, most of the simulations focus on the macroscopic grain structure, and the solute transport process has not been well demonstrated yet. In the present work, we develop a two-way fully coupled model based on cellular automaton and finite volume method to simulate the solute transport and dendritic structure evolution during the melting and solidification of the SLM process. The results reveal the microstructural evolution and solute transport during the melting, spreading, and smearing of the powder. The proposed model framework shows good potential to be applied to further numerical investigation on the solidification behaviours of the SLM process.

1. Introduction

In aerospace and medical fields, selective laser melting (SLM) is one of the most promising additive manufacturing methods for the precision forming of the components [1]. However, this developing technique still faces problems in providing parts of reliable working performance, such as the cracks in the built superalloys[2]. Grain structure and solute segregation are generally considered to be factors determining the crack susceptibility, and should be further understood and strictly controlled. Due to the difficulties in the in-situ observation of superalloys, numerical simulation is an important choice to study the cycled melting and solidification of the SLM process.

Recent decades, multi-scale models have been developed [3] and then applied to understand the grain structure of the built parts. Lian et al.[4], developed a parallelized 3D cellular automaton computational model to predict grain morphology for solidification of metal during the additive manufacturing process, where a one-way coupling strategy is used to account the temperature field. Yang et al.[5] established a phase field- thermal fluid flow model to simulate grain evolutions during powder-bed-fusion. Especially, the solid-state grain coarsening in the heat affected zone was reproduced. However, rather a few numerical simulations have taken the effect of melt convection on the solute transport into account. Shinjo et al.[6] demonstrated the solute transport of common solute elements in the metal pool on the macroscopic scale via computational fluid dynamics (CFD) simulations, but the solute distribution is not coupled with grain structure. Although Yu et al.[7] made attempts to evaluate the impact of fluid flow on the dendritic growth and the formation of new grain in additive manufacturing, their simulation is actually based on the pre-defined flow field. For the grain structure prediction, up till now, what most of the simulations focused on is the grain structure, but the solute distribution, including but not limited to the solute enrichment in the grain boundaries and interdendritic regions, was not well demonstrated. In the hard-to-print alloy systems, such as superalloys, the solute element distribution is also of great importance to estimate the crack susceptibility.

In this paper, based on cellular automaton and finite volume method, we develop a two-way fully coupled model to simulate the solute transport and dendritic structure evolution during the melting and solidification of the SLM process. Necessarily, the two-phase flow of gas and liquid metal as well as heat transfer are also included in the model framework. Except for the thermal-fluid dynamic aspects as well as the concomitant solute transport of the SLM process, the fusion of the powder and substrate, liquation of the interdendritic region, and the melting of the dendritic trunks are also well demonstrated. Thus, the proposed model framework provides a powerful tool to study the solutal segregation and alloying process in the SLM and the other additive manufacturing process.

2. Model description

Based on the combined cellular automaton and finite volume method, we develop a fully-coupled model describing the multi-physical fields (including temperature field, flow field, and solute element field) and evolution of dendritic structure on the same set of computational grids. The control volumes in finite volume framework are also employed as the cells in cellular automaton, and hence the multiple physics are two-way fully coupled with dendritic structure.

2.1. Multiphase flow

Generally, there are three phases considered in the CFD simulation of SLM process, liquid melt (fused substrate and powder), gas (air or protective gas such as argon), solid metal (including unfused substrate and powder, and solidified melt). In our model framework, we did not distinguish the liquid melt and solid metal as two separated fluid phases. Instead, we use solid volume fraction to characterize the local solidification or melting sequence. Considering the significant change in the surface morphology of the powder during the melting process, volume of fluid approach is used to track the interface due to its powerful robustness in handling two-phase interface. The continuity equation is as follows:

$$\frac{\partial \alpha}{\partial t} + (\vec{v} \cdot \nabla)\alpha = 0 \tag{1}$$

where α is the volume fraction of gas or metal, vector $\boldsymbol{v}$ is the velocity of fluid, t is time. Consequently, the physical properties in each control volume, such as viscosity and thermal physical properties, are the volume averaged ones of the two phases under the finite volume model framework. It should be noted that the sum of the volume fractions of gas and metal is unity.

Navier-Stokes equations is used to describe the fluid flow in the SLM process. The fluid of gas and metal is assumed to be incompressible, and the density dependence on the temperature and solute concentration is described using Bousinesq approximation, where the density difference is regarded as body forces in the momentum conservation.

$$\left[\frac{\partial(\rho\vec{v})}{\partial t} + \nabla \cdot (\rho\vec{v}\vec{v})\right] = -\nabla p + \nabla \cdot (\mu \nabla \vec{v}) + \vec{F}_T + \vec{F}_C + \vec{F}_M + \vec{F}_{st} \tag{2}$$

where ρ is density, μ is dynamic viscosity, p is pressure, and the four sources stand for thermal buoyancy, solutal buoyancy, momentum damp in mushy zone, and surface tension, respectively. Thermal expansion coefficient and solutal expansion coefficient are used to cover the deviation from the density under reference temperature and nominal composition. The momentum damp in the solid/liquid interface is described with Blake-Kozeny equation, where the resistance is the function of solid volume fraction. Surface tension is calculated using the continuum surface force model, and Marangoni effect is described with temperature dependent surface tension coefficient [8].

2.2. Heat transfer

Enthalpy equation is used to calculate the heat transfer during the SLM process, where enthalpy equals to the temperature integral of specific heat. Latent heat released during solidification or absorbed during melting is loaded as a source term. The latent heat absorbed by evaporated metal is similar, but the phase of metal vapor is neglected in the continuity and momentum equations. Besides, the thermal boundary conditions on the gas/metal interfaces are also loaded as source terms using the interfacial operator under the volume of fluid framework.

$$\frac{\partial \rho \vec{v} h}{\partial t} + \nabla \cdot (\rho \vec{v} h) = \nabla \cdot (\lambda \nabla T) + Q \tag{3}$$

where h is sensible enthalpy, λ is thermal conductivity, T is temperature, and Q is the heat sources, including latent heat of fusion and evaporation, convective and radiative gas/metal interfacial thermal dissipation, and laser power input [9].

2.3. Species transport

Two sets of solute conservation equations are employed for the solute elements in liquid melt and solidified metal. In the melt, except for the solutal diffusion and the species transfer caused by melting or solutal partition during solidification, the solute transport induced by melt convection is also included[10]. The species transfer caused by the solid-liquid transition is loaded as source terms in the two equations.

$$\frac{\partial \alpha_m \rho_m C_l}{\partial t} + \nabla \cdot (\alpha_m \rho_m \vec{v} C_l) = \nabla \cdot (D_l \alpha_m \rho_m \nabla C_l) + C_{ml} \tag{4}$$

$$\frac{\partial \alpha_m f_s C_s}{\partial t} = \nabla \cdot (D_s \alpha_m \rho_m f_s \nabla C_s) + C_{ms} \tag{5}$$

where C is the mass fraction of certain solute element, D is solute diffusivity, C_{ml} and C_{ms} are solutal sources induced by phase transformation, f_s is solid volume fraction, and the subscript l and s represent liquid and solid, respectively.

2.4. Dendritic evolution

Cellular automaton method is used to describe the preference growth of the dendritic structure. Two states of active and inactive are defined to govern the dendritic structural evolution. Active state means phase transformation as well as the concomitant heat and mass transfer would occur in the local control volume, while they are impossible to occur in the control volumes of inactive state. In other words, the inactive state means steady liquid or solid phase. Von-Neumann neighbour is used in the cellular automaton framework for the moment, and neighbour control volumes are captured by the active one when the dendritic tip touches their borders.

Except for being captured by active control volume, the inactive control volume is activated when the local undercooling reaches the critical nucleation undercooling, or the local temperature increases to the eutectic temperature. Conversely, the active state is converted to inactive when the local solid volume fraction decreases to zero, or the temperature drops under the eutectic temperature. It should be noted that only in the control volume whose metal volume fraction is larger than 0.5 can the inactive control volume be converted to the active state. Thermodynamical equilibrium condition is used to calculate the melting or solidification progressing[11], i.e., the solid volume fraction change:

$$\Delta f_s = f_l \frac{C_l^* - C_l}{C_l^*(1-k)} \tag{6}$$

where k is partition coefficient, f_l is liquid volume fraction, and the superscript * represent the equilibrium condition. It should be noted that the solid volume fraction here ranges from zero to unity, and the product of f_s and the volume fraction of metal α_m is the actual solid volume fraction of the control volume.

2.5. *Simulation parameters*

The SLM process of IN 718 is simulated, and the operating parameters and material properties are listed in Table 1. According to it is reported by Lee et al.[12], for the metal pool of SLM, such as SLM-ed nickel-based superalloy IN 718, the scale along the scanning direction (about 750 μm) is much longer than the scale perpendicular to the scanning path (180 μm or so). Therefore, though SLM process is actually three-dimensional for strictly speaking, it could still be sensible to perform two-dimensional simulation to capture certain characteristic of the multi-physical field transports in the process.

Figure 1 shows the computational domain and schematic diagram of laser heat source on the surface of the powder. The two-dimensional computational domain is the longitudinal section perpendicular to the scanning path, and the heat flux is calculated according to the section of the three-dimensional Gaussian distribution profile. The complicated reflection in the powder layer is simplified to be uniformly distributed on the gas/metal interfaces of both the powder and the substrate.

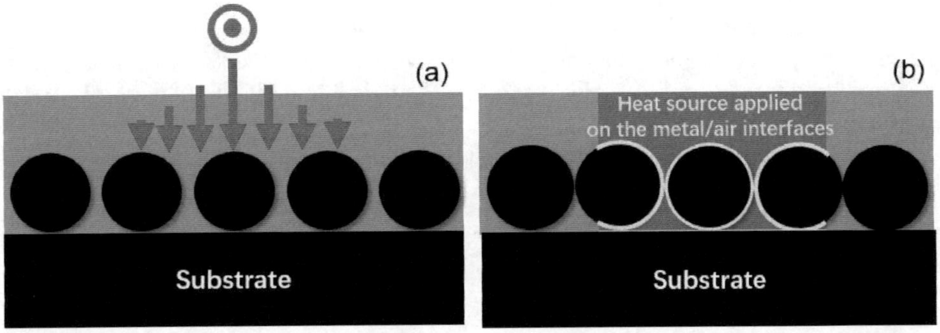

Figure 1. Schematic diagram of heat source induced by laser input in SLM.

Table 1. Processing parameters and material properties.

Parameters	Value
Laser scanning speed, m/s	0.96
Laser power, W	50
Diameter of laser beam spot, μm	100
Diameter of powder, μm	15
Convective heat transfer coefficient, W/m^2-K	80
Property	
Liquidus temperature, K	1609
Solidus temperature, K	1523
Surface tension coefficient, kg/s^2	1.882
Temperature coefficient of surface tension, kg/s^2-K	-1E-4

According to the reported solutal partition coefficients of the solute elements in the IN 718, Nb shows the most severe segregation tendency. Considering the great challenges in describing the solidification path and dynamical process in the multicomponent superalloys in numerical modelling, based on the assumption that the interactions of alloy elements on the solidification path could be neglected, only the Ni-Nb binary system in the superalloy IN 718 is focused on in the benchmark simulation case. The phase diagram parameters are listed in Table 2. For the stability of the solution process, the binary system is further linearised using constant partition coefficient and slope of liquidus.

Table 2. Phase-diagram parameters of the Ni-Nb binary system in IN 718

Parameters	Value
Partition coefficient	0.48
Slope of liquidus, K wt.$^{-1}$	-10.50
Critical nucleation undercooling degree, K	9.2
Standard deviation of Gaussian distribution, K	2.0
Maximum of nucleation density, m^{-3}	10^{10}
Solutal diffusion coefficient in liquid, m/s^2	2E-9

3. Results and discussion

Figure 2 shows the melting process of the powder and the substrate, as well as the liquation of the interdendritic region of the substrate. The grey scale in the left half of the contour stands for the mass fraction of solute elements, and the isolines coloured in rainbow colour map is isotherms. On the right half of the contour, rainbow colour map represents the local solid volume fraction, which indicates the sequence of melting and solidification, while red arrows are velocity vectors. The initial condition of the benchmark case of the SLM process is shown in Figure 1a, where the solute element is diluted in the dendritic trunks while enriched in the interdendritic region.

As the laser heat source is applied to the surface of the powder, the heat transfers from the surface to centre quite fast, and metal droplets are formed before the melt flows downward to the surface of the substrate. The interdendritic region is liquated first, that is the low melting point eutectic phases considered in our model framework. Actually, other phases of low melting point, such as Lave phases, carbides, and even borides in some kinds of superalloys, are also formed and growing in the interdendritic regions.

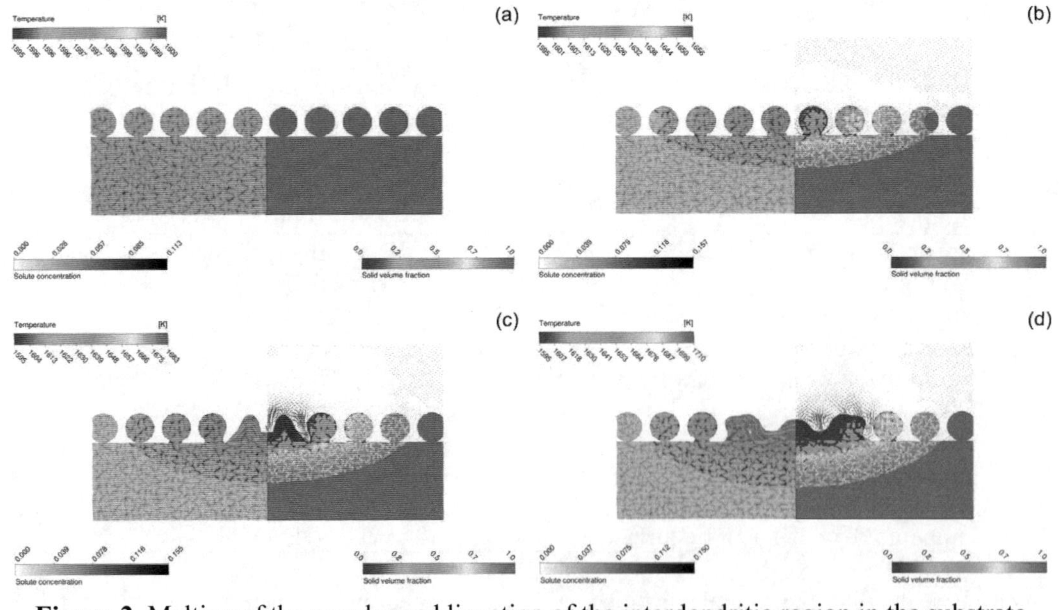

Figure 2. Melting of the powder and liquation of the interdendritic region in the substrate.

Figure 3 demonstrate the spreading and smearing of the fused powder. The metal droplets collapse, and the melt/gas interface spreads laterally. The spreading of the metal droplets and the fusion of the substrate leads to the formation of a shallow metal pool. The periphery powder is smeared by the free surface of the metal pool, and then heated to be fused. As soon as the part of the powder is fused, the local solute distribution is significantly changed under the melt convection. The enriched solute in the interdendritic region and dilute solute in the original dendritic trunks are well mixed, and thus there is an obvious boundary between the solute uniformly distributed metal pool and solute segregated powder and substrate.

In the heat affected zone, though the dendritic trunks are not completely remelted, the interdendritic region has already completely liquated. Therefore, the dendritic trunks are actually loosely connected with each other in the liquid films. Although the melt convection in the metal pool is rather intensive, the interdendritic flow is still rather weak and hardly influenced by the fully-fused bulk region. Consequently, the solute originally enriched in the interdendritic region could not be transferred away by convection. It can be deduced that the liquation in the heat affected zone would not lead to the change in the local composition. Furthermore, under a relatively lower temperature, the liquid film among the dendritic trunks suffers smaller thermal stress during the thermal cycles, and hence the risk of hot tearing could be much smaller.

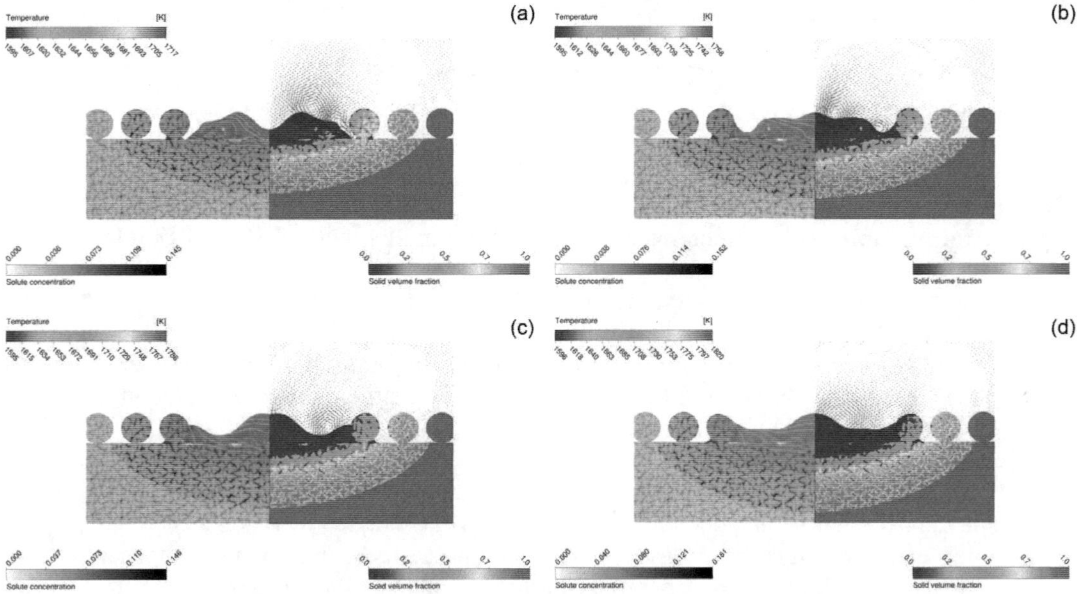

Figure 3. Spreading and smearing of the melted powder.

Figure 4 shows the oscillation of the free surface and the induced convection-driven solute transport in the metal pool. It should be noted that the temperature of the liquid metal does not reach the boiling temperature, and thus there is no recoil pressure acting on the free surface of the metal pool. Therefore, the metal pool is much shallower than the keyhole-mode SLM process, and consequently the melt flow is much weaker. Without the effect of the recoil pressure, the spreading and oscillating of the free-surface plays the dominating role in the expanding of the metal pool along both the horizontal and the building directions.

Besides, though the energy input is much smaller, a good metallurgical bonding still seems to be achieved between the fused powder and the substrate. However, the results need to be confirmed further in the actual powder layer of complicated particle distribution, where the fusion process shows more aspects, such as the lack of fusion, insufficient spreading of the free surface, and unexpected solidification of the border of the metal pool.

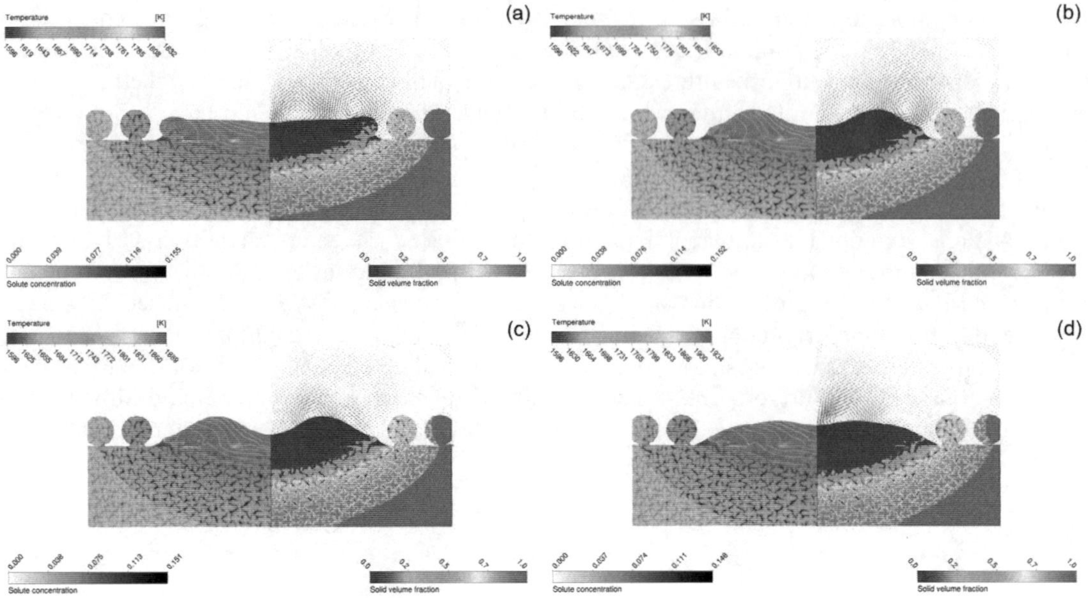

Figure 4. Oscillation of the gas/melt interface and solute transport in the metal pool.

4. Conclusions and perspective

A two-way fully coupled cellular automaton-finite volume model is developed to simulate the solute transport and dendritic structure evolution during the SLM process. The simulation results reveal the liquation in the interdendritic region, the fusion of dendritic trunks, and concomitant solute transport during the melting, spreading, and smearing of the powder. Under a relatively low heat input, when the peak of the temperature of the free-surface can not reach the boiling temperature, the free-surface dynamics plays the dominating role in the convection-driven solute mixing, and a good metallurgical bonding can still be achieved under this operating condition.

The presented model shows potential in predicting the solute transport on the scale of 10^{-7}~10^{-6} μm. The dendritic growth after the laser spot pass away is also expected to be demonstrated using the developed model in the further simulations. Besides, as it was discussed by Pinomaa et al.[13], the effect of rapid solidification can also be describe using the continuous growth model proposed by Aziz et al[14]. The local non-equilibrium solidification could be revealed by implicit solving the crystal growing rate, partition coefficient, and slope of liquidus.

Acknowledgements

This work was sponsored by National Natural Science Foundation of China (No. 52074182) and General fund of Shanghai Municipal Commission of science and technology (Grant No. 22ZR1430700).

References

[1] Panwisawas C, Tang Y T and Reed R C 2020 Metal 3D printing as a disruptive technology for superalloys *Nat. Commun.* **11** 2327

[2] Tang Y T, Panwisawas C, Ghoussoub J N, Gong Y, Clark J W G, Németh A A N, McCartney D G and Reed R C 2021 Alloys-by-design: Application to new superalloys for additive manufacturing *Acta Mater.* **202** 417–36

[3] Li J, Xu X, Ren N, Xia M and Li J 2022 A review on prediction of casting defects in steel ingots: from macrosegregation to multi-defect model *J. Iron Steel Res. Int.* **29** 1901–14

[4] Lian Y, Lin S, Yan W, Liu W K and Wagner G J 2018 A parallelized three-dimensional cellular automaton model for grain growth during additive manufacturing *Comput. Mech.* **61** 543–58
[5] Yang M, Wang L and Yan W 2021 Phase-field modeling of grain evolutions in additive manufacturing from nucleation, growth, to coarsening *npj Comput. Mater.* **7** 56
[6] Shinjo J and Panwisawas C 2021 Digital materials design by thermal-fluid science for multi-metal additive manufacturing *Acta Mater.* **210** 116825
[7] Yu Y, Wang L, Zhou J, Li H, Li Y, Yan W and Lin F 2022 Impact of fluid flow on the dendrite growth and the formation of new grains in additive manufacturing *Addit. Manuf.* **55** 102832
[8] Wang L and Yan W 2021 Thermoelectric Magnetohydrodynamic Model for Laser-Based Metal Additive Manufacturing *Phys. Rev. Appl.* **15** 64051
[9] Tang C, Tan J L and Wong C H 2018 A numerical investigation on the physical mechanisms of single track defects in selective laser melting *Int. J. Heat Mass Transf.* **126** 957–68
[10] Ren N, Li J, Panwisawas C, Xia M, Dong H and Li J 2021 Thermal-solutal-fluid flow of channel segregation during directional solidification of single-crystal nickel-based superalloys *Acta Mater.* **206** 116620
[11] Zhu M F and Stefanescu D M 2007 Virtual front tracking model for the quantitative modeling of dendritic growth in solidification of alloys *Acta Mater.* **55** 1741–55
[12] Lee Y S and Zhang W 2016 Modeling of heat transfer, fluid flow and solidification microstructure of nickel-base superalloy fabricated by laser powder bed fusion *Addit. Manuf.* **12** 178–88
[13] Pinomaa T, Lindroos M, Walbrühl M, Provatas N and Laukkanen A 2020 The significance of spatial length scales and solute segregation in strengthening rapid solidification microstructures of 316L stainless steel *Acta Mater.* **184** 1–16
[14] Aziz M J 1982 Model for solute redistribution during rapid solidification *J. Appl. Phys.* **53** 1158–68

IOP Conf. Series: Materials Science and Engineering 1281 (2023) 012004 doi:10.1088/1757-899X/1281/1/012004

Modelling and validation of laser based DED processes reinforced with nanoparticles in view of the effect of wetting

G Zheng[1], P Rey[2], R Laqua[1], M Salgueiro[2] and J Jakumeit[1,*]

[1] Access e.V. Intzestr. 5, 52072 Aachen, Germany

[2] Technology Centre AIMEN, Relva 27A - O Porriño, Pontevedra, 36410, Spain

* corresponding author: j.jakumeit@access-technolgy.de

Abstract. Although the observation of wetting phenomena has a long history both in science and in everyday life, nanoscale wetting properties leave open a multitude of questions. Improving the properties of Direct Energy Deposition (DED) products through reinforcement with nanoparticles (NPs) is attractive. Due to the involved NPs, the wetting behaviour of the deposition differs from that in macroscales. Modelling and validation of a laser-based DED process reinforced with NPs in view of the effect of wetting is demonstrated in this study. Metal alloys reinforced with NPs (Ti6Al4V- and Al7075-TiC) are melted by laser light and deposit on a substrate. Heating of the alloys by laser light is tackled using a volumetric heat source model. A three-phase (solid, liquid and gas) melting and solidification simulation methodology is established in a continuum approach. Nanoparticles are treated in the Lagrangian framework. Effect of nanoparticle reinforcement on the wetting behaviour of melt bead during the printing process is investigated numerically, taking into account the influence of nanoparticles on contact angle and surface tension. Simulation results are compared with the experimental deposition of Ti6Al4V and Al7075 alloys through dimensional analysis of the tracks by macrography.

1. Introduction

Nanoparticles reinforced metal matrix composites (MMnCs) demonstrate their advantages of enhanced mechanical properties [1,2,3], improved wear [4,5], or corrosion resistance [6,7] in various industrial fields. As a metal Additive Manufacturing (AM) process, Directed Energy Deposition (DED) is a feasible technique because of its high deposition rate, ability in manufacturing relative larger parts and flexibility in creating arbitrary shapes on even and uneven substrates [8]. Improving the properties of DED products through reinforcement with nanoparticles (NPs) is attractive. Due to the involved NPs, the wetting behavior of the deposition differs from that in macroscales. Although the observation of wetting phenomena has a long history both in science and in everyday life, nanoscale wetting properties leave open a multitude of questions. Modelling and validation of a laser-based DED process reinforced with NPs in view of the effect of wetting is demonstrated in this study.

Modelling of a laser based DED process reinforced with NPs was conducted in continuum approach, and the influence of the reinforcement of NPs on the deposition process are investigated by simulation and experiment in this study. A three-phase melting and solidification simulation methodology has been used to investigate the melting and solidification during DED printing of single line tracks. Metal alloys reinforced with nanoparticles TiC are melted by laser light and deposit on a substrate. Heating of the alloys by laser light is tackled using a volumetric heat source describing the specific absorption of the laser power by the metal alloys. Nanoparticles are treated in the Lagrangian framework. The wetting characteristics of the reinforced liquid metal are modelled in continuum approaches. Effect of nanoparticle reinforcement on the wetting behavior of melt bead during the printing process is investigated numerically, taking into account the influence of nanoparticles on contact angle and surface tension coefficient. Simulation results are compared with the experimental deposition of Ti6Al4V and Al7075 alloys through dimensional analysis of the tracks by macrography.

2. Modelling of nanoparticle reinforced DED processes

2.1. Modelling of the DED process

A three-phase melting and solidification simulation methodology, implemented in the multi-physics computational fluid dynamics solver STAR-CCM+, has been used to investigate the melting and solidification during DED. Latent heat release (absorption) during solidification (melting) of the metal is handled using an enthalpy formulation. A sharp interface between gas and melt phase is ensured by a high-resolution interface capturing (HRIC) [9]. A detailed description of the simulation approach can be found in [10]. Gas phase is treated as ideal gas with physical properties of air. Heating of the alloy by the laser is modelled as volumetric heat source with Gaussian distribution in radial direction for alloy and substrate regions. The volumetric heat source is active in the entire alloy region and in the substrate region limited to a depth of 0.1 mm. The effective laser power, i.e. the power absorbed by the reinforce alloy or substrate, is monitored separately for both regions.

2.2. Modelling of nanoparticles motion in reinforced alloys involved in the DED process

Particle motion results from various forces and torques acting on it, exerted by its surrounding medium. The Lagrangian approach computes the spatial position due to trajectory of each particle by integrating the forces and torques balance on each particle. Spin of nanoparticles in melts is not considered in this study and the integrating of forces is written as equation (1)

$$\frac{d}{dt}\left(\frac{dx_{i,pt}}{dt}\right) = \frac{d}{dt}\left(v_{i,pt}\right) = \sum f_i \ , \tag{1}$$

where x, v and f represents the centroids, velocities of each parcels and the force per unit particle mass, respectively, and the subscripts i = 1, 2, 3 denote the dimensions in a coordinate system, and pt represents particle. To investigate particle motion and distribution during melt fluid simulations, various forces, as listed in the Table 1 of [11], were used for enclosing the momentum equation. The forces of gravity, drag, pressure gradient, virtual mass, and shear fluid have already been implemented in the program STAR-CCM+. The thermophoretic force and the Brownian force were implemented and verified in [11].

Surface tension forces are added to model the motion of the Lagrangian phase in this study and the following formulae were implemented in the program STAR-CCM+. For a curved interface, the surface tension force density (i.e., the surface tension force per unit volume) can be resolved into two components in the normal and the tangential directions:

$$f_\gamma = f_{\gamma,n} + f_{\gamma,t} \tag{2}$$

The surface tension density in the normal direction can be determined by:

$$f_{\gamma,n} = \gamma\kappa\nabla\varphi_{melt} \tag{3}$$

and the tangential force is evaluated as:

$$f_{\gamma,t} = (\nabla\gamma)_t|\nabla\varphi_{melt}| \ , \tag{4}$$

where γ is the surface tension coefficient of liquid melt, $\nabla\varphi_{melt}$ is the gradient of the volume fraction of liquid melt, $(\nabla\gamma)_t$ is the gradient of the surface tension coefficient in the tangential direction. The curvature of the interface, κ, can be expressed in terms of the divergence of the unit normal vector:

$$\kappa = -\nabla \cdot \frac{\nabla\varphi_{melt}}{|\nabla\varphi_{melt}|} \tag{5}$$

As an example, Figure 1 demonstrates the computed volume fraction of Ti6Al4V (as denoted by the color bar in blue-red) and surface tension force (as denoted by the arrows in grayscale) in the cross section of a deposited bead. It is obvious that the surface tension force works at the liquid/gas interface and points normally towards the liquid phase.

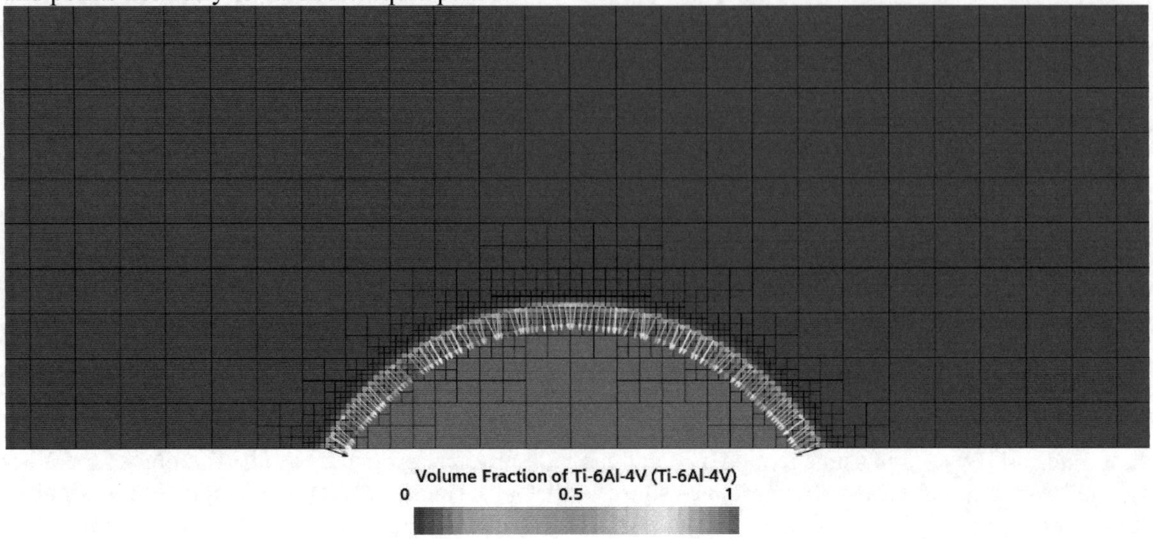

Figure 1. The computed surface tension force vectors of a melt bead of Ti6Al4V in the cross section.

Effects of Brownian force, thermophoretic force and surface tension force on the motion of nanoparticles in metal melts were evaluated in two ways: one way is to normalize the forces by gravitational force, as denoted by equation (6); the other way is to derive a critical velocity by dividing the friction factor based on the Stokes law, as given by equation (7):

$$F^* = \frac{F}{\rho_{pt} V_{pt} g} \tag{6}$$

$$v_c = \frac{F}{3\pi\mu d_{pt}} \quad , \tag{7}$$

where the symbol F represents various forces, μ denotes the dynamic viscosities of the applied melts, g is the gravitational acceleration, d_{pt}, V_{pt} and ρ_{pt} represents the diameter, the volume and the density of particle, respectively.

In Figure 2(a), the relationships between the normalized forces and particle diameter are shown to give an idea, how the relation between the forces are changing with increasing particle size. F_B, F_{tp} and F_σ represent the Brownian force, the thermophoretic force and the surface tension force, respectively. The dashed line denotes $F^* = 1$, i.e., the force effect is equal to the gravitational force. For nanoparticles, usually in the scale of 1 to 100 nm range, the effects of the Brownian force, the surface tension force and the thermophoretic force are dominant compared to the gravitational force. Figure 2(b) demonstrates the critical velocities of the Brownian force, the thermophoretic force and the surface tension force at different particle diameters. At the up-side of the lines, where the local velocities are larger than the critical velocities, the drag forces dominate; whereas, at the down-side of the lines, the corresponding forces dominate. It is obvious that under the usual velocities of DED processes, for nanoparticles, the Brownian force and the surface tension force dominate the drag force.

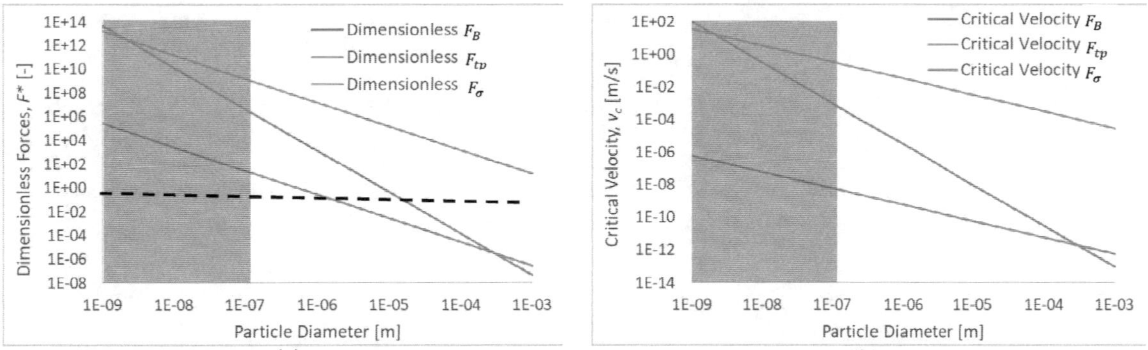

(a) (b)

Figure 2. Relationships between the normalized forces (a) and the critical velocities (b) as function of the particle diameter for TiC in Al7075 at the liquidus temperature of Al7075.

3. Influence of Nanoparticle on wettability

3.1. Surface tension

The surface tension coefficient of a liquid metallic alloy at high temperatures is very difficult to measure. A general trend of surface tension is not quite clear from the literature as both increasing and decreasing trends with nanoparticle concentration have been reported [12-18]. The temperature dependence of surface tension, $\partial\gamma/\partial T$, is a crucial characteristic, which is termed by Marangoni convection providing a driving force for fluid flow towards the region of high surface tension. In most materials, the surface tension coefficient decreases with an increase in temperature. Therefore, the surface tension gradients induce fluid flow from the high towards the low temperature positions.

The surface tension of a molten metal can be related to its heat of vaporization by the equation [19]:

$$\gamma = T\frac{\partial\gamma}{\partial T}\frac{1}{n} + \frac{\Delta h_{vap} - RT}{k\left(\frac{M_W}{\rho}\right)^{2/3} N_A{}^{1/3}} \quad . \tag{8}$$

The differential equation has the following solution to surface tension given as:

$$\gamma = \frac{\Delta h_{vap} - \left(\frac{n}{n-1}R - T^{n-1}\right)T}{k\left(\frac{M_W}{\rho}\right)^{2/3} N_A{}^{1/3}} \quad , \tag{9}$$

where (ρ) is the density, (M_w) the molar mass, (Δh_{vap}) the heat of vaporization, (R) the universal gas constant and (N_A) the Avogadro's constant. According to the surface tension values in the database of ACCESS e.V., the achieved parameters k and n for Ti6Al4V and Al7075 are listed in Table 1. The dependences of surface tension on temperature for the two applied alloys are plotted in Figure 3, together with the data from [20], for Ti6Al4V (a) and for Al7075 (b).

Table 1. The determined parameters k and n for alloys Ti6Al4V and Al7075.

	k	n
Ti6Al43	4.366	1.11
Al7075	4.159	1.076

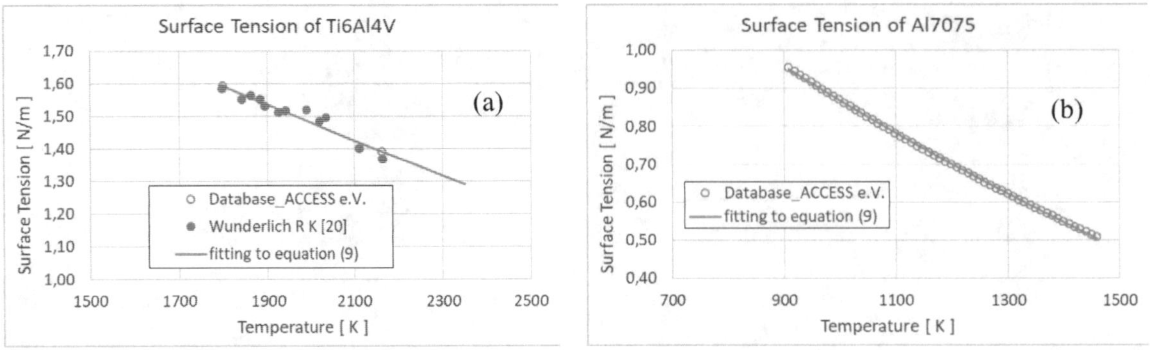

Figure 3. Dependences of surface tension on temperature for alloys Ti6Al4V (a) and Al7075 (b).

3.2. Contact angle

The contact angle can be considered as a measure of the surface tension of an applied liquid, to balance the surface energy of the solid on which the liquid is deposed. This consideration is well interpreted by the Young's equation [21]. As mentioned in the section 3.1, the effect of nanoparticle was not taken into account for the determination of the surface tensions of Ti6Al4V and Al7075. To compensate for this, this study assumes that its effect can be captured by the contact angle.

The following procedure was used to determine the contact angle in a self-consistent way: From a geometric analysis, as shown in Figure 4(a), the contact angle of a sessile droplet placed on the flat surface of a substrate can be determined by the following equation:

$$W = 2H\frac{\sin\theta}{1-\cos\theta} \quad , \tag{10}$$

where H and W are the height and the width of the droplet, θ is the contact angle. DED processes using Al7075 alloy reinforced with different concentrations of TiC nanoparticles were carry out at AIMEN and the dimensions (height, width) of the deposed tracks were determined by macrographs. Using equation (10) the corresponding contact angles for three nanoparticle TiC concentrations were determined and are plotted in Figure 4(b). The contact angle decreases with the addition of nanoparticles TiC for concentrations below 0.5 %. Similar behavior was observed for other nanofluids [22]. For concentrations above 0.5% the value seams to remain unchanged, a depends could not be deduced due to the absence of data in between 0,5 % and 2 %.

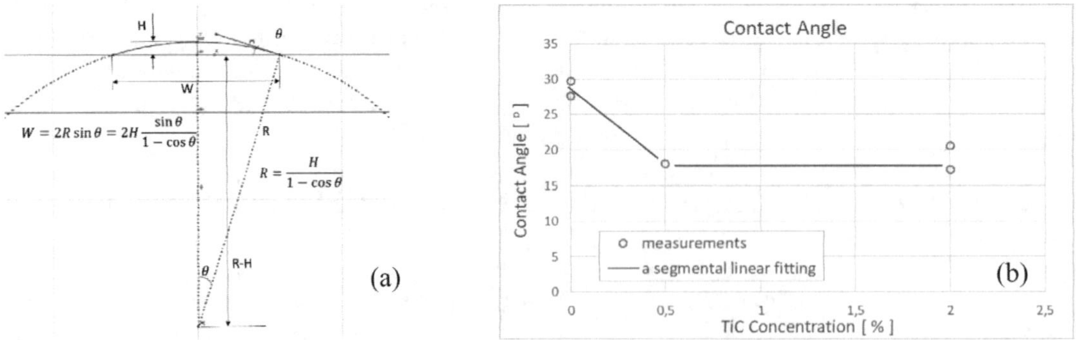

Figure 4. Sketch for the geometric analysis (a) and the dependence of contact angle on TiC concentration for Al7075.

The surface tension given by the equation (9) and the effect of TiC concentration on contact angle are then used for the simulation of the DED process and track width and height are determined from the simulation result. If the measured width and height of a printed track agreed with the values achieved by simulation for a certain addition amount of TiC, the dependency of contact angle on this TiC concentration is considered to be determined in a self-consistent way. A workflow of this assessment procedure is shown in figure 5.

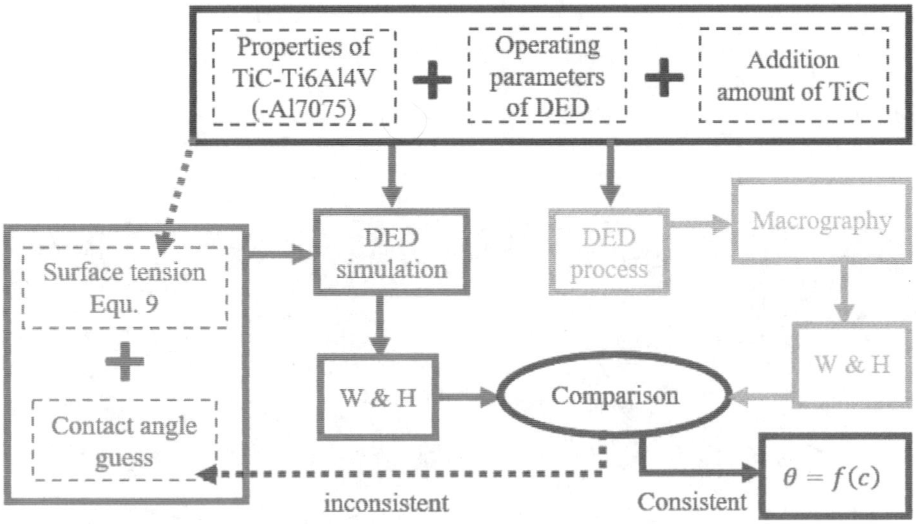

Figure 5. Workflow of a self-consistent determination of the nanoparticle concentration dependence of the contact angle.

4. Experimental and simulation results

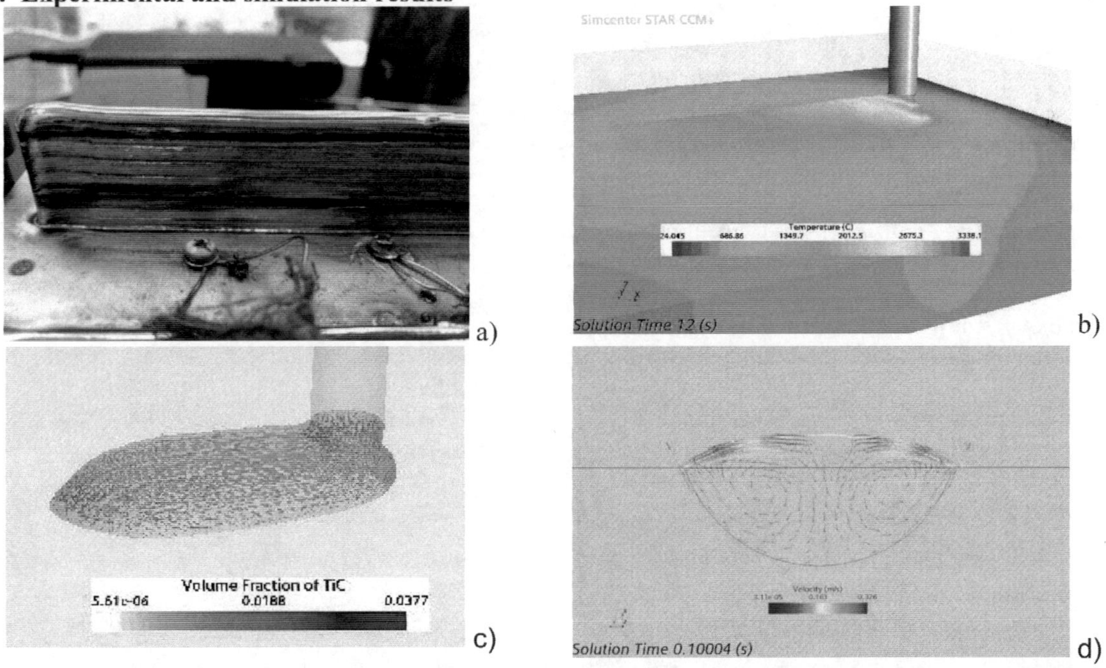

Figure 6. A multilayer track-form (a), and the simulated temperature distribution (b), nanoparticle distribution (c) and velocity vectors (d), respectively, of a Ti6Al4V DED process.

A multilayer track-form printed by a DED process at AIMEN is shown in Figure 6(a), where two thermocouples are installed at the top surface of the substrate to monitor the instant temperatures during printing. Simulations were carried out to verify the modelling and the implemented formulae. Particles are injected, together with alloy, from the inlet of the nozzle and accompany the whole melting, depositing and solidification procedure. NPs are added assuming a homogeneously initial distribution. The modelling of the DED process was validated by comparing the simulated temperature, as shown in Figure 6(b) with the measured temperatures of printing the 1st layer of single track [23]. The simulated particle distribution of a single track and the velocity vectors inside the melt pool at the cross section in the stream-wise direction is demonstrated in Figure 6(c) and (d), respectively.

The determined surface tension and contact angle were then used for DED simulations. Taking the 1st layer of the single tracks of Al7075 as examples, the simulation results were compared with the experimental results of different reinforcements of TiC nanoparticles, as demonstrated in Figure 7.

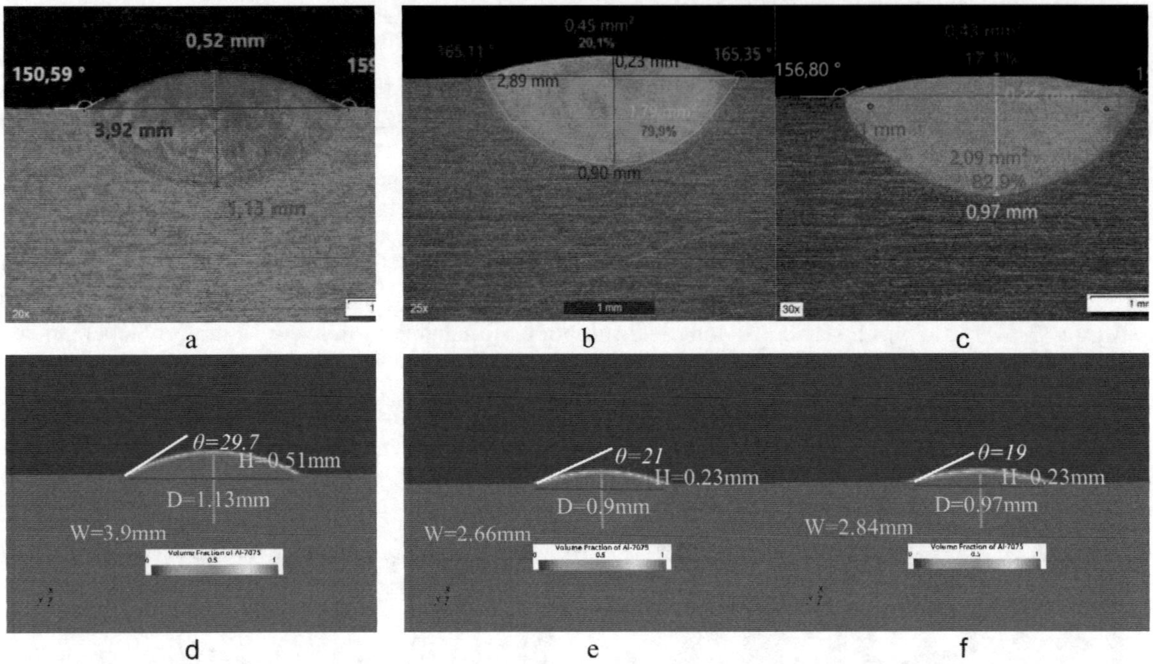

Figure 7. Single track profiles achieved by experiment (a-c) and simulation (d-f) for Al7075 reinforced with nanoparticle TiC concentrations at 0% (a,d), 0.5% (b,e) and 2.0% (c,f), respectively.

The computed track heights and widths for various reinforcement concentrations are in agreement with the experiment measurements with an error less than 5%.

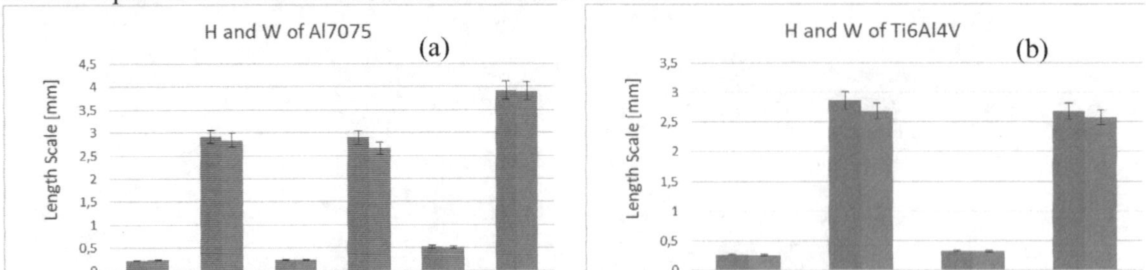

Figure 8. Comparisons of track heights and widths between simulation and experiments for Al7075 (a) and Ti6Al4V (b) reinforced with different concentrations of TiC nanoparticles.

5. Conclusions

Laser based DED processes of nanoparticles TiC reinforced Ti6Al4V and Al7075 alloys were modelled and simulated in this study. A Volume of Fluid (VoF) based-multiphase melting and solidification method of the computational dynamics solver STAR-CCM+, has been used to simulate the DED processes. Heating of the alloys by laser is treated as a volumetric heat source with Gaussian distribution in radial direction for reinforced alloys and substrate regions. A Lagrangian framework was utilized to capture nanoparticle tracking during the deposition and solidification stages of the DED processes. Brownian, thermophoretic, and surface tension forces were implemented to analyze the motion and the distribution of nanoparticles in DED simulations.

Track profile of the DED deposition was investigated in view of the effect of wetting. For alloys Ti6Al4V and Al7075 reinforced with different concentrations of TiC nanoparticles the temperature dependent surface angle was taken from the Access database without considering nanoparticle effects. The contact angle was determined using the TiC nanoparticle dependence of the contact angle determined from DED process results for Al7075 from AIMEN. Surface tension and contact angle, determined in this way, were applied to the simulation of the DED-process at AIMEN. The computed track profile (i.e. the heights and widths from converged simulations) was found to be in good agreement with the measurement, the determined surface tension and contact angle lead to a consistent simulation approach. With this procedure, the wetting characteristics of the TiC reinforced Ti6Al4V and Al7075 can be performed in DED processes without a direct measurement of surface tension and contact angle and their dependence on the nanoparticle content.

Acknowledgement
The authors acknowledge the financial support of this work from EU in relation to the project 814552-LightMe-H2020-NMBP-TO-IND-2018-2020/H2020-NMBP-HUBS-2018. The authors acknowledge the support by Siemens PLM, agreement 60068580, for providing STAR-CCM+ licenses.

References

[1] Fudger S et al 2017 *Material Science and Engineering A* Residual Stress Induced Mechanical Property Enhancement in Steel Encapsulated Light Metal Matrix Composites pp 10-17
[2] Hemanth J et al 2009 *Materials Science and Engineering A* Development and property evaluation of aluminum alloy reinforced with nano-ZrO2 metal matrix composites (NMMCs) pp 110-113
[3] Rajmohan T et al 2013 *Trans Nonferrous Met Soc China* Evaluation of mechanical and wear properties of hybrid aluminum matrix composites pp 2509-17
[4] Moorthy A et al 2012 *Int. J. Mod. Eng. Res.* Dry sliding wear and mechanical behavior of aluminum/fly ash/graphite hybrid metal matrix composite using taguchi method pp 1224-1230
[5] Selvam J D R et al 2013 *Energy Procedia* Synthesis and characterization of Al6061-Fly Ashp-SiCp composites by stir casting and compocasting methods pp 637-646
[6] Alaneme K K et al 2013 *Tribol. Ind.* Mechanical properties and corrosion behavior of aluminum hybrid composites reinforced with silicon carbide and bamboo leaf ash pp 25-35
[7] Escalera-Lozano R et al 2007 *Mat. Charact.* Corrosion characteristics of hybrid Al/SiCp/MgAl2O4 composites fabricated with fly ash and recycled aluminum pp 953-960
[8] Nagamatsu H Sasahara H 2022 Improvement of Cooling Effect and Dimensional Accuracy of Wire and Arc Additive Manufactured Magnesium Alloy by Active-Cooling-Based Contacting Copper Blocks *J. Manuf. Mat. Process.* **6** 27 (doi.org/10.3390/jmmp6020027)
[9] Teskeredžić A Demirdžić I Muzaferija S 2002 *Numerical Heat Transfer: Part B*: Fundamentals, Numerical method for heat transfer, fluid flow, and stress analysis in phase-change problems **42**(5) 437-459
[10] Jakumeit J Zheng G Laqua R Clark S J Zielinski J Schleifenbaum J H & Lee P D 2021 *Additive Manufacturing* Modelling the complex evaporated gas flow and its impact on particle spattering during laser powder bed fusion **47** 102332
[11] Zheng G Jakumeit J Pabel T Kneissl C Magagnin L 2021 *Light Metals* Computational Simulation

of Nanoparticle Distributions in Metal Matrix Composite Casting Processes pp 171-179

[12] Kumar R Milanova D 2009 Effect of surface tension on nanotube nanofluids *Appl. Physics Letters* **94** 073107

[13] Kim S J Bang I C Buongiorno J Hu L W Surface 2007 wettability change during pool boiling of nanofluids and its effect on critical heat flux *Intl. J. Heat Mass Transfer* **50** 4105–4116

[14] Ranjbar H Khosravi-Nikou M R Safiri A Bovard S Khazaei 2015 An experimental and theoretical investigation on nano-fluid surface tension *J. Natural Gas Sci. Eng.* **27** 1806–1813

[15] Vafaei S Purkayastha A Jain A Ramanath G Borca-Tasciuc T 2009 The effect of nanoparticles on the liquid–gas surface tension of Bi2Te3 nanofluids *Nanotechnology* **20** 185702

[16] Moosavi M Goharshadi E K Youssefi 2010 A Fabrication characterization and measurement of some physicochemical properties of ZnO nanofluids *Intl. J. Heat Fluid Flow* **31** 599–605

[17] Jeong Y H Chang W J Chang S H 2008 Wettability of heated surfaces under pool boiling using surfactant solutions and nano-fluids *Intl. J. Heat Mass Transfer* **51** 3025–3031

[18] Tanvir S and Qiao L 2012 Surface tension of nanofluid-type fuels containing suspended nanomaterials *Nanoscale Res. Letters* **7** 226–236

[19] Sageman D R 1972 Surface tension of molten metals using the sessile drop method *Theses and Dissertations* 6117

[20] Wunderlich R K 2008 Surface tension and viscosity of industrial Ti-alloys measured by the oscillating drop method on board parabolic flights High Temperature Materials and Processes Special Issue Vol **27** No 6

[21] Tadmor R 2004 Line Energy and the Relation between Advancing, Receding, and Young Contact Angles *Langmuir* **20** 7659-7664

[22] Harikrishnan A R Dhar P Agnihotri P K Gedupudi S Das S K 2017 Wettability of complex fluids and surfactant capped nanoparticleInduced quasi-universal wetting behavior *J. Phys. Chem. B* **121** 6081−6095

[23] Zheng G Laqua R Rey P Salgueiro M Gipperich M Riepe J Jakumeit J 2022 Influence of nanoparticles on melting and solidification during a Directed Energy Deposition process analysed by simulation *ICASP6* Le Bischenberg June 20-24

IOP Conf. Series: Materials Science and Engineering 1281 (2023) 012005 doi:10.1088/1757-899X/1281/1/012005

Thermo-mechanical modelling of additive manufacturing: activation of non-constructed domain and effect of kinematic hardening

J Keumo Tematio, L Ravix, Y Zhang*, M Bellet

Mines Paris, PSL University, CEMEF – Centre de Mise en Forme des Matériaux, CNRS UMR 7635, Sophia Antipolis, France. *E-mail: yancheng.zhang@minesparis.psl.eu

Abstract. A thermo-mechanical finite element model is developed for additive manufacturing by directed energy deposition (DED). The simulation is conducted at the part scale by modelling the progressive deposition in the way of the fraction of the layer. To incrementally resolve the displacement, strain, and stress fields, a theoretical formulation for the kinematic positioning is proposed to minimize the distortion of the non-constructed fraction by considering current displacement and strain in the constructed part, which can resolve the discontinuous problem at the interface for a patch of material depositions. Moreover, it is found that the kinematic hardening couldn't be ignored for the back-and-forth deposition mode of the DED process. The application for a turbine blade with a strong curvature is adopted, and the distributions of distortion and stress during mid-construction and final construction are studied.

1. Introduction

Among different additive manufacturing (AM) processes, directed energy deposition (DED) is widely adopted due to its efficiency, flexibility, and construction possibility of near-net-shape [1]. However, expensive and time-consuming experimental work is still needed to find the optimized parameters for the new structures and new materials. Recently, numerical simulation of DED processes has received great interest, especially with developments of computational resources and numerical techniques. To access the evolution of the part during its construction, such as distortions and stress, macro-scale thermo-mechanical numerical simulations are usually proposed.

In the available models, the quiet element method, inactive element method, and hybrid inactive/quiet element method are usually adopted to deal with the non-activated domain. Then two questions are posed: how to deal with the physical information consisting of thermal and mechanical aspects at the interface between the active and inactive (or quiet) elements? which type of material constitutive law should be adopted to simulate back-and-forth deposition mode?

Once the elements in the non-construction domain are activated, the physical information of the interface should be carefully handled. Regarding the thermal aspect, Michaleris [2] gives detailed discussions on numerical treatments of the quiet element method and inactive element method. No special treatment is needed for the quiet element method, but the scaling factors for the thermal conductivity and specific heat should be carefully verified to avoid artificial energy input. For the inactive element method, the activated elements and their nodes shared by active elements may not be at the initial temperature, which can result in artificial energy input, so resetting the new activated elements to the initial temperature for the numerical solver resolution is required.

However, from the mechanical point of view, few articles give detailed information on the element activation of the non-activated domain. For the method of the quiet element, the initial distortion existed for the new active elements by considering a small Young's modulus in the non-activated domain, the different magnitude of a small Young's modulus may lead to different modes of distortion as presented in the work of Biegler *et al.* [3]. For the method of inactive element (Baiges *et al.* [4]), the initial displacement field of the newly activated elements is induced by the nodes shared with the pre-existing active elements, and the inherited initial strain obtained by the initial displacement is removed from the following stress computation. Continuous displacement can be obtained by the above treatment, while the discontinuities of strain or stress may happen for activating a patch of elements.

Additive manufacturing is a high temperature accompanied process, the reasonable material laws at high temperatures and low temperatures should be adopted to simulate the process simulation, especially for a big number of layer depositions and long dwell time. Most of the numerical models employ the elastoplastic law with strain hardening or without strain hardening (Biegler *et al.* [3]). The elasto-viscoplastic law is also adopted by Zhang *et al.* [5], in which the viscoplastic flow rule is adopted for the solid and semi-solid states, and an incompressible Newtonian constitutive model is applied for the liquid state. Similarly, elasto-viscoplastic law is proposed by Stender *et al.* [6] for the solid and semi-solid states. Moreover, a recrystallization effect and static recovery are modeled. However, the kinematic hardening has not been yet considered, it may be a key factor in the accurate prediction of stress and strain, especially for the back-and-forth deposition mode of the DED process.

In this work, a fraction of layer-based thermo-mechanical finite element simulation is developed for DED additive manufacturing by inactive element method, where a theoretical formulation for the kinematic positioning is proposed to minimize the distortion of the non-constructed fraction by considering the current displacement and strain of the interface shared elements in the active constructed part. Simulation results of the proposed inactive element method are compared with those of the quiet element method. Moreover, it is found that it's essential to consider kinematic hardening for stress prediction in the back-and-forth deposition of the DED process.

2. Quasi-static mechanical analysis

Under the hypothesis of small deformation, the static equilibrium and mass conservation are resolved for the FE resolution by equation (1).

$$\begin{cases} \boldsymbol{\nabla}\cdot\boldsymbol{\sigma} & = & \mathbf{0} \\ \boldsymbol{\nabla}\cdot\boldsymbol{v}+\dfrac{\dot{p}}{\chi_b}-3\alpha\dot{T} & = & 0 \end{cases} \tag{1}$$

where $\boldsymbol{\sigma}$ is the stress tensor, p is the pressure equal to $(-\frac{1}{3})tr\boldsymbol{\sigma}$ and $\chi_b = \frac{E}{3(1-2v)}$ is the bulk elastic modulus., α is the linear thermal expansion coefficient, and $\dot{T}$ is the temperature rate.

One main objective of this paper is to propose a thermo-elastic-viscoplastic (EVP) constitutive model considering the isotropic and kinematic hardening. The general formulations of the EVP constitutive equations are given below:

$$\dot{\boldsymbol{\varepsilon}} = \dot{\boldsymbol{\varepsilon}}^{el} + \dot{\boldsymbol{\varepsilon}}^{vp} + \dot{\boldsymbol{\varepsilon}}^{th} \tag{2}$$

$$\dot{\boldsymbol{\varepsilon}}^{el} = \left(\mathbf{D}^{el}\right)^{-1}\dot{\boldsymbol{\sigma}} = \frac{1+\upsilon}{E}\dot{\boldsymbol{\sigma}} - \frac{\upsilon}{E}\mathrm{tr}(\dot{\boldsymbol{\sigma}})\mathbf{I} \tag{3}$$

$$\dot{\boldsymbol{\varepsilon}}^{th} = \alpha\dot{T}\mathbf{I} \tag{4}$$

$$\dot{\boldsymbol{\varepsilon}}^{vp} = \frac{3}{2\bar{\sigma}}\left[\frac{\bar{\sigma}(\mathbf{s}-\mathbf{X}) - (\sigma_Y + R(\bar{\varepsilon}))}{K_{vp}}\right]_+^{1/m}(\mathbf{s}-\mathbf{X}) \tag{5}$$

where the strain rate tensor $\dot{\boldsymbol{\varepsilon}}$ (equation (2)) is split into an elastic part, $\dot{\boldsymbol{\varepsilon}}^{el}$, a viscoplastic part, $\dot{\boldsymbol{\varepsilon}}^{vp}$, and a thermal part, $\dot{\boldsymbol{\varepsilon}}^{th}$. The latter consists of the thermal expansion rate (equation (4)), with α the one-dimensional thermal dilation coefficient. Equation (3) is the hypo-elastic Hooke's law where E denotes Young's modulus, and υ the Poisson's coefficient. $\dot{\boldsymbol{\sigma}}$ denotes the total time derivative of the stress tensor. Finally, equation (5) expresses the relation between the viscoplastic strain-rate tensor and the stress deviator $\mathbf{s}$ and back stress $\mathbf{X}$. Coefficient K_{vp} is the viscoplastic consistency, σ_Y denotes the initial yield stress. The function $[x]_+$ is equal to 0 when x is negative and to x otherwise. Coefficient m is the strain-rate sensitivity, and the strain hardening function R depends on the equivalent plastic strain $\bar{\varepsilon}$ which is the time-integral of the generalized strain-rate $\dot{\bar{\varepsilon}}$. From equation (5), when $\dot{\boldsymbol{\varepsilon}}^{vp} \neq \mathbf{0}$, and calculating $\dot{\bar{\varepsilon}}$, the following scalar relation can be easily obtained:

$$\bar{\sigma} = \sigma_Y + R(\bar{\varepsilon}) + k\dot{\bar{\varepsilon}}^m \tag{6}$$

Consequently, equation (5) can be straightforwardly rewritten:

$$\dot{\boldsymbol{\varepsilon}}^{vp} = \frac{3\dot{\bar{\varepsilon}}}{2\bar{\sigma}}(\mathbf{s}-\mathbf{X}) \tag{7}$$

Similar to the work of Smith *et al.* [7], the back stress **X** related to the kinematic hardening is defined as a non-linear form by ignoring the temperature and field variable dependencies.

3. Activation of the non-constructed domain

The activation of the non-constructed domain is based on the fraction of the layer along the deposited path, which contains a set of elements. The treatment of thermal aspects follows the work of Michelaris [2], which will not be elaborated on here. The main objective of this section will focus on the mechanical treatment in the following work.

Based on the inactive element method, it consists in describing the set of equations that determine the Lagrangian distortion $\boldsymbol{u}_{nb}$ related to the positioning of the non-built domain Ω_t^{nb}, which locates at t based on the built domain Ω_t. It will prepare for the deposition of a new fraction (dashed zone in **Figure 1**) at time interval $[t, t+\Delta t]$. The global principle of these equations will minimize the quantity of the initial distortions and strains within Ω_t^{nb} by equation (8), under the constraint of continuity of the global distortion $\boldsymbol{u}_t$ at the interface of activation Γ_t by equation (10), with the assumption of the incompressibility of volume by equation (9). The penalty coefficient η_p is introduced to control the weight of the minimization term of the global distortion $\boldsymbol{u}_t$, which can be determined by sensitivity analysis.

$$f(\boldsymbol{u}_t^{nb}) = \min\frac{1}{2}\int_{\Omega_t^{nb}}\left(\boldsymbol{\varepsilon}(\boldsymbol{\omega}_t):\boldsymbol{\varepsilon}(\boldsymbol{\omega}_t)+\eta_p\boldsymbol{\omega}_t:\boldsymbol{\omega}_t\right)\mathrm{dv} \tag{8}$$

$$\nabla.\boldsymbol{u}_t^{nb} = 0 \ in \ \Omega_t^{nb} \tag{9}$$

$$\boldsymbol{u}_t = \boldsymbol{u}_t^{nb} \text{ at } \Gamma_t \tag{10}$$

where $\boldsymbol{\omega}_t$ is the intermediate variable to determine $\boldsymbol{u}_t^{nb}$. The initialized distortion is based on the CAD configuration. Once the initialization step is finished, the physical deposition is performed during $[t, t + \Delta t]$ as described in section 1. The proposed method can resolve the discontinuities of deformation and strain (stress).

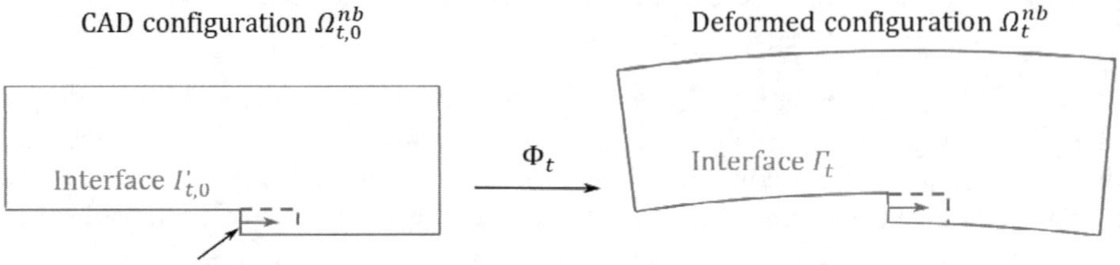

Figure 1. Initialization of no-built domain based on the CAD configuration

4. 3D Geometrical Model

In this paper, a thin-walled and curved DED geometry from the work of Biegler *et al.*[3] was chosen, as given in **Figure 2**, where the local curvilinear coordinate frame (t, n, z) is defined for demonstrating the local mechanical information (**Figure 2** (b)).

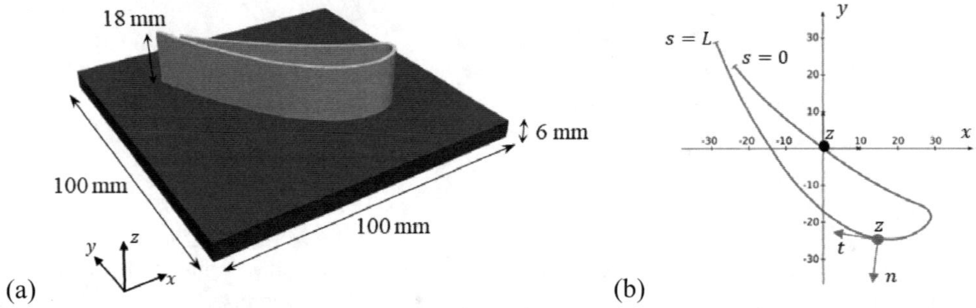

Figure 2. The geometry of the 30-layer turbine blade mock-up and defined coordinates: (a) 3D view; (b) deposition path in the horizontal plane, with the definition of global and local referential frames.

As the structure has varying curvature, high stiffness at the tip and low stiffness at the free ends, the numerical predictions of the distortion and stress will be very sensitive to different numerical methods and different material constitutive models. The same process parameters are also taken from the work of Biegler *et al.* [3], but only a regular dwell time of 30 s is considered, without the cooling break of 300s. The adopted thermal properties (Kim *et al.* [8]), the corresponding thermal boundary conditions, process parameters, and the dimension of the part are given in **Figure 2**. The non-linear thermal analysis is performed, more detailed information can be found in the PhD thesis of Keumo Tematio [9].

Table 1. Simulation data: material and process parameters, geometrical data of the turbine blade mock-up.

	Properties	**Values or reference**
Material properties (316L)	Thermophysics	Kim *et al.* [8]
	Initial temperature, T_0 [K]	298.15
	Ambient temperature, T_∞ [K]	298.15
Heat exchange	Convection coefficient, h_{conv} [W m^{-2} K^{-1}]	35
	Emissivity, ε	0.6
	Stefan–Boltzmann constant, σ [W m^{-2} K^{-4}]	5.670374×10^{-8}
	Nominal laser power, P_L [W]	400
Laser processing	Scan speed, v_L [mm s^{-1}]	10
(Biegler *et al.* [3])	Reflection coefficient, R	0.7
	Spot diameter, ϕ_L [mm]	0.6
	Curvilinear length, L [mm]	155.4
Dimension of part	Layer width, l_D [mm]	1.2
(Biegler *et al.* [3])	Layer thickness, h_D [mm]	0.6
	Total height, H [mm]	18

5. Effect of minimization of non-built domain

For the mechanical analysis, the values of plastic material properties are fitted by the work of Depradeux [10], in which only isotropic hardening is considered by the powder law $R(\bar{\varepsilon}) = H\bar{\varepsilon}^n$, where H is the hardening modulus and the n strain-hardening exponent. The elastic properties are given in Depradeux [10].

To study the method of minimization of the initial distortions and strains within the non-activated domain, a value of 0.25 is taken for the penalty coefficient η_p of the inactive element method, based on the equations (8-10). The simulation results are compared with those of the quiet element method by setting Young's modulus of 1GPa and 10GPa at mid-construction and final construction after dwell time.

5.1. During the mid-construction

Figure 3 presents the distributions of displacement (u_t) and stress (σ_{tt}) of the entire simulation domain for both inactive element method and quiet element method. The results are demonstrated at the instant close to half of the deposition of the 15th layer, at the mid-construction of the part. To better demonstrate the non-construction domain, the semi-transparent plot is adopted. By the displacement distributions, the value of u_t in the non-construction domain trends to be zero by the standard inactive element method except at interface, while non-negligible values are found for the quiet element method for both 1GPa and 10GPa of the Young's modulus in the non-construction domain. For the stress distribution of σ_{tt}, fewer differences are found for the two methods, while the σ_{tt} distribution of 1GPa is closer to that of the standard inactive element method.

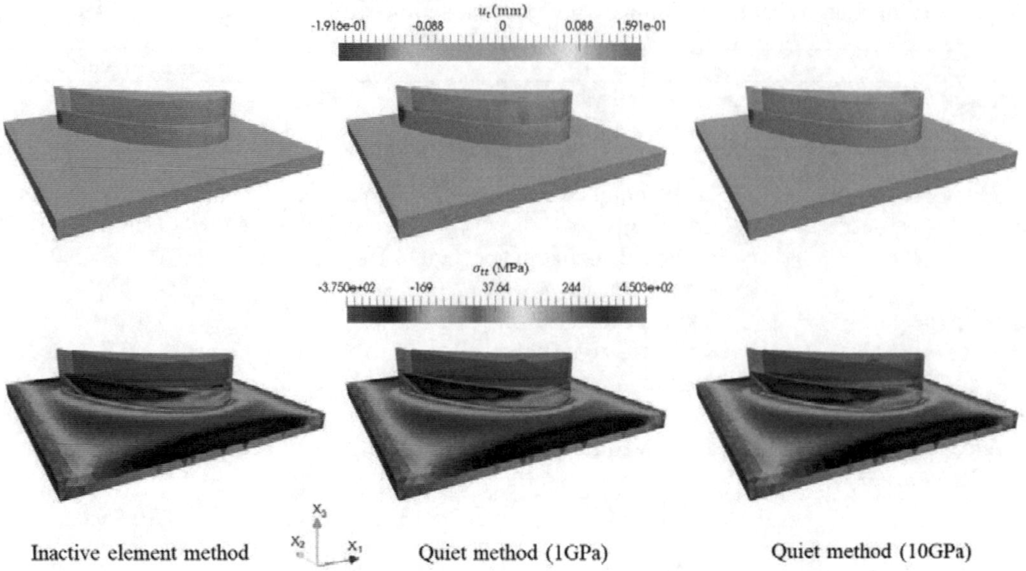

Figure 3. Displacement and stress distributions with inactive element method and quiet element method at the half deposition of the 15th layer. Construction domain: solid plot; non-construction domain: semi-transparent plot.

5.2. End of the construction

Compared with mid-construction, similar phenomena are found for both displacement and stress distribution at the end of construction. In **Figure 4**, the distortion predicted by the inactive element method lies between those estimated by the quiet element method for 1GPa and 10GPa, and the case of 1GPa overestimates the distortion and 10GPa underestimates the distortion. For the stress distribution, the two methods give similar results. Again, for the small Young's modulus, the stress distribution of 1 GPa nearly superposes with that of the inactive element method.

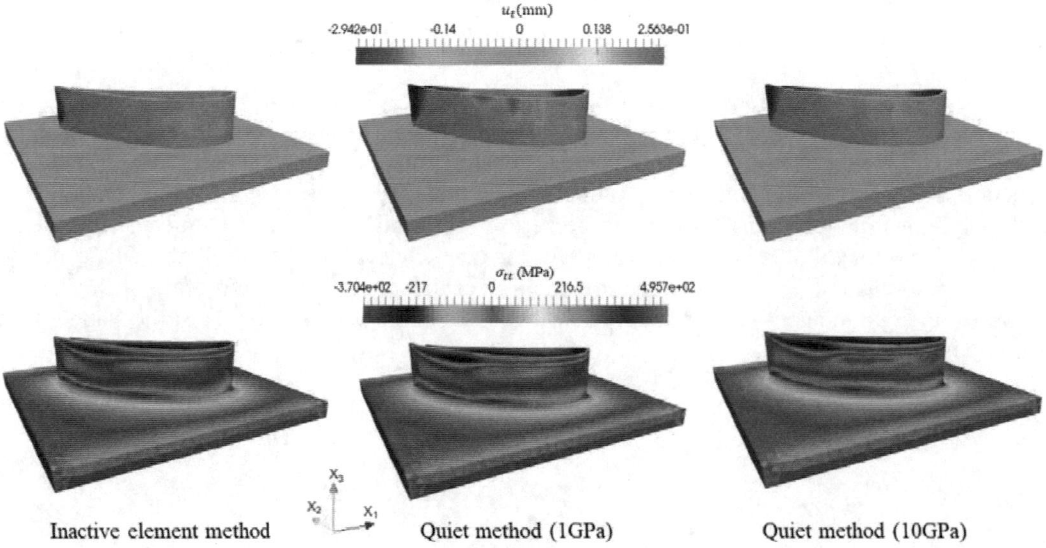

Figure 4. Displacement and stress distributions with inactive element method and quiet element method at the 30th layer.

6. Effect of kinematic hardening

To study the effect of kinematic hardening on the alternative deposition, the mixed isotropic-kinematic hardening simulation is adopted, where the material parameters of the non-linear form related to equation (6) are taken from the work of Smith *et al.* [7], which are obtained based on the traction-compression tests of two cycles. Parallelly, another group of parameters only with isotropic hardening is fitted based on the experimental traction curve in the first cycle. The elastic properties are adopted from Muransky *et al.* [11].

Based on the inactive element method, the following two sections will focus on the effect of kinematic hardening on results at mid-construction and end of the process.

6.1. During the process

As described above, Isotropic hardening and mixed isotropic-kinematic hardening are applied for the process simulation. **Figure 5** shows the comparison, in terms of predicted distortion and stress fields, between two different material laws. The results are demonstrated at the instant close to half of the deposition of the 15th layer, at the mid-construction of the part. It can be shown that the effect of kinematic hardening on the distortion u_t is quite limited, while the stress distribution of σ_{tt} is different for the substrate and the constructed wall.

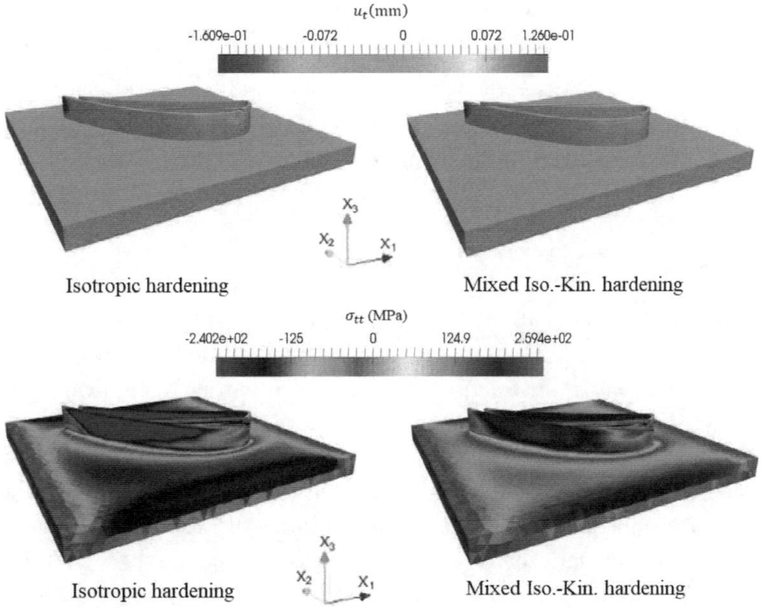

Figure 5.Comparison between isotropic hardening and mixed isotropic-kinematic (Mixed Iso.-Kin.) simulations: displacement field and stress field. The comparison is shown at about half the deposition of the 15th layer.

6.2. End of Construction

After the dwell time of the final construction of the 30th layer, the distortion and stress distributions are given in **Figure 6**. Like those observed at the mid-construction of the 15th layer, the influence of kinematic hardening on the σ_{tt} is obvious. While for the distortion distribution, the influence is not significant.

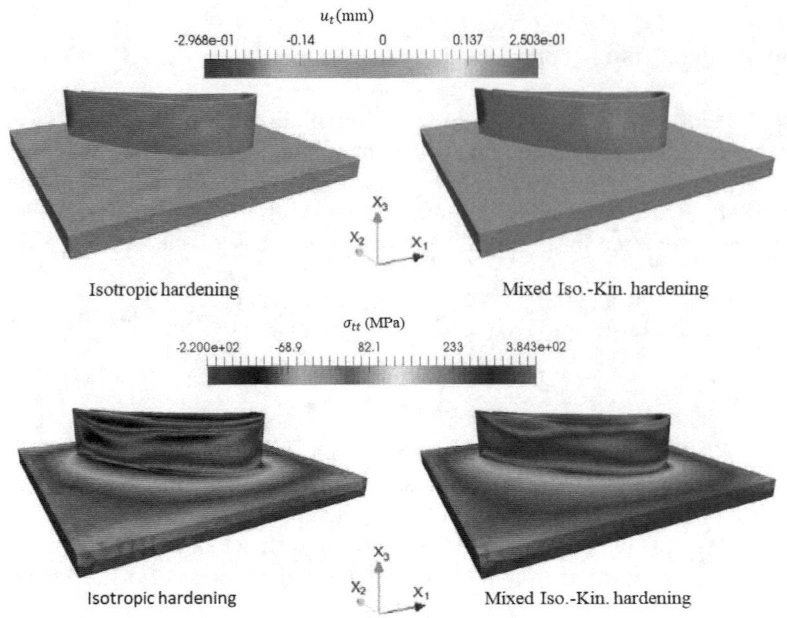

Figure 6. Comparison between isotropic hardening and mixed isotropic-kinematic (Mixed Iso.-Kin.) simulations: displacement field and stress field. The comparison is shown at the final deposition of the 30th layer.

By the above studies on the mid-construction and final construction, the kinematic hardening shows a significant influence on the stress distribution by present studies, while the effect on the distortion is quite limited.

7. Conclusion

In this paper, a thermo-mechanical finite element simulation is developed for additive manufacturing by directed energy deposition (DED). The simulation is conducted at the scale of the part, by modelling the progressive deposition by the way of the fraction of layer, where a group of elements is activated at the same time based on the method of inactive element method. To resolve the problem of discontinuities at the interface between active and inactive elements, a minimization of the quantity of the initial distortions and strains in the inactive domain is proposed. The comparison between the inactive element method and the quiet element method on displacement and stress distribution is discussed. In addition, another important discussion of the effect of kinematic hardening included in the material constitutive on DED process simulation is performed. All the discussions are presented for the mid-construction and the end of construction.

By the comparison of displacement and stress distributions between the inactive element method and the quiet element method, it is found that the inactive element method gives more reasonable results for the distortion prediction, and the stress distribution of the quiet element method with a set of small Young's modulus (1 GPa) in the non-construction domain is like those of the inactive element method.

The kinematic hardening shows a significant influence on the stress distribution, while its effect on the displacement distribution is slight.

IOP Conf. Series: Materials Science and Engineering 1281 (2023) 012005 doi:10.1088/1757-899X/1281/1/012005

By the above studies, the minimization method of the non-construction domain produces less uncertainty compared with the quiet element method. The kinematic hardening is necessary to be considered for the back-and-forth deposition mode of additive manufacturing.

Acknowledgments
The authors wish to acknowledge the financial support from the doctoral school of Mines Paris for the project of **AM-Multi**.

8. References

[1] Wilson J M, Piya C, Shin Y C, Zhao F and Ramani K 2014 *J. Clean. Prod.* **80** 170
[2] Michaleris P 2014. *Finite Elem. Anal. Des.* **86** 51
[3] Biegler M, Marko A, Graf B, Rethmeier M 2018 *Addit. Manuf.* **24** 264
[4] Baiges J, Chiumenti M, Moreira C A, Cervera M and Codina R 2021 *Addit. Manuf.* **37** 101650
[5] Zhang Y, Chen Q, Guillemot G, Gandin C A, Bellet M 2018 *C. R. Mecanique* **346** 1055
[6] Stender M E, Beghini L L, Sugar J D, Veilleux M G, Subia S R, Smith T R, San Marchi C W, Brown A A and Dagel D J 2018 *Addit. Manuf.* **21** 556
[7] Smith M C, Muransky O, Austin C, Bendeich P and Xiong Q 2018 *Int. J. Press. Vessel. Pip. INT J PRES VES PIP* **164** 93
[8] Kim C S 1975 *Technical Report Argonne National Laboratory*, Ill., USA
[9] Keumo Tematio J 2022 *PhD Thesis*, PSL University, Mines Paris
[10] Depradeux L 2004 *Ph.D. Thesis*, INSA Lyon
[11] Muransky O, Smith MC, Bendeich PJ, Edwards L 2011 *Comp. Mater. Sci.* **50** 2203

IOP Conf. Series: Materials Science and Engineering 1281 (2023) 012006 doi:10.1088/1757-899X/1281/1/012006

Optimization of powder catchment efficiency for industrial laser cladding processes

Mitchell Grams, Gentry Wood, Ata Kamyabi-Gol, Douglas Hamre
Apollo-Clad Laser Cladding, Leduc, AB T9E 0B3, CANADA

E-mail: mitchell.grams@apollomachine.com

Abstract. Laser clad overlays provide enhanced corrosion resistance and wear performance to components across a wide range of industries. Powder fed laser processes provide flexibility in material selection, however have traditionally yielded lower material usage efficiency (60–80%) compared to wire fed processes. A general model was developed to predict powder catchment efficiency based on the overlap between the molten pool and powder cloud distribution. Mathematical techniques of asymptotic analysis and blending were used to obtain closed-form expressions relating tabulated material properties and the key process inputs with the critical melt pool dimensions of leading length (x_{f}) and maximum width (y_{m}). An improved experimental technique was implemented to measure the powder cloud distribution, a key process input which previously was not well understood. For baseline cladding conditions, melt pool dimensions were smaller than the powder cloud leading to catchment inefficiency. A custom laser optics module was designed to enable adjustment of the relative position of the melt pool and powder cloud independently from other process parameter changes. Combining optimized parameters with a shift of 1 mm, catchment efficiency >90% has been achieved and sustained in production cladding operations. This work is part of a broader program at Apollo-Clad, using simple, accurate, and fast analytical modelling techniques to generate engineering design rules for laser-based processes.

1. Introduction

Laser cladding is a form of additive manufacturing that enables production and repair of components with improved surface characteristics compared to bulk material properties. The use of powder feed enables cladding with complex multi-component systems that offer superior quality and performance. A significant potential drawback of powder feed is the reduced (60–80%) material usage efficiency, commonly referred to as catchment efficiency, relative to wire fed processes (> 90%).

Improving powder catchment efficiency is of immediate concern for industrial laser cladding and additive manufacturing practitioners. The economics of cladding involve a combination of direct material costs and equipment time. Small improvements in material deposition efficiency can have large impacts on the overall process economics, due to the potential to reduce the direct costs of materials and simultaneously increase material deposition rate and productivity. Particularly for additive manufacturing applications, improved catchment is essential to enhancing the competitiveness with traditional manufacturing processes. Optimizing catchment efficiency is also of significant importance in laser hard-facing applications, with material systems such as nickel tungsten carbide (Ni-WC), where the direct cost of materials is responsibile for the majority of the total cost of production.

1.1. Methodology
There is an immediate need for an improved understanding of the relationships between catchment efficiency, material properties, and process variables. Industrial application requires this understanding to be in the form of engineering design rules, suitable for use in optimization of existing processes as well as new process qualification. Trial-and-error is a well-established empirical technique for optimizing industrial processes. This method is inefficient but generally effective for well-understood processes and material systems; however, no insight is given into the underlying physical mechanisms. Any process variation, beyond what is explicitly tested, requires a new experimental investigation. This work demonstrates the use of analytical methods as an alternative to traditional trial and error. The mathematical techniques of asymptotic approximation and scaling are used to produce guidelines which are simple, accurate, and fast to implement and thus capable of meeting the needs of industrial welding practitioners.

2. Analytical Modelling
Powder catchment efficiency (η) is determined by the proportion of process powders that are incorporated into the clad deposit expressed as a fraction (or wt. %) relative to the total mass of powder provided to the process. Calculation of the catchment efficiency involves integrating the projection of the powder distribution on the molten pool. For a first order approximation, attenuation of laser power due to interaction with the powder cloud is neglected. The essential variables involved in controlling the size and shape of the molten pool are assumed to be approximately independent of the variables which control the size of the powder distribution. The governing physics of heat and mass transfer may then be decoupled and analysed independently.

2.1. Mass Transfer
Process variables associated with the spatial powder distribution (i.e., mass transfer) include the powder nozzle geometry, powder carrier and shield gas flow rates, as well as density and particle size range of the powder. For coaxial nozzles, the projected powder distribution at the working distance (i.e., the powder focal plane) has the form of a circular gaussian distribution with a total mass flow rate ($\dot{m}$) and standard deviation (σ_p) [1, 2]. In this work, the effect of mass flow rate is assumed to be secondary and only the distribution parameter is used to fully characterize the powder incident on the cladding substrate. The fractions of powder residing within a radius of $1\,\sigma_\mathrm{p}$, $2\,\sigma_\mathrm{p}$, and $3\,\sigma_\mathrm{p}$ are 39%, 86%, and 99%.

2.2. Heat Transfer
The heat transfer aspects of laser beam processes are well approximated by considering a gaussian distributed heat source [3]. Essential variables for determining the molten pool geometry (i.e., heat transfer) include process parameters: laser power (q), distribution parameter of laser intensity (i.e., gaussian standard deviation, σ), travel speed (U), and initial or preheat temperature (T_0), as well as material properties of the substrate: thermal conductivity (k), thermal diffusivity (α), and melting temperature (T_m). It has been shown [4] that the relationship between the process parameters and the geometry of isotherms surrouding a moving heat source can be expressed in terms of only two dimensionless parameters Ry and σ^*.

The Rykalin number is a dimensionless quantity that describes the relative magnitude of advective to conductive heat transfer for a moving heat source on a thick plate, and is given by [5]:

$$\mathrm{Ry} = \frac{qU}{4\pi k\alpha(T_\mathrm{m} - T_0)} \tag{1}$$

For laser cladding processes, higher power necessitates higher travel speeds in order to avoid undesirable metallurgical effects that can result from excess heat input to the substrate material.

High Rykalin numbers are thus associated with higher cladding speeds or 'fast' processes. Operating points which produce low Rykalin numbers are comparatively 'slow'. Although it is theoretically possible to combine process variables to achieve values of Ry from 0–∞, practical considerations result in a far more limited domain. For industrial cladding operations, productivity is maximized by utilizing both high power and travel speed. Substrate materials such as stainless steels and Ni-based superalloys, give Ry on the order of 10^1–10^2 (i.e., fast processes).

The dimensionless distribution parameter describes the relative size of a gaussian distributed heat source, such as a laser beam, and is equivalent to a Peclet number [6]:

$$\sigma^* = \frac{\sigma U}{2\alpha} \tag{2}$$

Smaller values of the distribution parameter result in more concentrated heat sources and an increased risk of underdesirable metallurgical effects (e.g., porosity) due to excessive temperatures in the melt pool. A practical upper bound for the distribution parameter is given by the need for a sufficiently high peak temperature (T_p) to achieve melting of the clad substrate. The threshold value for $T_\mathrm{p} = T_\mathrm{m}$ is denoted σ^*_cr and may be expressed as a function of Ry [4]. Typical cladding practice involves σ^* in the range of 0.3–0.6 σ^*_cr.

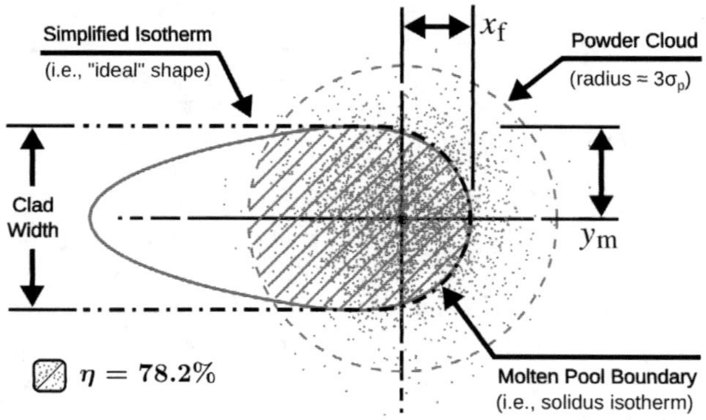

Figure 1: Monte-Carlo simulation representing the powder cloud projection (grey dashed line) overlayed on the molten pool of the laser cladding process (solid red line); catchment efficiency is approximated as the fraction process powders that land within the common area between these two regions, which for this example is calculated to be $\eta \approx 78\%$.

An example of the characteristic molten pool size and shape are shown in Fig. 1 alongside a representative powder cloud which is slightly large but of the same order as the melt pool width. The typical range of Ry and σ^* will tend to produce elongated melt isotherms where the back of the pool extends beyond the powder cloud boundary. For the purposes of computing catchment efficiency, the isotherm is approximated with the simplified geometry outlined in Fig. 1. This shape corresponds to the asymptotic case of $\sigma^* \gg 1$, Ry $\gg 1$, $\sigma^*/\sigma^*_\mathrm{cr} \to 0$ and consists of two parallel lines joined by a semicircle at the leading edge. Figure 1 also identifies the two critical dimensions for the idealized molten pool geometry: the maximum isotherm half-width (y_m) and the isotherm leading length (x_f).

Exact numerical solutions for isotherm geometry may be obtained using numerical techniques as outlined in Ref. [7]. To aid practical implementation of this work in industrial environments, closed-form expressions for each of these values are also presented in Appendix A. Detailed derivation of the asymptotic expressions and blending laws is beyond the focus of this work, but a brief summary is provided here. The expression for y_m was obtained through application of the techniques presented in Refs. [7, 8, 9, 10]. Ref. [7] proposes a two step solution for problems of blending with two independent parameters (e.g., Ry and σ^*). The first blending step produces expressions for the asymptotic cases of high and low Ry. The second blending step is used to combine these two expressions. Blending of x_f followed a similar approach but the second stage of blending combined the difference between the high and low Ry expressions with a specially selected function ($\mathrm{Ry}\,\sigma/\sigma_\mathrm{cr}$). Traditional blending techniques require that all input functions do not change sign, but this modification allows for consideration of both positive and negative values of x_f, which occur with highly distributed heat sources ($\sigma/\sigma_\mathrm{cr} \to 1$).

2.3. Catchment Efficiency

The primary source of laser cladding powder catchment inefficiency is a mismatch between the powder cloud size and melt pool size, such that powder is incident on the cladding substrate outside of the molten pool [3]. The catchment model presented in this work assumes all powder incident on melt pool is to be captured in the final deposit volume. Other potential sources of inefficiency (e.g., powder bouncing or reflecting off the molten material) are considered secondary and neglected.

Uncaptured powders can be classified into two categories based on where they land on the substrate relative to the melt pool. The first category includes powders landing on either side of the molten pool; the proportion of powder lost in this region depends primarily on the molten pool width, y_m. The second category includes the powder incident ahead of the molten pool; the proportion of powder lost in this region depends primarily on the leading isotherm length, x_f. The analytical expression for catchment efficiency is thus written as the product of two independent expressions:

$$\eta = \eta_\mathrm{max}\left(\frac{y_\mathrm{m}}{\sigma_\mathrm{p}}\right)\, f\left(\frac{x_\mathrm{f}}{\sigma_\mathrm{p}}\right) \tag{3}$$

The first expression, $\eta_\mathrm{max}(y_\mathrm{m}/\sigma_\mathrm{p})$, may be interpreted as the maximum possible catchment efficiency for a given melt pool width. In practice, the isotherm leading length will never exceed the isotherm width, and achieving this theoretical maximum catchment is not possible. The second expression, $f(x_\mathrm{f}/\sigma_\mathrm{p})$, is a factor that corrects for the reduction in catchment compared to the theoretical maximum value depending on the isotherm leading length. The MATLAB multivariate normal cumulative distribution function `mvncdf` [11] was used to compute an exact numerical value for η_max. The asymptotic expression for low values of $(y_\mathrm{m}/\sigma_\mathrm{p})$ was observed to take the form of a power law with an approximate exponent of -3. For high values of $(y_\mathrm{m}/\sigma_\mathrm{p})$ the maximum catchment efficiency is simply 1. Using blending techniques, an approximate closed-form solution was then obtained:

$$\eta_\mathrm{max}\left(\frac{y_\mathrm{m}}{\sigma_\mathrm{p}}\right) = \left[1 + \left(0.80\,\frac{y_\mathrm{m}}{\sigma_\mathrm{p}}\right)^{-3}\right]^{-\frac{1}{3}} \tag{4}$$

This expression has a maximum error of $< 3\%$ compared to the exact numerical solution.

Considering the idealized isotherm shape (see Fig. 1) a series of Monte-Carlo simulations were performed using MATLAB with a range of values of $(x_\mathrm{f}/\sigma_\mathrm{p})$. The resulting correction factor (f) was observed to have the form of an S-curve, and so it is possible to obtain an empirically fit curve using a logistic function:

$$f\left(\frac{x_\mathrm{f}}{\sigma_\mathrm{p}}\right) = \frac{1}{1 + \exp\left(-a\left(\frac{x_\mathrm{f}}{\sigma_\mathrm{p}}\right) + b\right)} \tag{5}$$

where the best fit coefficients are $a = 1.56$ and $b = 0.35$. A reasonable target for catchment is given by $\{y_\mathrm{m},\, x_\mathrm{f}\} \gtrsim 2.3\,\sigma_\mathrm{p}$ ($\eta_\mathrm{max} \gtrsim 0.95$ and $f \gtrsim 0.96$) which yields $\eta > 90\%$.

3. Process Optimization

The general model for catchment efficiency presented in this work satisfies a critical gap in current industrial knowledge. In almost all practical cases, optimization of catchment efficiency is not pursued in isolation, but is treated as one of many factors that contribute to a successful process. The following procedure highlights how the catchment model may be incorporated alongside additional considerations related to process quality and productivity. As a general outline, the steps for application of this optimization to an industrial laser cladding process are:

(i) Determine constraints on dimensional process inputs including: power (q), travel speed (U), laser power distribution parameter (σ), and powder distribution (σ_{p}); the primary industrial challenge in this step comes from practical measurement of the distribution parameters.

(ii) Compute target dimensionless output values (i.e., pool geometry) (y_{m}/σ) and (x_{f}/σ) in terms of required dimensionless input variables (i.e., Ry and σ^*).

(iii) Combine results of previous steps along with additional production goals (e.g., maximize productivity) and metallurgical considerations (e.g., limiting value on peak temperature) to establish operating point with optimized catchment efficiency.

3.1. Measurement of Process Inputs

Allowable ranges for process inputs in industrial laser cladding are given by the limitations of the equipment. These constraints include: maximum laser power (q), maximum travel speed (U), minimum laser power distribution parameter (σ), and minimum powder cloud distribution (σ_{p}). The laser power and travel speed are inherent to the laser energy source and motion system respectively; the acceptable range and operating point of both these variables are typically known quantities. The laser beam distribution and powder cloud distribution, depend on the configuration of laser optics and powder delivery system (i.e., powder nozzle), and are typically not well known. Measurement of these distribution parameters is critical to establishing an optimal operating point.

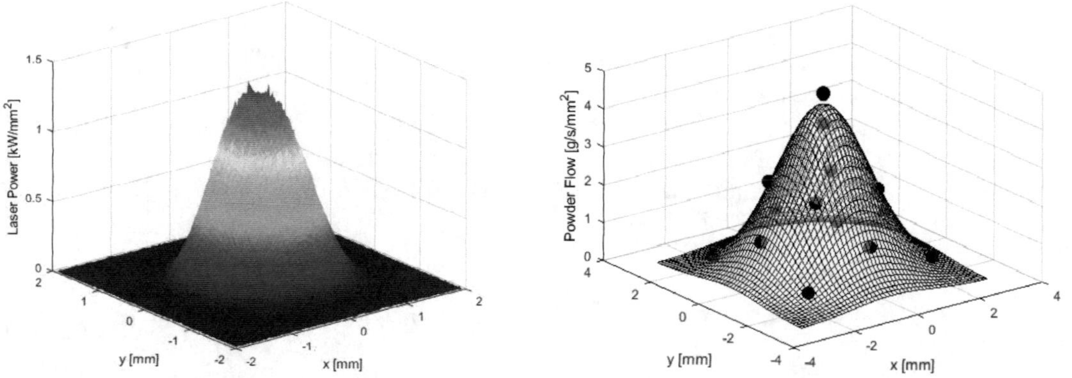

(a) (b)

Figure 2: Measured values for (a) laser power; and (b) powder flow illustrating approximate gaussian distribution.

Figure 2a provides an example of a typical laser beam distribution. This distribution was obtained with a beam profiling type device that samples the laser power at various points within a given working plane [12]. The measurements at each point are then used to fit a gaussian function and determine the effective laser power distribution parameter (σ). Guidelines for performing this analysis are given in ISO 11146-1 [13]. For industrial laser cladding, the minimum possible laser distribution parameter of a focused beam is on the order of 0.1 mm, however it is more common to operate beyond the minimum focus point using a value of σ in the range of 1–2 mm.

Characterization of the powder cloud distribution represents an intense area of recent work within the laser cladding research community. The validation carried out through the course of this work relied on a custom measurement apparatus which was purpose-built for determining the powder distribution parameter (σ_p) on a production laser cladding system. The measurement technique is based upon the continuous pinhole-orifice method presented in Ref. [14], but utilizes a more flexible and robust discrete sampling pattern. Eliminating the need for tight integration with the motion control improves system portability and is preferable in an industrial setting. An example of a typical powder cloud distribution is shown in Fig. 2b. Coaxial powder nozzles are designed for operation at a specific nozzle working distance, corresponding to the powder focus plane, providing less flexibility for adjustment compared to the laser power distribution. Practical values for σ_p are in the range of 0.5–1.5 mm.

3.2. Dimensionless Rules for Parameter Optimization

In prior work [4, 7, 15], isotherm dimensions were normalized by the characteristic length ($2\alpha/U$). For process optimization, it is more convenient to choose the characteristic length as the critical distribution parameter (σ_{cr}). Asymptotic expressions for fast and slow heat sources are then independent of Ry and only a function of the dimensionless distribution parameter (σ/σ_{cr}). Figure 3 illustrates the relationships between the dimensionless distribution parameter and critical melt pool geometry (y_m/σ_{cr}) and (x_f/σ_{cr}) given by Eqs. A.1–A.6 (see Appendix A).

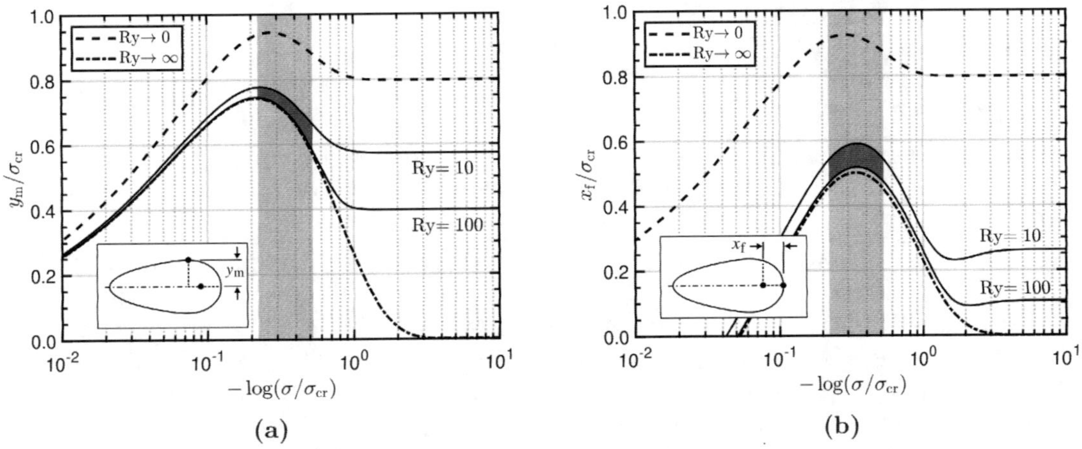

Figure 3: Plot of scaled values for isotherm: (a) maximum width (y_m); and (b) leading length (x_f) with the typical laser cladding operating windows Ry = 10–100, $\sigma/\sigma_{cr} = 0.3$–0.6 highlighted in grey.

Both the maximum width and leading isotherm length exhibit a local maximum which occurs within the typical cladding range. A physical meaning for this point is revealed in the asymptotic expression for the critical distribution parameter of fast heat sources [4]:

$$\sigma_{cr}|_{Ry\gg1} = 1.014 \left(\frac{2\alpha}{U}\right) Ry^{\frac{2}{3}} \tag{6}$$

Substituting Eq. 1 into Eq. 6 gives σ_{cr} as proportional to $(q^2/U)^{1/3}$. The local maximum then represents the largest possible isotherm dimensions for a given cladding power and travel speed. Operating at the corresponding value of σ will maximize the dimensions of the molten pool. Fig. 3a shows that the optimum molten pool width corresponds to $\sigma/\sigma_{\rm cr} \approx 0.6$ and $y_{\rm m} = 0.75\,\sigma_{\rm cr} = 1.3\,\sigma$.

Comparing Fig. 3a and Fig. 3b it is observed that $x_{\rm f} < y_{\rm m}$ throughout the typical operating range. If the process is optimized for $y_{\rm m}$ (i.e., $\sigma/\sigma_{\rm cr} = 0.6$), the leading length is then $x_{\rm f} \approx 0.5\,\sigma_{\rm cr} \approx 0.8\,\sigma$. The leading length is less than the optimized target by an amount $y_{\rm m} - x_{\rm f} \approx 0.5\,\sigma$. To compensate for this difference, a custom laser optics module has been designed to shift alignment of the laser power distribution relative to the alignment of the powder distribution independently from other process parameter changes. The lateral shift (x_0) introduced by of this module can effectively increase or decrease the apparent isotherm leading length such that $x_{\rm f,eff} = x_{\rm f} + x_0$. Figure 4 illustrates key aspects of the design and operation of the shift module. Considering typical values for σ, optimized laser cladding conditions of $x_{\rm f} + x_0 = y_{\rm m}$ requires a forward shift on the order of $x_0 = 1\,\rm mm$.

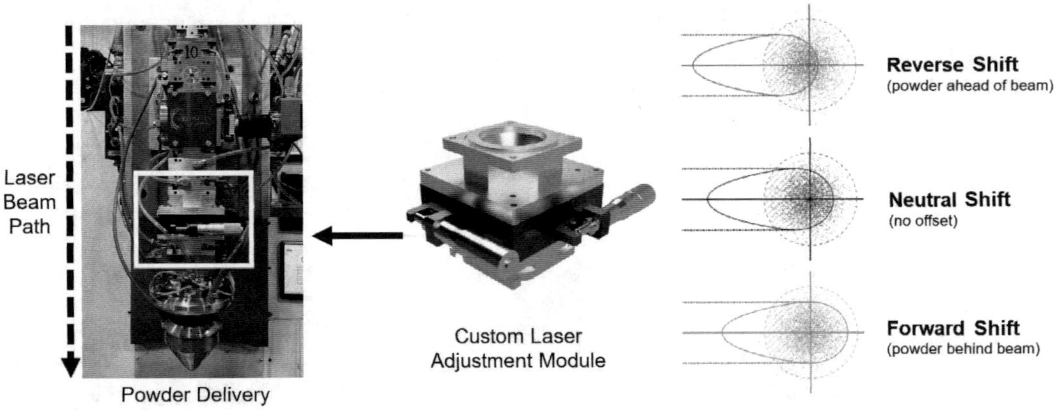

Figure 4: Install position and operating principles of the custom laser adjustment module that enables a precise relative shift between the powder cloud distribution and the molten pool.

Selection of the operating point at $\sigma/\sigma_{\rm cr} = 0.6$ is desirable to maximize productivity (i.e., $y_{\rm m}\,U$) when laser power is limited. Reduced shift values may be attained with an operating point that coincides with the local maximum for $x_{\rm f}$, which from Fig. 3b occurs at $\sigma/\sigma_{\rm cr} \approx 0.45$. The shift module can theoretically be used to achieve optimized catchment outside the typical operating range for laser cladding (e.g., with $\sigma/\sigma_{\rm cr} < 0.3$). This may be advantageous in some cases to overcome equipment limitations, however operating in this regime can introduce additional concerns beyond catchment. For example, the risk of metallurgical defects may be increased due to the inverse relationship between (σ/σ_{cr}) and the normalized peak temperature rise $(T_{\rm p} - T_0)/(T_{\rm m} - T_0)$ [4].

3.3. Selection of Operating Conditions

The dimensionless guidelines for optimized catchment may be translated into a straightforward procedure for selection of optimal dimensional operating values for q, U, σ, and $\sigma_{\rm p}$. Depending on the configuration of the laser system, the exact values chosen to achieve maximum power and travel speed may differ. In practice, the operating point is often the result of: (a) laser power limitations, (b) motion control system limitations, (c) process stability limitations (e.g., solidification cracking), (d) powder delivery limitations (higher speed necessitates larger powder flow to achieve a clad bead area).

Optimum melt pool dimensions $(y_{\rm m}, x_{\rm f})$ are related to the laser and powder distribution parameters. Implementation of the shift module permits the optimization to initially focus on

y_m and an appropriate shift can later be chosen to achieve an effective optimized value for $x_{f,eff}$. The target melt pool width must satisfy both the condition for optimized mass transfer (i.e., catchment) $y_m = (2.3)\,\sigma_p$, in addition to the condition for optimum heat transfer (i.e., productive utilization of the laser energy) $y_m = (1.3)\,\sigma$. More strict limitations on σ_p are often present and it is convenient to specify the target melt pool width relative to the measured powder distribution parameter (σ_p). The required laser distribution $\sigma/\sigma_p = (2.3)/(1.3) = (1.8)$ may then be verified to fall within the limitations of the equipment.

The range of suitable values for the laser power and travel speed to achieve the target melt pool width may be determined using the blended expression for σ_{cr} given in Ref. [4]. For the case of fast heat sources it is sufficient to use the asymptotic solution in Eq. 6 which yields:

$$y_m = (0.75)\,\sigma_{cr} = 0.28 \left(\frac{\alpha}{k^2(\Delta T_m)^2}\right)^{1/3} \left(\frac{q^2}{U}\right)^{1/3} \tag{7}$$

The relationship between power and travel speed is such that any number of operating points may now be selected for a given target melt pool width and material system. To optimize productivity, it is desirable to choose the operating point which results in the maximum power and travel speed. For fast heat sources this corresponds to the smallest possible melt pool width since, from Eq. 7, an incremental increase in speed produces a correspondingly incremental decrease in melt pool width and an overall net increase in area coverage (i.e., productivity).

4. Case Study

The utility of engineering design rules for process optimization is demonstrated with a case study for laser cladding of Ni-WC hardfacing material on a 4140 alloy steel substrate. Single bead cladding trials were performed to compare a set of baseline conditions to optimized cladding parameters obtained according to the procedure outlined in Section 3. The relevant process inputs and key dimensionless values for each case are given in Table 1.

Table 1: Process inputs and dimensionless values for baseline and optimized laser cladding of Ni-WC hardfacing material on 4140 alloy steel.

	σ_p [mm]	σ [mm]	y_m/σ_p	y_m/σ	Ry	σ/σ_{cr}	Shift [mm]
Baseline	1.4	1.1	1.4	1.8	18	0.4	–
Optimized	1.1	1.9	2.3	1.3	23	0.6	1

The optimized conditions used an improved powder nozzle with a reduced σ_p, selected by direct in-situ powder cloud measurement. The laser beam distribution and laser power were both adjusted to facilitate an increased melt pool width. Using Eq. 3–5, the catchment efficiency for the baseline operating point and optimized conditions were calculated to be $\eta = 67\%$ and $\eta = 93\%$ respectively. The resulting metallurgical cross sections for the two single bead trials are shown in Fig. 5.

The improved catchment results in a larger bead height with a substantial increase in the area of the clad deposit. Measured catchment efficiency for the optimized parameters was $\eta = 94\%$, representing an increase of nearly 20% compared to the baseline cladding conditions and almost perfectly matches the predicted value. The measured catchment efficiency for the baseline case was slightly higher than predicted ($\eta = 75\%$) but within the expected variation for industrial process equipment.

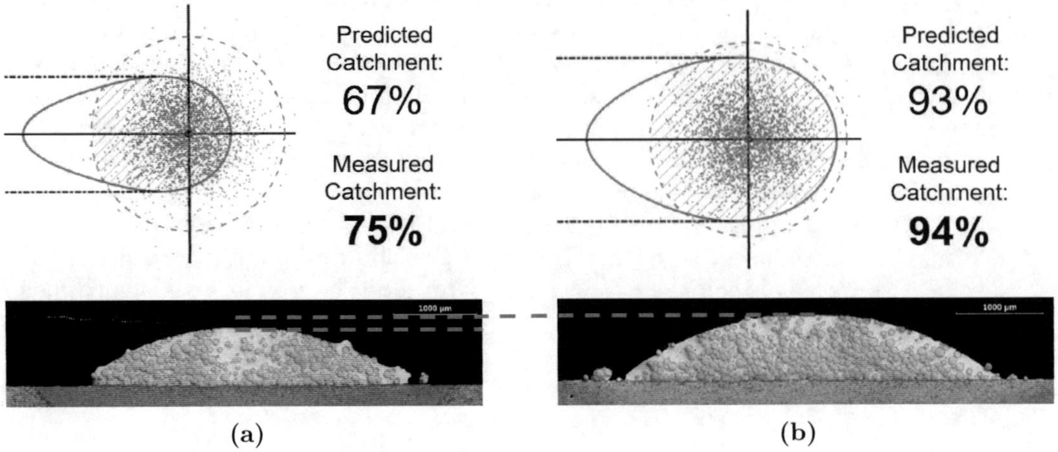

Figure 5: Validation of catchment predictions with single bead comparisons for (a) baseline operating parameters; and (b) optimized parameters (i.e., smallest possible σ_p, $y_\mathrm{m}/\sigma = 1.3$, $y_\mathrm{m}/\sigma_\mathrm{p} = 2.3$, and shift of 1mm).

The optimized parameters were further validated in an 8-hr production run and continuously produced catchment efficiency values of 92–93%. This represents a substantial improvement in both productivity and cost savings. A key benefit of this universal analytical solution is the potential for standardized application across industrial laser systems with different operational limits. Following the success of this case study, identification and implementation of optimized parameters for production laser cladding operations is an area of intense focus for future work.

5. Conclusions

This work provides practical engineering tools for the optimization of powder catchment efficiency in industrial laser cladding and additive manufacturing processes:

(i) A simple, general, and accurate model was established to relate the catchment efficiency with the overlapping geometry of the powder distribution and molten pool.

(ii) Dimensionless analysis and asymptotic blending were used to obtain closed-form expressions which relate the critical molten pool width (y_m), and leading length (x_f), with known material properties and process inputs.

(iii) A new and industrially robust technique was introduced for measurement of the powder cloud distribution parameter (σ_p).

(iv) A universally applicable step-by-step procedure was presented for simultaneously achieving high productivity and catchment efficiency by minimizing the powder distribution (σ_p), matching the melt pool width to the powder cloud size ($y_\mathrm{m} = 2.3\,\sigma_\mathrm{p}$), identifying parameters for optimized cladding conditions ($\sigma = y_\mathrm{m}/1.3$ and $\sigma_\mathrm{cr} = \sigma/0.6$), and shifting the alignment of laser and powder distributions by a value of $(2.3\,\sigma_\mathrm{p} - x_\mathrm{f})$.

(v) This methodology was demonstrated with an industrial laser cladding case study that resulted in a nearly 20% improvement over baseline conditions and a maximum catchment efficiency of 94%.

Acknowledgments

This study was partially supported by Mitacs under an Accelerate post-doctoral research grant No. IT25441. Thank you to A. Varghese and M. Salim for assistance with industrial implementation of the continuous powder cloud measurement technique. The authors also gratefully acknowledge Y. Lu, Y. Wang, and P.F. Mendez of the Canadian Centre for Welding at the University of Alberta, for their foundational work on scaling of welding heat sources and its contribution to the realization of this current work.

References

[1] Toyserkani E, Khajepour A and Corbin S F 2004 *Laser Cladding* 1st ed (Boca Raton: CRC Press)
[2] Balu P, Leggett P and Kovacevic R 2012 *Journal of Materials Processing Technology* **212** 1598–1610
[3] Wood G and Mendez P F 2016 *Mathematical Modelling of Weld Phenomena 11* 701–734
[4] Wang Y 2021 *Design Rules for Characteristics of Heat Flow in Welding on Thick Plates using Asymptotics and Blending* Ph.D. thesis Univ. of Alberta (https://doi.org/10.7939/r3-0ep4-gw27)
[5] Fuerschbach P W and Eisler G R 2002 *Proc. 6th Int. Trends Weld. Res.* (Pine Mountain, GA, Apr. 15-19) pp 782–786
[6] Eager T W and Tsai N S 1983 *Welding J.* **62** 346–355
[7] Lu Y 2021 *A Systematic Methodology to Develop Scaling Laws for Thermal Features of Temperature Field Induced by a Moving Heat Source* Ph.D. thesis Univ. of Alberta (https://doi.org/10.7939/r3-12bk-va06)
[8] Wang Y, Lu Y, Grams M R, Hintze Cesaro A and Mendez P F 2019 *Mathematical Modelling of Weld Phenomena 12* 907–932
[9] Mendez P F 2020 *IOP Conf. Ser.: Mater. Sci. Eng.* **861** 012003
[10] Churchill S W and Usagi R 1972 *AIChE Journal* **18** 1121–1128
[11] The MathWorks, Inc *MATLAB: Multivariate normal cumulative distribution function* (ver. 9.13, R2022b) Natick, MA URL `https://www.mathworks.com/help/stats/mvncdf.html`
[12] PRIMES GmbH 2021 *FocusMonitor FM+* URL `https://www.primes.de/en/products/beam-distribution/focus-measurement/focusmonitor-fmplus.html`
[13] ISO/TR 11146-1:2021 Lasers and laser-related equipment — Test methods for laser beam widths, divergence angles and beam propagation ratios — Part 1: Stigmatic and simple astigmatic beams
[14] Eisenbarth D, P E, Wirth F and Wegener K 2019 *Surf. Coat. Technol.* **362** 397—-408
[15] Wang Y and Mendez P F 2022 *Int. J. Therm. Sci.* **172** 107334

Appendix A. 2-D Blended Asymptotics for Isotherm Geometry

To simplify notation, the normalized laser power distribution parameter is defined as $s = \sigma/\sigma_{\mathrm{cr}}$. The asymptotic expression for maximum isotherm width of slow (i.e., small Ry) heat sources is:

$$\frac{y_{\mathrm{m,\,Ry.0}}(\mathrm{Ry}, s)}{\sigma_{\mathrm{cr}}} = (1 + 0.63\,\mathrm{Ry})^{-\frac{1}{6}} \left\{ \left[\sqrt{2/\pi}\,\left(1 - s^2\right) \right]^3 + \left[2\,s\sqrt{-\ln(s)} \right]^3 \right\}^{\frac{1}{3}} \tag{A.1}$$

The asymptotic formula for fast (i.e., large Ry) heat sources is:

$$\frac{y_{\mathrm{m,\,Ry.Inf}}(s)}{\sigma_{\mathrm{cr}}} = s\sqrt{-3\ln(s)} \tag{A.2}$$

The general blended, asymptotic expression for all values of Ry and s is:

$$y_{\mathrm{m}}^{+}(\mathrm{Ry}, s) = \left\{ \left[y_{\mathrm{m,\,Ry.0}}(\mathrm{Ry}, s) \right]^{7.45} + \left[y_{\mathrm{m,\,Ry.Inf}}(s) \right]^{7.45} \right\}^{1/7.45} \tag{A.3}$$

The asymptotic expression for leading isotherm length of slow heat sources is:

$$\frac{x_{\mathrm{f,\,Ry.0}}(\mathrm{Ry}, s)}{\sigma_{\mathrm{cr}}} = \left\{ \left[\frac{\ln(\mathrm{Ry} + \mathrm{e})}{(2.8880\,\mathrm{Ry} + 1.4031)^{2/3}} \left(1 - s^2\right) \right]^3 + \left[2\,s\sqrt{-\ln(s)} \right]^3 \right\}^{\frac{1}{3}} \tag{A.4}$$

The approximate asymptotic expression for fast heat sources is:

$$\frac{x_{\mathrm{f,\,Ry.Inf}}(s)}{\sigma_{\mathrm{cr}}} = \left(1.15 + \frac{0.15\,s}{1 + 0.15\,s}\right) s\sqrt{-3\ln(s)} - 0.7650\,s \tag{A.5}$$

The general blended, asymptotic expression for all values of Ry and s is:

$$x_{\mathrm{f}}^{+}(\mathrm{Ry}, s) = \left\{ \left[x_{\mathrm{f,\,Ry.0}}(\mathrm{Ry}, s) - x_{\mathrm{f,\,Ry.Inf}}(s) \right]^{-0.65} + \left[s\,\mathrm{Ry} \right]^{0.65} \right\}^{-1/0.65} + x_{\mathrm{f,\,Ry.Inf}}(s) \tag{A.6}$$

IOP Conf. Series: Materials Science and Engineering 1281 (2023) 012007 doi:10.1088/1757-899X/1281/1/012007

General expressions for heat distribution, pressure, and current density in argon welding arcs

R. Villarreal-Medina[1], A. Delgado-Álvarez[1], A. B. Murphy[2], P. F. Méndez[3], M. A. Ramirez-Árgaez[1]*

1 Metallurgical Engineering Department, School of Chemistry. UNAM. Mexico City, Mexico

2 CSIRO Manufacturing, PO Box 218, Lindfield NSW 2070, Australia

3 Department of Chemical and Materials Engineering, University of Alberta, Edmonton, Canada

*marco.ramirez@unam.mx

Abstract. In welding and wire-arc additive manufacturing (WAAM), a mobile arc is the heat source that enables the deposition of metals and the resulting properties of the final product. Because the arc involves temperatures of 20 000 K, and gas velocities of the order of 300 m/s, there are only a few experiments and models available to determine optimal, or at least acceptable, parameters for the operation such as current, voltage, and arc length. On the other hand, there is a lack of engineering guidance to optimize the processes resulting in costly and time-consuming trial-and-error optimization methods, which also involve wasteful use of energy and scrap parts. In this work, a numerical model of the gas-tungsten arc welding (GTAW) arc was created and validated against experiments. The model considers the arc interactions between a non-consumable electrode and the weld pool and accounts for multiple coupled heat transfer mechanisms: Joule heating, conduction, advection, radiation, and Thomson effect. The conditions considered cover the vast majority of GTAW welding operations. The results are generalized in the form of engineering expressions suitable to be embedded in metamodels, in which the heat source is just a part. Applications include penetration and width of welds and deposition rate in external-wire WAAM.

1. Introduction

Arc welding is a typical process in the manufacturing industry. Recently, wire-arc additive manufacturing made it possible to build complex shapes without machining. These two processes use a high-intensity direct current electric arc to provide enough heat to melt the filler metal in WAAM or the workpiece in GTAW welding.

The harsh conditions inside the arc column make it difficult to take direct measures of temperature, velocity, and pressure in the arc. Although there are measurements through spectroscopic techniques, thin probes, and electric potential to obtain the main parameters of welding arcs, these measurements are only a few [1]. For this reason, mathematical models are valuable tools for understanding the arcs in welding or WAAM processes.

Two approaches can be used to model the arc. The first is to consider it as a heat source into a substrate. This approach allows for defining the extent of the heat-affected zone; therefore, it can be used to predict the penetration depth of the arc, among other essential features. There are different ways to consider the heat source, and the most popular is the so-called double ellipsoidal model [2]. The second approach to modeling the arc column is through the numerical solution of the first principles governing arc physics, which is a more realistic way to understand the arc. When it is adopted the assumption of Local Thermal Equilibrium, the arc model involves the Magneto Hydro Dynamics (MHD) equations [3], [4]. There are different deterministic mathematical models of the arc reported in the literature. These were used with different goals: some assess the role of the shielding gas composition [5], [6], the impact that metal vapor plays on the arc characteristics[7], [8], and some models quantify the effect of the arc current and length on the arc-substrate interactions or determine the impact of the tip angle of the cathode and filler metal [9]–[11], among other studies. The complexity of the reported models also show marked differences; some models include multiphase modeling [12], other solve the non-equilibrium nature of the species present in the arc [13], and a few simulate transient arcs [14]. The set of equations solved in these models is a highly nonlinear and coupled system of partial differential equations. Therefore, these models represent a challenge to solve, and these tools are not always available for welding or WAAM practitioners.

Then, the asymptotic solutions with simple expressions can be used to take advantage of the information provided by the models and provide easy-to-use tools for welding and WAAM practitioners. These tools are provided by the minimal representation technique (MRCF) of the system of partial differential equations. A general procedure to obtain algebraic equations from the principal parameters has been applied in welding processes by Mendez et al. [15]–[17]. This procedure yielded simple equations describing the maximum arc pressure and arc velocity near the cathode.

Ramírez-Argáez et al. [18] developed an Electric Arc Furnace (EAF) dimensionless representation in a steelmaking process with simple equations. The physics of both arcs (welding and EAF) are almost the same, but the flow regime of the arcs in EAF is turbulent, while in welding, the flow regime is laminar. Thus, the algebraic equations of arcs obtained in EAF are not immediately applicable to welding arcs.

This work shows the existence of a universal and dimensionless shape for a GTAW welding arc. Under this hypothesis, dimensionless algebraic equations that provide the main arc characteristics and arc-weld pool interactions were obtained. These dimensionless equations can be used to model the arc as a heat source that can be applied to predict arc behavior quickly and accurately during welding and additive manufacturing processes with a hand calculator without using a fancy MHD numerical model.

2. Methodology

2.1. Arc model

In previous work, a numerical model of an atmospheric pressure argon arc with a non-consumable DC-negative electrode was developed, solved numerically through the CFD code PHOENICS, and validated against experimental results and other numerical models by comparing the main arc characteristics and arc-weld pool interactions extensively [19], [20]. The main assumptions to model the arc column include considering Local Thermodynamic Equilibrium (LTE), axisymmetry, and optically transparent plasma in a steady state.

Under the assumptions mentioned, the governing equations of arc physics are: Mass continuity, the axial and radial components of the Navier Stokes, and the energy conservation equations. Also, the electric potential conservation equation is solved to obtain the electromagnetic variables. This procedure is called the "potential approach" [21]. Next, Ohm's law is used to calculate the current density's axial and radial components, and Ampere's law allows for obtaining the azimuthal magnetic field. The coupling of the electromagnetic parameters and the mass, momentum, and heat conservation equations is through Lorentz's forces, the Joule heating, and the Thomson effect inside the conducting fluid. The model uses temperature-dependent thermophysical and transport properties reported by Murphy and Arundell [22], making converging the numerical problem difficult.

Boundary conditions are shown in Figure 1. The cathode and anode electrodes are considered non-deformable walls with no-slip condition. Current only flows through a cathode spot area of radius R_c, that can be obtained by using Equation (1).

$$R_c = \left(\frac{I}{\pi J_c}\right)^{0.5} \tag{1}$$

The current density flowing through this area is assumed to be constant with a value of $J_c=6.5\times10^7 \frac{A}{m^2}$. This value is consistent with other works [23] and corresponds to the current density for thermionic emission of a thoriated tungsten electrode with a work function of 3.5 V [24].

The commercial Computational Fluid Dynamics (CFD) code, Phoenics version 2017, was used to solve the governing equations numerically. Fortran subroutines were written to solve the electromagnetic problem to obtain the electromagnetic variables, like Lorentz's forces, Joule and Thomson heat sources, and the heat transferred from the arc to the weld pool. After a grid sensitivity study, a 60x60 nonuniform mesh was used to discretize the simulation domain. The model was used to simulate arcs burning in argon at three currents of 200 A, 250 A and 300 A, and three arc lengths of 5 mm, 7 mm and 10 mm. Each simulation took around 15,000 iterations and about an hour of CPU time to obtain the converged solution.

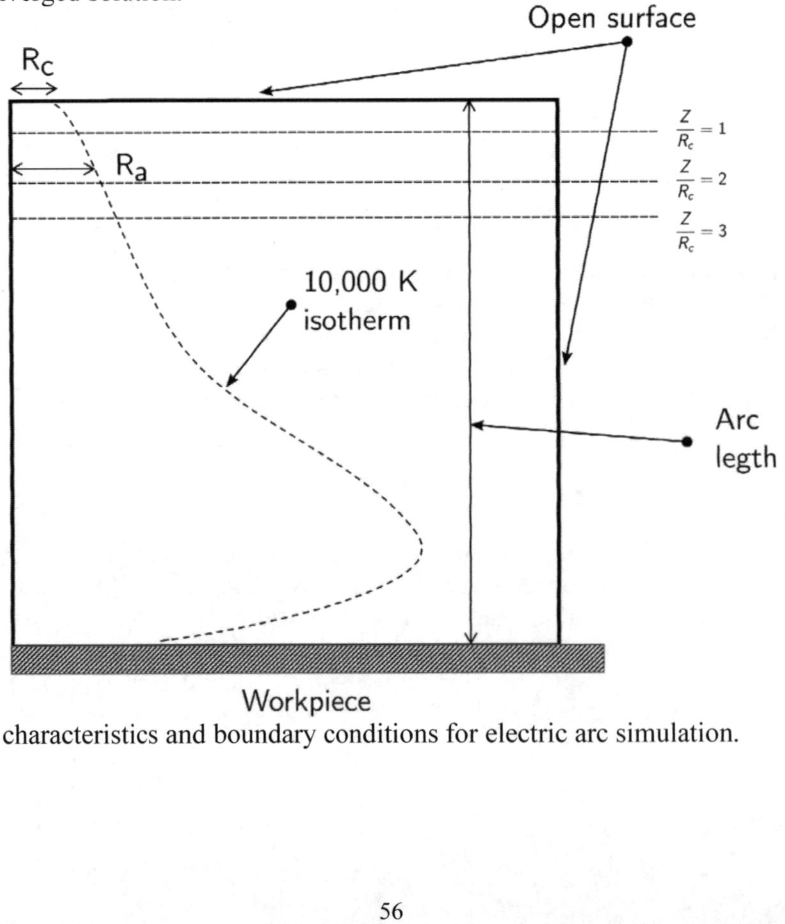

Figure 1. Main characteristics and boundary conditions for electric arc simulation.

2.2. *Universal representations*

Contour maps of temperature, velocity, and magnetic field were plotted for each arc simulated. Also, radial profiles of those fields were plotted at selected axial positions for an arc of 10 mm and 200 A (see Figure 1). The horizontal lines are drawn at dimensionless axial positions ($\frac{Z}{R_c}$), far from the impingement of the arc jet on the anode.

The arc radius (R_a) was defined by the loci of the 10,000 K isotherm (valid for arcs burning in an argon atmosphere). This value was selected because the electrical conductivity of the plasma is high enough to keep the current flowing between the electrodes.

3. Results

3.1. *Arc Shape*

Figure 2 shows the loci of the 10,000 K isotherm of all arcs simulated (nine simulations ranging from 5 mm and 200 A to 10 mm and 300 A) against the dimensionless axial and radial positions, defined by the division of each coordinate over the cathode spot radius (R_c). The region near the anode, where the isotherms take the bell-like shape, is not included in the plot. When the arc is graphed using these variables, the arc boundary coincides for all arcs regardless of arc length and current, as shown in Figure 2. Therefore, the arc radius in dimensionless form is the same; this means that the shape of the arc column is universal.

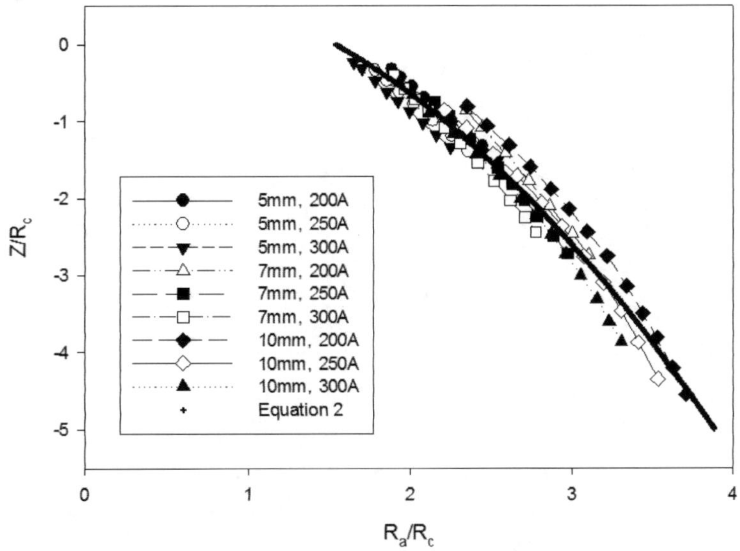

Figure 2. Dimensionless axial distance ($\frac{Z}{R_c}$) as a function of dimensionless arc radius ($\frac{R_a}{R_c}$) for all arc lengths and currents considered.

Equation (2) expresses the universal arc shape. It represents the arc radius increase proportionally to the axial distance's square root up to an offset value. The current spread over a more extensive area while the axial length increases from the cathode, so the current density decreases.

$$\frac{R_a}{R_c} = \left(a - b\frac{z}{R_c}\right)^{0.5} \quad R^2 = 0.959 \tag{2}$$

The coefficient $z = 0.9310R_c$ in Equation (2) is an offset value that places the vertex of the parabola formed by the universal shape formula inside the electrode at a depth almost equal to the radius of the cathode area.

3.2. Arc column characteristics

Having shown that the arc shape is unique, the arc column characteristics can be determined as a single universal profile. Universal profiles are obtained by plotting the velocity, magnetic field, and temperature along the dimensionless radius $(\frac{R}{R_a})$ at specific axial positions. Dimensionless characteristics are obtained by normalizing the radial profile with the maximum characteristic value at that axial position. The results for dimensionless radial profiles are shown in Figure 3. The shape of the radial velocity profile and temperature (Figure 3 a) and b), respectively) shows a Gaussian shape typical of jet flow behavior.

Meanwhile, the magnetic field shows a linearly increasing profile until it reaches a maximum at the boundary of the arc. After that, the magnetic field decreases. This behavior is consistent with Ampere's law.

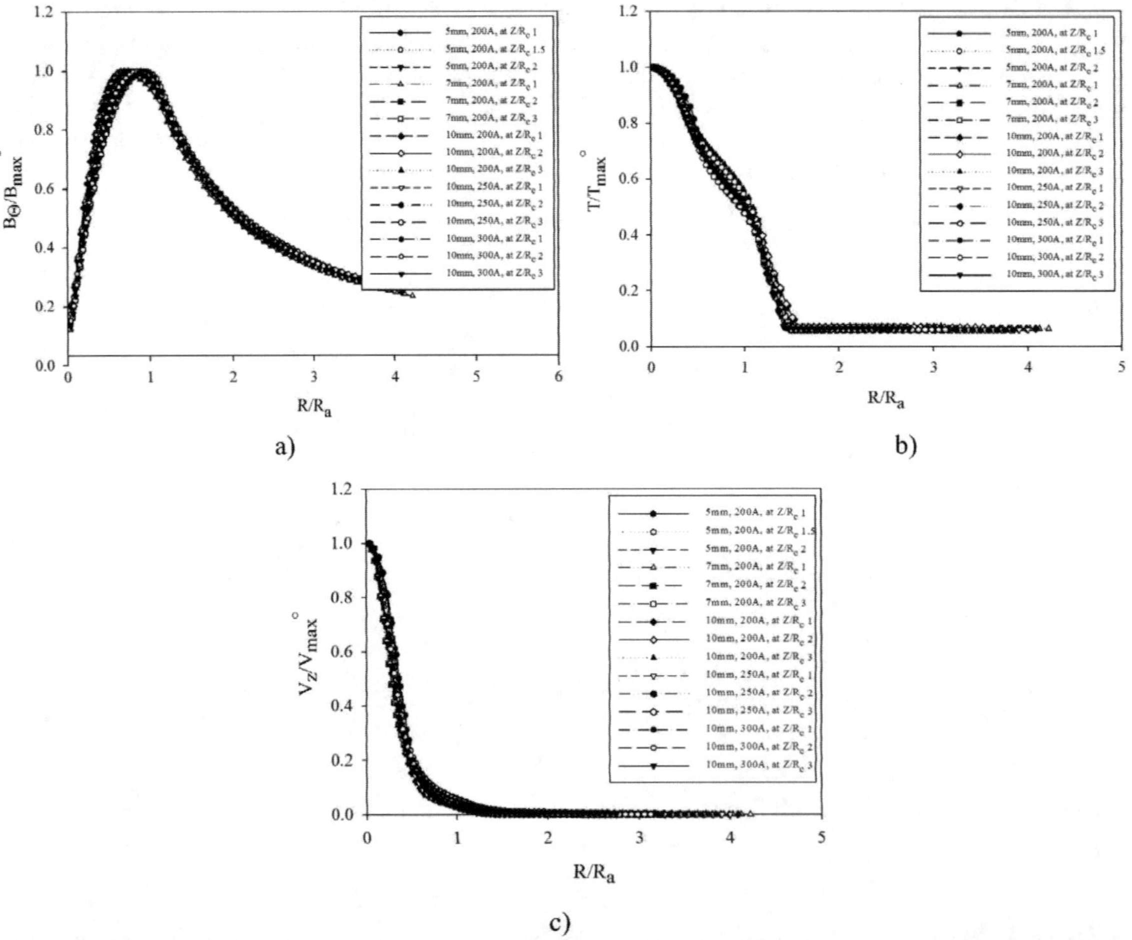

Figure 3. Dimensionless representations of arc characteristics as a function of dimensionless radius at various dimensionless axial positions, a) magnetic flux density, b) temperature, c) axial velocity

The radial profiles for these characteristics can be expressed by Equations (3)-(5), where R is the radial position. $\boldsymbol{B_\theta}$, T and $\boldsymbol{V_z}$ are the azimuthal magnetic field, temperature, and axial velocity, respectively. The correlations obtained need the maximum values of these quantities at specific axial locations ($\boldsymbol{B^\circ_{max}}$, $\boldsymbol{T^\circ_{max}}$, $\boldsymbol{V^\circ_{max}}$) that are calculated from the maximum values of the same variable in the whole domain ($\boldsymbol{B_{max}}$, $\boldsymbol{T_{max}}$, $\boldsymbol{V_{max}}$). It is worth mentioning that the maximum value of the magnetic field is at the arc hinge (the hinge of the arc is also the 10 000 K isotherm), while maximum temperature and velocity values are located at the symmetry axis.

$$\frac{B_\Theta}{B_{max}{}^{\circ}} = a + b\left(\frac{R}{R_a}\right) + c\left(\frac{R}{R_a}\right)^2 + d\left(\frac{R}{R_a}\right)^3 + e\left(\frac{R}{R_a}\right)^4 + f\left(\frac{R}{R_a}\right)^5 + g\left(\frac{R}{R_a}\right)^6 + h\left(\frac{R}{R_a}\right)^7 + i\left(\frac{R}{R_a}\right)^8 \quad (3)$$

$$\frac{T}{T_{max}{}^{\circ}} = a + b\left(\frac{R}{R_a}\right)^2 + c\left(\frac{R}{R_a}\right)^4 + d\left(\frac{R}{R_a}\right)^6 + e\left(\frac{R}{R_a}\right)^8 + f\left(\frac{R}{R_a}\right)^{10} + g\left(\frac{R}{R_a}\right)^{12} + h\left(\frac{R}{R_a}\right)^{14} + i\left(\frac{R}{R_a}\right)^{16} + j\left(\frac{R}{R_a}\right)^{18} \quad (4)$$

$$\frac{V_z}{V_{max}{}^{\circ}} = \frac{1}{\left(a + b\left(\frac{R}{R_a}\right) + c\left(\frac{R}{R_a}\right)^2 + d\left(\frac{R}{R_a}\right)^3\right)} \quad (5)$$

All arc characteristics can be easily obtained from the algebraic equations presented earlier. Table 1 shows mathematical expressions to get the maximum values of each field in the whole domain and at specific axial locations.

Table 1. Estimation of global and axial (marked with °) maximum values for arc characteristics.

Physical quantity	Dimensionless correlation	Eq.
$B_{max}{}^{\circ}$, at each Z.	$B_{max}{}^{\circ} = a + (b*I) + (c*L) + (d*B_{max}) + (e*Z)$	(6)
$T_{max}{}^{\circ}$, at each Z	$T_{max}{}^{\circ} = \left[a + b\left(\frac{Z}{(R_cL)^{0.5}}\right)^{0.5} - c\left(\frac{Z}{(R_cL)^{0.5}}\right) + d\left(\frac{Z}{(R_cL)^{0.5}}\right)^{1.5} - e\left(\frac{Z}{(R_cL)^{0.5}}\right)^2 + f\left(\frac{Z}{(R_cL)^{0.5}}\right)^{2.5} - g\left(\frac{Z}{(R_cL)^{0.5}}\right)^3\right] * T_{max}$	(7)
	$T_{max}{}^{\circ} = \left[a + bln\left(\frac{Z}{L}\right) + \frac{c}{ln\left(\frac{Z}{L}\right)} + d\left(ln\left(\frac{Z}{L}\right)\right)^2 + \frac{e}{\left(ln\left(\frac{Z}{L}\right)\right)^2} + f\left(ln\left(\frac{Z}{L}\right)\right)^3 + \frac{g}{\left(ln\left(\frac{Z}{L}\right)\right)^3} + h\left(ln\left(\frac{Z}{L}\right)\right)^4\right] * \left(\frac{R_c{}^{0.5} * I^{0.1}}{L^{0.2}}\right)$	(8)
$V_{max}{}^{\circ}$, at each Z.	$V_{max}{}^{\circ} = \left[\frac{\left(-a+b\left(\frac{Z}{(R_cL)^{0.1}}\right)^2\right)}{\left(1+c\left(\frac{Z}{(R_cL)^{0.1}}\right)^2+d\left(\frac{Z}{(R_cL)^{0.1}}\right)^4\right)}\right] * V_{max}$	(9)
	$V_{max}{}^{\circ} = \left[a\text{-}b\left(\frac{Z}{L}\right) - \frac{c}{\left(\frac{Z}{L}\right)} + d\left(\frac{Z}{L}\right)^2 + \frac{e}{\left(\frac{Z}{L}\right)^2} - f\left(\frac{Z}{L}\right)^3\right] * \left(\frac{R_c{}^{1.4} * I^{0.2}}{L^{0.1}}\right)$	(10)
B_{max}	$B_{max} = a + (b*I) + (c*L)$	(11)
T_{max}	$T_{max} = a + (b*I) + (c*L)$	(12)
V_{max}	$V_{max} = a + (b*I) + (c*L)$	(13)

3.3. *Arc-weld pool interactions*

Since the weld pool interaction with the arc highly influences the final quality of the welding joint and additive manufacturing processes, obtaining approximate values of those interactions is crucial. The interactions analyzed here are arc pressure (P), heat flux (q), current density (J), and shear stress (τ). Figure 4 shows the normalized interactions against a modified dimensionless radial position $\frac{R}{(R_aL)^{0.5}}$.

This expression was obtained using Equation (2) and Z=L to obtain the arc radius and length. The addition of these parameters in the dimensionless radial coordinate is explained by the dependence of the interaction value on the arc length. An increase in arc length is followed by a smaller interaction value but a more extensive propagation along the workpiece. In contrast, a decrease in arc length is followed by a larger interaction value but a smaller radial spread. Figure 4 also shows the result of applying the same methodology to other numerical and experimental results obtained by other authors. [6], [23], [25]–[27]. The dimensionless plot for all independent results and our correlations shows excellent agreement between them.

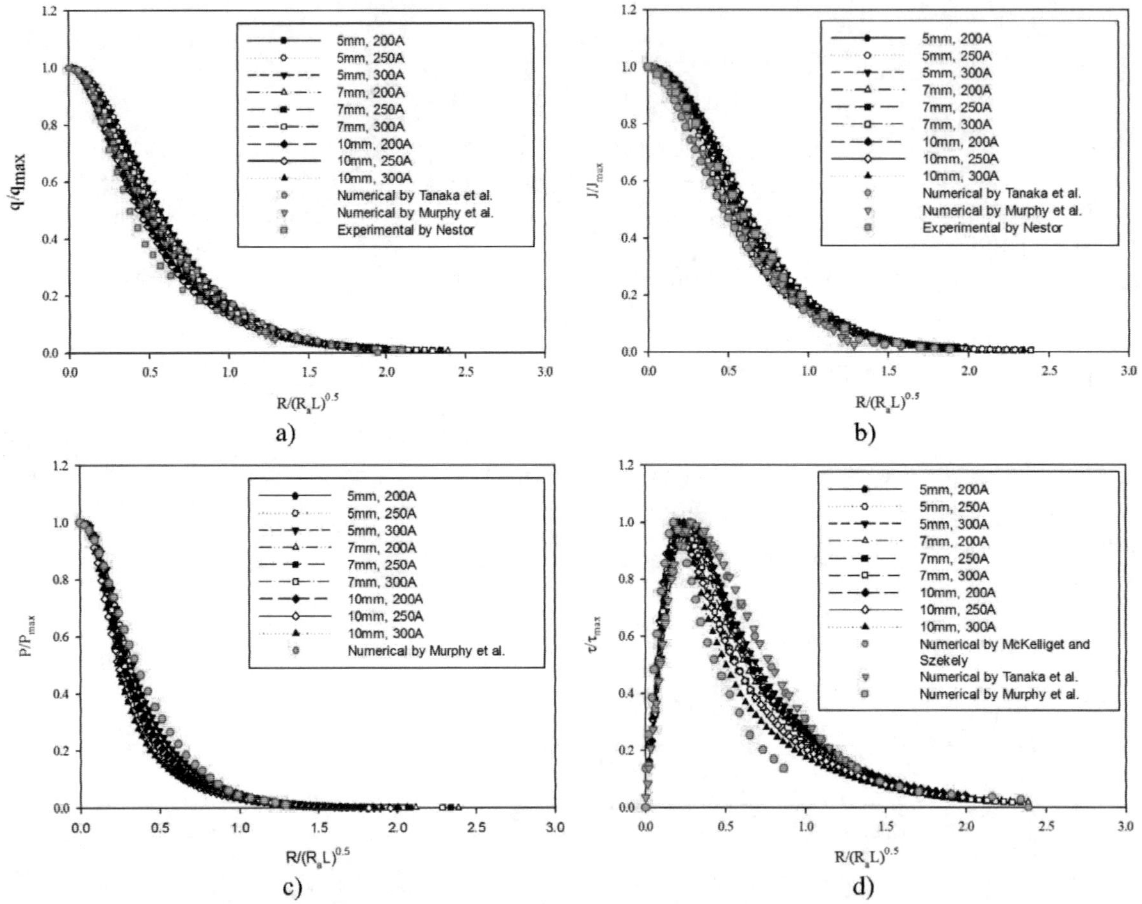

Figure 4. Dimensionless radial profiles of arc-bath interactions, a) heat flux, b) current density, c) current density, d) shear stress.

As expected, the profiles of dimensionless pressure, heat flux, and current density exhibit a Gaussian shape due to the nature of the charge distribution at the anode and the influence of the electronic and convective heat transfer mechanisms. The shear stress at the anode shows a more complex behavior, which can be explained by the boundary layer formed when the plasma jet impinges on the workpiece. While the arc pressure shows a maximum, the shear stress at the stagnation point is zero. When the plasma propagates in a radial direction, the velocity and the shear stress reach a maximum. Both interactions decrease when the plasma loses momentum in the radial coordinate.

Because pressure ($\frac{P}{P_{max}}$), heat flux ($\frac{q}{q_{max}}$) and current density ($\frac{J}{J_{max}}$) dimensionless profiles have a common shape, all of them can be expressed by Equation (14), where $\frac{Interaction}{Interaction_{max}}$ can be replaced by the desired interaction. Dimensionless shear profile ($\frac{\tau}{\tau_{max}}$) can be obtained by using Equation (15).

$$\frac{\text{Interaction}}{\text{Interaction}_{\text{max}}}=\frac{1}{\left(a+b\left(\frac{R}{(R_aL)^{0.5}}\right)+c\left(\frac{R}{(R_aL)^{0.5}}\right)^2+d\left(\frac{R}{(R_aL)^{0.5}}\right)^3+e\left(\frac{R}{(R_aL)^{0.5}}\right)^4+f\left(\frac{R}{(R_aL)^{0.5}}\right)^5\right)} \quad (14)$$

$$\frac{\tau}{\tau_{\text{max}}}=\frac{\left(a+b\ln\left(\frac{R}{(R_aL)^{0.5}}\right)+c\left(\ln\left(\frac{R}{(R_aL)^{0.5}}\right)\right)^2\right)}{\left(1+d\ln\left(\frac{R}{(R_aL)^{0.5}}\right)+e\left(\ln\left(\frac{R}{(R_aL)^{0.5}}\right)\right)^2\right)} \quad (15)$$

The maximum values of each interaction are needed to use Equations (14) and (15). Therefore, the maximum value of each interaction was correlated with the most common operating parameters in welding: Arc length (L) and current (I), as shown in Equation (16).

$$\text{Interaction}_{\text{max}} = a + bL + cI \quad (16)$$

The coefficients for Equation (16) and all other dimensionless equations can be obtained from Delgado-Alvarez et al. [20]. These equations can provide welding practitioners with a valuable set of relationships to help design and improve welding operations.

Conclusions

Relevant characteristic values of an axisymmetric GTAW argon arc can be accurately predicted using algebraic expressions. The algebraic expressions were obtained by numerical simulation and dimensionless scaling. The arc shape, temperature, axial velocity, magnetic field, and pressure fields in the arc show universal behavior as a function of arc length and input current. In addition, the results contain universal expressions for the interactions of the arc with the workpiece, such as heat density, current density, and shear stress. The obtained expressions were validated by comparing their predictions with independent experimental and numerical results obtained by other authors.

Further work addresses the effects of the electrode tip and the metal vapor released from the weld pool on these relationships. The goal is to provide welding professionals with easy-to-use algebraic expressions for planning and designing welding operations.

Acknowledgments

Rodrigo Villarreal Medina is a Ph.D. student in the "Programa de Doctorado en Ingeniería Química" at the Universidad Nacional Autónoma de México (UNAM) and is receiving fellowship CVU 1002868 from CONACYT.

References

[1] D. Uhrlandt, "Diagnostics of metal inert gas and metal active gas welding processes," *J Phys D Appl Phys*, vol. 49, no. 31, 2016, doi: 10.1088/0022-3727/49/31/313001.

[2] A. Das, A. Kumar, K. Shankhwar, and N. Gubeljak, "A review of heat source and resulting temperature distribution in arc welding," *J Therm Anal Calorim*, Nov. 2022, doi: 10.1007/s10973-022-11589-w.

[3] A. B. Murphy, “A Perspective on Arc Welding Research: The Importance of the Arc, Unresolved Questions and Future Directions,” *Plasma Chemistry and Plasma Processing*, vol. 35, no. 3, pp. 471–489, 2015, doi: 10.1007/s11090-015-9620-2.
[4] A. B. Murphy and J. J. Lowke, “Heat transfer in Arc Welding,” in *Handbook of Thermal Science and Engineering*, 2007, pp. 2657–2727.
[5] B. Silwal, N. Pudasaini, S. Roy, A. B. Murphy, A. Nycz, and M. W. Noakes, “Altering the Supply of Shielding Gases to Fabricate Distinct Geometry in GMA Additive Manufacturing,” *Applied Sciences (Switzerland)*, vol. 12, no. 7, Apr. 2022, doi: 10.3390/app12073679.
[6] A. B. Murphy, M. Tanaka, S. Tashiro, T. Sato, and J. J. Lowke, “A computational investigation of the effectiveness of different shielding gas mixtures for arc welding,” *J Phys D Appl Phys*, vol. 42, no. 11, 2009, doi: 10.1088/0022-3727/42/11/115205.
[7] M. Schnick, U. Fuessel, M. Hertel, M. Haessler, A. Spille-Kohoff, and A. B. Murphy, “Modelling of gas-metal arc welding taking into account metal vapour,” *J Phys D Appl Phys*, vol. 43, no. 43, 2010, doi: 10.1088/0022-3727/43/43/434008.
[8] A. B. Murphy, “The effects of metal vapour in arc welding,” *J Phys D Appl Phys*, vol. 43, no. 43, 2010, doi: 10.1088/0022-3727/43/43/434001.
[9] M. Goodarzi, R. Choo, and J. M. Toguri, “The effect of the cathode tip angle on the GTAW arc and weld pool: I. Mathematical model of the arc,” *J Phys D Appl Phys*, vol. 30, no. 19, pp. 2744–2756, 1997, doi: 10.1088/0022-3727/30/19/013.
[10] M. Goodarzi, R. Choo, T. Takasu, and J. M. Toguri, “The effect of the cathode tip angle on the gas tungsten arc welding arc and weld pool: II. The mathematical model for the weld pool,” *J Phys D Appl Phys*, vol. 31, no. 5, pp. 569–583, 1998, doi: 10.1088/0022-3727/31/5/014.
[11] M. Abid, S. Parvez, and D. H. Nash, “Effect of different electrode tip angles with tilted torch in stationary gas tungsten arc welding: A 3D simulation,” *International Journal of Pressure Vessels and Piping*, vol. 108–109, pp. 51–60, 2013, doi: 10.1016/j.ijpvp.2013.04.006.
[12] W. Zhao, H. Jin, X. Du, J. Chen, and Y. Wei, “A 3D arc-droplet-molten pool integrated model of Al alloy GMAW process: Heat transfer, fluid flow and the effect of external magnetic field,” *Vacuum*, vol. 202, no. April, p. 111129, 2022, doi: 10.1016/j.vacuum.2022.111129.
[13] M. Baeva, R. Kozakov, S. Gorchakov, and D. Uhrlandt, “Two-temperature chemically non-equilibrium modelling of transferred arcs,” *Plasma Sources Sci Technol*, vol. 21, no. 5, 2012, doi: 10.1088/0963-0252/21/5/055027.
[14] M. Tanaka *et al.*, “Time-dependent calculations of molten pool formation and thermal plasma with metal vapour in gas tungsten arc welding,” *J Phys D Appl Phys*, vol. 43, no. 43, 2010, doi: 10.1088/0022-3727/43/43/434009.
[15] Y. Wang and P. F. Mendez, “International Journal of Thermal Sciences Isotherm penetration depth under a moving Gaussian surface heat source on a thick substrate,” vol. 172, no. March 2021, 2022.
[16] M. R. Grams and P. F. Mendez, “Reduced order models for welding and solidification processes,” 2020, doi: 10.1088/1757-899X/861/1/012003.
[17] P. F. Mendez, “Scaling Laws in Welding Modeling Scaling Laws in Welding Modeling,” no. January, 2014.
[18] M. Ramírez and G. Trapaga, “Mathematical Modeling of a DC Electric Arc - Dimensionless Representation of a DC Arc,” *ISIJ International*, vol. 43, no. 8, pp. 1167–1176, 2003, doi: 10.2355/isijinternational.43.1167.
[19] M. A. M. A. Ramírez, G. Trapaga, and J. McKelliget, “A comparison between different numerical formulations for welding arc representations,” *J Mater Process Technol*, vol. 155–156, no. 1–3, pp. 1634–1640, 2004, doi: 10.1016/j.jmatprotec.2004.04.313.
[20] A. Delgado-Álvarez, P. F. Mendez, and M. A. Ramírez-Argáez, “Dimensionless representation of the column characteristics and weld pool interactions for a DC argon arc,” *Science and Technology of Welding and Joining*, vol. 24, no. 7, pp. 634–643, 2019, doi: 10.1080/13621718.2019.1584455.

[21] M. A. Ramírez, G. Trapaga, and J. McKelliget, “A comparison between different numerical formulations for welding arc representations,” *J Mater Process Technol*, vol. 155–156, no. 1–3, pp. 1634–1640, 2004, doi: 10.1016/j.jmatprotec.2004.04.313.

[22] A. B. Murphy and C. J. Arundell, “Transport Coefficients of Argon , Nitrogen , Oxygen , Argon-Nitrogen , and Argon-Oxygen Plasmas,” vol. 14, no. 4, 1994.

[23] J. McKelliget and J. Szekely, “Heat Transfer and Fluid Flow in the Welding Arc,” vol. 17, no. July, pp. 1139–1148, 1986.

[24] J. Lancaster, Ed., *The Physics of Welding*, 2nd ed. Elsevier, 1986. doi: 10.1016/C2013-0-03805-4.

[25] M. Tanaka, H. Terasaki, M. Ushio, and J. J. Lowke, “A unified numerical modeling of stationary tungsten-inert-gas welding process,” *Metall Mater Trans A Phys Metall Mater Sci*, vol. 33, no. 7, pp. 2043–2052, 2002, doi: 10.1007/s11661-002-0036-2.

[26] O. H. Nestor, “Heat Intensity and Current Density Distributions at the Anode of High Current, Inert Gas Arcs,” *J Appl Phys*, vol. 33, no. 5, pp. 1638–1648, 1962, doi: 10.1063/1.1728803.

[27] M. Tanaka, S. Tashiro, T. Satoh, A. B. Murphy, and J. J. Lowke, “Influence of shielding gas composition on arc properties in TIG welding,” *Science and Technology of Welding and Joining*, vol. 13, no. 3, pp. 225–231, 2008, doi: 10.1179/174329308X283929.

Phase-field simulation of the formation of new grains by fragmentation during melting of an ABD900 superalloy

B. Böttger, M. Apel

Access e.V. Aachen, Germany

b.boettger@access-technology.de

Abstract. Laser powder bed fusion (L-PBF) is an additive manufacturing method which involves local laser melting of powder particles, a partial remelting of previously deposited layers, and subsequent re-solidification under high thermal gradients and cooling rates. The transition between melting and re-solidification becomes visible as melt pool boundaries in optical micrographs and plays a crucial role: Apart from creating a strong segregation zone, the transition determines whether the microstructure is inherited and carried over to the next layer, or whether new grains with new orientations are formed. While heterogeneous nucleation is suppressed due to the lack of seeding particles at the small length scales inherent to L-PBF, alternatively, new grains can form via dendrite fragmentation, as demonstrated in this paper by phase-field simulations using the software MICRESS®. By strong coupling between the phase-field equation and a thermal 1D-cylinder approach for the long-range temperature field, consistency between latent heat and microstructure is ensured. To allow for a systematic variation of the orientation relationship between the dendrite growth direction and the respective temperature gradient, a two-step simulation procedure for two overlapping tracks with variable gradient directions is developed. Growth conditions which promote fragmentation and formation of new grains are analyzed and discussed.

1. Introduction

Phase-field modeling is a popular approach to simulate microstructure evolution during melting and solidification. In additive manufacturing (AM), phase-field simulation has been performed with a range of alloy types (simple binary to complex multicomponent and multiphase), and topics (micro-segregation or solute trapping to secondary phase precipitation or grain selection) [1-5]. However, the inherently transient character of laser-powder bed fusion (LPBF), starting with heating and partial melting of the previous layers, followed by stagnation of the solid-liquid front and dynamic resolidification, has not yet been simulated at the microscopic scale. Instead, quasi-stationary directional growth typically has been considered, assuming a constant temperature gradient and cooling rate, often estimated from analytical or process simulation results [6].

However, the boundary regions between the sequentially deposited layers formed during AM are of special interest. The microstructure parameters of one layer that are inherited by the next layer and whether nucleation or fragmentation leads to formation of new grains are determined in the transition zones between melting and solidification. The discontinuity of growth at this boundary often causes unwanted effects like segregation, melting of grain boundaries in the heat-affected zone, or formation of new grains and grain boundaries and thus can be responsible for the formation of pores and cracks as well as the deterioration of mechanical properties [7].

The complexity of the thermal boundary conditions that connect the temperature fields of the process scale with the microscopic scale have been discussed in a previous publication [8]. Here, we focus on the fragmentation associated with dendrite arm detachment. By floating freely in the melt, fragments can change orientation, leading to the formation of new grains. Grain refinement by fragmentation during solidification has been widely discussed in the literature as caused by ripening, recalescence, convection, or transient growth conditions [9-11]. However, fragmentation by melting during LPBF has been ignored.

To simulate temperature boundary conditions, a thermal 1D-cylinder was used, which is strongly coupled to the microstructure simulation model. It uses the specific process parameters of a laser-based additive process. Here, the term "strong coupling" (versus "weak coupling") stands for bi-directional interdependency within the model. This is achieved by using "Homoenthalpic Approximation" [12], which was developed for other casting processes.

In this work, microstructure formation during melting and solidification of subsequent layers is addressed for LPBF processing of the Ni-base-superalloy ABD®-900AM using the phase-field software, MICRESS® [13]. Simulations were carried out for various simulation geometries and orientation relationships between the temperature gradient and the crystallographic orientation for adjacent laser tracks. Growth conditions that promote fragmentation and new grain formation are analyzed and discussed.

2. Model description

2.1. Phase-field model with integrated 1D-temperature solver

MICRESS® [13] is based on a multiphase-field model. Thermodynamic and mobility databases are coupled via the Fortran TQ-interface of Thermo-Calc Software [14], thus permitting the application of MICRESS® to multicomponent technical alloys. See [15,16] for details on the model equations.

A 1D-macro model integrated into the software permits advanced coupling to macroscopic temperature fields on the process level [12,17]. This enables microstructure simulations with a strong coupling between the heat flow and local release of latent heat caused by phase transformations. With this aim, a 1D temperature field was placed in direction of the z-coordinate of the 2/3D-microstructure domain, such that it fully overlapped and could further extend above and below the microstructure domain (figure 1). In the overlapping region, latent heat is obtained directly using enthalpy and heat capacity data from the Calphad database, while in the extensions $H(T)$ and $C_P(T)$ are used from tabulated data. To solve the 1D-heat conduction problem, an explicit finite difference scheme is used that is based on the heat source method [18]. The heat conductivity, λ, is either approximated by volume averaging within the region overlapping with the microstructure domain as $\boldsymbol{\lambda} = \sum_\alpha \boldsymbol{\lambda}_\alpha$ or tabulated as $\lambda(T)$ outside the region.

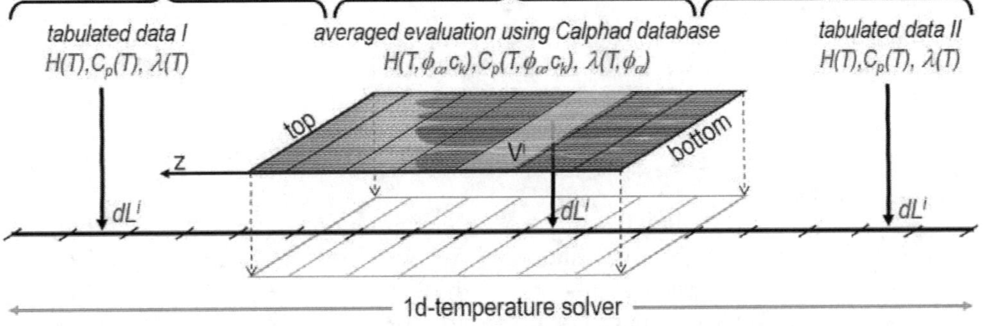

Figure 1. Coupling of the 1D-temperature solver with latent heat

2.2. Thermal 1D-cylinder approach

For the fast-moving laser heat source in LPBF, the temperature field within the processed material can approximately be described by its cross-section normal to the laser direction, which is usually used for experimental characterization of the melt pool shape as well. Here, the temperature field within this cross-section is further simplified to a 1D-cylinder with an inner surface of radius r, which is heated by the laser, and an outer spherical surface with radius $s+r$, which extends far into the material (figure 2). The radius is chosen such that the diameter d of the laser beam corresponds with the surface ($r=d/\pi$), thus also defining the position of the origin of the 1D-cylinder coordinate system (figure 2).

This simplified approach cannot replace a dedicated process model. Rather, it should be thought of as an extension to the microstructure model allowing for strong coupling between the macro- and micro-scale. In this way, latent heat is consistently taken into account. Further, the new approach allows for direct variation of process parameters like laser speed or laser power.

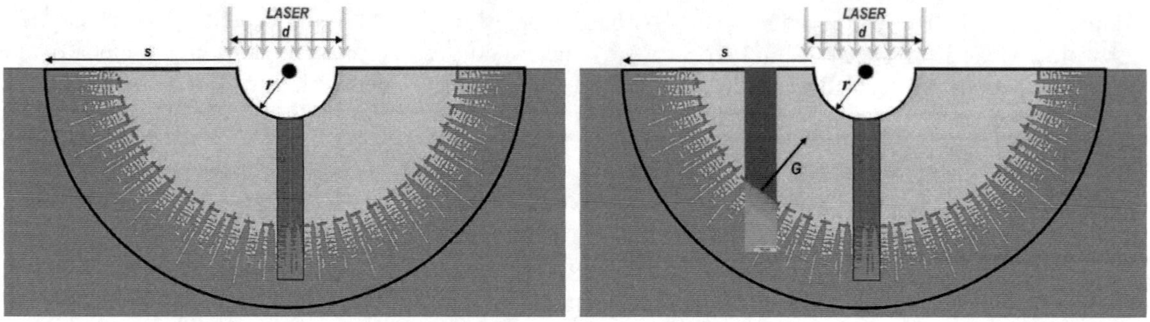

Figure 2. left: 1D-cylinder approach, with 1D-temperature field (black, with spherical inner and outer surface), the 2d-microstructure domain (green), the melt pool of depth l (red), and the laser beam with diameter $d=\pi r$. Right: Modified approach with arbitrary direction of the temperature gradient G.

To further permit taking into account the effect of changing directions of the temperature gradient G upon subsequent remelting of different layers or adjacent tracks, the approach was generalized by allowing for a free rotation of the direction of G (with rotation angles <90°). This generalization requires that integration of latent heat within the region of each 1D-temperature grid cell does not follow anymore the x/y-coordinate plane of the microstructure domain (see figure 1). Furthermore, the periodic boundary condition at the x/z- and y/z-planes needs to be replaced by a shifted periodic condition, which is progressively introduced as the liquid phase expands from the top during melting. Such a dynamic region-wise boundary condition is needed because the solid microstructure of the previous layer, which is used as initial microstructure for the second remelting, does not necessarily possess the same periodicity, so that isolating (i.e. non-periodic) boundary conditions must be used.

3. Phase-field simulation setup

3.1. Model parameters and data

Thermodynamic data were provided by online coupling to the database TCNI10 [14] using all alloying elements (table 1) and phases (table 2). Diffusion data for γ and γ' were obtained from the MOBNI5 mobility database [14]. Phase-related data like nucleation conditions and interface energy data (σ) are listed in table 3. Interface kinetics were assumed to be diffusion limited by correcting for artefacts of the finite interface thickness [19].
The 1D-temperature field had a total size, s, of 1 cm with a grid resolution of 0.25 µm and was tilted with respect to the z-axis of the microstructure domain by the angles α_x and α_y in the x- and y-directions, respectively. To ensure symmetry when changing from positive to negative angles, the origin of rotation was placed on the center line (half width, half thickness) close to the height of the melt pool border. A transient heat flux boundary condition (figure 3) was applied to the inner surface of the 1D-temperature field (figure 2), leading to a heating of the simulation domain from the top.

Different 2D- and 3D-setups and spatial resolution were chosen according to computational requirements (<2 weeks calculation time per simulation run on 14/28 CPU cores, physical memory usage <64 GB), while exploring only a small part of the possible combinations of crystallographic and temperature gradient orientations for occurrence of fragmentation. For simulation of the initial micro-structure, corresponding to the solidification microstructure of the previously deposited layer or track, the simulation domains were chosen with a height of 100 µm and a width of 15 µm with a grid spacing of Δx=0.01µm (2D), or with a width and thickness of 1,5 µm and a grid spacing of Δx=0.02µm (3D). Initially. the domain was filled by a single fcc-grain with crystallographic orientation φ which has been chosen as parallel to the z-direction (φ=0°) or 20° in x-direction (φ=20°). Correspondingly, the numerical grid dimension was 1500 x 10000 or 75 x 75 x 5000 cells, respectively. Given that the height of the simulation domain contained the full remolten area, temperature coupling was strong and bidirectional (without need of an iteration approach), directly leading to a representative set of *H(T)* and $C_p(T)$ data. Additionally, a 3D-variant with a smaller height and higher width (300 x 75 x 2500) was simulated, relying on the Homoenthalpic Approach [12] and already calculated latent heat data using the 75 x 75 x 5000 geometry. For all these geometries, fragmentation was not considered in this first simulation step, so that a single-crystalline dendritic microstructure was obtained.

The second simulation step was performed using the results of the first simulation as initial micro-structure. Reading of the structure was achieved without information loss by a technique using the restart output (.rest) of the former simulation [21,22]. The initial microstructure was shifted in negative z-direction according to the thickness of 40 µm of the new layer. In case of 2D-micro-structures, the full height of the simulation domain of 100 µm was retained and a new rectangular homogenous fcc-grain of nominal composition was placed at the top of the domain as a substitute for the powder layer. In case of the 3D-simulations, the domain consisted of only part of the original height of the initial microstructure, so that an addition of an extra layer at the top was not necessary.

Two different strategies were followed to get a bigger 3D-domain filled with representative initial microstructure data: Firstly, based on the simulation results of the first step with a 75 x 75 x 5000 cells domain (see figure 5c), an expanded domain of 300 x 75 x 1200 cells was filled by periodical continuation in x-direction, taking into account the proper displacement in z-direction of the shifted-periodic boundary condition (see section 2.2). Secondly, an already bigger (but shorter) domain of 300 x 75 x 2500 was used for the first simulation step as explained above (see figure 5c). In both cases, a domain size of 300 x 75 x 1200 cells was obtained focusing on the expected border of the melt pool, where the transition between melting and re-solidification is observed. The *H(T)* and $C_p(T)$ data from the first simulation was used for the temperature solver in the regions outside the shorter simulation domain. Thus, again, no iteration of the temperature field was required because the effect of the initial microstructure on H(T) turned out to be minimal.

Further, fragmentation was activated in the second simulation which implies actively checking for detached fragments of dendrite arms. If such a fragment was detected, a new phase-field parameter and a random orientation was attributed to this fragment, leading to the formation of a new grain. In order to avoid false fragmentation, fragments with contact to a non-periodic boundary condition were discarded. Also, for the second simulation step a different gradient direction (α_{x2}, α_{y2}) than in the first simulation was chosen to account for the effect of hedging or overlapping of adjacent tracks.

Table 1. Chemical compositions (in wt. %) of ABD®-900AM

	Ni	C	Cr	Co	Mo	W	Ti	Al	Ta	Nb
ABD®-900AM	bal.	0.06	17.0	20.0	2.0	3.0	2.0	2.0	2.0	2.0

Table 2. Phases and phase related parameters

No	name in database [12]	identity/ constitution	diffusion data	nucleation in phase	$\sigma_{0/x}$ / (J/cm²)	$\sigma_{1/x}$ / (J/cm²)
0	LIQUID	melt	MOBNI5	1 (bulk)	-	2,0 x 10^{-5}
1	FCC_L12	γ	MOBNI5	1 (split)	2.0 x 10^{-5}	-
2	FCC_L12#2	γ'	MOBNI5	0	2.0 x 10^{-5}	3.0 x 10^{-5}
3	FCC_L12#3	(Nb,Ti)C	-	0	1.8 x 10^{-5}	2.0 x 10^{-5}

Table 3. Process Parameters

T_0	v	P	d
200 °C	1.0 m/s	285 W	85 µm

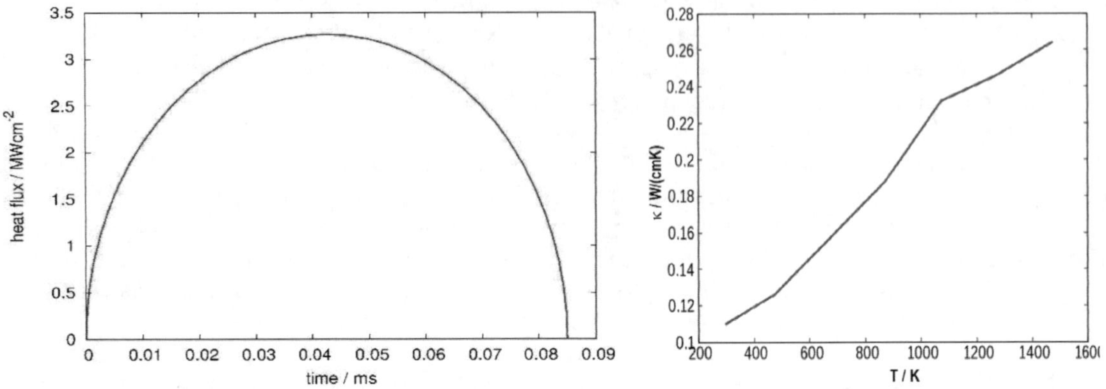

Figure 3. left: Heat flux density vs. time at the upper boundary condition of the 1D-temperature field, assuming a disc-shaped laser cross-section with diameter $d=\pi r$ and a homogeneous power distribution. Right: T-dependency of heat conductivity of solid ABD900 (according to OXMET Technologies)

3.2. Calibration of the thermal 1D-cylinder model

Given that the thermal boundary conditions can be directly linked to the process parameters by means of the initial temperature and the heat flux curve (figure 3), quantitatively correct results should be expected within the simplified assumption of a one-dimensional approach. However, two important factors are of high importance for the melting dynamics and thus the melting depth, which are outside the scope of the model. The first is the absorption efficiency η, which can be estimated by thermal process simulation or experimental measurements (heating of the sample). The second one is the effective heat conductivity κ of the melt which is strongly affected by convection in the melt pool. The effect of κ on the melting rate and melt pool depth is significant and can be seen in figure 4. Therefore, calibration has been performed using figure 4 and the measured cross-sectional area of the melt pool, which has been obtained from single-track experiments, finally leading to the parameter set $\eta=65$ % and $\kappa=0.75$ W/(cmK). More details on this calibration procedure can be found in [8].

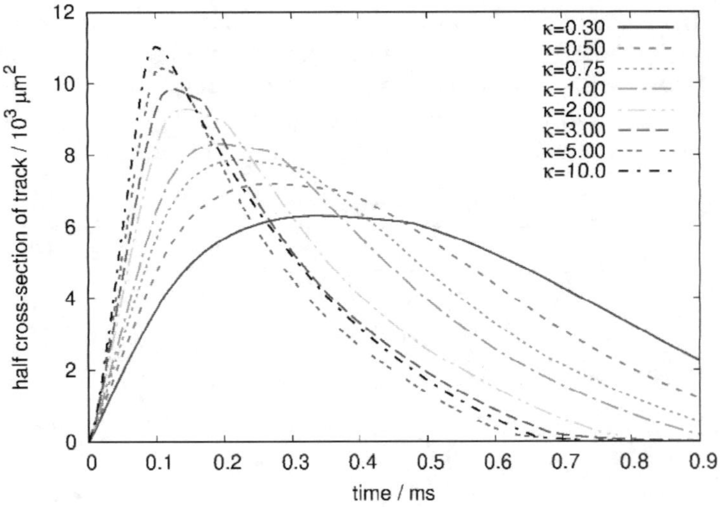

Figure 4. Development of molten half cross-section area with time for $\eta=65\%$ and varying thermal conductivity of the melt κ in W/(cmK).

4. Results

4.1. Simulation of initial microstructure

In figure 5 the different types of initial microstructures are shown which have been obtained for the different geometries of the simulation domain and different combinations of the orientations of the [110]-crystallographic direction and the temperature gradient. Displayed is the concentration distribution of Cr, which indicates a cellular-dendritic pattern in the upper remolten part, while in the lower part the initial homogeneous structure is essentially unchanged. Generally, an increase of the primary dendrite spacing (PDAS) and also of the development of side-branches is observed with increasing deviation of the dendrite orientation φ (=growth direction) and the orientation of the temperature gradient α. Specifically, for the 2D-simulations, the highest PDAS is observed for α=0° and φ=30° (figure 5a) and for α=40° and φ=0° (figure 5b).

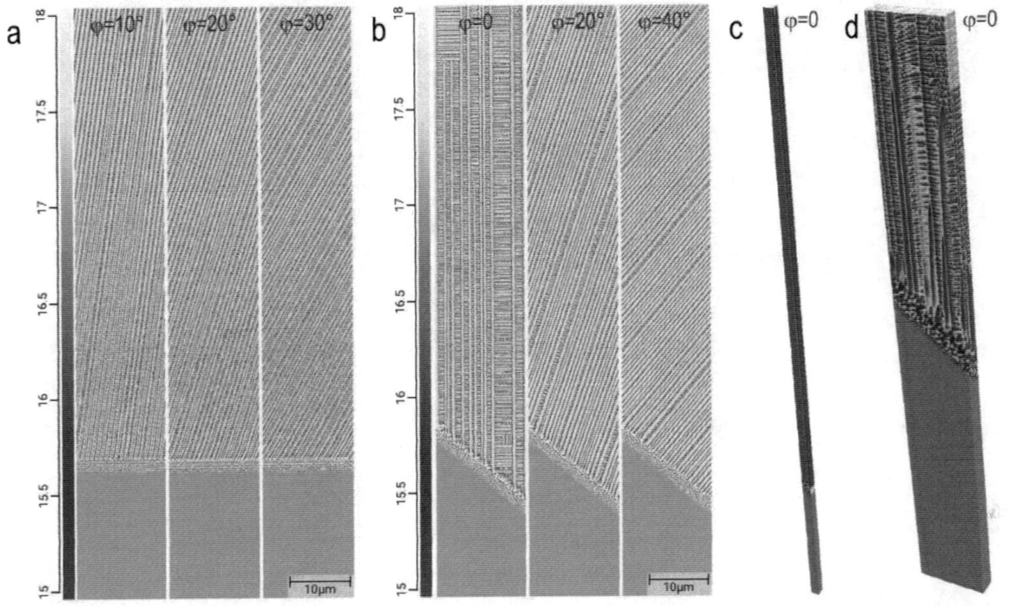

Figure 5. Initial microstructures obtained for different simulation domain geometries (shown is Cr-composition): a) 2D-domain 15 x 100 µm with T-gradient in z-direction (α_x=0°) and crystallographic orientations φ, b) 2D-domain 15 x 100 µm with α_x=40°, c) 3D-domain 1.5 x 1.5 x 100 µm with α_x=α_y=-40° and φ_x=φ_y=0°, d) 3D-domain 6.0 x 1.5 x 50 µm with α_x=α_y=-40° and φ_x=φ_y=0°

4.2. Simulation of remelting

In a second step, the initial microstructure as obtained in section 4.1, which represents a typical single-crystalline part of a previously deposited layer or track, was subject to remelting in the course of depositing a second subsequent layer. The general aim of this simulation was to explore microstructure formation during remelting with focus on the edge of the melt pool where melting changes to solidification. More specifically, the question was whether at this region fragmentation of the dendrite network could lead to the formation of new grains.

In contrary to a homogeneous grain, melting of a dendritic microstructure proceeds preferentially along the interdendritic regions, thus unveiling the dendritic mushy-zone which was formed during the prior solidification process. This is shown in figure 6 which illustrated the temporal development of the solid-liquid interface during remelting of the upper part of the microstructure (shown in figure 5b). While at the beginning of melting (left side of figure 6) fast melting and high temperature gradients lead to a relatively thin mushy zone, and fragments and carbide particles are released into the melt, at a later stage, when getting closer to the border of the melt pool, the mushy zone becomes deeper. Before resolidification starts, coarsening occurs, and the detached particles in the melt disappear. Resolidification essentially starts from the former dendrite tips, so that the prior PDAS is partly retained. This can be more easily seen in case of the 2D-simulations, e.g. in figure 8, where nearly all prior dendrites were successfully continued. Furthermore, the unsteady solidification velocity leaves a segregation band close to the border of the melt pool (see figure 9).

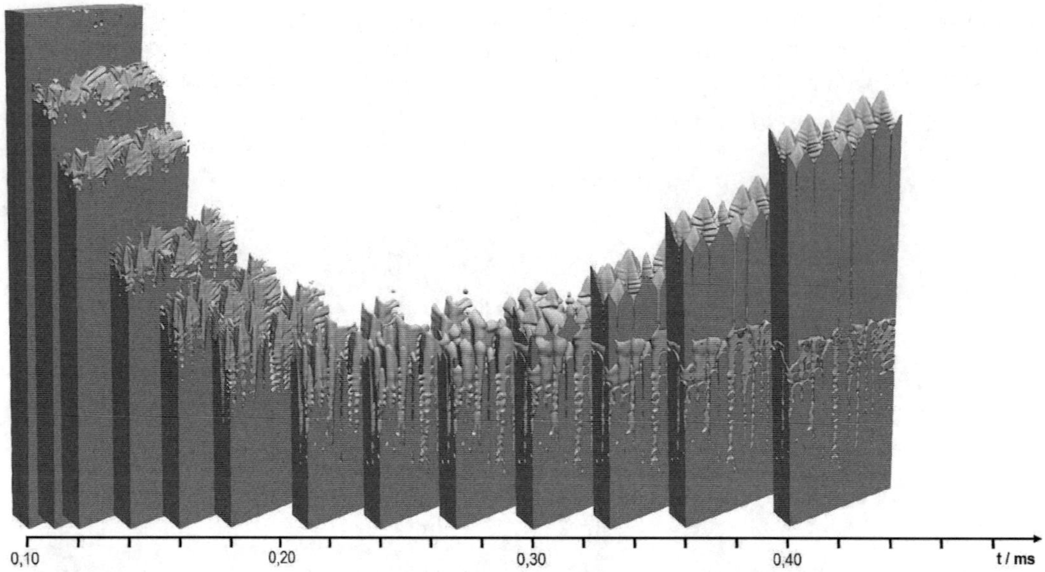

Figure 6. Time development of the solid-liquid interface during remelting of 3D-microstructure (top part of figure 5d) with α_{x2}=10° and α_{y2}=30°

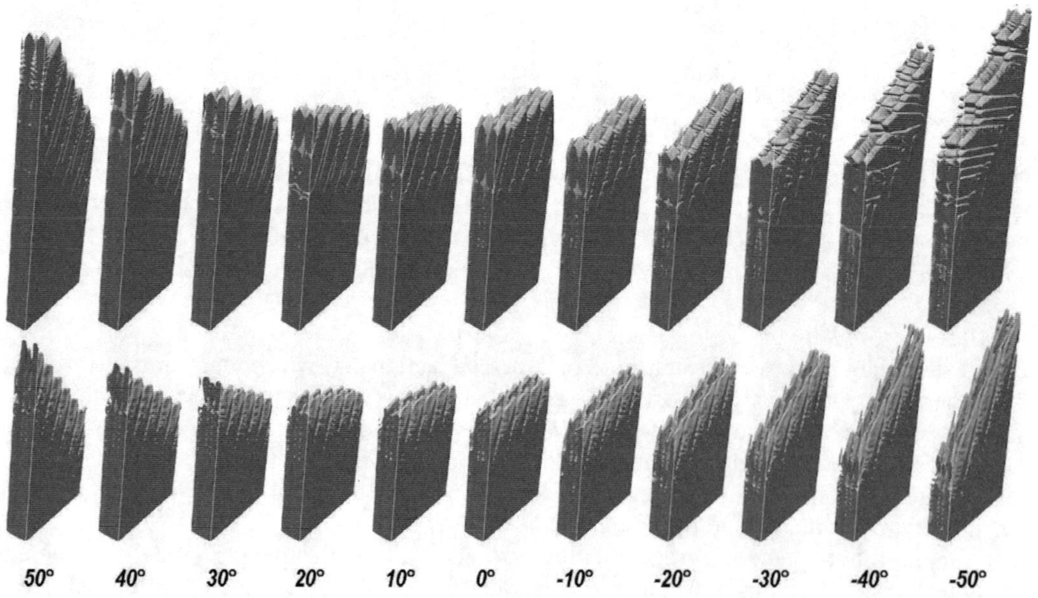

Figure 7. 3D-solid-liquid interfaces for remelting of a periodically enlarged initial microstructure using the parameter set α_{x1}=-20°, α_{y1}=0°, α_{y2}=0°, φ_x=20° and φ_y=0° and various values of α_{x1} between -50 and 50°. Bottom: at maximum melting (t=0.22 ms), top: t=0.34 ms.

However, at the time of this publication, fragmentation has only been observed in 2D-simulations, although a number of 3D-conditions has been tried, including variation of all parameters α_{x1}, α_{y1}, α_{x2}, α_{y2}, φ_x and φ_y. As example, figure 7 shows a systematic variation of α_{x2} between -50° and +50° for the parameter set α_{x1}=-20°, α_{y1}=0°, α_{y2}=0°, φ_x=20° and φ_y=0°. No sign of fragmentation is visible for all values of α_{x2}, possible reasons for that will be discussed in section 5. Notably, for α_{x2} above ~+30° the growth direction of the dendrites changed by 90° without formation of a grain boundary. This is caused by cubic symmetry, which allows side branches to develop to new dendrites if they are closer oriented to the T-gradient direction than the original trunk.

In figure 8, the fragmentation during remelting of the 2D-microstructures of figure 5a is shown. Under such conditions without inclined T-gradient in both steps (α_1=α_2=0°), fragmentation is found to be favored by larger misorientations between crystallographic [001] and T-gradient directions. Less clear is the situation in figure 9, where the T-gradient direction changed from +40° to -40°. Here, fragmentation was observed for φ=0° and φ=40° and not for φ=20°.

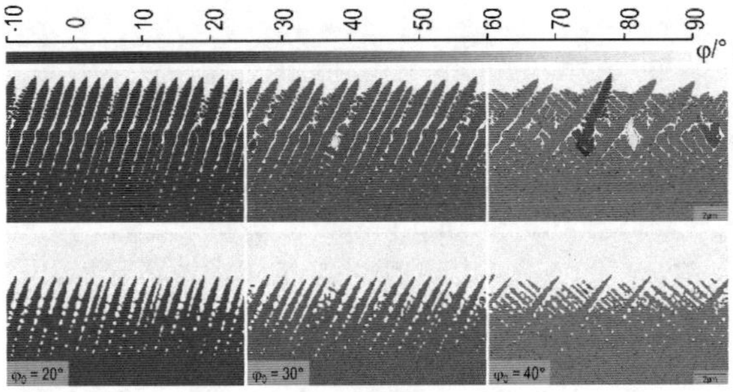

Figure 8. Detail of the microstructures of figure 5a during remelting. Bottom: end of melting (t=0.22 ms), top: t=0.34 ms.

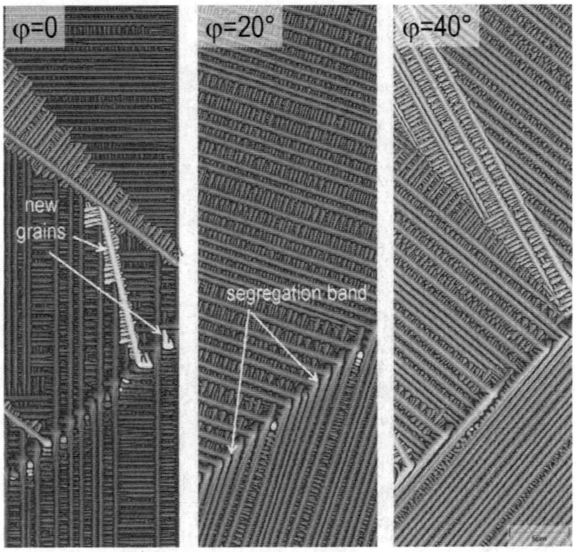

Figure 9. Detail of remolten microstructures of figure 5b, shown is the Cr composition distribution (modulated with 0.01φ to highlight new grains)

5. Discussion

The fragmentation phenomenon had been observed by the authors in 2D-simulations of alloy melting under various conditions [22,23], especially when melting homogeneous or homogenized materials, but also for remelting not homogenized material like in this study. However, our systematic approach to reproduce this behaviour with 3D-simulations in this work has failed so far.

Essentially, there are three conditions which must be fulfilled to allow for the formation of new grains via fragmentation of dendrite arms: Firstly, fragmentation must occur, i.e. some parts of the dendrite must be detached by the melting process. This is quite common during fast melting, but gets less probable when melting is slowing down close to the border of the melt pool. Breaking of dendrite arms should be more relevant for dendritic structures with pronounced side-branching as for cellular structures. This tendency is supported by the 2D-results in figures 8 and 9 and generally should favor fragmentation in microstructures which have grown under unfavourable T-gradient directions, and also should occur more easily in case of pre-existing grain boundaries. There should also be an effect of the relative orientation of the T-gradient to the dendrite growth directions during melting which probably cannot easily be predicted on purely theoretical considerations.

Secondly, the fragmented part of the dendrite must not melt before resolidification starts. This requires not only fragmentation to occur at the very end of melting, but also strongly depends on the temperature gradient in front of the dendrites, and the exact melting temperature and composition distribution of the dendrites and the fragment. Therefore, simulation approaches which intend to correctly predict formation of new grains in additive processes must be able to correctly describe the transient temperature-time profiles during the whole remelting process, and use a correct initial microstructure and a consistent description of latent heat at the micro-scale.

Thirdly, formation of new grains from broken fragments can only be successful, if the fragment can adopt a favorable orientation to the T-gradient during resolidification. This should be more easily fulfilled in cases where the original dendrites have an unfavorable orientation, so that they can be easily overgrown.

Given these three factors which must be fulfilled, and the large number of orientation variables which can be independently varied for the first and second simulation step, it is easily possible that favorable conditions just have not been found yet. However, there are also fundamental reasons which could make fragmentation more difficult in 3D. For example, dendrite selection is more difficult in 2D, often leading to higher PDAS values with exaggerated side-branch formation, which could falsely favor fragmentation in 2D. Similarly, random rotation of detached fragments in 2D has a higher probability to result in favorable orientations (close to the gradient direction). Indeed, in few cases, formation of detached fragments has been observed in the 3D-simulations, which however have been readily overgrown by the main dendrite.

On the other hand, the numerical grid resolution and the relatively large interface thickness ($\eta=2.85\Delta x$) could also be a problem, given that for sake of computational time Δx was chosen as 0.02 μm for the 3D-simulations, while 0.01 μm had been used for 2D. This could be important because the applied mechanism for fragmentation requires that the diffuse interfaces of the fragment and the main dendrite do not overlap anymore, which gets more difficult with higher values of η. Finally, the simulated 3D-geometries may have been simply too small to present a realistic probability for successful fragmentation.

6. Conclusion

In this work, simulation of microstructure formation during melting and solidification of subsequent layers has been simulated in 2D and 3D for LPBF processing of the Ni-base-superalloy ABD®-900AM using the software MICRESS®. With a new 1D-cylinder approach acting as strongly-coupled thermal process boundary condition, this contribution further focusses on the interlayer boundaries which are formed between sequentially deposited layers. For different simulation geometries and orientation relationships between the temperature gradient and the crystallographic orientation for adjacent laser tracks, simulations were carried out. While fragmentation has been found in some of the 2D-simulations, no such effects could be detected yet in 3D-simulation. Reasons for this behaviour could be fundamental differences between fragmentation in 2D and 3D, a too large parameter space which prevented finding suitable conditions so far, a lack of representativity of the relatively small 3D-domains, or numerical issues connected to a too large interface thickness. Future research in this field will focus on more representative and higher resolved 3D-simulations and also include the effects of pre-existing grain boundaries and intermediate (partial) homogenisation.

Acknowledgements

The authors acknowledge funding by the German Federal Ministry for Economic Affairs and Climate Action (BMWK) within the project "*SAPHIR – Simulationsmethoden zum additiven Prozessieren von Hochtemperaturlegierungen – Mikrostruktur, In-service-Eigenschaften und Reparatur*" under grant n° 03EE5049C.

References

[1] Tan JHK, Sing SL, Yeong WY. Microstructure modelling for metallic addi-tive manufacturing: a review. Virtual Phys Prototyp 2020; 15(1):87-105.

[2] Papazoglou EL, Karkalos NE, Karmiris-Obratański P et al. On the Modeling and Simulation of SLM and SLS for Metal and Polymer Powders: A Review. Arch Computat Methods Eng 2022; 29:941–973.

[3] Boussinot G, Döring M, Hemes S, Stryzhyboroda O, Apel M, Schmidt M. Laser powder bed fusion of eutectic Al–Ni alloys: Experimental and phase-field studies, Mater Des 2021; 198:109299.
[4] Kumara C, Ramanathan Balachandramurthi A, Goel A, Hanning F, Moverare J. Toward a better understanding of phase transformations in additive manufacturing of Alloy 718. Materialia 2020; 13:100862.
[5] Körner C, Markl M, Koepf JA. Modeling and Simulation of Microstructure Evolution for Additive Manufacturing of Metals: A Critical Review. Metall Mater Trans A 2020; 51:4970–4983.
[6] Zielinski J, Boussinot G, Laschet G et al. From Melt Pool to Microstructure to Mechanical Properties: A Simulative Approach to L-PBF Processed Material Behaviour. Berg Huettenmaenn Monatsh 2020; 165:175–180.
[7] Shifeng W, Shuai L, Qingsong W, Yan C, Sheng Z, Yusheng S. Effect of molten pool boundaries on the mechanical properties of selective laser melting parts. J Mater Process Technol 2014; 214(11):2660–2667.
[8] Böttger B, Apel M. On the choice of thermal boundary conditions for microstructure modelling of additive processes, Procedia CIRP - 12th CIRP Conference on Photonic Technologies [LANE 2022], pp. 291-295, https://doi.org/10.1016/j.procir.2022.08.024.
[9] Yasuda, H., Morishita, K., Nakatsuka, N. *et al.* Dendrite fragmentation induced by massive-like δ–γ transformation in Fe–C alloys. *Nat Commun* **10**, 3183 (2019). https://doi.org/10.1038/s41467-019-11079-y
[10] Neumann-Heyme H, Eckert K, Beckermann C. Dendrite fragmentation in alloy solidification due to sidearm pinch-off. Phys Rev E Stat Nonlin Soft Matter Phys. 2015 Dec;92(6):060401. doi: 10.1103/PhysRevE.92.060401.
[11] Cool T, Voorhees PW. Dendrite fragmentation: an experiment-driven simulation. Phil. Trans. R. Soc. A376: 20170213. http://dx.doi.org/10.1098/rsta.2017.0213.
[12] Böttger B, Eiken J, Apel M. Phase-field simulation of microstructure formation in technical castings – A self-consistent homoenthalpic approach to the micro–macro problem, J. Comput. Phys. 2009; 228:6784-6795.
[13] http://www.micress.rwth-aachen.de
[14] Themo-Calc Software, https://www.thermocalc.com
[15] Eiken J, Böttger B, Steinbach I. Multiphase-field approach for multicomponent alloys with extrapolation scheme for numerical application. Phys. Rev. E 2006; 73:066122.
[16] Böttger M, Apel M, Budnitzki M, Eiken J, Laschet G, Zhou B. Calphad coupled phase-field model with mechano-chemical contributions and its application to rafting of γ' in CMSX-4. Comput Mater Sci 2020; 184:109909.
[17] Viardin A, Böttger B, Apel, M. Phase field assisted analysis of a solidification based metal refinement process. Materials Theory (2022) 6:14, https://doi.org/10.1186/s41313-022-00042-y.
[18] Lewis W, Morgan K, Thomas HR, Seethar-amu KN. The Finite Element Method in Heat Transfer Analysis. Hoboken: Wiley, 1996; pp. 123–130.
[19] Carré A, Böttger B, Apel M. Implementation of an antitrapping current for a multicomponent multiphase-field ansatz. J Crystal Growth 2013; 380:5–13.
[20] Promoppatum P, Yao S-C, Pistorius P, Rollett A. A Comprehensive Comparison of the Analytical and Numerical Prediction of the Thermal History and Solidification Microstructure of Inconel 718 Products Made by Laser Powder-Bed Fusion. Engineering 2017; 3(5):685-694.
[21] Böttger B, Altenfeld R. Laschet G, Schmitz GJ, Stöhr B, Burbaum B. An ICME Process Chain for Diffusion Brazing of Alloy 247. Integr. Mater. Manuf. Innov 2018; 7:70-85.
[22] Böttger B, Apel M, Jokisch T, Senger A. Phase-field study on microstructure formation in Mar-M247 during electron beam welding and correlation to hot cracking susceptibility. IOP Conf Ser: Mater Sci Eng 2020; 861 012072.
[23] Böttger B, Apel M, Laux B, Piegert S. Detached Melt Nucleation during Diffusion Brazing of a Technical Ni-based Superalloy: A Phase-Field Study. Mater Sci Eng 2015; 84 012031.

IOP Conf. Series: Materials Science and Engineering 1281 (2023) 012009 doi:10.1088/1757-899X/1281/1/012009

A 3D simulation of grain structure evolution during laser rescanning process of powder bed fusion additive manufacturing

K Kang[1]**, L Yuan**[2] **and A B Phillion**[1,*]

[1] Department of Materials Science and Engineering, McMaster University, Hamilton, ON L8S 4L8, Canada
[2] Department of Mechanical Engineering, University of South Carolina, Columbia, SC 29208, USA

* corresponding author: philliab@mcmaster.ca

Abstract. Laser powder bed fusion (LPBF) is an extensively used additive manufacturing process that can build metal parts with complicated geometric designs. However, because of the rapid solidification conditions and the layer-by-layer building, its application is challenged by products having poor surface quality and reduced mechanical properties. The laser rescanning process is often used as a refinement method during LPBF to improve the quality of products. In this study, grain structure formation during the LPBF laser rescanning process is modelled by a 3D cellular automaton based microstructure model coupled with finite element analysis. The coupled model considers different nucleation mechanisms, including epitaxial growth, which are applicable to rapid solidification in melt pool. The model is adapted to reproduce the grain structure and evolution of an Al-Si alloy manufactured by LPBF utilizing a laser rescanning strategy. The effects of laser remelting on the characteristics of the grain structure (i.e., the grain size, aspect ratio and orientation) are evaluated. The mechanisms that enable unique grain structure (i.e., grain refinement) are discussed.

1. Introduction

Laser powder bed fusion (LPBF) is an extensively used additive manufacturing (AM) method, where the product is built in a layer-by-layer manner. This has become an emerging method to build metal parts with complicated 3D geometric designs that are difficult to build using conventional casting techniques [1]. However, LPBF is challenged by the requirements of engineering standards. The major drawbacks of LPBF include poor surface quality and residual porosity, which lead to limited mechanical strength as well as fatigue resistance [2].

To solve these problems, laser rescanning is sometimes adopted as an in-processing treatment to improve the product's mechanical properties. The basic strategy of laser rescanning in LPBF is that, after a powder layer has been melted and solidified, the same layer is rescanned by the laser beam before the addition of another layer of powder. The effects of this strategy in LPBF have been, via experiments, studied widely in recent years including its effects on: reducing residual stress, developing homogeneity, and increasing density of the as-printed part, as well as microstructure modification [3-5]. In addition to removing porosities and cracks at the macro scale, the microstructural changes caused by laser rescanning play a key role in other property improvements. Studies show that laser rescanning can further refine the grain structure, as the coarse columnar grains in the melted zone are seen to be replaced by small equiaxed grains [2]. However, the use of laser rescanning can also create a significant crystallographic texture owing to epitaxial growth [5]. Unfortunately, identifying the origins of these effects in laser rescanning by experimental monitoring methods is still challenging, as the formation of the microstructure involves a complex interplay between the thermal environment given by the heat source and the powder bed, rapid solidification conditions, and phase transformations—all within a small timescale and small length scale of 10^{-3} s and 10^{-6} m, respectively [6].

Cellular automata (CA) simulation is a well-suited computational technique for studying the complex, mesoscale phenomena involved in LPBF, such as the heat and mass transport processes that occur during laser melting and the resulting evolution of the grain structure. In recent years, the CA method has been widely used to study many LPBF-related phenomena, including the influence of process parameters and scanning strategies on the formation of grain structure as well as the overall quality of the final parts. The grain structure evolution that occurs in a CA simulation is based on solidification theories, suggesting that the grain nucleation and grain growth are controlled by local undercooling [7]. This necessitates knowledge of the evolving temperature field. Therefore, CA models are usually coupled with thermal analysis such as finite element analysis (FEA)—a method known as CAFE and employed by Gandin et al. [8]. The link between undercooling and grain growth rate has often been determined using the Kurz-Giovanola-Trivedi (KGT) [7] model based on interface stability analysis.

Although the CAFE approach has been used extensively for investigating LPBF processing, it would appear that such studies have not fully considered the remelting effects or any rescanning processes, and the origins of grain refinement phenomena during remelting remain unaddressed. As such, existing CA simulations are not sufficient to provide guidance for process parameter adjustment for grain refinement control during actual production. Besides adding grain refiners and altering alloy compositions, the variation of thermal conditions during melting and remelting plays a decisive role in grain growth as well as grain refinement [9].

In this study, a CA model developed from an open-source code μMatIC [10, 11], coupled with FEA via Abaqus and the KGT model to account for rapid solidification conditions, is utilized to study laser rescanning and the related remelting phenomena during LBPF of alloys where dendritic growth dominates. Using as an example the AlSi10 alloy, a simulation of a single track/single layer build is performed (first laser scan on the powder layer) followed by a laser rescan for remelting (second laser scan on the as-solidified layer). Owing to computational limits, only a small domain with a length of 100 μm on each side is simulated. The remelting effects on the characteristics of grain structure (i.e., grain size, aspect ratio and orientation) are evaluated. The thermal conditions that lead to the grain refinement are discussed.

2. Model description

The CA model is adapted to AM by extending the existing μMatIC CA model developed by Yuan and Lee for simulating dendritic growth during the directional solidification process in welding and casting [10, 11]. The model has been highly parallelized with MPI implementation [12] and is coupled with the commercial Abaqus FEA software obtains the simulated temperature history used for CA modelling of grain growth. The Goldak heat source model [13] is applied to determine the relevant heat input for melting and remelting. A function relating the dendrite tip velocity in terms of undercooling under rapid solidification conditions is derived from the KGT model to reduce computation time. This function is used as a grain growth rate control function in the CA simulation.

2.1 Heat source model

The temperature field data is required for the CA model to carry out a simulation. Briefly speaking, the CA model needs to update the temperature of each cell at every timestep, so that the undercooling, ΔT (which controls nucleation and grain growth), can be calculated. It is assumed that only heat conduction affects the temperature field during laser scanning. Also, the start stage (initial transient), as well as the end stage (final transient) of the single laser track are not considered. Thus, a steady-state temperature profile for a melt pool can be obtained with a moving heat source.

The spatial distribution of heat input by the laser is approximated by using two ellipsoid heat sources (i.e., the Goldak model). As defined, the front half of the source is the quadrant of one ellipsoidal source, while the rear half belongs to another. Meanwhile, each of the heat sources is distributed in a Gaussian manner over these ellipsoid areas. The power density distribution, q_f, inside the front quadrant is defined as:

$$q_{\mathrm{f}}(x,y,z,t) = \frac{6\sqrt{3}(f_{\mathrm{f}}P)\alpha_{\mathrm{R}}}{a_{\mathrm{f}}bc\pi\sqrt{\pi}}\exp\left(-\frac{3(x-Vt)^2}{a_{\mathrm{f}}^2}-\frac{3y^2}{b^2}-\frac{3z^2}{c^2}\right)(x \geq 0) \tag{1}$$

Similarly, the power density distribution q_{r} inside the rear quadrant is:

$$q_{\mathrm{r}}(x,y,z,t) = \frac{6\sqrt{3}(f_{\mathrm{r}}P)\alpha_{R}}{a_{\mathrm{r}}bc\pi\sqrt{\pi}}\exp\left(-\frac{3(x-Vt)^2}{a_{\mathrm{r}}^2}-\frac{3y^2}{b^2}-\frac{3z^2}{c^2}\right)(x < 0) \tag{2}$$

where α_R is the thermal absorptivity of the material, f_{f} and f_{r} are proportion coefficients representing energy distribution in the front and rear quadrant, and the sum of f_{f} and f_{r} is always equal to 2. P is the laser power, V is the scanning speed of laser, a_{r}, a_{f}, b and c are the ellipsoid parameters, and x, y, z are local coordinates of a point with respect to heat source.

The heat transfer during the moving heat source is calculated by using FEA to solve the heat conservation equation [14]:

$$Q = \rho C_p \frac{\partial T}{\partial t} - \nabla\cdot(k\nabla T) \tag{3}$$

where Q is the heat input, ρ is the density, C_p is the specific heat capacity, T is the temperature, t is the time and k is the thermal conductivity. As such, both temperature profiles in the 3D domain over time during the first laser scan for powder consolidation and the second laser scan for remelting can be obtained. Please note that the thermal properties of the base material influence the temperature field during processing. Hence, the differences in properties within the base materials during the two scans are considered, including thermal absorptivity, thermal conductivity and density, which leads to different melt pools in the first and second scans.

2.2 Grain growth model

The CA model simulates dendritic growth by developing octahedral-shaped grains in a grid of cells. In the case of simulation via µMatIC, the mesh is a regular grid of cubic cells. The evolution of each cell considers the states of its 26 neighbors. Once a grain nucleates in a cell (following a probabilistic nucleation model), an octahedral-shaped grain seed is placed at the center of the cell and grows. The growth rate of the seed, v, is a function of local undercooling, ΔT, at the cell, which controls how fast the seed captures the adjacent cells and grows. This growth function, $v(\Delta T)$, also known as the dendrite tip velocity function, is a crucial factor influencing the accuracy of the solidification simulation via CA modelling. An approximation of this grain growth function is used for computational efficiency, which has a form of power function suggested by Guillemot et al. [15]:

$$v(\Delta T) = A(\Delta T)^B \tag{4}$$

where A and B are alloy dependent constants. The foundation of this approximation is made up of (a) the stability analysis of the moving morphological dendritic interface, a.k.a. the dendrite tip radius selection criterion [16]:

$$v\frac{m_{\mathrm{L}}^{v}(1-k_v)C_{\mathrm{tip}}}{D\Gamma C_0}\xi_{\mathrm{c}} - G = \frac{1}{R^2\sigma^*} \tag{5}$$

as well as (b) a model for calculating undercooling at a parabolic dendrite tip interface [16] :

$$\Delta T_{\mathrm{t}} = -m_{\mathrm{L}}^{v}C_0\left[\frac{(1-k_v)Iv(Pe_{\mathrm{c}})}{1-(1-k_v)Iv(Pe_{\mathrm{c}})}\right] + C_0(m_{\mathrm{L}} - m_{\mathrm{L}}^{v}) + \frac{2\Gamma}{R} \tag{6}$$

where m_{L} is the slope of liquidus line at equilibrium given by the phase diagram, m_{L}^{v} and k_v are the velocity-dependent liquidus slope and partition coefficient, C_{tip} is the dendrite tip composition in terms of Ivantsov solution, D is the diffusion coefficient in the liquid, Γ is the Gibbs-Thomson coefficient for the solid phase, C_0 is the initial composition of alloy, ξ_c is the dendritic function of solutal Péclet number, G is the temperature gradient, R is the dendrite tip radius, σ^* is the tip selection parameter, Iv is the Ivantsov function, and Pe_c is the solutal Péclet number. The velocity-dependent liquids slope and partition ceoefficient, m_L^v and k_v are defined as [16]:

$$m_L^v = m_L \frac{1 - k_v\left(1 - \ln\frac{k_v}{k_0}\right)}{1 - k_0} \tag{7}$$

$$k_v = \frac{k_0 + Pe_i}{1 + Pe_i} \tag{8}$$

where k_0 is the equilibrium partition coefficient given by the phase diagram, Pe_i is the interfacial Péclet number for solute redistribution.

Equation (5) and equation (6) are solved together numerically to find a series of unique solution pairs of $(v, \Delta T)$. The collection of $(v, \Delta T)$ pairs is then fitted as a power function in the form of equation (4). As the result, the alloy-dependent values of A and B are determined, and the grain growth rate function of undercooling is finally obtained.

For the case of AlSi10, used in this study, the relevant materials properties are provided in table 1, and taken from the work of Mohammadpour et al. [18], who studied the application of solidification microstructure selection maps to AM of the AlSi10 alloy. With that information, the fitted grain growth rate function is obtained:

$$v = 5.95 \cdot 10^{-8} \cdot (\Delta T)^{3.77} \tag{9}$$

The formation of nuclei is driven by the undercooling in solidification, and the nucleus can survive and develop to a new grain only if the undercooling nearby is over a critical value. A probabilistic nucleation model with a form of Gaussian distribution is be used to determine if a nucleus can form in a cell [17]:

$$p(\Delta T) = \frac{n_{max}}{\sqrt{2\pi}} \exp\left[-\frac{(\Delta T - \Delta T_{mean})^2}{2\Delta T_\sigma}\right] \tag{10}$$

where $p(\Delta T)$ is the probability of the formation of a nucleus at the cell, n_{max} is the maximum nucleation density, ΔT_{mean} is the mean nucleation undercooling and ΔT_σ is the standard deviation of undercooling.

Additionally, during grain growth, each cell is characterized by its crystallographic orientation and grain index, defining which grain the cell belongs to. The crystallographic orientation is allocated randomly by applying the Euler rotation matrix characterizing the orientation in the crystal coordinate system with respect to the global coordinate. The application of crystallographic orientation is used to reflect the influence of the preferential growth direction of grains on grain geometry, which is named as decentered square growth algorithm in CA [17]. For cubic systems such as Al-Si the preferred dendritic growth direction is <100>.

Table 1 also summarizes additional parameters used in the CA simulation. The nucleation parameters were taken from Mohebbi et al. [19], who developed a CA model for additive manufacturing of AlSi10 alloy with integration of a particle-based nucleation mechanism (without rescanning), and verified the predicted grain structure against EBSD data. In that study, the nucleation density was tuned and set to match with previous experimental results from Thijs et al. [20]. Finally, the cell size was chosen to satisfy the Courant criterion for CA growth, as suggested by Gandin et al. [8].

2.3 Coupling cellular automata with FEA

The temperature profiles during the laser scan and the laser rescan in LBPF processing of AlSi10 are obtained by FEA via the heat conservation equation, equation (3). The computations were performed with ABAQUS 2021. The relevant processing parameters and temperature-dependent material properties are listed in table 2, and taken from the work of Mohebbi et al. [19]. Please note that the thermal conductivity of liquid AlSi10 has been multiplied by a factor of 3 in order to account for the effect of thermal convection. The material absorptivity, needed for the Goldak equation, is based on the research by Boley et al. [21] on metal powder absorptivity modeling. The ratio of heat source parameters a_r, a_f is set at 60:40 [22], resulting in proportional changes to the heat source dimensions based on their values used by Mohebbi et al. [19]. The temperature at the thermal simulation domain surfaces is maintained at ambient temperature, except the top surface. In the simulation, the laser scanning speed, laser power and scanning direction remain the same during the first laser scan and the laser rescan. Besides, experimental results suggest that the differences in thermal properties of the powder layer and consolidated grains cannot be neglected, where a shallower melt pool is generally found during the rescan [23]. As such, the base material properties during the first scan (powder) and the rescan (consolidated grains) are set differently.

Table 1. Parameters used in grain growth model and CA simulation [18, 19]

Parameter	Unit	Value
Initial composition of Si, C_0	wt.%	10
Melting temperature of pure Al	K	933.3
Liquidus slope of phase α in Al-Si, m_L	K/ wt.%	-6.83
Partition coefficient of phase α in Al-Si, k_0	-	0.131
Diffusion coefficient in the liquid, D	m^2/s	3.0×10^{-9}
Gibbs-Thomson coefficient, Γ	mK	2.36×10^{-7}
Dendrite tip selection parameter, σ^*	-	$(2\pi)^{-2}$
Cell size in CA simulation	m	5.0×10^{-7}
Mean nucleation undercooling, ΔT_{mean}	K	5
Standard deviation of nucleation distribution, ΔT_σ	K	1
Maximum nucleation density, n_{max}	m^{-3}	1.5×10^{15}

Table 2. Parameters used in thermal simulation by FEA [19, 21, 22]

Parameter	Unit	Value
Dimension of thermal simulation domain l_x,l_y,l_z	μm	1200, 800, 400
Laser scanning speed, V	m/s	0.2
Laser power, P	W	120
Heat source parameters a_r, a_f, b and c	μm	72, 48, 60, 120
Ambient temperature	K	573
Thermal absorptivity of powder AlSi10 α_p (1st scan)	-	0.53
Thermal absorptivity of solid AlSi10 α_s (rescan)	-	0.50
Latent heat of fusion	kJ/kg	474
Thermal conductivity of powder AlSi10 at 573K, 800K,865K	W/(mK)	160, 165, 250,
Mass density of powder AlSi10 at 573K, 800K,865K	kg/m^3	2600, 2500, 2300,
Thermal conductivity of solid AlSi10 at 573K, 800K,865K	W/(mK)	180, 185, 255
Mass density of solid AlSi10 at 573K, 800K,865K	kg/m^3	2680, 2530, 2300
Specific heat of AlSi10 at 573K, 800K,865K and 1050K	J/K	1000, 1100, 1150, 1150
Thermal conducitivity of liquid AlSi10 at 1050K	W/(mK)	270
Mass density of liquid AlSi10 at 1050K	kg/m^3	2200

The obtained temperature profiles during the first scan and the rescan are combined to make a sequential temperature history which is used as input for CA simulation at a later stage. In this simulation, the diffusion field is not calculated, and the effect of convection on composition distribution is neglected. In other words, the grain nucleation and grain growth are only determined by thermal conditions (i.e., local undercooling). Furthermore, the simulation assumes that the resulting grain structures from both laser scans are fully dense and free of porosity. The CA simulation is implemented in a cubic domain with a side length of 100 μm. The domain is placed at a position shown in figure 1. The domain for simulation (the square with blue mesh grid) is fully contained by the melt pool volume (the red dashed curve) during the first laser scan, ensuring all the cells in the cube to start solidification from liquid status when the first laser scan temperature profile is applied. The melt pool boundary during the laser rescan is shown by the yellow dashed line in figure 1, representing a shallower melt pool as compared to the first laser scan. The simulation domain is a little bit greater than the melt pool during the rescan, which means some of the grains formed in the first laser scan remain un-melted during the rescan. As such, the effects from as-solidified grains are considered, and the differences between grain structures formed in the first scan and the rescan can be easily compared.

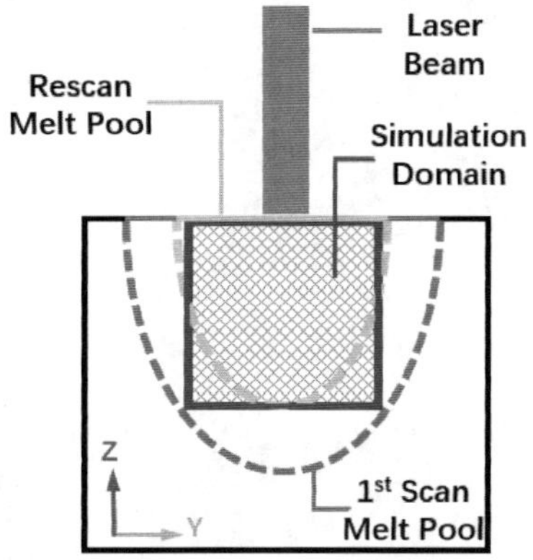

Figure 1. Schematic of the melt pool and simulation domain during the 1st scan and the rescan.

3. Results and discussion

3.1 Temperature profile

Figure 2 shows the predicted temperature field and melt pool during the first laser scan (figure 2(a) and (b)) and the laser rescan (figure 2(c) and (d)). In the simulation, the laser scans along X-axis, and the layer is built along Z-axis. Figure 2(a) and (c) show the top views, and figure 2(b) and (d) show the cross sections at Y-Z plane during laser scans. Assuming the temperature at the melt pool boundary equals the liquidus temperature of AlSi10 in Al-Si phase diagram (865K), the approximated melt pool depth during the first scan and the rescan is 110 μm and 90 μm respectively, while the approximate width is 160 μm and 125 μm respectively. The decreased melt pool dimension during the rescan results from the decreased thermal absorptivity, increased thermal conductivity and increased density of consolidated grains comparing to powder particles [23]. The obtained temperature profiles are then applied to the CA model to simulate grain structure evolution.

3.2 CA grain structure simulation results

The predicted grain structures after the first scan and the rescan with the given temperature profile are shown in figure 3. As can be seen, (a1)-(a4) show the grain structures after the first scan, while (b1)-(b4) show the grain structures after the rescan. The grain structures are shown from 3D view and three cross-sectional views (X-Y, Y-Z and X-Z). The yellow dashed line in (b2) and (b3) represent the approximated melt pool boundaries during the rescan. The green arrows in (b2)-(b4) identify the thermal gradient during the rescan. Compared to the grain structure formed in the first scan, grain elongation along the thermal gradient in the rescan is clearly evident in (b2)-(b4). This kind of elongation becomes more remarkable near the melt pool boundary during the rescan. The grain elongation phenomena are caused by the domination of epitaxial growth at the melt pool boundary. The great temperature gradient during LBPF processing (generally higher than 10^6 K/m) results in insufficient undercooling for nucleation near the melt pool boundary [9]. Hence, in the rescan, the liquid near the melt pool boundary is cooled to a temperature which enables solidification first, but new grain formation is suppressed there due to great temperature gradient. As the result, epitaxial growth occurs along the preferred growing direction of the partially melted grains. After the heat source passes, with the combined influences of cooled-down liquid and decreased thermal gradient, undercooling becomes sufficient for nucleation. Thus, epitaxial growth at the pool center is suppressed by the formation of new grains. Therefore, the post-rescan grain morphology at the melt pool boundary and the center of the melt pool are very different.

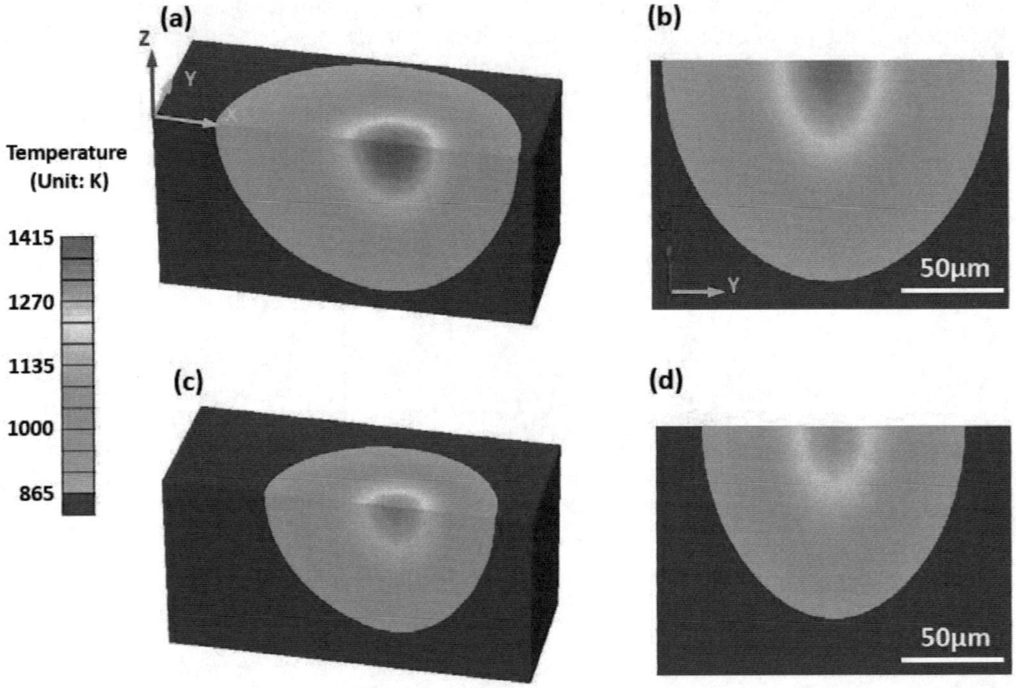

Figure 2. Temperature profiles calculated by FEA: (a) 3D temperature profile during the first scan; (b) Temperature profile at Y-Z cross section during the first scan; (c) 3D temperature profile during the rescan; (d) Temperature profile at Y-Z cross section during the rescan.

The effects of laser rescanning on grain structure are quantitively analyzed by comparing the volume of the grains, the dimension of the grains and their statistical distributions before and after the rescan. 1196 grains and 1208 grains are formed after the first scan and the rescan, with average volumes of 836.1 μm^3 and 827.8 μm^3, respectively. The nucleation of the grains is not limited by the maximum nucleation density applied to the simulation as the parameter is set as $1.5\times10^{15}/m^3$, which equals to 1500 in the simulation domain. Even though the difference in the average grain volume between the first scan and the rescan is not obvious, figure 4(a) reveals a significant difference in the grain volume distribution between the two scans. As shown in figure 4(a), there is a large number of grains with a grain volume of 200 to 600 μm^3 after the first scan, with a peak between 300 to 400 μm^3. In contrast, most of the grains formed in the rescan lie in the range of 0 to 400 μm^3, with a peak between 0 to 100 μm^3. These small grains are located in the middle of the melt pool, replacing the larger grains formed during the first scan. Additionally, compared with the first scan, the number of large grains with a volume over 100 μm^3 formed in the rescan has increased. This is due to epitaxially grown grains at the melt pool boundary during the rescan.

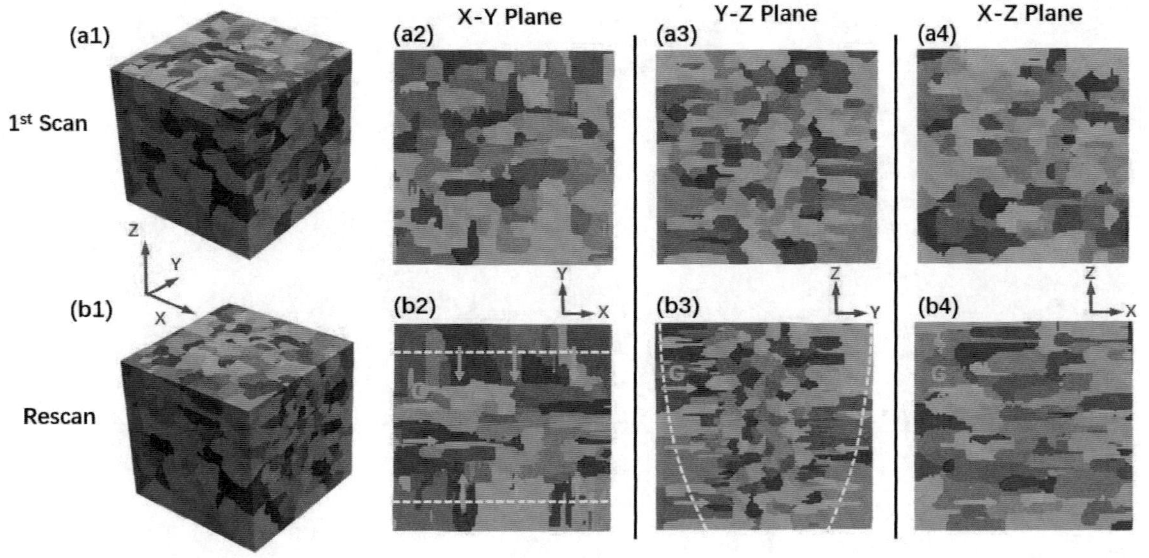

Figure 3. Simulated grain structure results by CA: (a1): 3D view of simulation domain after the first scan; (a2)~(a4):X-Y, Y-Z and X-Z cross-sectional views after the first scan; (b1): 3D view after the rescan; (b2)~(b4): X-Y, Y-Z and X-Z cross-sectional views after the rescan.

The dimensions of the grains along the X-axis (laser scanning direction), Y-axis (transverse direction) and Z-axis (build direction) are also measured to further reveal the morphology of the grain structure. The length of the grain along a certain coordinate axis is obtained by measuring the length of the straight line within the grain passing through the nucleation center and parallel to that coordinate axis. The average grain length along the three axes after the first scan and the rescan are (9.3±4.1 μm, 10.9±3.7 μm, 8.0±2.9 μm) and (10.7±8.6 μm, 11.2±7.1 μm, 6.3±4.0 μm), respectively. The grain size distributions along the axes are shown in figure 4(b)~(d). The increased grain size along the X-axis after the rescan is caused by the increased thermal gradient at the melt pool boundary and so caused grain elongation along the direction of laser scanning, where more grains are in the range of 15 to 30 μm compared to the first scan shown in figure 4(b). The average grain size change along the Y-axis is minor. The reason is that although a lot of small grains are formed in the middle of the melt pool, the grains that survive at the melt pool boundary during the rescan are enlarged along Y axis due to epitaxial growth, so that their effects on average size are cancelled. The decreased average grain size along the Z-axis is due to the formation of small equiaxed grains and weaken elongation effect along the build direction. Thus the distribution is shifted to the left as a whole compared to the first scan.

Finally, the collected crystallographic orientation information from the CA simulations has been processed by Mtex-MATLAB to calculate the pole figures (PF) images of the simulation results. Figure 5 gives the PF images of the grain structure formed in the first scan and the rescan. The impact of the laser rescan on crystallographic orientation characteristics cannot be ignored. As shown in figure 5(a1)~(a3), the grain structure after the first scan represents maximum polar densities of 8.2, 2.7 and 3.5 in the (100), (110) and (111) PF images, respectively. For comparison, the rescanned grain structure has the maximum densities of 10.0, 3.3 and 3.8 in the PF images for the same three directions, showing a much stronger crystallographic texture. This is caused by more competitive growth and elongation effects along the direction of the thermal gradient during the rescan.

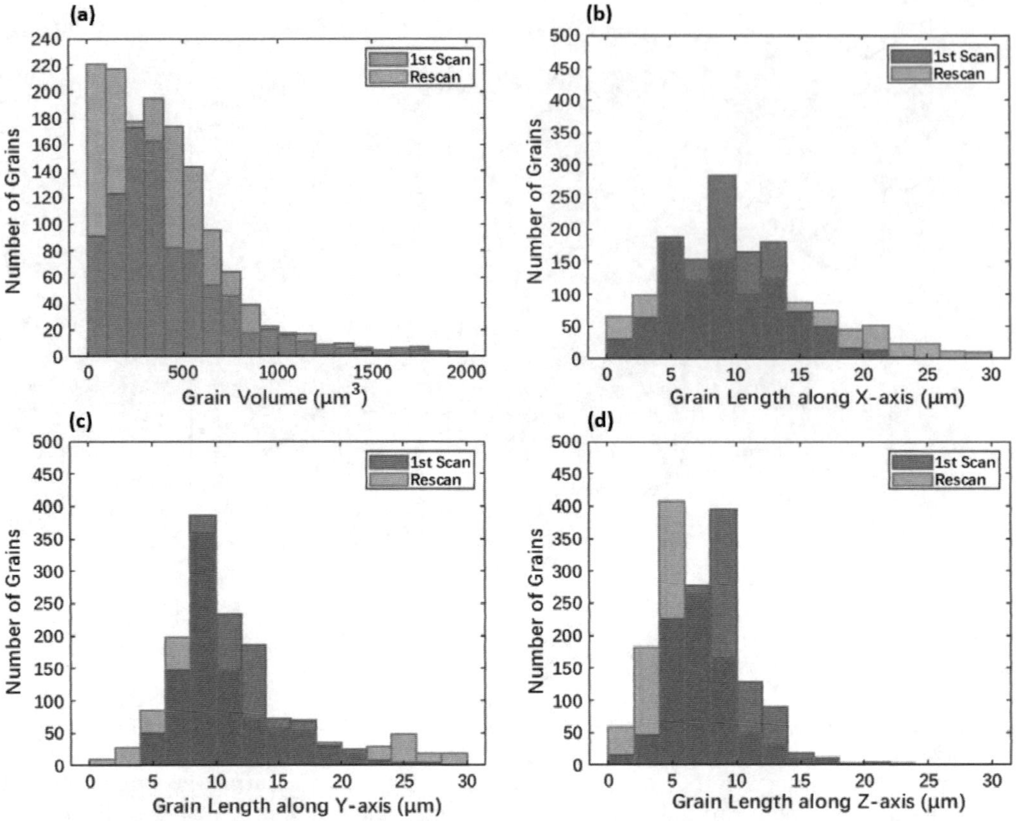

Figure 4. Grain volume and length statistical distributions after the first scan and the rescan: (a): Grain volume distribution; (b,c,d) Grain length distributions along the X, Y, and Z axes.

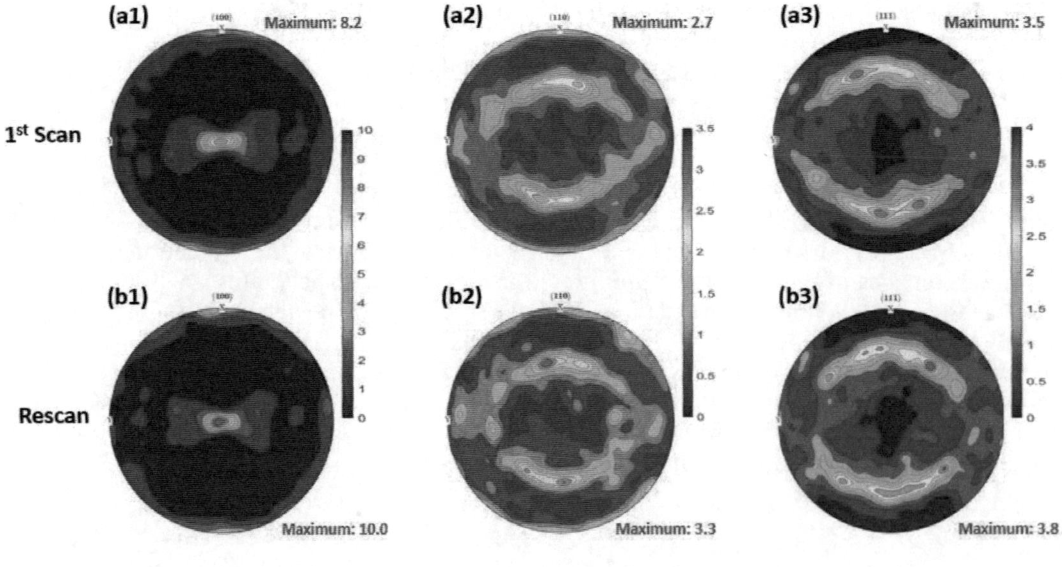

Figure 5. Pole figures for grain structure formed in 1st scan and the rescan: (a1)~(a3): Pole figures for the grain structure after the 1st scan in (100),(110) and (111), respectively; (b1)~(b3): Pole figures for the grain structure after the 1st scan in (100),(110)and (111), respectively.

4. Conclusions

The grain structure evolution in the laser rescanning process of LBPF has been studied numerically using a 3D cellular automata-finite element analysis simulation for the example case of the AlSi10 alloy. The model is shown to predict the formation of a mixed grain structure of columnar and equiaxed grains in the melt pool depending on the local thermal conditions. The following specific conclusions can be drawn:

1. Laser rescanning significantly changes the grain structure in the melt pool formed during the first scan due to altered thermal conditions, where thick and columnar grains appear at the new melt pool boundary, and fine columnar and small equiaxed grains form at the center of the new melt pool.
2. The distribution of grain volumes after the rescan suggests that the newly formed small grains driven by sufficient undercooling in the melt pool contribute to the grain refinement effects, impeding the epitaxial growth at the melt pool boundary when the molten is cooling down.
3. Laser rescanning creates a stronger crystallographic texture in the melt pool following the direction of temperature gradient, which is verified by incremental polar densities in PF figures.

The final aim of this study is to give meaningful suggestions for grain refinement control of laser rescanning strategy during the practical LBPF process. To further explore the laser remelting phenomena, future work will include considerations of different laser scanning strategies (i.e., different hatch styles and processing parameters), simulation of multi-tracks within multiple layers in the LPBF process and corresponding experimental validations of simulation results with tuned processing parameters.

References

[1] Khairallah S A, Anderson A T, Rubenchik A and King W E 2016 *Acta Mater.* **108** 36-45
[2] Liu B, Li B Q and Li Z 2019 *Results Phys.* **12** 982-88
[3] Guan J, Jiang Y, Zhang X and Chong X 2020 *Mater. Charact.* **161** 110079
[4] Shiomi M, Osakada K, Nakamura K, Yamashita T and Abe F 2004 *CIRP Ann.* **53** 195-98
[5] Zhou J, Han X, Li H, Liu S and Yi J 2021 *Mater. Des.* **210** 110092
[6] Körner C, Markl M and Koepf J A 2020 *Metall. Mater. Trans. A.* **51** 4970-83
[7] Kurz W I, Giovanola B and Trivedi R 1986 *Acta Metall.* **34** 823-30
[8] Gandin CA, Desbiolles J L, Rappaz M and Thévoz P 1999 *Metall. Mater. Trans. A.* **30** 3153-66
[9] Yuan L, Sabau A S, StJohn D, Prasad A and Lee P D 2020 *IOP Conf. Ser.: Mater. Sci. Eng.* **861** 012007
[10] Yuan L and Lee P D 2010 *Modell. Simul. Mater. Sci. Eng.* **18** 055008
[11] Yuan L and Lee P D 2012 *Acta Mater.* **60** 4917-26
[12] Sabau A S, Yuan L, Fattebert J L and Turner J A 2023 *Comput. Phys. Commun.* **284** 108605
[13] Goldak J, Chakravarti A and Bibby M 1984 *Metall. Trans. B.* **15** 299-305
[14] Chen S, Guillemot G and Gandin C A 2016 *Acta Mater.* **115** 448-67
[15] Guillemot G, Gandin C A, Combeau H and Heringer R 2004 *Modell. Simul. Mater. Sci. Eng.* **12** 545
[16] Trivedi R and Kurz W 1994 *Int. Mater. Rev.* **39** 49-74
[17] Gandin C A and Rappaz M 1994 *Acta Metall. Mater.* **42** 2233-46
[18] Mohammadpour P, Plotkowski A and Phillion A B 2020 *Addit. Manuf.* **31** 100936
[19] Mohebbi M S and Ploshikhin V 2020 *Addit. Manuf.* **36** 101726
[20] Thijs L, Kempen K, Kruth JP and Van Humbeeck J 2013 *Acta Mater.* **61** 1809-19
[21] Boley C D, Mitchell S C, Rubenchik A M and Wu S S 2016 *Appl. Opt.* **55** 6496-500
[22] Kik T 2020 Mater. **13** 608
[23] Griffiths S, Rossell M D, Croteau J, Vo N Q, Dunand D C and Leinenbach C 2019 *Mater. Charact.* **143** 34-42

IOP Conf. Series: Materials Science and Engineering 1281 (2023) 012010 doi:10.1088/1757-899X/1281/1/012010

Understanding conformal cooling in the diecasting of aluminum alloys

Xin He[1], Chenglu Tang[1], Corey Vian[2], Chi-Seng Wong[3], Shuei-Wan Juang[3], Wei Vian[1], and Xiaoming Wang[1*]

[1]School of Engineering Technology, Purdue University, West Lafayette, IN 47907 USA

[2]Stellantis Corporation, 3360 US-31, Kokomo, IN 46901 USA

[3]Department of Mechanical and Mechatronic Engineering, National Taiwan Ocean University, Taiwan

*Corresponding author's email: wang1747@purdue.edu

Abstract. Heat transfer and stress distribution are studied using computational fluid dynamic and structural simulation for a conformal cooling die insert made of Maraging M300 steel by selective laser melting. The phase transformation of cooling water is studied to understand the cooling effects and stress concentration on die life. The results of simulation are validated in production. The failure of the die insert has been revealed a result of enhanced stress concentration and corrosion directly related to the formation of vapor in the conformal cooling channel.

1. Introduction

The segregation of gasses and shrinkage porosity has a significant adverse effect on the mechanical properties of diecasting products [1-3]. The defects are formed in solidification related to cooling rate, requiring a proper thermal management. Therefore, cooling channels are introduced to control the temperature of a die and therefore the solidification of alloys in the die cavity [4, 5]. Traditional cooling channels are drilled holes in die bodies, which offer moderate and uneven cooling effects. In comparison, conformal cooling allows an effective heat extraction from precise locations, reducing the cycle time and producing high-quality products [6-8]. Additive manufacturing, such as selective laser melting (SLM), enables conformal cooling dies of complex geometries [9-12]. The layout and the shapes of a cooling channel affect heat transfer significantly [13]. An optimized cooling channel reduces cooling time for hot spots efficiently, and both the shape and layout of a conformal cooling channel (CCC) should be considered in design [14].

When a high temperature alloy melt is injected into a die cavity, a large temperature gradient is produced between the die surface and a cooling channel [15]. The stress concentration thus produced nucleates cracks primarily at the surface and propagate towards the interior of a die [16]. In addition, SLM steel die is anisotropy in microstructure and cracks form along the inter cellular regions, causing fatigue [17]. Cracks may also be a result of the dissolution of die material into the alloy melt as erosion damage [18].

Various commercial software is used to simulate diecasting processes, with a great success. However, cooling water is still treated as a single-phase liquid and no phase change is considered [19], which does not reflect the real situation. The temperature of an alloy melt that has been injected into a die cavity is much higher than the boiling point of the cooling water and the heat flux is high enough to boil the water in a cooling channel. It should also be noted that the operation temperature of a die surface is above 480K, high enough to vaporize the cooling water already. The cooling effects of the two-phase mixture are different from that of a single-phase liquid water. In a work of on metal injection molding [20], heat transfer between the two-phase mixture and a melt increased significantly accompanied by hydraulic instabilities. Meanwhile, active atoms in the vapor accelerate chemical reactions with the die material leading to an enhanced corrosion [21]. However, the influence of phase change from liquid to vapor on the failure of dies is still unclear.

To further understand conformal cooling, a failed SLM M300 insert with a CCC for A380 aluminum alloy is analyzed. Computational Fluid Dynamics (CFD) software, ANSYS Workbench with Fluent package, is used to evaluate the effects of phase change from water to vapor on the thermal and stress states of the die insert. The results are correlated to the microstructure near the cracks under high stress and accelerated corrosion.

2. Simulation and materials

2.1. CFD mathematical model

To reduce the complexity of simulation, several assumptions are made as follows:
(1) The microstructure of the insert is homogeneous and isotropic.
(2) The temperature of the vapor is constant in the simulation process.

The governing equations are defined as follows:

$$\frac{\partial}{\partial t}(\rho E) + \nabla \cdot \left(\vec{v}(\rho E + p)\right) = \nabla \cdot \left(k_{eff} \nabla T - \sum_j h_j \vec{J}_j + \left(\bar{\bar{\tau}}_{eff} \cdot \vec{v}\right)\right) \tag{1}$$

$$E = h - \frac{P}{\rho} + \frac{v^2}{2} \tag{2}$$

$$h = \sum_j Y_j h_j + \frac{P}{\rho} \tag{3}$$

$$h_j = \int_{T_{ref}}^{T} c_{p,j}\, dT \tag{4}$$

where k_{eff} is the effective conductivity, and $\vec{J}_j$ is the diffusion flux of species j, the $\nabla \cdot \left(k_{eff} \nabla T - \sum_j h_j \vec{J}_j + \left(\overline{\overline{\tau}}_{eff} \cdot \vec{v}\right)\right)$ represents energy transfer due to conduction, species diffusion, and viscous dissipation, respectively; the sensible enthalpy h is defined for incompressible flows, Y_j is the mass fraction of species j and the T_{ref} is 298.15K.

For the volume fraction equation, the interface between the phases is tracked by the solution of a continuity equation for the volume fraction of the phase, for the Nth phase, the equation has listed as follows:

$$\frac{1}{\rho_n}\left[\frac{\partial}{\partial t}(\alpha_n \rho_n) + \nabla \cdot (\alpha_n \rho_n \vec{v}_n) = S_{\alpha_n} + \sum_{p=1}^{N} (\dot{m}_{pn} - \dot{m}_{np})\right] \tag{5}$$

where $\dot{m}_{np}$ is the mass transfer from phase n to phase p and $\dot{m}_{pn}$ is the mass transfer from phase p to phase n. The primary-phase volume fraction is computed based on the following constraint:

$$\sum_{n=1}^{N} \alpha_n = 1 \tag{6}$$

2.2. Simulation
The partial view of the die insert with a CCC is shown in figure 1.

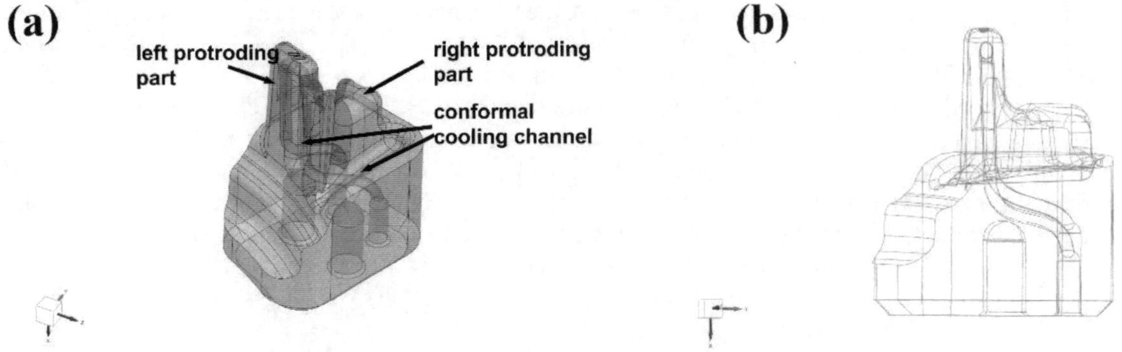

Figure 1. Geometry and dimensions of the insert: (a) isometric view and (b) front view.

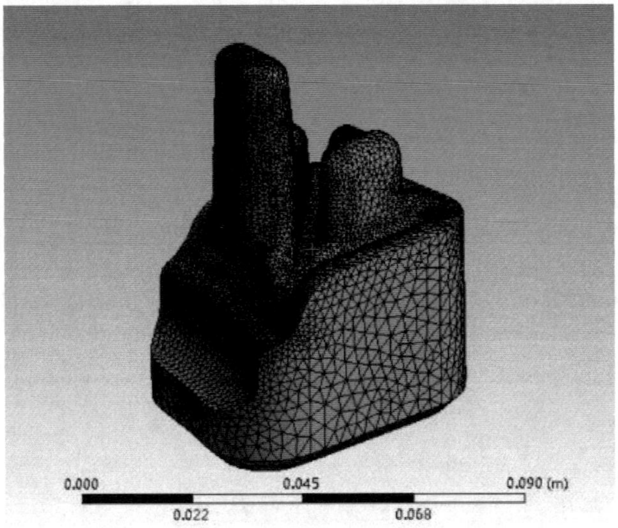

Figure 2. Solid model and mesh structure of the insert.

CFD simulation of the die insert was carried out using ANSYS Workbench with Fluent package. figure 2 shows the FE model and mesh structure of the die insert. Numerical mesh size and distribution are designed using automatic numerical mesh generator. The transient formulation is first-order implicit, and the SIMPLE scheme and direct iteration method is employed to solve the nonlinear differential governing equations. The calculation is considered converged when the energy residual is less than 10^{-6} and the velocity and continuity residual are less than 10^{-3} in each step.

2.3. Structural finite element analysis governing equation
The motion equation is shown as follows:

$$\sigma_{ji.j} + F_i = p\theta_{ii}u_i \tag{7}$$

This equation can be expanded into three dimensions,

$$\frac{\partial \sigma_x}{\partial x} + \frac{\partial \tau_{yx}}{\partial y} + \frac{\partial \tau_{zx}}{\partial z} + F_x = \rho \frac{\partial^2 u_x}{\partial t^2} \tag{8}$$

$$\frac{\partial \tau_{xy}}{\partial x} + \frac{\partial \sigma_y}{\partial y} + \frac{\partial \tau_{zy}}{\partial z} + F_y = \rho \frac{\partial^2 u_y}{\partial t^2} \tag{9}$$

$$\frac{\partial \tau_{xz}}{\partial x} + \frac{\partial \tau_{yz}}{\partial y} + \frac{\partial \sigma_z}{\partial z} + F_z = \rho \frac{\partial^2 u_z}{\partial t^2} \quad (10)$$

which are three equations with six independent unknowns (stresses).

The strain-displacement equations are given in the following, relating strains and displacements to nine independent unknowns (strains and displacements).

$$\varepsilon_{ij} = \frac{1}{2}(u_{j.i} + u_{i.j}),\ \epsilon_x = \frac{\partial u_x}{\partial x},\ \gamma_{xy} = \frac{\partial u_x}{\partial y} + \frac{\partial u_y}{\partial x} \quad (11)$$

$$\epsilon_y = \frac{\partial u_y}{\partial y},\ \gamma_{yz} = \frac{\partial u_y}{\partial z} + \frac{\partial u_z}{\partial y},\ \epsilon_z = \frac{\partial u_z}{\partial z},\ \gamma_{xz} = \frac{\partial u_z}{\partial x} + \frac{\partial u_x}{\partial z} \quad (12)$$

The continuous equation uses the Hook's equation.

$$\sigma_{ij} = C_{ijkl}\varepsilon_{kl} \quad (13)$$

where $\sigma_{ij} = \sigma_{ji}$ is the Cauchy stress tensor, F_i is the body force, $u_{j.i}$ is the displacement, C_{ijkl} is the stiffness tensor and $\varepsilon_{ij} = \varepsilon_{ji}$ is the strain.

2.4. Boundary and operation conditions

Table 1 gives the physical property of M300 tool steel, where $\mathrm{h_m}$, $\mathrm{h_i}$, $\mathrm{h_t}$, P_m and q_m are the thermal conductivities from melt to its surrounding (die), and from melt to insert and the insert to channel, the specific heat and coefficient of thermal expansion, respectively. $\mathrm{h_i}$ and $\mathrm{h_t}$ are system coupled (SC) and adopted from the software and updated in each step. The surroundings are assumed at constant temperatures. The operation condition is shown in table 2, where $\mathrm{T_m}$, $\mathrm{T_i}$, $\mathrm{T_t}$, $\mathrm{T_w}$ and $\mathrm{T_s}$ represent the initial temperature of the melt, insert, channel, and cooling water inlet, respectively. $\mathrm{V_w}$ represents the velocity of the cooling water inlet. The insert is in contact with the alloy melt (casting) while cooling water is flowing inside the insert for the simulation.

Table 1. Thermal physical property of M300 steel as the function of temperature [22, 23].

Temperature (K)	293	373	773	1273	1773
Density (kg m^{-3})	7750	7750	7750	7750	7750
Thermal conductivity (W m^{-1} K^{-1}), h_m	22.2	23.4	24.7	25.5	26.4
Thermal conductivity (W m^{-1} K^{-1}), h_i	SC	SC	SC	SC	SC
Thermal conductivity (W m^{-1} K^{-1}), h_t	SC	SC	SC	SC	SC
Specific heat (J kg^{-1} K^{-1}), P_m	501	511	532	548	574
Coefficient of Thermal expansion $\times$ 10^{-6} (K^{-1}), q_m	10.1	10.1	10.1	10.1	10.1

Table 2. Operation boundary conditions.

T_m (K)	T_i (K)	T_t (K)	T_W (K)	T_s (K)	V_W (m/s)
903	300.15	300.15	300.15	300.15	6.5

3. Results and discussions

3.1. Temperature gradient and phase change

A comparison of the temperature distributions across CCC with and without a phase change from water to vapor is given in figure 3. The temperature rises less for the cracking location when vapor forms in 0.5 s than that without vapor (i.e., a much better cooling). At the surface of the CCC, with vapor, the temperature and the temperature gradient near crack No. 1 are about 481.9 K and 129.22 K mm^{-1}, respectively. In comparison, without the formation of vapor, the temperature and the temperature gradient are 661.8 K and 73.85 K mm^{-1}, respectively.

Figure 4 shows the vapor boundary location with the same time step as in figure 3. Irregular vapor boundaries are generated first at two bending positions in the CCC, which is consistent with the observation in the section view (t=0.5 s). After the initial vaporization, the generation of vapor continues and flowing in the channel together with the water, which may cause minor flow-assisted corrosion (FAC) [24]. For comparison, the situation without the formation of vapor is shown in figure 4.

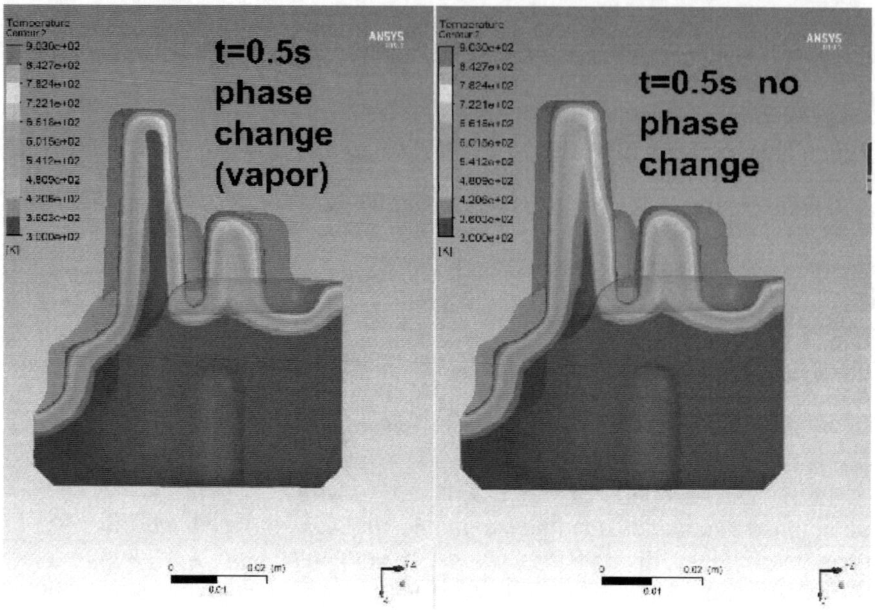

Figure 3. Temperature distribution in the die insert on the cross-section.

3.2. *Effect of phase change on heat transfer and pressure*

In figure 5, three yellow lines, locating on the right, left and middle of the CCC, are drawn near the major crack (will be given later), to determine the temperature with (figure 5a) and without vapor generation (figure 5b).

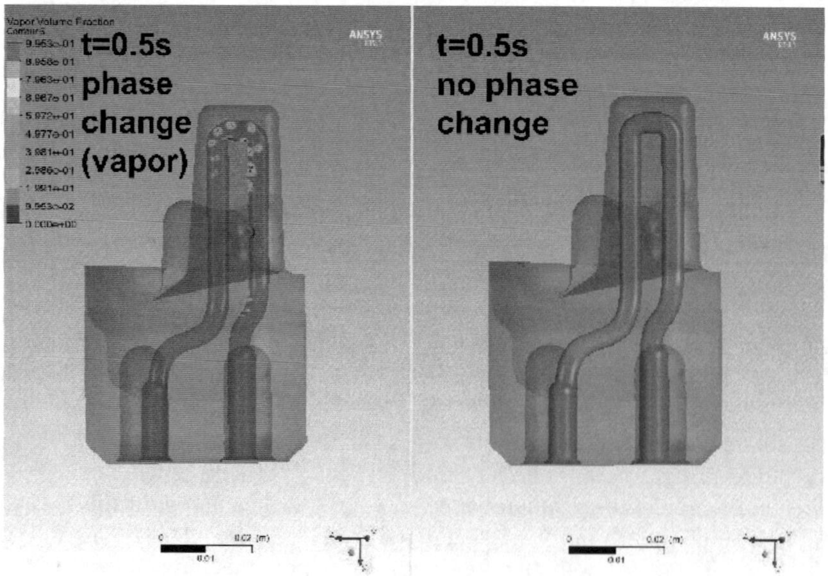

Figure 4. Volume fraction of vapor in the cooling channel on water domain surface.

Figure 6 shows temperatures along the three yellow lines. At t=0.1 s (figure 6c, and 6d), without vapor, the temperature of the cooling channel remains the same. When vapor forms at t= 0.5 s, the temperature in cooling channel doesn't change while the temperature in CCC with pure liquid water increases gradually. For CCC with vapor formation, a high temperature gradient than that without evaporation is resulted. And a large temperature gradient results in thermal stress concentration and thermal deformation leading to various defects [25]. With evaporation, the CCC is under large thermal stress, meaning that it has higher risk of fatigue than a channel that only has single liquid phase of water. In the left- (figure 6a and 6b) and right-side areas (figure 6e and 6f), as time increases, the formation of vapor in the CCC results in less temperature increase than that without vapor. For line 3, temperature is lower than line 1.

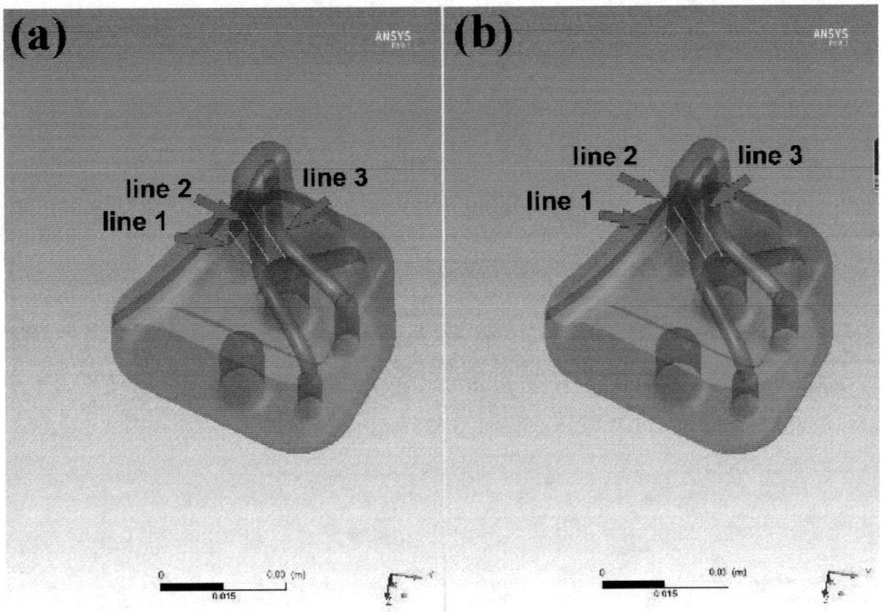

Figure 5. Yellow lines are drawn at the left (line 1) and right sides (line 3) of CCC, and across (line 2) CCC: top views (a) with and (b) without vapor.

Figure 7 shows a comparison of static pressure in the CCC with and without a vapor, with the same time step as in figure 3. A higher pressure in the first half of the CCC than the second half is shown. When vapor is generated at the two bending positions of the CCC, larger viscosity impedes water flow, resulting in greater forces acting on the first half of the channel. For the pressure in the CCC with single liquid phase of water, the pressure is a constant value.

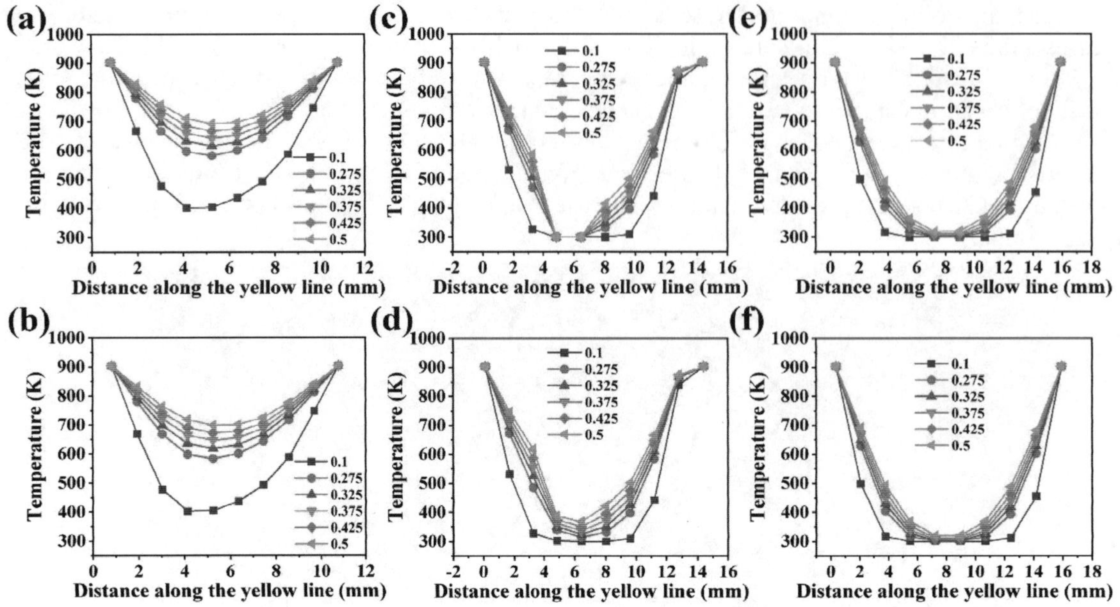

Figure 6. Temperature distribution along the yellow lines: (a) line 1, (c) line2 and (e) line 3 with and (b) line 1 (d) line 2, and (f) line 3 without vapor formation.

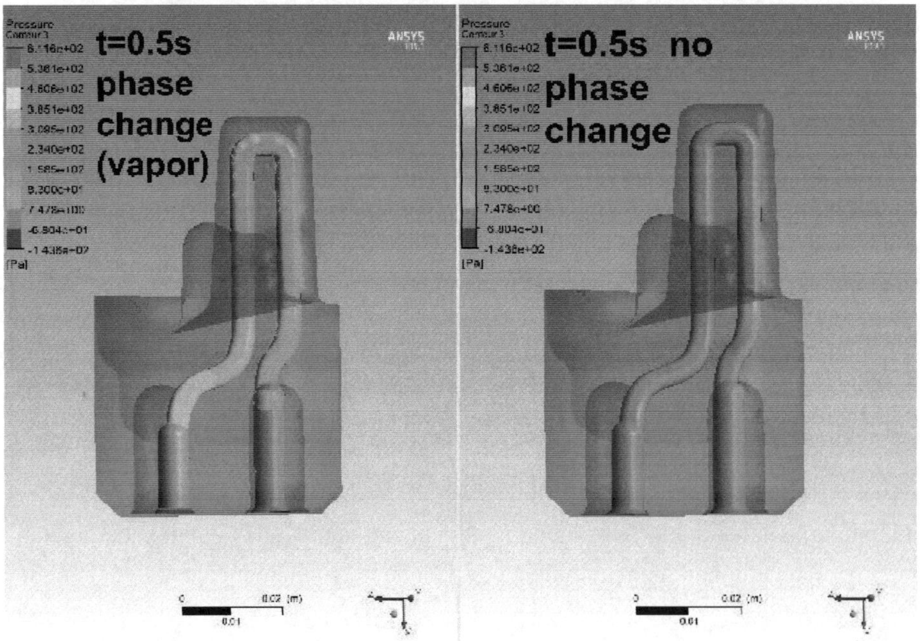

Figure 7. Comparison of pressure distribution in CCC with and without a phase change.

3.3. *Effect of phase change on the stress distribution*

The initial pressure value is 1 MPa on the wall of the CCC. The temperature of the cracking area on CCC wall was set to 480.9 K and 661.8 K for vapor generation and no vapor generation, respectively. This is taken from figure 3. The temperature of the die reaches 533.15 K, in a steady state.

Figure 8a shows the equivalent stress distribution in the die insert because of the vapor generation. It can be found that the maximum stress at the crack area is 1177.7 MPa which locates at the left protruding part exactly corresponding to the position of gas generation in the channel.

Figure 8b shows the internal temperature distribution without vapor. figure 9b shows the equivalent stress distribution in the die because of vapor generation. The maximum stress on the out surface is 1057.7 MPa which locates at the left protruding part. This stress is lower than that without vapor, meaning that vapor generation increases the stress in the channel.

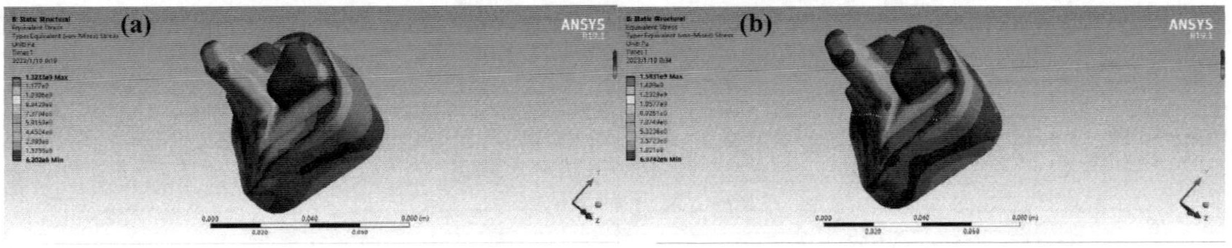

Figure 8. Stress distribution: (a) vapor generation, (b) without vapor generation.

3.4. Microstructural analysis

A major crack is observed on the surface of the die insert responsible for water leaking, as shown in figure 9. It appears longitudinally on the surface following the cooling channel.

Figure 9. Location of water leaking and crack.

3.5. Formation and propagation of cracks

Crack was initiated from the maximum stress location. And the maximum stress with vapor phase generation is much higher than that without vapor phase generation. A long-time exposure of the flow of cooling water and vapor will cause corrosion and it will promotes FAC and cracks. Vaporization increases heat transfer from the die, resulting in a larger temperature gradient and high thermal stress, favouring cracking. In addition, vapor generation increases the internal pressure due to the higher viscosity and the obstruction of the water flow.

4. Conclusions

CFD and structural analysis simulation were successfully applied to a SLM M300 die insert with CCC in predicting the initialization and propagation of cracks. The conclusions can be drawn as follows:
(1) A simple model was established for the prediction of flow fields and temperature distribution in a die insert.
(2) The predicted maximum von Mises stresses on the out surface of die insert and CCC wall are consistent with the location of cracking.
(3) CFD and structural simulation are effective in predicting the failure of the SLM manufactured M300 die insert with CCC.

References

[1] Inderjit and Sharma S, *Inter. J. Adv. Res. Innov. Ideas in Edu.* **3**
[2] Boeira A, Ferreira I, Garcia A, *Mater. Des.* **30** 2090
[3] Avalle M, *Inter. J. Fatigue.* **24** 1
[4] Shinde M and Ashtankar K, *Adv. Mech. Eng.* **9** 1
[5] Venkatesh G and Ravi Kumar Y, *Mater. Today: Proc.* **4** 2592
[6] Armillotta A, Baraggi R, Fasoli S, *Int. J. Adv. Manuf. Technol.* **71** 573
[7] Fiorentini F, Curcio P, Armentani E, Rosso C, Baldissera P, *Procedia Struct. Integr.* **24** 569
[8] Shayfull Z, Sharif S, Mohd A, Ghazali M, Saad R, *Adv. Polym. Technol.* **33** 1
[9] Kurtulus K, Bolatturk A, Coskun A, Gürel B, *Appl. Therm. Eng.* **194** 117105
[10] Park H and Dang X, *Procedia Manuf.* **10** 48
[11] Kuo C, Jiang Z, Yang M, You B, Zhong W, *Int. J. Adv. Manuf. Technol.* **114** 2697
[12] Mazur M, Brincat P, Leary M, Brandt M, *Int. J. Adv. Manuf. Technol.* **93** 881
[13] Hassan H, Regnier N, Lebot C, Defaye G, *Int. J. Therm. Sci.* **49** 161
[14] Hassan H, Regnier N, Pujos C, Arquis E, Defaye G, Appl. *Therm. Eng.* **30** 1547
[15] Shu L, Zhang Z, Ren Z, Zhang T, *J. Phys.: Conf. Ser.* **1939** 012002
[16] Bichler L, Elsayed A, Lee K, Ravindran C, *Int. J. Met.* **2** 43
[17] Damon J, Hanemann T, Dietrich S, Graf G, Lang K, Schulze V, *Inter. J. Fatigue.* **127** 395
[18] Persson A, Hogmark S, Bergström J, *J. Mater. Process. Technol.* **148** 108
[19] Wang G, Zhao G, Wang X, *Mater. Des.* **59** 310.
[20] Hestroni G, Mosyak A, Segal Z, *IEEE Trans. Comp. Packag. Technol.* **24** 16
[21] Tang Q, Hu Z, Cui X, Tao Z, Tang J, *Appl. Sci.* **12** 6813
[22] Yan L and Yan B, 2021 Math. Comput. Sci. *Preprint* 05.0114.v1
[23] Harvey P, American Society for Metals, eds., *Engineering properties of steel*, Metal Park, Ohio, 1982, **6** 477
[24] Kumar S and Balasubramaniam K, *J. Press. Vessel Technol.* **137** 061409
[25] Cao Z and Deng H, *Adv. Eng. Mater.* **22** 2101178

IOP Conf. Series: Materials Science and Engineering 1281 (2023) 012011 doi:10.1088/1757-899X/1281/1/012011

Phase-field study of polycrystalline growth and texture selection during melt pool solidification

Rouhollah Tavakoli and Damien Tourret

IMDEA Materials, Getafe, Madrid, Spain

E-mail: rouhollah.tavakoli@imdea.org, damien.tourret@imdea.org

Abstract. Grain growth competition during solidification determines microstructural features, such as dendritic arm spacings, segregation pattern, and grain texture, which have a key impact on the final mechanical properties. During metal additive manufacturing (AM), these features are highly sensitive to manufacturing conditions, such as laser power and scanning speed. The melt pool (MP) geometry is also expected to have a strong influence on microstructure selection. Here, taking advantage of a computationally efficient multi-GPU implementation of a quantitative phase-field model, we use two-dimensional cross-section simulations of a shrinking MP during metal AM, at the scale of the full MP, in order to explore the resulting mechanisms of grain growth competition and texture selection. We explore MPs of different aspect ratios, different initial (substrate) grain densities, and repeat each simulation several times with different random grain distributions and orientations along the fusion line in order to obtain a statistically relevant picture of grain texture selection mechanisms. Our results show a transition from a weak to a strong $\langle 10 \rangle$ texture when the aspect ratio of the melt pool deviates from unity. This is attributed to the shape and directions of thermal gradients during solidification, and seems more pronounced in the case of wide melt pools than in the case of a deeper one. The texture transition was not found to notably depend upon the initial grain density along the fusion line from which the melt pool solidifies epitaxially.

1. Introduction

The competition of columnar grains during directional solidification of metallic alloys occurs in many industrial processes like welding and additive manufacturing. It may promote the formation of a specific texture, and consequently the anisotropy in the mechanical properties of final products. In addition to melt convection and crystal nucleation ahead of the solid-liquid interface, the interactions between columnar grains mainly depend on the applied thermal field and physical properties of the material. According to the classical theory by Walton and Chalmers [1], dendritic grains with their preferred dendritic growth direction better aligned with the temperature gradient are favored during the growth competition. While this theory was applied to a wide range of applications with a reasonable outcome, experimental works [2, 3, 4] revealed that growth competition mechanisms are more complicated in practice. Recent phase-field (PF) studies on bi-crystal growth competition under a one-dimensional linear temperature field [5, 6, 7], have provided major insight in the fundamental understanding of grain growth competition mechanisms. However, simulations of dendritic grain growth competition in non one-dimensional temperature fields, with the level of accuracy afforded by PF, remain scarce [8, 9]. Here, we use parallel quantitative PF modeling to study columnar grain growth competition under a more general thermal condition. For this purpose, we consider the epitaxial polycrystalline solidification in the transverse cross-section of a melt pool corresponding to the selective laser melting (SLM) of a Nickel-based alloy. In particular, we explore the effect of the melt pool aspect ratio and the initial (substrate) grain size upon the selection of grain orientations and texture in the solidified melt pool.

2. Methods

While we explore a broader range of conditions, the models and simulations used here are similar in nature to those already presented and discussed in previous publications [8, 9]. Therefore, here we only remind some of the main features of the modeling framework. For more details, the reader is invited to refer to these prior works.

Essentially, we first calculate the alloy thermophysical data by computational thermodynamics using the CALculation of PHAse Diagram (CalPhaD) approach. Then, we perform the macro-scale thermal analysis of SLM process to determine the temperature field in and around the melt pool. Finally, we use the resulting thermal profile and history to simulate the solidification microstructure with phase-field simulations at the micro-scale. A one-way approach is used to link macro- and micro-scale simulations.

2.1. Computational alloy thermodynamics

The temperature-dependent density, $\rho(T)$, enthalpy density, $h(T)$, and effective heat capacity, $dh(T)/dT$ are computed by the CalPhaD approach (software ThermoCalc, database TCNI8). The phase diagram information is also computed by CalPhaD using a dilute pseudo-binary approximation of Inconel 718 (IN718) multi-component alloy. There are many different ways to perform a pseudo-binary approximation of a multi-component alloy [10]. Following [11], we consider Ni-5wt%Nb as the pseudo-binary surrogate for IN718. To make this approximation consistent with the transformation temperatures of the full alloy, first we compute the equilibrium liquidus temperature of IN718, T_L. Then, the partition coefficient of Nb, $k = c_s/c_l$, in IN718 is calculated at T_L, with c_s and c_l the Nb concentration in the solid and liquid phases, respectively. Finally, we calculate the slope of the liquidus line of the pseudo-binary phase diagram, m, at T_L as $m = -dT_L/dc$ where c denotes the concentration of Nb in IN718. The rest of the phase diagram data is readily obtained using the linear phase diagram assumption [8].

2.2. Macroscopic thermal simulation

Ignoring the effect of melt convection, the temperature history in the melt pool is computed by the numerical solution of the following heat transfer equation [12] in the macro-scale spatial domain, Ω_M,

$$\rho(T)\frac{dh(T)}{dT}\frac{\partial T}{\partial t} = \nabla \cdot \big(K(T)\nabla T\big) \tag{1}$$

The temperature-dependent thermal conductivity, $K(T)$, is tabulated from [13]. The spatial domain, Ω_M, is a parallelepiped of dimensions $5 \times 2 \times 2$ mm^3 along x, y, and z directions, respectively. Prior numerical convergence and size analysis [8] revealed that the resulting thermal field for a single track in its quasi-steady state was essentially independent of the domain size beyond these dimensions. The following boundary condition is applied on the top surface of Ω_M,

$$-K(T)\nabla T \cdot \boldsymbol{n} = h_a(T - T_a) + \epsilon_R\sigma_R(T^4 - T_a^4) + \frac{2\eta P}{\pi r_b^2}\exp\left(\frac{2d(x,z,t)^2}{r_b^2}\right) \tag{2}$$

where $\boldsymbol{n}$ denotes the unit outward vector on boundaries of Ω_M, h_a the convective heat transfer coefficient, T_a the ambient temperature, ϵ_R, the thermal radiation coefficient, σ_R the Stefan-Boltzmann constant, P the total laser power, η the power absorption coefficient, r_b the laser beam radius and d denotes the distance to the beam center. Assuming the initial location of the laser is denoted by $(x_0, z_0)^T$ on the top surface, then for every point $(x, z)^T$ on this surface, d is computed by $d(x, z, t) = \sqrt{(x - x_0 - Vt)^2 + (z - z_0)^2}$, where V denotes the scanning velocity. Importantly, neglecting fluid flow limits the applicability of our macroscopic analysis to low laser input energy density, i.e. conduction mode, and it may not be as adequate for keyhole regime. Another effect of fluid flow might be local temperature heterogeneities in the mushy zone, which are not represented with the elliptical temperature field approximation described further below.

Along the bottom surface, the powder bed is sitting on a substrate with thermal properties of stainless steel [8] and its inferior surface at $T = T_a$. The thermal properties of the powder bed are extrapolated from that of the dense material, considering average particle diameter, powder-bed compactness, inter-particle view factor, and properties of the surrounding gas (Argon) [8]. The temperature-dependent properties of the material switch from powder-bed to dense (fluid or solid) state when the local temperature first exceeds the alloy liquidus temperature. A zero heat flux condition is applied on the other boundaries. We solve Eq. (1) by the finite element method with a first-order Euler implicit scheme in time and trilinear hexahedral elements in space, using a uniform spatial mesh of element size 20 µm. Starting from the initial temperature T_a, the simulation runs until the temperature field reaches to the quasi steady-state conditions. Following Ref. [8], for our reference case, we used $P = 100$ W and $V = 0.1$ m/s, with $r_b = 35$ µm, $h_a = 15$ W/m^2/K, $T_a = 273$ K, $\epsilon_R = 0.3$, and $\eta = 0.55$.

Considering the location of the solidus isotherm, we denote x_c and x_e the x-coordinate of the deepest point and of the the tail of melt pool with respect to the frame moving with the scanning velocity. We then perform the two-dimensional phase-field simulation for (y, z) cross-sections of melt pool in the time interval $(0, t_f]$ where $t_f = |x_c - x_e|/V$. For the sake of computational performance, we fit an analytical function to the temperature field in the (y, z)-plane. For this purpose, the temperature is interpolated radially between the solidus temperature and the temperature at the center of melt pool $T(x_c, y_0, z_0)$. The solidus isotherm is approximated as a half-ellipse, and the central temperature is approximated using analytical functions — initially extrapolating radially to match reasonably the location of the liquidus isotherm, then transitioning toward a constant cooling rate as the melt pool center cools down below the liquidus temperature [9]. For the reference case, directly fitted to the finite element results, the initial (maximal) half-depth and half-width of the solidus isotherm are $d_S = 96$ µm and $w_S = 128$ µm, respectively. Figure 1 shows the considered evolution of the elliptically approximated solidus and liquidus isotherms in the (y, z) cross-section as the melt pool shrinks, considering these reference melt pool dimensions. In order to study the effect of melt pool geometry on the columnar grain growth competition, we simply multiply (w_S, d_S) by factors $(0.5, 2.0)$, $(1.0, 1.0)$, and $(2.0, 0.5)$ for the deep, reference, and wide melt pool scenarios, respectively.

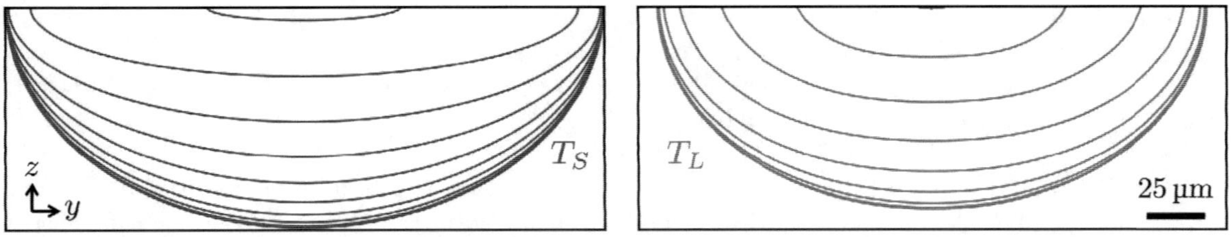

Figure 1. Time evolution of solidus (left) and liquidus (right) isotherms in the (y, z)-plane showing the shrinking of the melt pool. (Time increment between consecutive isotherms: 250 µs.)

2.3. Microscopic grain growth simulation

2.3.1. Phase-field model: We use a classical quantitative phase-field model for dilute binary alloys solidification introduced in Ref. [14], with a non-linear preconditioning of the phase-field [15] to reduce numerical mesh dependency. The resulting form of the equations can be written

$$\left(1-\frac{T-T_0}{mc_l^0}\right)a_s(\theta)^2\frac{\partial\psi}{\partial t}=\nabla\left[a_s(\theta)^2\right]\cdot\nabla\psi+a_s(\theta)^2\left[\nabla^2\psi-\phi\sqrt{2}\,|\nabla\psi|^2\right]-$$
$$-\frac{\partial}{\partial z}\left[a_s(\theta)\,a_s'(\theta)\,\frac{\partial\psi}{\partial y}\right]+\frac{\partial}{\partial y}\left[a_s(\theta)\,a_s'(\theta)\,\frac{\partial\psi}{\partial z}\right]$$
$$+\sqrt{2}\,\phi-\sqrt{2}\,\lambda(1-\phi^2)\left(U+\frac{T(z,y,t)-T_0}{mc_l^0(1-k)}\right) \tag{3}$$

$$\left(\frac{1+k}{2}-\frac{1-k}{2}\phi\right)\frac{\partial U}{\partial t}=\nabla\cdot\left(\tilde{D}\,\frac{1-\phi}{2}\,\nabla U+\left[1+(1-k)\,U\right]\frac{(1-\phi^2)}{4}\,\frac{\partial\psi}{\partial t}\,\frac{\nabla\psi}{|\nabla\psi|}\right)$$
$$+\left[1+(1-k)\,U\right]\frac{(1-\phi^2)}{2\sqrt{2}}\,\frac{\partial\psi}{\partial t} \tag{4}$$

where $T(z,y,t)$ denotes the temperature field that comes from the macro-scale simulation results, ϕ the phase-field variable, ψ the preconditioned phase-field variable, $\phi(z,y,t) = \tanh\left(\psi(z,y,t)/\sqrt{2}\right)$, $\theta = \arctan\left(\partial_y\psi/\partial_z\psi\right)$ the angle between the interface normal and the horizontal axis, U the dimensionless supersaturation, $U = \frac{1}{1-k}\left(\frac{2\,c/c_l^0}{1-\phi+k\,(1+\phi)} - 1\right)$, and c the solute concentration field, $c_l^0 = c_\infty/k$ the solute concentration of a flat interface at the reference (solidus) temperature T_0 for an alloy of nominal solute concentration c_∞. In Eqs (3)-(4), space is scaled in units of the diffuse interface width, W, and time is scaled in units of the relaxation time, τ_0, at T_0 [14]. The capillarity length, d_0, is computed at T_0 [14], $d_0 = \Gamma/\left(|m|c_\infty(1/k-1)\right)$, where Γ denotes the Gibbs-Thomson coefficient of the solid-liquid interface. The non-dimensional value for the liquid diffusion coefficient, $\tilde{D}$, and the coupling factor, λ, are computed according to the following identities: $\tilde{D} = D\tau_0/W^2 = a_1\,a_2\,W/d_0$, $\lambda = a_1\,W/d_0$, where D denotes the liquid diffusion coefficient, $a_1 = 5\sqrt{2}/8$ and $a_2 = 47/75$. We use a standard form of the fourfold anisotropy of the surface tension $\gamma(\bar{\theta}) = \bar{\gamma}a_s(\bar{\theta})$, with $a_s(\bar{\theta}) = 1 + \epsilon_4\cos(4\bar{\theta})$, where $\bar{\gamma}$ is the average surface tension in a $\langle 10\rangle$ plane, ϵ_4 is the strength of the surface tension anisotropy, and $\bar{\theta}$ is the angle between the normal to the interface and a fixed crystalline axis. For a crystal misorientation α_0 with respect to the coordinate axes, the anisotropy as function of θ between the interface normal and the x-axis follows $a_s(\theta) = 1 + \epsilon_4\cos\left(4\,(\theta-\alpha_0)\right)$. As the considered laser speed is relatively low ($V = 0.1$ m/s), the solid-liquid interface is assumed to remain under local equilibrium conditions. Therefore, the model does not consider kinetic undercooling or solute trapping, such that τ_0 is computed as $\tau_0 = a_2\,\lambda\,W^2/D$, and W is the only model parameter that should be appropriately chosen for the purpose of quantitative prediction. As in Refs [8, 9], the material properties for the Ni-Nb (IN718 surrogate) alloy used in PF simulations are $c_\infty = 5.0\,$wt%Nb, $k = 0.37$, $m = 9.0\,$K/wt%Nb, $\Gamma = 2.49\times10^{-7}$ Km, $D = 2.44\times10^{-9}$ m^2/s, $\epsilon_4 = 0.02$, with $T_M = 1670.43$ K and $T_0 = 1548.81$ K.

We solve Eqs. (3)-(4) under homogeneous Neumann (no-flux) boundary conditions along all directions. Moreover, the solid-liquid interface is initialized along the liquidus isotherm, with ψ initialized as the signed distance function to the liquidus isotherm (negative in the liquid region). The dimensionless supersaturation field is initialized based on the equilibrium concentration, i.e. $U(z,y,0) = -1.0$. The computational domain Ω_m is a rectangle of dimensions $2.01w_s \times 1.01d_s$, e.g. $\Omega_m = 257$ µm $\times$ 97 µm in the case of the reference melt pool configuration, with the center of the melt pool at the center of the top boundary (as represented in Fig. 1). The total simulated time is equal to 2.5 milliseconds in all cases.

2.3.2. Polycrystalline grain growth: To consider the columnar grain growth competition in the melt pool, the two-dimensional (2D) orientation of grains is stored in an auxiliary integer field. It assumes a value of 0 in the liquid phase and an integer in the range of $[1, 90]$ in the solid phase, which is also used as a discrete set of misorientation angles of the grains, in degrees. When $(1 - \phi^2)$ exceeds a threshold, here 0.01, the grain index is updated according to the local neighborhood. This method creates a halo of orientation field in the liquid around a grain. When a grid point is allocated a solid grain index value higher than 0, the index field no longer evolves, and neither does the solid-solid grain boundary (GB). While it does not account for solid-state microstructure evolution, this method is a computationally efficient alternative to multi phase-field models [16], since it relies on a single phase field. Its use is appropriate in the presence of well-developed dendritic structures, with GBs forming deep within the mushy zone, i.e. when the macroscopic orientation of resulting GBs depends primarily on the growth competition of primary and secondary dendrite tips in the the vicinity of the solidification front.

To initialize the grain index field, N_g number of points (Voronoi cell centers) are randomly distributed in Ω_m. A random integer value in the range of $[1, 90]$ is attributed to each point as the corresponding grain orientation. After the spatial discretization of Ω_m, the initial grain distribution is generated by a classic Voronoi tessellation algorithm. Finally, the grain orientation of computational cells located within the liquid region ($\psi < 0$) are reset to 0. The solution of Eqs (3)-(4) hence results in the epitaxial growth of columnar grains from the melt pool fusion line. Since we aim to assess the effect of initial grain size on the solidification microstructure, we considered different grain densities with N_g = 750, 1500, 3000 and 6000, corresponding to equivalent grain radii between 1.15 and 3.25 µm. Moreover, in order to gain a better statistical picture of final grain structure, each simulation is repeated five times with different initial grain distributions, hence resulting in 60 PF simulations in total.

2.3.3. Implementation: Phase-field Eqs (3)-(4) are solved in two dimensions by the finite difference method on a uniform spatiotemporal grid using the Euler explicit time integration scheme. The time step size, Δt, is considered as 0.3 of the maximum time step size based on the stability of Laplacian operators. A standard second-order five-point stencil is used to discretize Laplacian operators. The rest of terms in Eqs (3) and (4) are discretized by central difference schemes (see appendices of Ref. [6] for details).

The diffuse interface width $W = 0.8\,\Delta x$ and grid spacing Δx were determined based on a convergence study of the steady-state tip undercooling as a function of grid size in a unidirectional solidification with temperature gradient equal to its average in the mushy region and a pulling velocity equal to the scanning speed. Under conditions relevant to additive manufacturing, such a convergence study is quite limiting, since both the dendrite tip radius and the diffusion length are small. However, it is essential for the purpose of quantitative prediction of dendrite/cells growth kinetics and resulting grain structures. While the convergence study pointed at $\Delta x \approx 5$ nm for well-converged simulation in longitudinal simulations [8], here we can afford to use $\Delta x = 10$ nm, since isotherms and interface velocities are lower along the cross-section than they are along the longitudinal direction. Hence, the total number of spatial computational cells is equal to $25700 \times 9700 \approx 2.49 \times 10^8$ grid points in the case of the reference melt pool dimensions. While (y, z) dimensions are different, the total number of grid points is the same for the deep and wide configurations. The resulting time step is equal to 3×10^{-9} seconds, such that $O(10^5)$ time steps are required to complete the simulation.

Because of this grid size limitation for quantitative predictions, PF simulations at the scale of a full melt pool, even in 2D, are extremely computationally demanding. Therefore, advanced acceleration schemes are required. We implemented the model for massively parallel computing on multi-graphic processing units (Multi-GPU) with the computer unified device architecture (CUDA) programming language. Each simulation is performed on one computing node equipped with four Nvidia RTX-3090 GPUs. We use a layer-wise domain decomposition to distribute computation load among GPUs, with the computational domain divided into 4 almost equal layers along the y-direction (and the same number of grid points along x). An extra halo grid layer is added to the top and bottom rows of each domain to simplify the imposition of boundary conditions and inter-GPU data communication. Halo layer data is updated using direct GPU-GPU communication, so as to avoid expensive GPU-to-CPU and CPU-to-GPU data transfers.

3. Results and discussion

Figure 2 illustrates the evolution of the grain structure during competitive growth in a simulation for the reference melt pool size and a grain number $N_g = 750$. Figure 3 shows the final grain structures for one of the five simulation for each of the three considered melt pool sizes (reference, deep, and wide) and two considered initial grain numbers ($N_g = 750$ or 6000).

In all cases, the solid-liquid interface initially advances with a locally nearly-planar morphology at early stage. After a short time period, required for the development of a concentration boundary layer ahead of the interface, it destabilizes into cellular patterns. Several cells are eliminated early, leading to a rapid increase of the local primary spacing, while dendritic sidebranches appear after a short time. The average primary dendritic arm spacing (PDAS) was not found to markedly depend upon the initial grain density. Moreover, the iso-concentration line at $c = c_\infty$ in Fig. 2 (black line), which closely envelops the solidification front, shows that the diffusion length is of the same order or even smaller than the selected local primary spacings. While the diffusive conditions are an oversimplification of the transport regime in the melt pool — strong Marangoni convection is usually expected — this may provide an explanation for the typical lack of secondary sidebranches in additively manufactured metallic alloys.

A strong growth competition occurs between columnar grain in the melt pool. In qualitative agreement with the theory by Walton and Chalmers [1], grains with a $\langle 10 \rangle$ crystalline axis well aligned with the main heat transfer directions (i.e. perpendicular to the fusion line and isotherms) tend to prevail in the growth competition by eliminating the less favored grains with higher misorientation with the temperature gradient. In the case of the reference melt pool, the melt pool aspect ratio is close to unity (i.e. the fusion line is near circular). The resulting growth competition does not lead to a strong texture, as seen by the broad distribution of orientations present in the solidified melt pool (Fig. 3 a,b). In contrast, in the case of deep (Fig. 3 c,d) and wide (Fig. 3 e,f) melt pools, the growth competition results in a more noticeable $\langle 10 \rangle$ texture (apparent from the predominance of darker shades of red and blue). This $\langle 10 \rangle$ texture is primarily attributed to the large fraction of the melt pool solidifying under a mostly horizontal (Fig. 3 c,d) or vertical (Fig. 3 e,f) temperature gradient. The grain elimination mechanism proceeds via either dendrite impingement in the case of converging dendrites (grains) or side-branching in the case of diverging dendrites [6].

To further analyze texture selection, we quantified the orientation distributions in the different simulations. In Figure 4, histograms show normalized grain orientation (i.e. fraction of the melted area) with $N_{\text{bin}} = 10$ bins of width 9°, combining all grain densities (20 simulations per histogram). To illustrate the dependence upon initial grain densities, additional curves show idealized functions $H(\theta) = A\cos(4\theta) + 1/N_{\text{bin}}$, fitted to similar orientation distributions histograms within the five runs for each different initial grain density. The sole fitting parameter A may be interpreted as a measure of the strength of the $\langle 10 \rangle$ texture, with $A = 0$ corresponding to a flat distribution (no texture). This analysis confirm that deep ($A \approx 0.0749$) and wide ($A \approx 0.0855$) melt pools have a strong $\langle 10 \rangle$ texture (high population of grains with near 0 and 90 degrees orientations) in comparison to the reference melt pool geometry ($A \approx 0.0349$). Fitted curves for different N_g show that the effect of initial grain size on the final texture is marginal, with most values of A (all values for deep and wide cases) deviating by 20% or less from those obtained for the distributions combining all values of N_g (listed in parenthesis earlier).

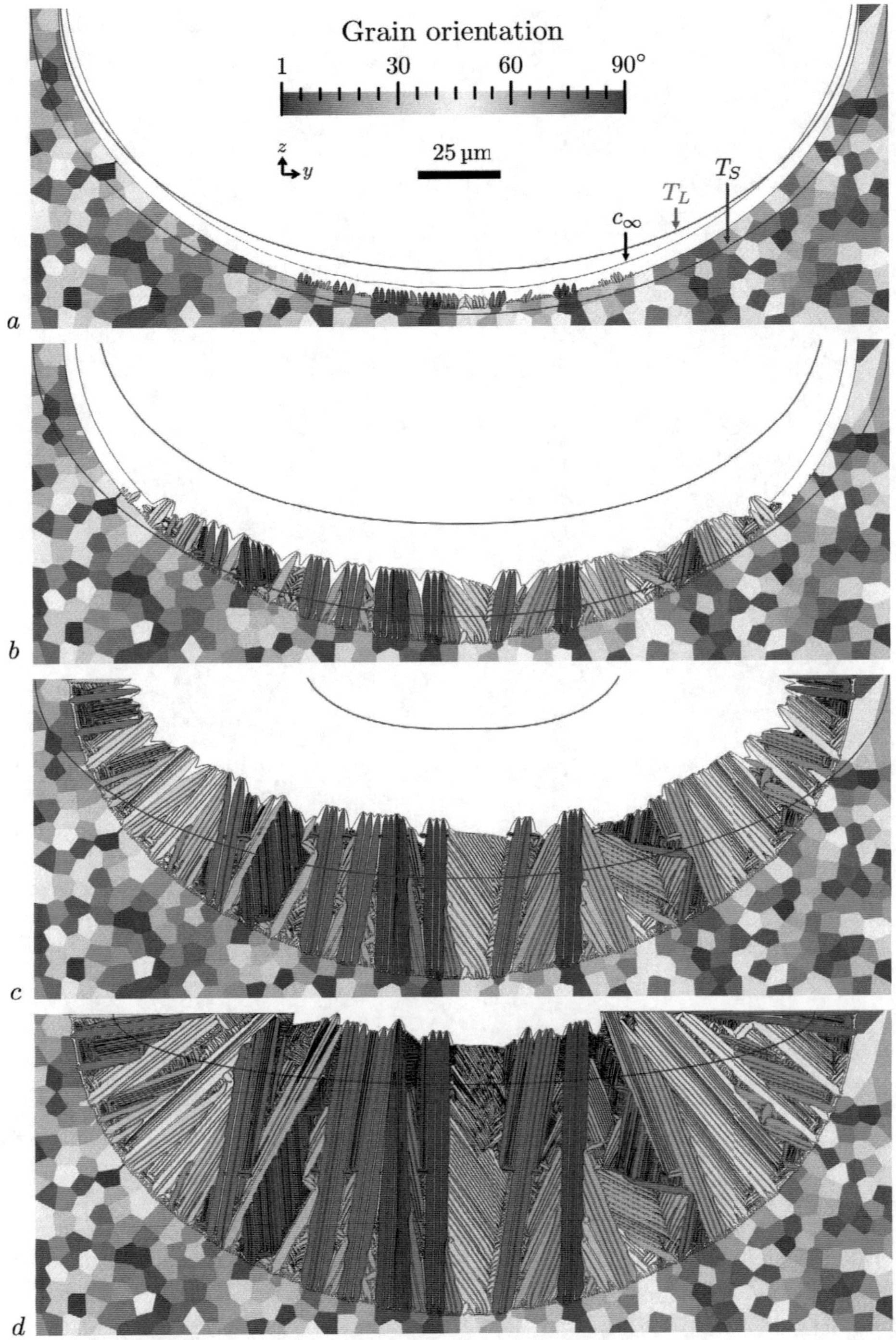

Figure 2. Evolution of grain structure with time for the reference melt pool size and $N_g = 750$, at t =(a) 0.75, (b) 1.25, (c) 1.75, and (d) 2.25 ms. Solidus and liquidus isotherms are marked in blue and red line, respectively, and the iso-concentration $c = c_\infty$ appears as black line.

IOP Conf. Series: Materials Science and Engineering 1281 (2023) 012011 doi:10.1088/1757-899X/1281/1/012011

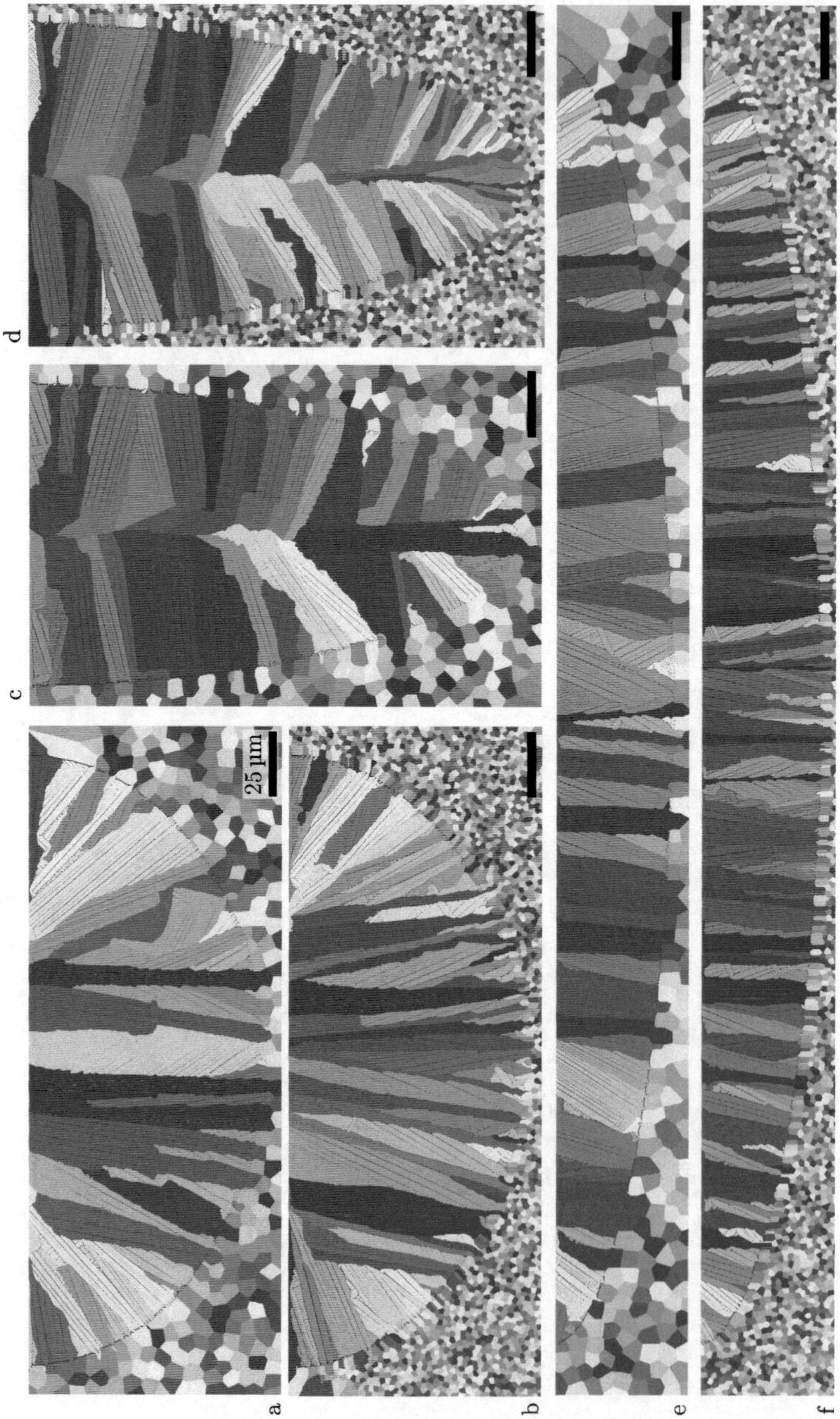

Figure 3. Representative final grain structures corresponding to reference (a,b), deep (c,d) and wide (e,f) melt pools with $N_g = 750$ (a, c, e) and $N_g = 6000$ (b, d, f). The grain orientation color map is similar to Fig. 2.

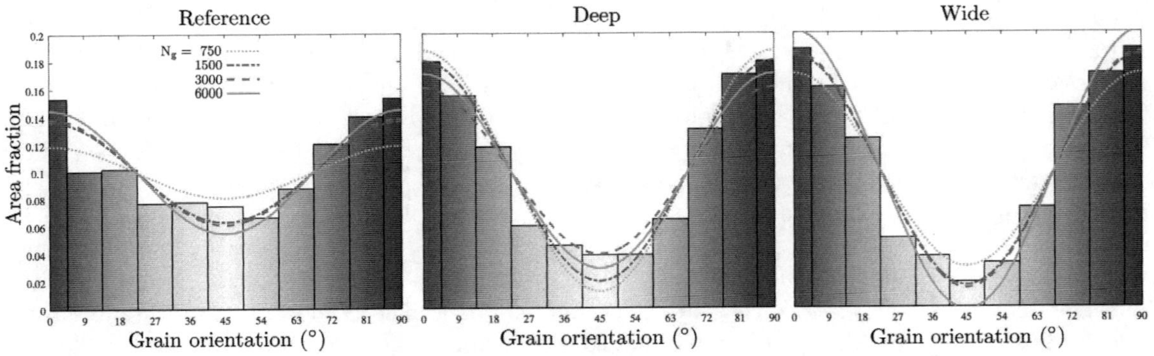

Figure 4. Histograms of grain orientations for reference (left), deep (center), and wide (right) melt pools. Each histogram shows the orientation distribution combining all four initial grain densities (20 simulations per histogram). Overlapped curves are the fitted cosine functions for the 4 different initial grain densities (5 simulations each). The grain orientation distributions are counted over each solid pixel and scaled by the total number of counted grid points, such that the ordinate axis corresponds to a fraction of the the melt pool area. Considering the fourfold symmetry of the crystal structure, histograms are periodic with respect to the orientation angle, such that the leftmost and rightmost bins are similar and include angles from $-4°$ (i.e. $86°$) to $4°$ (i.e. $94°$).

In the case of a deep melt pool, we can essentially divide the melt pool into two regions: a relatively texture-less bottom part with a circular/elliptical fusion line, and a textured upper part with a near vertical fusion line. As a result, the final $\langle 10 \rangle$ texture in the deep case is less strong compared to the wide melt pool, in which the main thermal gradient direction is essentially vertical. Interestingly, while these results show a clear texture transition from weak to strong $\langle 10 \rangle$ texture when the melt pool aspect ratio deviates from unity, they do not exhibit any indication of a transition toward other textures — such as, for instance, a possible $\langle 110 \rangle$ texture suggested from geometrical arguments and experimental hints [17, 18].

Finally, on the computational side, a preliminary scaling analysis shows a near linear scaling of the algorithm speedup with the number of GPUs. For instance, for the reference melt pool and $N_g = 6000$, the computational time is about 82, 43 and 22 hours using 1, 2 and 4 GPUs, respectively. Still, while the multi-GPU parallelization goes a long way in enabling full melt pool scale simulation in 2D, a similar 3D investigation remains unreasonable. Indeed, a single equivalent three-dimensional quantitative PF simulation at full melt pool scale would require a number of spatial grid points of $O(10^{13})$ and a similar number of time steps. Such a large-scale simulation, even using state-of-the-art petascale computing resources, would require O(day) to perform. It would therefore result in an enormous investment in time and resources for a relatively minor return in terms of physical insight into grain texture selection. For 3D simulations at this scale, coarse-grained approaches, e.g. cellular automaton, remain most convenient, while statistically capturing grain texture selection with reasonable accuracy [9].

4. Summary and perspectives

We performed quantitative phase-field simulation of solidification in thermal conditions relevant to SLM processing of a Nickel alloy. We limited our micro-scale simulations to two-dimensional cross-sections of the melt pool (perpendicular to the scanning direction), using a thermal history at quasi steady-state computed by 3D finite element analysis, using CalPhaD-computed alloy parameters [8, 9]. We studied the effect of the melt pool geometry (i.e. its aspect ratio) and of the initial grain density along the fusion line on the final microstructure. According to our results, a near-circular fusion line (i.e. an aspect ratio close to unity) does not lead to any notable texture, while relatively wider and/or deeper melt pools, which have more horizontally or vertically oriented fusion lines, promote the formation of a clear ⟨10⟩ texture.

Here, we only considered the single-track melting of a random Voronoi grain distribution. Therefore, the resulting grain texture is not completely representative of a realistic AM microstructure. Indeed, we decided to put the focus on the fundamental mechanisms of grain growth competition within a non-one-dimensional temperature field, which have received little attention so far. In order to produce a more realistic simulation of grain texture emergence from AM, one should account for multiple layers, multiple tracks, and nucleation events, all of which would be relatively straightforward to implement, as well as three-dimensional simulations, which, on the other hand, could be much more challenging computationally.

Perspectives following from this work are multiple. A three-dimensional quantitative phase-field study with the same level of accuracy (down to the level of individual dendrite) remains computationally prohibitive, and is thus unlikely to provide statistically-meaningful insight. However, even in two dimensions, the inclusion of additional physics into the simulations could certainly bring a fresh look onto grain growth competition within multidimensional temperature fields. For instance, including crystal nucleation in the melt pool would make it possible to study potential columnar-to-equiaxed transitions. The present study was also limited to the lower end of the velocity range relevant to SLM processing. In this low-velocity regime, the solid-liquid interface remains close to equilibrium, which is not always the case for typical SLM conditions. Ongoing work specifically focus on extending this kind of study to include kinetic undercooling and solute trapping using recent PF formulations [19, 20].

Acknowledgments

This work was supported by the Spanish Ministry of Science and the European Union NextGenerationEU (PRTR) through the MiMMoSA project (PCI2021-122023-2B) and a Ramón y Cajal fellowship (RYC2019-028233-I).

References

[1] Walton D and Chalmers B 1959 *Trans. Am. Inst. Min. Metal. Eng.* **215** 447–457
[2] D'Souza N, Ardakani M, Wagner A, Shollock B and McLean M 2002 *J. Mater. Sci.* **37** 481–487
[3] Wagner A, Shollock B and McLean M 2004 *Mater. Sci. Eng. A* **374** 270–279
[4] Zhou Y, Volek A and Green N 2008 *Acta Mater.* **56** 2631–2637
[5] Li J, Wang Z, Wang Y and Wang J 2012 *Acta Mater.* **60** 1478–1493
[6] Tourret D and Karma A 2015 *Acta Mater.* **82** 64–83
[7] Takaki T, Ohno M, Shibuta Y, Sakane S, Shimokawabe T and Aoki T 2016 *J. Cryst. Growth* **442** 14–24
[8] Elahi S, Tavakoli R, Boukellal A, Isensee T, Romero I and Tourret D 2022 *Comput. Mater. Sci.* **209** 111383
[9] Elahi S, Tavakoli R, Romero I and Tourret D 2023 *Comput. Mater. Sci.* **216** 111882
[10] Raghavan S, Singh G, Sondhi S and Srikanth S 2012 *Calphad* **38** 85–91
[11] Ghosh S, Ma L, Ofori-Opoku N and Guyer J 2017 *Modell. Simul. Mater. Sci. Eng.* **25** 065002
[12] Hong C 2019 *Computer modelling of heat and fluid flow in materials processing* (UK: CRC press)
[13] Mills K 2002 *Recommended values of thermophysical properties for selected commercial alloys* (England: Woodhead Publishing)
[14] Echebarria B, Folch R, Karma A and Plapp M 2004 *Phy. Rev. E* **70** 061604
[15] Glasner K 2001 *J. Comput. Phys.* **174** 695–711
[16] Steinbach I, Pezzolla F, Nestler B, Seeßelberg M, Prieler R, Schmitz G and Rezende J 1996 *Physica D* **94** 135–147
[17] Jadhav S D, Dadbakhsh S, Goossens L, Kruth J, Van Humbeeck J and Vanmeensel K 2019 *J. Mater. Process. Technol.* **270** 47–58
[18] Higashi M and Ozaki T 2020 *Mater. Des.* **191** 108588
[19] Pinomaa T and Provatas N 2019 *Acta Mater.* **168** 167–177
[20] Ji K, Dorari E, Clarke A J and Karma A 2023 *Phys. Rev. Lett.* **130** 026203

IOP Conf. Series: Materials Science and Engineering 1281 (2023) 012012 doi:10.1088/1757-899X/1281/1/012012

Modeling residual stresses of functionally graded deposits using the PTAAM

Geoffrey Bonias[1], Hani Henein[1], Tonya B. Wolfe[2]

[1]Department of Chemical and Materials Engineering, University of Alberta, Edmonton, AB, Canada
[2]Red Deer Polytechnic, Centre for Innovation in Manufacturing (CIMTAC), Red Deer, AB, Canada

E-mail: bonias@ualberta.ca, hhenein@ualberta.ca, tbwolfe@ualberta.ca

Abstract. High maintenance costs due to significant abrasive wear of components is experienced in the energy and mining sectors despite the current use of tough and hard coatings. During the coating process significant tensile residual stresses may build up and result in premature failure of a component. These tensile stresses can be reduced by adopting functionally graded structures of the composite. The main goal of the present study is to design an ideal material gradient in the additively manufactured composite using the plasma transferred arc (PTA) with a WC-Ni alloy graded in WC. To develop a comprehensive analysis of the functionally graded deposit, the thermal history of the WC and Ni alloy powders are first simulated as they travel through the plasma and deposit on the substrate. The initial temperature of each deposited bead depending on the WC content is deduced. These results are used as an input to predict the temperature and stress history of the solidifying deposit. Thermal stresses are computed and trends of residual stresses are obtained as a function of the WC gradient selected. The trends obtained are compared quantitatively before concluding on the most favorable gradient for this wear resistance application.

1. Introduction

The WC reinforced NiCrBSi or NiBSi Metal Matrix Composite (MMC) is a wear resistant coating in which the hardness of the tungsten carbides is combined with the toughness of the nickel matrix in order to withstand extreme wear conditions. The Direct Energy Deposition (DED) of that MMC using the Plasma Transferred Arc (PTA) has proven useful to protect mining tools against wear and abrasion [1]. The coating's mechanical and wear resistance, however, suffers from the development of large tensile residual stresses. One way to mitigate these residual stresses is to induce a smooth transition of the coating thermo-mechanical properties by gradually varying its material composition across its thickness; a method that fits into the concept of Functionally Graded Materials (FGMs). In this work the coating is deposited using an Additive Manufacturing (AM) method with its composition changing in a stepwise way (layer after layer), so the Functionally Graded coating is said to be a discontinuous one. One common gradient type is that of a power-law function: if the coating has two constituents A and B and the proportion of B in the mix (noted f_B) is increasing layer after layer according to that law, it will follow the following expression: $f_B(n) = f_B(N)(\frac{n}{N})^p$ with n being the current layer number and N the total number of layers. This equation can be applied in this work with f_B being the tungsten carbide content, negligible in the first layer but reaching up to 60wt% in the last layer (e.g. $f_{WC}(N) = 0.6$). With such a form, selecting an appropriate material gradient comes down to selecting the appropriate p exponent. It is then necessary to evaluate how the variation of this p exponent affects the resulting residual stresses in the graded coating. Predicting the thermal history of this PTA-AM deposition process is the preliminary step towards that goal. Such a prediction is not straightforward, as it requires a multi-scale understanding of the physics involved: first the interaction of the transferred plasma arc with the injected material to print (meso-scale level) and then the global heat transfer phenomena during the printing process (macro-scale level). The next section of the present paper focuses on the former, while the last section deals with the temperature history and the residual stresses predictions on a macro-scale level.

2. Thermal micro-modeling of the powder deposition process with the PTA

During the deposition process, the weldable material is introduced into the plasma transferred arc in the form of a powder by means of a carrier gas (Figure 1). The plasma gas is ionized by an electrode (typically the cathode) located inside the torch, with the nozzle having the effect of constricting the arc and increasing its energy density. The nozzle is continuously cooled by cold water and the weld is protected by a shielding gas also flowing from the same nozzle. To forecast the temperature history of the weldments and the accompanying thermal stresses, one must understand how these particles interact with the arc. Although these interactions have already been studied in the literature [2], the operating conditions specific to the present work differ from those considered by previous authors. These operating conditions may be listed as the Plasma Gas Flow Rate, Carrier Gas Flow Rate, Shielding Gas Flow Rate, Powder Flow Rate, Powder Composition, Powder Size Distribution, Powder Injection Angle and Initial Velocity Distribution, Nozzle to Substrate Standoff-Distance (or Voltage), Arc Current, Gases Composition, Torch Travel Speed, Electrode Composition and Tip Angle. The selected PTA operating parameters must be adapted to the powder size used and its material composition: here nickel-based and tungsten carbide particles are considered. The choice of these parameters will deeply impact how much these particles will be heated by the arc, and what their final temperature will be on average upon reaching the weld pool. In order to evaluate the impact of the operating conditions specific to this work on the powder particles final temperature, the characteristics of the arc must be studied, its interaction with the powder particles quantified, to be able to predict their temperature history inside the arc.

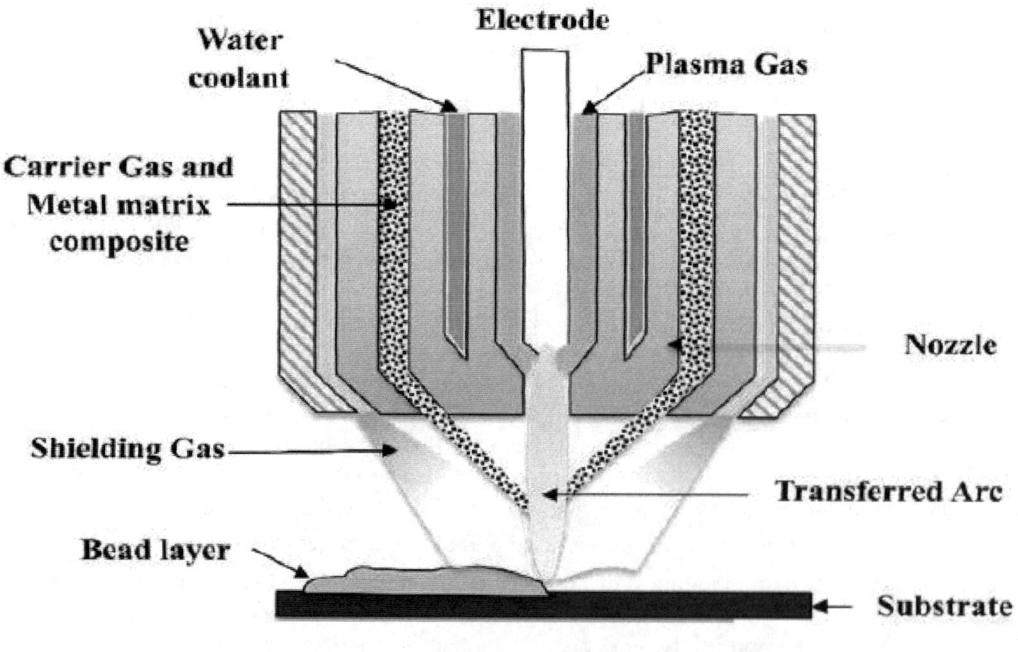

Figure 1: Schematic of the Transferred Plasma Arc (PTA) (Credit: [3])

2.1 Estimation of the velocity and average temperature of the plasma arc

A simplified modelling of the arc is conducted in order to get an estimate of its temperature and velocity. The main assumptions are the following:

- The plasma arc is stationary.
- Outside of the immediate vicinity of the cathode and anode, the plasma is in LTE (Local Thermodynamic Equilibrium), meaning that the electrons and heavy particles (atoms and ions) in this region of the plasma have the same temperature.
- The arc is axially symmetric and can be a cylinder shape with the same inner radius as the constricting nozzle (2mm). The arc length past the nozzle exit is 7mm.
- The current density is uniform and purely axial.
- There is a negligible amount of metallic vapor emanating from the powder and weld pool, therefore it does not affect the plasma temperature nor generate a thermal screening effect around the powder particles.
- The flow is laminar. This is verified by evaluating the Reynolds number.
- The plasma is optically thick, meaning that some radiations are re-absorbed within the plasma.
- The plasma can be modeled as a perfect fluid (negligible viscosity) everywhere except at the boundary with the powder particles or the electrodes.
- The carrier gas flow and shielding gas flow have no effect on the plasma temperature and velocity.
- The powder has a negligible influence on the plasma temperature and velocity. This is verified with calculations, considering the powder flow rate and size in use.

2.1.1. Plasma arc temperature. Additional assumptions are as follows. Conduction in the plasma and heat loss via the sides of the arc are negligible in comparison to radiation losses in the temperature range of 10,000–20,000K, and convection, not conduction, dominates axial heat transfer *[4]*. With the torch used in this work, the powder injection angles are such that the powder particles flow in the bottom half region (anode side) of the plasma arc. Axial convective heat transfer is less substantial on the anode side of the plasma arc than it is on the cathode side. It is assumed that the 1.5 lpm plasma gas flow rate utilized in this work is low enough for the upstream gas, which is typically hotter, not to significantly heat the plasma's anode zone.

Considering the aforementioned assumptions, the local heat balance equation in the central axis of the bottom area of the arc may be simplified such that the net radiation losses balance Joule heating. A comparative study with numerical model results shows this approximation yields sufficiently accurate predictions [5] when the operating conditions are similar to that of this work. The results obtained with this method are summarized in Table 1.

Table 1: Predicted average temperature of argon plasma at different currents

Plasma/Current	50A	57A	60A	63A	100A
Temperature	12 800K	13 100K	13 200K	13 400K	14 800K

2.1.2. Plasma velocity. Assuming a uniform current density, the radial pressure generated by the Lorentz forces can be calculated analytically (Maecker pressure) *[6]*. The electromagnetic pressure will be most elevated just below the cathode where the plasma arc is narrower. The radius of the arc when exiting the nozzle is assumed to match the nozzle inner radius. As the plasma arc becomes wider while flowing away from the nozzle, the electromagnetic pressure decreases, and this axial pressure gradient drives the flow. Using the previous assumptions on the plasma and by applying the Bernoulli theorem one can deduce the velocity profile of the arc *[5]*.

The velocity of the arc depends on the cathode tip radius, which can be calculated based on the current used and the cathode current density. The cathode current density is independent from the current and varies depending on the electrode shape and material composition. In this work a sharp thoriated tungsten electrode of 3.2mm diameter with 20° angle tip is used. For such an electrode, the value of current density is $1x10^8$ A/m^2 for a 5mm long arc [7]. In Table 2 are shown some central velocity predictions made using this calculation method.

Table 2: Central velocity of Ar & Ar/5%H_2 plasma at different currents

Plasma/Current	50A	57A	60A	63A	100A
Pure Argon	175 m/s	192 m/s	199 m/s	206 m/s	293 m/s

The plasma gas flow rate used in this work is 1.5 lpm. With the results given in Tables 1 and 2 and the calculated velocity profile of the arc, one can find that the cylindrical arc should indeed maintain a radius of about 2mm to satisfy the mass balance, at least for currents comprised between 50 and 63A [5]. Therefore, any additional acceleration effect due to the constriction of the thermally expanding plasma at the nozzle exit is ignored in this work. One can also find that the initial assumption of laminar flow is verified with the results obtained in Tables 1 and 2.

2.2 Interaction plasma-particles
In this section, a 60-150µm spherical nickel or tungsten carbide particle is considered, in an argon plasma in the temperature and velocity ranges presented in section 2.1.

2.2.1 Particle heating rate. For the range of particle sizes and plasma temperatures considered in this study, only the convective heat transfer was found to be significant [5]. It can be expressed through the classic Newton law, but the heat transfer coefficient is characterized using a modified Nusselt number. The modified Nusselt number takes into account the effect of varying plasma properties in the boundary layer around the particle (due to the extreme temperature gradient in that region); a suitable expression was proposed by Chen [8]. One should also factorize this expression with a corrective factor accounting for the Knudsen effect. The expression for this corrective factor can be found in [9] [10].

2.2.2 Particle acceleration. The forces applied on the powder particle include the drag force exerted by the plasma, the pressure gradient and gravity. Most of the driving forces, can be neglected compared to the viscous drag force for a powder particle under 100 µm diameter *[9]*. The standard expressions for the drag coefficient depending on the Reynolds number can be found in *[11]*. The Reynolds number can be evaluated using the integrally averaged plasma properties at the boundary layer *[5]*. Like for the Nusselt number, corrective factors must also be applied to the drag coefficient to account for varying plasma properties in the boundary layer and for the Knudsen effect. These corrections can be found in *[12]*.

2.3 Model validation and prediction of deposited particle temperature and velocity
The experimental results reported in [3] were compared to the model predictions. In the experimental results, 14 powder particles that were introduced into an argon plasma arc at 60A and 100A were tracked. The powder that was used was composed of tungsten carbide and nickel-based material, with particle sizes ranging roughly between 50 and 150 µm. The median particle diameter appeared to be 90 µm based on manufacturer data. The tracked particle sizes were not disclosed on an individual basis. In Table 3 are displayed the operating parameters used:

Table 3: Operating parameters used during the powder tracking experiments in [3]

Current	Powder flow (g/min)	Travel speed (mm/min)	Stand-off distance (mm)	Powder gas flow (slpm)	Shielding gas flow (slpm)	Plasma gas flow (slpm)	Electrode tip angle
60 / 100A	27	600	7	1.5	12	1.5	20°

On the basis of an examination of the particle tracking photographs shown in [3], initial velocities were found to vary between 1 and 1.5 m/s, and the usual range of injection angles was found to vary between 48° and 58° from the horizontal plane. For the 60A argon arc, multiple trajectories were analyzed in [3] and compared to the model results in this work. Among them, the trajectory #8 whose defining characteristics are listed in Table 4 represented an average path, and its final velocity was reported to be 3.7m/s. The proportion of plasma arc radius crossed by the particle at the end of the trajectory is mentioned, since the plasma velocity seen by that particle is considered in the model to be equal to the integrally averaged value of the velocity profile, in the region crossed by the particle. Table 5 shows the predictions made by the model for a spherical particle with the same boundary conditions as for the trajectory #8 (Cf. Table 4). The tracked particles individual sizes and their material compositions were not included in the experimental results, so different possibilities are considered in Table 5 to compare with the experimental result: a minimum size of 60 µm, a median size of 90 µm and a maximum size of 150 µm were selected based on the powder manufacturer's data for both the nickel-based and the tungsten carbide material.

Table 4: Boundary conditions of the tracked particle

Angle of injection	52°
Distance from the nozzle when entering the plasma	3.3 mm
Initial absolute velocity	1.5 m/s
Proportion of plasma arc radius crossed	75%

Table 5: Model predictions for a 60A Argon Plasma

Material	Particle size (µm)	NiBSi	WC
Final velocity (m/s)	60	9.4	7.5
	90	6.6	5.2
	150	4.4	3.5
Time spent in arc (ms)	60	0.72	0.89
	90	0.96	1.2
	150	1.3	1.6
Final Temperature (K)	60	2400	3140 (82% liq.)
	90	1770	2860
	150	1270	1840

The comparison for all studied trajectories is similar to how the model predictions in Table 5 compare to the observed final velocity of 3.7 m/s for trajectory #8. The conclusion is that if the investigated trajectories relate to larger-sized WC particles, the model predictions and the experimental results from [3] agree reasonably well. Indeed, the predicted final velocity for a 150 µm WC particle is consistently lower than the experimental values, whereas the predicted final velocity for a 90 µm WC particle is consistently greater. However, the predicted final velocity for a 150 µm NiBSi spherical particle is observed to be equivalent to or greater than the experimental measurements. The model predictions therefore rule out the likelihood that the tracked particles were made of NiBSi since the ultimate velocity decreases with the particle size. Unfortunately, the diameter and material composition of the particles tracked in [3] is unknown. Because they are easier to notice, one may suppose that the particles that were commonly tracked were the larger WC particles. But that assertion is unsupported by any evidence. To fully validate the model, more experimental data is required. However, the comparison in Table 5 and a comparison also made for the 100A Argon instance with more examined trajectories both seem to suggest that the model is reliable.

The model is now applied to predict the final temperature of an average sized (90 μm) NiBSi and WC particle respectively, under the assumption that these particles are travelling along a typical route in the 50-63A argon plasma arc. Table 6 lists the characteristics of a trajectory for such an arc (based on an analysis of the experimental data from [3]). Besides the current, the other operational parameters are set to be the same as in [3] (Cf. Table 3). The model results are displayed in Table 7.

Table 6: Characteristics of an average trajectory in a 50-63A argon plasma arc

Angle of injection (°)	Initial absolute velocity (m/s)	Initial vertical velocity (m/s)	Initial distance from the nozzle (mm)	Proportion of plasma arc radius crossed (%)
53	1.2	1.0	4	65

Table 7: Model predictions for the average particle path of a median sized (90 μm) powder particle for different powder material compositions & arc currents with pure argon plasma

Material	NiBSi	WC	NiBSi	WC	NiBSi	WC
Current (A)	50	50	57	57	63	63
Final velocity (m/s)	5.2	4.1	5.6	4.4	5.9	4.6
Time spent in the arc (ms)	0.98	1.2	0.92	1.1	0.88	1.1
Final temperature (K)	1590	2540	1640	2650	1700	2740

3. Functional gradient influence on the residual stresses of PTA additively manufactured walls

The main goal of this work, as indicated in the introduction, is to determine which material gradient in a WC-NiBSi coating is most likely to increase abrasion and wear resistance. It is believed that lowering the tensile residual stresses close to the coating surface will aid in boosting the coating's wear resistance.

The temperature of the particles at the end of their trajectory in the plasma will now be used as the initial temperature of the particles in the PTA deposit. Using a weighted average temperature based on the composition of the particles in the plasma, Table 8 is deduced from Table 7 for a 50 A plasma. When simulating the PTA-AM deposition of one of these mixed materials on a macro-scale with a finite element software, the initial temperature of the newly activated elements is set to be the final temperature displayed in Table 8.

Table 8: Powder mixtures average final temperature in the 50A argon plasma

Material Comp. (wt%)	100% Ni 0% WC	80% Ni 20% WC	60% Ni 40% WC	40% Ni 60% WC
50A Ar. Init. Temp. (K)	1590	1670	1770	1920

In this work, the coating is created by depositing a total of N=20 layers. The final layer should be composed of a metal matrix composite with 60% tungsten carbide in a nickel-based matrix. The tungsten carbide content needs to be a function of the layer number n. The weight % of WC will increase from 0 to 60 when additional layers are added. The power law-following set of gradation profiles is taken into consideration, as defined in Equation 1:

$$\text{WC wt\%} = 60.(\frac{n}{N})^p \quad (1)$$

Figure 2 displays the profiles taken into consideration for this study together with their exponent p. The gradient of WC content is higher near the first layers when p is lower than one, and higher near the last layers when p is higher than one. A p value of 0 corresponds to the case where the entire wall has 60 wt% WC (no gradient), and an infinite value corresponds to the case where only the last layer has 60 wt%

WC, but the other layers contain 0% WC. A p value of 0.01 or less is equivalent to 0, while a p value of 100 or more is equivalent to infinite. Abaqus was used to simulate the additive manufacturing of functionally graded walls using the PTA. This section compares the residual stress profiles that were obtained with each of the several gradation profiles that are taken into consideration, which are those in Figure 2.

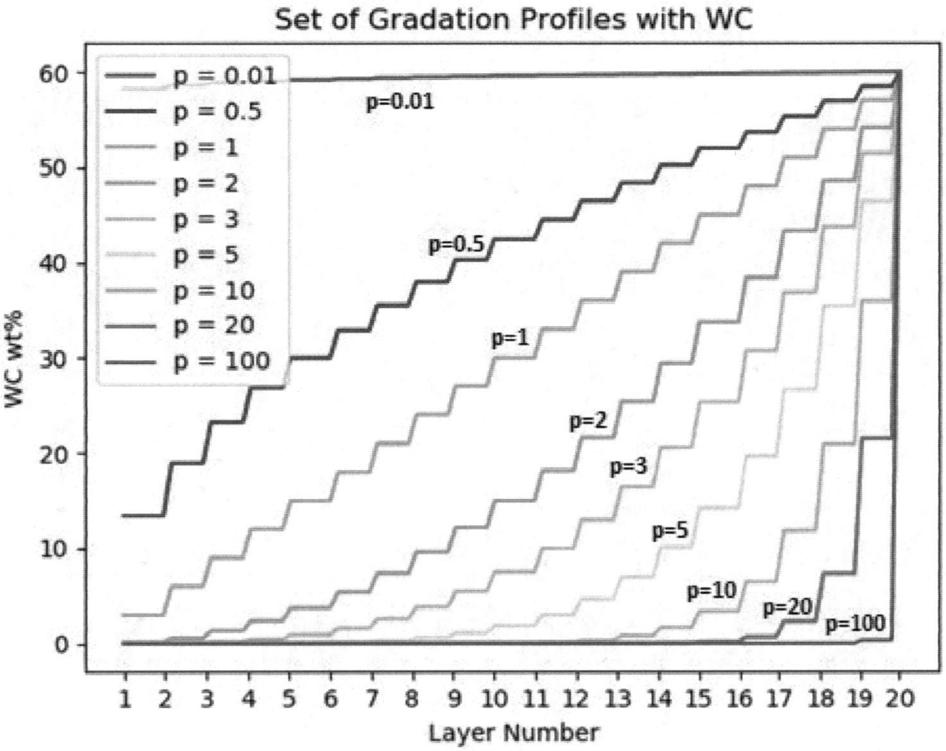

Figure 2: Set of tested gradation profiles

The simulation of the additive manufacturing of a functionally graded wall using the PTA with an argon gas was performed with Abaqus 2020 and its AM plug-in, assuming the operating parameters from Table 9. The graded wall comprised 20 layers printed on a substrate, each with a rectangular cross-section of 2.5x2.1 mm. Only half of the substrate and wall were modeled due to symmetry. The full dimensions of the substrate were 10x10x1cm and the length of the wall was 8 cm. A sensitivity analysis of the mesh was performed in which each layer was either divided into 50/100 elements in X direction, 1/2/3 elements in Y direction, and 3/5/10 elements in Z direction. The division by 50x2x5 elements was eventually selected, as the solidification time and residual stress results for each layer had sufficiently converged for that mesh size (accuracy < 0.2s/50MPa). In Figure 3 are displayed the residual stress results for one of the graded walls (p=5), and the mesh is visible with the following characteristics: 1.6x0.625x0.42mm layers element size, 13248 hexagonal elements (of linear geometric order, without reduced integration). One of the corners of the substrate surface is pinned to avoid rigid body motion (point A in Figure 3), and two other points located on the substrate surface 1 cm away from each of the wall extremities were assigned the YASYMM and ZSYMM constraints, (point B and C, respectively). The simulations were performed on a computer with an Intel Core i7-8700 (3.2 GHz) with 6 cores and 12 logical processors used in parallelization, and with 16GB of RAM. The thermal simulations took a few hours each to complete, while the mechanical simulations took about 2 days each.

Table 9: Selected PTA-AM Operating Parameters

Current	Powder flow (g/min)	Travel speed (mm/min)	Stand-off distance (mm)	Powder gas flow (slpm)	Shielding gas flow (slpm)	Plasma gas flow (slpm)	Electrode tip angle
50A	40	480	7	1.5	10	1.5	20°

Figure 3: Part Meshing

The initial temperature of each layer elements upon activation is derived from the conclusions of the previous section. The power transferred by the heat source was also estimated to be about 700W [5] (so 350W absorbed by the half part). The Goldak heat source was used to simulate the heat received from the arc; their dimensional parameters are summarized in Table 10 (mm units).

.

Table 10: Goldak dimensional parameters

Parameters	a	b	c_f	c_r	f_f	f_r
Value	2	3	2	4	0.67	1.33

An estimate of the air cooling was 15 $W.m^{-2}.K^{-1}$.To prevent the effect of heat buildup, an interpass time of 4 minutes was required following the deposition of each layer. Each layer was printed in the same direction, allowing for an equal length of time between each portion of a deposited layer and the deposit of the next one. Natural air convection causes the printed portion to gradually cool to room temperature (25 °C) after printing.

Each layer has been assigned a specific set of temperature-dependent material properties based on its WC content. More information is available in [5] regarding these properties (and the yield strength of the nickel matrix is here set at 493 MPa). The assigned mechanical properties for each layer are also described in the appendix. The substrate was assigned the temperature dependent properties of the 17-4PH stainless steel. It is important to state that the actual values of the elastic modulus and of the yield strength as a function of temperature for each layer are not known. But using the available information in the literature while not yielding actual residual stress values will yield useful trends as to which gradient will yield less residual stress than another. Using a sequentially coupled thermomechanical model, the projected temperature history was followed by the computation of the thermal strains and stresses. After the part has cooled, the residual stresses are obtained. In order to compare each set of result directly, for each wall with a specific gradient, the longitudinal residual stresses ("S11" in Figure 3) located in the middle cross-section of the wall were extracted as a function of the wall height. The extraction was made at equidistant intervals along the welding direction (dotted lines in Figure 3) and the obtained stress values were then averaged before being plotted against the height of the wall (Cf. Figure 4&5).

From Figure 4, it can be seen that the non-graded wall ($p=0$) has high tensile stresses near the surface (250-300 MPa). The WC rich gradient ($p<1$) corresponding to $p = 0.5$ has even higher tensile stresses near the surface (350MPa), which makes it a bad candidate for improving the coating abrasion resistance. Next, the results for some nickel rich gradients ($p>1$) are displayed in Figure 5 and compared to the wall without gradient ($p=0$). Unlike the WC rich gradients in which the WC content increases fast around the first layers (see Figure 2), the nickel rich gradients see their WC content increasing fast near the last layers.

From Figure 5, it can be seen that the residual stresses are similar across the wall for both of the displayed gradients except near the wall top/surface. This is natural since their WC content is similar for the first layers, but differ significantly for the last. Both of the nickel rich gradients ($p>1$) have lower tensile stresses at the last printed layer compared to the non-graded wall ($p=0$). This is due to the lower Coefficient of Thermal Expansion (CTE) of the tungsten carbide which reduces the shrinkage rate of the last layer compared to the previous ones upon cooling. Although only the results for a few gradients are displayed in Figure 4 and Figure 5, the comparison was made for all the gradients displayed in Figure 2 (p = 0; 0.5; 1; 2; 3; 5; 10; 20; infinity). It was found that the higher the p value, the lower the tensile stresses at the last layer. However, if the gradient is very high ($p>=10$), the tensile stresses at the previous layer (19th layer) exceed that of the non graded wall. For the other gradients ($1<p<5$), the tensile stresses at the previous layer reach a value similar to that of the non-graded wall. For most of the other layers, the tensile stresses in the walls with a gradient exceed that of the wall without gradient by a few dozens of MPa. The effect of applying a material gradient in the wall is to shift the location of these stresses from the most critical area (the coating's surface) to a less critical area.

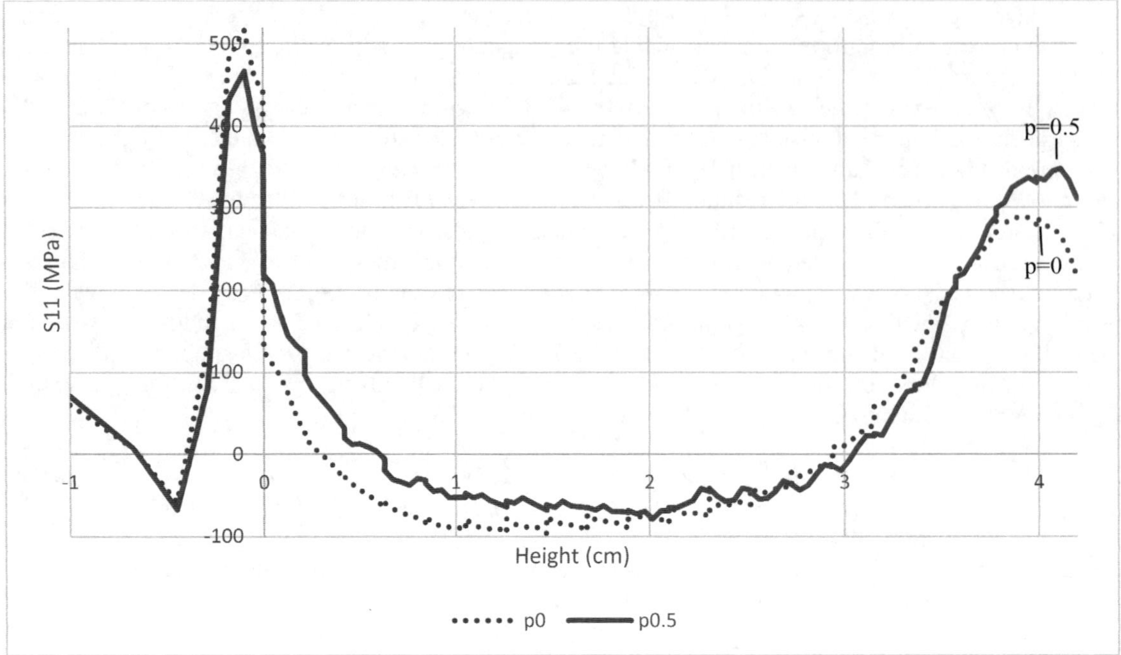

Figure 4: Longitudinal residual stresses in the middle of the wall, WC rich gradient

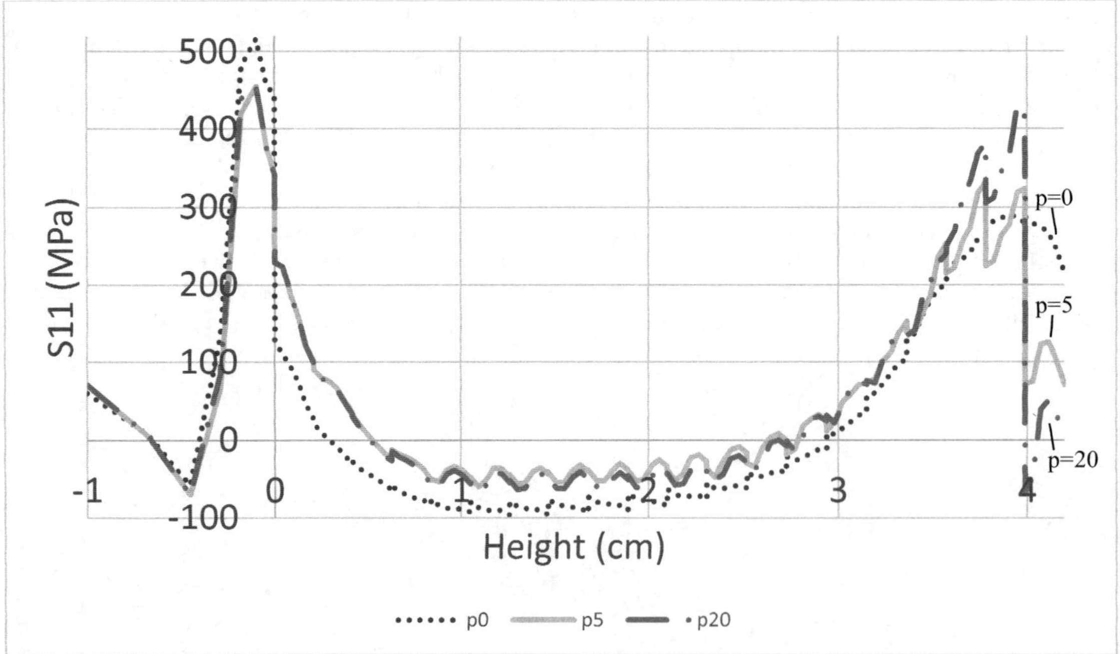

Figure 5: Longitudinal residual stresses in the middle of the wall, Ni rich gradients

4. Conclusion and discussion

With this comparison, it can be concluded that a WC rich gradient (p<1) is not recommended as the result would be worse than for the non graded wall. Any nickel rich gradient however (p>1) will have a positive effect and reduce the tensile stresses near the wall surface as long as the p value is not too high (does not exceed 5 or 10). In fact, based on these results a p value of about 5 seems to yield some of the best results as it significantly decreases the last layer tensile stresses (from 250 MPa to 100 MPa) while not significantly affecting the tensile stresses of the previous layers.

The observations made on suitable gradients of WC in the NiBSi matrix presented in this work are specific to the powder utilized and to the indicated operating conditions of the PTA-AM process. If a different distribution of powder particle sizes was assumed or if a different set of PTA operating parameters had been used, then this could result in different initial temperatures for each layer, and consequently different subsequent shrinkages, which could then result in a different residual stress profile. Nevertheless, the model can also be applied for a different set of operational parameters as long as the stipulated assumptions are still fulfilled, in which case it could provide the inputs for performing new finite element simulations in order to derive conclusions specific to these particular operating conditions.

It would be valuable to directly compare the predicted residual stress trends in this work with experimental measurements to validate the model. Such experimental measurements would involve some challenges because they would require the ability to measure stresses at a depth of about 1mm in a wall whose mechanical properties fluctuate with the height. Alternatively, directly measuring each wall's surface abrasion resistance would permit to verify the modeling strategy employed in this study.

References

[1] Wolfe T 2010 *Homogeneity of Metal Matrix Composites Deposited by Plasma Transferred Arc Welding* Doctoral dissertation (Edmonton: University of Alberta)

[2] Xibao W and Hua L 1998 *Surf. Coat. Technol.* **106** pp 156-61

[3] Gajbhiye R V 2022 *Image processing-based techniques for process parameter characterization in Plasma Transferred Arc Additive Manufacturing (PTA-AM) process* Master dissertation (Edmonton: University of Alberta)

[4] Dowden J and Kapadia P 1994 *J. Phys. D: Appl. Phys.* **27** pp 902-11

[5] Bonias G 2022 *Thermal and residual stress modeling of functionally graded deposits using the PTAAM* Master Dissertation (Edmonton: University of Alberta)

[6] Oh D S, Kim Y and Cho S M 2005 *Sci. Technol. Weld. Joining* **10** pp 442-6.

[7] Abid M, Parvez S and Nash D H 2013 *Int. J. Press. Vessels Pip.* **108-109** 51-60

[8] Chen X 1988 *Pure Appl. Chem.* **60** pp 651-62

[9] Wan Y P, Prasad V, Wang G X, Sampath S and Fincke J R 1999 *J. Heat Transfer* **121** pp 691-9

[10] Chen X 1999 *Thin Solid Films* **345** pp 140-5

[11] Vardelle M, Vardelle A, Fauchais P and Boulos M I 1983 *AIChE J.* **29** pp 236-43

[12] Pfender E and Lee Y C 1987 *Plasma Chem. Plasma Process.* **7**

[13] Anon 2022 Poisson's ratios metals *engineeringtoolbox.com*

[14] Li W, Kou H, Zhang X, Ma J, Li Y, Geng P, Wu X, Chen L and Fang D 2019 *Mech. Mater.* **139** 103194

[15] Farraro R and McLellan R 1977 *Metall. Trans. A* **8** pp 1563-5

[16] Zhu X K and Chao YJ. 2002 *Comput. Struct.* **80** pp 967-76

[17] Zhang X, Li W, Ma J, Geng P, Shao J and Wu X 2017 *Comput. Mater. Sci.* **129** pp 147-55

[18] Wang Z, Denlinger E, Michaleris P, Stoica AD, Ma D and Beese A M 2017 *Mater. Des.* **113** pp 169-77

[19] Toulouklan Y S, Kirby R K, Taylor R E and Desai P D 1975 *The TRPC Data Series* vol 12

[20] Pierson H O 1996 *Handbook of Refractory Carbides and Nitrides* chapter VI (Westwood: Noyes Publications)

[21] Weisbrook C M and Krawitz A D 1996 *Mater. Sci. Eng. A* **209** pp 318-28

[22] Anon 2021 Tungsten carbide: Density, strength, hardness, melting point *materials-properties.org*

[23] Ravichandran K S 1994 *J. Am. Ceram. Soc.* **77** pp 1178-84

[24] Serres N, Hlawka F, Costil S, Langlade C and Machi F 2011 *Appl. Surf. Sci.* **257** pp 5132-7

Appendix

Information on the mechanical characteristics of the MMC employed in this study is provided in this subsection. It is not required to aim for perfect accuracy in the results because the purpose of this study is only to demonstrate the trends in residual stresses that are produced from various material gradients. In the interest of keeping the study simple, an elastic-perfectly plastic model was chosen and the materials were approximated as isotropic.

A. Nickel alloy properties

Its young modulus and yield strength temperature dependence is not well known and little data is available in the literature, especially for high temperatures. Therefore, some correlations were employed in this work, which are displayed in this subsection.

A.1 Elastic Modulus and Poisson Coefficient. The Poisson coefficient of nickel is 0.31 [13] and is considered constant for all temperatures. Correlations for the temperature dependence of the elastic modulus for superalloys and for pure nickel were developed in [14] (Li's Correlation) and in [15] (Farraro's correlation) respectively. The application of these correlations, assuming a room temperature elastic modulus of 230 GPa, is shown in Figure 6. Both correlations yield similar results. In this work, Li's correlation was used.

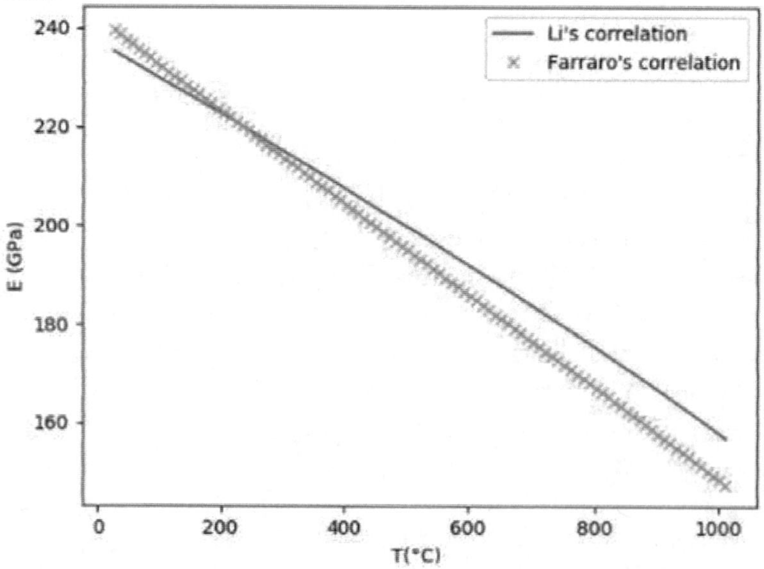

Figure 6: Estimations of the nickel superalloy elastic modulus dependency on temperature

A.2 Yield Strength. The temperature dependence of this property for the nickel alloy is not known. To cope with this type of data scarcity, an engineering approximation was proposed in [16] (Zhang's correlation). A model to predict the yield strength of a metallic alloy at any temperature based on other mechanical properties was also proposed in [17] (Zhu's correlation). Both of these correlations require the knowledge of the alloy room temperature yield strength. That property is also unknown, but the room temperature yield strength of another nickel superalloy (Inconel 625) was reported to be 493 MPa in [18]. Assuming a 493 MPa yield strength at room temperature, the respective correlations of Zhu and Zhang are compared in Figure 7.

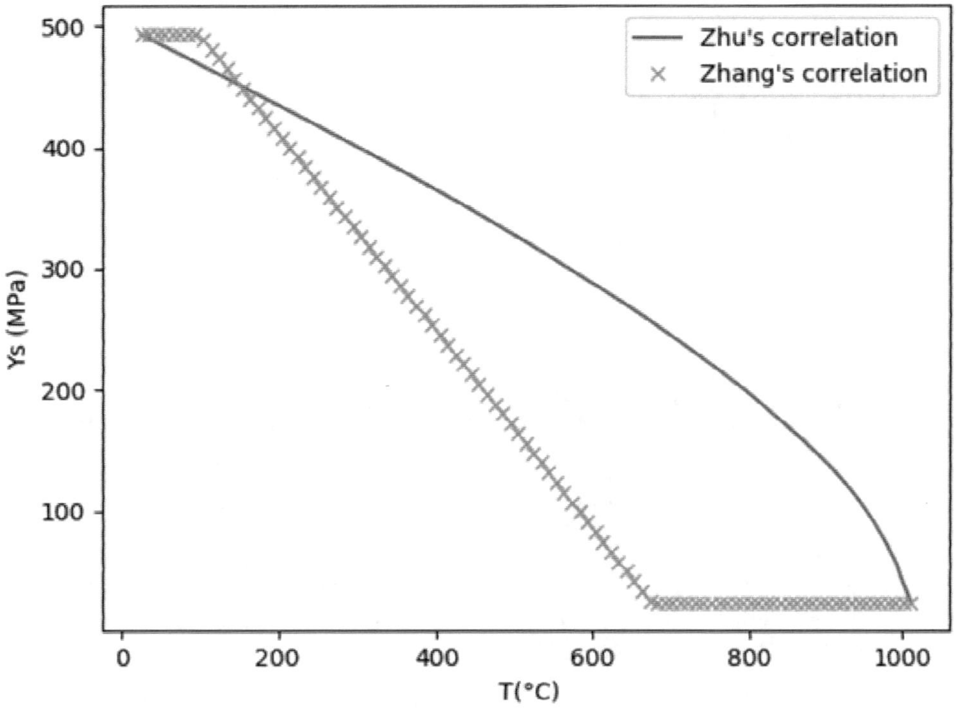

Figure 7: Estimations of the nickel superalloy yield strength dependency on temperature

In this work, the correlation of Zhang was used assuming a 493 MPa room temperature yield strength, and with a minimum yield strength of 50 MPa near melting temperature to allow for numerical convergence on Abaqus.

A.3 Linear Coefficient of Thermal Expansion. The temperature dependent CTE for Inconel 625 and for pure nickel were reported in [18] and in [19] respectively. Trendlines for these two materials were obtained on Excel from the extracted data, and they are compared in Figure 8. Although the CTE for the nickel alloy used in this study is not known, it can be seen from Figure 8 that the CTE difference between the Inconel 625 nickel superalloy and pure nickel is small (in the order of 1µm/m/K). Therefore, it is reasonable to assume that the CTE of the nickel alloy used in this study is also close from these values. The data for Inconel 625 was therefore used in this work.

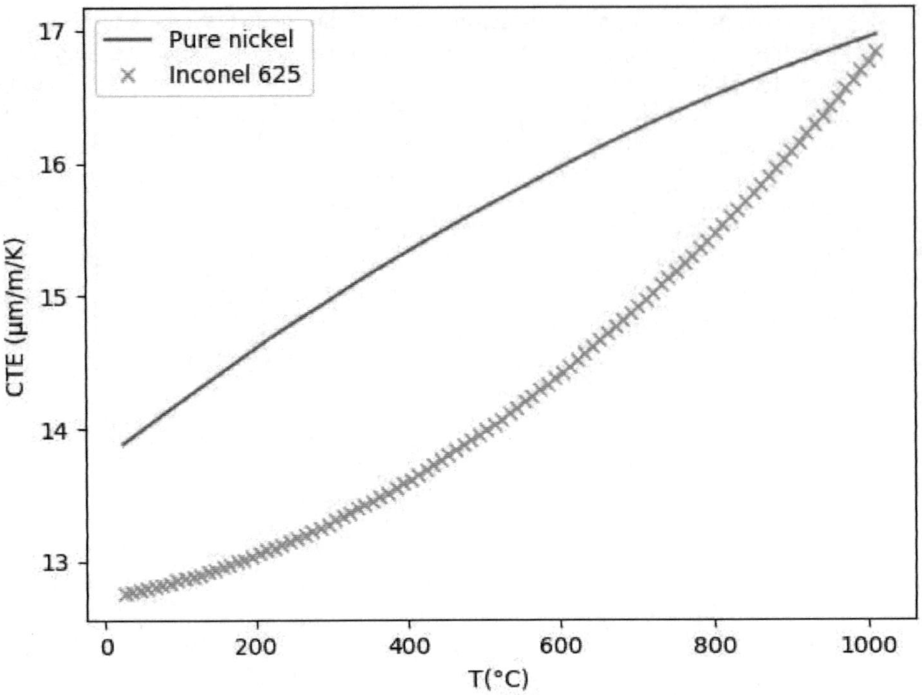

Figure 8: CTE for pure nickel and for Inconel 625

B. Tungsten carbide properties
For the tungsten carbide, all mechanical properties were considered to be temperature independent. The property data was sourced from [20], [21] and [22]. They are displayed in Table 11.

Table 11: Tungsten carbide mechanical properties

Elastic Modulus (GPa)	Yield Strength (MPa)	CTE (μm/m/K)	Poisson coefficient
620	330	5	0.25

C. Compositional properties
In a two-phase system with discontinuous reinforcements, mathematical formulas to evaluate the isotropic elastic properties were proposed [23]. These formulas were used in this study to estimate the temperature-dependent Poisson coefficients and elastic moduli at intermediate compositions in the scenario where WC is reinforcing the nickel-based matrix. Figures 9 and 10 display the results at room temperature as an illustration. Figure 9 includes a comparison with the experimental data from [24] for an alloy made of WC and NiCrBSi (with the min and max curves representing the experimental uncertainty range). As can be observed, Serre's experimental findings are more consistent with Ravichandran's correlation than the linear rule of mixture. As a result, Ravichandran's correlation was employed in the present work to evaluate the elastic properties of the composite. However, a linear rule of mixture was utilized for the estimation of the yield strength of the composite and the coefficient of thermal expansion (using the volume fraction of tungsten carbide).

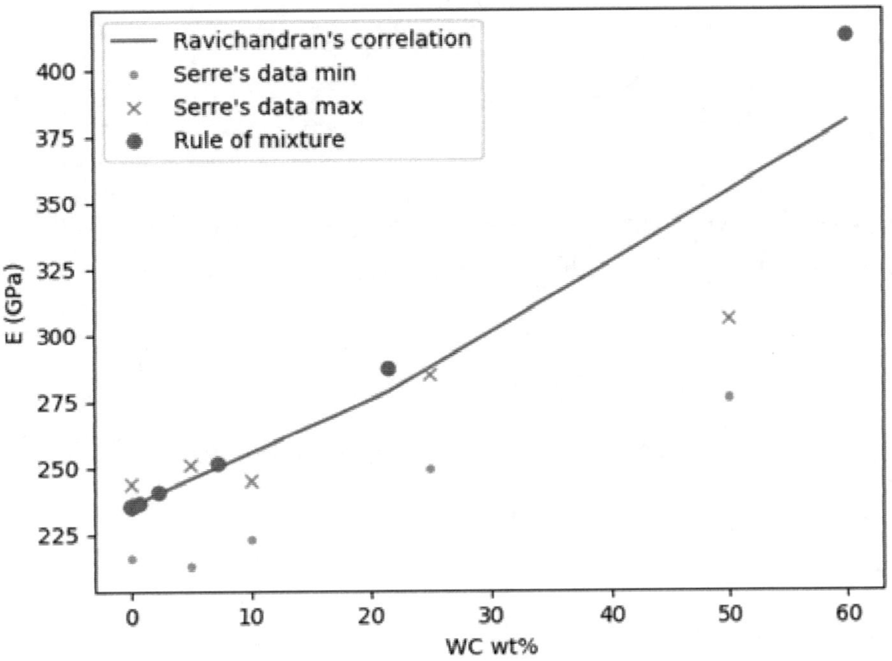

Figure 9: Estimations of the MMC elastic modulus dependency on the WC weight content

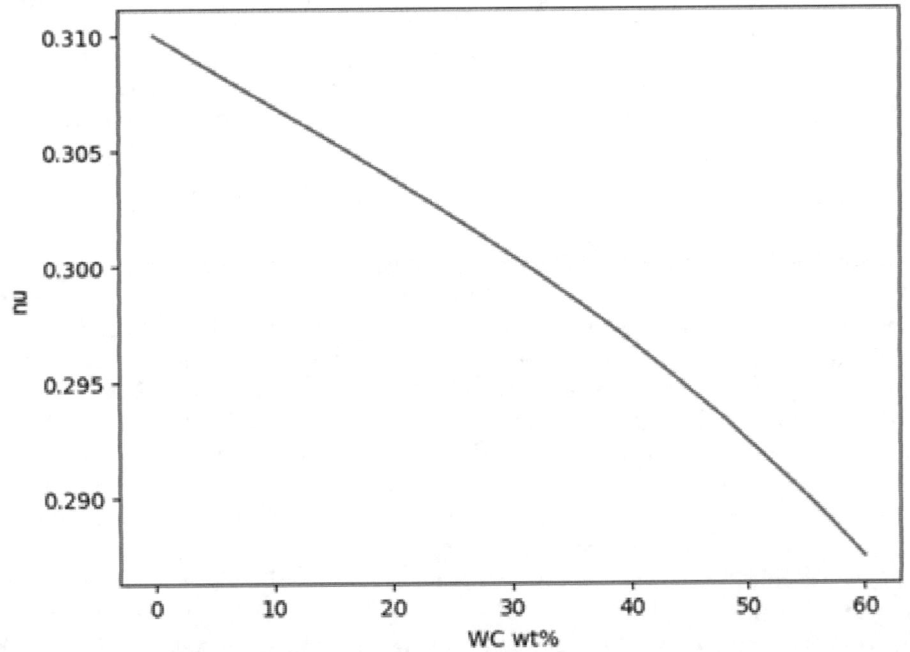

Figure 10: Estimations of the MMC poisson coefficient dependency on the WC weight content

IOP Conf. Series: Materials Science and Engineering 1281 (2023) 012013 doi:10.1088/1757-899X/1281/1/012013

Effect of powder properties on mesoscale thermal FEM simulations of powder bed additive manufacturing

Meet Upadhyay, Chad Sinclair, Daan Maijer

Department of Materials Engineering, University of British Columbia,

Vancouver, BC, Canada V6T 1Z4

Email: meet.upadhyay@ubc.ca

Abstract. Numerous studies have used FEM simulations to assess the effects of the heat source parameters on the melt pool volume during metal powder bed additive manufacturing. However, considerable debate still exists on how to incorporate the evolution of the thermophysical properties used to describe the powder as it undergoes heating, melting, consolidation and finally solidification. For single layer studies, since powder volume is much smaller compared to the substrate volume, highly detailed, computationally expensive powder property descriptions may not provide a commensurate increase in accuracy of simulation results. This study aims to quantify the effect of powder properties on the melt pool volume created during electron beam melting of Ti6Al4V powder using predictions from a FEM-based heat conduction model. The dependence of thermal conductivity, specific heat, and density of the powder on temperature and beam power density will be studied. Additionally, the relevance of the powder properties with changing layer height and beam power and speed will also be quantified.

1. Introduction

Metal powder-based additive manufacturing involves using a concentrated energy source, such as a laser or electron beam, to scan across a thin layer of metallic powder. The heat generated by the energy source melts the powder particles locally, creating a dense, solid component upon solidification. This process leads to the formation of spatially varying thermal fields within the component. To understand the effect of these thermal fields on material properties, thermal models are frequently used [1–3]. The finite element method (FEM) has gained widespread popularity for investigating the effects of additive manufacturing (AM) process parameters and their optimization due to its efficiency in resource usage [4]. By using FEM simulations, many trial-and-error experiments can be avoided, providing a useful technique for assessing over/under heating, residual stress, and distortion in parts produced through AM processes.

FEM simulations typically take one of two approaches to modelling powder particles; they either, explicitly resolve the particles [5] or treat them as a continuum body with adjusted properties [6]. Modeling the particles explicitly can provide more accurate and detailed results, but it also significantly increases the computation time and resources needed, limiting it to microscale, single-track studies. Hence, the continuum approach is more commonly employed. This approach is isotropic and homogenous, and while it may not offer the same level of microscopic insight, it is more practical for simulations involving larger numbers of particles [7].

The material properties of the powder change during the manufacturing process due to changes in temperature, consolidation, melting, and solidification. The density, thermal conductivity, and specific heat capacity of the powder may all affect the simulation results. The density of the powder bed, which is a function of the material density as well as the powder packing, influences the amount of latent heat absorbed/released during the phase changes, and together with specific heat will affect peak temperatures of the melt pool. Powder conductivity will affect how heat transmits through the powder layer and then into the substrate. Various approaches have been taken to incorporate the effect of the temperature and state-dependent behavior of these properties [4].

In previous studies, density has been considered constant [8], or different densities have been assigned for the powder and the solid substrate [9]. The powder density can vary from 40% to 60% of the solid density [10], based on a simple weight fraction or a detailed relationship between powder particle size and shape [11]. In a more detailed analysis, density was defined as varying with temperature and in some sintering based studies, sintering laws are applied to implement density as a function of temperature and level of sintering [12]. It is worth noting that in the domain of a thermal-only model, the volume remains constant and hence a change in density of the domain implies a change of mass in the domain which is not realistic.

Thermal conduction is the primary mode of heat transfer in the domain modelled by finite element-based approaches. Convection loss due to gas flow (in laser powder bed fusion) and radiation (in electron beam AM at high temperature) contribute minimally to the heat transfer and are rarely accounted for explicitly. The effective thermal conductivity of the powder is significantly lower than that of the solid, especially at lower temperatures as the contact area between adjacent particles is very small [13]. The effective thermal conductivity of powders can be modelled as function of multiple factors, such as particle size and shape, solid volume fraction, and the thermal conductivity of the gaseous media [14]. To reduce the complexity, authors have considered constant thermal conductivities, both for the powder and the solid state [9]. Others consider conductivity as temperature-dependent, defined at discrete points and interpolated in between [15]. In some cases, a constant conductivity is assumed for the powder and a temperature-dependent conductivity for the solidified material. Finally, it has been pointed out that a zero thermal conductivity for powders can be assumed (in a ray-tracing approach) as well without a noticeable effect on the simulation results, especially at high beam speeds [16].

As in the case of the thermal conductivity and density, a common approach is to use discrete values for the specific heat capacity and linearly interpolate between [17]. The specific heat capacity of a powder bed has been described by scaling the specific heat of the bulk material by the powder bulk density or for powder mixtures by applying the Kopp–Neumann law [18].

It is apparent that considerable debate remains in the literature on the best way to model the thermal properties of powders in FEM simulations of AM. Additional complexity may improve accuracy but with diminished returns due to increased computational cost. The influence of these complex definitions is further diminished for part scale simulations where initial and boundary conditions change rapidly from one track to the next. Even with detailed formulations, FEM analyses report errors in the range of 10% to even 50% when compared to experimental results [4]. Hence, this study aims to quantify the effect of powder on the melt pool size. It is hypothesized that the effect of powder properties is important, but importance depends on the melt pool size relative to the powder bed thickness. The variables in this study are the powder properties and the height of the powder layer. Melt pool width and depth between different cases will be considered to understand which variables in the study produce the greatest effect. If material property relationships can be simplified, computational requirements will be reduced making part-scale simulations more tractable.

2. Finite Element Model

To study the effect of the varying powder layer heights on the melt pool, a three-dimensional (3D) FE-based model was developed for a domain representing a volume of material within a component being processed. The 3D heat transport equation solved is given in Eq. 1.

$$\frac{\partial}{\partial x}\left(k\frac{\partial T}{\partial x}\right)+\frac{\partial}{\partial y}\left(k\frac{\partial T}{\partial y}\right)+\frac{\partial}{\partial z}\left(k\frac{\partial T}{\partial z}\right)-\rho C_p\frac{dT}{dt}+\dot{Q}=0 \qquad \ldots Equation\ 1$$

where x, y, and z (m) are the coordinates, k (W/m/°C) is the thermal conductivity, T (°C) is the temperature, ρ (kg/m^3) is the density, C_p (J/kg/°C) is the specific heat, t (s) is time, and $\dot{Q}$ (W/m^3) is a source term used to include the latent heat consumed and released during melting and solidification.

A thermal diffusion-only model with temperature dependent material properties was developed and implemented in the finite element software package ABAQUS™. A two-fold symmetry condition is assumed for the analysis along the centerline of the trajectory of the beam and the side and bottom surfaces are assumed to be adiabatic - see Figure 1. The domain was made sufficiently large such that the temperature change on the edges away from the symmetry surface was negligible.

The domain consists of a powder layer of varying thickness (50 µm to 750 µm) over a solid, fully dense substrate modelled using the properties of Ti6Al4V. Two additional cases where the full domain is assumed to have the properties of powder or solid were also performed. The region near the powder layer is meshed with hexahedral elements 50 µm in size to accurately capture the melt pool dimensions. Away from the top surface, the mesh is biased to be coarser to reduce computational time.

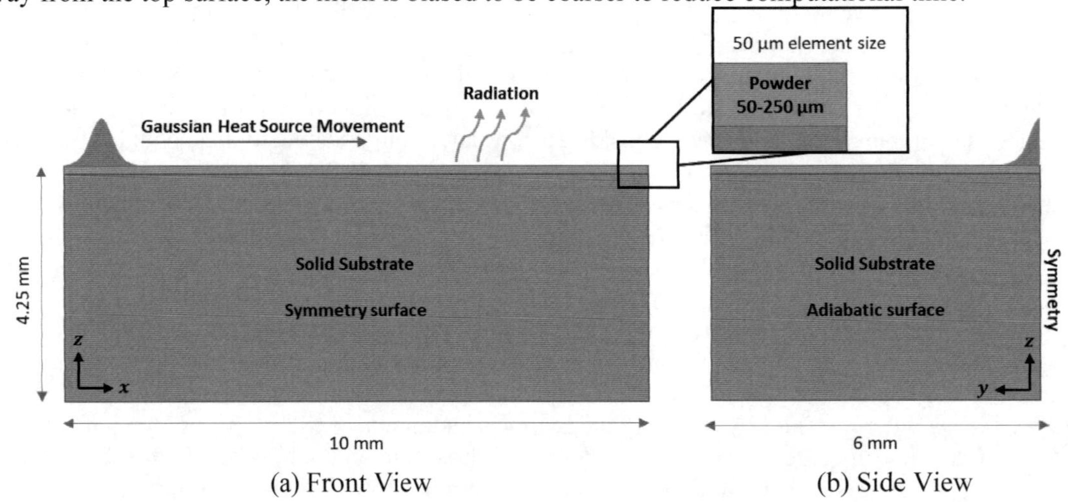

Figure 1: Schematic of finite element domain (a) front view and (b) side view. The direction of travel of the heat source, illustrated schematically as a half gaussian, shown by the arrow in (a).

2.1. *Heat flux boundary condition*

The heat deposited from the heat source was modelled as a normal distribution. Accordingly, the heat flux (W/m^2) at any location (x, y) on the heated surface at a distance $r(t)$ (m) from the center of the beam $(x_{beam}, 0)$ is:

$$q_r(x,y,t)=\frac{P\eta}{2\pi\sigma^2}e^{-\left(\frac{r(t)^2}{2\sigma^2}\right)} \qquad \ldots Equation\ 2$$

where $r(t)=\sqrt{(x-x_{beam}(t))^2+(y)^2)}$, P (W) is the beam power, η is the beam efficiency, and σ (m) is the standard deviation. For this study, the FEM model was run with beam parameters summarized in Table 1. The parameters selected represent a range of values used in EBAM processes [19]. The beam center moves with a velocity v (m/s), yielding the expression $x_{beam} = vt$. The moving heat flux associated with the beam is implemented in ABAQUS™ using the user-subroutine, DFLUX, which is applied to the top surface of the analysis domain within the sub-domain containing the finer mesh. Within DFLUX, the distributed heat flux is evaluated within a radius of 3σ of the beam centre at any time or is otherwise set to 0. The entire top surface is subjected to radiation and other surfaces are subjected to the adiabatic boundary conditions. A uniform initial temperature of 550°C is defined in the model as the starting temperature to represent the state of the powder layer following pre-heating.

Table 1: Beam parameters used in ABAQUS Model.

Parameter	Power, $\boldsymbol{P}$	Efficiency, $\boldsymbol{\eta}$	Beam Width, $\boldsymbol{2\sigma}$	Beam Speed, $\boldsymbol{v}$
Value	360 W	0.7	290 x 10^{-6} m	0.1 m/s

2.2. Material properties

The temperature dependent material properties used for this study are for a Ti-6Al-4V alloy and have been adapted from [20]. The thermal conductivity and specific heat implemented in the model are plotted in Figure 2 and 3, respectively. The material properties taken to be constant are described in Table 2. A constant, temperature independent density is used for the substrate and powder, ignoring shrinkage due to powder consolidation to prevent change of mass in the domain. The latent heat is linearly distributed over the solidus to liquidus temperature range of 1595°C to 1625°C. The conductivity is artificially increased in the liquid region to reduce large temperature gradients in the melt pool and to approximate the effects of mixing caused by fluid flow [19].

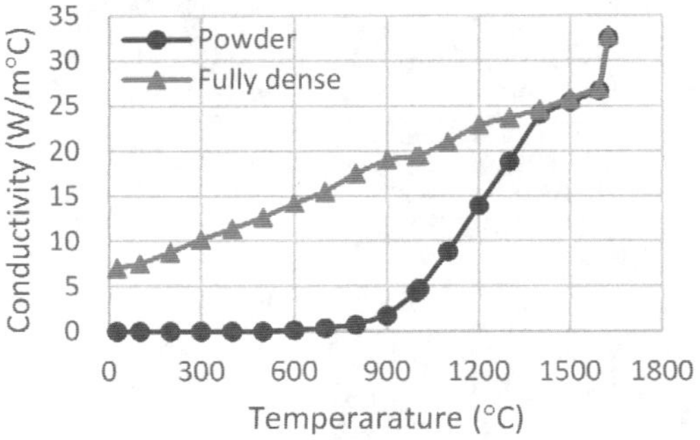

Figure 2: Powder and bulk thermal conductivity of Ti6Al4V varying with temperature.

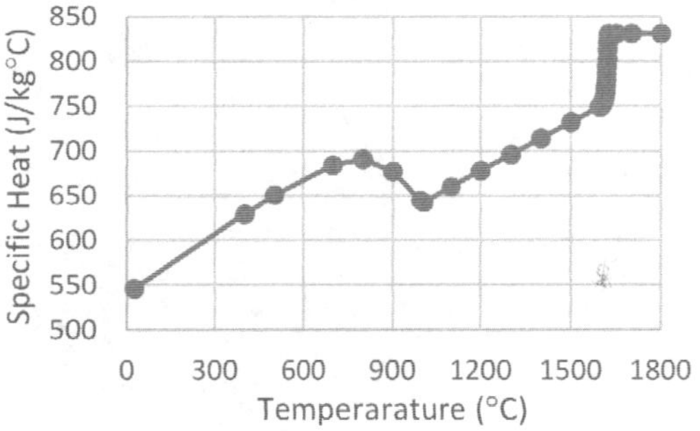

Figure 3: Specific heat of Ti6Al4V varying with temperature.

Table 2: Ti6Al4V material properties.

Material Property	Density (solid)	Latent heat	Liquidus	Solidus	Emissivity
Value	4198 kg/m^3	286 kJ/kg	1625 °C	1595 °C	0.7

Using a state variable and the USDFLD (user-defined field) subroutine in ABAQUS, the properties of the elements are switched from those of the powder to bulk when the melting point is reached. Four different combinations of thermal properties are considered, each having been taken from the literature, to understand their effect on melt pool geometry. The properties used for the different cases are described in Table 3. For Case 3, temperature dependent powder conductivity definition is applied, informed by experimental measurements of powder conductivity for powder beds subjected to holding at various temperatures. In this case, the powder conductivity remains negligible at lower temperatures and increases to the bulk properties nearer to the melting temperature based on the level of sintering of the powder [15]. For Case 1-4, a constant powder layer height of 250 μm is used. For the study of the effect of varying powder layer heights, properties corresponding to Case 3 are used.

Table 3: Powder properties for various cases.

Powder property	Thermal conductivity	Density	Specific heat capacity
Case 1	3.125 W/m°C	2098 kg/m^3	546 J/kg°C
Case 2	$f(T)$, 50% of solid	2098 kg/m^3	$f(T)$, same as solid
Case 3	Varying as Figure 2	2098 kg/m^3	$f(T)$, same as solid
Case 4	$f(T)$, 50% of solid	4198 kg/m^3	$f(T)$, 50% of solid

3. Results and Discussion

The predicted temperature contours on the top and front longitudinal surfaces and side section at maximum melt pool depth of the domain are shown in Figure 4. The temperature scale is limited to highlight the molten region. The powder layer height is 200 μm and powder properties correspond to those in Case 3. In the side section view, (Figure 4c) the effect of the lower powder conductivity is apparent. The temperatures are lower in the unconsolidated powder layer region, near the edge of the melt pool (red circle). Below the powder layer, the temperature contours are more uniform in the solid region. The powder layer acts as a barrier to heat conduction due to its low conductivity. The variation of temperature with time for a point on the domain edge (where beam intensity is maximum) was also extracted and is plotted in Figures 6 and 8 for different conditions. The melt depth and width were extracted by interpolating the distance between the domain edge and the iso-surface contour defining the melt pool edge, using the mesh as a scale. These are plotted in Figures 5 and 7.

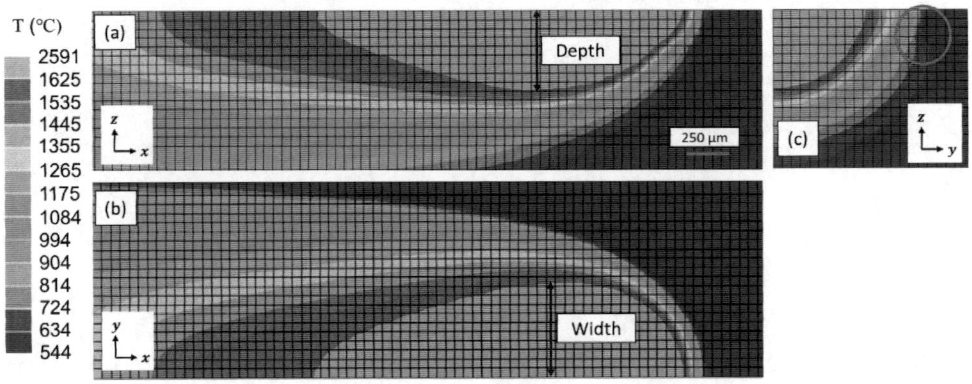

Figure 4: Temperature contours for powder layer height of 200 μm. (a) Front view showing the depth measurement. (b) Top view showing the width measurement. (c) Side section view showing the melt pool cross section at maximum melt pool depth. The elements in this region close to the melt pool are all 50 μm in size.

3.1. *Effect of powder property definition*

In Figure 5 and 6, the effect of the different property combinations (Table 3) can be seen. In Case 1, constant powder properties are used which do not vary with temperature. Thus, the conductivity and heat capacity are low even at high temperatures. This has the effect of increasing the size of the melt pool relative to the other cases. Case 2 and 3 are almost identical in Figure 6 and very close to one another in Figure 5. The higher conductivity in Case 2 allows more heat transfer into the powder layer, leading to a wider melt pool, while in Case 3 the near zero conductivity of the powder causes most heat to conduct vertically (as opposed to laterally) leading to a larger melt pool depth. Case 4 uses the same density as the bulk material with 50% of the bulk heat capacity and the same conductivity as Case 2 which means there is more powder mass available to absorb heat leading to the shallowest melt pool depth and a width similar to Case 2. The differences in these values are somewhat amplified as the layer height is 250 μm, almost 2.5x larger than the standard layer size of 80-100 μm used in electron beam additive manufacturing [19].

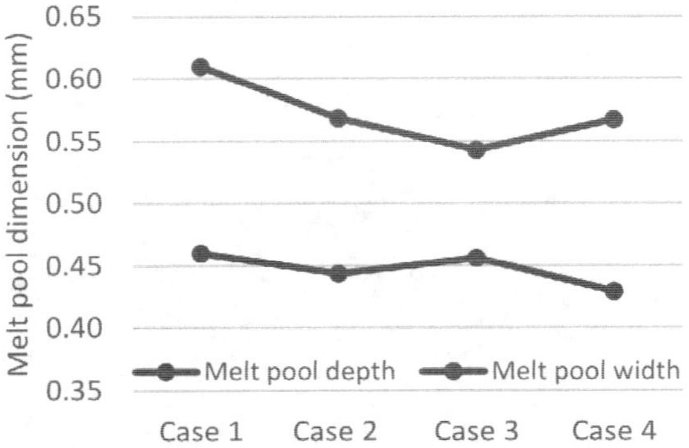

Figure 5: Variation of maximum melt pool width and depth for different powder property definitions (powder height = 250 μm).

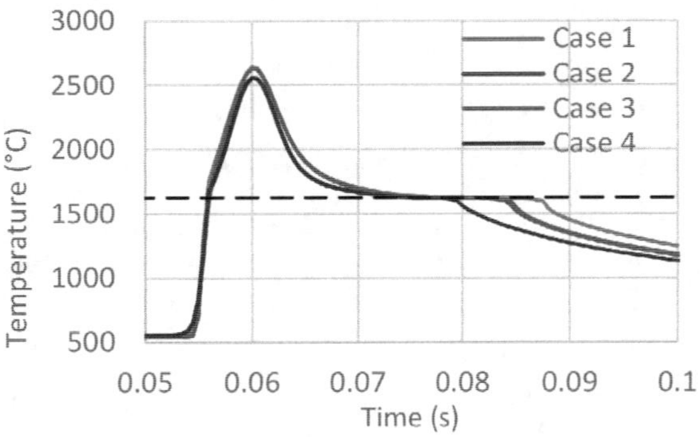

Figure 6: Variation of temperature with time when beam is centered on x = 6 mm for Case 1 – 4.

3.2. Effect of powder layer height

In Figure 7, the powder layer heights vary from 50 – 750 μm and the extreme ends of the graph represent the cases where the domain is completely solid (left) and completely powder (right). Powder properties corresponding to Case 3 are used. The error bars represent the size of the mesh element and place a bound on the maximum uncertainty in determining the dimensions from the FEM domain. Compared to Figure 5, the effect of powder layer height is much stronger. In Figure 8, as the powder layer height increases, the maximum temperature in the melt pool increases. The plateau region, where the node stays at the liquidus, also lengthens with increasing powder layer height. The powder layer's low thermal conductivity causes this effect. The heat flow into the substrate is restricted and a larger melt pool is formed, which means a higher amount of latent heat must be diffused until the melt pool solidifies.

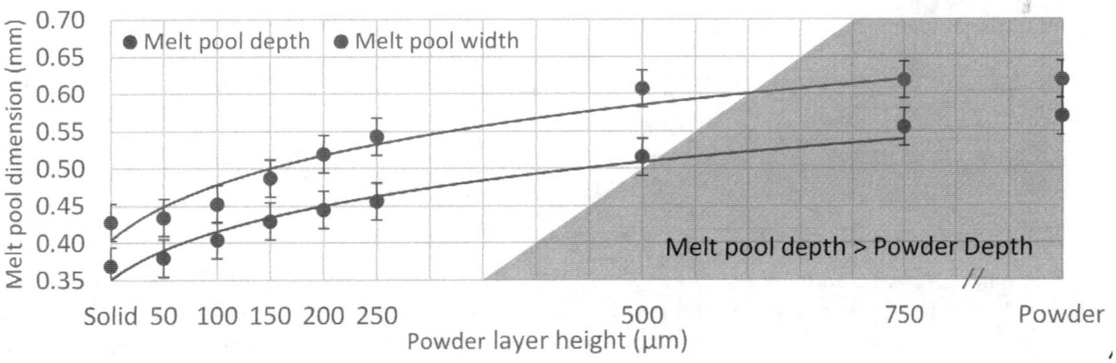

Figure 7: Variation of maximum melt pool width and depth for different powder layer heights.

Both the melt pool width and depth increase as the powder layer height increases. This effect slows as the powder layer height becomes comparable to the melt pool dimensions. Considering the heat loss by radiation is small, low powder conductivity implies that the primary mode of heat transfer is conduction vertically through the contact area of the melt pool and the solid substrate. Thus, the effect of the conduction is to reduce the size of the melt pool. As powder layer height increases, the conductive path for escape of heat from the melt pool reduces and melt pool size increases. Once the powder bed depth becomes of the order of the melt pool depth, the increase in melt pool size begins to saturate as the entire melt pool is contained within the low conductivity powder. Hence, the decrease in conduction at the solid substrate – melt pool interface leads to additional heat available to increase the size of the melt pool. However, these variations are not large (~15%) and fall within the error ranges reported for

FEM analyses in the literature. Additionally, this effect is further reduced in domains with shallower powder layers (80-100 um) typical of electron beam powder bed processes [21] as the conduction path to the solid substrate is shorter.

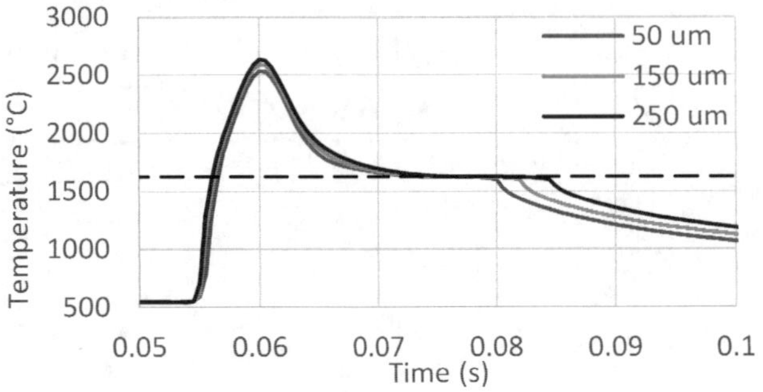

Figure 8: Variation of temperature with time when beam is centred on x = 6 mm for powder layer heights 50, 150 and 250 μm.

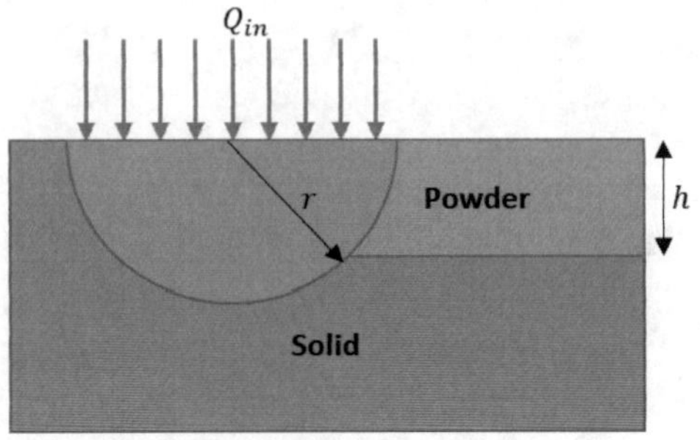

Figure 9: Schematic for the heat balance of a hemispherical melt pool.

A simplified heat balance has been performed on a hemispherical melt pool (radius r) in contact with a solid substrate and powder layer of thickness h (Figure 9). The heat flux into the melt pool from the beam is Q_{in}. Considering only the heat conduction into the solid, with zero conduction into the powder and that we can approximate the flux through the solid/liquid interface to be a constant (Q_c) then we have:

$$Q_{in}A_{melt\ surface} = \frac{\rho}{2\sigma/_v}(c_p\Delta T + L)V_{melt\ pool} + Q_c A_{melt-solid\ interface} \quad \ldots Equation\ (3)$$

where ρ is the density, $2\sigma/_v$ is a characteristic time - time taken for the beam to travel over a point (beam diameter / beam velocity), c_p is the specific heat capacity, ΔT is the temperature change from initial temperature of the metal to the liquidus temperature and L is the latent heat. This then gives:

$$Q_{in}(\pi r^2) = \frac{\rho v}{2\sigma}(c_p\Delta T + L)\frac{2}{3}\pi r^3 + Q_c(2\pi r(r-h))$$

which results in a quadratic relationship between melt pool dimensions and the powder depth for melt pool depths $r > h$:

$$c_1 r^2 = c_2 r^3 + c_3 r(r-h)$$
$$h = \frac{c_2}{c_3}r^2 + \frac{c_3 - c_1}{c_3}r$$

The parameters, c_i, were fit using the simulation data and the values of width and depth plotted in Figure 7. The results give the correct trends so long as $r > h$. Of course, as we can see in Figure 7, the melt pool is not hemispherical in shape and thus a more sophisticated model could be developed to account for changes in depth, width, and length. However, this simplified model highlights the important effect of reducing the conduction path by means of the powder bed depth.

4. Conclusions

In this study, the effect of powder property definitions and powder layer heights were studied. It was found that the powder layer height and consequently the conduction path to the solid substrate is more important than the exact property definition used.

The powder properties will affect the size of the melt pool formed based on how effectively heat is transferred in the domain. Higher powder conductivity allows for more lateral diffusion of heat leading to wider melt pools, while lower powder conductivity favors heat conduction through the depth towards the solid substrate leading to larger melt depths. These variations reduce as the powder layer height decreases.

IOP Conf. Series: Materials Science and Engineering 1281 (2023) 012013 doi:10.1088/1757-899X/1281/1/012013

The results of the simulations with varying powder layer height reveal that the larger the surface area of the melt pool in contact with solid substrate, the smaller are the resulting melt pools. As the layer height increases, the heat transported due to conduction decreases and more heat is available for the phase transformation of powder to liquid metal.

References

[1] Chen Q, Liang X, Hayduke D, Liu J, Cheng L, Oskin J, Whitmore R and To A C 2019 An inherent strain based multiscale modeling framework for simulating part-scale residual deformation for direct metal laser sintering *Addit. Manuf.* **28** 406–18

[2] Williams R J, Davies C M and Hooper P A 2018 A pragmatic part scale model for residual stress and distortion prediction in powder bed fusion *Addit. Manuf.* **22** 416–25

[3] Zinovieva O, Zinoviev A and Ploshikhin V 2018 Three-dimensional modeling of the microstructure evolution during metal additive manufacturing *Comput. Mater. Sci.* **141** 207–20

[4] Schoinochoritis B, Chantzis D and Salonitis K 2017 Simulation of metallic powder bed additive manufacturing processes with the finite element method: A critical review *Proc. Inst. Mech. Eng. Part B J. Eng. Manuf.* **231** 96–117

[5] Liu F R, Zhang Q, Zhou W P, Zhao J J and Chen J M 2012 Micro scale 3D FEM simulation on thermal evolution within the porous structure in selective laser sintering *J. Mater. Process. Technol.* **212** 2058–65

[6] Yin J, Zhu H, Ke L, Lei W, Dai C and Zuo D 2012 Simulation of temperature distribution in single metallic powder layer for laser micro-sintering *Comput. Mater. Sci.* **53** 333–9

[7] Nijhuis B, Geijselaers H J M and van den Boogaard A H 2021 Efficient thermal simulation of large-scale metal additive manufacturing using hot element addition *Comput. Struct.* **245** 106463

[8] Dai K and Shaw L 2004 Thermal and mechanical finite element modeling of laser forming from metal and ceramic powders *Acta Mater.* **52** 69–80

[9] Yadroitsev I, Gusarov A, Yadroitsava I and Smurov I 2010 Single track formation in selective laser melting of metal powders *J. Mater. Process. Technol.* **210** 1624–31

[10] Körner C, Bauereiß A and Attar E 2013 Fundamental consolidation mechanisms during selective beam melting of powders *Model. Simul. Mater. Sci. Eng.* **21**

[11] Moser D, Pannala S and Murthy J 2016 Computation of Effective Thermal Conductivity of Powders for Selective Laser Sintering Simulations *J. Heat Transfer* **138** 082002

[12] Childs T H G, Hauser G and Badrossamay M 2005 Selective laser sintering (melting) of stainless and tool steel powders: Experiments and modelling *Proc. Inst. Mech. Eng. Part B J. Eng. Manuf.* **219** 339–57

[13] Roberts I A 2012 *Investigation of residual stresses in the laser melting of metal powders in additive layer manufacturing* (University of Wolverhampton)

[14] Shapiro M, Dudko V, Royzen V, Krichevets Y, Lekhtmakher S, Grozubinsky V, Shapira M and Brill M 2004 Characterization of Powder Beds by Thermal Conductivity: Effect of Gas Pressure on the Thermal Resistance of Particle Contact Points *Part. Part. Syst. Charact.* **21** 268–75

[15] Raghavan N, Dehoff R, Pannala S, Simunovic S, Kirka M, Turner J, Carlson N and Babu S S 2016 Numerical modeling of heat-transfer and the influence of process parameters on tailoring the grain morphology of IN718 in electron beam additive manufacturing *Acta Mater.* **112** 303–14

[16] Wang X C, Laoui T, Bonse J, Kruth J P, Lauwers B and Froyen L 2002 Direct selective laser sintering of hard metal powders: Experimental study and simulation *Int. J. Adv. Manuf. Technol.* **19** 351–7

[17] Upadhyay M 2020 Fast to run model for thermal fields during metal additive manufacturing simulations

[18] Ren J, Liu J and Yin J 2011 Simulation of transient temperature field in the selective laser sintering process of W/Ni powder mixture *IFIP Adv. Inf. Commun. Technol.* **347 AICT** 494–503

[19] Debroy T, Wei H L, Zuback J S, Mukherjee T, Elmer J W, Milewski J O, Beese A M, Wilson-heid A, De A and Zhang W 2018 Progress in Materials Science Additive manufacturing of metallic components – Process , structure and properties **92** 112–224

[20] Zhao X, Reilly C, Yao L, Maijer D M, Cockcroft S L and Zhu J 2014 A three-dimensional steady state thermal fluid model of jumbo ingot casting during electron beam re-melting of Ti-6Al-4V *Appl. Math. Model.* **38** 3607–23

[21] Vock S, Klöden B, Kirchner A, Weißgärber T and Kieback B 2019 Powders for powder bed fusion: a review *Prog. Addit. Manuf.* **4** 383–97

IOP Conf. Series: Materials Science and Engineering 1281 (2023) 012014 doi:10.1088/1757-899X/1281/1/012014

Toward semantic standard and process ontology for Additive manufacturing

S Gouttebroze[1], J Friis[2], E W Hovig[1], K Boivie[3]

[1] SINTEF Industry, Forskningsveien 1, 0373 Oslo, Norway
[2] SINTEF Industry, Høgskoleringen 5, 7465, Trondheim, Norway
[3] SINTEF Manufacturing, S.P. Andersens vei 3, 7031 Trondheim, Norway

sylvain.gouttebroze@sintef.no

Abstract. Advanced modelling of additive manufacturing often requires the combination of models at multiple scales and multi-physics. Therefore, building the modelling workflow describing the process is complicated. The modelling is also only a part of the innovation process and must be connected to practices, experimental work, and characterisation. Efficient communication and data exchange between the different actors could quickly become a challenge. Recent developments in the frame of the EMMC (European Material Modelling Council) and in the EU project OntoTrans points toward the early integration of semantic description and the creation of dedicated domain ontologies. This require an unambiguous and consistent use of terms and definitions for various concepts within each field of technology, and international standards is an available source for structured technical terms and definitions. For additive manufacturing (AM) the international standard ISO/ASTM 52900 "Additive manufacturing - General principles - Fundamentals and vocabulary" is the internationally recognised source for terms and definitions. Basing the ontology on the AM terminology standard will greatly facilitate integration of AM processes as a part of an industrial manufacturing system. Therefore, the present work attempts to harmonise the standard terminology and the ontology concepts. Then, to improve the impact and connection to material science, the concepts will be connected to a microstructure domain ontology and to the top- and middle-level ontology EMMO. The conceptualisation and application of the ontologies will be illustrated through simple examples of process and material modelling.

1. Introduction

Additive manufacturing (AM) is becoming a reliable manufacturing route through the improvement of equipment, procedures, and feedstock. Normally a manufacturing process chain based on AM requires a series of operations and sub-processes besides the actual AM process. Preparations and post-processing operations can have a critical influence on the properties of the final product and therefore a combination of multiple models at multiple scales and multi-physics are necessary. This modern production method requires a high level of automation and integration of models and simulations to design the part, configure and automatise the process, and predict the performance. In industrial applications, processes such as laser-based powder bed fusion (PBF-LB) require both production time and high-quality feedstock material that can be quite expensive. Failure in production, or a failed product, can have a high impact on total production costs due to the wasted time and material, in particular for customised or small series manufacturing which is a common use for AM. Establishing reliable procedures and gathering expert knowledge has been a major focus on the last decade's work. The need to relate the component properties to process choices like support structures (internal and external), slicing, orientation, laser path, and laser parameters has led to multiple research studies and the development of numerical models. This knowledge and practical experience must be spread and reused to facilitate the industrial application of AM. Ideally this knowledge can be formalised and stored in a knowledge graph. This step is also required for the implementation of smart manufacturing where the next generation of manufacturing systems will have cognitive capabilities and the ability to control the execution of the task based on sensor measurements and reasoning on existing formal knowledge. The developments of ontologies for material, process and product are essential to convert existing knowledge into machine-readable, logical relations.

Per definition, international standards may contain product-performance requirements, describe recommended or required best practice testing procedures, or specify the content of services and how they should be performed. They can also include terminology. The aim of standards is to provide clear guidelines for consistent function and quality, improve processes, transparency, and comparison. Clarity in communication is critical, and therefore the use of vocabulary, including terms and their definitions, needs to be consistent and coherent throughout all standards on any given topic. This requires that a terminology standard is developed and widely accepted by consensus in the community. The development of international and industrial standards is mostly conducted through dedicated standard development organisations (SDOs) such as ISO, CEN/CENELEC, ASTM International, ASME, and other national or industry association-driven organisations, as reviewed by Kawalkar *et al.* [1].

On the other hand, an ontology represents knowledge as a map or graph of concepts within a domain and their relationships. The ontology is often expressed as an annotated machine-readable knowledge graph. The ontology is used to associate meaning to data (like measurements, procedure description, models, etc.), provides links between data, and allows for machine reasoning. Ontologies have proved themselves in various domains as being a valuable tool for solving knowledge and interaction problems. Still, in the manufacturing domain, ontologies are at an early development stage. Sanfilippo *et al.* [2] made a comparison of different ontologies and proposed a new one based on DOLCE (Descriptive Ontology for Linguistic and Cognitive Engineering) [3]. In this work, the ontology will be related to the EMMO [4]: "The Elementary Multiperspective Material Ontology (EMMO) is the result of a multidisciplinary effort within the European Material Modelling Council (EMMC), aimed at the development of a standard representational ontology framework based on current materials modelling and characterisation knowledge." EMMO development is based on description of the world anchored in physics and material sciences, integrating for example the concept of quantum of energy. A key advantage of EMMO is its inherent connection to material and process modelling that will facilitate the communication and integration of physics-based software.

The present work addresses AM technology and will therefore focus on extracting information only from the international standard for terminology in AM technology ISO/ASTM 52900 [5]. Building an ontology based on a published standard will require a parsing of the text to extract terms and definitions, and additional categorisation of the content. The aim is to develop a semi-automated procedure to build an ontology based on an online html version of the published terms and definitions. Then the ontology is linked to existing top-level and domain ontologies to provide a more comprehensive description. The meaning of and benefit for interoperability are described in the third section. Finally, the ontology is applied to describe existing research work.

2. Standardisation and ontology

The terms and definitions of ISO/ASTM 52900:2021 are available online through ISO's online browsing platform [6]. This is in the form of a structured html document, which enables automation of a part of the processing with the id and class of the html document. Automation is important to future application of the same methodology to other standards, and also to update the ontology following the upcoming revisions of the document. The main structure is presented in Figure 1. The terms are divided into three categories: general, processing, and parts. These sections are formed of the terms necessary to describe some key concepts but are not parenthood relations. For example, the term "3.12.4 final inspection" is a process and therefore not a subclass of the term "3.9.1 part" while "3.10.1 prototype" could be considered as a subclass of "3.9.1 part". The first step of the html document parsing is then to extract that section structure and at the same time the grammar information (as seen in Figure 2, the grammatical type is indicated).

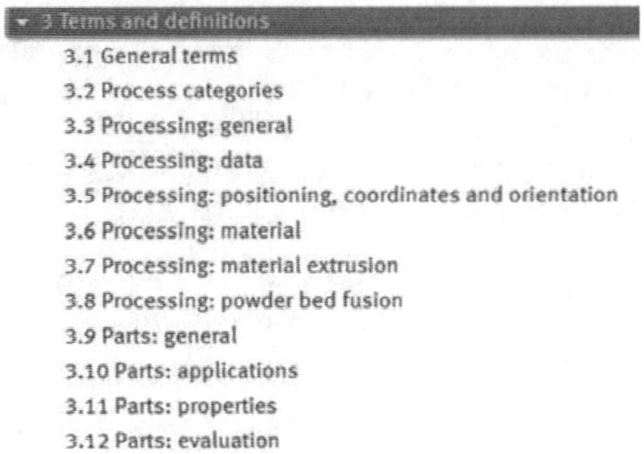

Figure 1. ISO/ASTM 52900:2021 structure for terms and definitions

The definition of a term also often includes references to other terms as illustrated in Figure 2. The procedure used to generate the ontology does not include language analysis and therefore will not extract the exact type of relation between the terms or the constraints. This data needs to be added manually later. For now, as illustrated in Figure 3, the procedure will create an ontology class that includes the definition, an id, a section number, and a list of term referenced in the definition. Please note that the terms have been replaced by the ontology concepts based on the section number. In the absence of defined relations, this information is kept as annotation of the class. Please note that as a convention, ontology classes will be written in Italic in the following sections.

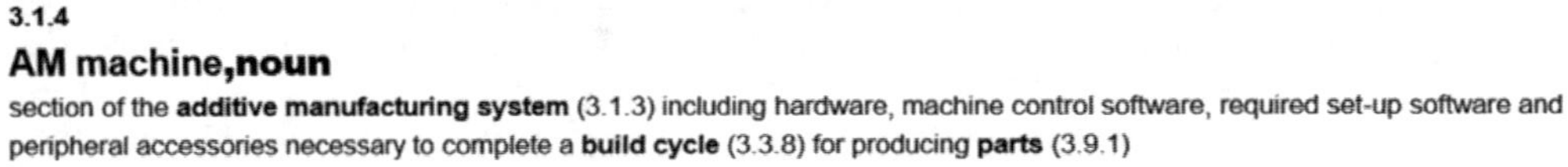

Figure 2. ISO/ASTM definition for AM machine.

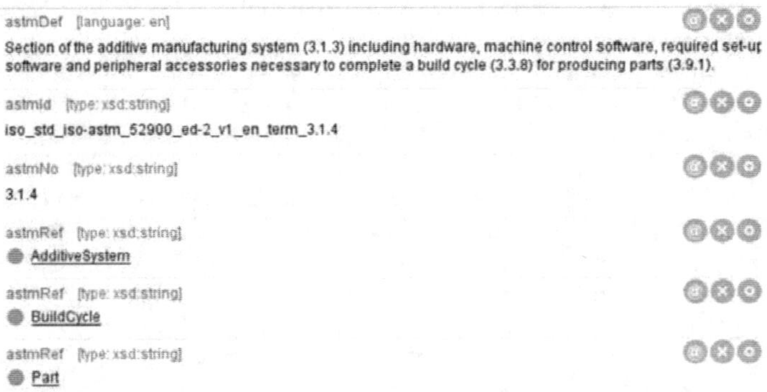

Figure 3. Ontologised definition of AM Machine

At this point the ontology structure reflects the grammatical information (*GrammaticalCategorisation*) and the paragraph structure (*ThematicCategorisation*). The next step is to categorise the concepts according to higher level concepts, called bridge concepts (as defined in

OntoCommons [7]). These higher-level concepts are ideally not new but related to established ontologies. We have defined 7 categories as presented in Figure 4.

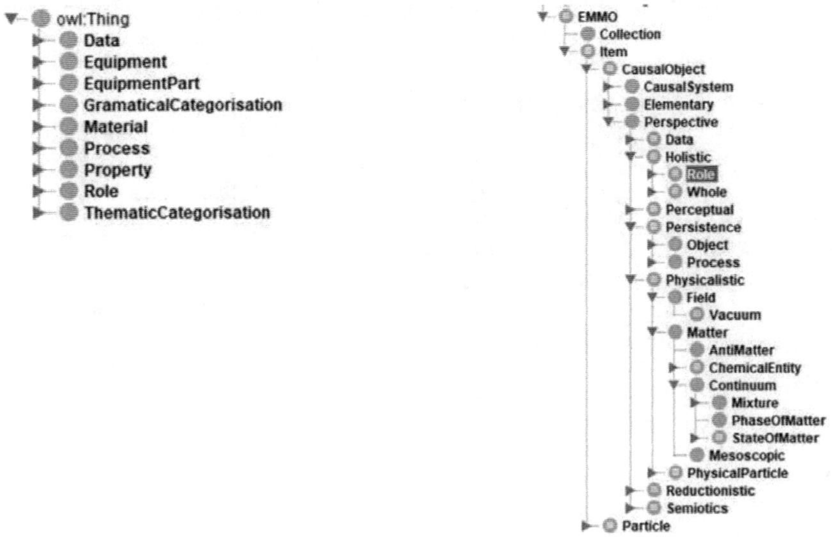

Figure 4. Main classes for the domain ontology for Additive manufacturing (left) and the top ontology EMMO (right)

All the previously parsed concepts are, manually, either defined as subclass of these categories or subclass of an existing class. For example, *Pellets* is defined as a subclass of *Feedstock* which is a subclass of *Material*. These categories represent different perspectives to describe the real world. It has been formalised in the *Perspective* class in EMMO. For example, *Role* is a subclass of *Holistic* perspective defines as "A holistic perspective considers each part of the whole as equally important, without the need to position the parts within a hierarchy (in time or space). The interest is on the whole object and on its parts (how they contribute to the whole, i.e., their roles), without going further into specifying the spatial hierarchy or the temporal position of each part." Similarly, *Process* is seen in the *Persistence* perspective: "The interest is on the 4D object as it extends in time (process) or as it persists in time (object): object (focus on spatial configuration) or process (focus on temporal evolution)." It is important to remember that the concept can be seen through different perspectives. For example, the laser is an equipment part that has a role (providing energy to melt the powder) but is also made of matter and is an assembly of components (with spatial positions and parthood relations).

The relation between the ISO/ASTM classes and the EMMO classes is important to harmonise the development and facilitate the extension of the capabilities. As illustrated in Figure 4, EMMO is still very generic, but domain ontologies have been built based on that logical representation. In the application section, we will see how part of the information used to describe an AM study is outside the scope of the ASTM standard (and so not included in the domain ontology generated from it), therefore requiring to add terms from a domain ontology for microstructure.

3. Data and interoperability

Interoperability (from latin, inter = between and operari = to work) is the ability of two or more systems to exchange information between them through a *common representational system* to perform a complex work that cannot be done by each single system alone. The presence of a common representation system provides the highest level of generalisation and replaceability and means that no privileged one-to-one connection between two system types should be implemented within the interfaces. In principle, in an interoperability scenario, one system can ignore the details about other systems. Figure 5 illustrate the difference between interoperability and compatibility. C*ompatibility* (from latin, cum = with and passus = to suffer) is the ability of two or more systems to establish a one-to-one connection between them, which is usually due to strong similarities in their internal representations that facilitate mutual understanding (e.g. for software this usually happens when systems are parts of a set of tools provided by a common developer) (Figure 5b). In a compatibility scenario, systems are fully aware of the type and identity of the other connected systems.

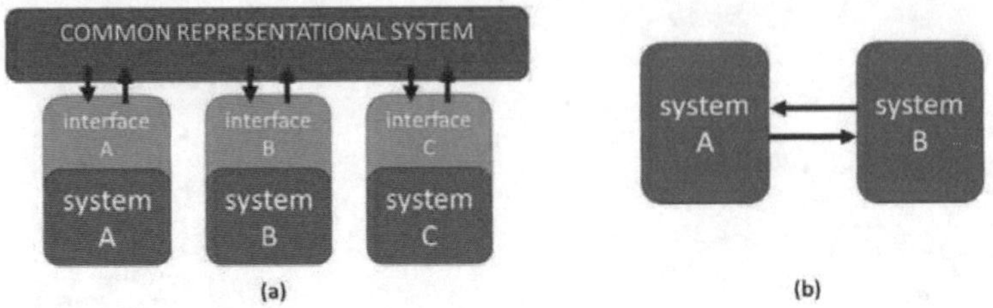

Figure 5. Interoperability (a) versus compatibility (b).

Interoperability can be at several semantic levels, such as scientific community level, user case level and numerical level. To support interoperability between experts from different scientific communities and different digital systems/simulation tools, the common representational system language must be understandable by both humans and machines. The ontological framework described in the last section serve exactly this purpose – to be a common representational system that is interpretable by both humans and machines.

There are different approaches for how to utilise ontologies in interoperability platforms. The SimPhoNy Open Simulation Platform [8] achieves interoperability by creating a representation using a set of connected classes in the Python programming language. This allows the user of SimPhoNy to seamlessly connect simulation engines, databases, and data repositories. However, for this to work, there must first exists a complete ontological description of the use case.

In the interoperability framework targeted by this work, we take another approach: we focus on the user and easy onboarding. The starting point is separation of concerns. Figure 6 shows an interoperability case where data from a database is used as input to a model. The database and model were developed completely independent of each other. The database provider knows the structure of the database but is not an expert on ontologies. The same is the case for the modeller, this person knows the input that the modelling tool expects, but not the details of the ontology. Instead of demanding these people to describe the data they provide or expect, they are only asked to describe their respective data in terms of simplistic, but formalised data models, whose structures can be made very similar to the database or model input. This makes it easy for the data provider/modeller or maybe even an external software engineer, to write a driver that can create an instance of data model A populated with data from the database or a driver that can serialise an instance of data model B to the form expected by the model. By mapping the properties of these data models to shared concepts in the ontology, it is now possible for the interoperability system to correctly create the input to the model from data in the database. It is only the ontologist that needs to know the details of the ontology and how properties can be converted from one concept to another. The low-level implementation of drivers for the database or modelling tool can be done by a software engineer. The ontologist can also help adding new concepts to the ontology as required by the domain experts.

A semantic interoperability framework based on the principles described above has been developed in a range of EMMC-related EU projects, especially OntoTrans [9] and OpenModel [10]. The main component of this framework is OTEAPI (Open Translation Environment Application Programming Interface) [11], which allows to document different data sources or data sinks in terms of reusable so-called partial pipelines, that can be stored in a knowledge base.

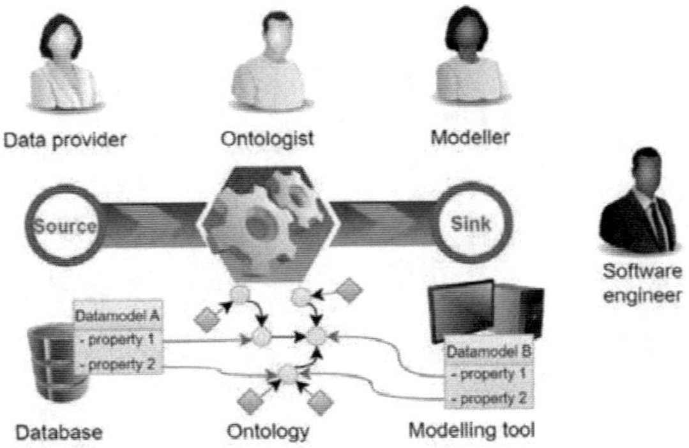

Figure 6. Achieving interoperability by separation of concerns.

A pipeline, like the one shown in Figure 6, can easily be created by connecting a partial pipeline for a data source with another partial pipeline for a data sink. This gives the user a high degree of freedom to mix and match data sources and modelling tools in a very flexible way. It is also possible to combine the pipelines into complex modelling workflows. Other important components of this semantic interoperability framework include the ontology, DLite [12] (an interoperability framework based on data models) and tripper [13] (a package that provide a common interface to the ontology regardless of how it is stored).

4. Application

The AM domain ontology [14] based on the ISO/ASTM 52900 standard can be applied to describe current research work on AM. As an illustration, we have selected three papers previously published by colleagues from SINTEF and NTNU (Norwegian University of Science and Technology) colleagues with slightly different perspectives. The first paper [15] focuses on process development and improvement. The AM description takes a significant part of the introduction section and the section on Materials and Methods. The second paper [16] is an intermediate paper linking the process to the properties. The focus on the AM process is smaller and seen more through the sample used for mechanical analysis. Finally, the third paper [17] focuses on the properties of the material produced with AM. The paragraph describing AM process parameters is even shorter and the second section is named only "Methods" (instead of "Materials and Methods"). It reflects the variations in the intention of the authors. For all papers, the interpretation of the results also includes references to AM terminology.

This case illustrates the importance of the common terminology and domain ontology as for all of them the description of the process should be understandable by all the actors to allow reproducible research. We will now attempt to convert the natural language description of the AM process into a semantic description relying on our new ontology. We will not completely cover all the data sets as it might require additional developments in the ontology and other application ontologies. Still, we will try to illustrate the necessary connection to other ontology in the second section.

4.1. Process and equipment

These studies are considering only laser powder bed fusion (PBF-LB) as the manufacturing process. The domain ontology used the term *PowderBedFusion* (see Figure 7). This term is referring to *AdditiveManufacturing* and *PowderBed*. Thus, it reflects that it is a manufacturing process, which could be inferred from the relations, and that it is applied to a material which is first considered as a geometrical domain as seen in Figure 8. This example illustrates as well that the ontology extracted from the standard using the current methodology is incomplete as it does not define explicit relations between the main component of the system, namely the powder material and the equipment.

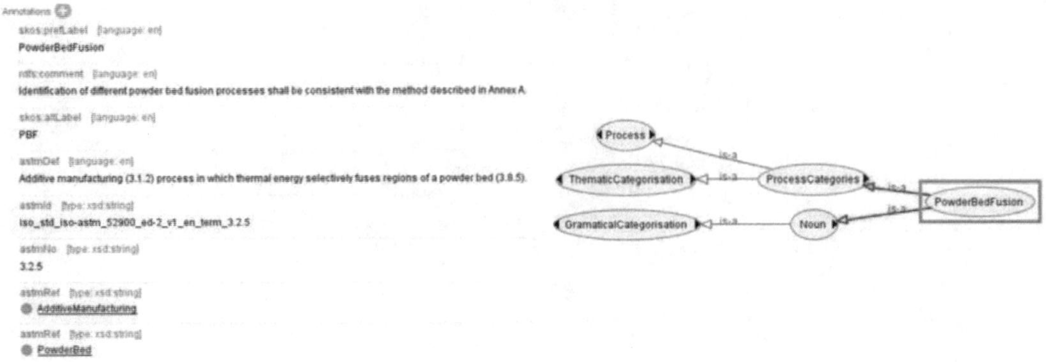

Figure 7. Definition and relations for the class PowderBedFusion

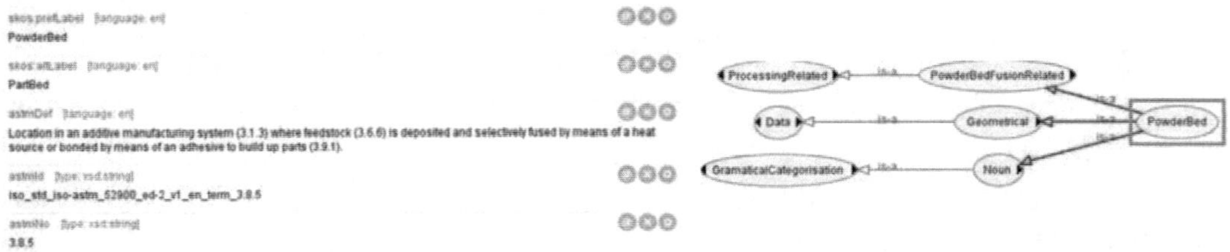

Figure 8. Definition and relations for the class PowderBed

The selected papers focus on the effect of anisotropy induced by the manufacturing process itself. From [15], "flat tensile specimens were produced in three sets, each consisting of 11 specimens built at different orientations. The sample orientation with respect to the build plate starts at horizontal (0°) increasing to vertical (90°) with 15° increments." The concept *BuildPlatform* is defined as an *Equipment,* but it has plural meanings in the previous description as it also refers to the *BuildSurface* (for the first layer) defined as a geometrical object. The proper definition of the orientation and position require as well as the definition of *BuildOrigin* and (*XAxis*, *YAxis*, *ZAxis*). Then the *PartPosition* and *PartReorientation* are defined and related to *PositionVector* defined in EMMO.

Important process parameters such as Laser power, Layer thickness, Hatch spacing, and Scan velocity are specified in the publication but not part of this standard. The terms "laser power" or "layer thickness" are considered as self-explanatory, and something that a qualified user of any AM standard would understand without the need for a definition, and ISO Directives states that they should not be defined in terminology standards. Only the generic concept of *ProcessParameters* is present. Nevertheless, EMMO already includes physical quantities. For example, *Power* is defined and associated with *PowerDimension* with the symbolic value "T-3 L+2 M+1 I0 Θ0 N0 J0" for Time - Length - Mass - Electric Intensity - Temperature - Amount - Luminous Intensity. *LayerThickness* would then be defined as a subclass of *Length* and inherit its properties and relations.

In addition, the samples are submitted to different post-process operations. In the standard, only *PostProcessing* is defined as a generic category. In the study, only heat treatment and HIP are used. These processes are generic and not specific to AM and therefore would naturally belong in another ontology.

4.2. Material and properties

The standard also allows us to describe the powder material. The main class is *Feedstock* that covers multiple AM processes. For PBF-LB, the powder characteristics are essential. The standard allows to specify the source of the powder (via *FeedstockManufacturer* and *FeedstockSupplier*) but more importantly indicate the possible reuse with *Virgin* and *UsedPowder*. In addition, the powder preparation is differentiated with *PowderBlend* and *PowderMix* as illustrated in Figure 9.

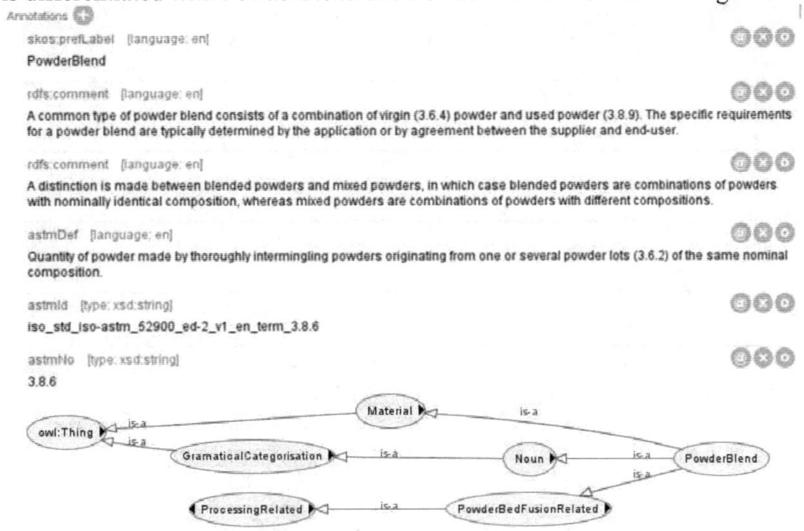

Figure 9. Definition and relations for the class PowderBlend

The description of the powder characteristics and part microstructure are not covered by the standard. Therefore, we include an additional microstructure ontology [18] based on EMMO. It allows to define the powder composition using the concept *ChemicalComposition*. The anisotropy of the microstructure also necessitates the concepts of matrix (*MicrostructureMatrix*), *Grain*, *Orientation* and *EulerAngles* to describe the texture. The microstructure ontology includes *PhasesFraction* (see Figure 10), *Dendrite*, *SecondaryArmSpacing*, *GrainBoundary* to specify the phase distribution and grain structure.

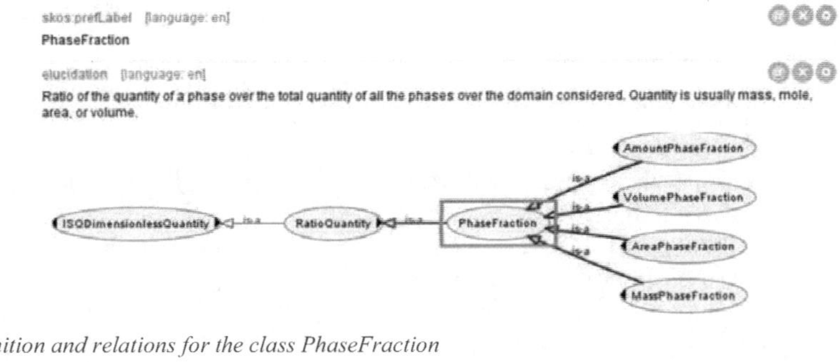

Figure 10. Definition and relations for the class PhaseFraction

The description of the mechanical properties would require an ontology as well which is not in place in the framework of EMMO.

4.3. Perspective on process and component modelling

In the previous sections, the experimental set-up and the process steps have been partly described using the AM ontology based on the standard and complementary ontologies. This must be seen as the first step to document semantically an existing process and component. The second step is to enable semantic interoperability for the simulation tools. For example, the definition of the laser power and path according to the standard should allow the automatic generation of the input file for the process simulation by the application of a wrapper layer that will connect the ontological concept to a variable in the model and its syntax for the input file. Similarly, the microstructure description could be correlated and compared with model prediction (like average model for grain size prediction or direct simulation with phase field). The systematic storage of data from measurements, experiments, and simulation as part of a knowledge graph would facilitate optimisation studies. The optimisation done by Azar et al. [19] requires linking surface characterisation and fatigue modelling by representing the surface roughness effect on crack initiation. Sharing common concepts and representing data in the appropriate format for the different communities is a key to enhance interdisciplinary collaboration.

4.4. Reasoning

As previously mentioned, this new ontology only includes information present in the standard and therefore does not pretend to generate a fully consistent ontology. Nevertheless, it allows us to identify the gap between standard documents and functioning ontologies. The semantic storage of all relevant information and measurements for the process, material, and part allows to build a massive knowledge graph. Such a knowledge graph could include as well other standards specifying the parameters for efficient production of specific components or structure. The extension with techno-economic data will also unlock production optimisation and decision support (see examples from Nagy *et al.* [20]).

Automated reasoning could be applied when semantic data is based on ontologies. Rules could be extracted to determine the ability to produce a component. For example, a rule could state that if a part is made of aluminium and its minimum wall thickness is greater than or equal to 3 mm, then it can be manufactured additively with these process parameters and these powder composition and sizing. These rules are deduced from the data stored in the knowledge and can later be applied to assess a design/production proposal automatically. A part that satisfies this rule will become a subclass of the *AMfeasible* class.

5. Concluding remarks

This paper presents the initial effort to partly automatise conversion of the ISO/ASTM AM terminology standard into a functioning ontology based on EMMO. The proper formatting of the online document allows us to extract important information and document structure. Nevertheless, the topics and categorisation of terms published in the present edition of this standard is not sufficient to build a complete ontology. Terminology standards and ontologies are developed to serve different purposes, and clearly this also means that the structures and functionalities are not perfectly matched. For building ontologies, it is necessary that the concepts defined in the standard are structured based on defined subclasses of high-level concepts. We have demonstrated that step and its linking to the top-level ontology EMMO. The AM ontology was then applied to describe existing research work. The focus on production of the selected standard does not cover all the concepts used by the referenced paper. It reveals the need for a broader development of ontologies dedicated to AM and materials science. The addition of the microstructure ontology developed in the frame of the EMMC enabled us to cover most of the microstructure concepts.

The growing integration of material and process modelling in component design and manufacturing will require the coupling of multiple models. The present work showed partly how the development of domain specific ontologies will enable semantic interoperability, therefore greatly accelerating the integration of new models in simulation workflows. In the future, it is expected that all numerical models will need to semantically describe their input and output to be broadly use and integrated in modelling software marketplaces.

The terms and definitions of ISO/ASTM 52900 are freely available through the ISO Online Browsing Platform, similarly the AM domain ontology is released on a public repository under the Creative Commons Attribution 4.0 International license (CC BY 4.0). It allows any manufacturer, designer, modeler to rely on the common vocabulary and integrates semantic at no cost. Standards are more than just terms definition, the fundamental need for clarity, consistency and coherency, as well as assuring the openness and transparency of the process, and a clear procedure for establishing consensus agreement and the publication of the standard requires a significant administration and clear directives, and this costs money. Currently the users pay for accessing the standards documents, so that the costs are covered by those who need, use and benefit from them. In the future, standards will need to be integrated in the digital world within knowledge graphs in order to enable smart manufacturing. This work can be seen as an early step in that direction where research work begins to be documented semantically based on ontologies covering production, material, processes. Later, the user could buy fully ontologised standards and ensure their application by computer reasoning on the stored data.

In the short term, we will aim at expanding the existing ontologies to cover the necessary measurements, simulations, and experimental set-up to completely describe our work. The objective is also to strengthen the connection to EMMO by exploiting the symbolic description of models and enhance interoperability.

Declaration of competing interest

The authors declare that they have no known competing financial interests or personal relationships that could have appeared to influence the work reported in this article.

Acknowledgments

This research work is funded by the SFI Manufacturing "Centre for competitive high value manufacturing in Norway" (grant number 237900) and SFI PhysMet "Center for sustainable and competitive metallurgical and manufacturing industry" (grant number 309584/F40) with support from the Research Council of Norway and the industrial partners. The interoperability work and development of EMMO was done through the project OntoTRANS, H2020-DT-NMBP-10-2019, with support from the EU commission (grant agreement 862136).

References

[1] Kawalkar R., *et al.*, 2022, Materials Today: Proceedings 50, pp. 1983–1990
[2] Sanfilippo E., *et al.*, 2019, Computers in Industry, 109, pp.182-194.
[3] Borgo S., and Masolo C. "Foundational choices in DOLCE". In Handbook on Ontologies, S.Staab and R.Studer, Eds. Springer Verlag Berlin Heidelberg, 2009.
[4] https://github.com/emmo-repo/EMMO
[5] "ISO/ASTM 52900:2021Additive manufacturing — General principles — Fundamentals and vocabulary", International Organization for Standardization (ISO) 2021, ASTM International 2021.
[6] https://www.iso.org/obp/ui/#iso:std:iso-astm:52900:ed-2:v1:en
[7] OntoCommons, https://ontocommons.eu/
[8] SimPhoNy, https://simphony.readthedocs.io/
[9] OntoTrans, https://ontotrans.eu/
[10] OpenModel, https://open-model.eu/
[11] OTEAPI, https://emmc-asbl.github.io/oteapi-core/
[12] DLite, https://github.com/SINTEF/dlite
[13] Tripper, https://emmc-asbl.github.io/tripper/
[14] https://raw.githubusercontent.com/emmo-repo/doam/main/astm52900.ttl
[15] Hovig E, *et al.*, 2018, Advances in Materials Science and Engineering, Article ID 7650303
[16] Hovig E, *et al.*, 2021, The International Journal of Advanced Manufacturing Technology, 114:1359–1372
[17] Solberg K, *et al.*,2021, International Journal of Fatigue 149, 106229
[18] https://github.com/emmo-repo/domain-microstructure
[19] Azar A., *et al.*, 2021, Rapid Prototyping Journal, Vol. 27 No. 2, pp. 256-267
[20] Nagy L., *et al.*, 2021, Complexity, Volume 2021, Article ID 8603515.

Understanding the inoculation of aluminium grains during welding under the influences of high cooling rate and high temperature gradient

Yijiang Xu[1], Geir Kvam-Langelandsvik[1], Ivan Bunaziv[1], Yanjun Li[2]

1. SINTEF industry, Richard Birkelands vei 3, 7034 Trondheim, Norway.

2. Department of Materials Science and Engineering, Norwegian University of Science and Technology (NTNU), N-7491 Trondheim, Norway

E-mail address: yanjun.li@ntnu.no (Yanjun Li)

Abstract. Inoculation is an important method to generate fine equiaxed grain structures in alloys during solidification, which is widely used in shape casting and direct chill casting processes of aluminium alloys. Different from the normal casting processes, welding and additive manufacturing can have ultrahigh cooling rate and temperature gradient during solidification. This makes the grain refinement by inoculation method more difficult. Therefore, a deeper understanding on the influences of cooling rate and temperature gradient on the nucleation and growth of grains is necessary. In this work, a numerical model is developed to quantitatively address the above influences. The model has been successfully applied to predict the grain size in the welding of aluminium alloys as a function of locations.

1. Introduction

Grain refinement by inoculation is widely used in shape casting and direct chill (DC) casting processes of aluminium alloys. Under such normal solidification conditions, the grain refinement behaviours including nucleation and grain growth, have been studied extensively through experimental studies and numerical modelling work in the past decades [1–11]. Among the experimental works, in-situ X-radiographic studies [10–13] have greatly contributed to the in-depth understanding of the nucleation and growth behaviours of grains in inoculated aluminium alloys. Based on the important free-growth criterion for heterogeneous nucleation of grains on potent inoculation particles proposed by Greer et al.[14], more advanced grain size prediction models for inoculated aluminium alloys have been developed in the last decade, in which the effects of addition level of grain refiners, solute concentration, cooling rate and temperature gradient have been well addressed [10,11,15,16]. The simulation results have a good agreement with the measured grain size of in-situ X-radiographic experiments.

Inoculation has also been explored in fusion welding and additive manufacturing of aluminium alloys to achieve fine equiaxed grain structure, which helps to reduce the solidification cracking and improve the mechanical properties such as strength, ductility, and toughness [17]. However, different from the normal casting, columnar grain structures are often obtained in welding and AM processes by adding normal grain refiner particles into the filler metals like wire and powder feedstock. The difficulty of grain refinement by inoculation in welding and additive manufacturing has been attributed to the much higher cooling rate and temperature gradient during solidification [18–22]. To date, deep understanding of the effect of such solidification conditions on heterogeneous nucleation and growth of grains has been limited.

In this work, a numerical model is developed to quantitatively address the influences of high cooling rate and high temperature gradient. The model will be applied to predict the grain size of the gas metal arc welding of aluminium alloys inoculated with Ti as a function of local solidification conditions.

2. Model for grain size prediction

The numerical model used in this work is an extension to a previous grain size prediction model developed for normal directional solidification conditions [11]. A schematic drawing of the 2D cross section of solidification domain used in the model is shown in Figure 1. Constant cooling rate and temperature gradient is applied in such local domain, while the local solidification parameter could be determined by finite element simulation. The interfacial non-equilibrium solidification effect on the dendrite tip growth is taken into account by implementing growth rate dependent kinetic partition coefficient k_v and kinetic liquidus slope m_v. The temperature gradient effect on the tip radius of dendrite was also included.

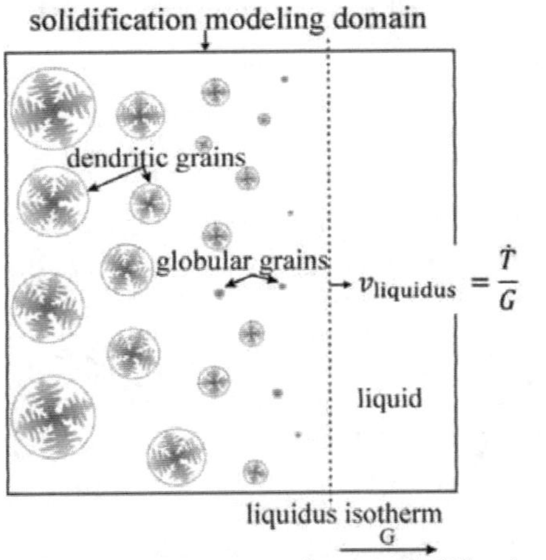

Figure 1. Schematic figure showing the 2D cross section of solidification modelling domain along the temperature gradient G direction. Nucleation and grain growth propagates from the left colder side towards the right hotter side.

The heterogeneous nucleation of grains on inoculant particles is based on the free growth criterion. The governing equations to determine the free growth undercooling and local undercooling experienced by inoculant particles, the size of inhibited nucleation zone around growing grains, and grain growth velocity were described in a previous model, and could be found in [11]. Here only some of the new equations are described.

During rapid solidification, the solute partition coefficient k_v is dependent of the moving velocity of solid/liquid interface V, and is estimated here based on [23]:

$$k_v = \frac{k + a_0 V/D_i}{1 + a_0 V/D_i} \tag{1}$$

where k is the equilibrium partition coefficient (could be assumed as constant or temperature dependent), a_0 is a characteristic length scale for solute trapping (of the order of interatomic distance), V is the growth velocity. D_i is the solute interface diffusion coefficient, which is often approximated to be the diffusion coefficient in the liquid D_l, given by the Arhenius type equation:

$$D_l = D_0 \exp\left(-Q/R_{gas}T\right) \tag{2}$$

And the velocity dependent liquidus slope m_v is calculated by the following equation [24]:

$$m_v = m\frac{1 - k_v[1 - \ln\left(\frac{k_v}{k}\right)]}{1 - k} \quad (3)$$

where m is the equilibrium liquidus slope.

For the spherical grain growth, the growth velocity is nearly the same as normal solidification condition, however, for dendritic growth, the temperature gradient will have a direct effect on the tip radius and growth rate. Here, the marginal stability criteria and Ivantsov solution [24] combining with the tip undercooling description are used to calculate the growth velocity, tip radius and tip concentration in the liquid.

Based on marginal stability criteria, the tip radius R_{tip} is [24]:

$$R_{tip} = [\frac{\Gamma}{\sigma^*(m_v G_c \xi_c - G)}]^{0.5} \quad (4)$$

Γ is the Gibbs-Thomson coefficient and σ^* is a stability constant given by $1/4\pi^2$, G is the temperature gradient. G_c is the solute concentration gradient in the liquid at the interface [25], and ξ_c is a stability parameter [24]:

$$\xi_c = 1 - \frac{2k_v}{\left[1+\left(\frac{2\pi}{Pe}\right)^2\right]^{0.5} - 1 + 2k_v} \quad (5)$$

3. Experiment

The Gas metal arc welding (GMAW) experiment was performed at SINTEF. The base metal for the investigation was an AA1070 aluminium alloy rolled plate of 3 mm in thickness. The welding wire was prepared by metal screw extrusion where elemental Ti powders were added into commercial purity Al (CP Al, 99.7 wt.%). The Ti composition in the welding filler wire was estimated around 0.85 wt.%. Welding was performed with I-groove geometry, with welding speed of 10 $mm \cdot s^{-1}$. Due to the dilution by base metal during welding, the final titanium content in the weld decreases relative to the original wire content. The chemical compositions of the weld zone were determined by Electron Probe Micro Analyser (EPMA) and listed in Table 1.

Table 1: Chemical compositions of different experimental materials, all in wt.%.

State	Al	Fe	Si	Ti	Ni	B
CP Al	99.83	0.12	0.04	0.00	0.003	0.00
Extruded Wire	98.98	0.12	0.04	0.85	0.003	0.00
Base metal AA1070	99.70	0.20	0.08	0.01	0.002	0.00
Weld	99.54	0.12	0.01	0.32	0.002	0.00

After welding, the weld part was cut transversally to the welding direction at the mid-length to obtain vertical cross section of metallographic samples. The sample was ground, polished mechanically, and etched anodically with a solution containing 5% HBF_4 and 95% H_2O to reveal the grain structure. Micrographs were made with an optical microscope using polarized light. The grain size at three locations was measured based on the circular intercept method described in ASTM E112-13.

4. Results and discussion

The grain structure in the cross section of welds can be seen in Figure 2. For the reference sample without adding Ti in the wire, the grain structure shown in Figure 2a shows a mixture of large columnar grains grow from the weld boundary and equiaxed grain in the weld centre. In comparison, the weld with adding Ti show mainly fine equiaxed grain structure, confirming that the Ti addition in the welding filler wire has a significant grain refinement effect. For grain size measurement, three locations were selected, i.e., at weld centerline, a mid-way section, and the transition area close to the weld boundary towards the base plate. These local areas are marked as '1', '2', and '3' in Figure 2, and the corresponding high magnification images are shown in Figure 3. As can be seen, the grains in area 1 show slightly elongated grain morphology, while they are equiaxed in the area 2 and area 3. The measured grain size from each area can be found in Table 2. It can seen that the grain size decreases from the weld boundary in area 1 to the weld centre in area 3.

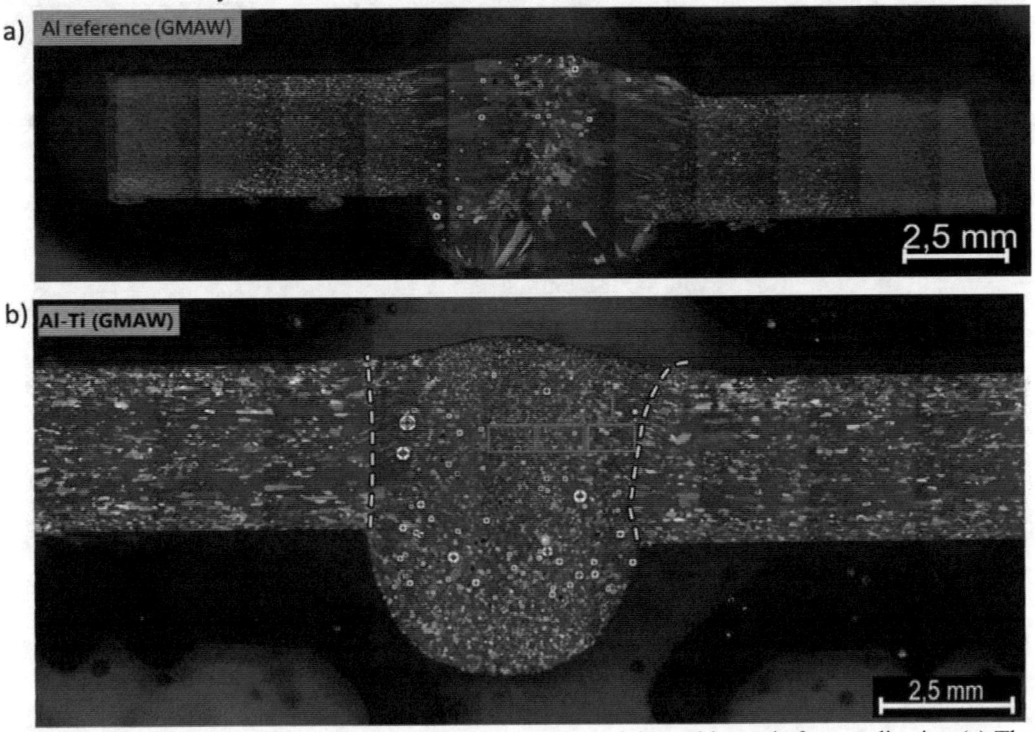

Figure 2. Optical microscope image of the cross section of the weld metal after anodization (a) The reference without adding Ti in the wire and (b) the one adding Ti in the wire. The three areas marked as 1, 2 and 3 in (b) were used for grain size measurement. The yellow dash lines indicate the melt pool boundary.

To obtain the local cooling rate and temperature gradient of the gas metal arc welding, temperature field simulation by the finite element modelling using the Abaqus software (2021 version) was performed. Such model has been validated with welding of steels [26,27]. The weld pool shape was modelled based on the experimental results of inoculated weld, as shown in Figure 2b with corresponding dimensions. The heat transfer modelling was done using DFLUX subroutine programmed in Fortran. The volumetric 3D heat source was the Goldak double ellipsoidal. The mesh was hex-dominated elements (DC3D8 type an 8-node linear heat transfer brick) of number 284 400. Temperature-dependent thermo-physical properties were applied: thermal conductivity for solid material was 210 $W \cdot m^{-1} \cdot K^{-1}$ and 100 for liquid $W \cdot m^{-1} \cdot K^{-1}$, density 2700 $kg \cdot m^{-3}$ (for liquid 2200 $kg \cdot m^{-3}$), latent heat of fusion 396 000 $J \cdot kg^{-1} \cdot K^{-1}$, and specific heat capacity for solid/liquid was 921/1050 $J \cdot kg^{-1} \cdot K^{-1}$ respectively. The temperature evolution under transient welding mode is shown in Figure 4.

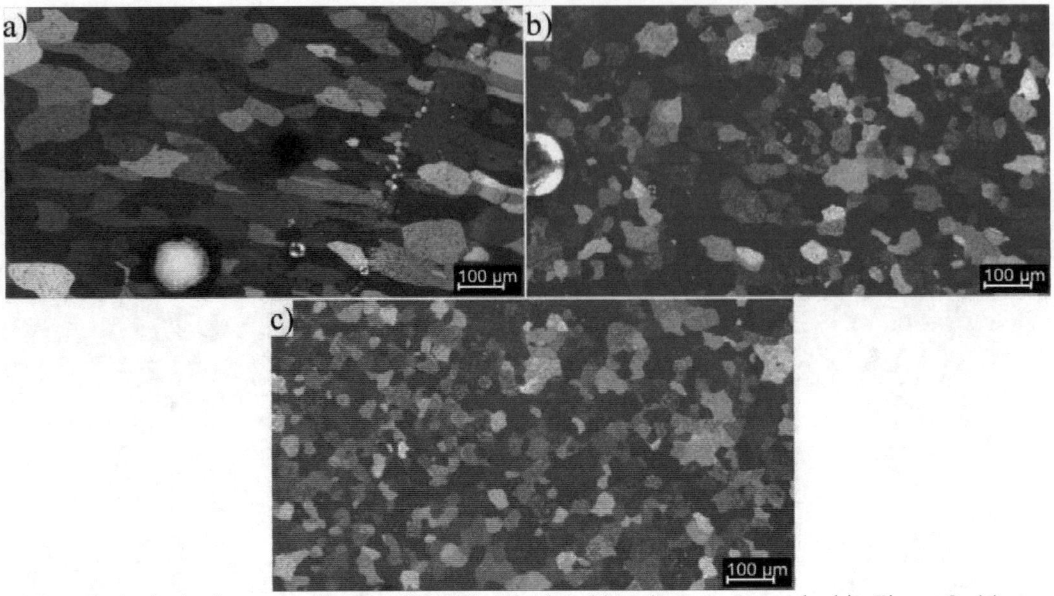

Figure 3. Optical microscope image of inoculated Al weld at three areas marked in Figure 2: (a) area 1, (b) area 2, and (c) area 3.

Table 2: Average grain size measured at three areas (marked as 1,2 and 3 in Figure 2) of inoculated aluminium weld.

Area	Average grain size [µm]	Standard deviation [µm]
1(weld boundary)	53.5	14.9
2 (midpoint of 1 and 3)	28.1	4.9
3 (weld centreline)	23.9	3.2

Figure 4. Temperature modelling and mesh outlook of the model during transient fusion welding.

The thermal cycle was extracted from the six locations of the weld pool at the cross section, as indicated in Figure 5a, with corresponding thermal cycle curves shown in Figure 5b. As can be seen, Point 1 close to the melt pool boundary has the lowest peak temperature, 679.4 °C, slightly higher than the melting point of the weld metal (~660 °C); while point 3 in the center reaches the highest peak temperature. Besides, there are plateaus in the cooling stage for some locations, such as point 2, 5 and 3, indicating the effect of latent heat release. Based on these thermal curves, the cooling rate close to the liquidus temperature and the local temperature gradient could be measured. The results are shown in Table 3. It shows that there is a big difference in cooling rate and temperature gradient at different locations in the weld and both decreases from the melt pool edge to the center of weld. These extracted data was used as input for the grain size prediction model.

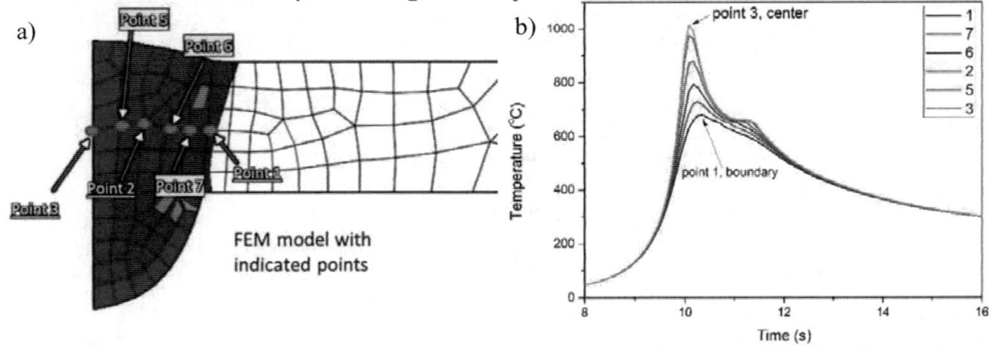

Figure 5. (a) Six node positions of weld pool cross section in Abaqus FEM model (from point 1 at the melt pool boundary to the center point 3) and (b) their corresponding thermal cycle curves as a function of time.

Table 3. Cooling rate close to the liquidus temperature and the local temperature gradient determined from the thermal cycle curves by Abaqus finite element simulation

Location	Cooling rate (K/s)	Temperature gradient (K/mm)
1	103	51.6
2	84	11.4
3	58	2.0

To simplify the influences of different solute contents in the weld on the growth kinetics of grains during solidification, the weld metal is simplified as a binary Al-0.322Ti (wt.%) alloy based on equivalent total growth restriction factor, $Q = \sum m_i Ci * (k_i - 1)$, where Ci is the concentration of solute element I, since Ti has the highest contribution to the Q value and the lowest diffusivity in liquid Al. The nucleant particles are in-situ formed Al_3Ti [28] before peritectic reaction. At the nucleation stage of Al grains, the total Ti solute is considered as the maximum solubility of Ti in liquid Al, 0.15 wt.%, which is used in the model for grain growth. The total volume fraction of the Al_3Ti particles is calculated as 0.362% based on the real Ti content in the weld. It is assumed that they have a log-normal particle size distribution [10], same as TiB_2 particles in Al-Ti-B refiner:

$$\frac{N(d)}{N_0} = \frac{\Delta d}{\sigma d\sqrt{2\pi}} exp\,(-\frac{(ln(d) - ln\,(d_0))^2}{2\sigma^2}) \tag{6}$$

where $N(d)$ is the number of particles with diameter between d and $d + \Delta d$, N_0 is the total number of TiB_2 particles, the ratio $\frac{N(d)}{N_0}$ is the relative population, d_0 =0.77 µm is the geometric mean of distribution and σ=0.5 is the geometric standard deviation. It is shown in Figure 6a. Besides, the total volume fraction, the size distribution and the total number density are considered independent of cooling rate studied here.

Figure 6b shows the grains size predicted by the current model in comparison to the experimental measurement shown in Table 2 at three locations of Al-Ti weld. The total number density of the Al_3Ti particles formed during solidification is an unknown parameter and thus has to be tuned based on grain size data at one of the locations, here location 1. Then this value is fixed and used as input for grain size simulation in location 2 and location 3. As can be seen, the grain size predicted by the model has a good agreement with the experimental results. From location 1 to location 3, the grain size decreases although the cooling rate (shown in Table 3) decreases as well. It reveals that the local temperature gradient has a greater influence than local cooling rate on the nucleation of Al grains in the weld pool. The larger grain size in location 1 is due to the larger temperature gradient, while smaller grain size in location 3 is promoted by smaller temperature gradient. Such agreement verifies that the model is able to simulate the effects of cooling rate and temperature gradient in welding of Al alloys on the grain size evolution and distribution.

Besides, the Al welding without adding Ti could also be simulated by the present model, assuming a small value of total number density of particles, for example typical amount of 0.1 wt.% addition of Al-Ti-B. The simulation results under solidification condition of location 1 show that the nucleation of new grains are not possible any more after first class nucleation at the melt boundary, indicating that columnar grains will form and grow towards the weld centre, which is in a good agreement with the experimental results presented in Figure 2.

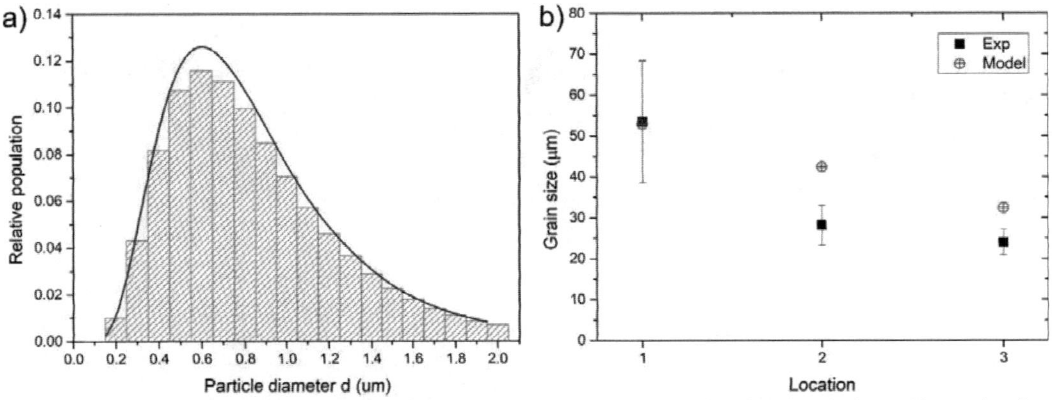

Figure 6. (a)Log-normal size distribution of Al_3Ti particles used for the simulation. (b) Grain size predicted by the current model, in comparison to the experimental results. The location should be referred to the ones marked in Fig. 2, where 1 is at the edge while 3 is at the centre of weld.

The possible change of non-equilibrium solute partition coefficient k_v and non-equilibrium liquidus slope m_v under current welding conditions have been investigated. Typical growth velocity evolution during solidification, for example, of the first class of grains nucleated in the melt at location 1, location 2 and location 3 is plotted in Figure 7a. As can be seen, the growth velocity will start from a very small value, then increase as solidification time increases (undercooling increase as well), and finally may reach a maximum value and then decreases again due to solute segregation effect of the residue melt. The corresponding non-equilibrium solute partition coefficient k_v, and non-equilibrium liquidus slope m_v at the growth interface are shown in Figure 7b and Figure 7c. It can be seen that k_v decrease slightly from 7 to about 6.98 at solidification condition of location 1, while in another two cases, it may decrease to 6.92. Nevertheless, the change of k_v is those conditions are not big. For mv shown in Figure 6c, the change is even less obvious. It could be concluded that k_v and m_v are not influenced much by the cooling rate and temperature gradient in common welding condition.

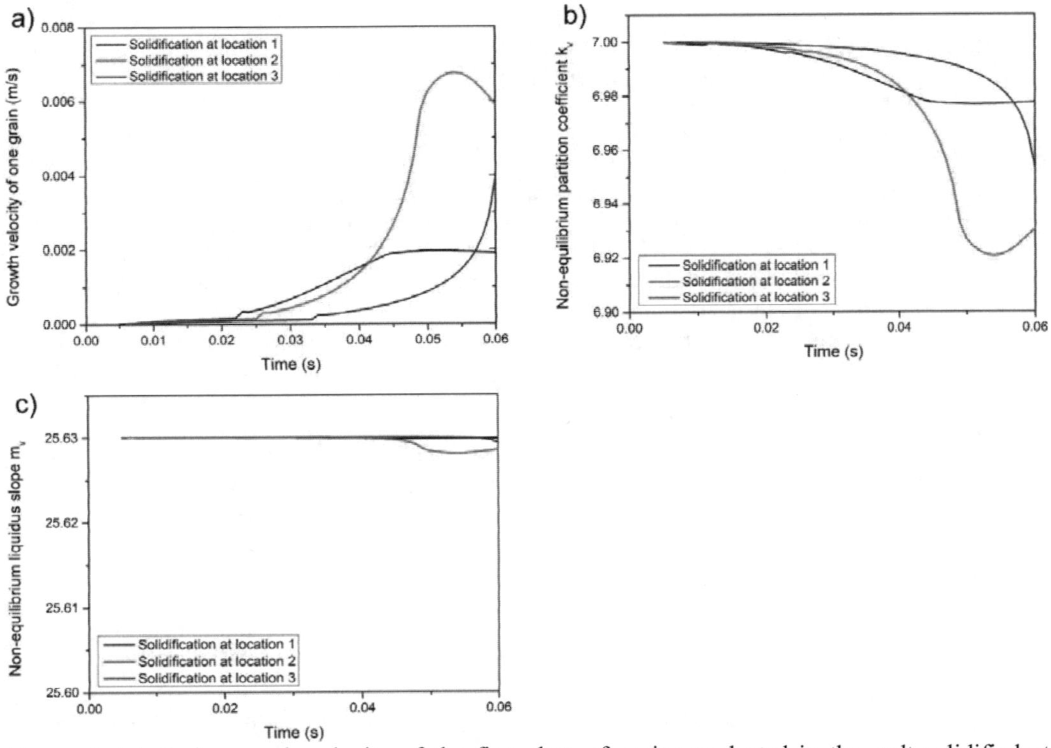

Figure 7. (a) Grain growth velocity of the first class of grains nucleated in the melt solidified at location 1, location 2 and location 3; (b) corresponding non-equilibrium solute partition coefficient k_v, (c) corresponding non-equilibrium liquidus slope m_v.

5. Conclusions

A numerical model was developed to quantitatively address the effect of high cooling rate and high temperature gradient on the heterogeneous nucleation and grain size of Al alloys during solidification. The model has been applied to the gas metal arc welding of aluminium alloys inoculated with Ti. By the finite element simulation of the temperature field, local cooling rate and temperature at different locations at the weld metal was obtained. The model has successfully predicted the grain size variation at those locations as a function of cooling rate and temperature gradient. It shows a good agreement with the measurement data of the welding experiment. It is found that, under the solidification condition of welding, the local temperature gradient in the welding pool has a more important influence on the heterogeneous nucleation of grains and therefore the final grain size. However, it can be expected that solidification conditions of additive manufacturing will have much more influences. It if worth of testing the validity of the model for the AM processes of aluminium alloys in a future work.

Acknowledgement
This research work has been funded by the SFI PhysMet, (Centre for Research-based Innovation, 309584). The authors gratefully acknowledge the financial support from the Research Council of Norway and the partners of the SFI PhysMet.

References

[1] Murty B S, Kori S A and Chakraborty M 2002 Grain refinement of aluminium and its alloys by heterogeneous nucleation and alloying *Int. Mater. Rev.* **47** 3–29
[2] Quested T E 2004 Understanding mechanisms of grain refinement of aluminium alloys by inoculation *Mater. Sci. Technol.* **20** 1357–69
[3] Easton M A, Qian M, Prasad A and StJohn D H 2AD Recent advances in grain refinement of light metals and alloys *Curr. Opin. Solid State Mater. Sci.* **20** 13–24
[4] Greer A L 2016 Overview: Application of heterogeneous nucleation in grain-refining of metals *J. Chem. Phys.* **145** 14
[5] Fan Z, Wang Y, Zhang Y, Qin T, Zhou X R, Thompson G E, Pennycook T and Hashimoto T 2AD Grain refining mechanism in the Al/Al–Ti–B system *Acta Mater.* **84** 292–304
[6] Iqbal N, van Dijk N H, Offerman S E, Moret M P, Katgerman L and Kearley G J 6AD Real-time observation of grain nucleation and growth during solidification of aluminium alloys *Acta Mater.* **53** 2875–80
[7] StJohn D H, Qian M, Easton M A and Cao P 7AD The Interdependence Theory: The relationship between grain formation and nucleant selection *Acta Mater.* **59** 4907–21
[8] Shu D, Sun B, Mi J and Grant P S 3AD A quantitative study of solute diffusion field effects on heterogeneous nucleation and the grain size of alloys *Acta Mater.* **59** 2135–44
[9] Du Q and Li Y J 2014 An extension of the Kampmann-Wagner numerical model towards as-cast grain size prediction of multicomponent aluminum alloys *Acta Mater.* **71** 380–9
[10] Xu Y, Casari D, Du Q, Mathiesen R H, Arnberg L and Li Y 2017 Heterogeneous nucleation and grain growth of inoculated aluminium alloys: An integrated study by in-situ X-radiography and numerical modelling *Acta Mater.* **140** 224–39
[11] Xu Y, Casari D, Mathiesen R H and Li Y 2018 Revealing the heterogeneous nucleation behavior of equiaxed grains of inoculated Al alloys during directional solidification *Acta Mater.* **149** 312–25
[12] Murphy A G, Mathiesen R H, Houltz Y, Li J, Lockowandt C, Henriksson K, Melville N and Browne D J 11AD Direct observation of spatially isothermal equiaxed solidification of an Al–Cu alloy in microgravity on board the MASER 13 sounding rocket *J. Cryst. Growth* **454** 96–104
[13] Murphy A G, Mirihanage W U, Browne D J and Mathiesen R H 8AD Equiaxed dendritic solidification and grain refiner potency characterised through in situ X-radiography *Acta Mater* **95** 83–9
[14] Greer A L, Bunn A M, Tronche A, Evans P V and Bristow D J 6AD Modelling of inoculation of metallic melts: application to grain refinement of aluminium by Al–Ti–B *Acta Mater.* **48** 2823–35
[15] Xu Y, Casari D, Mathiesen R H and Li Y 2019 Revealing the Heterogeneous Nucleation and Growth Behaviour of Grains in Inoculated Aluminium Alloys During Solidification TMS 2019 148th Annual Meeting & Exhibition Supplemental Proceedings (Cham: Springer International Publishing) pp 1665–75
[16] Xu Y, Du Q and Li Y 2015 Influence of Dendritic Growth of Equiaxed Grains on As-Cast Grain Size Prediction of Inoculated Aluminum Alloys *Trans. Indian Inst. Met.* **68** 1013–6
[17] Schempp P, Cross C E, Pittner A and Rethmeier M 2013 Influence of Solute Content and Solidification Parameters on Grain Refinement of Aluminum Weld Metal *Metall. Mater. Trans. A* **44** 3198–210
[18] Schempp P 2013 Grain refinement in aluminium GTA welds
[19] Li X P, Ji G, Chen Z, Addad A, Wu Y, Wang H W, Vleugels J, Van Humbeeck J and Kruth J P 2017 Selective laser melting of nano-TiB2 decorated AlSi10Mg alloy with high fracture strength and ductility *Acta Mater.* **129** 183–93

IOP Conf. Series: Materials Science and Engineering 1281 (2023) 012015 doi:10.1088/1757-899X/1281/1/012015

[20] Xiao Y K, Bian Z Y, Wu Y, Ji G, Li Y Q, Li M J, Lian Q, Chen Z, Addad A and Wang H W 2019 Effect of nano-TiB2 particles on the anisotropy in an AlSi10Mg alloy processed by selective laser melting *J. Alloys Compd.* **798** 644–55
[21] Tan Q, Zhang J, Sun Q, Fan Z, Li G, Yin Y, Liu Y and Zhang M-X 2020 Inoculation treatment of an additively manufactured 2024 aluminium alloy with titanium nanoparticles *Acta Mater.* **196** 1–16
[22] Tan Q, Zhang J, Mo N, Fan Z, Yin Y, Bermingham M, Liu Y, Huang H and Zhang M-X 2020 A novel method to 3D-print fine-grained AlSi10Mg alloy with isotropic properties via inoculation with LaB6 nanoparticles *Addit. Manuf.* **32** 101034
[23] Aziz M J 1982 Model for solute redistribution during rapid solidification *J Appl Phys* **53** 1158–68
[24] Kurz W, Giovanola B and Trivedi R 1986 Theory of microstructural development during rapid solidification *Acta Metall.* **34** 823–30
[25] Frenk A and Kurz W %J L in E 1992 Microstructure formation in laser materials processing **1** 193–212
[26] Bunaziv I, Akselsen O M, Frostevarg J and Kaplan A F H 2018 Laser-arc hybrid welding of thick HSLA steel *J. Mater. Process. Technol.* **259** 75–87
[27] Bunaziv I, Wenner S, Ren X, Frostevarg J, Kaplan A F H and Akselsen O M 2020 Filler metal distribution and processing stability in laser-arc hybrid welding of thick HSLA steel *J. Manuf. Process.* **54** 228–39
[28] Maxwell I and Hellawell A 2AD A simple model for grain refinement during solidification *Acta Met.* **23** 229–37

Computationally efficient model to predict the evolution in the thermal field in the EB-PBF and LP-DED processes

P Pourabdollah[1,2], F Farhang Mehr[1], S L Cockcroft[1] and D M Maijer[1]

[1] Department of Materials Engineering, University of British Columbia, 6350 Stores Rd, Vancouver, BC V6T 1Z4, Canada

[2] Corresponding author, E-mail address: pegah.pourabdollah@ubc.ca

Abstract. Understanding the development of a number of defects found in components fabricated by the metal Additive Manufacturing (AM) processes requires an understanding of the evolution in the thermal field within the component at both the macro- and meso-scales. As a first step, in this work, the agglomeration method was used in combination with a time-averaged input of energy to simulate the macro-scale evolution in temperature. Two example processes: 1) laser-based powder-fed directed energy deposition; and 2) electron beam powder bed fusion, are used to demonstrate the modelling methodology. The approach employed focuses on ensuring the conservation of heat and is applied using ABAQUS. The two applications have been validated by comparing the predicted thermal behaviour with process-derived data. The results indicate that this method is an efficient strategy to predict the thermal field at the scale of the component being fabricated.

1. Introduction

Additive Manufacturing (AM) is an advanced manufacturing method for producing components in which the feedstock material is consolidated into a metallic part in a layer-by-layer fashion [1]. The benefits of fabricating parts via AM include: the ability to produce geometrically complex near-net-shape structures; increased design flexibility; reduced part counts by reducing the need to assemble multiple components; and minimization of subtractive machining and material waste. For these reasons, AM is growing in acceptance as an alternative method for producing components for automotive, aerospace, medical, and energy applications [2,3]. However, despite these benefits, several challenges are associated with this process, including the high cost of the feedstock material in powder-based additive technologies, low productivity, porosity, surface roughness, and the development of residual stresses in the final component [4–6].

Metal AM processes include a wide range of methods, including Electron Beam Melting (EBM), Directed Energy Deposition (DED) and Selected Laser Melting (SLM). These AM techniques are based on selective melt/sintering of the build material, which may be powder or wire, using an intense, focused heat source. In these processes, the heat sources, such as an electron beam or laser, are rapidly moved over regions on a given layer consistent with the geometry of the part being fabricated [3,7,8].

In AM processes, large thermal stresses develop due to the temperature gradients associated with the rapidly moving intense heat source. Additionally, stresses may develop due to the thermal strain mismatch between the deposited material and the substrate or build plate. These strains can lead to residual stresses and component distortion. The post-fabrication removal from the substrate or the removal of support structures can also result in further deformation [9–11]. Predicting an accurate thermal history in the components fabricated by the AM processes at the macro-scale is crucial to estimating the Type-1 residual stresses, which are the macro-scale residual stresses responsible for component distortion. Various studies have been conducted to predict the temperature field and its effect during the AM processes [12–14].

Modelling the part-scale thermal field in melt-sintering AM processes is computationally intensive as it involves transient heat transfer over two significantly different time and length scales [10,15]. To facilitate practical AM process simulation, various studies have explored different techniques to reduce computational time. For instance, some researchers have used the agglomeration method, in which several physical deposition layers are lumped into one computational layer [16–18]. Another technique to reduce the computational time is the flash heating method, in which an equivalent uniform heat source is applied to each layer instead of modelling the precise details of the heat source-material interaction [19, 20]. Although the agglomeration and flash heating methods can reduce the computational cost, these methods cannot capture the thermal behaviour at high spatial and temporal resolution during the process – e.g., the size and location of the melt pool cannot be predicted.

In this work, the macro-scale evolution in temperature was simulated using a combination of the agglomeration and flash heat methods. Two applications were used to illustrate the analysis methodology: a Laser-based, Powder-fed Direct Energy Deposition (LP-DED) process and an Electron Beam Powder Bed Fusion (EB-PBF) process. The novelty of the approach presented is that it is formulated around the principle of heat conservation. In the LP-DED process, the time-averaged input of energy is corrected for the enthalpy of the deposited material. In the EB-PBF process, the time-averaged input of energy is corrected for the enthalpy of the powder following the powder layer preheat. The two applications were validated by comparing the predicted thermal behaviour to the experimentally derived data. The results show that this approach effectively predicts the thermal field at the part scale.

2. Model development

For the two applications, LP-DED and EB-PBF, separate 3D, transient thermal models were developed in ABAQUS[1].

2.1. Geometry, mesh and element activation

Figure 1 shows the computational domain in various views (a)-(c), which includes the baseplate, the component geometry and the fixturing used for the LP-DED process. The figure also includes mesh (d), which contained 12,621 nodes and 9208 elements. The part was partitioned into 11 super layers. Each super layer represents the agglomeration of 4 physical process layers. The thickness of each super layer is 1.96 mm = 4 layers × 0.49 mm/layer [21].

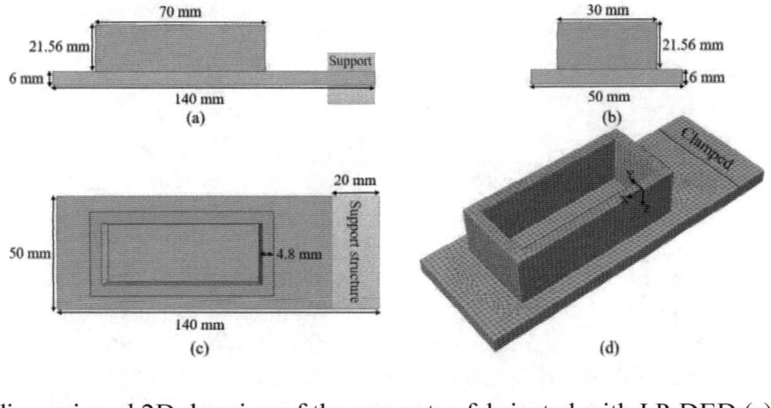

Figure 1. The dimensioned 2D drawing of the geometry fabricated with LP-DED (a) front view, (b) side view, (c) top view and (d) the analysis mesh [21].

[1] ABAQUS is the trademark for Dassault Systèmes.

Figure 2 shows the computational domain, including the part (a), baseplate and surrounding powder bed geometries (b) and mesh (c) for the EB-PBF process. The mesh contained 52,344 nodes and 44,700 elements. The powder bed and part were partitioned into 16 super layers. Each super layer represents the agglomeration of 6 physical process layers. The thickness of each super layer is 0.54 mm = 6 layers × 0.09 mm/layer.

For both applications, the number of physical layers per super layer was determined from a sensitivity analysis in which a balance was achieved between computational efficiency and accuracy. The deactivate/activate method was used to implement the sequential deposition of the layers. In this method, all elements in the domain are generated in the model before starting the analysis. Then, as the analysis starts, all elements of the part and/or powder bed are deactivated. During the analysis, the super layers of elements associated with the part and surrounding powder are then reactivated in a step-by-step sequence at the appropriate time.

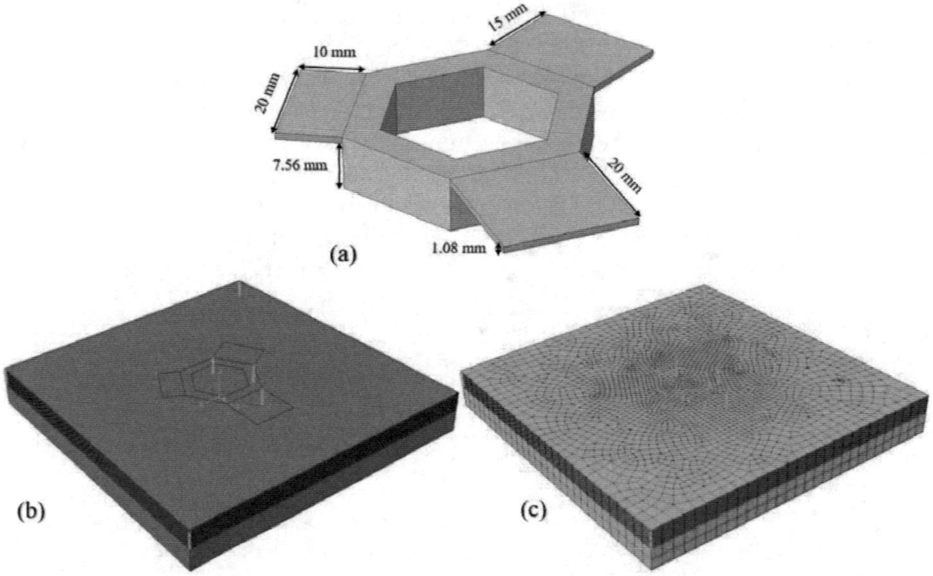

Figure 2. (a) fabricated component, (b) 3D computational domain including powder bed, baseplate, and component, and (c) mesh topography.

2.2. Thermal analysis

The governing equation for 3D transient heat balance is given in Eq. 1.

$$\frac{\partial}{\partial x}\left(k\frac{\partial T}{\partial x}\right)+\frac{\partial}{\partial y}\left(k\frac{\partial T}{\partial y}\right)+\frac{\partial}{\partial z}\left(k\frac{\partial T}{\partial z}\right)+\dot{Q}=\frac{\partial(\rho C_P T)}{\partial t} \tag{1}$$

where k is the thermal conductivity (W m^{-1} K^{-1}), T is the temperature (K), $\dot{Q}$ is the volumetric heat input rate (W m^{-3}), ρ is the density (kg m^{-3}), C_p is the specific heat capacity (J kg^{-1} K^{-1}), t is time (s), and x, y, z are the coordinates (m).

A time-averaged volumetric heat input methodology was implemented for each computational layer over the time associated with the deposition of the super layer. Thus, for example, in the EB-PBF process, the approach does not account for the process-related steps that would occur on a per powder layer basis – i.e., powder deposition, preheat-1, contouring, hatching and preheat-2 are not explicitly considered. Rather the timing of each is accounted for in the determination of the time-averaged energy input over the six process layers that are included in a single layer of elements.

In the approach used for both applications, the new layer of elements being added is activated at an elevated temperature. There are several reasons for doing this. In the LP-DED process, a portion of the laser energy is consumed by the powder stream entering the beam leading to this material being deposited at an elevated temperature. In the analysis of the LP-DED process, an assumption is made for the resulting temperature of the deposited material. In the case of the EB-PBF process, following the deposition of a powder layer, the new layer is preheated to a target temperature, which is reported to be ~950 °C, prior to beam contouring and hatching. In the context of the layer agglomeration method, these temperatures form the initial temperature for activation of the element layer, which must be considered in the overall system energy balance. This approach also benefits computational efficiency, and there are implications associated with capturing the generation of plastic strain at elevated temperatures, which are beyond the scope of this manuscript.

To correct the heat input for the enthalpy associated with activating the elements at the elevated temperature, the expressions shown in Eqs 2 and 3 are used.

$$\dot{Q}_{enthalpy} = \rho\big(C_p(T_{act} - T_{ref}\big)\frac{1}{t_{total}} \tag{2}$$

where ρ is the density (kg m^{-3}), C_p is the heat capacity (J kg^{-1} K^{-1}), T_{act} is the activation temperature of the new layer of elements (°C), T_{ref} is the reference temperature (°C) and t_{total} is the total time taken to process a super layer of elements.

$$\dot{Q}_{time\_avg} = \eta\dot{Q}_{beam}\frac{t_{beam}}{t_{total}} - \dot{Q}_{enthalpy} \tag{3}$$

where $\dot{Q}_{time\_avg}$ is the net time-averaged volumetric heat input rate (W m^{-3}) applied to each super layer, t_{beam} is the total time the beam is on during the deposition of the powder layers in a super layer, t_{total} is the total time taken to process powder layers in a super layer, $\dot{Q}_{beam}$ is the heat input rate (W m^{-3}), η is the absorptivity of the material being heated. In the case of the LP-DED process, the efficiency of heating, η, is set to 0.37, and for the EB-PBF process, it is set to 0.9.

For the LP-DED process, the beam power, $\dot{Q}_{beam}$, was 1500 W [18]. For the EB-PBF process, separate and time-varying heat inputs were applied to the component and the surrounding powder bed within each layer following activation. The energy input rates were extracted from data obtained from an ARCAM Q20Plus system, including the variation of voltage and current with time. The ARCAM Q20Plus system uses a series of different settings (including beam power, speed and focus) for what are referred to as process themes. The heat input for each process layer of the component is the sum of the energies from preheat-1, contour, hatch, and preheat-2 themes, whereas the heat input for the surrounding powder is the sum of the energies from preheat-1 and preheat-2 themes. The calculated heat inputs for each layer were then averaged over every six layers in the component and surrounding powder bed to calculate the heat input for each super layer of the component and surrounding powder bed in the analysis. The resulting time-averaged heat inputs, $\dot{Q}_{beam}$, are shown in Figure 3.

Referring to Figure 3, for the first few layers (up to the 12th layer), the time-averaged heat input increases due to the continuous application of the preheat-2 theme. The time-averaged heat input then decreases and oscillates around a constant value from the 12th to the 75th layer due to a reduction in the frequency of use of the preheat-2 theme. Toward the end of the build, the time-averaged heat input increases again during the deposition of the layers (84th layer to 96th layer) associated with the fabrication of the platforms on the component.

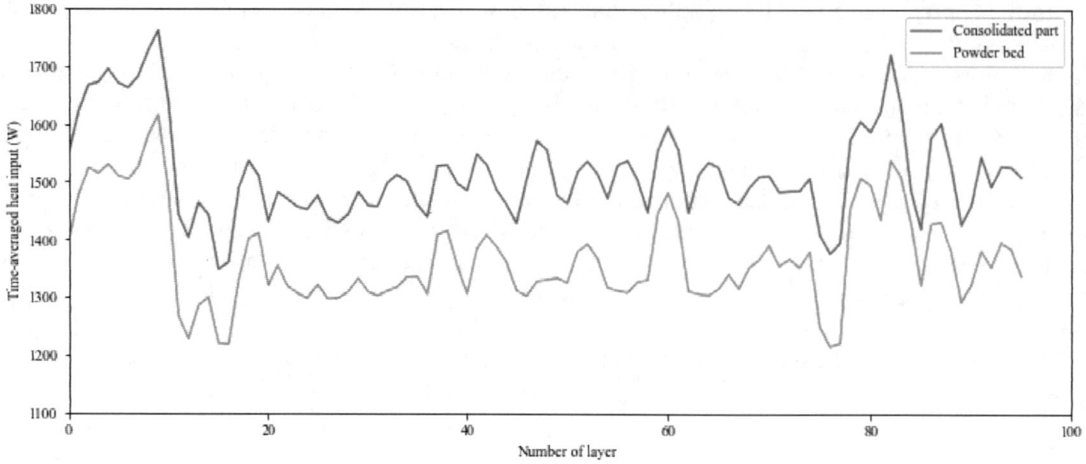

Figure 3. The time-averaged energy input for each layer of the component and surrounding powder bed.

2.2.1. *Boundary conditions*

During the LP-DED process, heat loss occurs via conduction to the baseplate, convection due to the flow of inert gas on the free surfaces (captured using a heat transfer coefficient of 5 W m^{-2} K^{-1}) and radiation from all the free surfaces on the component and baseplate to the surrounding environment. The baseplate also loses heat by conduction to the support structure that the baseplate is clamped to (refer to Figure 1). The ambient temperature was set to 25 °C. Conductive heat loss between the baseplate and the fixture was also defined via an increased heat transfer coefficient of 50 W m^{-2} K^{-1} [18].

During the EB-PBF process, heat loss occurs by conduction to the baseplate/surrounding powder bed and radiation from the top surface of the domain. The surface emissivity was set to 0.769 [22], and the ambient temperature was set to 350 °C. A contact boundary condition was defined to consider the conductive heat loss between the component and powder bed with a thermal conductance of 0.4 W m^{-2} K^{-1}.

2.2.2. *Initial conditions*

For the LP-DED process, the initial baseplate temperature was set to 25 °C. In addition, the activation temperature of the deposited material was set to 2200 °C [21]. For the EB-PBF process, the initial temperature of the baseplate was set to 750 °C, and the activation temperature of the deposited material for the component and powder bed was set to 950 °C.

2.3. *Material properties*

For the LP-DED process, the temperature-dependent thermo-physical properties of consolidated Ti-6Al-4V (Ti64) were used for both the baseplate and rectangular component [21]. For the EB-PBF process, the temperature-dependent thermo-physical properties of consolidated Ti64 were used for the component. The powder was treated as a bulk material, and the density and thermal conductivity were reduced by a factor of 0.5 relatives to bulk Ti64 to account for the void space between the discrete particles. The temperature-dependent thermo-physical properties of 304L stainless steel were used for the baseplate [23].

3. Results and discussion

The simulation results for both applications were validated by comparing the predicted thermal behaviour with process-derived data. Figure 4 shows a comparison between the predicted thermal history of the baseplate in the LP-DED process and the in-situ experimental measurements at two locations on the bottom surface of the baseplate reported by Lu *et al.* [18]. As can be seen, there is relatively good agreement between the simulated and measured thermal history of the baseplate, with a maximum error of approximately 10%. Furthermore, the execution time was less than 10 min using 12, 2.33 GHz, Intel Quad-core CPUs.

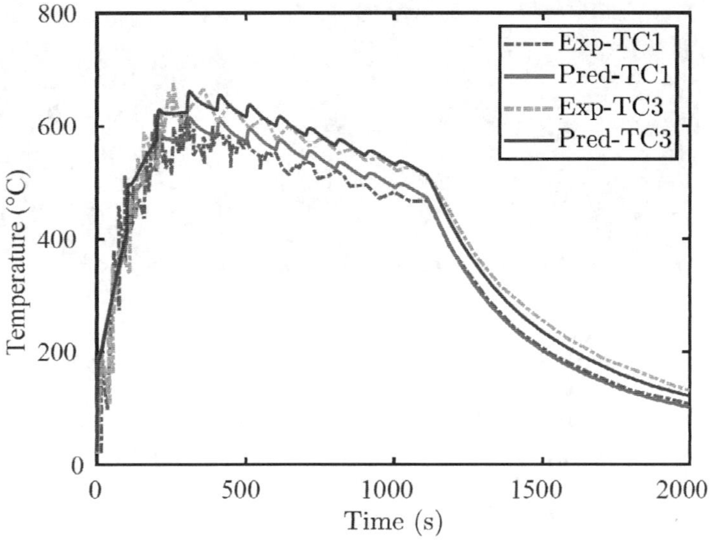

Figure 4. Measured [18] and predicted thermal history at two locations on the bottom surface of the baseplate [21].

Referring to Figure 4, the temperature of the baseplate increases rapidly during the addition of the first three super layers. The temperature then gradually reduces as the deposition continues. This reduction in temperature is due to the increased distance (and resistance to conduction) from the thermocouple locations to the layer of material being deposited. Finally, as the deposition of all layers is complete, the baseplate cools down. During the addition of each layer, it can also be seen that the predicted and measured thermal behaviour both show periodic, quick rises followed by gradual falls in temperature. This behaviour is associated with the deposition of hot material in a layer-by-layer process.

Figure 5 shows a qualitative comparison between the simulated temperature contour and a grey-scale image obtained by the ARCAM Q20Plus EB-PBF system during the manufacturing process. The image was captured after the deposition of all layers of the component (96 physical process layers, which corresponds to 16 super layers in the model). As can be seen, the model can qualitatively predict the difference in temperature observed between the inner support structure and the platforms. The higher temperature observed in the platforms is due to a combination of the increased area over which heat is input and their fabrication on un-consolidated powder resulting in a reduced path for conduction to the build plate. The execution time was approximately 2 hours using 12, 2.33 GHz, Intel Quad-core CPUs. Notes: 1) calibration of the grey-scale image from the ARCAM camera to temperature is ongoing; 2) the ABAQUS grey-scale shown is linear in temperature and must be corrected.

(a) (b)

Figure 5. A qualitative comparison between the predicted grey-scale temperature contour (a) and an image captured by the ARCAM Q20Plus EB-BPF system (b) at the end of the deposition of all layers.

As a second means of assessing the validity of the model predictions qualitatively, a comparison between a colour image captured during the EB-PBF manufacturing process and the predicted temperature contour is shown in Figure 6. Again, this comparison shows relatively good qualitative agreement between the model predictions and the process conditions.

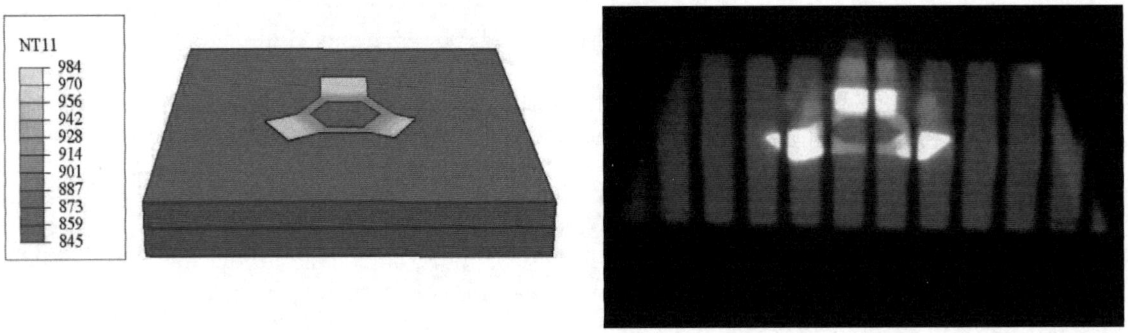

(a) (b)

Figure 6. A qualitative comparison between the simulated temperature contour (a) and an image captured during the EB-PBF manufacturing process (b).

4. Summary and conclusions

The accurate prediction of the temperature distribution at the macro-scale and its evolution during component fabrication in the metal AM processes is essential to be able to predict important phenomena developing at the meso-scale. These phenomena include pore formation, plastic strain accumulation, microstructure, and the evaporative losses of volatile alloy constituents. This work demonstrates a modelling methodology using the agglomeration method in combination with time-averaged heat input that is relatively fast to run and efficient. The method was implemented to simulate two process applications: an LP-DED process and an EB-PBF process. The conservation of heat was applied in the method employed in ABAQUS. The predicted thermal behaviour for both applications was compared against the process-derived data to validate the model. The results showed good quantitative agreement between predicted and experimental data in the case of the LP-DED process and good qualitative agreement in the case of the EB-PBF process.

5. Acknowledgement

The authors would like to acknowledge the NSERC Network for Holistic Innovation in Additive Manufacturing (HI-AM) for the financial support of this project.

References

[1] Gibson I, Rosen D, Stucker B. Additive manufacturing technologies: 3D printing, rapid prototyping, and direct digital manufacturing, second edition. 2015. https://doi.org/10.1007/978-1-4939-2113-3.

[2] DebRoy T, Wei HL, Zuback JS, Mukherjee T, Elmer JW, Milewski JO, et al. Additive manufacturing of metallic components – Process, structure and properties. Prog Mater Sci 2018;92. https://doi.org/10.1016/j.pmatsci.2017.10.001.

[3] Herzog D, Seyda V, Wycisk E, Emmelmann C. Additive manufacturing of metals. Acta Mater 2016;117. https://doi.org/10.1016/j.actamat.2016.07.019.

[4] Li C, Liu JF, Fang XY, Guo YB. Efficient predictive model of part distortion and residual stress in selective laser melting. Addit Manuf 2017;17. https://doi.org/10.1016/j.addma.2017.08.014.

[5] Aboulkhair NT, Everitt NM, Ashcroft I, Tuck C. Reducing porosity in AlSi10Mg parts processed by selective laser melting. Addit Manuf 2014;1. https://doi.org/10.1016/j.addma.2014.08.001.

[6] King WE, Barth HD, Castillo VM, Gallegos GF, Gibbs JW, Hahn DE, et al. Observation of keyhole-mode laser melting in laser powder-bed fusion additive manufacturing. J Mater Process Technol 2014;214. https://doi.org/10.1016/j.jmatprotec.2014.06.005.

[7] Körner C. Additive manufacturing of metallic components by selective electron beam melting - A review. International Materials Reviews 2016;61. https://doi.org/10.1080/09506608.2016.1176289.

[8] Li Y, Zhou K, Tan P, Tor SB, Chua CK, Leong KF. Modeling temperature and residual stress fields in selective laser melting. Int J Mech Sci 2018;136. https://doi.org/10.1016/j.ijmecsci.2017.12.001.

[9] Liang X, Cheng L, Chen Q, Yang Q, To AC. A modified method for estimating inherent strains from detailed process simulation for fast residual distortion prediction of single-walled structures fabricated by directed energy deposition. Addit Manuf 2018;23. https://doi.org/10.1016/j.addma.2018.08.029.

[10] Mukherjee T, Zhang W, DebRoy T. An improved prediction of residual stresses and distortion in additive manufacturing. Comput Mater Sci 2017;126. https://doi.org/10.1016/j.commatsci.2016.10.003.

[11] DebRoy T, Wei HL, Zuback JS, Mukherjee T, Elmer JW, Milewski JO, et al. Additive manufacturing of metallic components – Process, structure and properties. Prog Mater Sci 2018;92:112–224. https://doi.org/10.1016/J.PMATSCI.2017.10.001.

[12] Cao X, Ayalew B. Control-oriented MIMO modeling of laser-aided powder deposition processes. Proceedings of the American Control Conference 2015;2015-July:3637–42. https://doi.org/10.1109/ACC.2015.7171895.

[13] Toyserkani E, Khajepour A, Corbin S. 3-D finite element modeling of laser cladding by powder injection: effects of laser pulse shaping on the process. Opt Lasers Eng 2004;41:849–67. https://doi.org/10.1016/S0143-8166(03)00063-0.

[14] Hoadley AFA, Rappaz M. A thermal model of laser cladding by powder injection. Metallurgical Transactions B 1992 23:5 1992;23:631–42. https://doi.org/10.1007/BF02649723.

[15] Vastola G, Zhang G, Pei QX, Zhang YW. Controlling of residual stress in additive manufacturing of Ti6Al4V by finite element modeling. Addit Manuf 2016;12:231–9. https://doi.org/10.1016/J.ADDMA.2016.05.010.

[16] Ganeriwala RK, Strantza M, King WE, Clausen B, Phan TQ, Levine LE, et al. Evaluation of a thermomechanical model for prediction of residual stress during laser powder bed fusion of Ti-6Al-4V. Addit Manuf 2019;27. https://doi.org/10.1016/j.addma.2019.03.034.

[17] Hodge NE, Ferencz RM, Vignes RM. Experimental comparison of residual stresses for a thermomechanical model for the simulation of selective laser melting. Addit Manuf 2016;12. https://doi.org/10.1016/j.addma.2016.05.011.

[18] Lu X, Lin X, Chiumenti M, Cervera M, Hu Y, Ji X, et al. Residual stress and distortion of rectangular and S-shaped Ti-6Al-4V parts by Directed Energy Deposition: Modelling and experimental calibration. Addit Manuf 2019;26:166–79. https://doi.org/10.1016/j.addma.2019.02.001.

[19] Kiran A, Hodek J, Vavřík J, Urbánek M, Džugan J. Numerical simulation development and computational optimization for directed energy deposition additive manufacturing process. Materials 2020;13. https://doi.org/10.3390/ma13112666.

[20] Wang J, Zhang J, Liang L, Huang A, Yang G, Pang S. A line-based flash heating method for numerical modeling and prediction of directed energy deposition manufacturing process. J Manuf Process 2022;73:822–38. https://doi.org/10.1016/j.jmapro.2021.11.041.

[21] Pourabdollah P, Farhang Mehr F, Maijer DM, Cockcroft SL. A novel approach for the numerical analysis of in situ distortion in a component made by the directed energy deposition additive manufacturing process. The International Journal of Advanced Manufacturing Technology 2022 2022:1–14. https://doi.org/10.1007/S00170-022-10562-Z.

[22] Coppa P, Consorti A. Normal emissivity of samples surrounded by surfaces at diverse temperatures. Measurement 2005;38:124–31. https://doi.org/10.1016/J.MEASUREMENT.2005.05.001.

[23] ASM Handbook Volume 22B: Metals Process Simulation - ASM International n.d. https://www.asminternational.org/search/-/journal_content/56/10192/05281G/PUBLICATION (accessed December 18, 2022).

A phase field methodology for simulating the microstructure evolution during laser powder bed fusion in-situ alloying process

Z Li[1], M Greenwood[1,2] and A Phillion[1]

[1] Department of Materials Science and Engineering, McMaster University, Hamilton, ON, Canada

[2] Canmet MATERIALS, Natural Resources Canada, Hamilton, ON, Canada

E-mail: philliab@mcmaster.ca

Abstract. A recently-developed [1] multi-component phase field model has been utilized to investigate microstructure evolution during in-situ alloying of a blended elemental Ti-1Al-8V-5Fe alloy powder via the Laser Powder Bed Fusion process. The process of in-situ alloying, where elemental powder is used instead of pre-alloyed powder, was studied by performing two simulations having: (1) a uniform initial composition, and (2) a spatially varying initial composition to represent different powder particles. Specifically, the grain morphology, solute distribution, competitive growth and nucleation under the two different scenarios were simulated and compared. To assist the microstructure simulations, a macro-scale finite element model was developed to simulate the heat transfer during LPBF process. The thermal history data calculated by the finite element model was provided to the phase field model in order to simulate transient dendritic growth behaviour. The results show that a set of evenly-spaced columnar dendrites form in the uniform initial composition case, whereas when the initial composition is spatially varying, non-uniform dendrites having elongated shape can develop. It is also shown that competitive growth between dendrites is influenced by nucleation. For the spatially varying initial composition case, the results indicate that full alloying is difficult to achieve during the LPBF printing process; this incomplete alloying greatly influences the dendrite morphology and solute distribution.

1. Introduction

Additive Manufacturing (AM), which directly fabricates parts via layer-by-layer printing methods, has become a popular alternative to conventional subtractive manufacturing processes over the last 30 years [2, 3]. The major advantage of AM is that, after the AM process, only minimal post-processing machining is needed. This opens up significant possibilities in non-traditional markets for materials with poor machinability, e.g. β-Ti alloys [4].

β-Ti alloys are a class of Ti alloys where the predominant phase is β. This microstructure results in higher specific strength while still offering acceptable toughness and good fatigue resistance as compared to other Ti alloys [5]. However, β alloy possesses a number of significant limitations. First, due to the addition of Mo and Cr as β stabilizers, they are expensive. Second, because of their poor machinability, β alloys can be difficult to fabricate via conventional machining processes. To resolve these issues, Azizi et al. [3] proposed to utilize a low cost Fe-containing β alloy, Ti-1Al-8V-5Fe (wt %) alloy, and a Laser Powder Bed Fusion (LPBF) in-situ

IOP Conf. Series: Materials Science and Engineering 1281 (2023) 012017 doi:10.1088/1757-899X/1281/1/012017

alloying process, to make β-Ti components. Although the Ti-185 alloy was patented more than fifty years ago, the strong segregation tendency of Fe, and the resulting formation of brittle phases during casting [3] made this alloy not commercially feasible via conventionally processing. However, with the advancement of AM technologies, the utility of this alloy is being reassessed.

Laser Powder Bed Fusion (LPBF) is one of the most common metal AM processes as it offers high dimensional accuracy on small and medium size components. During LPBF, the molten pool undergoes rapid cooling whereby the cooling rates reaches values on the order of $10^5 - 10^6$ K/s, resulting in the formation of novel microstructures [6]. Traditionally, pre-alloyed powders have been used in LPBF, however, these can be costly especially for alloy development purposes.

Blended elemental powder are an alternative to pre-alloyed powders. This technique utilizes pure elemental powders as a substitute for prealloyed powders (e.g. Ti, Al and V powders instead of Ti-6Al-4V powders when building Ti-64 parts), and thus the alloying process occurs during the melting and subsequent reheating phase. Azizi et al. [3] built a Ti-1Al-8V-5Fe sample via LPBF of blended elemental powder of Ti, Al-V, and Fe, achieving in-situ alloying. Their Ti-185 sample showed the development of a fine microstructure, a remarkable combination of strength and plasticity, as well as homogeneity in alloy concentration. Ahmed et al. [7] performed a further investigation on the in-situ alloying process of Ti-185 using synchrotron X-ray diffraction. They found that the laser melting resulted in only partial alloying (i.e. alloyed and unmixed regions coexist), with the full alloying being achieved only during subsequent reheating cycles. However, it remains unclear how the alloying process proceeds and how the microstructure evolves during LPBF of blended elemental powders.

Over the last 30 years, the Phase Field (PF) method has become a popular approach to study the solidification process during conventional casting and emerging AM processes. In PF, a phase field ϕ is used to represent the liquid ($\phi = -1$) and solid ($\phi = 1$) phases, while a continuous function between $-1 < \phi < 1$ represents the mushy zone. The phase field will evolve along the direction of local free energy minimization, and thus the computational expense of interface tracking is not needed. Owing to the advantages of the PF method, it would seem perfectly suited to studying in-situ alloying during LPBF of blended elemental powders. Sahoo et al. [8] simulated microstructure evolution during electron beam AM processing of pre-alloyed Ti-64 powder in order to establish a relation between process parameters and grain size. However, nucleation was not considered in their model, and the multi-component system was simplified to a pseudo-binary alloy. Lenart et al. [9] simulated the Columnar to Equiaxed Transition (CET) of a pre-alloyed Ni-based superalloy during Directional Solidification, in order to study the influence of temperature gradient and solidification rate on the CET. Once again, a pseudo-binary assumption was applied.

While PF modelling using the pseudo binary approach improves the computational efficiency, and has been shown to simulate closely experimentally-produced microstructures, a homogeneous solute distribution is a necessary initial condition. Thus, it cannot be used to simulate in-situ alloying during LPBF of blended elemental powders. Multi-component PF models are highly accurate but computationally inefficient because they need to calculate the phase equilibrium of the multi-component system. Sun et al. [10] solved this problem for the case of dendritic growth of the Ti-64 alloy by considering multiple binary systems, specifically Ti-Al and Ti-V, each with their own partition coefficient k. However, the interaction between solutes was not considered and thus a true interfacial equilibrium was not achieved. Kundin et al. [11] adopted another multi-component PF model to simulate the rapid solidification processes. In their work the phase equilibrium was calculated at one concentration point, however the co-existence of multiple tie lines was ignored. Eiken et al. [12] improved the phase equilibrium prediction accuracy utilizing the quasi-equlibrium approach but at the cost of low computational efficiency since the quasi-equilibrium calculation was carried out at every node point and for each time step of the model [13]. Recently, Li et al.[1] proposed a new convex-hull based method to calculate the phase equilibrium of multi-component systems, and combined this with the grand potential PF model developed by Provatas [14] as well as a a dynamic adaptive mesh refinement method by Greenwood [15]. This combination of methods was shown to achieve a great combination of prediction accuracy and computational efficiency.

The other challenge when modelling microstructure development during LPBF via the PF method is the need to obtain the thermal history. While it is possible to simulate heat and mass transfer simultaneously within PF models, this is computationally expensive. Moreover, it is unnecessary since heat transfers much faster than mass [16, 17]. Simulating the thermal problem via finite element analysis (FEA) and then imposing the temperature history on the PF domain is a well-utilized substitution that provides computational efficiencies while still maintaining high accuracy in temperature evolution predictions [14, 8, 18]. Sahoo et al. [8] explored the relation between electron beam scanning speed and primary dendrite arming spacing during a Ti-64 alloy electron beam AM using this approach. However, the thermal conditions in the AM process were simplified to a directional solidification case where a single solidification rate was defined in their model. Thus, the complex transient thermal gradients and solidification rates that are known to occur during AM processing were ignored. Xiao et al. [18] overcome this issue by dividing the domain into different simulations to represent different regions within the molten pool. While more detailed knowledge of thermal gradients and solidification rates were obtained for the PF simulations, a Directional Solidification condition was still defined in each region and thus a panoramic picture of the whole molten pool was not obtained.

In the present study, a multi-component PF model is used to simulate microstructure development and in-situ alloying during LPBF of the Ti-185 alloy in blended elemental form. In-situ alloying is simulated by designing spatially varying initial composition fields to represent the elemental powders. Solidification of the same alloy but within a uniform initial concentration field, to represent pre-alloyed powder, is also simulated as a reference state. Finally, as input to the PF models, a finite element model is developed to simulate the transient heat transfer occurring during LPBF. These simulations are used to investigate grain morphology, solute distribution, and nucleation, and well as dendrite growth competition.

2. Numerical Methods

2.1. Phase Field Simulations

The multi-component PF model implemented in the present study was proposed by Provatas et al. [15, 19, 14, 1]. The formulation of this model begins with a quadratic description of the free energy of the solid and liquid phases,

$$F^{\theta}\left(c_1, c_2, ...c_n - 1\right) = \frac{1}{2}\sum_{i=1}^{n-1}\sum_{j=1}^{n-1} A_{ij}^{\theta}\left(c_i - \bar{c}_i^{\theta}\right)\left(c_j - \bar{c}_j^{\theta}\right) + \sum_{j=1}^{n-1} B_j^{\theta}\left(c_j - \bar{c}_j^{\theta}\right) + D^{\theta}, \tag{1}$$

where F^{θ} represents the free energy of phase θ, c_i and c_j are the concentration of component i and j, A_{ij}^{θ}, $\bar{c}_i^{\theta}$, B_j^{θ} and D^{θ} are fitting parameters. To get the fitting parameters, a data set $\{c_1, c_2, ..., c_{j-1}, F\}$ that contains a massive number of concentration-free energy data points is extracted from a thermodynamic database (e.g. Thermo-Calc), and then fit via a least square method. The fitting parameters will also be used in the phase field equations.

The evolution of the order parameter ϕ and chemical potential μ are expressed based on the free energy relaxation principle,

$$\tau(\vec{n})\frac{\partial \phi}{\partial t} = W_{\phi}(\vec{n})^2\nabla^2\phi + \phi - \phi^3 - (1-\phi^2)^2\left(\frac{(I-[K])^T}{2}\vec{U} + \hat{n}_c\right)^T [\lambda]\vec{U}, \tag{2}$$

and

$$[\chi]\frac{\partial\vec{\mu}}{\partial t}=\nabla\left[[M]\nabla\vec{\mu}+Wa(\phi)\,|\Delta\vec{c}_{eq}|\{\hat{n}_c+(I-[K])^T\vec{U}\}\frac{\partial\phi}{\partial t}\frac{\nabla\phi}{|\nabla\phi|}\right] + \frac{1}{2}|\Delta\vec{c}_{eq}|\{\hat{n}_c+(I-[K])^T\vec{U}\}\frac{\partial\phi}{\partial t}, \quad (3)$$

where $\vec{\mu}$ is a vector representing the chemical potentials of each solute component μ_i, I is the identity matrix, $\vec{n}\equiv\frac{\vec{\nabla}\phi}{|\vec{\nabla}\phi|}$ is the normal direction to the interface, $a_s(\vec{n})$ represents the 2D four-fold anisotropy, $W_\phi(\vec{n})=W_0a_s(\vec{n})$ sets the interface width [16], and

$$\left[\chi^\theta\right]^{-1}_{ij}=A^\theta_{ij}, \quad (4)$$

$$\vec{U}=\frac{[\chi^L]}{|\Delta\vec{c}_{eq}|}\left(\vec{\mu}-\vec{\mu}_{eq}\right), \quad (5)$$

$$\Delta\vec{c}_{eq}=\vec{c}^{\,L}_{eq}-\vec{c}^{\,S}_{eq}, \quad (6)$$

$$\hat{n}_c=\frac{\Delta\vec{c}_{eq}}{|\Delta\vec{c}_{eq}|}, \quad (7)$$

$$[K]=\left[\chi^L\right]^{-1}\left[\chi^S\right], \quad (8)$$

$$[\lambda]=\lambda_o\frac{|\Delta\vec{c}_{eq}|^2}{|\Delta\vec{c}_{eq,i}|^2}[\chi^L]^{-1}\chi^L_{ii}, \quad (9)$$

$$d_{0,i}=\frac{2\sigma_{SL}}{|\Delta\vec{c}_{eq,i}|^2}\chi_{ii} \quad (10)$$

$$\lambda_o=\frac{a_1W_0}{d_{0,i}} \quad (11)$$

$$[\chi]=\left[\chi^L\right]\{I-(I-[K])\,h(\phi)\}, \quad (12)$$

$$[M]=q(\phi)\left[D^S\right]\left[\chi^S\right]+(1-q(\phi))\left[D^L\right]\left[\chi^L\right]. \quad (13)$$

Note that in this set of equations, $h(\phi)$ and $q(\phi)$ are the interpolation equations commonly used in PF models, $d_{0,i}$ is the capillary length of the ith component, λ_o is the coupling coefficient [20], $a_1=0.8839$ is a constant, σ_{SL} is the solid-liquid interface energy, $\left[D^\theta\right]$ is the diffusion coefficient matrix that is related to the physical properties of the alloy system, matrices $[K]$ and $\left[\chi^\theta\right]$ are related to the free energy surface given in Eq. 1, and $\mu_{eq,i}$, $c^L_{eq,i}$ and $\vec{c}^{\,S}_{eq,i}$ represent the equilibrium chemical potential and the equilibrium concentration of the ith component in the liquid and solid phases. The last three terms, $\vec{\mu}_{eq}$, $\vec{c}^{\,L}_{eq}$ and c^S_{eq} all vary as a function of local temperature T; please refer [1] for the calculation methodology. Finally, the term $a_s(\vec{n})$ is calculated as

$$a_s(\vec{n})=1+\epsilon_4cos4(\theta+\theta_0), \quad (14)$$

where ϵ_4 is a material parameter characterizing the anisotropy strength, θ is the angle between the interface normal and pulling direction, and θ_0 is the misorientation angle between the preferred crystalline orientation and the global coordinate system [21].

In addition to the model described in [1], the model used in this study uses a nucleation module to take into account the formation of equiaxed grains. Assuming no solid grain movement, the nucleation probability P_n within an area $\Delta S = \Delta x \Delta y$ at a given time range Δt is calculated by [9, 22, 23]

$$P_n = -2\mu_N(\Delta T)\frac{\partial T}{\partial t}(1 - f_s)\Delta S \Delta t \tag{15}$$

where $\mu_N = 5 \times 10^6$ m^{-2}K^{-2} [23] is the nucleation possibility parameter, ΔT is the local undercooling, $\partial T/\partial t$ is the cooling rate and (1-f_s) is the local liquid fraction. Further details on this module's numerical implementation are described below.

2.2. Finite Element Thermal Analysis

The temperature evolution during LPBF of the Ti-185 alloy is simulated via the commercial ABAQUS software package. The model geometry includes one layer of deposited blended elemental powder on top of a substrate of uniform Ti-185. This approach was taken for computational efficient to approximate the steady-state printing of Ti-185. The dimensions of the substrate and deposited layer are: 3 mm × 1 mm × 1.06 mm and 3 mm × 1 mm × 0.06 mm, respectively. Heat transfer is assumed to be conduction-only, i.e.

$$\rho c_p \frac{\partial T(X, Y, Z, t)}{\partial t} = -\nabla \cdot \vec{q}(X, Y, Z, t) + Q(X, Y, Z, t) \tag{16}$$

where ρ is the density, c_p represents the specific heat capacity under constant pressure and T is the local temperature. Additionally, $\vec{q} = -k(T)\nabla T$ is the heat flux that results from the temperature gradient, which reprensents the heat conduction, $k(T)$ is the temperature dependent heat conductivity of the material, and Q represents other heat sources, which includes heat input from lasers and release of latent heat in this model.

In the present FEA model, the heat input from the laser beam is simulated by a Gaussian distributed moving heat source, written as [24]

$$Q(x, y) = \frac{2\lambda P}{\pi r_0^2} exp\{\frac{-2[(x - x_0)^2 + (y - y_0)^2]}{r_0^2}\} \tag{17}$$

where λ is the absorptivity of the material, P is the laser power, r_0 is the laser radius and (x_0, y_0) is the position of the laser center that varies with time according to the laser scanning speed, v. Convection and radiation are not considered. The preheat temperature, T_0 is used as the initial condition. Finally, the FEA domain is meshed with cuboid elements, 20×20×100 µm and 20×20×15 µm respectively in the substrate and the deposited layer.

2.3. Numerical Implementation of the PF, FEA, and Nucleation Modules

Fig. 1 shows a flowchart of the integration of the PF, FEA, and Nucleation modules. First, the FEA simulation is run to acquire the thermal history during LPBF processing. Second, the model parameters for the PF simulation are determined. Specifically, (1) the multi-component phase equilibrium is predicted using the method presented in [1] and (2) the capillary length of the material d_0 is calculated by calculating the $d_{0,i}$ for each component via Eq. 10 (Note - the reference component is chosen to be the one with the smallest capillary length). Third, the PF model is advanced from one time-step to the next. At each time-step the temperature values from the FEA simulation are linearly interpolated to the nodal points on the PF mesh. Please note that during the nucleation check stage, a random number between 0 and 1 is generated and compared with P_n. Nucleation will occur if P_n is larger than the random number. The

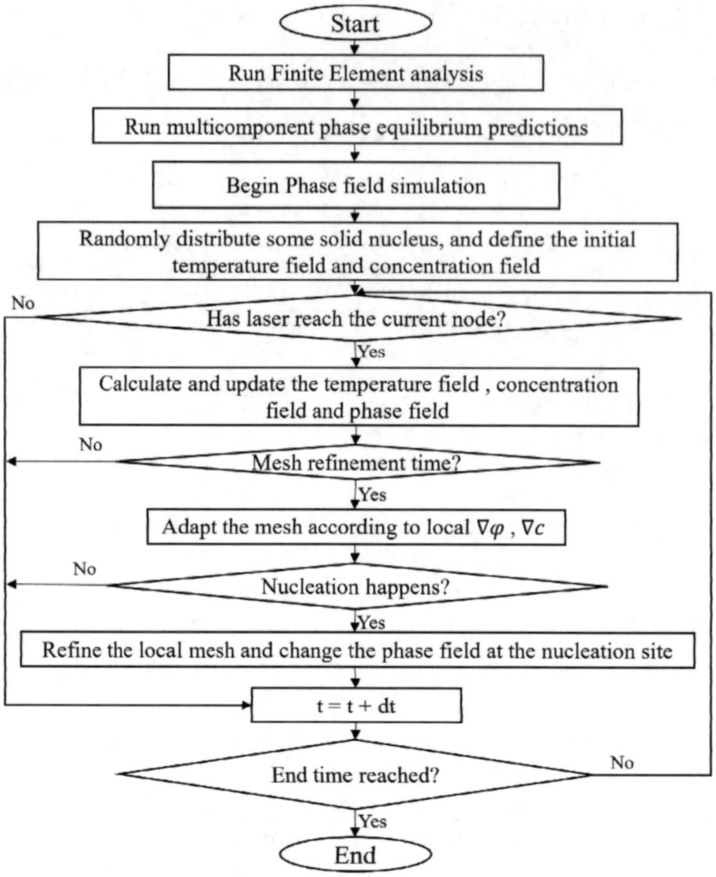

Figure 1. Flowchart showing the numerical implementation of the PF, FEA, and Nucleation Modules to simulate the LPBF process.

nucleation model, Eq. 15 is applied at each nodal point with the value of f_s being obtained from ϕ, and the values of δx and δy are assumed to be the grid spacing. Please note that the PF equations were solved using a bespoke C++ code via the finite volume method.

The relevant PF simulation parameters are listed in Table 1, while the relevant processing parameters for the thermal simulation are listed in Table 2. Given the high thermal gradient found in LPBF, the PF simulation contains a large number of nodes in order to reveal the fine microstructure details; the domain size, 280 μm $\times$ 70 μm was determined from the thermal simulation as discussed in Section 3. Note that the domain size and interface widths for the prealloyed powders and elemental powders were assumed to be the same. The thermodynamic free energy data of the Ti-alloy was extracted from the Thermo-Calc^{TM} TCTI 2 database. It should be noted that there are several assumptions with this model that must be acknowledged. Specifically, these assumptions include: (1) the local composition was only affected by diffusion without the consideration of convection in the melt pool; (2) the movement of the elemental powders and the new nucleated equiaxed grains were not considered; (3) only two stable phases exist throughout the simulation temperature range.

IOP Conf. Series: Materials Science and Engineering 1281 (2023) 012017 doi:10.1088/1757-899X/1281/1/012017

Table 1. Simulation parameters used in the phase field simulations [8][10]

PF parameters	Value
σ_{SL} (J/m^2)	0.046
ϵ_4	0.05
D_L (m^2/s)	9.5E-9
D_S (m^2/s)	5.0E-13
W ($\times 10^{-9}$m)	14.2
λ_o	11.76
grid points	6144×24576
domain size(µm)	70×280
Mesh spacing (dx)	0.8

Table 2. Process parameters of LPBF-Ti185 alloy used in the finite element analysis [3]

Laser power P	370 W
Scanning speed v	1035 mm/s
Laser spot size r_0	0.06 mm
Preheat temperature T_0	80 °C
Absorptivity λ	0.5

3. Results and discussions

3.1. Finite Element Analysis

The simulated temperature profile and molten pool geometry in XoY (top surface) and XoZ (side surface) planes are depicted in Fig. 2. These results show that the molten pool depth reaches about 70 µm, which exceeds the powder layer thickness and means that the substrate is also partially melted, thus a great connection between the newly deposited layer and prior layer (the substrate in this case) is expected.

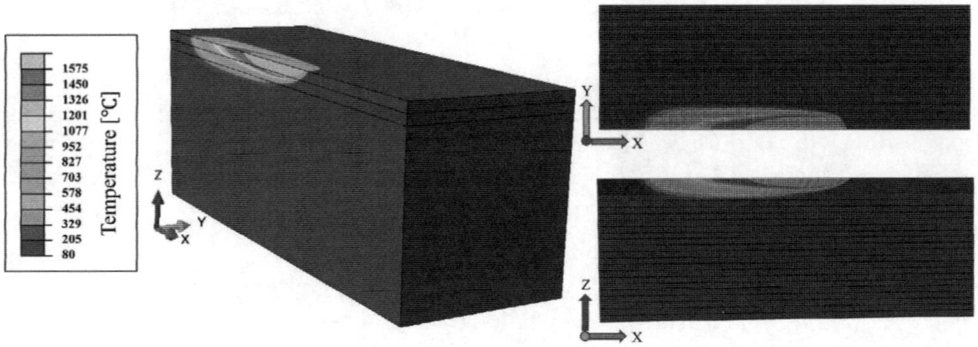

Figure 2. Contour plot and XoY and XoZ views of the temperature profile and molten pool geometry from the heat transfer finite element analysis

3.2. Initial condition for the Phase field models

Using the FEA results shown in Fig. 2 as a guide, a domain of 70 µm × 280 µm was utilized for the PF simulation. This was chosen to match the depth of the melt pool, while being four time larger in the direction of laser travel in order to simulate relevant phenomena. Fig. 3 provides (a) a schematic of the FEA temperature field applied to the PF domain, as well as (b) the corresponding uniform concentration map for the Ti-185 alloy to simulate the pre-alloyed condition, and (c) the spatially varying concentration map with particles of Ti (dark green), Al-V (light green), and Fe (yellow) to simulate the blended-alloy condition within a background of Ti-185 uniform composition. The number and radius of the Ti, Al-V and Fe powders were calculated based on their volume fraction.

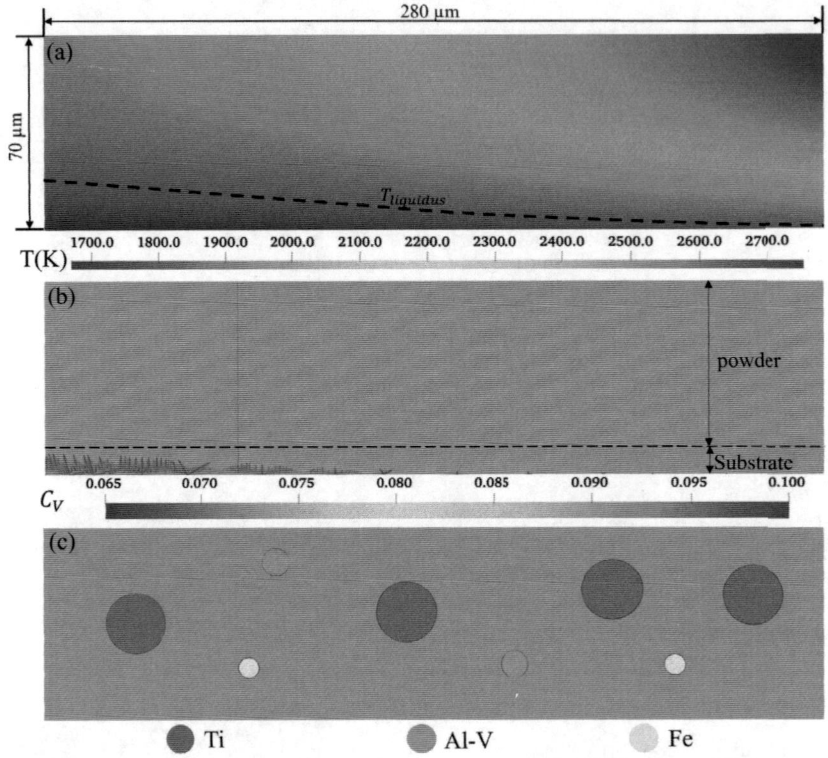

Figure 3. The initial condition for the phase field simulations, including (a) a snapshot of the temperature profile and domain geometries, (b) the uniform concentration map with initial microstructure, and (c) the spatially varying concentration map with a background of Ti-185.

As can be see in Fig. 3(a), the liquidus line at $T = 1848$ K is nearly parallel to the direction of laser beam travel with some incline angles with decreasing temperature. Analysis of the thermal data showed a temperature gradient along the Z direction G_z of about 1.4×10^7 K/m, and a temperature gradient along the X direction G_x of about 4.2×10^6 K/m. Referring now to Fig. 3(b), the initial microstructure is shown, with random crystal orientations placed at the bottom of the domain in the region of lowest temperature given by Fig. 3(a). Specifically, three different crystal orientations were imposed, $\phi_1 = 0$, $\phi_2 = \pi/12$ and $\phi_3 = \pi/6$, with respect to [001] direction. Fig. 3(c) shows the initial locations of the Ti, Al-V and Fe powders; the number and radius of the Ti, Al-V and Fe powders seen in the image were calculated based on their volume fraction. Note that the background phase was assumed to be Ti-185 and not pure Ti. This was done as a result of numerical limitations and real mass transfer condition. As the laser begins to scan over the blended elemental powders, the melting and alloying of different elemental powders has been initiated, thus a Ti-185 background at the beginning of solidification is a reasonable assumption. Additionally, the mesh spacing and time scale of the PF model is determined according to the thermodynamic data of Ti-185 alloy; carrying out the simulation with a background of pure Ti would result in significant numerical instabilities.

3.3. Phase field simulation during LPBF using pre-alloyed powders

The evolution of microstructure during LPBF in the pre-alloyed Ti-185 case is shown in Fig. 4. A differing solute redistribution phenomena is predicted by the PF model for the different solute components. Al (Fig. 4a) is enriched in the solid phase, while Fe (Fig. 4b)and V (Fig. 4c) are segregated in the liquid phase with a more evident segregation is predicted for Fe, which is consistent with thermodynamic predictions [1]. Furthermore, a fine columnar dendrite structure is seen, with competitive growth between dendrites of different orientation.

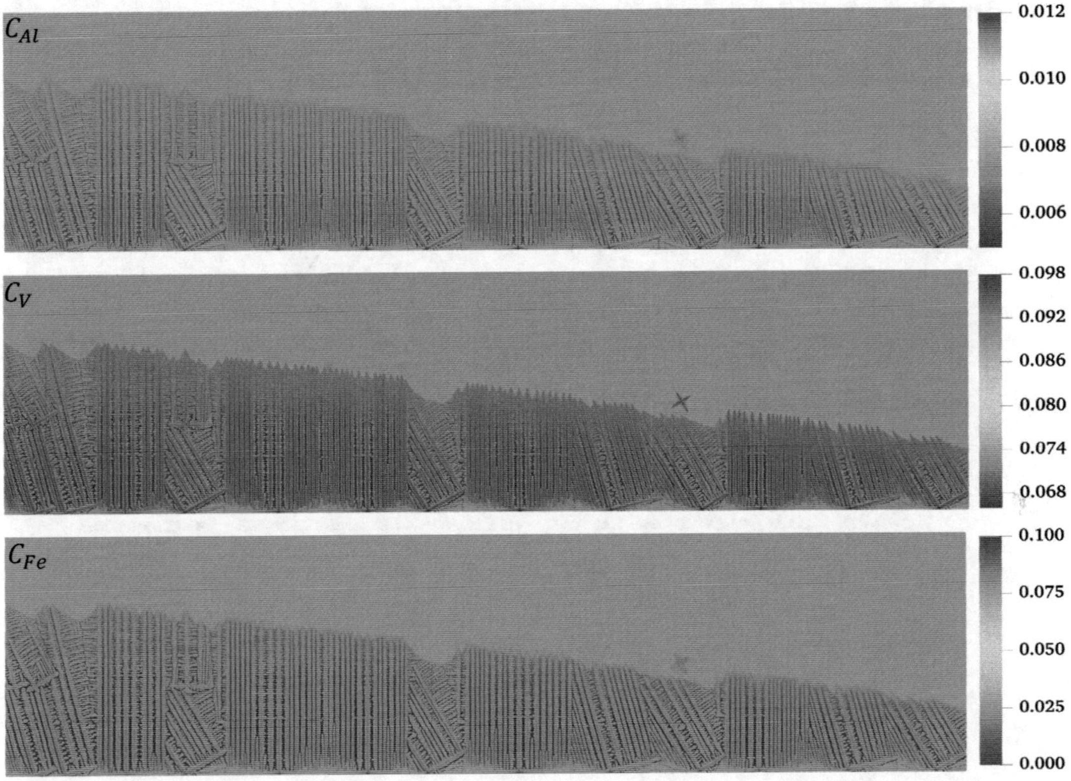

Figure 4. A snapshot at t = 0.65 ms of the concentration profile of Al, V and Fe during using pre-alloyed powder in laser powder bed fusion process

The simulation showed a strong [001] direction solidification texture, which is consistent with previous observations reported in the literature [2]. As shown in Fig. 4, the grains with their preferred grain growth direction aligned with the build (vertical) direction have a growth advantage, and are able to block growth of the adjacent grains. This growth advantage is well known, and is a result of the fact that for "inclined grains", a higher dendrite tip velocity and tip undercooling is needed when compared with the "vertical grains" to catch up to the pulling velocity [25]. This ultimately results in blockage by primary or even secondary dendrite arms of grains oriented in the build direction.

The PF simulation also shows several nucleation phenomena that may also influence the competitive growth process. Specifically, nucleation only occurred on the top of "inclined grains", and not the "vertical grains". This difference results from the difference in local undercooling between the two. As a high undercooling is needed to initiate the nucleation during the LPBF process, undercooling on the top of "inclined grains" tips is greater as compared with the "vertical grains". The new nucleated grains also block the growth of the "inclined grains", resulting in a further growth advantage of the "vertical grains".

3.4. Phase field simulation during LPBF using elemental powders

The evolution of microstructure during LPBF in the blended elemental Ti-185 case is shown in Fig. 5.

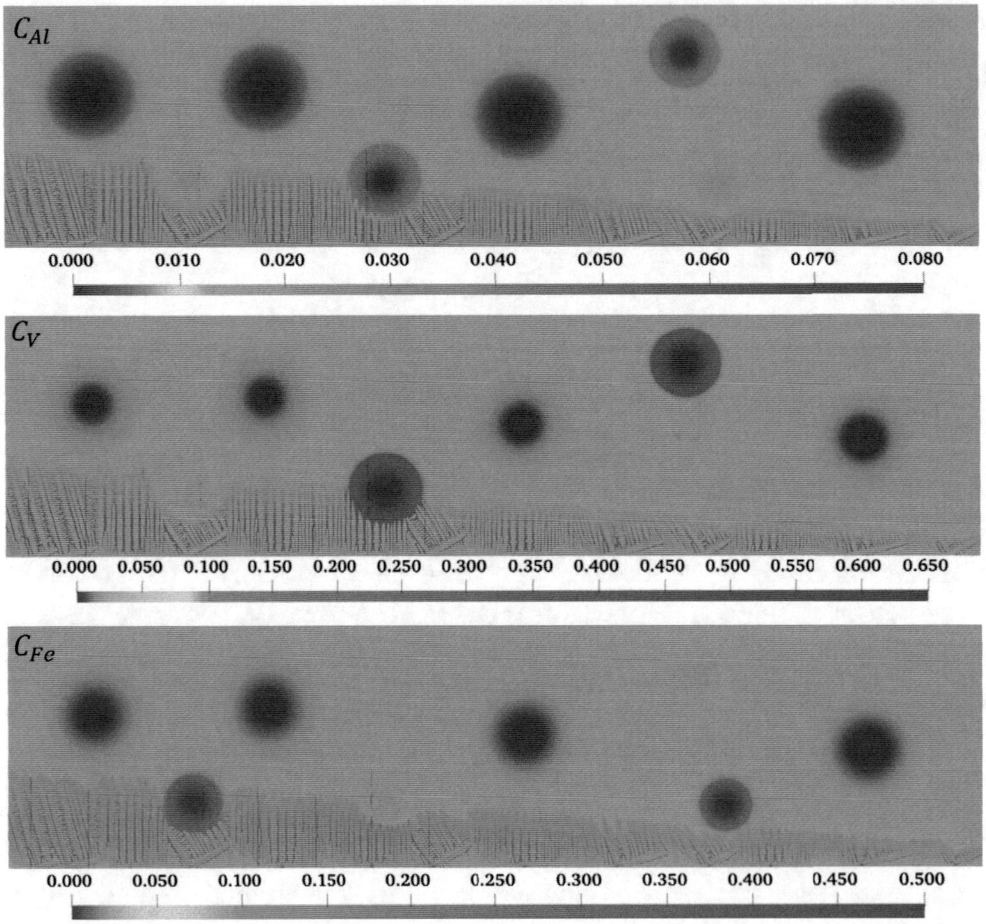

Figure 5. A snapshot at t = 0.45 ms of the concentration profile of Al, V and Fe during LPBF using pre-alloyed powders.

When using blended elemental powders, a significantly different microstructure evolution process is observed as compared to the case with pre-alloyed powders. First, the solute distribution is significantly altered due to the presence of elemental powders. Due to diffusion, these regions are centered at the initial locations of the elemental powders but are much larger in size. The highest solute concentrations are 8%, 65% and 50% for Al, V and Fe, respectively and reduce the further away from the powder centers. Dendrite growth begins far away from these solute rich regions and initially, the growth behavior is similar to that seen in the pre-alloyed powder case. However, once the dendrite tip reaches the regions of high solute concentration, further growth is inhibited. Dendrite growth is completely impeded by the local enrichment of Fe, and significantly slowed by local enrichment of Al and V.

The influence of solute enrichment on columnar dendrite growth can be explained from both thermodynamic and kinetic aspects. First, the enrichment of Fe and V would cause a decrease to the local liquidus and solidus according to the phase diagram. For a given temperature, this effectively decreases the undercooling and thereby reduces the driving force for dendrite growth. Second, Fe and V segregate into the liquid during solidification. However, local enrichment of Fe and V decreases the concentration gradient of these solutes, decreasing the diffusion rate of these elements. This increases the time taken for solute diffusion at the solid/liquid interface.

4. Conclusion

This study has examined how microstructure develops during LPBF of Ti-185 β titanium alloy. Using a multi-component PF model integrated with finite element thermal analysis and a model for grain nucleation, a comparison between pre-alloyed powder and a blended elemental powder has been made. The findings are as follows:

(i) The PF simulation during LPBF of Ti-185 using pre-alloyed powders shows a columnar dendrite growth: Fe and V segregate to the liquid phase while Al segregates to the solid. The grain with crystal orientation aligned to the build direction shows a considerable growth advantage. Nucleation is observed to occur above the "inclined grains", which blocks their growth of inclined dendrites.

(ii) The PF simulation during LPBF of Ti-185 using elemental powders shows that regions enriched in Fe significantly impede dendrite growth, while regions enriched in Al and V also slow the dendrite growth. This results in complete alloying not being achieved during a single layer deposition process, but, as shown experimentally by [7], requiring subsequent reheating cycles.

Acknowledgements

ZL and AP acknowledge the funding provided by the Natural Sciences and Engineering Council of Canada in support of this research. MG acknowledges funding provided the Natural Resources Canada's Office of Energy Research and Development (OERD).

References

[1] Li Z, Greenwood M and Phillion A 2022 *Computational Materials Science* **206** 111251
[2] Li J, Zhou X, Brochu M, Provatas N and Zhao Y F 2020 *Additive Manufacturing* **31** 100989
[3] Azizi H, Zurob H, Bose B, Ghiaasiaan S R, Wang X, Coulson S, Duz V and Phillion A 2018 *Additive Manufacturing* **21** 529–535
[4] Machado A and Wallbank J 1990 *Proceedings of the Institution of Mechanical Engineers, Part B: Journal of Engineering Manufacture* **204** 53–60
[5] Ng C, Bermingham M and Dargusch M 2022 *Additive Manufacturing* **59** 103176
[6] Liu Y, Xu L and Qiu C 2022 *Additive Manufacturing* **60** 103208
[7] Ahmed F F, Clark S J, Leung C L A, Stanger L, Willmott J, Marussi S, Honkimaki V, Haynes N, Zurob H S, Lee P D *et al.* 2021 *Materials & Design* **210** 110072
[8] Sahoo S and Chou K 2016 *Additive manufacturing* **9** 14–24
[9] Lenart R and Eshraghi M 2020 *Computational Materials Science* **172** 109374
[10] Sun W, Yan R, Zhang Y, Dong H and Jing T 2019 *Computational Materials Science* **160** 149–158
[11] Kundin J, Mushongera L and Emmerich H 2015 *Acta Materialia* **95** 343–356
[12] Eiken J, Apel M, Liang S M and Schmid-Fetzer R 2015 *Acta Materialia* **98** 152–163
[13] Eiken J, Böttger B and Steinbach I 2006 *Physical review E* **73** 066122
[14] Azizi H, Ebrahimi A, Ofori-Opoku N, Greenwood M, Provatas N and Mohammadi M 2021 *Acta Materialia* **214** 116983
[15] Greenwood M, Shampur K, Ofori-Opoku N, Pinomaa T, Wang L, Gurevich S and Provatas N 2018 *Computational Materials Science* **142** 153–171

[16] Provatas N and Elder K 2011 *Phase-field methods in materials science and engineering* (John Wiley & Sons)
[17] Echebarria B, Folch R, Karma A and Plapp M 2004 *Physical review E* **70** 061604
[18] Xiao W, Li S, Wang C, Shi Y, Mazumder J, Xing H and Song L 2019 *Materials & Design* **164** 107553
[19] Shampur K 2017 *A grand potential based multi-phase field model for alloy solidification* Master's thesis McGill University
[20] Fallah V, Amoorezaei M, Provatas N, Corbin S and Khajepour A 2012 *Acta Materialia* **60** 1633–1646
[21] Gaoyang M, Lingda X, Chunming W, Ping J and Guoli Z 2019 *Materials & Design* **181** 107980
[22] Nie P, Ojo O and Li Z 2014 *Acta Materialia* **77** 85–95
[23] Nastac L 1999 *Acta Materialia* **47** 4253–4262
[24] Promoppatum P, Yao S C, Pistorius P C and Rollett A D 2017 *Engineering* **3** 685–694
[25] Dantzig J A and Rappaz M 2016 *Solidification: -Revised & Expanded* (EPFL press)

Laser welding of dissimilar materials – simulation driven optimization of process parameters

M S Serdeczny

Flow Science Inc., 683 Harkle Road, Santa Fe, NM 87505, US

marcin.serdeczny@flow3d.com

Abstract. Laser welding can be used to join dissimilar materials to produce lightweight structures, and electric vehicle battery systems, which are important means of limiting the carbon emissions in the transport industry. Due to the differences in melting temperatures, thermal conductivities, and mutual solubility of dissimilar materials, it is still challenging to create defect-free joints with high mechanical strength, or low contact electrical resistance. In this work, we present a state-of-the-art numerical model of laser welding, developed within the Computational Fluids Dynamics (CFD) paradigm. The multi-physics model simulates melting, flow, and solidification of the alloys and accounts for the laser-material interactions, phase change, temperature and alloy-dependent thermophysical properties, recoil pressure, buoyancy force, and Marangoni effect. The simulation predicts weld penetration depth and width, alloy mixing, as well as the temperature gradient and cooling rate during the solidification that can be further fed into a micro-structure prediction model. The model is coupled with an optimization tool, which iterates over different process parameters to optimize the joint. The methodology is presented for steel to aluminium welding in lap configuration, but it can be used for other materials such as steel-copper, aluminium-copper, or steel-nickel, and in arbitrary geometrical configuration.

1. Introduction

Joining dissimilar materials efficiently and robustly is important for the automotive industry, as numerous parts have complex functions that require locally tailored material properties. Laser welding offers high efficiency, high precision, large throughput, and narrow heat affected zone. To this day, one of the main challenges of welding dissimilar materials are differences in the thermophysical properties. Among the widely used alloys, steel and aluminium couple is one of the most difficult to join via laser welding [1]. Steel has significantly higher melting point, lower thermal conductivity, and lower thermal expansion coefficient than aluminium. On the hand, aluminium has lower laser absorption rate that makes it difficult to select optimal parameters in the process. Moreover, an additional challenge is the formation of intermetallic phases (IMCs) that significantly reduce the electrical and mechanical performance of the joint [2]. IMCs formation depends on the dynamics of the dissimilar materials mixing as well as the temperature history of the weld. Thus, the optimization of laser welding parameters to ensure a proper bond while minimizing the formation of IMCs is an important research area.

Key tools in the analysis of laser welding of dissimilar materials are scanning electron microscopy (SEM), tensile tests, and measurements of the microhardness profile across the seam, which are all performed after the welding process is finished. In [3], Torkamany et al. looked at the dissimilar welding of carbon steel to aluminium with Nd:YAG pulsed laser. The influence of the laser peak power, pulse duration, and welding speed on the material mixing and IMCs formation was studied. It was found that too high line energy (high peak power, long pulse, or low welding speed) will cause significant mixing of the aluminium with steel and inevitable formation of the brittle IMCs as well as the crack propagation. This was found to be correlated with the low tensile strength of joints produced with high line energy input. On the other hand, when the line energy input was low, the shallow penetration of the laser resulted in low tensile strength, even though the IMCs content was minimal. Mathieu et al. [4] used a continuous wave Nd:YAG laser with 3.5 kW power to weld a 6016 T4 aluminium to a low carbon steel. They used an overlap joint configuration and applied a filler wire made of 85% Zn and 15% Al. They presented weld fracture locations and showed that the macroscopic geometry of the weld is equally important as the presence of brittle IMCs. Pereira et al. [1] demonstrated successful welding of DP1000 steel to Al050 in lap configuration but reported that in the butt configuration, the two materials were nearly impossible to weld. Sun et al. [5] added an aluminium ER4043 filler wire in the butt configuration and demonstrated successful welds of AA6013 with low cardon mild steel using laser welding-brazing mode. They investigated the influence of the laser power, wire speed, and the deviation of the laser spot with respect to the aluminium edge on the seam geometry, microstructure, IMCs layer thickness, as well as the tensile strength of the joint. With the increasing line energy density, they observed better wetting and penetration of the weld. Nevertheless, the specimens with higher line energy density yielded a smaller tensile strength due to a thicker layer of IMCs. The formation of intermetallic phases was attributed to the diffusion process of Fe into the melted filler material, as the diffusion coefficient of Fe increases with temperature. To sum up the experimental efforts, it can be noted that to obtain a high strength join the heat input needs to be optimized to ensure a thin layer of IMCs while the overall seam dimensions need to be maximized. For more experimental work on dissimilar materials welding, the reader is referred to the recent review by Sadeghian and Iqbal [2].

In contrast to most of the experimental methods, the numerical models allow for an in-situ analysis of the process, where the influence of physical phenomena on the material mixing can be studied individually and in detail, allowing for explanatory approach. Moreover, a validated numerical tool would allow for process parameters optimization without costly experimental trails. Several studies have focused on the modelling of the laser welding of dissimilar materials in the conduction mode [6, 7, 8, 9], where the surface of the material is typically assumed flat during the simulation. Keyhole simulations are more difficult to perform, as the free surface tracking algorithm is required, as well as the calculations of the evaporation pressure that drive the melt pool flow. Keyhole dynamics was studied using an-house Computational Fluid Dynamics (CFD) code by Kouraytem et al. [10] for a stainless steel 304. Esfahani et al. [6] simulated mixing of stainless steel with low carbon steel and found it to be influenced by the Marangoni flows that originate from the temperature dependent surface tension. In their model, the recoil pressure was neglected. Recently, Huang et al. [11] investigated numerically the mixing of aluminium and copper during the laser keyhole welding in a lap configuration. They showed how increasing the laser power relates to a larger recoil pressure and higher velocity inside the melt pool and therefore more efficient mixing of dissimilar materials. On the other hand, increasing the weld speed, reduces the time spent on accelerating the fluid by the recoil force, which limits the mixing capacity.

In this work, we study the pulsed laser welding of steel to aluminium using a CFD model. The pulsed laser is used to precisely control the heat input through the pulse duration, frequency, and peak power. First, a state-of-the-art model is developed and validated with experimental measurements. In the second part, the CFD model is coupled with an optimization algorithm that iterates over process parameters and minimizes the material mixing, while ensuring the maximum penetration of the material and good bonding. Section 2 describes the CFD model, simulation parameters, and the framework in which the model is coupled with the optimization tool. In Section 3, the results of the CFD simulations are shown. First, the agreement of the numerical results with the experimental measurements is discussed. Next, the influence of process parameters on the material mixing is investigated. At the end of Section 3, the optimization results are presented. Section 4 summarizes and concludes the study.

IOP Conf. Series: Materials Science and Engineering 1281 (2023) 012018 doi:10.1088/1757-899X/1281/1/012018

2. Methodology

2.1. Physics of the numerical model

In this work, the laser welding is simulated using a commercial CFD solver ***FLOW-3D* WELD** release 7 update 2 [12]. The core of the computational model is the solver for the mass, momentum and energy conservation equations that can be written in a simplified form as

$$\nabla \cdot \boldsymbol{u} = 0 \tag{1}$$

$$\rho \frac{D\boldsymbol{u}}{Dt} = -\nabla p + \nabla \cdot \left(\mu \overline{\overline{\boldsymbol{S}}}\right) + \rho g \tag{2}$$

$$\rho \frac{DI}{Dt} = \nabla \cdot (k \nabla T) + Q_L \tag{3}$$

$$\overline{\overline{\boldsymbol{S}}} = \nabla \boldsymbol{u} + \nabla \boldsymbol{u}^T \tag{4}$$

where $\boldsymbol{u}$ is the velocity vector, ρ is the material density, $D()/Dt$ is the notation of the material derivative, t is the time, p is the pressure, μ is the dynamic viscosity, $\overline{\overline{\boldsymbol{S}}}$ is the strain rate tensor, g is the gravitational acceleration, $I = C_v(T)$ is the internal energy of the material, C_v is the specific heat, T is the temperature, k is the thermal conductivity, and Q_L is the laser heat source.

The above system of equations is supplemented by the Volume-Of-Fluid (VOF) method that tracks the position of the material free surface with respect to the atmosphere [13]. The VOF method relies on solving an equation for variable F, called fraction of fluid.

$$\frac{\partial F}{\partial t} + \boldsymbol{u} \cdot \nabla F = 0 \tag{5}$$

Fraction of fluid takes values of 1 for the phase 1 (both liquid and solid matter), and value 0 for the phase 2, i.e. the atmosphere. In ***FLOW-3D* WELD**, the conservation equations (1) - (4) are solved only for phase 1. Phase 2 is approximated with a constant pressure and temperature and no velocity. This approach greatly enhances the computational time. The dissimilar alloys are tracked as a sub-phases of the phase 1, and their concentration is calculated by solving a scalar transport equation

$$\frac{\partial C}{\partial t} + \boldsymbol{u} \cdot \nabla C = \nabla \cdot (\phi \nabla C) \tag{6}$$

where C is the marker variable for alloys: 0 for steel and 1 for aluminium; ϕ is the diffusion coefficient that is calculated as

$$\phi = Sc \cdot \mu / \rho \tag{7}$$

where Sc is the Schmidt number that was assumed equal 1. Wherever the concentration C is between 0 and 1, the law of mixture is used to approximate the material properties (the viscosity, thermal conductivity, specific heat etc.).

Apart from the above, the following physical phenomena are included in the model:

- Density dependency on the temperature is included in the momentum conservation equation to capture buoyancy driven flows inside the melt pool.

- Surface Tension is modelled through an additional pressure term on the free surface

$$P_{ST} = \sigma(T) \cdot \kappa \tag{8}$$

  where σ is the surface tension coefficient that depends on the temperature. The dependency of σ on T allows to capture the Marangoni flows, where typically low temperature regions have higher surface tension and pull the fluid from the hot regions with low surface tension. κ is the curvature of the free surface at a given point on the material surface.

- Solidification phenomenon is included in the energy and momentum conservation by accounting for the latent heat of the phase change as well as the additional drag on the flow in the mushy zone (during the crystallization and melting). The velocity field is considered frozen in the solidified fluid cells.
- Laser absorption was approximated as constant, while in reality it is temperature and angle dependent. However, obtaining a reliable data is difficult. The starting values for low carbon steel and 5754 aluminum were $\varphi_{St14} = 0.35$ and $\varphi_{Al5754} = 0.25$, respectively. The values were further fine-tuned, when comparing the simulation results with experimental measurements, which is described in Section 3.1.
- A ray tracing method was used to track the reflections of the laser beam and subsequent absorption. The method allows to capture the effective increase of absorptivity in a keyhole scenario, i.e. when laser rays are trapped in the material due to multiple reflections. The ray tracing method works by dividing the laser beam into rays that are stored in the computational cells. When a ray interacts with the material, the incident and reflection angles are computed, as well as the portions of the absorbed and reflected energy. In the model, up to 7 reflections of a single ray are tracked, which is typically when the energy carried by a single ray becomes very small.
- The shield gas dynamic pressure was included in the simulation as

$$p_{sg} = \frac{1}{2}\rho_{sg}u_{sg}^2 \tag{9}$$

where ρ_{sg}, u_{sg} are the density and velocity of the shielding gas. Argon density and $u_{sg} = 1.0\frac{m}{s}$ were used to follow the experimental measurements used to calibrate the model.
- Liquid-Vapor phase change is modelled by including its effect in mass, energy, and momentum conservation of phase 1. Mass loss due to material evaporation is calculated according to the following equation [14]

$$\dot{m} = \alpha\sqrt{\frac{1}{2\pi R_{gas}T_{liq}^{inter}}}\left(P_{sat}(T) - P_{vap}\right) \tag{10}$$

where R_{gas} is the specific gas constant, T_{liq}^{inter} is the average temperature at the liquid-vapor interface, $P_{sat}(T)$ is the saturation pressure at the local temperature, P_{vap} is the partial pressure of the metal phase in the atmosphere (constant), α is the accommodation coefficient and it expresses the probability that the molecule escapes the liquid phase to the vapor phase. In practice, for alloys, this coefficient is unknown and must be chosen based on the experimental observations. In this work, we use a value of 0.01 based on the previous experience in numerical simulations of evaporation of steel and aluminum alloys.

Evaporation mass flux results in a change of the internal energy of the metal (cooling effect) due to the latent heat of evaporation

$$\Delta I = \Delta m \cdot \lambda_{evap} \tag{11}$$

ΔI is the change of internal energy, Δm is the change of mass due to evaporation of the material, λ_{evap} is the latent heat of evaporation.

The third implication of the liquid-vapor phase change is the evaporation (or recoil) pressure, which is due to the momentum of the molecules leaving the liquid state and rapidly expanding into the gaseous state. The evaporation pressure in ***FLOW-3D* WELD** is modelled with the following equation

$$P_r = AP_v \exp\left(B\left(1 - \frac{T_v}{T_s}\right)\right) \quad (12)$$

where P_v, T_v are coordinates of a single point on the saturation curve, T_s is the average temperature at the liquid-vapor interface, and A, B are the model parameters. In serval works [15, 16, 17], A is assumed to be equal 0.54, following the early works of Anisimov, which is used as a starting point in this work. The coefficient B can be estimated by comparing Eq. (12) to the Clausius-Clapeyron equation

$$B = \frac{\lambda_{evap}}{(\gamma - 1)c_v T_v} \quad (13)$$

where γ is the specific heat ratio.

2.2. Geometry of the model

Figure 1 depicts the geometry of the welding configuration, as implemented in the numerical model. The geometry follows the experimental measurements of Torkamany et al. [3] and Ghaini et al. [18], which will be reproduced as a validation case. The lap configuration consists of a low carbon steel St14 plate (blue block in Figure 1) on top of an aluminium 5754 plate (red block in Figure 1). The thickness of the steel plate is 0.8 mm, except the validation cases B-1 to B-5 (Section 3.1. where it is equal 0.7 mm. The thickness of aluminium plate is 2 mm, except cases B-1 to B-5 where only the steel plate is used. The width and length of the plates in the model are 20 mm and 25 mm, respectively, which is considerably smaller than in the actual measurements but necessary to make the simulation time feasible. The plates are stationary while the laser beam moves at a constant velocity V in x-direction.

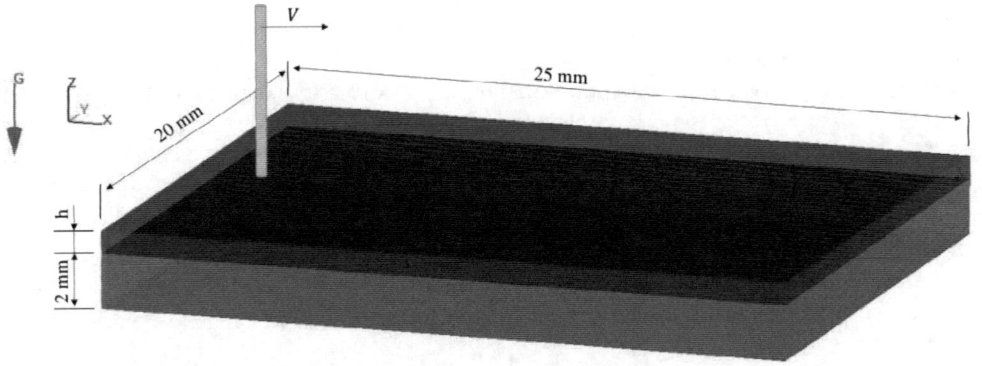

Figure 1. Geometry of the welding configuration implemented in the numerical model. Blue region is the low carbon steel St14, while the red region is the aluminium alloy 5754.

2.3. Mesh

Figure 2 shows the cartesian mesh that is used in the simulation. The mesh extends above and below the workpieces in z-direction by 0.5 mm and aligns with the plates dimensions in x and y. The mesh is made of two blocks: 1) Inner mesh block that serves to resolve the fluid flow inside the melt pool and heat transfer to the solidified regions in the vicinity of the melt pool; 2) Outer mesh block that serves to resolve the heat transfer in the vicinity of the domain boundaries. The details about the cell size and cell count are given in Table 1. Three levels of mesh refinement were tested (coarse, medium, and fine) and their influence on the melt pool dimensions and material mixing was checked. The results of the mesh sensitivity analysis are shown in Appendix A in Figure A.1., where the melt pool shape and dimensions (depth and width) in the YZ-cross-section can be seen. The melt pool dimensions were found to vary mildly between the medium and the fine mesh, while the computation time increases from approx. 5-10 hours to 2-3 days (depending on the case). There is, however, a difference in the alloy composition shown with the colouring variable, and for the fine mesh, more significant upward motion of the aluminium is observed. To limit the computational effort, we use the medium mesh for initial screening and for optimization, while the final validation cases are re-simulated with the fine mesh to confirm the findings.

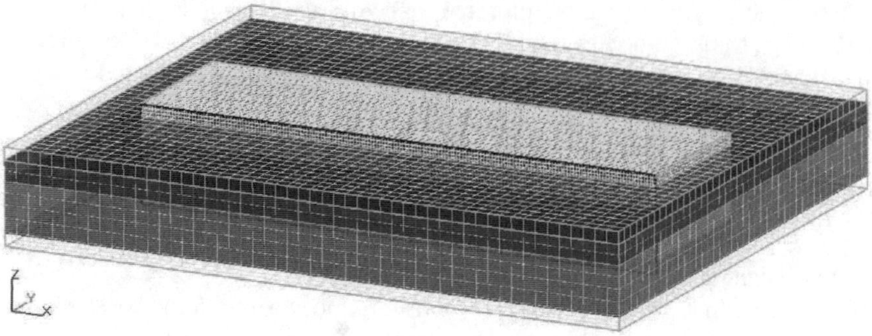

Figure 2. Cartesian mesh used in the simulations. The mesh consists of two blocks: inner – with small cell size, and outer that is coarser.

Table 1. Mesh properties.

	Outer Mesh Block			**Inner Mesh Block**		
	Coarse	Medium	Fine	Coarse	Medium	Fine
Cell size	960 μm	480 μm	240 μm	240 μm	120 μm	60 μm
Cell count	2.7e3	32e3	138e3	34.8e3	280e3	2.2e6

2.4. Initial and boundary conditions

The four vertical boundaries of the computational domain were prescribed a fixed temperature of 500 K. The initial temperature of the material as well as the ambient atmosphere temperature were also set to 500 K. A sensitivity analysis of the results with respect to this temperature was performed (at 400 K and 600 K) and no significant change in the melt pool dimensions, or mixing was observed. The top and bottom boundaries of the computational domain were prescribed a fixed pressure of one bar and the temperature of 500 K.

As the motion of the gas phase above the metal surface was not calculated explicitly, the convective heat loss was modelled by setting the following condition at the metal free surface

$$Q = h(T_s - T_\infty) \tag{14}$$

where h is the heat transfer coefficient (assumed 30 W/m^{2}K which is in the range of forced convection, that should be driven by the shielding gas [19]). T_s is the temperature at the free surface, T_∞ is the ambient gas temperature (assumed 500 K). The energy convected from the metal to the atmosphere is lost.

2.5. Laser parameters

The laser beam source was assumed to have a Gaussian intensity profile and have a cylindrical shape (no divergence or convergence of the beam)

$$H(r) = H_c \exp\left(-2\left(\frac{r}{W}\right)^2\right) \tag{15}$$

where H_c is the heat flux in the centre of the beam, r is the radial coordinate, W is the laser spot radius, defined as a point when the intensity (heat flux) falls below $1/e^2 \approx 0.135$ of the maximum intensity, which corresponds to two standard deviations of the normal distribution. For a Gaussian beam, this definition is equivalent to $D4\sigma$, where the beam diameter is measured by the 4 standard deviations.

IOP Conf. Series: Materials Science and Engineering 1281 (2023) 012018 doi:10.1088/1757-899X/1281/1/012018

2.6. Material parameters
The chemical compositions of the two alloys are presented in Table 2. The thermophysical properties were generated using JMatPro software [20] and included the temperature dependency of viscosity, density, specific heat, thermal conductivity and surface tension.

Table 2. Chemical compositions of the alloys used in this work, in wt% [3].

Element	Si	Cu	Mn	Mg	Cr	Ni	C	Fe	Al
Al 5754	0.12	0.04	0.25	3.009	0.08	0.006	-	0.37	Bal.
St 14	-	-	0.4	-	-	-	0.4	Bal.	-

2.7. Optimization tool
CFD model was coupled with an optimization tool ***FLOW-3D*(x)** 2022R1 that is a general-purpose optimization algorithm that employs hybrid and surrogate approach. The hybrid approach means that the algorithm will switch between the local or global minimum search depending on the number of iterations that are left till the end of the optimization (prescribed by the user). Surrogate model is particularly suitable for computationally expensive problems, such as CFD or Finite Element Method (FEM) and relies on reproducing the real objective function with an acceptable surrogate function that is cheap to evaluate. The surrogate function is used during the optimization process to select points at which the real objective function will be evaluated in the next iteration.

Figure 3 shows a simplified schema of a single iteration of the optimization tool. In Step 1, the guessed peak power and pulse duration of the laser are passed to a Python script that generates a .csv file, which contains a time series of square shape pulses. Next node, in Step 2, is the ***FLOW-3D* WELD** node, i.e. the CFD solver. The path of the csv file as well as the guessed value of welding speed are inputted to the CFD solver node, which runs the welding simulation. In Step 3, the solver node feeds the path of the simulation results to the execution node that runs a second Python script, which launches a CFD post-processor, ***FLOW-3D* POST**, and extracts the average melt pool penetration depth, as well as the average aluminium content. The average aluminium content is extracted in the weld region above the steel-aluminium interface, so it quantifies the amount of aluminium that flowed upwards due to the keyhole-driven mixing process. In Step 4, based on the surrogate objective function, algorithm guesses new values of peak power, peak duration, and welding speed, in order to maximize the penetration and minimize the aluminium content. The cycle repeats with the new input.

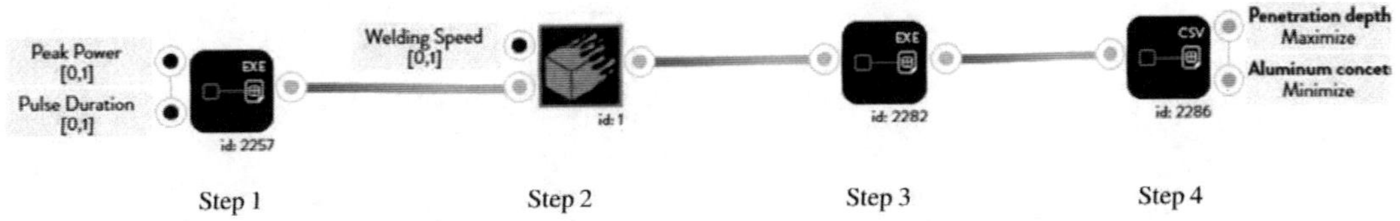

Figure 3. A diagram representing a single iteration of the optimization model in ***FLOW-3D (x).***

3. Results

3.1. Validation of the numerical model
We start by calibrating the model to the measurements of Torkmany and Ghaini et al. presented in the series of publications [21] [18] [3]. The first set of measurement data is summarized in Table 3 (B-1 to B-5, following the measurement numbering of [21] [18]) and consisted of a single pulsed laser scan on a 0.7 mm thick plate of low carbon steel St14. This is a simplified geometry with respect to Figure 1, as the there is no second material (aluminium plate) present yet in the simulation domain.

Table 3. Process conditions used in the validation of the numerical model. The conditions are taken from: B-1 to B-4 [20] [21], TN14 - TN19 [3]. In all cases, the pulse frequency was 20 Hz.

Simulation /Experiment	Peak power, P_p [kW]	Pulse duration, P_d [ms]	Welding speed, V [mm/s]	Laser spot diameter, $2W$ [mm]
B-1	1.33	6	8.3	1.2
B-2	1.50	6	8.3	1.2
B-3	1.67	6	8.3	1.2
B-4	1.83	6	8.3	1.2
B-5	2.00	6	8.3	1.2
TN-14	1.43	5	11.5	0.76
TN-15	1.43	5	10	0.76
TN-16	1.43	5	8	0.76
TN-17	1.43	5	6	0.76
TN-18	1.43	5	4	0.76
TN-19	1.43	5	2	0.76

The comparison of the simulated and measured width and depth of the melt pool is shown in Figure 4. Two simulation runs were performed – one with the nominal absorptivity of 0.35, and with increased absorptivity of the material to 0.45. The rationale of varying material absorptivity to laser radiation is due to unknown surface conditions of the material in the experiment (surface roughness, and oxides layer), which may significantly influence the reflectivity and absorptivity [22] [23]. Moreover, the absorptivity is approximated as constant in the model, while it varies with the angle of the incidence and temperature.

The measured melt pool width first increases with the laser peak power, and then remains constant. The simulations with nominal absorptivity $\varphi = 0.35$ capture this trend more accurately than when the absorptivity is increased. On the other hand, the melt pool depth is underestimated when $\varphi = 0.35$, hence the model was run with $\varphi = 0.45$. It is noted that in these simulations, the recoil pressure model coefficient A was increased to 2, as the default value of 0.54 (see Section 2.1.) yielded significantly smaller material penetration and no transitioning to the keyhole. Increase of coefficient A above 2 had small impact on the penetration depth, as the heat absorbed by the material is a major driver for the evaporation process and the recoil force. As the accurate prediction of the penetration depth is prioritized, absorptivity value of 0.45 is used in the following simulations.

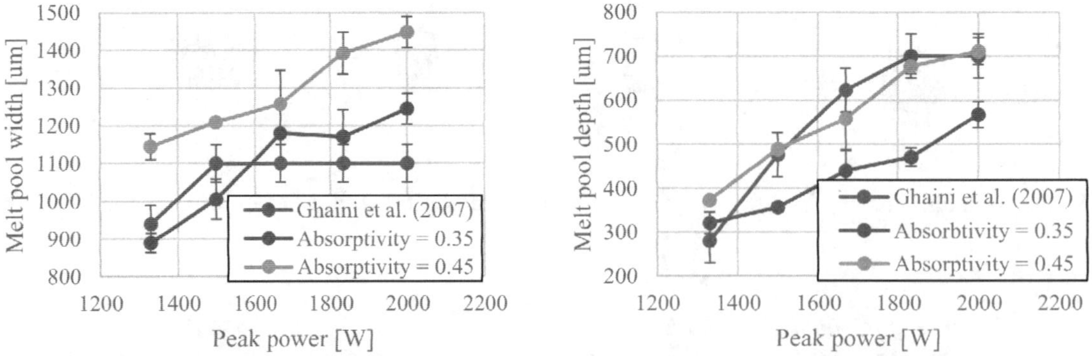

Figure 4. Validation of the numerical model to the experimental measurements of Ghaini et al. [21] Left: Melt pool width. Right: Melt pool depth. For each simulation point, the width and depth were calculated at three cross-sections along the weld (x = 5 mm, 7.5 mm and 10 mm) and averaged. The standard deviation from the 3 calculations is shown with the error bar. For the measurements of Ghaini et al. [21], 50 μm error bar was assumed.

Figure 5 shows the cross-sections of the melt pools obtained with the numerical model (light blue colour) for the absorptivity value of 0.45 and photographed by Torkamany et al. [21] (black dashed line). The overall dimensions of the simulation and measured melt pools are similar but there are differences in the shape. The simulated cross-sections feature a concave surface that indicates strong action of the recoil pressure, while the measured cross-sections are rather flat. Moreover, the measured melt pool shapes are more triangular and pointier. This could be due to the differences in the laser spot diameter between the experiment and the simulation that could stem out of either measurement error, or inconsistent definitions of the beam diameters. Here, we used four standard deviations of the Gaussian distribution to define the beam diameter, while the exact definition is not mentioned in [21]. Alternatively, the intensity profile of the laser beam could be different than the Gaussian that was assumed in the numerical model.

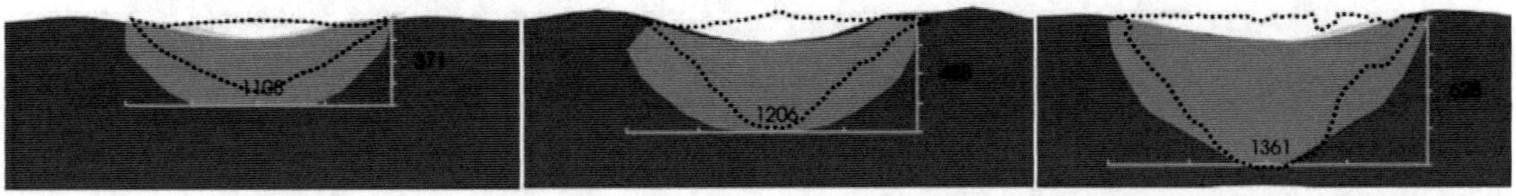

Figure 5. Comparison of the simulated (light blue) and measured [21] (black dashed line) melt pool cross-sections.

In the next step, the modelling results are compared to the melt pool measurements during welding of steel and aluminium [3] that are considered more complex to reproduce. Table 3 summarizes the welding parameters used in the experiment and to run the simulations in this round (TN-14 to TN-19). Figure 6 presents the comparison between the measured and simulated melt pool dimensions as a function of welding speed. First, the model is run with absorptivity value 0.45, that was used in the previous simulation round. It can be seen that the width of the melt pool is captured accurately at high welding speeds (low line energy density), but it is underestimated at lower welding speeds. At the same time, the melt pool depth is underestimated in the entire range of welding speeds indicating that eighter too little energy is absorbed by the material, or the recoil pressure is underestimated and does not lead to sufficient penetration of the material (in this simulation round the recoil pressure coefficient was increased to $A = 10$).

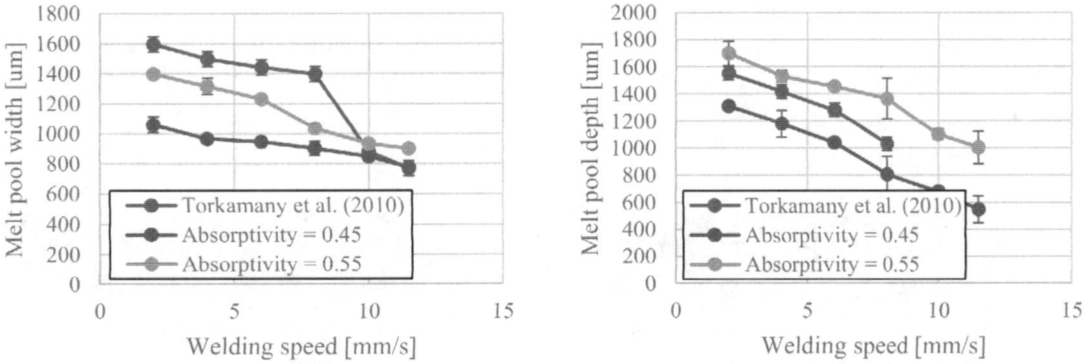

Figure 6. Validation of the numerical model to the experimental measurements of Torkamany et al. [3]. Left: Melt pool width. Right: Melt pool depth. For each simulation point, the width and depth were calculated at three cross-sections along the weld (x = 5 mm, 7.5 mm, and 10 mm) and averaged. The standard deviation of the 3 calculations is shown with the error bar. For the measurements of Torkamany et al. [3], 50 μm error bar was assumed.

The melt pool cross-sections simulated with absorptivity value of 0.45 and $A = 10$ are compared with the measurements in Figure 7. The measured weld becomes deeper with decreasing welding speed, as the heat input per length increases. The model captures this dependency although the penetration depth is smaller than in the experiment. Based on the measured profile, it can be seen that the weld shape becomes more irregular at the higher powers, which is due to unstable keyhole behaviour. The simulated weld shape features a widening, close to the interface between the two alloys. This widening is mildly seen in the experiment when $V = 4$ mm/s.

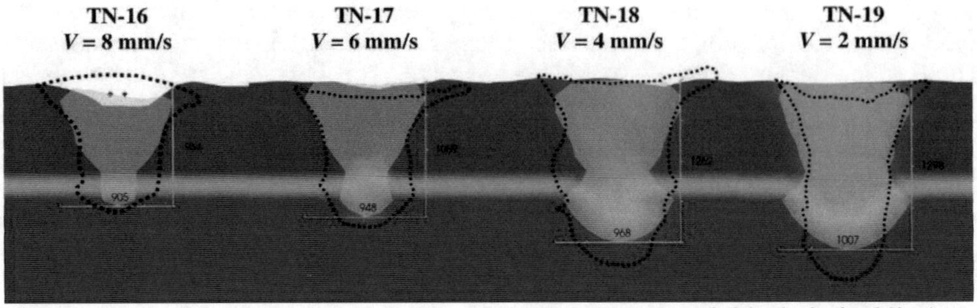

Figure 7. Comparison of the simulated (light blue region) and measured (black dashed line) melt pool cross-sections. Dark blue indicates steel, while red indicates aluminium. For simulations TN-14 and TN-15, the penetration was too shallow to create a joint (both in the simulation and experiment).

3.2. Analysis of the material mixing

In this section, we look on how the process conditions influence the mixing of the alloys. Figure 8 shows a cut through the weld in *XZ*-plane. On the left, simulation performed with welding speed of 2 mm/s is shown, while on the right with 6 mm/s. At low welding velocity, larger concentration of aluminium is observed in the weld, which agrees with the experimental observations in [3]. This is contributed to the higher energy density, and therefore larger recoil pressure and longer time that the laser acts on the material, which allows for larger accelerations of the molten aluminium. Clusters of aluminium are observed to be lifted upwards in the weld due to a vortex that forms behind the keyhole [11]. In Figure 9, aluminium volume fraction was plotted along the weld. The observed periodic spikes in the aluminium content can be regions of elevated probability of IMCs formations.

Figure 10 shows the aluminium volume fraction in the weld region above the steel-aluminium interface. In other words, it measures the aluminium volume that flowed upwards due to the keyhole mixing. Aluminium volume fraction falls almost linearly with the increasing welding speed until the heat input per length is insufficient to create a keyhole and the melt pool does not reach the Al plate.

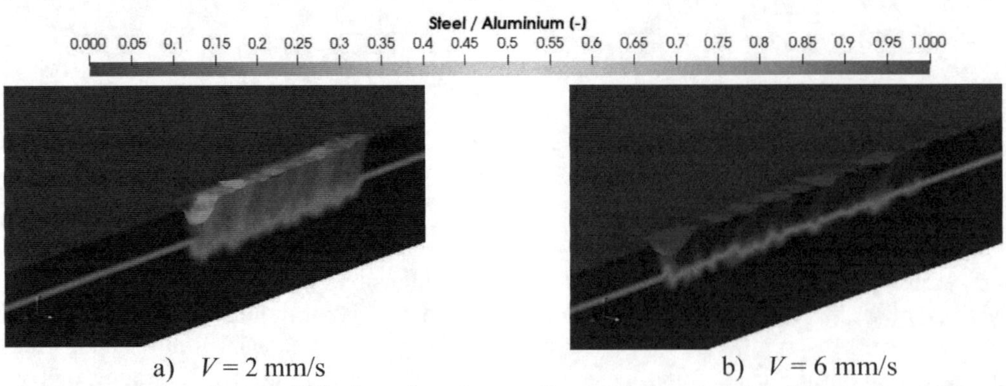

a) $V = 2$ mm/s b) $V = 6$ mm/s

Figure 8. Cross-section through XZ plane that shows alloys mixing simulated for two welding speeds. Other parameters were set as in Section 3.1.

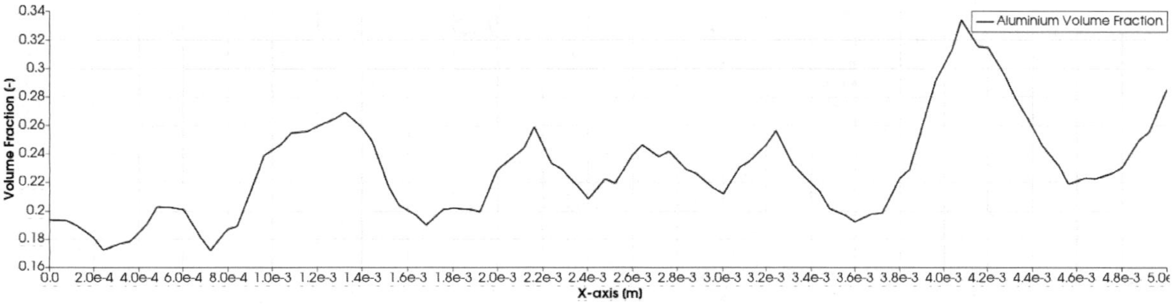

Figure 9. Variation in aluminium concentration along the weld at $Z = -500$ µm and $Y = 0$ µm for $V = 2$ mm/s.

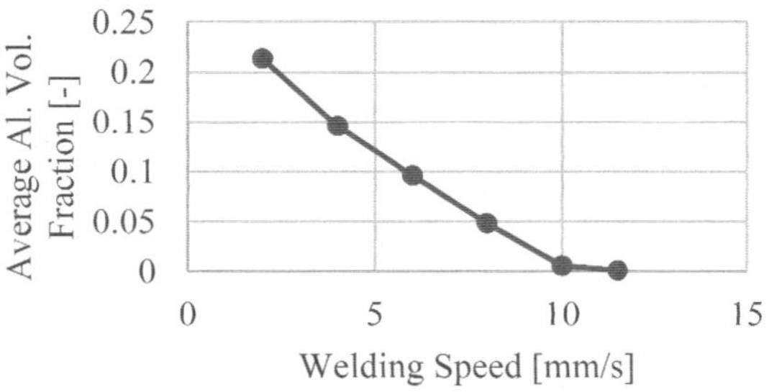

Figure 10. Dependency of the average aluminium volume fraction in the weld region above steel/aluminium plates' interface.

3.3. Laser welding optimization results

In this part, the calibrated model was run using the optimization tool described in Section 2.7. The algorithm was set to maximize the penetration depth and minimize the aluminium content in the upper part of the weld, while varying the peak power, pulse duration, and welding velocity. The model was allowed to run 7 iterations on a single 10 core machine, which took around 14 hrs. The preliminary results are shown in Figure 11, where each point is a single iteration. Y-axis shows the mean penetration depth, X-axis shows the laser peak power, while the bubble size indicates the average aluminium content in the weld region. Generally, the average aluminium content increases with the penetration depth, which makes it a very difficult optimization task. However, some of the trends in Figure 11 indicate that by a proper choice of process conditions, improvements are possible. For instance, as indicated by the black arrow, the mean penetration depth could be increased from approx. 1.2 mm to 1.5 mm, while keeping almost the same aluminium volume fraction in the weld, which yields larger joint area without increasing the risk of brittle IMCs formation.

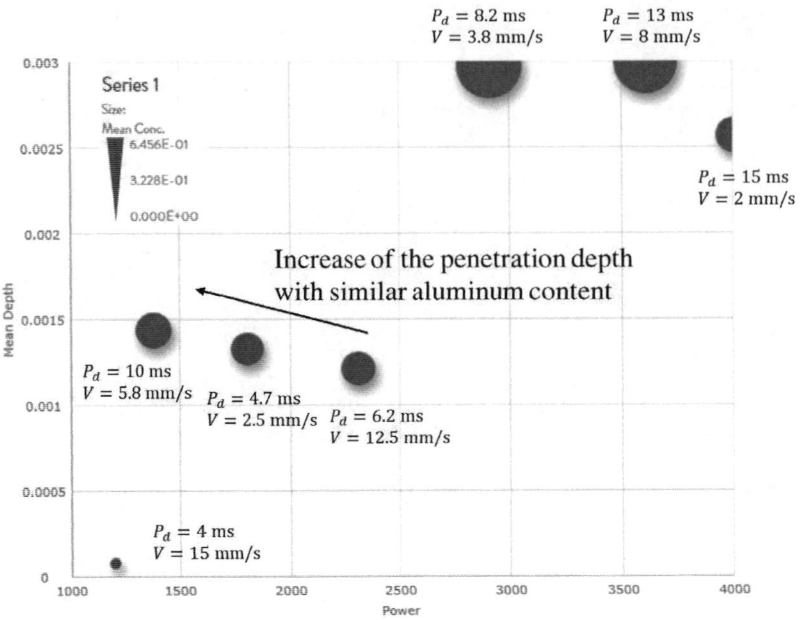

Figure 11. Preliminary optimization results, obtained for 7 iterations.

4. Summary and conclusions

This work documents a development of a computational model for simulating the pulsed laser welding of steel to aluminium. The model relies on solving the conservation equations of mass, momentum, and energy. Moreover, the method uses an empirical equation to predict the recoil pressure to capture the keyhole regime. The model was calibrated to the experimental measurements by comparing the simulated and measured melt pool cross-sections. As calibration parameters, material absorptivity and recoil pressure coefficients were used. However, for the chosen set of process conditions, the numerical results were primarily found to be sensitive to the absorptivity. The model correctly predicts the trends in the melt pool dimensions when the heat input per length is increased (increase in peak power or decrease in welding speed). After the calibration, the simulated melt pool shapes presented a good agreement with the measured melt-pool cross-sectional shapes, considering the complexity of the keyhole dynamics. The discrepancies were attributed to the uncertainty of the beam diameter definition in the experiment, unknown laser beam intensity profile, as well as simplistic recoil pressure model.

In the second part of the work, aluminium and steel mixing was observed as a function of welding speed. It was found that the clusters of aluminium flow upward due to the keyholing and lead to ununiform aluminium content in the weld with periodic spikes that can potentially be sources of brittle intermetallic phases formations.

In the third part, the calibrated CFD model was coupled with an optimization algorithm that iterates over the peak power, peak duration, and welding velocity to maximize the penetration depth, and minimize the aluminium content in the seam that increases the risk of IMCs formation. So far, the preliminary results indicated that the incremental improvements are possible, although the positive correlation between the aluminium content and the penetration depth, makes this optimization task difficult.

A. Appendix

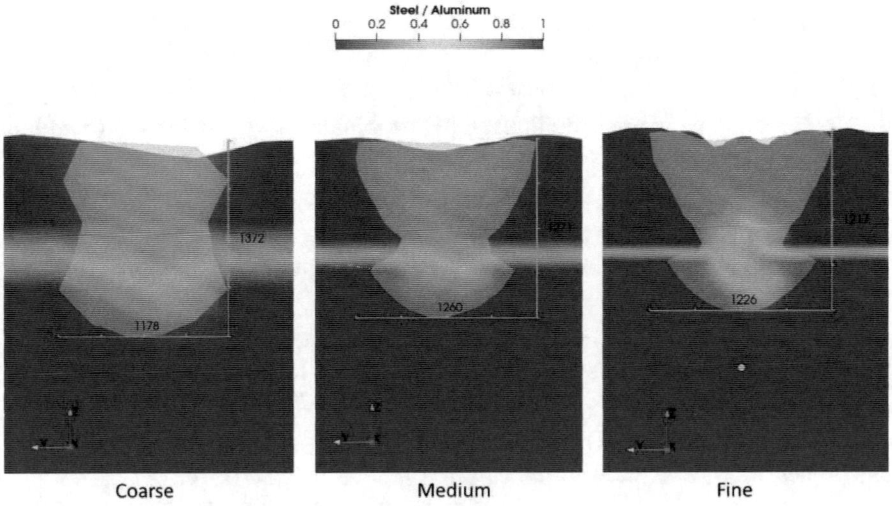

Figure A.1. Mesh convergence study results.

References

[1] A. B. Pereira, A. Cabrinha, F. Rocha, P. Marques, F. A. O. Fernandes and R. J. A. d. Sousa, "Dissimilar Metals LaserWelding between DP1000 Steel and Aluminum Alloy 1050," *Metals,* vol. 9, no. 102, 2019.

[2] A. Sadeghian and N. Iqbal, "A review on dissimilar laser welding of steel-copper, steel-aluminum, aluminum-copper, and steel-nickel for electric vehicle, battery manufacturing," *Optics and Laser Technology,* vol. 146, p. 107595, 2022.

[3] M. Torkamany, S. Tahamtan and J. Sabbaghzadeh, "Dissimilar welding of carbon steel to 5754 aluminum alloy by Nd:YAG pulsed laser," *Materials and Design,* vol. 31, p. 458–465, 2010.

[4] A. Mathieu, R. Shabadi, A. Deschamps, M. Suery, S. Matteı , D. Grevey and E. Cicala, "Dissimilar material joining using laser (aluminum to steel using zinc-based filler wire)," *Optics & Laser Technology,* vol. 39, pp. 652-661, 2007.

[5] J. Sun, Q. Yan, W. Gao and J. Huang, "Investigation of laser welding on butt joints of Al/steel dissimilar materials," *Materials & Design,* vol. 83, pp. 120-128, 2015.

[6] M. N. Esfahani, J. Coupland and S. Marimuthu, "Numerical simulation of alloy composition in dissimilar laser welding," *Journal of Materials Processing Technology,* vol. 224, pp. 135-142, 2015.

[7] Z. Xue, S. Hu, D. Zuo, W. Cai, D. Lee and E. K.-A. Jr, "Molten pool characterization of laser lap welded copper and aluminum," *J. Phys. D: Appl. Phys,* vol. 46, no. 495501, p. 9, 2013.

[8] Y. Hu, X. He, G. Yu, Z. Ge, C. Zheng and W. Ning, "Heat and mass transfer in laser dissimilar welding of stainless steel and nickel," *Applied Surface Science,* Vols. 5914-5922, p. 258, 2012.

[9] S. Zhao, G. Yu, X. He, Y. Zhang and W. Ning, "Numerical simulation and experimental investigation of laser overlap welding of Ti6Al4V and 42CrMo," *Journal of Materials Processing Technology,* vol. 211, pp. 530 - 537, 2011.

[10] N. Kouraytem, X. Li, R. Cunningham, C. Zhao, N. Parab, T. Sun, A. D. Rollett, A. D. Spear and a. W. Tan, "Effect of Laser-Matter Interaction on Molten Pool Flow and Keyhole Dynamics," *Phys. Rev. Applied*, vol. 11, p. 064054, 2019.

[11] W. Huang, H. Wang, T. Rinker and W. Tan, "Investigation of metal mixing in laser keyhole welding of dissimilar materials," *Materials and Design,* p. 109056, 2020.

[12] "FLOW-3D | Solving the world's toughest CFD problems," Flow Science Inc. , [Online]. Available: www.flow3d.com.

[13] C. Hirt and B. Nichols, "Volume of fluid (VOF) method for the dynamics of free boundaries," *Journal of Computational Physics,* vol. 39, no. 1, pp. 201-225, 1981.

[14] R. W. Schrage, A Theoretical Study of Interphase Mass Transfer, Columbia University Press, 1953.

[15] M. Bayat, S. Mohanty and J. H. Hattel, "Multiphysics modelling of lack-of-fusion voids formation and evolution in IN718 made by multi-track/multi-layer L-PBF," *International Journal of Heat and Mass Transfer,* vol. 139, pp. 95-114, 2019.

[16] V. Semak and A. Matsunawa, "The role of recoil pressure in energy balance during laser materials processing," *J. Phys. D: Appl. Phys,* vol. 30, pp. 2541-2552, 1997.

[17] W. E.King, H. D.Barth, V. M.Castillo, G. F.Gallegos, J. W.Gibbs, D. E.Hahn, C. Kamath and A. M.Rubenchik, "Observation of keyhole-mode laser melting in laser powder-bed fusion additive manufacturing," *Journal of Materials Processing Technology,* vol. 214, pp. 2915-2925, 2014.

[18] [Online]. Available: https://www.engineeringtoolbox.com/convective-heat-transfer-d_430.html.

[19] S. Software, "Sente Software - JMatPro," [Online]. Available: https://www.sentesoftware.co.uk/jmatpro. [Accessed 19 January 2023].

[20] M. J. Torkamany, M. J. Hamedi, F. Malek and J. Sabbaghzadeh, "The effect of process parameters on keyhole welding with a 400W Nd :YAG pulsed laser," *Journal of Physics D: Applied Physics,* vol. 39, no. 21, 2006.

[21] F. M. Ghaini, M. Hamedi and M. T. a. J. Sabbaghzadehb, "Weld metal microstructural characteristics in pulsed Nd: YAG laser welding," *Scripta Materialia 56,* vol. 56, pp. 955-958, 2007.

[22] D. Bergström, J. Powell and A. Kaplan, "The absorptance of non-ferrous alloys to Nd:YAG and Nd:YLF laser light at room temperature," *Applied Optics,* vol. 48, no. 8, pp. 1290-1301, 2007.

[23] D. Bergström, A. Kaplan and J. Powell, "The Absorptance of Steels to Nd:YLF and Nd:YAG Laser Light at Room Temperature," *Applied Surface Science,* vol. 253, no. 11, pp. 5017-5028, 2007.

IOP Conf. Series: Materials Science and Engineering 1281 (2023) 012019 doi:10.1088/1757-899X/1281/1/012019

Enhancement of a quasi-analytical solution for modelling additive manufacturing processes

S. Cooke, C. Sinclair, D. Maijer
Department of Materials Engineering, The University of British Columbia, Canada
E-mail: src@student.ubc.ca

Abstract. Numerical modelling methods (e.g. finite element) can provide accurate descriptions of long-range temperature fields in laser or electron-beam melting processes, however the high computational costs at part-scale make them unsuitable for process modelling in additive manufacturing (AM). Alternative methods such as semi-analytical solutions based on a moving heat source reduce the computational expense but at the cost of unrealistic assumptions. Radiation, temperature-dependent physical properties and latent heat are not considered in the semi-analytical approach but can have a significant effect on the thermal history. In this study, the error associated with each of these contributions are assessed against the conduction-only semi-analytical solution for a range of processing parameters for surface melting on solid Ti-6Al-4V. The semi-analytical model is then "enhanced" using results from finite element simulations to better account for the heat transfer in the AM process.

1. Introduction

Due to the high number of variables that influence fabricated component quality in metal additive manufacturing (AM), finding optimal printing conditions can be time consuming and costly [1, 2]. Previous studies have shown that printed part quality depends strongly on the selected processing parameters. The complex dependence on these parameters and part quality makes it difficult to know how to design build strategies *ab initio* [3, 4, 5]. In order to avoid resource intensive trial-and-error experiments to determine build parameters, computational models have been developed and deployed to help predict the thermal, mechanical and microstructural history of a part for a given set of AM processing conditions [3, 4, 6, 7].

Finite element (FE) and computational fluid dynamics (CFD) models are computationally expensive and cannot be used as a screening tool to mitigate defects in full-sized parts. Analytical and semi-analytical approaches have been explored in AM with a significant reduction in computation time [8, 9, 10]. Although computationally more efficient than FE, or CFD, analytical solutions to the heat equation that make use of superposition require that material properties are constant (e.g. temperature independent) and that conduction is the only heat transfer mechanism. Some efforts have been made to better address these limitations but their applications are limited [9, 10, 11, 12].

In this paper, a quasi-analytical model for surface melting is developed and enhanced to include the effects of radiative heat transfer, latent heat evolution and temperature dependent material properties by training with FE data sets generated for a variety of process conditions. Material properties consistent with those of solid Ti-6Al-4V (Ti64) are used here, though the model is generally applicable to any material. This study provides foundational work to enable the study of other aspects of AM processes, including the impact of powder properties and multiple layer builds. The final objective is to predict the long-range thermal history for any arbitrarily complex beam trajectory and processing parameters to identify defect-prone regions owing to over or under melting in the electron beam powder bed fusion (EB-PBF) process.

2. Methodology

2.1. Semi-analytical modelling of a moving heat source

The underlying assumptions for analytical models of moving heat sources require that: 1) the thermophysical properties are independent of temperature and space, 2) the domain is subject to semi-infinite conditions, and 3) heat transfer through convection, radiation heat loss and latent heat evolution due to phase transformations are not considered [13, 14]. A quasi-analytical (QA) transient model of heat conduction, originally proposed by Nguyen *et al* [15] for a moving volumetric heat source, has been used as the starting point in this study. Equation 1 presents Nguyen's solution.

$$T(t) - T_0 = \frac{2\eta Q}{\rho c_p (\pi/3)^{3/2}} \int_0^t \frac{1}{\sqrt{\phi_x \phi_y \phi_z}} \exp\left(\frac{-3x\,(t')^2}{\phi_x} - \frac{3y\,(t')^2}{\phi_y} - \frac{3z\,(t')^2}{\phi_z} \right) dt' \qquad (1)$$

where $\phi_i = 12\alpha(t - t') + \sigma_i^2$ for $i = x, y, z$. T_0 is the initial temperature, η is the absorption efficiency and Q is the nominal beam power. The material properties enter via the prefactor as density (ρ) and heat capacity (c_p) as well as through the thermal diffusivity (α) contained in ϕ_i. As noted above, this solution explicitly assumes constant ρ, c_p and α. The characteristics of the volumetric Gaussian heat source are determined by σ_x, σ_y and σ_z, and the coordinates $x(t')$, $y(t')$ and $z(t')$ are the relative distances from the point of interest to the beam center as a function of time. Built into the QA solution, a zero flux boundary condition is applied along the top surface of the evaluated area, mirroring the beam motion.

This model was implemented in C++ using a similar approach to Stump *et al* [9], making use of Gaussian-Legendre integration. Here, it was ensured that a smooth curve for any given beam velocity was achieved by discretizing the integration path into an appropriate number of segments. By doing so, the number of integration points required is also reduced. It should be noted that different integration schemes affect the calculated peak temperature. However, it has been determined that the calculated peak temperatures do not affect the meltpool size or the long-range temperatures after a smooth curve is achieved. Lastly, all constant properties used in the QA model are evaluated at the initial temperature, T_0.

2.2. Finite Element Model

Single beam thermal history analyses on a solid Ti64 substrate were conducted in ABAQUS in order to compare and assess the impact of the limiting assumptions in the QA model. Different combinations of initial substrate temperatures, beam power and beam velocity were considered and the effect on the thermal history was examined. Four separate FE analyses with varying levels of complexity were developed in order to better understand the effects of each contribution to the thermal history including: 1) conduction only with constant properties evaluated at the initial temperature, (FE1), 2) inclusion of radiative heat losses with constant properties (FE2), 3) radiative heat losses with the addition of temperature-dependent properties (FE3) and 4) accounting for the latent heat of fusion with the addition of radiative heat losses and varying thermal properties (FE4). All FE analyses with constant properties (FE1 and FE2) and the QA model employed material property values (specific heat and thermal conductivity) evaluated at the initial temperature, T_0. The density was assumed to be temperature independent for all FE models in this study with a value of $\rho = 4200$ kg/m^3. It is important to mention that the heat loss due to evaporation was not considered in this study, although it is acknowledged that this can have an effect on the thermal field at higher energy densities [16]. Table 1 summarizes the different FE models developed and used in this study.

Table 1. FE analyses used to assess contributions of different heat transfer mechanisms

FE analyses	Const. Props	Radiation	Var. Props	Latent heat
FE1	X			
FE2	X	X		
FE3		X	X	
FE4		X	X	X

A Gaussian surface heat flux distribution with a beam diameter of $\sigma_x = \sigma_y = 0.145$mm and an absorption efficiency of $\eta = 72\%$ was used. To compare different processing parameters, combinations of beam velocities (300, 500, 1000 and 3000 mm/min) and powers (100, 150, 300 and 500W) were evaluated at different initial temperatures of 35, 200, 600 and 1000°C. Thermo-physical properties from Ref. [17] were used for the solid Ti64 substrate in both the FE and QA models. For the FE models that incorporate radiative heat losses (FE2, FE3, FE4), Boltzmann radiation was assumed with an emissivity of 0.7 and an ambient temperature of 35°C. The FE mesh consisted of 157,500 DC3D8 elements and 172,172 nodes arranged into a rectangular domain (15mm x 6mm x 4.25mm (LxWxH)). The elements around the beam path have dimensions of 50μm x 100μm x 50μm. The mesh size is increased 1mm from the the beam center and 250μm below the surface as the thermal gradient decreases. A mesh sensitivity study was performed to ensure the fidelity of the model is not compromised with the current mesh size. Lastly, symmetry was assumed at the central plane of the domain to reduce the size of the problem and subsequently, the computation time. All exposed faces other than the top surface of the domain are subjected to an adiabatic condition. A large enough domain was chosen to achieve near-semi infinite conditions so that it may be compared to the QA model. Figure 1 below illustrates an example of the mesh and steady-state thermal profile from the FE model.

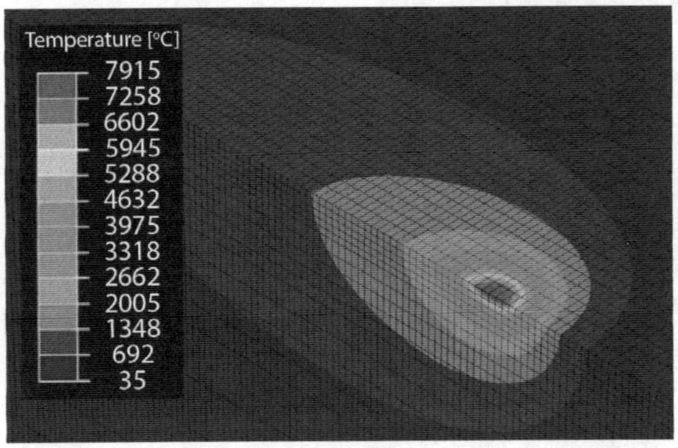

Figure 1. FE heat transfer model of single beam track steady-state profiles developed in ABAQUS used for this study.

3. Results and Discussion

One-dimensional steady-state temperature profiles from the top surface of the domain along the center-line of the beam trajectory from the FE and QA models are extracted and used for comparison, following the method of Steuben *et al* [10]. Since the long-range portions of the temperature profiles are of interest (locations outside of the meltpool), the difference in temperature calculated between the FE and QA models were calculated and used to correct the QA model. This was done for locations where the temperature profile is at or below 1595 °C (solidus temperature of Ti64). All differences reported in this paper are determined using the root mean squared error (RMSE) calculated with Equation 2 below:

$$\mathrm{RMSE} = \sqrt{\frac{\sum_{i=1}^{N}\left(T_i - \hat{T_i}\right)^2}{N}} \tag{2}$$

where T_i is the FE model temperature at the ith position being compared to the QA model, $\hat{T_i}$, and N is the number of points in the temperature data set. In this study, the QA model is "corrected" or scaled down using the input power, Q, within the pre-factor term in front of the integral in Equation 1 to minimize the RMSE between the FE model. The RMSE values in this study will also be presented as a factor of the conduction-only "uncorrected" QA model and the FE model in question where the minimized RMSE for the corrected QA model ($RMSE_{min}$) will be divided by the uncorrected base model ($RMSE_{base}$).

3.1. Conduction-only with constant properties comparison

First, it is necessary to ensure that the QA model can closely replicate the results given by the FE model under the same conditions (e.g. FE1). The temperature along the length of beam trajectory on the surface is compared to the FE solution in Figure 2. It was determined that for all conditions studied in this model the QA solution is in very good agreement with the FE1 model (constant properties and conduction only). With the validity of the QA model confirmed against its FE counterpart, the solution can then be optimized to account for other contributions to the thermal history such as radiation, temperature dependent properties, and, latent heat of fusion.

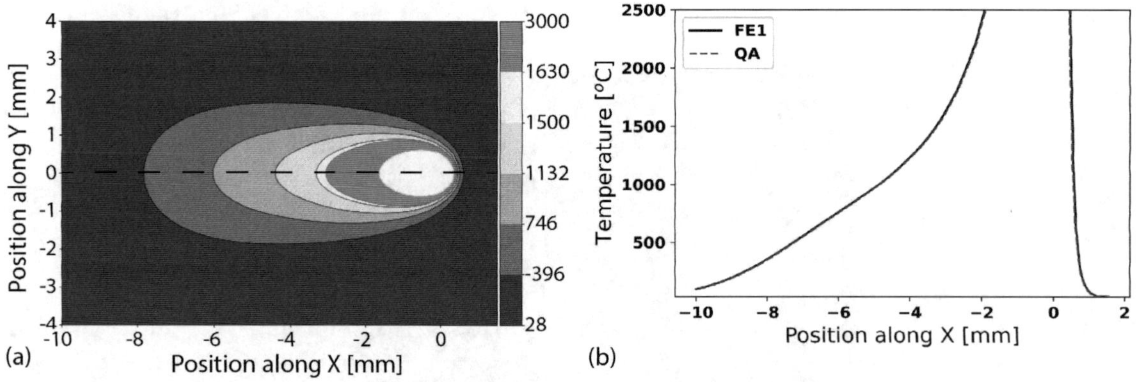

Figure 2. QA model at an initial temperature of 35°C, beam velocity of 1000mm/min and power of 300W showing a) temperature contour of top surface and the center line for the 1D temperature profile and b) 1D temperature profile comparison for the QA model and its FE counterpart evaluated under the same conditions and properties (FE1).

3.2. Radiation losses under constant properties

To account for the effect of radiation losses at the top surface of the sample, while assuming constant material properties, and varying process parameters we have taken a highly simplified approach similar to that taken by Shahabad *et al* [12]. We consider that the majority of losses occur from within the meltpool where the temperatures are largest. Given the short life-time of the meltpool, we therefore have approximated radiative losses as a reduction in the input beam power. To find the best matching "reduced power" for the QA simulation, the RMSE between the QA solution and the FE2 solution was minimized by adjusting the beam power Q for each set of velocity, power and initial temperature. The optimal "corrected" QA temperature profile calculated based on this reduced input power Q^* is compared to the FE2 solution in Figure 3 for different combinations of processing parameters and initial temperatures. Here, the RMSE of the corrected solution compared to FE2 ($RMSE_{min}$) is divided by the uncorrected RMSE compared to FE2 ($RMSE_{base}$) to visualize the trend in "improvement" as the QA solution is optimized. A smaller value in $RMSE_{min}/RMSE_{base}$ indicates a lower error after optimization compared to the uncorrected QA model. The shaded region shows the maximum and minimum RMSE determined for the different processing parameters (velocity, power) evaluated at the given initial temperature. It is evident that with increasing initial temperature, the model requires a smaller "correction" multiplier to the QA solution.

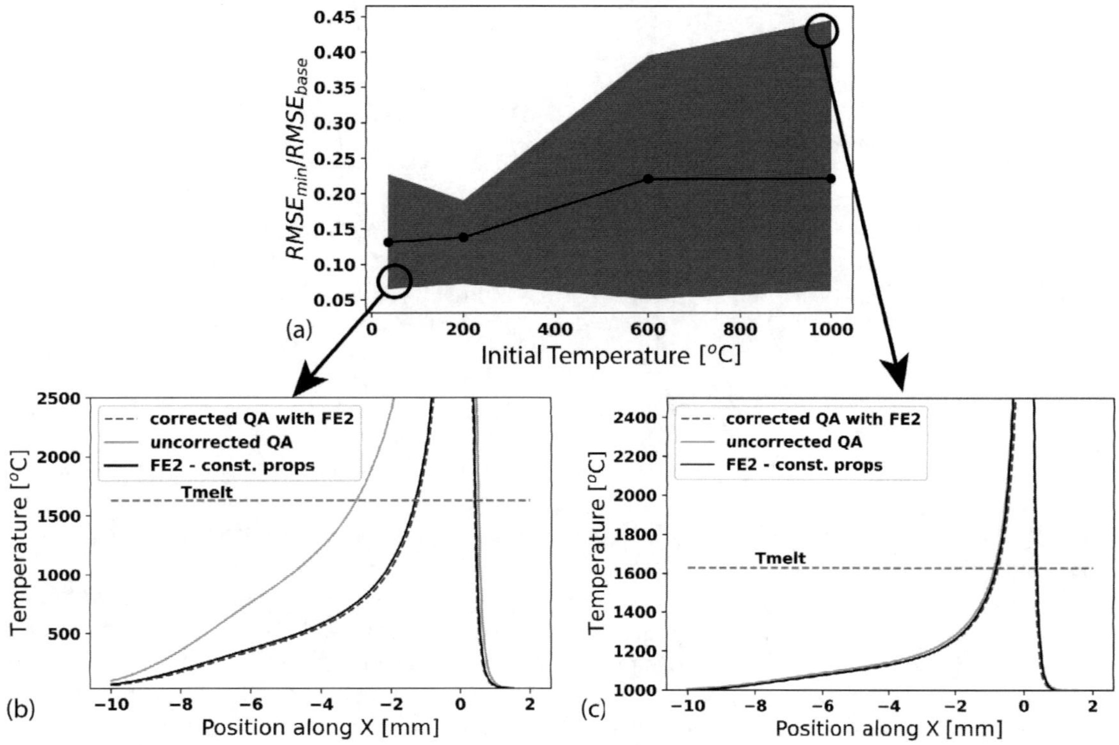

Figure 3. Comparison between the FE2 model and the QA model showing a) minimized RMSE of the corrected QA model against the FE2 model (radiation and constant properties) divided by the RMSE of the uncorrected QA compared to FE2, b) temperature profile comparison for a beam velocity of 1000mm/min and a power of 300W at 35°C, and c) temperature profile comparison for a beam velocity of 3000mm/min and a power of 100W at 1000°C. *Tmelt* represents the liquidus temperature of Ti64.

For different initial temperatures, and combinations of power and velocity, the equivalent radiative losses were determined using the "nominal input power reduction" (Q^*/Q) factor as a function of the FE1 (or equivalently, the uncorrected QA model) meltpool (MP) area. The results show that radiation is easily accounted for using this method and accurate temperatures are predicted for a wide range of power, velocity and initial temperature combinations by finding the optimal scaling parameter to reduce the input power. The relationship between the nominal power reduction factor (eg. Q*/Q) versus the MP area exhibits a power law relationship. Figure 4 illustrates the power law dependence on equivalent radiative losses through scaling the QA solution. The resulting power law relationship for the corrected heat input is shown in Equation 3 which is a function of the conduction-only melt pool area (MP_{area}), initial temperature (T_{base}) and the coefficient (a) and exponent (b) for the fitted power equation.

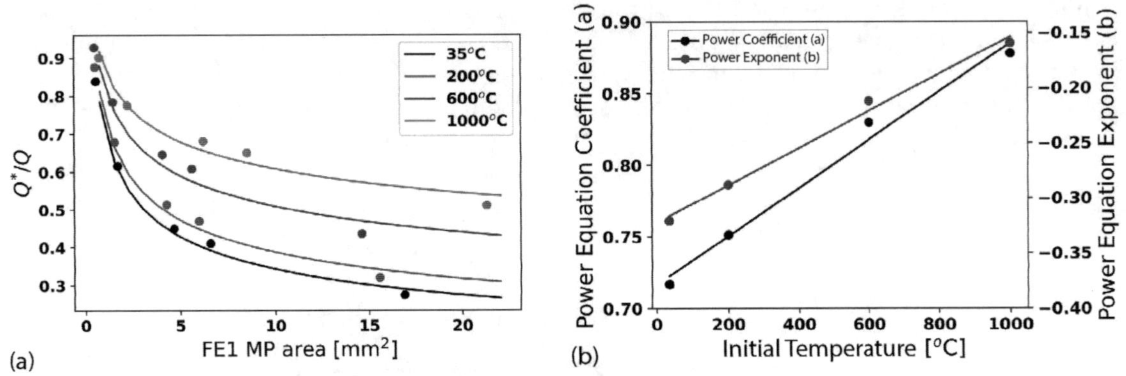

Figure 4. Input power reduction Q^*/Q relationship after minimizing the RMSE with FE2 model (with radiation and constant properties) against the FE1 melt pool area showing a) power equation fit b) power coefficient and exponent linear relationship with temperature.

$$\frac{Q^*}{Q} = a(T_{base}) \cdot MP_{area}{}^{b(T_{base})} \tag{3}$$

Previous studies [12, 18, 19, 20] on similar QA models attributed radiative losses at high energy density as the largest source of error. From these figures, it is evident that radiation is easily addressed by scaling the QA solution and reducing the input power in order to closely match the FE2 solution for different process parameters and initial temperatures. Further, there is a greater required correction to the input heat at lower initial temperatures compared to those at higher temperatures. At high initial temperatures and subsequently, greater thermal conductivities, more heat is able to dissipate from the meltpool. At lower initial temperatures with poorer conduction to the substrate, higher melt pool temperatures and thermal gradients are experienced. Thus, there are greater radiative losses as the initial temperature decreases, justifying the need for a greater correction to the input power for the QA solution.

3.3. Constant vs. temperature-dependent properties

With the effect of radiation with constant properties better understood, varying material properties with radiation but without latent heat will be assessed. First, it is best to compare the differences between FE2 and FE3 model results for different process parameters. Figure 5 illustrates the temperature profile comparison between constant (FE2) and varying material properties (FE3) for different processing parameters.

The RMSE between FE2 and FE3 was also determined for multiple combinations of power, velocity (linear energy density (power/velocity)[21]) and initial temperature and is illustrated in Figure 6. It is evident that the effect of temperature-dependent material properties decreases

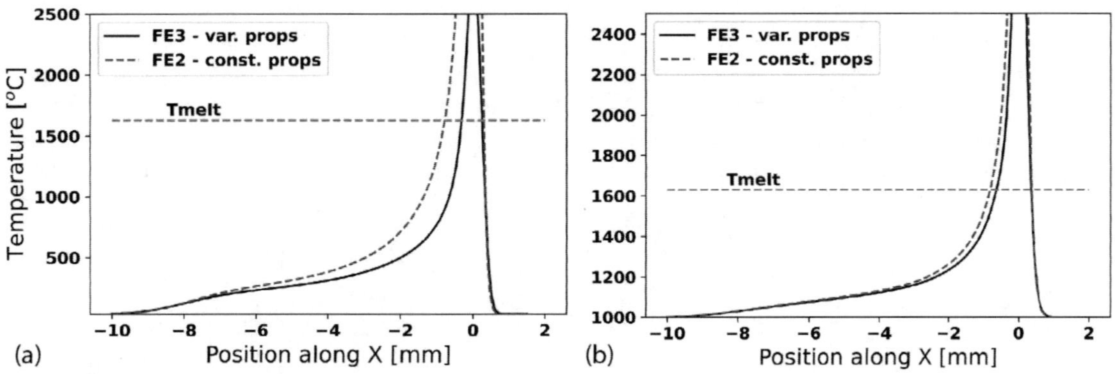

Figure 5. FE2 - constant properties and FE3 - variable properties comparison for 100W and 3000 mm/min at a) 35°C and b) 1000°C

with increasing initial temperature. Additionally, with an increase in energy density, the effect of temperature-dependent material properties increases. For high temperature AM processes, such as EB-PBF, one may find that constant properties are sufficient as the RMSE decreases with increasing initial temperature. For laser systems which are generally operated at much lower temperatures, the temperature-dependent properties have a more significant impact on the long-range thermal history. Depending on the objective, constant properties at low temperatures may still provide sufficient predictions of the thermal history to resolve macro-scale effects such as residual stress. However, when considering temperature sensitive phenomena such as phase transformations or recrystallization, the error between constant and varying properties will likely need to be considered.

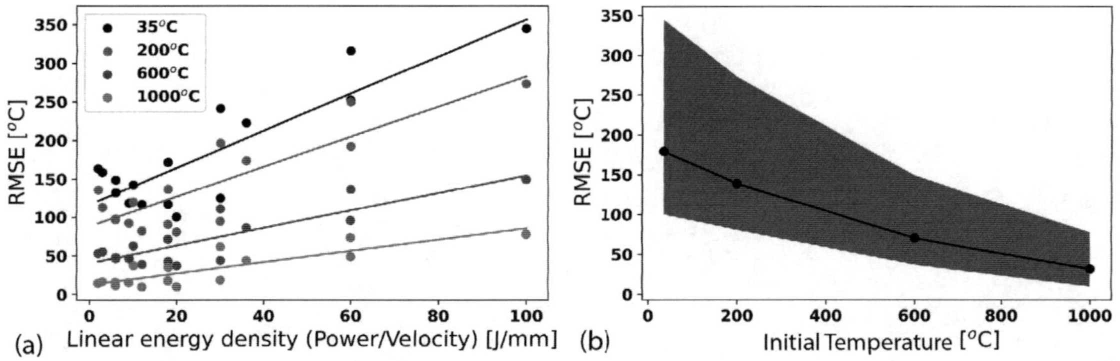

Figure 6. RMSE between FE2 and FE3 for constant and varying thermal properties showing a) dependence of linear energy density and initial temperature and b) average RMSE for all processing parameters as a function of initial temperature.

A similar approach for radiative losses in the previous section was employed here to account for the influence of temperature-dependent properties. The RMSE between the QA and FE3 (varying material properties) models were minimized for different combinations of initial temperature, velocity and power by adjusting the input power, Q. Figure 7 illustrates the comparison in temperature profile for the FE3 model and the relative RMSE comparison as a function of initial temperature. The RMSE in Figure 7(b) is plotted for the corrected QA model fit with FE2 (as shown in the previous section) and fit with FE3 (varying properties). Even though $RMSE_{min}$ for the FE2 fit remains the same as the previous section, the trend compared to Figure 3 changes with the new comparison of the uncorrected QA model to the FE3 solution ($RMSE_{base}$). The shaded regions provide the upper and lower RMSE limits compared with the varying properties FE3 model which arise from varying combinations of power and velocity.

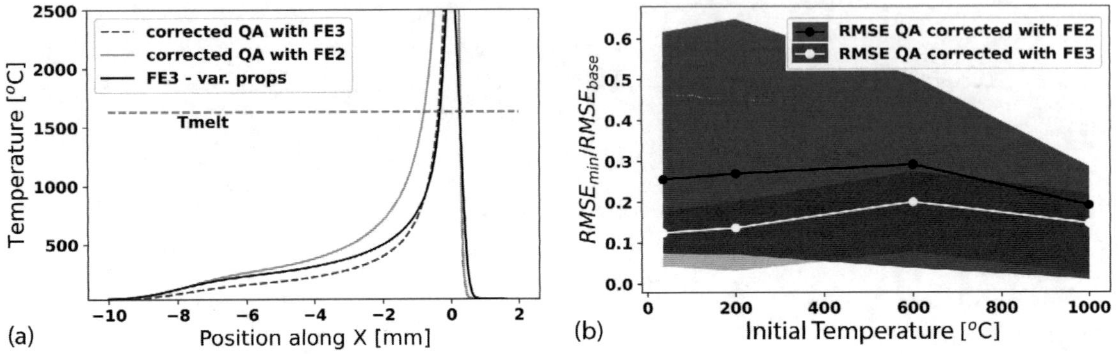

Figure 7. QA model minimized by the FE3 model with radiation and varying properties showing a) temperature profile comparison at 100W, 3000mm/min at 35°C for RMSE minimized against FE2 and FE3 to account for temperature-dependent properties and b) average RMSE for all processing parameters as a function of initial temperature.

From this, it is evident that the QA model can be further improved, specifically at lower initial temperatures where the effects of varying properties are significant. Furthermore, it can be seen that the maximum and minimum RMSE limits at varying energy densities are also truncated when correcting the QA solution to the FE3 model. A relationship between the power and meltpool area was determined similar to Equation 3 where Q^*/Q is a function of initial temperature and the uncorrected QA meltpool area. The power coefficient follows a linear relationship with increasing initial temperature as shown in Figure 8, however in contrast to Equation 3, the power exponent is a constant value at -0.06.

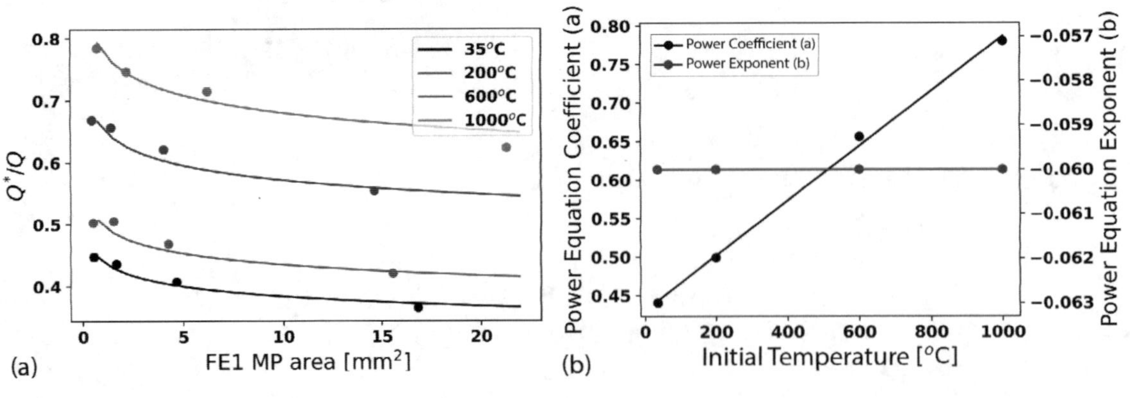

Figure 8. Input power reduction Q^*/Q relationship after minimizing the RMSE with FE3 model (with radiation and varying properties) against the FE1 melt pool area showing a) power equation fit b) power coefficient and exponent linear relationship with temperature.

IOP Conf. Series: Materials Science and Engineering 1281 (2023) 012019 doi:10.1088/1757-899X/1281/1/012019

3.4. Latent heat

Since the evolution of latent heat will affect the temperature distribution during both melting and solidification, it is an important contribution that must be considered. Neglecting latent heat will tend to over predict both the interface velocity and the thermal gradient at the solidification interface [11]. Thus, as discussed earlier, phenomena that are sensitive to small temperature changes will be affected by latent heat. To account for latent heat of fusion in the QA model, a secondary heat source with a similar Gaussian distribution, located behind the beam center, was used to simulate the heat released during solidification. With the FE model (FE4), the length and location of the latent heat "shoulder" from the beam center was determined by taking the spatial temperature gradient. To best fit the FE4 data, the second heat source location behind the beam center, its input power and intensity distribution was modified and added to the modified QA model fitted to the FE3 (radiation and varying properties) model from the previous section. Examples of incorporating latent heat for four different initial temperatures, each with varying effects of latent heat, are illustrated in Figure 9. In this figure, the effect of initial temperature at 300W and 1000mm/min on the latent heat was observed. The uncorrected and corrected QA solution (RMSE minimized with FE3 and including a latent heat source with FE4) is compared to the FE4 model and its RMSE evaluated. For these processing parameters, an overall RMSE reduction ($1 - RMSE_{min}/RMSE_{base}$) between 55-81% is achieved for varying degrees of latent heat "effect" on the temperature profile.

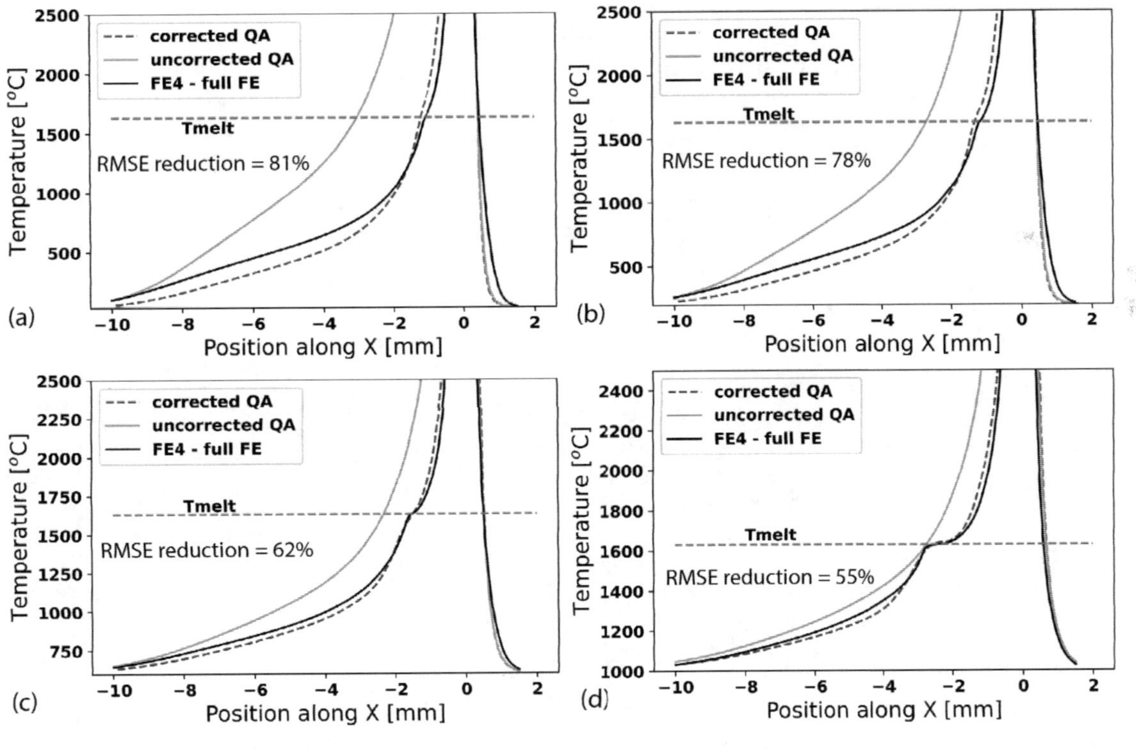

Figure 9. Latent heat comparison for 300W at 1000 mm/min at a) 35°C and b) 200°C c) 600°C and d) 1000°C

4. Conclusions

In this paper, a semi-analytical model for surface melting on solid Ti64 was developed and enhanced with FE data for application to additive manufacturing processes. First, single-track FE models for different beam powers, velocities and initial temperatures were developed. Here, different levels of complexity were introduced in order to study the effects of radiation, temperature-dependent properties and latent heat. In doing so, the importance of these contributions to the thermal history can be assessed and better understood. The effect of radiation losses was shown to have a power law relationship with increasing energy density and meltpool area. Thus, a predictive relationship between the conduction-only melt pool area and initial temperature could be determined to scale the quasi-analytical solution to best match with its FE counterpart. The effect of temperature-dependent properties was then assessed and it was shown that with increasing initial temperature and consequently, increasing conductive properties of Ti64, the RMSE between the two FE models with constant and varying properties decreased. For AM process that operate at lower bed or substrate temperatures (below 600°C), this error may need to be further decreased. However, for processes that operate at higher temperatures, like EB-PBF, constant properties may provide sufficient approximations for long-range thermal history predictions without compromising the fidelity of the model. Lastly, the effect of latent heat of fusion was implemented by incorporating a secondary heat source. The second heat source was placed a certain distance behind the beam center depending on the process parameters in the FE counterpart. The second heat source was then scaled down and its distribution was widened to best match the FE data. Overall, for a wide range of processing parameters and initial temperatures, the RMSE for the QA solution was able to be drastically improved compared to the full FE model with radiation, temperature-dependant properties and latent heat. Future work will work toward improving the model to be applicable to AM conditions by implementing powder properties and multi-layer build capabilities.

References

[1] Sames W J, List F, Pannala S, Dehoff R R and Babu S S 2016 *International materials reviews* **61** 315–360
[2] Fu Z and Körner C 2022 *European Journal of Materials* **2** 54–117
[3] Heigel J, Michaleris P and Reutzel E W 2015 *Additive manufacturing* **5** 9–19
[4] Denlinger E R and Michaleris P 2016 *Additive Manufacturing* **12** 51–59
[5] Oliveira J P, LaLonde A and Ma J 2020 *Materials & Design* **193** 108762
[6] Yang X, Barrett R A, Tong M, Harrison N M and Leen S B 2021 *Journal of Manufacturing Processes* **61** 428–439
[7] Cooke S, Sweet G, Ahmadi K, Bishop P and Herring R 2022 *Journal of Manufacturing Processes* **81** 537–561
[8] Pinkerton A and Li L 2004 *Proceedings of the Institution of Mechanical Engineers, Part C: Journal of Mechanical Engineering Science* **218** 531–541
[9] Stump B and Plotkowski A 2019 *Applied Mathematical Modelling* **75** 787–805
[10] Steuben J C, Birnbaum A J, Michopoulos J G and Iliopoulos A P 2019 *Additive Manufacturing* **25** 437–447
[11] Plotkowski A, Kirka M M and Babu S 2017 *Additive Manufacturing* **18** 256–268
[12] Imani Shahabad S, Karimi G and Toyserkani E 2021 *Lasers in Manufacturing and Materials Processing* **8** 288–311
[13] Rosenthal D 1941 *Welding journal* **20** 220–234
[14] Carslaw H S and Jaeger J C 1947 *Conduction of heat in solids*
[15] Nguyen N, Ohta A, Matsuoka K, Suzuki N and Maeda Y 1999 *WELDING JOURNAL-NEW YORK-* **78** 265–s
[16] Klassen A, Forster V E, Juechter V and Körner C 2017 *Journal of Materials Processing Technology* **247** 280–288 ISSN 0924-0136
[17] Zhao X, Reilly C, Yao L, Maijer D, Cockcroft S and Zhu J 2014 *Applied Mathematical Modelling* **38** 3607–3623
[18] Hekmatjou H, Zeng Z, Shen J, Oliveira J P and Naffakh-Moosavy H 2020 *Metals* **10** 436
[19] Promoppatum P, Yao S C, Pistorius P C and Rollett A D 2017 *Engineering* **3** 685–694
[20] Ji X, Wang Y and Liang S Y 2022 *Journal of Materials Processing Technology* **301** 117452
[21] Bertoli U S, Wolfer A J, Matthews M J, Delplanque J P R and Schoenung J M 2017 *Materials & Design* **113** 331–340

Measurements of deposition rate and droplet temperature in GMAW of aluminum alloys

Rishiekesh Ramgopal[1], Zhaoyang Yan[2], Kevin S. Scott[3], Shujun Chen[4] and Patricio F. Mendez[5]

[1,5] University of Alberta, Edmonton, Alberta, Canada
[2,4] Beijing University of Technology, Beijing, China
[3] Emerson Automation, Boulder, Colo

E-mail: [1] rishieke@ualberta.ca

Abstract. This article explores the reason behind the higher deposition rate in Mg-containing aluminum alloys in gas metal arc welding (GMAW). Experiments performed on ER1100 (Mg-free alloy) and ER5183 (high Mg alloys) measured current, wire feed speed, and droplet temperature. A non-linear analysis of Joule heating at the electrode extension and an analysis of the power needed to heat the consumable to the droplet temperature were performed. Analysis revealed that Joule heating is two orders of magnitude smaller than what was necessary for the higher wire feed speed. The droplet temperature measurements were done for the first time for both consumables under identical conditions, which revealed that ER5183 had a lower droplet temperature than ER1100 by almost 350 K.

1. Introduction

Deposition rate for Mg-containing aluminum alloys has been observed to be higher than aluminum alloys with other alloying elements in gas metal arc welding (GMAW) [1]. There was no knowledge on the droplet temperatures and deposition rates for ER1100 and Mg-containing ER5183 ahead of this work. The common hypothesis was that the larger resistivity of ER5183 causes higher joule heating, which leads to higher deposition rates in the Mg-containing alloy. To verify this hypothesis, two sets of calculations were carried out based on the wire feed speeds and current from Table 3. In the initial case, the energy generated through resistive heating at the electrode extension with time was measured for both consumables. In the latter case, the power provided to the wire to reach the droplet temperature was calculated for both consumables. The results obtained show that the impact of Joule heating was minimal and two orders of magnitude less than what was necessary to cause the higher deposition rate.

2. Materials

Lincoln Electric provided the materials required for this project. The composition of ER1100 was taken from the Lincoln Electric website, assuming identical Si and Fe content. Composition of ER5183 and wires for both ER1100 and ER5183 with 1.2 mm diameter was provided for the particular batch used. Table 1 contains the alloy compositions. The shielding gas utilized was industrial purity Argon (99.9995 vol%).

IOP Conf. Series: Materials Science and Engineering 1281 (2023) 012020 doi:10.1088/1757-899X/1281/1/012020

Table 1. Wire compositions used (wt%)

	Be	Mg	Si	Ti	Cr	Mn	Fe	Cu	Zn	Zr
ER5183	0.0002	4.93	0.05	0.08	0.08	0.63	0.13	0.002	0.001	0.00
ER1100	0	0	0.09	0	0	0	0.09	0.09	0	0

3. Experiments performed

Figure 1 illustrates the experimental layout, which involved a unique copper cathode [2] and an orifice that allowed molten droplets to fall through a solid state calorimeter [3]. The setup comprised of a cylindrical copper block of high purity, weighing 612 g with 35 mm height and 2 in. (50.8/ mm) diameter. Four K-type grounded thermocouples of diameter 1.57 mm measured the temperature across the cross-section of the calorimeter. Fluke 9173 Metrology well was used to calibrate the thermocouples between 50°Cand 300°C. Data acquisition for the experiment recorded the temperature simultaneously with time, using an OMEGA TC-08 board (±0.2% accuracy) at a rate of 10 Hz [4].

A Lincoln PowerWave S500 was utilized for all welds in constant voltage (CV) mode, using direct current electrode positive polarity (DCEP) with Program #5 in weld set reference Z153615. The weld setup included a Lincoln 84 Dual Feeder and a water-cooled Tregaskiss Tough I.C.E torch. Voltage calculation was done using a CHV-25P/100 data acquisition system with an accuracy of ±1%, in the middle of the cathode and contact tip. The current calculation was done using a Hall-effect sensor CHB-500T, which could measure upto ±0.5% accuracy.

Contact tip at the electrode was setup as reported in [3] for a dependable and easy to reference point of electrical contact. Contact tip to the plane of reference of cathode was consistently kept at 1 in. (25.4 mm). The distance from the arc attachment to the point of electrical contact was maintained at 0.5 in. (12.7 mm) by adjusting it in the previous test to each experiment. The uncertainty involved with setting the electrode extension was 0.6 mm, since it was set visually and could not help analyze arc attachment as in [5].

The experiment involved establishing and sustaining the arc between the copper cathode and the electrode. Thermal content of the molten droplet was estimated by letting the droplet fall through the cathode orifice. Arc was started when a small piece of aluminum foil covered the opening of the orifice, which connected the cathode and electrode ahead of each experiment. The arc ignition blows away the foil immediately. A movable flux collecting tray was situated under the cathode to trap the molten metal during the starting transient and then taken out after 3 seconds, thus only permitting the steady state droplets to the calorimeter.

Wire feed speeds are measured for calibrating machine panel readings. Calibration was done separately for ER5183 between 108 ipm to 241 ipm and for ER1100 between 150 ipm and 235 ipm (machine panel values). Calibration started by estimating the length of wire fed in 10s. Length of the wire was measured using a measuring tape with an accuracy of 1/16 in.,time was measured using a stopwatch app, considering the variable human reaction time of 0.1 s, which resulted in the calibration accuracy of the order of 1%. Differences in calibration among wires or differences in calibration at varied wire feed speeds, were statistically undistinguishable and are equivalent to an actual wire feed speed 5.1% faster than reported by the machine panel.

Figure 1. Experimental setup to record droplet temperature.

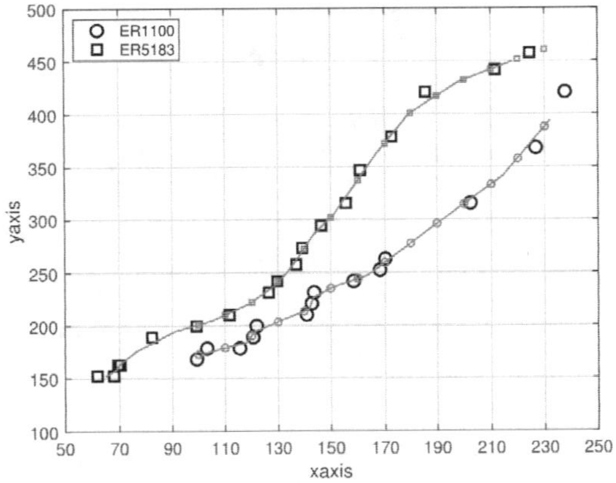

Figure 2. Wire feed speed of ER1100 and ER5183 with 1.2 mm wire diameter

High speed videos of every experiment were taken with a Phantom V210 camera with a Nikon 180mm f/2.8 lens and an 850nm long wave pass filter and were synchronized with the data acquisition. Table 3 indicates the mass of the droplets (m_d) after cooling, the scale used for measuring had accuracy of 0.01 g. The order of experiments was randomized for ER5183 to not cause systematic experimental errors. The order of experiments for ER1100 was done with increasing wire feed speeds, although it did not indicate any evidence of systematic error.

3.1. Droplet temperature measurement

The comparison of calibrated wire feed speed along with increasing current is shown in Figure 2, a similar trend was found in welding under actual conditions in the field [1]. Wire feed speed (i.e. deposition rate) of ER5183 was higher than ER1100 at any current tested, which indicates that Mg-containing consumable have a higher deposition rate. In the figure, the black squares and circles represent every experiment conducted, and the gray line which spans across the range represents the trendline to a moving average of two points. Small gray markers in the figures are added to provide a comparison between both wires at similar currents, and they are present every 10 A in between 100 A and 230 A on the trendline.

Figure 3 illustrates the temperature evolution of the calorimeter recorded by the thermocouples for one of the experiments. The gray lines in the figure indicate the individual evolution of the four thermocouples, one placed 5 mm from the surface, which causes the transient spike observed. The black line indicates the average temperature observed.

In every experiment conducted, the temperature of the calorimeter increases from its initial temperature $T_{\mathrm{c,i}}$ at time t_{i} till it reaches a maximum T_{f} at time t_{f}. After the calorimeter reaches its highest temperature, the temperature gradually reduces due to heat losses since the molten metal is cooling down. The calorimeter was calibrated using molten Sn [3] to find the heat losses involved in the calorimeter from its initial time to the time the calorimeter reached its maximum temperature in Equation 1. In Equation 1 the Q_{loss} term stands for the energy lost in J to the surroundings at T_{∞}(considered as 20°C) from the time the droplet enters the calorimeter till the time the calorimeter reaches its highest temperature.

$$Q_{\mathrm{loss}} = K_{\mathrm{loss}} \left(T_{\mathrm{f}} - T_{\infty}\right) \left(t_{\mathrm{f}} - t_{\mathrm{i}} - t_{\mathrm{trans}}\right) \quad (1)$$

IOP Conf. Series: Materials Science and Engineering 1281 (2023) 012020 doi:10.1088/1757-899X/1281/1/012020

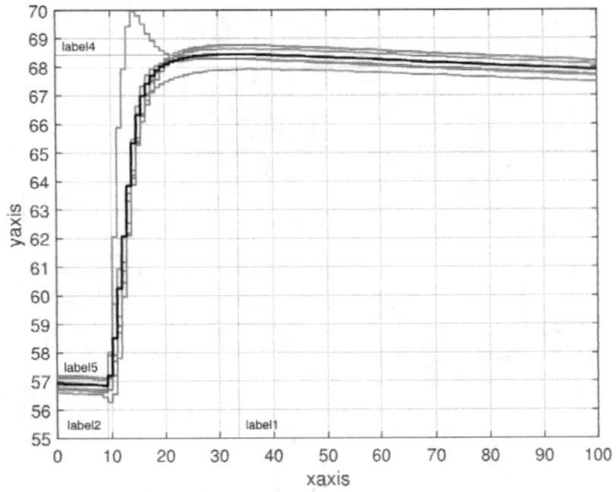

Figure 3. Typical temperature evolution for ER1100 at 250 A

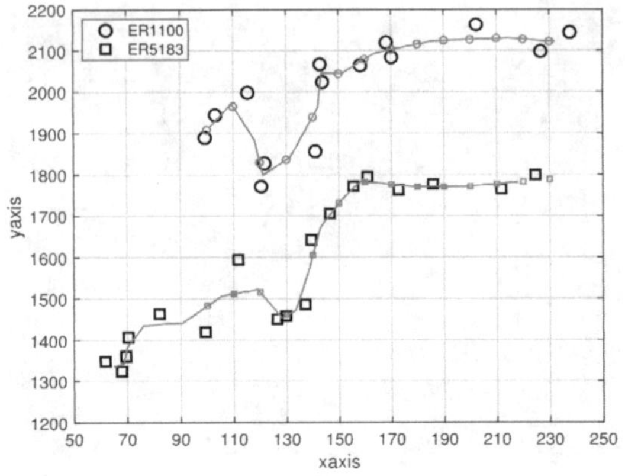

Figure 4. Droplet temperature of ER1100 and ER 5183 with 1.2 mm wire diameter

Curve fitting during the calibration of the calorimeter using molten Sn provided K_{loss} =5.372 10^{-2} W K^{-1} and t_{trans} =8.689 s. The values found during curve fitting were practical, with K_{loss} corresponding to an overall heat transfer coefficient of 5.6 W m^{-2} K^{-1} (based on the area of the copper block) and t_{trans} corresponding to a Fourier number of 0.65 (based on block radius and thermal diffusivity of copper of 1.11 10^{-4} m^2s^{-1}), the Fourier number value is consistent with a transient period of temperature distribution inside the block. The calorimeter showed a biot number of 3.5 10^{-4} according to its body size and thermal conductivity of copper. The low biot number value indicates a very high temperature uniformity in the calorimeter at the time corresponding to the maximum temperature attained in the calorimeter [6].

$$T_{\text{d}} = T_{\text{ref}_{\text{liq}}} + \frac{1}{c_{\text{ref}_{\text{liq}}}} \left[\frac{m_{\text{c}}}{m_{\text{d}}} c_{\text{c}} \left(T_{\text{f}} - T_{\text{c,i}} \right) - c_{\text{ref}_{\text{sol}}} \left(T_{\text{ref}_{\text{sol}}} - T_{\text{f}} \right) - \Delta h_{\text{ref}} + \frac{Q_{\text{loss}}}{m_{\text{d}}} \right] \tag{2}$$

Table 2. Thermophysical properties of alloys used

	$T_{\text{ref}_{\text{sol}}}$ (°C)	$T_{\text{ref}_{\text{liq}}}$ (°C)	$c_{\text{ref}_{\text{sol}}}$ (J kg^{-1}K^{-1})	$c_{\text{ref}_{\text{liq}}}$ (J kg^{-1}K^{-1})	Δh_{ref} (J kg^{-1})
ER1100	200.01	2497.4	1066	1172	3.028 10^6
ER5183	200.01	2000.4	1090	1173	2.464 10^6

Equation 2 indicates a mass and energy balance involving the calorimeter considering the early stage in which the molten droplet falls into the cold calorimeter to the end stage where the droplet and the calorimeter attain a unified maximum temperature. Table 2 illustrates the thermal properties of the alloys, which were ascertained using Thermocalc version 2022a. The specific heat of the copper calorimeter was c_{c} =333.5 J kg^{-1}K^{-1}, and the mass of the calorimeter m_{c} = 612.0 g. The reference temperature for molten and solid states for an alloy are $T_{\text{ref}_{\text{liq}}}$ and $T_{\text{ref}_{\text{sol}}}$, while the equivalent specific heats are $c_{\text{ref}_{\text{liq}}}$ and $c_{\text{ref}_{\text{sol}}}$. The enthalpy change when the alloy transitions from the solid reference state to the molten reference state is Δh_{ref} and was also calculated using Thermocalc. The enthalpy between the reference states was chosen instead of the solid and liquid states to avoid any error in calculation while using specific heats. The most significant source of uncertainty in the absolute measurement of droplet temperature was the specific heat of the calorimeter with an error of the order of 20% [3]. The relative error in the same calorimeter did not have a notable source of error. Current methods are aimed at improving this technique.

Figure 4 represents the droplet temperatures of the alloys that are listed in Table 3. The measured droplet temperature values are in accordance with the earlier results obtained by [2, 3, 7, 8], which had a local minimum at the transition from globular to spray transfer with increasing current. Figure 4 illustrates that both ER5183 and ER1100 have similar transition current ranges (around 110 A to 150 A), and both alloys have similar droplet temperature ranges to [2, 9]. ER1100 displayed very high droplet temperatures, which approached the boiling temperatures (2470°C) in spray transfer mode. At the same time ER5183 had slightly more than 300 K lower in spray transfer mode and around 400 K lower in globular mode.

4. Joule heating

Equation 3 represents the energy produced through resistive heating at the electrode extension through Joule heating(q_{EE}). Electrode extension for all experiments conducted was maintained constant at 0.5 in. Joule heating was calculated to understand the effect of change in resistivity through Mg addition. If Joule heating is high, then less energy must be provided to melt the wire. In Equation 3, A_{c} represents the cross-sectional area of the electrode, L represents electrode extension length, and $\rho_{\mathrm{e_{eff}}}$ represents the effective resistivity. Figure 5 illustrates that ER5183 has higher resistivity over ER1100 at any given temperature due to the Mg content. The resistivity difference among the alloys was significant at higher temperatures. During lower temperatures, Mg precipitates out of the solution, and the matrix is similar to ER1100, thus causing similar resistivities.

$$q_{\mathrm{EE}} = \frac{\rho_{\mathrm{e_{eff}}} L I^2}{A_{\mathrm{c}}} \tag{3}$$

$\rho_{\mathrm{e_{eff}}}$ was calculated precisely using equations from [10], where $\rho_{\mathrm{e_0}}$ represents electrical resistivity at the wire spool temperature (assumed 20°C). $\Delta\rho_{\mathrm{e_1}}$ is the variation in resistivity between the spool temperature and solidus, $\Delta\rho_{\mathrm{e_2}}$ is the maximum departure from linear in the variation in resistivity, indicated with vertical segments in Figure 5. The term ρ represents the wire density at room temperature, L is the length of electrode extension, $\Delta h_{0,\mathrm{sol}}$ is the variation in enthalpy between spool temperature and solidus, A_{c} is the cross sectional area of the wire, and wire diameter is d_{c}. Table 4 lists the resistive properties obtained through JMatPro® necessary for solving Equation 4.

$$\rho_{\mathrm{e_{eff}}} = \rho_{\mathrm{e_0}} M_1 \left[2 \frac{\gamma - 1}{b(1-\gamma) + d(1+\gamma)} \right] \tag{4}$$

$$M_1 = \frac{\rho \Delta h_{0,\mathrm{sol}} U_{\mathrm{c}} A_{\mathrm{c}}^2}{\rho_{\mathrm{e_0}} I^2 L} \tag{5}$$

$$\gamma = \exp(d/M_1) \tag{6}$$

$$d = \sqrt{b^2 - 4a} \tag{7}$$

$$a = -4\Delta\rho_{\mathrm{e_2}}/\rho_{\mathrm{e_0}} \tag{8}$$

$$b = (\Delta\rho_{\mathrm{e_1}} + 4\Delta\rho_{\mathrm{e_2}})/\rho_{\mathrm{e_0}} \tag{9}$$

Analysis of the effective resistivities, as listed in Table 3 shows very similar values to the resistivity at room temperatures. Joule heating causes only a small increase in the temperature of the wire, while the wire melts mainly due to the heat from the arc tip. The maximum joule heating value was 22 W, and the maximum joule heating difference between the alloys was 2.3 W, as listed in Table 5.

Table 3. Data obtained from droplet temperature measurements

U_c (ipm)	I (A)	m_d (g)	t_i (s)	t_f (s)	T_i (°C)	T_f (°C)	T_d (°C)	$\rho_{e_{eff}}$ (10^{-8} Ωm)
ER1100								
168.2	99.32	0.88	9.89	57.56	27.5	38.1	1890	3.11
178.7	103.22	0.49	8.25	28.95	45.3	51.3	1944	3.12
178.7	115.52	0.89	8.64	60.82	36.4	47.5	1998	3.13
189.2	120.48	1.48	7.64	42.73	48.0	64.6	1771	3.14
199.7	121.79	1.41	8.21	41.51	58.6	74.8	1828	3.13
210.2	141.04	0.81	8.44	33.63	68.2	77.5	1857	3.16
220.7	142.9	1.62	8.63	49.12	70.2	90.6	2067	3.16
231.2	143.83	2.64	10.11	52.42	79.5	112.0	2024	3.15
241.7	158.35	1.06	12.53	44.93	51.5	65.1	2064	3.17
252.2	168.1	1.47	9.41	57.08	41.4	60.7	2119	3.18
262.7	170.15	0.89	9.25	29.95	56.7	68.3	2084	3.18
315.3	202.38	1.86	9.05	41.45	54.3	79.2	2163	3.21
367.8	226.73	4.05	9.52	52.73	73.1	124.8	2097	3.22
420.4	237.71	3.35	9.58	36.58	112.5	155.7	2144	3.21
ER5183								
152.4	61.81	1.65	9.00	90.00	37.6	52.0	1348	3.26
152.4	67.93	1.53	21.90	111.90	22.6	36.2	1324	3.28
162.9	69.55	1.41	8.55	72.43	22.7	35.5	1360	3.28
162.9	70.37	1.15	11.12	66.91	35.0	45.6	1407	3.28
189.2	82.16	1.12	7.86	88.83	45.9	56.2	1464	3.30
199.7	99.27	0.72	8.89	61.98	49.5	55.9	1420	3.35
210.2	111.72	0.31	8.16	34.26	59.7	62.8	1594	3.38
231.2	126.45	0.47	11.01	40.70	52.7	57.0	1450	3.41
241.7	129.69	0.78	10.62	63.71	42.5	49.8	1458	3.42
257.5	137.07	0.56	10.67	31.37	77.3	82.6	1486	3.43
273.2	139.36	0.83	8.78	44.79	54.2	62.8	1642	3.42
294.3	146.53	1.51	9.66	92.40	52.6	68.3	1705	3.43
315.3	155.66	1.21	10.30	87.67	33.7	47.2	1772	3.44
346.8	160.92	1.11	10.28	84.96	58.4	70.4	1795	3.43
378.3	172.73	1.88	9.28	65.93	63.7	84.2	1763	3.45
420.4	185.83	2.41	8.63	68.86	75.1	101.3	1777	3.46
441.4	211.7	2.31	10.47	94.74	75.4	99.8	1764	3.53
457.2	224.58	1.42	10.39	46.36	94.4	109.9	1800	3.56

5. Energy provided to the wire

Equation 10 denotes the energy required to melt the consumable and is derived from the enthalpy of the wire and mass flow. The term q_{d} represents the energy required per second to melt the wire, T_0 represents the initial temperature of the wire, $T_{\mathrm{ref_{sol}}}$ and $T_{\mathrm{ref_{liq}}}$ stand for the solid and molten state reference temperature.

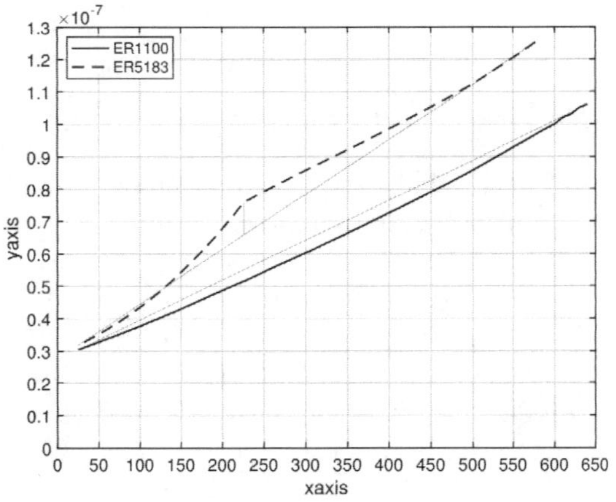

Figure 5. Electrical resistivity of ER1100 and ER5183 with 1.2 mm wire diameter

Table 4. Joule heating properties for the alloys used

	ρ_{e_0} (Ωm)	$\Delta\rho_{e_1}$ (Ωm)	$\Delta\rho_{e_2}$ (Ωm)	ρ (kg m^{-3})	$\Delta h_{0,sol}$ (J kg^{-1})	d_c (mm)
ER1100	$3.040\,10^{-8}$	$7.560\,10^{-8}$	$-4.051\,10^{-9}$	2710	$6.44\,10^{5}$	1.190
ER5183	$3.182\,10^{-8}$	$9.341\,10^{-8}$	$1.004\,10^{-8}$	2660	$6.03\,10^{5}$	1.185

Table 5. Joule heating and power required to melt for ER1100 and ER5183

I (A)	U_c (ipm)		T_d (°C)		q_{EE} (W)		Δq_{EE} (W)	q_d (W)		Δq_d (W)
	1100	5183	1100	5183	1100	5183		1100	5183	
100	172.6	201.0	1908	1483	3.5	3.8	0.3	559	512	-48
110	179.0	210.2	1965	1512	4.3	4.6	0.4	595	544	-51
120	190.6	222.3	1830	1515	5.1	5.6	0.5	596	577	-19
130	203.5	241.2	1836	1461	6.0	6.6	0.6	638	607	-31
140	213.5	272.3	1939	1605	7.0	7.6	0.7	702	742	40
150	235.0	301.9	2044	1730	8.0	8.8	0.8	809	877	68
160	244.2	337.4	2079	1782	9.1	10.0	0.8	854	1006	153
170	259.1	372.0	2102	1777	10.3	11.3	1.0	915	1106	191
180	277.5	400.5	2115	1770	11.6	12.7	1.1	985	1187	202
190	295.9	416.7	2124	1771	13.0	14.2	1.3	1055	1236	181
200	314.5	432.1	2126	1772	14.4	15.9	1.5	1122	1282	160
210	333.1	441.6	2129	1777	15.9	17.6	1.7	1189	1314	124
220	357.8	451.0	2127	1783	17.5	19.5	2.0	1276	1346	69
230	387.5	460.5	2122	1789	19.1	21.5	2.3	1379	1378	-2

$$q_{\mathrm{d}} = \rho U_{\mathrm{c}} A_{\mathrm{c}} \left[c_{\mathrm{ref_{sol}}} \left(T_{\mathrm{ref_{sol}}} - T_0\right)) + \Delta h_{\mathrm{ref}} + c_{\mathrm{ref_{liq}}} \left(T_{\mathrm{d}} - T_{\mathrm{ref_{liq}}}\right)\right] \quad (10)$$

6. Discussion

Magnesium droplets evaporate at a lower temperature when compared to aluminum, this could make the ER5183 droplets attain a thermal balance around the arc at lower temperatures in comparison to ER1100, which does not contain Mg.

The influence of energy loss through evaporation and the effect of metal vapors in anode fall voltage requires additional research to be validated. The influence of droplets falling through the arc before reaching the calorimeter might affect the measured droplet temperature [11]. Droplet temperature measurements could be further improved to eliminate the error during the experiment. In all cases of this experiment, the errors identified are constant, systematic, and in a similar direction, thus creating minimal impact on the comparison among alloys.

7. Conclusions

Droplet temperature in GMAW of aluminum alloys has been measured through calorimetry, with the ER5183 (Mg-containing) droplets being 300 K to 400 K lower than ER1100 (no Mg) droplets. The results are consistent with previous works. Analysis of Joule heating at electrode extension and droplet enthalpy content implied that Joule heating could not explain the enthalpy difference in the droplets. . The droplet temperature difference could explain the higher deposition rate in ER5183 over ER1100 since the lower droplet temperature of ER5183 enables faster wire feed speed for a given current.

Acknowledgments

The authors would like to acknowledge the donation of consumables and equipment from Lincoln Electric, shielding gas flow meters from Alicat.

References

[1] Electric L 2014 Gas metal arc welding. product and procedure selection Report C4.200 Lincoln Electric
[2] Soderstrom E, Scott K M and Mendez P F 2011 *Welding Journal* **90** 77s–84s
[3] Scott K M 2011 *Heat Transfer and Calorimetry of Tubular Ni/WC Wires Deposited with GMAW* MSc University of Alberta
[4] Yan Z, Scott K, Chen S and Mendez P 2022 *Welding Journal* **101**
[5] Egerland S 2015 *Soldagem & Inspecao* **20** 367–380
[6] Gibbs J W, Schlachter C, Mayr P, Kamyabi-Gol A and Mendez P F 2015 *Materials Transactions A* **46A** 148–155
[7] McIntosh C, Chapuis J and Mendez P F 2016 *Welding Journal* **95** 273s–279s
[8] Bin M S, Titinan M, Shinichi T and Manabu T 2017 *Yosetsu Gakkai Ronbunshu/Quarterly Journal of the Japan Welding Society* **35** 160s–164s ISSN 0288-4771
[9] McIntosh C and Mendez P F 2017 *Welding Journal* **96** 354s–366s
[10] Lehnhoff G and Mendez P F 2011 *International Journal of Heat and Mass Transfer* **54** 2651–2660
[11] Zhang G, Goett G and Uhrlandt D 2020 *Journal of Physics D: Applied Physics* **53** 395202 ISSN 0022-3727

IOP Conf. Series: Materials Science and Engineering 1281 (2023) 012020 doi:10.1088/1757-899X/1281/1/012020

Appendix A. Notation

Variable	Unit	Description
m_{d}	kg	Mass of the droplet
T_{d}	°C	Droplet temperature
$T_{\mathrm{c,i}}$	°C	Initial temperature of Calorimeter
T_{f}	°C	Maximum temperature of Calorimeter and droplet
$T_{\mathrm{ref,sol}}$	°C	Reference temperature of Sn in solid state
$T_{\mathrm{ref,liq}}$	°C	Reference temperature of Sn in molten state
T_0	°C	Initial temperature of the wire
T_{∞}	°C	Reference temperature for surroundings is 20°C
t_{i}	s	Time at initial temperature of calorimeter
t_{f}	s	Time at maximum temperature of Calorimeter
Q_{loss}	J	Energy lost by the calorimeter
q_{EE}	W	Power produced through Joule heating
q_{d}	W	Power required to melt the consumable
m_{c}	kg	Mass of the calorimeter
Δh_{ref}	J kg^{-1}	Variation in enthalpy between the two reference temperatures
$c_{\mathrm{ref,sol}}$	J kg^{-1} K^{-1}	Specific heat of Sn at solid state
$c_{\mathrm{ref,liq}}$	J kg^{-1} K^{-1}	Specific heat of Sn at liquid state
ρ	kg m^{-3}	Wire density at room temperature
ρ_{e_0}	Ωm	Electrical resistivity at the wire spool temperature (Assumed 20 °C)
$\rho_{e_{\mathrm{eff}}}$	Ωm	Effective resistivity

IOP Conf. Series: Materials Science and Engineering 1281 (2023) 012021 doi:10.1088/1757-899X/1281/1/012021

Calculation of thermal features in welding and additive manufacturing

Patricio F. Mendez

University of Alberta, Edmonton, Canada

E-mail: pmendez@ualberta.ca

Abstract. This paper summarizes multiple engineering expressions that enable the prediction of thermal magnitudes of interest associated with moving heat sources. The expressions use only fundamental parameters available before performing any experiments, and their calculation is algebraic, without the need for numerical methods. All expressions are based on the fundamental governing equations of heat transfer in the solid. The magnitudes predicted include maximum width and its location, maximum penetration, thickness of HAZ, maximum temperature and its location, leading and trailing edge of an isotherm, heating and cooling rate, aspect ratio of an isotherm, melting efficiency, cooling time from 800°C to 500°C, solidification time, and maximum distribution of a heat source to reach a target temperature. Parameters involved include heat source power and speed, thermal conductivity and diffusivity of the substrate material, temperature of interest and preheat or interpass temperature. Temperature-dependent properties are accounted for by the use of effective properties. The expressions proposed can be extended into sophisticated geometries for welding and specific additive manufacturing cases.

1. Introduction

Both welding and additive manufacturing are characterized by the use of moving heat sources, such that the workpiece experiences localized high temperature (typically melting), while the rest of the component is unaffected. The thermal features have a significant effect on the resulting phase transformations and metallurgical properties. Practical questions associated with moving heat sources include the cooling rate after the heat source passed, thickness of the heat affected zone, desired preheat to reach a target cooling rate, and solidification time. Thermal features can be estimated from a large variety of numerical, analytical and semi-analytical models. In all cases, the results are a temperature as a function of location and time, and the thermal features are calculated from these results. Even for the simplest of models, the process can be tedious, for example to determine the width of a weld, it is necessary to solve the implicit equation giving the isotherm, and then find its maximum.

The practical application of mathematical models in welding and additive manufacturing often requires fast answers that can be obtained by a practitioner of welding. The solutions need not be more accurate than the ability to measure the results; however, the expressions used must be trusted to never give unrealistic results. Analytical expressions (closed-form, such as Rosenthal's solutions) can be faster and more reliable than numerical models, at the expense of necessary simplifications, which can affect the accuracy. Semi-analytical models use numerical integration (e.g. Green's functions) based on analytical models. Even for analytical models, numerical methods might be necessary. For example, obtaining the the width of an isotherm from Rosenthal's solutions is not possible using algebraic operations, and requires a root finding operation to determine the width of the isotherm at different longitudinal coordinates, and then an optimization algorithm to find the maximum of all widths. Although the two numerical operations can be performed with single commands in high-level packages such as Matlab, the small hurdle is still significant for welding and additive manufacturing practitioners.

This work reviews explicit forms of target magnitudes of interest in welding and additive manufacturing, greatly expending the pioneering work of [1–4] in aiming to develop predictive fromulas from existing analytical solutions. The methodology followed is the MRCF described in [5]. The results presented are based on fundamental concepts and closed-form solutions. The simplifications involved include the consideration of the heat source as a point, the consideration of very thick or thin substrates, and particular shapes for the distribution of the heat source. The analysis relies on the foundations laid in [1–14], and the analysis developed in [15–21].

2. Magnitudes of interest

In welding and additive manufacturing, practitioners are typically more interested in values at particular locations or values at particular temperatures more than on a complete field description. The magnitudes explored in this paper are illustrated in Figure 1.

The maximum isotherm width y_{max} is of practical value because it can be measured from cross sections of welds or AM parts. Relevant temperatures considered include the melting temperature and austenitization temperature ($\mathrm{A_{c3}}$) for the HAZ in steels. The trailing length of isotherm x_{b} is the length of the "hot tail" trailing the heat source such as the length of the molten tail (which is often protected with inert gases of limited reach). The penetration z_{max} is a crucial parameter in welding cladding to understand dilution, and laser heat treatment to determine case depth. The centerline cooling rate, $\dot{T}_{\mathrm{b}}$, is representative of the cooling rate of the whole area affected by the heat source [2, 22]. The cooling rate is crucial to determine the microstructure and mechanical properties such as hardness and strength after a pass [23, 24]. The leading length of an isotherm, x_{f}, is an indication of how much the molten bead "leads" the heat source [25, 26]. The centerline heating rate $\dot{T}_{\mathrm{f}}$ is relevant to determine the austenization temperatures $\mathrm{A_{c1}}$ and $\mathrm{A_{c3}}$ [27]. The maximum temperature T_{max} reached at a given point in a cross section gives an indication of possible phase changes and transformations. The gradient of maximum temperature dT_{max}/dy can be used to estimate the thickness of the HAZ. For distributed heat sources, T_{max} is the maximum surface temperature, and $x_{T_{\mathrm{max}}}$ its location; also σ_{max} or L_{max} are the maximum distribution parameter with which a target temperature is reached.

Important magnitudes can be derived from the primary magnitudes described above. The aspect ratio AR is a metric of how elongated the isotherm is, the melting efficiency η_{m} captures the superheat above melting is used to create a bead, and is related to dilution [28]. The characteristic value $t_{8/5}$ is a metric of cooling rate in the welding and AM of steels, and it is fairly insensitive to the location in the vicinity of weld centerline [29,30]. The solidification time at the centerline t_{sl} is useful when studying the microstructure resulting from solidification. The thickness of the HAZ Δy_{HAZ} can also be observed directly in cross sections.

3. Minimal representations

Minimal representations are asymptotic cases when all secondary phenomena are considered negligible, and are the foundation of the the MRCF (Minimal Representation and Correction Factors) methodology described in [5]. In this work the minimal representation will consider point, gaussian, and rectangular top hat heat sources. Point heat sources will be based on Rosenthal's hypothesis and solutions explored in detail in [15–18]. For gaussian heat sources the analysis will follow [19–21], and for a square top hat source, the treatment in [4, 31]. With the notation used in this work, Rosenthal's 2D and 3D problems are illustrated in Figure 2, with

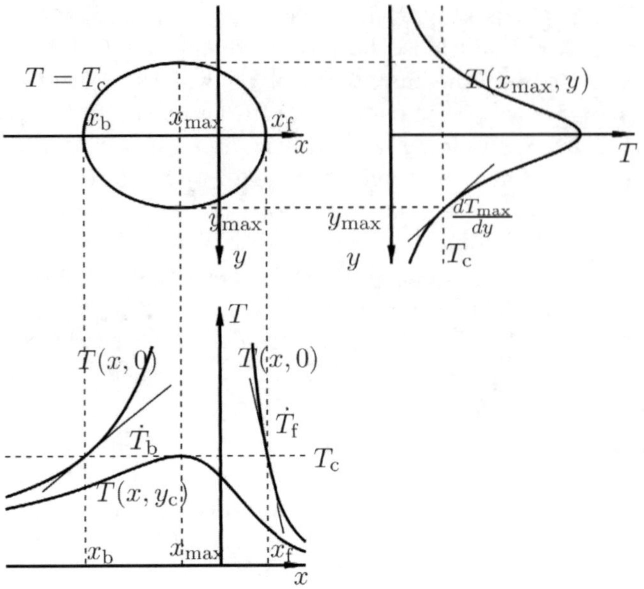

Figure 1: Magnitudes of interest for the thermal field induced by a moving heat source [16]

the corresponding solutions:

$$T(x, y, z) = T_0 + \frac{q}{2\pi k d} \exp\left(-\frac{Ux}{2\alpha}\right) K_0 \left[r\sqrt{\left(\frac{U}{2\alpha}\right)^2 + \frac{h_{\text{top}} + h_{\text{bot}}}{kd}} \right] \tag{1}$$

$$T(x, y, z) = T_0 + \frac{q}{2\pi k r} \exp\left[-\frac{U}{2\alpha}(r + x)\right] \tag{2}$$

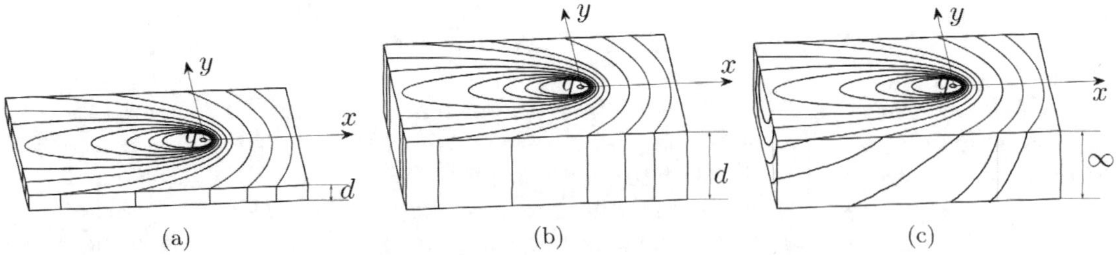

Figure 2: Point heat sources in an infinite domain. (a) 2D problem on a thin substrate (b) 2D problem on a thick substrate (line heat source) (c) 3D problem (adapted from [16]

The 2D solution includes the effect of surface heat losses through the convection coefficients, and is applicable for point heat sources on a thin substrate. The 2D solution can also account for the welding of thick substrates when the heat source is a line penetrating through the thickness, for example full-penetration welding in keyhole mode. The 3D solution corresponds to a point heat source in a very thick substrate. For the case of the 2D solution penetration is the whole thickness of the substrate, while for the 3D solution, because it is axisymmetric around the

IOP Conf. Series: Materials Science and Engineering 1281 (2023) 012021 doi:10.1088/1757-899X/1281/1/012021

x-axis, the penetration of an isotherm is the same as its half-width. Figure 2 (c) shows how the isotherms are axisymmetric for a 3D, with semi-circular intersections with the left surface of the block illustrated, while for the 2D problem in (b) the intersections with the left surface are vertical straight lines.

For the case of a distributed gaussian heat source on a thick substrate, the solution used is [10, 11]:

$$T(x,y,z) = T_0 + \frac{q\alpha^{\frac{1}{2}}}{2k\pi^{\frac{3}{2}}} \int_0^\infty \frac{t^{-\frac{1}{2}}}{2\alpha t + \sigma^2} \exp\left(-\frac{x^2 + 2xtU + U^2t^2 + y^2}{4\alpha t + 2\sigma^2} - \frac{z^2}{4\alpha t}\right) dt \tag{3}$$

while for the case of a fast and broad top hat source on a thick substrate the solution is illustrated in Figure 3 with the following solution [4]:

$$T(x,z) = T_0 + T_1(x,z) + T_2(x,z) \tag{4}$$

$$T_1(x,z) = \begin{cases} 0 & \text{if } x > 0 \\ \frac{2}{\sqrt{\pi}} \frac{q}{kLW} \sqrt{-\frac{\alpha x}{U}} \exp\left(\frac{z^2 U}{4\alpha x}\right) - \frac{qz}{kLW} \operatorname{erfc}\left(\frac{z}{2\sqrt{-\frac{\alpha x}{U}}}\right) & \text{if } x \le 0 \end{cases} \tag{5}$$

$$T_2(x,z) = \begin{cases} 0 & \text{if } x > -L \\ \frac{2}{\sqrt{\pi}} \frac{q}{kLW} \sqrt{-\frac{\alpha(x+L)}{U}} \exp\left[\frac{z^2 U}{4\alpha(x+L)}\right] - \frac{qz}{kLW} \operatorname{erfc}\left[\frac{z}{2\sqrt{-\frac{\alpha(x+L)}{U}}}\right] & \text{if } x \le -L \end{cases} \tag{6}$$

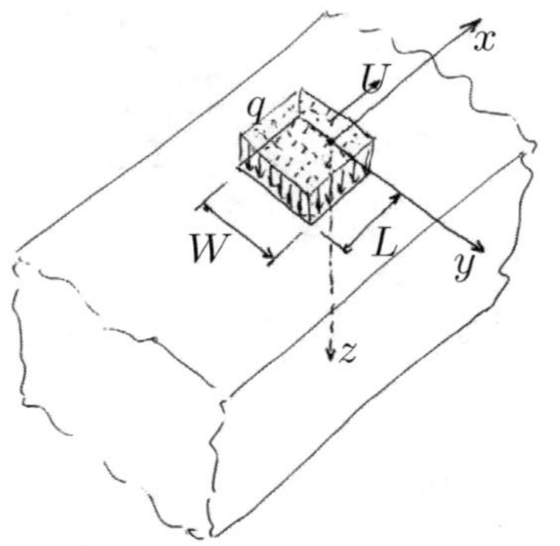

Figure 3: Rectangular top hat heat source on an infinite domain

Even a system as simplified as a point heat source in an infinite domain, there is a further simplification, which is considering whether the heat source is moving fast or slow. This analysis is presented in [15, 17]. The consideration of fast or slow motion cannot be based in a Peclet number, as traditionally done, because there is no length reference in the description of the problem. The metric of speed for moving heat sources is expressed in dimensionless form as the Rosenthal number (Ro, for the 2D model) and the Rykalyn number (Ry, for the 3D problem).

$$\text{Ro} = \frac{q}{2\pi k d\,(T_c - T_0)} \tag{7}$$

$$\text{Ry} = \frac{qU}{4\pi k\alpha\,(T_c - T_0)} = d^*\text{Ro} \tag{8}$$

IOP Conf. Series: Materials Science and Engineering 1281 (2023) 012021 doi:10.1088/1757-899X/1281/1/012021

where $d^* = Ud/(2\alpha)$ is a dimensionless substrate thickness. These dimensionless groups were introduced in [32], and are fundamental for the understanding and tabulation of moving heat sources that can be approximated as a point in a 2D or 3D space. The dimensional analysis justifying the relevance of these numbers is detailed in [5]. In all cases, the parameter q is the amount of heat entering the substrate, which is smaller than the heat generated by the source because the thermal efficiency η of all processes is always lesser than 1; thus, $q = \eta\, q_{\text{source}}$.

4. Regimes

Regimes are combinations of parameters such that the system resembles a given minimal representation over others. For point heat sources on a substrate of finite thickness, there are then 4 possible combinations of minimal representations: fast heat source on a thick substrate (Regime I), slow heat source on a thick substrate (Regime II), fast heat source on a thin substrate (Regime III), and slow heat source on a thin substrate (Regime IV). These regimes can be identified in the map of Figure 4, which is consistent with the maps of [12].

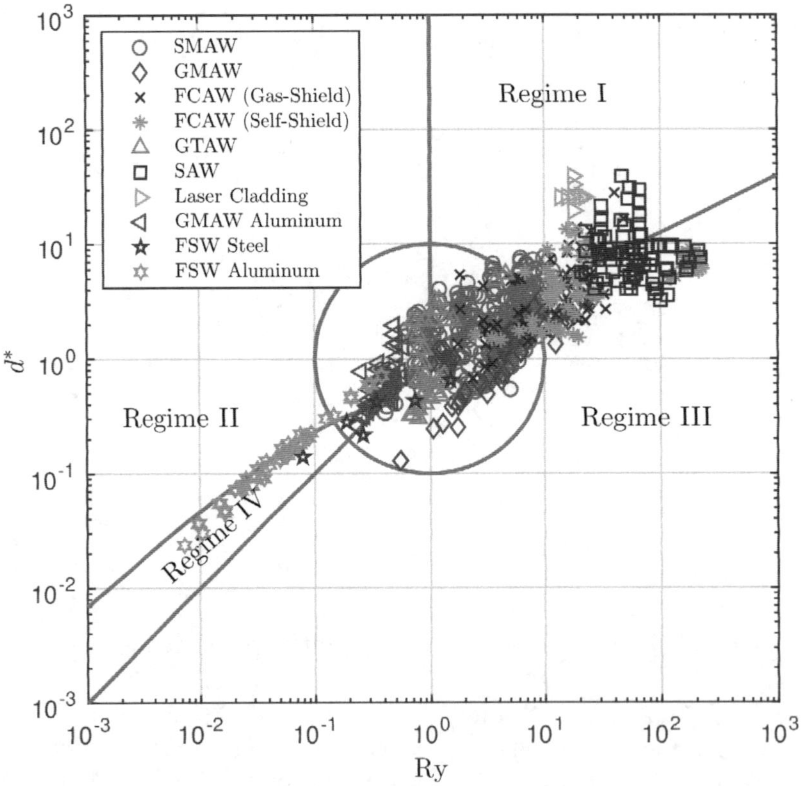

Figure 4: Domains of minimal representations for point heat sources. SMAW: Shielded Metal Arc Welding, GMAW: Gas Metal Arc Welding, FCAW: Flux Core Arc Welding, GTAW: Gas Tungsten Arc Welding, SAW: Submerged Arc Welding, FSW: Friction Stir Welding

The limit between regimes is gradual, not sharp; however, it is practical to establish a conventional boundary between contiguous regimes. Contiguous regimes are those that share at least one feature in common. For example, for point heat sources, those corresponding to the 3D model have the contiguous regimes I (3D fast) and II (3D slow), and fast heat sources have the contiguous regimes I (3D) and III (2D). Regimes I and IV are not contiguous because they do not share their geometry or speed range.

IOP Conf. Series: Materials Science and Engineering 1281 (2023) 012021 doi:10.1088/1757-899X/1281/1/012021

One practical approach to setting a boundary between contiguous regimes is to consider the dimensionless groups under which the predictions of a given magnitude of interest yields the same result with the minimal representations of both contiguous regimes. The boundaries set this way typically vary slightly with the magnitude of interest considered. For the case of point heat sources, if we consider x_{b} (length of the tail) as the magnitude of interest, the resulting boundaries are:

$$\mathrm{Ry}_{\text{I-II}} = 1 \tag{9}$$

$$d^*_{\text{I-III}} = \sqrt{\pi/2}\mathrm{Ry}^{1/2} \tag{10}$$

$$d^*_{\text{III-IV}} \approx \mathrm{Ry} \tag{11}$$

$$d^*_{\text{II-IV}} \approx -\mathrm{Ry}\ln\mathrm{Ry} \tag{12}$$

For the case of x_{b} in particular, regimes I and II always predict the same value, and the boundary is set at Ry=1 based on the limit observed for other magnitudes of interest.

The locations near the regime boundaries are intermediate regimes. Typically, when a condition is a factor of 20 away from the boundary, the error incurred in considering the minimum representation instead of the exact solution is less than 10% [20, 21]. Intermediate regimes require special treatment, typically the blending techniques described in [5].

In Figure 4, the markers correspond to the Ry and d^* based on the substrate thickness, travel speed, current, measured or estimated voltage of the welding passes listed in [33], supplemented with FSW data from [34]. The thermophysical properties and efficiencies used are detailed in [4]. The target temperature for steels is 727°C, representative of the A_1 temperature at the edge of the HAZ. The circle around the origin indicates an intermediate region in which all regimes need to be corrected. It can be observed that most welds in practice cluster around the edge between Regimes I and III, corresponding to fast welds on substrates of intermediate thicknesses, and within those, the majority is within the circle indicating both intermediate thicknesses and intermediate velocities. Some of the welding passes considered as intermediate thickness are full-penetration (e.g. FSW), and in that case, the 2D solution is applicable, regardless the value of d^*. Large or small values of d^* are not by themselves an indication problems, Figure 4 and knowledge of whether the weld is full penetration are necessary to determine the regime. For the case of AM of a wall of only one pass in width, the heat source straddles the thickness between both faces of a wall, and the 2D model ("full penetration") should be used.

When considering finite heat sources, new parameters are needed to account for their distribution. For the case of axisymmetric gaussian heat sources, the parameter σ (standard deviation of the heat source) appears in Equation 3, with its dimensionless counterpart

$$\sigma^* = \frac{\sigma U}{2\alpha} \tag{13}$$

The appearance of the new dimensionless group σ^* creates new regimes. Although it is tempting to view large values of σ^* as "broadly distributed" the true criterion for broad distribution is when the temperature of interest reaches depths much shallower than the size of the heat source. For a thick substrate, a fast broadly distributed heat source will be Regime V, and a slow distributed heat source will be Regime VI, studied in [19–21]. Narrowly distributed heat sources on a thick substrate converge to Regimes I and II respectively. For a thin plate a fast moving broadly distributed heat source will be Regime VII, and its slow counterpart will be Regime VIII. These last two regimes are not explored in this work, but their foundation is developed in the study of gaussian heat sources on substrates of intermediate thickness in [35]. A rectangular top hat source would require two new dimensionless groups. If the source is considered square, then a single additional parameter is necessary, similarly to the case of the axisymmetric gaussian heat source.

Rosenthal's 2D solution (Eq. 1) allows for accounting surface heat losses. If the surface heat losses at the top and bottom are combined in a single coefficient h, a new dimensionless group h^* capturing the effect of surface heat losses appears:

$$h^* = \frac{4\alpha^2 h}{kdU^2} \tag{14}$$

This problem has been explored in [20], and is relevant for cases such as underwater wet welding, in-service repairs, and when the temperature of interest is low, such as the yield temperature when calculating residual stresses [36–38]. In this case, two new regimes are created when surface heat losses are dominant: Regime IIIa for fast heat sources and Regime IVa for slow heat sources. The various regimes mentioned are summarized in Table 1.

Table 1: Regimes corresponding to minimal representations of heat sources

			no surface losses	surface losses
point	3D	fast	I	
		slow	II	
	2D	fast	III	IIIa
		slow	IV	IVa
distributed	3D	fast	V	
		slow	VI	
	2D	fast	VII	
		slow	VIII	

5. Prediction of magnitudes of interest

Table 2 summarizes the expressions to predict the magnitudes of interest for concentrated sources without heat losses corresponding to regimes I and II [15, 16] and regimes III and IV [17, 18]. Table 3 contains predictive expressions for concentrated sources on a thin substrate with heat losses, corresponding to regimes IIIa and IVa [20], and Table 4 contains predictive expressions for distributed sources on a thick substrate, corresponding to regimes V and VI [19–21, 31]. These references also contain correction factors based on blending for intermediate regimes, dimensionless expressions of the magnitudes of interest, and regime boundaries.

In these tables, the function W_0 is the upper branch of the Lambert function [39], with an asymptotic behavior $W_0(x) \approx x$ for small x, and $W_0(x)/\ln x \approx 1$ for large x. For the case of melting, an intermediate temperature is considered between solidus and liquidus, and for the case of $t_{8/5}$, an intermediate temperature between 800°C and 500°C. A suggested choice of intermediate temperature between T_1 and T_2 is $T_\mathrm{i} = T_0 + \sqrt{(T_1 - T_0)(T_2 - T_0)}$ [16], where the case of $t_{8/5}$ on a starting temperature of 20°C yields T_i =632°C.

6. Extension to other geometries

The 3D and 2D point heat source solutions can be extended to many practical geometries other than those described above. The extension is based on the use of an effective heat q_eff in the formulas presented in tables 2 and 3. For regimes I and II, $q_\mathrm{eff} = (180°/\phi)\,q$, where $\phi = \phi_1 + \phi_2$ as shown in Figure 5(a). For thin plates with no losses (regimes III, IV), the ratio q/d in the

Table 2: Expressions to predict magnitudes of interest with point heat sources

Magnitude	Regime I	Regime II	Regime III	Regime IV
y_{max}	$\sqrt{\frac{0.234\,\alpha q}{kU(T_c-T_0)}}$	$\frac{0.159\,q}{k(T_c-T_0)}$	$\frac{0.242\,\alpha q}{Udk(T_c-T_0)}$	$\frac{4\,\alpha}{U\exp\left(\gamma+\frac{1}{Ro}\right)}$
$-x_{max}$	$\frac{5.85\,10^{-2}\,q}{k(T_c-T_0)}$	$\frac{U}{\alpha}\left[\frac{0.113\,q}{k(T_c-T_0)}\right]^2$	$\frac{\alpha}{U}\left[\frac{0.171\,q}{dk(T_c-T_0)}\right]^2$	$\frac{50.3\,\alpha dk(T_c-T_0)}{Uq\left[\exp\left(\gamma+\frac{1}{Ro}\right)\right]^2}$
$-x_b$	$\frac{0.159\,q}{k(T_c-T_0)}$		$\frac{\alpha}{U}\left[\frac{0.282\,q}{dk(T_c-T_0)}\right]^2$	$\frac{4\alpha}{U\exp\left(\gamma+\frac{1}{Ro}\right)}$
$-\dot{T}_b$	$\frac{kU(T_c-T_0)^2}{0.159\,q}$		$\frac{U^2d^2k^2(T_c-T_0)^3}{0.159\,\alpha q^2}$	$\frac{U^2q\exp\left(\gamma+\frac{1}{Ro}\right)}{25.1\,\alpha dk}$
x_f	$\frac{\alpha W_0(2Ry)}{U}$	$\frac{0.159\,q}{k(T_c-T_0)}$	$\frac{\alpha W_0\left(6.28Ro^2\right)}{2U}$	$\frac{4\alpha}{U\exp\left(\gamma+\frac{1}{Ro}\right)}$
z_{max}	$\sqrt{\frac{0.234\,\alpha q}{kU(T_c-T_0)}}$	$\frac{0.159\,q}{k(T_c-T_0)}$	d	
$\dot{T}_f$	$\frac{U^2(T_c-T_0)}{\alpha}\left[\frac{1}{W_0(2Ry)}+1\right]$	$\frac{kU(T_c-T_0)^2}{0.159\,q}$	$\frac{U^2(T_c-T_0)}{\alpha}$	$\frac{U^2q\exp\left(\gamma+\frac{1}{Ro}\right)}{25.1\,\alpha dk}$
$T_{max}(y_c)-T_0$	$\frac{0.234\,\alpha q}{kU{y_c}^2}$	$\frac{0.159\,q}{ky_c}$	$\frac{0.242\,\alpha q}{Udky_c}$	$\frac{q\ln\left(2+\frac{2\alpha}{Uy_c}\right)}{6.28\,dk}$
$-\left.\frac{dT_{max}}{dy}\right\rvert_{y_{max}}$	$\sqrt{\frac{17.1\,kU(T_c-T_0)^3}{q\alpha}}$	$\frac{6.28k(T_c-T_0)^2}{q}$	$\frac{Udk(T_c-T_0)^2}{0.242\,\alpha q}$	$\frac{Uq\exp\left(\gamma+\frac{1}{Ro}\right)}{25.1\,\alpha dk}$
Æ	$\sqrt{\frac{2.70\,10^{-2}\,qU}{k\alpha(T-T_0)}}$	1	$\frac{0.164\,q}{dk(T_c-T_0)}$	1
η_m	0.368	$\frac{3.98\,10^{-2}qU}{k\alpha(T_m-T_0)}$	0.484	$\frac{8dk(T_m-T_0)}{q\exp\left(\gamma+\frac{1}{Ro\,m}\right)}$
$t_{8/5}$	$\frac{0.159\,q(T_8-T_5)}{2\pi kU(T_i-T_0)^2}$		$\frac{0.159\,\alpha q^2(T_8-T_5)}{2\pi U^2d^2k^2(T_i-T_0)^3}$	$\frac{25.1\,\alpha dk(T_8-T_5)}{qU^2\exp\left(\gamma+\frac{1}{Ro_i}\right)}$
t_{sl}	$\frac{0.159\,qh_{sl}}{kcU(T_m-T_0)^2}$		$\frac{0.159\,\alpha q^2h_{sl}}{U^2d^2k^2c(T_m-T_0)^3}$	$\frac{25.1\,\alpha dkh_{sl}}{qU^2c\exp\left(\gamma+\frac{1}{Ro_m}\right)}$
Δy_{HAZ}	$\sqrt{\frac{5.85\,10^{-2}\,q\alpha}{kU(T_i-T_0)^3}}(T_m-T_{HAZ})$	$\frac{q(T_m-T_{HAZ})}{6.28\,k(T_i-T_0)^2}$	$\frac{\alpha q(T_m-T_{HAZ})}{4.13\,Ukd(T_i-T_0)^2}$	$\frac{25.1\,\alpha dk(T_m-T_{HAZ})}{qU\exp\left(\gamma+\frac{1}{Ro_i}\right)}$

Table 3: Expressions to predict magnitudes of interest in point heat sources with heat losses

Magnitude	Regime IIIa	Regime IVa
y_{max}	$\sqrt{\frac{kd}{h}}W_0\left(\pi Ro^2\right)$	$\frac{2\sqrt{kd/h}}{\exp(\gamma+1/Ro)}$
$-x_{max}$	$\frac{kdU}{4h\alpha}W_0\left(\pi Ro^2\right)$	$\frac{12.6\,Uk^2d^2(T_c-T_0)}{\alpha hq[\exp(\gamma+1/Ro)]^2}$
$-x_b$	$0.5\sqrt{\frac{kd}{h}}W_0\left(\pi Ro^2\right)$	$\frac{2\sqrt{kd/h}}{\exp(\gamma+1/Ro)}$
z_{max}	d	
$-\dot{T}_b$	$U\sqrt{\frac{h}{kd}}\left(T_c-T_0\right)$	$\frac{7.96\,10^{-2}\,qU\sqrt{h/(kd)^3}}{\exp(\gamma+1/Ro)}$

Table 4: Expressions to predict magnitudes of interest with distributed heat sources

Magnitude	Regime V (top hat)	Regime V (gaussian)	Regime VI (gaussian)
$y_{\max}$	W	$\frac{\alpha \mathrm{Ry}^{2/3}}{0.285\,U}\frac{\sigma}{\sigma_{\max}}\sqrt{\ln\frac{\sigma_{\max}}{\sigma}}$	$\frac{0.399\,q}{k(T_c-T_0)}$
$z_{\max}$	$1.13\sqrt{\frac{\alpha L}{U}}-\frac{k}{qLW}(T_c-T_0)$	$\frac{\alpha \mathrm{Ry}^{\frac{1}{3}}}{0.248\,U}\ln\frac{\sigma_{\max}}{\sigma}$	$\frac{0.25\,q}{k(T_c-T_0)}\ln\frac{\sigma_{\max}}{\sigma}$
$T_{\max}-T_0$	$\frac{1.13\,q}{kW}\sqrt{\frac{\alpha}{UL}}$	$\frac{0.230\,q}{k}\sqrt{\frac{\alpha}{\sigma^3 U}}$	$\frac{0.199\,q}{k\sigma}$
$-x_{T_{\max}}$	L	$0.765\,\sigma$	$\frac{\sigma^2 U}{2\alpha}$
$\sigma_{\max}$, $L_{\max}$†	$\left[\frac{1.27\,q^2\alpha}{Uk^2(T_c-T_0)^2}\right]^{1/3}$	$\left[\frac{5.28\,10^{-2}\,q^2\alpha}{Uk^2(T_c-T_0)^2}\right]^{1/3}$	$\frac{0.199\,q}{k(T_c-T_0)}$

†: For a square spot ($W = L$)

equations should be replaced by q'_{eff} as indicated in Figure 5. In these extensions, it is assumed that the fraction of heat going on each wedge or thin sheet is proportional to their corresponding angle or thickness such that $y_{\max}$ is the same for all members. For regimes I and II, $y_{\max}$ must be smaller than the size of the bevel; for regimes III and IV, $y_{\max}$ must be larger than the thickness of the plate.

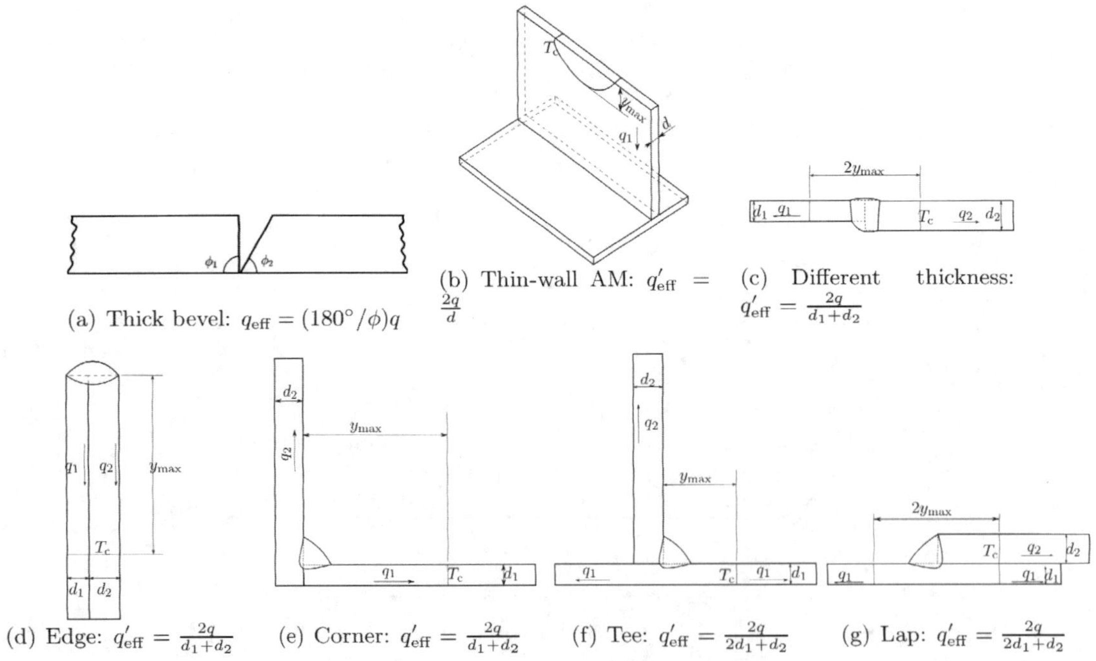

(a) Thick bevel: $q_{\text{eff}} = (180°/\phi)q$

(b) Thin-wall AM: $q'_{\text{eff}} = \frac{2q}{d}$

(c) Different thickness: $q'_{\text{eff}} = \frac{2q}{d_1+d_2}$

(d) Edge: $q'_{\text{eff}} = \frac{2q}{d_1+d_2}$

(e) Corner: $q'_{\text{eff}} = \frac{2q}{d_1+d_2}$

(f) Tee: $q'_{\text{eff}} = \frac{2q}{2d_1+d_2}$

(g) Lap: $q'_{\text{eff}} = \frac{2q}{2d_1+d_2}$

Figure 5: Effective heat source for regimes I and II (a), and regimes III and IV (all others)

7. Calculation of effective properties

The predictive expressions provided involve thermophysical properties of the substrate such as thermal conductivity, which are known to depend on temperature. Although the derivation of the root equations consider constant material properties, it is possible to account for variations in an approximate way by using effective properties based on the known temperature behavior of the properties of interest between the initial substrate temperature T_0 and the temperature of interest. For magnitudes that involve two temperatures such as $t_{8/5}$ and thickness of HAZ, those two temperatures should be used as the range considered for an effective value. Recommended effective parameters are:

$$k_{\text{eff}} = \frac{1}{\Delta T}\int_{T_1}^{T_2} k(T)dT \tag{15}$$

$$(\rho c)_{\text{eff}} = \left.\frac{\Delta(\rho h)}{\Delta T}\right|_{T_1}^{T_2} \tag{16}$$

$$\alpha_{\text{eff}} = \frac{k_{\text{eff}}}{(\rho c)_{\text{eff}}} \tag{17}$$

8. Example of application

This example was provided by Dr. T. Lienert, and was first discussed in [40]. Prisms approximately cubical of 10 mm side were made using powder bed laser additive manufacturing (LENS), as shown in Figure 6(a). The alloy used was a steel containing 9% Cr and 1% Mo, comparable to the well known ASTM A335 P91alloy. This alloy is known to result in a martensitic microstructure under all welding conditions; however, under AM conditions the resulting microstructure is not martensitic. The absence of martensitic is unexpected, as cooling rate in AM is faster than the welds that already result in martensite. It has been speculated that an excessively fast cooling rate might possibly prevent martensite formation, and the key question is then, "What is the cooling rate in this case?"

Two approaches can be taken to assess cooling rate. One is by the analysis of the solidification microstructure to estimate cooling rate during solidification, and the other is by using the expression for solidification time in Table 2. To use the predictive expression for solidification time, the following process parameters were used: T_0 = 20°C, U =0.70 m s^{-1}, d = 0.01 m, q_{laser} = 270 W, thermal efficiency η =0.2, resulting in q=54 W. The materials properties were determined from JMatPro h_{sl} = 2.58 10^5 J kg^{-1}, T_{solidus} = 1420°C, T_{liquidus} = 1502°C, with effective properties k = 26.77 W m^{-1}K^{-1}, $(\rho c)_{\text{eff}}$ =4.912 10^6 J m^{-3}K^{-1}. These parameters correspond to Ry=14.3 and d^* =385, deeply into the Regime I region in Figure 4. Using the formula for t_{sl} in Table 2, a time of 89 μs is obtained, and combining this with the freezing range of the alloy, the average cooling rate during solidification obtained is 0.93 10^6 K s^{-1}, which is much faster than what is common in welding, and in an unexplored range where unexpected effects are possible.

The theoretical prediction above was compared to the estimated solidification rate obtained from the microstructure in Figure 6(b), in which solidification cells are spaced at 0.63μm. Two correlations used in [41] are of the form $\lambda_1 = AR^{-n}$ where λ_1 is the cell spacing (μm), R is the cooling rate during solidification (Ks^{-1}), and A and n are empirical parameters for stainless steels listed in [41]: A =80, n =0.33(Katayama), and A =54, n =0.32: R =1.08 10^6 K s^{-1} (Nakao). The two correlations yield 2.3 10^6 K s^{-1} (Katayama), and 1.1 10^6 K s^{-1} (Nakao), very close to the theoretical prediction within the uncertainties involved, and validating the proposed expression.

IOP Conf. Series: Materials Science and Engineering 1281 (2023) 012021 doi:10.1088/1757-899X/1281/1/012021

(a)

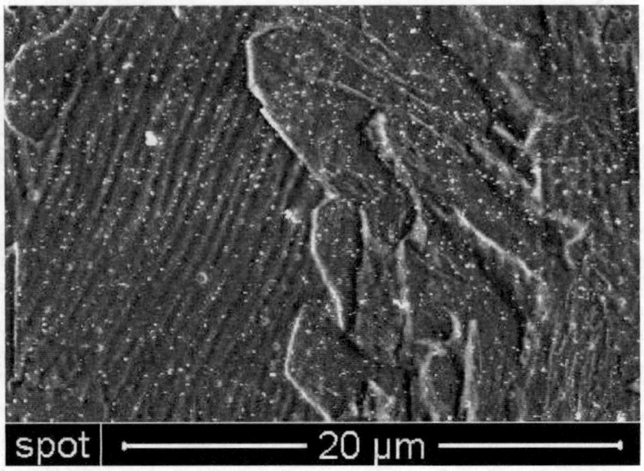

(b)

Figure 6: Prisms of alloy similar to P91 manufactured with powder bed laser AM (left), and resulting microstructure showing solidification cells(right)

9. Discussion

The predictions are consistent with the few that were provided in earlier work for the simplest cases, e.g. peak temperature and cooling rates for fast moving point heat sources in [2]. For some cases, they provide the proper behavior that was missing in previous work, such as width of an isotherm in Regime IV [1]. The case of volumetric heat sources, such as electron beams is analyzed in detail in [42–44]

The effect of molten metal flow is outside the scope of this paper, and it is often secondary to the other parameters considered here, but not always negligible. The level of difficulty in the treatment increases, but a rigorous scaling approach has been developed in [45], based on [46].

Some predictions of tables 2-4 are almost always applicable, while some are seldom applicable. Fortunately, important predictions such as $t_{8/5}$ and Δy_{HAZ} in regimes I-IV are often applicable to welding and AM of all metals under normal conditions. This is because because the location of the magnitude of interest is far enough from the heat source such that the exact shape of the heat source is secondary. Other prediction, such as x_{f} in regimes I-IV have a location often within the heat source distribution, and are thus of less practical value. The prediction of width under a gaussian heat source involves subtleties not discussed, such as the presence of two local maxima under some conditions in the width of an isotherm under a gaussian heat source [20].

When the conditions correspond to an intermediate regime, the technique of blending is useful, but it is not fully developed yet. When all but one dimensionless groups are in an asymptotic regime (1D blending), the effect of the single remaining parameter can be captured accurately with the blending techniques described in [5], some progress has been accomplished for the case in which all but two dimensionless groups are asymptotic (2D blending). No formal approach has been developed for more than two dimensionless groups.

Perhaps the most important aspect to be developed is a practical prediction of interpass temperature, especially for additive manufacturing. While the equations presented are always applicable, the variations in interpass temperature during additive manufacturing are very large and have a noticeable effect that should be anticipated, for the equations to have their best applicability.

10. Conclusions

Thermal features in welding and additive manufacturing can be predicted reliably with practical formulas listed in tables 2 to 4. The magnitudes predicted include maximum width and its location, maximum penetration, thickness of HAZ, maximum temperature and its location, leading and trailing edge of an isotherm, heating and cooling rate, aspect ratio of an isotherm, melting efficiency, cooling time from 800°C to 500°C, solidification time, and maximum distribution of a heat source to reach a target temperature. The expressions proposed can be extended into sophisticated geometries for welding and specific additive manufacturing cases as illustrated in Figure 5.

Parameters involved include heat source power and speed, thermal conductivity and diffusivity of the substrate material, temperature of interest and preheat or interpass temperature. Temperature-dependent properties are accounted for by the use of effective properties detailed in equations 15 to 17.

For additive manufacturing, the formulas are of immediate application; however, a proper estimation of interpass temperature is needed, and is the focus of current research.

Acknowledgements

The many formulas presented were developed over many years with the enormous effort talented and hard-working graduate students, in particular Dr. Lu Yi, Dr. Wang Ying, and Dr. Gentry Wood. Special thanks to Dr. Tom Lienert for suggesting and supplying information for the example of application.

References

[1] Wells A 1952 *Welding Journal* **31** 263s–267s
[2] Adams C M 1958 *Welding Journal* **37** 210–215
[3] Christensen N, Davies, V de L and Gjermundsen K 1965 *British Welding Journal* **12** 54–75
[4] Grong Ø 1994 *Metallurgical Modelling of Welding* 1st ed (Cambridge, Great Britain: Institute of Materials)
[5] Mendez P F 2020 Reduced order models for welding and solidification processes *15th International Conference on Modelling of Casting, Welding and Advanced Solidification Processes, MCWASP* vol 861 p 012003
[6] Rosenthal D 1935 Etude theorique du regime thermique pendant la soudure a l'arc *Comptes Rendus. (2eme Congres National des Sciences)* pp 1277–1292
[7] Rosenthal D 1941 *The Welding Journal* **20** 220–234
[8] Rosenthal D 1946 *Transactions of the A.S.M.E.* 849–866
[9] Rykalin N N 1951 *Calculation of Heat Flow in Welding* (Moscow: Mashgis)
[10] Cline H E and Anthony T R 1977 *Journal of Applied Physics* **48** 3895–3900
[11] Eagar T W and Tsai N S 1983 *Welding Journal* 346s–355s
[12] Myhr O R and Grong O 1990 *Acta Metallurgica Et Materialia* **38** 449–460
[13] Wilson H A 1904 *Proceedings of the Cambridge Philosophical Society* **12** 406–423
[14] Roberts O 1923 *Proceedings of the Royal Society of London. Series A* **104** 640–654
[15] Mendez P F, Wang Y and Lu Y 2018 *Journal of Heat Transfer* **140** 081301–9
[16] Wang Y, Lu Y and Mendez P F 2019 *International Journal of Heat and Mass Transfer* **135** 1118–1129
[17] Lu Y, Wang Y and Mendez P F 2020 *International Journal of Heat and Mass Transfer* **156** 119793, 1–14
[18] Lu Y and Mendez P F 2021 *International Journal of Heat and Mass Transfer* **166** 120671, 1–15
[19] Wang Y and Mendez P F 2022 *International Journal of Thermal Sciences* **172** 107334
[20] Lu Y 2021 *A Systematic Methodology to Develop Scaling Laws for Thermal Features of Temperature Field Induced by a Moving Heat Source* PhD University of Alberta
[21] Wang Y 2021 *Design Rules for Characteristics of Heat Flow in Welding on Thick Plates using Asymptotics and Blending* PhD University of Alberta
[22] Hess W F, Merrill L L, Nippes E F and Bunk A P 1943 *Welding Journal* **22** 377–422
[23] Inagaki M, Nakamura H and Okada A 1965 *Journal of the Japan Welding Society* **34** 1064–1075
[24] Graville B A 1973 *Welding Journal, Research Supplement* **52** 377s–385s
[25] Pinkerton A J and Li L 2004 *Journal of Physics D: Applied Physics* **37** 1885–1895
[26] Peyre P, Aubry P, Fabbro R, Neveu R and Longuet A 2008 *Journal of Physics D: Applied Physics* **41** 025403
[27] Hintze Cesaro A and Mendez P F 2021 *Welding Journal* **100** 338s–347s
[28] Borle S D, Le Gall I and Mendez P F 2015 *Welding Journal* **94** 1–7

[29] Inagaki M 1958 *Journal of the Japan Welding Society* **27** 716–722
[30] Kasuya T and Yurioka N 1993 *Welding Journal* **72** 107–115
[31] Mendez P F 2021 Thermal analysis of laser heat treatment Report PGS AP1-C02-R01-V03 PGS Technical Consulting
[32] Fuerschbach P W and Eisler G R 2002 Determination of Material Properties for Welding Models by Means of Arc Weld Experiments *6th Intl. Trends in Welding Research* (Pine Mountain, Georgia)
[33] Lincoln Electric 2000 *The Procedure Handbook of Arc Welding* 14th ed (Cleveland: The James F. Lincoln Arc Welding Foundation)
[34] Mendez P, Tello K and Lienert T J 2010 *Acta Materialia* **58** 6012–6026
[35] Kwon Y and Weckman D C 2008 *Science and Technology of Welding and Joining* **13** 539–549
[36] Grams M and Mendez P F 2022 *Journal of Applied Mechanics* **89** 021004
[37] Grams M R and Mendez P F 2021 *Journal of Applied Mechanics* **88** 011001, 1–10
[38] Grams M R and Mendez P F 2021 *Journal of Manufacturing Science and Engineering* **143** 121002–1, 121002–12
[39] Corless R M, Gonnet G H, Hare D E G, Jeffrey D J and Knuth D E 1996 *Advances in Computational Mathematics* **5** 329–359
[40] Mendez P F 2021 Design of procedures in welding and additive manufacturing of metals *AWS Annual Meeting*
[41] Lienert T J and Lippold J C 2003 *Science and Technology of Welding and Joining* **8** 1–9
[42] Gajapathi S S, Mitra S K and Mendez P F 2011 *International Journal of Heat and Mass Transfer* **54** 5545–5553
[43] Gajapathi S S, Mitra S K and Mendez P F 2012 *Science And Technology Of Welding And Joining* **17** 429–434
[44] Gajapathi S S, Mitra S K and Mendez P F 2012 *Science and Technology of Welding and Joining* **17** 435–440
[45] Havrylov D 2021 *Hydrodynamic regimes in autogenous fusion welding* PhD University of Alberta
[46] Rivas D and Ostrach S 1992 *International Journal of Heat and Mass Transfer* **35** 1469–1479

IOP Conf. Series: Materials Science and Engineering 1281 (2023) 012021 doi:10.1088/1757-899X/1281/1/012021

Appendix: Notation

Variable	Units	Description
$\mathcal{R}$	dimensionless	Aspect ratio of isotherm
c	$\mathrm{J\,kg^{-1}\,K^{-1}}$	Effective specific heat
d	m	Substrate thickness
h	$\mathrm{J\ kg^{-1}}$	Enthalpy
	$\mathrm{W\ m^{-2}K^{-1}}$	convection coefficient
k	$\mathrm{W\,m^{-1}\,K^{-1}}$	Effective thermal conductivity
K_0	dimensionless	Modified Bessel function of second kind and zero order
q	W	Power absorbed by substrate
Ro	dimensionless	Rosenthal number
Ry	dimensionless	Rykalin number
t	s	Time
t_{sl}	s	Solidification time at centerline
$t_{8/5}$	s	Cooling time from 800°C to 500°C
T	K	Temperature
T_0	K	Initial/preheat/interpass temperature
T_5	K	500°C
T_8	K	800°C
T_{c}	K	Temperature of interest
T_{HAZ}	K	Temperature of edge of HAZ
$T_{\max}$	K	Maximum temperature
$\dot{T}_{\mathrm{b}}$	$\mathrm{K\,s^{-1}}$	Centerline cooling rate
$\dot{T}_{\mathrm{f}}$	$\mathrm{K\,s^{-1}}$	Centerline heating rate
$dT_{\max}/dy$	$\mathrm{K\,m^{-1}}$	Gradient of maximum temperature
U	$\mathrm{m\,s^{-1}}$	Travel speed of moving heat source
W_0	dimensionless	Lambert function (upper branch)
x_{b}	m	Trailing length of isotherm
x_{f}	m	Leading length of isotherm
$x_{\max}$	m	Location of maximum isotherm width
$y_{\max}$	m	Maximum isotherm width
Δy_{HAZ}	m	Thickness of the affected zone
Greek symbols		
α	$\mathrm{m^2\,s^{-1}}$	Thermal diffusivity
γ	dimensionless	Euler-Mascheroni constant (0.5772...)
η_{m}	dimensionless	Melting efficiency
ρ	$\mathrm{kg\,m^{-3}}$	Density
σ	m	Standard distribution of gaussian heat source
ϕ	rad	Actual heat flow angle
Subscripts		
I, II, III, etc.		Corresponding to Regimes I, II, II, etc.
eff		Effective value
i		Intermediate temperature
m		Melting temperature
Decorations		
(symbol)*		Dimensionless value
$\overline{\text{(symbol)}}$		Average
$\dot{\text{(symbol)}}$		Time derivative

IOP Conf. Series: Materials Science and Engineering 1281 (2023) 012022 doi:10.1088/1757-899X/1281/1/012022

Modulating Meltpool Dynamics and Microstructure using Thermoelectric Magnetohydrodynamics in Additive Manufacturing

A Kao[1], C Tonry[1], P Soar[1], I Krastins[1,2], X Fan[3], PD Lee[3] and K Pericleous[1]

[1]Computational Science and Engineering Group, University of Greenwich, UK
[2]Institute of Physics, University of Latvia, Jelgavas iela 3, Riga, LV-1004, Latvia
[3]Department of Mechanical Engineering, UCL, UK

Email: a.kao@gre.ac.uk

Abstract. Meltpool modulation in Selective Laser Remelting Additive Manufacturing via an oscillating magnetic field generates Thermoelectric Magnetohydrodynamics (TEMHD) flow. Numerical predictions show that the resulting microstructure can be significantly altered. A multi-scale numerical model captures the meso-scale melt pool dynamics coupled to microscale solidification showing the microstructure evolution and solute redistribution. The results highlight the complex interaction of the various physical phenomena and also show the method's potential to disrupt the epitaxial growth defect. The model predictions are supported by preliminary experimental results that demonstrate the dependency of the melt pool depth on magnetic field orientation. The results highlight how a time-dependent field has the potential to provide an independent control mechanism to tailor microstructures.

1. Introduction

Additive Manufacturing (AM), also known as 3D printing of alloys is an emerging technology with tremendous potential in a wide range of sectors including energy, biomedical implants and aerospace [1-3]. However, unlike traditional casting methods with centuries of development, AM has yet to reach the same level of maturity and suffers from microstructural defects such as porosity and epitaxial growth [4-7]. The latter, leading to the formation of long grains and unwanted anisotropic properties. These microstructural features are related to the melt pool dynamics and heat transfer, which in turn are governed by complex physics such as radiation[8], evaporation[9] and intense convection from temperature dependent Marangoni flow[10]. The thermal Péclet number is generally much greater than unity in AM processes, so that the heat transfer and corresponding melt pool morphology are strongly dependent on the fluid dynamics in the melt pool. Controlling the melt pool dynamics is therefore critical to beneficially tailor the resulting microstructure and mitigate against defects. One method that is receiving increased attention is the application of an external magnetic field [11]. Magnetic fields are often used to introduce the electromagnetic damping force on liquid flow, however, in many AM and related welding processes, large thermal gradients are present and conducive to the formation of Thermoelectric (TE) currents [12]. These currents interact with the magnetic field generating a TE Lorentz force that drives fluid flow in the melt pool alongside other inherent fluid drivers such as buoyancy and Marangoni stress. The TE Lorentz force is dependent on the time-dependent orientation and magnitude of the magnetic field and therefore, the introduction of a magnetic field strategy in conjunction with the scanning strategy brings a truly independent mechanism to control AM processes and the resulting microstructure.

Previous work by the authors [13,14] showed this strong dependency of the orientation of the magnetic field on the melt pool depth and solute redistribution. In these previous works quasi-steady state solutions for the thermofluid problem were first solved and used as an input to predict microstructure evolution. Nie et al. [15] concluded that static magnetic field causes EM braking which then suppresses meltpool convection and allows the melt to redistribute between dendrites during remelting therefore reducing residual stresses. Yeung and Yan [16] demonstrated a successful meltpool size control by laser power tuning that is based on a data-driven predictive meltpool model. Mondal et al. [17] developed a hybrid framework that uses machine learning to model and optimize the meltpool geometry control. Wang and Yan [18] considered a free surface and induced magnetic field in their TEMHD model simulations and concluded that the induced magnetic field impact is negligible compared to an externally applied magnetic field. Gruzd et al. [19] presented a theoretical assessment of the effect of an alternating EM field on meltpool convection for the purpose of stainless-steel laser annealing. They concluded that the TE effect is much smaller for stainless steel than it is for aluminium alloys. Du et al. [20] studied the effects of an external magnetic field in laser powder bed fusion and reaffirmed the positive impact on material physical properties and the importance of TE Lorentz force on the meltpool flow dynamics.

In this paper we further expand on this work by coupling time dependent solutions between the mesoscopic thermofluid problem and microstructure evolution, to investigate the influence of a time-dependent magnetic field on the modulation of the melt pool.

2. Methodology

A numerical model representing the Selective Laser Melting (SLM) AM process has been developed. The model uses a multi-scale approach where on the mesoscale an Enthalpy based approach captures solidification and heat transfer, from this a staggered finite difference method resolves the thermoelectric currents, which are then used to calculate the thermoelectric force that appears as a source term in the Lattice Boltzmann method to capture TEMHD flow. Time dependent solutions of the temperature and velocity are then coupled to a microscale model that uses a Cellular Automata method based on the code µMatIC [21-23], to resolve solute redistribution and microstructure evolution. The mesoscale solutions have a resolution of 5 µm, while the microscale is resolved at 1 µm. A more detailed description of the method can be found in Kao *et al.* [13], however a summary of the governing equations begins with modified form of Ohm's law accounting for TE currents

$$\mathbf{J} = \sigma(\mathbf{E} - S\nabla T + \mathbf{u} \times \mathbf{B}), \tag{1}$$

where $\mathbf{J}$ is the current density, σ is the electrical conductivity, S is the Seebeck coefficient, T is the temperature, $\mathbf{u}$ the fluid velocity and $\mathbf{B}$ the magnetic field. The Lorentz force, $\mathbf{F} = \mathbf{J} \times \mathbf{B}$ then appears as a source term in the Navier-Stokes equation governing fluid flow

$$\rho\left(\frac{\partial \mathbf{u}}{\partial t} + \mathbf{u} \cdot \nabla \mathbf{u}\right) = -\nabla p + \mu \nabla^2 \mathbf{u} + \mathbf{J} \times \mathbf{B}, \tag{2}$$

where ρ is the density, t is time, p is pressure and μ is the dynamic viscosity. The solutions to (1) and (2) also satisfy continuity of mass and charge, $\nabla \cdot \mathbf{u} = \nabla \cdot \mathbf{J} = 0$. In this work the magnetic field is oriented in the $\hat{y}$-direction, transverse to the scan, $\hat{x}$-direction, and build, $\hat{z}$-direction, and varies sinusoidally in time as $B_y = B_0 \sin(2\pi t f)\, \hat{y}$, where B_0 is the amplitude of the magnetic field and f is the frequency of its oscillation. Velocity is used to calculate time-dependent convection for both temperature, T, and concentration, C. The heat transfer is given by

$$\frac{\partial T}{\partial t} + \mathbf{u} \cdot \nabla T = \nabla \cdot \alpha \nabla T, \tag{3}$$

IOP Conf. Series: Materials Science and Engineering 1281 (2023) 012022 doi:10.1088/1757-899X/1281/1/012022

where α is the thermal diffusivity. In the mesoscale Enthalpy model the latent heat of fusion L_f is related to the temperature using the volumetric enthalpy H by

$$H = (1 - \phi_s)L_f + c_p T, \tag{4}$$

where ϕ_s is the solid fraction and $c_p$the specific heat capacity. For the CA method the solid fraction concentration of solute in the solid, C_s, and liquid, C_l are governed by $C_s = kC_l$, where k is the partition coefficient of the solute. The concept of solute trapping is used to account for both the physical non-equilibrium effect, but also a numerical effect where the computational resolution is much coarser than the interdendritic spacing. Each cell is considered to be the average of both high concentration eutectic structures and dendrites. The Continuous Growth Model [24] corrects for this by proposing a modification of k given by

$$k = \frac{k_e + V/V_D}{1 + V/V_D}, \tag{5}$$

where V is the growth velocity and $V_D$the 'diffusion' velocity. In this case $V = u_s$is the scan velocity and $V_D \approx D_l/L$ where L is a characteristic length, typically the interface thickness, but to account for numerical trapping assumed much larger, close to the cell size, giving a value of k = 0.95 [13]. At the solid-liquid interface the solid fraction rate of change is calculated by

$$C_l(1 - k)\frac{\partial \phi_s}{\partial t} = -\nabla \cdot (D_e \nabla C_l) + [1 - (1 - k)\phi_s]\frac{\partial C_l}{dt}, \tag{6}$$

where $D_e = (1 - \phi_s)D_l + \phi_s D_s$is the equivalent diffusion coefficient and $T_i = m_l(C_l - C_0)$ is the interface equilibrium temperature with m_l the liquidus slope and $C_0$the reference concentration. The equivalent $C_e = (1 - \phi_s)C_l + \phi_s C_s$

$$\frac{\partial C_e}{\partial t} + \mathbf{u} \cdot \nabla C_l = \nabla \cdot D_e \nabla C_l, \tag{7}$$

with zero mass flux prescribed at the boundaries.

The Navier-Stokes equation (2) are solved using the Lattice Boltzmann Method (LBM) [14, 25,26]. The other equations (1), (3-4) and (6-7), are solved using finite difference method schemes [27, 28]. The equations are solved throughout the whole domain.

The multiple pass layers of the AM process are captured by using offset periodic boundary conditions at each edge. A moving mesh approach is used with solidified material leaving the domain at the back and entering the layer below. A new powder layer is then introduced at the front of the first layer. A schematic of this can be seen in figure 1. The computational domain is cuboid with dimensions 800 × 600 × 300 μm and the incoming powder layer has a height of 30 μm. With a scan speed of 0.1 m/s and a domain length of 800 μm incoming material will be re-introduced into the layer below after 8 ms and

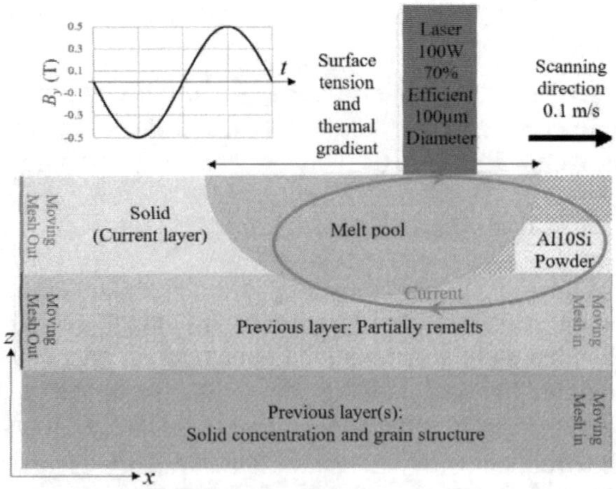

Figure 1. Schematic of the AM process, highlighting the TE current circulation. the moving mesh and the oscillating B_y field.

corresponds to a characteristic magnetic field frequency of ~100 Hz, assuming an oscillation occurs within a single layer. While this may be impractical experimentally and the induced currents from the time dependent field are neglected in the model, it is necessary to see the influence on successive layers without modelling a domain that is 100 to 1000 times larger. Newton's second law with a predicted Lorentz force as high as 10^8 N/m^3, gives ~10 μs as an estimate of the characteristic time for acceleration, several orders of magnitude smaller than the time scale for material to be re-introduced into the domain.

3. Results

In our previous work [13,14], numerical models predicted that application of a magnetic field aligned transverse to the build and scan directions, would interact with the dominant current density, in the scan direction, to generate a Lorentz force in the build direction i.e. $\mathbf{J} \times \mathbf{B} \sim J_x B_y \hat{z}$. This force was greatest near the laser spot and depending on the sign of the magnetic field it would either be directed upwards towards the surface of the melt pool reinforcing Marangoni flow or downwards competing against it. The resultant melt pool morphology then became either shallower (and wider) or deeper (and narrower).

Recent experiments conducted on the related Directed Energy Deposition (DED) process have shown strong evidence of the predicted mechanisms. Figure 2 shows bi-directional scanning for 5 layers of AlSi10 in the DED processes with a laser power of 200 W and a scan speed of 1 mm/s. Further details of the underlying experimental set up are reported by Chen *et. al* [29]. In this case the magnetic field is fixed, however the bi-directional scanning (left to right followed by right to left etc.) shares similarities to changing the sign of the magnetic field per layer. By alternating the scan direction, the thermoelectric

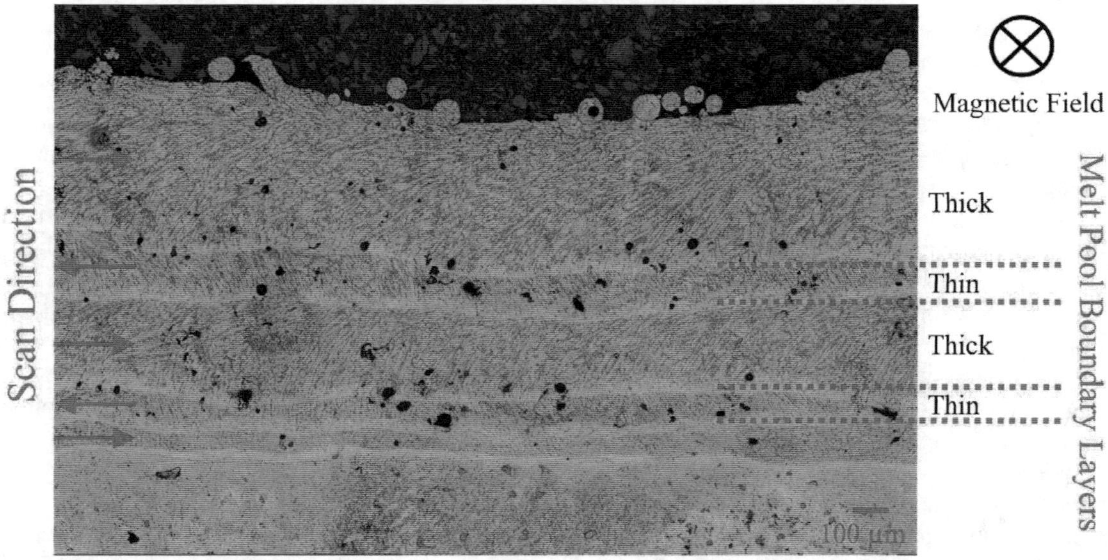

Figure 2. Experimental results demonstrating the effect of bidirectional scanning in the presence of a transverse magnetic field.

currents change sign, hence the Lorentz force changes sign. The melt pool boundary layers are clearly visible and indicated in the figure by regions of relatively parallel lighter lines and provide an indication of the depth of the melt pool. Except for the first layer that is printing onto the substrate, the experiment shows how the distance between successive melt pool boundary layers alternates between thick – thin – thick etc., giving the same predicted behaviour as if the melt pool was alternating between deep – shallow etc. for each layer.

The numerical modelling presented in this work uses an oscillating magnetic field which causes an oscillating force and modulation of TEMHD in the melt pool. Figure 3 shows the temperature field and melt pool morphology at different phases of the magnetic field. Figure 3a shows when the magnetic field is at a minimum, $B_y = -0.5$ T and the force is negative giving a deep meltpool, figure 3b is when $B_y = 0$ T and figure 3c is when $B_y = +0.5$ T and at a maximum with a shallow meltpool.

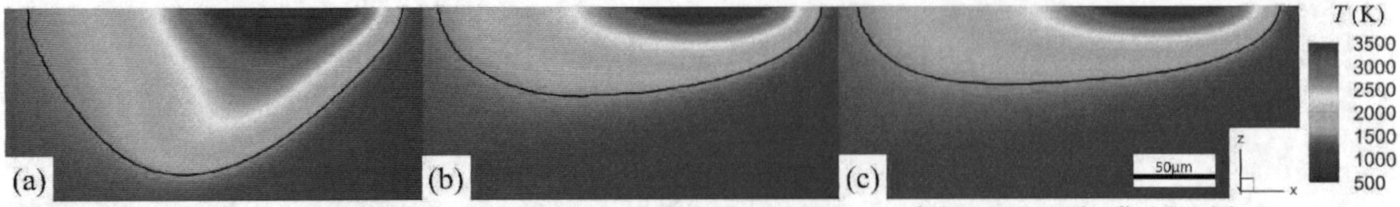

Figure 3. Melt pool temperature cross sections for various phases of the magnetic field with B_y intensities of a) 0.5 T, b) 0 T and c) - 0.5 T.

Figures 4, 5 and 6 show results for an oscillating magnetic field with a frequency matching the period of two layers, 62.5 Hz. Each figure shows the concentration and grain structure in different planes. Figure 4 shows the scan-build plane (x-z), figure 5 the scan-transverse plane (x-y) and figure 6 the transverse-build plane (y-z). The location of the slices taken for figures 5 and 6 are indicated in figure 4. In this case the region where the magnetic field is positive and negative overlap with the shallow positive magnetic field being remelted by the deeper negative magnetic field regions. Similar to the results in our previous work [13], in Figure 4 the low concentration melt pool boundary layer is evident showing the history over successive layers. The layers are uneven in thickness, since when the melt pool is deep it remelts through previous layers, while at the edges of the figure the magnetic field is tending toward zero leaving a shallower and hence thinner region between the melt pool boundary layers. The solute concentration in the vicinity of the melt pool boundary layer increases as the magnetic field tends to zero, while this effect is not seen in this region where the magnetic field is applied. This variation in concentration is due to TEMHD flow competing with Marangoni flow, at the edges, where the magnetic field is low, Marangoni flow dominates sweeping the flow towards the centre of the melt pool and hence the high concentration regions. In the centre of figure 4, TEMHD flow dominates and flow is pushed to the bottom of the melt pool and swept to the sides. This effect can be seen in the melt pool in figure 6a, but also in figure 5a, where high concentration regions in the solidified structure can been seen off the centre line, an indication that the solute was captured at the sides of the melt pool. Figure 5a also shows three meltpool boundary layers, starting from the centre of the figure and moving outwards these are: an intersection with the bottom of the deep melt pool of the topmost layer, the second is the boundary from previous time the melt pool was deep and thinner (two layers before), while the furthest is from the previous layer when the magnetic field causes a shallow wide melt pool.

Figure 4b shows long epitaxial grains can be seen in the regions where the magnetic field is applied, central to the figure, while at the edges there is a 'wavey' like structure. Grains compete and typically those whose orientation aligns with the thermal field will win this competition. However, as the thermal field is also oscillating the preferred orientation changes with time giving rise to complex competition that may be able to prevent epitaxial growth in certain regions. This disruption to epitaxial growth is also evident in the scan-transverse section given in figure 5a and 5b, where due to remelting through successive layers the thermal field and hence gradient during re-solidification is very different and favours different grain orientation. There is a clear disruption to epitaxial growth, in the region of the superimposed 'oval' in figure 5b, that matches the meltpool boundary layer of figure 5a. In the region where the magnetic field is applied, Figure 4a, shows long grains formed from epitaxial growth. However, figure 6b shows that these grains are quite 'thin' and are a consequence of scanning in the same location between each layer.

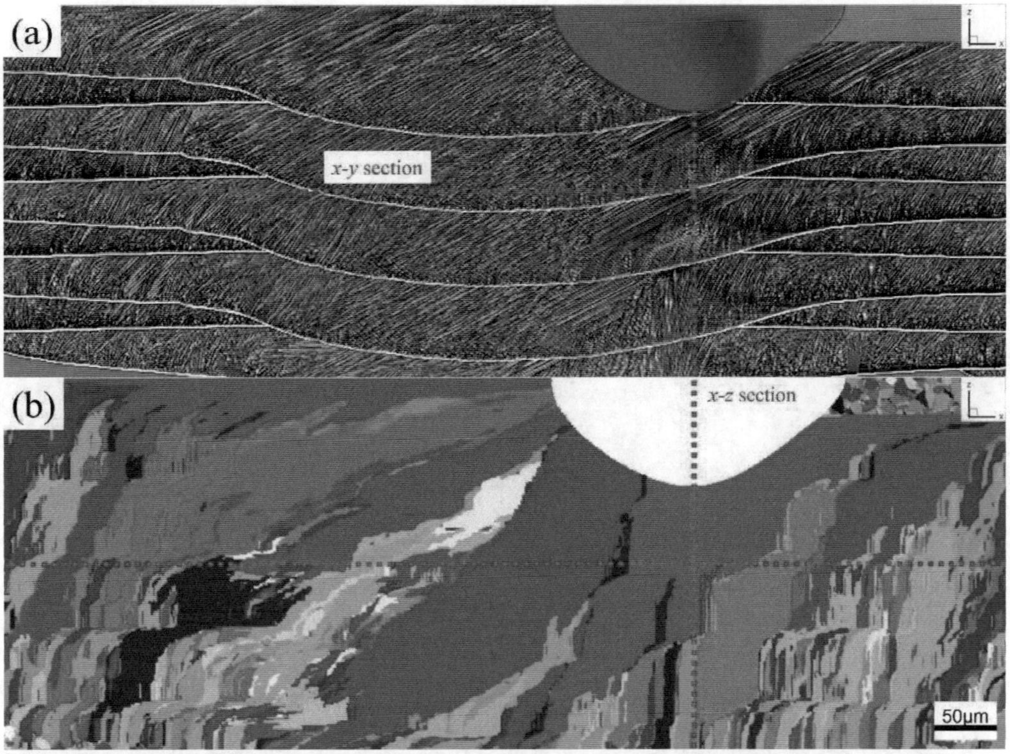

Figure 4. Concentration – dark/light regions are solute enriched/depleted respectively (a) and Grain Structure (b) for the scan-build plane (x-z) for a B_y field frequency 62.5 Hz.

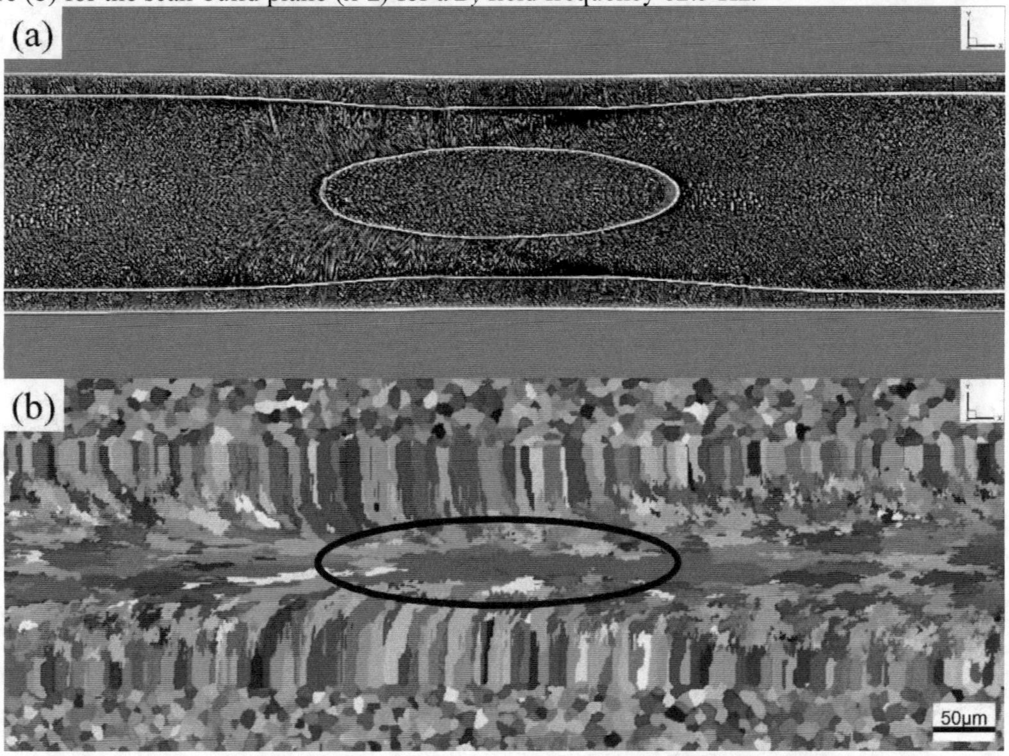

Figure 5. Concentration – dark/light regions are solute enriched/depleted respectively (a) and Grain Structure (b) for the scan-transverse plane (x-y) for a B_y field frequency 62.5 Hz.

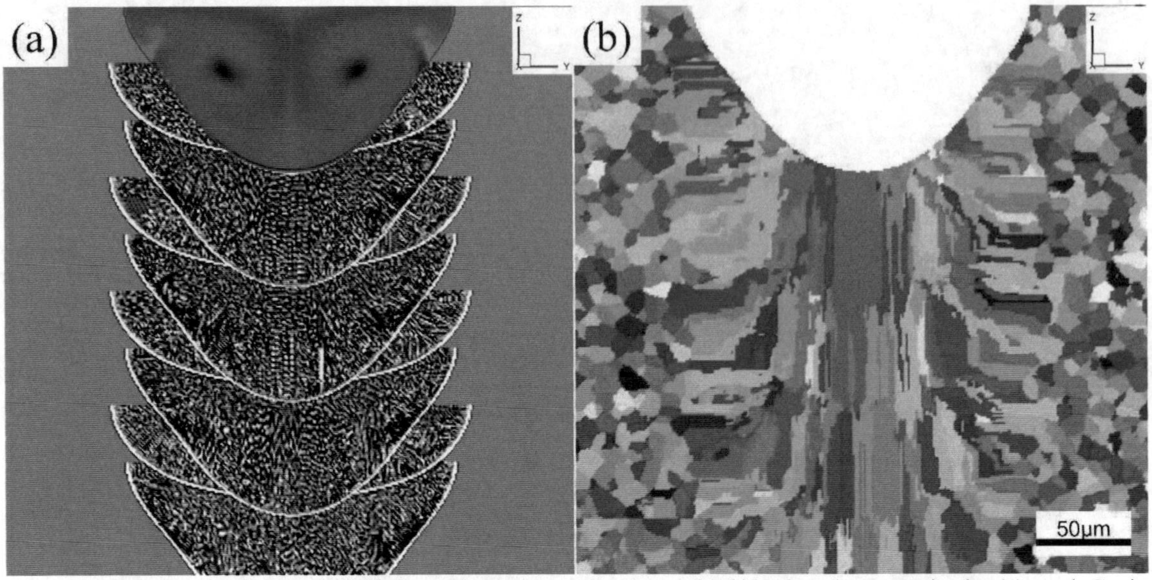

Figure 6. Concentration – dark/light regions are solute enriched/depleted respectively (a) and Grain Structure (b) for the transverse-build plane (y-z) for a B_y field frequency 62.5 Hz.

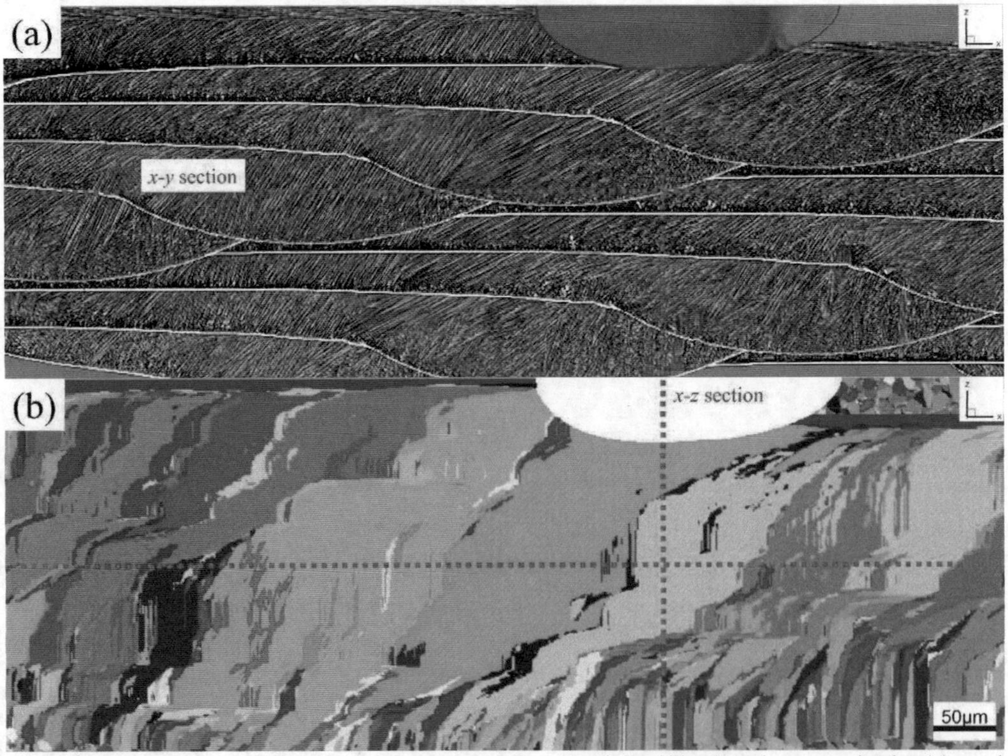

Figure 7. Concentration – dark/light regions are solute enriched/depleted respectively (a) and Grain Structure (b) for the scan-build plane (x-z) for a B_y field frequency 100 Hz

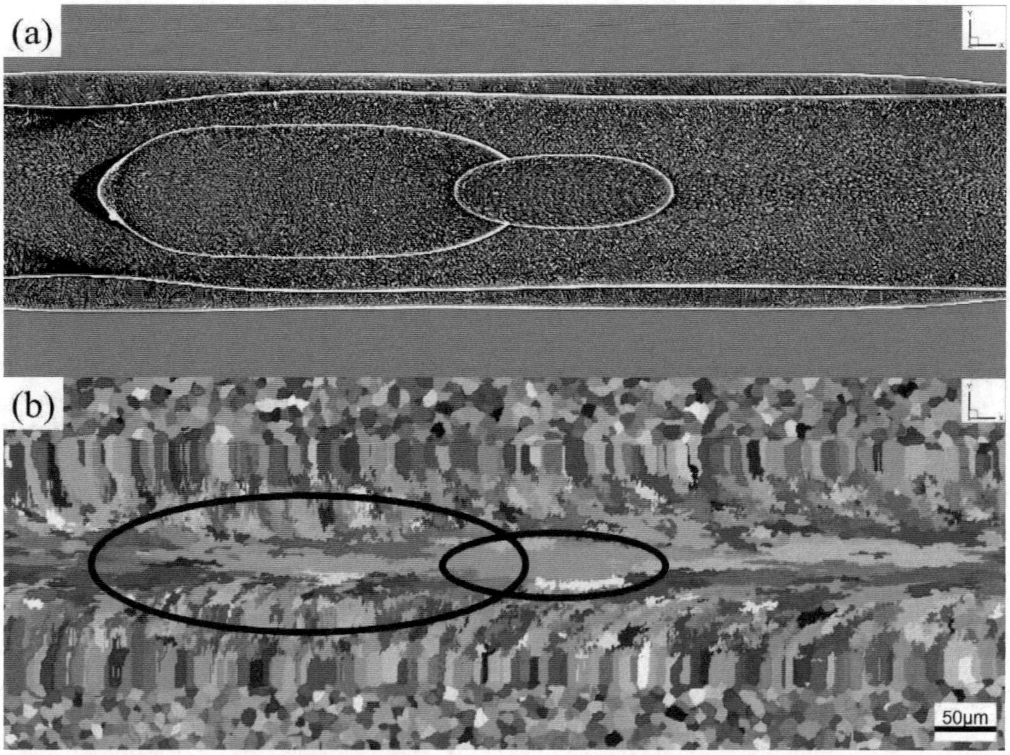

Figure 8. Concentration – dark/light regions are solute enriched/depleted respectively (a) and Grain Structure (b) for the scan-transverse plane (x-y) for a B_y field frequency 100 Hz

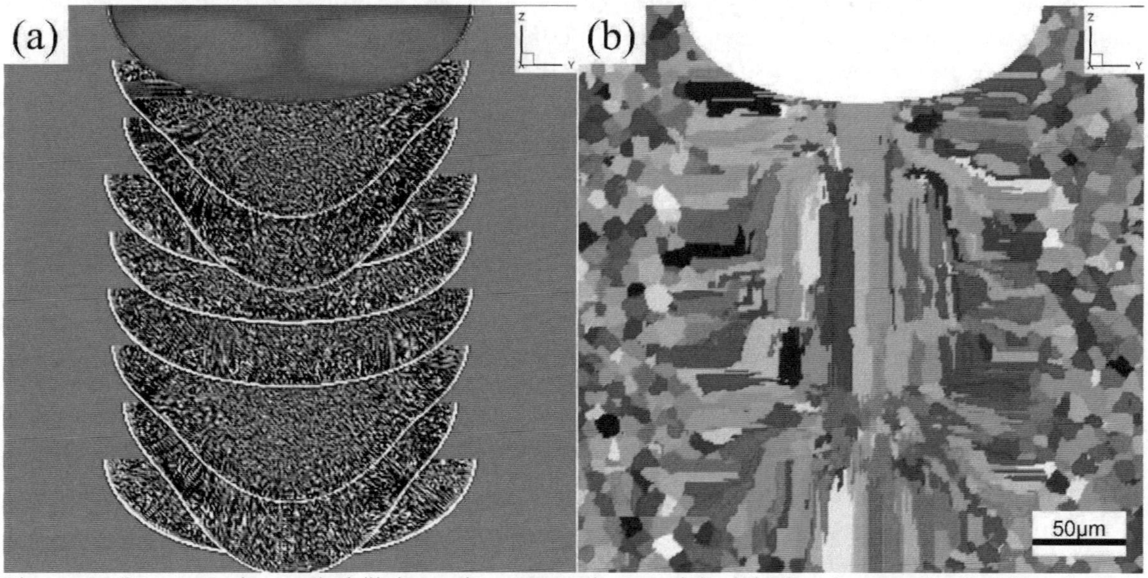

Figure 9. Concentration – dark/light regions are solute enriched/depleted respectively (a) and Grain Structure (b) for the transverse-build plane (y-z) for a B_y field frequency 100 Hz

The second case shown is with a high frequency of 100 Hz such that the positive and negative magnetic fields are not in the same relative location. Figures 7,8 and 9 are relevant to this case where again each figure shows the concentration and grain structure in different planes. Figures 7, 8 and 9 shows the scan-build plane (x-z), the scan-transverse plane (x-y) and the transverse-build plane (y-z), respectively and the locations are again indicated in figure 7. Due to the shift in position where the deep meltpool acts, figure 7a shows how the meltpool boundary layers take a much more complex structure, where intersections due to remelting of previous layers leads to a much more disparate distance between melt pool boundary layers. Regions of high concentration are more distributed throughout the domain due to the varying position where Marangoni flow dominates and solute is drawn to the centre of the melt pool, highlighted in the solute redistribution in the melt pool of figure 9a. On the grain structure, figure 7b, shows how the epitaxial growth appears to follow the shift in position of regions of high magnetic field, however looking in the other planes there are additional mechanisms for disruption of epitaxial growth. An example is in the scan-transverse section shown in figure 8, where the slice cuts through the bottom of two successive deep melt pools, indicated by the melt pool boundary layers in figure 8a and the two superimposed ovals in figure 8b. In a similar way to the previous case, the thermal field is contrasting during remelting and re-solidification, leading to a complex grain competition and the disruption of epitaxial growth. Another example is in the transverse-build plane, shown in figure 10b, where in the lower portion of the figure there is a region of small grains dissecting the longer grains. This is again due to the difference in thermal gradient during solidification. When the thermal gradient is more vertical, the preferred orientation would also be vertical and likewise when the thermal gradient is horizontal the preferred orientation would be horizontal. This change in grain morphology is essentially where epitaxial growth is disrupted, oriented more vertically to those further up that are more horizontal. This case highlights how a more complex and interesting behaviour of the microstructure evolution can be achieved by modulating the magnetic field and hence the thermofluid problem.

4. Conclusions

Experimental results in the related DED system have been shown to verify early predictions of how melt pool depth can be changed with a magnetic field. Further numerical predictions in this work have then gone to investigate the effect of a low frequency oscillating transverse magnetic field on the melt pool dynamics and grain morphology of SLM AM aluminium. The time-dependent magnetic field interacts with the TE currents in the melt pool leading to a modulation of TEMHD and ultimately modulation of the meltpool morphology through convective transport. Due to the synchronised spatial and temporal nature of the cases presented, the time-dependent melt pool morphology leads to spatial variations in the layer thickness and solute redistribution, as TEMHD competes with or enhances Marangoni flow. A host of other microstructure effects such as complex grain competition due to the varying thermal gradient at solidification leads to regions of enhanced or disrupted epitaxial growth. By simply changing the frequency of the magnetic field, so that the intensity is not in the same relative place at each layer, created an even more complex final microstructure, that warrants further analysis. This study also demonstrates that the parameter space for the magnetic field strategy is large and therefore could have a huge potential in tailoring microstructures by optimising a combined magnetic field and scan strategy.

Acknowledgements

The authors acknowledge financial support from the UK Engineering and Physical Sciences Research Council (EPSRC) through grants EP/W032147/1 and EP/W031167/1.

References

[1] Singh R, Lee P D, Dashwood R J and Lindley T C 2010. *Mater. Technol.* **25** 127–36
[2] Ibrahim K A, Wu B and Brandon N P 2016 *Mat. Des.* **106** 51–9
[3] Murr L E 2016 J. *Mat. Sci. Tech.* **32** 987–95
[4] Zhang B, Li Y and Bai Q 2017 *Chinese J. Mech. Eng.* **30** 515–27
[5] Leung C L A, Marussi S, Atwood R C, Towrie M, Withers P J, Lee P D 2018 *Nat. Comm.* **9** 1355

[6] Li R, Shi Y, Wang Z, Wang L, Liu J and Jiang W 2010 *Appl. Surf. Sci.* **256** 4350–6
[7] Shifeng W, Shuai L, Qingsong W, Yan C, Sheng Z and Yusheng S 2014 *J. Mater. Process. Technol.* **214** 2660–7
[8] Gusarov A, Yadroitsev I, Bertrand P and Smurov I 2009 *J. Heat Transfer* **131** 072101
[9] Geiger M, Leitz K H, Koch H and Otto A 2009 *Prod. Eng.* **3** 127–36
[10] Yuan P and Gu D 2015 *J. Phys. D: Appl. Phys.* **48** 035303
[11] Du D, Haley J C, Dong A, Fautrelle Y, Shu D, Zhu G, Li X, Sun B and Lavernia E J 2019 *Mat. Des.* **181** 107923
[12] Kern M, Berger P, Hugel H. 2000 Magneto-fluid dynamic control of seam quality in CO2 laserbeam welding. Weld. J. Res. Suppl. 79, 72s.
[13] Kao A, Gan T, Tonry C, Krastins I and Pericleous K 2020 *Philos. Trans. Royal Soc. A* **378** 2171
[14] Kao A, Gan T, Tonry C, Krastins I and Pericleous K 2020 IOP Conf. Series: Mat. Sci. and Eng. **861** 012009
[15] Nie J, Chen C, Shuai S, Lui X, Zhao R, Wang J, Liao H and Ren Z 2020 *J. Therm. Spray Technol.* **29** 1410–1423
[16] Yeung H, Yang Z and Yan L 2020 *Addit. Manuf.* **35** 101383
[17] Mondal S, Gwynn D, Ray A and Basak A 2020 *Metals* **10(5)** 683
[18] Wang L and Yan W 2021 *Phys. Rev. Applied* **15** 064051
[19] Gruzd S, Lomaev S, Simakov N, Gordeev G, Bychkov A, Gapeev A, Cherepetskaya E, Krivilyov M, Ivanov I 2022 *Metals* **12(9)** 1540
[20] Du D, Wang L, Dong A, Yan W, Zhu G, Sun B 2022 *Int. J. Mach. Tools Manuf.* **183** 103965
[21] Wang W, Lee P and McLean M 2003 *Acta Mater.* **51** 2971–87
[22] Yuan L and Lee P D 2010 *Modell. Simul. Mater. Sci. Eng.* **18** 055008
[23] Yuan L and Lee P D 2012 *Acta Mater.* **60** 4917–26
[24] Aziz M J and Kaplan T 1988 *Acta Metall.* **36** 2335–47
[25] Kao A, Krastins I, Alexandrakis M, Shevchenko N, Eckert S, Pericleous K 2018 *JOM* **71** 48–58.
[26] Kao A, Shevchenko N, Alexandrakis M, Krastins I, Eckert S and Pericleous K 2019 *Philos. Trans. Royal Soc. A* **377** 20180206
[27] Kao A, Cai B, Lee P and Pericleous K 2017 *J. Cryst. Growth* **457** 270–4
[28] Voller V R, Cross M and Markatos N C 1987 *Int. J. Numer. Methods Eng.* **24** 271–84
[29] Chen Y, Clark S J, Sinclair L, Leung CLA, Marussi S, Connolley T, Atwood RC, Baxter GJ, Jones MA, Todd I and Lee PD 2021 Addit. Manuf. 41 101969.

IOP Conf. Series: Materials Science and Engineering 1281 (2023) 012023 doi:10.1088/1757-899X/1281/1/012023

Arc length behaviour and voltage predictions for GMAW-Sp aluminum procedures in the Lincoln Procedure Hand Book

J. Eduardo Alvarez Rocha, and Patricio F. Mendez

Canadian Centre for Welding and Joining, Chemical and Materials Engineering, University of Alberta, Donadeo ICE 12-332, 9211 116st, Edmonton, AB T6G 1H9, Canada

E-mail: jealvare@ualberta.ca, pmendez@ualberta.ca

Abstract. Heat input is a key component of a welding procedure, which is dependent on the several fall voltages composing a total voltage loss. As current practice is primarily reliant on trial and error to determine voltage settings for a desired heat input, a means to understand and predict the voltage loss is of interest to welding engineers. Voltage, amperage, and arc length measurements using ER4043 1.2 mm at varying voltages were used to break down fall voltage constituents for a given weld with GMAW-Sp. Arc length was defined as the distance from the weld pool to the point where the metal vapour and ionized gas boundary attach to the consumable, and measured over 5 droplet cycles to obtain a time average. Aluminum procedures in the Lincoln Procedure Hand Book (LPHB) with 1.2, 1.6, and 2.4 mm consumables were analyzed to predict the individual fall voltage constituents using experimental results. Expected arc lengths ranged from 13-26 mm depending on voltage prescribed in the procedure, and agreed with comparative experiments. Combined anode/cathode, and the arc column were major contributors to overall fall voltage with 64±3%, and 34±3%, respectively. Contact tip, electrode extension, and lead cables were minor contributors, each contributing less than 1.2% to the overall voltage loss.

1 Introduction

Fall voltages are important characteristics that researchers and engineers seek to understand for applications including power source design, and welding output prediction (e.g. heat input and bead width). Nearly 80 years after gas metal arc welding (GMAW) was introduced [1], there is limited understanding of how the voltage set at the machine is isolated to distinct fall voltages that make up the total voltage loss (V_{Tot}) in the GMAW system as shown in figure 1. This has led to a costly trial and error approach to achieve desired settings for a successful weld. In a typical welding system, the anode and cathode fall voltages (V_{an} and V_{cath}) are the main contributors for heat input [2]. However, all fall voltages including the contact tip (V_{CT}), electrode extension (V_{EE}), arc column (V_{AC}), and welding leads (V_{WL}) must be accounted for to understand heat input and the required voltage set at the machine to generate a successful weld as defined by:

$$V_{\mathrm{Tot}} = V_{\mathrm{cath}} + V_{\mathrm{an}} + V_{\mathrm{CT}} + V_{\mathrm{EE}} + V_{\mathrm{AC}} + V_{\mathrm{WL}} \tag{1}$$

IOP Conf. Series: Materials Science and Engineering 1281 (2023) 012023 doi:10.1088/1757-899X/1281/1/012023

In particular, previous work has posed important questions around the definition of arc length (L_{AC}) [3,4], which is required to determine V_{AC}. Understanding the relationship between voltage and L_{AC} is paramount to controlling the GMAW process. The self-regulating arc in GMAW described by Phillips [5] allows for the quantification of effects to L_{AC} from changing variables such as V_{Tot}, welding current (I_{w}), and contact tip to work piece distance (L_{CTWD}).

The purpose of this paper is to: understand the relationship between V_{M} (measured voltage excluding V_{WL}) vs. L_{AC}, and L_{AC} vs. L_{CTWD}; determine a measurement for V_{F} (sum of V_{an} and V_{cath}); and apply experimental findings to predict fall voltage contributions for each fall voltage constituent from a set of aluminum procedures in the Lincoln Procedure Hand Book (LPHB) [6]. It is noted that the aluminum procedures in the LPHB are in direct current electrode positive (DCEP), which is consistent with the anode and cathode placement in figure 1. Per Lincoln Electric, globular transfer is not one of the recommended GMAW modes of metal transfer for aluminum. As such, this study excludes those procedures below the transition current listed by Lincoln Electric, which are deemed to be in globular transfer mode [1].

This paper covers background on fall voltages in section 2, experimental set up in section 3, equations used for experimental and prediction analysis in section 4, experimental and prediction results in section 5, discussion in section 6, and conclusion including outlook in section 7.

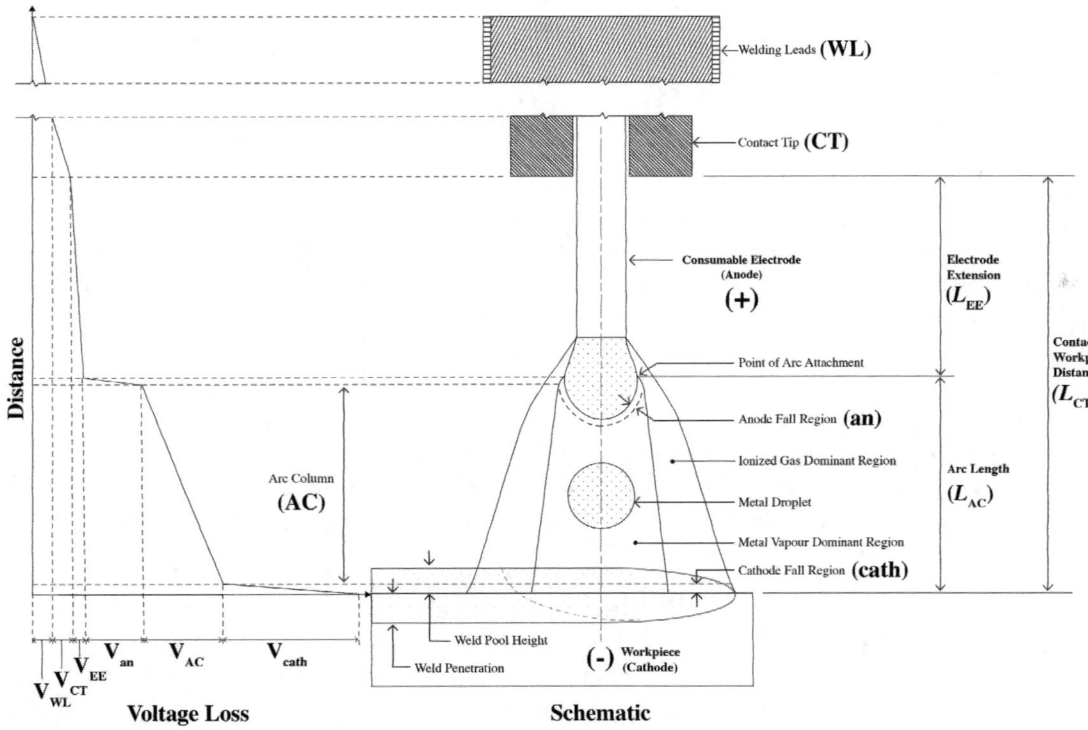

Figure 1. GMAW arc region schematic including: fall voltage (voltage loss) constituents for cathode (V_{cath}), anode (V_{an}), contact tip (V_{CT}), electrode extension (V_{EE}), arc column (V_{AC}), and welding leads (V_{WL}); metal vapour and ionized gas dominant regions; metal droplet; weld pool characteristics such as weld pool height and weld penetration; and associated dimensional references such as contact tip to work piece distance (L_{CTWD}), electrode extension (L_{EE}), and arc length (L_{AC}).

IOP Conf. Series: Materials Science and Engineering 1281 (2023) 012023 doi:10.1088/1757-899X/1281/1/012023

2 Background

2.1 Arc column

The correlation between welding voltage and L_{AC} is commonly used in the welding industry and often considered to be directly proportional [5]. Here, a distinction is made between a generic "voltage" and V_{AC}, which is exclusively proportional to the arc column electric field (V'_{AC}) in V mm^{-1}. To understand the respective V_{AC} for a given weld, the L_{AC} must also be known or anticipated for a given voltage setting (V_{Tot}).

Previous work has raised valid questions around a definition for L_{AC}. More specifically around where the arc attaches to the electrode. Egerland identified distinct points of attachment for a given weld, which were also dependent on mode of metal transfer [3]. Egerland suggested L_{AC} to be defined as the distance between the beginning of electrode tapering and the work piece surface, which also requires to account for either penetration or weld pool height as shown in figure 1. In a visualization study of the GMAW arc, Dos Santos et al. pointed out that the electric arc in GMAW assumes a conical shape and is divided into two distinct regions: 1) an outer cone composed of mostly ionized gas, and 2) an inner cone with a high concentration of metal vapour [4] as shown in figure 1. Lastly, Zhang et al. concluded a definition for L_{AC} by considering the length of the current path for the metal vapour dominant region [7]. Therefore, the point of arc attachment is defined and considered in this study to be the point where the boundary between the ionized gas and metal vapour dominant regions come in contact with the electrode consumable (figure 1).

In a study with aluminum (ER4043), McIntosh and Mendez determined V'_{AC} to be 0.77±0.19 V mm^{-1}. The value reported for V'_{AC} appeared to increase with current and accounted for ~32% of the total voltage loss [8].

2.2 Electrode extension (joule heating)

Voltage loss in the electrode extension is a function of the current running through the wire and the associated resistance along the electrode extension (L_{EE}). This V_{EE} will result in joule heating proportional to the L_{EE} [8]. Work by Lehnhoff and Mendez allows for calculation of the effective resistivity by considering the non-linearity in resistance along the L_{EE} [9]. For aluminum, V_{EE} is deemed a minor contributor to V_{Tot} with reported values of $< 0.5\%$ [8].

2.3 Contact tip

The contact between the welding consumable and the contact tip produces a resistance that is dependent on the contact force (the larger the force, the smaller the resistance). A constant resistance will contribute to stable metal transfer in GMAW [10]. Values for contact tip resistance (R_{CT}) have been reported in literature ranging from 1 to 2 mΩ and are considered to be minor contributors to the overall voltage loss in GMAW with less than 2% contribution [7, 8, 11].

2.4 Anode and cathode

The anode and cathode are recognized to be the main contributors to overall voltage loss in arc welding [5]. A study performed with aluminum reported a combined contribution of over 60% with values of 6.13±0.24 V (V_{an}) and 10.86±0.8 V (V_{cath}). Also, V_{an} and V_{cath} were found to be approximately independent of I_{w} [8].

3 Experimental set up

Voltage and amperage measurements were performed using the experimental set up depicted in figure 2 consisting of synchronized high-speed cinematography with data acquisition (DAQ), and welding equipment.

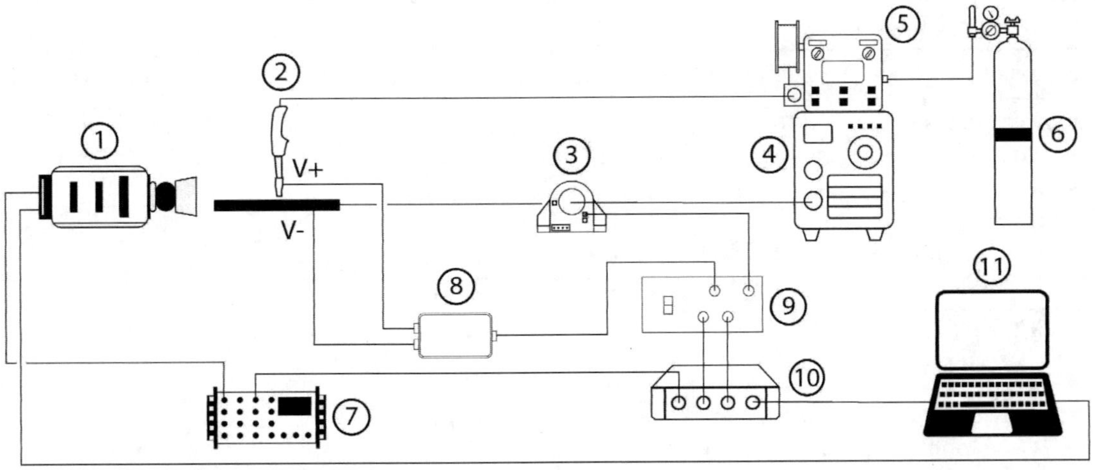

Figure 2. Experimental set up included: (1) high speed camera, (2) welding torch, (3) current transducer, (4) welding power supply, (5) wire feed unit, (6) gas cylinder, (7) signal synchronization terminal, (8) voltage transducer, (9) voltage and current transducer power supply, (10) DAQ, and (11) computer.

3.1 High-speed videography and data acquisition
A Phantom v210 camera was used to capture high speed videography and synchronized with an external DAQ system (National Instruments USB-6351) to collect video, current, and voltage data at 8.1 kHz over 0.5 s. The camera was placed perpendicular to the direction of travel and centered to capture the space between the gas cup and the bottom of the weld pool. An 800 nm narrow band pass filter with an aperture set at f/22, and an exposure time of 120 μs was used to view the metal vapour and ionized gas dominant regions shown in figure 1, which facilitated identifying the point of arc attachment to the consumable. Voltage sensor probes had electrical contact at the contact tip and ground terminal (placed within 150 mm of the weld zone) to exclude welding lead and ground cable voltage losses from the measurements.

3.2 Welding equipment
All welding was performed using a Lincoln Power Wave S500 power source and a Lincoln 84 Dual Power Feed unit set to program 5 for Gas Shield Standard Constant Voltage. A Lincoln Magnum PRO Al G450A K3355-2 gun was used with a 8 m #2 (35 mm) lead cable and a 3 m 4/0 grounding cable.

3.3 Experimental weld sets
Four sets of experiments with aluminum (AlS1, AlS2, AlS3, and AlS4) were performed (each set with up to 6 tests, denoted Tn, n=1,2,3,4,5,and 6). All welds were performed with direct current electrode positive (DCEP) in the 1G position with a 10° push angle. Constant welding parameters for all experimental sets are listed in table 1. Sets AlS1 and AlS3 varied L_{CTWD}, and sets AlS2 and AlS4 varied voltage setting for each weld trial. Table 2 lists the non-constant variables including wire feed speeds (U_{c}) used for each experimental set.

Table 1. Constant welding parameters for all experimental sets.

Description	Designation
Consumable Material	ER4043
Consumable Diameter	1.2 mm
Base Maerial	6061-T651
Shielding gas	Ar
Shielding gas flow	40 CFH*
Travel speed	55.1 cm min^{-1}

*Cubic Feet Per Hour

Table 2. Non-constant welding variables between all experimental sets.

Set ID	Fixed Variable	Variable Value	U_c (m min^{-1})	Changing Variable	Variable Range
AlS1	V_{Tot}	24.5 V	7.5	L_{CTWD}	12-16 mm
AlS2	L_{CTWD}	14 mm	7.5	V_{Tot}	20.0-25.0 V
AlS3	V_{Tot}	27.0 V	8.1	L_{CTWD}	16-20 mm
AlS4	L_{CTWD}	17 mm	8.1	V_{Tot}	20.0-26.0 V

4 Analysis

4.1 Arc column

Arc length measurements were performed on each weld using high-speed video analysis with the Phantom Camera Control (PCC) software. To measure the arc length, the datum and arc attachment locations were identified for each video as shown in figure 3. The datum was defined as the point were the droplet came into contact with the weld pool, and the attachment was identified as the point where the boundary between the metal vapour and ionized gas dominant regions contacted the consumable as shown in figure 1. Measurements were taken manually at each frame over 5 droplet transfer cycles to obtain a time averaged arc length.

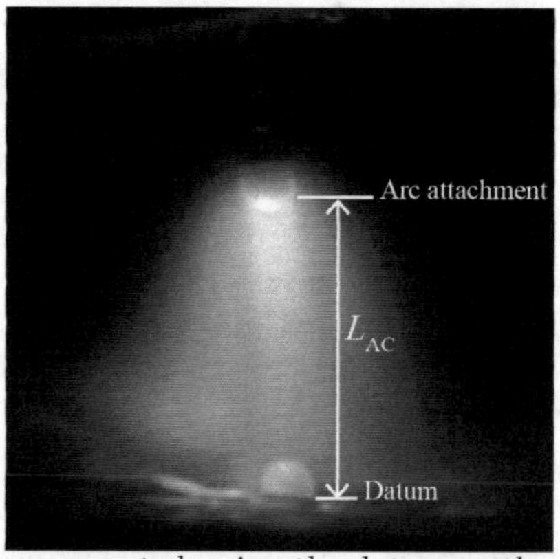

Figure 3. Typical arc length measurement showing the datum as the interface where the droplet comes into contact with the weld pool and the arc attachment as the point where the consumable contacts the boundary between the ionized gas and metal vapour dominant regions.

With a known or expected L_{AC}, the V_{AC} was found using:

$$V_{AC} = V'_{AC} L_{AC} \tag{2}$$

where V'_{AC} was identified by an empirical relationship between voltage measured (V_M) and L_{AC} for the varying voltage experiments with AlS2 and AlS4. The same relationship was used to determine an expected L_{AC} for the LPHB procedures based on the voltage settings prescribed.

4.2 Electrode extension

The electrode extension fall voltage was found using:

$$V_{\mathrm{EE}} = I_{\mathrm{w}} R_{\mathrm{EE}} = I_{\mathrm{w}} \frac{\rho_{\mathrm{c_{eff}}} L_{\mathrm{EE}}}{A_{\mathrm{c}}} \tag{3}$$

where the electrode extension resistance (R_{EE}) is composed of the effective resistivity of the consumable ($\rho_{\mathrm{c_{eff}}}$), L_{EE}, and the consumable's cross sectional area (A_{c}). A value for $\rho_{\mathrm{c_{eff}}}$ was calculated using expressions by Lehnhoff and Mendez [9], and is considered to be a constant value of 2.55x10^{-8} Ωm in this study as the negligent difference in $\rho_{\mathrm{c_{eff}}}$ over a 150 A range is on the order of magnitude of 10^{-9}. Values for I_{w} were measured from the experimental trials, and provided in the LPHB for the voltage prediction analysis.

The value for L_{EE} is determined by equation 4. The values for L_{CTWD} are not provided for aluminum procedures in the LPHB and were therefore based on the voltage settings to prevent burn back of the electrode into the contact tip.

$$L_{\mathrm{EE}} = L_{\mathrm{CTWD}} - L_{\mathrm{AC}} \tag{4}$$

4.3 Contact tip
The contact tip fall voltage was calculated using equation 5, where R_{CT} is the resistance between the contact tip and the consumable. The resistance from the contact tip was not measured as part of this study but is considered to be 1 mΩ per findings from Zhang et. al. [11].

$$V_{\mathrm{CT}} = I_{\mathrm{w}} R_{\mathrm{CT}} \tag{5}$$

4.4 Welding leads
For the LPHB voltage predictions, the welding leads fall votlage was calculated using:

$$V_{\mathrm{WL}} = I_{\mathrm{w}} R_{\mathrm{WL}} = I_{\mathrm{w}} \frac{\rho_{\mathrm{eff_{Cu}}} L_{\mathrm{WL}}}{A_{\mathrm{WL}}} \tag{6}$$

where R_{WL} is the resistance of the welding leads, $\rho_{\mathrm{eff_{Cu}}}$ is the resistivity of copper (considered to be 1.72x10^{-8} Ωm at 26°C [12]), L_{WL} is the length of the welding lead and ground cables (assumed to be 4 m), and A_{WL} is the cross sectional area of the welding leads. The values for A_{WL} were assigned based on what would be expected to be used for the range of parameters in the LPHB using literature from Lincoln Electric as guidance [13].

4.5 Anode and cathode fall voltage
For the experimental data, V_{F} was obtained using:

$$V_{\mathrm{F}} = V_{\mathrm{cath}} + V_{\mathrm{an}} = V_{\mathrm{M}} - (V_{\mathrm{AC}} + V_{\mathrm{CT}} + V_{\mathrm{EE}}) \tag{7}$$

The experimental results for V_{F} (discussed in section 5) were applied to predict the mid-point voltage settings (considered to be V_{Tot}) for the LPHB procedures using equation 1.

5 Results
5.1 Experimental arc length measurements
Time averaged L_{AC} measurements ranged from 1.6 to 12.2 mm and had a consistent standard deviation (σ) range of 0.2 to 0.5 mm between all weld sets. The consistent range of σ values shows that time averaged L_{AC} measurements are more precise than single point measurements. This is critical for welds with smaller L_{AC} since the arc becomes more erratic as the transfer mode approaches short circuit. It is noted that short circuits began to be intermittently identified for voltages ≤21.0 V. Sample images of AlS2T4 (V_{M}=23.0 V) show arc attachment point behaviour during one droplet cycle in figure 4. Respective L_{AC} measurements (average L_{AC}=7.7 mm) are shown in figure 5, where the arc attachment is shown to fluctuate approximately 1.1 mm throughout the droplet formation cycle until detachment.

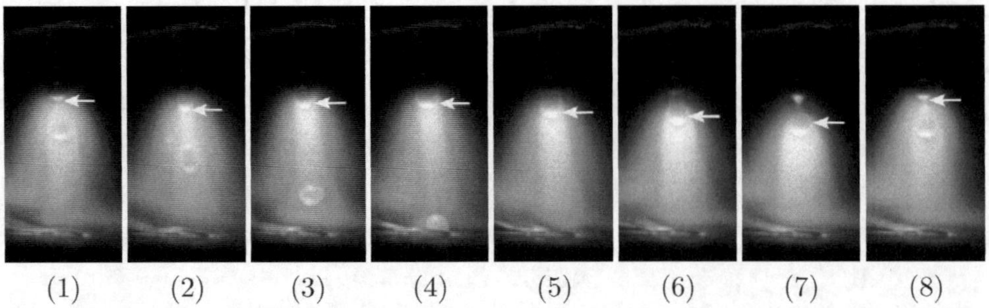

Figure 4. Images from AlS2T4 showing arc attachment point behaviour during a droplet cycle.

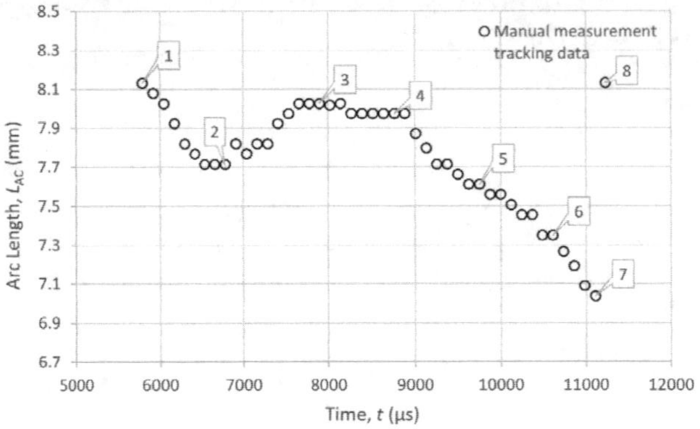

Figure 5. Sample of manually tracked measurements of 1 droplet cycle corresponding to images from figure 4 showing a fluctuation in arc length of approximately 1.1 mm for an average L_{AC} measurement of 7.7 mm for AlS2T4.

5.2 Experimental arc length behaviour

Arc length measurements allowed for a determination of V'_{AC} by plotting V_M vs. L_{AC} for AlS2 and AlS4, which consisted of varying voltage settings with a constant L_{CTWD} and U_c (refer to table 2). Considering V'_{AC} as the slope of the linear fit (least sum of squares) from figure 6, then $V'_{AC} = 0.62$ V mm^{-1}. From the limited data analyzed in figure 7, a consistent correlation of -0.07 mm A^{-1} was observed between L_{AC} and I_w. Varying L_{CTWD} was found to have a negligible effect on L_{AC} considering the operable range of L_{CTWD} (figure 8). Per the linear correlation expressions from figure 8, there is a 0.01 to 0.1 mm change in L_{AC} for every 1 mm change in L_{CTWD}.

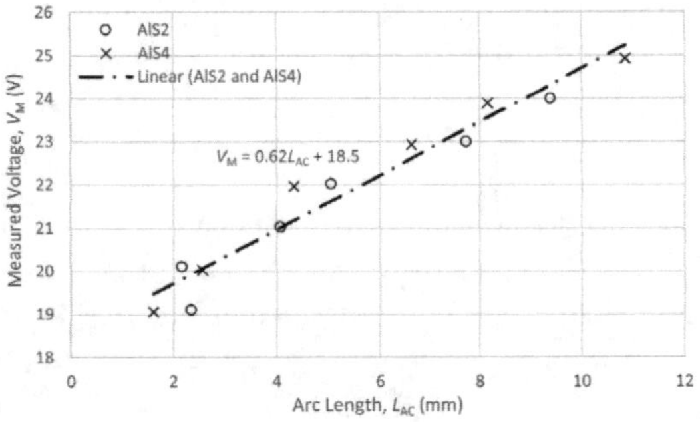

Figure 6. Measured voltage vs. arc length for AlS2 and AlS4 to determine the arc column electric field to be 0.62 V mm^{-1} from the linear fit expression.

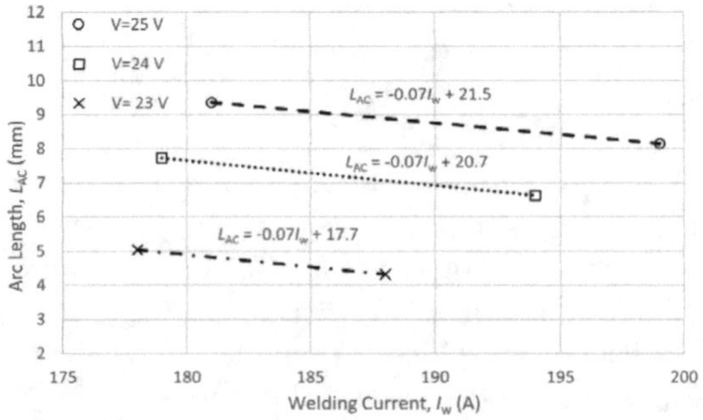

Figure 7. Arc length vs. welding current measured at different voltages showing a decrease in arc length of approx. -0.07 mm A^{-1}.

Table 3. Percentage contribution of fall voltages from experimental data.

	V_F (V)		V_{CT} (V)		V_{EE}(V)		V_{AC}(V)	
Label	Avg	(+/-)	Avg	(+/-)	Avg	(+/-)	Avg	(+/-)
AlS1	77%	1%	0.76%	0.01%	0.06%	0.02%	22%	1%
AlS2	85%	6%	0.82%	0.04%	0.08%	0.03%	14%	6%
AlS3	70.3%	0.3%	0.77%	0.01%	0.09%	0.04%	29.0%	0.3%
AlS4	84%	8%	0.85%	0.04%	0.11%	0.03%	15%	8%

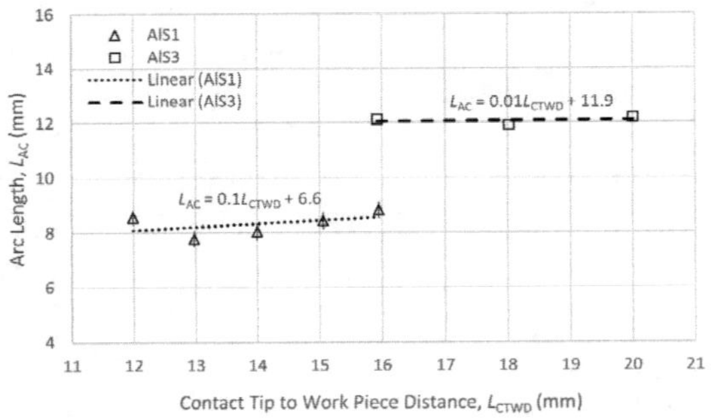

Figure 8. Effect on L_{AC} with changing L_{CTWD} showing a negligible increase in L_{AC} over an operable range for L_{CTWD}.

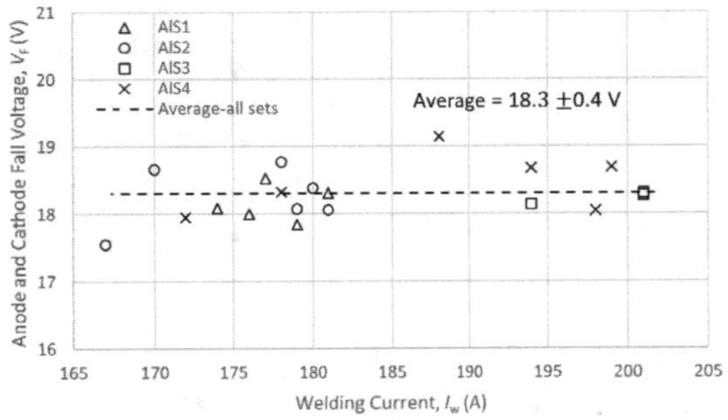

Figure 9. The sum of anode and cathode fall voltages for experimental data showing an average of 18.3±0.4 V for all data sets.

5.3 Experimental voltage contributions

Percent contributions to V_M from experimental data are listed in table 3 (respective error obtained from 1σ). The average value of V_F was found to be 18.3±0.4 V and approximately independent of I_w as shown in figure 9. The percent contribution for V_F ranged from ~70 to 85%, and attributed to the range of voltages used in the experiments. It was confirmed that V_{AC} is a major contributor to V_M with ~14 to 29% contribution that varied between sets depending on the voltage setting. Minor contributors to V_M were V_{EE} and V_{CT}, each contributing $< 1\%$.

5.4 Prediction results for procedures from LPHB

The expected L_{AC} values for the LPHB procedures were obtained by using the empirical relationship identified in figure 6. To align the empirical relationship to the procedures, V_{WL} was subtracted from the midpoint voltage setting (V_{Tot}) on the procedures since the experimental measurements excluded the welding leads. Thus, $L_{AC} = \frac{V_{Tot} - V_{WL} - 18.5}{0.62}$. The results are shown in figure 10, and range from 13 to 26 mm depending on the midpoint V_{Tot} from the procedure.

Considering the values for V'_{AC} and V_F obtained experimentally, midpoint V_{Tot} was predicted for each procedure using equation 1. Voltage prediction results are shown in figure 11 and were within 0.3% of the midpoint voltage settings on the procedures.

The main contributors to overall voltage loss were V_F and V_{AC} with an overall average of 64±3%, and 34±3% as shown in table 4. All other constituents were found to be minor contributors to overall voltage loss, with averages of 0.8±0.1% (V_{CT}), 0.02±0.01% (V_{EE}), and 1.1±0.1% (V_{WL}).

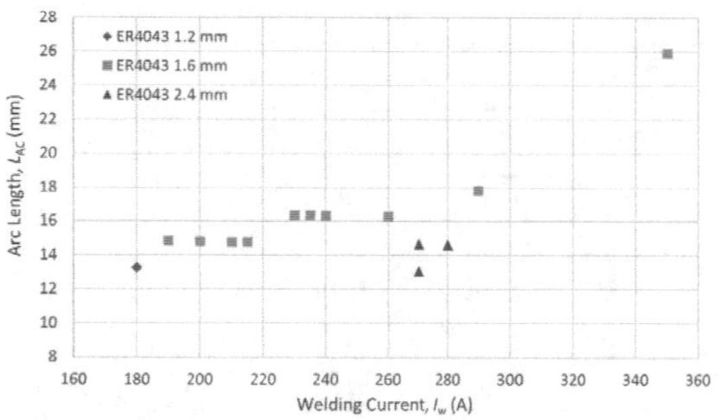

Figure 10. Expected arc length for procedures in the LPHB assuming the value of V_F from figure 9 and the V_M vs. L_{AC} relationship from figure 6.

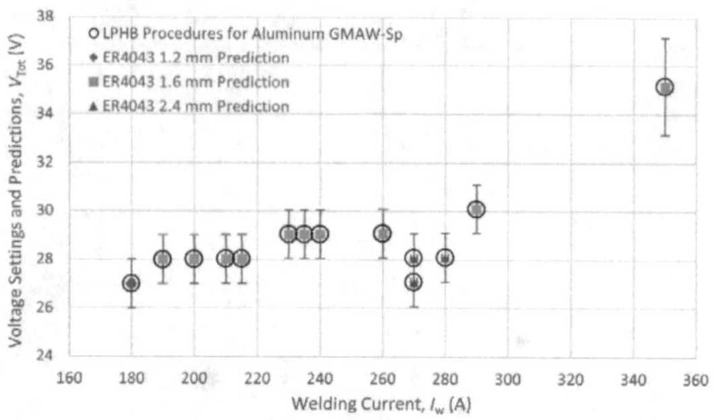

Figure 11. Midpoint voltage (V_{Tot}) predictions for procedures in the LPHB with error bars showing the range of acceptable voltage settings from the procedures.

Table 4. Constituents of total voltage and respective contribution to total voltage loss.

	V_F (V)		V_{CT} (V)		V_{EE}(V)		V_{AC}(V)		V_{WL} (V)	
Label	Avg	(+/-)	Avg	(+/-)	Avg	(+/-)	Avg	(+/-)	Avg	(+/-)
All Data	64%	3%	0.8%	0.1%	0.02%	0.01%	34%	3%	1.1%	0.1%
ER4043 1.2 mm	68%	0%	1%	0%	0.1%	0.0%	31%	0%	1%	0%
ER4043 1.6 mm	63%	3%	0.8%	0.1%	0.02%	0.01%	35%	3%	1.1%	0.1%
ER4043 2.4 mm	66%	1%	0.99%	0.02%	0.019%	0.004%	32%	1%	1.36%	0.02%

6 Discussion

6.1 *Experimental*

Arc attachment at the boundary between metal vapour and ionized gas considered in this study was a bigger challenge to trace due to the time dependency of the attachment point location. The time dependency is attributed to fluctuation of the arc attachment point as the metal vapour region is concentrated at the tip of the unmelted electrode immediately after droplet detachment, and begins to shift as the metal vapour engulfs the droplet as it grows in diameter until detachment. In contrast, the attachment at the beginning of taper region proposed by Egerland in [3], and considered in L_{AC} measurements by McIntosh and Mendez in [8] has less time dependency (i.e. the attachment point is approximately constant).

The consistency in σ (0.2 - 0.5 mm) from the time averaged L_{AC} measurements show that taking the time average of L_{AC} has greater impact for shorter arc lengths (eg. L_{AC} <3mm) than for longer arc lengths (e.g. L_{AC} >10 mm). Since L_{AC} measurements are subject to individual ability, future work would benefit from the application of a machine learning algorithm to standardize and expedite L_{AC} measurements over a greater number of droplet cycles.

Although the relationship between L_{AC} and I_w is not the focus of this paper, the consistency in the relationships observed in figure 7 drive interest for future work. The effect to L_{AC} from I_w (-0.07 mm A^{-1}) cannot be ignored with an operable range from 190 to 360 A (for 1.6 mm as an example from LPHB). On the contrary, the effect to L_{AC} from changing L_{CTWD} (0.01 to 0.1 mm mm^{-1}) can be considered negligible with an operable range for L_{CTWD} typically limited by gas shielding capability (range of 4 mm change to L_{CTWD} for welds in this study). Also, the negligible effect on L_{AC} from changing L_{CTWD} is supported by the self regulating arc concept discussed in [5].

Results show that L_{AC} is not directly proportional to V_{Tot}. More specifically, L_{AC} is directly proportional to V_{AC} by a factor of 0.62 V mm^{-1} for ER4043 as determined by this study. The 0.62 V mm^{-1} is lower than the average value reported by McIntosh and Mendez (0.77 V mm^{-1}) but comparable in order of magnitude considering the different methodology for measuring L_{AC}, and the larger current range (150-230 A) studied in [8]. Also, V_F obtained in [8] was slightly lower (17.0±1.0 V) than the 18.3±0.4 V obtained in this study. Thus explaining the fall voltage not accounted for in V_{AC} from this study was accounted for in the V_F in comparison to [8].

The average value for V_F (18.3±0.4 V) shown in figure 9 was smaller than the y-intercept (18.5 V) from the linear fit equation shown in figure 6. This is an important finding since the difference between these two values corresponds to the sum of V_{CT} and V_{EE}, which in the experimental trials summed up to an average of 0.2 V. This implies that approximately 18.5 V is required to obtain an L_{AC} >0 mm.

From the percent contributions for experimental data listed in table 3, it is noted that AlS2 and AlS4 were composed of varying voltages and therefore the respective percent contributions for V_F and V_{AC} have a large variance and are not to be considered as representative of what would be expected for ideal GMAW-Sp for aluminum. The settings for AlS3 are the most representative of what would be expected for GMAW-Sp for ER4043 1.2 mm per comparative procedure in the LPHB (e.g. with V_{AC} contribution ~30%). Thus, showing that AlS1 settings are below what would be desirable with 22% contribution to V_{Tot}.

6.2 *Application to LPHB procedures*

In this study, an empirical relation was used to predict L_{AC} for a given voltage setting for aluminum. Future work could consider an arc length formula from first principles by taking into account parameters that can be prescribed in a welding procedure such as: electrode diameter (d_c), material grade, V_{Tot}, U_c, and L_{CTWD}. This would allow to predict heat input and bead width for a given voltage setting by estimating the expected L_{AC} and associated thermal contribution from the arc. Also, the boundary between spray and short circuit transfer modes could be defined analytically as the L_{AC} approaches zero.

The shielding gas for the experiments was pure Ar as prescribed by the aluminum procedures in the LPHB. Further study is required to understand respective magnitudes of arc length with different gas mixtures and their respective ionizing energies.

It is noted that no material grades are specified in the LPHB. However, the results obtained with ER4043 are a starting point approximation for other aluminum alloys. Due to the low resistivity of aluminum on the order of 10^{-8}, and the low resistance from the contact tip (1-2 mΩ), it is expected that V_{AC}, V_{cath}, and V_{an} will always be the main contributors to total voltage loss for aluminum. If V_F is approximately constant over the operable current range, then any change in voltage setting will primarily manifest as V_{AC} resulting in a respective L_{AC}.

The predictions for percent contribution to overall voltage loss had a small σ ($\leq$ 3%) for the major contributors V_F, and V_{AC} (table 4). This consistency between the different electrodes shows an expected acceptable percent contribution for each fall voltage constituent as the settings in the procedures are considered to be a good starting point for welding aluminum.

7 Conclusions

7.1 *Experimental*

- Arc column electrical field (V'_{AC}) was 0.62 V mm^{-1} from an empirical relationship ($V_M = 0.62 L_{AC} + 18.5$) considering arc length (L_{AC}) and a measured voltage (V_M). The value of 0.62 V mm^{-1} was considered to be independent of welding current (I_w) when applied to LPHB procedure predictions. Future work will include experiments with different material grades and consumable diameters at varying I_w ranges to verify the impact of I_w on V'_{AC}.

- Time weighted L_{AC} measurements ranged from 1.6 to 12.2 mm with a standard deviation between 0.2 to 0.5 mm. Time weighted measurements are critical for welds with smaller L_{AC} as the arc becomes more erratic approaching short circuit transfer mode.
- A consistent decrease in L_{AC} was observed with increasing I_w (-0.07 mm A^{-1}). Results were consistent with limited data for three voltage settings and are subject for further study.
- L_{AC} was approximately independent of change in contact tip to work piece distance (L_{CTWD}) with an increase between 0.01 and 0.1 mm for every 1 mm increase in L_{CTWD}. Thus, considered negligible for the operable range of L_{CTWD} (4 mm range in this study).
- The sum of the anode and cathode (V_F) had an average of 18.3±0.4 V and was approximately independent of current.

7.2 Application to LPHB procedures:

- L_{AC} predictions were performed by using the empirical experimental correlation between V_M and L_{AC}. Expected arc lengths in the range of 13 to 26 mm were obtained, which depended on the midpoint voltage setting prescribed in the procedure. An arc length equation from first principles that is applicable to all materials and settings is subject for future work.
- Total voltage loss predictions were made within 0.3% of the midpoint voltage values suggested in the LPHB procedures using experimental values for V'_{AC} and V_F.
- The greatest contributors to overall voltage loss were V_F with an average contribution of 64±3%, and V_{AC} with 34±3%. The fall voltages associated with the contact tip, electrode extension, and welding leads were minor contributors with averages of 0.8±0.1% (V_{CT}), 0.02±0.01% (V_{EE}), and 1.1±0.1% (V_{WL}).

Acknowledgments
The author would like to thank: Lincoln Electric for supplying all welding equipment and consumables; Vicente Núñez Sánchez for assisting with experiments, and pulling data from JMatPro used for resistivity calculations; Carter Trautmann for assisting with experiments; Juan Pablo Romero Campos for assisting in processing the DAQ data, Viraj Patel for transcribing all LPHB procedures to Excel, and Dixie Albanez for assisting with diagrams.

References
[1] Lincoln Electric 2014 Gas metal arc welding guidelines *Product and Procedure Selection* Report C4.200 (Ohio: Lincoln Global Inc.)
[2] McIntosh C and Mendez P F 2017 *Welding Journal* **96** 121–132
[3] Egerland S 2015 *Soldagem & Inspeção* **20** 367–380
[4] Dos Santos E B, Kuroiwa L H, Ferreira A F C, Pistor R and Gerlich A P 2017 *Applied Sciences* **7** 503
[5] Phillips D H 2016 *Welding Engineering - An Introduction* (The Atrium, Southern Gate, Chichester, West Sussex, PO19 8SQ, United Kingdom: Wiley & Sons, Ltd.)
[6] Lincoln Electric 2000 *The Procedure Handbook of Arc Welding* 14th ed (Cleveland: The James F. Lincoln Arc Welding Foundation)
[7] Zhang G, Goett G, Kozakov R, Uhrlandt D, Reisgen U, Willms K, Sharma R, Mann S and Lozano P 2018 *Journal of Physics D: Applied Physics* **52** 085202
[8] McIntosh C and Mendez P F 2017 *Welding Journal* **96** 354S–366S
[9] Lehnhoff G R and Mendez P F 2011 *International Journal of Heat and Mass Transfer* **54** 2651–2660
[10] Luijendijk T 1996 *Int. J. for the Joining of Materials* **8** 1–4
[11] Zhang G, Goett G, Kozakov R, Uhrlandt D, Reisgen U, Willms K, Sharma R, Mann S and Lozano P 2018 *Journal of Physics D: Applied Physics* **52** 085201
[12] Matula R A 1979 *Journal of Physical and Chemical Reference Data* **8** 1147–1298
[13] Lincoln Electric Accessed on January 24, 2023 *https://www.lincolnelectric.com/en-us/support/welding-solutions/Pages/selecting-proper-size-welding-cables.aspx*

IOP Conf. Series: Materials Science and Engineering 1281 (2023) 012024 doi:10.1088/1757-899X/1281/1/012024

Uncertainty quantification for computational modelling of laser powder bed fusion

S Wells[1,4], A Plotkowski[2], J Coleman[2], M Rolchigo[2], R Carson[3], and M J M Krane[1]

[1]Purdue Center of Metal Casting Research; School of Materials Engineering, Purdue University, West Lafayette, IN 47907, USA
[2]Oak Ridge National Laboratory, Oak Ridge, TN 37830, USA
[3]Lawrence Livermore National Laboratory, Livermore, CA 94550, USA
[4]currently Continuous Casting Consortium; Department of Mechanical Engineering, Colorado School of Mines, Golden, CO 80401, USA

Wells130@purdue.edu

Abstract. Additive manufacturing (AM) may have many advantages over traditional casting and wrought methods, but our understanding of the various processes is still limited. Computational models are useful to study and isolate underlying physics and improve our understanding of the AM process-microstructure-property relations. However, these models necessarily rely on simplifications and parameters of uncertain value. These assumptions reduce the overall reliability of the predictive capabilities of these models, so it is important to estimate the uncertainty in model output. In doing so, we quantify the effect of model limitations and identify potential areas of improvement, a procedure made possible by uncertainty quantification (UQ). Here we highlight recent work which coupled and propagated statistical and systematic uncertainties from a melt pool transport model based in OpenFOAM, through a grain scale cellular automaton code. We demonstrate how a UQ framework can identify model parameters which most significantly impact the reliability of model predictions through both models and thus provide insight for future improvements in the models and suggest measurements to reduce output uncertainty.

1. Introduction

Additive manufacturing of metal alloys such as Ti-6Al-4V, 316 SS, IN625, and IN718 [1] is a class of metallurgical processes whereby components are printed in a layer-by-layer manner directed by a digital model of the part. Laser powder bed fusion (L-PBF) involves laser melting and localized consolidation of metal powders at sub-millimeter length scales, allowing for the production of complex components difficult to produce via more traditional casting or deformation processing. As such, metal additive manufacturing has found applications in aerospace, automotive, and biomedical fields [2]; yet detrimental defects and inconsistent build quality and performance [3-7] make certification of critical components difficult and limit industrial adaptation. Given the potential upsides of AM, reducing these defects and thus broadening the applicability of the process is at the forefront of ongoing research efforts. However, process complexity, laser system costs, and material costs mean experimental studies are difficult. Computational models offer a supplementary means of investigating the process while allowing for reproducibility and isolation of underlying physics.

Unfortunately, the complexity and scale of influential physics involved in metal additive manufacturing is too great to be captured by a single computational model. Instead, multiple models which specialize in a particular physical phenomenon must be developed and coupled to fully capture the process. Examples of computational models for AM research include melt pool transport models such as those used in [8-12], which provide temperature history data and melt pool geometries to aid in the development of scan strategies [13]. Also, microstructure evolution models [14-17] are vital for understanding as-built microstructures and crystal plasticity codes [18-21] predict the constitutive mechanical behavior. While computational models such as these are powerful tools to investigate and provide insight into metal additive manufacturing, their practical use relies on trade-offs between computational efficiency and model fidelity which reduce overall reliability of their predictions. Additionally, the data used to drive simulations (e.g., thermophysical properties) have inherent uncertainty that propagate to simulation results. It is therefore important to quantify the reliability of model predictions and the uncertainty therein, a process made possible via uncertainty quantification (UQ).

Two types of uncertainty are present in computational models such as those used for AM research, both of which must be considered. The first is statistical uncertainty, referred to as *aleatoric uncertainty*, resulting from natural variations in observations or measurements, oftentimes resulting from the equipment making the measurements. Aleatoric uncertainty is typically easy to quantify and therefore the propagation to model predictions can be readily determined. The second type of uncertainty is systematic uncertainty, referred to as *epistemic uncertainty*, resulting from a lack of knowledge due to data scarcity or model form error due to simplifications or non-physical parameters. Epistemic uncertainty can only be inferred, and correlations must be drawn. Given the role of simulation in AM research, it is vital that both forms of uncertainty and their propagation be accounted for, and their influences understood to quantify model limitations as well as identify potential areas of improvement.

The present work highlights a few key results from recent studies where an uncertainty quantification framework developed using TASMANIAN [22-26] was applied to two coupled models, a finite volume transport model based in OpenFOAM [9] and ExaCA [27], a cellular automata model for grain scale solidification, to demonstrate how both aleatoric and epistemic uncertainties can be studied and their influence quantified.

2. Methodology

The present work focuses on key findings from two separate studies. The first looks at uncertainty in solidification dynamics resulting from the role of solidification pathway selection. Second, the uncertainty in solidification dynamics is propagated to the as-built microstructure, specifically the grain morphology. The results presented here summarize a few key observations which are more thoroughly detailed in [29].

2.1. Melt Pool Transport Model

The melt pool transport model detailed in this work is based in OpenFOAM for simulations of laser powder bed fusion of Inconel 625 [9]. This model treats the domain as a continuum with volume-weighted thermophysical properties for all phases present in a computational cell.

2.1.1. Conservation Equations. The transport model solves the conservation of mass, momentum, and energy equations. In the present studies, fluid flow driven by Marangoni effects are disregarded and therefore the conservation of mass and momentum can be disregarded leaving only the conservation of energy given by Eqn. (1),

$$\rho C_p \frac{\partial T}{\partial t} = \nabla \cdot (k \nabla T) + S_T + \dot{Q} \tag{1}$$

where ρ is density, t is time, C_P is the specific heat capacity, T is temperature in Kelvin, and k is the thermal conductivity of IN625. S_T is an energy source term given by Eqn. (2) which accounts for the evolution of latent heat release due to phase change.

$$S_T = -\rho L_f \frac{\partial f_l}{\partial t} \tag{2}$$

$\dot{Q}$ is a volumetric laser heat source given by Eqn. (3) where η is the laser absorption efficiency, P is laser power, r is the laser radius and d is the penetration depth over which the laser energy is integrated. The x and y coordinates are the in plane scan directions with x_b being the laser center starting point, and z is the build direction over which multiple layers can be added.

$$\dot{Q}(x,y,z) = \frac{2\eta P}{r^2 d(\pi/3)^{3/2}} \exp\left(\frac{-3\left((x-x_b)+y\right)^2}{r^2} + \frac{-3z^2}{d^2}\right) \tag{3}$$

2.1.2. Boundary Conditions. For the present model, radiative heat transfer along the top boundary is governed by a heat transfer coefficient h given by Eqn. (4), and vaporization by heat flux (q''_v) given

$$h = \varepsilon\sigma\left(T^4 - T_\infty^4\right) \tag{4}$$

$$q''_v = A \cdot \frac{\exp\left(B - \frac{C}{T}\right)}{\sqrt{T}} \tag{5}$$

by Eqn. (5). Here ε is emissivity, σ is the Stefan-Boltzmann coefficient, T_∞ is the far field temperature taken to be 300K, and A, B, and C are algebraic coefficients. Input parameter values are provided in Table 1.

2.1.3. Simulation Domain, Time Step, and Grid Spacing. In the presented results two different simulation domains were used. A single line scan was used in Section 3.1 with a domain size of 5.0 x 0.8 x 0.4 mm to ensure a steady state melt pool was established at the probe locations while still approximating a semi-infinite domain surrounding the melt pool. In Section 3.2 and 3.3 a multi-pass, multi-layer simulation based on the AMB2018-01 test bed series [30] was used to create a build volume over which the microstructure was analyzed. This consists of alternating even/odd scan pattern totalling 65 layers. For both simulation domains a uniform grid spacing of 5 μm was used to ensure sufficient spatial resolution needed to capture the melt pool curvature. A 0.1 μs timestep was used for the single line scan, while an adaptive timestep was used for the multi-pass multi-layer simulation to reduce simulation time when the laser is off during dwell periods.

2.1.4. Outputs of Interest. For the single line scan study, the output of interest is the cooling rate through the tail of the melt pool along the top surface defined by a relative position through the mushy zone (x_{rel}) where $x_{rel} = 0$ at the liquidus isotherm and $x_{rel} = 1$ at the solidus isotherm. For the coupled models study, the outputs of interest were selected based on the inputs required by the weakly coupled CA grain scale solidification model. Here, the time at which each computational cell drops below the liquidus temperature and the instantaneous cooling rate at the liquidus isotherm are recorded and passed to ExaCA.

2.2. CA Grain Growth Model

2.2.1. Initialization. The grid spacing used by the transport model is too coarse to adequately resolve grain competition and nucleation, therefore the OpenFOAM temperature data was interpolated down to the CA grid spacing (2.5 μm here). Additionally, a substrate was simulated using a substrate grain size S_0 which was then used to calculate the number of grains (and thus grain seeds) in the substrate. Each grain seed was assigned to a random cell in the substrate with 1 of 10,000 possible cubic grain orientations. Cells not assigned a seed were grouped to the nearest seed to develop the final substrate grain structure.

2.2.2. Grain Growth. The decentered octahedron method is leveraged by ExaCA [27] which tracks the solidification front as it grows. An octahedral grain envelope half-diagonal direction, each half-diagonal aligned with one of the <001> crystallographic directions) is assigned to each active cell. During each time step where an active cell has a temperature below the liquidus, the half-diagonal length L of each octahedron is updated using Eqns. (6) and (7).

$$\Delta L = \frac{\Delta t}{\Delta x}\left(V(\Delta T)\right) \tag{6}$$

$$V(\Delta T) = \mathbf{A}(\Delta T)^3 + \mathbf{B}(\Delta T)^2 + \mathbf{C}(\Delta T) \tag{7}$$

Here, ΔL is the increment in octahedron half-diagonal length, Δt is the CA timestep, Δx is the grid spacing, and ΔT is the cell undercooling relative to the liquidus temperature. **A**, **B**, and **C** are fitting constants for a polynomial fit to experimental dendrite growth velocities $V(\Delta T)$ based on work by [30]. The octahedron half-diagonal length governed by Eqn. (6) was limited to 0.05 to prevent an order of iteration dependence growth. A timestep of 0.1μs was used to ensure the growth limit is not reached by the advancing melt pool periphery during steady state.

2.2.3. Heterogeneous Nucleation. In addition to grain growth, heterogeneous nucleation ahead of the front is modelled by ExaCA. All cells that will undergo solidification were assigned a random number $R_{nucleation}$ between 0 and 1. In Eqn. (8) N_0 is a heterogeneous nucleation site density. If Eqn. (8) is true, then the cell is a potential nucleation site.

$$R_{nucleation} < N_0(\Delta x)^3 \tag{8}$$

Each cell is then assigned a nucleation undercooling value sampled from a Gaussian distribution centered around ΔT_N with a standard deviation of ΔT_σ. If the nucleation undercooling value of a cell is reached prior to the cell becoming active via a cell capture event associated with grain growth, the nucleation event is successful, a new grain envelope with a random orientation is assigned to the cell, and the cell becomes active. All input parameter values used by the CA model are provided in Table 2.

2.2.4. Outputs of Interest. The grain growth model was used to produce as-built microstructures resulting from temperature data produced by the transport model. The presented work was focused on grain size and shape and thus the outputs of interest were 1) grain misorientation relative to the positive z axis, 2) average grain count, 3) volume weighted aspect ratio, and 4) average grain diameter of a sphere with the same volume. All these values were calculated in a representative value element (RVE) taken out of the middle of the as-built microstructure as depicted in Figure 1.

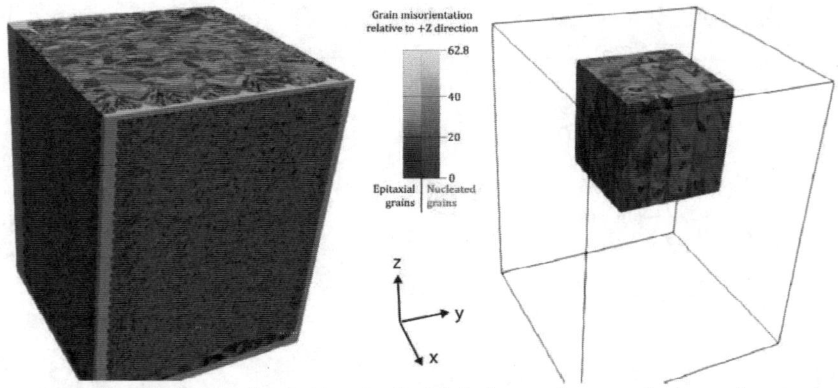

Figure 1. Representative volume element out of which the grain morphology data is drawn. The RVE avoids simulation boundaries and is taken near the top of the build when the microstructure is assumed to fully evolved. Gray indicates epitaxial grains and red grains are nucleated.

2.3. *Uncertainty Quantification*

To determine the influences of aleatoric and epistemic uncertainties on model reliability, the uncertainty quantification framework detailed in [11] was leveraged here for the analysis of solidification pathway selection using TASMANIAN[21-25]. This framework involves first establishing uncertainty bounds for each input parameters from the literature where possible, or by inference when not. A sensitivity analysis then revealed how the uncertainty in each parameter propagates and manifests as uncertainty in the outputs of interest. In doing so, not only can the most significant contributing uncertainties be determined, but so too can those which are negligible and can be disregarded to simplify the uncertainty space. Next, a sparse grid, which only features the parameters with the highest sensitivity, was constructed where grid points correspond to sets of simulation input parameters. Simulation results corresponding to these points are then used to construct a polynomial response surface. This response surface acts as a surrogate model which approximates the behavior of the computational model/models but can be interrogated rapidly. By sampling the surrogate 10 million times via a Monte Carlo based method, probability density functions (PDFs) are formed. These PDFs represent the uncertainty distribution for each output of interest due to the uncertainty in the considered input parameters.

Table 1. Transport model input parameters and associated uncertainties.

Parameter	Most likely value	Uncertainty	Units
$\rho_{s,l}$	7670.0	1.0%	kg/m^3
$C_{p,s}$	579.28	4.0%	J/(kg K)
$C_{p,l}$	750.65	4.0%	J/(kg K)
k_s	8.275 + 0.01472T	5.0%	J/(m s K)
k_l	4.889 + 0.01474T	5.0%	J/(m s K)
P	195.0	4.5 W	W
η	0.30	15.0%	---
d	20	25.0%	μm
r	85.7	---	μm
L_f	2.18×10^5	4.0%	J/kg
L_v	6.10×10^6	4.0%	J/kg
ε	0.40	---	---
A	2.121×10^{10}	---	W K$^{1/2}$/m^2
B	14.29	---	---
C	44,756	---	K

Table 2. Input parameter values and associated uncertainties for the CA grain growth model.

Parameter	Most likely value	Uncertainty	Units
S_0	25.0	10.0μm	μm
ΔT_N	20.0	75%	K
ΔT_σ	2	2K	K
A	-1.0302×10^{-7}	50%	m/(s K^3)
B	1.0533×10^{-4}	50%	m/(s K^2)
C	2.2196×10^{-3}	50%	m/(s K)
N_0	10^{15}	---	m^{-3}

3. Results and Discussion

3.1. Influence of Solidification Pathway Selection on Solidification Dynamics

One common assumption made in additive manufacturing modelling is that, due to the cooling rates and melt pool size, transport models lack the necessary resolution to resolve the release of the latent heat profile in the mushy zone and therefore any $f_s(T)$ function used in Eqn. (2) can be used if the entirety of the latent heat is accounted for. For this reason, in addition to thermodynamic models such as equilibrium or Scheil, non-physical but numerically convenient linear or sigmoidal functions (shown for IN625 in Figure 2) are oftentimes used. If the 'lack of resolution' assumption does not hold, then selection of the solidification pathway and latent heat release profile will impact model prediction reliability by increasing the uncertainty therein.

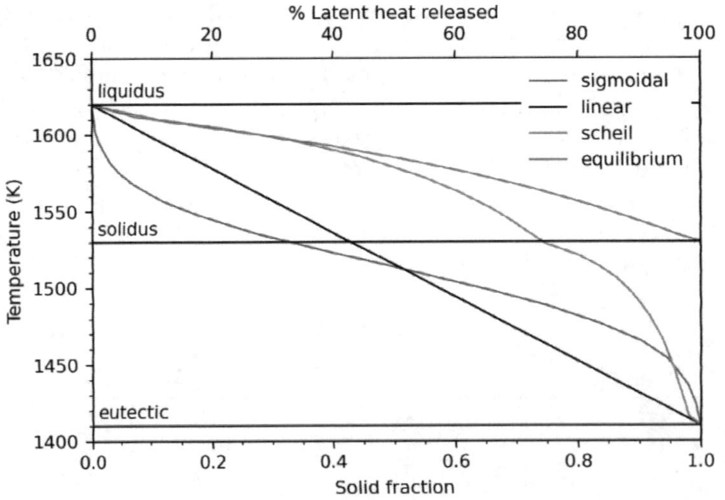

Figure 2. Common solidification pathways which dictate the release of latent heat during solidification.

Using the UQ framework and sensitivity analysis detailed [11], a 3-dimensional sparse grid was constructed for each of the solidification pathways which accounts for the uncertainty in laser absorption efficiency, and solid and liquid specific heat capacities of IN625 (the three parameters model predictions had the highest sensitivity to). 19 simulations were run using each of the pathways and used to construct polynomial response surfaces which act as surrogate models. All four surrogate models were sampled 10-million times to generate the statistics presented in Table 3. These statistics represent the uncertainty in model predictions due to the uncertainty in input parameters. Comparing results for each model illustrates the epistemic uncertainty associated with the solidification pathway selection. Differences between the pathways indicates that the lack of resolution assumption does not hold for the melt pool length or cooling rate at the liquidus isotherm.

Table 3. Probability density function statistics generated by sampling polynomial surrogate models 10 million times. Differences between the model indicates the lack of resolution assumption does not hold for melt pool length and cooling rate.

Output	Solidification pathway	Mean	2σ
Width [μm]	Scheil	140.91	2.10
	Equilibrium	134.01	2.15
	Linear	140.35	2.10
	Sigmoidal	140.42	2.08
Depth [μm]	Scheil	46.21	1.28
	Equilibrium	42.48	1.28
	Linear	45.84	1.26
	Sigmoidal	45.93	1.26
Length [μm]	Scheil	404.91	14.72
	Equilibrium	370.85	13.86
	Linear	418.22	15.38
	Sigmoidal	417.91	15.34
Cooling rate [K/s]	Scheil	418,751	27,015
	Equilibrium	360,329	19,112
	Linear	923,000	56,035
	Sigmoidal	1,632,218	76,071

In addition to the statistic provided above, Figure 3 depicts the cooling rate through the mushy zone at the tail of the melt pool. If the latent heat release profile were not adequate resolved, these curves would overlap significantly. However, this figure shows how the linear (black) and sigmoidal (blue) predict larger cooling rates near the liquidus isotherm ($x_{rel} = 1$) relative to the equilibrium and Scheil thermodynamic relations, but lower cooling rates near the end of solidification. Figure 2 shows how sigmoidal and linear release a significant amount of latent heat near the onset of solidification, but much less later, whereas sigmoidal and linear are more gradual in their latent heat release throughout the freezing range. These observations agree with those drawn from the surrogate model sampling as the latent heat release profile influences the solidification dynamics.

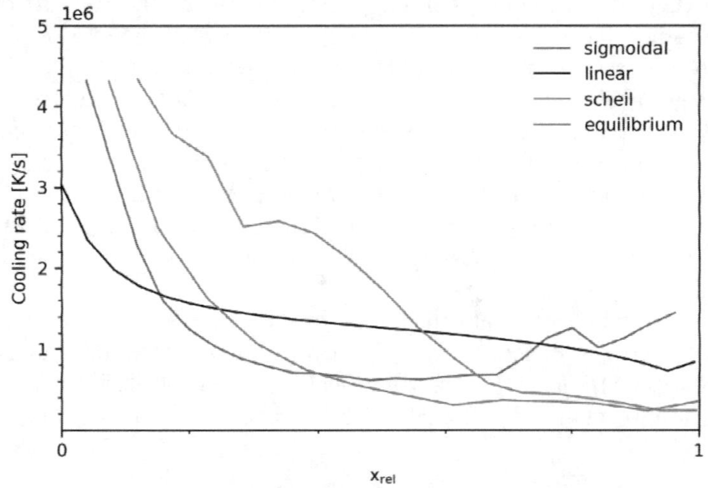

Figure 3. Cooling rate throughout the mushy zone for each of the solidification pathways. Sigmoidal and linear predict higher cooling rates early on, but lower near the end of solidification.

3.2. Grain Morphology Sensitivity Analysis

To begin the uncertainty propagation, a sensitivity analysis was conducted where input parameters for each model were individually adjusted by the uncertainties reported in Tables 1 and 2. These sensitivities are depicted in Figure 4 for each of the outputs of interest. Uncertainty in the interfacial response function coefficients (V(T)), the mean nucleation undercooling temperature, and the laser absorption efficiency showed the highest sensitivities and therefore were included in the surrogate model construction. All other input parameter uncertainties were neglected.

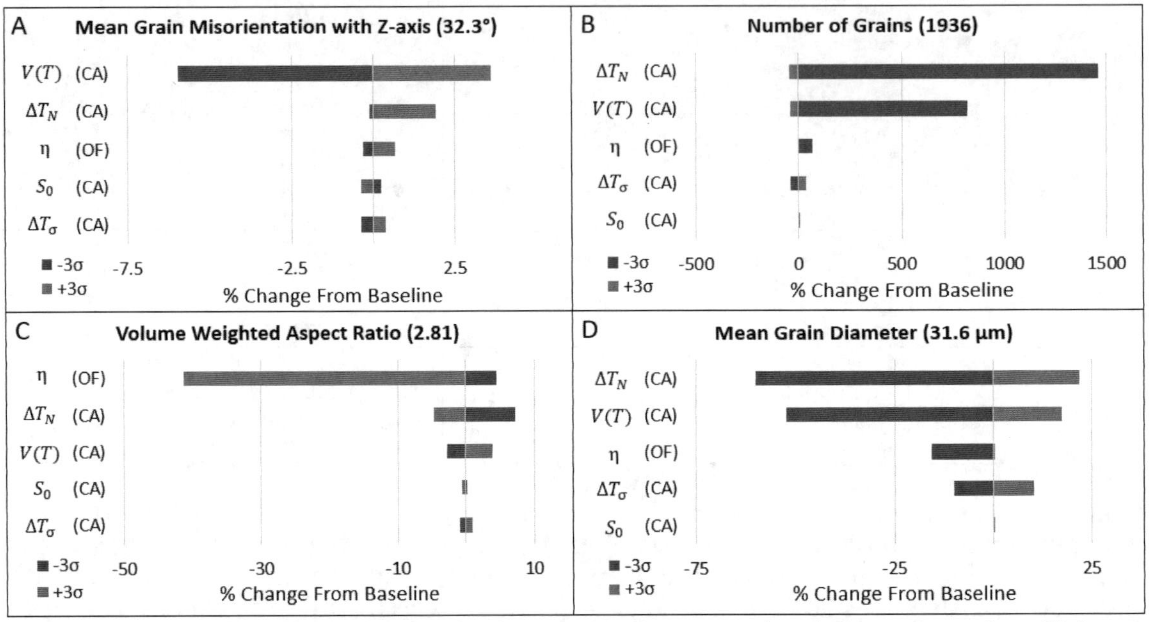

Figure 4. Sensitivity analysis for grain morphology showing uncertainty in laser absorption efficiency, IRF, and the mean nucleation undercooling temperature propagate significantly through the models.

3.3. Influence of Solidification Pathway on Grain Morphology

To determine the extent of uncertainty in grain morphology resulting from the solidification pathway assumption, temperature data from the OpenFOAM transport model for the linear and Scheil solidification pathways was propagated to ExaCA to generate an explicit microstructure. The equilibrium and sigmoidal relations were not carried forward to reduce the total number of simulations. A new 3D sparse grid was generated which included the uncertainty in laser absorption efficiency, mean undercooling for heterogeneous nucleation, and the coefficients dictating the interfacial response function. Surrogate models were generated for the linear and Scheil relations for each of the grain morphology outputs of interest discussed above. Sampling these surrogate models 10-million times produced the statistics presented in Table 4 and show a dependence of the microstructure on the solidification pathway.

Additionally, the probability density functions for the two latent heat release profiles are given in Figure 5. Differences between the curves can be attributed to the latent heat release profile through the mushy zone. These results indicate that the solidification pathway selection does impact the predicted as-built grain morphology, yet the significant overlap indicates that the influence is fairly small.

Table 4. Statistical generated by sampling the polynomial surrogate models 10-million times.

	Mean misorientation with z-Axis [°]	2σ	Number of grains	2σ	Volume weighted aspect ratio	2σ	Mean grain diameter [μm]	2σ
Scheil	33.2	1.69	3,910	12,100	2.75	0.124	31.4	11.1
Linear	32.3	1.10	4,290	12,700	2.76	0.343	30.2	11.4

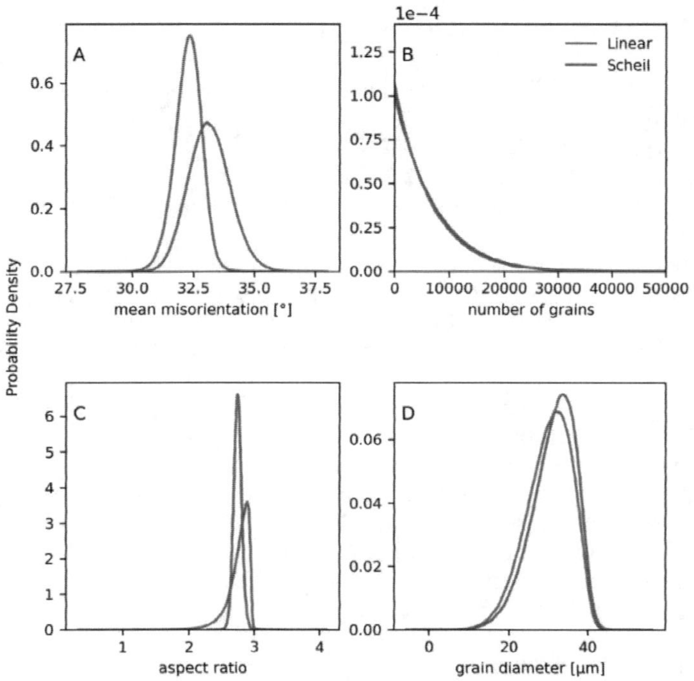

Figure 5. Probability density functions for grain morphology data showing a dependence on the solidification pathway selected.

4. Concluding Remarks

The results presented here are from a series of studies which investigated how statistical and systemic uncertainties impact the reliability of computational model predictions. Key observations from this work include:

- Uncertainty quantification frameworks such as that discussed in [11] allow for critical examination of modelling assumptions and their impact on model performance.
- The lack of resolution assumption often made in additive manufacturing modelling may not hold in the tail of the melt pool where mushy zone thickness is the greatest. When the assumption does not hold, selection of the solidification pathway may impact solidification dynamics and therefore the reliability of model predictions.
- Uncertainty in solidification dynamic predictions arising from the solidification pathway selection were shown to propagate to the predictions of as-built microstructure thus further reducing the reliability of model predictions.
- Lastly, while the results here only include a melt pool transport and cellular automaton grain growth model, it will be important to include a crystal plasticity code such as ExaConstit [27] to fully encompass the process-microstructure-property relations in future propagation studies.

This work was supported by the Exascale Computing Project (17-SC-20-SC), a collaborative effort of the U.S. DOE Office of Science and the NNSA. This manuscript has been authored by UT-Battelle, LLC under Contract No. DE-AC05-00OR22725 with the U.S. Department of Energy. The United States Government retains and the publisher, by accepting the article for publication, acknowledges that the United States Government retains a non-exclusive, paid-up, irrevocable, world-wide license to publish or reproduce the published form of this manuscript, or allow others to do so, for United States Government purposes. The Department of Energy will provide public access to these results of federally sponsored research in accordance with the DOE Public Access Plan (http://energy.gov/downloads/doe-public-access-plan).

References

[1] T DebRoy, H Wei, J Zuback, T Mukherjee, J Elmer, J Milewski, A Beese, A Wilson-Heid, A De, W Zhang 2018 *Prog. Mater. Sci.* **92** 112-224
[2] T Horn, O Harrysson 2012 *Sci. Prog.* **95** 255-282
[3] I Yadroitsev, I Smurov 2011 *Phys. Proc.* **12** 264-270
[4] W King, H King, V Castillo, G Gallegos, J Gibbs, D Hahn, C Kamath, A Rubenchik 2014 *J. Mater. Proc. Tech.* **214** 2915-2925
[5] A Martin, N Calta, S Khairallah, J Wang, P Depond, A Fong, V Thampy, G Guss, A Kiss, K Stone, C Tassone, J Nelson Weker, M Toney, T van Buuren, M Matthews 2019 *Nat. Comm.* **10** 1-10
[6] S Coeck, M Bisht, J Plas, F Verbist 2019 *Add. Manuf.* **25** 347-356
[7] T Mukherjee, W Zhang, T DebRoy 2017 *Comput. Mater. Sci.* **126** 360-372
[8] N Contuzzi, S Campanelli, A Ludovico 2011 *Intern. J. Simul. Mod.* **10** 113-121
[9] J Coleman, A Plotkowski, B Stump, N Raghavan, A Sabau, M Krane, J Heigel, R Ricker, L Levine, S Babu 2020 *J. Heat Transf.* **142** 122201
[10] N Hodge, R Ferencz, J Solberg 2014 *Comput. Mech.* **54** 33-51
[11] S Wells, A Plotkowski, M Krane 2021 *Metall. Mater. Trans. B* **52** 3016-3031
[12] S Khairallah, A Anderson 2014 *J. Mater. Proc. Tech.* **214** 2627-2636
[13] W Yan, W Ge, Y Qian, S Lin, B Zhao, W Liu, F Lin, G Wagner 2017 *Acta Mater* **134**
[14] O Zinovieva, A Zinoviev, V Ploshikhin 2018 *Comput. Mater. Sci.* **141** 207-220
[15] Y Zhang, J Zhang 2019 *Add. Manuf.* **28** 750-765
[16] M Rolchigo, R Carson, J Belak 2022 *Metals* **12** 1-26
[17] X Gong, K Chou 2015 *JOM* **67** 1176-1182
[18] Y Aoyagi, K Shizawa 2007 *Intern. J. Plast.* **23** 1022-1040
[19] M Zhang, J Zhang, DL McDowell, R Neu 2006 *Ninth Intern. Fatigue Congress.*
[20] H Lim, F Abdeljawad, S Owen, B Hanks, J Foulk, C Battaile: (2016), *Modell and Simul in Mater Sci and Engin* **24**
[21] T W Heo, S Khairallah, R Shi, J Berry, A Perron, N Calta, A Martin, N Barton, J Roehling, T Roehling, J Fattebert, A Anderson, A Nichols, S Wopschall, W King, J McKeown, M Matthews 2021 *J. Phys.: Mater.* **4**
[22] M Stoyanov, C. Webster 2016 *Comput. Math. Appl.* **71**
[23] M. Stoyanov 2018 *Lect. Note. Comput. Mater. Sci. Eng.* **123**
[24] M. Stoyanov 2015 *User Manual: TASMANIN Sparse Grids*
[25] Z Morrow M Stoyanov 2020 *SIAM J. Sci. Comput.* **42**
[26] M Stoyanov, D Munster, D Lebrun-Grandie, J Burkardt 2013 https://github.com/ORNL/Tasmanian
[27] M Rolchigo, S Reeve, B Stump, G Knapp, J Coleman, A Plotkowski, J Belak 2022 *Comput. Mater. Sci.* **214** 111692
[28] R Carson, S Wopschall, J Bramwell 2019 https://github.com/LLNL/ExaConstit
[29] S Wells 2022 Uncertainty Quantification of Laser Powder Bed Fusion Computational Models Ph.D. Purdue University Graduate School
[30] J Heigel, B Lane, L Levine, T Phan, J Whiting 2020 *J. Res. Nation. Instit. Stand. Tech.* **125** 1-17
[31] A Rai, M Markl, C Körner 2016 *Comput. Mater. Sci.* **124** 37-48

Transient simulation of melt flow, clogging, and clog fragmentation inside SEN during steel continuous casting

H Barati[1,2*], M Wu[2*], A Kharicha[2,3], A Ludiwg[2]

[1] K1-MET, Franz-Josef Street 18, 8700 Leoben, Austria
[2] Chair for Modeling and Simulation of Metall. Processes, Dept. Metall., Montanuniversitaet Leoben, Franz-Josef Street 18, 8700 Leoben, Austria
[3] Christian-Doppler Laboratory for Metallurgical Applications of Magnetohydrodynamics, Dept. Metall., Montanuniversitaet Leoben, Franz-Josef Street 18, 8700 Leoben, Austria

E-mail: hadi.barati@k1-met.com, menghuai.wu@unleoben.ac.at

Abstract. Clogging of submerged entry nozzle (SEN) during continuous casting of steel is an undesirable phenomenon leading to different problems like flow blockage, slag entrainment, nonuniform solidification, etc. A transient numerical model for nozzle clogging based on an Eulerian-Lagrangian approach was developed and it covers the main steps of clogging: (a) formation of the first oxide layer by chemical reactions on the steel-refractory interface; (b) motion of non-metallic inclusions (NMIs) due to the turbulent melt flow towards the SEN wall; (c) interactions between the melt, the NMI, and the wall; (d) formation and growth of the clog by the deposition of NMIs on the clog front and the flow-clog interactions; and (e) detachment/fragmentation of a part of clog due to the flow drag force. Clogging in an industrial scale SEN was simulated. The simulated clog front was compared with real as-clogged SENs. The modeling results have successfully explained the SEN clogging induced transient flow phenomenon in the mold region, i.e. the transition from the stable to an unstable and non-symmetrical flow.

1. Introduction

Clogging of submerged entry nozzle (SEN) during continuous casting (CC) of steel appears due to the build-up of solid materials on the inner wall. SEN clogging may lead to decreased productivity of the CC process and lower quality of the final product. Various clogging mechanisms have been proposed: attachment of non-metallic inclusions (NMIs) [1–3], reaction at the SEN wall [4–6], and temperature drop leading to solidification of steel or precipitation of alumina [7–9]. Among them, the deposition of NMIs on the SEN wall is still the primary cause of clogging [10].

Clogging has been studied experimentally and numerically for decades. In experimental studies, the main focus was on the analysis of clog [6,11], non-metallic inclusions [12,13], and effect of the steel composition [14,15]. The characteristics of these materials were correlated to clogging phenomena. Clogging can be generally considered in five steps, as depicted in Figure 1: (a) formation of the first oxide layer by chemical reactions on the steel-refractory interface; (b) transport of non-metallic inclusions (NMIs) by turbulent fluid flow towards the wall; (c) interactions between the fluid and the wall, and adhesion of the NMI on the wall; (d) formation and growth of the clog by the NMI deposition on the clog front and the flow-clog interactions; (e) detachment/fragmentation of a part of clog due to the flow drag force.

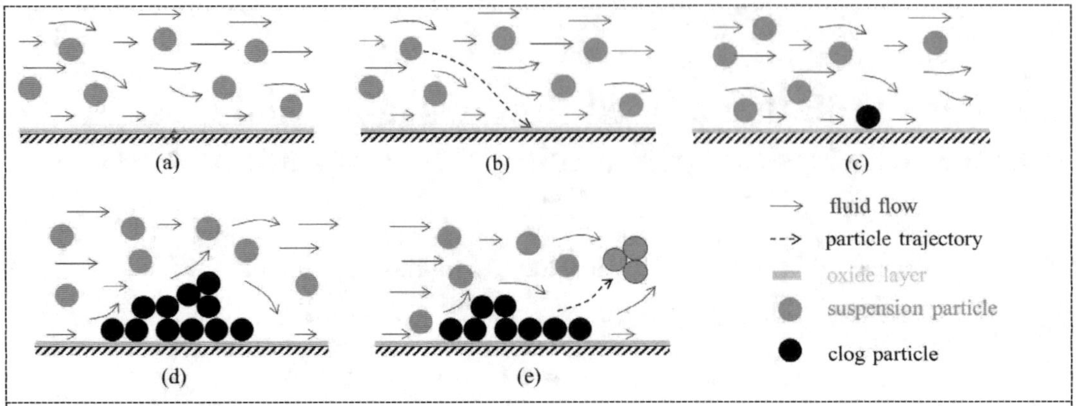

Figure 1. Schematic of main steps of clogging. (a) formation of the first oxide layer by chemical reactions; (b) transport of NMIs; (c) adhesion of the NMI on the SEN wall; (d) growth of the clog; and (e) detachment/fragmentation.

Different numerical models have been developed by considering one or more main steps of clogging. Single-phase-based Eulerian approach [10,16], Eulerian-Lagrangian approach [17–20], Eulerian-Eulerian two-phase approach [21,22] were tried to simulate clogging phenomena. Computational thermodynamics has also been adopted to investigate reactions and inclusion formation which are responsible for clogging [1,23,24]. Recently water models have also been considered to study the effects of as-clog SEN on the flow pattern inside the mold region [25].

The current authors have developed a transient model for nozzle clogging in steel CC that covers chemical reactions on the SEN refractory called 'early stage' of clogging [26] , i.e. step (a), along with clog growth due to NMI deposition called 'late stage' of clogging [27,28], i.e. steps (b) – (d) in Figure 1. The model was validated with laboratory experiments [27] and its accuracy and efficiency were verified for industry SENs [29]. In this work, the model is extended to consider one more sub-model for the detachment/fragmentation of a part of clog due to the flow drag force, i.e. step (e) in Figure 1.

2. Modeling

Eulerian–Lagrangian and volume-averaged approaches were adopted to simulate melt flow, NMI tracking, clogging, and fragmentation in a coupled way. The Eulerian approach was employed to calculate the turbulent flow and the Lagrangian approach was used to track the motion of the NMIs. A volume-average scheme is used to define the clog as a porous medium and track the growth of the clog. Details of this model were explained previously [27].

Modeling concept of clog fragmentation is schematically described in Figure 2. In the clogging model, clog is supposed to be a porous medium formed by random deposition of solid NMIs due to the turbulent flow, like Figure 2(a). This assumption is compatible with NMI networks observed in metallographic images of clog [16,30]. It is further assumed that the flow-induced drag force leads to mechanical fracture of the clog network, hence fragmentation of a part of the clog. To calculate the stress on the clog, the clog network is treated as a bunch of cylinders uniformly distributed, as shown in Figure 2 (b). Each cylinder called 'clog finger' is made out of sintered spheres, as shown in Figure 2 (c). The maximum stress occurs at the base of a clog finger. A similar idea was used by Pilling and Hellawell [31] to model the fragmentation mechanism for dendritic arms during solidification. They considered the dendrite as a cylinder with different radii between the root and main section. Similarly, clog finger was assumed to be a cylinder but with different radius at the base and flow is perpendicular to the clog finger, as shown in Figure 2(c). At the base of the clog finger, neck radius (r_n), increases over time due to the sintering. According to the calculations of Pilling and Hellawell, the maximum stress on the base of a clog finger (σ_n) is

$$\sigma_n = \frac{2FL_c^2}{\pi r_n^3}, \tag{1}$$

where F is the total force acting on the clog finger by the melt flow. L_c and r_n are length of clog finger and neck radius due to sintering at the base of clog finger, respectively, as shown in Figure 2. Neck radius can be calculated by the theory of the kinetic growth of sintering neck of spheres,

$$\frac{r_n^m}{r_p^n} = A_{(T)}\, t \tag{2}$$

where r_p is radius of sphere and $A_{(T)}$ is a temperature-dependent constant expressed as

$$A_{(T)} = \frac{K_1 \gamma V_0 D}{RT} \tag{3}$$

In this equation, K_1 is a constant; γ is the interfacial tension; V_0 is the molar volume; D is the self-diffusion coefficient of the material; R is the gas constant; and T is the temperature. The exponents m and n in Equation (2) define the sintering mechanism. Here, $m = 5$ and $n = 2$ according to the experimental observation of sintering alumina spheres in steel melt [32].

If the calculated σ_n is larger than the strength of alumina, the clogged part in the computational cell is removed and a partially-clogged cell turns into a free cell. The strength of alumina was considered as 300 MPa, as reported in the literature [9,33,34].

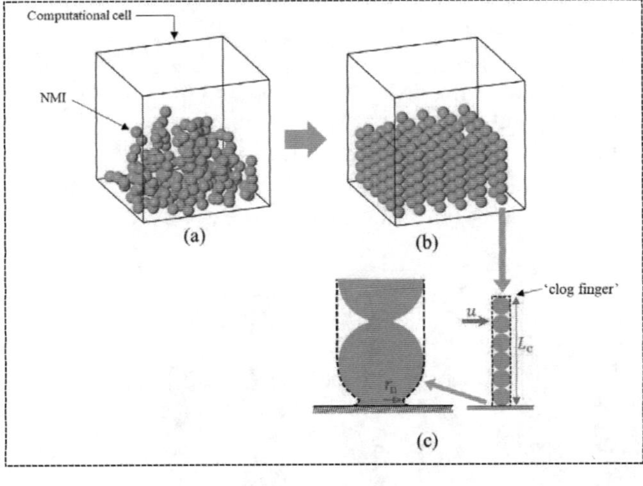

Figure 2. Schematic of fragmentation concept. (a) clog network constructed by random deposition of NMIs, (b) simplification of the clog network as a bunch of uniformly-distributed 'clog finger', (c) graphical description of the 'clog finger' as cylinder.

To test the model, a computational domain with boundaries is shown in Figure 3. The main objective of this work is to simulate melt flow and clogging inside SEN; however, a part of the CC mold was also included to have a reasonable flow pattern near the ports of SEN. In this case, accurate flow in the mold and cooling regions was not of interest. Hence, only a small length of mold and cooling part was included in the domain. At inlet, the mass flow rate of steel melt was 53.82 kg/s. Spherical alumina particles were injected with zero velocity and a rate of 0.0034 kg/s corresponding to 30 ppm oxygen content of the steel melt in the tundish before entering SEN. The particle size was 10 μm. Non-slip flow boundary conditions were considered for walls of SEN and CC mold and a shear-free wall was set for the free surface. The pressure outlet was set at the outlet. A clean SEN without any clogging is set as initial condition.

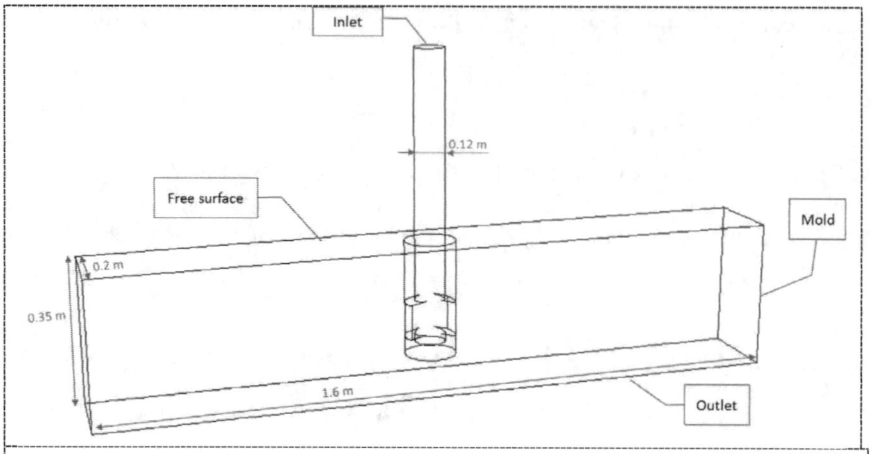

Figure 3. Computational domain and boundary conditions.

3. Results and discussions

In Figure 4, the transient growth of clog inside the SEN is plotted. For better visualization, the SEN is cut into two halves. It shows that clogging occurred initially at the bottom and ports. Then clogging continued on the tube part of the SEN. The random deposition of NMIs due to the turbulent melt flow resulted in random growth of the clog. The total clog fragmentation volume in each computational cell is shown in form of spheres in Figure 5. Because of continuous deposition and fragmentation, the total fragmentation volume in each cell could be larger than the cell volume. The major fragmentation occurred in the port area. Fragmentation in the bottom part was very small compared to that of the port area. With the evolution of clogging the major fragmentation area changed: in 9 min and 15 min, fragmentation was dominant on the side and bottom of the port area, but in 21 min and 27 min, the main fragmentation area was only the bottom of the ports. Fragmentation in the tube part was nearly uniform during the time. Another point is that the fragmentation in two halves of SEN was noticeably different after 9 min.

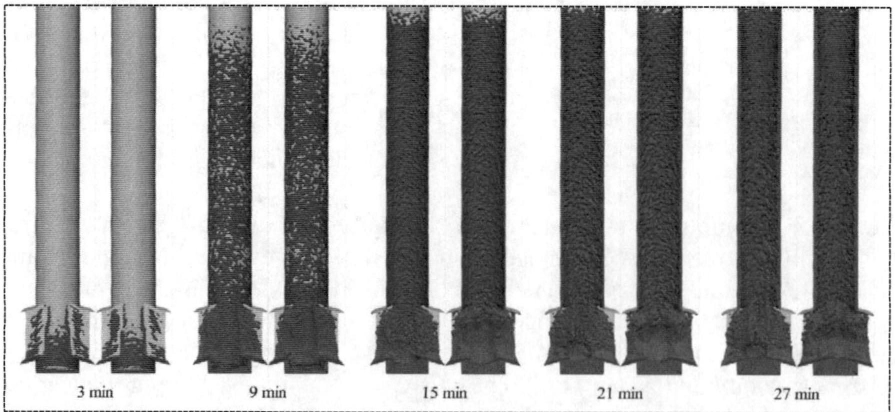

Figure 4. Growth of clog during the time.

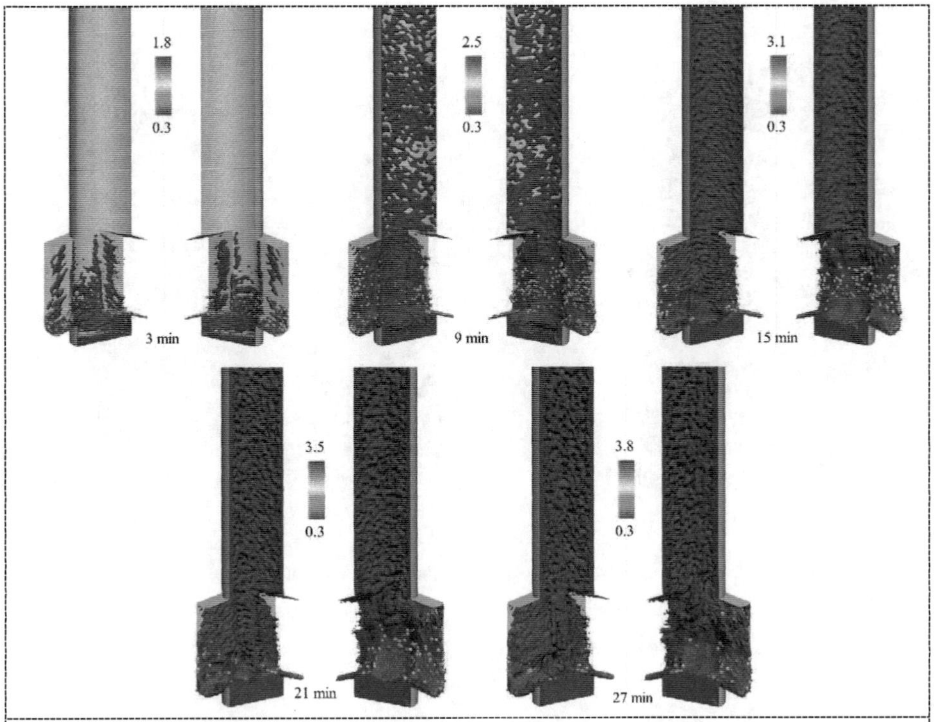

Figure 5. Clog front (grey) and total fragmentation volume (in form of colored spheres) during the time. Color bars show radius of spheres in mm.

In Figure 6, the flow inside the SEN is depicted. This figure shows that before 9 min, the flow in the ports was still conventional straight jets similar to that of non-clogged SEN. However, after 9 min, clog growth changed the flow, and a big vortex formed in the port area which lasted until the end of the simulation. This vortex explains why clog growth in the two halves of the SEN was very different.

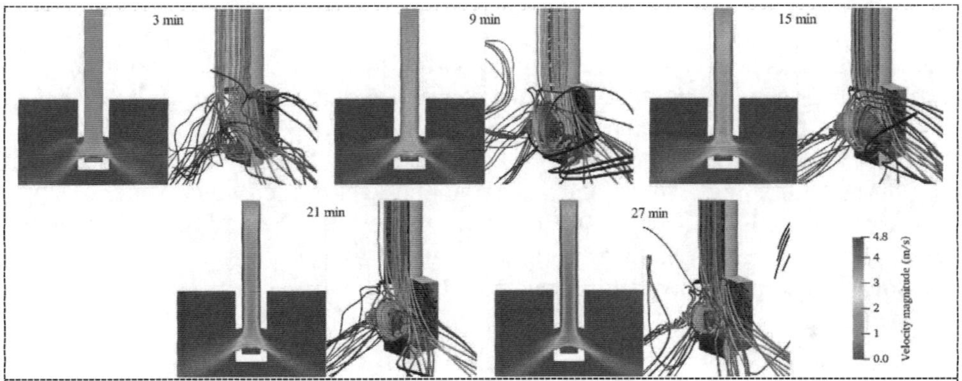

Figure 6. Melt flow on a symmetrical plane and flow streamlines inside the SEN during the time.

A qualitative comparison of as-clogged SENs from literature with the current simulation results, as shown in Figure 7, indicated that the current model has the potential to reproduce the clogging in real-size SENs. However, the proposed fragmentation sub-model requires still further refinements and improvements, like the treatment of turbulence in partially-clogged cells and the implementation of size and shape of NMIs.

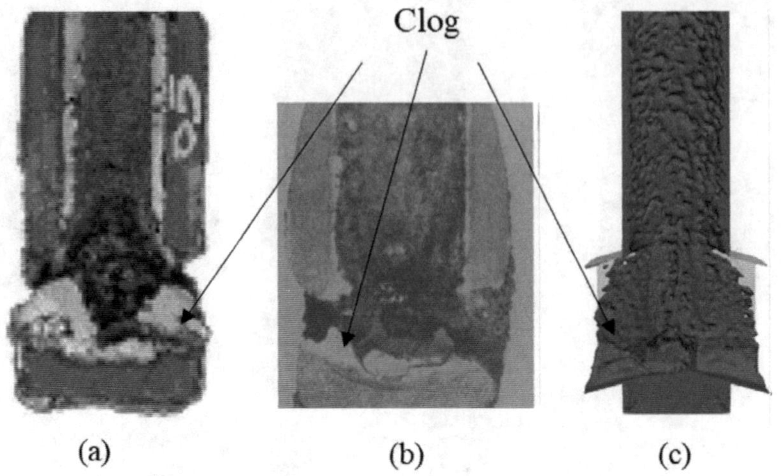

Figure 7. Qualitative comparison of as-clogged SENs, (a) from reference [8] and (b) from reference [16] with (c) the current simulation results.

4. Conclusions

A sub-model was developed for fragmentation during clogging in the continuous casting of steel. The sub-model was implemented in a transient clogging model. The simulation results of an industrial real-size SEN showed that after a while (in this case-study 9 min), the melt flow inside the SEN changed dramatically. This led to non-symmetrical growth of the clog, especially in the port area, and consequently non-symmetrical flow in the mold region. The addition of this feature to the developed clogging model made the model more realistic and more practical to study clogging in steel continuous casting.

Acknowledgements

The authors gratefully acknowledge the funding support of K1-MET GmbH, metallurgical competence center. The research program of the K1-MET competence center is supported by COMET (Competence Center for Excellent Technologies), the Austrian program for competence centers. COMET is funded by the Federal Ministry for Climate Action, Environment, Energy, Mobility, Innovation, and Technology; the Federal Ministry for Digital and Economic Affairs; the Federal States of Upper Austria, Tyrol, and Styria; and the Styrian Business Promotion Agency (SFG). In addition to the public funding from COMET, this research project was partially financed by scientific partners (Montanuniversität Leoben and Johannes Kepler University Linz) and industrial partners (voestalpine Stahl Linz GmbH, voestalpine Stahl Donawitz GmbH, and RHI Magnesita GmbH).

References

[1] Lee J-H, Kang M-H, Kim S-K and Kang Y-B 2018 Oxidation of Ti Added ULC Steel by CO Gas Simulating Interfacial Reaction between the Steel and SEN during Continuous Casting *ISIJ International* **58** 1257–66

[2] Vermeulen Y, Coletti B, Blanpain B, Wollants P and Vleugels J 2002 Material evaluation to prevent nozzle clogging during continuous casting of Al killed steels *ISIJ International* **42** 1234–40

[3] Sasai K and Mizukami Y 1994 Reaction Mechanism between Alumina Graphite Immersion Nozzle and Low Carbon Steel. *ISIJ International* **34** 802–9
[4] Miki Y, Kitaoka H, Sakuraya T and Fujii T 1992 Mechanism for separating inclusions from molten steel stirred with a rotating magnetic field *ISIJ International* **Vol. 32** 142–9
[5] Zhang L and Thomas B G 2006 State of the art in the control of inclusions during steel ingot casting *Metallurgical and Materials Transactions B* **37** 733–61
[6] Basu S, Choudhary S K and Girase N U 2004 Nozzle Clogging Behaviour of Ti-bearing Al-killed Ultra Low Carbon Steel *ISIJ International* **44** 1653–60
[7] Duderstadt G C, Iyengar R K and Matesa J M 1968 Tundish nozzle blockage in continuous casting *JOM* **20** 89–94
[8] Rödl S, Schuster H, Ekerot S, Xia G, Veneri N, Ferro F, Baragiola S, Rossi P, Fera S, Colla V and others 2012 *New strategies for clogging prevention for improved productivity and steel quality*
[9] Rackers K G and Thomas B G 1995 Clogging in Continuous Casting Nozzles *78th Steelmaking Conference Proceedings* vol 78 pp 723–34
[10] Bai H and Thomas B G 2001 Turbulent flow of liquid steel and argon bubbles in slide-gate tundish nozzles: Part I. model development and validation *Metallurgical and Materials Transactions B* **32** 253–67
[11] Karnasiewicz B and Zinngrebe E 2019 Observational Study of Clogging Specimens from the Tundish Well Showing Origin and Growth of a Clog in an Al-Killed Ti-Alloyed Steel Cast *Metallurgical and Materials Transactions B* **50** 1704–17
[12] Michelic S K and Bernhard C 2022 Significance of Nonmetallic Inclusions for the Clogging Phenomenon in Continuous Casting of Steel—A Review *Steel Res Int* **93** 2200086
[13] Zhang L F and Thomas B G 2003 State of the Art in Evaluation and Control of Steel Cleanliness *ISIJ International* **43** 271–91
[14] Dorrer P, Michelic S K, Bernhard C, Penz A and Rössler R 2019 Study on the Influence of FeTi-Addition on the Inclusion Population in Ti-Stabilized ULC Steels and Its Consequences for SEN-Clogging *Steel Res Int* **90** 1800635
[15] Deng X, Ji C, Cui Y, Tian Z, Yin X, Shao X, Yang Y and McLean A 2017 Formation and evolution of macro inclusions in IF steels during continuous casting *Ironmaking & Steelmaking* **44** 739–49
[16] Zhang L, Wang Y and Zuo X 2008 Flow Transport and Inclusion Motion in Steel Continuous-Casting Mold under Submerged Entry Nozzle Clogging Condition *Metallurgical and Materials Transactions B* **39** 534–50
[17] Hua C, Bao Y and Wang M 2021 Numerical simulation and industrial application of nozzle clogging in bilateral-port nozzle *Powder Technol* **393** 405–20
[18] Thomas B G, Yuan Q, Mahmood S, Liu R and Chaudhary R 2014 Transport and Entrapment of Particles in Steel Continuous Casting *Metallurgical and Materials Transactions B* **45** 22–35
[19] Pfeiler C, Thomas B G, Wu M, Ludwig A and Kharicha A 2008 Solidification and Particle Entrapment during Continuous Casting of Steel *Steel Res Int* **79** 599–607
[20] Yuan Q, Thomas B G and Vanka S P 2004 Study of transient flow and particle transport in continuous steel caster molds: Part II. Particle transport *Metallurgical and Materials Transactions B* **35** 703–14
[21] Ni P, Jonsson L T I, Ersson M and Jönsson P G 2014 The Use of an Enhanced Eulerian Deposition Model to Investigate Nozzle Clogging During Continuous Casting of Steel *Metallurgical and Materials Transactions B* **45** 2414–24
[22] Ni P, Jonsson L T I, Ersson M and Jönsson P G 2014 On the deposition of particles in liquid metals onto vertical ceramic walls *International Journal of Multiphase Flow* **62** 152–60
[23] Botelho T, Medeiros G, Ramos G L and Costa e Silva A 2017 The Application of Computational Thermodynamics in the Understanding and Control of Clogging and Scum in Continuous Casting of Steel *J Phase Equilibria Diffus* **38** 201–7

[24] Lee J-H and Kang Y-B 2020 Kinetics of CO Gas Dissolution into Stirred Liquid Fe at 1823 K and Its Impact on Nozzle Clogging during Continuous Casting *ISIJ International* **60** 258–66
[25] Li G, Lu C, Gan M, Wang Q and He S 2022 Influence of Submerged Entry Nozzle Clogging on the Flow Field and Slag Entrainment in the Continuous Casting Mold by the Physical Model *Metallurgical and Materials Transactions B* **53** 1436–45
[26] Barati H, Wu M, Michelic S, Ilie S, Kharicha A, Ludwig A and Kang Y-B 2021 Mathematical Modeling of the Early Stage of Clogging of the SEN During Continuous Casting of Ti-ULC Steel *Metallurgical and Materials Transactions B* **52** 4167–78
[27] Barati H, Wu M, Kharicha A and Ludwig A 2018 A transient model for nozzle clogging *Powder Technol* **329** 181–98
[28] Barati H, Wu M, Holzmann T, Kharicha A and Ludwig A 2018 Simulation of Non-metallic Inclusion Deposition and Clogging of Nozzle *CFD Modeling and Simulation in Materials Processing 2018* ed L Nastac, K Pericleous, A S Sabau, L Zhang and B G Thomas (Cham: Springer International Publishing) pp 149–58
[29] Barati H, Wu M, Kharicha A and Ludwig A 2019 Calculation Accuracy and Efficiency of a Transient Model for Submerged Entry Nozzle Clogging *Metallurgical and Materials Transactions B* **50** 1428–43
[30] Janis D, Karasev A, Inoue R and Jönsson P G 2015 A Study of Cluster Characteristics in Liquid Stainless Steel and in a Clogged Nozzle *Steel Res Int* **86** 1271–8
[31] Pilling J and Hellawell A 1996 Mechanical deformation of dendrites by fluid flow *Metallurgical and Materials Transactions A* **27** 229–32
[32] Nakamoto M, Tanaka T, Suzuki M, Taguchi K, Tsukaguchi Y and Yamamoto T 2014 Effects of Interfacial Properties between Molten Iron and Alumina on Neck Growth of Alumina Balls at Sintering in Molten Iron *ISIJ International* **54** 1195–203
[33] Uemura K, Takahashi M, Koyama S and Nitta M 1992 Filtration Mechamism of Non-metallic Inclusions in Steel by Ceramic Loop Filter. *ISIJ International* **32** 150–6
[34] Sambasivam R 2006 Clogging resistant submerged entry nozzle design through mathematical modelling *Ironmaking & Steelmaking* **33** 439–53

Influence of the adjustable EMBr on the asymmetric flow in a thin slab caster with a misaligned SEN

A Vakhrushev[1], E Karimi-Sibaki[1], M Wu[2], A Ludwig[2], G Nitzl[3], Y Tang[4], G Hackl[4], J Watzinger[5], A Kharicha[1]

[1]Christian-Doppler Lab for Metallurgical Applications of Magnetohydrodynamics, University of Leoben, Franz-Josef-Str. 18, 8700 Leoben, Austria
[2]Chair of Simulation and Modeling of Metallurgical Processes, University of Leoben, Franz-Josef-Str. 18, 8700 Leoben, Austria
[3]RHI Magnesita GmbH, Kranichberggasse 6, 1120, Vienna, Austria
[4]RHI Magnesita GmbH, Technolgy Centre, Magnesitstrasse 2, 8700, Leoben, Austria
[5]Primetals Technologies, Turmstrasse 44, 4031, Linz, Austria

E-mail: abdellah.kharicha@unileoben.ac.at

Abstract. The thin slab casting (TSC) of steel is a type of the continuous casting (CC) with a narrow funnel-shaped mold, characterized by the rapid solidification and fast production rates. A highly turbulent flow impacts on a growing solid shell due to the constant feeding of the fresh hot melt. That strongly affects the solidification profiles and final quality of the TSC slabs. The presented work numerically investigates the solidification inside the TSC mold with the asymmetric flow pattern caused by the misalignment (tilting) of the submerged entry nozzle (SEN). These effects were considered with and without the applied electromagnetic brake (EMBr). The influence of the adjustable EMBr on the asymmetric flow and solidification profiles including turbulent and magnetohydrodynamic (MHD) effects were studied. During consistent series of simulations, the EMBr was varied between the magnetic poles and the time-averaged velocity and temperature fields were collected. The results showed that symmetric EMBr of a local type could partially improve the asymmetry. An optimal braking scenario was found for the casing speed of 5.5 m/min and maximum EMBr value of 180 mT. The solidification and MHD models including turbulence were developed using OpenFOAM®.

1. Introduction

The thin slab casting (TSC) is known by the close to the net-shape profile, causing a strong turbulence generated in a funnel type mold. The turbulent jets come into the strong interaction with the solidifying shell after they are fed through the submerged entry nozzle (SEN) [1–3]. Thus, the impinging hot melt influences the mushy zone development and due to the partial remelting of the solid shell enhances the breakout risks [4–6]. To control the mold flow an efficient technique was developed through last decades by applying the magnetic field in the form of the electromagnetic braking (EMBr) [7].

The consideration of the Lorentz force in the EMBr-related research is typically limited to the flow damping effects. However, a series of recent studies has revealed the details of its complex topology, which leads to the entrainment of the quiescent melt into the formation of the reverse zones near the jets and in the upper region of the CC mold [8–10].

The flow pattern in the mold, the superheat distribution and the melt solidification are significantly altered under the EMBr during the continuous casting (CC) process [11–17].
The influence of the applied magnetic field on the asymmetric flow inside a TSC mold due to the partial SEN blockage was recently presented by the authors including effects on the solidification profiles [18]. The SEN clogging can occur during deposition of the non-metallic inclusions [19,20] or due to the combined effect with the parasitic solidification inside the SEN bore because of the heat losses [21,22]. A large eddy simulation was performed to investigate the transient asymmetric fluid flow, solidification, and inclusion transport in a slab CC under different EMBr power by Lu et al. [23].
The initial misalignment of the SEN, mechanical distortions due to the thermal expansion of the CC tundish or during the SEN ramping are known to be common issues in the TSC process. The presented work numerically investigates the solidification inside the TSC mold with the asymmetric flow pattern caused by the tilting of the SEN. These effects are considered with and without the applied EMBr.
A previously developed model is applied to simulate the turbulent flow including solidification [24]. The magnetohydrodynamic (MHD) effects are incorporated by using the electric potential method [25]. The influence of the adjustable EMBr on the asymmetric flow and solidification profiles is studied by varying the peak values between the magnetic poles. The time-averaged velocity and temperature fields are collected to define the optimal braking strategy.

2. Numerical model

A mixture volume averaging was used to model the turbulent melt flow accounting for the solidification and the MHD forces. The previously verified mixture approach [24] shows a high performance to model a complex real-size CC process. The corresponding momentum equation of the developed model is:

$$\rho\left[\frac{\partial \vec{u}}{\partial t} + \nabla \bullet (\vec{u} \otimes \vec{u})\right] = \nabla \bullet [\mathrm{dev}(\boldsymbol{\sigma}_{\mathrm{tot}}) - p\mathbf{I}] + \vec{F}_{\mathrm{mush}} + \vec{F}_{\mathrm{MHD}}\,, \tag{1}$$

where the stress $\boldsymbol{\sigma}_{\mathrm{tot}}$ includes molecular viscous forces and the turbulent stress. The large eddy simulation (LES) method was applied using the wall-adapting local eddy-viscosity (WALE) sub-grid scale (SGS) model [26]. The Darcy force $\vec{F}_{\mathrm{mush}}$ represents the drag between the mushy zone and the liquid melt [24]. The Lorentz force $\vec{F}_{\mathrm{MHD}}$ is calculated using the electric potential method [25]. The MHD model implementation can be found elsewhere [8,10]. A conservative formulation was used for the induced electric current density and for the Lorentz force, derived by Ni et al. [27].
The evolution of the temperature field T is obtained by solving corresponding energy equation [24]:

$$\rho C_{\mathrm{p}}\left[\frac{\partial T}{\partial t} + \nabla \bullet (\vec{u}T)\right] = \nabla \bullet \lambda \nabla T + \rho L\left[\frac{\partial f_{\mathrm{s}}}{\partial t} + \nabla \bullet (f_{\mathrm{s}} \cdot \vec{u}_{\mathrm{s}})\right], \tag{2}$$

with the specific heat C_{p}, latent heat L and solid fraction f_{s}. The distribution of the solid velocities $\vec{u}_{\mathrm{s}}$ follows the TSC funnel shape [24,28], defining the advection of the released latent heat [3]. The numerical model was implemented in the OpenFOAM® finite volume method (FVM) framework [29].

Table 1. Material properties and casting parameters of the low carbon DD11 steel.

Properties	Symbols	Units	Quantities
Density	ρ	$\mathrm{kg \cdot m^{-3}}$	7850
Specific heat	C_{p}	$\mathrm{J \cdot kg^{-1} \cdot K^{-1}}$	838.2
Thermal conductivity	λ	$\mathrm{W \cdot m^{-1} \cdot K^{-1}}$	35
Liquid dynamic viscosity	μ_{ℓ}	Pa· s	0.0061
Latent heat of fusion	L	$\mathrm{J \cdot kg^{-1}}$	243000
Casting velocity	$\vec{u}_{\mathrm{pull}}$	m/min	5.5
Solidus / liquidus temperature	T_{sol} / T_{liq}	K	1788.15 / 1805.15
Casting temperature	T_{cast}	K	1833.15

The detailed alloy properties calculated using a thermodynamic-kinetic software package IDS [30], casting temperature and pulling velocity are gathered in table 1. The simulation domain overlay is presented in figure 1-a. It includes a funnel-type TSC mold, a strand part, and a SEN; all corresponding dimensions, such as mold width and height, simulated strand length, are marked in figure 1-a for the 100 mm thick slab. The schematics of the SEN tilting is displayed in figure 1-b: the nozzle deviates from the mold symmetry line by an angle $\alpha = 2°$ in the wide face plane. The temperature dependent solid fraction curve is plotted in figure 1-c.

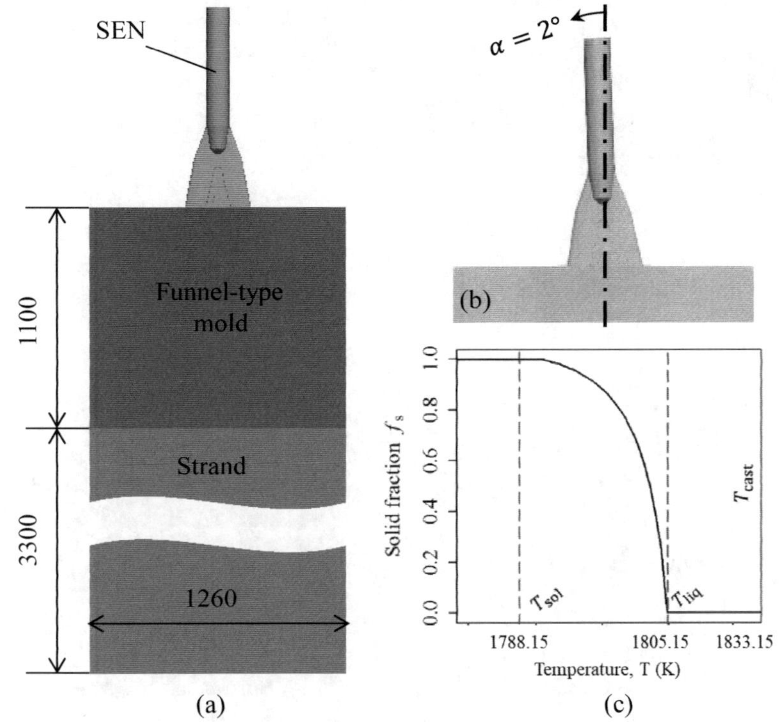

Figure 1. Setting for the TSC modelling: (a) domain outline; (b) schematics of the SEN tilting in the wide face plane; (c) temperature dependent solid fraction curve.

A mold heat flux, which is commonly taken from literature, evaluated from the plant measurements or modeled directly [24,31–33], is defined along the casting direction as a boundary condition in correspondence to figure 2 in Vakhrushev et al. [18]. The secondary cooling zone is described by an effective HTC of 1100 W/m^{2}K, obtained by solving the inverse conduction problem [34].

3. Simulation results

The presented numerical simulations of the melt flow and solidification inside the TSC mold with the tilted SEN (see figure 1-b) were performed according to the following scenarios: (i) without applied magnetic field; (ii) under the symmetric double-pole EMBr of 180 mT; (iii) with the varied EMBr by reducing its power for the left or right pole. The corresponding magnetic field distribution is in figure 2. The direction of the SEN tilting is marked by an arrow. The left brake corresponds to the case, when the maximum EMBr power is on the left side of the mold, as shown in figure 2-b. The similar principle applies for the scenario with the right-side brake in figure 2-c. The percentage marks the amount of the magnetic field reduction on the opposite site. The peak value is 180 mT for all scenarios.
The simulation results for the non-EMBr case and for the symmetric braking are detailed in figure 3. The time-averaged velocity field and temperature distribution were collected during the modelling for the comparison. Velocity magnitude in figure 3-a shows, that without magnetic field the jet on the right side of the mold (tilting-wise direction) goes deeper into the melt pool. The flow stagnation zone is

formed at the meniscus region. On the left side the upper roll remains close to the top surface. That results in the colder region on the right side in figure 3-c with more superheat transported to the left.

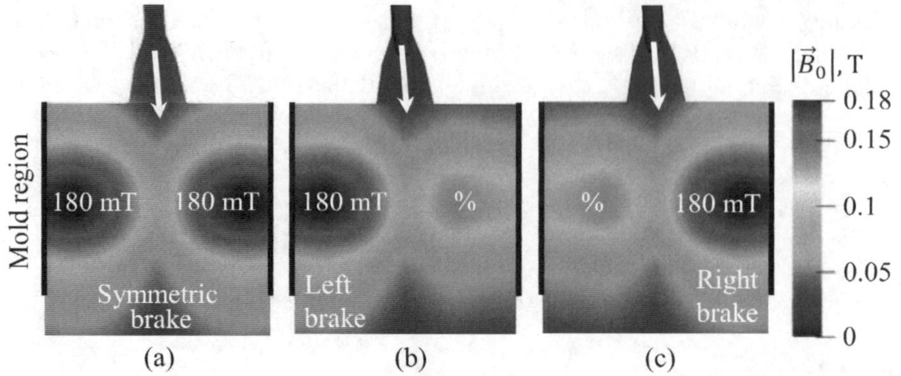

Figure 2. Distribution of the applied magnetic field for the cases (a) with the symmetric EMBr; adjustable braking with the peak value (b) on the left and (c) on the right side.

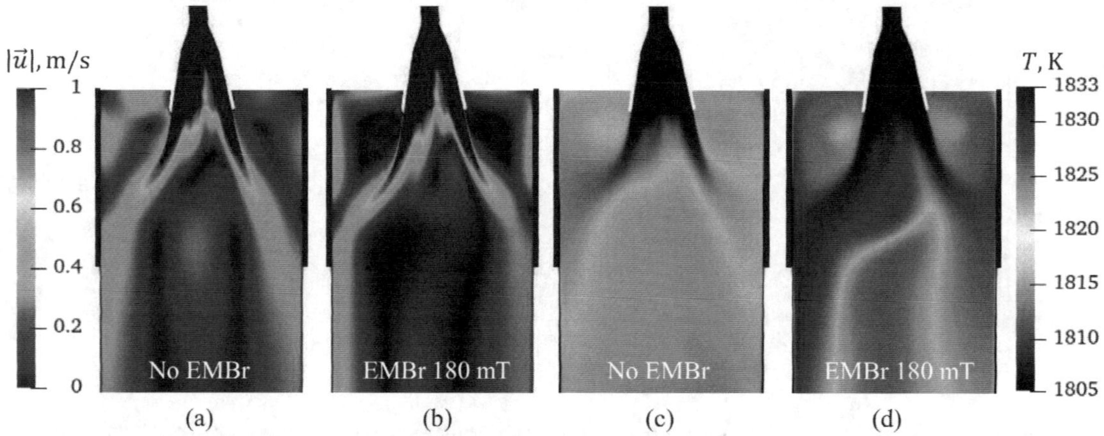

Figure 3. Time-averaged velocity magnitude for (a) non-EMBr case and (b) with the EMB; time-averaged temperature field (c) without and (d) under the applied magnetic field.

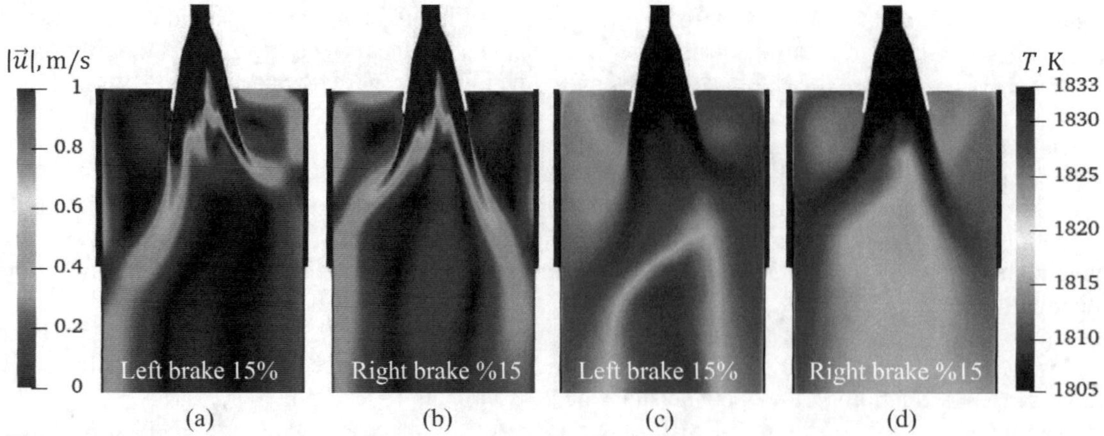

Figure 4. Time-averaged velocity distribution with the 15% more braking (a) on the left and (b) right side; time-averaged temperature field for the corresponding (c) left and (d) right 15% brake.

When the EMBr of 180 mT is applied the flow becomes more symmetric with a slightly slower upper roll on the right in figure 3-b. The uniformity of the temperature field in figure 3-d is improved as well. However, more superheat is brought to the narrow walls with a colder melt zone formed in the center. Next, the adjustable EMBr technique was applied and analyzed by varying the magnetic field between the poles by 15% of its maximum value of 180 mT. It should be reminded for the readers, that a linear variation of the EMBr power causes a quadratic change of the Lorentz force [10]. Thereby, a change of the applied magnetic field by 15% results in 32.2% for the acting MHD force.
The modelling results in figure 4 distinctively show, that a jet goes deeper on the dominant EMBr side. The superheat transport shifts correspondingly towards the opposite side with a weaker magnetic field.

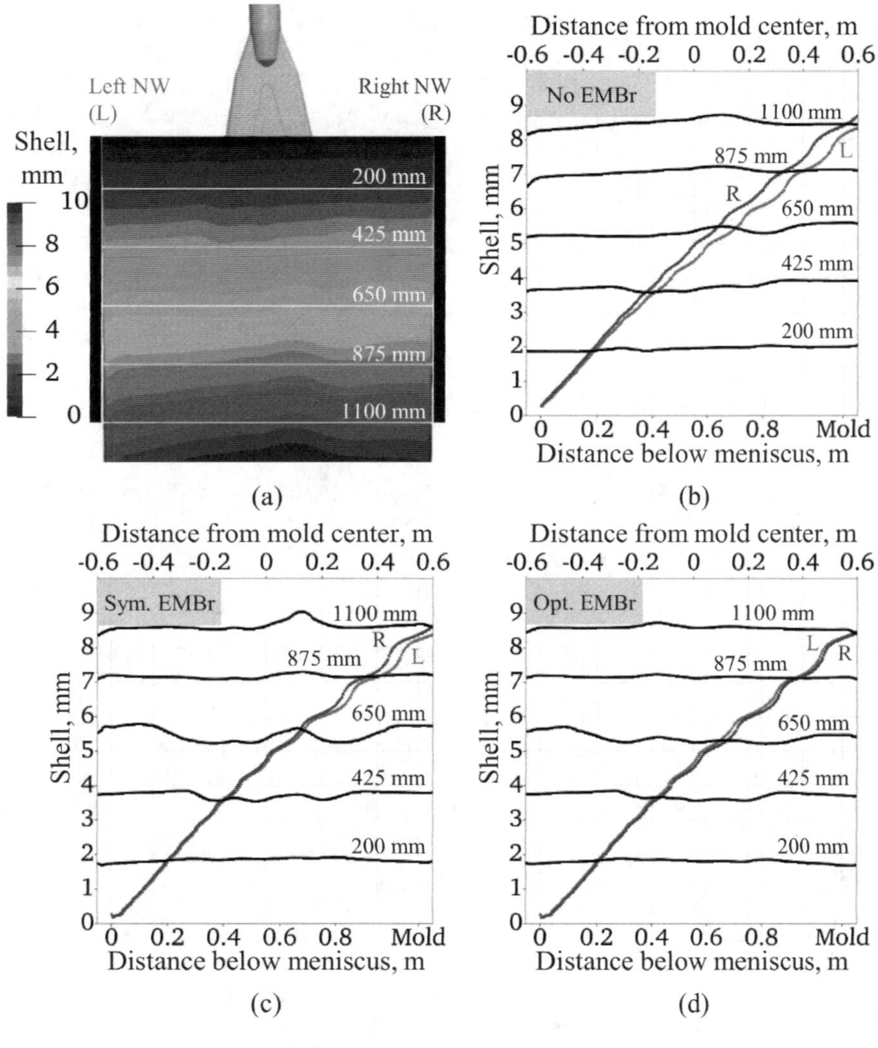

Figure 5. Adjustable EMBr simulation results: (a) shell thickness (front wide face) for the non-EMBr case; shell thickness profiles along the narrow sides and at the transversal lines for the scenarios (b) without magnetic field; (c) symmetric brake (180 mT); (d) optimized EMBr with 4% more power on the left side.

The simulations of the symmetric EMBr showed, that the application of the magnetic field leads itself to the improvements in the flow pattern with a slightly slower upper roll (figure 3-b) and colder region (figure 3-d) on the right side of the domain. Applying this knowledge along with the conclusions for the 15% left and right brake in figure 4, the optimal brake of 4% more power on the left side was consequently found through the manual simulation series.

The centerlines of the left and the right narrow sides are marked in figure 5-a with the red L and blue R letters and text respectively. They are specified for solid shell profiles along the casting direction. The transversal thickness is analyzed along the horizontal lines, which are at 200 mm, 425 mm, 650 mm, 875 mm, and 1100 mm (mold exit) below the meniscus in figure 5-a. Vertical and horizontal shell thickness profiles are presented on the corresponding subplots of figure 5.

Figure 5-a shows the distribution of the solid shell thickness across the front wide face of the TSC mold for the non-EMBr case. The remelting on the left and a thicker shell on the right are detected. Such an asymmetry can cause the mechanical stresses, tearing of the solid shell and leads to the costly breakout. The significant deviations, especially at the mold exit, are detected on the narrow sides in figure 5-b.

The uniform magnetic field improves shell thickness on the narrow sides in figure 5-c: deviation of the shell thickness between left and right is reduced. The solidified shell on the narrow sides in figure 5-c is slightly thinner than in the non-EMBr case in figure 5-b. That is explained by the enhanced superheat transport under magnetic field, if referred to figure 3-d in comparison to figure 3-c. The thickness kinks in the transversal shell profiles in figure 5-c are due to the colder central zone in figure 3-d. The remelting sites correspond to the location of the stabilized jets, that bring a fresh melt.

The resulting shell distribution under the optimized brake scenario (4% more EMBr power on the left) is detailed in figure 5-d. A good fit achieved between the narrow side profiles. The horizontal thickness distribution became more uniform and symmetric. Some insignificant deviations still appear; however, the main issues and the unevenness of the shell thickness are removed. More precise solution and the automation of the adjustable EMBr optimization are the topics for the future research.

4. Conclusions

In the presented work the solidification inside the TSC mold with the asymmetric flow pattern caused by the misalignment (tilting by an angle of 2°) of the SEN was numerically investigated. These effects were considered without and with the applied electromagnetic brake.

The developed solidification and MHD models were used to reveal the influence of the adjustable EMBr on the solid shell growth considering interaction with the asymmetric turbulent flow. The EMBr variation was done by increasing or decreasing the magnetic field between the left and right poles.

During consistent series of simulations, the time-averaged velocity and temperature fields were collected for the different magnetic field settings to perform a comparative study. The results showed that the symmetric EMBr of a local type could partially remove the asymmetry in the flow patter, superheat distribution and solidification. An optimal braking scenario was found for the casing speed of 5.5 m/min and maximum EMBr value of 180 mT by having the magnetic field stronger by 4% on the left side.

5. Acknowledgments

The authors acknowledge the financial support by the Austrian Federal Ministry of Economy, Family and Youth and the National Foundation for Research, Technology and Development within the framework of the Christian Doppler Laboratory for Metallurgical Applications of Magnetohydrodynamics.

References

[1] Wu M, Vakhrushev A, Nummer G, Pfeiler C, Kharicha A and Ludwig A 2010 *TOTPJ* **2** 16–23

[2] Vakhrushev A, Wu M, Ludwig A, Tang Y, Hackl G and Nitzl G 2012 *IOP Conf. Ser.: Mater. Sci. Eng.* **33** 012014

[3] Wu M, Vakhrushev A, Ludwig A and Kharicha A 2016 *IOP Conf. Ser.: Mater. Sci. Eng.* **117** 012045

[4] Iwasaki J and Thomas B G 2012 *Supplemental Proceedings* ed Tms (Hoboken, NJ, USA: John Wiley & Sons, Inc.) pp 355–62

[5] Luk'yanov S I, Suspitsyn E S, Krasilnikov S S and Shvidchenko D V 2015 *Int J Adv Manuf Technol* **79** 1861–8
[6] Zappulla M L S, Cho S and Thomas B G 2019 *Steel Res. Int.* **90** 1800540
[7] Thomas B G and Cho S M 2018 *IOP Conf. Ser.: Mater. Sci. Eng.* **424** 012027
[8] Vakhrushev A, Kharicha A, Liu Z, Wu M, Ludwig A, Nitzl G, Tang Y, Hackl G and Watzinger J 2020 *Metall. Mater. Trans. B* **51** 2811–28
[9] Kharicha A, Vakhrushev A, Karimi-Sibaki E, Wu M and Ludwig A 2021 *Phys. Rev. Fluids* **6** 123701
[10] Vakhrushev A, Kharicha A, Karimi-Sibaki E, Wu M, Ludwig A, Nitzl G, Tang Y, Hackl G, Watzinger J and Eckert S 2021 *Metall. Mater. Trans. B* **52** 3193–207
[11] Ha M Y, Lee H G and Seong S H 2003 *Journal of Materials Processing Technology* 18
[12] Chaudhary R, Thomas B G and Vanka S P 2012 *Metall and Materi Trans B* **43** 532–53
[13] Thomas B G, Singh R, Vanka S P, Timmel K, Eckert S and Gerbeth G 2015 *Journal for Manufacturing Science and Production* **15** 93–104
[14] Vakhrushev A, Kharicha A, Liu Z, Wu M, Ludwig A, Nitzl G, Tang Y, Hackl G and Watzinger J 2019 *The 8th International SteelSim Conference* SteelSim 2019 (AIST) pp 615–9
[15] Garcia-Hernandez S, Gonzalez-Guzman C H, Morales Davila R, Barreto J de J, Gutierrez E and Calderon-Ramos I 2020 *Crystals* **10** 958
[16] Vakhrushev A, Kharicha A, Wu M, Ludwig A, Nitzl G, Tang Y, Hackl G, Watzinger J and Rodrigues C M G 2020 *IOP Conf. Ser.: Mater. Sci. Eng.* **861** 012015
[17] Wang C, Liu Z and Li B 2021 *Metals* **11** 948
[18] Vakhrushev A, Kharicha A, Karimi-Sibaki E, Wu M, Ludwig A, Nitzl G, Tang Y, Hackl G and Watzinger J 2022 *steel research int.* **93** 2200088
[19] Barati H, Wu M, Kharicha A and Ludwig A 2018 *Powder Technology* **329** 181–98
[20] Barati H, Wu M, Michelic S, Ilie S, Kharicha A, Ludwig A and Kang Y-B 2021 *Metall Mater Trans B* **52** 4167–78
[21] Barati H, Wu M, Kharicha A and Ludwig A 2020 *Steel Res. Int.* **91** 2000230
[22] Vakhrushev A, Kharicha A, Wu M, Ludwig A, Tang Y, Hackl G, Nitzl G, Watzinger J and Bohacek J 2021 *Metals* **11** 1375
[23] Lu H, Zhong Y, Ren W, Ren Z and Lei Z 2022 *steel research int.* **93** 2200518
[24] Vakhrushev A, Wu M, Ludwig A, Tang Y, Hackl G and Nitzl G 2014 *Metall. Mater. Trans. B* **45** 1024–37
[25] Davidson P A 2001 *An Introduction to Magnetohydrodynamics* (Cambridge University Press)
[26] Nicoud F and Ducros F 1999 *Flow, Turbulence and Combustion* **62** 183–200
[27] Ni M-J, Munipalli R, Huang P, Morley N B and Abdou M A 2007 *Journal of Computational Physics* **227** 205–28
[28] Vakhrushev A, Kharicha A, Wu M, Ludwig A, Nitzl G, Tang Y, Hackl G, Watzinger J and Rodrigues C M G 2022 *J. Iron Steel Res. Int.* **29** 88–102
[29] Weller H G, Tabor G, Jasak H and Fureby C 1998 *Comput. Phys.* **12** 620
[30] Miettinen J, Somani M, Visuri V-V, Koskenniska S, Louhenkilpi S, Fabritius T and Kömi J 2022 *steel research int.* srin.202200120
[31] Camporredondo S. J E, Castillejos E. A H, Acosta G. F A, Gutiérrez M. E P and Herrera G. M A 2004 *Metall. Mater. Trans. B* **35** 541–60
[32] Zhang H and Wang W 2017 *Metall. Mater. Trans. B* **48** 779–93
[33] Niu Z, Cai Z and Zhu M 2020 *Ironmaking & Steelmaking* **47** 1135–47
[34] Bohacek J, Kominek J, Vakhrushev A, Karimi-Sibaki E and Lee T-W 2021 *OpenFOAM Journal* **1** 27–46

Combined defect prediction for large-area die-cast components using high-resolution multi-phase simulation

J. Jakumeit[1], H Behnken[1], R Laqua[1], J. Gänz[2], L. Becker[2], M. Fehlbier[3], O. Middelmann[4]

[1]Access e.V., Intzestr. 5, 52072 Aachen, Germany
[2]Siemens Industry Software GmbH, Nordostpark 3, 90411 Nürnberg, Germany
[3]GTK, University Kassel, Kurt-Wolters-Str. 3, 34125 Kassel, Germany
[4]KSM Castings Group GmbH, Cheruskerring 38, 31137 Hildesheim, Germany

E-mail: j.jakumeit@access-technology.de

Abstract. In the automotive industry there is a trend towards large components that are manufactured using high pressure die casting process (HPDC), in particular when many previously welded individual parts are to be replaced with one cast part to save costs and energy. In the case of large components, incipient solidification during mold filling cannot be ruled out. Casting defects due to cold running, air pockets and porosity cannot be separated spatially and temporally and influence each other. This stronger linking of the defects requires a realistic depiction of the casting process in the simulation, since the interactions between effects are more difficult to depict through approximations. The multi-phase approach presented here offers various options for this: The air is calculated as a compressible gas that is separated from the melt by a sharp boundary surface. Reduced melt flow due to solidification is represented by a porous media approach, followed by a complete flow stop at high solids. The formation of porosity due to volume shrinkage is coupled with a gas evaporation model. The multi-phase approach was validated by casting trials using a specially designed test geometry for thin-walled aluminum HPDC applications. First results of an industrial application are shown.

1. Introduction

Of all casting processes, the die casting process is the one with the highest plant costs as well as being technologically complex and demanding. It enables the production of complex, thin-walled and large cast components in large quantities and good reproducibility with high mechanical properties and high dimensional accuracy at the same time. The number of process parameters that are decisive for the quality of the casting is high: the controllable movement of the casting piston, the design of the casting mold and its temperature control, the extraction of air from the casting mold before it is filled, the spraying of the opened casting mold and the cycle time for the entire process cycle for a cast component. This results in a high optimization potential for the die casting process in relation to the process development time, the casting quality and the energy consumption. Computer simulation is used to accelerate process optimization. For years, several software packages have been established on the market that offer simulation models for die casting [1-4]. In order to limit the computing time to a few hours up to a few days, approximations are increasingly used. The influence of the air and the stopping of the melt during solidification are represented by models, some of which are greatly simplified. Also, the thin-walled areas can often only be resolved with a few mesh cells.

A three-phase simulation of casting processes is the basis for the simulation approach in this work: Liquid melt and gas are treated as compressible fluids separated by a sharp interface. The air is compressed and can thus build up a counterforce to the melt shooting into the mold, which has a decisive influence on the flow of the melt. Reduced melt flow due to solidification with increasing fraction solid is first calculated by a temperature dependent viscosity followed by a porous media approach before the melt flow is frozen entirely at high solids content. The multi-phase simulation approach was implemented in the flow solver STAR-CCM+ [5,6].
A comprehensive review of porosity modelling for casting processes was given by Stefanescu [7]. As mentioned in this review, one of the first porosity models validated by experiment was that of Beech [8]: based on the temperature dependent density and fraction solid of the melt the volume deficit is calculated for each time step and the amount of melt removed from the top region of the casting in respect to the gravity vector. The model assumes, that due to gravity the remaining melt fills the lower part of the cavity and the void gathers at the highest point. Modifications of this model are widely used in commercial casting simulation software. In fast solidification process like high pressure die casting (HPDC) with not negligible melt flow during solidification, the assumption that void gathers at the highest point is found to be not valid [9]. The Niyama criterion is a standard approach for micro-porosity prediction [10]. It was extended to the dimensionless Niyama criterion by Carlson and Beckermann [11] and the Böttiger-Laqua-Jakumeit criterion [12] by one of the authors. The Niyama criterion were found to be not applicable for the porosity prediction in HPDC-applications [9]. Beckermann und Monroe [13] came to the same conclusion when developing a porosity model for HPDC-applications. Their focus was the interplay between gas and shrinkage porosity, which was modelled by a sophisticated gas porosity model [14]. Recently, Khalajzadeh and Beckermann developed a new shrinkage porosity model [15,16] based on the observation, that porosity forms in regions where the solid fraction is the lowest.
The main focus of this work is the combination of air entrainment and shrinkage porosity formation in HPDC applications. To address this topic a coupled porosity model was developed based on the pressure calculation in melt and gas and the compressibility of the melt: If feeding is no longer possible due to solidification, the compressible melt compensates for volume shrinkage in areas that are still liquid by reducing pressure, which leads to melt expansion. If the pressure drops below a critical level, pores are formed by a gas evaporation model to keep the pressure at a defined level. When the velocities in the melt fall below a critical value, the flow simulation is stopped and porosity formation until complete solidification is calculated using a purely thermal shrinkage porosity model following the model of Beech [8] but forming the void at the location of lowest solid fraction [15,16].
Details of the simulation approach are given in the following chapter, followed by a description of the test casting geometry and the validation of the temperature distribution calculated by the simulation in chapter 3. Chapter 4 focuses on the porosity prediction by the coupled porosity model for a part of the full test geometry, the three-cam geometry. Finally, the model is applied to an industrial casting in chapter 5 before chapter 6 summarises the findings.

2. Theory

Simulation of the HPDC process is based on a multiphase simulation, which correctly treats both air and liquid melt as compressible fluid. The treatment of the air as a compressible gas phase enables the flow solver STAR-CCM+ [6]. The interface between air and melt is captured by a volume of fluid approach, a standard approach for describing interfaces in turbulent flow processes [17-19]. With the methods used, a special numerical trick, the High-Resolution-Interface-Capturing (HRIC) [19,20], is used to sharply image the interface even with very dynamic flow processes, such as mold filling in HPDC, a prerequisite for the correct calculation of the forces between air and melt as well as the surface tension. In the case of large-area components in particular, it cannot be ruled out that the melt will begin to solidify before the component is completely filled. The associated deceleration of the melt until it comes to a complete stop is described by a temperature-dependent viscosity model when the solids content is low. With a higher proportion of solids, the forces slowing down the melt are calculated using a model that describes the solidified melt as a porous medium with a permeability calculated using the Carman-Kozeny equation [21,22]. Above a solids content of 70%, the flow of the melt is completely stopped.

The multi-phase approach used has been validated in various applications by comparison with casting results [9,23].

The coupled calculation of the melt and gas phase was used to develop a fully coupled porosity model: During mold-filling, pressure build-up and starting solidification, flow and pressure are calculated coupled in the melt and gas phase. Both phases are treated as compressible fluids. If a region is separated from the piston chamber and inlet, volume shrinkage during solidification leads to a pressure drop in this region, which is determined by the pressure dependence $d\rho/dP$ of the compressible melt ($\rho(P)$ is the density of the melt). If the pressure P drops below a critical value P_{crit}, the pressure is kept close to the critical value by the compensation of the volume shrinkage of the melt by the evaporation of an additional pore gas phase. The volume increase of the pore gas phase ΔV is given by:

$$\frac{\Delta V}{V} = \frac{(P_{crit} - P) * \partial\rho/\partial P}{\rho + (P_{crit} - P) * \partial\rho/\partial P}$$

The pore gas phase is treated as a new ideal gas phase, which is separated from the melt phase by a VoF approach. The concept of this coupled porosity model was demonstrated for a cylindrical geometry in [9].

3. Validation castings

A special HPDC test geometry was designed to investigate misrun, gas entrainment and porosity in one geometry (see [9] for more details). Figure 1a gives a photo of the moving die of the test geometry and figure 1b the whole geometry with the four test geometries: two-hole geometry to investigate misrun and cold shuts, steps geometry for air entrainment and porosity analysis, three-cam geometry to enforce shrinkage porosity and U-profile geometry to analyze gas entrainment.

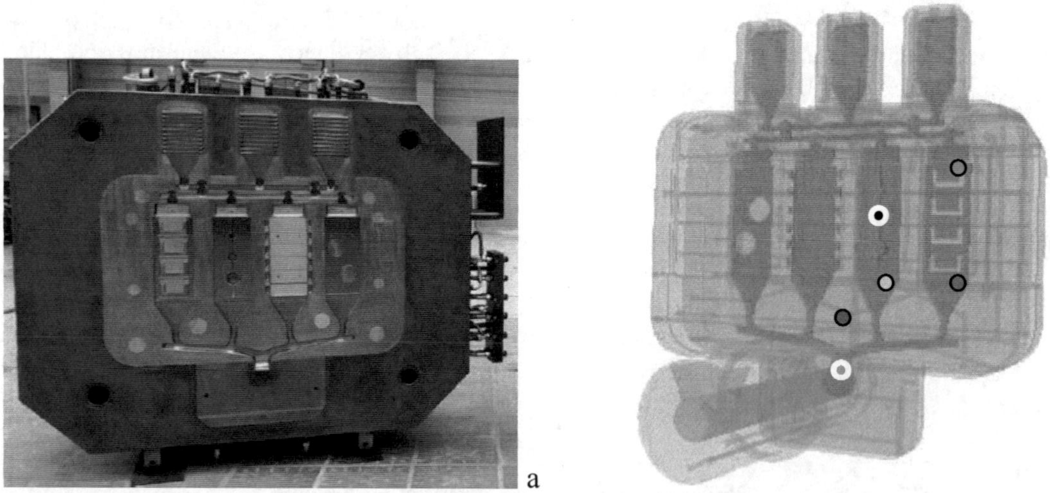

a b

Figure 1: Inner side of the moving die (a) and the geometry with four test cases two-holes, steps, three-can and u-profiles (b) and the shown in a)

Six thermocouples were placed in the geometry to validate the thermal settings of the simulation (see figure 1b). Several casting cycles including dosing, shot, intensification, solidification, die opening, part removal, spraying, drying and die closing were simulated with a pure thermal approach to get the temperature distribution of the cyclic process. Figure 2a compares the calculated temperature curves at the thermocouple position with the measurement results. The casting trials were performed on the GTK cold chamber machine (Bühler Carat 1400 t) using alloy AlSi9Cu3, a filling temperature of 680 °C, a maximal piston speed of 3.4 m/s, a preheating temperature of 250 °C, an intensification pressure of 500 bar and an vacuum of 50 mbar. Material data for AlSi9Cu3 and the steel die was taken from the Access material database AMat.pro. By adjusting the heat transfer coefficients (HTC, figure 2b) of the simulation and the cooling and heating of the die a good agreement between measured and calculated temperature curves could be found (see figure 2a).

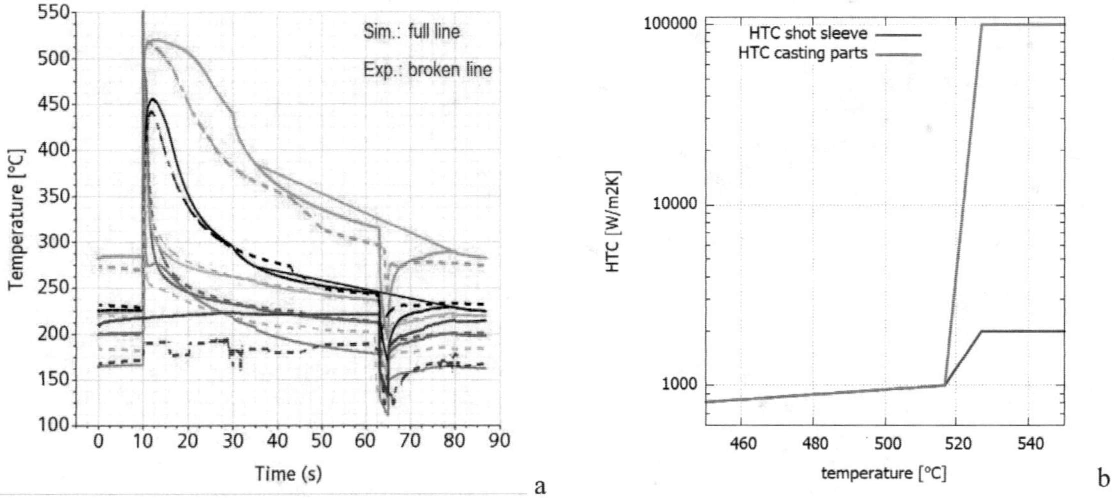

Figure 2: Comparison of measured and calculated temperature curves at the 6 measurement points (a) (see figure 1b) and the calibrated temperature dependent HTC (b)

4. Simulation of the three cam porosity geometry

In order to investigate the coupled porosity model with a high resolution simulation in acceptable computation times, a separate three cam porosity geometry was developed for simulation. The three cam geometry was extracted from the full test geometry and a simple inlet and outlet were added to form a reduced test case. Figure 3 shows the geometry and high resolution mesh with 2.4 Mio cells, 1.8 Mio in part region. 3 prism layer were used for a high resolution of the velocity and temperature gradients at the melt/mold interface. While in the full test geometry the piston movement is simulated by a moving mesh approach, a velocity inlet is used for the three cam porosity test geometry. An inlet velocity of 2 m/s results in velocities at the ingate of the geometry of ~ 100m/s, similar to the velocities calculated for the full geometry. The melt temperature at the inlet was set to 650 °C corresponding to the melt temperature before the shot calculated by the simulation of the full test geometry. The calibrated HTC-values (figure 2b) were used for all simulations.

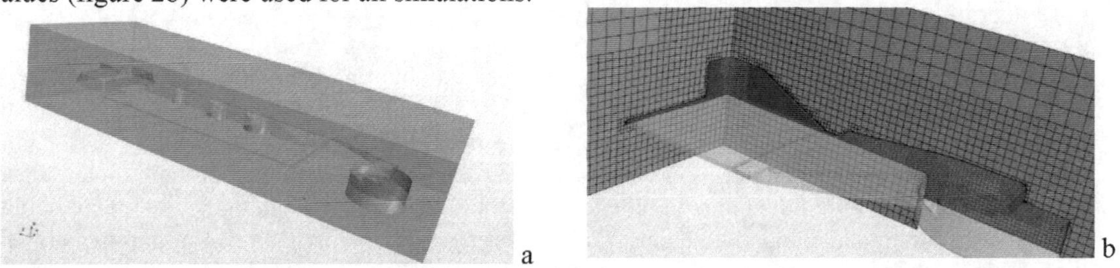

Figure 3: Three cam test geometry with simple inlet and outlet (a) and the high resolution mesh with 2.4 Mio cells, 1.8 Mio in part region. 3 prism layer at the melt/mold interface were used for a high resolution of the velocity and temperature gradients.

The simulation can be divided in to four steps:

1. The first step is the filling of the geometry with melt. Figure 4 shows the velocity (a) and entrained air distribution (b) after 0.055 s at the end of the filling step. Velocities up to 100 m/s are still found at the narrow connection between three cam geometry and overflow region. Velocities at the ingate are around 50 m/s. Air is still present in large parts of the cavity with volume fractions of up to 20

IOP Conf. Series: Materials Science and Engineering 1281 (2023) 012027 doi:10.1088/1757-899X/1281/1/012027

%. Small spots in the three camp geometry and larger regions in the overflow region and outlet are still filled with air.

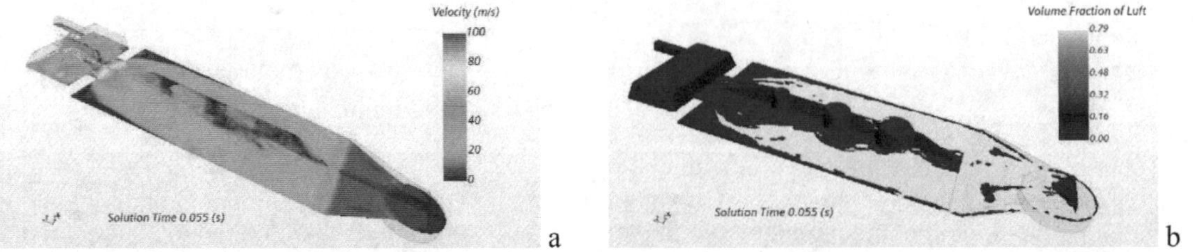

Figure 4: Velocity (a) and entrained air (b) distribution after 0.055 s at the end of the filling step.

2. The second step is the pressure intensification. When melt reaches the outlet, the pressure outlet is replaced by a wall boundary in the simulation to depict the closing of the exhaust valve. The inlet velocity is controlled to increase the pressure up to the intensification pressure of 500 bar and to keep the pressure at this level. As figure 5a shows, velocities up to 10 m/s can still be found in the cavity. The high pressure compresses the trapped air to less than 1% volume fraction for most of the cavity (see Figure 5b). Up to 5% air inclusions can only be found at the plate edge of the three-cam geometry and in the cam and overflow region.

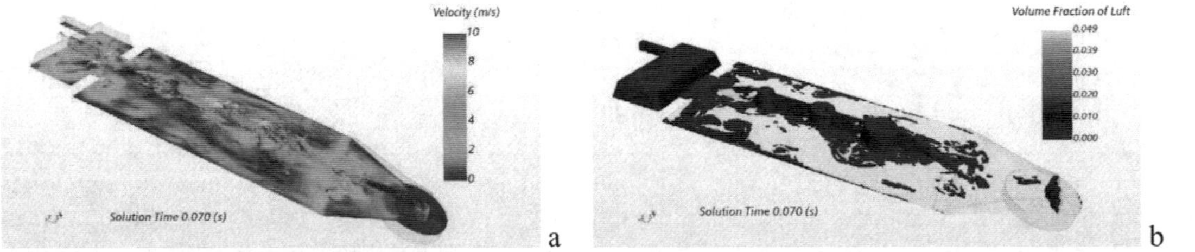

Figure 5: Velocity (a) and entrained air (b) distribution after 0.1 s at the end of the intensification step.

3. The third step is the coupled simulation of flow and solidification until the maximal velocities in the cavity drop to values below 0.3 m/s. In principle the step could be prolonged until the part is completely solidified. Since the simulation of flow demands small time steps, a coupled flow and solidification simulation of many seconds leads to simulation times of weeks. As figure 6a shows most parts of the geometry are already solidified after 1.1 s and flow with velocities above 0.1 m/s are found only closed to the ingate (figure 6b). Figure 6c gives the pressure distribution after 1.1s. The flow stop model freezes the pressure when the volume fraction of solidified melt is above the flow stop threshold and the model becomes active. Therefore, the low pressure at the top of the bridge between the two cams close to the outlet and the bridge in direction outlet is still visible, which leads to the evaporation of the pore gas. Figure 6d shows the pore gas distribution at the top if the bridges as a result of the reduced pressure and the coupled porosity model in good agreement with the experimental findings (see figure 7). The average pressure in the still liquid region is used as critical pressure for the gas evaporation assuming that dissolved hydrogen or air through gaps in the die can easily fill the void region. In the cams, the pressure is still not below the average pressure and pore gas is not evaporated. Looking at the distribution of solidified melt in figure 6a, it becomes clear that the cams are still connected by a small link of liquid melt with the inlet or outlet near overflow region, preventing a pressure reduction due to volume shrinkage of the melt.

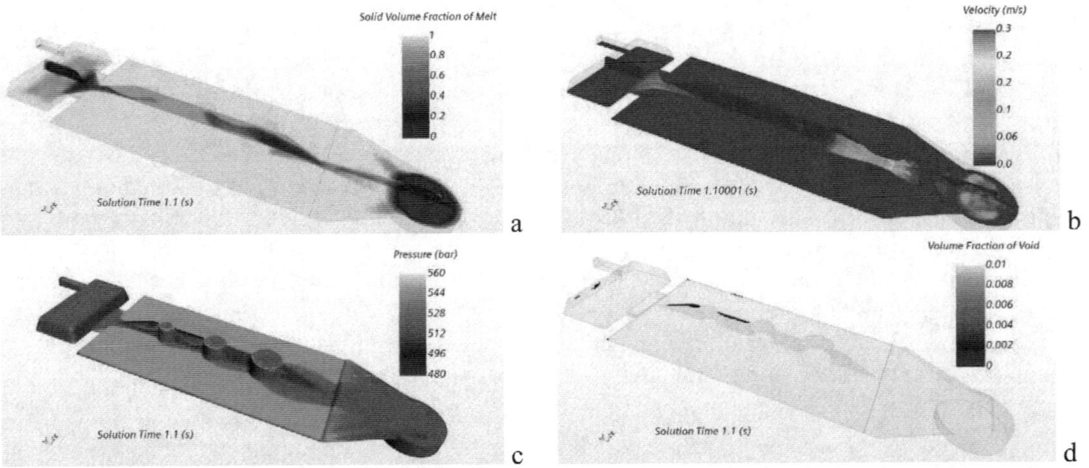

Figure 6: Fraction solid (a), velocity (b), pressure (c) and porosity (d) distributions after 1.1 s at the end of the coupled flow and solidification step.

4. The final step is the pure thermal calculation of the complete solidification while neglecting the flow. Figure 7 shows the final porosity distribution after all 4 steps. The pure thermal porosity model gives pores close to the middle of the cams, when the volume deficit is compensated by pore gas phase at the region of highest temperature or highest liquid fraction.

Along with the pore gas distribution figure 7 gives the porosity found in the real casting. The porosity prediction by the coupled porosity model agrees well with the experimental findings, even when the simplified three cam geometry is used. The filling of the cavity seems to have only a small effect on the shrinkage pore formation after complete filling of the cavity and the simplifications for the reduced test case do not effect a correct porosity prediction.

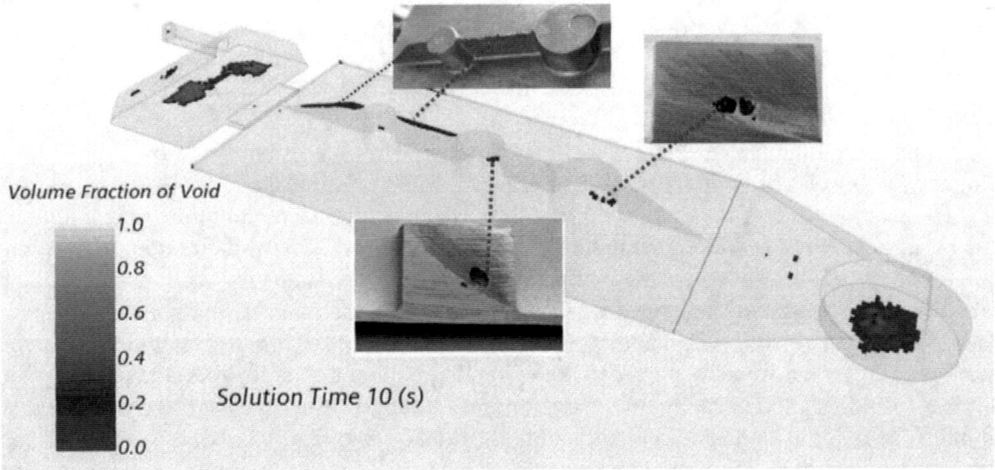

Figure 7: Comparison of the porosity prediction by the coupled porosity model and the experimental findings.

5. Industrial application

The coupled porosity model was applied to a structural part of the industrial partner KSM Castings Group GmbH. Figure 8 gives the geometry (a) and polyhedral mesh (b) together with the temperature and melt distribution during filling after 0.75 s (c) and the air entrainment after complete filling and pressure intensification (d). Shrinkage porosity was identified at two locations (C5, B2, see figure 9a and 9b) by CT-scans (figure 9c and 9d). Simulation was so far not able to predict these porosity locations.

The fully coupled mold filling simulation shows, that the critical regions C5 and B2 are filled after 0.75 s (figure 8c) and no entrained air can be found in the part after complete filling and pressure intensification (figure 8d). The entrained air is pushed into the overflow regions. Shrinkage porosity was indeed indicated by the pure thermal porosity model at the end of step 4 at the two locations C5 and B2 (figure 9e and 9f). The porosity criterion also marked other locations, where no porosity was found in the real casting. These wrong indications might be caused by a hypersensitivity of the model. Beckermann in [13] named several effects, which are not addressed by porosity models, and my cause a lower porosity in the real casting compared to the prediction by simulation.

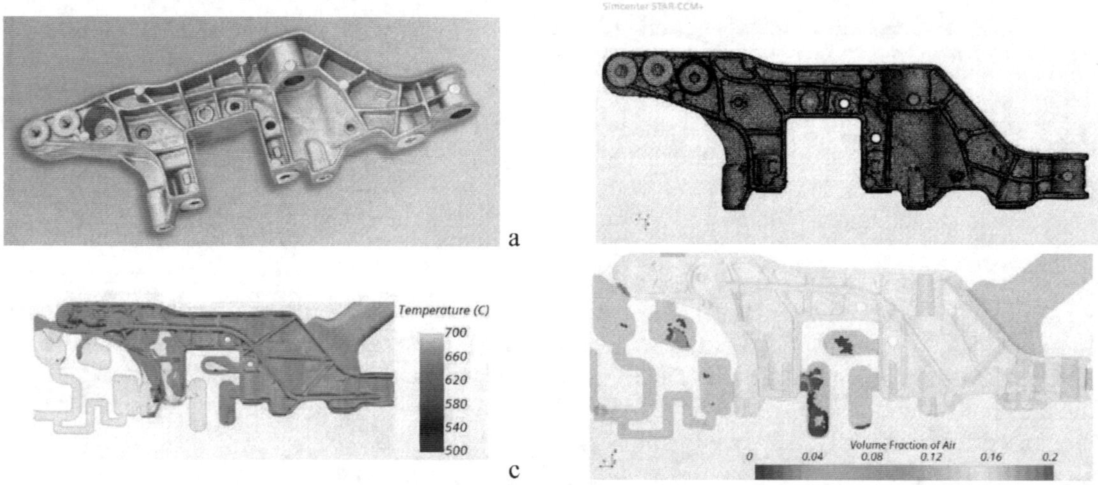

Figure 8: Geometry a), polyhedral mesh b), temperature and melt distribution during filling after 0.75 s c) and the air entrainment after complete filling and pressure intensification d).

6. Conclusion

Porosity formation in high pressure die casting (HPDC) applications is analyzed by a multi-phase flow and solidification simulation: Liquid melt and gas are treated as compressible fluids separated by a sharp interface. Reduction of melt flow due to solidification is achieved by a temperature dependent viscosity, followed by a mushy-zone model and a flow-stop function for high solid fractions. If feeding is no longer possible due to solidification, the compressible melt compensates for volume shrinkage in areas that are still liquid by reducing pressure, which leads to melt expansion. If the pressure drops below a critical level, pores are formed by a gas evaporation model to keep the pressure at a defined level. When the velocities in the melt are below a critical value, the flow simulation is stopped and porosity formation until complete solidification is calculated using a purely thermal shrinkage porosity model.

To examine misrun, cold shut, gas entrainment and porosity in HPDC casting a special test geometry was developed. The simulation of the test geometry was validated by a comparison of measured and calculated temperature curves at six points in the geometry. The calculated time dependent temperatures of a casting cycle at the measurement points agree well with the measured results of the real casting process.

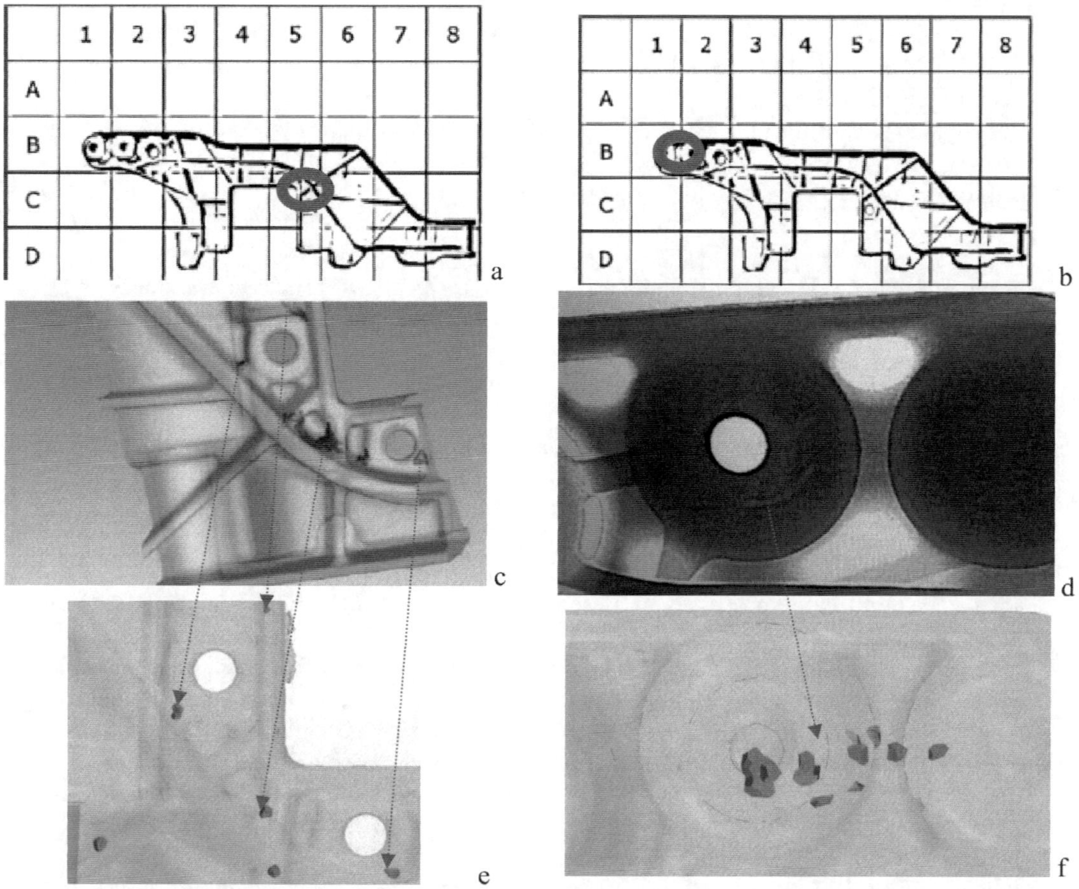

Figure 9: Industrial HPDC application showing porosity at two positions C5 (a) and B2 (b) of the casting revealed by CT-analysis (c) and (d) and the porosity prediction by the coupled simulation model for the position C5 (e) and B2 (f).

The high-resolution simulation of the three-cam geometry, specifically developed to represent shrinkage porosity, proves the predictive quality of the coupled porosity modeling. The intensification pressure reduces the volume of entrained air to a small volume fraction of less than 1%. Early solidification at the tip of the bridges connecting the cams results in shrinkage porosity caused by local pressure reduction. Late solidification in the cams leads to shrinkage pores that are closed toward the center of the cams in good agreement with the experimental results.

Finally, the coupled porosity model was applied to an industrial casting. The simulation reveals that entrained air is collect inside the overflow regions and shrinkage porosity forms at a late stage of solidification, identified by the pure thermal porosity model. The porosity, found in the real castings, is correctly predicted by the coupled porosity model.

Looking at the porosity prediction for the test cases and the industrial application, the shrinkage porosity dominates the porosity due to entrained air. In well-designed gating systems the entrained air is mainly pressed to the overflow regions and the remaining air volume in the part is low. In general, the results demonstrate the ability of the coupled porosity model to describe both effects and their interaction.

Acknowledgments

The work was supported by BMWK (Bundesministerium für Wirtschaft und Klimaschutz) joint project *SPEED*, funding contract 03ET1516A, and INNOKOM project *Fast Solid*, funding contract 49VF170031. Siemens PLM provides STAR-CCM+ licenses within agreement 60068580.

References

[1] Duan H Z, Shen J N, & Li Y P (2012). Comparative analysis of HPDC process of an auto part with ProCAST and FLOW-3D. In *Applied Mechanics and Materials* (Vol. 184, pp. 90-94). Trans Tech Publications Ltd.

[2] Bodhayana M R and Ramesha N 2014 *Int. J. Research in Engineering and Technology* **03** (11) 1163

[3] Aghion E, Moscovitch N and Arnon A (2007). The correlation between wall thickness and properties of HPDC Magnesium alloys. *Materials Science and Engineering: A*, *447*(1-2), 341-346.

[4] Wang T, Huang J, Fu H, Yu K and Yao S (2022). Influence of Process Parameters on Filling and Feeding Capacity during High-Pressure Die-Casting Process. *Applied Sciences*, *12*(9), 4757.

[5] Simcenter STAR-CCM+: https://www.plm.automation.siemens.com/global/de/products/simcenter/STAR-CCM.html

[6] Jana S, Kättlitz O, Hediger F, Jakumeit J and Aguilar J (2012). MCWASP XIII, IOP Publishing, IOP Conf. Series: Materials Science and Engineering.

[7] Stefanescu, D. M. (2005). Computer simulation of shrinkage related defects in metal castings–a review. *International Journal of Cast Metals Research*, *18*(3), 129-143.

[8] Beech J, Barkhudarov M, Changands K, Chin B (1998) in 'Modelling of casting, welding and advanced solidificationprocesses VIII', (ed. B. G. Thomas and C. Beckerman), 1071; Warrendale, PA, TMS.

[9] Jakumeit J, Behnken H, Laqua R, Mbewou S, Fehlbier M, Gänz J, & Becker, L. (2021). Coupled Modeling of Misrun, Cold Shut, Air Entrainment, and Porosity for High-Pressure Die Casting Applications. In *Light Metals 2021* (pp. 865-870). Springer, Cham.

[10] Niyama T, Uchida M, Morika W A, Saito S (1982) *Int. cast metals journal* **7**:52-63

[11] Carlson K D and Beckermann C (2009). *Metallurgical and Materials Transactions A*, *40*(1), 163-175.

[12] Jakumeit, J., Jana, S., Böttger, B., Laqua, R., Jouani, M. Y., & Bührig-Polaczek, A. (2012, January). In *IOP Conference Series: Materials Science and Engineering* (**27**, No. 1, p. 012066). IOP Publishing.

[13] Beckermann C and Monroe A 2009 Porosity simulation for high pressure die castings *Die Casting Engineer* (The North American Die Casting Association) ISSN 0012-25JX

[14] Carlson K D, Lin Z and Beckermann C (2007). Modeling the effect of finite-rate hydrogen diffusion on porosity formation in aluminum alloys. *Metallurgical and Materials Transactions B*, *38*(4), 541-555.

[15] Khalajzadeh V, Goettsch, D D and Beckermann C (2016). In *Shape Casting: 6th International Symposium* (pp. 35-42). Springer, Cham

[16] Khalajzadeh V and Beckermann C (2020). Simulation of shrinkage porosity formation during alloy solidification. *Metallurgical and Materials Transactions A*, **51**(5), 2239-2254.

[17] Ferziger J H and Peric M (ed) 2002 *Computational Methods for Fluid Dynamics* (McGraw-Hill) ISBN 2-540-42074-6

[18] Teskeredzic A, Demirdzic I and Muzaferija S 2002 *Num. Heat Trans*. B **42** 43

[19] Demirdzic I and Muzaferija S 1995 *Computer Methods in Appl. Mech. and Eng.* **125** 235-255.

[20] Muzaferija S and Peric M 1999 Computation of free surface flows using interface-tracking and interface-capturing methods. *Nonlinear Water Wave Interaction*, ed O Mahrenholtz and M. Markiewicz, (Southampton: WIP Press) 59-100

[21] Sabau S and Viswanath S 2002 *Met. Mat. Trans.* B **33B** 131

[22] Poirier D R (1987). Permeability for flow of interdendritic liquid in columnar-dendritic alloys. *Metallurgical Transactions B*, *18*(1), 245-255.Poirier D R, Yeum K and Mapples A L 1987 *Metall. Mater. Trans.* A **18** 1979

[23] Jakumeit, J., Subasic, E., & Bünck, M. (2016). Prediction of misruns in thin wall castings using computational simulation. In Shape Casting: 5th International Symposium 2014 (pp. 253-260). Springer International Publishing.

IOP Conf. Series: Materials Science and Engineering 1281 (2023) 012028 doi:10.1088/1757-899X/1281/1/012028

Model of multiphase flow, superheat transport, and particle capture in a steel continuous caster

Mingyi Liang[1], Seong-Mook Cho[2], Xiaoming Ruan[3], and Brian G Thomas[4]

[1]Convergent Science, Madison, WI, 53719, USA
[2]Metallurgical Engineering, Pukyong Nat. Univ., Nam-gu, Busan, 48513, S. Korea
[3]Research Inst. Metallurgical Processes, Baosteel, Shanghai, 201900, China
[4]Mechanical Engineering, Colorado School of Mines, Golden, CO, 80401, USA

bgthomas@mines.edu

Abstract. A model of turbulent multiphase flow, heat transfer, and particle entrapment during continuous casting of steel is presented. The model includes the top 7m of the vertical and curved strand, and considers the effects of argon gas injection and thermal buoyancy. RANS model flow results are compared with LES. Lagrangian particle transport through the Eulerian-Eulerian multiphase flow field from the k-ε RANS model is based on random walk and features advanced particle capture criteria, improved from previous work by including anisotropic turbulent velocity fluctuations near the walls. Particle capture via 3 mechanisms is included: capture by solidified hooks at the meniscus, entrapment between dendrites, and engulfment by surrounding large particles. The fluid flow and bubble/inclusion capture results are validated with plant measurements, including nail board dipping tests and ultrasonic tests of particle locations, and good agreement is seen. The superheat has negligible effect on flow in the mold region but causes complex flow in the lower strand by creating multiple recirculation zones due to the thermal buoyancy. With high (30 K) superheat, this leads to less penetration, and slightly fewer and shallower capture of particles. Lower (10 K) superheat may enable significant top surface freezing, leading to very large internal defect clusters. Lower superheat also leads to deeper meniscus hooks, and more surface capture. Capture bands occur near the transition from vertical to curved, where downward liquid flow velocity balances the particle terminal velocity.

1. Introduction

The entrapment of inclusions, bubbles, and other particles is one of the most important and chronic sources of defects in continuous casting of steel slabs. Inclusion sources include insufficient steel refining, refractory erosion, slag entrainment, reoxidation, and nozzle clogging. Argon gas bubbles are often injected to lessen nozzle clogging and may become covered with small inclusions as they are transported through the molten steel. If any of these particles become entrapped into the strand during continuous casting, this can lead to serious product defects. Meniscus hooks can capture particles, leading to surface defects such as blisters. Internal defects arise where particles are captured deeper in the caster. Small particles can be entrapped between dendrites, while larger particles can be surrounded by engulfment, leading to slivers in the rolled product [1]. Large particles are especially problematic.

Particle defect problems can be mitigated by improved upstream steelmaking and refining operations, careful control of refractories to avoid erosion, leaks leading to reoxidation, and control of turbulent fluid flow in the mold and strand. Flow depends on the nozzle and strand geometry, casting speed, argon gas injection, electromagnetic systems, stopper or slide-gate positioning, nozzle submergence depth, and superheat. In order to optimize flow parameters, it is crucial to have a quantitative understanding of how these parameters affect both the turbulent flow pattern and particle entrapment.

Plant nail-board measurements can quantify surface velocity in the mold, but cannot measure velocity inside the strand. Plant measurements can quantify inclusion content in molten steel samples and final product, but are very difficult to obtain during the process. Thus, significant effort has gone into understanding and quantifying flow and particle capture using advanced computational models, which is reviewed elsewhere [2-4]. In this work, turbulent flow and particle entrapment in a typical commercial steel slab caster are investigated using a system of computational models, and sample results focus on the effect of superheat.

2. Computational Models

A system of computational models was applied in this work to simulate turbulent multiphase flow, heat transfer, and particle entrapment during continuous casting of steel, as described in recent work by the same authors [1]. A flowchart of the submodels making up this system is shown in figure 1 [1]. The flow model considered the effects of argon gas injection and thermal buoyancy on the flow.

First, the mean bubble size needed by the multiphase flow model was predicted using a model of pressure distribution and flow rate within a slide-gate flow delivery system, PFSG [5], to find the local pressure, and a two-stage bubble size model for gas injection into downward flowing metal [6]. The results were implemented into a Rosin-Rammler distribution, based on water model measurements, in order to estimate the bubble size distribution for the particle transport model in the steel caster [1].

Next, the solidification heat-transfer model CON1D was calibrated to match plant measurements and applied to determine the shell thickness profile that defines the liquid pool domain, and the primary dendrite arm spacing (PDAS) profile [1]. The shell thickness profile was then implemented into the multiphase thermal - fluid flow model via UDFs for mass, momentum, and energy sinks.

Multiphase flow was computed with a steady Eulerian-Eulerian (EE) model of the Reynolds-Averaged Navier Stokes (RANS) equations for the molten steel and gas phases in Ansys Fluent, including the standard k-ε model for turbulence with standard wall function boundary conditions [1]. Effects of thermal buoyancy were included with a fully-coupled heat transfer model of superheat transport.

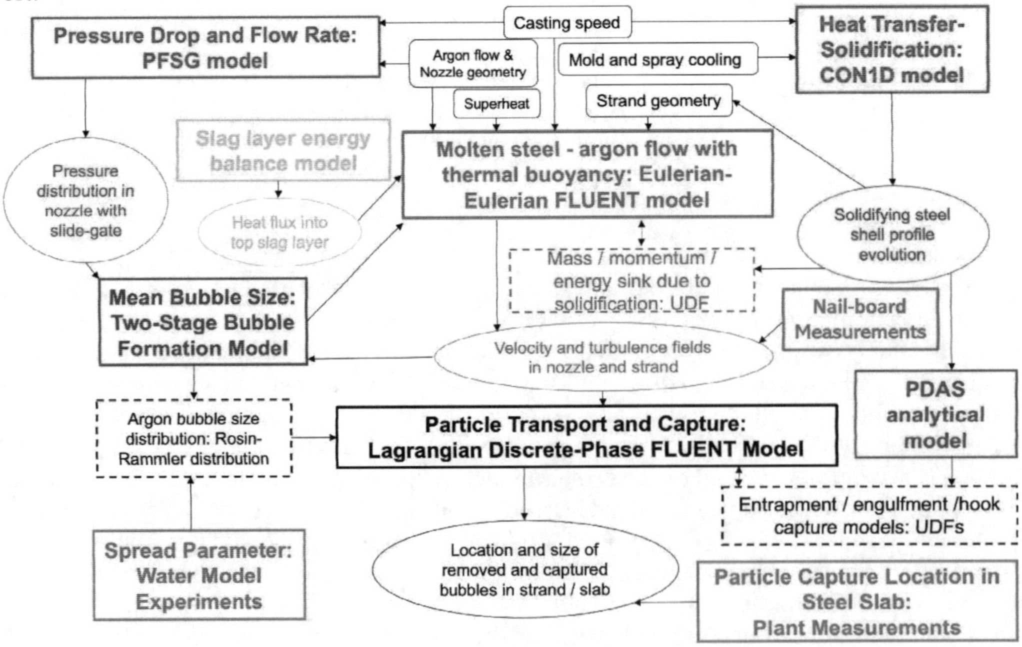

Figure 1. Model system flowchart

The model domain is shown in figure 2. Exploiting symmetry between the narrow faces, it included half of the nozzle, slide-gate flow control system, and 7m of the vertical and curved strand. The mesh included ~20,000 cells for the nozzle domain and ~1.1 million cells for the liquid pool in the strand. It was constructed entirely of hexahedral cells, which in the mold region were aligned with a 15° downward angle to align with the flow direction, and thereby reduce false diffusion. Taking the solidification front as the domain boundaries in the strand greatly aided accuracy and convergence, as wall laws could be applied, resolution of the steep gradients across the solidification front was not needed, and external heat transfer conditions in the mold and secondary cooling were not needed. Instead, in the thermal model, these boundaries were naturally fixed at the liquidus temperature of the typical low-carbon steel alloy. A simple energy balance model of the top slag layer was developed and verified with previous models to calculate the heat flux lost from the top surface of the molten steel [1].

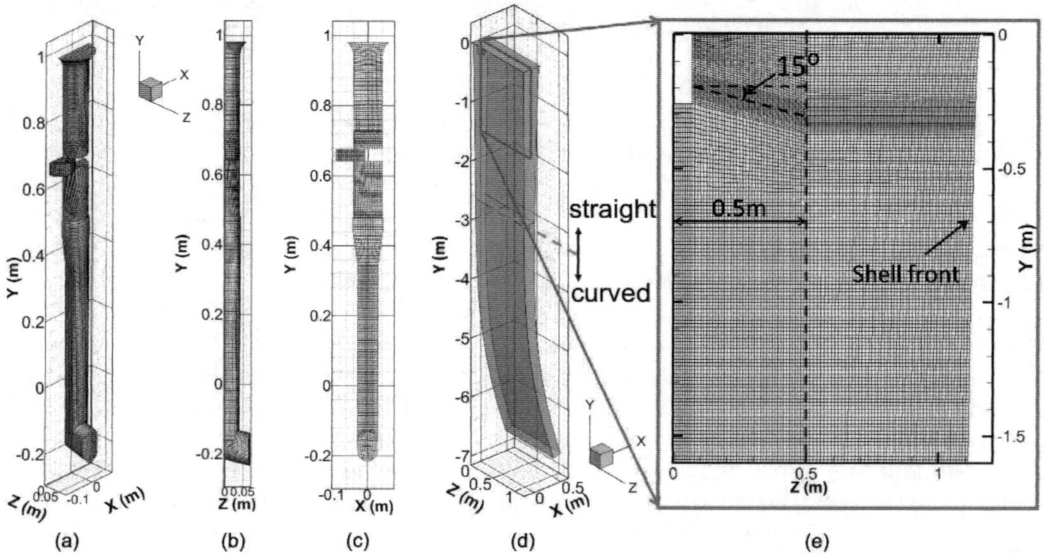

Figure 2. Flow domain and mesh: (a) nozzle 3D view, (b) nozzle front view, (c) nozzle center plane, side view, (d) liquid pool, and (e) liquid pool center plane

Next, large groups of the different bubble sizes were injected into the strand flow field and tracked using a Lagrangian discrete phase model (DPM) transport model in Ansys Fluent. Considering the EE RANS model, this means that the bubbles were modeled in two independent ways, and the results were manually compared for consistency. The DPM particle transport model adopts the Discrete Random Walk model (DRW) to calculate velocity fluctuations for each particle, based on the local turbulent kinetic energy, k, and assuming isotropic turbulence:

$$u_i' = \lambda_i \sqrt{\overline{u_i^2}} = \lambda_i \sqrt{\frac{2}{3}k} \quad (1)$$

where u_i' are the turbulent fluctuation velocities in x, y, and z directions. λ_i are Gaussian-distributed random numbers. This isotropic assumption is reasonable in the bulk flow region, far from the walls.

The particle capture submodel features advanced particle capture criteria, including capture by solidified meniscus hooks, entrapment, and engulfment between dendrites, as illustrated in figure 3. These three different capture mechanisms were implemented as user-defined functions (UDFs) in Fluent to evaluate the capture of each argon bubble touching the solidification front during the simulation [1]. A particle entering a small flow region beneath the meniscus hook is considered to be captured by the hook, if it stays longer than the critical time in that region. Elsewhere, the empirical relation for increasing primary dendrite arm spacing (PDAS) down the caster is applied to determine 1) if each particle contacting the shell is small enough to be entrapped between dendrites, or 2) if it requires a force balance to determine if the drag from cross-flow velocities and 5 other forces (buoyancy, lift, lubrication, van der Waals, and surface tension gradient forces) are small enough to allow it to become suspended long enough for capture by engulfment, giving dendrites time to grow around, surround, and capture it [1].

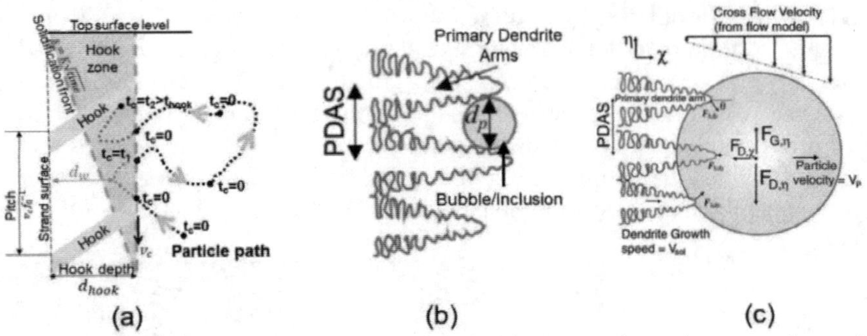

Figure 3. Particle capture mechanisms modeled: (a) capture by a downward-moving hook; (b) entrapment between primary dendrites arms; (c) engulfment by the growing solidification shell

When a particle larger than PDAS touches the solidification front, (domain wall), the velocity difference between the particle and the molten steel is used to calculate the drag and lift forces on this particle, which determines its capture fate.

In the "old UDF" used in previous work [1,7], the particle velocity is roughly the sum of the tangential components of the fluid velocity and the velocity fluctuation due to turbulence, based on the Random Walk Model, which depends on the local level of isotropic turbulence, k. The velocity components are calculated at the center of the near-wall cell. This overpredicts the particle velocity, which is supposed to decrease to moving at the casting speed in the casting direction, at the instant it touches the solidification front. In addition, the fluid velocity, which should be taken close to the wall according to the particle radius (< 1mm), is also overpredicted when taken at the cell center, with near-wall cell size ~10mm. The overpredictions of both particle and fluid velocity roughly cancel, which explains why the old UDF gives reasonable predictions of particle capture locations and fractions.

To improve the calculation of the tangential drag force, based on the difference between particle and fluid velocity near the wall, a "new UDF" is developed here. When a particle touches the solidification front, its velocity is set as the casting speed in the casting direction. The fluid flow velocity at the particle center is interpolated from the standard wall function [8,9]. Furthermore, calculation of the fluctuation velocity components due to the turbulence is improved by considering the anisotropic nature of turbulence near the wall. The normal component of the fluctuation velocity is dampened by the wall, while the lost momentum causes the parallel components to increase, according to empirical equations based on advanced DNS analysis of internal channel flow [12]. Specifically, the local isotropic turbulent fluctuation velocity predicted using the random walk model in the EE RANS flow model at the cell center is scaled into components in the new UDF according to

$$u_i' = f_i \lambda_i \sqrt{\frac{2}{3}k} \tag{2}$$

where u_n' is the turbulent fluctuation velocity component normal to the wall (where i=n), f_n is the corresponding empirical scaling factor, decreased to 0.07 according to the empirical equations for velocity fluctuations in turbulent channel flow [13], λ_i are random Gaussian-distributed numbers for each of the 3 velocity components, and k is the local isotropic turbulence from the EE RANS flow model. The average turbulent fluctuation velocity component parallel to the wall, u_p', is increased by setting f_p= 1.4 in this work. These f_i values also require making assumptions for the dimensionless distance from the wall, $y^+ = \frac{u_\tau y}{\mu_s} = 1$, the local friction velocity, $u_\tau = \sqrt{\frac{\tau_\omega}{\rho_s}} = 0.4$ Pa, and the local near-wall turbulence level, k=0.006 m^2/s^2, which is reasonable for 0.6mm particles. The difference between this flow velocity near the wall and the modified particle velocity is used to calculate the drag and lift forces in the new UDF. The capture locations and fraction of bubbles captured by both methodologies (old and new UDF) are compared in the results of this study.

Finally, a new post-processing method was applied to adjust the particle capture fractions to properly compare them with measurements in slab samples. First, a statistically-relevant number of particles (30,000 in this work) of each size is injected at the same time and simulated with the DPM model and capture criteria. Then the injection rate of the real particle size distribution entering the caster is factored to determine the number of particles captured per unit time, and (knowing the casting speed), per unit length of the slab product. Further details of these computational models are provided elsewhere [1,7].

3. Results

Two cases with different superheats (10 K and 30 K) were simulated on the same commercial caster, for 300 mm thick x 2300 mm wide slabs of low-carbon (0.055 %C) steel (1796 K liquidus), with density (kg/m^3) of 8278.2 - 0.7 T(K), cast at 0.6 m/min with 3.7 SLPM argon injection (7.2% hot gas volume fraction, with negligible density of 0.42 kg/m^3). The slide gate was 62% open (linear distance ratio), nozzle port angle 15° down, and submergence depth of 140 mm. A transient large eddy simulation (LES) of the turbulent flow for this caster took ~ 6 months to complete 71s of coupled LES-DPM simulation time on a 96-node high-performance computer, but was less accurate than the current steady EE-DPM multiphase thermal-flow model, which used a coupled algebraic multigrid solver on a staggered grid and took only ~5.5 h [1]. This was due to 71s being insufficient time for the transient turbulence to penetrate sufficiently deep into the caster, and perhaps also to the symmetry assumption.

The EE-DPM RANS model flow pattern is shown in figure 4 for both superheat cases in two vertical sections through the strand.

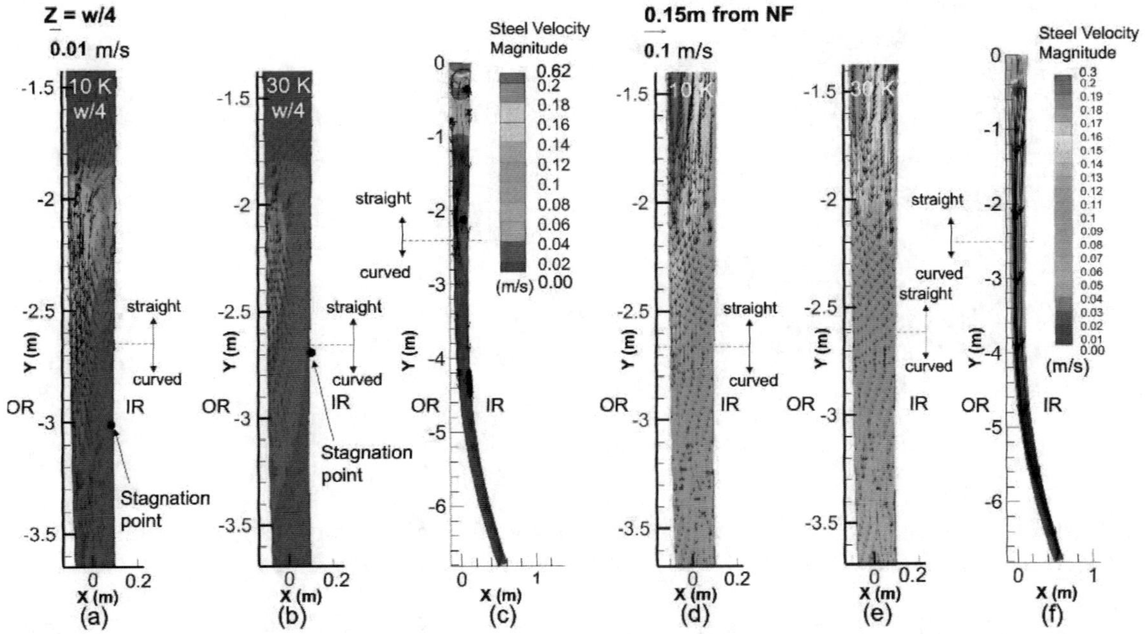

Figure 4. Molten steel velocity distribution—side views: on width/4 plane: (a) 10 K superheat close-up view, (b) 30 K superheat close-up view, and (c) 30 K superheat full view; on plane 0.15 m from NF: (d) 10 K superheat close-up view, (e) 30 K superheat close-up view, and (f) 30 K superheat

Flow in the mold has a stable double-roll flow pattern, with surface velocity < 0.4m/s, as shown in figure 5 (a), which is the best practice to avoid slag entrainment. The superheat has negligible effect on flow in the mold region, but causes more complex flow in the lower strand by creating multiple recirculation zones due to the thermal buoyancy. With high (30 K) superheat, the jets down the narrow face are slower, comparing 4(a) with 4(b), so have less penetration deep into the strand. This is due to the stronger extra recirculation zones in the lower strand. This leads to slightly fewer and shallower capture of larger particles, as shown in figure 5(b). Lower (10 K) superheat also has lower temperature near the meniscus (around the mold perimeter at the top surface), which leads to deeper surface hook formation, and more particle capture at the surface. In extreme cases, low superheat might enable significant top surface freezing, leading to very large internal defect clusters.

Most small particles are captured in the strand, as shown in figure 5 (b), due to their negligible buoyancy and easy entrapment between dendrites. In contrast, most large particles escape into the slag layer. The capture fraction experiences a large drop between 0.2 mm and 1mm, which corresponds to the PDAS, and a transition from particle entrapment to the more difficult mechanism of particle engulfment. Over 99% (old UDF) and 97% (new UDF) of the 0.6 mm bubbles escape into the slag layer. Even though only 0.6% (old UDF) and ~2% (new UDF) of these particles are captured into the steel product, this still represents several hundred or a few thousand large particles captured per meter of cast slab length. Larger particles (>= 1 mm diameter) are even more rare to be captured, representing only <0.02% of those entering the process. This is still important, however, because even one such large particle can cause a serious defect in the rolled product, resulting in downgrading or scrapping of the coil, for stringent finished products, such as exposed automotive surfaces.

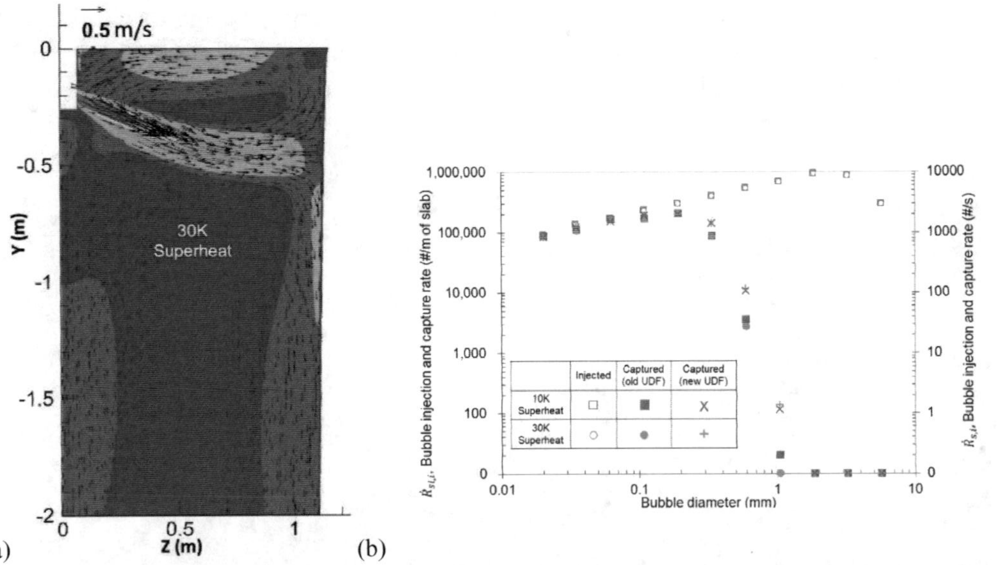

(a) (b)

Figure 5. (a) Molten steel velocity distribution—front view with 30 K superheat; (b) Rates of injected and captured particles, comparing 10 K and 30 K superheat

Figure 6 shows the capture locations of 0.6 mm bubbles on the inner radius and outer radius sides of the wide face (WFIR and WFOR). In figure 5, even though the new UDF predicted more 0.6mm bubbles captured, the bubbles are captured shallower in the caster, as shown in figure 6. More bubbles are captured near the SEN region, especially with the new UDF. Where the molten steel flows generally downwards, it is more likely to balance the upward bubble terminal velocity, and facilitate engulfment.

Figure 7 shows the distribution of final locations where particles were either captured in the strand during casting or escaped into the top-surface slag layer, for two different sizes, predicted with the old UDF. Most of the small and intermediate (0.1 mm) size particles are captured, figure 7 (a), as explained already. Most of these are captured in the mold region, leading to surface or slightly subsurface defects. Below the vertical portion of the strand, only ~7% of these particles are captured, leading to internal defects. More of these (~5%) are found on the inside radius of the curved shell. As they circulate within the lower recirculation zones, these particles gradually spiral upwards towards the inside radius of the curved strand, owing to their buoyancy. Smaller particles (not shown) are almost equally distributed between inside and outside radius, as their buoyancy is too small to affect their motion, relative to the turbulent flow.

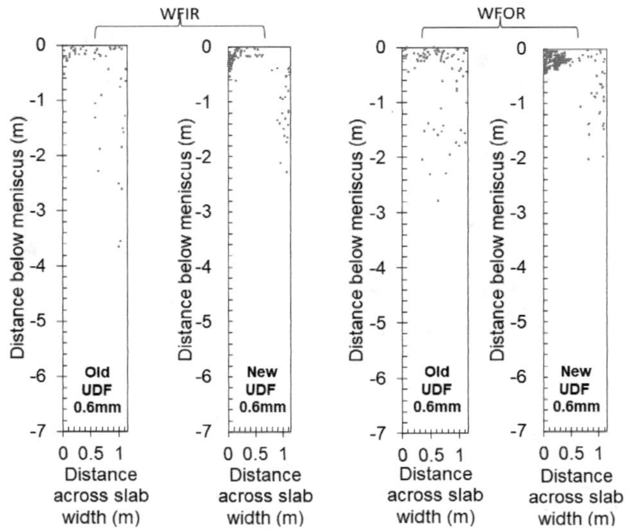

Figure 6. Capture locations of 0.6mm bubbles on the inner radius and outer radius sides of wide face, predicted with the old and new UDFs

Figure 7 (b) shows the fate of larger (0.6 mm) particles. Most (99.4%) of these bubbles escape into the top slag layer. A few (0.46%) of them are captured in the vertical portion of the strand, especially towards the outside radius in this case. Deeper in the strand, only a tiny fraction (~0.01%) of these particles are captured, with more towards the inside radius, owing to their buoyancy and due to flow recirculation that arises at the transition from vertical to curved.

4. Model Validation

Plant measurements were conducted at Baosteel to validate the simulation results of this study, including the flow pattern and particle capture predictions. Nailboard dipping test measurements revealed the velocity profile across the top surface in the mold, shown in figure 8 (a) RHS as circular symbols (flow towards SEN) and square symbols (cross flow velocity). The current (EE-DPM RANS) thermal-flow model predictions are roughly consistent with the measurements, with velocities towards the SEN reaching ~0.25 m/s. The LES model results show significant reverse flow near the SEN and crossflow towards the outer radius of the mold, which were not observed in the measurements.

Ultrasonic (UT) tests were conducted on sixteen plates at Baosteel, measuring 0.21 × 3.9 × 36.2 m (thickness × width × length) rolled from 0.3 × 2.3 × 4.31 m slabs samples. Figure 8 (b) shows an example of the UT tests for a case with 29 K superheat. Particle capture bands occur at ~40 – 60 mm beneath the strand surface. This corresponds to ~1.6 – 3.5m down the caster, which includes the transition from vertical to curved. In this region, the downward liquid velocity roughly balances the particle terminal rising velocity, which increases the chance of suspending large particles at the solidification front long enough to be captured by the engulfment mechanism. Similar capture bands have been observed in other recent measurements conducted on commercial continuous-cast slabs [10,11].

Figure 8 (b) and (c) also show the raw model predictions using the old and new UDF. Only the capture locations of large (0.6 mm) bubbles are shown, because the UT measurements cannot detect small defects. Most of the predicted particles captured are near the surface, where the UT measurements are not accurate. The predicted number of inclusions in the capture band is much less than measured. However, this is due to the injected number of 0.6 mm bubbles (30,000) being far less than actually injected during the appropriate time interval in the commercial process. After accounting for this via post-processing, both the measurements and calculations show ~200 particles within the capture band (which is indicated by red dashed lines in the figure).

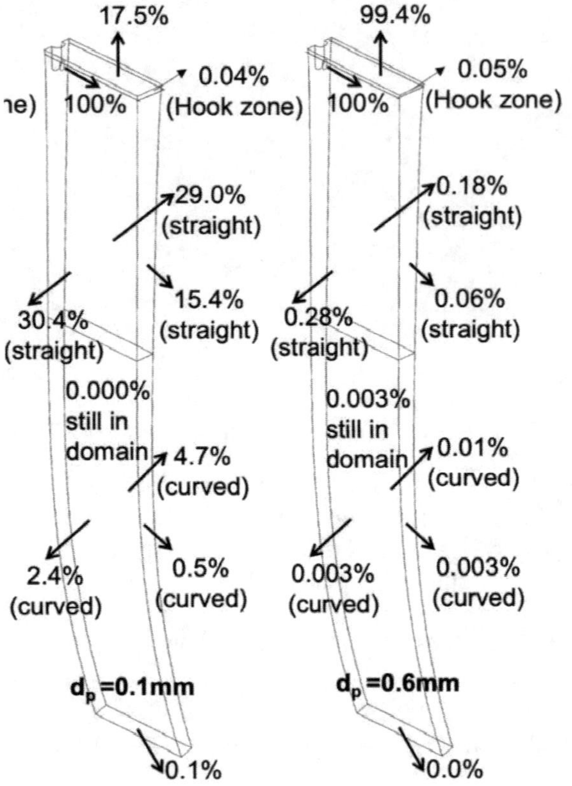

Figure 7. Final location of entrapped particles with 10 K superheat predicted with the old UDF: (a) intermediate (0.1mm) particles and (b) large (0.6 mm) particles

Finally, several particles are captured in the central region of the slab (UT measurements). These correspond to simulated particles that exit the domain, which were very few.

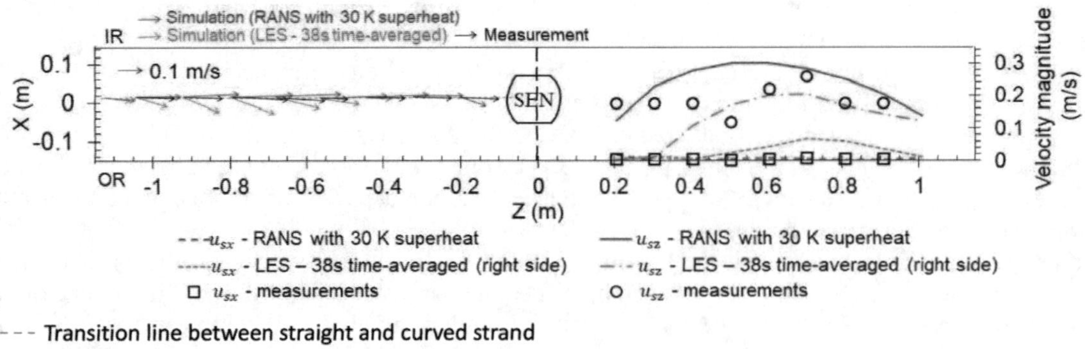

(a)

----- Transition line between straight and curved strand
——— Domain outlet
----- Capture band boundary • UT-measured defect dots

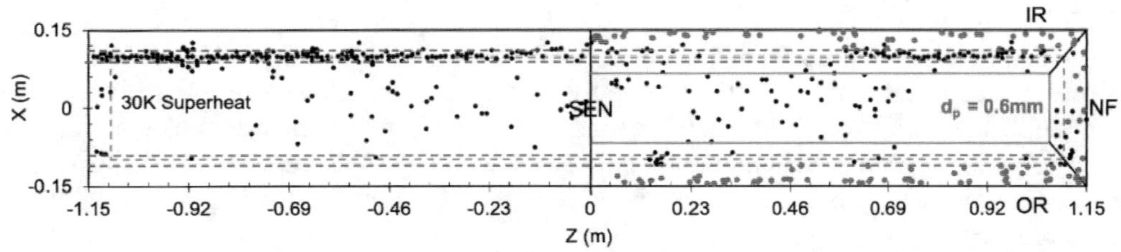

(b)

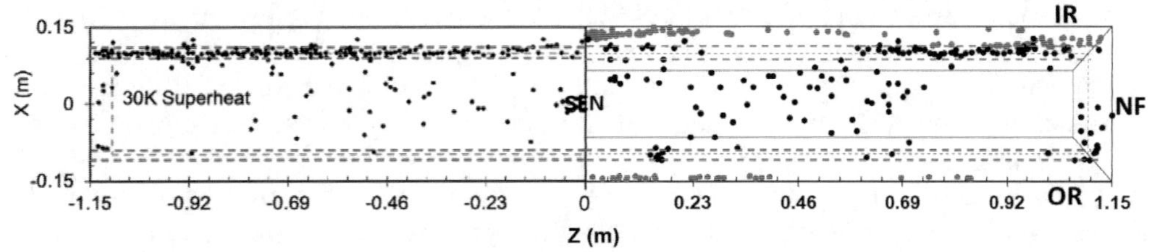

(c)

Figure 8. Model validation—top view (a) surface velocity comparing simulations with nail board measurements (black); location of entrapped particles in transverse plane to the casting direction, comparing simulated (red dots on RHS) with measured (black dots) for 30 K superheat case with (b) the old UDF (c) the new UDF

Figure 9 shows the number of 0.6 mm bubbles captured inside the capture band on WFIR and WFOR, comparing the UT measurements with simulation results using both the old and new UDFs. At 30 K superheat, both UDFs predict more 0.6 mm bubbles captured on WFIR than WFOR, which matches the measurements both qualitatively and quantitatively. Zero 0.6 mm bubbles are captured on WFOR. Both simulations predict ~200 bubbles captured on WFIR at 30 K superheat, which lies in the range of the measurements. At 10 K superheat, the old UDF predicts more capture on WFOR than WFIR. The new UDF predicts a consistent trend at both superheats that more bubbles are captured on the WFIR, which matches better with logical expectations from bubble buoyancy towards the inner radius in the lower strand, with the plant measurements, and with previous work.

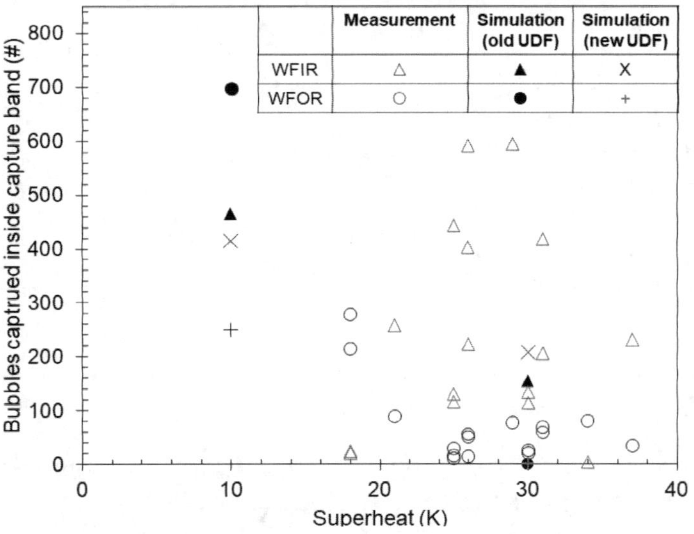

Figure 9. Model validation – number of 0.6mm bubbles captured inside the capture band, comparing UT measurements and simulations with the old and new UDFs

5. Conclusions

A system of computational models of turbulent multiphase flow, superheat transport, and particle capture is presented and applied to simulate a commercial continuous slab caster with vertical bending. The flow model features simulation of particle capture via three different mechanisms including capture by solidified hooks at the meniscus, entrapment between dendrites, and engulfment by the growing shell. It also has a post-processing feature to enable sufficient particles for statistical accuracy, while scaling the results to enable fair comparisons with the different-size particles in the real commercial process.

The capture model used in previous work (old UDF) is improved by modifying the particle velocity and the molten steel velocity at the particle center. In the new UDF, non-isotropic turbulent fluctuation velocity components are implemented by scaling the local turbulence k to dampen the component normal to the solidification front (wall), and to transfer that momentum to increase the component parallel to the solidification front.

Thermal-flow model results show that superheat has a negligible effect on flow in the mold region, where inertia is dominant over thermal buoyancy, but is significant in the lower strand, where thermal buoyancy creates multiple complicated recirculation regions, which influence particle capture slightly.

The particle transport and capture results show that most small (<0.1 mm diameter) particles are captured in the strand by entrapment between the primary dendrite arms. Only a few large (>0.6 mm) particles are captured, especially by frozen meniscus hooks near the top surface in the mold, especially at lower superheat when the meniscus hooks are deeper. Very few large particles penetrate deep into the caster. More particles are captured on the WF inner radius, (consistently with the new UDF), due to their buoyancy, which matches plant observations and previous work.

Comparison with plant measurements is promising, but further work is needed to evaluate and improve the computational models in this system to include other phenomena, such as slag entrainment. In addition, the effects of upstream processing, such as tundish flow, should be investigated to improve the initial particle size distributions entering the mold. Then, extensive parametric studies are needed to quantify the effects of process parameters such as nozzle and caster design, casting speed, electromagnetics, and gas injection on flow-related defects in the process.

6. Acknowledgements

The authors thank the National Science Foundation (Grant No. CMMI 18-08731) and the Continuous Casting Center, Colorado School of Mines, for support of this project. Thanks are also given to Baosteel, Shanghai, P.R. China, for providing the casting conditions and plant measurements.

References

[1] Liang M, Cho S-M, Ruan X, and Thomas B G, 2022, *Processes*, 10, No. 1429
[2] Thomas B G, 2018, *Steel Research International,* 1700312, 89 (1), No. 1700312, pp 1-21
[3] Yang H, Vanka S P and Thomas B G, 2019, *ISIJ Internat.*, 59 (6), pp 956-972
[4] Cho S-M and Thomas B G, 2019, *Metals*, 9, No. 471, pp 1-38
[5] Olia H, Yang H, Cho S M, Liang M, Das L, and Thomas B G, 2022, *Metall. Mater. Trans. B*, 53B (3), pp 1661-1680
[6] Bai H and Thomas B G, *Metall. Mater. Trans. B*, 2001, 32, pp 1143–1159
[7] Liang M, 2022, PhD Thesis, Mechanical Engineering, Colorado School of Mines, Golden, CO
[8] Launder B E and Spalding D B, 1974, *Computer Meths. in Applied Mech. & Eng.*, 3, pp 269-289
[9] ANSYS FLUENT 21.0 User Manual
[10] Chen W, Zhang L, Wang Y, Ren Y, Ren Q, Yang W, 2022, *Int. J. Heat Mass Transf.* 190, 122789
[11] Cho S-M, Thomas B G, Hwang J-Y, Bang J-G, Bae I-S, 2021, *Metals*, 11, No. 654, pp 1-23
[12] Dehbi A, 2006, *Work Proc benchmarking CFD codes Appl to Nucl React Saf.*, pp 703-720

Numerical modelling of macrosegregation in three-dimensional continuous casting of steel billets

K Mramor[1], R Vertnik[1,2] and B Šarler[1,3]

[1] Department of Fluid Dynamics and Thermodynamics, Faculty of Mechanical Engineering, University of Ljubljana, Aškerčeva 6, Ljubljana, SI-1000, Slovenia
[2] Štore Steel d.o.o., Železarska cesta 3, SI-3220 Štore, Slovenia
[3] Laboratory for Simulation of Materials and Processes, Institute of Metals and Technology, Lepi pot 11, SI-1000 Ljubljana, Slovenia

* corresponding author: bozidar.sarler@fs.uni-lj.si

Abstract. Macrosegregation presents a considerable defect in the continuous casting of billets and can critically affect the final properties of the product. The numerical modelling can help to predict and better understand the segregation and flow patterns inside the mould. The process is modelled with a physical model described by a set of conservation equations describing the t heat transfer, turbulence, fluid flow, solidification and segregation. A two-equation low-Re k-epsilon model and Abe-Kondoh-Nagano closures are used to close governing equations in this incompressible fluid flow example. The Boussinesq approximation is applied to account for the thermo-solutal buoyancy effects, and the Darcy approximation is applied for the description of the flow through the porous mushy zone. On a microscale, a lever rule solidification model is used to couple liquid fraction, temperature and concentration. The three-dimensional model is solved with the method based on local collocation with multiquadric radial basis functions on seven-nodded subdomains. The aim of this contribution is to explore the three-dimensional macrosegregation patterns of 0.51 wt% carbon steel in the solidified shell of the steel in the mould.

1. Introduction

The solidification and macrosegregation of steel in the continuous casting (CC) process is on a macroscale, affected by interconnected complex physical phenomena such as the turbulent motion of molten steel, energy and solute transfer. The consequent variations in the composition, known as macrosegregation, can significantly affect the condition of the product.

The solidification in the CC process, together with the macrosegregation, has been previously extensively studied with a wide variety of numerical methods. The methods considered ranged from classical finite volume [1–6] and finite differences [7] to modern meshless methods [8–11]. In this paper, we apply a local radial basis function collocation method (LRBFCM) [12], for which a three-dimensional (3D) numerical model was constructed to explore the influence of various process parameters on final macrosegregation. The presented model considers the turbulence of incompressible fluid and macrosegregation of 0.51% carbon steel. Among various available turbulence models [13–16], the two-equation low-Re eddy-viscosity model proposed by Abe-Kondoh-Nagano [17] was chosen. This selection is chosen due to its stability and relatively low computational times.

This article explores the effect of input parameters, precisely the casting temperature, velocity, and submerged entry nozzle (SEN) depth, on the final macrosegregation patterns and is a follow-up publication of [9,11].

2. Numerical model

The physical model is posed in the Cartesian coordinate system and formulated within the mixture continuum framework [18], in which steel densities in liquid and solid phases are equal and constant $\rho = \rho_S = \rho_L$. The model describes a simplified 3D CC process and considers the mass

$$\nabla \cdot \boldsymbol{u} = 0; \qquad \boldsymbol{u} = f_S \boldsymbol{u}_S + f_L \boldsymbol{u}_L, \tag{1}$$

momentum

$$\frac{\partial(\rho \boldsymbol{u})}{\partial t} + \nabla \cdot (\rho \boldsymbol{u}\boldsymbol{u}) = -\nabla p + \nabla \cdot [(\mu_L + \mu_t)(\nabla \boldsymbol{u} + (\nabla \boldsymbol{u})^T)] - \boldsymbol{F}_D + \boldsymbol{F}_B, \tag{2}$$

energy

$$\frac{\partial(\rho h)}{\partial t} + \nabla \cdot (\rho \boldsymbol{u} h) = \nabla \cdot (\lambda \nabla T) + \nabla \cdot [\rho f_S (h_L - h_S)(\boldsymbol{u} - \boldsymbol{u}_S)] + \nabla \cdot (f_L \frac{\mu_t}{\mathrm{Pr}_t} \nabla h_L), \tag{3}$$

and species conservation equations

$$\frac{\partial(\rho C)}{\partial t} + \nabla \cdot (\rho C \boldsymbol{u}) = \nabla \cdot (D \nabla C) + \nabla \cdot [\rho (C_L - C)(\boldsymbol{u} - \boldsymbol{u}_S)] + \nabla \cdot (f_L \frac{\mu_t}{\sigma_C} \nabla C_L). \tag{4}$$

The ψ_ℓ ($\psi = \boldsymbol{u}, C, h, f$) is a general phase variable standing for $\ell = \mathrm{L}$ liquid and $\ell = \mathrm{S}$ solid phases. The variables used in equations are mixture velocity $\boldsymbol{u}$, volume fraction f, time t, pressure p, dynamic viscosity μ_L, turbulent viscosity μ_t, Darcy term $\boldsymbol{F}_D = \mu_L[(1-f_L)^2/(K_0 f_L^3)](\boldsymbol{u} - \boldsymbol{u}_S)$ describing flow through porous media, permeability K_0, Boussinesq approximation $\boldsymbol{F}_B = \rho \boldsymbol{g}[\beta_T(T - T_{ref}) + \beta_C(C_L - C_{ref})]$ describing the effect of the buoyancy on the melt, gravitational acceleration $\boldsymbol{g}$, temperature T, reference temperature T_{ref}, thermal expansion coefficient β_T, solute expansion coefficient β_C, the concentration of alloying element C, the nominal concentration of the alloying element C_{ref}, mass-specific mixture enthalpy h, thermal conductivity λ, turbulent Prandtl number Pr_t, diffusivity D, and closure coefficient σ_C. The temperature-enthalpy relation is taken from [18].

The lever rule microsegregation model is given as

$$f_L = 1 - \frac{1}{1-k_p}\frac{T - T_L}{T - T_m}; \qquad C_L = \frac{C}{1 + (1 - f_L)(k_p - 1)}; \qquad T_L = T_m + (T_e - T_m)\frac{C}{C_e}, \tag{5}$$

where the partition coefficient is $k_p = C_S / C_L$, T_m are melting and T_e eutectic temperatures, and C_e is eutectic concentration obtained from the iron-carbon phase diagram.

The Low-Re k-epsilon model is used to model the turbulent flow. The equations for turbulent kinetic energy k is defined as

$$\frac{\partial(\rho k)}{\partial t} + \nabla \cdot (\rho \mathbf{u} k) = \nabla \cdot \left[\left(\mu_L + \frac{\mu_t}{\sigma_k}\right) \nabla k \right] + P_k + G_k - \rho \varepsilon + \rho D_k - \mu_L \frac{(1-f_L)^2}{K_0 f_L^3} k \tag{6}$$

and dissipation rate ε as

$$\frac{\partial(\rho \varepsilon)}{\partial t} + \nabla \cdot (\rho \mathbf{u} \varepsilon) = \nabla \cdot \left[\left(\mu_L + \frac{\mu_t}{\sigma_\varepsilon}\right) \nabla \varepsilon \right] + (c_{1,\varepsilon} f_1 (P_k + c_{3,\varepsilon} G_k) - c_{2,\varepsilon} f_2 \rho \varepsilon) + \rho E - \mu_L \frac{(1-f_L)^2}{K_0 f_L^3} \varepsilon \tag{7}$$

where σ_k, σ_ε, $c_{1\varepsilon}$, c_μ, $c_{2,\varepsilon}$, D_k, E, f_μ, f_1, and f_2 are Abe-Kondoh-Nagano closures and damping functions [17]. The shear production of turbulent kinetic energy and generation of turbulence due to buoyancy terms P_k and G_k are the same as in [10].

2.1. Numerical implementation

The modern Fortran is used to code the LRBFCM solution. The code is compiled with Intel® Fortran Compiler Classic 2021.7.0 for 64-bit Windows applications. The details of the numerical implementation can be found in [10].

3. Numerical method and solution procedure

3.1. Local radial basis function collocation method

The explicit Euler method is used for time discretisation, and the LRBFCM applied on a local seven-nodded sub-domain is used for space discretisation. The radial basis functions are augmented by a constant, and the shape parameter is set to 32. The incompressibility of the fluid flow is closed by the fractional step method pressure-velocity coupling. A detailed description of the numerical method, its implementation and the solution procedure can be found in [10].

4. Numerical example

4.1. Geometry

The present article's simplified illustration of a CC process is considered as a numerical example. It is represented in simplified 3D geometry and consists of a vertical longitudinal billet of 1.5 m height. The computational domain thickness is set to 0.18 m and accounts for the whole width and depth of the billet. The SEN is set at the top of the billet so that the nozzle entry is 0.12 m below the free surface of the meniscus. Its' inner diameter is 0.035 m and outer diameter is 0.065 m. The height of the mould is 0.8 m. The described geometry was constructed on 473081 nodes and was calculated with time step 10^{-3} s.

4.2. Material Properties

The simplified material properties [19], presented for carbon steel are listed in table 1. In the present model, the carbon content in steel is set to $C_0 = 0.51$ wt%.

Table 1. Material properties and process parameters

	Value	Unit
ρ	7300	kg/m^3
λ	33	W/(m K)
c_{pL}	698	J/(kg K)
c_{pS}	804	J/(kg K)
μ	0.006	kg/(m s)
β_T	10^{-4}	K^{-1}
β_C	$4.0\cdot 10^{-3}$	wt%$^{-1}$
K_0	10^8	m^2
D_S	$1.6\cdot 10^{-4}$	m^2/s
D_L	10^{-8}	m^2/s
T_{cast}	1525	°C
T_S	1400	°C
T_L	1482	°C
T_m	1539	°C
u_{cast}	1.6	m/min

4.3. Boundary conditions

The boundary conditions are carefully chosen to solve the problem's strongly coupled governing equations. In the present case, inlet, free surface, moving wall, and outlet boundary condition types are set, as shown in figure 1.

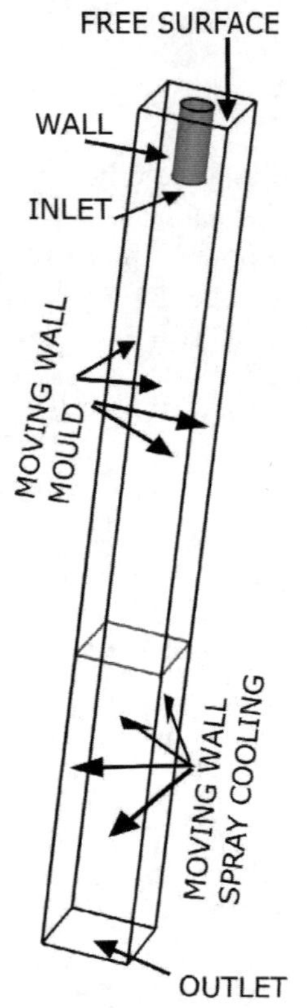

Figure 1. Domain scheme with boundary conditions.

The inlet boundary conditions are prescribed as Dirichlet boundary conditions for temperature $T = T_{\text{cast}}$, velocity $\boldsymbol{u}(x,y,z) = (0 \text{ m/s}, \ u_{\text{cast}}\, 4a_{\text{b}}^{2} / (\pi d_{2}^{2}), \ 0 \text{ m/s})$, concentration $C = C_{\text{ref}}$, turbulent kinetic energy $k = 1.5(I_t u_y)^2$; where a semi-empirical correlation for turbulent intensity is calculated as $I_t = 0.16\text{Re}^{-1/8}$ and $\text{Re} = \rho u_y d_2 / \mu$ is Reynolds number, and dissipation rate $\varepsilon = c_{\mu}^{0.75} k^{1.5} / (0.07 d_2)$. The Neumann boundary condition is defined for pressure $\partial p / \partial \mathbf{n} = 0$ Pa/m. The SEN a_{b} and billet width d_2 are used for rescaling the casting velocity to maintain a constant flux. The boundary conditions at the free surface are

$$\begin{aligned} &\partial u_x / \partial \boldsymbol{n} = 0 \text{ s}^{-1}; \quad u_y = u_z = 0 \text{ s}^{-1}; \quad \partial T / \partial \boldsymbol{n} = 0 \text{ K/m}; \quad \partial C / \partial \boldsymbol{n} = 0 \text{ m}^{-1}; \quad \partial p / \partial \boldsymbol{n} = 0 \text{ Pa/m}; \\ &\partial k / \partial \boldsymbol{n} = 0 \text{ m/s}^2; \quad \partial \varepsilon / \partial \boldsymbol{n} = 0 \text{ m/s}^3 \end{aligned} \quad . \tag{8}$$

The boundary conditions on the outlet are

$$\begin{aligned} &\partial \boldsymbol{u} / \partial \boldsymbol{n} = 0 \text{ s}^{-1}; \quad \partial T / \partial \boldsymbol{n} = 0 \text{ K/m}; \quad \partial C / \partial \boldsymbol{n} = 0 \text{ m}^{-1}; \quad p = 0 \text{ Pa}; \\ &\partial k / \partial \boldsymbol{n} = 0 \text{ m/s}^2; \quad \partial \varepsilon / \partial \boldsymbol{n} = 0 \text{ m/s}^3 \end{aligned} \quad . \tag{9}$$

The boundary conditions on the moving walls are

$$\begin{aligned} &u_x = u_z = 0 \text{ m/s}; \quad u_y = u_{\text{cast}}; \quad q_{\text{cz}} = h_{\text{cz}}(T - T_{\text{cw}}); \quad \partial C / \partial \boldsymbol{n} = 0 \text{ m}^{-1}; \quad \partial p / \partial \boldsymbol{n} = 0 \text{ Pa/m}; \\ &k = 0; \quad \varepsilon = 2\mu k_w / (\rho y^2) \end{aligned} \quad . \tag{10}$$

The heat flux q_{cz}. is calculated as a combination of heat transfer coefficient h_{cz}, which is set to 2000 W/(m^2 K) at the mould and to 800 W/(m^2 K) at the secondary cooling system and temperature of the

IOP Conf. Series: Materials Science and Engineering 1281 (2023) 012029 doi:10.1088/1757-899X/1281/1/012029

cooling water T_{cw} set to 293 K. Turbulent kinetic energy is k_w and y is the normal distance to the nearest domain node from the boundary.

5. Results

The concentration fields (C/C_0) are presented for four different casting temperatures at vertical and horizontal cross-sections. The vertical cross-section is shown for an x-y plane in the middle of the billet, and the horizontal cross-section is presented for an x-z plane at the height of 0.19 m. The casting temperatures under observation are 1505°C, 1515°C, 1525°C, and 1535°C. The concentration field is shown in figure 2 and shows that lower casting temperatures result in a more homogeneous concentration field. In general, the concentration in the totally fluid part is close to the nominal concentration. In the mushy region, the concentration is elevated. The most notable differences in the concentration fields can be observed in the totally solid region.

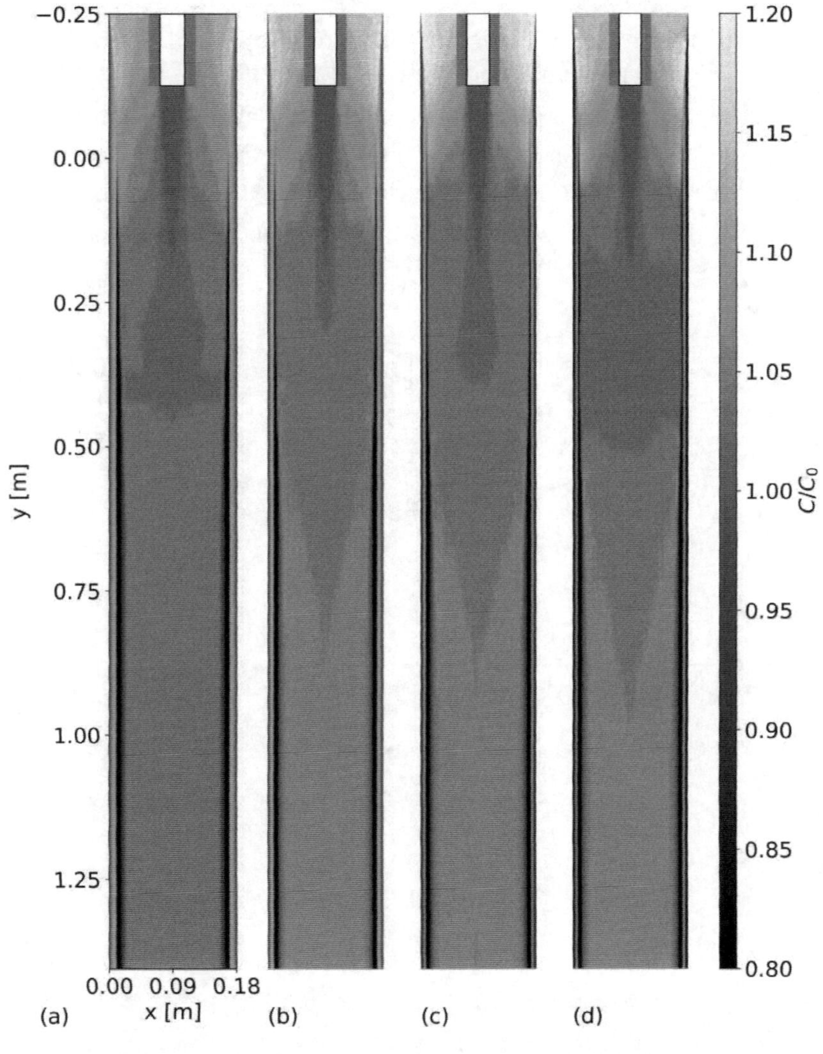

Figure 2. Comparison of concentration fields for different casting temperatures: a) 1505°C, b) 1515°C, c) 1525°C, and d) 1535°C, for a vertical cross-section at the centre of the billet.

Next, the horizontal cross-section placed at $y = 0.19\,\text{m}$ is presented in figure 3. Graphs for all temperatures confirm the observation from vertical cross-sections, where the concentration in the totally liquid region at the centre of the billet is close to the nominal concentration. In contrast, the concentration in the mushy zone is depleted. The solidified edges show the most significant difference between the concentration fields.

Figure 3. Comparison of concentration fields for different casting temperatures: a) 1505°C, b) 1515°C, c) 1525°C, and d) 1535°C, for a horizontal cross-section at 0.19 m, casting velocity 1.6 m/min and SEN depth 0.12 m. The green line presents a solidus temperature of 1400°C.

The concentration is then compared for different casting temperatures (1505°C, 1515°C, 1525°C, and 1535°C at SEN depth 0.12 m and $u_{cast} = 1.6$ m/min) in figure 4 and for different casting velocities (1.5 m/min,1.6 m/min and 1.7 m/min at SEN depth 0.12 m and $T_{cast} = 1525\,°C$) and different SEN depths (0.10 m, 0.12 m, and 0.13 m at $u_{cast} = 1.6$ m/min and $T_{cast} = 1525\,°C$) in figure 5, both at cross-section $y = 0.19$ m and $z = 0.09$ m. The concentration in the liquid is enriched in all of the calculated examples. In the mushy and solid part, the concentration is first depleted, then enriched and again depleted. Higher casting temperatures result in a lower concentration near the boundary.

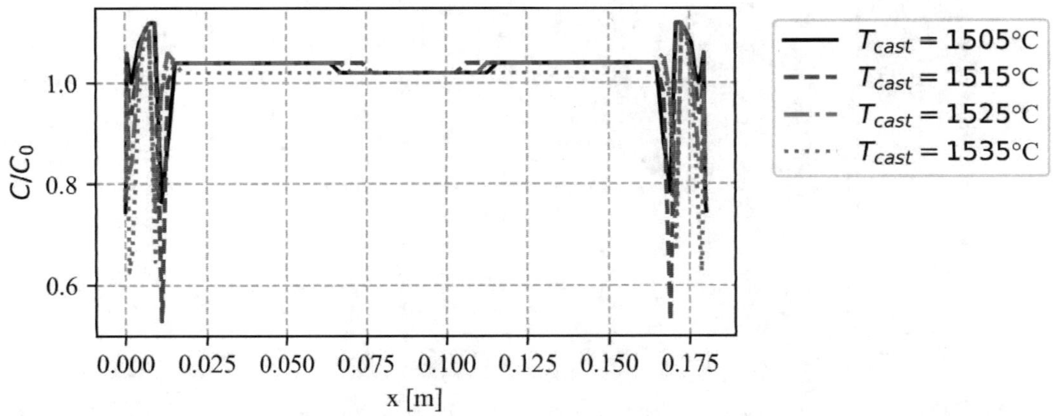

Figure 4. Comparison of concentration cross-sections for different casting temperatures.

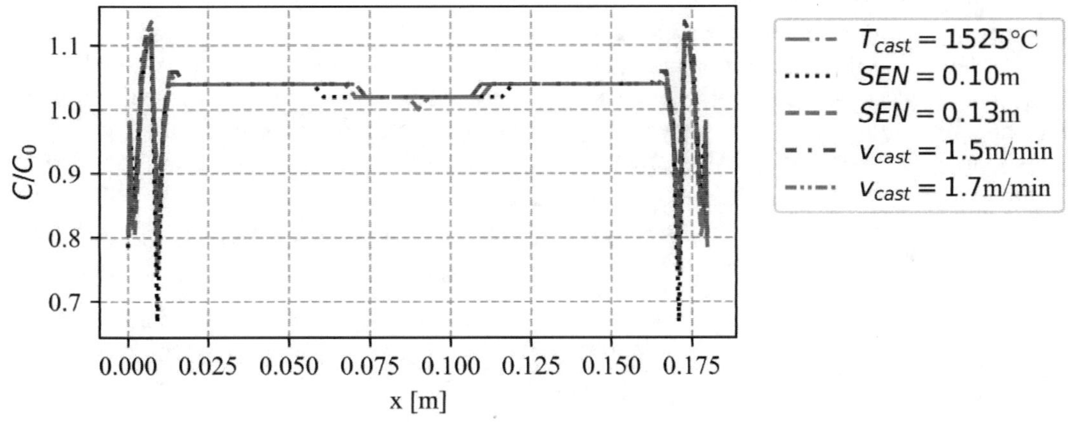

Figure 5. Comparison of concentration cross-sections for different casting velocities and SEN depths.

Figure 6 presents temperature fields at horizontal cross-section $y = 0.19$ m for four different casting temperatures. Higher casting temperatures return higher overall temperatures, which is most visible at the centre of the billet. At the boundary, higher temperature gradients result in a higher concentration differences.

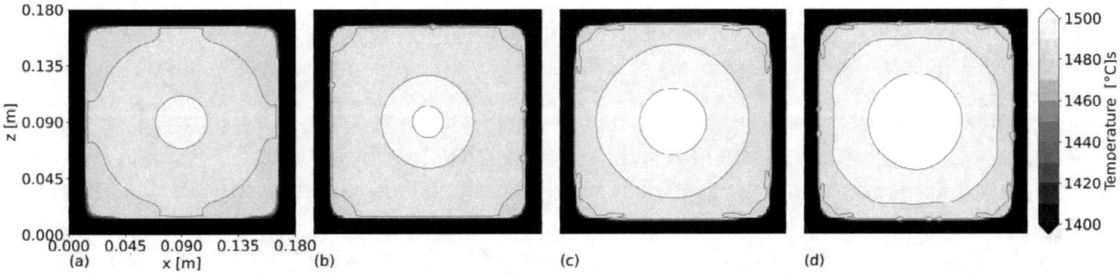

Figure 6. Comparison of temperature fields for different casting temperatures: a) 1505°C, b) 1515°C, c) 1525°C, and d) 1535°C, for a horizontal cross-section at 0.19 m, casting velocity 1.6 m/min and SEN depth 0.12 m.

The velocity field at horizontal cross-section $y = 0.19$ m is presented next. Figure 7 shows velocity magnitude fields for different casting temperatures. Higher casting temperatures result in marginaly higher velocities, which can be visible in the centre of the billet. At the boundary, velocities are constant and equal to casting velocity.

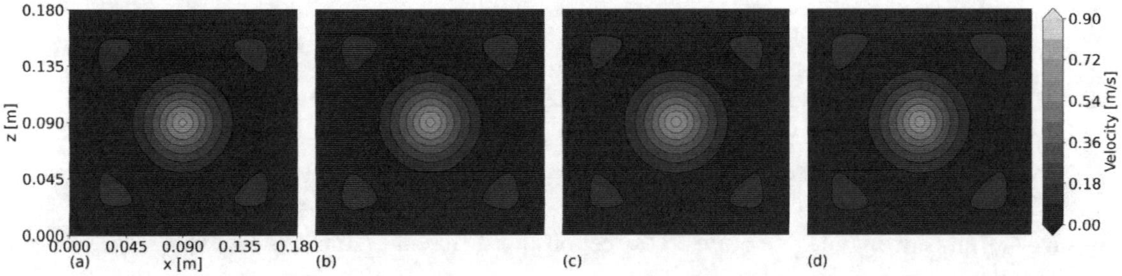

Figure 7. Comparison of velocity magnitude fields for different casting temperatures: a) 1505°C, b) 1515°C, c) 1525°C, and d) 1535°C, for a horizontal cross-section at 0.19 m, casting velocity 1.6 m/min and SEN depth 0.12 m.

The horizontal cross-section at $y = 0.19$ m and $z = 0.09$ m of velocity magnitude is shown in figure 8. The effect of casting temperature on velocity is negligible. We can observe small increase in the velocity with increasing casting temperature, however the differences are very small. A more pronounced effect can be seen when changing the casting velocity and SEN depth. The increase in SEN depth has similar effect as the increase in casting velocity, both of which result in slight increase of velocity magnitude in the centre of the billet.

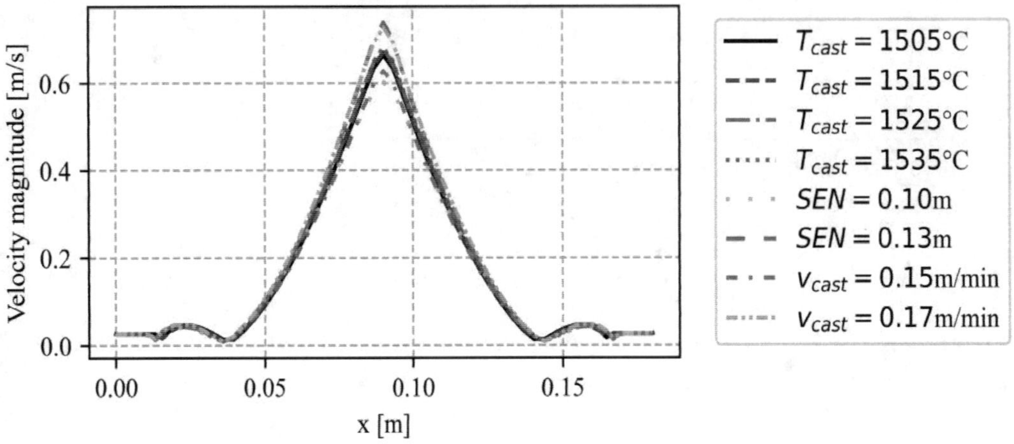

Figure 8. Comparison of velocity magnitude cross-sections for different casting temperatures, velocities and SEN depths.

Temperature cross-sections at $y = 0.19$ m and $z = 0.09$ m are presented in figure 9. Lower casting temperatures result in lower overall temperature in the central part of the billet. The change in central temperature at this cross-section due to the change in casting velocity is small. Bigger differncs can be observed due to the change in the SEN depth.

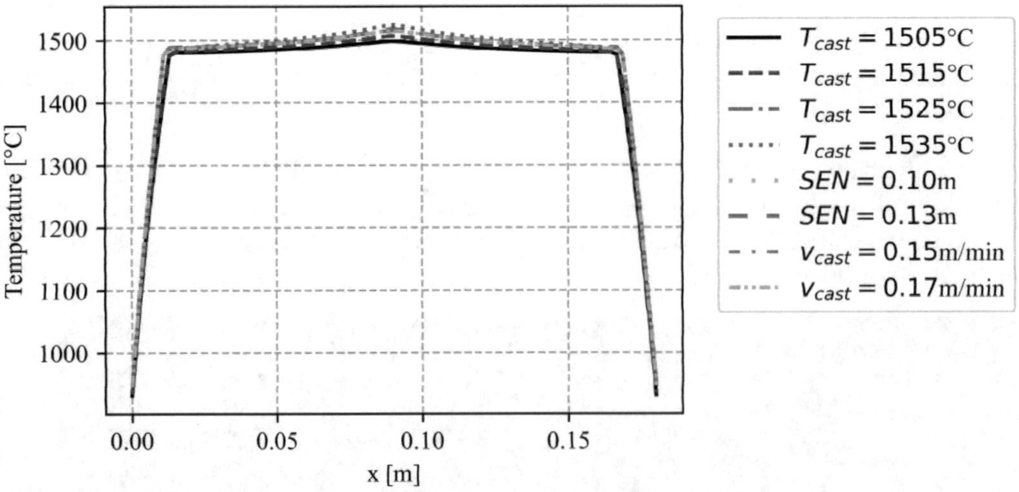

Figure 9. Comparison of temperature cross-sections for different casting temperatures, velocities and SEN depths.

This parametric study of our numerical model shows that the increase or decrease in the casting temperature has a more significant effect on the final solidification than the changes in the casting velocity or SEN depth.

6. Conclusions

The present study presents the macrosegregation in a 3D continuous casting mould solved with LRBFCM. The numerical example is evaluated for the Low-Re k-epsilon turbulence model. The results, calculated on 7-noded local subdomains with MQ RBFs, show a successful application of the RANS model for predicting macrosegregation in the 3D example of the CC of steel. Although lower casting temperatures result in a more homogenous concentration, slightly higher casting temperatures are used in real-life casting due to technological difficulties. The simulations will be compared with 3D LES model in the perspective paper [19].

Acknowledgements

This work was funded by the Slovenian Research Agency (ARRS) and Štore-Steel company in the framework of applied research project L2-3173 Advanced Simulation and Optimization of the Entire Process Route for Production of Topmost Steels, and program P2-0162 Multiphase Systems.

References

[1] Vakhrushev A, Wu M, Ludwig A, Tang Y, Hackl G and Nitzl G 2012 *IOP Conf. Ser. Mater. Sci. Eng.* **33**
[2] Aboutalebi M R, Hasan M and Guthrie R I L 1995 *Metall. Mater. Trans. B* **26** 731–44
[3] Schneider M C and Beckermann C 1995 *Metall. Mater. Trans. A* **26** 2373–88
[4] Zhang H, Nagaumi H, Zuo Y and Cui J 2007 *Mater. Sci. Eng. A* **448** 189–203
[5] Gu J P P and Beckermann C 1999 *Metall. Mater. Trans. A Phys. Metall. Mater. Sci.* **30** 1357–66
[6] Wu M, Ludwig A and Kharicha A 2019 *Metals.* **9**
[7] Bozzini B, Antonetti L, Boniardi M, Serra M and Fanigliulo A 2002 *Int. J. Comput. Appl. Technol.* **15** 186–94
[8] Mramor K, Vertnik R and Šarler B 2014 *3rd International Conference on Computational Methods for Thermal Problems, ThermaComp 2014* pp 283–6
[9] Šarler B, Vertnik R and Mramor K 2012 *IOP Conf. Ser. Mater. Sci. Eng.* **33**
[10] Vertnik R, Mramor K and Šarler B 2019 *Eng. Anal. Bound. Elem.* **104** 347–63
[11] Vertnik R and Šarler B 2016 *IOP Conf. Ser. Mater. Sci. Eng.* **117**
[12] Šarler B and Vertnik R 2006 *Comput. Math. with Appl.* **51** 1269–82
[13] Bredberg J 2001 *On Two-equation Eddy Viscosity Models* (Goteborg)
[14] Launder B 1974 *Int. Commun. Heat Mass Transf.* **1** 131–7
[15] Steffen J R C 1993 *29th AIAA Joint Propulsion Conference and Exhibit* pp 1–13
[16] Lilly D K 1966 *Natl. Cent. Atmos. Res.* **123**
[17] Abe K, Kondoh T and Nagano Y 1995 *Int. J. Heat Mass Transf.* **38** 1467–81
[18] Vertnik R and Šarler B 2014 *Eng. Anal. Bound. Elem.* **45** 45–61
[19] Mramor K, Vertnik R and Šarler B 2022 *Metals (Basel).* **12** 1–19

IOP Conf. Series: Materials Science and Engineering 1281 (2023) 012030 doi:10.1088/1757-899X/1281/1/012030

How electromagnetic stirring influences fluid flow in continuous casting of steel

Jakob Fainberg, Erik Hepp and Evgenii Shvydkii

MAGMA Gießereitechnologie GmbH, Kackertstr. 16-18, 52072 Aachen, Germany

E-mail: J.Fainberg@magmasoft.de, E.Hepp@magmasoft.de, E.Shvydkii@magmasoft.de

Abstract. State-of-art simulation models provide quantitative insights into flow, solidification and stress formation for continuous casting processes. This includes the entire process, from the tundish and the flow into the mold to the solidifying strand, which is withdrawn through various cooling zones. Process simulation and optimization provides important information about quality and productivity to evaluate process alternatives. This is only be possible if all relevant process parameters can be taken into account. The use of electromagnetic stirring is a technology which plays a significant role in the majority of continuous casting processes worldwide and its effects cannot be neglected in simulation models. This paper will discuss the modeling of electromagnetic stirring (EMS) and its impact on steel slab continuous casting.

Two cases with and without EMS are presented. The theoretical background to calculate the Lorentz force are described. The EMS calculation described here works with traveling (linear) magnetic fields. The effect of the EMS on the flow behavior, solidification and macrosegregation is shown on an industrial-scale slab casting.

This information leads to a better understanding of the EMS process in industrial applications to avoid casting defects, improve the quality of the final product, and increase the efficiency of the casting process.

1. Introduction

Continuous casting is a crucial process for steel production. Over 96% of the globally produced raw steel is continuously-cast steel [1]. At the same time steel continuous casting is a sophisticated technological process. It involves a complex system of inter-related phenomena, such as for example turbulent liquid metal flow, heat transfer issues, solidification and microstructure formation. This multiphysical nature in combination with a large scale of industrial casters makes this process challenging for computational modelling. The state of the art of continuous casting numerical simulation is summarized in [2–4]. Although much progress is made, accurate defect prediction is still in its beginning.

Macrosegregation is one of these defects in continuous-cast steel. Moreover it is the most serious because macrosegregation cannot be removed by heat treatment as diffusion in solids is slow even for elevated temperatures and large diffusion distances might result in impracticably long holding times for homogenization. [5]. Modeling of macrosegregation is expensive due to the need to compute the transport of multiple chemical species. Most macrosegregation models for continuous casting use simplified binary alloy systems. However cast steels are multicomponent systems [6].

Other models are needed to include electromagnetics. Implementation of electromagnetic fields has several applications in continuous casting [7]. It enables a contactless control of the fluid flow to improve quality and avoid defects in continuous cast steel slabs. Electromagnetic stirring inserted into the mold (M-EMS) was initially introduced to improve surface and subsurface quality. Control of the shell thickness allows to increase the casting speed as well. Moreover M-EMS prevents the superheat penetration deep down into the strand, thereby prolongs the mushy zone and promotes equiaxed crystal growth [8, 9].

Coupling together of different phenomena as macrosegregation and electromagnetics into multiphysics models remains a challenge [2]. The aim of this work is to illustrate the use of combined macrosegregation and EMS simulation by means of ready to use commercially available software. This work mirrors the actual state of continuous casting numerical simulation in the framework of MAGMASOFT® and MAGMA CC [10].

2. Method

2.1. Electromagnetic formulation

The Lorentz force appears in electric conductors imposed into a time dependent electromagnetic field. The flow in the metallic melt is driven among other things by the Lorentz force. This volumetric force results basically from interaction between the induced electric eddy current in the melt and the external magnetic field, both time dependent in the considered case. The Lorentz force is superimposed with other effects such as the thermal buoyancy and the forced flow due to the mass inflow and due to the strand withdrawal. In this work, the Lorentz force is computed from the solution of the complete Maxwell equation set. The Maxwell equations are formulated in terms of vector and scalar potentials for the frequency, or quasi-static domain and follow mainly the procedure described in [11]. It is assumed that the imposed electric currents in the inductors are harmonic with a prescribed amplitude and phase shift. Then, it is assumed that all field quantities Φ have the harmonic time dependence with the cyclic frequency of inductor currents due to linearity of the Maxwell equations.

$$\Phi = \Phi^{Re} cos(\omega t) + i\Phi^{Im} sin(\omega t) \tag{1}$$

Φ is a complex amplitude with two components (Φ^{Re}, Φ^{Im}), which are the in phase and out of phase components. The Maxwell equations are formulated in the frequency domain, which means that the time derivative is eliminated from them.

A vector potential A and a scalar potential ϕ are introduced as primary variables.

$$A = \vec{A}^{Re} + i\vec{A}^{Im} \tag{2}$$

$$\phi = \phi^{Re} + i\phi^{Im} \tag{3}$$

The considered Maxwell equations in the frequency domain are

$$\nabla \times E - i\omega\mu\mu_0 H = j_0 \tag{4}$$

$$\nabla \times H - (\sigma - i\omega\varepsilon)E = 0 \tag{5}$$

The local amplitude of the current density j_0 in the inductors is given by the total electric current amplitude defined for each inductor. It is assumed that there is no electric charge accumulation, and the magnetic permeability within a material is constant. Additionally to the equations (4) and (5), the definition of the magnetic induction B (6) is used:

$$B = \mu\mu_0 H \tag{6}$$

IOP Conf. Series: Materials Science and Engineering 1281 (2023) 012030 doi:10.1088/1757-899X/1281/1/012030

The complex potentials are defined in the unusual form as taken from [11]

$$E = A\nabla\phi \tag{7}$$

$$\nabla \times A = i\omega\mu H \tag{8}$$

The potential definitions in form (7) and (8) are inserted into the Maxwell equations (4)-(5) formulated in frequency domain (4) and (5). Then they are transformed with usage of the Coulomb gauge condition for the vector potential $\nabla A = 0$, and the resulting elliptic equations for unknown potentials in the frequency domain are obtained, in the electrically conducting domain

$$\nabla^2 A + i\omega\mu\mu_0\sigma(A + \nabla\phi) = -i\omega\mu j_0 \tag{9}$$

$$\nabla(\sigma A) + \nabla(\sigma\nabla\phi) = 0 \tag{10}$$

and else

$$\nabla^2 A = -i\omega\mu j_0 \tag{11}$$

The equations (9) and (10) describe the unknown complex vector A with 6 scalar components and the unknown complex scalar ϕ with 2 scalar components. 8 scalar equations and 8 unknowns result from (9) and (10). The electrically conducting domain consists of the solidified metal, the melt and optionally the permanent mold, all other materials are treated being not electrically conducting and are described by equation (11).

The boundary condition at the external boundary of the computational domain for all particular scalar equations are of Neumann type, i. e. no flux. There is a significant formulation drawback, the magnetic permeability μ cannot be discontinuous. In reality, the piecewise constant distribution of the permeability is more appropriate because of the jumps at the boundary between magnetic and non-magnetic materials. These jumps could not be taken into account in the actual formulation, hence bunching and amplifications of the magnetic field has to be approximated by the adjustment of electric current in the inductors. In order to overcome this limitation and allow a spatial jump of the magnetic permeability, the formulation has to be modified in future.

The equations are discretized according to the finite volume method on a Cartesian grid. The matrix of the total equation system consists of the 8×8 blocks associated with the particular elliptic equations and with the cross-terms proportional to the coupling parameter.

$$\nabla^2\Phi = 0 \tag{12}$$

The potential equation is a Laplace one for the computational domain consisting of the single inductor. The boundary conditions are symmetry everywhere excepted at the contact areas with electrodes. The Dirichlet boundary conditions are applied at both contact areas with cathode and anode. The Dirichlet value unity is prescribed at the cathode and zero at the anode.

The local current density is computed from the solution of (12). The gradient of the obtained potential is scaled with a constant factor K to obtain the current density accordingly to the input of the inductor current. The current is the surface integral of the unknown current density over the contact area between the inductor and cathode. Hence

$$K = \frac{I}{S|\nabla\Phi|} \tag{13}$$

IOP Conf. Series: Materials Science and Engineering 1281 (2023) 012030 doi:10.1088/1757-899X/1281/1/012030

with I the prescribed inductor current, S the contact area between the inductor and cathode and $|\nabla\Phi|$ the absolute value of potential gradient at the contact with the cathode. The obtained current density is then

$$j = K\nabla\Phi \tag{14}$$

2.1.1. Electromagnetic computational domain and the solution procedure. The calculation of the continuous casting process with electromagnetic stirring is split internally into two steps. In the first step, the Lorentz force associated with each stirring device is computed and stored. In the second one, the Lorentz forces from all available active stirrers are applied to the computational fluid dynamics within the melt. The requirements on the grid resolution of electromagnetic and thermally coupled CFD computations can be strongly different, and the region affected by the electromagnetic field in vicinity of the stirrer takes usually only a small part of the total computational domain. Hence the trial to compute both problems on the same grid would lead to a huge numerical grid and extensive computational time of the continuous casting process.

Therefore, the electromagnetic computation is completely separated from the continuous casting computational process and computed on a separate grid.

The CAD preprocessor allows to define an arbitrary number of stirrers which includes a number of elements such as inductors, electrodes and magnetic cores. The definition of a single stirring region is illustrated in Fig. 1.

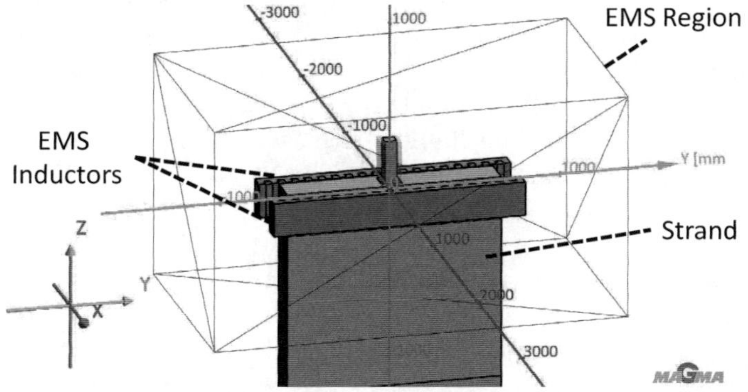

Figure 1. Definition of a single stirring region. The geometrical components, the size of the stirrer region and its location either in absolute coordinates or with respect to the strand are prescribed.

The computational process starts from creation of the project subversion for electromagnetic computation, the grid generation for the considered EMS region and the solution of equations (12) and (9)-(11). All these steps are executed for each available active electromagnetic region automatically based on the geometric and other definitions done in the basic continuous casting setup. The equation system can be effectively solved using the algebraic multigrid technique.

2.1.2. Lorentz force. The stored results are the time averaged Lorentz Force density and auxiliary results such as eddy current densities, magnetic induction and components of complex scalar and vector potentials. After results for all EMS regions are obtained, the computation is switched to the main part of the continuous casting process. The Lorentz force densities are read from the stored files of the particular computed EMS subversions and interpolated onto the actual grid used for the continuous casting fluid flow simulation. The interpolation is done in such a way that the volume integral over each component of the Lorentz force is preserved.

The Lorentz force is given as a cross product of the current density and the magnetic field, or

$$F_L = \sigma(E \times B) \tag{15}$$

Taking in account that each of E and B vectors changes harmonically with time, one obtains a harmonic and a constant components of (15). The harmonic component has a double frequency of the inductor current and is not accounted for in the acceleration of the flow. The constant component of the Lorentz force, returned by time averaging is applied as a momentum source term.

$$\langle F_L \rangle = \sigma \frac{1}{T} \int (E \times B) dt \tag{16}$$

The averaging via the integral (16) is applied for one period of the inductor current oscillation.

One obtains the following expression for Cartesian components of the Lorentz force after execution of integral (16)

$$\langle F_L \rangle = \frac{1}{2}\sigma \begin{pmatrix} E_y^{Re} B_z^{Re} + E_y^{Im} B_z^{Im} - E_z^{Re} B_y^{Re} - E_z^{Im} B_y^{Im} \\ E_z^{Re} B_x^{Re} + E_z^{Im} B_x^{Im} - E_x^{Re} B_z^{Re} - E_x^{Re} B_z^{Re} \\ E_x^{Re} B_y^{Re} + E_x^{Re} B_y^{Re} - E_y^{Re} B_x^{Re} - E_y^{Re} B_x^{Re} \end{pmatrix} \tag{17}$$

The real and imaginary components of the electric and magnetic fields E and B that are complex amplitudes are obtained directly from solution of the potential problem (9)-(11) according to definitions (6)-(8).

2.2. Species transport

The species transport in both liquid and solid phases has to be considered in the continuous casting configuration. The transport in the liquid phase differs from that in the solid phase by the macroscopic diffusion but its relative contribution is typically small in comparison with the convective transport. No relative velocity of the liquid and solid phase is considered, both phases are moved together. For the melt temperature above the predefined freezing temperature, the movement of the equiaxed crystals is assumed with the liquid phase velocity, no flotation or sedimentation is considered. Below the freezing temperature both phases are moved with the prescribed withdrawal velocity.

The formulation is based on the work of Schneider and Beckermann [6]. Species transport with segregation is considered for an arbitrary number of chemical elements in two phases, liquid and solid. Two transport equations are solved for each chemical element.

$$g_s \rho_s \frac{\partial C_s}{\partial t} + g_s \rho_s V_s \nabla C_s = \Gamma_{sl}(\kappa C_l - C_s) + S_v \rho_s D_s \frac{\kappa C_l - C_s}{l_{sl}} \tag{18}$$

$$g_l \rho_l \frac{\partial C_l}{\partial t} + g_l \rho_s V_l \nabla C_l = \nabla(g_l \rho_l D_l \nabla C_l) - \Gamma_{sl}(\kappa C_l - C_l) - S_v \rho_s D_s \frac{\kappa C_l - C_s}{l_{sl}} \tag{19}$$

The volume averaged species concentration in the liquid and solid phases are denoted as C_l and C_s, the volume liquid and solid phase fractions correspondingly g_l and g_s. The index of the considered chemical element is omitted in (18)-(19) for simplicity. The flow velocities in the liquid and solid phases are V_l and V_s. The temperature dependent partition coefficient is denoted as κ and the temperature dependent diffusion coefficients in two phases as D_l and D_s.

Γ_{sl} is the local solidification rate, computed via the total time derivative as

$$\Gamma_{sl} = \frac{\partial}{\partial t}(g_l \rho_s) + \nabla(g_s \rho_s V_s) \tag{20}$$

The ratio of the diffusion area to the diffusion length in the back diffusion term is approximated as

$$\frac{S_v}{l_{sl}} = \frac{12}{d_2^2 g_s} \tag{21}$$

with d_2^2 quadrat of secondary dendrite arm spacing.

The transport and segregation of 7 chemical elements was considered in this study. Chemical elements and their properties are shown in Table 1.

Table 1. Physical properties of the 18Cr-8Ni alloy components used for the simulations.

	[wt.%]	κ_s	κ_l	β_s $[\frac{1}{wt.\%} \cdot 10^{-3}]$	β_l $[\frac{1}{wt.\%} \cdot 10^{-3}]$	D_s $[\frac{m^2}{s} \cdot 10^{-14}]$	D_l $[\frac{m^2}{s} \cdot 10^{-13}]$
C	0.1	0.28	0.12	11.0	11.0	59000	10300
Si	1.0	1.21	0.97	17.9	18.1	630	160
Mn	1.0	0.84	0.73	2.7	3.1	4.8	1.8
Cr	18.0	0.74	1.00	0.74	0.63	2800	1100
Cu	1.0	0.68	0.43	−0.76	−0.89	10	1.0
Mo	0.5	0.51	0.95	−2.8	−3.1	7.8	22
Ni	8.0	1.14	0.81	−0.1	−0.87	3.2	1.1

Table 1 contains properties for 7 considered chemical elements in the 18Cr-8Ni alloy. C_{in} is the initial element concentration. Partitioning coefficients κ, volumetric expansion coefficient β and solid state diffusion coefficient are given at liquidus and solidus temperatures. Hence, the partitioning coefficient is given at the liquidus (κ_l) and solidus (κ_s) temperatures. The coupling with fluid flow is given via solutal buoyancy by means of the volumetric expansion coefficient β.

2.3. Reference case

As a reference case the work of Kihara et al. [12] is chosen. The authors simulated a slab continuous casting for austenitic stainless steel and experimentally validated it. The caster is supplemented by a mold electromagnetic stirrer (M-EMS) which is aimed to avoid a negative segregation underneath the slab surface.

Geometry of the chosen case is partially shown in figure 1. The 35.8 m long strand has a cross section of 200 × 1258 mm and a bow radius of 10 m. M-EMS consists of two travelling magnetic field inductors placed at the strand wide sides near the meniscus.

The applied numerical mesh consists of totally 1,243,702 cartesian cells. The mesh is refined in z-direction at the mold area and in the expected position of the wall boundary layers along the whole strand.

Heat transfer coefficients are set according to [12]. Primary cooling coefficient at the mold area h_{mold} is 1200 W/m^{2}K, secondary cooling $h_{secondary}$ = 418 W/m^{2}K and the meniscus $h_{meniscus}$ = 52 W/m^{2}K. The start melt temperature T_{in} = 1492°C which means a superheat of 45°C. The casting speed v_{cast} = 1.2 m/min, this corresponds to a maximum velocity in the SEN of $v_{SEN} \approx 0.96$ m/s

3. Results

The Lorentz force is calculated before the thermal and fluid flow simulation starts. Figure 2 shows the distribution of these induced forces on the slab cross section. Two travelling magnetic field inductors generate forces along the slab wide faces. Due to the skin effect these forces are located near the surface and do not penetrate into the middle of the slab. Direction of forces on lower surface in figure 2 is opposite to the upper, thus a counterclockwise torque is produced. The maximum value is about 5000 N/m^3 which corresponds to the reference case in [12].

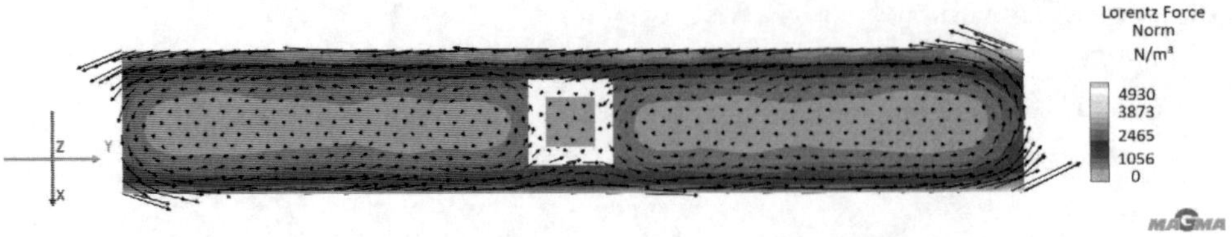

Figure 2. Lorentz forces

Figure 3. Velocity y-component at the 100 mm from the meniscus for 1400 s.

The velocity distribution at the middle height of the EMS inductors is shown in figure 3. The EMS forces the fluid flow in the horizontal plane of the strand. The EMS generates a positive y-component velocity at the bottom face as shown in figure 3 and negative at the opposite face. Thus a counterclockwise flow pattern occurs. Without EMS, the flow pattern is dominated by jets from the submerged entry nozzle (SEN) and is symmetric.

Figure 4 shows a flow pattern by streamlines. There is a conventional four-vortex flow pattern controlled by the SEN jets without EMS. Whereas electromagnetic forces generate the superimposed rotation about the vertical axis z. Electromagnetically driven flow appears near the meniscus where the EMS is placed.

This has an effect on the temperature field in the strand. The strand surface temperature is shown in figure 5. The temperature gradually decreases in the withdrawal direction and reaches $\sim$500°C at the surface of the horizontal part due to primary and secondary cooling. EMS shifts isotherms which results in a non-symmetric temperature distribution comparing with the reference case without EMS.

Next important parameter is the solidified shell thickness. Liquid fraction in the slab cross section is shown in figure 6. The solidified shell stays symmetrical without EMS. It became thicker in the slab corners and in the middle slab faces with EMS. The applied EMS deforms the solidified shell. It thins the shell at upper right and lower left positions of the slab cross section by means of superheat advection. The metallurgical length is 16.96 m without EMS case and 17.34 m for the case with EMS.

The distribution of the averaged slab shell thickness along the strand length is shown in the figure 7. Results with and without EMS match well within the first strand interval of 300 mm. Below this interval, EMS produces a thinner shell compared with the case without EMS. Intensive stirring keeps superheat in the upper mold part preventing its penetration deep down into the strand. The difference in the obtained shell thickness is about 3 mm and remains

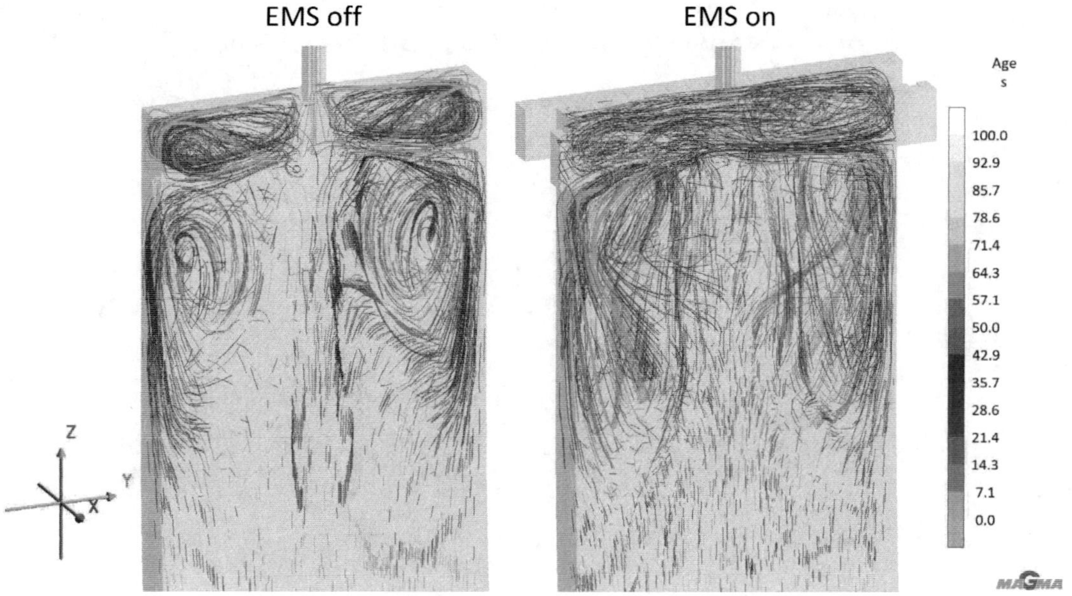

Figure 4. Flow pattern streamlines

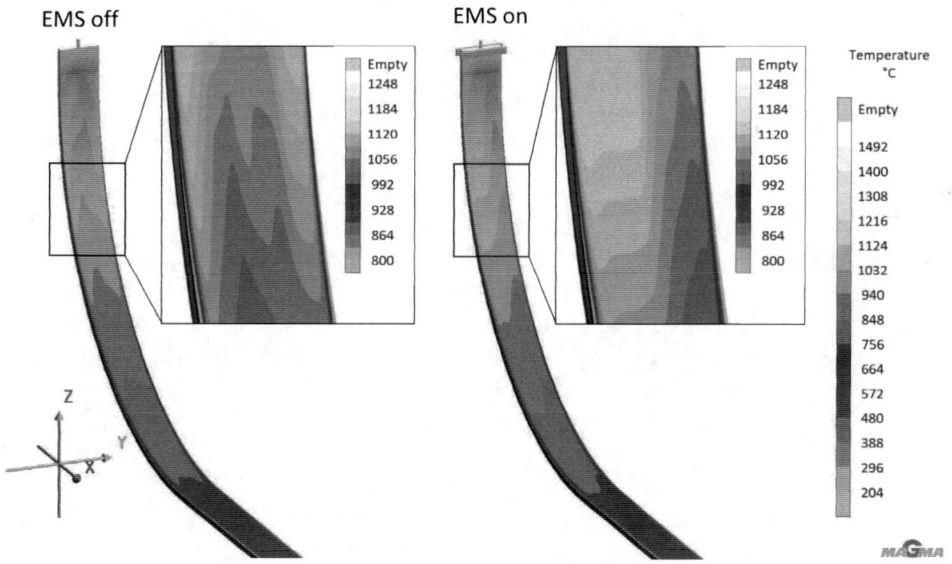

Figure 5. Strand surface temperature.

constant along the strand length.

Figure 8 shows concentration distribution in the slab cross section for the considered chemical alloy elements. The results are evaluated at the end of simulation at 30 minutes of the process time, in the strand cross section located just after the metallurgical length position at 18 m of strand length. Left column is without EMS and right one with EMS. The most prominent segregation is observed for the Carbon, due to the lower value of its partition coefficient (see

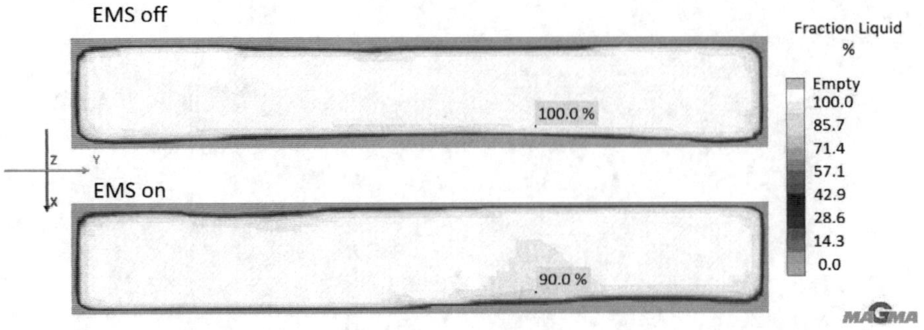

Figure 6. Liquid fraction at the 2300 mm from the meniscus.

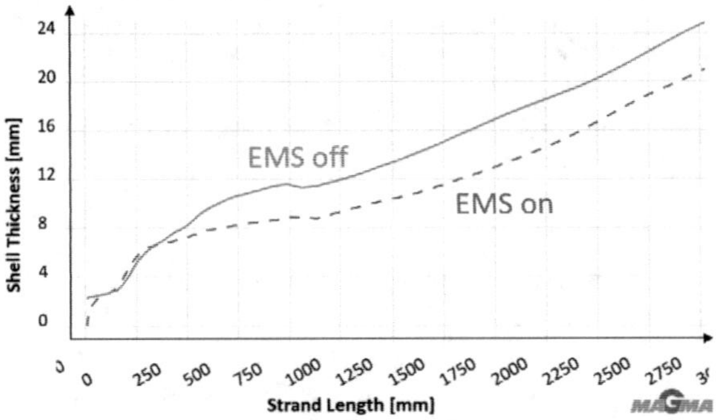

Figure 7. Averaged shell thickness over the strand length.

table 1). Almost no macrosegregation for Si is obtained in opposite to Carbon.

Electromagnetic stirring significantly affects the C and Cr distribution. The stirring in the plane of the strand cross section transports the solute to the upper right and the lower left corners and increases the Carbon concentration up to 0.15% there. The concentration pattern remains more symmetric without EMS. In addition, EMS eliminates the negative segregation spot at the center of the slab for Mn, Mo and Ni elements.

4. Conclusion

A global numerical model for stainless steel slab continuous casting is presented. Two cases with and without EMS are considered. The influence of EMS forced flow on temperature field, solidified shell and macrosegregation is shown. The longitudinal temperature distribution along the strand length changes by EMS. This is due to the effect of the enhanced superheat advection. EMS deforms the solid shell, breaks its symmetry and makes it thinner as well. The explicit effect of EMS is noticeable for the Carbon macrosegregation mainly. Other elements shows minor effects but for e.g. Cu, Mo and Ni concentration distribution is more uniform in the EMS case.

The presented model is a commercially available tool to simulate the continuous casting process at industrial scale. The temperature field, fluid flow, solidification and magrosegregation are computed simultaneously, in 3D and transient. Moreover, electromagnetic field calculation is incorporated into the computational workflow in an automatic way to enable an easy calculation of the electromagnetic stirring. Demonstrated cases are computed with standard PC in parallel mode with 32 cores. The computational time was about 32 hours.

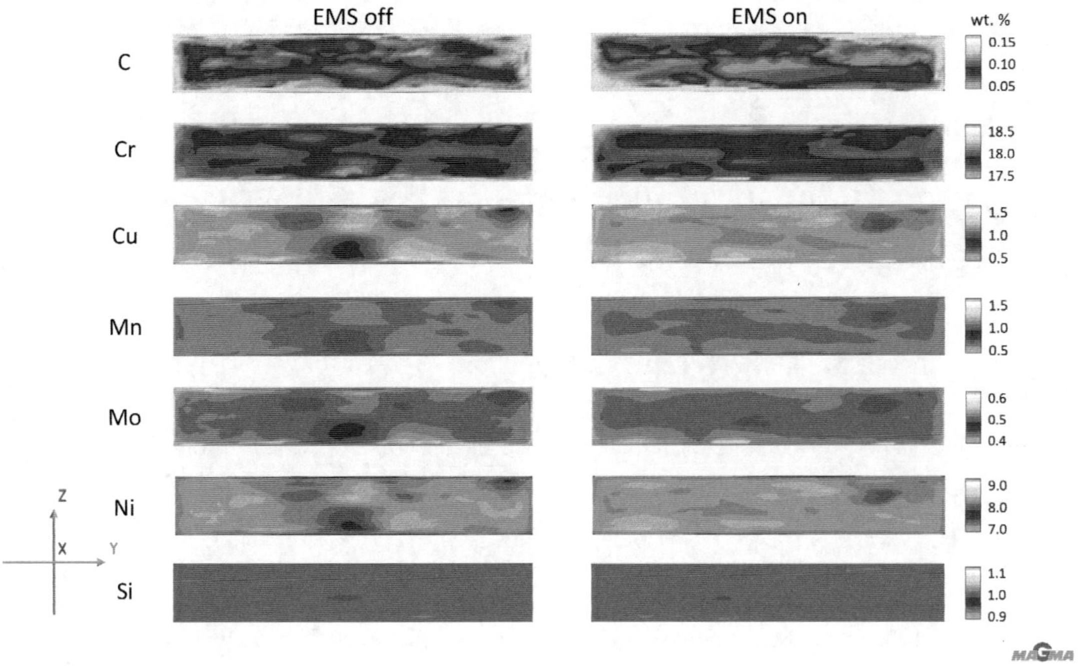

Figure 8. Distribution of the concentration for 7 chemical elements in the slab cross section at the position of 18 m of the strand length.

References

[1] Steel statistical yearbook 2018 https://worldsteel.org/wp-content/uploads/Steel-Statistical-Yearbook-2018.pdf
[2] Thomas B G 2018 *steel research international* **89** 1700312 URL https://doi.org/10.1002/srin.201700312
[3] Miłkowska-Piszczek K and Falkus J 2018 *Metals* **8** ISSN 2075-4701 URL https://www.mdpi.com/2075-4701/8/8/591
[4] Vynnycky M 2018 *Metals* **8** ISSN 2075-4701 URL https://www.mdpi.com/2075-4701/8/11/928
[5] Ludwig A, Wu M and Kharicha A 2015 *Metallurgical and Materials Transactions A* **46** 4854–4867 URL https://doi.org/10.1007/s11661-015-2959-4
[6] Schneider M C and Beckermann C 1995 *Metallurgical and Materials Transactions A* **26** 2373–2388 URL https://doi.org/10.1007/bf02671251
[7] Cho S M and Thomas B G 2019 *Metals* **9** ISSN 2075-4701 URL https://www.mdpi.com/2075-4701/9/4/471
[8] Kunstreich S 2003 *Revue de Métallurgie* **100** 395–408 URL https://doi.org/10.1051/metal:2003198
[9] Kunstreich S 2003 *Revue de Métallurgie* **100** 1043–1061 URL https://doi.org/10.1051/metal:2003113
[10] Fainberg J, Hepp E, Schäfer W and Krafthöfer C 2020 *IOP Conference Series: Materials Science and Engineering* **861** 012017 URL https://dx.doi.org/10.1088/1757-899X/861/1/012017
[11] Aruliah D A, Ascher U M, Haber E and Oldenburg D 2001 *Mathematical Models and Methods in Applied Sciences* **11** 1–21 URL https://doi.org/10.1142/s0218202501000702
[12] Kihara K, Okada N, Saito S and Kawashima K 2022 *ISIJ International* **62** 1862–1873 URL https://doi.org/10.2355/isijinternational.ISIJINT-2022-120

Non-metallic inclusion (NMI) deposition in a slide-gate submerged entry nozzle (SEN)

K U R M Shibly[1], S Tullis[2], A B Phillion[1]

[1]Dept. of Materials Science and Engineering, McMaster University, Hamilton, ON
[2]Dept. of Mechanical Engineering, McMaster University, Hamilton, ON
E-mail: philliab@mcmaster.ca (A B Phillion)

Abstract. Clogging of the submerged entry nozzle (SEN) is a common and persistent issue in the continuous casting process for steel alloys. The dominant clogging mechanism has been attributed to the deposition and accumulation of non-metallic inclusions (NMI) in the steel melt. Prior studies of NMI deposition assume that every collision between an inclusion and the nozzle wall results in adhesion, which is unrealistic. In this study, a macroscopic transport model for fluid and NMI motion is combined with a microscale model for NMI adhesion and applied to a slide-gate nozzle. Eight NMI sticking probabilities (S) and three slide-gate linear opening positions are explored. Simulation results indicated that the more closed the slide-gate, the greater the total deposition of NMI. When the slide-gate was partially open, the particle area number density was highest above, within and just below the slide-gate. But when the slide-gate was fully opened the deposition was concentrated in the upper tundish nozzle. Deposition behaviour fell into two regimes based on S. When $S \leq 0.05$, the particle deposition was low and increased rapidly with sticking probability. When $S \geq 0.05$, the particle deposition was high but changed very little with sticking probability. Changes to sticking probability did not significantly affect the deposition locations or relative distribution of particles within the nozzle.

1. Introduction

Nozzle clogging is a notorious issue in the continuous casting process where solid material can accumulate within the submerged entry nozzle (SEN) and disrupt the flow of molten steel. Clogging can have several deleterious effects on the casting process, including uneven flow conditions [1,2] and decreased production rates [3]. Clogging may also cause issues with the quality of the finished product due to incorporation of pieces of the clog into the steel [4–6]. In severe cases, the flow of molten steel can be blocked and production halted, until the nozzle is replaced [7].

Various methods have been proposed to solve the problem of nozzle clogging [8–10], with the two most common being calcium treatment [11] and argon injection [12]. These methods have been helpful but have not completely eliminated the problem. While different mechanisms for clogging have been proposed in the scientific literature [13,14], the majority of research indicates that clogging occurs due to the deposition and accumulation of inclusions on the nozzle wall [15–17].

Given that NMI deposition is a significant factor in clogging, researchers have used different modelling methods to study how particles deposit in the nozzle. The most common approach is the

IOP Conf. Series: Materials Science and Engineering 1281 (2023) 012031 doi:10.1088/1757-899X/1281/1/012031

Eulerian-Lagrangian method in which the fluid flow is modelled using the Eulerian method as a continuous phase, while the particles are modelled as a discrete phase using the Lagrangian frame of reference. The fluid motion is modelled using one of several standard models such as RANS (Reynolds-averaged Navier-Stockes simulation), LES (large-eddy simulation) or DES (detached-eddy simulation). Particle motion is modelled using a force balance. While there are minor variations in which forces are modelled, the most common forces considered are drag, buoyancy, lift, virtual mass and pressure gradient forces. Of those forces modelled, drag is typically reported to be the dominant force [18].

Pfeiller [19] and Thomas [18] used a RANS model to study particle entrapment in a solidifying shell in the mould. Yuan et al. modelled fluid motion in the mould using LES, then inserted particles to study the inclusion behaviour [20]. Mohammadi-Ghaleni et al. [21] used DES to study inclusion motion in a slide-gate nozzle, while Zhang et al. [2] used a RANS approach to study inclusion motion in a slide-gate nozzle and a coupled mould. Both Mohammadi-Ghaleni and Zhang showed qualitatively that inclusion entrapment locations are found around the slide-gate and exit ports of the nozzle. Additionally, Zhang showed that the flow pattern around the exit ports of the nozzle are inaccurate, unless the mould is also simulated.

Previous research in studying inclusion deposition has a limitation in that it assumed that all inclusions stick to the SEN wall if they come into contact with it. Inclusion rebound or deflection after collision with the nozzle wall are not considered. This study aims to address this limitation by examining the adhesion of particles to the SEN wall as a stochastic process. The fluid transport and particle motion in a slide-gate SEN is considered with three slide-gate opening positions and eight sticking probabilities. The effects on the degree of particle deposition and deposition location are explored.

2. Modelling

The overall model consists of a $k-\omega$ shear-stress transport (SST) model for fluid flow, a particle force balance for particle motion and a stochastic model for particle-wall interaction. The melt was assumed to be of uniform temperature and the particles were assumed to be spherical. Due to their small size, one way coupling between the particles and fluid was used. Additionally, particle-particle interaction in the melt were considered negligible for the simulation. The details of the model are below in their respective sections.

2.1. Fluid Model

The molten steel is modelled using a control volume approach and by solving the standard equations for the conservation of mass and momentum. The unsteady turbulent flow is modelled using a well-known RANS approach, the $k-\omega$ SST model proposed by Menter [22], which is a combination of the $k-\varepsilon$ and $k-\omega$ models.

2.2. Particle Force Model

Particle motion is calculated by solving the transport equation in a Lagrangian frame of reference. The solid non-metallic inclusions are modelled as spherical particles and are treated as a discrete secondary phase. The particle transport model, is essentially a force balance consisting of five forces,

$$m_p \frac{\mathrm{d}\vec{u}_p}{\mathrm{d}t} = \vec{F}_{Buoyancy} + \vec{F}_{Drag} + \vec{F}_{Lift} + \vec{F}_{Virtual\,Mass} + \vec{F}_{Pressure} \tag{1}$$

where the subscript p denotes a particle parameter and subscript f denotes a fluid parameter.

The buoyancy force occurs due to the difference in density between the particle (ρ_p) and fluid (ρ_f). For a particle of diameter (d_p) subject to a gravitational field $(\vec{g})$, the buoyancy force is:

$$\vec{F}_{Buoyancy} = \frac{(\rho_p - \rho_f)\pi d_p^3}{6}\vec{g} \tag{2}$$

IOP Conf. Series: Materials Science and Engineering 1281 (2023) 012031 doi:10.1088/1757-899X/1281/1/012031

The drag force results from the friction of the fluid moving across the surface of the particle, and is expressed as:

$$\vec{F}_{Drag} = \frac{1}{8}\pi d_p^2 \rho_f C_D \left|\vec{u}'_f - \vec{u}_p\right| \left(\vec{u}'_f - \vec{u}_p\right) \tag{3}$$

with the particle Reynolds Number (Re_p) and Drag Coefficient (C_D) and being given by:

$$Re_p = \frac{\left|\vec{u}_f - \vec{u}_p\right| d_p \rho_f}{\mu} \tag{4}$$

$$C_D = \begin{cases} \dfrac{24}{Re_p} & Re_p < 0.1 \\ \dfrac{24}{Re_p}\left(1 + 0.15 Re_p^{0.687}\right) & 0.1 \leq Re_p \leq 10^3 \\ 0.44 & Re_p \geq 10^3 \end{cases} \tag{5}$$

As the fluid flow is modelled using a RANS model, turbulent eddies are not simulated. The effect of eddies on particle motion is accounted for using a random walk model:

$$\vec{u}'_f = \vec{u}_f + \xi \sqrt{\frac{2k}{3}} \tag{6}$$

where $\vec{u}'_f$ is the instantaneous fluid velocity and is implemented in the drag force term (Equation 3), $\vec{u}_f$ is the time average velocity from the momentum equation, ξ is a Gaussian distributed random number and k is the turbulent kinetic energy.

A velocity gradient across the body of the particle results in the lift force. The lift force model proposed by Saffman and extended by Mei [23] is used and is described by the following series of equations:

$$\vec{F}_{Lift} = 1.61 C_L d_p^2 \left(\mu \rho_f\right)^{0.5} \left|\vec{G}\right|^{-0.5} \left(\vec{U}_s \times \vec{G}\right) \tag{7}$$

$$\vec{G} = \nabla \times \vec{u}_f \tag{8}$$

$$\vec{U}_s = \vec{u}_f - \vec{u}_p \tag{9}$$

$$\beta = \frac{d_p \left|\vec{G}\right|}{2\left|\vec{U}_s\right|} \tag{10}$$

$$C_L = \begin{cases} (1 - 0.3314\beta)\exp\left(-0.1 Re_p\right) + 0.3314\beta^{0.5} & Re_p \leq 40 \\ 0.052\left(\beta Re_p\right)^{0.5} & Re_p > 40 \end{cases} \tag{11}$$

$$Re_p = \frac{\left|\vec{U}_s\right| d_p \rho_f}{\mu} \tag{12}$$

where G is the shear rate of the mean flow, $\overrightarrow{U_s}$ is the difference between the streamwise velocities of the fluid and particle, Re_p is the particle Reynolds number and C_L is a correction factor proposed by Mei.

As the particle accelerates, it displaces the surrounding fluid. The inertia added to the system is called the virtual mass force and can be expressed as:

$$\vec{F}_{Virtual\,Mass} = \frac{\rho_f \pi d_p^3}{12} \cdot \frac{\mathrm{d}}{\mathrm{d}t}(\vec{u}_f - \vec{u}_p) \tag{13}$$

Finally, the difference in pressure across the surface of the particle generates the pressure gradient force, which is expressed as:

$$F_{Pressure} = \frac{\rho_f \pi d_p^3}{6} \cdot \frac{\mathrm{D}\vec{u}_f}{\mathrm{D}t} \tag{14}$$

2.3. Particle-Wall Interaction Model

The particle-wall interaction is described using a simple stochastic model. If the particle collides with the wall, there are two possible outcomes. Either the particle sticks to the wall or rebounds. A user defined sticking probability (S) is used to calculate the outcomes, where the value of S lies between 0 and 1. A value of 0 means the particle never sticks to the wall (i.e. the particle always rebounds) and a value of 1 means the particle always sticks to the wall (i.e. the particle never rebounds). For values between 0 and 1, the particle behaviour is determined at run-time using a random number generated on collision. In the case of rebound, the coefficient of elasticity is set to 1.

2.4. Numerical Implementation

The model was applied to a slide-gate submerged entry nozzle, where part of the tundish is also simulated. During the casting process, the slide-gate position is automatically adjusted. This results in different flow behaviour within the nozzle. To understand the effects of this behaviour on particle deposition, three slide-gate opening positions were tested, varying from 20% to 100%, where 100% represents the maximum degree to which the slide-gate can be opened. The slide-gate linear opening is defined as the length of overlap between the slide-gate and nozzle bore, divided by the nozzle bore diameter. The deposition behaviour of inclusions was tested using alumina particles with eight sticking probabilities. The particles were inserted at the nozzle entrance and allowed to pass through the nozzle. At the end of the simulation, only the particles which adhered to the nozzle remained. The particle deposition behaviour was then quantified and the deposition locations identified.

As indicated in Figure 1, the parts of the industrial nozzle have been labelled upper-tundish nozzle (UTN), slide-gate, nozzle length and exit ports, for ease of reference. The nozzle has a length of 1.17 m, with an entrance diameter of 110 mm which narrows to a bore of 78 mm. The exit ports have a width and height of 80 mm and 78 mm respectively and are inclined downwards at an angle of 15°. The slide-gate is oriented 90° to the exit ports (i.e. if the exit ports point left and right, the slide-gate moves back to front). In order to accurately simulate the flow in the nozzle it is necessary to simulate the flow in the tundish. In reality, the tundish is very large, but simulating only a part of it was considered sufficient to accurately capture the flow field at the nozzle entrance. A tundish base of 500 x 500 mm was considered sufficient for this purpose as it is ~ 5x the nozzle inlet diameter. A melt height of 500 mm was used, which is within the range of tundish bath depths reported in literature [24].

An unstructured hex-dominant mesh was used for the simulation. Of the ~1.8 million cells, 96% of them were hexahedra. The rest were prisms, tetrahedral wedges and polyhedral cells. The use of an unstructured mesh allowed for the simulation domain to follow the contours of the complex geometry of the SEN while the hex-dominant nature of the mesh ensured that the simulation would not run too slowly due to the use of an unstructured mesh. The cells within the nozzle were smaller than the cells in the tundish by a factor of 4. Typical cell size throughout the nozzle and close to the nozzle wall was 2 mm. Due to the fineness of the mesh, refinement of wall cells was considered unnecessary.

A uniform pressure boundary condition was applied to the tank inlet and nozzle outlet. A uniform pressure of $\rho g h$ was applied to the tank inlet where h is the height of the tank surface above the bottom of the nozzle (which was 1.76 m). The nozzle outlet was assigned a pressure of 0 Pa. For velocity a Neumann boundary condition was applied at the inlet with a gradient of zero. At the exit ports, an outlet boundary condition was used (i.e. any inflow velocity was set to 0 ms^{-1}). As previously noted by Zhang, the flow around the exit ports cannot be accurately simulated without also coupling the nozzle geometry to the mould during the simulation [2]. As this would be computationally too expensive, the mould is neglected for this work. Instead the focus was on flow transport and particle deposition in the upper parts of the nozzle. A no-slip condition was applied to the walls.

The model was implemented using custom code in OpenFOAM, a free and open-source software. The simulations were run on the Digital Research Alliance of Canada's computing cluster using 128 cores and took about two weeks each. The solver convergence was set to 10^{-6} for pressure and 10^{-5} for velocity, turbulence kinetic energy (k) and specific dissipation rate (ω). The initial velocity, pressure, k and ω were set to zero. The maximum Courant number was set to 0.3 and the solver dynamically adjusted the timestep to ensure this value. This typically resulted in a timestep of $\sim 10^{-5}$ s.

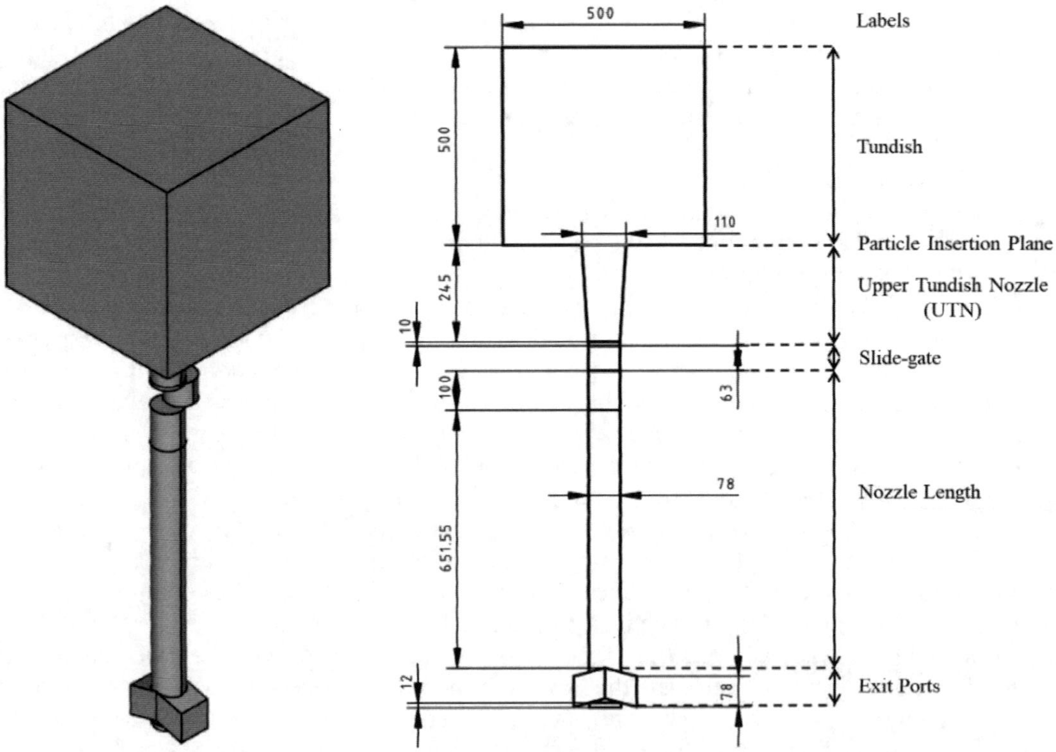

Figure 1. Left: Isometric view of simulation domain. Right: tundish and nozzle dimensions. All dimensions are in mm.

Alumina inclusions were used in the simulation, due to the ubiquity with which clogging occurs in Al killed steels. The inclusions were modelled as spherical particles with a diameter of 10 μm and a density of 3700 kg m^{-3}. For collision with the nozzle wall, a coefficient of elasticity of 1.0 and a coefficient of restitution of 0.0 were used. For the steel melt, typical values of density and kinematic viscosity were used which were 7020 kg m^{-3} and 0.006 kg $\mathrm{m}^{-1}\mathrm{s}^{-1}$, respectively.

The total simulation time was 20 s and the simulation procedure was as follows. First, the simulation was run for 5 s to allow the flow field to 'settle.' While the velocity fluctuations reduce over the first 5 s, they never reach steady state when the slide-gate is partially opened. Secondly, particles are inserted from the insertion plane atop the UTN. Particles enter the simulation with a velocity matching the surrounding fluid. Over 5 s a total of 1,000,000 particles are inserted (an injection rate of 200,000 particles per second). Finally, another 10 s was allowed for particles to travel through and exit the nozzle. At the end of the simulation, only particles which have adhered to the nozzle wall remain.

Eight sticking probabilities (S = 1.0, 0.75, 0.50, 0.25, 0.05, 1E-2, 1E-3, 1E-4) and three slide-gate linear opening positions (20%, 50% and 100%) were tested, resulting in a total of 24 simulations.

3. Results and Discussion

Figure 2 shows the velocity field within the tank, UTN and slide-gate. The glyphs represent only the direction of the flow while the background colour represents the velocity magnitude. In all three cases, the fluid entered the nozzle with a considerable degree of lateral velocity. When the slide-gate was partially open (the 20% and 50% opening cases), the fluid formed a 'jet' upon entering the slide-gate. This jet had a much higher velocity than the surrounding fluid. A great deal of recirculation occured within the slide-gate itself and the velocity fluctuations were high as the fluid exited the region. For the fully opened case (the 100% opening case), there were far less fluctuations in the velocity field. The fluid simply entered the nozzle and accelerated as the nozzle diameter narrowed. As the pressure head was the same for all three cases, the greater the slide-gate opening, the larger the flow rate. The flow rates for the 20%, 50% and 100% opening cases were $1.17 \times 10^{-3}\ m^3s^{-1}$, $6.98 \times 10^{-3}\ m^3s^{-1}$ and $2.26 \times 10^{-2}\ m^3s^{-1}$.

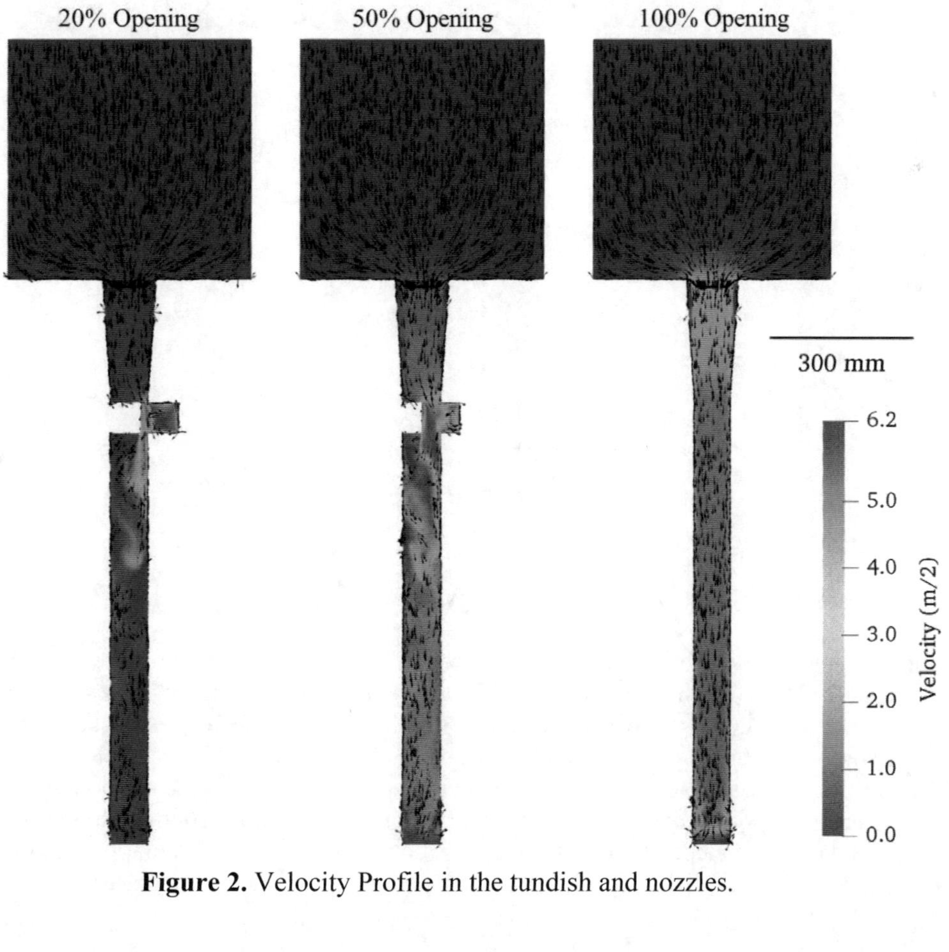

Figure 2. Velocity Profile in the tundish and nozzles.

Figure 3 shows the distribution of turbulence kinetic energy within the nozzle. As the turbulence within the tank was negligible and did not significantly affect particle motion, the tank has not been visualised. When the slide-gate was partially open, the turbulence within the nozzle was much higher. Flow around and within the slide-gate is unsteady and the use of an unsteady RANS model allows for the calculation of large-scale flow oscillations. The highest levels of both unsteady flow and turbulence were found within the slide-gate, followed by the flow downstream of it. Flow in the fully opened case was much less turbulent and the turbulence kinetic energy with significant levels only near the walls of the nozzle.

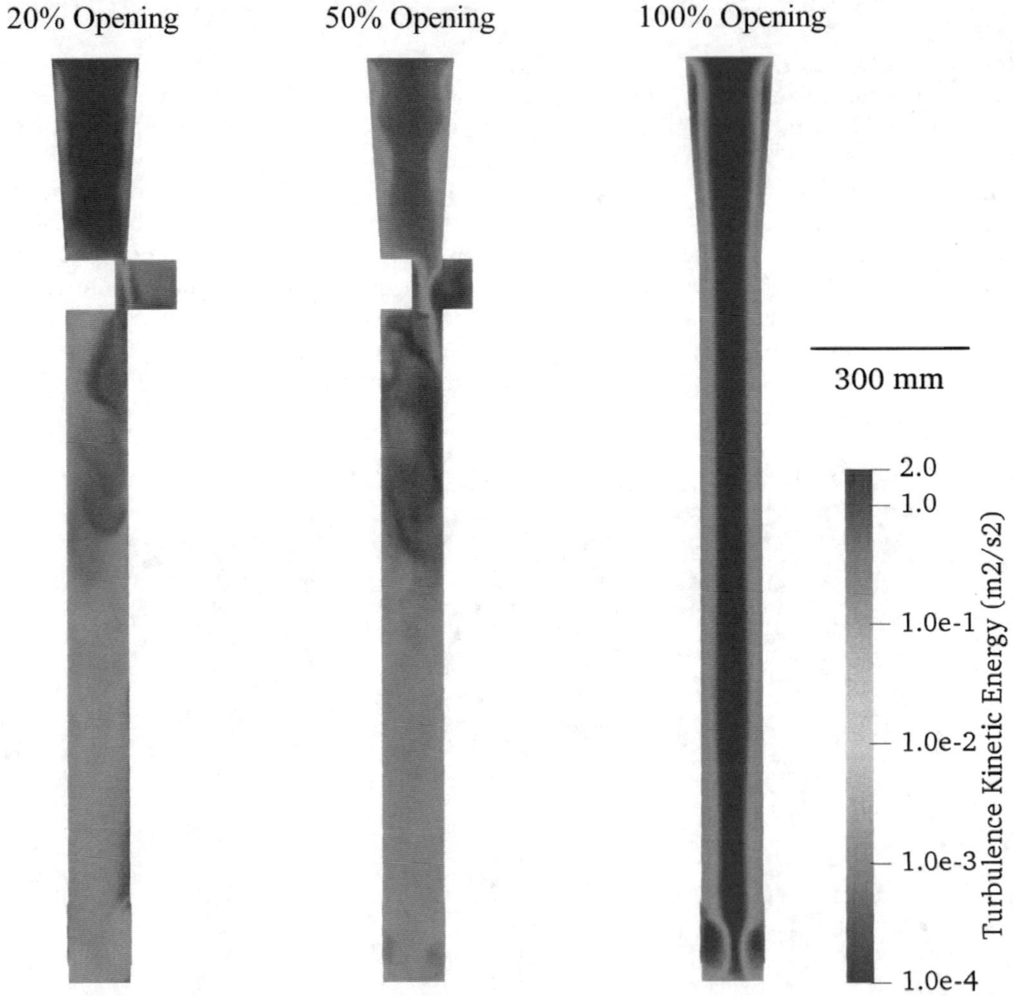

Figure 3. Turbulence Kinetic Energy in the nozzles. Note the use of a log scale.

Figure 4 shows the deposition pattern of the particles within the nozzle cross section. The deposition was highest in the 20% opening case, followed by the 50% and 100% opening cases. For the cases where the slide-gate was partially open, the particle area number density was highest just above the slide-gate, within the slide-gate and part of the nozzle wall below the slide-gate. Deposition at the bottom of the UTN and bottom of the slide-gate is mainly due to direct impingement of the flow on the surface. Within the slide-gate, deposition is due to a combination of recirculation and higher k. Along the nozzle length, just below the slide-gate, deposition is due to a combination of the jet, recirculation and increased k. The deposition pattern is very different for the 100% opening case. The particle area number density was highest in the UTN and gradually decreased deeper into the nozzle. While Figure 4 only shows the deposition locations when $S = 1$, this overall deposition pattern holds true even at lower values of S. Changes to sticking probability did not significantly alter deposition locations.

Figure 5 quantifies the deposition of particles within the nozzle. First, it must be noted that deposition of particles around the exit ports of the nozzle was neglected. This is because in order to properly simulate the flow in the outlet ports, the nozzle must be coupled with a mould. The flow pattern has a significant effect on particle deposition behaviour but as simulating the mould would drastically increase the computational cost of the study, this was not done. Hence, the study focused on deposition on the UTN, slide-gate and along the length of the nozzle. Total deposition (%) was calculated by dividing the number of particles attached to the nozzle by the total number of particles inserted (in this case, 1 million).

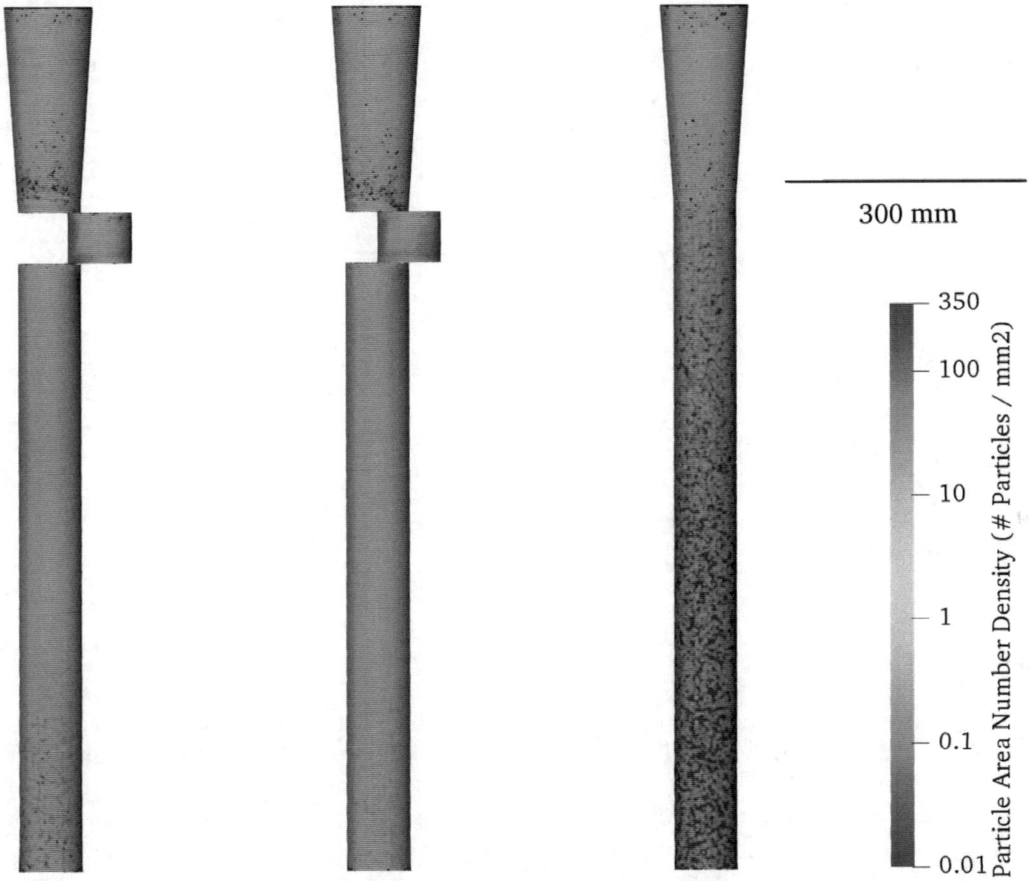

Figure 4. Cross section of the nozzle showing particle area number density. S = 1.00. Note the use of a log scale. The exit ports have been neglected.

As can be seen from Figure 5, the deposition rate was very high when $S = 1$ and varies from 85% for the 20% opening case to 30% for the 100% opening case. A clear trend can be seen where the deposition increases as the slide-gate opening decreases. This can be attributed to the increase in circulation within the slide-gate, as well as the increase in fluid turbulence at lower slide-gate opening positions. The circulation and increase in turbulence kinetic energy allowed for increased particle-wall collisions, enhancing deposition.

The effect of S on particle deposition fell into two regimes. Initially, as S increased, there was a rapid increase in deposition until $S = 0.05$. As S increased beyond 0.05, there was very little increase in particle deposition, although the deposition does remain high. The relative insensitivity to sticking probability when $S \geq 0.05$ can be attributed to the fact that inclusions collided with the wall multiple times while travelling through the nozzle. Hence, if the inclusion did not adhere to the wall on the first collision, there were several subsequent collisions where adhesion could occur. It was only when the sticking probability fell below 0.05, that a significant drop in deposition occurred. Hence, it is no surprise that for the partially open slide-gate cases, the majority of deposition occurred around and within the slide-gate. This was the region in which the largest velocity fluctuations and highest turbulence kinetic energy were located.

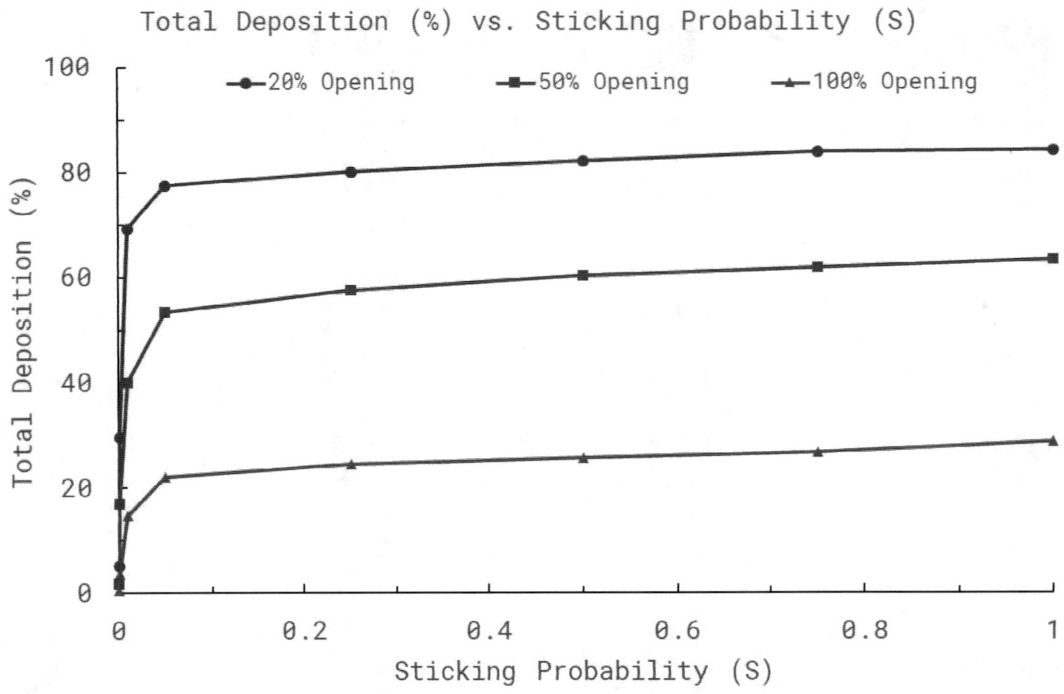

Figure 5. Deposition of particles in the nozzles. Deposited particles are the percentage of inserted particles left in the nozzle at the end of the simulation. Deposition on the exit ports is neglected.

In Figure 6, the relative deposition (%) of particles has been quantified. The relative deposition was calculated by dividing the number of particles in a region by the total number of particles deposited within the nozzle. For the partially opened cases, the largest proportion of particles was deposited on the nozzle length, followed by the slide-gate. The lowest relative deposition occurred in the UTN. It must be noted that as the slide-gate was a region with a smaller area than the nozzle length, the particle deposition was more concentrated (higher particle area number density), as shown previously in Figure 5. However, the results were different when the slide-gate was fully opened. The majority of particle deposition (72%) occurred on the UTN and only 5% deposited on the slide-gate.

During a typical casting process, the slide-gate begins with a more closed position and gradually opens throughout the process. This indicates that initially, particle deposition and clog formation is favoured around and within the slide-gate. Later in the casting process, clog formation is more likely to occur in the UTN. Varying the initial slide-gate linear opening position can affect the nascent location of clog formation. Depending on if the limiting factor for production is clog formation below the slide-gate, or in the UTN, the initial slide-gate position can influence casting time prior to nozzle replacement.

However, the effect of slide-gate opening on other production factors, such as the effect on casting speed, will need to be considered as well.

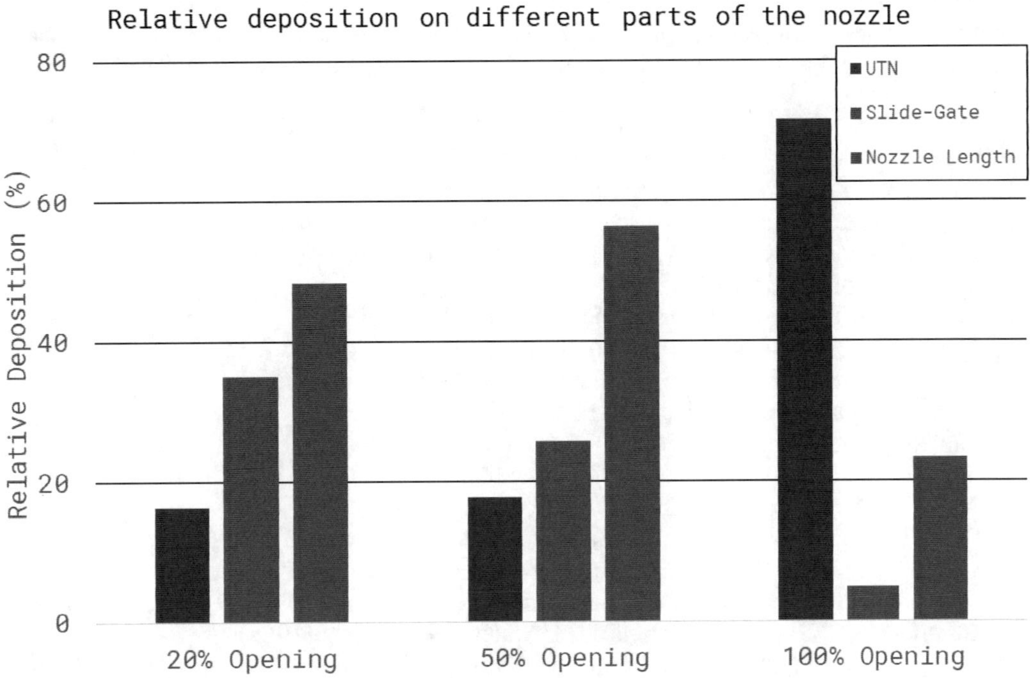

Figure 6. Relative deposition on different parts of the nozzle when $S = 1$. The relative deposition is the number of particles in a region divided by the total number of particles deposited in the nozzle.

4. Conclusion

A study was conducted on the effects of particle sticking probability on particle deposition behaviour inside a slide-gate SEN. The flow was modelled using a $k - \omega$ SST model, particle motion was modelled using a force balance and a simple stochastic approach was used to model particle adhesion behaviour. A set of 24 simulations were conducted with three slide-gate linear opening positions (20%, 50% and 100%) and 8 levels of sticking probability with S ranging from 1.00 to 1E-4.

A significant portion of the particles travelling through a nozzle collide with the nozzle wall, varying from 85% for the 20% opening case to 30% for the 100% opening case. The total deposition increased as the linear slide-gate opening position decreased. When the slide-gate was partially opened (20% and 50% opening), the particle deposition was concentrated just above, within and below the slide-gate. When the slide-gate was fully opened (100% opening), particle deposition was more concentrated in the UTN and decreased further into the nozzle. Examining relative deposition showed that the lower the slide-gate opening, the more deposition occurred within the slide-gate. With the slide-gate fully opened, 72% of deposition occurred in the UTN, for $S = 1$. Changes to S did not significantly alter deposition location.

The percentage of deposited particles decreased sharply when $S < 0.05$ and increased modestly yet linearly when $S \geq 0.05$. This was likely because in the case of particle rebound, the flow turbulence pushed the particle to repeatedly collide with the wall. When $S \geq 0.05$, most particles underwent a sufficient number of collisions to eventually adhere to the wall. When < 0.05, the number of particle-wall collisions were small enough to begin to affect deposition.

The use of a stochastic model has demonstrated that even a small amount of sticking probability can result in large effective deposition rates. However, this prediction is only true for the initial deposition behaviour. As the clog builds up and causes changes to the flow field, the deposition behaviour may change throughout the casting process.

Acknowledgements
The authors are grateful for the support and funding provided by the members of the McMaster Steel Research Centre as well as the Natural Sciences and Engineering Research Council of Canada.

References

[1] E. Szekeres, in *4th International Conference on Clean Steel,,(Balatonszeplak, Hungary)* (1992).
[2] L. Zhang, Y. Wang, and X. Zuo, Metall and Materi Trans B **39**, 534 (2008).
[3] S. Basu, S. K. Choudhary, and N. U. Girase, ISIJ International **44**, 1653 (2004).
[4] G. C. Duderstadt, R. K. Iyengar, and J. M. Matesa, JOM **20**, 89 (1968).
[5] W. Fix, H. Jacobi, and K. Wünnenberg, Steel Research **64**, 71 (1993).
[6] S. Ogibayashi, Taikabutsu (Refractories);(Japan) **46**, (1994).
[7] H. Cui, Y. Bao, M. Wang, and W. Wu, Int J Miner Metall Mater **17**, 154 (2010).
[8] J. K. S. Svensson, F. Larsson, A. Memarpour, S. Ekerot, V. Brabie, and P. G. Jönsson, Ironmaking and Steelmaking **45**, 105 (2018).
[9] R. Sambasivam, Ironmaking and Steelmaking **33**, 439 (2006).
[10] W. Bin Dai, X. L. Zhou, X. Yang, G. P. Tang, D. Bin Jia, N. L. Cheng, and J. K. Yu, Acta Metallurgica Sinica (English Letters) **29**, 500 (2016).
[11] Z. Deng, M. Zhu, B. Zhong, and D. Sichen, ISIJ International **54**, 2813 (2014).
[12] K. G. Rackers and B. G. Thomas, 78th Steelmaking Conference Proceedings **78**, 723 (1995).
[13] J. W. Farrell and D. C. Hilty, in *Electric Furnace Proceedings.*, Vol. 29 (1971), pp. 31–46.
[14] G. C. Duderstadt, R. K. Iyengar, and J. M. Matesa, Jom **20**, 89 (1968).
[15] L. Zhang, J. Aoki, and B. G. Thomas, Metallurgical and Materials Transactions B: Process Metallurgy and Materials Processing Science **37**, 361 (2006).
[16] Y. Miki, H. Kitaoka, T. Sakuraya, and T. Fujii, ISIJ International **32**, 142 (1992).
[17] S. Basu, S. K. Choudhary, and N. U. Girase, ISIJ International **44**, 1653 (2004).
[18] B. G. Thomas, Q. Yuan, S. Mahmood, R. Liu, and R. Chaudhary, Metallurgical and Materials Transactions B: Process Metallurgy and Materials Processing Science **45**, 22 (2014).
[19] C. Pfeiler, B. G. Thomas, M. Wu, A. Ludwig, and A. Kharicha, Steel Research International **79**, 599 (2008).
[20] Q. Yuan, Transient Study of Turbulent Flow and Particle Transport During Continuous Casting of Steel Slabs, University of Illinois at Urbana-Champaign, 2004.
[21] M. Mohammadi-Ghaleni, M. Asle Zaeem, J. D. Smith, and R. O'Malley, Metallurgical and Materials Transactions B: Process Metallurgy and Materials Processing Science **47**, 3056 (2016).
[22] F. R. Menter, AIAA Journal **32**, 1598 (1994).
[23] C. T. Crowe, *Multiphase Flows with Droplets and Particles*, 2nd ed. (CRC Press, Boca Raton, FL, 2012).
[24] H. Bai and B. G. Thomas, Metall and Materi Trans B **32**, 707 (2001).

IOP Conf. Series: Materials Science and Engineering 1281 (2023) 012032 doi:10.1088/1757-899X/1281/1/012032

In situ observation of solidification in peritectic steels and evaluation of the strains induced in the solidifying shell

H Yasuda[1], T Narumi[1], R Katsube[1], Y Nanri[1]

[1] Department of Materials Science and Engineering, Kyoto University, Sakyo, Kyoto 6068501, Japan

yasuda.hideyuki.6s@kyoto-u.ac.jp

Abstract. In the continuous casting of peritectic steels, unevenness and cracks of the solidifying shell are typical casting defects. This study focuses on the ferrite–austenite transformation and the influence of the transformation on the solidifying shell deformation. In situ observations revealed that the austenite hardly nucleated and, consequently, the undercooled ferrite massively transformed to the austenite in the solid state. Simultaneously, the austenite rapidly solidified in the remaining liquid. The volume changes due to the ferrite–austenite transformation and the solidification shrinkage were measured using time-resolved tomography. On the basis of the observations and measurements, a model that includes the ferrite–austenite transformation and the austenite solidification was proposed to evaluate the volumetric strains in the solidifying shell. There was a sharp peak in the volumetric strains derived from the model at hypo-peritectic compositions. This work demonstrates the need to include the ferrite–austenite transformation in the solid state and the rapid austenite solidification for understanding the solidifying shell deformation in peritectic steels.

1. Introduction

In the continuous casting of peritectic steels (i.e., approximately 0.1–0.18 mass% C), the surface quality is crucial [1-7]. Surface depressions (i.e., unevenness of the solidifying shell) are known to arise during the early stages of solidification within molds for peritectic carbon steel [1, 2]. These depressions have also been found for Fe–18mass%Cr–8mass%Ni stainless steel, Fe–11.6mass%Mn alloy, and Fe–5.6mass%Si alloy [1]. In Ref. [1], it was noted that the unevenness could not be explained only by considering the volume change during the peritectic solidification in which a transformation from the δ phase (i.e. ferrite, bcc) and the liquid phase to the γ phase (i.e., austenite, fcc) occurs during solidification, as shown in Fig. 1. The surface depressions in the peritectic steels have been extensively studied [1-20] because they can cause casting defects.

A mathematical model that considered the peritectic reaction ($L+\delta\rightarrow\gamma$) and the δ–γ transformation was proposed by Matsumiya et al. in 1982 [16]. In this model, the tensile strength induced in the solidifying shell reached a maximum value at a hypo-peritectic composition (0.14 mass% carbon). The induced forces could cause an uplift of the solidifying shell (i.e., the detachment of the solidifying shell from the mold). The uplift could be induced if the heat extraction from the casting to the mold is locally reduced (for example, by creating a groove on the mold surface, which increases the resistance for heat extraction). The model explained the unevenness of the solidifying shell at hypo-peritectic compositions. The deformations of the solidifying shell could significantly enhance the unevenness.

X-ray imaging studies have revealed an alternative transformation mode in FeC alloys [21-24] and FeCrNi alloys [25, 26]. In the solidification in FeC steels (<0.45 mass% carbon), the γ nucleation

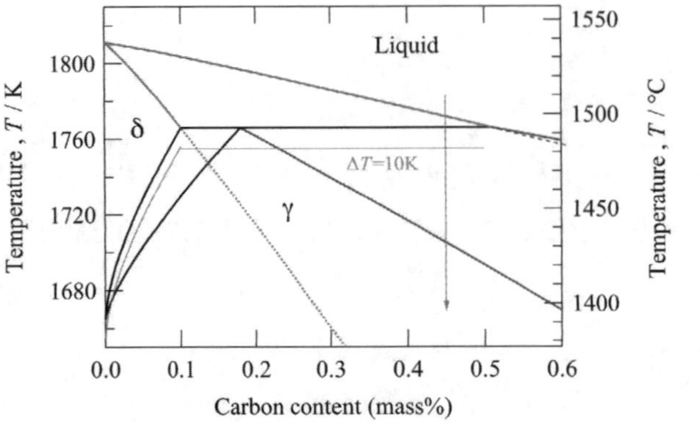

Figure 1. Peritectic reaction in the Fe–C binary system. The green line indicates the γ nucleation used in the calculation.

requires a relatively large undercooling from the peritectic temperature. Consequently, the δ phase continues to grow below the peritectic temperature. According to the Fe–C equilibrium phase diagram (Fig. 1), the solid-state transformation from the δ phase to the γ phase is thermodynamically admissible when the δ phase is cooled below the T_0 temperature (approximately 10 K below the peritectic temperature). The solid-state transformation can occur without solute partitions between the solid phases. In fact, the undercooled δ phase transforms to the γ phase rapidly in the solid state (referred to as a massive-like transformation). To the best of our knowledge, the unevenness of the solidifying shell has not previously been considered in terms of the massive-like transformation. Thus, it is of interest to examine the strains induced in the solidifying shell by the massive-like transformation.

In this paper, we present results of X-ray imaging (transmission imaging and tomography) to characterize the transformation. We then introduce a model for evaluating the strains and discuss the relationship between the massive-like transformation and the strains caused in the solidifying shell.

2. X-ray imaging

We review the δ–γ transformation to characterize the peritectic steel solidification and to highlight the novelty of the model. The typical δ-γ transformation mode during solidification in the Fe–0.45mass%C–0.6mass%Mn–0.3mass%Si alloy (referred to hereafter as Fe–0.45C–0.6Mn–0.3Si steel) is shown in Fig. 2 [21, 22, 27]. The difficulty of the γ nucleation resulted in the growth of δ dendrites at 0 s, nearly 100 K below the peritectic temperature. Once the γ phase nucleated between 0 s and 1 s, the undercooled δ phase transformed into the γ phase within 1 s. As indicated by the gray arrow at 0.45C in Fig. 1, the specimen was located in the single γ region of the equilibrium phase diagram. Thus, the γ solidification in the remaining liquid phase immediately followed the δ–γ transformation. The many dark spots (yellow arrows at 125 s) indicate the refined γ grain formation in a δ dendrite arm. In addition, the white regions (orange arrows) show the shrinkage grooves in the interdendritic part due to the rapid γ solidification. The δ–γ transformation and the rapid γ solidification are the primary features of the

Figure 2. Transmission images of solidification at 50 K/min in Fe–0.45C–0.6Mn–0.3Si steel [21,22,27].

Figure 3. Transmission image of unidirectional solidification at 50 μm/s in Fe–0.3C–0.6Mn–0.3Si steel [21,28].

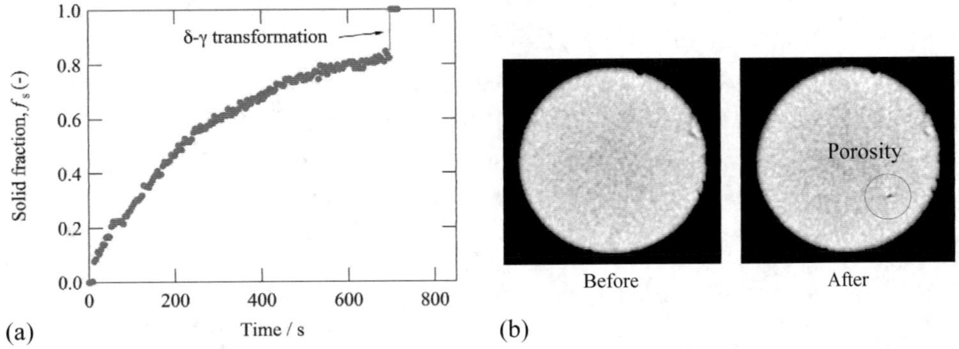

Figure 4. (a) Change in the solid fraction of Fe–0.45–C0.6Mn–0.3Si steel estimated by 4D-CT [30] and (b) pore formation induced by the massive-like transformation.

massive-like transformation [21, 22]. The results proved that the massive-like transformation in a non-equilibrium state must be included for understanding the microstructure evolution.

The transformation mode is also observed in the steady-state unidirectional solidification, as shown in Fig. 3 [21, 28]. The dark spots showed that refined γ grains were formed by the δ-γ transformation. The refined γ grain formation was also followed by the coarsened γ grains. The diffusion-controlled peritectic reaction was observed when the growth rate was less than 10 μm/s [21, 28]. Accordingly, the solidification structure in peritectic steels must be formed through the massive-like transformation mode in practical casting processes. On the other hand, the geometrical configuration of the three phases in the massive-like transformation is similar to that expected in the peritectic reaction. Thus, it will be possible to estimate the strains of the solidifying shell during steady-state solidification based on previous studies of the peritectic reaction [1-20]. It should be noted that the massive-like transformation contributes to the unique strain formation in the surface region because of non-steady-state solidification and transformation.

Time-resolved CT (4D-CT) has allowed observations of the solidification structure and measurements of volume change during heating and cooling for 3d transition-metal alloys, including steels [29, 30]. Figure 4(a) shows the solid fraction during solidification in carbon steel. The rapid increase in the solid fraction at 700 s was mainly caused by the rapid γ solidification. Figure 4(b) shows the porosity formed by the volume shrinkage due to the γ rapid solidification. The formation proves that the volume reduction induced by the rapid transformations should be included as a driving force for casting defects formation.

The volume changes required to evaluate the strains in the solidifying shell were measured using 4D-CT. Figure 5 shows the volume change during cooling at 20 K/min for Fe–0.05C–0.6Mn–0.3Si steel. The solidification shrinkage of the δ phase was −2.5% and the volume reduction due to the δ–γ transformation was −0.56%. The volume changes for Fe–0.45C–0.6Mn–0.3Si steel were also reported

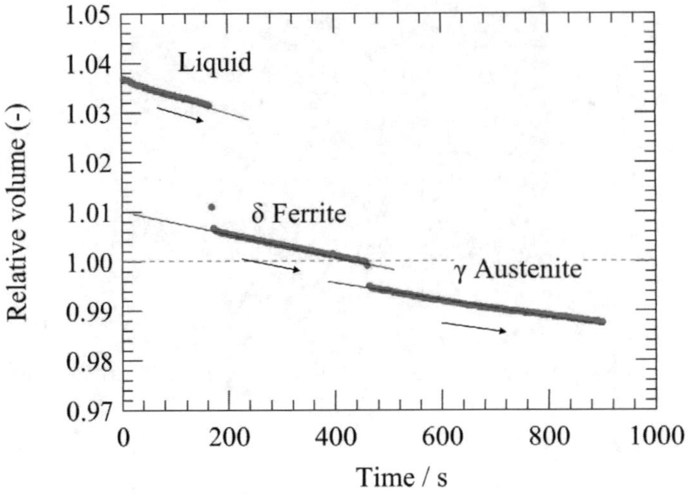

Figure 5. Relative volume measured during cooling at 20 K/min for Fe–0.05C–0.6Mn–0.3Si steel. The volume of the δ phase before the δ–γ transformation was used for the normalization.

to be −3.0% and −0.5%, respectively [27]. We note that an earlier paper reported that the solidification shrinkage of pure iron was −3.5% [31]. The carbon content may change the solidification shrinkage. Although further study is required to clarify this discrepancy, the small discrepancy does not intrinsically influence the conclusions of the present study. In the present study, the volume expansion coefficients are measured to be 80×10^{-6} K^{-1} (L), 61×10^{-6} K^{-1} (δ), and 65×10^{-6} K^{-1} (γ), respectively. The solidification shrinkage (L→γ) was estimated to be −3.5% by assuming it to be equal to the sum of the shrinkages due to δ-phase solidification and the δ–γ transformation. These measurement details will be published elsewhere [32].

3. Modeling

Figure 6 depicts the solidification sequence of the surface region in the peritectic steels based on the in situ observations. The δ dendrites grow along the temperature gradient from the mold side. If the γ phase

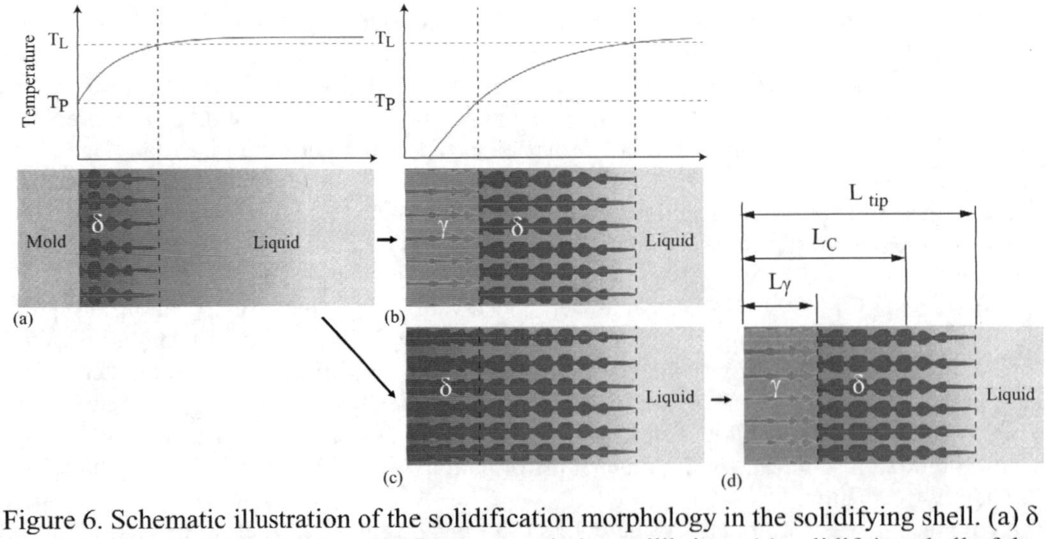

Figure 6. Schematic illustration of the solidification morphology in the solidifying shell. (a) δ dendritic growth, (b) peritectic solidification nearly in equilibrium, (c) solidifying shell of the undercooled δ phase without γ nucleation, and (d) solidifying shell of the δ and γ phases when the δ phase massively transforms to γ phase in the surface region.

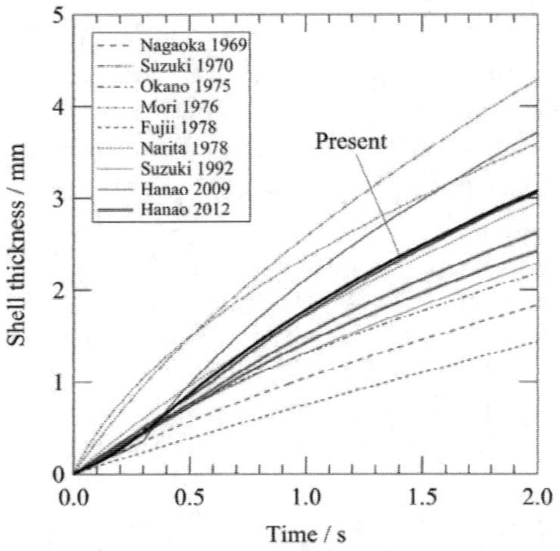

Figure 7. Shell thickness and solidification time [4,5,33-39]. The black line (present work) was used for evaluating the volumetric strains.

Table 1. Physical properties used in the calculation

Superheat	30 K
Latent heat	271×10^3 J/kg
Hear capacity	8.24×10^2 J/kgK
Density	7.3×10^3 kg/m^3
Thermal conductivity	40.3 J/mK
Thickness of casting	0.2 mm
Solidification shrinkage (δ phase)	-3.5%
Volume change (δ-γ)	-0.5%

nucleates at the surface in equilibrium, the δ solidification is followed by γ growth (Fig. 6(b)). There is no significant difference in the volume change in the solidifying shell between the massive-like transformation and the peritectic reaction. Thus, the strains induced in the solidifying shell can be considered based on the results of the previous works [1-20]. On the other hand, the induced strains in the surface region can completely differ from those induced by the peritectic solidification if the γ nucleation requires an undercooling from the peritectic temperature (approximately 10 K or more). This is the case we consider in this study. The γ phase does not nucleate and the δ dendrites continuously grow, as shown in Fig. 6(c). Once the γ nucleation occurs at the surface region below the peritectic temperature, the δ phase in the surface region immediately transforms to the γ phase, which simultaneously and rapidly solidifies in the remaining liquid phase. Because the δ–γ transformation and the γ solidification are completed quickly, the strain rate induced by the transformations will be much higher than that expected for near-equilibrium solidification and transformation. The strain and strain rate are crucial factors for the deformation of the viscoelastic solidifying shell. This study considers the strain in the solidifying shell.

Figure 7 shows the thickness of the solidification shell in continuous casting reported in the literature [4, 5, 33-39] and that calculated in the present work. We simulated the development of the δ-solidifying shell to evaluate the volume changes in continuous casting. For simplicity, equilibrium δ solidification in the binary Fe–C system was assumed because of the high carbon diffusivity at high temperatures. Table 1 lists the physical properties used in the calculation. The heat extraction from the casting to the mold was adjusted to fit the reported values [4, 5, 33-39]. The calculated shell thickness for Fe–0.18C steel (black line in Fig. 7) is within the range of the reported thickness [4, 5, 33-39]. Thus, the volume changes calculated in this study are expected to simulate the δ–γ transformation in continuous casting.

The undercooling of the γ-phase nucleation should be considered for evaluating the strains. As shown in the transmission images in Fig. 2, the undercooling was often larger than 100 K. The typical specimen dimensions for transmission images were 5 mm × 5 mm × 0.1 mm (2.5 mm^3). A small specimen with a volume of several mm^3 has a potential for large undercooling for γ-phase nucleation. Because the nucleation rate should be proportional to the volume, the undercooling in continuous casting must be much smaller than in these specimens. A simple calculation was performed to consider the undercooling of γ nucleation. A γ-nucleation event was assumed to be 2 mm from other nucleation events (the same order of magnitude as observed in previous X-ray studies [21]). If the two events occur at the same time, the δ/γ interfaces move toward each other. Here, we assumed that the surface cooling rate was 100 K/s. The growth velocity of the massive transformation was typically on the order of 10^{-2} m/s. The temperature at the midpoint between the nucleation positions decreased 20 K until the γ-growing interfaces reached the midpoint. In other words, the δ–γ transformation at the midpoint occurred at 20 K below the peritectic temperature. Although more-detailed investigations are necessary, an undercooling of 10 K for the γ nucleation in the surface region of the solidifying shell was selected to avoid overestimation of the induced strains. The nucleation temperature used in the calculation is indicated by the green line in Fig. 1.

Figure 8(a) shows the temperature distribution in Fe–0.05C steel casting when the surface temperature reaches the nucleation temperature (10 K below the peritectic temperature). The fractions of the liquid and the δ phases before the γ nucleation are shown in Fig. 8(b). The tip position of the δ dendrites was located at 2.5 mm and the δ solidification was completed in the surface region (<1.5 mm). The phase fractions after the γ phase had massively developed in the undercooled region are shown in Fig. 8(c). The thickness of the γ phase after development was only in the surface region (<0.3 mm), and the thickness of the δ shell was 1.5 mm. To evaluate the strains induced by the transformation and the solidification, the local volumetric strains, $\varepsilon_V^{(\delta-\gamma)}$ and $\varepsilon_V^{(L-\gamma)}$ are defined as

$$\varepsilon_V^{(\delta-\gamma)} = f_{\delta-\gamma} \cdot \xi_{\delta-\gamma} \tag{1}$$

$$\varepsilon_V^{(L-\gamma)} = f_{L-\gamma} \cdot \xi_{L-\gamma} \tag{2}$$

where $f_{\delta-\gamma}$ and $f_{L-\gamma}$ are the volume fraction changes due to the δ–γ transformation and γ solidification, respectively. The volume change due to the δ–γ transformation, $\xi_{\delta-\gamma}$, was −0.5%, and the solidification shrinkage, $\xi_{L-\gamma}$, was −3.5% in this evaluation. The total strain is simply given by

$$\varepsilon_V = \varepsilon_V^{(\delta-\gamma)} + \varepsilon_V^{(L-\gamma)} \tag{3}$$

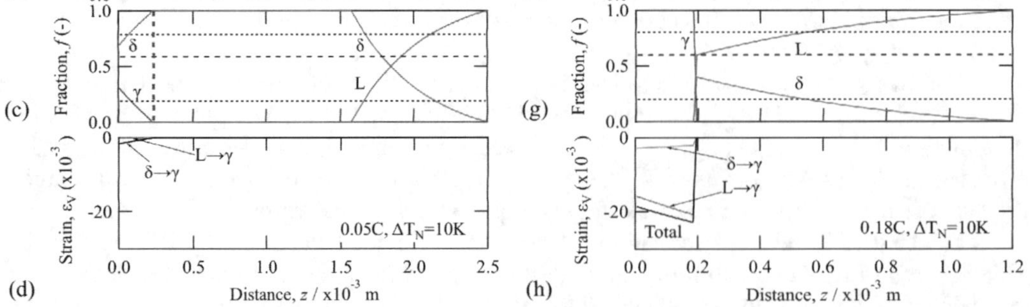

Figure 8. Temperature profile, phase fractions before and after the massive-like transformation, and the local volumetric strains in the solidifying shell. The undercooling of the γ phase was 10 K.

Figure 8(d) shows the local strains induced by the transformation. The total strain was as small as 10^{-3}, because the rapid solidification of the γ phase did not occur in Fe–0.05C steel. Figure 8(e)–(h) shows the calculated results for Fe–0.18C steel. When the undercooling at the surface reached 10 K, the fraction of the δ phase was slightly lower than 0.6. In other words, the δ phase transformed to the γ phase in the semisolid. As shown in Fig. 8(g), the fraction of the γ phase reached unity in the surface region (<0.2 mm). As a result, large strains were induced by the massive-like transformation. It is noted that the γ solidification contributed to 80% of the local strains.

The local strains did not directly contribute to the deformation of the solidifying shell. For example, the thickness of the γ phase produced by the massive-like transformation was only 0.3 mm, and the remaining δ phase was more than five times thicker than the γ phase for Fe–0.05C steel. Thus, the impact of the massive-like transformation on shell deformation was not large. For Fe–0.18C steel, the solid fraction was less than 0.6 immediately before the transformation. If the solidifying shell does not have mechanical strength, it is not necessary to consider its deformation. To estimate the impact of volume change on shell deformation, the average volumetric strains were defined as:

$$\tilde{\varepsilon}_V^{(\delta-\gamma)} = \int_0^{L_C} \varepsilon_V^{(\delta-\gamma)} dz \Big/ \int_0^{L_C} dz \tag{4}$$

$$\tilde{\varepsilon}_V^{(L-\gamma)} = \int_0^{L_C} \varepsilon_V^{(L-\gamma)} dz \Big/ \int_0^{L_C} dz \tag{5}$$

$$\tilde{\varepsilon}_V = \tilde{\varepsilon}_V^{(\delta-\gamma)} + \tilde{\varepsilon}_V^{(L-\gamma)} \tag{6}$$

The length, L_C, indicated in Fig. 6(d), corresponds to the thickness of a solidifying shell that has mechanical strength. Thus, the mechanical strength, especially the tensile strength, of the solidifying shell should be considered here. The tensile strength of semisolid carbon steel (carbon content of 0.005–0.55 mass%) was measured by Mizukami et al. [40]. The partially melted specimen was held without a crucible in the tensile test. The measured tensile strength was 0 for $f_s < 0.6$ and increased with increasing f_s. We consider the thickness of the solidifying shell with $f_s > 0.6$ as the length, L_C. This study evaluated the average volumetric strain using $f_s > 0.6$ and $f_s > 0.8$.

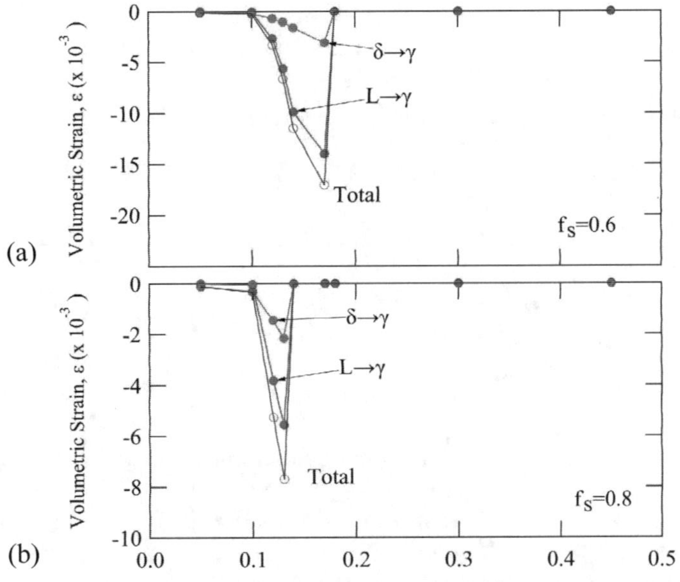

Figure 9. Average volumetric strain and carbon content. The thickness of the solidifying shell was defined as (a) $f_s > 0.6$ and (b) $f_s > 0.8$.

Figure 9 shows the average volumetric strains induced by the massive-like transformation. For f_s > 0.6, negative strains were observed for carbon contents ranging from 0.1 to 0.17 mass%. For a lower carbon content (<0.1 mass%), a larger thickness of the δ phase than that of the γ phase resulted in negligibly small strains. At higher carbon content (>0.18 mass%), the solidifying shell had zero tensile strength and the local strains did not contribute to shell deformation. As a result, the average strains had a sharp peak at a hypo-peritectic composition. For f_s > 0.8, the peak was sharper and shifted to lower carbon content. This evaluation indicates the strong dependence of the average volumetric strains on the carbon content. The massive-like transformation could contribute to the unevenness of solidifying shell. Further investigations are still needed on the γ-nucleation undercooling, the mechanical properties of semisolid steels, and the growth velocity of the massive-like transformation.

4. Summary

In situ observations revealed that the δ phase massively transforms to the γ phase in peritectic steels. The δ–γ massive-like transformation and the subsequent γ solidification in the undercooled melt characterized the development of the solidifying shell. This transformation can induce a larger strain and strain rate in the solidifying shell than peritectic solidification. We considered the impacts of the δ–γ transformation and γ solidification on the solidifying shell (f_s > 0.6 or 0.8). The average volumetric strains, which are a driving force for shell deformation, had a sharp peak at a hypo-peritectic composition when the undercooling for the γ nucleation was assumed to be 10 K. These results suggest that the massive-like transformation and the rapid γ solidification influence the unevenness of the solidifying shell in peritectic steels. Thus, surface defects in peritectic steels should be considered in future research in terms of the massive-like transformation.

Acknowledgments
X-ray imaging experiments were performed as general projects in a synchrotron radiation facility SPring-8/JASRI (2018B1521, 2019B1482, 2020A1564, 2020A0769, 2021A1499, 2021A1485, 2021B1560). The observation and the modeling were supported by JSPS KAKENHI Grant Numbers 17H06155, 22H00264, 22H04963. We thank Adam Brotchie, PhD, from Edanz (https://jp.edanz.com/ac) for editing a draft of this manuscript.

References
[1] Sugitani Y, Nakamura M 1979 *Tetsu-to-Hagane* **65** 1702-11
[2] Murakami H, Suzuki M, Kitagawa T, Miyahara S 1992 *Tetsu-to-Hagane* **78** 105-12
[3] Mizukami H, Yamanaka A 2010 *ISIJ Int* **50** 435-44
[4] Hanao M, Kawamoto M, Yamanaka A 2009 *ISIJ Int* **49** 365-74
[5] Hanao M, Kawamoto M, Yamanaka A 2012 *ISIJ Int* **52** 1310-9
[6] Presoly P, Pierer R, Bernhard C 2013 *Metall Mater Trans A* **44** 5377-88
[7] Singh S N, Blazek K E 1974 *JOM* **26** 17-27
[8] Won Y M, Yeo T J, Seol D J, Kyu Hwan O H 2000 *Metall Mater Trans B* **31** 779-94
[9] Thomas B G 2002 *Metall Mater Trans B* **33** 795-812
[10] Sengupta J, Shin H J, Thomas B G, Kim S H 2006 *Acta Mater* **54** 1165-73
[11] Sengupta J, Thomas B G, Shin H-J, Lee G-G, Kim S-H 2006 *Metall Mater Trans A* **37** 1597-611
[12] Matsumiya T 2006 *ISIJ Int* **46** 1800-4
[13] Maehara Y, Yasumoto K, Tomono H, Nagamichi T, Ohmori Y 1990 *Mater Sci Technol* **6** 793-806
[14] Saeki T, Ooguchi S, Mizoguchi S, Yamamoto T, Misumi H, Tsuneoka A 1982 *Tetsu-to-Hagane* **68** 1773-81
[15] Clyne T W, Wolf M, Kurz W 1982 *Metall Trans B* **13** 259-66
[16] Matsumiya T, Saeki T, Tanaka J, Ariyoshi T 1982 *Tetsu-to-Hagane* **68** 1782-91
[17] Ueshima Y, Mizoguchi S, Matsumiya T, Kajioka H 1986 *Metall Trans B* **17** 845-59
[18] Miyazawa K, Schwerdtfeger K 1979 *Ironmak Steelmak* **6** 68-74
[19] Brimacombe J K, Sorimachi K 1977 *Metall Trans B* **8** 489-505
[20] Yamanaka A, Nakajima K, Okamura K 1995 *Ironmak Steelmak* **22** 508-12

[21] Yasuda H, Morishita K, Yoshiya M, Narumi T 2020 *ISIJ Int* **60** 2755-64
[22] Yasuda H, Morishita K, Nakatsuka N, Nishimura T, Yoshiya M, Sugiyama A, Uesugi K, Takeuchi A 2019 *Nat Comm* **10** 3183-
[23] Yasuda H, Nagira T, Yoshiya M, Uesugi M, Nakatsuka N, Kiire M, Sugiyama A, Uesugi K, Umetani K 2011 *IOP Conf Ser Mater Sci Eng* **27** 012084-
[24] Yasuda H, Nagira T, Yoshiya M, Nakatsuka N, Sugiyama A, Uesugi K, Umetani K 2011 *ISIJ Int* **51** 402-8
[25] Nishimura T, Matsubayashi R, Morishita K, Yoshiya M, Nagira T, Yasuda H 2021 *ISIJ Int* **61** 1053-
[26] Nishimura T, Matsubayashi R, Morishita K, Yoshiya M, Nagira T, Yasuda H 2019 *ISIJ Int* **59** 459-65
[27] Yasuda H, Hashimoto T, Sei N, Morishita K, Yoshiya M 2019 *IOP Conf Ser Mater Sci Eng* **529** 012013-
[28] Nishimura T, Morishita K, Yoshiya M, Nagira T, Yasuda H 2020 *ISIJ Int* **60** 930-8
[29] Nakano K, Narumi T, Morishita K, Yasuda H 2020 *Mater Trans* **61** 596-604
[30] Yasuda H, Kawarasaki T, Tomiyori Y, Kato Y, Morishita K 2019 *IOP Conf Ser Mater Sci Eng* **529** 012023-
[31] Jablonka A, Harste K, Schwerdtfeger K, Schwerddeger K 1991 *Steel Research* **62** 24-33
[32] Nanri Y, Yasuda H, Narumi T, Katsube R 2023 *to be submitted to ISIJ international*
[33] Nagaoka N, Inamoto K, Shirai G, Nemoto S, Ohkawa Y 1969 *Tetsu-to-Hagane* **55** S109
[34] Suzuki A, Suzuki T, Shouzou M 1970 *Tetsu-to-Hagane* **56** S427
[35] Okano S, Nishimura T, Ooi H, Chino T 1975 *Tetsu-to-Hagane* **61** 2982-90
[36] Mori T, Ayata K, Fujimaki M, Soejima T, Kawahara M 1976 *Tetsu-to-Hagane* **62** S129-S36
[37] Fujii H, Ohash T 1978 *Tetsu-to-Hagane* **64** S646-S53
[38] Narita K, Mori T, Ayata K, Miyazaki J, Fujimaki M, Siomi T 1978 *Tetsu-to-Hagane* **64** S659-S
[39] Suzuki M, Miyahara S, Kitagawa T, Uchida S, Mori T, Okimoto K 1992 *Tetsu-to-Hagane* **78** 113-20
[40] Mizukami H, Hiraki S, Kawamoto M, Watanabe T 1998 *Tetsu-to-Hagane* **84** 763-9

IOP Conf. Series: Materials Science and Engineering 1281 (2023) 012033 doi:10.1088/1757-899X/1281/1/012033

Quantifying X-ray diffractometer precision for measurement of dislocation density and the evolution of defect structures during solidification of additive manufacturing powders

C Azersky[1], S Jeon[2], C Settens[3], M Fakhrul[4], and D Matson[4]

[1] NASA Goddard Space Flight Center, Greenbelt, Maryland, MD 20771, USA
[2] Deutsches Elektronen-Synchrotron (DESY), Notkestraße 85 D-22607 Hamburg, Germany
[3] X-ray Diffraction Shared Experimental Facility, Massachusetts Institute of Technology, Cambridge, MA 02139, USA
[4] Department of Mechanical Engineering, Tufts University, Medford, MA 02155, USA

ColbyAzersky@gmail.com

Abstract. In this work, the Phillips PW 3020 X'Pert Diffractometer and PANalytical X'Pert Pro MPD X-ray Diffractometer were used to perform in-situ monitoring of quenched microstructures to define the defect evolution of Inconel 718 additive manufacturing powder. Dislocation density analysis based on X-ray Diffraction (XRD) measurement is sensitive to XRD peak broadening. The X-ray line profile is affected by the instrumental effects of the diffractometer, so quantifying and analyzing XRD instrument performance is important for accurate dislocation density analysis. The average measurement bias of the Phillips diffractometer was found to be -1.5490·10^{-3} Å while the precision was found to be 6.1000·10^{-5} Å. The average measurement bias of the PANalytical diffractometer was found to be -1.7633·10^{-4} Å while the precision was found to be 7.6917·10^{-5} Å. The average dislocation density calculated from the data was 3.93·10^{14} m^{-2} for the smaller particle size range and 2.58·10^{14} m^{-2} for the larger particle size range. The number of diffraction peaks utilized in dislocation density analysis was found to be a more significant factor than the instrumental differences. This work confirmed that differences in defect structure density can be observed across differently sized particles sourced from additive manufacturing powder.

1. Introduction

Line profile analysis of X-ray diffraction (XRD) data requires accurate estimation of instrumental broadening effects because the physical profile of a Bragg reflection is a convolution of microstrain, crystallite size, and instrumental effects. Understanding and quantifying each of these contributions are vital to accurate estimation of dislocation density, dislocation character of edge and screw types, and the crystallize size of nanocrystalline materials. Controlling the microstructure of metallic powders used in additive manufacturing (AM) is crucial to the production of AM parts with desired mechanical properties. In this series of experiments, a Philips PW 3020 X'Pert diffractometer and a PANalytical X'Pert Pro MPD X-ray diffractometer were used to study instrumental effects on dislocation density analysis. In both experiments, Inconel 718 additive manufacturing powder was used because it is an industrially relevant superalloy. This gas atomized powder was readily available and easily isolated into distinct particle size ranges by sieving the as-received material. Thus, the powder was separated into distinct particle size ranges of 20-25 μm and 38-45 μm particle diameter using a sieve shaker. These particle size ranges were confirmed using scanning electron microscopy.

For nanocrystalline materials, atomic dislocations in the otherwise regularly ordered atomic lattice can generate a strain field around the dislocation[1]. Differences in interplanar spacing between the atoms can be measured using X-ray diffractometry because the associated diffraction will occur at certain angles based on the Bragg law, resulting in a broadening of diffraction peaks caused by the strain and crystallite size effects[2]. The Modified Williamson-Hall and Warren-Averbach methods[3-10] separate the line profiles by crystallite size and microstrain, based on their different dependence on the diffraction vector. Thus, these methods were used to evaluate the dislocation density of the Inconel 718 additive manufacturing powder samples.

The two diffractometers (λ = 1.5406 Å) used in this series of experiments had differences in their observable angular ranges. The Philips diffractometer had a 2θ angular range of 20° to 90°, whereas the PANalytical diffractometer had an angular range of 20° to 120°. Both diffractometers, can measure slightly above and below these limits, but these angles were the chosen parameters for this experiment. This difference in angular range for each facility impacts the number of diffraction peaks that may be evaluated during both calibration and powder characterization.

After the calibration process, the Inconel 718 powders were measured using a full 2θ angular scan. Due to the differences in angular range for the two diffractometers, the PANalytical diffractometer could access five diffraction peaks for the Inconel 718 samples while the Philips diffractometer could only access three diffraction peaks. To account for this difference, the two diffractometers were compared once using only the first three diffraction peaks, then once using all available diffraction peaks. By calibrating and comparing the devices using the mutually accessible Inconel 718 diffraction peaks, a true comparison of the accuracy and precision achieved by each of the diffractometers can be evaluated.

Modified Williamson-Hall[6] analysis uses diffraction data where the instrumental broadening effects have been previously removed. The diffraction peak broadening is further separated into the contributions by crystallite size and microstrain. The classical Williamson-Hall analysis was modified to account for microstrain anisotropy at each crystal plane which induced error in the original analysis and the modified version is given as follows:

$$\Delta K \cong \frac{0.9}{D} + \left(\frac{\pi M^2 b^2}{2}\right)^{1/2} \rho^{1/2} K \bar{C}^{1/2} + O(K^2 \bar{C}) \tag{1}$$

where K is the magnitude of the diffraction vector and defined by $K = 2\sin\theta/\lambda$, where θ and λ are the diffraction angle and the wavelength. ΔK is the FWHM (full-width-at-half-maximum) in K and defined by $\Delta K = 2\cos\theta(\Delta\theta)/\lambda$. D is the coherent domain size (or crystallite size), M is a dimensionless parameter depending on dislocation density and outer cut-off radius of dislocations, b is the magnitude of Burgers vector, ρ is the dislocation density, $\overline{C}$ is the average contrast factor of dislocations, and O is the higher order terms.

Modified Warren-Averbach[9,10] analysis utilizes results from the Modified Williamson-Hall method to determine dislocation density using Fourier transformations. The signal from X-rays penetrating to different depths of the atomic crystal lattice combines effects from many different atomic planes with different associated microstrains. A Fourier transformation can be used on this combined signal to evaluate the total density of dislocations for the entire sample.

$$\ln\big(A(L)\big) \cong \ln\big(A^s(L)\big) - \frac{\rho\pi b^2 L^2}{2} \ln\left(\frac{R_e}{L}\right)(K^2\bar{C}) + Q(K^4\bar{C}^2) \tag{2}$$

where A is the real part of the Fourier coefficient, A^s is the size Fourier coefficient[3], L is the Fourier length defined as $L = na_3$, where a_3 is $\lambda / [2(\sin\theta_2 - \sin\theta_1)]$, n are integers starting from zero, and $(\theta_2 - \theta_1)$ is the theta range for the XRD profile, ρ is the dislocation density, b is the magnitude of Burgers vector, R_e is the effective outer cut-off radius of dislocations, K is the centroid of the diffraction peak, C is the contrast factor, and Q is the collection of quadratic terms.

2. Experimental Methods

The gas atomized Inconel 718 powder was isolated into two distinct particle size ranges utilized in this series of experiments. The as-received Inconel powder was sieved using a DUAL Manufacturing Co. Inc. Model # D-4326 sieve shaker and DUAL Manufacturing Co. Inc. A.S.T.M. E-11 specification mesh screens. This particle isolation was using validated Pro X PHENOM Model 800-07334 SEM. The material analysis, as shown in table 1, was performed by Luvak Inc. under the ASTM E 1019-18 and ASTM 1097-12 standards.

Table 1. Material analysis of Inconel 718.

Element	Weight %
Oxygen	0.016
Carbon	0.029
Sulfur	0.003
Nickel	52.1
Chromium	19.16
Iron	*Balance*
Niobium	5.01
Molybdenum	2.96
Cobalt	0.12
Titanium	0.90
Manganese	0.042
Silicon	0.055
Copper	0.031
Aluminum	0.51
Tantalum	<0.02
Phosphorus	0.004
Boron	0.0037
Magnesium	0.0010

XRD data were collected in one session at the Philips diffractometer while the data was collected at the PANalytical diffractometer in three different sessions. In all cases, a lanthanum hexaboride sample was measured for calibration before any data was collected. Due to the nature of the diffractometers, the powder sample was loaded into the machines using different methods. At the Philips diffractometer facility, a thin layer of Inconel powder was mounted on Kapton tape. At the PANalytical diffractometer facility, the powder was loaded into the well of a sample holder which was inserted into the diffractometer.

The data points for diffraction peaks for both the lanthanum hexaboride calibration samples and the experimental samples were fitted using a Lorentzian curve fit. These fits provided both the peak center location and the FWHM for each peak. The diffraction peaks, which had a majority Lorentzian character, were fitted using Lorentzian profiles:

$$y = y_0 + \frac{2A}{\pi}\frac{w}{4(x - x_c)^2 + w^2} \tag{3}$$

where y_0 is the baseline height, A is the amplitude, w is the width, and x_c is the peak center.

The apparent peak centroid shift due to instrument effects was corrected using a curve fit generated from the Wilson formula[13] to the calibration sample data points.

$$\Delta_{2\theta}(\theta) = axtan^{-1}(\theta) + bx + cx\tan(\theta) + dxtan^2(\theta) + extan^3(\theta) \quad (4)$$

where $\Delta_{2\theta}$ is the change in peak centroid position from the expected value, θ is the 2-theta angle of X-ray incidence, and a, b, c, d, e, are fitting parameters.

The peak broadening for each of the lanthanum hexaboride diffraction peaks was used to estimate the instrumental broadening effects. A curve fit using the Caglioti formula[11] was used to determine he instrumental broadening as a function of incident angle, theta. The instrumental broadening effect was subtracted from the measured breadth to obtain the actual breadth.

$$FWHM(\theta) = [W + Vtan(\theta) + Utan^2(\theta)]^{1/2} \quad (5)$$

The parameters used for the Philips PW 3020 X'Pert diffractometer were scan range of 2θ values from a minimum of 20° to a maximum of 90° at a resolution of 0.01°. This allowed the dataset to capture three diffraction peaks for Inconel 718: FCC (111), (200), and (220) occurring at angles of 43.57°, 50.72°, and 74.58°. The PANalytical X'Pert Pro MPD X-ray diffractometer had a scan range of 2θ values from a minimum of 20° to a maximum of 120°. The scans were performed at a resolution of 0.008°. This allowed the dataset to capture the five diffraction peaks for Inconel 718: (111), (200), (220), (311), and (222) occurring at angles of 43.57°, 50.72°, 74.58°, 90.49°, and 95.84°. For the same exposure time of X-ray and the same amount of power samples, relative intensity of diffraction peaks for data collected at the Philips facility was roughly 7 times less intense than then intensities of similar diffraction peaks at the PANalytical facility.

3. Results

XRD patterns of the NIST Standard lanthanum hexaboride samples were collected in a 2θ angular range from 40° to 90° at both the Phillips and PANalytical facilities. Each of the diffraction peaks was analyzed using a Gaussian curve fit. The Caglioti formula was applied to this data to find the instrumental width fit as shown in figure 1.

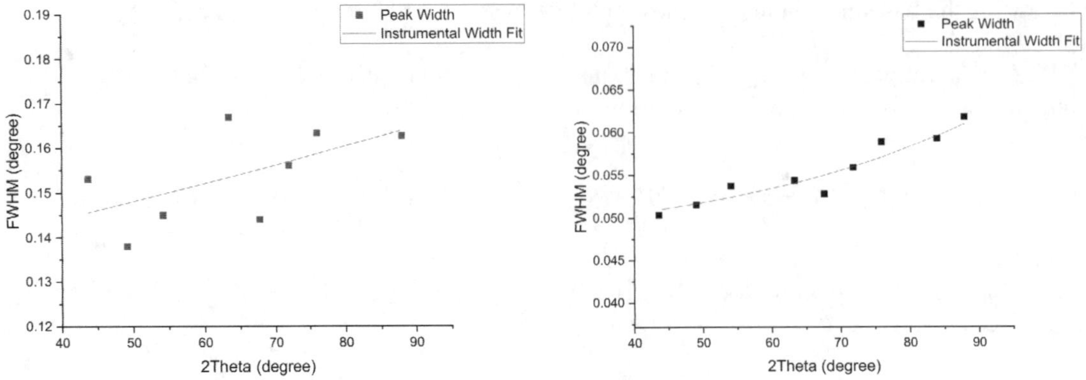

Figure 1. Instrumental width fit versus actual peak width for lanthanum hexaboride powder calibration samples. Results are shown for the Philips PW 3020 X'Pert diffractometer (left) and the PANalytical X'Pert Pro MPD X-ray diffractometer (right) for angle ranges between 40° and 90°.

Rietveld refinement was performed on the lanthanum hexaboride powder calibration sample data sets from both facilities. The lattice parameter precision was extracted from each of these fits and an average, bias, and standard deviation were found and provided in table 2.

Table 2. Diffraction peak rietveld refinement

Facility	Average Lattice Parameter (Å)	Diffractometer Measurement Bias (Å)	Standard Deviation [Precision] (Å)
NIST Standard	4.1568	N/A	N/A
Phillips	4.1566	$-1.5490 \cdot 10^{-3}$	$6.1000 \cdot 10^{-5}$
PANalytical	4.1553	$-1.7633 \cdot 10^{-4}$	$7.6917 \cdot 10^{-5}$

The instrumental broadening influence identified using the calibration samples was removed from the datasets and the microstrain was deduced from the breadth infromation from the raw specimens for each particle size range and provided in table 3.

Table 3. Calculated microstrain results

Facility	Particle Diameter (μm)	Microstrain (%)
Phillips	20 – 25	0.28
Phillips	38 – 45	0.21
PANalytical	20 – 25	0.35
PANalytical	38 – 45	0.33

Using the Modified Williamson Hall and Warren Averbach methods, the dislocation density was evaluated for the 20 – 25 μm and 38 – 45 μm particle diameter Inconel 718 samples. While the dataset collected from the PANalytical facility measured five diffraction peaks, the dataset was first evaluated using the three available diffraction peaks and compared to the Philips facility to evaluate both instruments using the same number of peaks. Then, the dataset collected from the PANalytical facility was also evaluated using the full five diffraction peaks. This data is shown in figure 2.

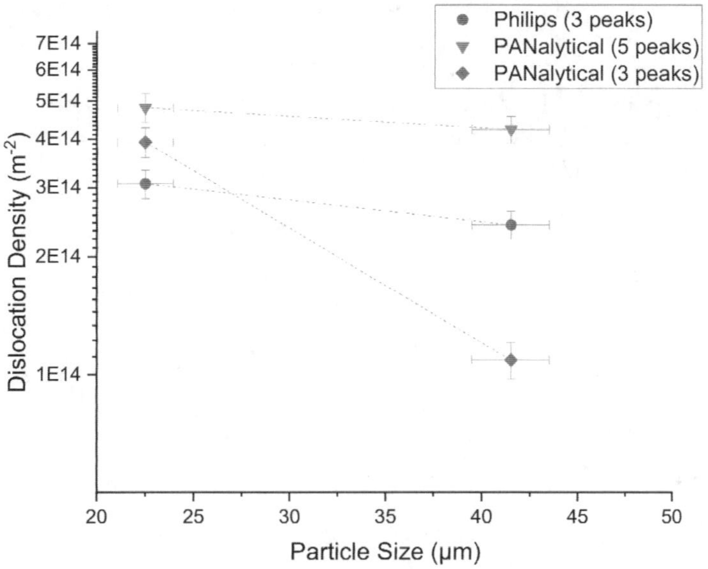

Figure 2. Dislocation density results for the Inconel 718 additive manufacturing powder samples at particle diameter ranges 20 – 25 μm and 38 – 45 μm. Results are shown for the PANalytical X'Pert Pro MPD X-Ray diffractometer using 5 diffraction peaks and 3 diffraction peaks and the Philips PW 3020 X'Pert diffractometer using 3 diffraction peaks.

Field emission scanning electron microscopy was used to image the surfaces of individual particles as shown in figure 3. Then, individual grains were identified and analyzed to determine dendrite arm spacing. This information was used to calculate cooling rate as a function of particle diameter.

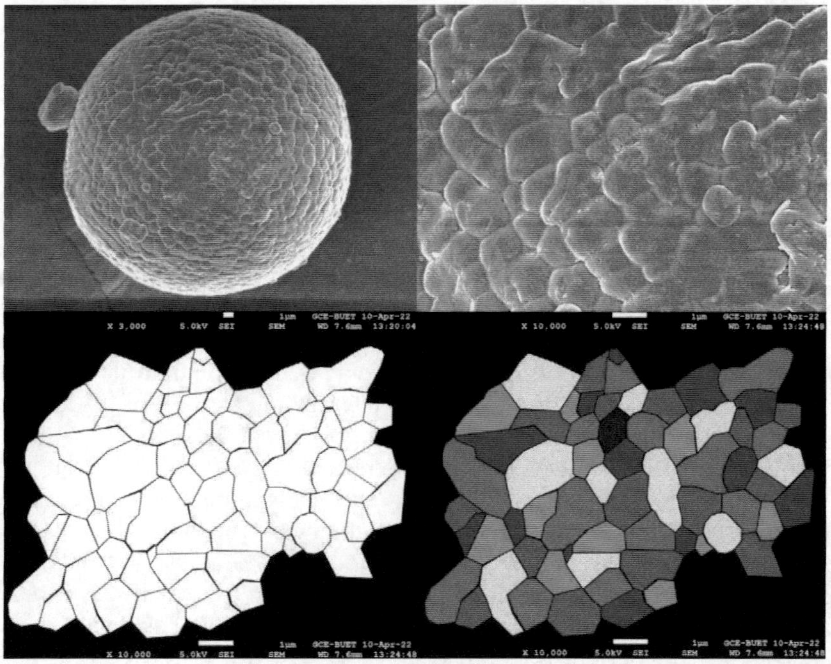

Figure 3. Field emission scanning electron microscope images and subsequent grain partitioning for a single Inconel 718 particle.

The relationship between primary dendrite arm spacing (d.a.s.) and cooling rate for Inconel 718 reported by Shohoji[14]. Primary d.a.s. was analyzed using the grain size data from the scanning electron microscope images. The primary d.a.s. was used to calculate the cooling rate for several individual particles as shown in figure 4. The results were plotted with a 95% confidence interval and an exponential fit was found with the following parameters:

$$CR = 25050 + 1977000 \cdot e^{-0.0821D}$$

where CR is the cooling rate of the particle (K/s), and D is the particle diameter (μm).

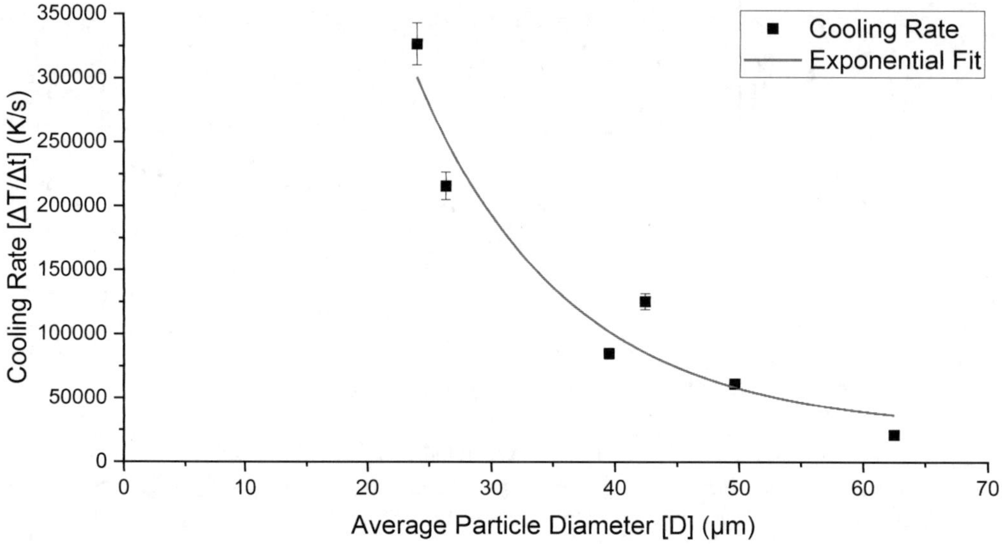

Figure 4. Cooling rate results derived from dendrite arm spacing for the Inconel 718 additive manufacturing powder samples at several different particle diameters.

4. Conclusions and Outlook

The data from both facilities resulted in dislocation densities which were both within the same order of magnitude and consistent with values published in existing literature[15]. All results showed a trend of decreasing dislocation density with increasing particle size: the average dislocation density calculated from the data across both facilities was $3.93 \cdot 10^{14}$ m^{-2} for the smaller particle size range and $2.58 \cdot 10^{14}$ m^{-2} for the larger particle size range. This trend was previously demonstrated for $AlCrFe_2Ni_2$ eutectic high entropy alloy (HEA) powders[12]. This is also consistent with the findings from the microstrain analysis. An increase in dislocation density would generate an increased microstrain in the atomic lattice. The smaller particle size range was calculated to have an average microstrain of 0.315% while the larger particle size range was calculated to have an average microstrain of 0.270%.

The magnitude of the diffractometer bias was identified to be significantly larger at the Phillips facility when compared to the PANalytical facility. While the magnitude of the precision of the Phillips facility was calculated to be smaller than the PANalytical facility, it should be noted that funding for this research only encompassed one calibration data set from the Phillips facility and three data sets from the PANalytical facility. Further research should be conducted on the precision of the Phillips facility as the reported precision may be artificially low due to a lack of comparison data.

It was found that the number of diffraction peaks used during the dislocation density analysis had a larger impact on the associated calculation than the instrumental differences in the diffractometers. The PANalytical diffractometer had a larger angular scan range than the Phillips facility meaning that five diffraction peaks were captured in the data while only three peaks were captured from the Phillips facility. Analysis of the PANalytical facility data using five peaks resulted in a dislocation density of $4.81 \cdot 10^{14}$ m^{-2} for the smaller particle size and $4.23 \cdot 10^{14}$ m^{-2} for the larger particle size. Analysis of the PANalytical facility data using three peaks resulted in a dislocation density of $3.93 \cdot 10^{14}$ m^{-2} for the smaller particle size and $1.09 \cdot 10^{14}$ m^{-2} for the larger particle size. Analysis from the Phillips facility data resulted in a dislocation density of $3.08 \cdot 10^{14}$ m^{-2} for the smaller particle size and $2.41 \cdot 10^{14}$ m^{-2} for the larger particle size. It should be noted that the dislocation density calculations can be further extrapolated to find the corresponding misorientation angle and subsequently the dislocation energy; however, this is beyond the scope of this research. When evaluating the respective accuracy of the methods, using less diffraction peaks generated a larger average variation than using a different diffractometer.

Differences in mounting techniques may have produced differences in datasets which were not due to the instruments themselves. At the facility with the Phillips diffractometer, the powder sample was mounted directly to Kapton tape. Whereas, at the facility with the PANalytical diffractometer, the powder was loaded directly into the sample holder. Diffractometer setup, which can include divergence slits, anti-scatter slits, masks, Soller slits, and filters, also may play a role in differences between datasets. It should also be noted the input parameters, angular range, resolution, and intensity for both diffractometers can be altered. While all these factors may affect the resulting data, this research aimed to compare the nominal operation of these diffractometers.

The Philips PW 3020 X'Pert diffractometer is fast and accessible, so it is best for feasibility experiments to determine which materials should be further studied. The PANalytical diffractometer is relatively slower and the cost for access is larger, but it provides a higher fidelity signal resulting in increased data quality. Differences in the number of diffraction peaks, a result of differences in maximum scan angle and diffractometer intensity, was found to be a significant factor in dislocation density determination. Therefore, further experimentation should utilize facilities that can access the largest number of diffraction peaks possible.

Dislocation density and rapid solidification of metallic powders are important factors that can affect the properties and performance of parts produced through additive manufacturing. The rapid solidification that additive manufacturing powders experience during production generates high densities of dislocations which will affect the mechanical properties of the final part. Therefore, it is important to carefully control the processing conditions during the manufacturing of metallic powders used in additive manufacturing to achieve the desired properties in the final product. This may involve optimizing the processing parameters to control the dislocation density and achieve the desired solidification.

References

[1] Zhang Z, Ódor E, Farkas D, Jóni B, Ribárik G, Tichy G, Nandam S, Ivanisenko J, Preuss M and Ungár T 2020 *Metallurgical and Materials Transactions* **51** 513
[2] Scardi P, Ermrich M, Fitch A, Huang E, Jardin R, Kuzel R, Leineweber A, Mendoza-Cuevas A, Misture S T, Rebuffi L and Schimpf C 2018 *Applied Crystallography* **51** 831
[3] Hajyakbary F, Sietsma J, Bottger A J and Santofimia M 2015 *Materials Science and Engineering* **603** 208
[4] Shintani T and Murata Y 2011 *Acta Materialia* **59** 4314
[5] Martinez-Garcia J, Leoni M and Scardi P 2009 *Acta Crystallographica* **65**, 109
[6] Ungár T and Borbély A1996 *Applied Physics Letters* **69** 3173
[7] Ungár T, Gubicza J, Ribárik G and Borbély A 1999 *Applied Crystallography* **32** 992
[8] Ungár T, Gubicza J, Ribárik G and Borbély A 2001 *Applied Crystallography* **34** 298
[9] Ungár T 2001*Materials Science and Engineering* **309** 14
[10] Ungár T, Ott S, Sanders P G, Borbély A and Weertman J R 1998 *Acta Materialia* **46** 3693
[11] Caglioti G, Paoletti A and Ricci F P 1958 *Nuclear Instruments* **3** 223
[12] Jeon S, Liu X, Azersky C, Ren J, Zhang S, Chen W, Hyers R W, Costa K, Kolbe M and Matson D M 2021 *Materialia* **18** 101161

[13] Wilson A J C 1963 *Mathematical theory of X-ray powder diffractometry*
[14] Shohoji N 2014 *Surfaces and Interfaces of Materials* **182** 2
[15] Shia J J, Lia X, Zhang Z X, Cao G H, Russell A M, Zhou Z J, Li C P and Chen G F 2019 *Materials Science and Engineering* **765** 138282

Acknowledgments

Portions of this work were funded by NASA under grants NNX17AH41G and 80NSSC19K0256. This work made use of the MIT Materials Research Laboratory X-ray facility that is supported by the NSF under Grant No. DMR-1419807. The authors declare no conflict of interest.

Effects of instabilities on the properties of ingots made by the ESR and VAR processes

S Hans[1*] and A Mitchell[2]

[1] Aubert Duval, 63770 Les Ancizes, France
[2] Department of Materials Engineering, University of British Columbia, Vancouver BC, V6T1Z4, Canada

* corresponding author: stephane.hans@aubertduval.com

Abstract. The ESR and VAR processes are very often carried out under less-than-ideal conditions where the procedure involves inadvertent instabilities such as interruptions in metal flow or abrupt changes in the process heat balance. In this work we demonstrate the effects of typical examples of such instability as related to the remelting processes. It is suggested that a major direction in modelling development should be towards an understanding of the extent to which instability can be tolerated relative to the product requirements. Such an understanding would assist in the establishment of suitable quality standards governing the production of high-quality alloys.

1. Introduction

The ESR (Electro Slag Remelting) and VAR (Vacuum Arc Remelting) processes are widely used to produce special steels, Ni-base superalloys, and titanium alloys to meet the needs of the aeronautics, energy and nuclear industrial sectors for materials with high added value, intended for critical applications. Excellent models of these processes have been developed over the past few years [1-4]. They all rest on the basic premise that the processes are axisymmetric, and that the energy balance is constant during the steady-state melting cycle. The output from the computations is a thermal picture in which the steady-state ingot temperature regime is constant, producing a symmetrical liquidus isotherm and constant solidification rate profile. The computed solidification result is shown as a temperature pattern in which it is assumed that solidification will result in a symmetrical pattern of columnar dendrites whose spacings are dependent on the radial position [1]. Models of this format have been used by the industry for some time to design process parameters in order to obtain a particular ingot structure The solidification growth of the ingot can be modeled using different software, for example by using the CAFE concepts, in which the dendrites progressively form as the crucible is gradually filled. A transition to the equiax mode may take place if the solidification rate is low [5] and has been considered in some of the models [5].

2. Irregularities in the Practical ESR Procedures

The ESR process heat balance is determined principally by the balance of heat generation in the slag against heat extraction by the water-cooled mold. It is generally assumed that these two processes are axisymmetric and constant. However, if there is a change in the current distribution in the slag leading to an equivalent change in the slag temperature distribution, there is a consequent effect on the ingot solidification pattern. Such changes can occur in regular production. Examples are:

- metal in the ingot pool breaking through the surrounding solid slag film and contacting the mold
- variations in the electrode position control causing variations in slag temperature
- progressive changes in slag composition due to oxidation of the metal

In principle, such changes in the energy balance can be modeled with the existing software, except for the necessity to change the base from axisymmetric to 3D, since the process changes are not necessarily radially uniform. All of the effects, unfortunately, are exacerbated by the low melting rates normally used for the production of high-quality alloys due to the lower stability of the operations. Some irregularities are axisymmetric and can, in principle, be modeled with existing software. Examples are changes in the melting rate, power interruptions or the electrode changing process [6 – 10]. However, the thermal effects which would be predicted lack the necessary quantitative connection to the resulting solidification effects. The effects on ingot solidification are strongly dependent on the alloy composition. In alloys solidifying as single phase, such as the austenitic stainless steels, the changes in dendrite formation are generally not significant after the product is mechanically worked. If the alloy forms primary precipitates, as in IN718, the effect can lead to the formation of precipitate size variations, tree-ring segregation bands or freckling which have significant effects on mechanical properties in the final product. Examples of these effects are shown in figure 1.

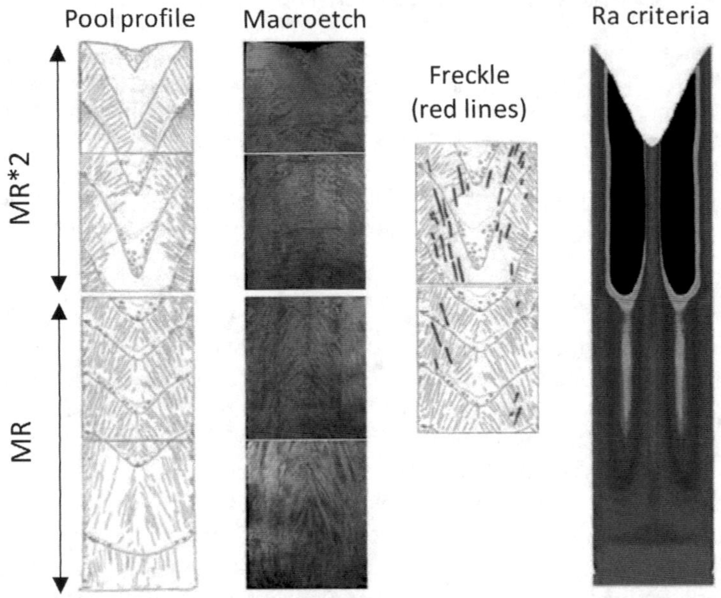

Figure 1. Freckles in an ESR ingot, Nimonic 80, influenced by an increased melting rate. The potential freckling area is indicated by the black Ra region, “SOLECS” computation [2].

If the melting rate is close to the maximum level permitted by the model-predicted analysis a small positive deviation can cause the solidification rate at the liquidus central region to fall below the value for the columnar/equiax transition and so make a segregated region in the ingot centre. In the case of a smaller short-duration change of energy input the consequent change in solidification rate can convert the dendritic structure into a central region of equiax grains. The equiax grain structure is associated with enhanced microsegregation [19]. Models have been developed to demonstrate the effect [5], but to date they do not appear to have been used as a quality control tool.

A common process issue on both ESR and VAR is that of electrode cracking before or during the melt [11]. The high-alloy electrodes can contain large internal stresses after cooling from the VIM casting process. The stresses can produce cracks both before remelting and during the process itself. The effect is a considerable perturbation of the process, which is not necessarily axisymmetric and which significantly perturbs the solidification pattern

3. Irregularities in the VAR Process

The VAR process does not possess the intrinsic thermal capacity of the ESR process and so is more sensitive to process perturbations; the situation is exacerbated by the low melting rates generally used in VAR due to lower arc stability and reduced liquid pool depth. In addition, since the energy input is through a very small, high energy-density arc with a rapidly changing radial position and length variations in the arc behaviour are transmitted strongly to the solidifying region of the ingot.

The arc is nominally controlled by the "drip-short" method [12], intended to keep the arc length constant; but as the recent studies using the Varmetric technique have shown [13], this method does not provide an effective control of the arc energy distribution. In addition, changes in the arc energy transfer to the ingot surface are necessarily 3-dimensional and not axisymmetric. Nonetheless, alloy ingots which show solidification resembling the predicted output from CAFE computations can be made and are regularly found in industrial practice (figure 2).

Figure. 2 "Ideal" VAR ingot structure, 500mm diameter, Waspaloy.

The transition between a regular dendritic structure and the equiax region has been modeled in VAR for both nickel-base alloys [5] and for titanium alloys [14]. However, due to the difficulty of exact and consistent arc control, difficulties remain in respect of producing consistent ingot structures. In addition, heat transfer from the ingot to the mold takes place through the direct contact of the two surfaces and the ruling heat transfer coefficient varies widely and not necessarily axisymmetrically, as a function of the ingot contact surface roughness, contact pressure and the gas pressure in the gap between the surfaces if He injection is used. As a result, the solidification pattern of the ingot can vary considerably from ingot-to-ingot even with the imposition of a "fixed practice".

The most discussed feature resulting from solidification irregularities is freckling; the formation of solute-rich channels in the mushy zone arising in the movement of segregated liquid through the dendrite network. The effect has been widely studied and mechanisms postulated which can be reduced to a "possibility of freckling" for a given process condition by the computation of the local Rayleigh Number (Ra) in the mushy zone using the software developed to describe the thermal/mechanical forces operating in that volume of the ingot. The critical value of the Ra number which would indicate a possibility of freckling depends on the precise formulation used in the Ra model. In the example of figure 3, the critical value would lie between 0.4 and 0.8 using the method of Valdes [15].

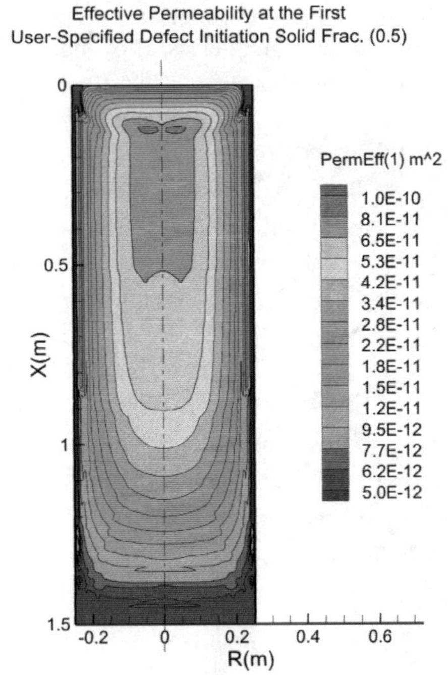

Figure 3. "MeltflowVAR" [1, 16] analysis of VAR ingot of IN718, 500mm diameter melted at 3.5 Kg/min

Studies of the practical incidence of freckling, however, indicate that although the Ra computation forms a basis for the freckle mechanism in an "ideal" ingot, the incidence of freckles in production ingots is closely related to the level of stability in the process. It is suggested that the initiation of freckle channel formation is the result of local irregularities in the dendrite network which lead to local increases in the network permeability. The random nature of the irregularities accounts for the formation of a small number of freckles in an ingot with no indicated regions of high Ra number [18]; also for the demonstrated possibility of producing an ingot without freckles under process parameters which would produce a high Ra value in the ingot [17]. The irregularities are the result of perturbations in the heat balance of the solidifying dendritic front brought about principally by lack of control of the arc heating process. In some alloys, the perturbations do not lead to the full development of freckles. Instead, the minor local increases in permeability lead to the formation of small segregation "spots" in which larger-than-normal primary carbide precipitation can take place. The mechanical effect of these spots on the alloy properties is varied. In the Nb-containing superalloys the effect appears to be negligible but in tool steels the result is a significant decrease in fatigue life. Modeling of this aspect of VAR requires a 3-D approach. Some progress is being made in this direction with the prediction of the effect of electrode off-centering on the complex fluid flow in the VAR liquid pool and associated local change in pool inclination which increases the risk to generate freckles (figure 4).

IOP Conf. Series: Materials Science and Engineering 1281 (2023) 012034 doi:10.1088/1757-899X/1281/1/012034

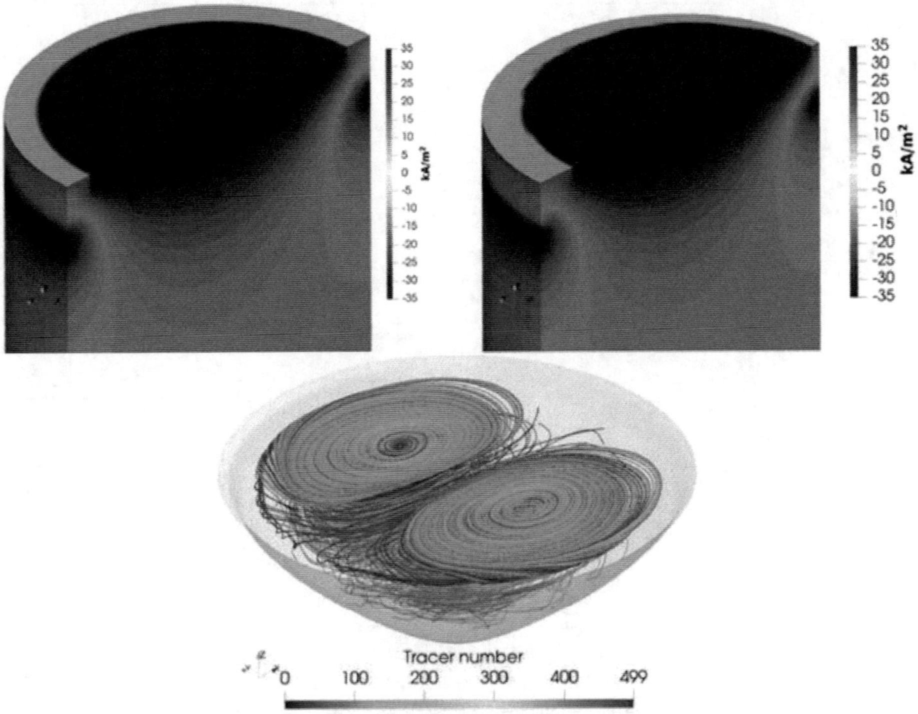

Figure 4. " RAVEL" model, 3D VAR prediction – electrode off-centering study for 500 mm IN718 (top ; vertical current density centered (left) and off centered electrode (right) – down : streamlines for off centered electrode) [21].

A further aspect of the industrial issues in VAR lies in macro-segregation leading to composition variation in both the radial and axial directions in the ingot. In many alloy systems processed by VAR the permitted composition range for the alloy elements is very small. Titanium alloys in particular (for example Grade 5, "Ti6/4") are very sensitive to this problem; small variations in 6/4 composition require the user to make small adjustments to the heat treatment schedule depending on the exact billet composition since the variation of alloy element concentrations affects the transus temperature. Macro-segregation during ingot formation has been modeled using various software formulations [1-4, 16]. The models, however, must take into account the potential periodic asymmetry of the operation due to arc movement and its effect on the fluid dynamics of the liquid pool at the head of the ingot. Some success is reported in this direction with 2D software taking into account the orthoradial stirring movement of the liquid due to Lorentz forces. Modelling results are very sensitive to the dendritic microstructure and permeability of the mushy zone which drive the fluid flow of the inter-dendritic liquid. For the right permeability coefficient, radial macro-segregation is well predicted for all alloying elements compared to the average composition measured on ten industrial ingots (figure 5). However, if the average segregation behavior is well predicted, the scatter of the individual measurements can be of the same order of magnitude as the average radial segregation. In order to reduce this scatter very stable melting conditions as well as very uniform electrode chemistry are necessary [22].

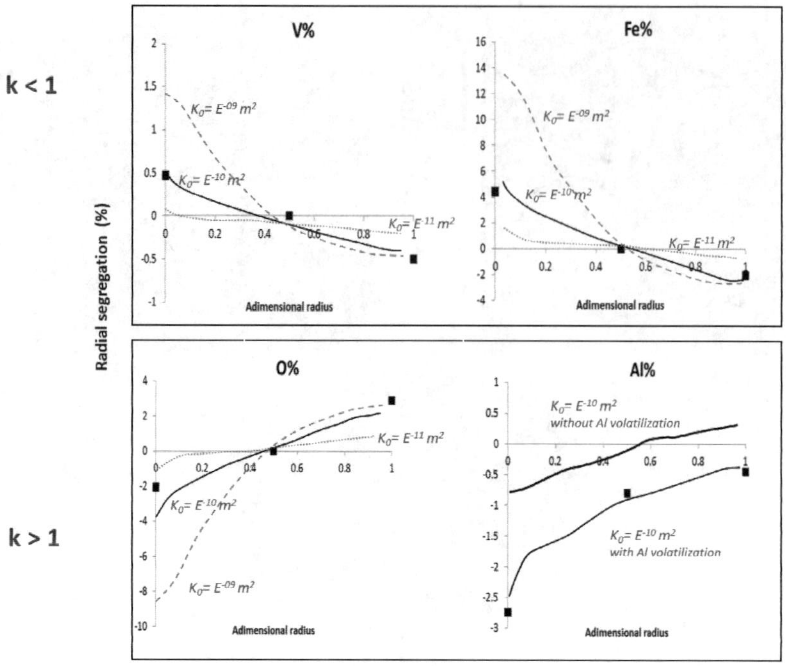

Figure. 5 : Comparison between measured average radial segregation on Ti64 ingots (black dots) and SOLAR modelling results (dotted line) for different permeability coefficients K_0 [22]

4. Relationship to Process and Solidification Modeling

Models, by their nature, must involve some level of assumption in the input values and in the model structure itself. The value of the output must be judged in relation to these assumptions. The existing models of the remelting processes are excellent general descriptions of the assumed stable, controlled behaviour of the systems. They have indicated valuable general rules for the process parameters such as melt-rate ranges, optimum ingot/electrode geometry and in the case of ESR the relationships between the slag volume, its properties and the overall heat balance. The models, however, do not often represent the detailed situation in practice. Moreover, many of them have been validated only on the results obtained from small-scale laboratory furnaces in which the processes have a different comportment than that in large industrial furnaces [20].

In order for the models to be useful at the production level they must have the capability to describe and include the process variations which are seen in production which implies 3-dimensional models with a time resolution that can describe the short-term solidification effects. The value of such models would not only be in better assessing the melt-to-melt variations experienced, but also in the aspect of quality control related to the product. The producer and end-user must generally agree on a "fixed practice" process procedure which will ensure the required level of properties in the product. This agreement always includes two levels of potential process variations; one level of severity which requires additional product testing and one of greater severity which renders the product unacceptable. In some applications, the restriction goes further to the disqualification of the producer if the variation has led to a serious defect in the end-product. The precise establishment of these levels by experimental melting of multiple ingots is extremely expensive and is also a major cause of the long time required for the introduction of a new alloy or new process into industrial use. The attribution of "process robustness" is in effect the tolerance of the process to variations which affect the product quality. Process models which could predict the relation between process variations and product quality would greatly assist this aspect of remelting. They would require at least 3-D geometry and a capability in small time steps; also the necessary steps to relate the results to the details of ingot structures.

The use of the remelting processes themselves introduces a difficulty of quality assurance through the feature of progressive melting. The product is worked into billet in which any defects are identified by either ultrasonic testing (UST) or macro-etching of selected sections. The segregation defects outlined above are not reliably detected by UST and are found only by macro-etching, which necessarily does not cover the entire ingot and detects the defects only on a small number of surfaces. As a result, the product quality relies principally on process control rather than inspection. In turn, this aspect makes the knowledge of the relation of process irregularities to ingot structure a critical requirement so emphasizing the need for process models which describe the practical process.

5. Conclusions

The present models of the remelting processes have been useful to establish the general operating parameters of the industrial operations. However, the practical details of the operations lead to many situations where the solidification does not result in the assumed ideal structure. The resulting irregularities require further modeling efforts directed towards 3-dimensional representations with sufficient resolution in time that they can represent the corresponding solidification variations. Such efforts would be valuable to help define the required operational windows for satisfactory product quality.

Acknowledgements

The authors gratefully acknowledge the contributions from colleagues in the industry.

Conflict of Interest

On behalf of the authors the corresponding author states that there is no conflict of interest.

References

[1] Kelkar K M, Patankar S V, Srivatsa S K, Minisandram R S, Evans D G, deBarbadillo J J, Smith R H, Helmink R C, Mitchell A, H A Sizek ; 2013, Computational modeling of ESR, *Proc. Liquid Metal and Casting Conference 2013, eds. M J M Krane and A Jardy, publ.TMS New York 2013*, pp 11 – 21.

[2] Weber V, Jardy A, Dussoubs B, Ablitzer D, Ryberon S , Schmitt V, Hans S, and H. Poisson H, 2009. A comprehensive model of the electroslag remelting process description and validation, *Met Trans* **40**, pp 271 – 280.

[3] Hans S, Jardy A and Ablitzer D, 1994, A numerical model for the prediction of transient turbulent flow, heat transfer and solidification during vacuum arc melting, *Proc Liquid Metal processing and Casting Conference 1994 eds Mitchell A and Fernihough J, publ AVS New York 1994,*p 143 – 152.

[4] Eikhoff M, Ruckert A, Pfeiffer H, Tewes J, Hoffmann Tand Klower J,2017, Numerical simulation of the vacuum arc remelting process with Ansysfluent, *Proc Liquid metal Processing and Casting Conference 2017, eds Krans J M, Ward M R, Rudoler S, Elliot A J, Patel A, publ TMS New York 2017,* p 1 – 7.

[5] Kermanpur A, Evans D G, Sidall R J, Lee P D and McLean M, 2003, Effect of process parameters on grain structure formation during VAR of Inconel alloy 718, *Proc Liquid Metal Processing and Casting Conference. 2003, eds Lee P D, Mitchell A, Bellot J-P, Jardy A, publ SFM2 Paris France 2003,* p 39 – 49.

[6] Jackson R O, 1972, A study of the electrode change process in ESR, 1972, M.A.Sc. Thesis, University of British Columbia, Vancouver, Canada

[7] Jackson R O, Mitchell A, Luchok J, 2000, An examination of electrode change practice in electroslag remelting, *Journal of Vacuum Science and Technology* ***9**, 1972 https://doi.org/10.1116/1.1317026*

IOP Conf. Series: Materials Science and Engineering 1281 (2023) 012034 doi:10.1088/1757-899X/1281/1/012034

[8] Wang W, Kermanpur A, Lee P D and McLean M, 2001, An investigation of the effect of perturbations on dendritic growth in nickel-base superalloys,, *Proc Liquid Metal Processing and Casting Conference 2001, eds Mitchell A, Van Den Ayvle J, publ AVS New York 2001,* p 267 – 280.
[9] Ward R M, Xu X, Lee P D, McLean M and Jacobs M H, 2001, Process/structure relationships during VAR of Inconel 718, *Proc Liquid Metal processing and Casting Conference 2001, eds Mitchell A, Van Den Ayvle J, publ AVS New York 2001,* p 187 – 200.
[10] Karimi-Sibaki E, Kharicha A, Wu M, Ludwig A, Holzgruber H, Ofner B, Scheriau A Kubin M and Ramprecht M, 2017, CFD Modeling of the electrode change during the ESR process, *Proc Liquid metal Processing and Casting Conference 2017, eds Krane M J M, Ward R M, Rudoler S, Elliot A J and Patel A, publ TMS New York 2017,* p218 – 216.
[11] Williamson R L, Beaman J J, Melgard D K, Shelmidine G J, Patel A D and Adasczik C B, 2003, A demonstration of melt-rate control during VAR of cracked electrodes, *Proc Liquid Metal Processing and Casting Conference. 2003, eds Lee P D, Mitchell A, Bellot J-P, Jardy A, publ SFM2 Paris France 2003,* p 29 – 39.
[12] Zanner F J, Williamson R I, Harrison R P, Flanders H D, Thompson R D, Szeto W C, 1989, Vacuum arc melting of alloy 718, *Proc Conference Superalloy Metallurgy and Applications 1989, ed Loria E, publ TMS New York 1989.* p 17 – 32.
[13] McCulley D, Motley J, Cibula M and King P, 2022, Elucidating the relationship between arc behaviour and solidification defects during VAR, *Proc Conference on Liquid Metals Processing and Casting 2022, eds Krane M, Kharicha A, Ward R M , publ TMS New York, 2022,* p 1 – 9.
[14] Zagrebelnyy D V, 2007, Modeling macro-segregation during the vacuum arc remelting of Ti-10v-2Fe-3Al alloy, Ph.D.Thesis, Purdue University, West Lafayette, Indiana, USA.
[15] Valdes I, King P, and Liu X, 2010, On the formulation of a freckling criterion for nickel-base superalloys in VAR, *Met Trans A, 2010,* ***41(4)***, p 2408 – 2416.
[16] MeltFlow-ESR Technical Background and Users Manual, Innovative Research, LLC, Plymouth, MN 55446
[17] Bond B J, Jackman L A and Ballantyne A S, 2002, Method of manufacture of large ingots of alloy 718, US Patents 6,416,564 (2002), 6,719,858 (2004).
[18] Morita, T. Suzuki, T. Taketsuru, Evans D G, and Yang W, 2001, The tendency for freckle formation in alloy 718, *Proc. Superalloys 718 625, 706 and Various Derivatives, ed. E Loria, publ TMS New York, 2001*, pp 149 – 160
[19] Zagrebelnyy D and Krane M J M, 2009 Segregation Development in Multiple Melt Vacuum Arc Remelting, *Metals and Materials Transactions,* 2009, **40B**, p 281 – 288.
[20] Mitchell A and Sjoqvist-Persson Ewa, 2022, Comments on oxide inclusion formation/removal in the ESR process, *J Ironmaking and Steelmaking,* https://doi.org/10.1080/03019233.2022.2069998
[21] Matveichev A, 2022, RAVEL software and electrode off-centering study, O2M Solutions - SOLAR Club, private communication
[22] Doridot E, Hans S, Jardy A and Bellot J-P, 2020, Industrial applications of modelling tools to simulate the PAMCHR casting and VAR process for Ti64, *14th World Conference on Titanium (Ti 2019) MATEC Web of Conferences 2020,***321**, https://doi.org/10.1051/matecconf/202032110011

IOP Conf. Series: Materials Science and Engineering 1281 (2023) 012035 doi:10.1088/1757-899X/1281/1/012035

Modeling of oxide inclusions in steel casting

R Donahue, R Hardin, and C Beckermann

Department of Mechanical Engineering, University of Iowa, 3131 Seamans Center, Iowa City, IA 52242, USA

becker@engineering.uiowa.edu

Abstract. A model is presented that predicts the amount and location of oxide inclusions in steel castings. A number and size distribution of inclusions, defined about a mean diameter, enters the casting system at its inlet during the filling process and are transported to their final locations in the casting. Model parameters for inclusion density, drag and wall friction are used to calculate the motion and locations of the oxide particles. Model results are presented to study the effects of casting shape and surface orientation on the final inclusion locations and distributions within castings. These results are compared with inclusion tracking experiments where the geometry of the gating system and orientation of casting cope surfaces affect the final distribution of inclusions in the castings. Measured and simulated inclusion area percent coverage, inclusion count and mean diameter are compared for a range of modelling parameters and inclusion size distributions. The size and number distribution at the casting system inlet, and other model parameters, are determined which provide the best agreement between measured and simulated inclusion area, count, and size.

1. Introduction

Oxide inclusions in steel castings are estimated to contribute 20% to the cost of a casting due to the costs of removing them and repairing the casting [1]. They are also a frequent cause of premature failure of steel castings when not detected during production. There are numerous sources of oxides such as the ladle lining and poorly deoxidized melt, and many casting process variables can affect the levels of oxides in steel castings. The cleanliness of the melt can vary from heat to heat, so called dirty heat versus clean heat, due to poor control of the melt practice. Considering all the sources of oxide inclusions, reoxidation inclusions, formed during pouring of the metal into the mold, are a common cause of inclusion defects in steel castings, if not the most common. Reoxidation inclusions form when deoxidized steel comes into contact with oxygen during mold filling. They are reported to make up 83% of oxide inclusions in low-alloy steel castings and 48% of inclusions found in high-alloy steel castings [2]. Reoxidation of the steel during pouring can be minimized by employing well designed gating systems. Much research has been performed for over 50 years to establish rules for gating castings. However, the design of gating systems is still more of an art than a science. Instead of investigating source of the inclusions, like reoxidation, the present paper focuses on modeling the transport of inclusions throughout gating and casting systems and their final locations.

One of the authors has previously modeled reoxidation inclusions originating on the free melt surface with their growth controlled by oxygen transfer from the atmosphere and their motion and agglomeration [3]. This paper presents the results from a model where the number and size distribution of inclusions, defined about a mean diameter, enters the casting system at its inlet during the filling. Generation of inclusions in the casting system by reoxidation is not considered. Inclusions enter the casting system from the ladle and are transported to their final locations in the casting using the inclusion motion model presented in [3]. Model results are compared to a series of casting experiments. The experiments are designed to investigate the control of oxide inclusions in steel casting using simple geometries to affect their transport and final locations. It is assumed that inclusions measured in the experiments either came directly from the ladle or were generated in the pouring cup. This assumption is based on an experimental investigation [4] using the melt and pouring practice used in the experiments presented here. In [4] ceramic foam filters were positioned at several locations in the casting system: at the inlet to the pouring cups filtering the stream, at the exit of the pouring cups (top of down sprue), at the base of the downsprue, and at the ingate to the casting. By pouring repetitions of experiments for the various filter location cases, it was found that a negligible number of inclusions was generated in the gating system. A range of "good" and "poor" gating systems were used as described in [4]. Inclusions measured in the experiments either came directly from the ladle or were generated in the pouring cup.

Therefore the model applied here assumes a distribution of inclusions entering the casting system only at the pouring cup with no growth or agglomeration. The model predicts the amount of area coverage and final locations of oxide inclusions in steel castings. The primary objectives of the present study are to determine the number of inclusions, their size distribution and other model parameters that give best agreement between the model and the measurements. Percent area coverage of inclusions, average inclusion indication size and number of indications on the casting surfaces are used to determine the agreement.

2. Description of Model and Analysis of Simulation Results

The model used in this paper was developed based on previous work [3,5]. It is referred to as a "dross" model as it applies to liquid metals forming dross inclusions (i.e., ductile iron [5], nickel-aluminum bronze and aluminum alloys) when the liquid metal surface is exposed to oxygen. In the case of this model, inclusions were determined to primarily come from the ladle used to fill the casting. In all experiments discussed and simulated here, ASTM A216 WCB steel was poured at the University of Northern Iowa Metal Casting Center using a 300-pound lip-pour ladle. During the holding of the steel prior to pouring, inclusions accumulated on the surface of the ladle due to reaction with the air, and convection and buoyancy effects bring them to the top surface. While efforts were made to de-slag and remove the inclusions from the melt surface prior to pouring, inclusions remained on the melt surface and entered the castings via the inlet stream. Upon entering the casting with the filling flow, the inclusions are transported with the flow under the effects of buoyancy and drag to their final locations in the solidifying casting.

Because of the observations discussed above, the current model assumes the sole source of inclusions is at the inlet to the casting system having; 1) a prescribed total number of inclusions N_{inc} entering the casting system during filling, and 2) a size distribution ranging from 50% to 150% about a prescribed mean inclusion diameter d_{inc}. The inclusions are assumed to be spheres. All other aspects of the model for calculating the motion and final locations of the inclusions are as described in [3] and [5] except that agglomeration is not considered here. Therefore, additional parameters used in the inclusion equation of motion [3] are the inclusion density ρ_{inc} and the wall slip coefficient λ_{inc}. As in [3] and [5], the model has been developed within the framework of a commercial casting simulation code.

The lip pour ladle filling stream was simulated as realistically as possible. As the lip pour ladle is tilted progressively to pour castings, the inlet stream falls at an angle from the lip to the inlet of the casting system. In the software used, an inlet is defined in an orientation aligned with the gravitational vector (the vertical direction) in order to define the inlet pouring stream. An angled stream such as that from a lip pour ladle is difficult to define in the software. To generate a more realistic lip pour stream than using the vertically aligned inlet, the inlet is aligned vertically, and the inlet filling stream flows through a channel designed to generate a stream falling at an angle before entering the pouring cup at the top of the downsprue.

The inclusion area coverage and number are measured in the experiments. To compare the simulation results to the experiments, the inclusion simulation results must be quantitatively evaluated to determine the inclusion area coverage and inclusion number on the casting surfaces as well. The inclusions on the experiment casting surfaces are observed to be pancake-shaped, while the inclusions formed in the simulations are spheres. To make the comparison to the measurements, the spherical inclusions on the surfaces of the simulated castings are converted to flattened disks having the same volume. A flattened inclusion has a larger diameter than a sphere of the same volume. Based on an experimental study of inclusions removed from the surfaces of castings, it was found that the actual flattened inclusions had an average diameter 1.7 times the diameter of a sphere having the same volume. To analyze the simulated inclusion size distributions, an image of the inclusion distribution result on the surface analyzed is made in the software's post-processor with the inclusions scaled to their true size. To model the flattened shape of the inclusions observed in the experiments, these spherical results are converted to flat disk diameters by increasing their size using a factor of 1.7 in the post-processor. Then the image of the flattened inclusions is converted to a binary image and analyzed using the software *ImageJ* [6]. In *ImageJ* the particle analysis feature is used to calculate the inclusion area coverage on the casting surface, inclusion count/number and the average diameter.

3. Experimental Inclusion Measurements

The objective of the experiments was to generate inclusions and produce flow patterns in various casting geometries to study the effects of casting shape and surface orientation on the final inclusion locations and distribution within the casting. This experimental data is generated to compare results with the model described above, and will provide data for calibration and validation of the model.

The experiments were quantitatively analyzed, and inclusions were measured. The as-cast cope surfaces of the experiments are first cleaned with a wire brush and then media blasted. The cleaning exposes inclusion pits since the inclusions are removed by media blasting. The casting surfaces are inspected and the inclusion pits are marked. The marked areas of inclusions are digitally photographed. These digital images are converted to binary images. The binary images are quantitatively analyzed using image analysis software [6] in the same way the simulation results are analyzed to measure the inclusion area, inclusion count and average inclusion diameter on the casting surfaces.

The six experiment cases presented here are shown in Figure 1. Dimensions of the experimental castings and other geometric features are provided. The first experiment (case A) shown in Figure 1(a) has two inclined surfaces at angles of 55° and 30° and is referred to as an "Angled Cope" casting. The second experiment case B is shown in Figure 1(b) and is an experiment with flow around a circular core. The next two cases C1 and C2 are plates inclined by 30° and 60° as seen in Figure 1(c). These cases are similar to the "Angled Cope" experiment case except that the inclination angles are split into separate cases. The last two experiments, having a ring flow pattern during filling are shown in Figure 1(d). In both experiments the filling flow enters the ingate and circles counter clockwise as viewed from above. In one "Ring Flow" experiment a recessed notch is positioned on top of the casting (case D1) and in the other a dirt trap is positioned on top of the casting. The dirt trap feature is essentially the opposite of the notch case, creating a cavity above the casting's cope surface. The notch and the dirt trap have similar dimensions. Chills are used in all cases as discussed below.

The experiments were designed so that the castings would capture any and all inclusions possible for measurement. In steel castings, inclusions can be removed from castings during filling, if the flow carries them into feeders. It was determined in the experiment design process that feeders should be avoided for this reason. However, it was determined that without feeding the solidification shrinkage, the casting cope surface would form a sump, and suck into the casting. This made the measurements difficult. To avoid this, chill plates, shown in blue in Figure 1, were used at the cope surfaces to create a uniformly flat surface through rapid directional solidification. This resulted in the surface being in excellent condition for inclusion measurements.

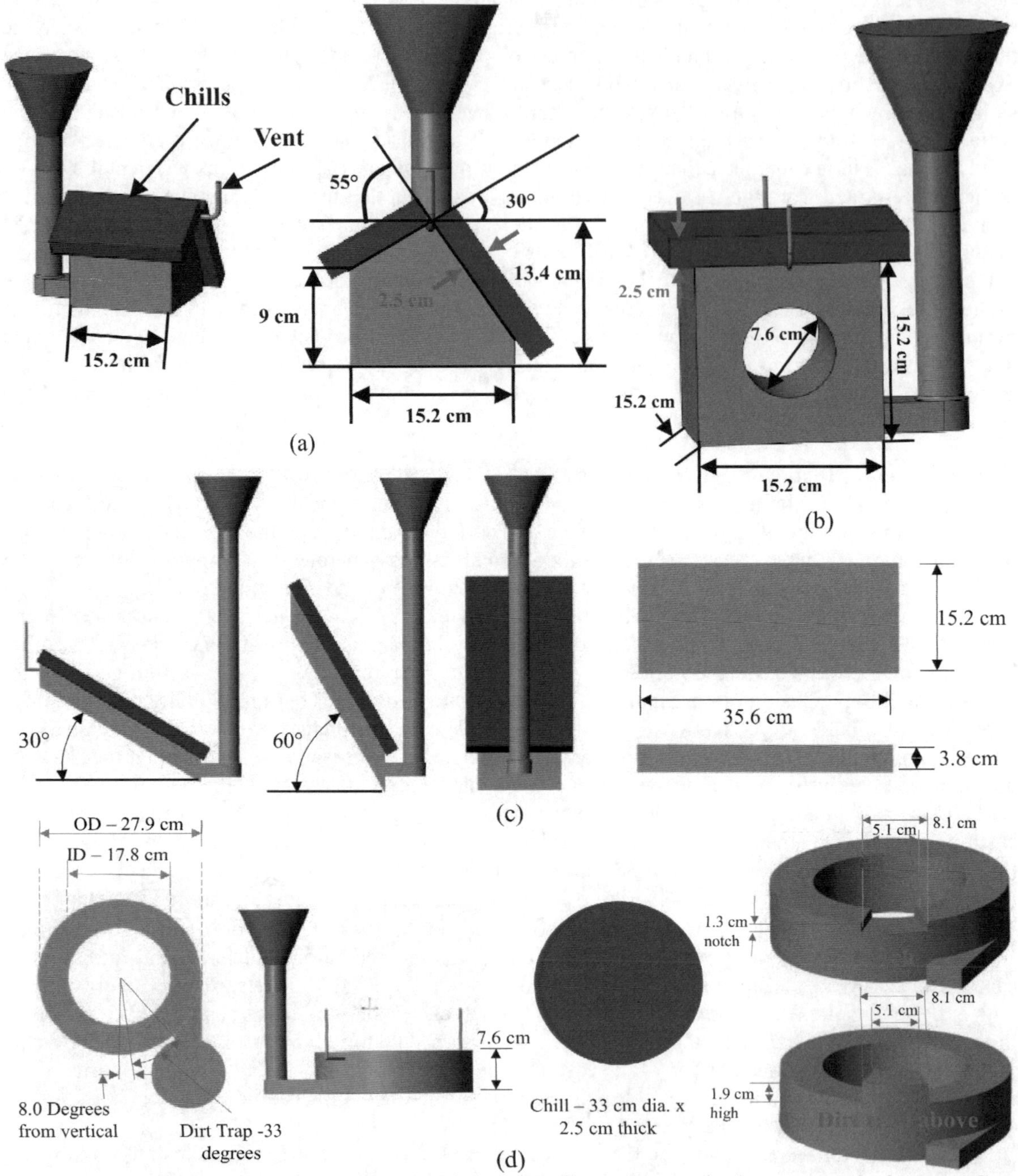

Figure 1. The six experiment cases simulated showing dimensions and other geometric features. Case A(a) has two inclined surfaces, case B (b) is an experiment with flow around a circular core, cases C1 and C2 (c) are plates inclined by 30° and 60°, and cases D1 and D2 have a ring flow pattern with a notch and dirt trap, respectively.

4. Comparison Between Simulations and Experiments

The images of, and data from, the inclusion analysis on the surfaces of the experimental and simulated castings are given in Figures 2 to 5. In these figures, the binary/black and white images of the inclusion sizes and locations on the chilled cope surfaces, and core surfaces for case B, are shown. Inclusion locations are black. The red rectangles in the images indicate the ingate locations. The results of the image analyses of the inclusion distributions are given below the images. These results are inclusion area percentage (inclusion area/surface area), and inclusion count (number of individual/discrete inclusions) and the average equivalent diameter of the individual inclusions. For the simulation results, three of the four model parameters are constant; the mean inclusion diameter d_{inc} is 0.912 mm, the inclusion density ρ_{inc} is 3,230 kg/m^3 and the wall slip coefficient λ_{inc} is 0.025 cm. The total number of inclusions N_{inc} entering via the inlet was varied. The rational for varying N_{inc} was an observed effect of pouring order on the cleanliness of the melt for the lip pour ladle used. The pouring temperature and time simulated for each case are taken from thermocouple measurements and videos of the experiments, respectively. Model results for case A are shown in Figure 2(b) where the N_{inc} is 2000. For both measured and simulated results in Figure 2, case A has inclusions located at the upper ends of the surfaces at the "peak of the casting." The lower part of the casting was much cleaner by comparison for both results. The inclusion area percentages agree well for both cope surfaces. Inclusion counts are higher and sizes lower for the measurements compared to simulations. For the cored casting in case B, N_{inc} is again 2000. For case B in Figures 3(a) and 3(b), both measurements and simulations have a large amount of inclusions on the cope surface of the block with the inclusion locations biased on the ingate side of the casting. In Figures 3(b) and 3(d) the drag side at the core (top surface of the core) was almost entirely clean for both results. For the cope surface of the core (bottom side of core) in Figures 3(e) and 3(f), both results have numerous inclusions and the locations are biased at the mid-length of the core. For the two inclined plates (case C1 and C2) in Figure 4, both results have inclusion locations towards the bottom/ingate end of the plates. For case C1 N_{inc} is 4000, and N_{inc} is 2000 for C2. For the steeper inclined plate C2, both results have observably more inclusions distributed at the upper end of the plate in Figures 4(c) and 4(d). For both inclined plates the melt appears to be "cleaned" progressively as inclusions stick to the cope surface during filling. The ring flow cases D1 and D2 are shown in Figure 5. For case D1 N_{inc} is 4000, and N_{inc} is 500 for D2. The measurements showed that the dirt trap case D2 in Figure 5(c) had much fewer inclusions than the notch case D1 shown in Figure 5(a). Both the

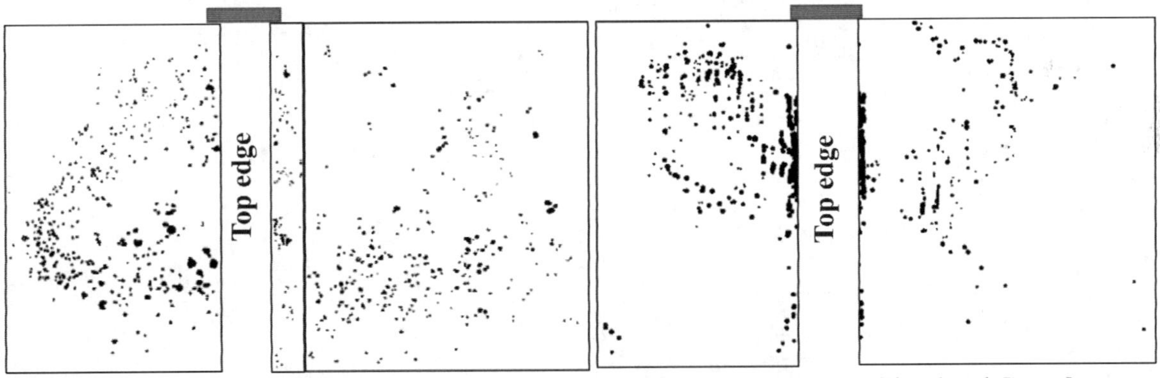

Measured Cope 1	Measured Cope 2	Simulated Cope 1	Simulated Cope 2
Inclusion area 4.0%	Inclusion area 2.0%	Inclusion area 4.39%	Inclusion area 2.17%
Inclusion count 345	Inclusion count 280	Inclusion count 156	Inclusion count 129
Average diam. 1.27 mm	Average diam. 1.13 mm	Average diam. 1.97 mm	Average diam. 1.75 mm

(a) (b)

Figure 2. Comparison between measured inclusion data in case A (a) and simulation results (b).

measured dirt trap and the notch cases were cleaner upstream of the notch and trap features compared to downstream of them. The simulations in Figures 5(b) and 5(d) had the opposite distribution with more inclusions upstream than downstream.

A summary of the comparisons between measured and simulated inclusion data is given for the inclusion area percentage in Figure 6(a) and also for the inclusion count in Figure 6(b). The nine points in each plot correspond to the surfaces plotted in Figures 2 to 5. In Figure 6(a) the inclusion area

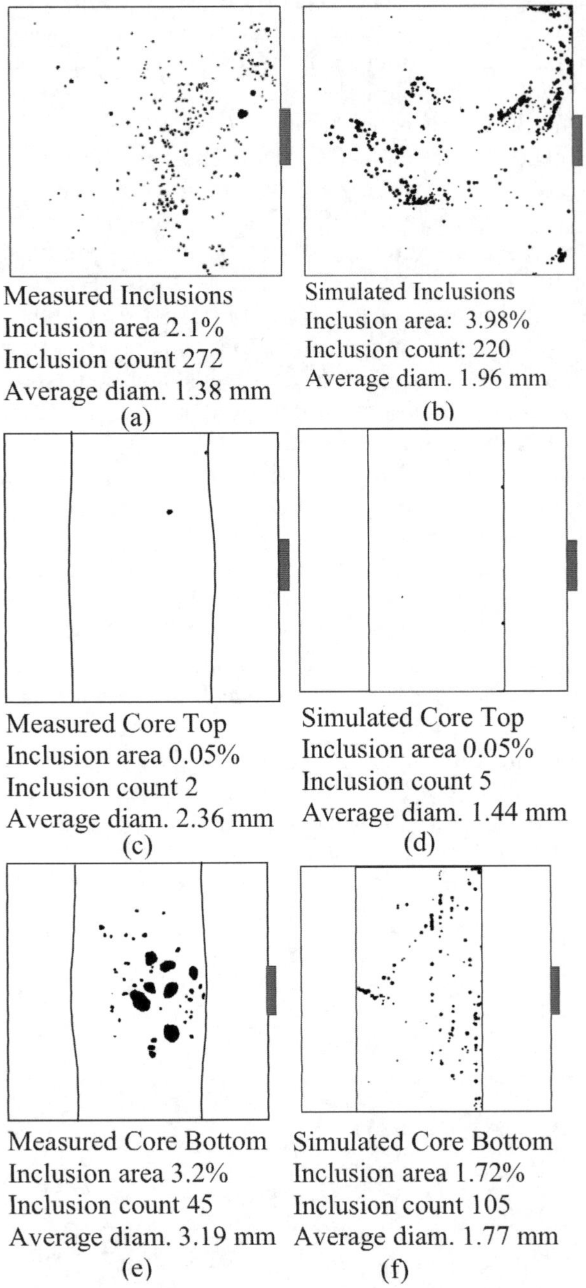

Figure 3. Comparison between measured and simulated inclusion data for case B. Measured cope surface (a) and simulation results (b). Measured core top surface (c) and simulation (d). Measured core bottom surface (e) and simulation (f).

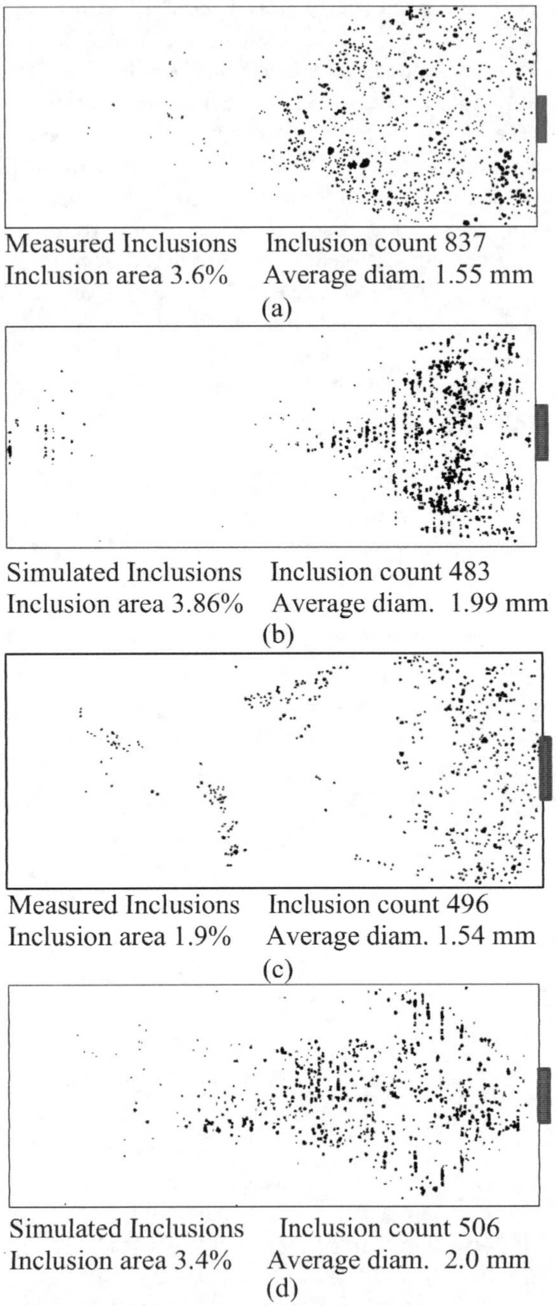

Figure 4. Comparison between measured and simulated inclusion data for cases C1 and C2. Measured case C1 cope surface (a) and simulation results (b). Measured case C2 cope surface (c) and simulation (d) results.

percentage shows an excellent trend over the range of values and remarkable agreement between measurement and simulation. The inclusion count in Figure 6(b) shows a good trend between the experiment and model, but the simulations appear to consistently under predict the inclusion count. The mean inclusion diameter did not predict a trend with the simulation diameter range being from 1.4 to 2.1 mm and the measured range from 1.1 to 3.2 mm.

5. Conclusions

Results from a model are presented that predicts the motion and final locations of oxide inclusions in steel castings. The simulations use a distribution of inclusions entering the casting system via the inlet stream, ranging in size from 50% to 150% of a prescribed mean diameter. Upon entering the casting system with the filling flow, the inclusions are transported with the flow under the effects of buoyancy and drag to their final locations in the solidifying casting. Model results are compared to experimental measurements results on nine surfaces of five casting experiments. A mean inclusion diameter of 0.91 mm is used in the simulation results given here. The inclusion density used is 3,230 kg/m^3 and the wall slip coefficient is 0.025 cm. The total number of inclusions entering via the inlet was varied from 500 to 4000 inclusions entering the casting during the filling process. Comparisons are made between the measured and simulated inclusion area coverage, inclusion count/number and the average diameter on the casting surfaces, and the standard errors in the model results are 1.0%, 179,

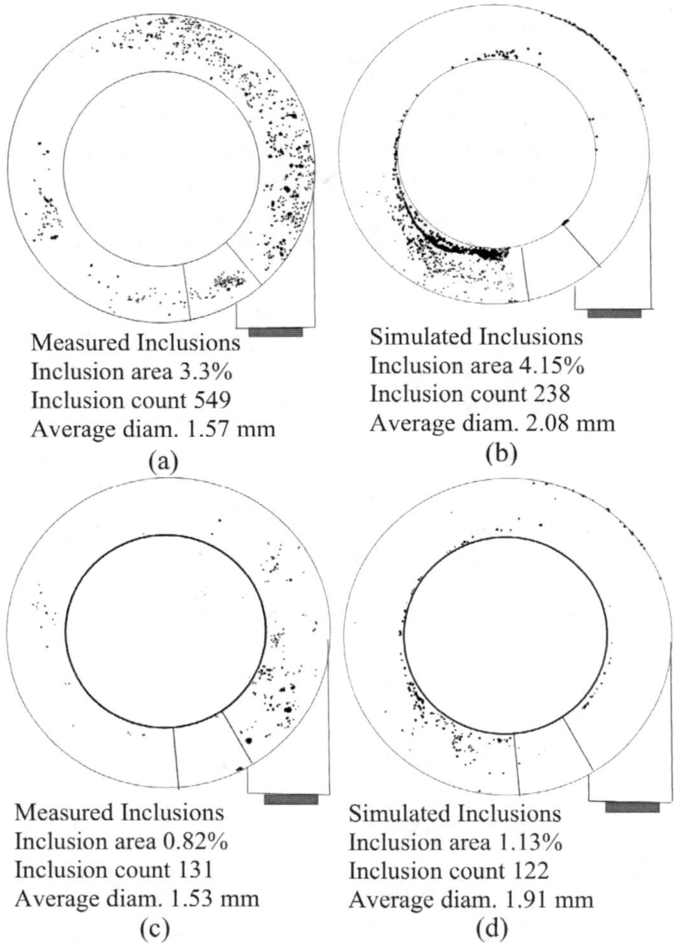

Figure 5. Comparison between measured and simulated inclusion data for cases D1 and D2. Measured case D1 cope surface (a) and simulation results (b). Measured case D2 cope surface (c) and simulation (d) results.

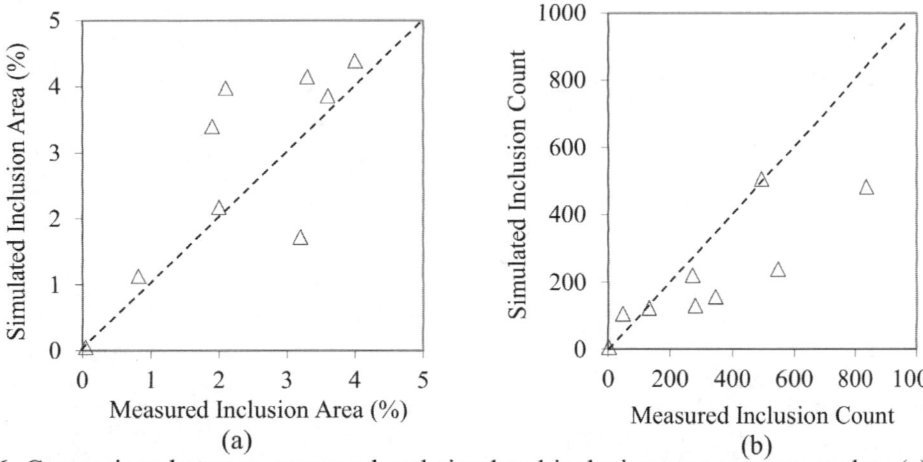

Figure 6. Comparison between measured and simulated inclusion area percentage data (a) and count (b) for all cases.

and 0.67 mm, respectively, compared to the measurements. For the 30 and 60 degree inclined plate cases, the measurements and simulations showed more inclusions near the lower end of the plates. The inclusions entering the casting system were cleaned from the melt as these cases filled. For the cored casting, both measured and simulation results had clean core top surfaces and dirty core bottom surfaces. In order to verify the modeling capability of predicting final inclusion locations, many repeated experiments of the cases presented here would be needed. However, the model results presented compare well with the experiments overall in predicting the area coverage on the casting surfaces. This demonstrates the model's usefulness in designing filling systems that provide cleaner melt flow by transporting inclusions to less harmful locations in the casting system.

Acknowledgements

This research is sponsored by the DLA-Troop Support, Philadelphia, PA and the Defense Logistics Agency Information Operations, J68, Research & Development, Ft. Belvoir, VA.

References

[1] Svoboda J, Monroe R, Bates C and Griffin J 1987 *AFS Trans.* **95** 187-202
[2] Griffin J and Bates C 1991 *Ladle Treating Pouring and Gating for the Production of Clean Steel Castings* (Steel Founders' Society of America Research Report No. 104)
[3] Melendez A, Carlson K, and Beckermann C 2010 *Int. J. Cast Metals Res.* **23** 278-88
[4] Donahue R, Hardin R and Beckermann C 2021 *Proc. of the 75th SFSA Technical and Operating Conf.* (Chicago IL: Steel Founders' Society of America) Paper No. 4.11
[5] Majidi S, Beckermann C, Fainberg, J, Schaefer W, and Bodenburg M 2018 *Mat. Sci. Forum* **925** 419-26
[6] Schindelin J, Arganda-Carreras I and Frise E 2012 *Nature Methods* **9** 676-82

IOP Conf. Series: Materials Science and Engineering 1281 (2023) 012036 doi:10.1088/1757-899X/1281/1/012036

Flow-enhanced remelting of settling/floating globular crystals during mixed columnar-equiaxed solidification

M Wu*, Z Zhang, H Zhang, A Ludwig, A Kharicha
Chair for Simulation and Modeling of Metallurgy Processes, Department of Metallurgy, University of Leoben, A-8700 Leoben, Austria

E-mail: menghuai.wu@unileoben.ac.at

Abstract. Previously, the authors have used a mixed columnar-equiaxed solidification model, successfully 'reproduced' the solidification benchmark experiments on the Sn-10wt.%Pb alloy under natural/forced convections (travelling magnetic stirring) as performed at SIMAP laboratory [*Int. J. Heat Mass Transf.* **85** (2015) 438-54]. The current contribution is to address the flow-effect on the remelting of settling/floating crystals during the mixed columnar-equiaxed solidification. The remelting or growth is controlled by diffusion of solute in the liquid boundary layer. The diffusion length due to the flow-effect is modelled as a function of Schmidt and Reynolds numbers. The modelling results show that remelting rate of the floating/settling crystals, which originate from fragmentation and then brought to the superheated region by the forced flow, can be enhanced by the flow. In turn the released latent heat can reduce the temperature locally (even globally), hence to speed up the solidification of the columnar structure. Additionally, the solidification-migration-remelting of equiaxed grains present an important macrosegregation mechanism. By solidification of a crystal in the cold region it rejects solute, while by remelting of the crystal it dilutes the surrounding melt. These phenomena are found critical in many engineering castings with mixed columnar-equiaxed solidification.

1. Introduction

Solidification and remelting occur concurrently during solidification of many engineering castings. As shown in figure 1(a), a crystal with certain *mass*, which forms near the mold wall, would release the latent heat $mass \cdot L$ and reject solute $mass \cdot c_\ell^* \cdot (1-k)$ in its surrounding melt. When this crystal is transported to a superheat region and re-melted there, it would take the same amount of energy $-mass \cdot L$ from melt of its new surrounding and dilute the melt there $-mass \cdot c_\ell^* \cdot (1-k)$. This concurrent solidification-remelting phenomenon, together with grain transport, is critical for a numerical model [1]. This conference contribution is (1) to quantify the importance of the remelting effect in the alloy solidification and (2) to address the flow-effect on the remelting. As shown in figure 1(b) and (c), the solidification/remelting rate, i.e. the formed or re-melted *mass* is proportional to growth or shrinkage velocity of the crystal v_R, with + or – sign of its value indicating the growth or shrinkage of the radius of the globular grain.

$$v_\mathrm{R} = \frac{D_\ell}{l_\ell} \cdot \frac{\left(c_\ell^* - c_\ell\right)}{\left(c_\ell^* - c_\mathrm{e}^{\mathrm{interface}}\right)}, \tag{1}$$

where D_ℓ is the diffusion coefficient of solute in the melt. c_ℓ and c_ℓ^* are bulk and thermodynamic equilibrium concentrations. $c_\mathrm{e}^{\mathrm{interface}}$ is the so-called interface concentration at the crystal surface, which is assumed as the equilibrium concentration of the crystal c_e^* for solidification and the average concentration of the crystal c_e for remelting. Here we ignore the back diffusion in the solid crystal. The flow effect is described by the adapted diffusion length in the liquid boundary layer l_ℓ, which is a function of the Reynolds and Schmidt numbers. For solidification [2],

$$l_\ell = \min\left(\frac{D_\ell}{v_R},\ R\left[\frac{1}{1-f_e^{1/3}} + \frac{\mathrm{Sc}^{1/3}\,\mathrm{Re}^n}{3f_\ell}\right]^{-1}\right), \tag{2}$$

for remelting [3],

$$l_\ell = \frac{R}{1.0 + 0.3\cdot \mathrm{Sc}^{1/3}\,\mathrm{Re}^{1/2}}, \tag{3}$$

where $n = \frac{1}{3} + \frac{\mathrm{Re}^{0.28}}{3\left(\mathrm{Re}^{0.28}+4.65\right)}$, $\mathrm{Re} = \frac{d_e \rho_\ell \left|\vec{u}_\ell - \vec{u}_e\right|}{\mu}$ and $\mathrm{Sc} = \frac{\mu}{\rho_\ell D_\ell}$.

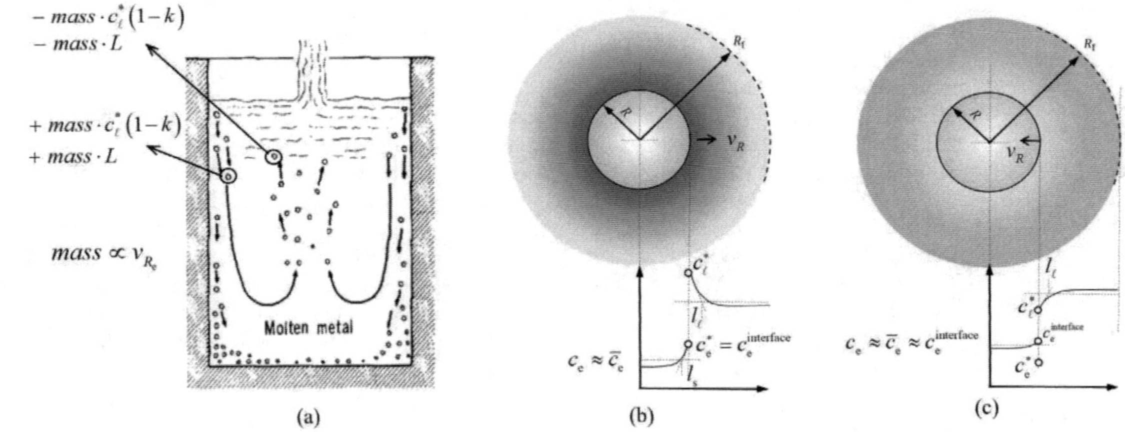

Figure 1. Schematics of (a) solidification, grain motion and remelting during solidification of an ingot casting; schematic of the diffusion-governed growth (b) and remelting (c) of a globular crystal. The flow affects the diffusion length l_ℓ during solidification/remelting.

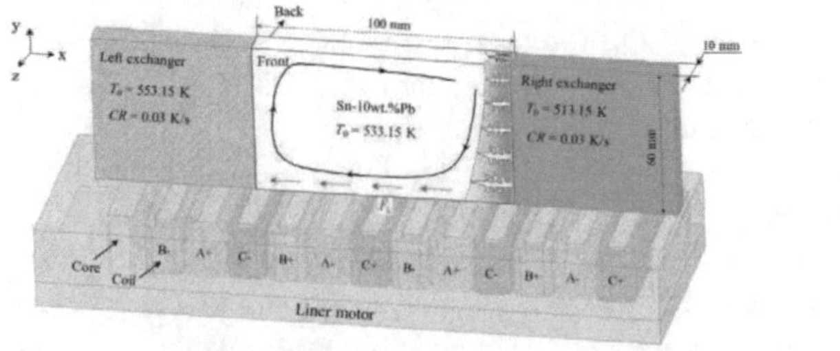

Figure 2. Geometry configuration of the experiment facility of the solidification benchmark with TMF [6].

2. Methods

A three-phase mixed columnar-equiaxed solidification model [4, 5] is used to study a solidification benchmark (100 × 60 × 10 mm^3) on the Sn-10wt.%Pb alloy as performed at SIMAP laboratory [6]. This conference contribution focuses on the flow-effect on the remelting/solidification of settling/floating globular

IOP Conf. Series: Materials Science and Engineering 1281 (2023) 012036 doi:10.1088/1757-899X/1281/1/012036

equiaxed crystals. Therefore, only details of the corresponding modelling part are described by equations (1) - (3), the rest refers to [4, 5]. Four benchmark experiments at the SIMAP were performed [6]: solidification under natural convection condition, and solidification under forced convections (TMF: travelling magnetic field) with the Lorentz force acting in different directions. As shown in figure 2, only one experiment with TMF acting in the direction from-right-to-left in the bottom of the benchmark is considered in the current study. Two heat exchangers were applied to control the heating/cooling rate on both lateral walls of the sample. The solidified sample was metallographically analysed for the as-cast structure and macrosegregation. Settings for the numerical model, including boundary and initial conditions, material properties, were presented elsewhere [7, 8]. 3 cases are simulated: Case A considers all solidification/remelting and flow effect; Case B ignores the remelting; Case C considers the remelting, but it ignores the flow-effect.

3. Results

3.1. Numerical reproduction of the experimental benchmark

Case A represents the most complete scenario. Its solidification sequence (time = 180, 420, 1020, 1920 s) is shown in figure 3. As the benchmark is cooled from right side, solidification starts with columnar structure from right wall, as seen in figure 3 (a.1) - (f.1) at 180 s. The applied TMF (Lorentz force) acts in the same direction as the thermo-solutal buoyancy; hence a strong clockwise convection loop form. The cooler melt from the right side is brought along the sample bottom towards the hotter left side. Equiaxed crystals form by the mechanism of crystal fragmentation [7] in the front of columnar mushy zone. The volume averaged solidification rate is very high ($M_{\ell e} \sim 80$ kg/m^3/s) just ahead of the columnar tip front. The equiaxed crystals are also brought by the melt flow to the hotter region (bottom), where remelting ($M_{\ell e} \sim -50$, even reaches -80 kg/m^3/s) occurs. It means that during the initial stage of the solidification (180 s), equiaxed grains, which form in the cold region, are almost re-melted in the hotter region. To a later moment (figure 3 (a.2) – (f.2) at 420 s), the same phenomena continue. Some equiaxed grains are brought to the left side of the sample and remelt there. The columnar tip front starts to tilt, and channel segregates start to form in the lower part of the columnar mushy zone. At 1020 s (figure 3 (a.3) – (f.3)), the superheat region disappears. Fragmentation and solidification of equiaxed grains continue with less intensity, but no remelting occurs. Finally, most equiaxed grains are brought to the left side of the sample, and the right side is occupied by columnar structure. Relatively strong channel segregates in the lower-right region of the sample. Globally, relative strong negative segregation occurs in the upper region; very strong positive segregation occurs in the lower left region.

The final as-cast structure and macrosegregation pattern are shown in figure 4, and compared with the experimental results. Excellent qualitative agreement was obtained. The as-cast structure of the experiment cannot be quantified. Quantitative result of macrosegregation shows that the negative segregation in the upper wall and the positive segregation in the bottom wall seem to be overestimated by the model.

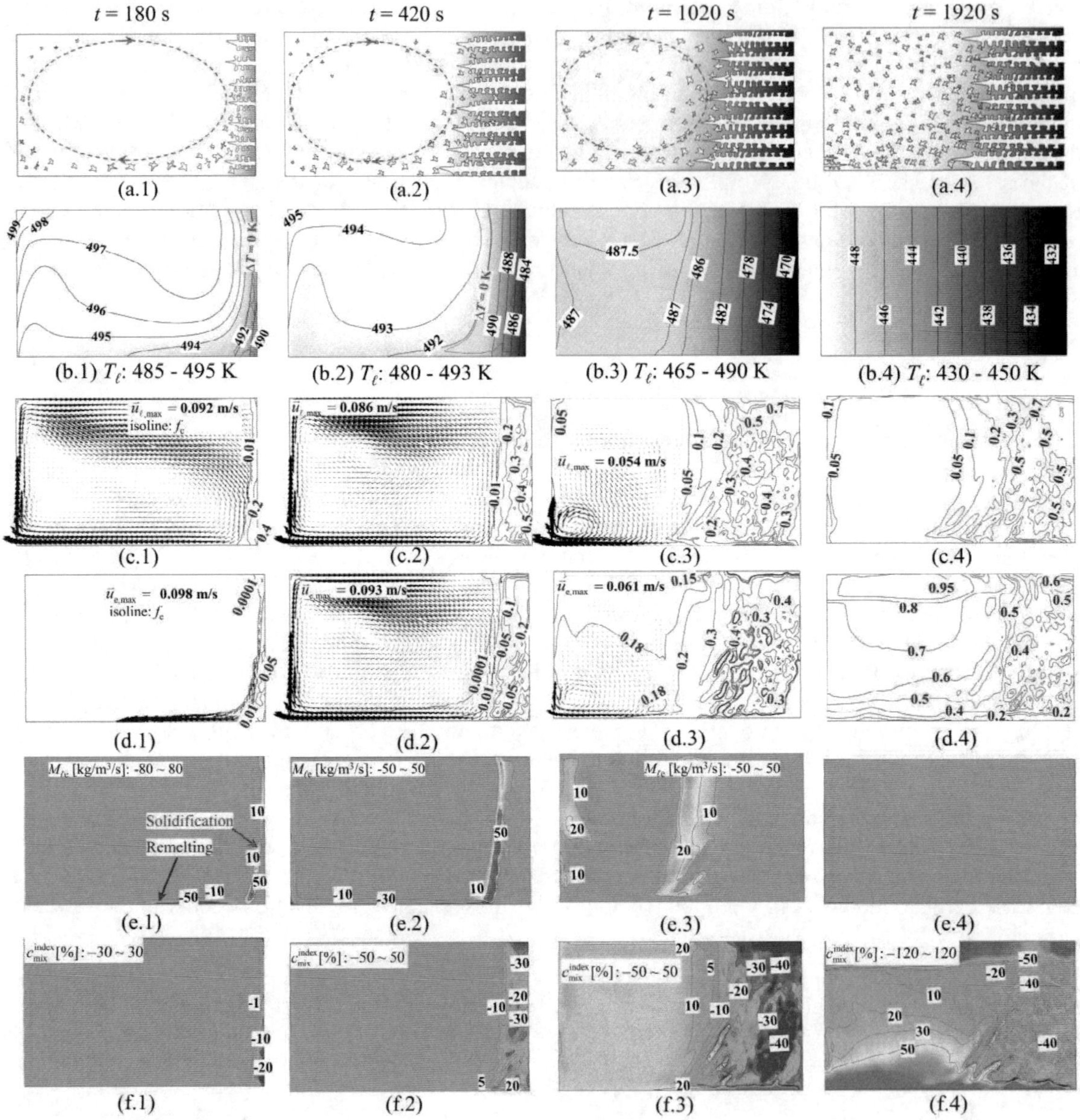

Figure 3. Solidification sequence (on the middle vertical plane) of the benchmark (Case A): (a.x) schematic views of the liquid flow and solidification pattern; (b.x) contour and isolines (in black) of T (in K) overlaid with the red isoline of liquidus (identical to the undercooling $\Delta T = 0$ K); (c.x) liquid velocity overlaid with the isolines of f_c; (d.x) velocity of equiaxed grains overlaid with the isolines of f_e; (e.x) contours of $M_{\ell e}$, yellow/red regions for solidification ($M_{\ell e} > 0.0$) and blue regions for remelting ($M_{\ell e} < 0.0$); (f.x) contours of the c_{mix}^{index}[%], yellow/red regions for the positive segregation ($c_{mix}^{index} > 0.0$) and blue regions for negative segregation ($c_{mix}^{index} < 0.0$).

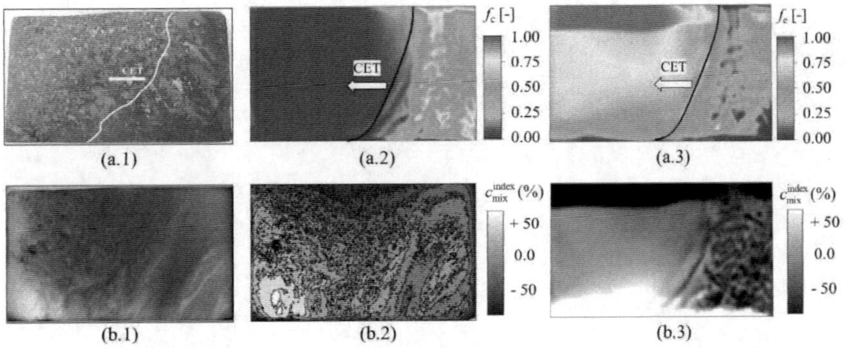

Figure 4. Comparison of the simulation result (Case A) with the experiment. (a.1) as-cast structure of the experiment; (a.2) simulated f_e distribution; (a.3) simulated f_c distribution; (b.1) X-radiography of the as-cast sample, which indicates the intensity of macrosegregation through the sample thickness; (b.2) macrosegregation map digitally processed from (b.1); (b.3) simulation result of c_{mix}^{index} through the sample thickness.

3.2. Importance of the remelting phenomenon

Case B has ignored the remelting of equiaxed phase, i.e. equiaxed grains which are brought to the left side do not re-melt despite of the superheat. Its modelling result for the moment at 420 s is shown figure 5, and compared with that of Case A (complete scenario with remelting). Differences between two cases are observed. (1) The average temperature in the sample of Case B is ca. 1 ~ 2 K higher than that of Case A. Remelting of equiaxed crystals in the left-side superheat region in Case A consumes energy (latent heat), leading to a lower-down of the local temperature. This effect will further influence the temperature field in the rest domain due to convective heat transfer. It means that the ignorance of the remelting phenomenon in Case B has overestimated the temperature in the sample. (2) the remelting of equiaxed crystals does not only influence the solidification of equiaxed phase itself, but it also influences the solidification of columnar structure from the right side. The solidification of columnar structure is also slowed-down by the ignorance of the remelting of equiaxed phase in Case B. This effect is indirect, because Case B has overestimated the temperature everywhere and the growth of columnar structure becomes slower. (3) The velocity is also different, i.e. the maximum velocities (appear at the bottom left corner) of the melt and the equiaxed phase are underestimated in Case B. This might be due to the accumulation of the equiaxed grains (due to the ignorance of the remelting function) in the bottom left corner, which resists the flow in Case B. (4) The remelting-induced dilution of solute in the superheat region is ignored in Case B. This result is not presented in this paper due to limited space. (5) Another interesting phenomenon is also observed: the ignorance of remelting of equiaxed crystals in Case B does not necessarily mean more equiaxed structure to form. It is true that the as-created equiaxed crystals in Case B would survive forever (as the remelting function is completely ‘switched-off’), even they are shortly exposed in a superheat region. However, the higher temperature in Case B would supress the formation of growth of both columnar and equiaxed structures. The final as-cast structure shows that the total equiaxed amount is overestimated only by ca. 13% in Case B. (6) More equiaxed phase is found in the columnar-dominant region along the right-wall region in Case B. Those equiaxed phase comes from the equiaxed crystals, which survive the superheat (due to the ignorance the remelting function) and finally are brought back and entrapped in the columnar structure along the upper part of the right-side wall. This phenomenon occurs in the early stage of solidification. However, it does not happen to Case A. Almost all equiaxed crystals created during the early stage of solidification are mostly (almost all) re-melted by the superheat, and no (almost no) equiaxed crystals can survive the superheat and be brought back in the columnar structure along the right-side wall.

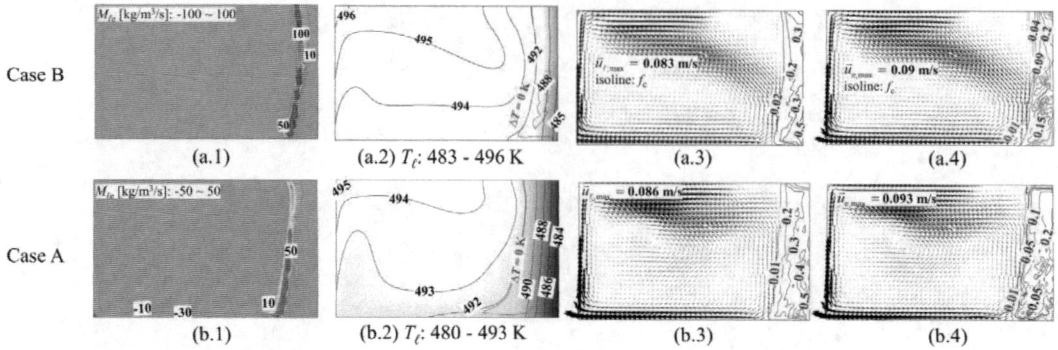

Figure 5. Importance of remelting phenomenon: comparison of Case B with Case A for the moment at 420 s. (x.1) contours of $M_{\ell e}$, yellow/red regions for solidification ($M_{\ell e} > 0.0$) and blue regions for remelting ($M_{\ell e} < 0.0$); (x.2) contour and isolines (in black) of T (in K) overlaid with the red isoline of liquidus (identical to the undercooling $\Delta T = 0$ K); (x.3) liquid velocity overlaid with the isolines of f_c; (x.4) velocity of equiaxed grains overlaid with the isolines of f_e.

3.3. Flow-effect on the remelting

Case C has considered the remelting of equiaxed grains, but the flow-effect is not treated properly. The diffusion length (equation (3)) is simply set as the radius of the grain (R), i.e. the Reynolds number (Re) is set zero. As shown in figure 6, the comparisons of the as-cast structure between the three cases are made. The difference between Case A and Case B is significant. The calculated as-cast structure of Case C is closer to the Case A than to Case B, e.g. the final amount of equiaxed phase of Case C is only about 0.04% overestimated in comparison to Case A.

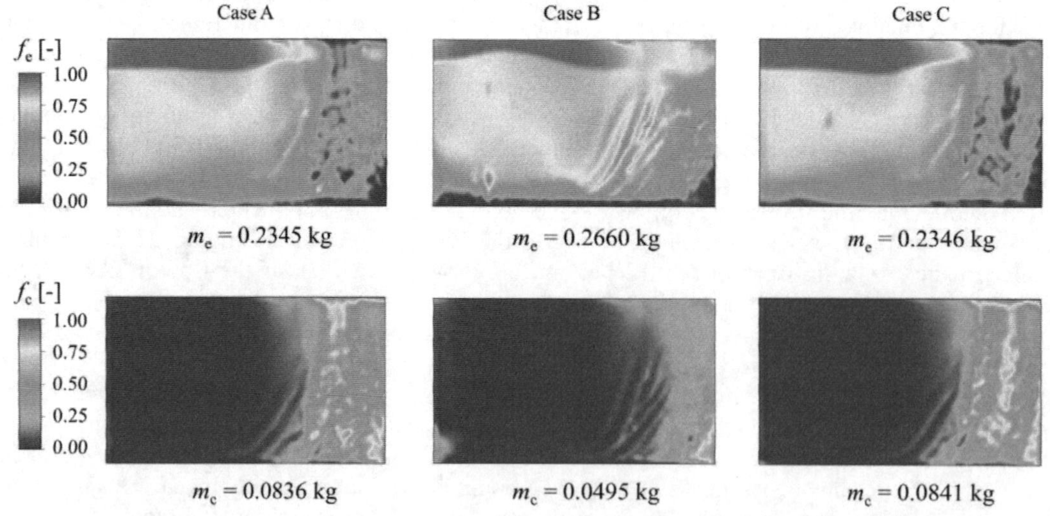

Figure 6. Comparisons of the as-cast structure between the three cases.

To further investigate the flow-effect on remelting phenomenon, some relevant solidification parameters at the bottom-left corner of the sample at the moment of 320 s (where/when remelting is significant) are analysed for the Case A. The relative velocity, $|\Delta u| = |\vec{u}_\ell - \vec{u}_e|$, is 0.2 mm/s; the averaged grain side d_e (= $2R$) is 50.0 μm; the Re is 0.09; according to equations (3) and (1), the calculated diffusion length l_ℓ and shrinkage (remelting) velocity v_R are 18.0 µm and –20.0 µm/s, correspondingly. If we ignored the flow-effect by imposing Re to zero, the calculated l_ℓ would be 25.0 µm (38.9% overestimation) and v_R would be -17.0 µm/s (15.0 % underestimation). The 15.0 % underestimation of v_R means ca. 38.6 % underestimation of mass remelting rate. In other words, the mass remelting rate as estimated by ignoring the flow-effect is only about 3/5 of what it should be.

4. Conclusion and discussions

To explain the as-cast structure and macrosegregation in the solidification benchmark at SIMAP laboratory [6], one needs to understand the origin of equiaxed crystals by crystal fragmentation, growth, transport and remelting of the equiaxed crystals. As a unidirectional heat transfer condition (cooling from right-side and heating from left-side) and forced convection (TMF) are imposed to the solidification sample, the equiaxed crystals as-formed from the cooler side can be brought to the hotter side, where it is initially superheated, and remelting of the equiaxed crystals occurs. This conference contribution has demonstrated the importance of such remelting phenomena and forced flow effect. A numerical model ignoring the remelting of the floating/settling crystals would lead to following consequences: overestimation of the temperature, delayed growth of columnar structure, overestimation of the amount of equiaxed, underestimation of flow intensity, error estimation of the macrosegregation and structure distribution. The forced flow interacts with the floating/settling equiaxed grains, and it can enhance the remelting rate by 1.7 times in the bottom-left corner region at a certain moment (might be even more in other regions or moments). It is anticipated that such concurrent phenomena of solidification-migration-remelting of equiaxed crystals should occur in engineering processes such as continuous, semi-continuous castings, ingot and shape castings. To what extend do these phenomena impact the as-cast structure formation in the above engineering processes? Further study is demanded. It might depend on the size of the casting when it is solidified under natural convection condition, or depends on the flow intensity when special flow control measures are implemented such as using electromagnetic stirring etc. A numerical model targeting the as-cast structure and macrosegregation must treat those phenomena properly.

5. Acknowledgements
The authors acknowledge the financial support from the Austrian Research Promotion Agency (FFG) through the project Bridge I (No. 868070) and the Austrian Science Fund in the framework of the FWF-NKFIN joint project (FWF, I 4278-N36).

References

[1] Campbell J 1991 *Castings* (Londen: Butterworth-Heinemann Ltd)
[2] Rowe P N and Claxton K T 1965 *Trans. Instin. Chem. Engrs.* **43** T321-T331
[3] Gu J P, Beckermann C and Giamei A F 1997 *Metall. Mater. Trans.* **28A** 1533-1542
[4] Wu M and Ludwig A 2006 *Metall. Mater. Trans.* A **37A** 1613-1631
[5] Wu M, Ludwig A and Kharicha A 2019 *Metals* **9** 229
[6] Hachani L, Zaidat K and Fautrelle Y 2015 *Int. J. Heat Mass Transf.* **85** 438-454
[7] Zhang Z, Wu M, Zhang H, Ludwig A and Kharicha A 2023 *Int. J. Heat Mass Transf.* **208**, 124050.
[8] Zheng Y., Wu M., Karimi-Sibaki E., Kharicha A and Ludwig A 2018 *Int. J. Heat Mass Transf.* **122** 939-953.

Predicting the local solidification time using spherical neural networks

Maximilian Erber[1,°,*], Tobias Rosnitschek[2,°], Constantin Bauer[1], Muhammet Ali Güldali[1], Bettina Alber-Laukant[2], Stephan Tremmel[2], Wolfram Volk[1] and Christoph Hartmann[1]

[1] Chair of Metal Forming and Casting, Technical University Munich, Walther-Meissner-Strasse 4, 85748 Garching, Germany
[2] Chair of Engineering Design and CAD, University of Bayreuth, Universitaetsstr. 30, 95447 Bayreuth, Germany
° These authors contributed equally.
* Corresponding author.

E-mail: maximilian.erber@tum.de

Abstract. Castings are predestined for the application of structural optimization, but to date, the integration of process simulation into structural optimization is limited due to high computational cost and is therefore often neglected at the beginning of the design process. This leads to the need for surrogate models, which allow a fast and simplified evaluation of design proposals during the optimization in order to improve the integration. This article introduces a novel approach that estimates the solidification time of randomly created geometries solely based on the casting geometry. The approach uses ray-tracing methods to calculate the distance function along preset directions. The estimated solidification time is calculated using a Spherical Convolutional Neural Network (CNN). The training data is obtained by several thousand solidification simulations using the optimization toolkit of a commercial casting simulation software combined with further data augmentation. The model is experimentally validated for five different geometries in the sand casting process.

1. Introduction

Structural optimization plays a prominent role in the seek for lowering energy consumption and improving resource efficiency in product design and manufacturing process. Casting processes are often used to manufacture such lightweight structures, as those processes allow a high degree of freedom in the design of the components. In general, numerical structural optimization leads to design proposals that are often non-manufacturable [1–3]. Hence, to improve the capability of the optimization as well as the design proposals a lot of effort is put into the coupling of process knowledge and topology optimization. In particular, so-called manufacturing constraints, such as symmetry, extrusion, or minimum length scale, were developed [4–7]. A more sophisticated approach is to combine numerical process simulation with numerical structural optimization [8, 9]. Although numerical methods for accurately describing the filling and solidification in the casting process by well-known physical laws have been developed for decades [10–13], their integration into structural optimization is still limited due to the remaining computational cost [1, 14]. Accordingly, there is a strong desire for methods leading to results with comparable

IOP Conf. Series: Materials Science and Engineering 1281 (2023) 012037 doi:10.1088/1757-899X/1281/1/012037

accuracy at less computational cost [15]. Therefore, this article presents an approach to predict the local solidification time for arbitrary complex geometries by artificial neural networks. This method comprises a uniform description of arbitrary geometries to enable evaluation using deep learning, the data acquisition of a sufficient data basis as well as the neural network approaches specified for the current application. Finally, our method is validated by casting experiments as well as numerical analysis.

2. Background

The solidification process is limited by the rate at which the heat of the liquid metal diffuses out of the liquid metal regions. The rate of solidification is mainly influenced by five main resistances that impede heat flow [16]:

- The liquid
- The solidified metal
- The metal-mould interface
- The mould
- The surroundings of the mold

In the sand casting process, the resistances in the liquid as well as the surroundings of the mold are negligible. While for light metals the resistance in regions of solidified metal has minor influence due to high thermal conductivity, the insulating effect of the sand mold predominantly influences the solidification process [16]. This led to surrogate models that estimate the solidification time based on the casting geometry. Most models are based on Chvorinov's rule [17], where the total solidification time t_f of a casting is determined by the ratio between the volume V and the surface A:

$$t_f = k\left(\frac{V}{A}\right)^2 \tag{1}$$

Flemings proposed a more generalized expression to calculate t_f using thermal parameters of the sand mold and the melt. For this the heat flow problem for simple sand castings is solved by solely considering the heat diffusion in the mold and assuming a constant metal temperature [18],

$$\frac{V}{A} = \frac{T_M - T_0}{\rho_s H'}\left(\frac{2}{\sqrt{\pi}}\sqrt{k_m\rho_m c_m}\sqrt{t_f} + \frac{nk_m t_f}{2r}\right) \tag{2}$$

where r is the casting radius and $n = 1$ for cylinder, 2 for sphere. This rule approximates the total solidification time t_f based on the ratio of a heat content volume V and its surrounding heat transferring area A by considering the thermal properties of the melt and the mold. This ratio is called the casting modulus. The heat H' is generated by the latent heat of the melt H and the superheat $c_{melt}\Delta T$ at constant temperature T_M and density ρ_s. The heat diffusivity, defined by the product of the heat conductivity k_m, density ρ_m, and specific heat capacity c_m of the mold, is a quantitative measure for the absorption of heat in the mold. For the complete solidification of a sphere ($n = 2$) or a cylinder ($n = 1$), the radius can be substituted by the modulus M using $r = 3M$ for $n = 2$ and $r = 2M$ for $n = 1$.

By solving equation 2 for t_f and replacing r we get the total solidification time in dependence on the modulus M and the material parameters for the cases of the cylinder and the sphere, see equation 3, as well as for the plate shown in equation 4.

$$t_f = \frac{2c_g c_s + c_s^2 c_\alpha^2 \pm c_\alpha \sqrt{4c_g c_s^3 + c_s^4 c_\alpha^2}}{2c_s^2 c_g^2} M^2, \quad \text{for } n = 1, 2 \tag{3}$$

$$t_f = \frac{1}{c_s^2 c_\alpha^2} M^2, \quad \text{for } n = 0 \tag{4}$$

$$\begin{aligned} \text{with: } M &= \frac{V}{A} \\ c_s &= \frac{T_M - T_0}{\rho_s H'} \\ c_\alpha &= \frac{2}{\sqrt{\pi}} \sqrt{k_m \rho_m c_m} \\ c_g &= \begin{cases} \frac{n k_m}{4}, & \text{for } n = 1 \\ \frac{n k_m}{6}, & \text{for } n = 2 \end{cases} \\ H' &= H + c_{melt} \Delta T \end{aligned} \tag{5}$$

Equations 3 and 4 are valid for the simple geometries, where the constant k in equation 2 is adapted to the respective geometry. Nevertheless when it comes to more complex arbitrary geometries and the need for comparing different areas of a single component, the determination of the corresponding volume and heat-transferring area becomes difficult and prone to errors. Therefore, methods have been developed that approximate the casting modulus [19, 20]. One approach also known as section modulus was introduced by Neises et al. [19]:

$$M_{Section} = \frac{2}{\sum_{i=1}^{N} \frac{1}{r_i}} \tag{6}$$

In this method, the section modulus $M_{Section}$ at a given point is determined by the shortest euclidean distance r_i towards the casting surface along given directions. As mentioned above Chvorinov's rule neglects the resistance of heat flow in the solidified metal as well as at the metal-mould interface. Further, material parameters are constant and temperature dependence is not taken into account. In standard casting software tools, these factors are considered. We use the section modulus as the state of the art for geometrically estimating the local solidification time t_s. Therefore we define the dependency of the local solidification time with the section modulus, in mm, by fitting a second-order polynomial curve, see equation 7, which showed acceptable accuracy for the data basis used in the current work. We prefer this approach for our work because it allows the estimation of t_s at arbitrary positions in a casting and is valid for complex castings, for which the casting surface is not describable with simple geometries like those found at topology optimised geometries. This approximated formulation includes thermal material parameters in its constants and is, therefore, usable to benchmark our proposed models to already known models.

$$t_s = 32,549 \cdot M_{Section}^2 + 815 \cdot M_{Section} \tag{7}$$

Previous work on integrating geometrical methods in structural optimization is scarce. Some preliminary work focused on the identification of solidification paths [21–23] and the integration in the topology optimization cycle [15]. There the directed solidification was determined solely based on the euclidean distance toward the closest casting surface.

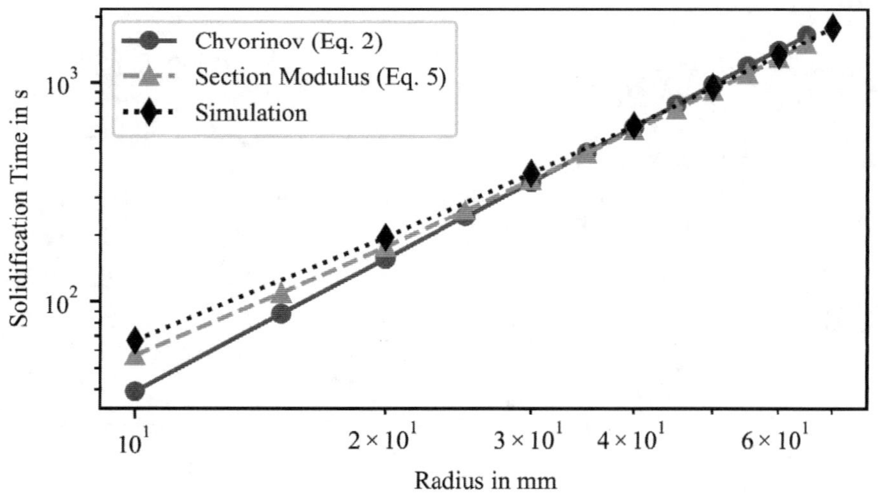

Figure 1. Approximating the solidification time for spheres with varying radius using Chvorinov's equation (Eq. 2) and using section modulus (Eq. 7). The results are compared to simulation results. For equation 3 following parameters are used: T_M: 580 °C, T_0: 20 °C, ρ_s: 2570 kg/m^3, H': 570 MJ/kg, k_m: 0.7 W/mK, ρ_m: 1280 kg/m^3, c_m: 0.8 kJ/kgK, n: 2.

3. Uniform Geometrical Description

Since we propose a model using an artificial neural network, a method is needed to integrate the casting geometry into the data-based model. Input parameters describing the geometry, have to be structured consistently for different arbitrary geometries [24]. We assume that the local solidification time is invariant to rotations of the surrounding casting. This assumption, combined with the need of an uniformly structured input data set, limits the possible descriptions of the surface. The use of Platonic solids enables a rotational invariant and equidistant sampling along a sphere's surface. We propose the use of the icosahedron, the Platonic solid with the highest number of faces, shown in Fig. 2. Some recent research showed excellent performance for the classification of projected images, climate patterns as well as acoustic sound source detection [25, 26]. This solid has twelve equally spaced vertices and consists of 20 congruent triangular faces and, therefore, results in 60 equivariant rotational symmetries. This mesh can be recursive refined leading to higher sample sizes and a more detailed description of the spherical shape, see Fig. 2. Thereby every face is subdivided into four smaller faces and new vertices are defined at the center of every edge. The number of refinement steps r results in a sample size of $N = 5 \cdot 2^{2r+1} + 2$ and refined samples retain their rotational symmetry.

One advantage of the icosahedral mesh is that its mesh is describable by standard file formats like STL (Standard Triangle Language). With this the triangulated surface mesh is defined by the position of its vertices in the three-dimensional space and its faces defined by their bounding vertices. The coordinates of the vertices define the displacement of the vertex towards a fixed central point. The coordinates of a vertex v_i can be also described by its norm $|v_i|$ times the normalized direction vector d_i, see equation 8. For the three-dimensional space d_i represents the position of a point on the surface of a unit sphere.

$$v_i = |v_i| \cdot d_i \tag{8}$$

In this work, a refined icosahedron with $r = 3$ and therefore 642 vertices, respectively, sample

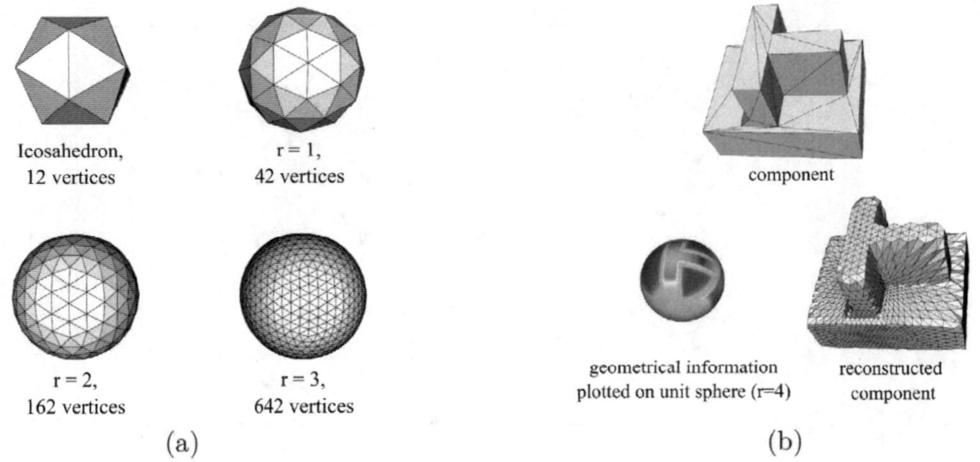

Figure 2. (a) Icosahedron with different refinement steps r. In every refining new vertices are created at the center of every edge. With increasing refinement, the sphere's surface is mapped more accurately. (b) Sampling of exemplary component plotted on the surface of an unit sphere and it's reconstruction.

points is used. Further, we define our geometry around the central point. Arbitrary geometries are created by repositioning the vertices along the directions d_i. Therefore, we must only take into account the norm $|v_i|$ of every vertex, which leads to an input vector with a size of 642, see Fig 2b. The position of a value in this input vector implies the position on the unit sphere d_i as well as neighborhood connections, because the faces of the surface mesh are defined by the corresponding vertices. In actual casting geometries, the norm is calculated using ray tracing. Hence, we call those values ray lengths in the following.

4. Data Acquisition and Augmentation

During the topology optimization of components, complex geometries and structures are generated as intermediate results in the individual iteration steps. These structures are not bound to the common description by means of simple volumes, as they mostly occur in the design of cast components. As a consequence, for a local evaluation of the castability, which means the comparison of different positions in the component, for example, to identify hot spots, surrogate models are necessary that can handle these complex structures. To the best of our knowledge, this generalization for the evaluation of arbitrary casting geometries is so far only possible with sufficient accuracy by discretization and numerical methods. The quality of results and generalization of data-based models requires a sufficient data basis. In this work, we generated 3000 randomized geometries and determined the local solidification time in the center point of those geometries by performing a solidification simulation to ensure this requirement. The solidification time at this center point does not need to be the maximum solidification time, as this work's objective is to determine the solidification time at a defined location.

The geometry is defined by randomly choosing N points out of the 642 vertices. The exact number N differs for every geometry in a range between 30 and 50 points and is randomly defined within these limits. For every geometry, we define a random mean ray length μ between 10 mm and 70 mm and a standard deviation σ in the range of 5 mm to 30 mm. Then for every vertex of the set N a ray length is generated, where the set for all points of N results in a Gaussian distribution defined by μ and σ. To ensure geometries with reasonable dimensions newly generated lengths that exceed a maximum length of 70 mm or fall below the minimum length of 10 mm are adapted to the maximal or respective minimal length. A triangulated mesh of an icosahedron with $r = 3$ along the surface of a unit sphere is created. While retaining the directions of the vertices, their distance towards the center points $(0, 0, 0)$ is adjusted by the randomly selected ray lengths. Interstitial vertices that do not belong to the set N are interpolated. This results in a complex morphed geometry but ensures that the surface mesh of this geometry is watertight and rays from the center point in each direction intersect the surface mesh at only a single point. For further evaluation, the created surface mesh is saved as STL as well as a compressed format saving only the ray lengths. A reconstruction of the geometry is ensured because the directions of the vertices were not changed in the creation process. The set of created geometries is shown in Fig. 3, where every geometry is represented by its mean ray length and its standard deviation.

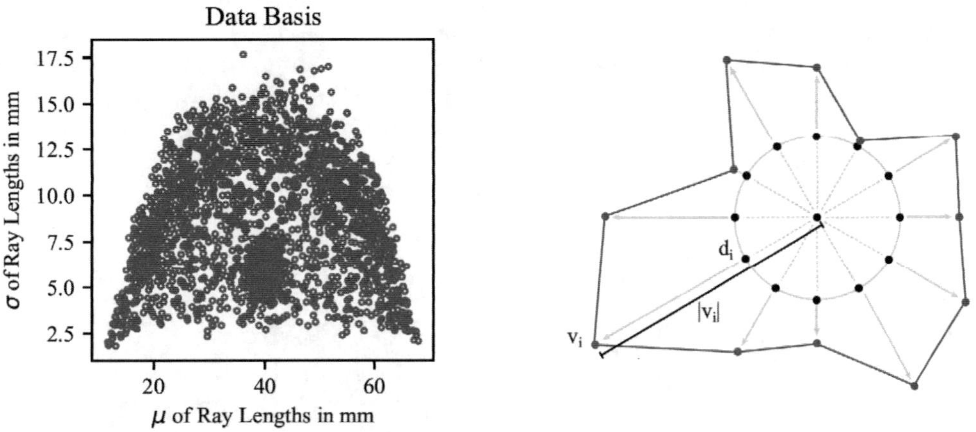

(a) (b)

Figure 3. (a) The data set of randomized geometries is represented by the respective mean value and standard deviation. (b) Deformation exemplary shown for a 2D geometry. The position on the unit sphere d_i is predefined and will not be changed at the deformation. The vertex v_i can therefore be defined only by changing $|v_i|$.

For every geometry, a solidification simulation is performed using the optimization toolbox of the commercial software MAGMAsoft (v5.4.1, MAGMA, Aachen, Germany). More detailed information is given in section 6. For our purpose for every geometry, the local solidification time at the center point of the geometry is calculated. We propose that the local solidification time is rotational independent. Therefore, rotations as well as reflections of our geometry have the same local solidification time at the center point. To augment our data set we rotated and flipped each geometry 30 times. This results in a total data set of 183,000 samples.

5. Neural Network Approaches

5.1. Multi Layer Perceptron

The MLP (Multi-Layer Perceptron) model was built using the python libraries NumPy [27], pandas [28, 29], TensorFlow [30, 31] and Auto-Keras [32]. We used Bayesian Optimization to identify a Multi-Layer Perceptron architecture and hyperparameters to build the model. Such approaches are also known as automated machine learning, since the costly search for the Neural Network architecture and its adjacent hyperparameters are systematized and automatized [32–34]. In this work, the objective was to minimize the model's mean squared error. Within the optimization process, each suggested model was trained for 100 epochs and evaluated subsequently. In total, 100 iterations were used. The best model after the optimization process was trained for 1000 epochs, using a validation split of 0.2 and five-fold cross-validation.

5.2. *Spherical Convolutional Neural Network*

Although the MLP approach allows a universal description of complex problems, an increase in the number of input parameters results in an exponential increase of the neural network's trainable parameters, and information about connections between the input parameters gets lost. As our uniform geometrical description in section 3 is defined by ray lengths pointing in defined directions and the geometrical neighborhood is known by the icosahedral's surface mesh, a neural network for this purpose is desired. We propose the use of icosahedral neural networks presented by Cohen et al. [25]. Thereby, the surface of the refined icosahedron is unfolded and subdivided into an atlas with five charts. This unfolding allows the transformation of the sphere's surface of a r-times refined icosahedron as a padded array of shape $5 \cdot (2^r + 2) \times (2^{r+1} + 2)$. This matrix preserves neighborhood relationships and allows a regression analysis with Convolutional Neural Networks (CNNs). We used the model shown in Fig. 4, which was built using the python libraries NumPy [27], pandas [28, 29], pyTorch [35] and icoCNN [26]. The final model was trained for 1000 epochs while minimizing the model's mean squared error.

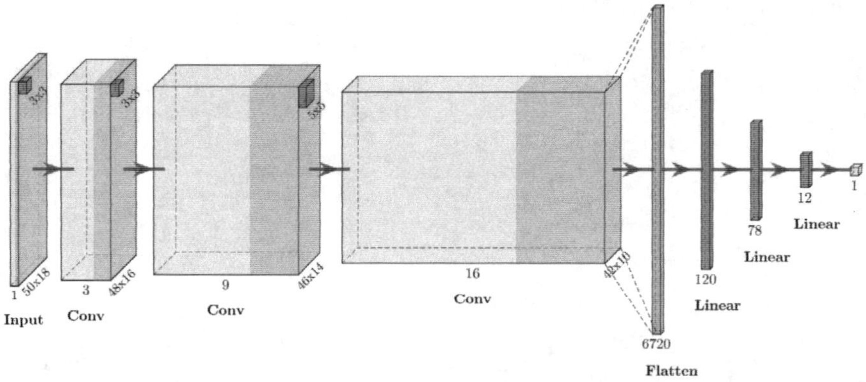

Figure 4. Network architecture of the CNN used for regression analysis of the unfolded icosahedron's surface with a refinement $r = 3$.

5.3. *Neural Network Evaluation*

We used two metrics to compare our Neural Networks, the root mean squared error (RMSE) and coefficient of determination (R^2), see equation 9 and equation 10. The prediction of the solidification time by section modulus is rotational independent. With our presented neural networks this equivariance is broken due to the use of fully connected layers, which leads to different results in dependence of the geometries orientation. To improve our prediction results we propose to calculate multiple orientations and use the mean of those predictions. In our work, we predicted every geometry in 62 different orientations.

$$RMSE = \sqrt{\frac{1}{N} \sum_{i}^{N} (y_i - \hat{y}_i)^2} \tag{9}$$

$$R^2 = 1 - \frac{\sum_i^N (y_i - \hat{y}_i)^2}{\sum_i^N (y_i - \bar{y}_i)^2} \tag{10}$$

6. Simulation

The commercial software MAGMAsoft (v 5.4.1, MAGMA, Aachen, Germany) was used for creating the training data set, as well as for designing and validating the experiments. The material used in the sand casting process is aluminum AlSi12. The mold material was set to furan bonded silica sand. As in the experimental setup two thermocouples were defined, one at the center of the sphere and one placed at the center point of the randomized geometry. The heat transfer coefficient was determined by performing an optimization with the objective to match the simulated and measured temperature profile at the thermocouple position of the sphere. For the optimization only the solidification process of the experimental geometry including the ingate system was investigated. The initial melt temperature was predefined to 650 °C. Then a process simulation was performed including the mold filling with a filling time of five seconds and a pouring temperature of 685 °C.

For the complete casting simulation 2,000,000 elements were used. For the generation of the training data, the simulation time had to be minimized in order to create as many different geometries as possible in a reasonable time. Therefore the filling simulation was neglected and the number of elements was reduced to 500,000. The initial temperature of those simulations was set to 650 °C, which was in good agreement with the measured temperature of the experiments. The aim of these simulations was to determine the local solidification time in the center point of the randomized geometries. Those times were later used as the targets during the training of the neural networks. For that the solidus to liquidus temperature at the center of the geometries was set as a objective and sequential simulations each containing 100 geometries were performed.

7. Validation Experiments

For evaluating the prediction and accuracy of the proposed method in the real casting process validation experiments have been performed. The objective of these gravity casting process experiments was to prove that the input training data correlates with the real casting process. Five different geometries, shown in Fig. 5, with comparable volumes but significantly differing solidification times were chosen out of the randomly created geometries. The selected volumes were almost equal to the volume of a sphere with a diameter of 50 mm. This led to geometries with a simulated local solidification time in a range from 134.53 seconds for the most deformed geometry 1244 and 244.77 seconds for geometry 1404, see table 1.

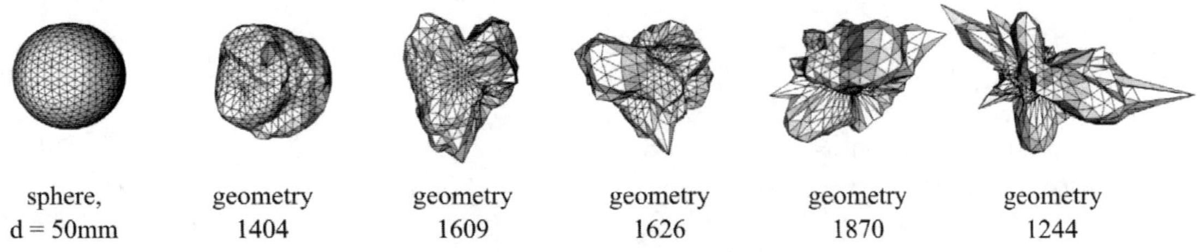

Figure 5. Randomly created validation geometries, used in the sand casting experiments

The experiments were performed using the near-eutectic aluminum alloy AlSi12 with a casting temperature of 690 +/- 10 °C. The sand mold was produced by 3D printing. Furan resin was the

IOP Conf. Series: Materials Science and Engineering 1281 (2023) 012037 doi:10.1088/1757-899X/1281/1/012037

Table 1. Geometrical information about the specimen used for validation.

Geometry	V in mm^3	A in mm^2	M in mm	t_s in s
sphere Ø50 mm	65,417	7,850	0.0779	278.96
geometry 1404	63,340	9,383	0.0736	244.77
geometry 1609	61,650	10,652	0.0683	212.03
geometry 1626	64,349	11,271	0.0670	209.30
geometry 1870	68,668	13,683	0.0581	173.15
geometry 1244	67,586	16,025	0.0565	134.53

bonding agent for the sand. To validate the simulation results, the temperature was measured with a thermocouple (Typ K, class I) at the center point of the random geometry and the sphere's center point. Filling the molted material into a pouring basin in the middle of the sand mold ensured the comparability between different castings experiments. From the pouring basin, two running systems led on one side to the randomized geometry and on the other side to a sphere with a diameter of 50 mm. The cavity of the test geometry is filled from bottom to top. Every experiment was repeated four times. The final casting for geometry 1244 is exemplarily shown in Fig. 6. By evaluating the measured time-temperature curve the local solidification time, defined as the time span during the cooling of the part between liquidus temperature T_{liq} and solidus temperature T_{sol}, was determined. We predefined T_{liq} to 580 °C and T_{sol} to 570 °C.

Figure 6. Casting experiments including the runner system of geometry 1244.

8. Results

8.1. Experimental Results and Simulation Accuracy

The validation geometries show local solidification times t_s in a range between 130 and 270 s and different complexity. For every experiment a sphere with a diameter of 50 mm was casted, which had a mean measured t_s of 269.8 +/- 11.53 s over all experiments. The randomized geometries showed an even smaller standard deviation in the range between 2.77 s for geometry 1244 and 11.34 s for geometry 1626. This shows a good comparability for the experimental setup. The comparison of those results with the simulation reveals that the simulation approximates t_s with a maximum absolute error of 20 s, see Fig. 7. The simulation of the full setup including a filling simulation rather underestimates the local solidification time, whereas the simulation of the geometries without ingate system and neglecting the filling slightly overestimates it.

The assumption that all rotations of casting do have the same local solidification time is a simplification to enlarge the data set and will induce an inaccuracy. We evaluated this error by performing simulations with the geometries 1244 and 1870 at different rotation variants, see 2. With a standard deviation of about 1.5 to 2 percent of t_s it is acceptable.

Table 2. Results of variations in the local solidification time due to rotating the geometry.

	μ_{t_s} in s	σ_{t_s} in s	$\max(t_s)$ in s	$\min(t_s)$ in s
geometry 1870	170.6	2.8	173.9	165.8
geometry 1244	135.7	2.7	141.3	132.2

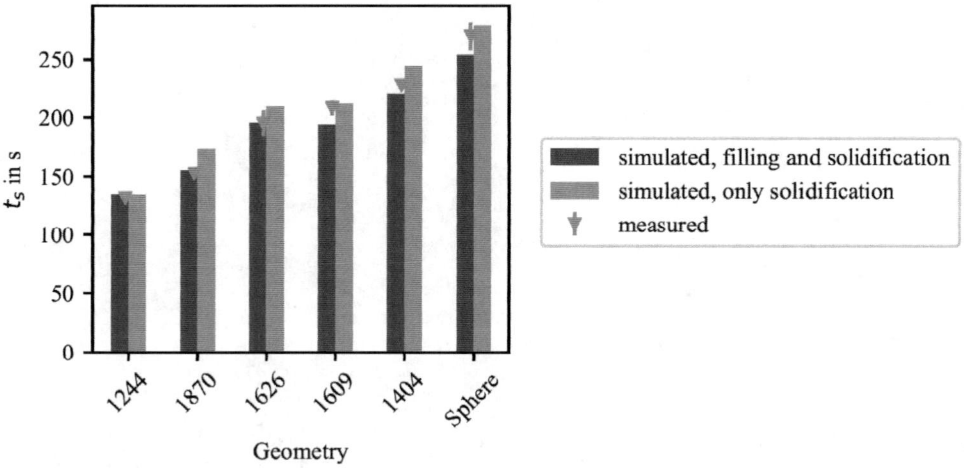

Figure 7. Experimentally measured local solidification times and the corresponding simulation results. The experimental setup was simulated using a filling and solidification simulation. The light grey results of the validation geometries, shown in figure 5 have been simulated neglecting the ingate system.

8.2. Bayesian Architecture Optimization and Neural Network Predictions

Estimating the solidification times using the polynomial approximation of the section modulus by equation 7 has a R^2 score of 0.993 and a RMSE of 23.97 s. Still as shown in Figure 8 some geometries show a higher deviation. The MLP with the optimized architecture, shown in table 4, results in a higher accuracy with a R^2 score of 0.996 an a RMSE of 19.55 s. The optimization of the MLP model including the section modulus as input resulted in the architecture shown in table 5. After training, this model had a R^2 score of 0.998 for the validation data set. The most accurate prediction is achieved by using the spherical neural network with a R^2 score of 0.999 and a RMSE of 7.33 s. By calculating the mean predictions of multiple rotations of one geometry the accuracy can be further improved, with a minimal RMSE of 5.10 s using the spherical network. For the best model this results in a relative error reduction of about 70 percent for estimating the geometry. All metrics are summarized in table 3.

Table 3. Regression metrics for the validation data set.

Model	R^2	R^2 Mean of Rotations	RMSE in s	RMSE Mean of Rotations in s
Section Modulus, p=2	0.9933	-	23.968	-
MLP	0.9956	0.9984	19.553	11.425
MLP incl. section modulus	0.9979	0.9986	13.359	10.914
Spherical CNN	0.9993	0.9997	7.334	5.102

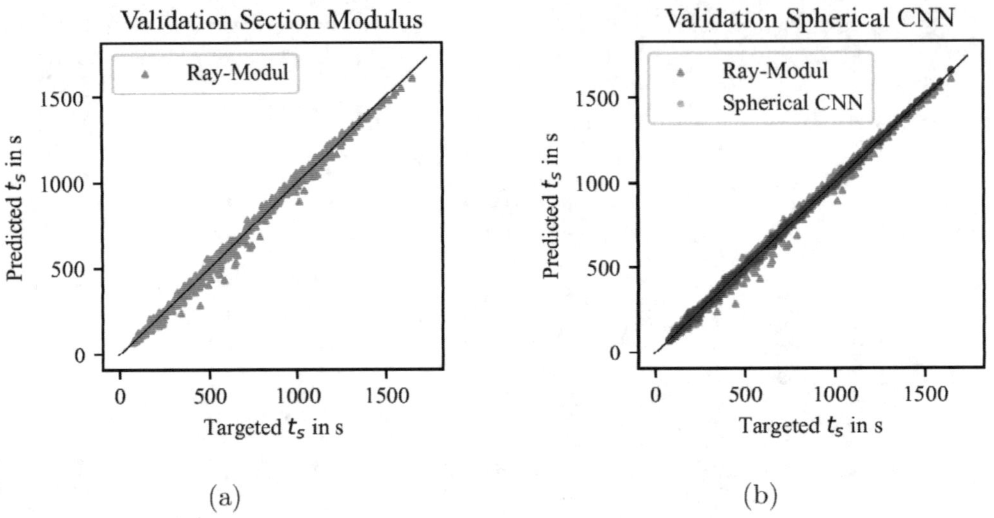

Figure 8. (a) Comparative data of local solidification time predicted by polynomial fitting with section modulus. (b) Predictions of validation data with icosahedral CNN. The predictions of the section modulus are shown for comparison.

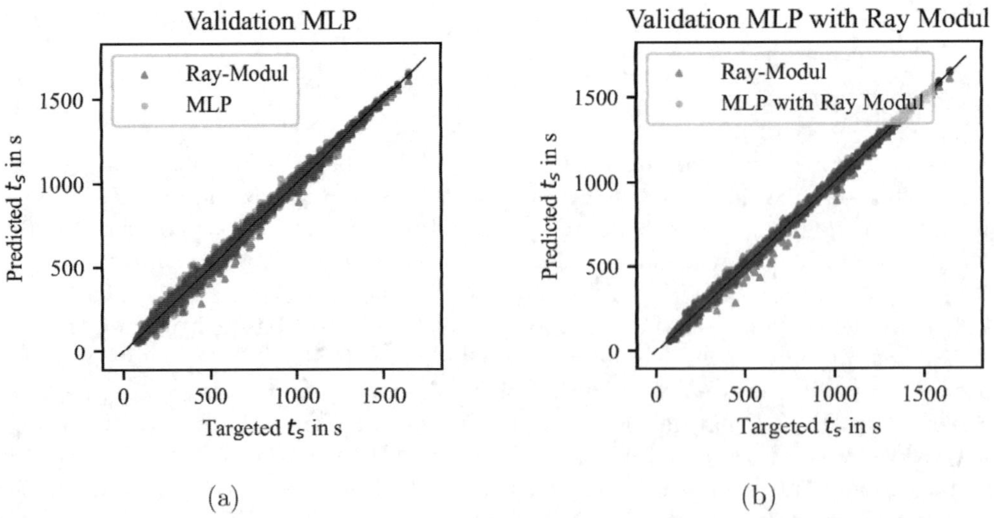

Figure 9. Predictions of validation data with MLP (a) and the MLP taking the section modulus as input in account (b). The predictions of the section modulus are shown for comparison.

Table 4. Architecture of the Multi-Layer Perceptron model after Bayesian Optimization.

Layer	Output Shape	Number of Parameters
Input Layer	642	0
Multi-Category Encoding	642	0
Linear + ReLu	1024	658432
Linear + ReLu	1024	1049600
Linear + ReLu	16	16400
Regression Head	1	17

Table 5. Architecture of the Multi-Layer Perceptron model including the section modulus as input after Bayesian Optimization.

Layer	Output Shape	Number of Parameters
Input Layer	643	0
Multi-Category Encoding	643	0
Normalization	643	0
Linear + ReLu	1024	659456
Linear + ReLu	1024	1049600
Linear + ReLu	1024	1049600
Regression Head	1	1025

8.3. Experimental Validation of Neural Network Predictions

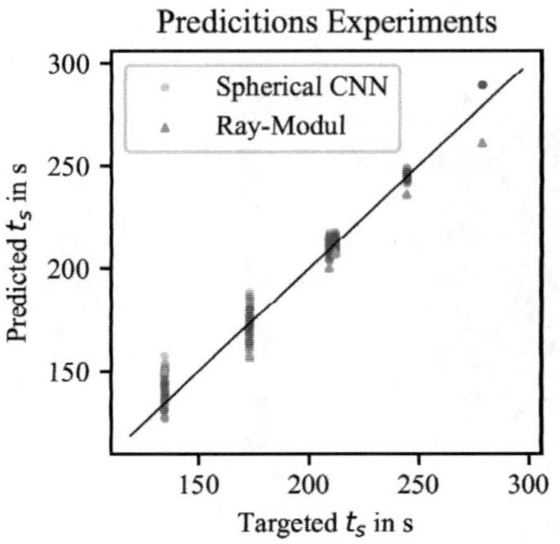

Figure 10. Predictions with MLP including the section modulus as input. The predictions of the section modulus are shown for comparison.

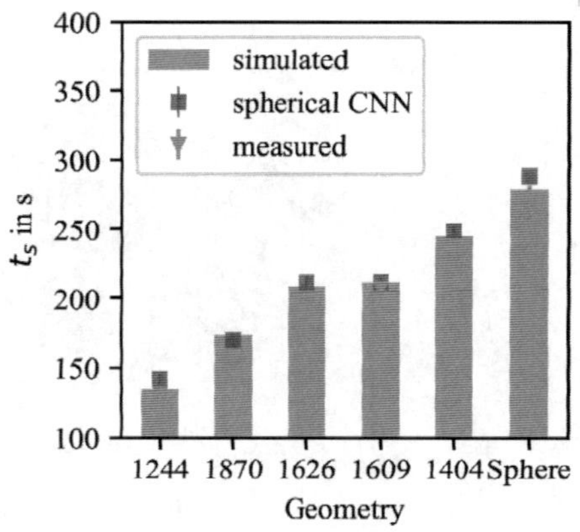

Figure 11. Comparison of the simulated and measured solidification times with the predicted t_s using the spherical CNN.

The prediction of the neural networks is further investigated using the casted validation geometries. Those geometries are not used for training the spherical neural network and are therefore unknown to the model. Figure 10 shows that for the five validation geometries as well as

IOP Conf. Series: Materials Science and Engineering 1281 (2023) 012037 doi:10.1088/1757-899X/1281/1/012037

for the sphere with a diameter of 50 mm the error compared to the section modulus is reduced with regard to the targeted respective simulated t_s. By comparison with the experimentally measured t_s, it is evident that the prediction tends to overestimate the measured t_s. However, the simulated results without gating system is used for training the network, which overestimates t_s compared to the experiments as well and the experimental results are unknown to the network. Therefore the targeted t_s is predicted accurately and the error with regard to the experiments is mainly due to deviations of training and validation data basis.

9. Conclusion

In this work we show a way to enable the description of arbitrary geometries in a uniform representation and a fast way to predict their local solidification time t_s. This representation allows the generation of randomized geometries and providing a good database for training data-based models. The validity of this database was evaluated by experimental investigations. As models the determination by polynomial approximation of the section modulus, two different multi-layer perceptron models as well as a spherical convolutional neural network, which can represent geometric neighborhoods, are applied. The spherical CNN shows the best results and achieved a 70 percent accuracy improvement in comparison to the section modulus. Compared to the experimental data, a higher error is obtained. However, this error is due to the simulation model setup for generating the input data. In our work we did not investigate different process parameters or thermal material parameters. Nevertheless the used material properties, including temperature-dependant thermal properties, are implicitly incorporated in the neural network approaches because they are as well considered in their training data. Further integration of process parameters requires more training data, but implementation can be done without any problems. By now, only the center point of the geometries are evaluated, from which rays pointing outward toward the casting surface intersect the component surface only once. Therefore the future goal is to adapt our model to predict t_s for any position in the casting geometry, as examplary shown for a simple bar in figure 12.

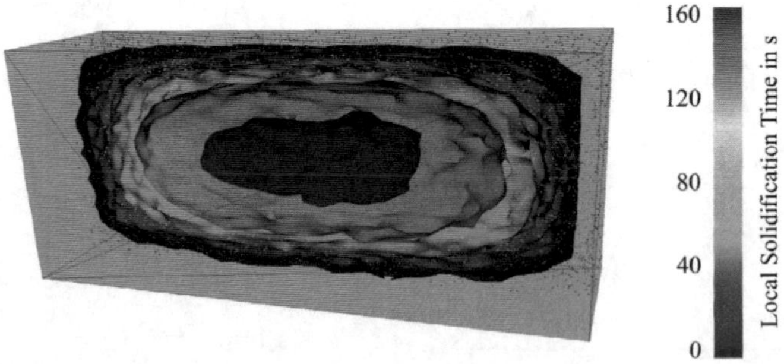

Figure 12. Exemplary results for a bar with the dimensions of 70 mm x 30 mm x 20 mm predicted by the spherical CNN. For better visibility isosurfaces for local solidification times of 50 s, 75 s, 100 s, 125 s and 150 s are shown.

Acknowledgments

The authors would like to thank the German Research Foundation for financial support under grant number 434348474.

References

[1] Heilmeier F, Goller D, Opritescu D, Thoma C, Rieg F, Volk W. Support for Ingate Design by Analysing the Geometry of High Pressure Die Cast Geometries Using Dijkstra's Shortest Path Algorithm. Advanced Materials Research. 2016 Aug;1140:400-7. Available from: https://www.scientific.net/AMR.1140.400.

[2] Harzheim L, Graf G. A review of optimization of cast parts using topology optimization: I - Topology optimization without manufacturing constraints. Structural and Multidisciplinary Optimization. 2005 Dec;30(6):491-7. Available from: http://link.springer.com/10.1007/s00158-005-0553-x.

[3] Harzheim L, Graf G. A review of optimization of cast parts using topology optimization: II-Topology optimization with manufacturing constraints. Structural and Multidisciplinary Optimization. 2006 May;31(5):388-99. Available from: http://link.springer.com/10.1007/s00158-005-0554-9.

[4] Vatanabe SL, Lippi TN, Lima CRd, Paulino GH, Silva ECN. Topology optimization with manufacturing constraints: A unified projection-based approach. Advances in Engineering Software. 2016 Oct;100:97-112. Available from: https://linkinghub.elsevier.com/retrieve/pii/S0965997816301703.

[5] Wang Y, Kang Z. Structural shape and topology optimization of cast parts using level set method: Structural shape and topology optimization of cast parts using level set method. International Journal for Numerical Methods in Engineering. 2017 Sep;111(13):1252-73. Available from: http://doi.wiley.com/10.1002/nme.5503.

[6] Li Q, Chen W, Liu S, Fan H. Topology optimization design of cast parts based on virtual temperature method. Computer-Aided Design. 2018 Jan;94:28-40. Available from: https://linkinghub.elsevier.com/retrieve/pii/S0010448517301434.

[7] Sato Y, Yamada T, Izui K, Nishiwaki S. Manufacturability evaluation for molded parts using fictitious physical models, and its application in topology optimization. The International Journal of Advanced Manufacturing Technology. 2017 Sep;92(1-4):1391-409. Available from: http://link.springer.com/10.1007/s00170-017-0218-0.

[8] Franke T, Fiebig S, Bartz R, Vietor T, Hage J, vom Hofe A. Adaptive Topology and Shape Optimization with Integrated Casting Simulation. In: Rodrigues HC, Herskovits J, Mota Soares CM, Araújo AL, Guedes JM, Folgado JO, et al., editors. EngOpt 2018 Proceedings of the 6th International Conference on Engineering Optimization. Cham: Springer International Publishing; 2019. p. 1263-77. Available from: http://link.springer.com/10.1007/978-3-319-97773-7_109.

[9] Franke T, Fiebig S, Paul K, Vietor T, Sellschopp J. Topology Optimization with Integrated Casting Simulation and Parallel Manufacturing Process Improvement. In: Schumacher A, Vietor T, Fiebig S, Bletzinger KU, Maute K, editors. Advances in Structural and Multidisciplinary Optimization. Cham: Springer International Publishing; 2018. p. 1815-30. Available from: http://link.springer.com/10.1007/978-3-319-67988-4_136.

[10] Flender E, Sturm J. Thirty Years of Casting Process Simulation. International Journal of Metalcasting. 2010 Apr;4(2):7-23. Available from: http://link.springer.com/10.1007/BF03355463.

[11] Šabík V, Futáš P, Pribulová A. Failure analysis of a clutch wheel for wind turbines with the use of casting process simulation. Engineering Failure Analysis. 2022 May;135:106159. Available from: https://linkinghub.elsevier.com/retrieve/pii/S1350630722001339.

[12] Sowa L, Skrzypczak T, Kwiatoń P. Influence of Basic Physical Phenomena of the Casting-Riser System Solidification Process on Defects Formation in The Casting. Acta Physica

Polonica A. 2022 Jul;142(1):48-51. Available from: http://przyrbwn.icm.edu.pl/APP/PDF/142/app142z1p11.pdf.

[13] Khalajzadeh V, Beckermann C. Simulation of Shrinkage Porosity Formation During Alloy Solidification. Metallurgical and Materials Transactions A. 2020 May;51(5):2239-54. Available from: http://link.springer.com/10.1007/s11661-020-05699-z.

[14] Rosnitschek T, Erber M, Hartmann C, Volk W, Rieg F, Tremmel S. Combining Structural Optimization and Process Assurance in Implicit Modelling for Casting Parts. Materials. 2021 Jul;14(13):3715. Available from: https://www.mdpi.com/1996-1944/14/13/3715.

[15] Erber M, Rosnitschek T, Hartmann C, Alber-Laukant B, Tremmel S, Volk W. Geometry-based Assurance of Directional Solidification for Complex Topology-optimized Castings using the Medial Axis Transform. Computer-Aided Design. 2022:103394. Available from: https://www.sciencedirect.com/science/article/pii/S0010448522001324.

[16] Campbell J. Complete casting handbook: Metal casting processes, techniques and design. 2nd ed. Amsterdam: Butterworth-Heinemann; 2015.

[17] Chvorinov N. Theorie der Erstarrung von Gussstücken. Giesserei. 1940:177-86.

[18] Flemings MC. Solidification processing. McGraw-Hill series in materials science and engineering. New York: McGraw-Hill; 1974.

[19] Neises SJ, Uicker JJ, Heine RW. Geometric Modeling of Directional Solidification based on Section Modulus. AFS Transactions. 1987;95:25-30.

[20] Ravi B, Srinivasan MN. Casting solidification analysis by modulus vector method. International Journal of Cast Metals Research. 1996;9(1):1-7.

[21] Pao WKS, Ransing RS, Lewis RW, Lin C. A medial-axes-based interpolation method for solidification simulation. Finite Elements in Analysis and Design. 2004;40(5-6):577-93.

[22] Ransing RS, Sood MP, Pao WKS. Computer implementation of Heuvers' circle method for thermal optimisation in castings. International Journal of Cast Metals Research. 2005;18(2):119-28.

[23] Sood MP. A novel geometry based approach for casting process optimisation [Dissertation]. Swansea: Swansea University; 2006.

[24] Hartmann C, Opritescu D, Volk W. An artificial neural network approach for tool path generation in incremental sheet metal free-forming. Journal of Intelligent Manufacturing. 2019;30(2):757-70.

[25] Cohen TS, Weiler M, Kicanaoglu B, Welling M. Gauge Equivariant Convolutional Networks and the Icosahedral CNN. Proceedings of the 36th International Conference on Machine Learning. 2019. Available from: http://arxiv.org/pdf/1902.04615v3.

[26] Diaz-Guerra D, Miguel A, Beltran JR. Direction of Arrival Estimation of Sound Sources Using Icosahedral CNNs. IEEE/ACM Transactions on Audio, Speech, and Language Processing. 2023;31:313-21. Available from: http://arxiv.org/pdf/2203.16940v2.

[27] Harris CR, Millman KJ, van der Walt SJ, Gommers R, Virtanen P, Cournapeau D, et al. Array programming with NumPy. Nature. 2020 Sep;585(7825):357-62. Available from: https://www.nature.com/articles/s41586-020-2649-2.

[28] Reback J, McKinney W, Jbrockmendel, Bossche JVD, Augspurger T, Cloud P, et al.. pandas-dev/pandas: Pandas 1.0.3. Zenodo; 2020. Available from: https://zenodo.org/record/3715232.

[29] McKinney W. Data Structures for Statistical Computing in Python. In: Proceedings of the 9th Python in Science Conference. Austin, Texas; 2010. p. 56-61. Available from: https://conference.scipy.org/proceedings/scipy2010/mckinney.html.

IOP Conf. Series: Materials Science and Engineering 1281 (2023) 012037 doi:10.1088/1757-899X/1281/1/012037

[30] Martín Abadi, Ashish Agarwal, Paul Barham, Eugene Brevdo, Zhifeng Chen, Craig Citro, et al.. TensorFlow: Large-Scale Machine Learning on Heterogeneous Systems; 2015. Available from: https://www.tensorflow.org/.

[31] Developers T. TensorFlow. Zenodo; 2021. Available from: https://zenodo.org/record/5799851.

[32] Jin H, Song Q, Hu X. Auto-Keras: An Efficient Neural Architecture Search System. In: Proceedings of the 25th ACM SIGKDD International Conference on Knowledge Discovery & Data Mining. Anchorage AK USA: ACM; 2019. p. 1946-56. Available from: https://dl.acm.org/doi/10.1145/3292500.3330648.

[33] Yao Q, Wang M, Chen Y, Dai W, Li YF, Tu WW, et al. Taking Human out of Learning Applications: A Survey on Automated Machine Learning. arXiv:181013306 [cs, stat]. 2019 Dec. ArXiv: 1810.13306. Available from: http://arxiv.org/abs/1810.13306.

[34] Victoria AH, Maragatham G. Automatic tuning of hyperparameters using Bayesian optimization. Evolving Systems. 2021 Mar;12(1):217-23. Available from: https://link.springer.com/10.1007/s12530-020-09345-2.

[35] Paszke A, Gross S, Massa F, Lerer A, Bradbury J, Chanan G, et al. PyTorch: An Imperative Style, High-Performance Deep Learning Library. In: Advances in Neural Information Processing Systems 32. Curran Associates, Inc.; 2019. p. 8024-35. Available from: http://papers.neurips.cc/paper/9015-pytorch-an-imperative-style-high-performance-deep-learning-library.pdf.

Simulation of dendritic deformation induced by convection in alloy solidification

Luwei Yang[1,a], Neng Ren[1], Jun Li[1,2,b], Mingxu Xia[1,2], Hongbiao Dong[3], Jianguo Li[1]

[1] School of Materials Science and Engineering, Shanghai Jiao Tong University, Shanghai 200240, China
[2] Shanghai Key Laboratory of Advanced High-temperature Materials and Precision Forming, Shanghai Jiao Tong University, Shanghai 200240, China
[3] School of Engineering, University of Leicester, Leicester LE1 7RH, United Kingdom

Email: [a] yangluwei@sjtu.edu.cn, [b] li.jun@sjtu.edu.cn

Abstract. Dendritic deformation induced by convection of thermal fluid is one of the factors leading to dendrite fragmentation and plays a crucial role in grain structure of alloy, which lacks in-depth understanding. In this paper, we simulate the flow-induced mechanical deformation of dendrites during the solidification of Al-4.5wt.%Cu alloy by combining the cellular automaton - finite volume method (CA-FVM) for the dendrite growth and the finite element method (FEM) for handling the dendritic deformation with the complex boundary conditions given by CA-FVM results. It shows that the dendritic deformation strongly depends on the flow velocities of melt and dendritic morphology. The dendrites can undergo visible bending above the critical flow velocities for dendrite yield (ranging from 0.023 m/s to 0.126 m/s as the inlet velocity increases) and the von Mises stress increases as the flow is enhanced during the growth process.

1. Introduction

Melt convection is a common phenomenon during alloy solidification and plays an important role in macrosegregation, solidification defects and grain structure (columnar/equiaxed grains) [1-3]. The melt flow velocity under external forces such as electromagnetic stirring and ultrasonic vibrations can reach ~0.4 m/s while the thermal-solutal-fluid flow during solidification of large steel ingots can reach ~0.05 m/s [4-6], which causes strong inter-dendrite fluid flow and even dendrite fragmentations. It is generally accepted that high flow velocities under melt convection, especially in electromagnetic stirring and ultrasonic melt processing, promote dendritic fragmentation and lead to grain refinement and the transitions of columnar to equiaxed dendrites [7-10]. However, there are still some unclear points regarding flow-induced dendrite fragmentation.

Previous studies have well revealed numerous mechanisms of dendrite fragmentation such as solute enrichment remelt [11], thermal remelt [12], pinch-off [13-15], etc. There is some controversy with regard to dendrite deformation and mechanical fracture. Zhang *et al* [16] suggested that thermal perturbation coupled with mechanical fracture is the main mechanism for dendrite fragmentation under ultrasound and concluded that the fracture strength of dendrites near the melting point is less than 1 MPa. Wang *et al* [17] found that under ultrasonic melt processing, fluid flow can cause mechanical fracture of intermetallic dendrites. Nevertheless, the experiments do not provide direct evidence of dendrite deformation, especially the generation of flow-induced dendrite deformation stresses, the simulations can be used as an advantageous means to capture dendrite deformation and possible fracture. Piling *et al* [18] developed an analytical model of flow-induced dendrite deformation in 1996, which was a fundamental work and the model that was adopted by many later researchers. But the model ignored the interaction between dendrite growth and fluid flow, the flow-induced dendrite deformation stress under complex flow fields and the very low yield strength of the dendrite near the melting point. In this work, we combine cellular automaton - finite volume method (CA-FVM) and displacement-based finite element method (FEM) to simulate flow-induced dendrite deformation in the solidification of Al-4.5wt.%Cu alloy under a parallel fluid flow and to shed light on the factors influencing flow-induced dendrite deformation.

2. Model description

Our model framework consists of two main parts, one is the cellular automaton-finite volume method and the other is the displacement-based finite element method. The overall model is in a two-dimensional approach and the framework is shown in figure 1.

One of the CA-FVM models describing dendrite growth under fluid flow is based on the Eulerian multiphase flow framework. In this model we consider the liquid phase and the solid phase as two different hydrodynamic phases. The liquid phase is treated as an incompressible Newtonian fluid, while the solid phase is treated as a pseudo-fluid phase with an infinite viscosity and is considered stationary. Cellular automaton is used to model dendrite growth, and the evolution rules for cells use the von Neumann capture rule of four nearest neighbors. The details of the model can be found in the references [19].

The other part of the model is displacement finite element method model that simulates the deformation of the dendrites. In this model we consider that only the solid phase is subject to flow-induced mechanical deformation and that the fluid flow force acts only at the liquid-solid interface (e.g. taking the solid phase iso-lines as the liquid-solid interface). Several assumptions are listed as follows: (1). The solid is considered as a deformed body in plane stress state and the solid constitutive laws are considered to be elastic. (2). Thermal stresses are neglected as the solidification is at isothermal conditions. (3). The elastic modulus (E) is constant at the undercooled temperature and calculated by JMatPro software.

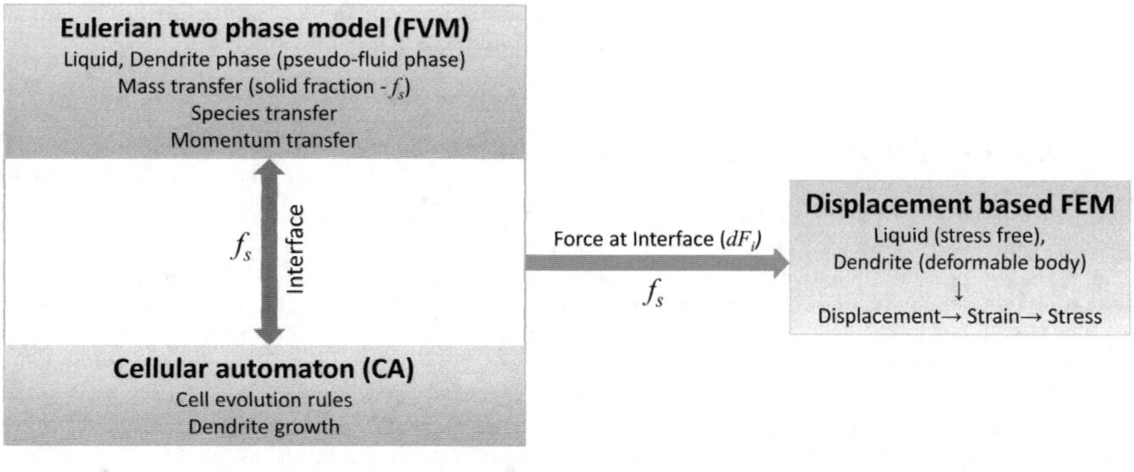

Figure 1. The overall approach of combining CA-FVM and FEM model.

The forces imposed on the solid/liquid interfaces due to fluid flow are based on the follows [20]:

$$\boldsymbol{\sigma}^f \cdot \boldsymbol{n}^f = -\boldsymbol{\sigma}^s \cdot \boldsymbol{n}^s \tag{1}$$

where $\boldsymbol{\sigma}^f$ and $\boldsymbol{\sigma}^s$ are the stress tensor of the fluid and solid in the interface, respectively. $\boldsymbol{n}^f$ and $\boldsymbol{n}^s$ are the normal vectors of the interface in the fluid and solid domain, respectively. The stress tensor of liquid is obtained by solving the Navier-Stokes equations in CA-FVM model. The stress tensor of liquid (incompressible Newtonian fluid) can be expressed as:

$$\sigma_{ij}^f = \left(-p\delta_{ij} + \mu^f\left(\frac{\partial u_i^l}{\partial x_j} + \frac{\partial u_j^l}{\partial x_i}\right)\right) \tag{2}$$

where p is the pressure in the fluid, δ_{ij} is the Dirac delta function, μ^f is dynamic viscosity of liquid, u_i^l is the velocity of liquid in the x_i direction. The forces act on the interface are given by:

$$dF_i = \sigma_{ij}{}^s \cdot n_j \cdot d\boldsymbol{S} \tag{3}$$

where F_i is the force in the x_i direction, n_j is the directional cosine of the normal vector of the interface $\boldsymbol{S}$ with respect to the x_j direction. The calculated stress tensor of fluid in the CA-FVM model are re-interpolated to the nodal points at the fluid-solid interface and used as the load condition of the FEM model.

The stress balance equation, strain equation and constitutive equation are used to describe the mechanical deformation of dendrites under the impact of fluid. The stress balance equations are given by:

$$\nabla \cdot \boldsymbol{\sigma} + \boldsymbol{b} = 0 \tag{4}$$

where $\boldsymbol{b}$ is the body force within an element (per unit volume), $\boldsymbol{\sigma}$ is the Cauchy stress in the current configuration. For a two-dimensional approach, there are two partial differential equations in two directions. The strain equation is as follows:

$$\boldsymbol{\varepsilon} = \frac{1}{2}\left(\frac{\partial \boldsymbol{u}}{\partial \boldsymbol{X}} + \left[\frac{\partial \boldsymbol{u}}{\partial \boldsymbol{X}}\right]^T\right) \tag{5}$$

where $\boldsymbol{u}$ is the displacement of a material point, $\boldsymbol{X}$ is Lagrangian coordinate. The linear elastic constitutive laws are given by:

$$\boldsymbol{\sigma} = \boldsymbol{D}^{el} : \boldsymbol{\varepsilon} \tag{6}$$

where $\boldsymbol{D}^{el}$ represents the fourth-order elasticity tensor. Therefore, the deformation of dendrite induced by fluid flow is solved with finite element method based on these governing equations.

3. Simulation parameters and benchmark configurations

The benchmark configuration is shown in figure 2. The velocities of the inlet are from 0.01 m/s to 0.08 m/s and the undercooling of the melt is set to 5 K to keep isothermal conditions. The right boundary is set as outflow boundary conditions, and the other two boundaries are set as no slip wall boundary conditions. The initial nucleus is placed in the centre of computational domain in CA-FVM, while the nucleation locations with several grids are considered as an encastre boundary conditions to fix the displacements in FEM. The domain with 1mm × 1mm is meshed into 640,000 (800 × 800) structured square grids. The mesh size is 1.25 μm. The thermodynamic, physical and mechanical properties of Al-4.5wt.%Cu alloy are shown in Table 1. The governing equations in the CA-FVM model are solved with phase coupled SIMPLE algorithm. The time step is set as 2×10^{-5} s to ensure that the equations can converge under the fine meshes. The governing equations of mechanical deformation model are directly solved for an input specific physical time result from CA-FVM under linear constitutive laws. Numerical implementation is used a combination of in-house developed code and commercial CFD and FEM software.

IOP Conf. Series: Materials Science and Engineering 1281 (2023) 012038 doi:10.1088/1757-899X/1281/1/012038

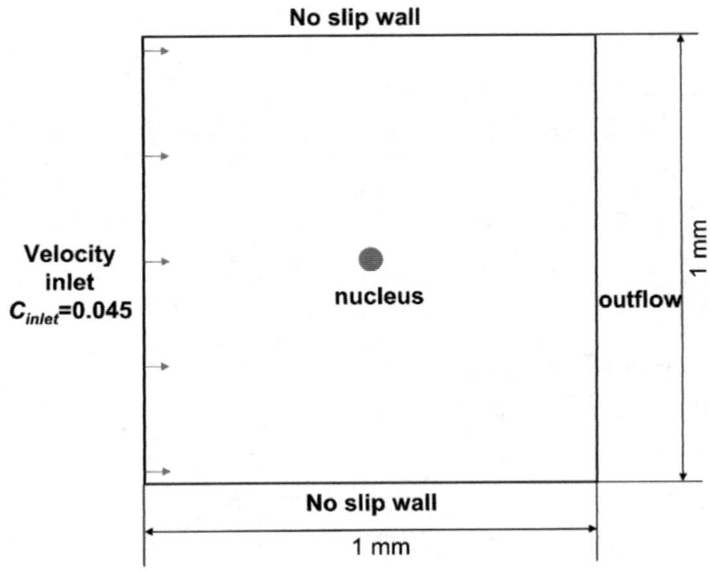

Figure 2. Benchmark geometry with boundary and initial conditions, and the nucleation locations in the center was applied an encastre boundary conditions in FEM.

Table 1. Thermodynamic, physical and mechanical properties of Al-4.5wt.%Cu alloy

Property	Symbol	Value	Ref.
Thermodynamic properties			
Melting point of pure metal (K)	T_f	933	[21]
Liquidus slope ($K\cdot(wt.\%)^{-1}$)	m	-3.36	[21]
Equilibrium partition coefficient	k	0.17	[21]
Specific heat ($J\cdot kg^{-1}\cdot K^{-1}$)	c_p	1086	[22]
Gibbs - Thompson coefficient ($m\cdot K^{-1}$)	Γ	2.4×10^{-7}	[23]
Solute diffusion coefficient in liquid ($m^2\cdot s^{-1}$)	D_l	3.4×10^{-9}	[24]
Solute diffusion coefficient in solid ($m^2\cdot s^{-1}$)	D_s	3×10^{-13}	[24]
Thermal expansion coefficient (K^{-1})	β_T	1×10^{-4}	[25]
Solutal expansion coefficient ($(wt.\%)^{-1}$)	β_c	-0.92	[25]
Latent heat ($J\cdot kg^{-1}$)	Δh_f	389320	[22]
Physical properties			
Dynamic viscosity ($kg\cdot m^{-1}\cdot s^{-1}$)	μ^f	0.0013	[22]
Density ($kg\cdot m^{-3}$)	ρ_l, ρ_c	2460	[22]
Thermal conductivity of liquid ($W\cdot m^{-1}\cdot K^{-1}$)	k_l	95	[22]
Thermal conductivity of solid ($W\cdot m^{-1}\cdot K^{-1}$)	k_s	180	[22]
Mechanical properties			
Elastic Modulus at undercooled temperature (GPa)	E	0.13	*Cal.
Poisson's ratio	v	0.33	[26]

* Calculated by JMatPro

4. Results and discussion

Figure 3 shows the flow field, von Mises stress and total displacements evolution with elastic models as the dendrite grows. The von Mises stress and deformation displacements increase with the enhancements of flow velocities. There are stress concentrations along the longer primary dendrite trunks. As the dendrite grows the locations of stress concentration propagate from the centre of the primary dendrite (roots of primary dendrite) to the sides (the regions of secondary dendrites intersecting with primary dendrite trunks). At 0.052 s, the dendrite undergoes a visible deformation and the maximum displacements is 22 μm. The stress curve along the path line near the surface of longer primary dendrite trunks is shown in figure 4. It indicates that there are many peaks in the curve (stress concentrations). As time varies, the stress concentration disappears in some locations and enhances in most locations, although the stresses are continued to increase.

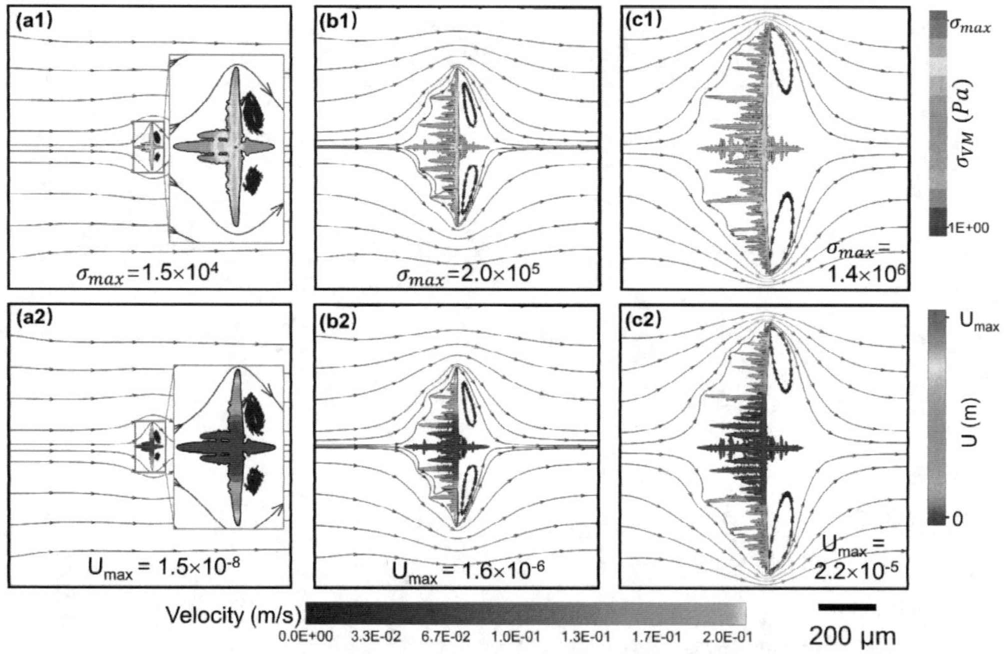

Figure 3. Flow field, von Mises stress (σ_{VM}) and displacements (U) evolution with elastic models as the dendrite grows under the inlet velocity of 0.02 m/s at (a) 0.010 s, (b) 0.034 s, (c) 0.052 s.

Figure 5 is the maximum von Mises stress (Max-MS) at each moment during the dendrite grows, which is the σ_{max} variation in Figure 3 at each moment. It can be seen that the Max-MS increases with the time and exceed the yield stress (*E/1000*) at 0.026 s. The Max-MS varies from ~10^3 Pa to ~10^6 Pa. Figure 6 is the critical maximum flow velocity (Cr-Ve) at the solid/liquid interface when the Max-MS exceed *E/1000* at different inlet velocities. The Cr-Ve increases with the inlet velocities ranging from 0.023 m/s to 0.126 m/s due to the deformation of the dendrites is hindered by the interconnected secondary dendritic arms. It means that the lowest velocities corresponding to dendrite yielding may be lower than 0.023 m/s when the inlet velocity is lower than 0.01 m/s, which is associated with the specific flow field and dendrite morphology. It is a further indication that the importance of flow-induced dendrite deformation should not be ignored at a wide range of melt flow velocities.

However, the present work is only focused on a two dimensional approach. The dendrite growth in the experiments is in three dimensional and there are fluid flow and more secondary dendrites in the third direction. Although there are the CA methods in three dimensional approach [27], it is a challenge to couple dendrite growth, fluid flow and dendrite deformation in three dimensional due to the complex shapes of dendrites are difficult to reconstruct in FEM. Nevertheless, the current simulation results are favorable for flow-induced dendrite deformation during the solidification of alloys in thinner samples due to the weak fluid flow in the third direction.

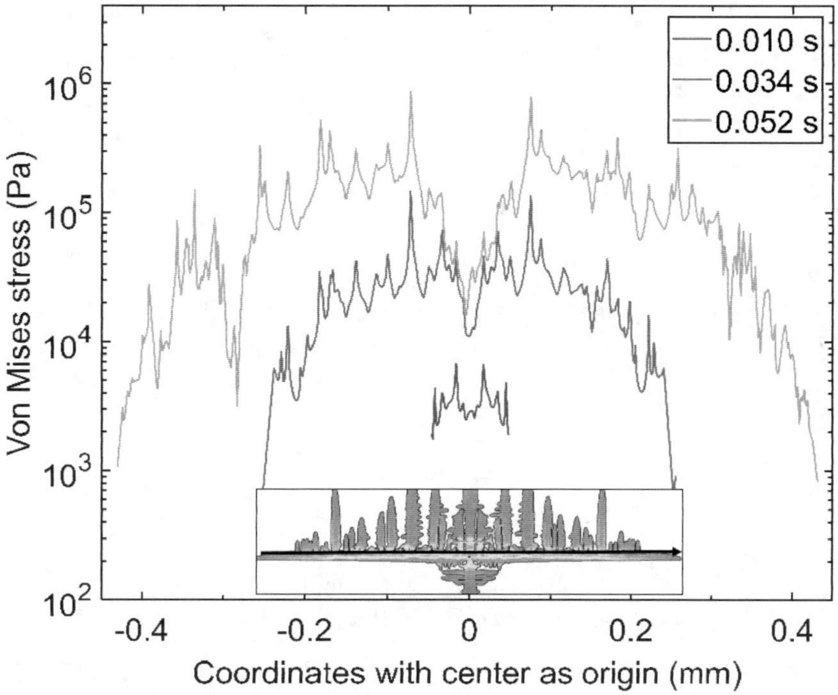

Figure 4. Von Mises stress distribution along the path line in the inset as the dendrite grows under the inlet velocity of 0.02 m/s.

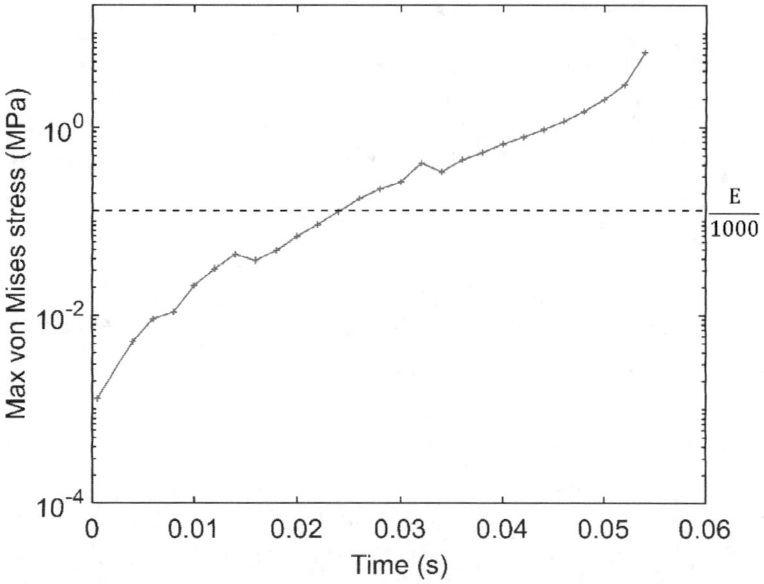

Figure 5. Maximum von Mises stress variation with dendrite growth using elastic models at the inlet velocity of 0.02 m/s.

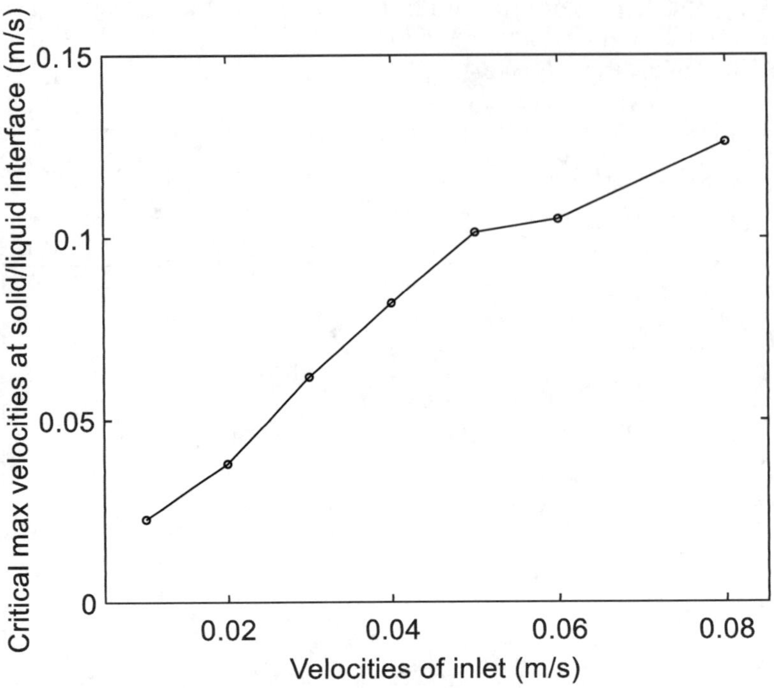

Figure 6. Critical maximum liquid velocity at the solid/liquid interface at the moment corresponding to exceeding the yield strength (*E/1000* in figure 5) at different inlet velocities.

5. Summary

A combination of CA-FVM and displacement-based FEM model was developed to simulate the dendrite growth and flow-induced mechanical deformation of the dendrite under a parallel inlet fluid flow. The results show that the von Mises stress and deformation displacements increase as the dendrite grows, and the dendrite can undergo a visible deformation under fluid flow. There are many stress concentrations on the primary dendrite trunks. The locations of stress concentrations can be enhanced at most locations and eliminated at a few locations. The critical flow velocities corresponding to dendrite yield ranges from 0.026 m/s at the inlet velocity of 0.01 m/s to 0.126 m/s at the inlet velocity of 0.08 m/s.

Acknowledgements

This work was sponsored by National Natural Science Foundation of China (No. 52074182 and No. 91860121) and Natural Science Foundation of Shanghai (Grant No. 22ZR1430700).

References

[1] C.J. Todaro, M.A. Easton, D. Qiu, D. Zhang, M.J. Bermingham, E.W. Lui, M. Brandt, D.H. StJohn, M. Qian, *Nat. Commun.* 11(1) (2020) 142.

[2] J. Li, X.-w. Xu, N. Ren, M.-x. Xia, J.-g. Li, *J. Iron. Steel Res. Int.* 29(12) (2022) 1901-1914.

[3] S.-k. Yin, S. Luo, W.-j. Zhang, W.-l. Wang, M.-y. Zhu, *J. Iron. Steel Res. Int.* 28(4) (2020) 424-436.

[4] E. Liotti, A. Lui, R. Vincent, et al., *Acta Mater.* 70 (2014) 228-239.

[5] Z. Chen, H.-f. Shen, *Int. J. Miner. Metall. Mater.* 27(2) (2020) 200-209.

[6] H. Ge, F. Ren, J. Li, X. Han, M. Xia, J. Li, *Metall. Mater. Trans. A* 48(3) (2017) 1139-1150.

[7] E. Liotti, A. Lui, S. Kumar, Z. Guo, C. Bi, T. Connolley, P.S. Grant, *Acta Mater.* 121 (2016) 384-395.
[8] D. Räbiger, Y. Zhang, V. Galindo, S. Franke, B. Willers, S. Eckert, *Acta Mater.* 79 (2014) 327-338.
[9] J.C. Jie, S.P. Yue, J. Liu, D.H. StJohn, Y.B. Zhang, E.Y. Guo, T.M. Wang, T.J. Li, *Acta Mater.* 208 (2021).
[10] N. Balasubramani, G. Wang, D.H. StJohn, M.S. Dargusch, *J. Mater. Sci. Technol.* 65 (2021) 38-53.
[11] N. Ren, C. Panwisawas, J. Li, M. Xia, H. Dong, J. Li, *Acta Mater.* 215 (2021).
[12] M. Rettenmayr, *Int. Mater. Rev.* 54(1) (2013) 1-17.
[13] L.K. Aagesen, A.E. Johnson, J.L. Fife, P.W. Voorhees, M.J. Miksis, S.O. Poulsen, E.M. Lauridsen, F. Marone, M. Stampanoni, *Nat. Phys.* 6(10) (2010) 796-800.
[14] L.K. Aagesen, A.E. Johnson, J.L. Fife, P.W. Voorhees, M.J. Miksis, S.O. Poulsen, E.M. Lauridsen, F. Marone, M. Stampanoni, *Acta Mater.* 59(12) (2011) 4922-4932.
[15] H. Neumann-Heyme, K. Eckert, C. Beckermann, *Phys. Rev. E* 92(6) (2015) 060401.
[16] Z. Zhang, C. Wang, B. Koe, C.M. Schlepütz, S. Irvine, J. Mi, *Acta Mater.* 209 (2021).
[17] F. Wang, D. Eskin, J. Mi, C. Wang, B. Koe, A. King, C. Reinhard, T. Connolley, *Acta Mater.* 141 (2017) 142-153.
[18] J. Pilling, A. Hellawell, *Metall. Mater. Trans. A* 27(1) (1996) 229-232.
[19] N. Ren, J. Li, N. Bogdan, M. Xia, J. Li, *Comput. Mater. Sci* 180 (2020).
[20] K. Sugiyama, S. Ii, S. Takeuchi, S. Takagi, Y. Matsumoto, *J. Comput. Phys.* 230(3) (2011) 596-627.
[21] N. Ponweiser, C.L. Lengauer, K.W. Richter, *Intermetallics* 19(11) (2011) 1737-1746.
[22] D.G. Eskin, Q. Du, L. Katgerman, *Metall. Mater. Trans. A* 39(5) (2008) 1206-1212.
[23] W. Lee, Y. Jeong, J.-W. Lee, H. Lee, S.-h. Kang, Y.-M. Kim, J. Yoon, *J. Mater. Sci. Technol.* 49 (2020) 15-24.
[24] H.B. Dong, P.D. Lee, *Acta Mater.* 53(3) (2005) 659-668.
[25] M. Ahmadein, M. Wu, A. Ludwig, *J. Cryst. Growth* 417 (2015) 65-74.
[26] D. Fuloria, P.D. Lee, D. Bernard, *Mater. Sci. Eng. A* 494(1-2) (2008) 3-9.
[27] C.A. Gandin, M. Rappaz, *Acta Mater.* 45(5) (1997) 2187-2195.

IOP Conf. Series: Materials Science and Engineering 1281 (2023) 012039 doi:10.1088/1757-899X/1281/1/012039

Meshless simulation of a macrosegregation benchmark considering the solid motion

Viktor Govže[1], **Igor Vušanović**[2] **and Božidar Šarler**[1,3,*]

[1] Institute of Metals and Technology, Ljubljana, Slovenia
[2] Faculty of Mechanical Engineering, University of Montenegro, Podgorica, Montenegro
[3] Faculty of Mechanical Engineering, University of Ljubljana, Ljubljana, Slovenia

E-mail: viktor.govze@imt.si, igorvus@ucg.ac.me, *bozidar.sarler@fs.uni-lj.si

Abstract. We have extended the existing two-dimensional rigid solid phase benchmark for binary substance with the solid phase motion in the present paper. Incompressible laminar Newtonian flow is assumed, and a standard mixture formulation is used for the mass, momentum, energy, and solute transport. A coherency solid motion model accounts for the free-floating grains, assuming that the solid velocity is proportional to the mixture velocity and the liquid fraction. The lever rule is used to describe the mass fractions of the phases. A two-dimensional benchmark is solved using the semi-implicit meshless diffuse approximate method with an adaptive subdomain upwinding strategy. The results of the meshless method are compared to the finite volume method results with a reasonable agreement. The new benchmark results show that the solid motion has an essential effect on the macrosegregation pattern.

1. Introduction

The incorporation of the movement of free-floating grains into solidification simulation results in improved fidelity of the results. For example, it was observed that the solid motion might cause inverse segregation in the billet centre during the direct chill casting of aluminium alloys [1]. Several benchmark cases were proposed in [2] to test and verify the implementations of the solidification models. In the present paper, we aim to extend one of the proposed benchmarks with a model considering the solid motion on a reduced scale, like in [3]. So far, different models for treating free-floating grains were developed, which differ in complexity. For example, in [4], a two-region approach has been developed, which divides the solidification region into the slurry and the rigid region. In the slurry region, the solid is allowed to move with the convective currents, and in the rigid region, it is assumed to behave as a rigid structure, with a critical solid fraction being the slurry-mushy partitioning parameter. More recently, an attempt to reduce the complexity of such models was presented [5]. This enables to include the solid motion in the existing mixture continuum equations developed in the eighties [6]. In the present benchmark, a simple non-parametric coherency solid motion model is used, which assumes that the solid velocity is proportional to the mixture velocity through the liquid fraction. Previously, the benchmark with the solid motion model, based on the rigid solid assumption, was successfully solved with different strong-form meshless methods [3, 7]. In the present extended benchmark, the diffuse approximate method (DAM) is used to obtain the numerical solution, and the results are compared to the finite volume method (FVM) with reasonable agreement.

IOP Conf. Series: Materials Science and Engineering 1281 (2023) 012039 doi:10.1088/1757-899X/1281/1/012039

2. Problem definition

The physical model is derived based on the mixture continuum assumption for a multi-component substance elaborated in [6]. The coupled mass, energy, species, and momentum conservation equations are formulated for an incompressible Newtonian fluid. The solidification process of a binary alloy is considered, and the lever rule is used for the mass fractions of the phases. A linearized binary phase diagram is used to describe the local equilibrium conditions.

The governing equations describing the macroscopic model are given as

$$\nabla \cdot \mathbf{v} = 0, \tag{1}$$

$$\rho \frac{\partial h}{\partial t} + \rho (\mathbf{v} \cdot \nabla) h = k \nabla^2 T - \rho \nabla \cdot [(h_l - h)(\mathbf{v} - \mathbf{v_s})], \tag{2}$$

$$\frac{\partial C}{\partial t} + (\mathbf{v} \cdot \nabla) C = -\nabla \cdot [(C_l - C)(\mathbf{v} - \mathbf{v_s})], \tag{3}$$

$$\rho \frac{\partial \mathbf{v}}{\partial t} + \rho (\mathbf{v} \cdot \nabla) \mathbf{v} = -\nabla p + \mu \nabla^2 \mathbf{v} + \mathbf{b}_D + \mathbf{b}_B, \tag{4}$$

where ρ, $\mathbf{v}$, h, k and C are the mixture density, velocity, enthalpy, thermal conductivity, and species concentration, μ is the liquid viscosity, T is the temperature, and p is the pressure. The individual phase variables are denoted with indexes l for liquid and s for solid, therefore h_l is the liquid enthalpy, C_l is the liquid species concentration, and $\mathbf{v}_s$ is the solid velocity. The additional body forces $\mathbf{b}_D$ and $\mathbf{b}_B$ exerted on the mixture are the hydraulic drag force in the solidifying region modelled with Darcy law and the buoyancy force modelled with the Boussinesq approximation. Their definitions are

$$\mathbf{b}_D = -\frac{\mu}{K} (\mathbf{v} - \mathbf{v}_s), \tag{5}$$

$$\mathbf{b}_B = \rho_{ref} [1 - \beta_T (T - T_{ref}) - \beta_C (C_l - C_{ref})] \mathbf{g}, \tag{6}$$

where K is the isotropic hydraulic permeability, ρ_{ref}, T_{ref}, and C_{ref} represent the reference density, temperature and species concentration, β_T and β_C are the thermal and species volumetric expansion coefficient, and $\mathbf{g}$ is the gravitational acceleration. The hydraulic permeability is modelled with the Carman-Kozeny relation, defined as

$$K = K_0 \frac{g_l^3}{(1 - g_l)^2}, \tag{7}$$

where K_0 is the Kozeny constant and g is the volumetric fraction. The mixture quantities are defined as

$$\mathbf{v} = g_l \mathbf{v}_l + g_s \mathbf{v}_s, \tag{8}$$

$$h = c_p T + g_l L, \tag{9}$$

$$C = [g_l + (1 - g_l) k_p] C_l, \tag{10}$$

where c_p is the mass specific heat at a constant pressure of the liquid and the solid phase, L is the latent heat, and k_p is the partition coefficient. The equilibrium liquidus temperature T_{liq} is defined as

$$T_{liq} = T_f + m C_l, \tag{11}$$

where T_f is the pure substance temperature of fusion and m is the slope of the liquidus line.

In the present paper, we compare the results of the rigid and the coherency solid motion models. The former is defined as

$$\mathbf{v}_s = \mathbf{v}_{sys}, \tag{12}$$

where $\mathbf{v}_{sys}$ is the velocity of the system, and the latter is defined as

$$\mathbf{v}_s = g_l \mathbf{v} + (1 - g_l)\mathbf{v}_{sys}, \tag{13}$$

where the liquid and solid fractions measure the intensity of the solid phase movement. In the present article, $\mathbf{v}_{sys} = \mathbf{0}\ \mathrm{m\,s^{-1}}$, since the system is stationary.

The problem geometry is shown in figure 1 together with the mathematical description of the boundary conditions, initial conditions, and the reference points with cross-sections used in the analysis of the results in tables 1-3.

A two-dimensional cartesian coordinate frame with basis vectors $\mathbf{i}_x$ and $\mathbf{i}_y$ and the position vector $\mathbf{p} = x\mathbf{i}_x + y\mathbf{i}_y$ is used to describe the problem. Points A, B, C and D define the corner points of the rectangular cavity and segments AB, BC, CD, and AD represent the boundaries of the cavity. The boundary is denoted with Γ and the interior is denoted with Ω.

The problem is symmetric about the left boundary. The cavity is impermeable for species transfer on the top, right, and bottom and thermally insulated at the top and bottom. Heat is extracted from the right boundary through forced convection with a heat transfer coefficient $\tilde{h}$ to the surroundings with temperature T_{env}. A sticking boundary condition is used on all non-symmetry boundaries. Initially, the melt is at rest with the temperature and the species concentration uniformly distributed and set to T_0 and C_0. The numerical values of the parameters used in the simulation are given in table 4.

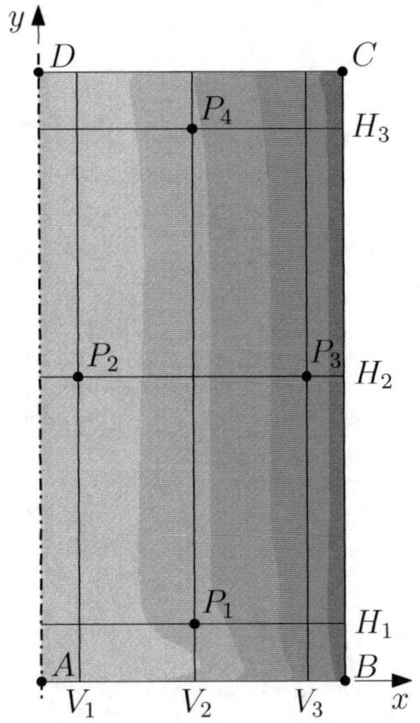

Figure 1: The solidifying domain.

Table 1: The reference points.

	A	B	C	D	P_1	P_2	P_3	P_4
x [mm]	0	10	10	0	5	1	9	5
y [mm]	0	0	20	20	2	10	10	18

Table 2: The cross sections.

	H_1	H_2	H_3	V_1	V_2	V_3
x [mm]	x	x	x	P_2	P_1	P_3
y [mm]	P_1	P_2	P_4	y	y	y

Table 3: The boundary and initial conditions.

$\mathbf{p} \in \Gamma_{AB}, t > 0$ s
$C = C_0,\ \mathbf{v} = \mathbf{v}_{sys},\ \frac{dT}{d\mathbf{n}} = 0\ \mathrm{K\,m^{-1}}$

$\mathbf{p} \in \Gamma_{BC}, t > 0$ s
$C = C_0,\ \mathbf{v} = \mathbf{v}_{sys},\ \frac{dT}{d\mathbf{n}} = \tilde{h}\,(T - T_{env})$

$\mathbf{p} \in \Gamma_{CD}, t > 0$ s
$C = C_0,\ \mathbf{v} = \mathbf{v}_{sys},\ \frac{dT}{d\mathbf{n}} = 0\ \mathrm{K\,m^{-1}}$

$\mathbf{p} \in \Gamma_{DA}, t > 0$ s
$\frac{dC}{d\mathbf{n}} = 0\ \mathrm{m^{-1}},\ v_x = 0\ \mathrm{m\,s^{-1}},\ \frac{dv_y}{d\mathbf{n}} = 0\ \mathrm{s^{-1}},\ \frac{dT}{d\mathbf{n}} = 0\ \mathrm{K\,m^{-1}}$

$\mathbf{p} \in \Omega, t = 0$ s
$C = C_0,\ \mathbf{v} = \mathbf{v}_{sys},\ T = T_0$

IOP Conf. Series: Materials Science and Engineering 1281 (2023) 012039 doi:10.1088/1757-899X/1281/1/012039

3. Solution procedure

The solution to the problem stated is calculated by using DAM, which belongs to local strong-form meshless methods. This method uses weighted least squares to determine a locally smooth solution and has been used to solve several solidification problems before [8, 7, 9, 10]. The shape functions Ψ_k used in this work are Gaussian function weighted monomials defined as

$$\Psi_k(\mathbf{r}) = e^{-W\left(\frac{\|\mathbf{r}\|}{D}\right)^2} \psi_k\left(\frac{\mathbf{r}}{D}\right), \qquad \mathbf{r} = \mathbf{p} - \mathbf{p}_0, \quad (14)$$

where W is the Gaussian weighting function sharpness parameter, $\|\cdot\|$ represents the Euclidian norm, $\mathbf{r}$ represents the relative position vector, D is a normalization constant, ψ_k is the k-th monomial from the set of monomials $\{1, r_x, r_y, r_x^2, r_y^2, r_x r_y\}$, and $\mathbf{p}_0$ is the position of the node of interest.

The governing equations are propagated through time by using a first-order accurate scheme. The energy (2) and species (3) equation are solved in a fully explicit manner, whilst the momentum equation is solved in a semi-implicit manner.

Table 4: Thermophysical parameters.

ρ	2450	$\mathrm{kg\,m^{-3}}$
c_p	1000	$\mathrm{J\,kg^{-1}\,K^{-1}}$
k	192	$\mathrm{W\,m^{-1}\,K^{-1}}$
L	4×10^5	$\mathrm{J\,kg^{-1}}$
μ	1.2×10^{-3}	$\mathrm{Pa\,s}$
β_T	1.3×10^{-4}	$\mathrm{K^{-1}}$
β_C	-7.3×10^{-3}	$\mathrm{wt\%^{-1}}$
ρ_{ref}	2450	$\mathrm{kg\,m^{-3}}$
T_{ref}	465	°C
C_{ref}	4.5	wt%
$\mathbf{g}$	$-9.8\mathbf{i}_y$	$\mathrm{m\,s^{-2}}$
K_0	5.56×10^{-11}	$\mathrm{m^2}$
T_f	660	°C
m	-3.43	$\mathrm{°C\,wt\%^{-1}}$
k_p	0.173	/
$\tilde{h}$	500	$\mathrm{W\,m^{-2}\,K^{-1}}$
T_{env}	20	°C
T_0	700	°C
C_0	4.5	wt%
$\mathbf{v}_{sys}$	$\mathbf{0}$	$\mathrm{m\,s^{-1}}$

During solidification, the Darcy term (5) in the mixture momentum equation poses a serious numerical stability issue if a fully explicit approach is used. For this reason, the part of the Darcy term associated with the mixture velocity is treated implicitly. To avoid singularity when the liquid fraction approaches 0, the permeability is calculated up to a small liquid fraction ϵ and $K(g_l = \epsilon)$ is used at $g_l < \epsilon$. In this way, no error is introduced in the permeability calculation when $g_l \geq \epsilon$. In the present paper, $\epsilon = 1 \times 10^{-2}$, where the Darcy drag force is practically infinite and the mixture is stationary.

The momentum and continuity equations are directly coupled to satisfy the discrete incompressibility constraint exactly. To prevent numerical instabilities resulting from the advective terms in the governing equations, an upwinding stabilization method is used similar to the one in [11].

The support subdomains include $N_{loc} = 9$ nearest nodes and the chosen Gaussian weighting parameter is $W = 5$. The normalization constant is defined as

$$D = \max\left(\|\mathbf{r}_j\|\right), \qquad j = 1, \ldots, N_{loc}, \quad (15)$$

where j represents the local node index. The time step is $\Delta t = 4 \times 10^{-5}$ s. To test the convergence of the solution, grids of 20×40, 25×50, 40×80, 50×100, and 80×160 nodes were used with the same time step and numerical parameters. The meshless simulation is performed on an Intel Xeon Gold 6146 CPU and is parallelized to use 6 cores. The wall time of the simulation run for the 80×160 set of nodes was 34 h.

The reference results are calculated using FVM. The momentum equation is solved in a fully implicit manner and the pressure-velocity coupling is made with the SIMPLE algorithm as proposed by Patankar (1980). The energy and species concentration equations are solved in a fully explicit manner. The advection terms are treated by a first-order upwinding stabilization method. The results shown are calculated on a regular 80×160 mesh with the time step $\Delta t = 8 \times 10^{-6}$ s.

4. Results

As can be seen in the comparison of the final macrosegregation in figure 2, very different results are obtained just by using a different solid velocity constitutive relation. Since the species transport is governed exclusively by advection in both models, whatever influences the mixture velocity will also influence the final macrosegregation. Compared to the rigid model, the species concentration is more evenly distributed throughout the cavity when using the coherency model. Even though its peak is still at the bottom, it is significantly lower than in the rigid model. The convergence of the solutions is monitored in terms of the species concentration field standard deviation and the minimum and maximum absolute value of segregation $S_{\min\,|\,\max}$ defined as

$$\operatorname{std}(C) = \sqrt{\frac{1}{N-1}\sum_{n=1}^{N}\left(C\left(\mathbf{p}_n\right) - C_0\right)}, \quad S_{\min\,|\,\max} = \min|\max\left(\left|C\left(\mathbf{p}_n\right) - C_0\right|\right), \quad n = 1, \ldots, N, \tag{16}$$

where N is the number of nodes used in the simulation. The convergence is measured against the relative grid spacing. For both the rigid and the coherency model, the convergence of the chosen parameters is shown in figure 3.

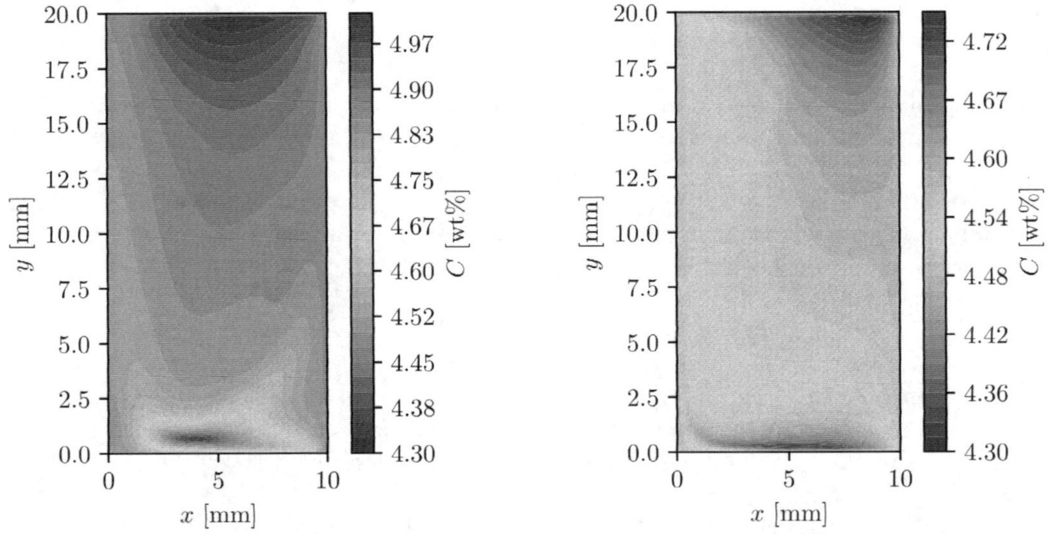

Figure 2: Macrosegregation comparison of the DAM solutions of the (a) rigid and the (b) coherency model.

Even though a relatively fine regular grid of nodes is used, the solution is still underresolved in the vicinity of the boundaries for the coherency model both in the FVM and DAM solutions of the problem. Compared to the rigid model's solution, the coherency model's solution exhibits sharper species concentration peaks, which require a dense node arrangement to be properly resolved. Even more nodes than used would be required, especially near the boundaries, which is impractical to do with a regular grid node arrangement. A node set biased towards the boundaries of the cavity would be better suited. The results converge slower than the rigid model results but are nevertheless slowly convergent concerning std(C). A parametric study with denser node arrangements would have to be used to determine more accurate results.

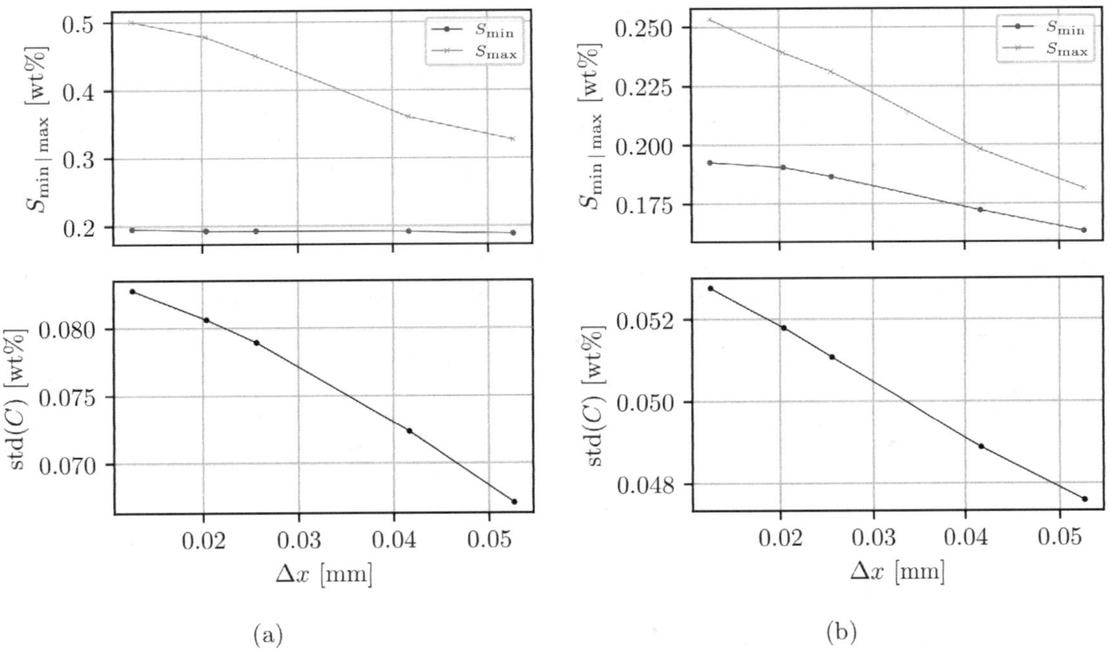

Figure 3: DAM solutions convergence for (a) rigid and (b) coherency model.

Practically speaking, the macrosegregation on the 80 × 160 node arrangement yields a very similar solution as the 40 × 80, and the trend is unlikely to change with denser nodes.

To analyze the results more thoroughly, horizontal and vertical cross-section results are provided in figure 4. The DAM solution of the rigid model was verified using the results presented in [3] and the comparison is for clarity not shown in the figure. A qualitative agreement between the results of the FVM and DAM is obtained. One can attribute the quantitative disagreement to the first-order upwind method in FVM, which is known to introduce numerical diffusion. In the meshless solution procedure, the upwinding method is analogous to the second-order finite difference upwinding method, and therefore, less numerical diffusion is introduced as in FVM solution.

From the time evolution of the magnitude of the mixture velocity for each of the reference points in figure 5a the mixture velocity in the coherency model is equal to or larger than its rigid counterpart. That is because the moving solid incorporated through the coherency model imposes less drag on the mixture flow in the solidification region than the rigid model. The coherency model progressively builds up the drag by a factor of $1 - g_l$ since $\mathbf{v}_{sys} = \mathbf{0}\ \mathrm{m\,s^{-1}}$. As a result, the flow is faster in the solidifying region allowing for more solute advection. Consequently, this leads to more evenly distributed macrosegregation, as seen in figure 2b.

Both models are similar in terms of the solidification start and end times, which are at 5 s and 45 s, respectively. As expected, both models behave the same in the fully liquid region but quite differently during the solidification. The time it takes for the macrosegregation to appear like the final macrosegregation is very different. In the rigid model, it takes 4 s after the solidification start, whilst in the coherency model, it takes 12 s. This can be attributed to the flow being less constrained in the slurry, allowing the solute to be more freely distributed, resulting in changing species concentration field. This effect can be more clearly seen in figure 5b, where a small peak in species concentration is reached in P_2 before decreasing and increasing again. The temperature distribution remains approximately the same, since the maximum local Péclet

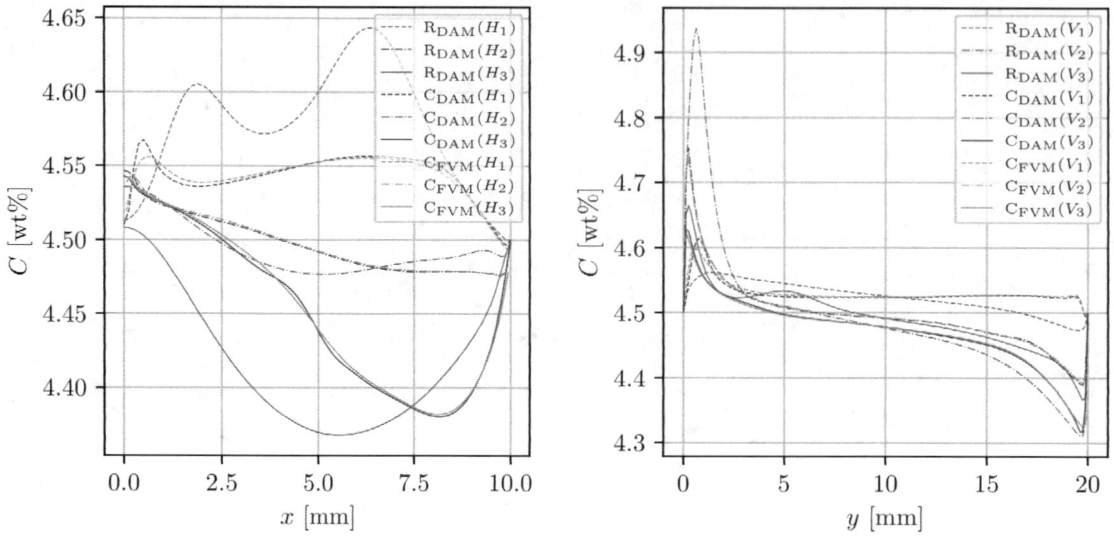

(a) (b)

Figure 4: A comparison of the solution of the rigid (denoted as R) and the coherency (denoted as C) models with DAM and FVM. Horizontal (a) and vertical (b) cross section.

number is on the order of magnitude of 2×10^{-1} and the diffusive heat transport dominates in the energy transport equation.

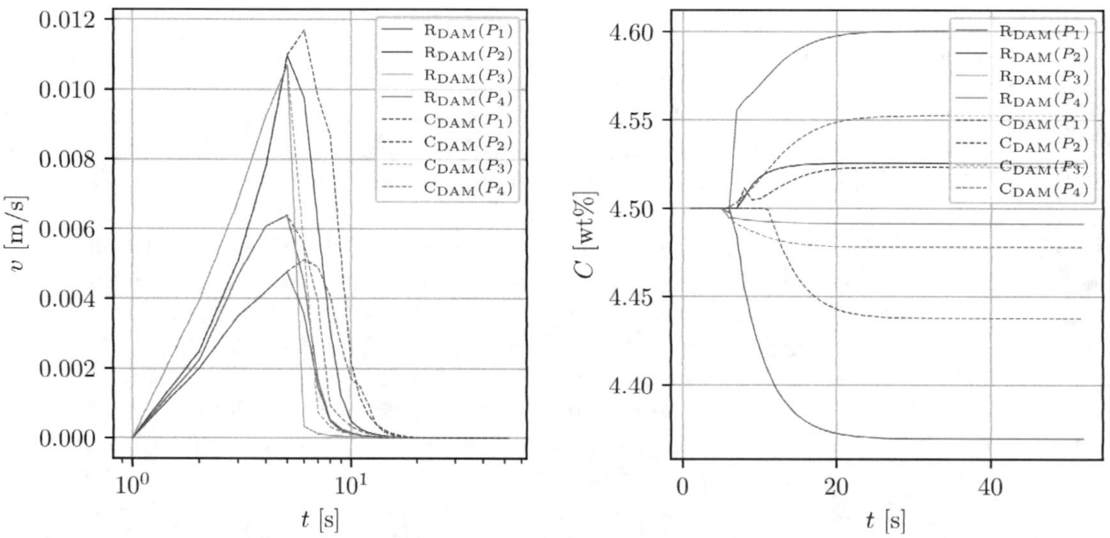

(a) (b)

Figure 5: Concentration and velocity magnitude at the reference points for the rigid (denoted as R) and the coherency (denoted as C) models of a meshless DAM solution.

IOP Conf. Series: Materials Science and Engineering 1281 (2023) 012039 doi:10.1088/1757-899X/1281/1/012039

5. Conclusion

In the present paper, the macrosegregation benchmark [3] has been extended to consider the solid motion through a simple coherency model, and the results were compared to the original benchmark case. Different numerical approaches, namely the DAM and FVM have been used to implement the physical model, and a qualitative agreement was reached between the two methods. The elaboration of the meshless solution procedure will be presented in a follow-up publication. Convergence of the results of the meshless method is shown where it can be seen that DAM can be used to get accurate predictions of the macrosegregation. The effect of the solid motion model is demonstrated, producing significantly different results than the rigid model case.

Acknowledgments

This work was supported by the Slovenian Research Agency (grant numbers J2-4477, P2-0162) and bilateral Slovenian-Montenegrin project (grant number BI-ME/23-24-009).

References

[1] Nadella R, Eskin D G, Du Q and Katgerman L **53** 421–480 ISSN 0079-6425 URL https://www.sciencedirect.com/science/article/pii/S007964250700059X

[2] Bellet M, Combeau H, Fautrelle Y, Gobin D, Rady M, Arquis E, Budenkova O, Dussoubs B, Duterrail Y, Kumar A, Gandin C, Goyeau B, Mosbah S and Založnik M **48** 2013–2016 ISSN 12900729 URL https://linkinghub.elsevier.com/retrieve/pii/S129007290900177X

[3] Kosec G, Založnik M, Šarler B and Combeau H **22**(2) 169–196 ISSN 1546-2218, 1546-2218 URL https://www.techscience.com/cmc/v22n2/22619

[4] Ni J and Incropera F P **38** 1285–1296 ISSN 0017-9310 URL https://www.sciencedirect.com/science/article/pii/001793109400237P

[5] Vušanović I and Voller V **169** 120923 ISSN 00179310 URL https://linkinghub.elsevier.com/retrieve/pii/S0017931021000260

[6] Bennon W D and Incropera F P **30** 2161–2170 ISSN 0017-9310 URL https://www.sciencedirect.com/science/article/pii/0017931087900949

[7] Hatić V, Mavrič B and Šarler B **28** 361–380 ISSN 0961-5539 URL https://www.emerald.com/insight/content/doi/10.1108/HFF-04-2017-0143/full/html

[8] Hatić V, Mavrič B, Košnik N and Šarler B **54** 170–188 ISSN 0307904X URL https://linkinghub.elsevier.com/retrieve/pii/S0307904X17305863

[9] Hatić V, Cisternas Fernández M, Mavrič B, Založnik M, Combeau H and Šarler B **142** 121–133 ISSN 12900729 URL https://linkinghub.elsevier.com/retrieve/pii/S1290072918319197

[10] Hatič V, Mavrič B and Šarler B **113** 191–203 ISSN 09557997 URL https://linkinghub.elsevier.com/retrieve/pii/S0955799719306678

[11] Chan Y, Shen L, Wu C and Young D **89** 157–166 ISSN 00457930 URL https://linkinghub.elsevier.com/retrieve/pii/S0045793013004155

IOP Conf. Series: Materials Science and Engineering 1281 (2023) 012040 doi:10.1088/1757-899X/1281/1/012040

A mathematical examination of the impact of mould transparency to infrared radiation on solidification during the investment casting process

C A Jones[1], A E W Jarfors[2], E Rogers[3], P Vrethed[4], P Silva[4], M R Jolly[1]

[1] Sustainable Manufacturing System Centre, School of Aerospace, Transport and Manufacturing, Cranfield University, Bedfordshire, MK43 0AL, UK
[2] Jönköping University, School of Engineering, Box 1026, 551 11 Jönköping, Sweden
[3] Surface Engineering and Precision Centre, School of Aerospace, Transport and Manufacturing, Cranfield University, Bedfordshire, MK43 0AL
[4]TPC Components AB, Brånstaleden 2, 734 32 Hallstahammer, Sweden

Email: c.a.jones@cranfield.ac.uk

Abstract. Investment casting is a highly dynamic process during which multiple competing physical phenomena are at work. Those seeking to understand and simulate such processes computationally are confronted with a considerable task, balancing accuracy with efficiency. Approximations and models based on well-understood and documented fundamental physics are powerful tools in a modeller's arsenal. Driven by observed discrepancies between experimental thermocouple measurements and simulation predictions of casting temperatures, this work explores the additional alloy cooling mechanism of mould transparency to infrared radiation, targeting a new mathematical approximation applicable in such situations. Direct attenuation, scattering from coarse sand, sand distribution in the mould and material temperatures play a role in the extent of radiation transparency that must be considered. From this model, estimation of the additional cooling rate resulting from expected mould transparency can be determined and applied as a corrective measure to computation fluid dynamics (CFD) simulation results that do not capture this phenomenon.

1. Introduction

Managing heat transfer in any casting process is crucial in controlling the generation of many casting defects such as misruns, undesirable microstructures and residual stresses [1]. Such defects are particularly prolific when producing aerospace turbine blades via investment casting where the component's large area, thin cross-sectional regions are a principal source of such issues. Management of these thermal flows is generally controlled by applying insulating wool fibre to the mould which prevents excessive heat loss, alternatively through increasing mould porosity or adjusting mould composition to improve the overall insulating capability of the mould. These solutions are admirable at controlling heat transfer by thermal conductivity, and in general for radiation heat transfer but rely on radiation interacting with the mould. Previous research conducted into understanding and validating the thermophysical properties of an investment casting mould [2] succeeded, in addition to those aims, in identifying an alloy cooling in experimentation greater than that predicted using CFD simulations. This suggests that a quantity of thermal energy escaping from the casting, in reality, has been omitted from heat transfer consideration within the simulation calculations. It is hypothesised that the relatively high porosity, Silica based investment casting mould used in the manufacture of Nickel-based superalloys could result in a mould system with non-negligible transparency to infrared radiation leading to an additional, uncontrolled cooling of the alloy. It is this hypothesis that this work seeks to investigate through discussion of the interaction between electromagnetic (EM) radiation and matter in comparison to experimental transmission measurements conducted on the mould material.

2. Theoretical Model

EM radiation passing through solid matter has a common mathematical root defined by Maxwell's equations and applies irrespective of the solid material properties or composition (although such details are required for a solution). As such, this represents an admirable starting point for constructing a generic, mathematical description of radiation semi-transparency. It is common knowledge that EM radiation consists of a combination of orthogonal electric and magnetic fields, but the magnetic component is generally negligible compared to the electric field element during interactions with matter [3]. Equation 1 shows the expanded form of such an electric field of an arbitrary EM wave based on Maxwell's equations that describes the electric field strength at a position, x, within a material at a time, t. This expansion is based on a substitution for the velocity, v, in terms of the complex refractive index, $1/_v = n/_c - ik/_c$. Here n is the real portion and k is the imaginary part of the complex refractive index with c representing the velocity of light.

$$E = E_0 e^{i2\pi f[t-(x/_v)]} \rightarrow E = E_0 e^{(i2\pi ft)} e^{-(i2\pi xn/_c)} e^{-(2\pi fkx/_c)}$$

Equation 1. Mathematical wave equation for the electric field component of a generic electromagnetic wave propagating through a solid.

Developing this theory, let us consider the associated power of an EM wave which can be defined by the square of the electric field strength multiplied by the electrical conductivity of the solid, σ_e. Taking the ratio between the power at a generic position, x, in the material and the initial power, as shown in Equation 2, reduces the expression considerably to a single exponential governing the damping of radiation passing through a material. The exponent of this exponential can be expressed in terms of either frequency, f, or wavelength, λ_w, and are related through the velocity of light, $f = c/_{\lambda_w}$.

$$\frac{P(x)}{P(0)} = \frac{\sigma_e E(x)^2}{\sigma_e E(0)^2} = e^{-(4\pi fkx/_c)} = e^{-(4\pi kx/_{\lambda_w})} = e^{-\alpha x}$$

Equation 2. The ratio of the EM wave power at position x to the initial EM wave power at position zero

The salient result of these equations is the definition of the absorption coefficient, α, from this remaining exponential which introduces the material specifics into the problem through the extinction coefficient, k (imaginary part of the complex refractive index). This wavelength-dependent parameter governs the attenuation per metre as EM radiation passes through a material. The wavelength-dependent makes sense as it would not be expected that materials would respond similarly when exposed to different radiation sources. Absorption in different atomic structures and atoms would give rise to variance in the extinction coefficients for composite solids.

Developing this equation further, the subject of the formula can be translated analogously into radiation intensity in terms of the absorption coefficient – a far more practical form as shown in Equation 3. It should be made clear that the same exponential term derived previously appears in this expression but here as a one minus value (bracketed term). This is an essential alteration as this allows the expression to predict the intensity of radiation at position x that survives to propagate further into the solid.

$$I = I_0(1 - e^{-\alpha x}) = \varepsilon_a \sigma T^4 (1 - e^{-\alpha x})$$

Equation 3. Electromagnetic wave intensity that successfully penetrates a distance x into a material. Valid for a uniform composition material.

The initial intensity of the incident radiation, I_0, has been elaborated on by considering the emitting radiation source. It can be assumed that any hot object will emit energy as a blackbody. The Stefan-Boltzmann law [3] defines the emission intensity based on the fourth power of the emitter's surface temperature and the Stefan-Boltzmann constant, σ. The alloy emissivity, ε_a, is also a consideration as no object truly emits as a perfect radiator; an emission inefficiency must be considered. A study of the blackbody spectrum provides a second piece of essential information to introduce; the wavelength at which the radiation will be emitted depending on the source temperature, which is required for absorption coefficient calculation. The peak of a blackbody spectrum defines the most probable wavelength of emission dependent on the surface temperature, which changes inversely with temperature following Wein's displacement law [3], Equation 4.

$$\lambda_w = \frac{2.898 \times 10^{-3}}{T}$$

Equation 4. Wien's displacement law for calculation of the most probable wavelength of emission from an object based on surface temperature in Kelvin

Equation 3 represents the most basic understanding of the attenuation of EM radiation during interaction with matter, as it only considers a single homogeneous material. Setting the value of x to the corresponding total thickness of a casting mould would grant a basic transmittance estimate for a single composition, homogeneous material. For more realistic composite materials, there will be multiple absorption coefficients to consider, addressed by multiplying the corresponding thickness of each absorber with the absorption coefficient and summing the results in the exponent. Considering that emission will occur from all surfaces of the casting, and because the mould completely encases the casting, it can be assumed that radiation emitted from the casting will pass through the mould system without exception. Hence the volume fraction of each material can be used in place of the corresponding material thickness that an incident beam would encounter. Generally, this is easier to determine compared with physical distances and has the major advantage of independence from the angle of incidence, so it will not be impacted by scattering angles and boundary reflection phenomena.
It is not only direct attenuation that affects the passage of radiation in solids, but also surface reflections and scattering interactions with large particulates and defects. Firstly, because investment casting moulds are constructed layered, escaping radiation must cross each interface irrespective of the surface emission point. Should the successive layers have significant differences in refractive index, reflections will occur that will reduce the intensity of the propagating radiation as per Equation 5 [4]. The calculated reflectance, ranging between 0 and 1, would appear in mathematics as a multiplicative factor that reduces the initial intensity.

$$R = \left(\frac{n_1 - n_2}{n_1 + n_2}\right)^2$$

Equation 5. Surface reflectance in terms of the refractive index of medium 1, n_1, and medium 2, n_1

Scattering is a complex phenomenon dependent on multiple factors that must be considered on a case-by-case basis. Scattering behaviour depends on the relative disparity between the particle size (d) and the radiation wavelength interacting with them. Broadly, scattering can be segregated into three regimes; Rayleigh ($d \ll \lambda_w$), Mie ($d{\sim}\lambda_w$) and Optical ($d \gg \lambda_w$) scattering [5]. In investment casting, even the fine stucco particles present in the mould prime coat (90 mesh) will generally be several orders of magnitude larger than the radiation wavelength. As such, scattering behaviour in these systems will be driven by optical scattering descriptions. Several models for optical scattering can be relevant when the underlying assumptions are in line with the facts of the casting system. Arguably more helpful than an assumed model are phase functions, which are probability distributions defining the angular distribution of intensity around a particle following a scattering event. An example of such a phase function employed in this model is the Henyey-Greenstein (HG) phase function [6], detailed in Equation 6, which can be applied to particles of nanometres to millimetres scales. The parameter g is known as the anisotropy factor spanning the range zero to one and is used to tune the angular distribution to a specific system. Depending on the selection, the intensity distribution can be altered from uniform scattering ($g = 0$), to preferentially backwards ($g < 0$) or forwards ($g > 0$) scattering. This parameter is not selected arbitrarily but instead calculated from the cosine expectation value, $g = \langle cos(\theta) \rangle$ [7].

$$P(g,\theta) = \frac{1}{4\pi}\frac{(1-g^2)}{[1+g^2-2g\,cos(\theta)]^{3/2}}$$

Equation 6. Henyey-Greenstein (HG) phase function in terms of scattering angle, θ, and anisotropy factor, g

This phase function can be integrated before unification with the wider theory which, although not essential, reduces future calculations. Defining interest as the quantity of intensity that is not able to escape from the solid, the upper and lower integration limits can be set at π and $\pi/2$ respectively. Equation 7 shows the results of performing this integration by substitution ($\mu = cos(\theta)$), where the additional factor of 2π is included for normalisation. The result of this integration, the backscatter fraction ($S_{BS}(g)$) ranges from zero to one and now depends solely on the anisotropy factor as the angle has been integrated out through defined limits.

$$S_{BS}(g) = 2\pi\int_{\pi/2}^{\pi}\frac{1}{4\pi}\frac{(1-g^2)}{[1+g^2-2g\,cos(\theta)]^{3/2}}d\theta = \int_{0}^{-1}\frac{1}{2}\frac{(1-g^2)}{[1+g^2-2g\mu]^{3/2}}d\mu$$

$$S_{BS}(g) = \frac{1-g}{2g}\left(\frac{g+1}{\sqrt{g^2+1}}-1\right)$$

Equation 7. Fraction of backscattered intensity calculated from the integration of the HG function via substitution method

This simplified expression can be used directly in the theoretical model as a multiplicative factor acting to reduce the outbound intensity. In systems with a low concentration of scattering elements, the backscatter fraction can be scaled by the ratio of the total scattering cross-section to the emitting surface area, equivalent to the volume fraction. As with the boundary reflections, this backscatter fraction will appear one minus the value in the final expression as it is the fraction of escaping radiation sought. The backscatter angle was chosen for the integration to allow consideration of an effective thermal conductivity parameter in future work. Integration has a secondary importance; literature suggests estimation of backscattering using the HG function specifying both g and θ gives poor results [6]. Integrating across a range of angles improves accuracy by broadening the field of interest. Using the HG function without integration remains possible through an alternative form, the double HG function [6], which consists of summing two weighted HG functions where one has a positive anisotropy factor and the second has a negative anisotropy factor. The weighting represents the relevant dominance of the associated forward and backscattering. Given that the integrated single HG function reports satisfactory values when considering backscattering, the choice was made to use this simpler variant in the final expression over the more complex double function that has the potential to introduce additional uncertainty through a second anisotropy factor and weightings.

Equation 8 shows the final expression for the escaping radiation intensity after consolidating all aspects of consideration previously discussed. As mentioned, the backscatter fraction has been scaled by stucco volume fraction as some systems might exhibit lower concentrations of particles that would influence the predictions. Here indices n, m and i are positive, real numbers governing the number of boundary interfaces, the number of different scattering elements and the number of absorbing elements respectively. Although not explicitly mentioned up until this point, it should be made clear that mould porosity should be included among the absorbing factors with the appropriate volume fraction.

$$I = F\varepsilon_a\sigma T^4\sum_{1}^{n}(1-R_n)\sum_{1}^{m}(1-V_mS_{BS_m})\left(1-e^{-\sum_1^i\alpha_iV_i}\right)$$

Equation 8. The final expression for escaping radiation intensity through an arbitrary solid

A closing consideration included in the final expression is the concept of view factors, F. This does not generally apply to investment casting as the mould structure is always parallel with the casting surface by the nature of its manufacture. This may be applicable in other casting processes, or when considering subsequent radiation interactions with other parts of the casting that present an angle to the emitter. If the absorbing media present an angle to the emitting surface, only a fraction of the surface area will participate in absorption. This fraction is defined by the ratio of the product of the area and cosine of the angle between the absorbing material and the surface normal of the emitter, as shown in Equation 9 where subscripts r denotes the receiver (absorbing object) and e denotes the emitter.

$$F = \frac{q_r}{q_e} = \frac{A_r \cos(\theta_r)}{A_e \cos(\theta_e)}$$

Equation 9. The equation for the view factor affecting the extent of radiation intensity interaction between two surfaces

3. Experimental Investigation

Mathematical models of phenomena are useful tools for understanding the impacts on the final cast component. Comparing model predictions against experimental measurements is an essential development step in any theory to assess the agreement quality and accuracy. To this end, a typical mould used in producing Ni-based superalloy castings consisting of approximately nine layers was selected for assessment. The three compositionally distinct regions of the mould, defined as materials 1, 2 and 3, (Table 1) were investigated independently using a Jasco FT/IR-6200 Fourier transform infrared spectroscopy (FTIR) equipped with an integrating sphere.

Table 1. Details of the experimental mould composition

Material Identity	Layer	Composition		
		Name	Details	
Material 1	1	Cobalt Aluminate with 90 mesh Zircon sand	Chilches Microzir Flour 200M IC	$ZrSiO_2 > 63.5\%$ $TiO_2 < 0.2\%$ $Fe_2O_3 < 0.15\%$ $Al_2O_3 < 1.3\%$ $SiO_2 < 33\%$
			Cobalt Aluminate	$CoAl_2O_4$
			Primecote® Plus	Binder
Material 2	2	Backup with 30/80 mesh Mullite sand	Ranco-Sil 140F	SiO_2
			Matrixol 30	Binder
			MXC Excel X2	Fibre Reinforcement
			30/80 Sand	-
Material 3	3-9	Backup with 16/30 mesh Mullite sand	Ranco-Sil 140F	SiO_2
			Matrixol 30	Binder
			MXC Excel X2	Fibre Reinforcement
			16/30 sand	-

The integrating sphere is an important FTIR attachment when measuring semi-transparent samples, as it allows the detector to remain offset from the incident source beam. A Gold reference standard was used to block the upper aperture during the baseline measurement, ensuring no light escapes. Both reflection and transmission data are required to obtain information on the absorption properties of a material. For reflection data, samples are positioned in place of the Gold standard at the top of the sphere and covered with an opaque lid to prevent the escape of light. For transmission measurements, the Gold standard remains in place while samples are placed directly in the path of the incident beam by covering the entry point at the left-hand side of the integrating sphere.

Producing samples suitable for testing was highly challenging, as the low penetrating power of the FTIR demanded correspondingly thin samples to investigate transmission. Although the production of thin films is possible, the fragility of such films makes multiple loading and unloading of samples without damage a challenge. A solution exists in the form of a transparent substrate on which to build the films. Sapphire is an ideal material for this application as it has optical transparency in the infrared region of the EM spectrum, preventing adverse impact on material transmission measurements. Circular Sapphire microscope slides with a thickness of 0.5 mm were used as the base for preparing all samples, where the diameter was selected to ensure complete coverage of the FTIR apertures. Three samples were produced on individual Sapphire slides, one for each material (1, 2 and 3) with each receiving a rainfall sanding of stucco (in line with foundry application processes). Before experimentation, all samples were fired in a muffle furnace to ensure they were representative of true production mould components. The heating and cooling ramps were intentionally selected to be gradual to minimise the risk of cracking the films, with several pauses during the increase made for the same reason as outlined in Table 2.

Table 2. The stages of sample heating during the firing of the thin films before the FTIR assessment

Stage	Temperature range (C)	Rate (K min^{-1})
1	21 → 300	20
2	300	Hold for 5 minutes
3	300 → 600	20
4	600	Hold for 5 minutes
5	600 → 900	20
6	900	Hold for 5 minutes
7	900 → 1050	20
8	1050	Hold for 90 minutes
9	1050 → 21	20

Conducting the FTIR assessment on the samples gives reflectance and transmission results illustrated in the upper panels of Figure 1. The reflectance and transmittance results are not a single measurement but are an average from five repeat measurements. Each measurement was taken targeting a different area of the sample, ensuring subtle variations in surface finish or thickness would not negatively impact the final result. The effect of such variations does not appear to notable impact the measurement as the uncertainty bound remains tightly constrained around the mean. There is a marginally higher uncertainty in the absorbance as the uncertainty in both reflectance and transmission propagates into absorbance mathematically. Absorbance calculation is exceedingly simple, relying on energy conservation; the sum of reflectance, transmittance and absorbance must equal 100% at any wavelength.

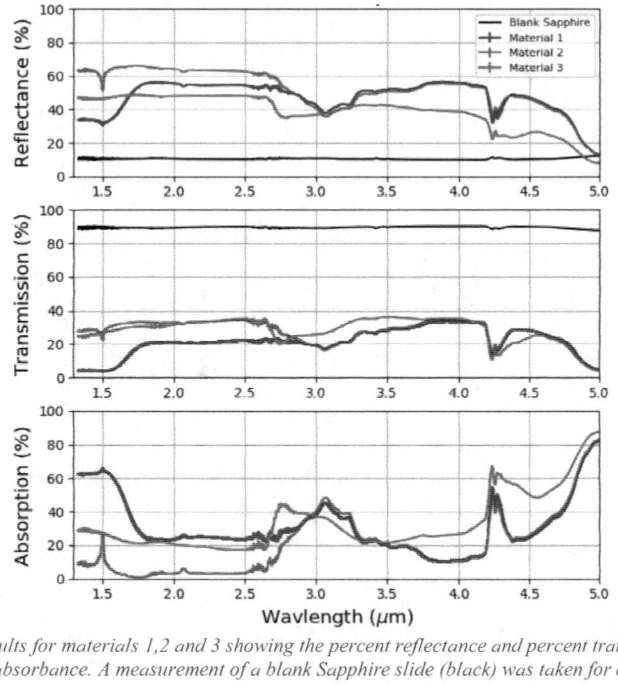

Figure 1. FTIR results for materials 1,2 and 3 showing the percent reflectance and percent transmittance alongside calculated percent absorbance. A measurement of a blank Sapphire slide (black) was taken for comparative purposes

In general, the reflectance of the samples in all cases dominates the optical properties with the greatest percentage contribution. Several troughs are present in the profile with corresponding peaks in absorbance, indicating elemental absorption bands present in the materials. Interestingly, there is a non-

zero transmittance observed in all samples across all wavelengths; an observation suggesting transparency may not be limited to defined optical windows but may operate more broadly.
Alongside the sample measurements, reflectance and transmittance measurements for a blank Sapphire slide were taken to confirm its desirable optical properties. Measurements were taken between 1.3 – 20 μm, although, above 5 μm, the Sapphire reflectance begins to rise rapidly with a corresponding decline in transmittance. This behaviour limits the useful range of experimental results to the displayed range of 1.3 – 5 μm. However, this range is sufficient. Returning to Wein's displacement law, the corresponding range of alloy surface temperatures can be estimated as 306.6 – 1956.2 °C, which comfortably covers the alloy melting, solidification interval and a significant part of subsequent cooling. The absorption coefficient can be calculated based on the measured reflectance and transmittance data using Equation 10 [8]. Knowledge of the sample thickness is needed for this calculation, which can readily be calculated based on the total sample thickness subtracting the Sapphire slide thickness. The sample thickness for materials 1, 2 and 3 were 0.44 mm, 0.47 mm and 1.42 mm respectively.

$$\alpha = \left(1 - \frac{R}{100} - \frac{T}{100}\right)\left(\frac{ln(10)}{1 \times 10^{-3}d}\right) = \frac{A}{100}\left(\frac{ln(10)}{1 \times 10^{-3}d}\right)$$

Equation 10. Method for calculation of the absorption coefficient from experimental reflectance and transmittance data with the addition of the sample thickness, d

Figure 2 shows the results of performing this calculation for the absorption coefficient in the left-hand column as a function of both wavelength and alloy surface temperature. The extinction coefficients in the right-hand panel have also been determined using the relation introduced in Equation 2. These results are considerably more relevant than the raw measurement results discussed previously as they feature thickness information; the per metre nature of these results means they are more representative of the true absorbing potential of these materials. Considering the thickness leads to enhanced absorption of material 1, which would be expected given it is the densest of the materials with a high Zircon and Cobalt concentration. As materials 2 and 3 have similar compositions, with the difference being the coarseness of the stucco, the magnitude of the absorption is similar. However, there is a considerable peak in material 2 at 3 μm which given the comparable response in material 1 is potentially a result of increased scattering of the finer stucco particles in these materials.

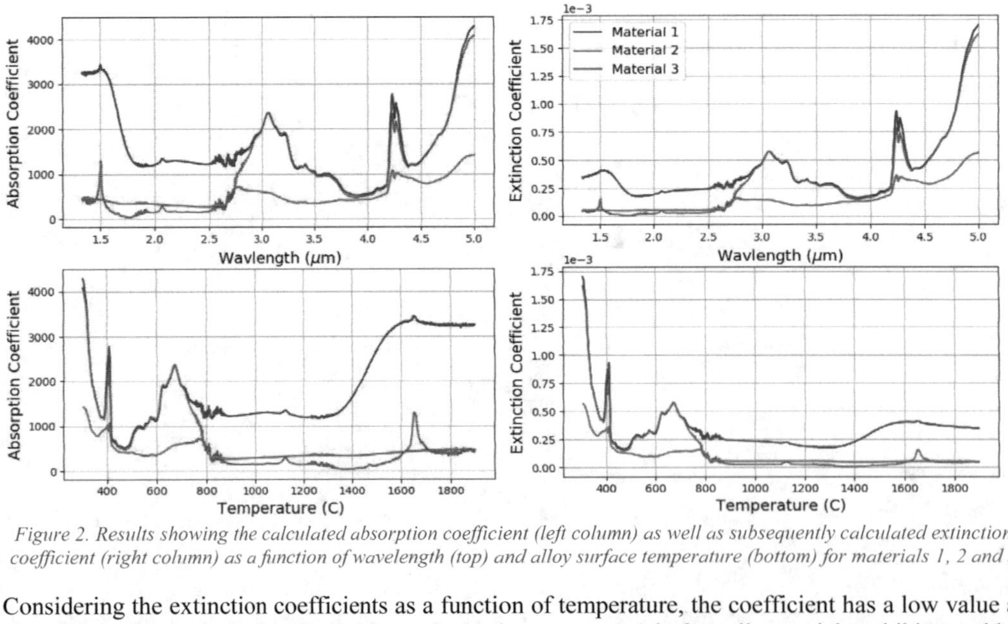

Figure 2. Results showing the calculated absorption coefficient (left column) as well as subsequently calculated extinction coefficient (right column) as a function of wavelength (top) and alloy surface temperature (bottom) for materials 1, 2 and 3

Considering the extinction coefficients as a function of temperature, the coefficient has a low value at elevated casting temperature (indicating enhanced transparency) before all materials exhibit a sudden increase in opacity around an alloy temperature of 800 C. This result is exciting as it implies significant transparency across alloy melting, solidification and several hundred degrees of subsequent cooling. This has the potential to give rise to greater cooling than expected during a period of critical microstructural and residual stress formations that may not be accurately understood or captured by computational models of the casting process.

The final consideration is to combine the calculated extinction coefficients from FTIR analysis with the correct material thicknesses of a true production mould to assess the extent of mould transparency in a real situation. Because the absorption coefficient obtained experimentally already contains information on the boundary reflection and scattering features, the true behaviour of a production mould can be evaluated using a far simpler expression (Equation 11). This expression is related to the information discussed during the construction of the theory, but only concerns the exponential decay of initial radiation intensity. A factor of 100 has been added and the expression divided by I_0 to allow the percentage of incident radiation that will escape the mould system to be calculated.

$$\frac{I}{I_0} = 100e^{-\alpha_1 x_1 - \alpha_2 x_2 - \alpha_3 x_3} \qquad [\%]$$

Equation 11. Expression for the percentage radiation escape applied to the experimental measurements to determine real behaviour

Figure 3 shows the final comparison between the measured experimental transparency and that predicted by the model introduced in this work. Considering the experimental results initially, across the range of wavelengths presented, there are two distinct windows of transmission separated by a deep absorbance band centred ~3 μm. This is likely the result of atmospheric water [9] as experiments were not conducted in a vacuum. Both transmission peaks represent a significant amount of radiation escape with peak values of approximately 7.5 % and 4.5 %. Applying the conversion between wavelength and temperature and narrowing the range of interest to 750 – 1450 °C (because the vast majority of interesting solidification behaviour occurs across this range), it was found that only the first peak of the wavelength plot is present. It appears in the temperature-dependent plots as a mirror image due to the inverse relationship between temperature and wavelength. It is fascinating that radiation transparency is non-zero across the entire duration of mould filling, alloy solidification and initial cooling to 800 °C, a feature of such ceramic mould that is likely poorly understood within the industry. Such transparency would have the effect of steepening the cooling in the alloy in susceptible systems that potentially introduce unexpected or undesirable solidification behaviour into cast components. Moreover, this phenomenon is highly influenced by variations in mould thickness that can give rise to local enhancements or reductions in the alloy cooling rate.

The predictions from the theoretical model were generated using data from several literature sources, as detailed in Table 3. Due to materials 2 and 3 being compositionally identical and stucco size being the only dividing factor, the decision was made to amalgamate these materials into a single entity, defined as the backup region, to simplify the calculation. Elements with a 'calculation' source were determined during the investigation of the mould materials during previous unpublished work, which will be summarised herein. Consulting scanning electron microscope images of the mould cross-section using Fiji image analysis software [10], converting the backscatter electron image to a binary format allows identification of the percent porosity in the backup region. This method identifies a 21 % porosity where the remainder is split approximately evenly between the Silica matrix and Aluminosilicate sand. Material 1, known as the prime coat, was more complex as the volume fractions could not be determined from image analysis alone. Instead, it was estimated based on the mass of materials in the slurry, assuming the volatile binders would burn away during firing to leave a 0.075 % porosity. The fractions of Zircon and Cobalt Aluminate were estimated directly based on masses against the total mass in the slurry.

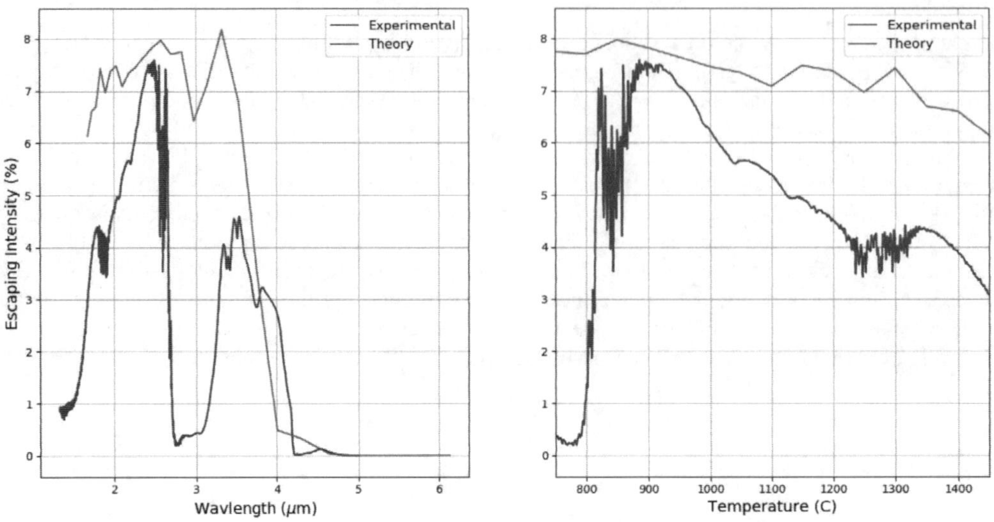

Figure 3. Comparison plots of theoretical model prediction (green) and calculation from experimental measurements (blue) as a function of wavelength (left) and temperature (right)

Estimations from the theoretical model have a reasonable agreement with the experimental measures, in which critical features such as the double peak in transmission and the gradual reduction in transmission at low wavelengths. However, the extent of these drops in transmission is underestimated. As minimal consideration of different crystallographic structures forming during sample firing was given, it is possible that increased absorption or scattering from these different structures may be responsible for this discrepancy – this will require additional investigations to confirm. In general, the theory accurately predicts the magnitude and approximate profile of transmitted radiation although further refinement will be needed to capture the nuances of radiation transparency.

Table 3. Catalogue of data sources used in the determination of theoretical model predictions

Parameter	Value	Source
Alloy emissivity	Tabular data	[11]
Refractive index (prime matrix)	1.7	[12]
Refractive index (prime stucco)	1.94	[12]
Refractive index (backup matrix)	1.44	[12]
Refractive index (backup stucco)	1.65	[12]
Extinction Coefficient (prime)	Tabular data	[13]
Extinction Coefficient (backup)	Tabular data	[14]
Extinction Coefficient (air)	Tabular data	[15]
Volume fractions (prime)	$ZrSiO_4$–91.25, $CoAl_2O_3$–8% Porosity–0.075%	Calculated
.Volume fractions (backup)	SiO_2–39.5%, Al_2O_3–39.5%, Porosity–21%	Calculated

4. Conclusions

It has been shown through an experimental assessment conducted on a representative ceramic mould that transparency to EM radiation presents a non-negligible mechanism of heat loss that will likely impact alloy cooling. Multiple solidification mechanisms will likely be impacted with magnitudes of 7.5% at its maximum that occurs at elevated alloy temperature. Applying general EM theory has shown that a representative model of semi-transparent materials, evaluated with reasonable literature data, can be successfully implemented. Although the fine details of this model require further investigation to tune the theory, the predictions correctly capture the key features and magnitude of radiation transparency. As this model is founded in physics common to all casting processes, its applicability is universal across multiple casting disciplines, requiring only input of the relevant material properties to ascertain the possible importance of IR transparency in a system.

High emissivity coating has been a point of active research for some time where complex geometry castings are concerned, but when considering semi-transparent materials, such materials only represent a partial solution. Specially designed coatings intended to absorb infrared radiation at selected wavelengths offers foundries the ability to prevent the unmonitored escape of radiation from the casting. Equally, this research offers a potential re-focus to existing efforts into high emissivity coatings by restricting identifying narrower wavelength ranges in which coatings should be effective. Should a stronger cooling trend be desirable in a specified region of a casting, high emissivity coating should consider enhancing effectiveness in regions of the spectrum where the natural transmittance of the mould is poorer.

Acknowledgements
The authors wish to thank the Engineering and Physical Sciences Research Council (EPSRC) [Grant EP/L016389/1] and TPC Components AB for funding this research.
The author would like to thank Cranfield University for providing access to the scientific equipment used in this investigation and TPC Components AB for providing the necessary materials for sample construction.
Data associated with this work is stored on the Cranfield Online Research Data repository, available to access through 10.17862/cranfield.rd.22153778.

References
[1] A. A. Keste, S. H. Gawande, and C. Sarkar, "Design optimization of precision casting for residual stress reduction," *J. Comput. Des. Eng.*, vol. 3, no. 2, pp. 140–150, 2016.
[2] C. A. Jones *et al.*, "A verification of thermophysical properties of a porous ceramic investment casting mould using commercial computational fluid dynamics software," *IOP Conf. Ser. Mater. Sci. Eng.*, vol. 861, 2020.
[3] R. M. Eisberg, *Fundamentals of Modern Physics*. John Wiley & Sons, 1961.
[4] X. U. Zhang, D. J. Faber, A. L. Post, T. G. van Leeuwen, and H. J. C. M. Sterenborg, "Refractive index measurement using single fiber reflectance spectroscopy," *J. Biophotonics*, 2019.
[5] O. Boucher, *Atmospheric Aerosols*. Springer, 2015.
[6] F. Zhang and J. Li, "A note on double Henyey–Greenstein phase function," *J. Quant. Spectrosc. Radiat. Transf.*, vol. 184, pp. 40–43, 2016.
[7] T. Binzoni, T. S. Leung, A. H. Gandjbakhche, D. Rüfenacht, and D. T. Delpy, "The use of the Henyey-Greenstein phase function in Monte Carlo simulations in biomedical optics," *Phys. Med. Biol.*, vol. 51, no. 17, 2006.
[8] M. T. Abdullah, L. M. Raoof, M. H. Hasan, A. N. Abd, and I. M. Mohammed, "The Effect of Different Thickness on the Optical and Electrical Properties of TiO2Thin Films," in *Journal of Physics: Conference Series*, 2021, vol. 1999.
[9] National Institute of Standards and Technology, "Water Infrared Spectrum," *NIST Chemistry WebBook*, 1969. [Online]. Available: https://webbook.nist.gov/cgi/cbook.cgi?ID=C7732185&Type=IR-SPEC&Index=1. [Accessed: 02-Jan-2023].
[10] J. Schindelin, I. Arganda-Carreras, and E. Frise, "Fiji: an open-source platform for biological-image analysis," *Nat. Methods*, vol. 9, no. 7, pp. 676–682, 2012.
[11] B. P. Keller, S. E. Nelson, K. L. Walton, T. K. Ghosh, R. V. Tompson, and S. K. Loyalka, "Total hemispherical emissivity of Inconel 718," *Nucl. Eng. Des.*, vol. 287, pp. 11–18, 2015.
[12] Malvern Instruments, "Sample dispersion and refractive index guide - Mastersizer 2000." 2007.
[13] E. D. Palik and R. Khana, "Zircon (ZrSiO4)," in *Handbook of Optical Constants of Solids III*, 1998, pp. 987–999.
[14] R. Kitamura, L. Pilon, and M. Jonasz, "Optical Constants of Silica Glass for Extreme Ultraviolet to Far Infrared at Near Room Temperature," *Water Res.*, vol. 46, no. 33, 2007.
[15] R. W. Fenn, S. A. Clough, W. O. Gallery, R. E. Good, F. X. Kneizys, J. D. Mill, L. S. Rotherman, E. P. Shettle, F. E. Volz, "Chapter 18: Optical and Infrared Properties of the Atmosphere," in *Handbook of Geophysics and the Space Environment*, 1985.

IOP Conf. Series: Materials Science and Engineering 1281 (2023) 012041 doi:10.1088/1757-899X/1281/1/012041

Droplet temperature measurement in GMAW

Vicente Núñez Sánchez, Zhaoyang Yan, Shujun Chen and Patricio F. Mendez

MSc. Student of Welding Engineering, Canadian Centre for Welding and Joining (CCWJ), University of Alberta

E-mail: nunezsan@ualberta.ca

Abstract. This work presents measurements of the droplet temperature for the Gas Metal Arc Welding process. Four different aluminium electrodes were used for experiments performed under the same conditions. The approach used yielded higher droplet temperatures than were obtained in previous experiments with the same raw data because the energy balance approach differed. However, trends were similar, with a minimum value close to the transition current and a tendency to increase with the current. Anode fall voltages behaved similarly, increasing with the current. Differences in temperature were found between the four electrodes. This can be explained by the different alloying elements present on the electrodes, which evaporate and take energy out of the droplet.

1. Introduction

The measurement of the droplet temperature in welding has been of interest to researchers, in line with the development of new alloys. The fumes generated during welding can be harmful to workers when inhaled. Measurements of the droplet temperature have been performed for years, with the same physics: measuring the heat content of the droplet. In the gas metal arc welding (GMAW) process, the droplets deposited are directly from the rod of the consumable, the electrode.

The heat content of droplets has been studied since the 1960. Pokhodyna and Subtel [1] concluded that the temperature decreases when changing from electrode positive to electrode negative polarity. Another method of measuring the droplet temperature or heat content is with cameras. Zhang *et al.* [2] performed this study with two-colour pyrometry. Droplet temperature has been studied respect to the transition current by Soderstrom and Mendez [3], having a small decay close to it. The same raw data were used by Yan *et al.* [4], but only with ER1100 and ER5183 alloys. With the calculations of droplet temperature, it is possible to develop calculations for the anode fall voltage and evaporations.

2. Calculations

2.1. Energy balance

An energy balance was performed on the calorimeter at initial and final conditions. The initial condition is when the calorimeter is at room temperature and droplets separated from it at initial temperature. The final condition is when the calorimeter and droplets are together at the same temperature and start to cool down by losing heat. The control volume can be seen in red in figure 1 for the two points studied.

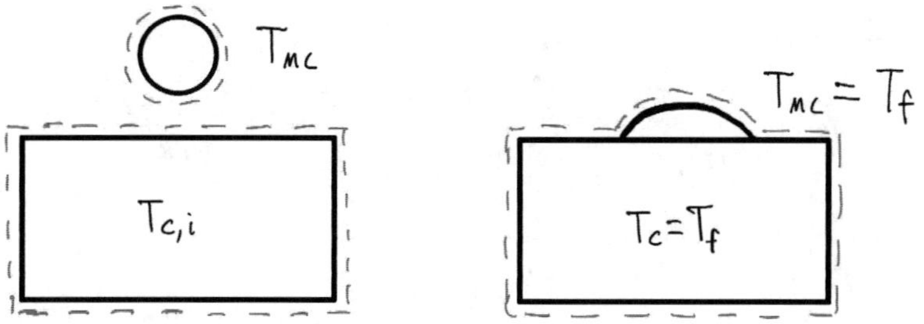

Figure 1. Control volume studied

The initial heat balance in equation 1 does not consider heat generated or stored in the system.

$$Q_{\text{in}} = Q_{\text{out}} + Q_{\text{loss}} \tag{1}$$

It is assumed that the droplet will not evaporate during the time that it touches the calorimeter and cools down. The heat radiating from the droplet to the calorimeter is negligible, as well as the heat of the arc that was on the upper part of this system, which is shown later. With these assumptions, in equation 2 is developed the energy balance.

$$m_{\text{MC}}h_{\text{MC,i}} - m_{\text{C}}h_{\text{C,i}} = m_{\text{MC}}h_{\text{MC,f}} + m_{\text{C}}h_{\text{C,f}} + Q_{\text{loss}} \tag{2}$$

The mass of the molten consumable and the copper calorimeter are m_{MC} and m_{C}, with enthalpies h_{MC} and h_{C}. The heat loss Q_{loss}, only considered when the calorimeter with droplets starts to cool down.

$$Q_{\text{loss}} = m_{\text{MC}}(h_{\text{MC,i}} - h_{\text{MC,f}}) - m_{\text{C}}(h_{\text{C,f}} - h_{\text{C,i}}) \tag{3}$$

The enthalpies $h_{\text{MC,i}}$ and $h_{\text{MC,f}}$ are the enthalpy of the molten consumable at the beginning of the process, when it leaves the electrode and at the end, when the temperature of the molten consumable is the same as the calorimeter. The enthalpies $h_{\text{C,i}}$ and $h_{\text{C,f}}$ are for the calorimeter, at the initial and final temperature, the final the same as the final molten consumable. The different enthalpies are defined in equation 4, 5, 6 and 7 for the molten consumable and calorimeter:

$$h_{\text{MC,i}} = h_{\text{ref,liq}} + c_{\text{ref,liq}}(T_{\text{MC,i}} - T_{\text{ref,liq}}) \tag{4}$$

$$h_{\text{MC,f}} = h_{\text{ref,sol}} + c_{\text{ref,sol}}(T_{\text{f}} - T_{\text{ref,sol}}) \tag{5}$$

$$h_{\text{MC,i}} - h_{\text{MC,f}} = c_{\text{ref,sol}}(T_{\text{ref,sol}} - T_{\text{f}}) + \Delta h_{\text{ref,MC}} + c_{\text{ref,liq}}(T_{\text{MC,i}} - T_{\text{ref,liq}}) \tag{6}$$

$$h_{\text{C,f}} - h_{\text{C,i}} = c_{\text{ref,hi}}(T_{\text{f}} - T_{\text{C,i}}) \tag{7}$$

The equation is performed with $T_{\text{ref,sol}}$ and $T_{\text{ref,liq}}$ as reference temperatures in the solid and liquid state. The references temperatures are at room temperature for solid state and the average of previous measurements of molten consumable for the liquid reference. The enthalpy difference between these two temperatures is added with $\Delta h_{\text{ref,MC}}$ for the molten consumable. The specific heats are $c_{\text{ref,sol}}$ and $c_{\text{ref,liq}}$ and obtained from the reference temperatures in solid and liquid state.

Combining all equations leads to equation 8 for heat loss in terms of parameters from the molten consumable and calorimeter, including the mass, enthalpy, temperatures and specific heat.

IOP Conf. Series: Materials Science and Engineering 1281 (2023) 012041 doi:10.1088/1757-899X/1281/1/012041

$$Q_{\text{loss}} = \Big[m_{\text{MC}}\left[c_{\text{ref,sol}}(T_{\text{ref,sol}} - T_{\text{f}}) + \Delta h_{\text{ref,MC}} + c_{\text{ref,liq}}(T_{\text{MC,i}} - T_{\text{ref,liq}})\right] - m_{\text{C}}\left[c_{\text{ref,hi}}(T_{\text{f}} - T_{\text{C,i}})\right]\Big] \tag{8}$$

Equation 8 can be written in two terms, given by the mass from the molten consumable and the calorimeter. These two terms are shown in equations 9 and 10.

$$Q_{\text{MC}} = m_{\text{MC}}\left[c_{\text{ref,sol}}(T_{\text{ref,sol}} - T_{\text{f}}) + \Delta h_{\text{ref,MC}} + c_{\text{ref,liq}}(T_{\text{MC,i}} - T_{\text{ref,liq}})\right] \tag{9}$$

$$Q_{\text{C}} = m_{\text{C}}\left[c_{\text{ref,hi}}(T_{\text{f}} - T_{\text{C,i}})\right] \tag{10}$$

2.2. Heat loss

Heat losses are studied and defined by the calorimeter design. For this solid state calorimeter, the approach by Scott [5] is due with a previous calibration with tin. The equation 11 describes that approach, which is fitted as in equation 12 for values K in W s^{-1} K^{-1} and t_{trans} in s.

$$\widehat{Q}_{\text{loss}} = K\left(T_{\text{f}} - T_{\infty}\right)\left(t_{\text{f}} - t_0 - t_{\text{trans}}\right) \tag{11}$$

$$min\left(Q_{\text{loss}} - \widehat{Q}_{\text{loss}}\right)^2 \tag{12}$$

Where T_∞ and T_{f} are temperatures in infinity, or room temperature and for maximum temperature, both in K. Times t_{f} and t_0 in s are for the maximum and minimum temperature.

The heat loss by Scott [5] is given by equation 13:

$$\widehat{q}_{\text{loss}} = K\left(T_{\text{f}} - T_{\infty}\right) \tag{13}$$

The definition of this calculations, does not accounts for the heat losses in the heating stage of the calorimeter due to the heat from the droplet. The model presented in equation 14 describes the heat losses when cooling down, by using the slope of the cooling curve.

Energies in equation 13 and 14 are compared beforehand with the values of the Tin calibration. It is possible to see in figure 2 a similar behaviour between both approaches for the Tin calibration.

$$q_{\text{loss}} = \dot{T}\left(c_{\text{ref,sol}} m_{\text{d}} + c_{\text{ref,low}} m_{\text{c}}\right) \tag{14}$$

The characteristic time for equation 14 is obtained with a projection of the cooling slope at the moment t_{MC} that is the average between t_{f} and t_0, as the point where the droplet touches the calorimeter and begins to lose heat to the environment. As an average time between both temperatures, it is possible to obtain realistic values comparable with previous ones.

$$Q_{\text{loss}} = \dot{T}\left(c_{\text{ref,sol}} m_{\text{d}} + c_{\text{ref,low}} m_{\text{c}}\right) t_{\text{MC}} \tag{15}$$

2.3. Droplet temperature

With the definition of the heat losses in equation 8 and in Eq. 15, it is possible to obtain the initial droplet temperature, as equation 16 shows.

$$T_{\text{MC,i}} = \frac{1}{c_{\text{ref,liq}}}\left[\frac{1}{m_{\text{MC}}}\left[Q_{\text{loss}} + m_{\text{C}} c_{\text{ref,hi}}(T_{\text{f}} - T_{\text{C,i}})\right] - \Big(\left[c_{\text{ref,sol}}(T_{\text{ref,sol}} - T_{\text{f}}) + \Delta h_{\text{ref,MC}}\Big)\right] + T_{\text{ref,liq}} \tag{16}$$

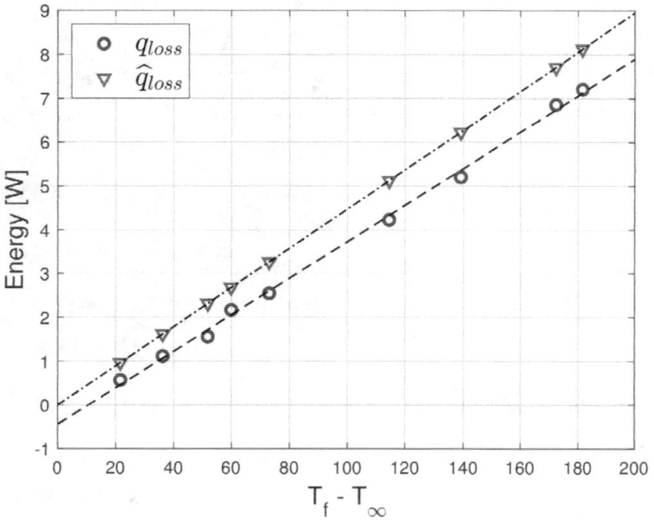

Figure 2. Comparison of two energy loses methods

2.4. Anode fall voltage
Values of droplet temperature were previously used in the laboratory but obtained with a different method. This new values of droplet temperature can be applied to previous works [6], as equation 17 shows to obtain the anode fall voltage.

$$V_{\text{anode}} = \frac{1}{I}\Big[U_{\text{c}} m'_{\text{c}} \left(\Delta h_{\text{ref}} + c_{\text{MC,liq}} \left(T_{\text{MC}} - T_{\text{ref,liq}}\right)\right) + \dot{m}_{\text{g}} \left(\Delta h_{\text{liq,gas}} + c_{\text{ref,gas}} \left(T_{\text{g}} - T_{\text{ref,liq}}\right)\right) - R_{\text{ct}} I^2 - \frac{I^2 L \rho_{\text{eff}}}{A_{\text{e}}} - q_{\text{arc}}\Big] \tag{17}$$

3. Experiments
The experimental setup comprises an anode (electrode) and water-cooled copper cathode (figures 3 and 4). The droplets of wire are allowed to fall onto the calorimeter through the cathode orifice. The calorimeter has 10 type K thermocouple inside, which give the temperatures reading through the *Omega TC-8 Module.* The current and voltage were acquired between the torch and the base plate with an *NI-DAQ USB 6351* synchronised with a *Phantom v210* high speed video camera. The welding machine used is a *Power Wave S500* by *Lincoln Electric.*

3.1. Calorimeter design
For the measurement of the temperature of the calorimeter, six thermocouples from ten were used to read temperatures. The sample rate of the temperatures reading was 0.0156 seconds per sample. The mass of the calorimeter was 633.72 g. The Biot number for the calorimeter was $3.5\ 10^{-4}$, that means that the calorimeter presents small variations of temperature gradients.

3.2. Experimental procedure
The experiments performed were in DCEP with a 100% of Argon as shielding gas. The electrode diameter was 1.2 mm and the cathode was a copper water cooled. Experiments were performed for WFS between 100 and 450 ipm, while maintaining the electrode extension in 12.5 mm all the time. When set the WFS, the voltage was changed until the electrode extension was the

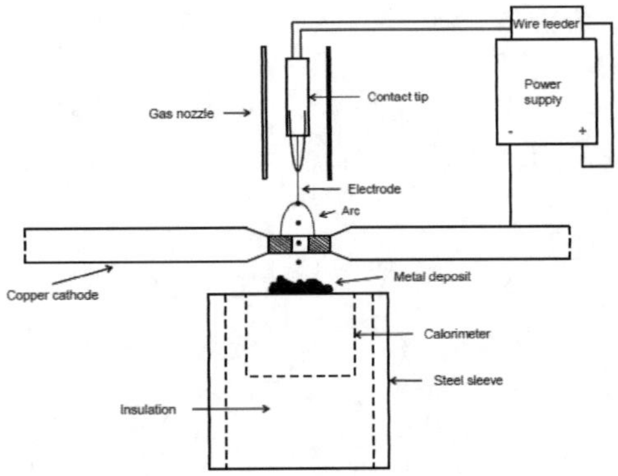

Figure 3. Same setup as from Scott [5]

Figure 4. Anode and Cathode Setup

same as described. The droplets were allowed to fall into the flux tray until the parameters were obtained. When everything was running, the flux tray was removed and the droplets allowed to fall into the calorimeter during two to six seconds. After this time, the weld was stopped and the acquisition of temperature run for the next three minutes to allow the calorimeter to heat until a maximum and cool down. The mass of the droplets deposited were between 0.5 and 3 g.

The values measured during the experiments were the voltage and current of it, 6 different temperatures inside the calorimeter and two inside the water cooled cathode, the flow of water inside the cathode, and the mass of the droplet that felt into the calorimeter. With the temperatures before and after the arc in the cathode, is possible to measure the heat of it. This calculations, were not performed for this publication.

4. Calculations

4.1. Thermocouple reading and correction

Each thermocouple was calibrated before being used in the experiments, using known temperatures. The following calibration method for each thermocouple was used.

$$T = T_{\text{read}} + \Delta T_{\text{corr}} \tag{18}$$

When obtaining T in °C for each reading, a fitting as in equation 19 in Kelvin is performed to obtain the values. The values for A, B, C, D and E are listed in table 1.

$$\Delta T_{\text{corr}} = AT_{\text{read}}^4 + BT_{\text{read}}^3 + CT_{\text{read}}^2 + DT_{\text{read}} + E \tag{19}$$

4.1.1. Key values The values obtained for temperature and time, are obtained after having the mean of the six thermocouples. All temperatures are a characteristic of the average of the first six thermocouples. Each value has a criteria to be searched that is applied in the Matlab code, being these values:

- T_0: is the minimum value.
- T_{f}: is determined as the highest point measured.
- $T_{\text{MC,slope}}$: is determined as the average between T_{f} and T_0.
- t_0: time corresponding to T_0.
- t_{f}: time corresponding to T_{f}.

Table 1. Values obtained for each thermocouple calibrated

	A	B	C	D	E
	K^{-3}	K^{-2}	K^{-1}	-	K
TC1	$-2.39\,10^{-9}$	$4.11\,10^{-6}$	$-2.57\,10^{-3}$	$6.85\,10^{-1}$	$-6.56\,10^{1}$
TC2	$-2.40\,10^{-9}$	$4.20\,10^{-6}$	$-2.68\,10^{-3}$	$7.39\,10^{-1}$	$-7.43\,10^{1}$
TC3	$-2.05\,10^{-9}$	$3.51\,10^{-6}$	$-2.17\,10^{-3}$	$5.69\,10^{-1}$	$-5.32\,10^{1}$
TC4	$-2.05\,10^{-9}$	$3.56\,10^{-6}$	$-2.23\,10^{-3}$	$5.95\,10^{-1}$	$-5.71\,10^{1}$
TC5	$-3.64\,10^{-10}$	$5.56\,10^{-7}$	$-2.66\,10^{-4}$	$3.99\,10^{-2}$	$6.44\,10^{-1}$
TC6	$-7.04\,10^{-10}$	$1.19\,10^{-6}$	$-6.93\,10^{-4}$	$1.65\,10^{-1}$	$-1.27\,10^{1}$
TC7	$-1.73\,10^{-9}$	$2.99\,10^{-6}$	$-1.87\,10^{-3}$	$5.02\,10^{-1}$	$-4.82\,10^{1}$
TC8	$-1.37\,10^{-9}$	$2.40\,10^{-6}$	$-1.50\,10^{-3}$	$3.99\,10^{-1}$	$-3.75\,10^{1}$
TC9	$-2.06\,10^{-9}$	$3.49\,10^{-6}$	$-2.13\,10^{-3}$	$5.48\,10^{-1}$	$-4.96\,10^{1}$
TC10	$-2.40\,10^{-9}$	$4.05\,10^{-6}$	$-2.48\,10^{-3}$	$6.44\,10^{-1}$	$-5.95\,10^{1}$

- t_{MC}: time when $T_{MC,i}$ starts cooling.

The values explained are shown in figure 5, a plot of the calorimeter heating up and cooling down.

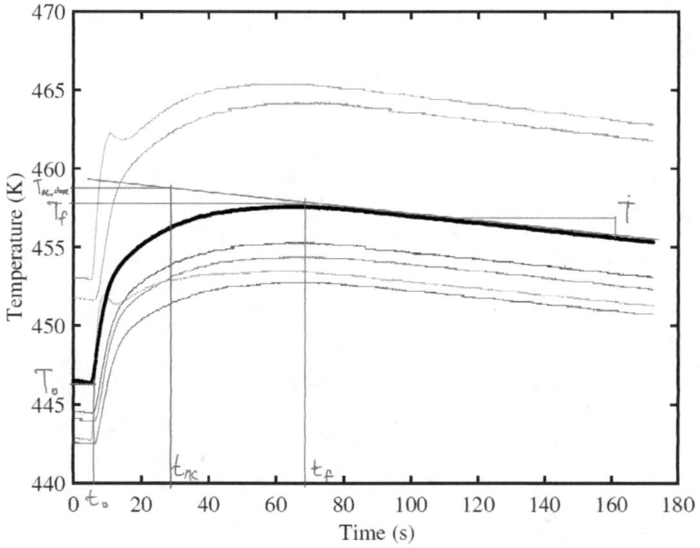

Figure 5. Curve with cooling ratio and characteristic temperatures

4.2. Electrode composition

Four different electrodes were used for these experiments: ER1100, ER4043, ER5554 and ER5183, as shown in Table 2. ER1100 can be considered as pure aluminium and is compared with ER4043, that have a similar composition but with a 5% of Si in it [7]. ER5183 has a value of 5% of Magnesium and is compared with ER5554 that has the half of Mg content. The electrodes were decided in a way, to understand better the effects of the vapours that the electrode generates.

IOP Conf. Series: Materials Science and Engineering 1281 (2023) 012041 doi:10.1088/1757-899X/1281/1/012041

Table 2. Electrode composition by the supplier

Electrode	%Mg	%Al	%Si	%Ti	%Cr	%Mn	%Fe	%Cu	%Zn
ER1100	0.00	99.730	0.09	0.00	0.00	0.00	0.09	0.09	0.000
ER4043	0.03	94.529	5.26	0.01	0.00	0.01	0.15	0.01	0.001
ER5554	2.66	96.288	0.06	0.08	0.08	0.69	0.14	0.00	0.002
ER5183	4.93	94.079	0.05	0.08	0.08	0.63	0.13	0.02	0.001

4.3. Material properties of calorimeter and electrodes

All the materials properties were obtained from *JMatPro* [8] and before used compared with the properties from *Thermocalc* [9]. The enthalpy of copper, which is 99.999% pure copper, is shown in table 3 for the different temperatures of the calorimeter.

Table 3. Thermodynamic properties of Cu

	Temperature °C	Density kg m^{-3}	Specific Heat J kg^{-1}K^{-1}	Enthalpy Jkg^{-1}	$\Delta h_{\mathrm{ref,Cu}}$ J kg^{-1}
T_{room}	20	8.96 10^3	3.84 10^2	-1.88 10^3	
$T_{\mathrm{ref,lo}}$	100	8.93 10^3	3.97 10^2	2.93 10^4	
$T_{\mathrm{ref,hi}}$	115	8.92 10^3	3.99 10^2	3.53 10^4	5.97 10^3

Table 4. Material properties for the electrodes

Temperature K	Enthalpy J kg^{-1}	Specific heat J kg^{-1} K^{-1}
ER1100		
293.15	-7.304 10^3	8.968 10^2
2383.15	2.771 10^6	1.175 10^3
ER4043		
293.15	-1.181 10^4	8.871 10^2
2273.15	2.665 10^6	1.165 10^3
ER5554		
293.15	-2.876 10^4	8.956 10^2
2083.15	2.399 10^6	1.180 10^3
ER5183		
293.15	-2.414 10^4	8.928 10^2
1893.15	2.178 10^6	1.174 10^3

5. Results

All the values obtained for aluminium electrodes are possible to see in figure 6 and 7 with the average of two points and characteristic values in table 5.

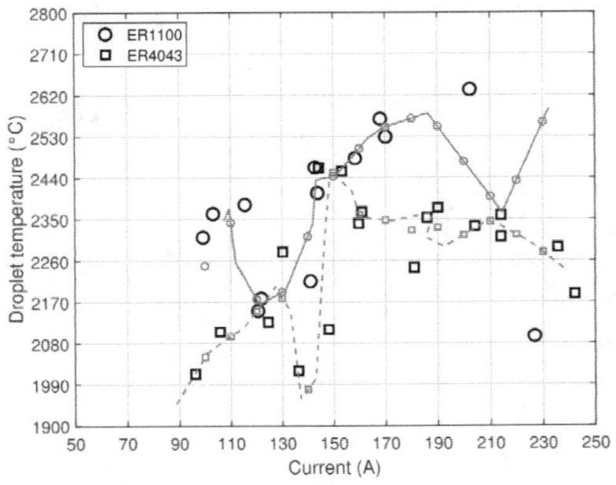

Figure 6. Droplet temperatures for electrodes ER1100 and ER4043

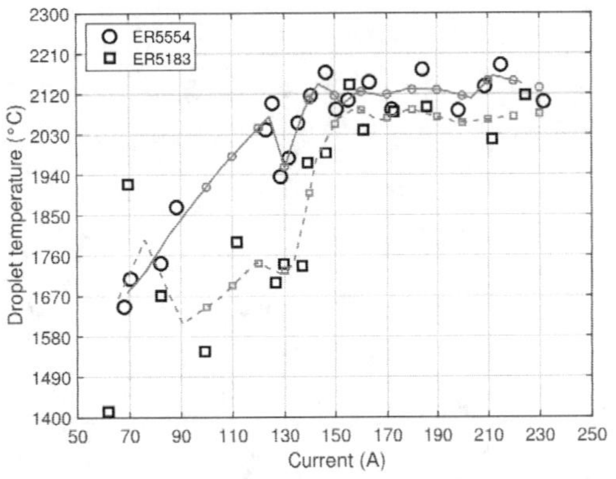

Figure 7. Droplet temperatures for electrodes ER5554 and ER5183

Table 5. Droplet temperatures for the different electrodes

Electrode -	Slope K A^{-1}	Intercept at 200 A K	Average K
ER1110	3.8108	2889.6	2640.5
ER4043	3.4360	2665.4	2494.5
ER5554	2.8027	2447.9	2296.4
ER5183	3.6137	2368.1	2145.6

Anode fall voltages have the tendency to increase with current, as can be seen in table 6, that shows the slope of the tendency line, the intercept of it at 200 A and the average of the values. As in equation 17 an increase in droplet temperature should also increase in a small amount the anode fall voltage, as can be seen in table 7 and figure 8, that shows the slope of the tendency line and the average of them.

Table 6. Anodes fall voltages for the different electrodes

Electrode -	Slope V A^{-1}	Intercept at 200 A V	Average V
ER1110	0.004543	1.066126	0.758919
ER4043	0.004292	0.604506	0.404771
ER5554	0.003267	1.010291	0.83359
ER5183	0.011291	1.802143	1.107038

6. Analysis

It is possible to see in figure 6 and 7 how the droplet temperature increases when increasing the current. With the fitting curve it is possible to see a small decay over the 100 A and below the 150 A, that is where the transition current is for this alloys and diameter. A higher curve was obtained for ER1100, after ER4043, then for ER5554 and by last ER5183 that has more alloying elements. It is important to consider the amount of alloying elements, their boiling and melting

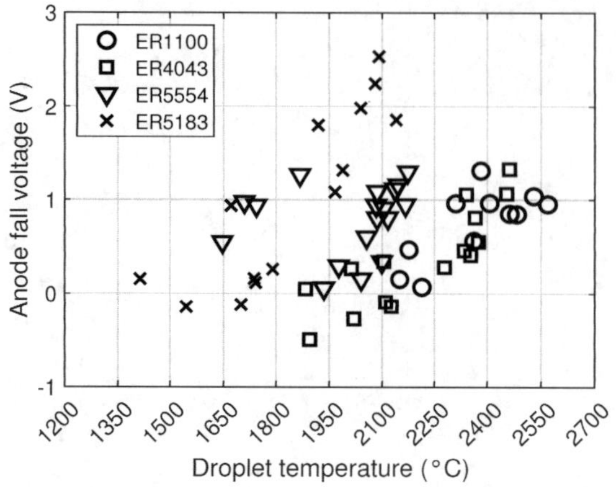

Figure 8. Anode fall voltages and droplet temperatures for aluminium electrodes

Table 7. Anode fall voltages for the different electrodes compared with temperatures

Electrode -	Slope V K^{-1}	Average V
ER1110	0.002083	0.758919
ER4043	0.002244	0.404771
ER5554	0.000588	0.83359
ER5183	0.003619	1.107038

temperatures. If the droplet is at 2500 K and the Magnesium has a boiling temperature of 1363 K, it is very likely that it evaporates at a higher rate than other element with higher boiling temperature. This evaporation takes an important amount of heat from the droplet, making it reduce its temperature. As can be seen in table 5 the average of the droplet temperature decreases for alloys ER5554 and ER5183. This two alloys have alloying elements with lower melting points than aluminium. Compared directly with the values obtained from Yan *et al.* [4] it is possible to see an increase in droplet temperature.

The anode fall voltage increases around 0.02 V per A with the increase of the current, for all of the electrodes. Even though the increase is small, the anode energy in W, increases with the WFS, meaning that the system has more energy. The anode fall voltages shown in figure 8 and in table 7 tend to increase with droplet temperature.

This approach to obtain droplet temperatures has still details to configure, but it reduces the error propagation compared with the Tin calibration. Times of data collection are a new source of error. If the time of cooling down were higher, the final cooling slope could be more accurate.

7. Conclusions

This work proposes a reliable method for the calculation of the droplet temperature with a solid state calorimeter. This solution gives new values that by now are accurate in the magnitude and behaviour of the temperature. The different alloying elements affect the behaviour of the droplet heat content. The temperature of the droplet is still a function of current as shown in previous works, as well as the alloying elements. Future work will consider more electrode compositions and diameters.

IOP Conf. Series: Materials Science and Engineering 1281 (2023) 012041 doi:10.1088/1757-899X/1281/1/012041

References

[1] Pokhodnya I and Suptel A 1967 *Automatic Welding USSR* 16–24
[2] Zhang G, Gött G and Uhrlandt D 2020
[3] Soderstrom E J 2009 *Gas Metal Arc-Welding Electrode Heat-and-Mass Transfer Mechanisms* Doctor of philosophy Colorado School of Mines
[4] Zhaoyang Z, Scott K M, Chen S and Mendez P 2022 *Welding Journal*
[5] Scott K M 2011 *Heat Transfer and Calorimetry of Tubular Ni/WC Wires Deposited with GMAW* Master of science University of Alberta Edmonton, Canada
[6] Núñez Sánchez V 2021 *Energy Balance in Gas Metal Arc Welding* Bsc. thesis Universidad de Chile - Facultad de Ciencias Físicas y Matemáticas
[7] The James F Lincoln Arc Welding Foundation 2000 The Procedure Handbook of Arc Welding (Cleveland)
[8] JMatPro *v12.4*
[9] Thermocalc *2022a*

Multiphase flow model of semi solid solidification of novel Al-15Mg_2Si-Si composite

Prosenjit Das[1]

[1]Advanced Manufacturing and Materials Processing Laboratory, Department of Materials Engineering, Indian Institute of Science Bangalore, Bengaluru-560012, India.

*Email: prosenjitdas@iisc.ac.in

Abstract. In this work, a volume averaging technique based Eulerian multiphase flow CFD (Computational Fluid Dynamics) model is developed to simulate the semi-solid solidification of the cooling slope generated Al-15Mg_2Si-4.5Si composite slurry. The four phases considered here are liquid (*l*), primary solid (Mg_2Si), secondary solid (α-Al) and air (*a*). Whereas latent heat release during solidification is tackled via a temperature correction scheme. The effect of recalescence is also observed in the present solidification model. The simulation findings include shape and size of the α-Al and primary Mg_2Si grains, solid fraction, temperature, melt viscosity etc. The developed multiphase flow model of cooling slope casting (semi solid solidification) is applied to a test case having melt pouring temperature (into the slope) of 650°C (923K). The simulation findings are then validated with the help of measurements of microstructural features observed in the experimental (optical) micrographs of the solidified samples.

1. Introduction

Semisolid metal processing (SSM) has emerged as a potential manufacturing process, in recent times, to develop near net shape engineering components. In the last 50 years, the research activities in the field of SSM has been intensified, as depicted by a number of comprehensive reviews [1-3]. In recent times, aluminum matrix composites (AMCs) grabs center stage among the family of light engineering materials. Whereas among others, Mg_2Si reinforced aluminium matrix composites offers noticeable improvements in mechanical and physical properties such as improved high temperature stability, higher wear resistance, enhanced specific strength etc. [4-6].

There is a handful of literature available on CFD based mathematical model of semisolid processing and solidification. Wang and Beckerman developed CFD model of solidification, at the onset of melt flow and grain sedimentation [7-10], in case of A356 alloy. Whereas Wu. et al [11-15] employed CFD model to investigate macro segregation, temporal temperature distribution, and grain evolution during solidification of Al-Si alloy. One of the pioneer studies on Eulerian multiphase flow model of semisolid slurry preparation during cooling slope processing is reported by Wang et al [16]. Whereas CFD based two phase flow numerical model is developed by Das et al [17, 18] to optimize the process variables of cooling slope slurry preparation technique such as slope inclination angle, degree of slope preheat as well as melt pouring temperature into the slope, in case of A356 aluminium alloy.

However, there is hardly any CFD based numerical study is available on semi solid slurry generation and subsequent solidification of an aluminum matrix composite (AMC). In the above context, the present study discusses the development of an experimentally validated multiphase flow model (CFD) of semi solid slurry preparation and subsequent solidification of novel Al-15Mg_2Si-4.5Si composite. Here, four hydrodynamic phases such as liquid (*l*), primary solid (Mg_2Si), secondary solid (α-Al) and air (*a*) are considered, where two nucleation and two mass transfer equations are solved. The developed model is capable of studying the solidification transport phenomena as well as remelting of phases of the present composite (Table 1).

Table 1: Chemical constituents of the Al-15Mg_2Si-4.5Si composite.

Si	Mg	Al
0.12	0.1	0.78

Table 2: Nomenclature of the above mentioned CFD model.

C_D	drag coefficient	c_p	specific heat, $J.kg^{-1}.K^{-1}$
d	diameter, m	m	liquidus slope K/wt%
$f_{s,\max}$	maximum solids fraction	F_l	liquid volume fraction
g	gravity acceleration, $m.s^{-2}$	f	volume fraction
K	momentum exchange coefficient, $kg.m^{-3}.s^{-1}$	h	enthalpy, $J.kg^{-1}$
U	momentum exchange , kg $m^{-2}s^{-2}$	k	thermal conductivity, $W.m^{-1}.K^{-1}$
Pr	Prandtl Number	H	heat transfer coefficient, $W.m^{-2}.K^{-1}$
P	pressure, Pa	L	latent heat, J/kg
T	Temperature, K	Q	energy exchange by heat transfer, $J.m^3.s^{-1}$
T^s	solidus temperature of alloy, K	k_p	Partition coefficient,
T^m	melting temperature of pure Al, K	T^L	liquidus temperature of alloy, K
ρ	density, kg m-3	v	grain growth rate
μ	viscosity, kg m^{-1} s^{-1}	g_a	grain growth $m.s^{-1}$
$\vec{\mathbf{u}}$	velocity vector, m. s^{-1}	$\bar{\bar{\tau}}$	stress-strain tensors, kg. m^{-1} $.s^{-2}$
D	diffusion coefficient, m^2s^{-1}	$\vec{\mathbf{u}}^*$	interface velocity, m/s
N	grain production rate, $m^{-3}s^{-1}$	M	mass transfer rate, $kgs^{-1}m^{-3}$
C	species exchange rate, $kgm^{-3}s^{-1}$	n	grain density, m^{-3}
C_{mix}	mixture concentration	c^*	interface species concentration
t	time, s	Δt	time step, s
Superscripts	p stands for phase transfer related part		
	d stands drag related part		
l,s,p,a	Stands for liquid metal, solid α-Al grain, Mg_2Si air and liquid metal-α-Al solid grain mixture (Semi solid slurry) respectively		

2. Mathematical model

In the present investigation four hydrodynamic phases are considered, wherein liquid matrix is the primary phase (l) and evolving primary solids i.e., Mg_2Si (p) and α-Al (s) as well as air (a) are the three secondary phases. Mass, energy, grain transport, and momentum conservation equations are solved for individual phases, as per methodologies of Eulerian multiphase flow technique. Same pressure field for each phase is considered. In this model, solid movement is allowed within the bulk melt. Whereas a mixture model of viscosity is adopted to calculate the flow behaviour of the solidified grains. The semi solid slurry is produced via cooling slope processing, so, both type of solid grains i.e., Mg_2Si and α-Al are considered as globular. It is to be noted here that the shear flow experienced by the composite melt during cooling slope processing facilitates dendrite to globular transformation of primary solid grains. Independent nucleation and mass transfer models are employed here for the primary solid (Mg_2Si) as well as secondary solid (α-Al). Although solidified grain is actually solid but here it is considered as fluid. However, the drag co-efficient is determined via appropriate exchange model relevant to the fluid-solid exchange. Further, exchange models relevant to the fluid-solid exchange and fluid-fluid exchange are considered for the solid-air and liquid-air interfaces, respectively. The developed multiphase flow CFD model is found to be capable of predicting the volume fraction of constituent phases, diameter/size of the α-Al and primary Mg_2Si grains, solid fraction, temperature, melt viscosity etc. The thermo solutal convection is also considered in this model. Model description us elaborated in our earlier publications [19, 20].

2.1 Governing equations

Mass conservation equation:

$$\frac{\partial}{\partial t}(f_i \rho_i) + \nabla \cdot (f_i \rho_i \vec{\mathbf{u}}_i) = M_{ji} + M_{ki} \text{ where, } \sum_{i=1}^{4} f_i = 1 \quad (i = l, s, p, a) \tag{1}$$

Momentum conservation equation:

$$\frac{\partial}{\partial t}(f_i \rho_i \vec{\mathbf{u}}_i) + \nabla \cdot (f_i \rho_i \vec{\mathbf{u}}_i \otimes \vec{\mathbf{u}}_i) = -f_i \nabla p + \nabla \cdot \bar{\bar{\tau}}_i + f_i \rho_i \mathbf{g} + \mathbf{U}_{ji} + \mathbf{U}_{ki} \text{ where, } \bar{\bar{\tau}}_i = \mu_i f_i \times \left[\nabla \cdot u_i + (\nabla \cdot u_i)^T\right] \tag{2}$$

Energy conservation equation:

$$\frac{\partial}{\partial t}(f_i \rho_i h_i) + \nabla \cdot (f_i \rho_i \vec{\mathbf{u}}_i h_i) = \nabla \cdot (f_i k_i \nabla T_i) + Q_{ji} + Q_{ki} \text{ where, } h_i = \int_{Tref}^{T_i} c_{p(i)} dT + h_i^{ref} \tag{3}$$

Species transport equation:

$$\frac{\partial}{\partial t}(f_i \rho_i c_i) + \nabla \cdot (f_i \rho_i \vec{\mathbf{u}}_i c_i) = \nabla \cdot (f_i D_i \nabla c_i) + C_{ij} \tag{4}$$

Grain transport equation:

$$\frac{\partial}{\partial t}(n) + \nabla \cdot (u_i n) = N \tag{5}$$

Source term and exchange term

$$\text{Nucleation- } \frac{dn}{d\Delta T} = \frac{n_{\max}}{\sqrt{2\pi} \cdot \Delta T_\sigma} \exp\left(-\frac{1}{2}\left(\frac{\Delta T - \Delta T_N}{\Delta T_\sigma}\right)^2\right) \tag{6}$$

Grain growth and mass transfer-

$$M_{ls} = -M_{sl} = v \times (n \times d_s^2) \times \rho_s \times f_l \text{ and } M_{lp} = -M_{pl} = g_{a2} \times \frac{c_l^* - c_l}{c_l^*} \times (n_p \times 6.324 \times a_p^2) \times \rho_p \times f \tag{7}$$

$$\text{Momentum transfer- } U_{ji} = -U_{ij} = U_{ij}^p + U_{ji}^d \text{ where, } U_{ij}^p = u^* \times M_{ij} \text{ and } U_{ij}^d = K_{ij}(u_i - u_j) \tag{8}$$

$$\text{Enthalpy transfer- } Q_{ls} = -Q_{sl} = Q_{ls}^d + Q_{ls}^p, \; Q_{ls}^p = h^* \cdot M_{ls}, \; Q_{ls}^d = H_{ls}^*(T_l - T_s) \tag{9}$$

$$\text{Species transfer- } C_{ls} = -C_{sl} = C_{ls}^p = c^* \times M_{ls}, \text{ where, } c^* = k \times c_l \tag{10}$$

Auxiliary term

$$\text{Viscosity: } \mu_{mix} = \mu_l \left(1 - \frac{f_p + f_s}{f^{crit}}\right)^{-2.5 f^{crit}} \quad \text{Diameter: } d_s = \sqrt[3]{\frac{6 f_s}{\pi n_s}} \tag{11}$$

IOP Conf. Series: Materials Science and Engineering 1281 (2023) 012042 doi:10.1088/1757-899X/1281/1/012042

2.2 Numerical procedure

In this study, Ansys Fluent software is used as a solver platform to find solutions to the above depicted governing equations. Control volume technique is adopted to find numerical solutions of the governing equations, employing Phase Coupled SIMPLE algorithm. Here, cooling slope slurry generation and subsequent solidification process of the said composite is simulated via Eulerian multiphase flow approach. C based codes are written and compiled with the FLUENT solver to define the nucleation model, viscosity evolution, absorption and release of latent heat and associated scheme of temperature correction as well as source terms relevant to the conservation equations. Here, time step (Δt) value is chosen as 0.001s based on trial numerical experiments, wherein individual time steps are divided into a set of 20 iterations. The chosen time step size and the number of iterations are found to be successful to yield the optimum simulation results/findings as well as convergence [19, 20]. The conditions prevail at the beginning of the simulation study as well as at the boundaries of the computational domain are depicted below in Fig. 1.

Table 3: Material properties and nucleation parameters considered for the Al-15Mg_2Si-4.5Si composite.

	Nucleation parameter	
$\rho_l = \rho_0[1-\beta_T(T-T_0)-\beta_c(c_l-c_0)]$	**For αAl phase:**	**For Mg_2Si phase:**
$\rho_0 = 2606$ kg m^{-3}	$n_{max} = 4\times10^{14}$	$n_{max} = 3.6\times10^{14}$
$\beta_T = 1\times10^{-4}$ K^{-1}, $\beta_c = -9.2\times10^{-3}$ (wt%)$^{-1}$	$\Delta T_\sigma = 8$K	$\Delta T_\sigma = 4$K
ρ_s, ρ_p, ρ_a = 2700,1998,1.225 kg.m^{-3}	$\Delta T_N = 20$K	$\Delta T_N = 10$K
k_l, k_s, k_p, k_a = 167, 167, 9.71, 0.0242 W/m.K	$T^L = 863$K	$T^L = 906$K
$c_{p(l)}, c_{p(s)}, c_{p(p)}, c_{p(a)}$ =1082, 963, 863, 1006.43 J/kgK	$T^S = 848$K	$T^S = 863$K
$D_l = 5\times10^{-9}$ m^{2}s^{-1}, $d_a = 10^{-5}$m, $f^{crit} = 0.74$	m = -187.5K	m =-115K
μ_l, μ_a = 1.3×10^{-3}, 1.8×10^{-5} kg m^{-1} s^{-1}	k=0.145	$g_{a2} = 1\times10^{-4}$

IOP Conf. Series: Materials Science and Engineering 1281 (2023) 012042 doi:10.1088/1757-899X/1281/1/012042

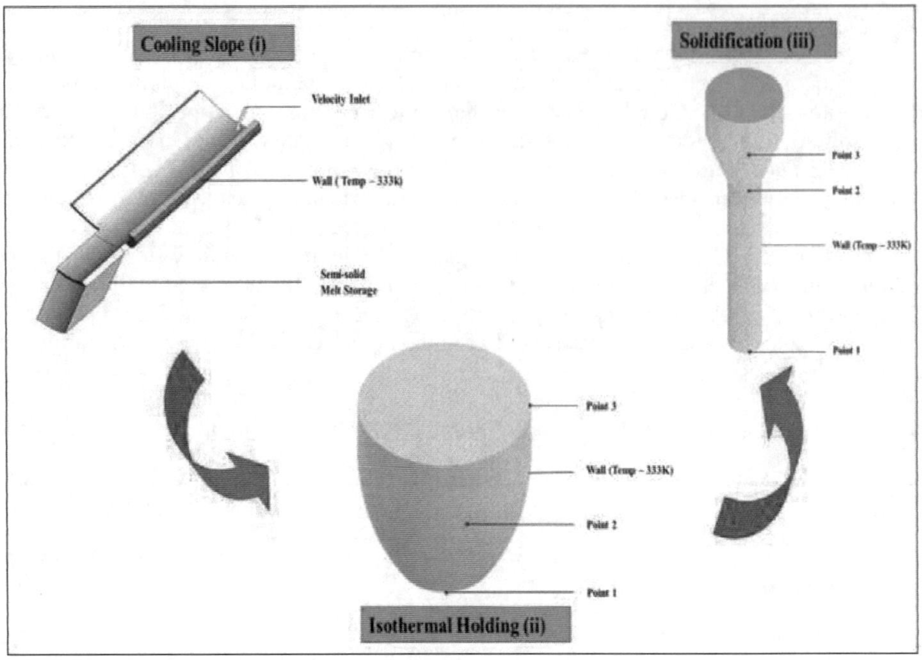

Figure 1. Computational domain showing boundary and initial conditions.

3. Semi solid slurry generation using cooling slope

The developed multiphase flow model of cooling slope casting (semi solid solidification) is applied to a test case where melt flow length, melt inlet velocity as well as melt pouring temperatures are assumed as 500 mm (over the cooling slope), 1.25 m/s (calculated from experiment), and 650°C (923K), respectively. Due to the sudden chilling at the contact zone of the impinging composite melt and cooling slope, a lot of nuclei are generated. The nuclei's are then transported downstream along the slope via freshly poured upstream melt. The above discussed crystal separation mechanism facilitates globular grain structure of primary Mg_2Si as well as α-Al grains, instead of dendritic morphology. After formation, nuclei's experiences ripening driven growth and yields maximum average grain size and solid content in the slurry, at slope leaving state. So, the cooling slope slurry generation technique is a periodic process i.e., pouring - nucleation – growth – flow -pouring, wherein temperature reduces, whereas solid fraction and viscosity of the composite slurry increases till the cooling slope exit.

3.1 Temperature and solid fraction:

The superheated melt experiences forced convection driven cooling during its travel down the slope, which leads to its temperature drop to 861K starting from the melt pouring temperature of 923 K, as shown in Fig. 2. Relatively cooler slope free surface (333 K) causes burst nucleation within the impinging melt, whereas forced convection driven cooling/temperature drop causes progressive nucleation and growth of solid grains within the composite melt. The above mechanisms causes maximization of slurry solid content up to the slope exit. Temperature and solid evolution curves of the flowing melt is shown below in Fig. 3. Figure 3 confirms near maximization of primary Mg_2Si content of 0.18 till the inception of α-Al formation within the composite melt. Accordingly, the total slurry solid content of 0.36 is equally distributed between α-Al and primary Mg_2Si phases, as evident from Fig. 4.

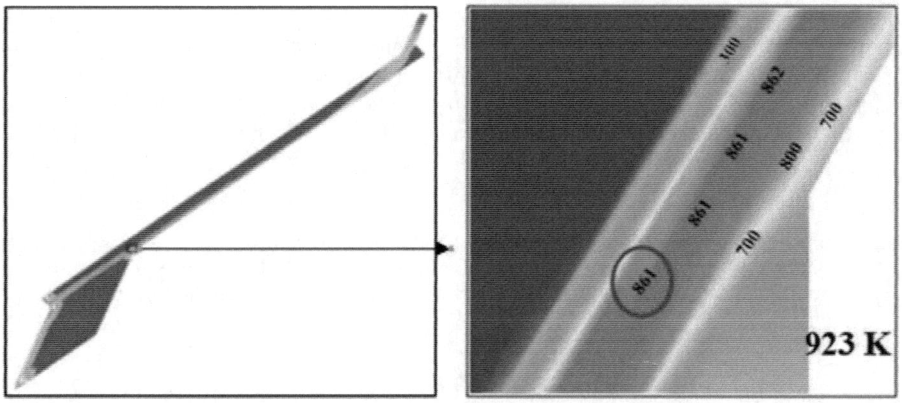

Figure 2. Temperature distribution at the slope exit.

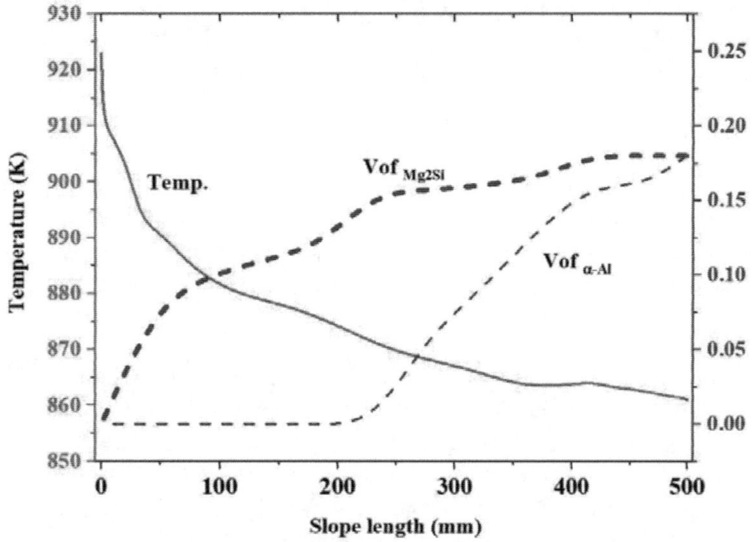

Figure 3. Temperature and solid evolution curves of the composite slurry during its downstream transport over the slope free surface.

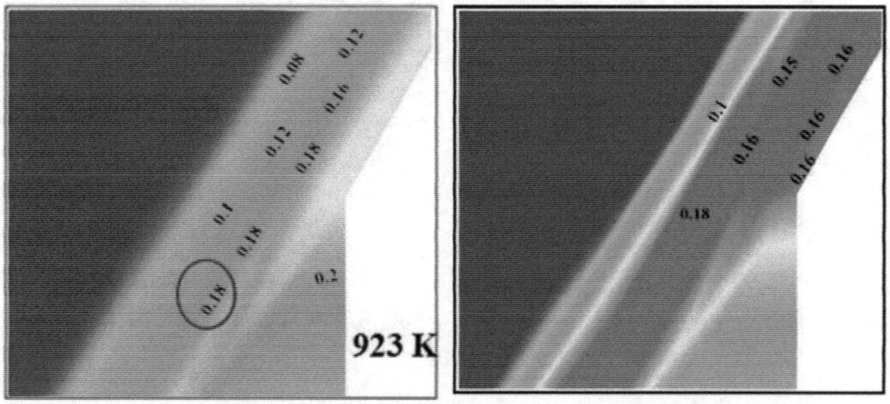

Figure 4. Solid fraction (αAl and Mg_2Si) distribution at the slope exit.

3.2 Velocity and Segregation

Decreasing melt temperature and increasing solid content is found to be accompanied with the increasing melt velocity during slurry flow towards the slope exit. Dominant effect of gravitational acceleration over melt viscosity increase is found to be responsible for increasing slurry velocity along the slope. The slope exit slurry velocity is found to be 2.91 m/s, which is significantly higher than the melt inlet velocity of 1.25 m/s (Fig. 5a)

Figure 5b shows the evolution of micro segregation of solutal Si during melt flow over the slope. Here, micro segregation is triggered by solute portioning at the solidification interface, which reaches its maxima till the slope exit. Progressive increase in solid content of the melt (Fig. 3) triggers increasing interfacial (solid-liquid interface) accumulation of solutes and facilitates its maxima at the slope exit.

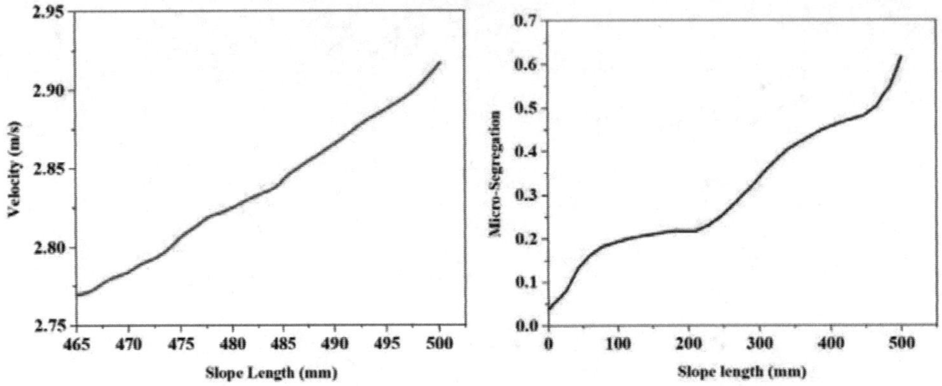

Figure 5. (a) Velocity plot for the rear part of melt flow length, (b) Evolution of micro-segregation within the liquid phase during cooling slope slurry processing.

3.3 Grain Structure

Post nucleation at the melt contact zone (primary Mg_2Si) and at approx. 220 mm slope length (α-Al), the nuclei's experiences diffusion driven growth up to the slope exit. Here, Hexahedral and equiaxed morphology is assumed for the evolving primary Mg_2Si and α-Al grains, respectively, at their inception/formation. However, they experiences shear driven spherodization up to the slope exit. Sphericity evolution for both the primary grains are shown in Fig. 6a and 6b, respectively, for the rear part of melt flow length. Whereas their grain size distribution is compared at slope exit in Fig. 7.

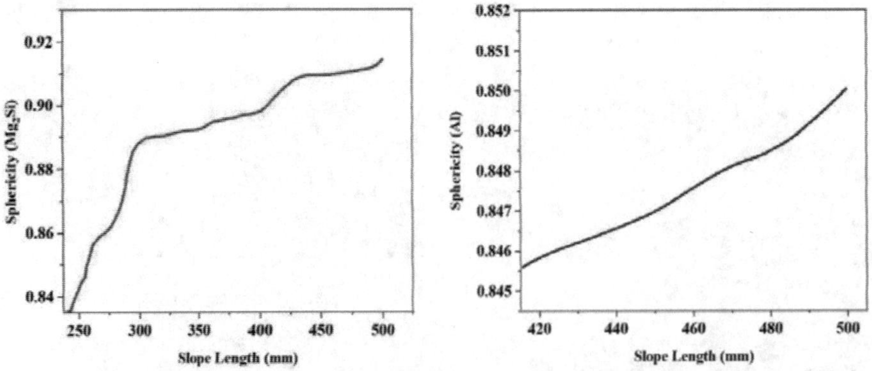

Figure 6. Sphericity plot at the slope exit region: (a) primary Mg_2Si and (b) α-Al grains.

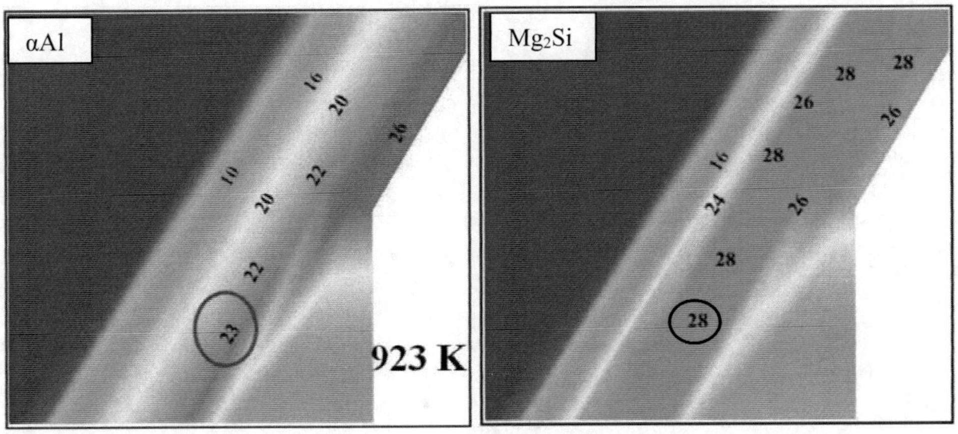

Figure 7. Grain size distribution of (a) α-Al and (b) primary Mg_2Si grains at slope exit region.

3.4 Isothermal holding

Post slurry preparation using cooling slope, grain coarsening is studied in the present CFD model by invoking the modified Lifshitz-Slyozov Wagner (LSW) equation [19, 20]. The modified LSW equation (given below) quantifies the individual effect of Ostwald ripening as well as coalescence on grain coarsening behaviour of the generated slurry.

$$d^n - d_0^n = at^b \tag{12}$$

where d_0, d, t, n, a, and b denote the initial grain dia of respective primary phases, grain dia at any given time, coarsening exponent (n=3 as per convention of LSW theory), coefficient for Ostwald ripening, coefficient for coalescence, respectively. In this study, experimentally determined values of Ostwald ripening as well as coalescence coefficients of individual phases (α-Al and primary Mg_2Si) are invoked in the multiphase flow CFD model. Morphological evolution of primary grains (α-Al and primary Mg_2Si) in terms of average grain dia. as well as degree of sphericity, after 8 minutes of slurry holding, are found to be 31.41μm, 0.85 and 28 μm, 0.86, respectively.

4. Solidification

In this case, instantaneous filling of semi solid melt is assumed within the cylindrical mould, after undergoing isothermal holding. Due to the chilling effect at the contact zone of liquid melt and mould wall a no. of new nuclei's (mostly α-Al) are generated near the mould wall. Thereafter, heat of the semi solid melt is extracted continuously by the mould wall (523K), which facilitates its solidification. Here, grain transport and melt convection is considered for the simulation purpose, wherein no-slip boundary condition is defined at the mould wall-composite melt interface. Temperature distribution of the solidified billet is shown at the final instance of simulation i.e., at 2.5s (Fig. 8a). Whereas distribution of primary Mg_2Si content within the solidified billet of the afore mentioned composite is shown below in Fig. 8b. Finally, temporal evolution of α-Al content is shown graphically for three chosen points of the simulation domain (Fig.1). Figure 8b also demonstrates the effect of simultaneously occurring remelting as well as solidification phenomenon on α-Al volume fraction.

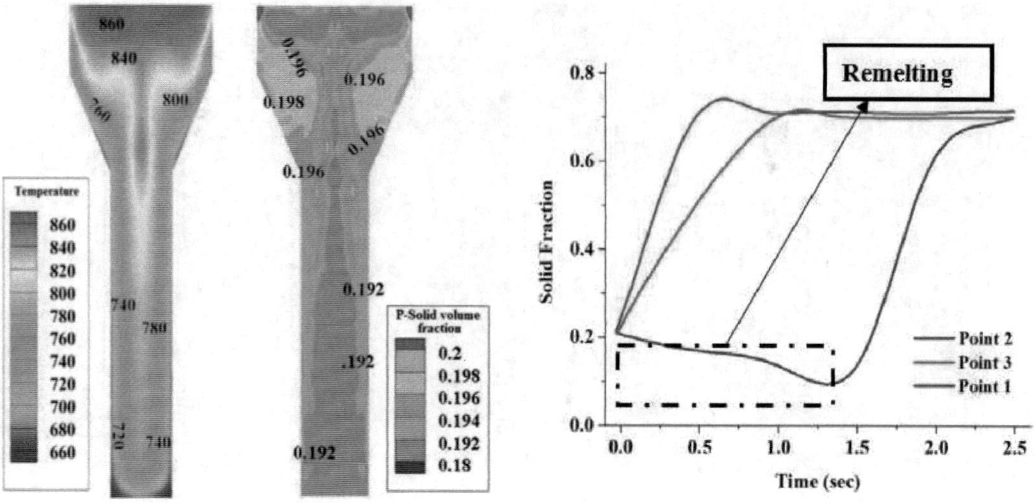

Figure 8. Distribution of (a) temperature and (b) Mg_2Si volume fraction within the solidified billet, and (c) temporal evolution of α-Al volume fraction within the composite melt.

4.1 Grain evolution

As solidified grains moves to casting surface, so the grain diameter in the surface area is bigger at initial time steps. Whereas the situation reverses up to the end of solidification, wherein grain diameter at the central region of the solidified billet is found to be bigger than the surface region for both the primary phases (Fig. 9). Slight decrease invα-Al grain dia. is noted at the time of remelting, since outer surface of α-Al grain melts first during remelting. This happens due to the lesser solidus temperature of α-Al phase than that of primary Mg_2Si. At the end of solidification, noted avg. grain dia. and degree of sphericity values are found to be 50μm, 0.8 and 29 μm, 0.9, for α-Al and primary Mg_2Si grains, respectively.

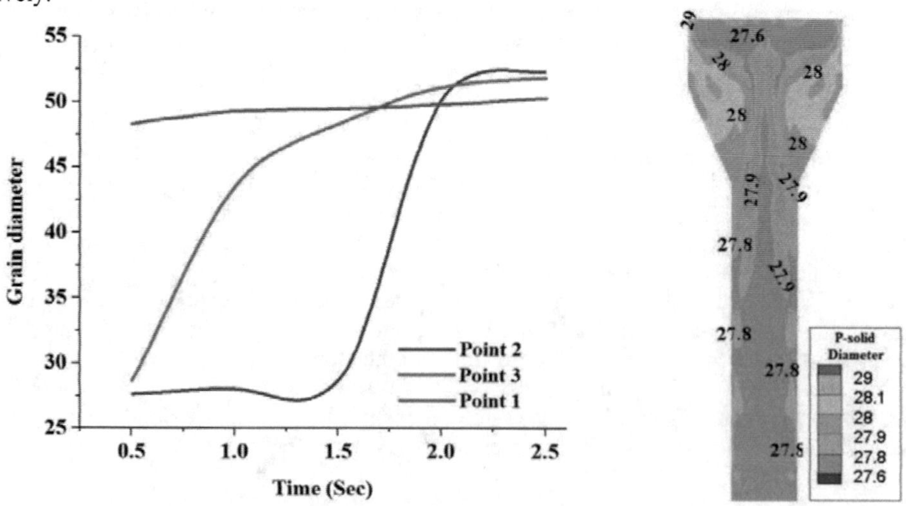

Figure 9. (a) Temporal evolution of α-Al grain diameter within the composite melt, and (b) grain size distribution of primary Mg_2Si within the solidified billet.

4.2 Experimental validation

Accuracy of the simulation findings are validated by performing cooling slope casting experiment of the novel Al-15Mg₂Si-4.5Si composite. Post solidification, metallographic samples are obtained from representative location of the composite billet and the resulting optical micrograph is shown below in

Fig. 10. The experimentally measured grain dia. and sphericity values are noted as 50μm, 0.79 and 26.64 μm, 0.88, or α-Al and primary Mg_2Si grains, respectively. The experimental findings are found to be in close agreement with the above mentioned simulation results.

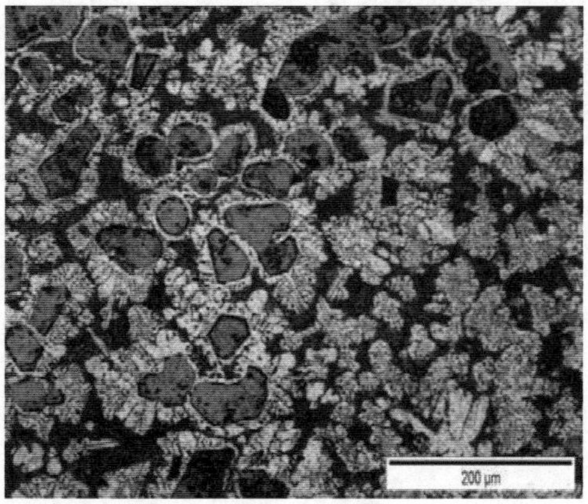

Figure 10. Optical microstructure of the Rheo/semi solid cast Al-15Mg_2Si-4.5Si composite.

5. Conclusion

The concluding remarks are listed below.
i. The developed Eulerian multiphase flow solidification model of cooling slope casting process of novel Al-15Mg_2Si-4.5Si composite is found to be capable of accurate prediction of microstructural morphology of the solidified composite.
ii. The developed semi solid solidification model predicts the slurry morphology with reasonable accuracy, at slope exit state as well as after undergoing isothermal coarsening for a finite time.
iii. The simulations findings successfully captures the recalescence and remelting of α-Al grains, happens during solidification of the composite, starting semi solid state.
iv. Close agreement between the experimental measurements of slurry microstructure in terms of avg. grain dia. and sphericity of α-Al and primary Mg_2Si grains (50μm, 0.79 and 26.64 μm, 0.88) to that of simulation findings (50μm, 0.8 and 28 μm, 0.9) establishes applicability of the developed semi solid solidification model of the novel Al-15Mg_2Si-4.5Si composite as a process control tool.

Acknowledgement
The author would like to thank SERB, DST, Government of India for their financial support to this work vide grant no. TTR/2021/000113. Author also like to acknowledge former Project JRF Mr. Chayan Das for his support with the data analysis. Finally, the author would like to express his gratitude to Director, IISc for his encouragements.

References:

[1] Flemings MC. “Behavior of metal alloys in the semisolid state.” Metallurgical Transactions A, vol. 22, pp. 957-981, 1991.
[2] Atkinson HV. “Modelling the semisolid processing of metallic alloys.” Progress in Materials Science, vol. 50, pp. 341-412, 2005.
[3] Fan Z. “Semisolid metal processing.” Int. Mater. Rev., vol. 47(2), pp. 49–85, 2002.
[4] Zhang J, Fan Z, Wang YQ, Zhou BL. “Microstructural development of Al–15wt.%Mg_2Si in situ composite with mischmetal addition” Mater Sc and Eng A, vol 281, pp. 104-112, 2000.
[5] Qin QD, Zhao YG, Zhou W, Cong PJ. "Effect of phosphorus on microstructure and growth manner of primary Mg_2Si crystal in Mg_2Si/Al composite" Mater Sc and Eng A, vol 447, pp. 186-191, 2007.
[6] Li C, Wu Y, Li H, Liu X. "Microstructural formation in hypereutectic Al–Mg_2Si with extra Si" Journals of alloys and compounds, vol 477, pp. 212-216, 2009.
[7] Wang CY, Beckermann C. "Equiaxed Dendritic Solidification with Convection: Part 1. Multi-Scale /-Phase Modeling" Metall. Mater. Trans. A, vol. 27, pp. 2754-2764, 1996.

[8] Wang CY, Beckermann C. "Equiaxed Dendritic Solidification with Convection: Part 2. Numerical Simulations for an Al-4wt% Cu Alloy" Metall. Mater. Trans. A, vol. 27, pp. 2765-2783, 1996.
[9] Beckermann C, Ni J. "Simulations of Sedimentation in Equiaxed Alloy Solidification" Int. Comm. Heat Mass Transfer, vol. 23, pp. 315-324, 1996.
[10] Wang T, Wu M, Ludwig A, Abondano M, Pustal B, Buhring-Polaczek A. "Modelling the thermo solutal convection, shrinkage flow and grain movement of globular equiaxed solidification using a three phase model" International Journal of Cast metals research, vol.18(4), pp. 221-228, 2005.
[11] Wu M, Ludwig A, Buhring-Polaczek A, Fehlbier M, Sahm PR. "Influence of convection and grain movement on globular equiaxed solidification" International journal of heat and mass transfer, vol. 46, pp. 2819-2832, 2003.
[12] Wu M, Ludwig A. "A Three-Phase Model for Mixed Columnar Equiaxed Beckermann C Solidification" Metallurgical and Materials Transactions, vol. 37, pp. 1613-1631, 2006.
[13] Wu M, Ludwig A. "Using a Three-Phase Deterministic Model for the Columnar-to- Equiaxed Transition" Metallurgical and Materials Transactions A, vol. 38, pp. 1465-1475, 2007.
[14] Wu M, Ludwig A. "Modeling equiaxed solidification with melt convection and grain sedimentation-I. Model description" Acta Materialia, vol. 57, pp. 5621-5631, 2010.
[15] Wu M, Ludwig A. "Modeling equiaxed solidification with melt convection and grain sedimentation-II. Model verification" Acta Materialia, vol. 57, pp. 5632-5644, 2010.
[16] Wang T, Pustal B, Abondano M, Grimming T, Buhring-Polaczek A, Wu M, Ludwig A. "Simulation of cooling channel rheocasting process of A356 aluminum alloy using three phase volume averaging model" Trans. Nonferrous Met. Soc. China, vol. 15(2), pp. 389-394, 2005.
[17] Das P, Samanta SK, Chattopadhyay H, Dutta P. "Effect of pouring temperature on cooling slope casting of semi-solid Al-Si-Mg alloy" Acta Metall. Sin. (Engl. Lett.), vol. 25(5), pp. 329-339, 2012.
[18] Das P, Samanta SK, Chattopadhyay H, Sharma BB, Dutta P. "Eulerian two-phase flow simulation and experimental validation of semisolid slurry generation process using cooling slope" Materials Science and Technology, vol. 29(1), pp. 83-92, 2013.
[19] Das P, Samanta SK, Mondal B, Dutta P. "Multiphase Model of Semisolid Slurry Generation and Isothermal Holding During Cooling Slope Rheoprocessing of A356 Al Alloy." Metallurgical and Materials Transactions B, vol. 49 (2018), pp. 1925 -1944, , 2013.
[20] Das P. "Effect of process variables and melt treatment conditions during cooling slope semi solid slurry generation process of Al-7Si-0.3 Mg alloy." Materials Today Communications, vol. 33, 104944, 2022.

Integrated property predictions during casting and heat treatment of Al-Si-Cu-Mg alloys by precipitation simulation

Fengxin Mao[1*], Moritz Weidt[1,2], Frank Roeser[1], Jesper Thorborg[1], Luis Barrales-Mora[3]

1 MAGMA Gießereitechnologie GmbH, Kackertstraße 16-18, 52072, Aachen, Germany

2 MAGMA Engineering (Suzhou) Co., Ltd, No. 98 Suhui Road, Suzhou Industrial Park, Jiangsu Province, 215021 China

3 Georgia Institute of Technology—CNRS 2958, George W. Woodruff School of Mechanical Engineering, Metz Technopole, 2 Rue Marconi, 57070 Metz, France

f.mao@magmasoft.de

Abstract. The production and manufacture of aluminium castings typically involves multiple heat treatment process steps. Simulation tools are very useful to help predict the evolution of mechanical properties, to improve process parameters and as a guideline for casting design. In this study, a kinetic model, based on Nucleation and Growth Theory (CNGT), was developed and implemented for modelling heat treatment in a casting process simulation tool. Starting from information generated from the simulation of solidification and cooling of the casting, simulation was carried out to predict further precipitation behaviour of multiple Cu or Mg-Si -containing phases during solution treatment, quenching, and artificial ageing processes in heat treatment of Al alloys. Model validation was carried out within the research cluster "Advance Metals and Processes" in Aachen, Germany. Heat treatments were performed on two different Al-Si-Cu-Mg alloys with different quench methods and at different ageing temperatures. Experimental investigations including atom probe, wavelength dispersive X-ray spectroscopy, and tensile tests were conducted to calibrate and validate the model. The simulations showed good agreement for the studied conditions and a promising modelling approach for integrated simulation of casting and heat treatment for industrial applications.

1. Introduction

Cast aluminum alloys have a wide application in various industrial products. The manufacturing of castings typically involves solidification and multiple heat treatment process steps, during which the microstructure undergoes complex phase transformations [1]. The property evolution of heat treatable castings during solution treatment, quenching, and ageing is a complex process that has motivated many detailed investigations with the objective to determine optimal industrial parameters as well as alloying composition while maximizing mechanical properties. In Al-Si-Cu-Mg alloys, precipitates have been found by various experimental methods [2–6] to be Al_2Cu, Al_2CuMg, Mg_2Si, and their metastable forms.

IOP Conf. Series: Materials Science and Engineering 1281 (2023) 012043 doi:10.1088/1757-899X/1281/1/012043

It has also been observed in these investigations that the precipitation sequence is mostly influenced by the alloying content of Mg and Cu [2, 7, 8].

The simulation of solid-state precipitation is an important tool in industry that is used to predict the evolution of the microstructure and thus, the mechanical properties of structural materials. Intensive works have been made by applying different modelling methods to simulate precipitation and microstructure evolution in aluminum alloys. Among the various methods, classical nucleation and growth theory (CNGT) based approaches has rendered itself as an excellent option owing to its accuracy and computational performance in predicting particle growth in various aluminum alloys [4, 9–13]. In this study, a CNGT based kinetic precipitation model has been developed to simulate the dissolution and formation of particles during solution treatment, quenching, and artificial ageing treatment of Al-Si-Cu-Mg alloys. The evolution of chemical composition and yield strength has been predicted and compared to experimental investigations. The simulation has shown promising results and good agreement for the studied cases.

2. Experimental

2.1. Material and heat treatments

The investigation was conducted in two commercial Al-Si-Cu-Mg cast aluminum alloys, designated A1 and A2. Both alloys were provided by Nemak Dilligen GmbH, Dillingen, Germany. Their alloying composition is shown in Table 1. The alloys were cast into a stepped wedge mold in a permanent mold tilt casting process. The casting procedure was carried out by the Gießerei Institut (GI), RWTH Aachen University, Aachen, Germany. Heat treatments involved in this study are summarized in Table 2. Solution treatment was conducted in an air furnace followed by water and air quenching, resulting in quench rates of 150 Ks^{-1} and 1Ks^{-1}, respectively. The samples were then aged at 200 °C for up to 4 hours in an air furnace.

Table 1: Alloying composition of the investigated alloys

	Si (wt. %)	Mg (wt. %)	Cu (wt. %)
A1	~8.0	<0.2	3.0
A2	~7.0	0.4	0.5

Table 2: Heat treatments of the investigated alloys

	Solution treatment	Quenching	Artificial ageing
A1	485 °C for 4h	water and air	200 °C
A2	530 °C for 2h		

2.2. Validation experiments

2.2.1. Tensile tests. To determine the mechanical properties of the alloys, tensile tests were conducted at the Gesellschaft für Werkstoffprüfung mbH (GWP), branch office Saarland, Dilligen, Germany, on the quenched and aged specimens. Three tests, at room temperature, were performed at each condition using a strain rate of $10^{-3/s}$. In turn, the yield strength was measured using the $\sigma_{0.2}$ characteristic stress on the flow curve.

2.2.2. Wavelength dispersive X-ray spectroscopy (WDX). WDX measurements were carried out at the Gemeinschaftslabor für Elektronenmikroskopie (GFE), RWTH Aachen University, Aachen, Germany.

Line scans over a range of 400 μm in length for Mg and Cu content were carried out for the quenched samples.

2.2.3. Atom probe tomography (APT). APT experiments were performed at Materials Chemistry Institute (MCh), RWTH Aachen University, Aachen, Germany. Specimens of both alloys were quenched and stored in liquid Nitrogen. Times from taking sample out to APT measurements were approximately 7–8 hours. The sample tip with a radius of 35–50 nm were made and a scan over a cylinder area of 10 nm in radius and 100 nm in length was carried out. Local composition of Mg and Cu were evaluated.

3. Model description

For modelling precipitation, the algorithm introduced in ref. [13] was further developed and extended to consider the precipitation behavior during solution treatment and quenching. Particles of multiple co-existing phases were discretized into bins by an "Euler-like" method [14]. In the model, particles were assumed to have spherical shapes. Heterogeneous nucleation and elastic distortion of the Al–matrix was neglected. The algorithm is shown in Figure 1 as a flow chart.

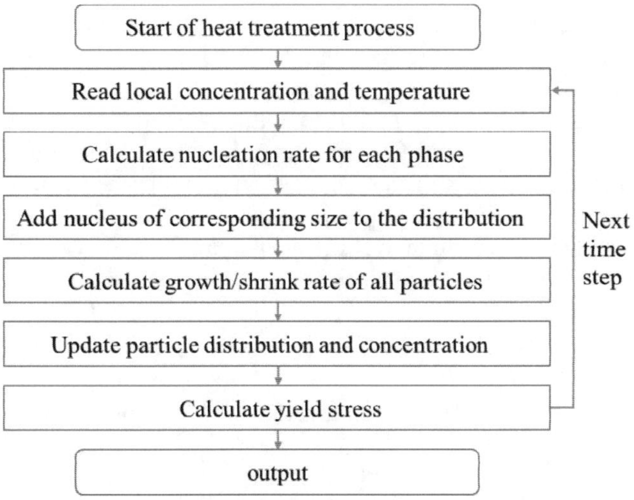

Figure 1 Flowchart of the applied model

The precipitate evolution with time is determined essentially by the nucleation and the growth rate. The particles are initialized with a critical size at a temperature and concentration dependent rate and the gradient of the concentration is calculated to drive the precipitate to either grow or shrink. The concentration and temperature is updated at every step, thus the time step should be adapted to a fine enough range to prevent numerical disturbance to the kinetics. The yield stress evolution was calculated by applying the approach in previous work [13], assuming a linearly additive contribution of hardening effects of all particles.

The nucleation rate $\dot{N}$ of all possible particle phases is calculated by Equation (1):

$$\dot{N}(t) = N_0 exp\left(-\frac{16\pi}{3}\frac{f_0\gamma^3}{kT}\left(\frac{V_m}{\Delta g_T}\right)^2\right) \quad (1)$$

where N_0 is the total number of all possible homogeneous nucleation sites in an aluminum matrix, V_m is the molar volume of the precipitate, f_0 is the effective homogeneous nucleation factor, γ is the interface energy between particle and matrix, and Δg_T is the chemical driving force of the corresponding phase.

Growth/shrinkage of the particles is driven by the concentration gradient between the matrix and particle–matrix interface of all alloying elements that contribute to phase formation. This diffusion–controlled rate $\dot{r}^k$ of element j in phase k is described by Equation (2):

$$\dot{r}^k = \left(r \sum_j \left[D_j \frac{c_j^m - c_j^{i,k}}{c_j^k - c_j^{i,k}} \right]^{-1} \right)^{-1} \tag{2}$$

where r is the radius of the particle (assumed spherical for simplification), D is the diffusion coefficient. Matrix concentration, particle-matrix interface concentration and particle concentrations are denoted, respectively, as c^m, c^i and c^k. Finally, the contribution of yield stress by precipitation hardening of phase k can be calculated from Equation (3):

$$\sigma_k = \frac{ME_{disl}}{b} \left(\frac{6f_k}{\pi} \right)^{1/2} \frac{f_s}{r_k} \tag{3}$$

where M is the Taylor factor, E_{disl} is the dislocation energy, b is the Burgers vector, r is the particle size, f_k is the particle volume, and f_s is the shearing factor which equals to 1 or r/r_s when the particle is larger or smaller than the shearable radius r_s, respectively.

The model, as explained above, was applied in solution treatment to determine the dissolution of particles that were formed during solidification. In order to simulate the clustering effect and the depletion of solute atoms in the matrix during quenching, two phases, namely Mg_xSi_y and Al_xCu_y were added into the model and allowed to nucleate with a small size around one nanometer. Ageing was then simulated at 200 °C to evaluate the influence of the quench rates on the subsequent age hardening effect. Parameters of each phase calculated in this work are listed in Table 3.

Table 3: Applied phase parameter in this study

	γ (J m^{-2})	f_0 (–)	r_s (nm)
Mg_xSi_y	0.20	0.10	5.0
Al_xCu_y	0.20	0.05	5.0
Mg_2Si	0.20	0.10	10.0
Mg_5Si_6	0.16	0.14	4.0
Mg_9Si_5	0.17	0.12	5.0
Al_2Cu	0.20	0.08	10.0
Al_2CuMg	0.16	0.05	5.0

4. Result and discussion

4.1. Chemical composition evolution

As an example, the result of the WDX line scan of the water quenched A2 sample is shown in Figure 2. Several "valleys" can be observed for both Mg and Cu concentration. This is believed to be a consequence of particles that were removed from the sample surface during sample preparation, which results in a drop of concentration when these regions are scanned. The evaluation of the matrix concentration suggested a matrix concentration of 0.38 wt.% Mg and 0.50 wt.% Cu, indicating a relatively full solutionized condition.

Results of the APT measurements conducted to the specimen of Figure 2 are shown in Figure 3. The scan was focused on a local area and aimed for a detailed atom distribution in this particular sample tip.

The 40 nm scan indicated an average concentration of ~0.19 at.% Mg and ~0.43 at.% Cu (0.39 wt.% Mg and 0.45 wt.% Cu, respectively) which agrees well with the WDX measurements.

The evaluated concentration from both WDX and APT (with standard deviation) are plotted and compared against simulation results in Figure 4 & 5. Both measurements indicate a significant concentration loss in the matrix for air–quenched conditions compared to water–quenched samples. APT results deliver slightly less concentration compared to WDX, especially in the case of air quenched Alloy 1 samples. The relative high concentration of Cu in Alloy 1 caused a higher scattering of WDX measurements, resulting in larger standard deviations. Future WDX experiments should be carried out on a different position of the castings as well as at faster quench rates to minimize the scattering in data, and obtaining a better calibration of the model.

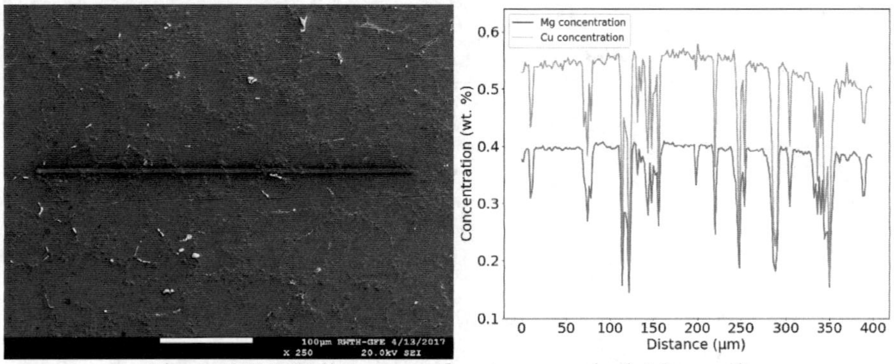

Figure 2 WDX line scan of water quenched A2 sample

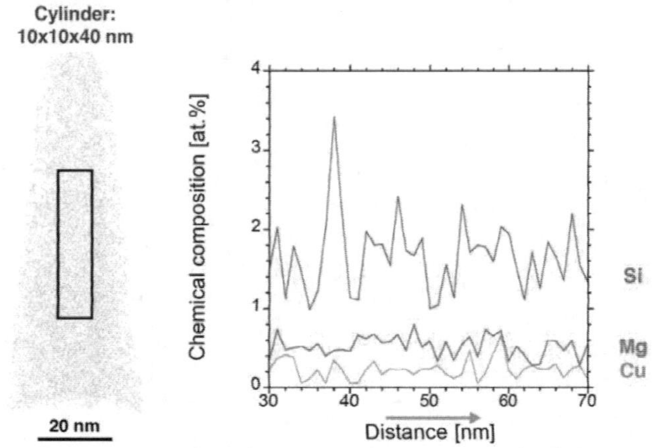

Figure 3 APT concentration scan of water quenched A2 sample

Overall, the simulated concentrations agreed well with most of the experimental findings from WDX and APT. An increase of the concentration occurred during solution treatment because of the dissolution of particles formed during solidification that enriched the matrix.

During quenching, the concentration decreased because of the clustering of particles effect and the formation of metastable particles. This decrease is also a quench rate dependent phenomenon. In fact, it is believed that there exists a critical quench rate that controls and determines the formation of clusters or precipitates during quenching [15]. By applying a CNGT approach in this case, the simulation always gives a certain temperature range for the particle to nucleate and grow, resulting in a more pronounced concentration loss at lower quench rates (e.g. air–quenched), as the system is exposed longer times at those temperatures that exponentiate nucleation events. In general, the prediction shows a good potential for simulating the concentration evolution during a quenching process. Nevertheless, further validation with more quench rates should be performed for a better understanding of the kinetics.

The concentrations of Mg decreased further during ageing, indicating a strong precipitation of Mg–containing phases. Cu, on the other hand, only decreased during ageing when its concentration was above approximately 2 wt.% in the matrix. We assume that Cu–containing phases are less favored energetically compared to Mg phases.

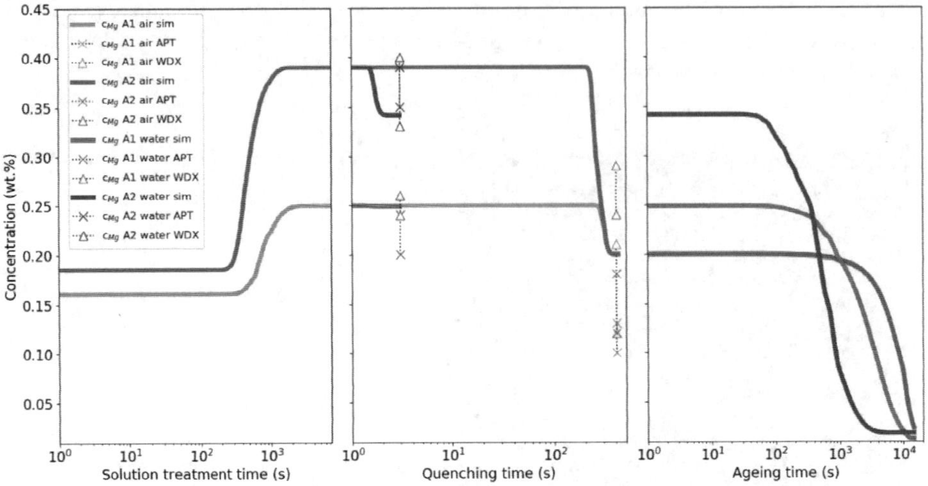

Figure 4 Evolution of chemical concentration of Mg in the matrix

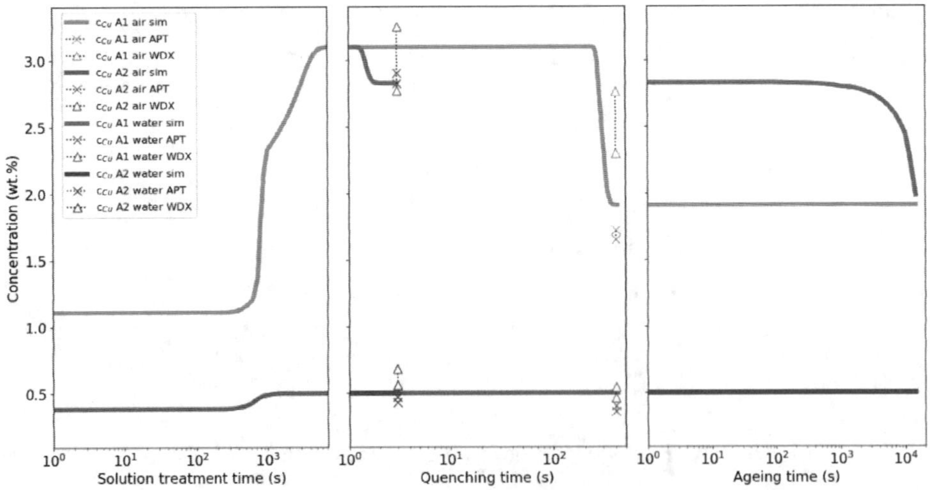

Figure 5 Evolution of chemical concentration of Cu in the matrix

4.2. Yield stress predictions

Figure 6 shows the predicted yield stress evolution of both alloys after different quench methods and compared with tensile test results. The solution hardening effect is calibrated for non–aged conditions so that all four cases are starting from a reasonable calculated stress level. The simulated water–quenched conditions match reasonably well with the experimental values, and the air–quenched curves shows that the kinetics are properly represented. The stress level for the alloy A1, air–quenched case is slightly over predicted by a value of ~20 MPa, while the stress for the alloy A2, air–quenched case is about 30 MPa too low.

In general, the slight mismatch in both air–quenched samples could be explained by a lack of information on the exact particle phase that formed during quenching and a possible dissolution during the early phase in artificial ageing. The stress level of air–quenched non–aged samples suggests that the concentration loss simulated during quenching is still not high enough, which also agrees with the APT findings. This could be improved by considering more accurately the formation of clusters or metastable phases during quenching. Though, this task requires extensive experimental research. Meanwhile, the precipitated particles during quenching are also believed to be partially involved in the dissolution behavior of precipitates in the subsequent ageing process [16, 17]. This behavior is currently neglected in the model, and should be considered in future work.

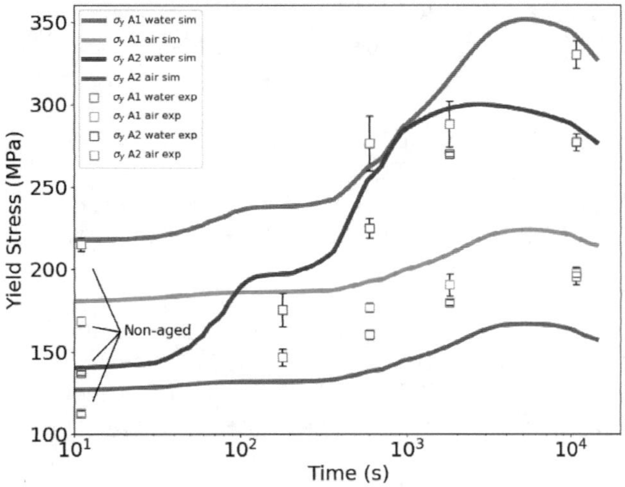

Figure 6 Comparison of predicted yield stress evolution to tensile testing results during ageing

5. Conclusions

The implemented CNGT model was employed to simulate the solution treatment, quenching and ageing throughout the heat treatment process of Al-Si-Cu-Mg cast alloys. The results showed a good agreement with experimental findings. The model is uncomplicated, efficient, and has already been applied in a commercial software tool for industrial application. Simulation of precipitation during quenching and corresponding concentration evolution agreed well with experimental observations from WDX and APT results. Predictions of the yield stress of different quenched cases matched accurately experimental measurements. Based on the results of this investigation, the model shows a good potential for its application on complex industrial heat treatments of Al-Si-Mg-Cu cast alloys. Subsequent investigations will be focused on investigating further details of quenching phases and corresponding validation cases for understanding this physical phenomenon and improve the accuracy of the model.

6. Acknowledgement

The research work has been carried out within the framework of the AMAP (Advanced Metals and Processes) research cluster. The supply of the investigated materials from project partner NEMAK Europe is gratefully appreciated. Support and help from the scientific staff of GI, GWP, GFE, and MCh on corresponding experimental works are warmly appreciated.

References

[1] M.V. Glazoff, A.V. Khvan, S.V. Zolotorevsky, A.N. Belov, A.T. Dinsdale, Casting Aluminum

Alloys; Butterworth-Heinemann: United Kingdom, 2018
[2] A.R. Farkoosh, M. Pekguleryuz, Enhanced mechanical properties of an Al–Si–Cu–Mg alloy at 300°C: Effects of Mg and the Q-precipitate phase. Materials Science and Engineering: A, 2015. 621: p. 277−286.
[3] S. Roy, L.F. Allard, A. Rodriguez, T.R. Watkins, A. Shyam, Comparative Evaluation of Cast Aluminum Alloys for Automotive Cylinder Heads: Part I—Microstructure Evolution. Metallurgical and Materials Transactions A, 2017. 48(5): p. 2529−2542.
[4] X. Larráyoz Izcara, A. Guirao Blank, F. Pyczak, P. Staron, S. Schumann, N. Huber, Characterization and modeling of the influence of artificial aging on the microstructural evolution of age-hardenable AlSi10Mg(Cu) aluminum alloys. Materials Science and Engineering: A, 2014. 610: p. 46−53.
[5] P.A. Rometsch, L.F. Cao, X.Y. Xiong, B.C. Muddle, Atom probe analysis of early-stage strengthening behaviour in an Al-Mg-Si-Cu alloy. Ultramicroscopy, 2011. 111(6): p. 690−694.
[6] E. Sjölander, S. Seifeddine, Artificial ageing of Al–Si–Cu–Mg casting alloys. Materials Science and Engineering: A, 2011. 528(24): p. 7402−7409.
[7] S.W. Choi, H.S. Cho, C.S. Kang, S. Kumai, Precipitation dependence of thermal properties for Al–Si–Mg–Cu–(Ti) alloy with various heat treatment. Journal of Alloys and Compounds, 2015. 647: p. 1091−1097.
[8] S, Beroual, Z, Boumerzoug, P, Paillard, Y, Borjon-Piron, Effects of heat treatment and addition of small amounts of Cu and Mg on the microstructure and mechanical properties of Al-Si-Cu and Al-Si-Mg cast alloys. Journal of Alloys and Compounds, 2019. 784: p. 1026−1035.
[9] A. Deschamps, Y. Brechet, Influence of predeformation and ageing of an Al–Zn–Mg alloy—II. Modeling of precipitation kinetics and yield stress. Acta Materialia, 1998. 47: p. 293−305.
[10] O.R. Myhr, Ø. Grong, Modelling of non-isothermal transformations in alloys containing a particle distribution, Acta Materialia, 2000. 48: p. 1605−1615.
[11] Q. Du, K. Tang, C.D. Marioara, S.J. Andersen, B. Holmedal, R. Holmestad, Modeling over-ageing in Al-Mg-Si alloys by a multi-phase CALPHAD-coupled Kampmann-Wagner Numerical model. Acta Materialia, 2017. 122: p. 178−186.
[12] F. Mao, C. Bollmann, T. Brüggemann, Z. Liang, H. Jiang, V. Mohles, Modelling of the Age-Hardening Behavior in AA6xxx within a Through-Process Modelling Framework. Materials Science Forum, 2016. 877: p. 640−646.
[13] F. Mao, M. Weidt, P. Staron, S. Busch, A. Stark, L. Barrales-Mora, Investigation on the age-hardening behavior of Al–Si–Cu–Mg cast alloys by mechanical testing, microstructural characterization, and a CNGT based precipitation model, International Journal of Material Research, 2022.
[14] M. Perez, M. Dumont, D. Acevedo-Reyes, Implementation of classical nucleation and growth theories for precipitation. Acta Materialia, 2008. 56(9): p. 2119−2132.
[15] B. Milkereit, M.J. Starink, P.A. Rometsch, C. Schick, O. Kessler, Review of the Quench Sensitivity of Aluminium Alloys: Analysis of the Kinetics and Nature of Quench-Induced Precipitation. Materials 2019, 12, 4083.
[16] O. Engler, C.D. Marioara, Y. Aruga, M. Kozuka, O.R. Myhr, Effect of natural ageing or pre-ageing on the evolution of precipitate structure and strength during age hardening of Al–Mg–Si alloy AA 6016, Materials Science and Engineering: A, 2019. 759: p. 520−529
[17] S.N. Khangholi, M. Javidani, A. Maltais, X.-G. Chen, Effects of natural aging and pre-aging on the strength and electrical conductivity in Al-Mg-Si AA6201 conductor alloys, Materials Science and Engineering: A, 2021

Modelling microstructure formation in TiAl by isomorphic inoculation

*** J.R. Kennedy[1,2], A.K. Boukellal[1], M. Založnik[1,2], D. Daloz[1,2], J. Zollinger[1,2]**

1 Université de Lorraine, Institut Jean Lamour, Department of Metallurgy and Materials Science and Engineering, Nancy, France

2 Laboratory of Excellence on Design of Alloy Metals for low-mAss Structures ('LabEx DAMAS'), France

Jacob.kennedy@univ-lorraine.fr

Abstract. Ti-Al alloys have replaced Ni-based superalloys in the last stages of some aircraft engines to improve fuel efficiency. In order to improve their properties, grain refinement has been investigated via isomorphic inoculation with Ti-Al-Nb particles. This inoculation method is orders of magnitude more efficient on a particle-by-particle basis than traditional inoculation, rather than multiple inoculant particles added to form a solidified bulk phase grain, in isomorphic inoculation each particle added results in the formation of multiple grains. As the particles are indistinguishable from the matrix after solidification, a model was used to elucidate this mechanism. Two phenomena were considered to calculate the number of particles acting during solidification: particle breakup along grain boundaries and complete particle dissolution. The grain size of the particles was calculated with an empirical model from initial TKD analysis of the particles and high temperature molten salt heat treatments. Particle dissolution was estimated via mass transport of the slowest diffusing Nb species. This showed the population of isomorphic inoculant particles which can act during solidification is near a 1:1 ratio with the number of grains formed, confirming the mechanism of grain refinement by direct epitaxial growth from the particles.

1. Introduction

The high specific strength and oxidation resistance of TiAl alloys make them particularly appealing for use in aerospace applications [1]. Currently they have been introduced in the last low pressure stage of the GEnx turbine engine as part of its goal to increase fuel efficiency. In order to better utilize these alloys grain refinement is needed, however, current methods rely on the introduction of brittle ceramic phases which act as inoculants [2]. Nucleation plays a key role when using such traditional inoculants for grain refinement. Inoculants facilitate heterogeneous nucleation as they reduce the activation energy required [3]. Models of heterogenous nucleation are broken into three main categories, spherical cap, free growth or constitutional supercooling [4]. In the first model, the effectiveness of the nucleant depends essentially on its characteristics. The kinetics are then determined by the formation of the nucleus and the adsorption rate of atoms into them. In the second model, heterogeneous nucleation is considered athermal and the active nucleants are determined by their size distribution. In the third, it is the constitutional undercooling at the front of growing crystals that triggers the active nucleants. In each case there is an energy barrier to the formation of new solid from the nucleant particles. In the case of slow solidification rates, predicting the grain size relies on linking both nucleation and crystal growth in a complex manner, highly dependent on experimental conditions and alloy composition. Recent developments introduced isomorphic inoculants (ISI) as potential grain refiners for cast Ti-Al [5–7], these particles have a coherent interface with the solidifying bulk, as they solidify by epitaxial growth rather than nucleation of new solid, and are of the same phase, eliminating the need for brittle second phase particles. Such grain refiners also have potential to simplify prediction of as-cast grain size due to the suppression of the nucleation of new solid.

ISI particles are engineered alloy powders to grain refine during solidification without embrittlement. The ISI alloy is chosen such that at the temperature of a given melt they are of the same phase that will solidify while also being thermally stable [5]. The ISI alloy is then only effective when used for solidification of the phase it was designed for. The thermal stability of the ISI particles is achieved through refractory metal additions, in the case of the Ti-46Al (wt%) alloy investigated here the ISI alloy is Ti-10Al-25Nb (wt%). This alloy has a melting point above that of the bulk, 1800°C compared to ~1540°C, while existing as β-Ti, the TiAl alloy solidification phase, at high temperature. It is also important that the ISI particle density remains near that of the parent alloy or excess flotation or settling may occur preventing their active participation in solidification [6]. The mechanism of solidification from ISI particles differs from that of traditional inoculants which act as heterogeneous nucleation sites, but rather solidification occurs by epitaxial growth [8]. This results in solidification without a detectable interface between the particles and solidified bulk, and as such particles which participate in solidification cannot be detected ex-situ [7]. This is particularly interesting as since solidification is not occurring by nucleation of new solid, the energy barrier of nucleation is not encountered and as such particles of different sizes should induce solidification all at the same, minimal, undercooling as the only barrier to free growth is the strain from the difference in lattice parameters between the ISI alloy and bulk. This has interesting implications for validation of solidification models as each ISI particle added participates in solidification and should be responsible for the formation of an as-cast grain.

Casting trials of TiAl using TiAlNb ISI particles were carried out which showed each ISI particle added was responsible for the formation of multiple as-cast grains. Simple models were developed to account for the thermal and chemical effects of the ISI/melt interaction on the number of particles present during solidification. Taking into account these effects, the number of active ISI particles was used to predict the as-cast grain size assuming each particle formed a single grain.

2. Casting Trials

In order to test the efficacy of ISI particles as nucleant particles casting trials were carried out [5–7]. ISI particles were produced from commercially pure metals in an induction heated cold crucible apparatus as a bulk ingot. Drillings taken from the ingot were cryomilled for 3h under Ar and 3, 6, 9, or 11h under air to produce particles of different sizes. Particle size distributions had median (D50) sizes ranging from 233 μm (3h under Ar) to 48 μm (11h under air) as measured by SEM image analysis. There exists no simple way to add particles to the cold crucible apparatus used for the casting trials as melting and solidification are conducted under a high purity Ar atmosphere to prevent oxidation of the TiAl. In order to add the particles to the melt they were then mixed with pure Al powder in a 1:1 ratio and compressed to form a pellet which could held by means of a quartz vacuum tube until the TiAl was fully molten at which point the vacuum could be released and the pellet would fall into the melt. A target mass of 1g of ISI particles was added during each trial to TiAl ingots of roughly 40g. An interaction time of 20 s between the particles and melt was used in each trial to ensure there was no cooling effect from the particles on solidification. In each casting trial the grain size was reduced and equiaxed fraction increased compared to a reference ingot cast without ISI particles or accounting for the solutal effects of the Nb [5].

The as-cast grain size and equiaxed fraction, assuming spherical grains, can be used to calculate the number of equiaxed grains formed during solidification, while the size distribution of particles and mass of particles added yields the number ISI particles added. Figure 1 shows how these values compare across all casting trials. In conventional inoculation, where the inoculant particles act as heterogeneous nucleation sites, and as such an energy barrier exists in order for solid to grow from the particles, the size distribution of inoculants is of critical importance, as the largest particles require the smallest undercooling to nucleate new solid they act first during solidification and the smallest particles are not active participants in solidification [9]. As such, the ratio of particles added to new grains formed is always less than 1:1, i.e., often 100 to 1000 or more particles need to be added to form a single new grain. This is in stark contrast to Figure 1 where in each trial more than 1 grain is formed by each particle added. It is then evident that further phenomena are occurring while the ISI particles are interacting with the melt to make this possible. While interacting with the melt the particles are subjected to both thermal and chemical conditions which will affect their properties. Both these factors must be accounted for in order to use ISI particles to predict the as-cast microstructure.

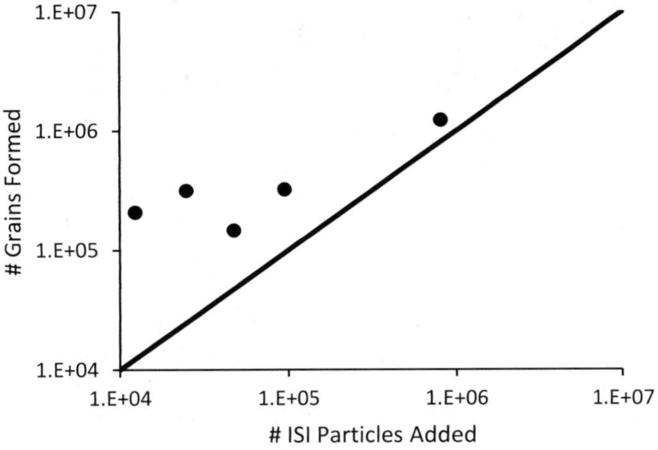

Figure 1: Relationship between the number of grains formed and number of ISI particles added

3. Thermal Effect: Particle Microstructure

Casting trials involving more highly stable TiTa particles which could survive prolonged contact with a TiAl melt revealed that dissolution was occurring preferentially along grain boundaries within the particles [6]. This is then likely also true of the TiAlNb particles used here which were processed in the same manner. The internal particle microstructures are then of critical importance. Initial investigations of the TiAlNb ISI particles by optical or SEM investigation did not have the resolution to reveal the internal microstructure of the particles. In order to evaluate the microstructure electron transparent samples was removed from 3h and 9h milled particles to be analysed by TEM and TKD. A TEM micrograph is shown in
Figure 2a from a 3h milled particle. The process of cryomilling resulted in a superplastically deformed nano-scale microstructure consisting of elongated grains after only 3h of milling. Further milling did not significantly change the grain size, however TKD analysis showed the fraction of low angle grain boundaries decreased with further milling, while high angle grain boundaries increased. The median grain size within the as milled particles was 35 nm [7].

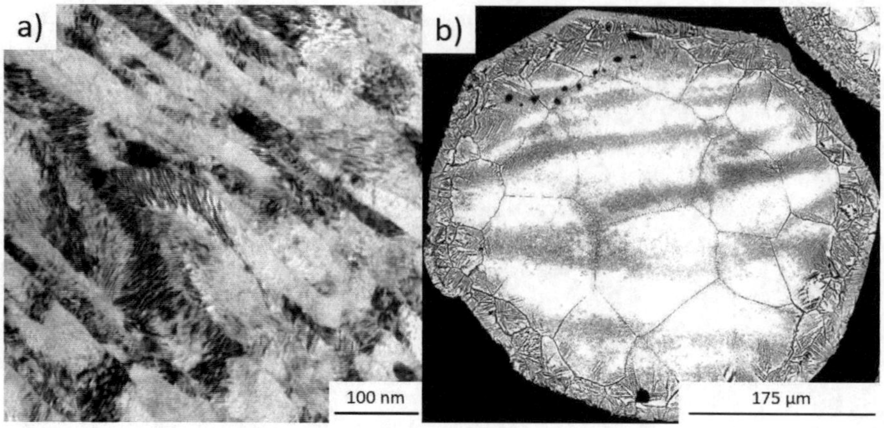

Figure 2: Microstructure of 3h cryomilled TiAlNb particles a) TEM micrograph of as milled particles and b) SEM BSE micrograph after 20s of interaction with molten CaF at 1600°C

The thermal effects of the TiAl melt on the particle microstructures cannot be evaluated from the casting trials as the epitaxial growth mechanism made location of the original particles impossible. Furthermore, during the casting trials the particles were also subjected to the chemical effect of the melt. In order to evaluate the microstructure of the particles during solidification the thermal effect must be separated from the chemical effect. To that end a high temperature molten salt heat treatment was carried out on representative samples of 3h and 9h milled particles. A high temperature molten salt was chosen which did not react with the particles, CaF, and was stable at the melt temperature of 1600 °C. In order to replicate the casting conditions alumina crucibles containing CaF were heated to 1600 °C until the salt was completely melted then a representative sample of particles added to the CaF for 20s then quenched in water. After this heat treatment there was significant grain growth,

Figure 2b shows an SEM BSE micrograph of a 3h milled particle where the grain size has increased to be nearly 40 µm. These easily delineated interior grains were measured across multiple particles, both 3h and 9h milled, the results of which are shown in Figure 3. While the 9h milled particles tended to have smaller grain sizes, and the 3h larger, this corresponded to the 9h particles themselves also being smaller. The heat treated grain size was found to depend more on the size of the particles than their milling time, particles of similar size with either milling time had similar grain sizes. The average grain size across all the particles was found to be 37 µm, three orders of magnitude greater than the average as milled microstructure. A linear trend was observed between the particle size and interior grain size after heat treatment independent of milling time:

$$Gs=0.1069(Ps)+1.773$$

Where Gs is the resultant interior grain size and Ps the size of the milled particles. This trend could then be used to predict the grain size of particles after interaction with the melt.

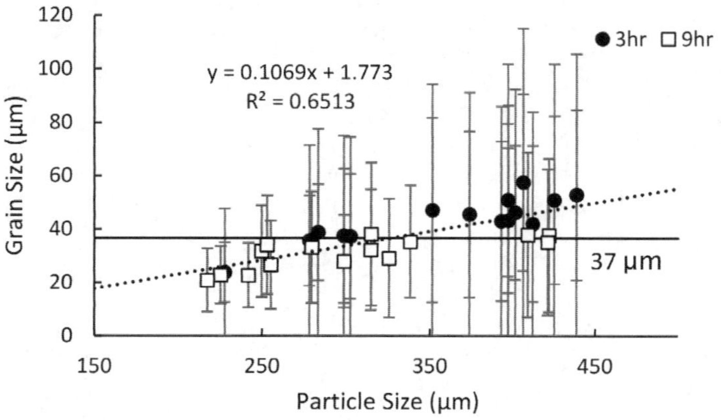

Figure 3: Measured grain sizes within particles of different sizes after replicating interaction with the melt.

4. Chemical Effect: Solid Dissolution

The design criteria of an ISI particle requires that they be thermally stable in the melt, i.e., their melting temperature is above that of the melt, however while the particles do not melt there is also the chemical effect of the molten TiAl which must be considered. Since preferential dissolution was observed in TiTa particles [6], it stands to reason that the particles also undergo partial or complete dissolution. As with the thermal effect, this cannot be evaluated from the casting trials alone. In order to evaluate the chemical dissolution of the particles a simple dissolution model was developed based on the work of Hsu and Lin [10], and validated through a rod dipping experiment.

4.1. Dissolution Model

A simplified first calculation model was developed to evaluate how dissolution is affecting the ISI particles. The casting trials were conducted in a complex environment, involving 3D fluid flow induced by the induction field, and the entire population of particles, these first calculations simplify the flow to 1D and the particle distributions are simplified to ten particle sizes each which do not interact with one another. In these calculations dissolution was according to film theory, i.e., the rate limiting step is the transport of the solute away from the interface [11]. This theory considers there is no solute buildup effect in the liquid, that it is an infinite medium. The model developed by Hsu and Lin builds upon film theory by also incorporating the volume of liquid the particles are interacting with, if solute buildup is a factor in the dissolution kinetics it is then considered [10]. Their model operates under the assumption that the dissolution rate is controlled by the transport of solute atoms through a boundary layer between the particles and the bulk liquid. This model was adapted for use with the ISI particles as shown in the schematic, Figure 4.

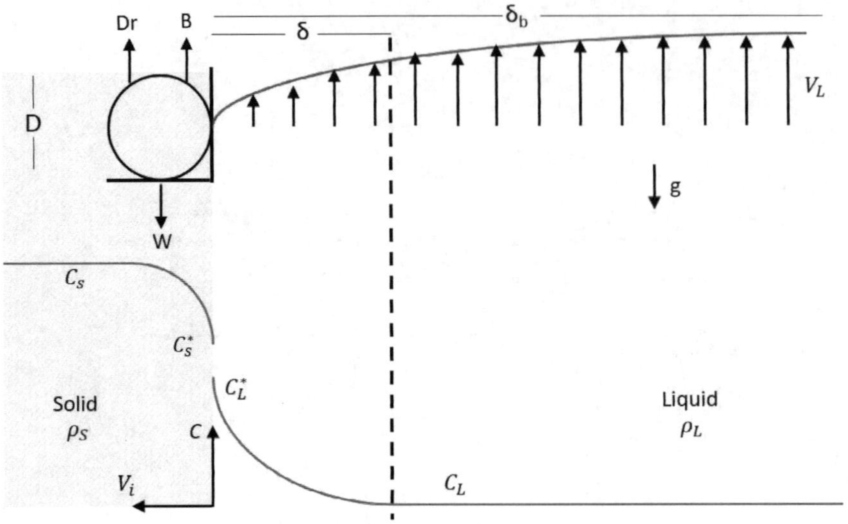

Figure 4: Schematic of dissolution model

The particles are not stationary in the melt, as such in order to calculate the width of the boundary layer their velocity relative to the liquid must be known. To simplify the 3D fluid flow only the maximum relative velocity, when the particle settling velocity is opposite that of the the liquid velocity, was considered. This was calculated by a force balance on the particle, shown in Figure 4, where the drag force (Dr) and buoyancy (B) act opposing the weight (W) of the particle. The relative velocity (V_{rel}) is then:

$$V_{rel} = \sqrt{4gD(\rho_s - \rho_L)/3C_D\rho_L}$$

where g is acceleration due to gravity, D is the particle diameter, ρ_s and ρ_L the particle and liquid density respectively and C_D the drag coefficient, which is dependent on the Reynolds number. As the particles remain in a laminar flow regime C_D can be taken as [12]:

$$C_D = 24/Re \quad Re < 53$$
$$C_D = 0.45 \quad 53 < Re$$

The initial velocity of the particles was taken to be the liquid velocity (assuming stationary particles) then the velocity, Reynolds number and drag coefficient calculations were iterated ten times, until variation was less than 1%. These values are needed to calculate the width of the fluid velocity boundary layer δ_b which is the distance from the particle where the fluid velocity returns to 99% of its free-flowing velocity. In reality this boundary layer around a sphere is non-uniform, compressed at the front of the particle and elongated in the rear. To simplify the boundary layer calculations the particles are instead considered to be cuboid and the boundary layer only considered on the faces parallel to V_{rel}, this is shown schematically in Figure 4. The mass transfer boundary layer taken to be the layer used for the dissolution calculations can be found by dividing the boundary layer by the Schmidt number yielding:

$$\delta = (D_L 5r)/(v\sqrt{Re})$$

Where δ is the mass transfer boundary layer, D_L the liquid diffusion rate of Nb, r the particle radius, ν the kinematic viscosity of the melt and Re the Reynolds number. The Nb diffusion rate in the liquid was calculated using the stokes-einstein relation as there is no literature data available. The thermophysical properties of our case required to implement the method of Hsu and Lin are given in Table 1.

Table 1: Thermophysical Properties for Dissolution Calculations

Variable	Symbol	Units	Value	Ref
Solute Liquid Diffusion	D_L	m^2/s	2.14×10^{-9}	PW
Particle Density	ρ_s	g/m^3	5.01×10^6	[7]
Liquid Density	ρ_L	g/m^3	3.82×10^6	[13]
Solute Molecular Weight	M	g/mol	93	[14]
Solute Liquid Concentration	C_L	mol/m^3	0	PW
Solute Interface Concentration	C_i	mol/m^3	8164	[15]
Kinematic Viscosity	ν	m^2/s	4.93×10^{-7}	[16]
Dynamic Viscosity	η	Pa s	1.88×10^{-3}	[16]
Temperature	T	°C	1600	PW

4.2. Experimental Validation

A further experiment was needed to validate the dissolution of TiAlNb in molten TiAl using the above model. As with the thermal effect the addition of particles to the melt could not validate the model as the particles could not be located after solidification. A larger sample of TiAlNb which could be inserted into the melt, be easily located after solidification and would survive prolonged contact with the melt was needed. In order to fabricate such a sample a further bulk TiAlNb ingot was produced from pure elements in the same manner as for the ISI particles, a rod could then be produced by vacuum pulling. A quartz tube was inserted into the molten TiAlNb and a vacuum applied, the molten TiAlNb was pulled up the tube and solidified as a rod 5 mm in diameter and ~10 cm long. This rod could then be inserted into molten TiAl in a controllable manner under the same conditions as the casting trials, i.e., temperature and interaction time. The resultant rod/ingot assembly could be cross sectioned in order to measure the change in the rod diameter after interaction with the melt. An optical micrograph of the rod and measured radii along its length are shown in Figure 5. The average measured dissolution length after 20s of interaction with the melt at 1600°C was 94 μm.

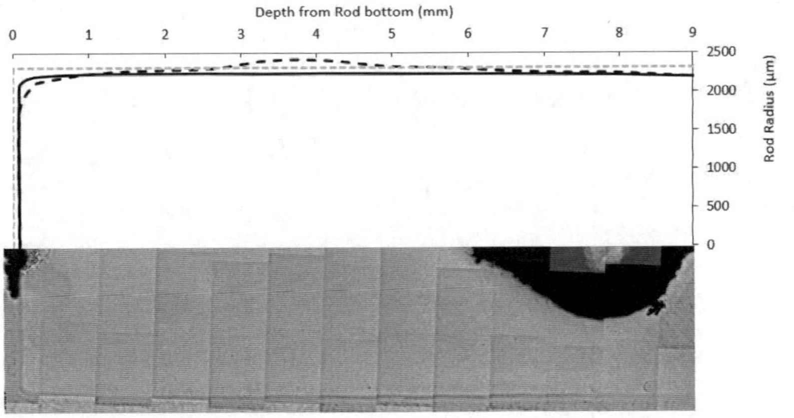

Figure 5: Optical micrograph and measured radii of TiAlNb rod after 20s interaction with melt. The grey line represents the initial radii and the black radii of either side of the rod after dissolution.

The details of the rod dipping were used in the dissolution model, namely the starting diameter of 5 mm, a length of 1 cm dipped into the melt and an interaction time of 20 s. The model then calculates the dissolution length over time as shown in Figure 6. After 20s of interaction with the melt the dissolution length was 97 μm, close to the experimental value of 94 μm. This means the model and assumptions are reasonable for the given experimental conditions and can be applied to the real particle size distributions used in the casting trials.

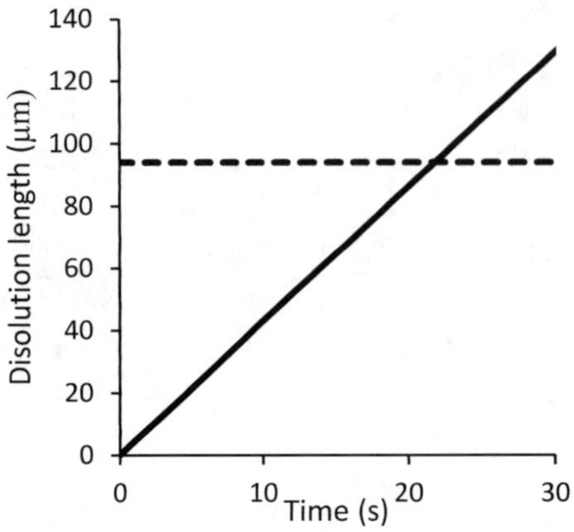

Figure 6: Calculated dissolution over time for TiAlNb rod in TiAl at 1600°C

5. Chemical and Thermal Effects on ISI Particles

The adapted model of Hsu for the chemical, dissolution, effect and the analytical grain size relation for the thermal, grain growth, effect can both then be applied to the ISI particle size distributions used in the casting trials. The known number of ISI particles added to the melt is increased by the thermal effect of grain growth/grain boundary dissolution, and simultaneously decreased by the chemical dissolution of some particles. Each of these effects is shown separately in Figure 7. Accounting for only thermal effects greatly increases the supposed number of active ISI particles, moving the distributions away from the 1:1 ratio of ISI: grains formed and into the regime of traditional inoculation [9]. In contrast accounting for only the thermal effect of particle dissolution moves the distributions further from 1:1 in the other direction with each particle responsible for even more new grains than previously. However, when both effects are considered, by firstly applying the chemical effect of particle dissolution then the thermal effect of grain growth on the dissolved particle sizes the ratio approaches 1:1, with the exception of the 11h milled sample. The median size of these particles, ~50 μm, is significantly smaller than that of the other distributions, ~110-230 μm, additionally in the case of the 11h milled particles the dissolution model predicts over 96% of the particles dissolving completely, this is likely an overestimation and results in fewer ISI particles predicted during solidification.

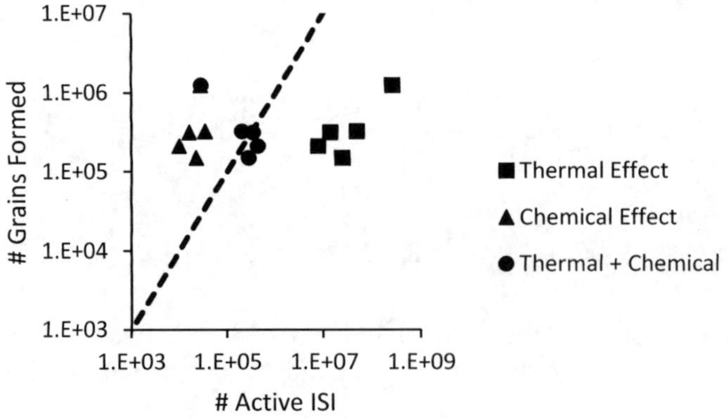

Figure 7: Relationship between the number of grains formed and number of ISI particles added considering different interaction phenomena

The calculated number of active ISI particles taking into account the chemical, thermal effects or both can then also be used to predict the as-cast grain size. Since ISI particles have been shown to induce solidification by epitaxial growth a 1:1 particle to grain relation can be inferred [8]. In the case of the casting trials the equiaxed volume of each ingot is known, so dividing this by the number of active ISI particles gives a grain volume which, assuming spherical grains, yields an estimated grain size. This is shown for each size distribution in Figure 8a considering: no interactions with the melt (particles introduced), only a thermal effect (grains introduced), only a chemical effect (particles remaining) or both a chemical and thermal effect (grains remaining). With the exception of the smallest introduced particle distribution considering both thermal and chemical effects predicts grain sizes most similar to those measured. However, the trend of predicted grain sizes increases with decreasing particle size while the experimental trials decrease grain size with decreasing particle size. This also points to an overprediction of the dissolution of small particles, leading to fewer predicted active ISI particles and larger predicted grain size. Even considering this it can be seen in Figure 8b that considering both thermal and chemical effects results in predictions closer to the experimental values. It is also interesting to note that when considering only the thermal, grain growth/ particle breakup, effect the predicted as-cast grain sizes are consistently 60-70% smaller than the experimental across all particle distributions. This indicates that there is a strong relationship between the internal grains of the particles and the as-cast grain size and that further work is needed to better estimate the dissolution of the particles to predict the grain size more accurately.

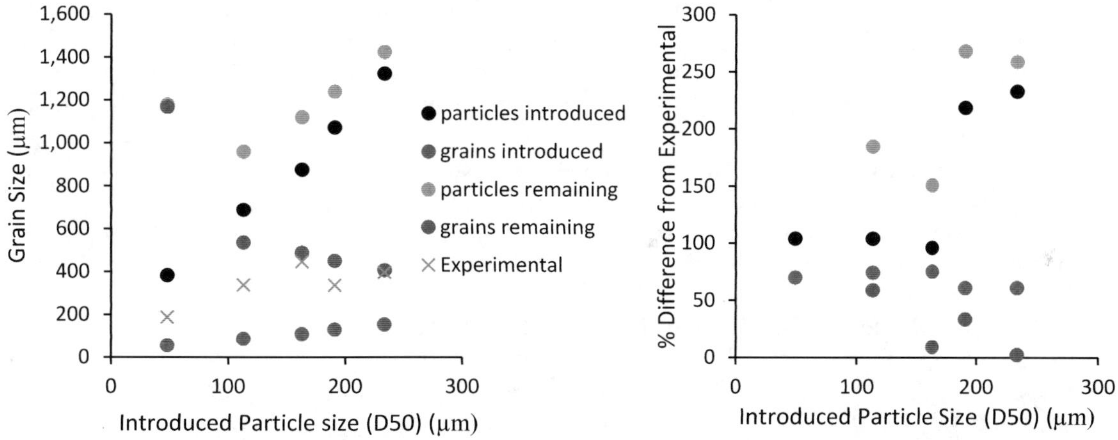

Figure 8: Predicted grain size and difference from experimental grain size for given introduced size distributions

6. Conclusions

ISI particles show significant potential as a tool to model as-cast grain size as they bypass the nucleation energy barrier by inducing solidification by epitaxial growth. However, the number of particles added to the melt does not correspond to the number of grains formed. This is due to both chemical and thermal effects of interaction between the melt and particles. The thermal effect of the interaction results in grain growth within the particles, which break up along their grain boundaries, while the chemical effect results in dissolution of some particles. Both these factors work together to change the number of active ISI particles present during solidification; dissolution decreases their number while particle breakup increases them. Taking into account both these effects using an empirical model for grain growth and a dissolution model based on the work of Hsu and Lin [10], the number of ISI particles present during solidification can be calculated. If the smallest particle size distribution is disregarded due to an overestimation of the dissolution of small particles, the predicted grain sizes are relatively close to those measured from experimental trials. In order to apply ISI particles to improve grain size modelling a dissolution model is needed which can account for dissolution of these small particles more accurately.

Additionally, the calculations could be simplified by manufacturing the particles using atomisation so each particle has a single grain rather than cryomilling which resulted in nano scale superplastically deformed grains.

References:

[1] Progress in the understanding of gamma titanium aluminides | SpringerLink, (n.d.). https://link.springer.com/article/10.1007/BF03221103 (accessed January 5, 2023).
[2] U. Hecht, V. Witusiewicz, A. Drevermann, J. Zollinger, Grain refinement by low boron additions in niobium-rich TiAl-based alloys, Intermetallics. 16 (2008) 969–978. https://doi.org/10.1016/j.intermet.2008.04.019.
[3] M. Rappaz, J.A. Dantzig, Solidification, EPFL Press, 2016.
[4] S.B. Park, Heterogeneous nucleation models to predict grain size in solidification, Prog. Mater. Sci. 123 (2022) 100822. https://doi.org/10.1016/j.pmatsci.2021.100822.
[5] J.R. Kennedy, D. Daloz, B. Rouat, E. Bouzy, J. Zollinger, Grain refinement of TiAl alloys by isomorphic self-inoculation, Intermetallics. 95 (2018) 89–93. https://doi.org/10.1016/j.intermet.2018.02.001.
[6] J.R. Kennedy, B. Rouat, D. Daloz, E. Bouzy, J. Zollinger, Effect of Inoculant Alloy Selection and Particle Size on Efficiency of Isomorphic Inoculation of Ti-Al, Materials. 11 (2018) 666. https://doi.org/10.3390/ma11050666.
[7] J.R. Kennedy, Development of a New Generation of Inoculants for Ti-Al Alloys, thesis, Université de Lorraine, 2018. http://www.theses.fr/2018LORR0101 (accessed April 20, 2020).
[8] J.R. Kennedy, A.E. Davis, A. Caballero, A. Garner, J. Donoghue, S. Williams, J. Zollinger, E. Bouzy, E.J. Pickering, P.B. Prangnell, Isomorphic grain inoculation in Ti-6Al-4V during additive manufacturing, Mater. Lett. X. 8 (2020) 100057. https://doi.org/10.1016/j.mlblux.2020.100057.
[9] T.E. Quested, A.L. Greer, The effect of the size distribution of inoculant particles on as-cast grain size in aluminium alloys, Acta Mater. 52 (2004) 3859–3868. https://doi.org/10.1016/j.actamat.2004.04.035.
[10] J.-P. Hsu, M.-J. Lin, Dissolution of solid particles in liquids, J. Colloid Interface Sci. 141 (1991) 60–66. https://doi.org/10.1016/0021-9797(91)90302-O.
[11] T.K. Sherwood, R.L. Pigford, C.R. Wilke, Mass transfer, McGraw-Hill, 1975.
[12] C.T. Crowe, Multiphase flows with droplets and particles, CRC Press, 2012.
[13] I. Egry, D. Holland-Moritz, R. Novakovic, E. Ricci, R. Wunderlich, N. Sobczak, Thermophysical Properties of Liquid AlTi-Based Alloys, Int. J. Thermophys. 31 (2010) 949–965. https://doi.org/10.1007/s10765-010-0704-1.
[14] P.T.B. Shaffer, Plenum Press Handbooks of High-Temperature Materials, Springer US, Boston, MA, 1963. https://doi.org/10.1007/978-1-4899-6405-2.
[15] V.T. Witusiewicz, A.A. Bondar, U. Hecht, T.Ya. Velikanova, The Al–B–Nb–Ti system, J. Alloys Compd. 472 (2009) 133–161. https://doi.org/10.1016/j.jallcom.2008.05.008.
[16] K. Zhou, H.P. Wang, J. Chang, B. Wei, Surface tension measurement of metastable liquid Ti–Al–Nb alloys, Appl. Phys. A. 105 (2011) 211. https://doi.org/10.1007/s00339-011-6491-0.

IOP Conf. Series: Materials Science and Engineering 1281 (2023) 012045 doi:10.1088/1757-899X/1281/1/012045

A study of the complex dynamics of dendrite solidification coupled to structural mechanics.

P Soar, A Kao, G Djambazov and K A Pericleous

Centre for Numerical Modelling and Process Analysis, University of Greenwich, Old Royal Naval College, Park Row, London, SE109LS, United Kingdom

E-mail: P.Soar@gre.ac.uk

Abstract. The impact of structural mechanics is often overlooked when modelling the solidification of dendritic microstructures, despite experimental observations that the interaction between these processes can be a factor leading to the development of crystal mosaicity throughout the microstructure which can itself lead to more serious defects. When considered at all, the structural mechanical behaviour of columnar dendrites is often considered as being analogous to a cantilever beam both in interpretations of experimental results and in existing numerical modelling. While this is not an unreasonable assumption when considering a dendrite in isolation, this is a scenario that infrequently occurs. In this paper a parametric study is presented using a Cellular Automata solidification solver coupled to a Finite Volume Structural Mechanics solver. These results highlight the complex non-linear behaviour that arises when considering dendrite interaction, demonstrating the significantly different microstructures that can be obtained by varying only the force experienced by the system.

1. Introduction

Structural mechanical effects have been widely observed in experimental results as one of the factors than can influence the development of solidifying dendritic microstructures, with examples identifying structural effects as the root cause of dendrites observed bending [1], fracturing [2] and leading to the formation of serious casting defects such as slivers [3]. Even in less extreme instances, the structural mechanical influences on the microstructure development can be observed in more gradual dendrite misorientations [4], which are not uncommon in any dendritic microstructure and can lead to the substructures of misoriented dendrites to occur even in single crystal castings in a phenomenon known as crystal mosaicity [5].

The interaction between structural mechanics and microstructure solidification is a complex, interdependent process that defies easy characterisation when considering the development of larger systems comprising multiple interacting dendrites. For this reason, when interpreting results the analogy has often been made that a single columnar dendrite behaves in a manner analogous to that of a cantilever beam anchored to either the nucleation site at the edge of the sample [6], by the solidifying interdendritic front [3, 7, 8, 4] or when considering a secondary arm anchored to a primary dendrite trunk [1]. This assumption of beam like behaviour has also been made when modelling the process, where beam models have been used to make assumptions about dendrite fragmentation [2]. Indeed, investigations using the Finite Element Method have shown that while a single dendrite will experience higher stress than a comparable

IOP Conf. Series: Materials Science and Engineering 1281 (2023) 012045 doi:10.1088/1757-899X/1281/1/012045

simple cantilever due to the morphology around the dendrite neck, the behaviour of the modelled dendrite still follows a comparable power law in its behaviour [6].

While the assumption of cantilever beam like behaviour appears appropriate when considering the mechanical behaviour of a single dendrite, this overlooks a great deal of complexity when considering the full solidifying system. There is the issue of in situ observations being difficult for many realistic systems, leading to questions of when during the solidification process deformation occurs, with some experiments positing thermal contraction to cause cantilever beam like deformation of dendrites in an already significantly developed microstructure [3, 8, 4, 9]. In other cases it has been proposed that misorientation can occur near the tips of dendrites which are still in the mushy zone, before having any deformation frozen in place by the advancing solidification of the interdendritic material [7], meaning that the region which is free to deform like a cantilever will remain limited to the length of the mushy zone rather than the total length of the dendrite in question. Even within the mushy zone, the mechanical behaviour of dendrites may be further constrained by the convergence of the secondary arms, which allows load to pass between dendrites [7]. This can potentially lead to regions of high stress where a dendrite deform like a cantilever causing secondary arms to impact [10], while in other cases these impinging secondary arms lead to successive dendrites forming a strip which acts together in a cantilever beam like manner [11].

The impact of structural mechanics is often neglected when numerically modelling the microstructure solidification process, with limited examples of fully coupled modelling [12] and no 3D modelling of the process other than the prior work by the authors [13] where some of the more complex structural behaviours which can only be represented in 3D such as out of plane bending and torsion were investigated. However, in this work these behaviours will not be considered, focussing instead on the relatively simpler case of a constant body force acting perpendicular to the dendrite growth orientation. An experimental example of a system analogous to this case which demonstrated dendrites becoming misoriented in the direction of gravity can be found in figure 1, which was previously compared to an earlier 2D version of this modelling in Soar et al. (2022) [14]. Consequently, a parametric study will be conducted which varies this constant body force to explore how this can cause significant differences in both growth behaviour and final morphology of the dendritic structures, these results will then be used to consider how well the assumption holds that columnar dendrites behave like cantilever beams throughout the simulations.

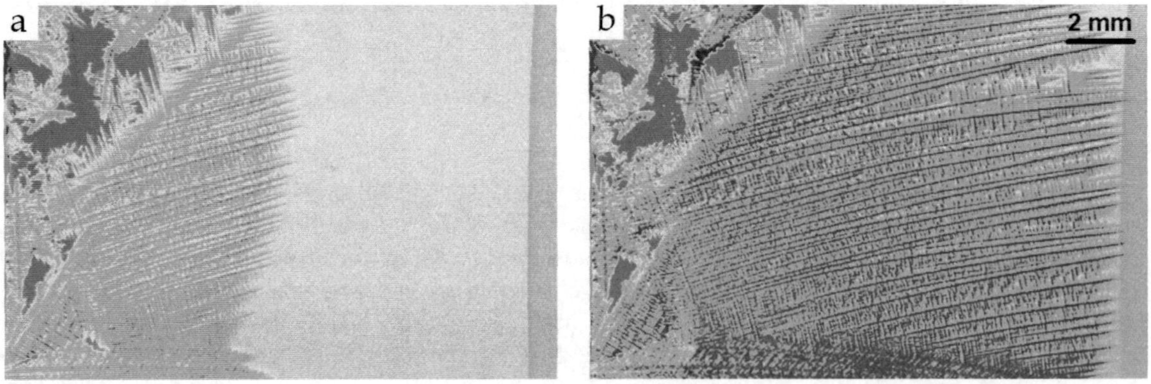

Figure 1: Experimental results from Soar et al. (2022) [14] showing the accumulation of misorientation in columnar dendrites during the solidification of a Ga − 25wt%In alloy.

IOP Conf. Series: Materials Science and Engineering 1281 (2023) 012045 doi:10.1088/1757-899X/1281/1/012045

2. Methodology

This section will provide a quick summary of the methods used to obtain the results. Other than the problem setup, the numerical modelling used is identical to that presented in previous papers by the authors [14, 13] which provide more in depth descriptions of the methodology used.

The model for microstructure alloy evolution uses a finite difference decentred octahedron Cellular Automata (CA) method based on the Imperial College µMatIC code [15] which has been further developed into the parallel multi-physics code Thermo Electric Solidification Algorithm (TESA) [16]. This solidification method uses the following governing equation for the transient phase change and solute partitioning

$$C_l\left(1-k\right)\frac{\partial \phi_s}{\partial t} = -\nabla\cdot\left(D_e\nabla C_l\right) + \left[1-\left(1-k\right)\phi_s\right]\frac{\partial C_l}{\partial t}, \tag{1}$$

where C_l and C_s are respectively the liquid and solid concentration, $k = C_l/C_s$ is the partition coefficient, $0 \leq \phi_s \leq 1$ is the solid fraction, D_e is the mass diffusivity and t is time. The diffusive transport of this system is given by

$$\frac{\partial C_e}{\partial t} = \nabla\cdot\left(D_e\nabla C_l\right), \tag{2}$$

defining the equivalent concentration as $C_e = (1-\phi_s)\,C_l + \phi_s C_s$. A frozen temperature approximation has been taken with a constant thermal gradient in the primary growth direction and a constant cooling rate. As the system has a Lewis number of approximately 10000, the thermal field drives to quasi-steady equilibrium much faster than the solute allowing this convenient assumption to be made to simplify the system being solved. The equilibrium temperature at the interface given as

$$T_i = T_0 + m_l\left(C_l - C_0\right), \tag{3}$$

where C_0 is the initial concentration, T_0 is the initial liquidus temperature and m_l is the liquidus slope.

A Linear Elasticity material model has been used to represent the three-dimensional structural mechanical behaviour of the system. This material model requires assumptions that stress and strain are linearly related, that the displacements are 'small' and that there is no yielding behaviour. While these are certainly limiting assumptions that cannot hold for all cases where microstructure solidification is influenced by structural mechanics, they are sufficient for scenarios where the gradient of the displacements, and thereby orientation change, have a larger impact on the microstructure development than the absolute displacements. The displacements are governed by the following three sets of linked partial differential equations which are solved simultaneously using a quasi-stationary Finite Volume solver where the body forces act transiently so that the change in force since the last time step rather than the total force is considered.

$$(\lambda+\mu)\left(\frac{\partial^2 u}{\partial x^2}+\frac{\partial^2 v}{\partial x\partial y}+\frac{\partial^2 w}{\partial x\partial z}\right)+\mu\left(\frac{\partial^2 u}{\partial x^2}+\frac{\partial^2 u}{\partial y^2}+\frac{\partial^2 u}{\partial z^2}\right)+F_x=0 \tag{4}$$

$$(\lambda+\mu)\left(\frac{\partial^2 v}{\partial y^2}+\frac{\partial^2 u}{\partial y\partial x}+\frac{\partial^2 w}{\partial y\partial z}\right)+\mu\left(\frac{\partial^2 v}{\partial x^2}+\frac{\partial^2 v}{\partial y^2}+\frac{\partial^2 v}{\partial z^2}\right)+F_y=0 \tag{5}$$

$$(\lambda+\mu)\left(\frac{\partial^2 w}{\partial z^2}+\frac{\partial^2 u}{\partial z\partial x}+\frac{\partial^2 v}{\partial z\partial y}\right)+\mu\left(\frac{\partial^2 w}{\partial x^2}+\frac{\partial^2 w}{\partial y^2}+\frac{\partial^2 w}{\partial z^2}\right)+F_z=0 \tag{6}$$

IOP Conf. Series: Materials Science and Engineering 1281 (2023) 012045 doi:10.1088/1757-899X/1281/1/012045

where u, v and w are the displacements in x, y and z, F_x, F_y and F_z are the corresponding body forces and μ and λ are the Lamé constants:

$$\lambda = \frac{\eta E}{(1+\eta)(1-2\eta)} \tag{7}$$

$$\mu = \frac{E}{2(1+\eta)} \tag{8}$$

where E and η to be the Young's Modulus and Poisson's ratio of the material respectively. Material properties may vary both spatially and temporally as solidification progresses, allowing the Young's modulus of the dendrites to vary according to a linear relationship with the solid fraction. The impact of thermal contraction has been neglected for the cases examined, with the material properties assumed to be temperature independent. The crystallographic orientation of the dendrites is used as a means of coupling solidification to the structural mechanics, where the displacements are used to alter the local orientation by obtaining three extrinsic rotations θ_x, θ_y and θ_z around their corresponding axes which can be combined into a single 3D orientation change for each computational cell comprising the dendrite, though it should be noted that for the cases examined in this paper the orientation change in the direction of the constant body force (θ_z) is significantly larger and dominates the orientation change behaviour such that any out of plane bending or torsion is negligible.

3. Results

The numerical model has been used to run a parametric study to examine the changing behaviour of a dendrite growing under a constant body force, where all cases in the study are identical other than the magnitude of the body force. The dendrite was seeded on the west wall of a 2mm × 2mm × 12mm domain with an initial orientation of $\theta_z = 20°$ and $\theta_x = \theta_y = 0°$ and a constant thermal gradient G along the x axis which cools at a constant rate Q throughout the simulation. Periodic boundaries were used for the North, South, High and Low boundaries so that rather than modelling a single dendrite growing in a 'box', the modelling instead represents a subsection of an infinite field of identical dendrites. Whilst still an idealised scenario, this setup was chosen as it serves to highlight the impact of dendrite interaction as well as preventing interaction with the wall from limiting both the growth and displacement behaviour of the dendrite. All relevant variables related to the material properties and general model setup can be found in table 1, as representative of those used in experiments for solidifying a Ga – 25wt%In alloy, chosen for easier comparison with earlier work which used this material [14, 13]. The only properties of note are the Young's Modulus (E) and Poisson's Ratio (η), with accurate values for these properties being unclear due to the lack of information in the literature for the mechanical behaviour of semi-solid dendrites, with $\eta = 0.3$ being chosen as a the standard value for the fully solidified alloy while $E = 10$MPa was chosen arbitrarily so as to be low enough to allow significant orientation change to occur during the largest magnitude case examined. It should be noted that as the Young's Modulus and the force are related quantities such any problem with the same ratio of $E : F$ will have identical results, meaning that the results presented can be considered valid for an entire family of solutions using a given ratio.

This study will examine five cases where the constant body force in y is changed, taking the values $F_y = 100$ N/m^3, 1000 N/m^3, 2500 N/m^3, 5000 N/m^3 and 10000 N/m^3, where the body forces in the other directions are fixed as $F_x = F_z = 0$ N/m^3 . These forces act on the solid material only and any other forces that may realistically arise in this scenario are neglected. For the duration of the paper, these five cases will be identified by a number corresponding to what multiple of the initial 100 N/m^3 case, making the examined cases respectively 1, 10, 25, 50 and 100. The simulations for all cases have been run for 250000 timesteps, meaning the final state of the simulation occurs 1250 seconds after the initial seeding, with this final state for all cases being visible in figure 2. For cases 1 and 10 little visible orientation change has occurred along the length of the dendrite arm, though the cumulative orientation changes throughout the simulation have caused the tip of the dendrite in case 10 to have a significantly different position and orientation to case 1, along with various minor changes in the secondary arm development. There is some clear orientation change across the primary dendrite trunk observable in case 25, with the dendrite orientation flattening off and almost approaching 0° by the end of the simulation, while in case 50 this change is even further exaggerated such that the total orientation at the tip is now clearly negative. By case 100 it becomes difficult making direct comparisons to the other cases due to the impact of the significant misorientation on the morphology of the microstructure, where the tip of the main dendrite arm is approaching an orientation parallel to gravity and being outcompeted by a secondary arm that would likely become a new primary arm if the simulation was continued.

Throughout the simulations the transient behaviour of the orientation change and displacements have also been recorded to see if there are any obvious patterns in growth behaviour to be observed across the different cases. The first structural behaviour examined is the orientation change (such that any values presented are the change relative to the seeded orientation of $\theta_z = 20°$ rather than the current orientation at the tip), where values at regular 50000 timestep intervals the maximum observed orientation change for all cases have been recorded in the left section of table 2, with graphs showing more granular orientation change information from the simulation for easier visual interpretation in figure 3.

The graphs show both the raw values indicating the absolute difference in values in figure 3a and a 'scaled' orientation change in figure 3b for easier comparison of their relative behaviour. In this context 'scaled' means the final observed orientation is divided by the force multiplier corresponding to the case number, which means that if there was a linear relationship between force increase and the orientation change then the orientation change lines for case 1 and all the scaled cases should be identical.

By utilising a combination of the outputs in table 2 and figure 3 some clear patterns in the development of the orientation change emerge. Considering only the simulation up to step $t = 100000$, there appears to be an essentially linear relationship between the force and

Property	Variable	Value	Unit
Density Ga	ρ_{Ga}	6095	kgm^{-3}
Density In	ρ_{In}	7020	kgm^{-3}
Young's Modulus	E	10	MPa
Poisson's Ratio	η	0.3	–
Liquidus Slope	m_l	−2.9375	$K\%wt.^{-1}$
Partition Coefficient	k	0.5	–
Liquid Mass Diffusivity	D_l	2×10^{-9}	m^2s^{-1}
Solid Mass Diffusivity	D_s	1×10^{-12}	m^2s^{-1}
Cooling Rate	Q	0.01	Ks^{-1}
Thermal Gradient	G	1	Kmm^{-1}
Cell Size	Δx	10	µm
Time step	Δt	5	ms

Table 1: Material properties and problem setup used for all cases.

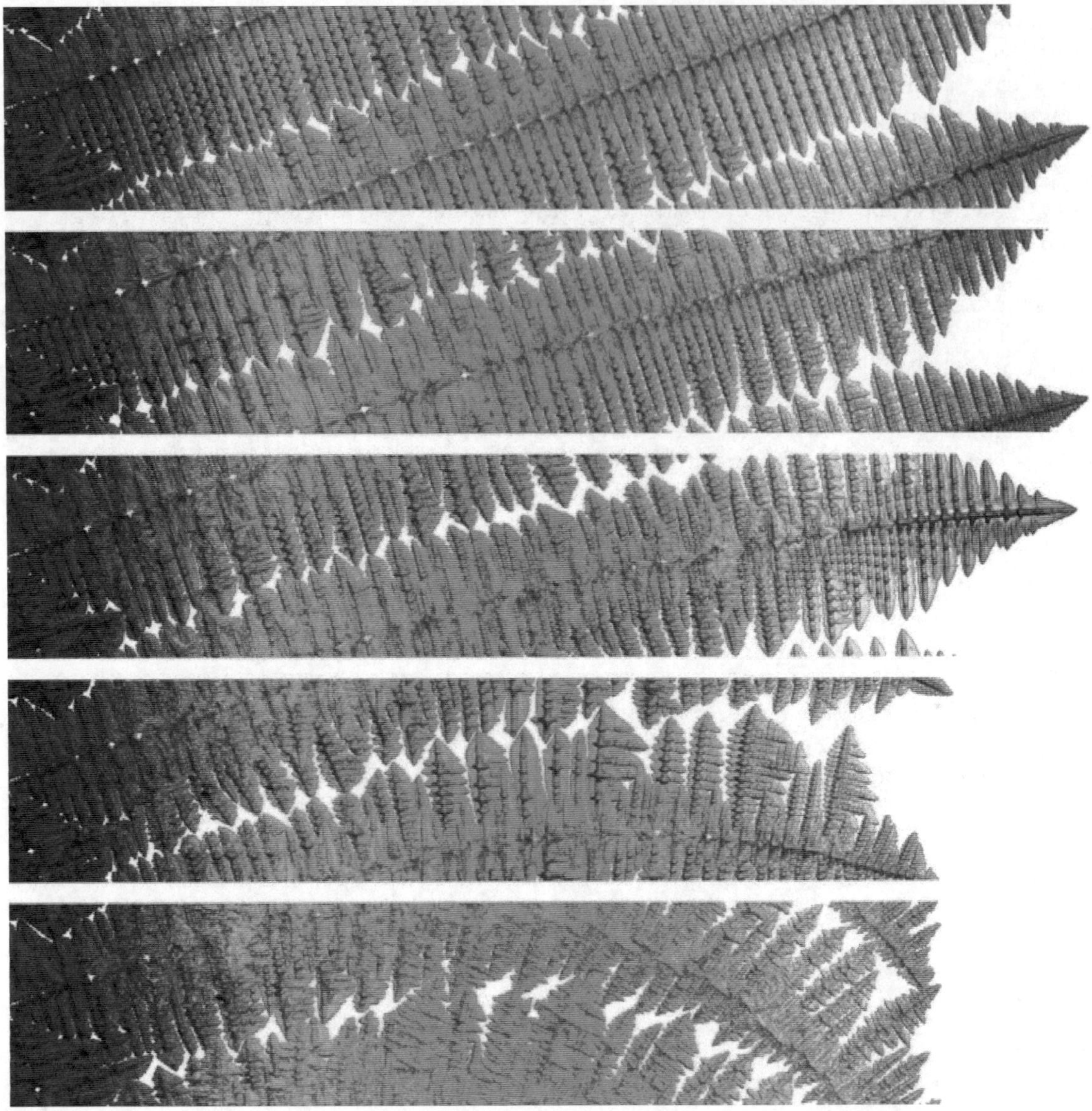

Figure 2: Dendrite structures for cases 1, 10, 25, 50 and 100 in descending order at final time step $t = 250000$.

orientation change. Consequently, the orientation change observed for case 100 seems to be approximately 100 times larger than the orientation observed for case 1, with the intermediate cases behaving in the same manner. For cases 1 and 10 this almost linear relationship between force and orientation change holds for the entire simulation, with very similar behaviour and scaled values being observed. However, the other three cases experiencing larger forces quickly diverge significantly from any linear relationship between force size and orientation change, with this disparity likely being caused by the impact of orientation changes on the morphology of the dendrite which does not significantly impact cases 1 and 10. Up until step 100000 all of the dendrites are morphologically quite similar, as the orientation change has been relatively small even for case experiencing larger forces, but after this point cases 25, 50, and 100 begin to show radically different microstructural developments and hence structural behaviour.

t	(a) Orientation change ($\theta°$)					(b) Accumulated v deformation (µm)				
	1	10	25	50	100	1	10	25	50	100
50000	-0.0364	-0.367	-0.905	-1.85	-3.64	-0.0612	-0.604	-1.50	-3.17	-6.66
100000	-0.132	-1.39	-3.54	-7.38	-14.4	-0.503	-5.49	-13.8	-32.7	-65.1
150000	-0.256	-2.79	-7.33	-17.6	-32.3	-0.998	-11.6	-32.9	-130	-200
200000	-0.383	-4.14	-12.0	-26.9	-60.1	-1.39	-18.9	-57.6	-230	-504
250000	-0.507	-5.49	-18.5	-34.0	-78.1	-2.00	-27.0	-144	-282	-687

Table 2: Comparison of (a) orientation change ($\theta°$) and (b) maximum accumulated v deformation (µm) values for the 5 cases throughout the simulation.

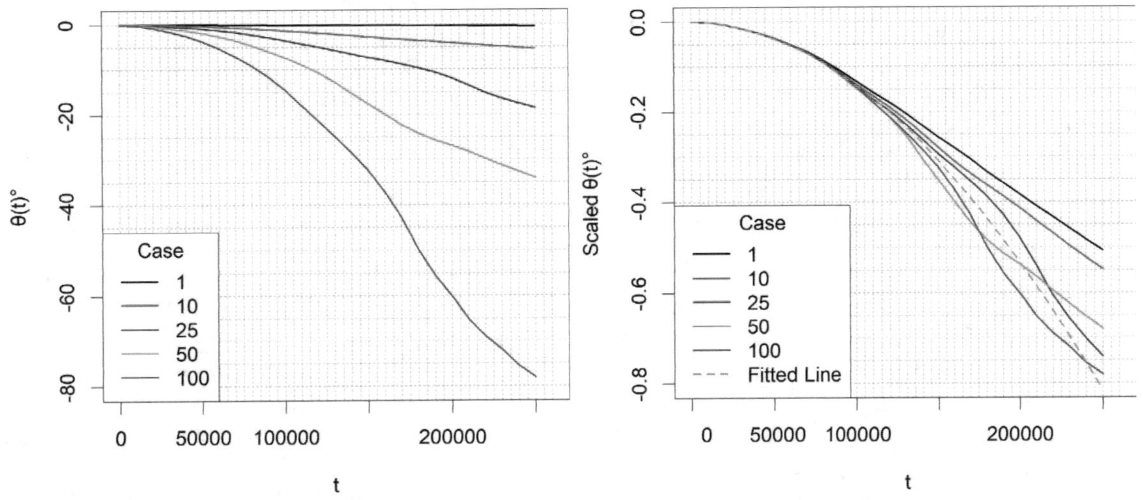

(a) Real values.

(b) Scaled values

Figure 3: Case comparison plots for largest observed orientation change θ at time step t. Scaled orientation change showing full simulation including a fitted power law of $\theta(t) = -4 \times 10^{-11} t^{1.91}$.

When considering the transient orientation change behaviour for the first 100000 steps it can be seen that all cases approximately follow a power law of $\theta(t) = -4 \times 10^{-11} t^{1.91}$, but after this point the orientation change behaviour of all cases stops following this distribution and begins to vary dramatically between cases. The rate of change for cases 1 and 10 becomes essentially linear, this is likely as the dendrites transition from being a generally freely moving dendrite (which could be considered as being analogous to a cantilever beam fixed to a wall) to being limited by interaction with surrounding dendrites. Consequently, a relatively consistently sized region near the tip of the dendrite ahead of the arm interaction remains free for the duration of the simulation, causing the rate of orientation change to become almost constant. In cases 25, 50 and 100 the behaviour after step 100000 becomes more complicated to characterise, with signifiant behavioural differences between cases and no clear fixed pattern. From this point onwards the rate of orientation change both increases and decreases at seemingly unrelated time steps and while the cases using larger forces always have a correspondingly larger orientation change in an absolute sense, when considering the scaled outputs it can been seen that in relative terms the cases experiencing the larger forces do not always correspond to a larger scaled orientation change. For example, if considering the scaled solutions for case 50 this experiences the largest relative orientation changes of all the cases between $t = 125000$ to 175000, yet by the end of the simulation it has a relative orientation change lower than that observed in case 25. This seems to highlight that predicting the orientation change becomes increasingly difficult in a coupled system once the impact of structural mechanics begins changing the morphology of the dendrite and when interaction with other dendrites begins.

Similar conclusions can be drawn when examining the transient behaviour of the accumulated displacements v across the different cases (the displacements in u and w have been neglected due to being significantly smaller in magnitude, having little impact on microstructure development). The largest accumulated deformation recorded every 50000 time steps can be found in the right half of table 2 and graphs presenting this information at more frequent intervals can be seen in figure 4, again presenting both the real and 'scaled' versions of the data relationships.

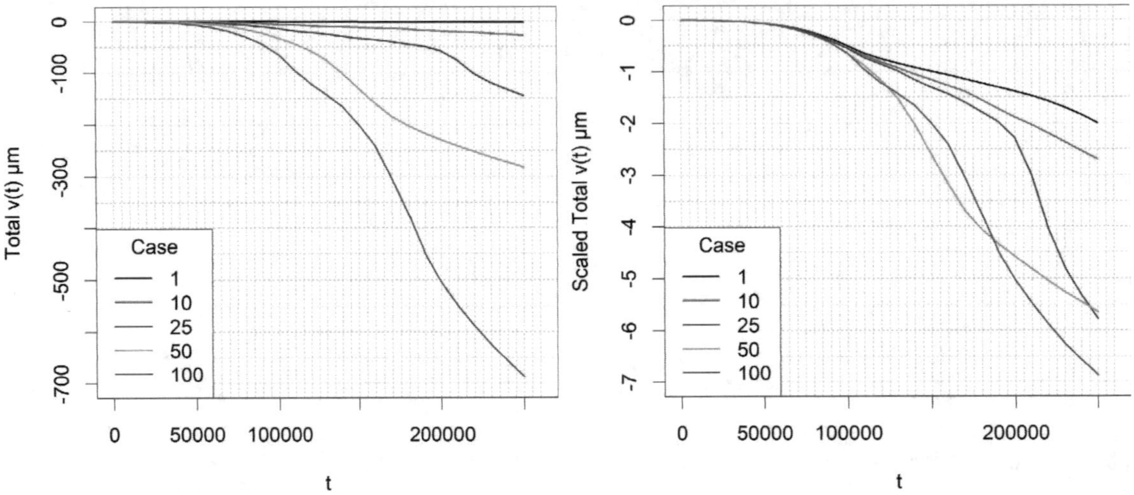

(a) Real values. (b) Scaled values.

Figure 4: Case comparison plots for largest observed cumulative v displacement at time step t.

When considering the transient behaviour of the scaled values in figure 4b, the accumulated behaviour seems to match the general relationships previously described for the orientations, which makes intuitive sense when considering that in this framework the orientation change is a quantity derived from the obtained displacements. Consequently, much as with the orientations, the relative scaled values up to step 100000 are similar for all cases, with cases 1 and 10 remaining similar in behaviour throughout the simulation while the other 3 cases diverge significantly. Considering the real values highlights the non-linear increase in displacement magnitude as the force increases, as when considering cases 1 and 100, the displacements are nearly 350 times larger that those observed in case 1, where a linear relationship would anticipate that they would be approximately 100 times larger.

Finally the maximum step displacement will be examined, which is the single largest v displacement calculated anywhere in the dendrite at a given timestep, to be contrasted with the cumulative displacements considered above which aggregate all the displacements experienced by the dendrite throughout the simulation. Values of this quantity have not been included in a table like the cumulative displacement and orientation as the step displacement has a high variability depending on the state of the system at a given time, such that only by considering the evolution of the step displacements in more granular detail can anything useful be inferred, where graphs of the real an scaled values showing this for all cases throughout the simulation are provided in figure 5.

When considering the scaled values there again seems to be a generally similar step

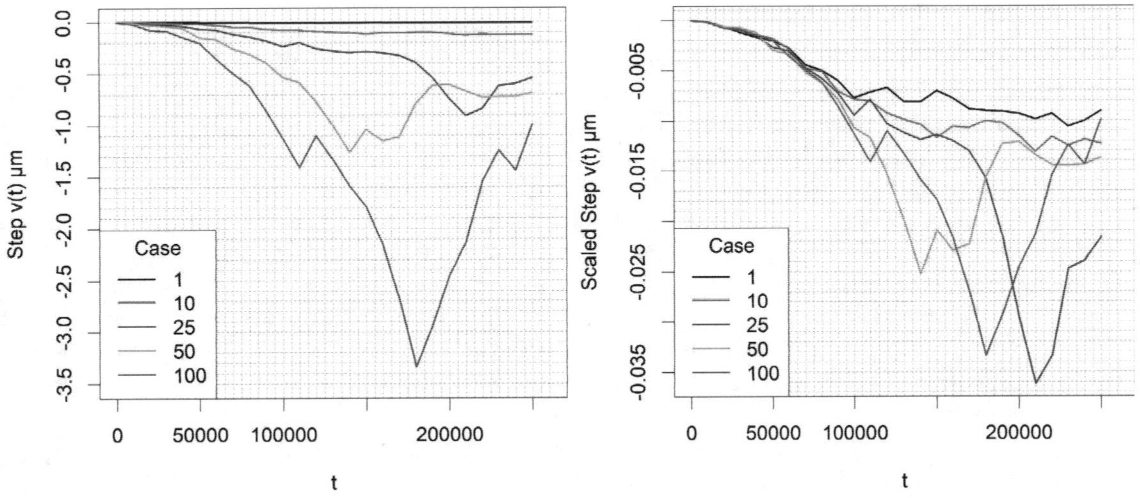

(a) Real values.

(b) Scaled values.

Figure 5: Case comparison plots for largest observed single v displacement at time step t.

displacement behaviour for all cases up to around step $t = 100000$, broadly corresponding with the power law behaviour described when considering the orientation, at which point the behaviour becomes unique for each of the cases. For cases 1 and 10 the step displacements almost plateau off to a constant rate of displacement, with only a slight increase throughout the rest of the simulation. This seems to support the idea that only the region of the tip ahead of the secondary arm interaction is free to move, with this length staying approximately the same for the entirety of the simulation and the slight increase observed is likely due to the small deformations in the growing region constrained by dendrite arm interaction. Case 25 sees a steady increase broadly in line with that observed in cases 1 and 10 until a very sharp increase occurs around $t = 17500$ to 210000, where it obtains the largest observed scaled displacement, at which point the displacements continue to decrease until the end of the simulation (though still having the highest scaled displacement by $t = 250000$). It seems probable that there was a relatively long period where the region near the tip achieved a relatively long length without constriction from secondary arm interaction. This was curtailed significantly around step 210000 as an increasing length of the dendrite becomes constrained by this arm interaction. Case 50 also demonstrates a sharp increase in step displacement between $t = 100000$ and 140000, followed by an equally sharp fall until 190000 at which point the displacements level off to being almost constant. This would again seem to be a case where the altered growth behaviour allows a relatively large region near the tip to remain free ahead of secondary arm interaction for a short period, however once the interaction begins the displacements shrink and become effectively constant due to this free length remaining comparable for the rest of the simulation. For case 100 there is an increase in step displacements until around step $t = 180000$, at which point it beings to decrease at such a precipitous rate that by the end of the simulation even the real step displacement values are comparable to those observed in cases 25 and 50, while the scaled step displacements are comparable to those seen in case 1. This occurs as the tip goes from having the freedom to deform without secondary arm interaction before experiencing this along almost the entire length of the dendrite, finally having the tip become trapped by the end of the simulation. Here the microstructure of the system changes from a free dendrite to an almost continuous block with a significantly lower capacity for displacement.

The assumption made in the literature treating a columnar dendrite as a structure analogous to a cantilever beam is generally reasonable when considering the behaviour of a single dendrite at a single point in time. However, this study highlights the limitations of this assumption in more realistic scenarios where structural mechanics and microstructure solidification are being resolved concurrently such that the growth behaviour of the microstructure is altered. The cantilever analogy holds broadly true for growing dendrites while they don't experience significant orientation changes and are free from interaction with other dendrites, such as could be observed in approximately the first 100000 steps of all cases. However, even in cases 1 and 10 where orientation changes are relatively small, the simulations still reach a point of secondary arm interaction where only the small region ahead of the arm interaction is free to behave like a beam. As this free region stays approximately the same length for the duration of the simulation, while the behaviour of this region at any given step is similar to that of a deforming beam, the transient displacements essentially progress at a constant rate. For the cases experiencing larger forces, aside from the still existing issue of secondary arm interaction limiting beam like behaviour, these cases will ultimately grow into a structure with a more complex curved morphology that makes any comparisons with a cantilever beam increasingly dubious.

4. Conclusion

A parametric study was undertaken varying the constant body force applied to a solidifying system using a coupled CA solidification and Finite Volume structural mechanics solver. The varying force was found to significantly alter both the growth behaviour and final morphology of the solidifying dendritic system. These results were then considered in the context of the analogy of cantilever beam-like structural behaviour when describing the mechanical behaviour of columnar dendrite. While this analogy can be helpful for understanding the behaviour of a single dendrite, it generally fails to describe the transient dynamics of the true interdependent system, which requires fully coupled modelling to truly understand the behaviour even for relatively simple cases such as those examined here.

Acknowledgments

Acknowledgments P.S. gratefully acknowledges the financial assistance offered by the University of Greenwich VC scholarship for his PhD studies which provided the foundation for this work.

References

[1] Reinhart G, Nguyen-Thi H, Mangelinck-Noël N, Baruchel J and Billia B 2014 *JOM* **66** 1408–1414
[2] Ananiev S, Nikrityuk P and Eckert K 2009 *Acta Mater.* **57** 657–665
[3] Aveson J, Tennant P, Foss B, Shollock B, Stone H and D'Souza N 2013 *Acta Mater.* **61** 5162–5171
[4] Hu S, Liu L, Yang W, Sun D, Huo M, Huang T, Zhang J, Su H and Fu H 2019 *Metall. Mater. Trans. A* **50** 1607–1610
[5] Hallensleben P, Scholz F, Thome P, Schaar H, Steinbach I, Eggeler G and Frenzel J 2019 *Crystals* **9** 149
[6] Takaki T and Kashima H 2011 *J. Cryst. Growth* **337** 97–101
[7] Aveson J, Reinhart G, Nguyen-Thi H, Mangelinck-Noël N, d'Souza N and Stone H 2014 *MATEC Web of Conferences* vol 14 (EDP Sciences) p 05003
[8] Sun D, Liu L, Huang T, Yang W, Li Y, Yue Q, Zhang J and Fu H 2018 *Prog. Nat. Sci.: Mater. Int.* **28** 489–495
[9] Huo M, Liu L, Yang W, Hu S, Sun D, Su H, Zhang J and Fu H 2019 *Vacuum* **161** 29–36
[10] Billia B, Bergeon N, Thi H N, Jamgotchian H, Gastaldi J and Grange G 2004 *Phys. Rev. Lett.* **93** 126105
[11] Chen C, Sun J, Diao A, Yang Y, Li J and Zhou Y 2022 *J. Alloys Compd.* **891** 161949
[12] Yamaguchi M and Beckermann C 2013 *Acta Mater.* **61** 4053–4065
[13] Soar P, Kao A, Djambazov G and Pericleous K 2022 *JOM* 1–9
[14] Soar P, Kao A, Shevchenko N, Eckert S, Djambazov G and Pericleous K 2022 *Philos. Trans. R. Soc. London, Ser. A* **380** 20210149
[15] Wang W, Lee P D and Mclean M 2003 *Acta Mater.* **51** 2971–2987
[16] Kao A, Krastins I, Alexandrakis M, Shevchenko N, Eckert S and Pericleous K 2019 *JOM* **71** 48–58

IOP Conf. Series: Materials Science and Engineering 1281 (2023) 012046 doi:10.1088/1757-899X/1281/1/012046

CA-LBM studies on the grain formation process of aluminum alloy under convection

Yanying Zhang [1], Ruifeng Dou [1], Junsheng Wang [2], Yuzhu Tian [1], Xunliang Liu [1] and Zhi Wen [1]

1 School of Energy and Environmental Engineering (SEEE), University of Science and Technology Beijing (USTB), Beijing, 100083, China

2 Advanced Research Institute of Multidisciplinary Science (ARIM), Beijing Institute of Technology (BIT), Beijing, 100081, China

*Corresponding author: douruifeng@126.com, douruifeng@ustb.edu.cn

Abstract. A cellular automata–lattice Boltzmann coupled model was established to simulate the grain formation process of aluminum alloy under convection conditions. The influence of flow on crystal grain growth was simulated. Results show that the existence of convection prevents the crystal from growing symmetrically compared with the pure diffusion condition. Moreover, the growth of dendrite at the upstream side is promoted, whereas that at the downstream side is inhibited because of solute segregation. The greater the flow velocity is, the stronger the asymmetry of the grain is. The greater the undercooling is, the greater the dendritic growth driving force is， the faster the dendritic growth is, the dendrite arms coarsen, and the secondary dendrite arms become increasingly developed, thereby increasing the solute concentration at the solid–liquid interface at the tip of the dendrite and aggravating the solute segregation.

1. Introduction

Cast aluminum alloy has become an important alloy material in the industry because of its high strength-to-weight ratio, good ductility and corrosion resistance, low cost, and excellent casting performance[1-2]. The macro properties of metal alloys are mainly determined by their microstructure. Various methods of improving material properties start from controlling the solidification structure. The morphology, size, and distribution of solidification structure affect the final properties of aluminum alloys. Research and control of the solidification structure of metal materials have become hot issues in materials science[3-4]. As a multiscale phenomenon, solidification involves heat and mass transfer on the macroscale, growth of dendrites and grains on the mesoscale, and nucleation and growth on the microscale. The nucleation and growth of dendrites determine the morphology of the microstructure. After nucleation, the shapes of these grains are columnar or equiaxed. In the process of metal solidification, the convection of liquid metal directly affects the transfer of heat and solute and plays a decisive role in the distribution of alloy solidification structure[5-6].

Directly observing the convection phenomenon in the solidification process is complex. Thus, this study uses a numerical method to simulate the convection phenomenon in the solidification process. The phase field method is often applied to the problem of dendrite growth under 2D or 3D melt convection[7-12]. However, the calculation amount is very large, the calculation efficiency is low, and the computational domain is small. The cellular automata (CA) model, which has the advantages of simplicity, rapidity, and high computational efficiency, has attracted much attention in microstructure simulation[13-15]. In addition, the lattice Boltzmann method (LBM), with fast computation and natural parallelism, can easily simulate the flow of complex geometry under various boundary conditions; this method can be adapted for large-scale parallel computations[16-18].

In the present research, the CA method for calculating dendrite growth was coupled with the LBM for calculating the flow, concentration, and temperature fields. A 2D CA-LBM coupled model was established. The solidification growth processes of Al-4%Cu dendrites under convection were simulated. The effects of dendritic morphology, flow field, temperature field, solute field and undercooling on dendritic growth were analyzed.

2. Materials and methods

In this study, Al-4%Cu alloy was selected as the research material. The physical property parameters are presented in Table 1.

Table 1. Physical property parameters of Al-4%Cu alloy.

Physical Parameter	Symbol	Value	Unit
Liquidus temperature	T_l	648	K
Solid diffusivity	D_s	3×10^{-13}	m^2/s
Liquid diffusivity	D_l	3×10^{-9}	m^2/s
Liquidus slope	m_l	-2.6	$m \cdot K/\%$
Kinematic viscosity	ϑ	1.4×10^{-3}	Pa· s
Thermal diffusivity	α	4.313×10^{-5}	m^2/s
Partition coefficient	k	0.17	-
Liquid density	ρ	2600	Kg/m^3
Gibbs–Thomson constant	Γ	$2.4*10^{-7}$	m·K
Interfacial energy anisotropy	ϵ	0.0267	-
Specific heat of liquid phase	c_p	1070.0	J/(Kg·K)
Latent heat	L	397.0	KJ/Kg

2.1. Nucleate model

Nucleation occurs in a series of nucleation positions. The nucleation model is a continuous rather than discrete distribution function. Grain nucleation density $n(\Delta T)$ is essentially a function of undercooling ΔT. The density of the new crystal nucleus[19] can be expressed as

$$n(\Delta T) = \int_0^{\Delta T} \frac{dn}{d(\Delta T)} d(\Delta T) \tag{1}$$

The relationship between the grain nucleation density and the undercooling is as follows:

$$\frac{dn}{d(\Delta T)} = \frac{n_{max}}{\sqrt{2\pi} \cdot \Delta T_\sigma} exp\left(-\frac{1}{2}\left(\frac{\Delta T - \Delta T_{max}}{\Delta T_\sigma}\right)^2\right) \tag{2}$$

where n_{max} is the maximum grain density; ΔT_{max} is the maximum nucleation undercooling; ΔT_σ is the standard deviation of undercooling.

2.2. CA model

The interface growth kinetics is determined according to the local solute balance:

$$V_n C_l^*(1-k) = D_s \frac{\partial C_s}{\partial n}|^* - D_l \frac{\partial C_l}{\partial n}|^* \tag{3}$$

$$C_s^* = k C_l^* \tag{4}$$

where V_n is the normal growth velocity of the solidification interface; k is the equilibrium partition coefficient of the solute; C_s^* and C_l^* are equilibrium solute concentrations of solid and liquid phases at the solidification interface, respectively. C_l^* is calculated according to the local temperature and curvature at the solidification interface:

$$C_l^* = C_0 + \frac{1}{m_l}(T^* - T_l + \Gamma\kappa f(\theta,\theta_0)) \tag{5}$$

where C_0 is the initial content of the solute; m_l is the liquidus slope; Γ is the Gibbs–Thomson coefficient of the alloy; ϵ is the anisotropy parameter; $f(\theta,\theta_0)$ is the interface anisotropy function[20], which is defined as

$$f(\theta,\theta_0) = 1 - 15\epsilon cos\,[4(\theta - \theta_0)] \tag{6}$$

where θ_0 is the preferential growth orientation; θ is the angle between the interface norm and the x axis, and it can be calculated by

$$\theta = arccos\,(-\frac{\partial_x f_s}{\sqrt{(\partial_x f_s)^2 + (\partial_y f_s)^2}}) \tag{7}$$

As the interface curvature, κ is obtained by Eq. (8).

$$\kappa = \frac{1}{dx}[1 - \frac{2}{N+1}(f_s + \sum_{k=1}^{N} f_s^k)] \tag{8}$$

where dx is the space step; N is the number of adjacent cells of the current cell, and for this 2D model calculation field, $N = 9$; f_s^k is the solid fraction of the kth adjacent cell.

The decentered square algorithm is used in the capturing process of cells. Figure 1 shows the schematic diagram of the decentered square algorithm. The solid fraction of this cell increases continuously when a cell in the calculation domain nucleates. A square with an inclined angle θ_0 is arranged at the center of the nucleating cell according to the solid fraction of the cell. The diagonal length of the square increases with grain growth, and the semidiagonal length of the square can be calculated by the following formula[21]:

$$L(t + dt) = L(t) + V_n \cdot dt \tag{9}$$

The grain growth process is equivalent to the increase in the solid fraction. The calculation formula for the solid fraction is as follows:

$$f_s(t + dt) = \frac{L(t + dt)}{dx(|sin\theta| + cos\theta)} \tag{10}$$

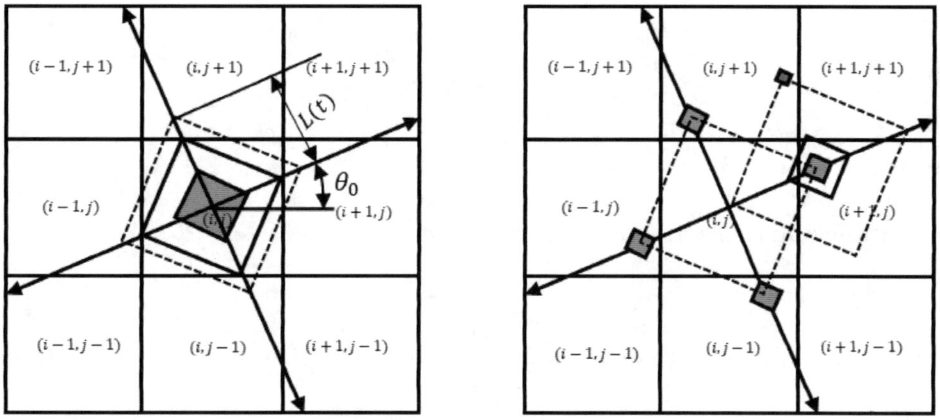

Figure 1. Decentered square algorithm.

2.3. LBM model

LBM[22-23] has been widely used to calculate fluid flow for its simplicity and amenability to parallelization. In the present study, the D2Q9 model, namely, the LGK (Lipton-Glicksman-Kurz) model[24], was used to calculate the flow, temperature, and solute fields in the 2D dendritic growth process.

The evolution equation of the flow field in LBM is

$$f_\alpha(r + e_\alpha, t + dt) - f_\alpha(r,t) = -\frac{1}{\tau_f}[f_\alpha(r,t) - f_\alpha^{eq}(r,t)] \tag{11}$$

where r is the position vector; t is the time; dt is the time step; $\alpha = 0,1,\ldots,8$ represent 9 discrete velocity directions; τ_f is the relaxation time related to the kinematic fluid viscosity ϑ.

$$\tau_f = 3\frac{\vartheta}{c^2 dt} + 0.5 \tag{12}$$

f_α^{eq} is the equilibrium distribution function derived from a Maxwell–Boltzmann distribution:

$$f_\alpha^{eq} = w_\alpha \rho(x,t)(1 + \frac{e_\alpha \cdot u}{c_s^2} + \frac{(e_\alpha \cdot u)^2}{2c_s^4} - \frac{u^2}{2c_s^2}) \tag{13}$$

where $\rho(x,t)$ represents the lattice density in the LBM model at time *t*; c_s is the sound speed of particles, and $c_s = \frac{c}{\sqrt{3}}$, $c = \frac{dx}{dt}$; dx is the space step; u is the fluid velocity of the node; w_α is a lattice constant depending on the chosen lattice scheme, and it can be calculated by

$$w_\alpha = \begin{cases} \frac{4}{9} & \alpha = 0 \\ \frac{1}{9} & \alpha = 1-4 \\ \frac{1}{36} & \alpha = 5-8 \end{cases} \tag{14}$$

The grid boundary adopts the nonequilibrium extrapolation format. The corresponding macro density and speed can be determined by the following formulas:

$$\rho = \sum_{\alpha=0}^{8} f_\alpha \tag{15}$$

$$u = \sum_{\alpha=0}^{8} f_\alpha \, e_\alpha / \rho \tag{16}$$

Similarly, the evolution of solute and temperature fields can also be calculated by the corresponding lattice Boltzmann equation:

$$g_\alpha(r + e_\alpha, t + dt) - g_\alpha(r,t) = -\frac{1}{\tau_g}[g_\alpha(r,t) - g_\alpha^{eq}(r,t)] + w_\alpha C_l(1-k)\Delta f_s \tag{17}$$

$$h_\alpha(r + e_\alpha, t + dt) - h_\alpha(r,t) = -\frac{1}{\tau_h}[f_\alpha(r,t) - h_\alpha^{eq}(r,t)] + w_\alpha \Delta f_s L / c_p \tag{18}$$

where Δf_s is the solid phase increment of the interface cell in a time step; C_l is the liquid solute concentration; L is the latent heat; c_p is the specific heat of the liquid phase.

The equilibrium distribution functions of the concentration and temperature fields are

$$g_\alpha^{eq} = w_\alpha C(x,t)(1 + \frac{e_\alpha \cdot u}{c_s^2} + \frac{(e_\alpha \cdot u)^2}{2c_s^4} - \frac{u^2}{2c_s^2}) \tag{19}$$

$$h_\alpha^{eq} = w_\alpha T(x,t)(1 + \frac{e_\alpha \cdot u}{c_s^2} + \frac{(e_\alpha \cdot u)^2}{2c_s^4} - \frac{u^2}{2c_s^2}) \tag{20}$$

where $C(x,t)$ represents the lattice concentration in the LBM model at time *t*; $T(x,t)$ represents the lattice temperature in the LBM model at time *t*.

The relaxation time of the concentration field τ_g and temperature field τ_h is a function of solute diffusion coefficient *D* and thermal diffusion coefficient α:

$$\tau_g = 3\frac{D}{c^2 dt} + 0.5 \tag{21}$$

$$\tau_h = 3\frac{\alpha}{c^2 dt} + 0.5 \tag{22}$$

The corresponding macro concentration and temperature can be determined by the following formulas:

$$C = \sum_{\alpha=0}^{8} g_\alpha \tag{23}$$

$$T = \sum_{\alpha=0}^{8} h_\alpha \tag{24}$$

2.4. CA and LBM coupling

In the initial state, one or more crystal nuclei are placed as nucleation points in the domain. According to the above capture rules, the solid phase grid captures the surrounding liquid phase grid, making it the solid–liquid interface grid. The velocity, temperature, and concentration distributions of the flow field are obtained by the LBM model, which determines the concentration and temperature of the solid–liquid interface in this time step. According to the temperature and concentration of the interface, the dendrite growth in this time step is calculated by the CA method. The latent heat and solute released from solidification calculated by the CA model are added to the LBM evolution equation as source terms. The flow chart of the simulation program is shown in Figure 2.

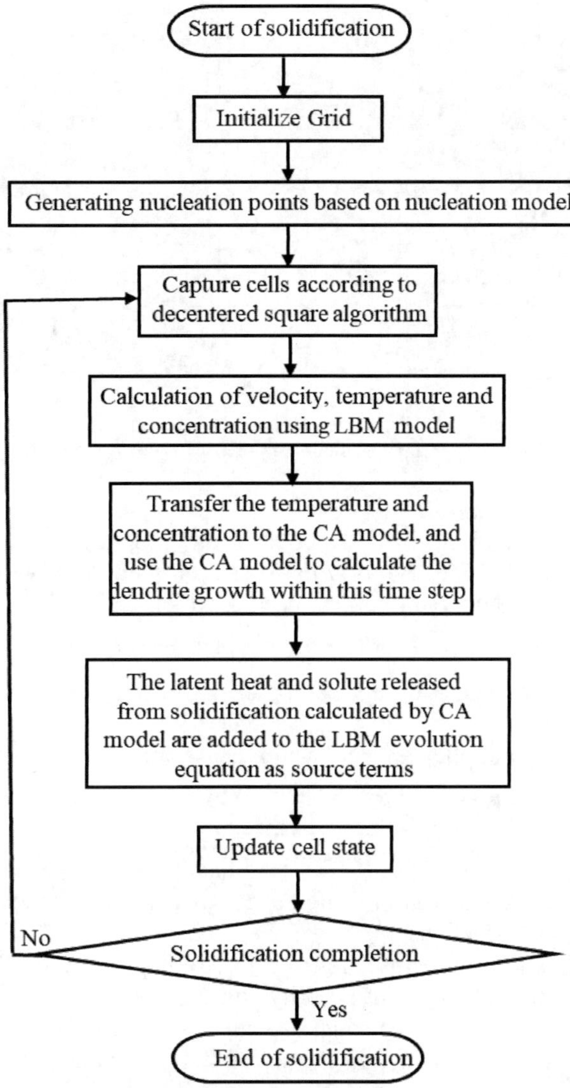

Figure 2. LBM-CA simulation flow chart.

3. Model validation

3.1. Dendrite morphology under pure diffusion condition

The domain is divided into 300 × 300 grids, with a grid size of $dx = 2.5$ μm. At $t = 0$, an initial preferred orientation of $\theta_0 = 45^\circ$ is set at the center of the calculation area. The initial undercooling is 10 K, and the cooling rate is 10 K/s. Under the pure diffusion condition, the solute precipitated from solidification and the latent heat released only depend on the diffusion effect. Figure 3 shows the concentration and temperature distribution of the single dendrite growth under the pure diffusion condition. The figure shows that during the growth process, the dendrite morphology is completely symmetrical. The solute and temperature distribution are also symmetrical. The simulation results are identical to solidification theory[25], thereby proving that the CA-LBM coupled model is correct.

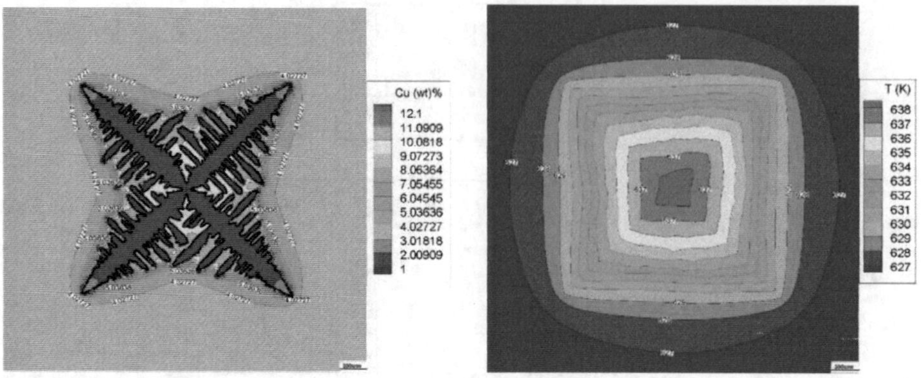

(a) (b)

Figure 3. Distribution diagram of concentration (a) and temperature (b) of single dendrite growth under pure diffusion condition.

3.2. Comparison with the LGK model

Liption et al.[26] proposed an analytical LGK model for the growth of equiaxed crystals in undercooled melts based on critical stability. The total driving forces for dendrite growth are thermal undercooling, composition undercooling, and curvature undercooling, which are expressed by the following formula:

$$\Delta T = \frac{\Delta H}{C_p} Iv(P_t) + |m| C_0 \left\{1 - \frac{1}{1-(1-K)Iv(P_c)}\right\} + \frac{2\Gamma}{R} \tag{25}$$

where P_c and P_t are the solute Peclet number and thermal Peclet number; $P_c = RV/2\text{D}$; $P_t = \text{RV}/2\alpha$; Iv (P) is an Ivantsov function.

R is the tip radius of the parabolic dendrite, which is assumed to be equal to the shortest wavelength perturbation according to the marginal stability criterion:

$$R = \frac{\Gamma/\sigma^*}{\frac{P_t L}{C_p} \frac{2P_c m_l c_0 (1-k_0)}{1-(1-k_0)I_v(P_c)}} \tag{26}$$

The LGK theoretical analytical solution of the tip growth rate and tip radius of dendrite can be obtained from the above two equations.

For quantitative verification, the simulation results are compared with the LGK model. Figure 4 shows the relationship between tip growth velocity and time under two grid sizes (2.5 and 5.0 μm). In the early stage of simulation, the dendritic growth speed is fast because of the low interface concentration. The precipitation of solute during solidification increases the solute concentration in the front of the interface, thereby inhibiting dendritic growth. The growth rate also tends to stabilize when the release and diffusion of solute reach equilibrium. Figure 4 shows that the equilibrium dendrite growth rate under two grid sizes is close to the theoretical value of LGK. The model is reliable in simulating aluminum alloy solidification, and the mesh size has a slight effect on the results.

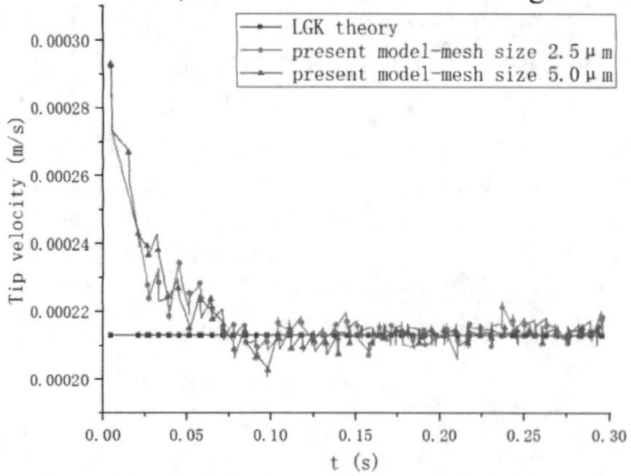

Figure 4. Comparison of the dendrite tip growth rate of Al-4%Cu alloy with the LGK model.

4. Numerical simulations

4.1. Grain growth under forced convection

Figure 5 shows the distribution of (a) flow field, (b) concentration field, and (c) temperature field during the solidification of a single grain at different times (t = 0.15 s, t = 0.40 s, t = 0.90 s) under convection (u = 0.6 mm/s). Compared with the case under pure diffusion, the grain morphology has changed significantly under the effect of the flow field. This finding shows that the dendrite on the upstream side grows fast, whereas the dendrite on the downstream side grows slowly. The reason is that the flow washes away the solute near the upstream primary arm under the condition of forced flow. This phenomenon increases the effective undercooling at the dendrite tip, thereby significantly promoting the growth of the upstream dendrite. The growth of the downstream-side dendrite is inhibited by solute accumulation. With the progress of solidification, the solid fraction increases, accompanied by a large amount of solute precipitation. The solute distribution difference between the solid–liquid interface of the upstream and downstream dendrites becomes increasingly obvious under the action of melt convection.

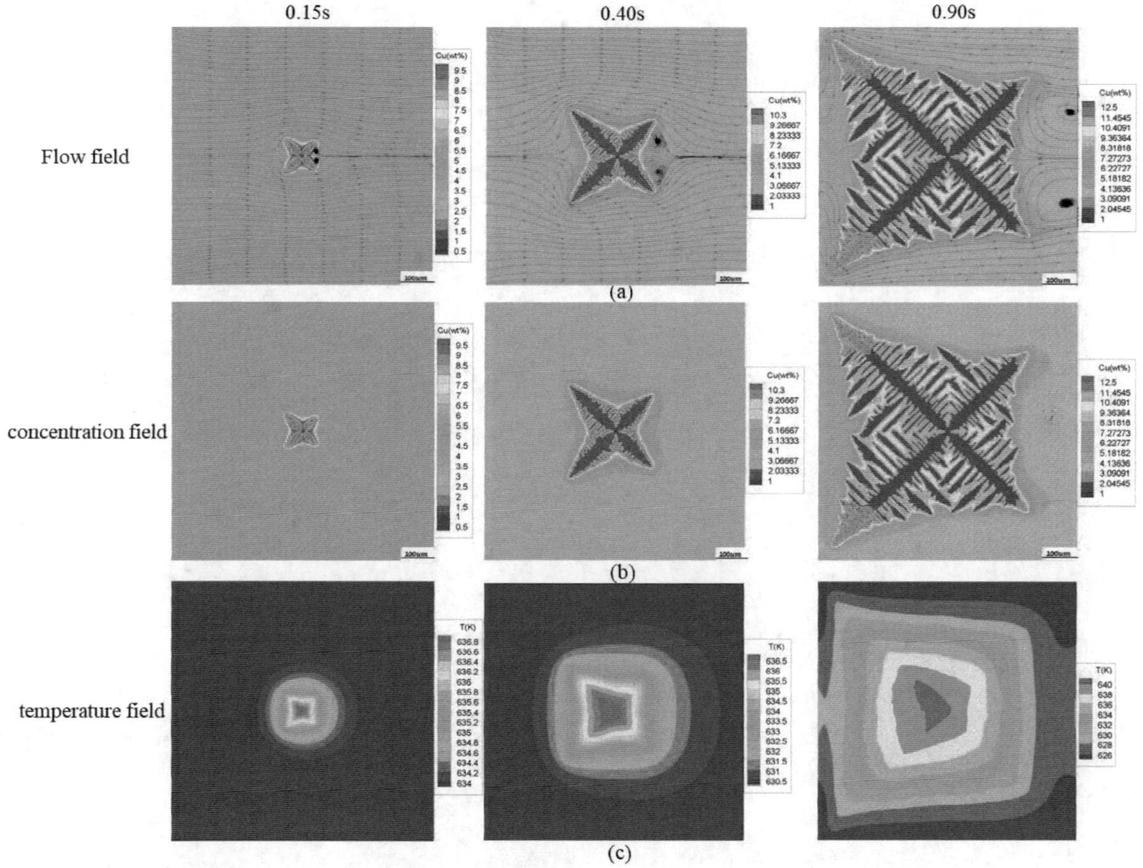

Figure 5. (a) Flow field, (b) concentration field, and (c) temperature field distributions at different times (t = 0.15 s, t = 0.40 s, t = 0.90 s) under the inflow velocity of u = 0.6 mm/s.

4.2. Influence of flow velocity

Figure 6 shows the grain morphology at different flow velocities: (a) u = 0.3 mm/s, (b) u = 0.45 mm/s, and (c) u = 0.6 mm/s. The figures indicate that the increase in the flow velocity strengthens the promotion effect on the upstream side and the inhibition effect on the downstream side, leading to solid crystal asymmetry. This scenario occurs because the convection intensity is enhanced with the increase in the flow rate. Moreover, the solute and heat are easily carried downstream by the melt. Thus, the dendritic growth driving force on the upstream side is great, and the dendrite growth speed is high. Much solute and heat are gathered on the downstream side. Thus, the growth on the downstream side is inhibited, and the growth asymmetry is obvious.

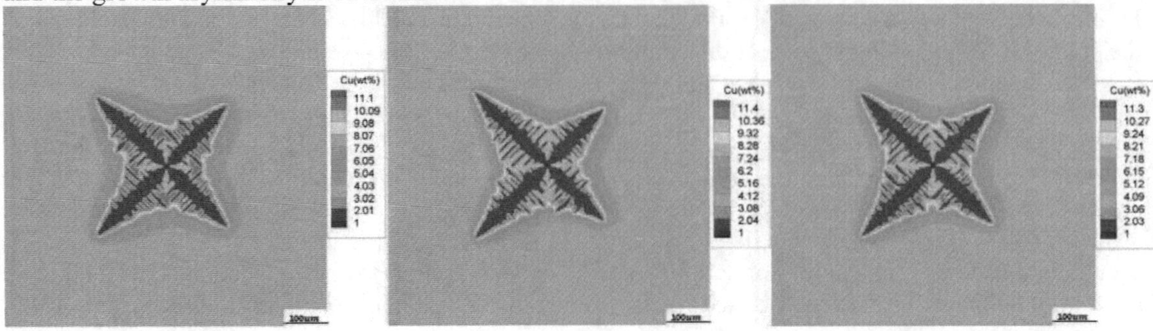

(a) (b) (c)

Figure 6. Dendrite morphology at different flow velocities: (a) $u = 0.3$ mm/s, (b) $u = 0.45$ mm/s, (c) $u = 0.6$ mm/s.

4.3. *Influence of undercooling under forced convection*

Figure 7 shows the grain morphology at different undercooling states, (a) $\Delta T = 2$ K, (b) $\Delta T = 6$ K, (c) $\Delta T = 12$ K, at the same time when the inflow speed is 0.6 mm/s. The greater the undercooling is, the faster the dendrite growth is, and the coarser the dendrite arm is. The reason is that the large initial undercooling strengthens the instability of the dendrite solid–liquid interface. The solid–liquid interface instability drives the formation of secondary dendrites. Finally, it increases the solute segregation phenomenon.

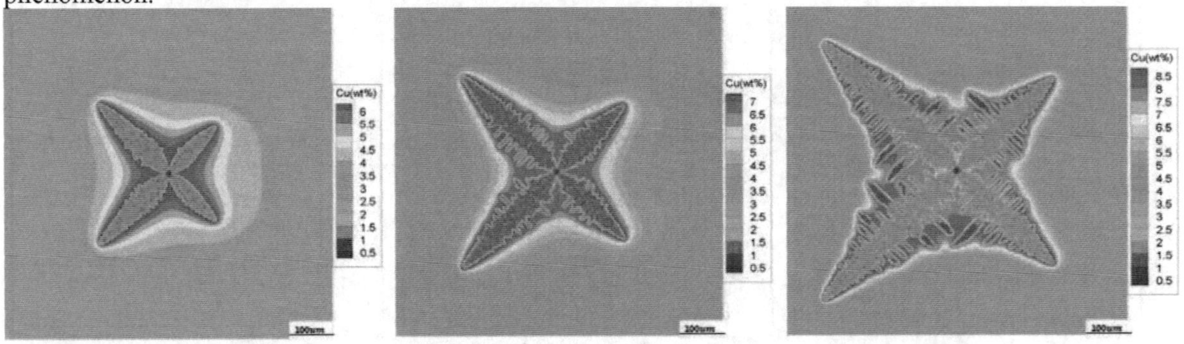

(a) (b) (c)

Figure 7. Dendrite morphology at different undercooling states: (a) $\Delta T = 2$ K, (b) $\Delta T = 6$ K, and (c) $\Delta T = 12$ K.

4.4. *Simulation of multigrain growth under forced convection*

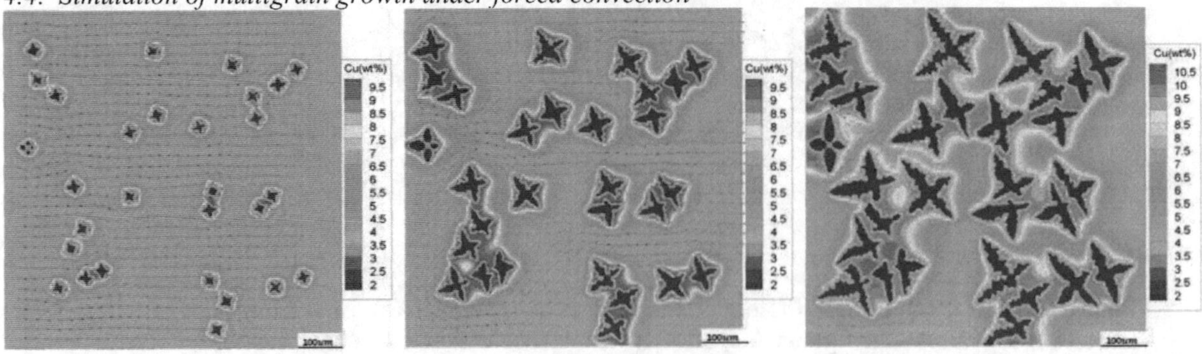

(a) (b) (c)

Figure 8. Morphological distribution of dendrites under forced convection at different times: (a) 0.015 s, (b) 0.05 s, (c) 0.1 s.

IOP Conf. Series: Materials Science and Engineering 1281 (2023) 012046 doi:10.1088/1757-899X/1281/1/012046

Figure 8 shows the distribution of grain morphology at different times, (a) 0.015 s, (b) 0.05 s, (c) 0.1 s, during the solidification process under the inlet velocity of 0.8 mm/s. The primary dendrite arm gradually thickens with the solidification process. Moreover, the secondary dendrite arm gradually appears. The growth of the primary and secondary arms on the upstream side is promoted under the effect of inlet velocity. The interdendritic fluid flow is inhibited when the dendrites are close to one another.

5. Conclusions

A CA-LBM coupled model was established to simulate the flow, temperature, grain, and solute fields simultaneously. The decentered square algorithm was adopted. The CA-LBM coupled model was applied to simulate the grain formation process of aluminum alloy in the convection process and reveal the growth process of dendrites under the convection of the superheated melt during the solidification of the grains. The main conclusions are summarized as follows:

1) The model was based on the principle of solidification thermodynamics by considering the growth factors of nucleation. The model was compared with the LGK model in the literature to verify the reliability of the former. The simulation results showed that the CA model could reproduce the morphology of the single crystal and polycrystal in the presence of a flow field.

2) The influence of flow on the solidification process of aluminum alloy was studied through numerical simulation. The results showed that the existence of convection prevented the crystal from growing symmetrically. Moreover, convection promoted the growth of upstream-side dendrites, and the growth of downstream-side dendrites was inhibited because of solute segregation.

3) The effects of flow velocityand undercooling on grain growth were studied. The results showed that the greater the flow velocity is, the stronger the asymmetry of the dendrites is. Moreover, the greater the undercooling is, the greater the dendrite driving force is, the faster the dendrite growth is, the dendrite arms coarsen, and the secondary dendrite arms become increasingly developed, thereby increasing the solute concentration at the solid–liquid interface at the tip of the dendrite and intensifying the solute segregation.

References

[1] Crum K A, Mcmichael J and Novak M 2012 *NEJ.* **124** 105-114

[2] Cahoon J R, Lennart B and Sigworth G K 1992 *Metall Trans.* **23** 3399–3404

[3] Mats J, Lennart B and Sigworth G K 1993 *Metall Trans.* **24** 481–491

[4] Eskin G I. 2001 *ULTRASON SONOCHEM.* **8** 319–325

[5] J A Dantzig and M Rappaz 2009 *Solidification* Italy

[6] M Rappaz. 1989 *Int. Mater. Rev.* **34** 93–123

[7] M Asta, C Beckermann, A Karma, W Kurz, R Napolitano, M Plapp, G Purdy, M Rappaz and R Trivedi 2009 *Acta Mater.* **57** 941–971

[8] Z Guo, J Mi, S Xiong and P S 2014 *J.Comput. Phys.* **257** 278–297

[9] Z Guo, J Mi, S Xiong and P S.2013 *Metall. Mater. Trans.* **44** 924–937

[10] I Steinbach 2009 *Acta Mater.* **57** 2640–2645

[11] R Tönhardt and G Amberg 2000 *J. Cryst. Growth.* **213** 161–187

[12] R Siquieri, J Rezende, J Kundin and H Emmerich 2009 *Eur. Phys. J. Spec. Top.* **177** 193–205

[13] Y Natsume, K Ohsasa and T Narita 2002 *Mater. Trans.* **43** 2228–2234

[14] W L Wang, S Luo and M Y Zhu 2014 *Comput. Mater. Sci.* **95** 136–148

[15] S Luo, W L Wang and M Y Zhu 2016 *ISIJ Int.* **56** 564–573

[16] L Yuan and P D Lee 2012 *Acta Mater.* **60** 4917–4926

[17] J Zudrop, K Masilamani, S Roller and P Asinari 2017 *Comput. & Fluids.* **153** 20–33

[18] E Calore, A Gabbana, J Kraus, E Pellegrini, S F Schifano and R Tripiccione 2016 *Parallel Comput.* **58** 1–24

[19] A Randles and E Kaxiras 2014 *J. Comput. Phys.* **270** 577–586
[20] Thévoz P,Desbiolles J L and Rappaz M 1989 *MMTA* **2** 311-322
[21] Beltran Sanchez L and Stefanescu D M 2004 *MMTA*. **35** 2471-2485
[22] KangWei Wang, MengWu Wu,BingHui Tian and ShouMei Xiong 2022 *Chin. Phys. B*. **31** 1-12
[23] Qian YH, Humie`res DD and Lallemand P 1992 *Europhys Lett.* **17** 479
[24] Kru¨ger T, Kusumaatmaja H, Kuzmin A, Shardt O, Silva G and Viggen EM 2017 *Basel: Springer*. **61** 1-13
[25] Qian Y H 1992 *EL*. **17** 479-484
[26] Kurz W, Fisher D J and Trivedi R 2019 *Int. Mater. Rev.* **64** 311-354
[27] Liption J,Glicksman M E and Kurz W 1984 *MSE*. **65** 57-63

Phase-field modeling of austenitic steels used in turbines

Salvador Valtierra Rodriguez[1], Michael Greenwood[3], Delin Li[3], Jean-Benoît Lévesque[4], Vladimir Timoshevskii[4], Daniel Paquet[4], Nikolas Provatas[2],

[1] Department of Mining and Materials Engineering, McGill University, Montreal, QC H3A 2T8, Canada
[2] Department of Physics, McGill University, Montreal, QC H3A 2T8, Canada
[3] Natural Resources Canada, CanmetMATERIALS, 183 Longwood Road, Hamilton, Ontario, Canada, L8P 0A5
[4] Hydro Quebec Research Institute (IREQ), 1800 Bd Lionel-Boulet, Varennes, Quebec, Canada, J3X 1S1

E-mail: salvador.valtierra.rodriguez@gmail.com

Abstract. The performance in hydro-electric turbine casting and repair requires understanding of how process parameters and chemistry selection affect solidification microstructures. The aim of this study is to provide a quantitative phase-field formulation for process-microstructure relationships that seeks to model stainless steels. We have developed a phase-field model to simulate austenitic stainless steel solidification under experimental thermal histories. To this end we look at a pseudo-binary approximations for numerical efficiency. The pseudo-binary formulation is underpinned by the alloying element equivalent value, a metallurgical tool used to analyze the microstructural impact of "minor" alloying elements in stainless steels. For model validation we develop thin wall casting experiments to measure the thermal history and chemistry controlled microstructure. The models incorporate a thermodynamic parameterization and are linked to a thermal-phase transformation model which represents the experimentally measured thermal history. The results display a good agreement with the primary branch spacing and cellular to dendritic transition of the casting experiments. These models and software provide the basis for future expansion to include more complex microstructures.

1. Introduction

Materials fatigue has a huge negative impact upon the lifetime performance and cost-effectiveness of turbine equipment. Hydraulic turbines are fabricated and repaired using different types of steels (martensitic, austenitic, and ferritic-pearlitic). Their microstructure varies according to their chemical composition and the processes employed for their fabrication (ex. welding, casting, rolling, heat treatment) [1–3]. The methodology used for designing and repairing modern hydraulic turbines must consider the quintessential role that microstructure mechanisms play in driving fatigue damage. Pristine turbines that develop cracks or other features indicating imminent failure are repaired through a welding process [2, 4, 5].

During the welding procedure, different phase transformations (solidification, austenitic-martensitic transformation, etc.) take place, which leads to the formation of distinct (reformed)

IOP Conf. Series: Materials Science and Engineering 1281 (2023) 012047 doi:10.1088/1757-899X/1281/1/012047

Table 1. Experimental composition modified to obtain a purely austenitic stainless steel.

Cr	Ni	Si	Mo	Mn	C	N	P	S	Fe
21.96	15.64	0.67	0.3	1.87	0.036	0.009	0.001	0.04	Bal

microstructure in the welding regions. Thus, the impact of microstructural properties in both the repaired and pristine turbine regions upon the micromechanical and macroscopic properties of turbine steels must be considered in any predictive model of turbine failure. Throughout the reparation process, the welding filler (which is an austenitic steel in this particular case) solidifies during said reparation process. This region is the one where liquid-solid and subsequently other phase transformations take place.

In order to simulate this complex welding process, a modeling technique capable of effectively simulating these processes in the welding filler region is needed. To this end, we employ a phase-field theoretical approach to simulate the solidification of an austenitic steel near the "chill wall" region and to predict its resulting microstructure. A great success of the phase-field (PF) theoretical approach in modeling solidification processes is due to the extreme computational effectiveness of this method and to its fundamental physics background [6–10]. Among others, it is capable of accurately capturing morphological instabilities that yield cellular-dendritic regions, as well as grain boundary energies, constitutional undercooling, solute-trapping and directional solidification [11–15].

In this article, Sec. 2 explains the composition selected for the analysis, its corresponding chill-castings experiments for said composition and the microstructural analysis. In Sec. 3, the pseudobinary phase-field model is explained as well as the thermokinetic parameterization. Afterwards, we proceed to explain the thermal-model used to obtain the thermal history data for this alloy in Sec. 4, to obtain the thermal gradients (G) and solidification front velocities (V). In Sec. 5 we present our phase-field results as well as a discussion of the results. Finally, in Sec. 6 we present the conclusions of this research and future work.

2. Casting experiments and microstructural analysis

Given the importance of stainless steel alloys E309L, this chemical composition was chosen as the reference point. However, this alloy normally displays a two-phase microstructure composed of FCC austenite and BCC delta ferrite [16, 17]. In order to study a simpler one-phase system as a first step before proceeding with more complex systems, the chemical composition of E309L was modified by increasing the amount of nickel and reducing the amount of chromium. The former being an austenite stabilizer and the latter a delta ferrite stabilizer. This can be inferred by analyzing a Schaeffler diagram where the amounts of Cr and Ni modify the amounts of austenite and delta-ferrite [18]. This new chemical composition was then used in the experiments henceforth presented and is presented in table 1.

A series of chill-castings were designed, cast and then analyzed utilizing optical microscopy, scanning electron microscopy (SEM). The optical analysis observed in Fig. 1 reveals a fully cellular/dendritic, austenitic microstructure where a symmetry line can be drawn at the center, given that directional solidification took place from the walls of the chill walls in the casting. From this microstructure, at an approximate distance of 1.4 mm from the chill-wall, primary branch spacing was calculated for several grains throughout the sample. On average, a primary branch spacing of 20.7μm is observed in this region. This will be the main length-scale we analyze

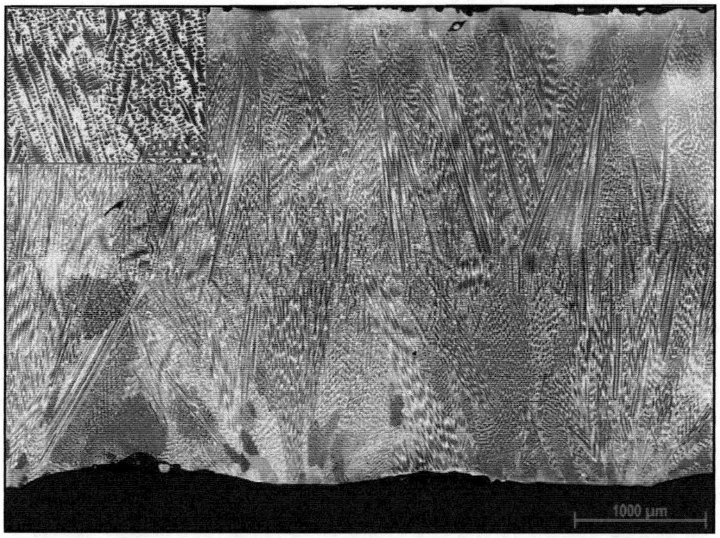

Figure 1. Optical microscope image displaying the full casting. A fully cellular-dendritic structure can be observed, where very fine cells emerge from the chill-regions and become dendritic as the thermal-gradient across the sample decreases. An inset displaying more clearly the dendritic features is included and denoted by the red block.

in this work, further analysis of other characteristic length-scales will be explored in future work.

EBSD analysis was also performed in order to obtain relevant information about grain direction and grain density across the system as observed in Fig. 2. On average a density of 13.3 $grains/mm$ was found at the chill-wall region and 7.25 $grains/mm$ at the 1.4 mm mark. It can be inferred that slightly more than half the grains that originate and develop at the chill region survive until they encounter the incoming solidification front from the other wall. The purpose of this data will be relevant as it provides the necessary information to generate the initial condition from which the phase-field model shall proceed.

3. Pseudo-binary phase-field model

We begin from a multi-order, pseudo-binary phase-field model for solidification [19–21]. Multiple orientations can be simulated and are depicted by α where each one represents a different crystallographic orientation [10]. The order parameter ϕ_α takes a value from -1 to 1, which determines the interpolation functions to be used for the model. The evolution of each order parameter is described through the following equation:

$$\tilde{\tau}_\alpha \frac{\partial \phi_\alpha}{\partial t} = \tilde{\nabla}^2 \phi_\alpha - f'_{DW}(\phi_\alpha) - \omega_{obs}\phi_\alpha \sum_{\beta \neq \alpha}^{N} \phi_\beta^2 - \lambda U D_\Lambda(\phi_\alpha)\left(1 + \frac{(1-k^{ef})}{2} U\right) g'(\phi_\alpha), \quad (1)$$

where $f'_{DW}(\phi_\alpha)$ represents the double-well free energy for a given order parameter, ω_{obs} represents the repulsion/grain-boundary energy coefficient, λ is the coupling constant derived from the interface width W_o and the solutal capillary length d_o, U represents the local chemical supersaturation factor that provides the driving force for alloy solidification, $D_\Lambda(\phi_\alpha)$ is an overlap interpolation function that reduces the solidification (liquid-to-solid) driving force close to zero where order parameters overlap to form a complete solid, k^{ef} is the effective partition function

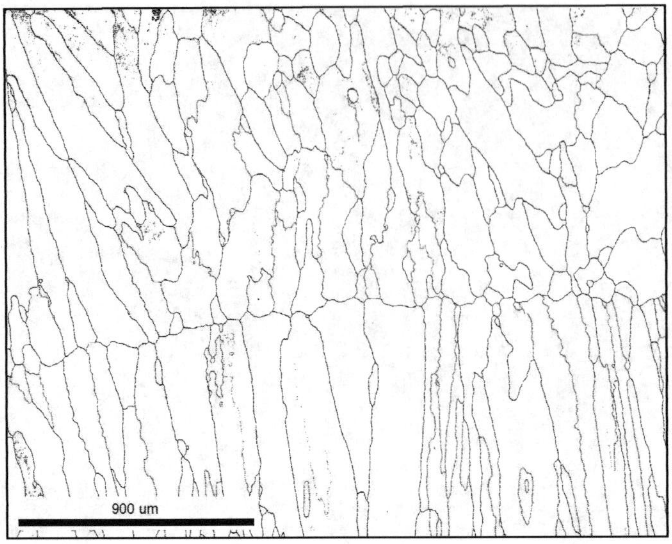

Figure 2. EBSD analysis showing the grain structure of the sample. From this image, an average grain density and orientations are extracted for the initial conditions of the model.

defined as the ratio between the free energy curvatures of the solid and liquid, and finally $g'(\phi_\alpha)$ is an interpolation function that mediates the imbalance between the free energy wells. These interpolation functions can be found in previous phase-field modeling literature [10, 19, 21, 22].

It is to be noted that space has been renormalized as $\tilde{x} = x/W_o$ as well as time $\tilde{t} = t/\tau_o$. The second relevant dynamics equation for concentration diffusion is:

$$\chi(\phi)\frac{\partial \mu}{\partial t} = \tilde{\nabla}\left[\tilde{D}q(\phi)\tilde{\nabla}\mu + a_t\Delta c^{eq}\{1+(1-k^{ef})U\}\sum_{\alpha}^{N}\frac{\partial \phi_\alpha}{\partial t}\hat{n}_\phi\right] + \Delta c^{eq}\frac{1+(1-k^{ef})U}{2}\sum_{\alpha}^{N}\frac{\partial \phi_\alpha}{\partial t}, \quad (2)$$

where $\chi(\phi)$ is the susceptibility function, $\tilde{D}$ is the renormalized, dimensionless liquid diffusion coefficient, $q(\phi)$ is an interpolation function mediating diffusion across order parameters, Δc^{eq} is the difference in equilibrium concentrations between liquid and solid, and $\tilde{n}_\phi$ is the normal vector of the order parameter's gradient. As with the order parameter evolution equation, these functions can be found in the phase-field literature [21, 22].

We map our multi-component system onto an "effective" binary alloy. To achieve this, we firstly use the Schaeffler diagram, which allows us to express the effect of the chemistry of an alloy in terms of an equivalent value of nickel and chromium. This, in turn, makes it possible to determine the expected equilibrium microstructure based on its chemical composition [18]. In this way we gather information about the requirements we will need for the model. The compositions obtained in terms of this new effective values yields $Cr_{eq} = 23.265$, $Ni_{eq} = 18.255$ and $Fe = 58.8$ in wt%. We then must turn this to molar fractions which approximately yields $XNi = 0.1722$, $XCr = 0.2477$, and $XFe = 0.5799$ in order to produce our phase diagram and use in our model. This phase diagram can be observed in Fig. 3.

To obtain a quantitative phase-field model, we must obtain the necessary free energies from a thermodynamic database. The CALPHAD method offers an excellent reference point to obtain

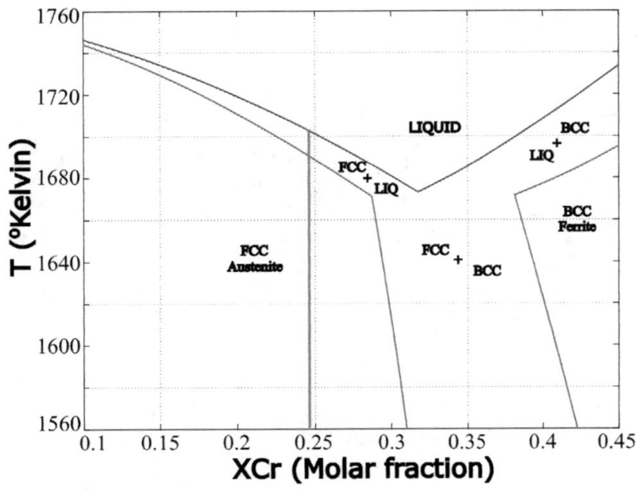

Figure 3. Phase Diagram showing the FCC, BCC and Liquid (LIQ) phases for a system with $XNi = 0.1722$ obtained from the CALPHAD free energies.

the necessary free energies to construct phase diagrams [23]. In this work, we focus on the literature applied to ultralow carbon stainless steels, where the most critical components are Fe, Cr and Ni. The bulk of the thermodynamic data we employ can be found in the work of Dinsdale, Andersson-Sundman, Hillert-Qiu, Lee and Miettinen [23–29]. We then proceed to use the CALPHAD method free energies (G-HSER), defining the pure-component free energies according to the data provided by Dinsdale for pure Fe, Cr and Ni [23]. Afterwards, the binary and ternary excess enthalpy interaction terms are taken from the work of Lee and Hillert-Qiu [25, 26]. In order to ascertain the data is accurate, we take the binary excess terms and produce their respective binary phase diagram (Fe-Cr, Fe-Ni, Cr-Ni). Finally, we produce a phase diagram using a constant amount of Ni molar fraction XNi, which permits setting $XFe = 1 - XCr - XNi$, in order to generate a pseudo-binary phase diagram representation T vs XCr. The reason it is possible to set XNi as a constant is due to the fact that the concentration tie-line for Ni is observed in the full ternary model, there is barely any change in this value across the solidification regime.

The aim of this procedure is the pursuit of a pseudo-binary phase-field model where only the concentration of Cr is used while employing 3-component free energies for thermodynamic parameterization. In other words, we use a 3-component (Fe,Cr,Ni) CALPHAD free energy, fix the amount of XNi as it was mentioned before in order to only explicitly handle XCr, produce free energies as a function of XCr and use these as reference to produce our own phase-field free energies. As already stated beforehand, the phase-field model employs parabolic free energies [19]. Thus, we must perform a parabolic fit of the obtained CALPHAD free energy curves of the FCC and LIQ phases. Firstly however, the molar volume (m^3/mol) is required to convert the CALPHAD molar free energies (J/mol) to free energy densities (J/m^3). This molar volume is obtained through Thermocalc and is assumed to be constant, given that its variation in the concentration range of interest is smaller than 5%. A parabolic function of the form $f_i = \frac{A_i(T)}{2}[c - c_i^{min}(T)]^2 + B_i(T)$ is employed where i represents the given phase (FCC, LIQ) and min represents the minima of the parabolic curve. Thus, by performing a parabolic fit of the CALPHAD LIQ and FCC free energies we obtain the coefficients A_{FCC}, A_L, c_{FCC}^{min}, c_L^{min}, B_{FCC} and B_L. From these parameters we can then compute the equilibrium concentrations $c_i^{eq} = c_i^{min} + \mu_i^{eq}/A_i$, where

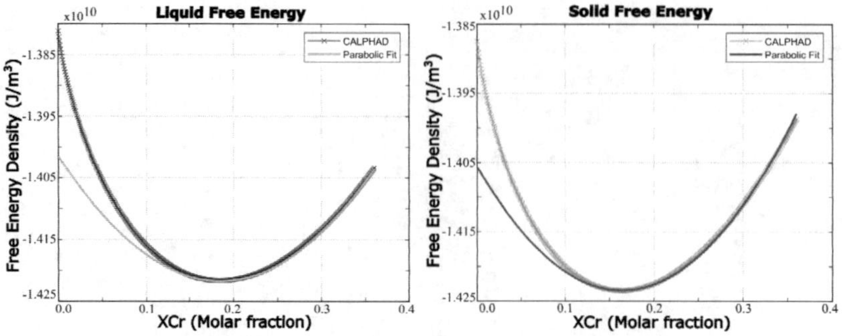

Figure 4. Comparison between the CALPHAD free energy density and the parabolic fit for a system with $XNi = 0.1722$ and $T = 1694$ for both liquid and FCC phases. The fit for the FCC free erergy is adjusted based on the constraint that the equilibrium concentrations predicted by the phase diagram must be achieved based on the parabolic approximation.

$$\mu^{eq} = \frac{\Delta c}{\chi^{l(eq)}(1-k^{ef})} + \left(\sqrt{1 + 2\frac{\chi^{l(eq)}(1-k^{ef})\Delta B}{(\Delta c)^2}} - 1\right). \quad (3)$$

The remaining model parameters are calculated through $\chi^{l(eq)} = 1/A_L$, $\chi^{FCC(eq)} = 1/A_{FCC}$, $\Delta B = B_L - B_{FCC}$, $\Delta c = c_L^{min} - c_{FCC}^{min}$, $\Delta c^{eq} = c_L^{eq} - c_{FCC}^{eq}$ and $k^{ef} = \chi^{FCC(eq)}/\chi^{l(eq)}$. The model is implemented in dimensionless form, thus, the molar volume is employed again since the 2nd-order parabolic coefficients have units of J/m^3, we may then proceed to normalize our free energies by RT where R is the gas constant in units of J/mol^oK. The plots showing the CALPHAD free energy densities and the parabolic fit performed in this work can be observed in Figs. 4a) and 4b). It can be appreciated how both figures reproduce to a fair degree the CALPHAD free energy within the most critical concentration range to replicate essential features.

In order to obtain this fitting, a nonlinear least-squares solver function is used. The relevant constraints are the following: The obtained c_L^{eq} and c_{FCC}^{eq} calculated from the fitting must match those of the phase diagram within a 1% margin of error. The free energies curvatures in the relevant concentration range $XCr \approx 0.24$ must also be within a small margin of error. Once all the thermodynamic parameters have been obtained, we may then proceed to calculate characteristic length-scales and time-scales; Respectively, these are the capillary length d_o and the kinetic time-scale τ_o, where $d_o = 2\sigma_{sl}/(\Delta c_{eq})^2 f''(c)|_{c_l}$ and $\tau_o = a_2\lambda W_o^2/D_L$ [10, 20].

To complete the model parameterization, the solid-liquid surface tension and the liquid diffusion coefficient are still necessary. These are extracted from the work of Mas, Brechet et.al. and set to $D_L = 1.5\text{x}10^{-9}m^2/s$ and $\sigma_{s_l} = 0.358J/m^2$ [30]. Finally, $\lambda = a_1 W_o/d_o$ and W_o is the solid-liquid interface width which is a phase-field model parameter, set to 360 nm here. Thus, based on this parameter choice, $d_o = 66.5333\ nm$ and $\tau_o = 258.9642\ \mu s$. As an example, if a grid spacing $\Delta x = (1.0/3.0)$ is chosen and $\Delta t = 0.008$ is used, a system size of 983 μm x 1986 μm ($\sim$ 1 mm x 2 mm along the chill-wall and growth directions, respectively) and if simulated for half a million timesteps, a total simulation time of 1.0358 seconds is obtained. The above parameter relations between W, tau and lambda are selected to map the solid-liquid interfaces of phase field model onto the sharp interface model of solidification [10, 31].The initial condition is handled through a separate phase-field employing the same equations and thermokinetic parameters except that $W_o = 240$ and the time-spacing $dt = 0.0002$ instead of the other values.

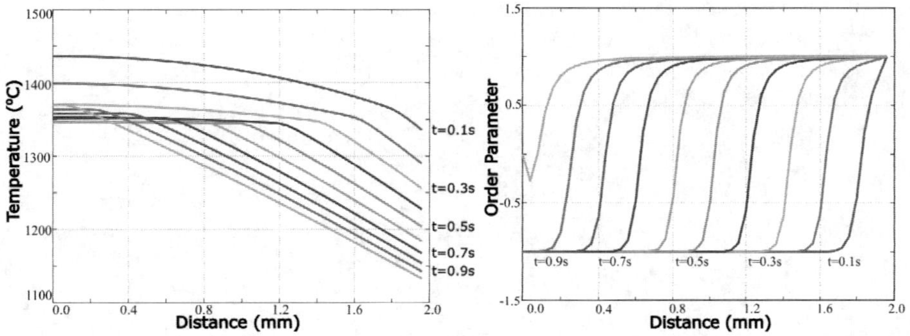

Figure 5. Temperature and order parameter profiles for a 4mm-long sample where the solidification time is approximately 1s.

These are set in this manner to maintain stability during the nucleation stage, which is the main goal of this stage. This homogeneous-grid phase field model employs a ghost-field in order to simulate "nucleation" in a random-unbiased manner [32]. A heterogeneous lambda λ_{het} is employed, where the wetting angle for heterogeneous nucleation follows a non-zero mean, random Gaussian distribution. The parameters of said Gaussian distribution used in the λ_{het} are parameterized in order to yield approximately the number of seeds observed in the experiments by Dr. Greenwood and IREQ, which is one solid grain every 70 μm, approximately. This λ_{het} in conjunction with the noise fluctuations present in the ghost-field trigger nucleation events. Each seed is then randomly assigned an order parameter $\phi_i = \{\phi_1, \phi_2, \phi_3\}$.

Once enough seeds to replicate the number of grains observed in the EBSD have nucleated, their position in the system is determined for later use. However, for our purposes, we merely use this latter model to determine the seeding and orientation of these seeds which will comprise the initial condition for the adaptive mesh refinement (AMR) model [21]. The positions of the seeds are then registered and an AMR phase-field model simulation code is employed to substantially decrease the total simulation time with the seed information used as initial condition. Fourteen seeds are obtained and their approximate position along the chill-wall direction (1mm long) is input into the AMR phase-model. It is assumed that all seeds start off with the smallest possible radius for them to undergo growth, which is approximately 1.4 μm.

4. Thermal model for casting runs

A thermal model was developed by Prof. Greenwood capable of reproducing the thermal measurements observed in the chill-casting experiments. This thermal model allows us to approximate the thermal profile as a gradient G and the solidification front velocity V at the solidification interface. These are implemented in the form of a thermal gradient G and a pulling velocity V_p commonly implemented in phase-field simulations of directional solidification [8, 21]. The mathematical form is to modify the chemical potential driving force and phase-field time-scale as $U \to U + G(x - V_p t)/\Delta T$ and $\tau_\alpha \to \tau_o[1 - (1 - k_{eff})G(x - V_p t)/\Delta T]$, where ΔT is the solidification range of the alloy [31]. Time and space are also renormalized in these expressions based on the aforementioned scalings, which leads to a dimensionless thermal gradient $\tilde{G}\tilde{x} = (GW_o/\Delta T)\tilde{x}$ and dimensionless pulling velocity $\tilde{V_p}\tilde{t} = (GV_p\tau_o/\Delta T)\tilde{t}$, where G is a function of time based on experimental data. The temperature and solidification front (order parameter) are plotted.

The thermal gradient is fitted as $G(t) = G_0[G_1 + G_2 \exp(-G_3 t^{G_4})]$, where $G_0 = 73.0298$,

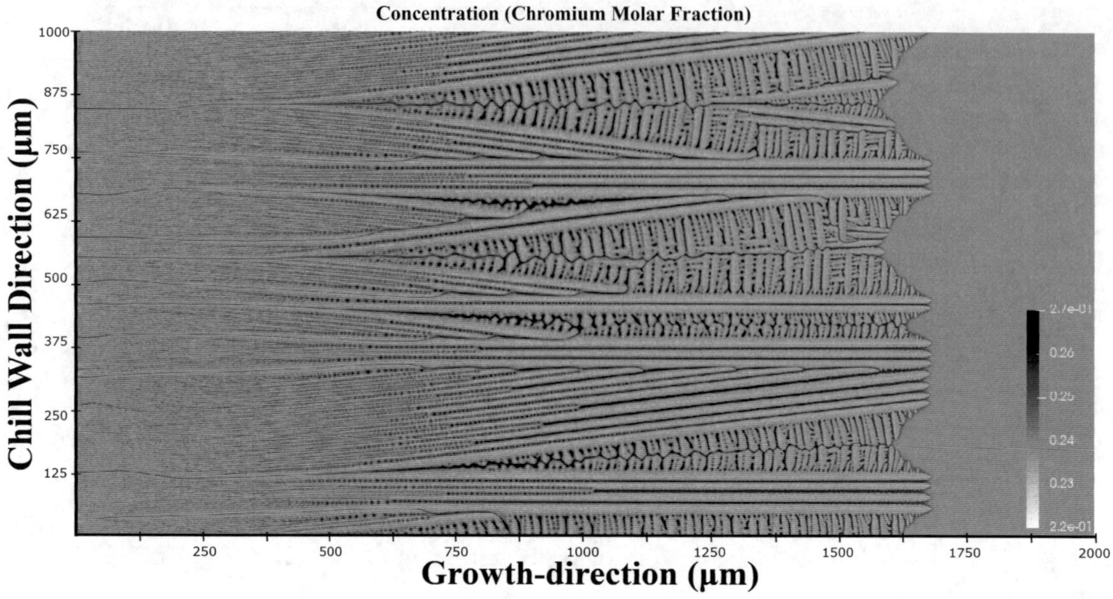

Figure 6. Phase-field simulated solidification which starts from the left side where a chill region is placed, and proceeds to the right. We plot the concentration field representing the chromium molar fraction The picture shows the system after a total simulation time of 0.9 seconds.

$G_1 = 0.13857$, $G_2 = 0.97362$, $G_3 = 24.5982$, and $G_4 = 2.2937$, where the units of G are in K/mm. Also, when $t > 0.4s$ it is maintained to a constant value of 11.6619. However, for completeness, the entire time evolution of the thermal model is presented. Afterwards, two modifications are implemented into the model in order to apply the thermal gradient time dependence and the chill-quench conditions. In terms of dimensionless implementation, it is as follows: $G(x - V_p t)/\Delta T = \tilde{G}\tilde{x} - \tilde{V}_p\tilde{t}$ where $\tilde{G} = GW_o/\Delta T$ and $\tilde{V}_p = GV_p\tau_o/\Delta T$, note that $\tilde{x} = i\Delta x$ where i is the grid-point along the x-direction and $\tilde{t} = t_s\Delta t$ where t_s is the timestep. Finally, chill quench conditions are applied by modifying the temperature profile at the wall through the following expression $Gx \rightarrow G(x - d)$ where d can be computed from $G(x - d) = T - T_o$, if x = 0 and $T - T_o = \Delta T$ is the solidification range (set as $18K$), then $d = 18K/73.0298$. Since $\tilde{d} = d/W_o$, we can numerically determine where this position is on the grid through $\tilde{d} = i_f\Delta x$, where i_f is the coordinate that reflects the distance d (given that $\Delta x = 1/3$, in this case $i_f = 2054$). This modification makes physical sense as it is expected that contact with the chill-wall drops the liquid metal's temperature in direct contact with the wall.

5. Phase-field Model Dynamics and Results

An adaptive-mesh-refinement (AMR) technique is utilized with the phase-field model, permitting us to simulate length-scales and time-scales that may be considered infeasible with a regular homogeneous grid [21]. The initial condition is constituted by a liquid with fourteen solid seeds placed in the chill region. Several simulations were performed with different initial conditions and get a clear average for our observations/measurements. Some of these results can be observed in Fig. 6. Firstly, it is interesting to compare the extent/length of each of the transition regions in both the microstructure (Fig. 1) and the simulation. In the microstructure, the immediate region to the chill-wall can be on average determined to be very-fine cells that emerge and later become wider cells. The length of this region can approximately be estimated to be around

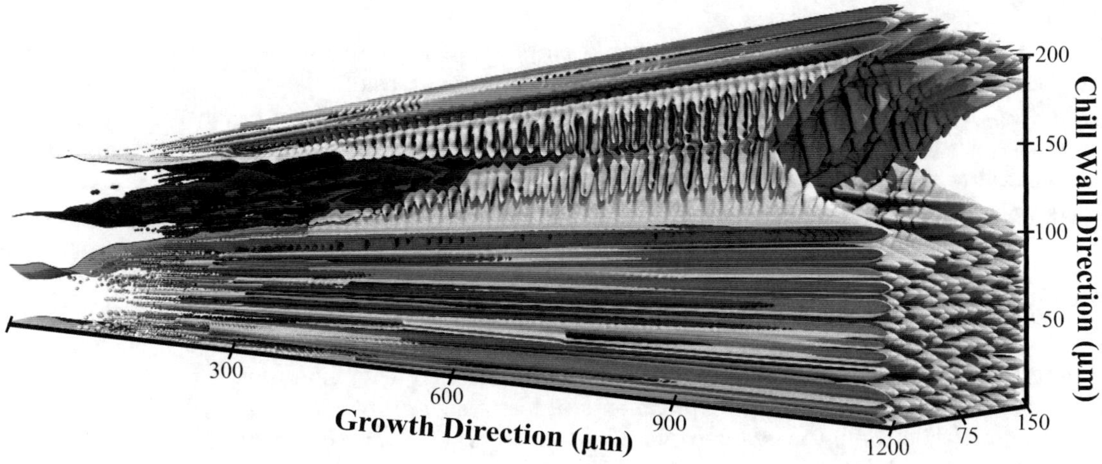

Figure 7. 3D Phase-field simulated solidification which starts from the left side where a chill region is placed, and proceeds to the right. We illustrate the order parameters here to clearly display the grain competition. The red, blue and white indicate different grains (crystallographic directions). The gray order parameter representing the least misoriented grain with respect to the thermal gradient is more favorable and therefore quickly outcompetes the other two grains.

$500\mu m$. Afterwards, a very noticeable directional cellular region can be observed throughout the sample. This region's length is approximately $800\mu m$, the rest being a cellular/dendritic microstructure before colliding with the solidification front coming from the other wall, which is approximately $400\mu m$.

If we compare this to the simulation, we can also observe that the very fine cellular region emerging from the chill-wall before clear cellular can be identified takes place at around $600\mu m$. A clear cellular region extends for further $700\mu m$, at which point the strong secondary branching from the dendritic growth starts yield cellular dendrites that now compete with the solidification front. Thus, the cellular/dendritic region would approximately measure correspondingly $400\mu m$. Modifying the thermal gradient and pulling velocity results in different transition region lengths which allows this phase-field model to replicate different morphological points based on the thermal history being introduced. These results indicate that the model is fairly capturing the average length at which morphological transitions occur yielding different cellular and dendritic microstructures in the experiments.

We then proceed to compute the average primary branch-spacing across multiple simulations with different initial conditions (such as seed placement). We find that the simulations yield a primary branch spacing of $18.35\mu m$, which is in fair agreement with the observed primary branching from the casting experiments. Noteworthy, is the fact that throughout these battery of simulations, the starting number of seeds is 14 at the chill-wall, but the final number of grains is approximately 10 grains at the estimated merger length between solidification fronts. We attribute this to a limitation of 2D models where the cell/grain competition is more limited than the 3D case. To this end, we display some preliminary results of the grain competition mechanisms that take place in 3D.

IOP Conf. Series: Materials Science and Engineering 1281 (2023) 012047 doi:10.1088/1757-899X/1281/1/012047

5.1. 3D Dynamics

In this brief subsection, we display the grain competition mechanisms that takes place in 3D as observed in Fig. 7. Three different grains with different degrees of misorientation are seeded in a system size of 153.6 μm x 204.8 μm x 1683.4 μm. As different grains interact and compete, two different effects can observed. The first involves how one grain can be slightly or considerably more favorable than the rest of the grains due to a lesser degree of misorientation with respect to the thermal gradient [33]. The second one involves how any grain with a misorientation along the height of the system can move away from a given region and effectively "phase out" of the picture. Both of these effects can either decrease the number of grains across the growth direction due to competition, or generate the impression of a reduced number of grains due to a grain "vanishing" out of the domain through the third dimension when a slice or cut is extracted in either simulation or experiment. This explains why the number of grains observed from start to finish in the casting experiments differs considerably from that observed in the 2D simulations. While the latter simulations took on average 4 days to completion using 256 cores, 3D simulations of this kind of system took nearly three times as much, in spite of making the chill-wall and width dimensions smaller.

6. Summary, Conclusions and Future Work

We presented a brief formulation of the phase-field model, which is currently intended to simulate the solidification process of AISI E309L stainless steel and applied to study the solidification of an austenitic stainless steel similar in composition. We developed a phase-diagram from the available literature (CALPHAD) for a stainless steel with a composition similar to that of AISI E309L (with slight differences to promote a complete austenite phase formation from the liquid). Through the use of the alloying element equivalent value, we simplify the effect of the chemistry on the microstructure in order to obtain a computationally tractable model. It is however noted that, using this approach, the amount of chromium and nickel in the system will be higher compared to that of the real system. Notably, segregation of chromium might also be higher by an offset than those observed in experiments. Afterwards, a phase-field model parameterization was performed by extracting the parabola coefficients from the phase diagram fitting, where a two component pseudo-binary phase-field model is employed. Thermo-kinetic parameters were determined as either certain experimental constants from the literature or based on the size of the simulated system. The thermal history of the solidification process was determined by a combined experimental-theoretical study of the casting process of a real stainless steel. Finally, we presented preliminary phase-field solidification simulations of the 2000 $\mu m \times 1000$ μm region near the chill wall of the system, where a fair agreement with experiment is observed for the primary branch spacing. Good agreement is also observed for the lengths at which instability-driven morphological transitions occur yielding cellular/dendritic microstructures. Further work in 3D simulations aids in the explanation of the grain density throughout different sections of the system. Future work will seek to reinforce some of these findings while also elucidating in a more clear manner how length-scales, segregation patterns and competitive side-branching take place for both 2D and 3D systems. In particular, segregation pools will be explored to observe differences arising from the use of the equivalent value and identify how the approximations used in this work impact these regions.

7. Acknowledgments

We wish to acknowledge the financial support that made this research possible from MITACS, IREQ-Hydro-Quebec and the office of energy research and development. Computational support was provided by the Digital Alliance of Canada and Calcul Quebec. We also wish to acknowledge the Canada Research Chair (CRC) Program for funding.

References

[1] Klueh R and King J 1982 *Weld. Res. (Miami); (United States)* **61**
[2] Bhaduri A, Gill T, Albert S, Shanmugam K and Iyer D 2001 *Nuclear Engineering and Design* **206** 249–259 ISSN 0029-5493
[3] Dhib Z, Guermazi N, Gaspérini M and Haddar N 2016 *Materials Science and Engineering: A* **656** 130–141 ISSN 0921-5093
[4] Balitskii A I, Dmytryk V V, Ivaskevich L M, Balitskii O A, Glushko A V, Medovar L B, Abramek K F, Stovpchenko G P, Eliasz J J and Krolikowski M A 2022 *Energies* **15** ISSN 1996-1073
[5] Trudel A, Sabourin M, Lévesque M and Brochu M 2014 *International Journal of Fatigue* **66** 39–46 ISSN 0142-1123
[6] Drolet F m c, Elder K R, Grant M and Kosterlitz J M 2000 *Phys. Rev. E* **61**(6) 6705–6720
[7] Elder K R, Grant M, Provatas N and Kosterlitz J M 2001 *Phys. Rev. E* **64**(2) 021604
[8] Pinomaa T, Lindroos M, Walbrühl M, Provatas N and Laukkanen A 2020 *Acta Materialia* **184** 1–16 ISSN 1359-6454
[9] Pinomaa T, Ofori-Opoku N, Laukkanen A and Provatas N 2021 *Phys. Rev. E* **103**(5) 053310
[10] Provatas N and Elder K 2010 *Phase-Field Methods in Materials Science and Engineering* 1st ed (Wiley-VCH) ISBN 9783527407477
[11] Badillo A and Beckermann C 2006 *Acta Materialia* **54** 2015–2026 ISSN 1359-6454
[12] Ofori-Opoku N and Provatas N 2010 *Acta Materialia* **58** 2155–2164 ISSN 1359-6454
[13] Ahmad N A, Wheeler A A, Boettinger W J and McFadden G B 1998 *Phys. Rev. E* **58**(3) 3436–3450
[14] Pinomaa T and Provatas N 2019 *Acta Materialia* **168** 167–177 ISSN 1359-6454
[15] Galenko P K, Abramova E V, Jou D, Danilov D A, Lebedev V G and Herlach D M 2011 *Phys. Rev. E* **84**(4) 041143
[16] Brooks J, Williams J and Thompson A 1983 *Metallurgical Transactions A* **14** 1271–1281
[17] Koseki T and Flemings M C 1997 *Metallurgical and Materials Transactions A* **28** 2385–2395
[18] Porter D A and Easterling K E 1992 *Phase Transformations in Metals and Alloys* 2nd ed (Chapman & Hall)
[19] Nikolas Provatas Tatu Pinomaa N O O 2021 *Quantitative Phase Field Modelling of Solidification* 1st ed (CRC Press)
[20] Plapp M 2011 *Phys. Rev. E* **84**(3) 031601
[21] Greenwood M, Shampur K, Ofori-Opoku N, Pinomaa T, Wang L, Gurevich S and Provatas N 2018 *Computational Materials Science* **142** 153–171 ISSN 0927-0256
[22] Yang L, Dong Z, Wang L and Provatas N 2022 *Journal of Materials Science and Technology* **101** 217–225 ISSN 1005-0302
[23] Dinsdale A 1991 *Calphad* **15** 317–425 ISSN 0364-5916
[24] Hertzman S and Sundman B 1985 *Scandinavian Journal of Metallurgy* **14** 94–102 ISSN 0371-0459
[25] Hillert M and Qiu C 1990 *Metallurgical Transactions A* **21** 1673–1680
[26] Hillert M and Qiu C 1991 *Metallurgical Transactions A* **22** 2187–2198
[27] Miettinen J 1999 *Calphad* **23** 231–248 ISSN 0364-5916
[28] Lee B J 1992 *Calphad* **16** 121–149 ISSN 0364-5916
[29] Byeong-Joo L 1993 *Calphad* **17** 251–268 ISSN 0364-5916
[30] Mas F, Tassin C, Roch F, Yescas M, Todeschini P and Bréchet Y 2018 *Metals* **8** 284 ISSN 2075-4701
[31] Echebarria B, Folch R, Karma A and Plapp M 2004 *Phys. Rev. E* **70**(6) 061604
[32] Azizi H, Ebrahimi A, Ofori-Opoku N, Greenwood M, Provatas N and Mohammadi M 2021 *Acta Materialia* **214** 116983 ISSN 1359-6454
[33] Dorari E, Ji K, Guillemot G, Gandin C A and Karma A 2022 *Acta Materialia* **223** 117395 ISSN 1359-6454

IOP Conf. Series: Materials Science and Engineering 1281 (2023) 012048 doi:10.1088/1757-899X/1281/1/012048

Multi-scale modeling of equiaxed dendritic solidification of Al-Cu at constant cooling rate

***A. K. Boukellal[1], M. Založnik[1,2], J.-M. Debierre[3]**
1Université de Lorraine, CNRS, IJL, F-54000 Nancy, France
2 Laboratory of Excellence on Design of Alloy Metals for low-mAss Structures ('LabEx DAMAS'), France
3 Aix-Marseille Université, CNRS, Université de Toulon, IM2NP UMR 7334, 13397 Marseille, France

E-mail: ahmed.boukellal@univ-lorraine.fr

Abstract. To provide quantitative predictions, multiscale models of dendritic solidification (e.g., GEM, DNN, CAFE) need to be validated and require model parameters, which can be calculated by phase-field simulations. We report on a multiscale modeling of dendritic solidification in samples that are cooled homogeneously at a constant rate. We consider three Al-Cu alloys and samples from thin to bulk thickness. We investigate how the alloy composition, the distance between the equiaxed dendrites and the sample thickness influence the transient growth velocity of the primary tips. Using 3D phase-field simulations, we calculate the tip selection parameter based on the microsolvability theory. We show that the selection parameter depends principally on the ratio between the sample thickness and the smallest tip diffusion length during the transient growth (D/v_m, where v_m is the maximum tip velocity). The extracted tip selection parameters are then used as inputs for three-dimensional grain envelope model (GEM) simulations. The comparison between TIPF and GEM shows that the GEM can reproduce transient growth of interacting equiaxed dendrites during cooling and can account for sample confinement effects.

1. Introduction

Aluminium (Al) is widely used in industry for its lightness (3 times lighter than iron) and its mechanical properties when alloyed to other elements like copper (Cu) [1]. These properties do not only depend on the chemical composition of the alloy, but also on the microstructures that develop during the solidification process. It is well known that isotropy of mechanical properties is enhanced by equiaxed microstructures, more precisely equiaxed dendrites that develop during isothermal solidification. For these reasons, solidification has been studied for several decades[2, 3]. Most of the theoretical studies have been carried out on 2D systems, with the extension to 3D systems with simple geometries [4, 5, 6]. Consequently, theoretical studies of real cases with complex geometries and in confined systems are still lacking.

To overcome this problem, Integrated Computational Materials Engineering (ICME) has emerged. Besides the power of dealing with the aforementioned problems, ICME can tackle solidification problems over a wide range of length scales. At the dendritic scale, the thin interface phase-field method (TIPF) is a method of choice to determine with high accuracy the operating state of the dendrite tip, i.e, its radius of curvature ρ and its growth velocity v [7, 8]. With the introduction of parallel programming techniques on graphical unit processors

(GPU), TIPF allows quantitative simulations over large domain sizes, up to 10 solutal diffusion lengths $l_s = D/v$ (where D is the diffusion coefficient). Larger domains cannot be simulated because of the requirement for fine mesh spacing (of the order of 0.1 ρ). Larger scales can be tackled with so-called mesoscopic models: the dendrite needle network model (DNN) [9], the cellular automaton–finite element model (CAFE) [10] or the grain envelope model (GEM) [11, 12]. In these models the description of dendrites is simplified. In the GEM, used in this work, the dendritic grains are described by envelopes, omitting local details of the solid-liquid (S-L) interface. Quantitative simulations can be made with much coarser meshes and on much larger domains than with TIPF models, provided the the GEM is well calibrated. One way to calibrate the GEM is to use the TIPF and microsolvability theory results as reference [13].

In the present work, we use TIPF and GEM for a multiscale modeling of isothermal solidification of Al-Cu samples cooled homogeneously at a constant rate, with a special attention to the confinement effects. To this aim, we use the TIPF model as a reference to calculate the operating state of the dendrite tips. This is an extension of the prior study of 3D TIPF simulations in thin samples [14] with additional simulations for different values of the sample thickness. The TIPF results are used to calculate the tip selection parameters σ^*, that are then used as inputs to perform GEM simulations. In the Section 2, we give a brief overview of the both models. Then, we explain the simulation procedure in Section 3. The results are presented in Section 4 together with our discussions. Finally, we summarize the main results of the work and suggest new perspectives for the future.

2. Methods

2.1. Phase-field model

We use the quantitative TIPF [7, 8] adapted to the study of isothermal solidification at a constant cooling rate [14]. The solid-liquid interface width, W_0, and the relaxation time, τ_0, are used as the unit length and time, respectively [15, 8]. Because Al-Cu has a rough solid-liquid interface, $\tau_0 = a_0 W_0^3/Dd_0$ is imposed to cancel the kinetic coefficient [15, 8]. Here, d_0 is the chemical capillary length, D the copper diffusion coefficient in the liquid phase, and $a_0 \approx 0.5539$. Diffusion in solid is assumed to be negligible (the one-sided model is considered here). The classical phase-field $\varphi \in [-1, 1]$ is replaced by the preconditioned one [16]

$$\psi = \sqrt{2}\tanh^{-1}(\varphi), \tag{1}$$

and the phase-field evolution equation is rewritten according to [17],

$$\begin{aligned}\Big(1-\frac{Rt}{mc_0}\Big)a_s^2\frac{\partial\psi}{\partial t} &= a_s^2\Big[\nabla^2\psi-\sqrt{2}\varphi(\vec{\nabla}\psi)^2\Big]+2a_s\nabla a_s\nabla\psi\\ &+\sqrt{2}\Big[\varphi-\lambda(1-\varphi^2)\Big(U+\frac{Rt}{\beta_k mc_0}\Big)\Big]\\ &+\frac{\sqrt{2}}{(1-\varphi^2)}\nabla.\vec{A}.\end{aligned} \tag{2}$$

Here, $\lambda = a_1\xi$, with $a_1 \approx 0.8839$, is the coupling parameter between the phase-field and the dimensionless concentration U, $m < 0$ is the liquidus slope.

$$a_s = (1-3\varepsilon_4)+4\varepsilon_4(n_x^4+n_y^4+n_z^4) \tag{3}$$

is the anisotropy function of surface energy with ε_4 the anisotropy strength and $n_{x,y,z}$ the components of the normal vector to the S-L intreface pointing toward the liquid. In addition,

$$A_{x,y,z} = 16\varepsilon_4\frac{(1-\varphi^2)}{\sqrt{2}}|\vec{\nabla}\psi|a_s n_{x,y,z}\Big[(n_x^4+n_y^4+n_z^4)-n_{x,y,z}^2\Big] \tag{4}$$

represents the components of the anisotropy vector of surface energy, and

$$c = c_0\Big(1+\beta_k U\Big)\Big(\frac{\alpha_k - \beta_k U}{2}\Big) \tag{5}$$

expresses the relationship between the dimensionless concentration U and the physical one c, with $\alpha_k = 1 + k$, $\beta_k = 1 - k$
and c_0 the nominal concentration of the alloy. The U evolution equation reads

$$\begin{aligned}(\alpha_k - \beta_k\varphi)\frac{\partial U}{\partial t} &= \tilde{D}(1-\varphi)\nabla^2 U - \frac{(1-\varphi^2)}{\sqrt{2}}\tilde{D}\nabla\psi\nabla U \\ &- \frac{1}{\sqrt{2}}\vec{n}\nabla G + G\Big(1 - \frac{\nabla.\vec{n}}{\sqrt{2}}\Big),\end{aligned} \tag{6}$$

where $\tilde{D} = D\tau_0/W_0^2$ is the dimensionless diffusion coefficient in the liquid and

$$\vec{j} = \frac{1}{2\sqrt{2}}\beta_k c_\ell^0 W_0 G\vec{n} \tag{7}$$

is the anti-trapping current, with

$$G = \Big(1+\beta_k U\Big)\frac{\partial\varphi}{\partial t} = \frac{1-\varphi^2}{\sqrt{2}}\Big(1+\beta_k U\Big)\frac{\partial\psi}{\partial t}. \tag{8}$$

2.2. *Tip selection parameter*

When capillarity effects are entirely neglected, Ivantsov demonstrated that a paraboloid of revolution is an exact solution of the steady state growth equations of a pure material growing in a infinite space [18]. He derived a relationship between the undercooling and the thermal Péclet number. Similarly, for binary alloys [19]

$$\Omega = \mathrm{Iv}_{3D}(\mathrm{P}) = \mathrm{P}\exp(\mathrm{P})\int_{\mathrm{P}}^{\infty} du\frac{\exp(-u)}{u} = \mathrm{P}\exp(\mathrm{P})\,\mathrm{E}_1\mathrm{P}. \tag{9}$$

links the tip supersaturation

$$\Omega = (c_l^0 - c_0)/[(1-k)c_l^0], \tag{10}$$

to the solutal Péclet number

$$\mathrm{P} = \rho_{\mathrm{Iv}}v/(2D). \tag{11}$$

for a paraboloid with a tip radius ρ_{Iv} and a tip velocity v. Here, c_l^0 is the equilibrium concentration on the liquid side at the S-L interface at a given temperature $T_0 < T_L$, c_0 is the far-field concentration of the alloy and D is the solute diffusion coefficient in the liquid phase. One can see that Eq.(9) leads to an infinite combinations of ρ_{Iv} and v for a given Ω, while experiments show that the dendrite has a unique operating state (ρ_{Iv}, v) at a given Ω [20]. The microsolvability theory shows that the anisotropy of surface energy acts as a singluar perturbation that selects a unique operating state at a given supersaturation [21, 22, 23, 24]. According to this theory, the scaling parameter

$$\sigma^* = \frac{2d_0 D}{\rho_{Iv}^2 v} = \frac{d_0 v}{2D\mathrm{P}^2}, \tag{12}$$

is constant at low supersaturations and depends solely on the anisotropy strength of the surface energy. If the temperature is not constant, but varies with time as $T(t)$, this induces time variations of the local equilibrium concentration at the S-L interface by $\partial c_l/\partial t = (1/m)\partial T/\partial t$. Consequently, the supersaturation, the capillary length, the tip radius of curvature, the tip growth velocity, and the Péclet number also vary with time. Thus, eqs. (9) to (12) become

$$\Omega(t) = \mathrm{Iv}_{3D}[\mathrm{P}(t)] = \mathrm{P}(t)\exp[\mathrm{P}(t)]\,\mathrm{E}_1\mathrm{P(t)}, \tag{13}$$

$$\Omega(t) = [c_l(t) - c_0]/[(1-k)c_l(t)], \tag{14}$$

$$\mathrm{P}(t) = \rho_{\mathrm{Iv}}(t)v(t)/(2D), \tag{15}$$

and

$$\sigma^* = \frac{d(t)v}{2D\mathrm{P}(t)^2} \tag{16}$$

where the instantaneous chemical capillary length is given by

$$d(t) = \frac{\Gamma}{|m|(1-k)c_l(t)}, \tag{17}$$

with
Γ the Gibbs-Thomson coefficient. Using eqs. (13), (14) and (17), eq. (16) becomes

$$\sigma^* = \frac{d_0[1-(1-k)\mathrm{Iv}_{3D}(\mathrm{P(t)})]v}{2D\mathrm{P(t)}^2} \tag{18}$$

with $d_0 = \Gamma/[m(k-1)c_0]$, where c_0 is the far-field concentration

2.3. Grain envelope model
The grain envelope method (GEM) [11, 25] is a mesoscopic model where the dendritic grain is described by its envelope. The envelope is a smooth surface that links the tips of the actively growing dendrite branches. Its shape and growth rate are determined by the tip growth kinetics. Each tip velocity (v) depends on the solute profile built up in the surrounding of the envelope during solidification. To account for the surroundings, the tip velocities are modeled with the stagnant-film formulation of the Ivantsov solution [26]

$$\Omega_\delta = \mathrm{P}\exp\left\{E_1(\mathrm{P}) - E_1\left(\mathrm{P}\left[1+\frac{2\delta}{\rho}\right]\right)\right\} \tag{19}$$

where Ω_δ is the supersaturation across the stagnant film. It is obtained from the numerically resolved concentration field around the envelope. The stagnant film width δ is a model parameter. This equation is supplemented by the tip selection criterion, eq. (16), to calculate the tip velocity: $v = 4\sigma^* D\mathrm{P}^2/d(t)$. The tips are supposed to grow in six perpendicular $\langle 100 \rangle$ growth directions. The normal speed of the envelope is thus given by

$$v_n = \vec{v}\cdot\vec{n} = v\cos(\theta) \tag{20}$$

where θ is the angle between the outward drawn normal to the envelope $\vec{n}$, and the tip growth direction that forms the smallest angle with $\vec{n}$. To propagate the envelope on a numerical mesh, the phase-field sharp-interface capturing method [27] is used as in previous works [28, 29].

Solute transport in the liquid at the mesoscopic scale is described by a volume averaged equation valid in the whole domain, i.e., both inside and outside the envelopes:

$$g_l\frac{\partial c_l}{\partial t} = D\nabla\cdot(g_l\nabla c_l) + c_l(1-k)\frac{\partial g_l}{\partial t} \tag{21}$$

The solid and liquid phases are assumed to be stationary. Outside the envelope, the material is fully liquid, i.e., the liquid volume fraction $g_l = 1$, and Eq. (21) reduces to a single phase diffusion equation. Inside the envelope, solidification follows the Scheil assumptions [30]: thermodynamic equilibrium in the liquid and no diffusion in the solid. This implies that the concentration of the binary liquid inside the envelope is linked to the temperature field by $c_l = (T - T_m)/m$, where T_m is the melting temperature of the pure solvent. With c_l known, Eq. (21) gives the evolution of the liquid fraction inside the envelope.

3. Simulations

We consider Al-Cu alloys undergoing isothermal equiaxed solidification at a constant cooling rate, R. We are interested in growth of solutally interacting equiaxed dendrites in domains confined by lateral walls. This is representative of growth regimes observed in in-situ radiography experiments [31, 32]. In order to study the lateral confinement by the sample walls, We first performed TIPF simulations of Al-1wt.%Cu in samples with increasing thickness, from 100 μm to 1000 μm, to study the confinement effects. These simulations complete those already published in [14]. To do so, eqs. (2) and (6) are discretized using the finite differences method on a cubic grid with a uniform mesh step $\delta\tilde{x} = 1$ and integrated in time using the explicit Euler scheme with a uniform time step $\Delta t = 0.9\delta\tilde{x}^2/6\tilde{D}$. Time simulations are optimised using the CUDA parallel programming techniques on a single GPU. The numerical interface width is set to $\xi = W_0/d_0 = 24$ [14]. Consequently, the physical mesh step $\delta x = W_0\delta\tilde{x}$ decreases with the alloy concentration (see table 2). The chosen values of δx allow to solve accurately the equations at the scale of the smallest radius of curvature during the simulations. The physical parameters are gathered in table 1 (same as in [14]). Simulations start with bringing an initial spherical

Table 1. Physical parameters used in the simulations.

Parameter	physical meaning	Value	unit
D	copper diffusion coefficient in the liquid	3000.0	μm^2/s
k	partition coefficient	0.14	–
m	Liquidus slope	−2.6	K/wt.% Cu
T_m	pure aluminum melting temperature	933	K
c_0	copper concentration	1.0, 1.5, 2.0	wt.% Cu
ε_4	crystal cubic anisotropy	0.01	–
Γ	Gibbs-Thomson coefficient	0.236	Kμm

solid seed of radius ≈ 50 μm to its equilibrium at temperature T_0.

Thanks to mirror boundary conditions and symmetry properties, only one eighth of the grain centered on the domain origin (0,0,0) is considered. Once equilibrium is reached, the sample is cooled at a constant rate $R = 0.5$ K/min and the growth velocity (v) and radius of curvature (ρ) of the primary tip growing along the $x-$axis are calculated. We emphasize that $\rho = 2\frac{\rho_{xy}\rho_{xz}}{\rho_{xy}+\rho_{xz}}$ is the local radius of curvature, with $1/\rho_{xy(xz)} = \nabla.\vec{n}_{xy(xz)}$ the local curvature in the $xy(xz)$plane. The selection parameter must be determined from the Ivantsov radius, which is different [33] and is linked to the local radius by $\rho_{Iv} = \rho/(1 - 15\varepsilon_4)$, with ε_4 the anisotropy strength for a material with a fourfold symmetry [33]. The parameters v and ρ_{Iv} are used to calculate the Péclet number in eq. (15), then to calculate the tip selection parameter at each time step using eq. (18). For each case in table 2, we retain the value $\sigma^* = \frac{1}{\Delta t}\int_{t'_m}^{t_m} \sigma(t)dt$ averaged over the fast growth regime $\Delta t = t_m - t'_m$ between the maximum growth velocity v_m and $v_m/2$.

The retained values of σ^* are used as inputs to perform the grain envelope model (GEM) simulations in identical conditions as in the phase-field ones. The solute conservation equation (21) and the interface-capturing method equation are solved by finite volume method, and are integrated in time using the implicit and explicit methods, respectively. The code is based on the OpenFOAM® [34] toolbox. The stagnant film thickness, δ, and the spatial mesh step, δx, are set relative to the smallest tip diffusion length, l_s^{min}, following recommendations from [13]. $l_s^{min} = D/v_m$ is determined from the maximum tip velocity v_m predicted by the phase-field simulations for each case. The initial condition consists of a spherical seed of radius $6.1\sqrt{2}\delta x$, centered on the domain origin (0,0,0). As in the phase-field simulations, only one eighth of the solid is considered and the tip velocity, v, is calculated along the $x-$axis. The physical and simulation parameters are gathered in table 1 and table 2 respectively.

Table 2. Simulation parameters for different copper concentrations and different domain sizes.

					Phase-field	Grain envelope model	
	Case	L_x (μm)	c_0 (wt.% Cu)	Ω_m	δx (μm)	δ / l_s^{min}	δx (μm)
L_z =100 μm	1	800	1	0.175	2.54	0.4	10
	2		1.5	0.214	1.69	0.6	10
	3		2	0.252	1.27	0.6	10
	4	1000	1	0.138	2.54	0.2	8
	5		1.5	0.166	1.69	0.4	8
	6		2	0.196	1.27	0.4	10
	7	1200	1	0.117	2.54	0.2	10
	8		1.5	0.138	1.69	0.2	8
	9		2	0.161	1.27	0.4	8
		L_z (μm)					
L_x =1000 μm	10	200	1	0.138	2.54	0.2	8
	11	300					
	12	400					
	13	500					
	14	1000					

4. Results and discussion

4.1. Tip selection parameter

Fig. 1(a) shows the tip selection parameter, σ^*, in thin samples ($L_z = 100\mu$m) as a function of the final grain size (800–1200 μm) and concentration (c_0 = 1–2 wt.% Cu). Three Cu concentrations and three domain lengths are considered. In thin samples, σ^* is rather constant at the two higher concentrations (in cases 2, 5, 8, and 3, 6, 9 it varies at most from 0.074 to 0.069). However, at c_0 = 1 wt.% Cu it decreases from 0.073 to 0.057 (cases 1, 4, and 7). These variations are related to the characteristic solutal diffusion length at the primary tip $l_s^{min} = D/v_m$ that the system reaches at maximum tip speed in each case. Indeed l_s^{min} decreases from 200 μm in case 3 to 50 μm in case 7. One can notice that σ^* is rather constant for cases with strong confinement in the lateral (z) direction, i.e., where the minimum diffusion length at the primary tip, l_s^{min}, exceeds the sample thickness, L_z: cases 1, 2, 3, 5, 6, 9. Strong variations are observed for cases where where $l_s^{min} \geq L_z$: cases 4, 7, 8. To obtain a representative value, we calculate an average σ_{av} of cases 1–9 and use it as the selection parameter in thin samples (L_z = 100 μm).

Fig. 1(b) shows σ^* as a function of sample thickness, from strongly confined ($L_z = 100\mu$m) to bulk ($L_z = 1000\mu$m) domains, for a single grain size and concentration (L_x = 1000 μm, c_0 = 1

Figure 1. Variations of the tip selection parameter with the domain size (a) variable side length $L_x = L_y$ in thin samples ($L_z = 100\mu$m), for three different solute concentrations ($c_0 = 1.0, 1.5, 2.0$ wt.% Cu) and (b) thickness for $c_0 = 1$ wt.% Cu and $L_x = 1000\ \mu m$

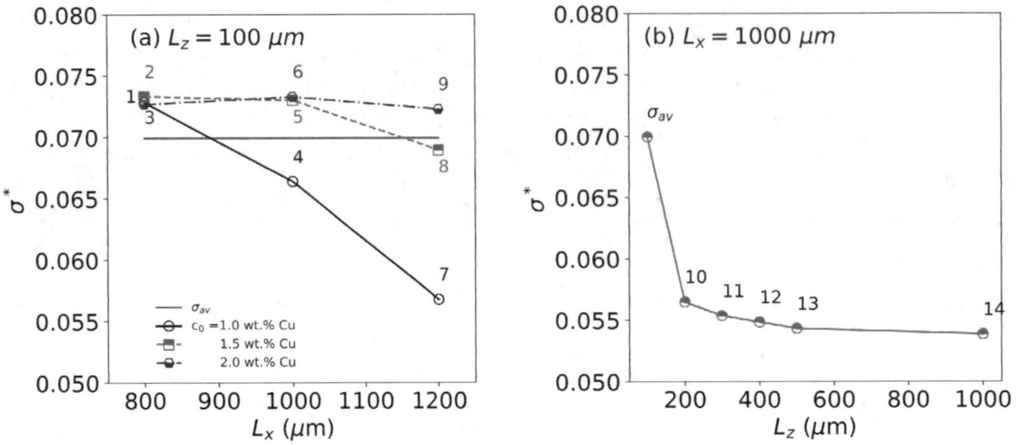

wt.% Cu). It confirms the dependence of σ^* on L_z. There is a sharp decrease of σ^* from 0.07 to 0.056 when L_z increases from 100 μm to 200 μm, that is to say when L_z/l_s^{min} increases from roughly 1.3 to 2 (see fig. 3 for v_m values). After that, σ^* decreases slowly from 0.056 to 0.054 when L_z/l_s^{min} increases from 2 to 5, then it remains almost constant. The trend of σ^* increasing with the confinement agrees with the microsolvability theory [35], that shows that σ^* is higher in 2D than in 3D for a given material with the same anisotropy strength of surface energy ε_4. To quantify these confinement effects, it is more convenient to measure the ratio L_z/l_s^{min} rather than L_z alone, as just discussed. This ratio provides more insight into the interactions of a dendrite tip with its surrounding because it takes into account its operating state (through v in l_s^{min}). Similar observations were made previously in [36].

In the fully 3D system, i.e $L_z = 1000\ \mu$m, $\sigma^* \approx 0.054$ (see fig. 1.b). Microsolvability theory [35] predicts $\sigma^*_{th} \approx 0.046$ for the same anisotropy strength. The small deviation of σ^* from σ^*_{th} may arise from the method we use to estimate the operating state. This method that could be improved as in [37], nonetheless this value is reasonably acceptable. It is worth mentioning that, to our knowledge, there is no theoretical method to calculate σ^* in confined systems. According to the above discussions, one only can expect it to be higher than in 3D systems. Our calculations provide a computational methodology to determine σ^* through TIPF simulations.

4.2. Tip growth velocity

Figs. 2 and 3 show the time variation of the tip growth velocities (v) calculated by the thin interface phase-field (lines) and the grain envelope (symbols) models in thin samples and in samples of increasing thickness respectively. Both methods lead to a similar behaviour of v: it increases from 0 to a maximum v_m then decreases sharply until it cancels out when the dendrite tip reaches the domain boundaries. On one hand the quantitative scaling laws already derived for v_m in thin samples [14] are recovered here and apply as well for the GEM calculations : v_m increases with the simulation domain length L_x and the inverse of the alloy concentration $1/c_0$ (fig. 2). On the other hand, v_m increases with the domain thickness, L_z, until this latter reaches a threshold value. This value is roughly five (5) times the aforementioned minimum solutal length l_s^{min} (fig. 2). Beyond the threshold thickness the tip does not feel the lateral confinement any more. A thorough study, which is beyond the scope of the present work, is necessary to

Figure 2. Tip growth velocities in thin samples ($L_z = 100\mu$m) calculated with phase-field (lines) and grain envelope (symbols) models. Three different alloy concentrations ($c_0 = 1.0, 1.5, 2.0\%$ wt.Cu) and three different box lengths ($L_x = 800, 1000, 1200$ μm) are considered.

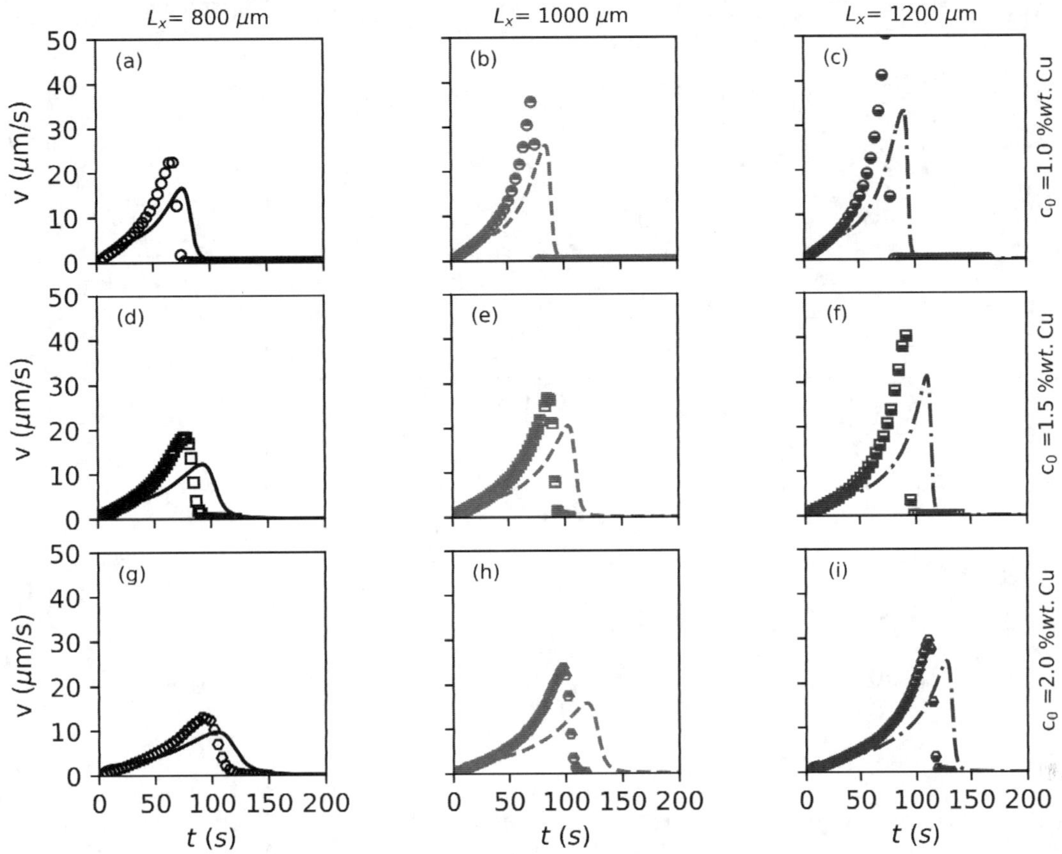

derive a quantitative scaling law for v_m with L_z.

The GEM gives a reasonably accurate description of the growth transient. The tip velocity is overestimated in all cases. This may originate from different reasons. Firstly, the initial evolution of the solid seed in TIPF and GEM are definitely different. In TIPF, the initial seed consists in an almost spherical solid that turns slowly into a dendrite-like shape after a quite long time, while in GEM, the initial spherical seed is larger and turns into a dendrite-like shape almost instantaneously. This allows the envelope tips to advance quickly and reach higher v_m earlier and further away from the boundaries of the domain. Secondly, the same average σ^* was used for all GEM simulations. Note that the tip velocity $v \sim \sigma^*$, thus variations resulting in up to up to 30% of relative error. Temporal variations of σ^* during the initial stages of growth were also neglected.

Finally, in the fully 3D system, the σ^* predicted by the TIPF is 10 % higher than the one predicted by the microsolvability theory [35] (as discussed in section 4.1).

5. Summary and conclusion

We presented a multiscale numerical study of isothermal solidification of Al-Cu alloys at a constant cooling rate. We first complemented the results of previously published TIPF

Figure 3. Tip growth velocities calculated with phase-field (lines) and grain envelope (symbols) models for Al-1 wt.% Cu for $L_x = 1000\ \mu$m. Six values of domain thickness L_z are considered.

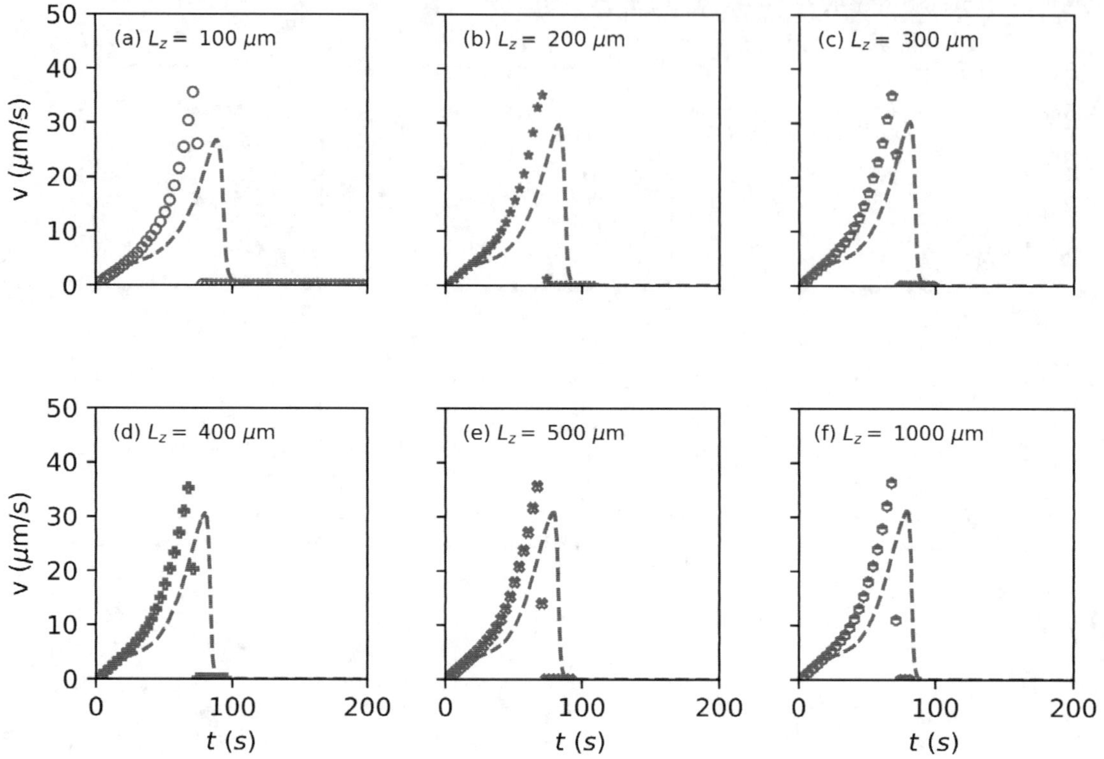

simulations forthin samples [14] by a study of the the sample thickness, from strongly confined to fully 3D systems. From the obtained results, we extracted the instantaneous operating state (v, ρ) of the dendrite tip to calculate the tip selection parameter σ^* during the fast growth regime. The values of σ^* were then used as inputs to perform grain envelope simulations (GEM) in the same conditions as the ones of the TIPF. The GEM results agree reasonably well with the TIPF. They show the robustness of the GEM to simulate accurately transient dendritic growth at mesoscales. Similar validations were recently performed for 2D directional solidification [29] and 3D free solidification [13].

Some points should be improved in the future. For the phase-field model, more 3D simulations should be performed to account for the thickness effects in different conditions with the aim of extending the scaling laws [14]. Moreover, those simulations, all together with the numerical method improvement to calculate the operating state, should lead to extract the tip selection parameter with higher accuracy. Finally, the study could be carried out on different materials to come up with general scaling laws in confined systems. Such laws would be useful to calibrate the tip growth velocities in the grain envelope model.

References

[1] Davis J R *et al.* 1993 *Aluminum and aluminum alloys* (ASM international)
[2] Dantzig J A and Rappaz M 2016 *Solidification: -Revised & Expanded* (EPFL press)
[3] Kurz W and Fisher D J 1984 *Fundamentals of solidification*
[4] Godrèche C, Che C G *et al.* 1991 *Solids far from equilibrium* vol 1 (Cambridge University Press)
[5] Kassner K 1996 *Pattern formation in diffusion-limited crystal growth* (World Scientific)

[6] Pelcé P and Libchaber A 2012 *Dynamics of curved fronts* (Elsevier)
[7] Karma A 2001 *Phys. Rev. Lett.* **87** 115701
[8] Echebarria B, Folch R, Karma A and Plapp M (2004) *Phys. Rev. E* **70** 061604
[9] Tourret D and Karma A 2013 *Acta materialia* **61** 6474–6491
[10] Gandin C A and Rappaz M 1994 *Acta metallurgica et materialia* **42** 2233–2246
[11] Steinbach I, Beckermann C, Kauerauf B, Li Q and Guo J 1999 *Acta Mater.* **47** 971–982
[12] Steinbach I, Diepers H J and Beckermann C 2005 *Journal of crystal growth* **275** 624–638
[13] Tourret D, Sturz L, Viardin A and Založnik M 2020 Comparing mesoscopic models for dendritic growth *IOP Conference Series: Materials Science and Engineering* vol 861 (IOP Publishing) p 012002
[14] Boukellal A K, Debierre J M, Nguyen-Thi H and Reinhart G (2018) *Materialia* **1** 62
[15] Karma A and Rappel W J (1998) *Phys. Rev. E* **57** 4323
[16] Glasner K (2001) *Journal of Computational Physics* **174** 695–711
[17] Boukellal A K, Rouby M and Debierre J M (2021) *Computational Materials Science* **186** 110051
[18] Ivantsov G 1947 *Dokl Akad Nauk SSSR* **58** 567
[19] Temkin D 1962 *Sov Phys Crystallogr* **7** 354–357
[20] Glicksman M, Schaefer R and Ayers J 1976 *Metallurgical Transactions A* **7** 1747–1759
[21] Langer J and Müller-Krumbhaar J 1977 *J. Cryst. Growth* **42** 11 – 14
[22] Langer J and Müller-Krumbhaar H 1978 *Acta Metall.* **26** 1681 – 1687
[23] Langer J and Müller-Krumbhaar H 1988 Theory of dendritic growth—ii. instabilities in the limit of vanishing surface tension**research supported by afosr grant f 44620–76-c-0103 and also in part by the cmu center for the joining of materials. *Dynamics of Curved Fronts* ed Pelcé P (San Diego: Academic Press) pp 289 – 295
[24] Müller-Krumbhaar H and Langer J 1978 *Acta Metall.* **26** 1697 – 1708
[25] Delaleau P, Beckermann C, Mathiesen R H and Arnberg L 2010 *ISIJ international* **50** 1886–1894
[26] Cantor B and Vogel A 1977 *Journal of Crystal Growth* **41** 109–123
[27] Sun Y and Beckermann C 2007 *Journal of Computational Physics* **220** 626–653
[28] Souhar Y, De Felice V F, Beckermann C, Combeau H and Založnik M 2016 *Computational Materials Science* **112** 304–317
[29] Viardin A, Založnik M, Souhar Y, Apel M and Combeau H 2017 *Acta Mater.* **122** 386–399
[30] Scheil E 1942 *n* **134** 70
[31] Bogno A, Nguyen-Thi H, Reinhart G, Billia B and Baruchel J 2013 *Acta Mater.* **61** 1303–1315
[32] Olmedilla A, Založnik M and Combeau H 2019 *Acta Materialia* **173** 249–261
[33] Karma A, Lee Y H and Plapp M 2000 *Phys. Rev. E* **61** 3996
[34] OpenFOAM, version 2.2.2, 2014, www.openfoam.org URL `www.openfoam.org`
[35] Barbieri A and Langer J S 1989 *Phys. Rev. A* **39**(10) 5314–5325
[36] Athreya B P, Dantzig J A, Liu S and Trivedi R 2006 *Philosophical Magazine* **86** 3739–3756
[37] Bellón B, Boukellal A, Isensee T, Wellborn O, Trumble K, Krane M, Titus M, Tourret D and Llorca J 2021 *Acta Mater.* **207** 116686

IOP Conf. Series: Materials Science and Engineering 1281 (2023) 012049 doi:10.1088/1757-899X/1281/1/012049

Development of a data assimilation system for the investigation of the dendrite solidification process by integrating in situ X-ray imaging and phase-field simulation

A Yamamura[1], S Sakane[2], M Ohno[3], H Yasuda[4] and T Takaki[2, *]

[1] Graduate School of Science and Technology, Kyoto Institute of Technology, Matsugasaki, Sakyo-ku, Kyoto 606-8585, Japan.
[2] Faculty of Mechanical Engineering, Kyoto Institute of Technology, Matsugasaki, Sakyo-ku, Kyoto 606-8585, Japan.
[3] Division of Materials Science and Engineering, Faculty of Engineering, Hokkaido University, Kita 13 Nishi 8, Kita-ku, Sapporo, Hokkaido 060-8628, Japan.
[4] Department of Materials Science and Engineering, Kyoto University, Yoshidahonmachi, Sakyo-ku, Kyoto 606-8501, Japan.

* E-mail: takaki@kit.ac.jp

Abstract. The dendrite solidification process has been observed and simulated using state-of-the-art techniques, such as time-resolved X-ray tomography (4D-CT) and high-performance phase-field (PF) simulations. 4D-CT has enabled the direct observation of the 3D dendrite growth in opaque alloys. However, the spatiotemporal resolution is not sufficient for investigating fast phenomena because a 3D solidification structure is obtained using hundreds of transmission images during the 180° rotation of a sample. High-performance PF simulations have enabled the simulation of multiple 3D dendrite growth phenomena. However, the material properties required in PF solutions of alloys are often unavailable. Therefore, integrating in situ X-ray observations with PF simulations using data assimilation is a promising approach for simultaneously solving these issues. In this study, we developed a data assimilation system with an ensemble Kalman filter, in which the solid fraction along the thickness of a sample was used as observation data to enable data assimilation using X-ray transmission images. The performance of the developed data assimilation system was evaluated via twin experiments for columnar dendrite growth during the directional solidification of a binary alloy in a thin film. The results showed that data assimilation using the solid fraction as observation data estimated the material properties and solidification morphologies with reasonable accuracy.

1. Introduction

Solidification microstructures of alloys are generally formed through dendrite growth; hence, a highly accurate prediction of dendrite growth is essential to enhance the quality of casting products [1].

The real-time and direct observation of solidification in metallic alloys was impossible owing to their opacity and high melting temperature. Therefore, the solidification of transparent organic alloys with cubic lattices, such as SCN–acetone, has been observed since the 1970s [2-4]. Nevertheless, since around 2000, in situ X-ray imaging of the dendrite solidification of metallic alloys in thin films became feasible using high-energy X-rays [5-7]. Furthermore, in around 2010, time-resolved X-ray computed tomography (4D-CT) was enabled to investigate the 3D dendrite growth process in real-time [8-10]. In

4D-CT, a 3D dendrite structure is reconstructed using hundreds of X-ray transmission images captured over a 180° rotation of a cylindrical sample. Recently, a reconstructed 3D image was obtained every 0.5–4.0 s and 4.0–8.0 s for Al-based [10] and Fe-based alloys [8], respectively. In addition, Yasuda et al. developed phase-field (PF) filtering for 4D-CT [8] to obtain a smooth solid–liquid surface morphology. However, despite the success of the 4D-CT application, the spatiotemporal resolution is insufficient for investigating the fast dendrite growth phenomena.

The PF method is the most accurate simulation model for predicting dendrite solidification [11-14]. However, because the PF method is a diffuse interface method, dendrite growth simulation using the PF method is computationally expensive. The development of high-performance computing schemes, such as the adaptive mesh refinement method [15] and parallel computing using multiple graphics processing units [16, 17], enabled the competitive growth simulation of multiple 3D dendrites [14, 18]. Nevertheless, the material properties required in PF simulations are often unavailable, particularly for alloys.

Data assimilation, originally developed from meteorology and oceanography, is a promising method for integrating experiments and simulations [19, 20]. Recently, data assimilation was applied to PF simulations of grain growth [21, 22], dendrite solidification [23-26], and sintering [27, 28]. The introduction of data assimilation to integrate in situ X-ray observations and PF simulations is expected to enable the highly accurate prediction of dendrite growth by overcoming the limitations of both methods. In this methodology, PF simulation can serve as a 3D reconstruction method with a high spatiotemporal resolution based on thermodynamics. Simultaneously, the experimental results can provide the material properties required in PF simulations. However, the sequential data assimilation performed in this study requires numerous simultaneous simulations, leading to high computational costs. Therefore, in our previous study, a data assimilation system using only the dendrite tip was developed for columnar dendrite growth [24].

The outputs of X-ray observations are 2D X-ray transmission images; hence, the development of a data assimilation system that uses transmission images as observation data is necessary. In this study, we developed a data assimilation system that uses the solid fraction along the thickness of a sample as observation data, which corresponds to the luminance distribution of the X-ray transmission images. The performance of the data assimilation system was evaluated using twin experiments of the columnar dendrite-growth problem in thin films.

The rest of this paper is structured as follows. Section 2 describes the data assimilation method that uses the solid fraction as observation data. Section 3 presents the preparation of observation data used in the twin experiments described in Section 4. Section 4 presents the twin experiments performed using the observation data whose preparation is presented in Section 3. Section 5 discusses the influence of the selection of the region and observation data used for data assimilation on the estimation results. Finally, Section 6 presents the conclusions and perspectives for future work.

2. Data assimilation

The ensemble Kalman filter (EnKF) [26, 29], a sequential data assimilation method, was employed in this study because of its relatively easy implementation with an existing PF simulation program. The data assimilation method used in this study comprises two processes: "forecasting" corresponding to PF simulations of the ensemble number and "filtering" to modify the simulation results closer to the observation data. Forecasting and filtering are alternately performed in data assimilation. In this study, PF simulations of columnar dendrite growth were performed during the directional solidification of a binary alloy as the forecasting process, the details of which are provided in the literature [24-26]. The state vector calculated in the forecasting process, $\mathbf{x}_{t|t-1}^{(i)}$, was updated to $\mathbf{x}_{t|t}^{(i)}$ during filtering using the following equation [26, 29]:

$$\mathbf{x}_{t|t}^{(i)} = \mathbf{x}_{t|t-1}^{(i)} + \mathbf{K}_t \left[\mathbf{y}_t + \breve{\mathbf{w}}_t^{(i)} - \boldsymbol{h}_t\left(\mathbf{x}_{t|t-1}^{(i)}\right) \right] \tag{1}$$

where $\boldsymbol{h}_t$ is an observation operator in the form $\boldsymbol{h}_t(\mathbf{x}_t^{(i)}) = \mathbf{H}_t\mathbf{x}_t^{(i)}$, which relates the state vector $\mathbf{x}_t^{(i)}$ to the observation data $\mathbf{y}_t$, $\mathbf{K}_t$ is the Kalman gain [26, 29], and $\mathbf{y}_t + \breve{\mathbf{w}}_t^{(i)}$ is a perturbed observation. The state vector $\mathbf{x}_t^{(i)}$ is defined as $\mathbf{x}_t^{(i)} = \{\boldsymbol{\phi}_t^{(i)}\ \mathbf{u}_t^{(i)}\ \mathbf{z}^{(i)}\}^T$, where $\boldsymbol{\phi}_t^{(i)}$ represents the vector of PF variables, $\mathbf{u}_t^{(i)}$ is the dimensionless supersaturation, and $\mathbf{z}^{(i)}$ represents certain estimated material properties. Here, the number of components of $\boldsymbol{\phi}_t^{(i)}$ and $\mathbf{u}_t^{(i)}$ is the number of grid points, and the superscript (i) indicates the i-th ensemble member. The PF variable ϕ is defined as $\phi = +1$ in the solid phase and $\phi = -1$ in the liquid phase [13].

In the twin experiment (described in Section 4), when the observation data $\mathbf{y}_t$ was substituted as $\mathbf{y}_t = \boldsymbol{\phi}^{obs}{}_t$, where $\boldsymbol{\phi}^{obs}{}_t$ is a vector with PF variables at all grid points, the number of components of the state vector $\mathbf{x}_t^{(i)}$ and observation data $\mathbf{y}_t$ became $2N_xN_yN_z + n$ and $N_xN_yN_z$, respectively, where N_x, N_y, and N_z are the numbers of grid points in the x-, y-, and z-directions, respectively, and n is the number of estimated material properties $\mathbf{z}$. The observation matrix $\mathbf{H}_t$ of size $N_xN_yN_z \times (2N_xN_yN_z + n)$ is expressed as follows:

$$\mathbf{H}_t = \left\{\mathbf{I}_{N_xN_yN_z} \quad \mathbf{0}\right\} = \left\{\begin{matrix} \mathbf{I}_{N_xN_y} & \mathbf{0} & \cdots & \mathbf{0} & \mathbf{0} \\ \mathbf{0} & \mathbf{I}_{N_xN_y} & \cdots & \mathbf{0} & \mathbf{0} \\ \vdots & \vdots & \ddots & \vdots & \vdots \\ \mathbf{0} & \mathbf{0} & \cdots & \mathbf{I}_{N_xN_y} & \mathbf{0} \end{matrix}\right\}, \tag{2}$$

where $\mathbf{I}_N$ is the identity matrix with $N \times N$ components.

Upon setting the observation vector $\mathbf{y}_t$ to the solid fraction s along the thickness of a sample, or the z-direction in this study, which is in the range [0, 1], $\mathbf{y}_t$ is expressed as

$$\mathbf{y}_t = \mathbf{s}_t^{obs} = \frac{1}{2}\left(\frac{1}{N_z}\sum_{k=0}^{N_z-1}\boldsymbol{\phi}_t^{obs} + 1\right). \tag{3}$$

In this case, $\boldsymbol{h}_t$ is expressed as

$$\boldsymbol{h}_t\left(\mathbf{x}_t^{(i)}\right) = \frac{1}{2}\left(\frac{1}{N_z}\mathbf{H}_t'\mathbf{x}_t^{(i)} + 1\right), \tag{4}$$

where

$$\mathbf{H}_t' = \left\{\mathbf{I}_{N_xN_y} \quad \mathbf{I}_{N_xN_y} \quad \cdots \quad \mathbf{I}_{N_xN_y} \quad \mathbf{0}\right\} \tag{5}$$

is an $N_xN_y \times (2N_xN_yN_z + n)$ matrix. Here, $\boldsymbol{h}_t$ sums up the PF variables along the thickness of the sample and converts it to a solid fraction.

3. Preparation of observation data

To validate the developed data assimilation system, twin experiments were performed to estimate the anisotropic strength of the interfacial energy, ε, and solute diffusion coefficient of the liquid, D_l [24]. ε expresses the orientation dependence of the interfacial energy [30] and is identical to ε_4 in previous studies [24-26]. The observation data were prepared via the PF simulation of columnar dendrite growth during directional solidification in a thin film of Fe–4 wt% Si. The material properties and computational conditions are listed in Table 1. The boundary conditions of ϕ and u were set to zero flux boundary condition for all boundaries ($d\phi/dx = du/dx = 0$ for x-direction, $d\phi/dy = du/dy = 0$ for y-direction and $d\phi/dz = du/dz = 0$ for z-direction). Figure 1 shows the time evolution of the 3D morphologies and solid fraction in the thickness direction. The simulation was performed in a domain with $N_x \times N_y \times N_z = 128 \times 1280 \times 16$ grids and a grid size $\Delta x = 0.781$ μm, which is one-fourth of the domain shown in Figure 1 considering the symmetry in the x- and z-directions. The typical columnar dendrite growth process is shown in Figure 1(a). The 3D PF distribution (Figure 1(a)) and solid fraction in the thickness direction (Figure 1(b)) were used as the observation data in the twin experiments.

Table 1. Material properties of Fe–4 wt% Si and computational conditions.

Property	Symbol	Value	Reference
Melting temperature of Fe	T_m	1811.0 K	[31]
Gibbs–Thomson coefficient	Γ	2.88×10^{-1} K·μm	[31]
Strength of anisotropy	$\varepsilon = \varepsilon_{true}$	0.02	*assumed*
Solute diffusion coefficient of liquid	$D_l = D_{l\,true}$	3.0×10^{-9} m²/s	*assumed*
Solute diffusion coefficient of solid	D_s	3.44×10^{-11} m²/s	[32]
Liquidus slope	m	–884 K/mol frac.	[33]
Partition coefficient	k	0.79	[33]
Initial temperature at $y = 0$	T_0	1741.2 K	
Initial supersaturation	u_0	–0.15	
Temperature gradient	G	10 K/mm	
Cooling rate	R	1 K/s	
Pulling velocity	V_p	0.1 mm/s	
Number of grid points	$N_x \times N_y \times N_z$	128 × 1280 × 16	
Grid size	Δx	0.78125 μm	
Interface thickness	W_0	0.977 μm	
Time increment	Δt	1.45×10^{-5} s	
Total time	t_{total}	10.0 s	

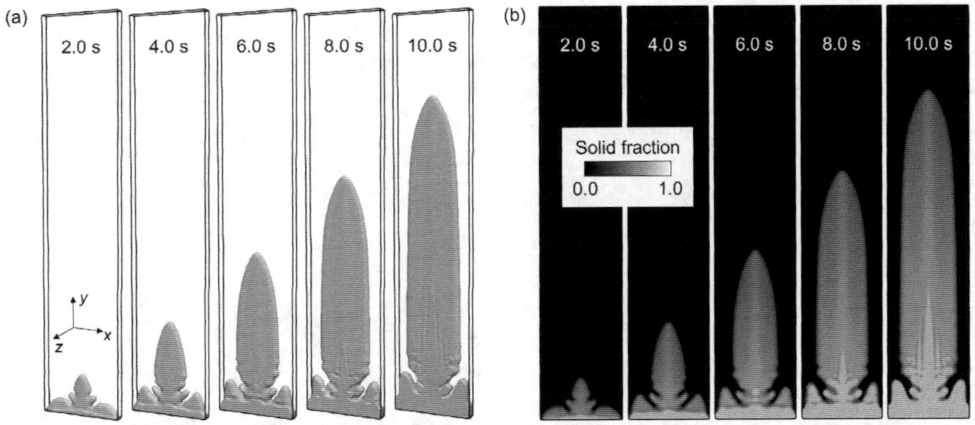

Figure 1. Time evolution of the (a) 3D morphologies and (b) solid fraction s in the thickness direction.

4. Twin experiment

Using the PF variable and solid fraction shown in Figure 1 as observation data, twin experiments were performed to confirm the possibility of data assimilation using the solid fraction. The twin experiments were also performed by changing the domain size used for the data assimilation. Two material properties, ε and D_l, were estimated in the twin experiments, and the reproducibility of the dendrite morphology was evaluated. Hyperparameters required in the data assimilation were determined in advance as follows: ensemble number = 128, time interval of filtering = 0.05 s, variance of system noise for $\varepsilon = (\varepsilon_{true}/1000)^2$, variance of system noise for $D_l = (D_{l\,true}/1000)^2$, variance of observation noise = 1.0^2, and system noise of ϕ and u = **0**. To avoid negative values for ε and D_l, the minimum values were set as $\varepsilon_{min} = \varepsilon_{true}/100$ and $D_{l\,min} = D_{l\,true}/10$ for ε and D_l, respectively. Here, ε_{true} and $D_{l\,true}$ are the predefined true values of ε and D_l, respectively, as listed in Table 1, implying that ε_{true} and $D_{l\,true}$ are the solutions of ε and D_l, respectively, in the twin experiments. Additionally, the maximum value of ε was set as $\varepsilon_{max} = 1/15 = 0.0667$ [30]. The maximum value of D_l was set as $D_{l\,max} = 2D_{l\,true}$ and used to determine the time increment Δt to explicitly solve the time-evolution equations of ϕ and u. The initial distributions of $\varepsilon^{(i)}$

and $D_l^{(i)}$ were set as Gaussian distributions with mean values $\varepsilon_{\text{true}}/2$ and $D_{l\,\text{true}}/2$ and variances $(\varepsilon_{\text{true}}/4)^2$ and $(D_{l\,\text{true}}/4)^2$, respectively.

4.1. Data assimilation using the entire domain

Figures 2 and 3 show the results of data assimilation using the PF variables and solid fraction as observation data, respectively, the estimation results of (a) ε and (b) D_l, and (c) cross-sectional morphologies of the solid–liquid interface at $z = N_z/2$ at $t = 3.3$, 6.7, and 10.0 s. The black circular marks and the error bars in Figures 2 and 3 represent the average values and standard deviations of $\varepsilon^{(i)}$ and $D_l^{(i)}$, respectively, whereas the red line represents the true values $\varepsilon_{\text{true}}$ and $D_{l\,\text{true}}$. In Figures 2(c) and 3(c), the black and red lines represent the morphologies of an ensemble member after filtering and the observation data, respectively.

Figures 2(a) and 2(b) show that ε and D_l rapidly approach the true value immediately after starting the data assimilation, and the estimation accuracy of D_l is higher than that of ε. This implies that the influence of D_l on the solidification morphology is greater than that of ε. In Figure 2(c), the estimated morphology at $t = 3.3$ s is almost the same as the observation data, implying that the material properties and morphologies can be accurately estimated. After the rapid approach to the true value, D_l decreases monotonically with time, while ε briefly maintains the same accuracy. After approximately $t = 4$ s, ε deviates from the true value and is significantly disturbed, thereby making the estimation of ε impossible. The estimated solidification morphology at $t = 6.7$ s slightly deviates from the observation data, and at $t = 10.0$ s, the morphology is observed to be completely different. The estimation results shown in Figure 3, where a solid fraction was used as the observation data, are similar to those of Figure 2. Meanwhile, the time required to reach the true value is shorter, and the estimation accuracies of ε and D_l are better in Figure 2 than in Figure 3. The estimated solidification morphology at $t = 6.7$ and 10.0 s in Figure 3(c) is more chaotic than those in Figure 2(c). Nevertheless, in the early stage ($t < 4.0$ s) of the data assimilation, the estimation of ε and D_l and the solidification morphology are possible with reasonable accuracy in Figure 3 when the solid fraction is used as the observation data. This indicates that data assimilation using the solid fraction is possible.

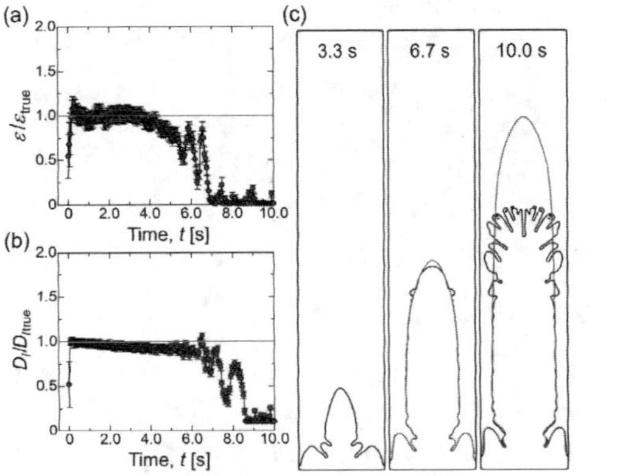

Figure 2. Results estimated from the data assimilation of (a) ε, (b) D_l, and (c) cross-sectional morphologies of an ensemble member (black) and observation data (red) using PF variables of the entire domain shown in Figure 1(a) as observation data.

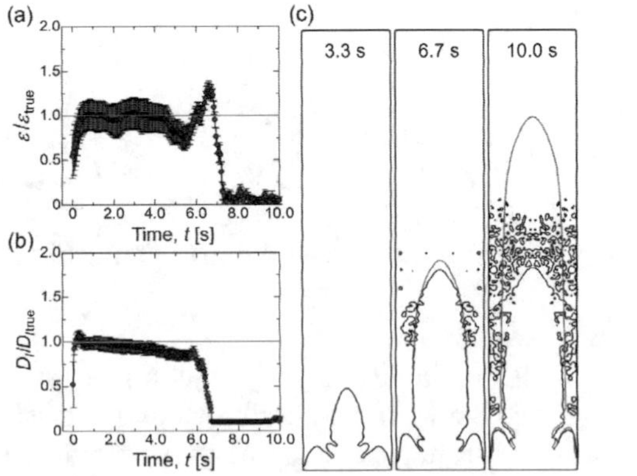

Figure 3. Results estimated from the data assimilation of (a) ε, (b) D_l, and (c) cross-sectional morphologies of an ensemble member (black) and observation data (red) using a solid fraction of the entire domain shown in Figure 1(b) as observation data.

The reduction in the estimation accuracy with time (as shown in Figures 2 and 3) can be attributed to the increase in the ratio of the immobile interface with the growth of the columnar structure. The sides of the cellular structure stop the migration at approximately $t = 6.0$ s, as shown in Figure 1. In such a situation, the effects of ε and D_l on the solidification morphology are reduced, making the estimation of ε and D_l difficult. Therefore, data assimilation was attempted using only the dendrite tip region.

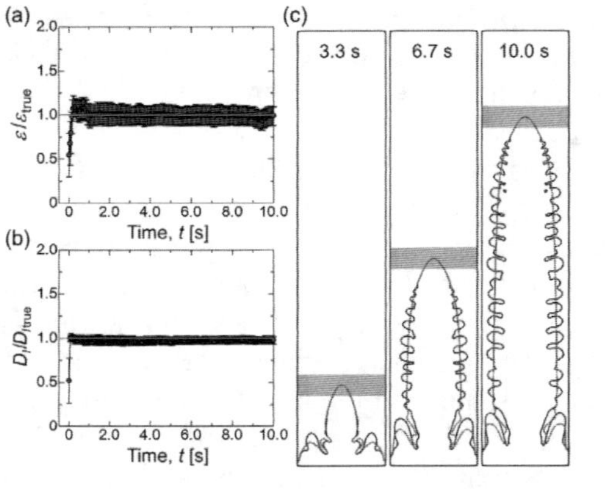

Figure 4. Results estimated from the data assimilation of (a) ε, (b) D_l, and (c) cross-sectional morphologies of an ensemble member (black) and observation data (red) using PF variables of the gray area with l_y = 64 grids as observation data.

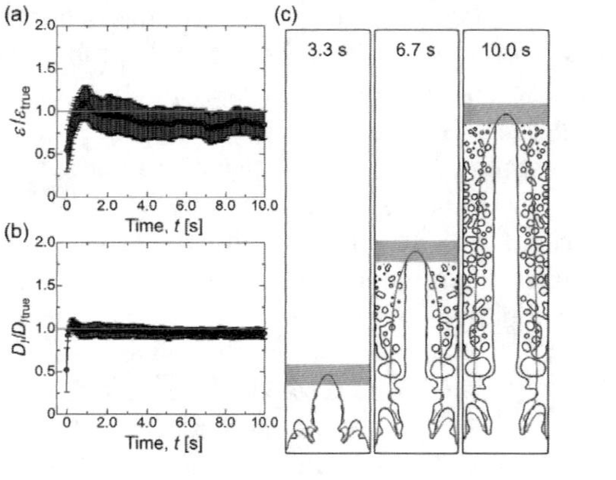

Figure 5. Results estimated from the data assimilation of (a) ε, (b) D_l, and (c) cross-sectional morphologies of an ensemble member (black) and observation data (red) using a solid fraction of the gray area with l_y = 64 grids as observation data.

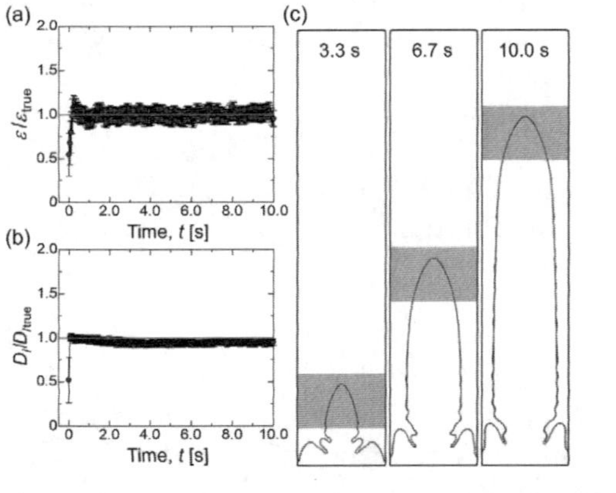

Figure 6. Results estimated from the data assimilation of (a) ε, (b) D_l, and (c) cross-sectional morphologies of an ensemble member (black) and observation data (red) using PF variables of the gray area with l_y = 160 grids as observation data.

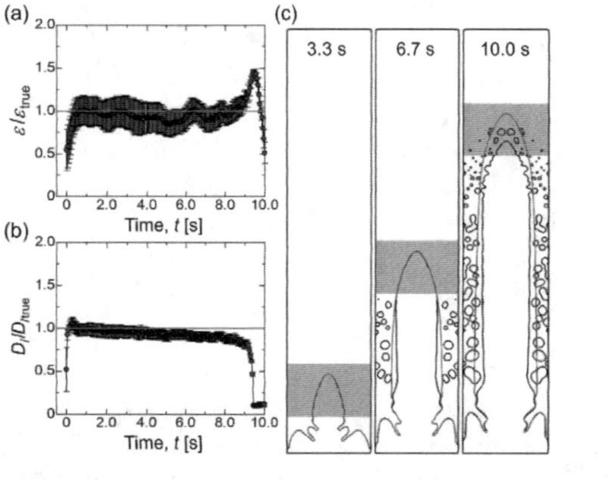

Figure 7. Results estimated from the data assimilation of (a) ε, (b) D_l, and (c) cross-sectional morphologies of an ensemble member (black) and observation data (red) using a solid fraction of the gray area with l_y = 160 grids as observation data.

4.2. Data assimilation using the region around the dendrite tip

Twin experiments using the gray area illustrated in Figures 4–7(c), where the size in the y-direction from $32\Delta x$ (= 25 μm) above the dendrite tip is $l_y = 64$ (Figures 4 and 5) and 160 grids (Figures 6 and 7) ($l_y\Delta x$ = 50 μm and 125 μm), were performed to investigate the effect of the domain size (used in data assimilation) on the estimation accuracy. The number of state-vector components $\mathbf{x}_t^{(i)}$ was $2N_xl_yN_z + n$, and one of the observation data $\mathbf{y}_t$ was $N_xl_yN_z$ and N_xl_y when the PF variables and solid fraction were used as observation data, respectively.

Figures 4 and 5 show the estimation results with l_y = 64 grids when the PF variables and solid fraction are used, respectively. Figures 6 and 7 show the estimation results for l_y = 160 grids. The estimation results of (a) ε, (b) D_l, and (c) the cross-sectional morphologies of the solid–liquid interface at $z = N_z/2$ and $t = 3.3$, 6.7, and 10.0 s are shown in the same manner as those in Figures 2 and 3. All the estimation results shown in Figures 4–7 become more stable (than those in Figures 2 and 3) by restricting the region used in the data assimilation only to the dendrite tip area. In Figures 4 and 6, where PF variables are used as observation data, the material properties and morphologies in the gray region are estimated accurately and stably. For the estimation of D_l, the result shown in Figure 4(b) is accurate and independent of time; however, in Figure 6(b), the accuracy decreases slightly with time. This implies that to estimate the material properties in dendrite growth, data assimilation should be performed for the region where the solid–liquid interface evolves. For solidification morphologies, Figure 6(c) with l_y = 160 grids shows accurate morphology reproduction for the entire region, whereas in Figure 4(c) with l_y = 64 grids, the morphology, except for the gray region, deviates from the observation data.

A comparison of Figures 3, 5, and 7 using the solid fraction as observation data shows that the smaller the area for data assimilation, the longer is the period during which material properties can be stably estimated. However, the standard deviations are almost the same, and the effect of the size of the area for data assimilation on estimation accuracy cannot be observed. The morphologies shown in Figures 3(c), 5(c), and 7(c) indicate that the smaller the area of data assimilation, the longer is the period during which the morphologies in the gray region can be stably estimated, whereas the morphologies outside the area cannot be estimated for all cases.

5. Discussion

Data assimilation was observed to become more stable and accurate when only the dendrite tip region was used rather than the entire domain. In a previous study [24], the region for data assimilation was restricted to the dendrite tip area to reduce the computational cost. The method [24] contributed not only to the reduction of the computational cost but also to the improvement of the estimation accuracy of the material parameters. Data assimilation using the solid fraction as observation data was observed to be possible for the estimation of material properties and morphologies with reasonable accuracy, although the accuracy was lower than that in the case in which PF variables were used. Thus, the possibility of this method being used as a new 3D reconstruction method for X-ray imaging was confirmed. However, all regions should be subjected to filtering to investigate the entire morphology via data assimilation.

6. Conclusions

To develop a highly accurate prediction method for dendrite morphology during alloy solidification by integrating in situ X-ray observations with the PF method, a data assimilation system was developed using the solid fraction along the thickness of a sample to enable data assimilation using X-ray transmission images as observation data. Twin experiments were then performed for the growth of columnar dendrites in thin films with relatively simple morphologies along their thickness to validate the developed data assimilation system. Data assimilation using the solid fraction as observation data estimated the material properties and solidification morphologies with reasonable accuracy, although the accuracy was lower than that when using 3D morphologies as observation data. Additionally, the estimation accuracy decreased as the ratio of the immobile interface in the total solid–liquid interface increased with dendrite growth and increased when the estimation region was limited to only around the dendrite tip. However, filtering the entire domain is necessary to apply data assimilation for the 3D reconstruction scheme of X-ray imaging. In future studies, a data assimilation method must be developed to simultaneously obtain material properties only around the dendrite tip region and dendritic morphologies by filtering the entire region.

Acknowledgements

This work was supported by KAKENHI Grant-in-Aid for Scientific Research (A) 20H00217 and (S) 22H04963. This study also used the TSUBAME3.0 computational resources provided by the Tokyo Institute of Technology through the HPCI System Research Project (Project ID: hp220089).

References

[1] Dantzig J A and Rappaz M 2016 *Solidification: 2nd Edition - Revised & Expanded* (Lausanne: CRC Press LLC)
[2] Glicksman M E, Schaefer R J and Ayers J D 1976 *Metall. Trans. A* **7** 1747-59
[3] Esaka H and Kurz W 1985 *J. Cryst. Growth* **72** 578-84
[4] Beckermann C and Wang C Y 1996 *Metall. Mater. Trans. A* **27** 2784-95
[5] Mathiesen R H, Arnberg L, Mo F, Weitkamp T and Snigirev A 1999 *Phys. Rev. Lett.* **83** 5062-5
[6] Nguyen-Thi H, Bogno A, Reinhart G, Billia B, Mathiesen R H, Zimmermann G, Houltz Y, Löth K, Voss D, Verga A and Pascale F D 2011 *J. Phys. Conf. Ser.* **327**
[7] Yasuda H, Yamamoto Y, Nakatsuka N, Yoshiya M, Nagira T, Sugiyama A, Ohnaka I, Uesugi K and Umetani K 2009 *Int. J. Cast Met. Res.* **22** 15-21
[8] Yasuda H, Kawarasaki T, Tomiyori Y, Kato Y and Morishita K 2019 *IOP Conf. Ser.: Mater. Sci. Eng.* **529** 012023
[9] Nakano K, Narumi T, Morishita K and Yasuda H 2020 *Mater. Trans.* **61** 596-604
[10] Garcia-Moreno F, Kamm P H, Neu T R, Bulk F, Noack M A, Wegener M, von der Eltz N, Schleputz C M, Stampanoni M and Banhart J 2021 *Adv Mater* **33** 2104659
[11] Kurz W, Fisher D J and Trivedi R 2019 *Int. Mater. Rev.* **64** 311-54
[12] Kurz W, Rappaz M and Trivedi R 2021 *Int. Mater. Rev.* **66** 30-76
[13] Ohno M and Matsuura K 2009 *Phys. Rev. E* **79** 031603
[14] Takaki T 2014 *ISIJ Int.* **54** 437-44
[15] Sakane S, Takaki T and Aoki T 2022 *Mater. Theory* **6** 3
[16] Sakane S, Takaki T, Ohno M, Shibuta Y and Aoki T 2020 *Comput. Mater. Sci.* **178** 109639
[17] Yamanaka N, Sakane S and Takaki T 2021 *Comput. Mater. Sci.* **197** 110658
[18] Takaki T, Sakane S, Ohno M, Shibuta Y and Aoki T 2020 *Comput. Mater. Sci.* **171** 109209
[19] Evensen G 2009 *Data assimilation: the ensemble Kalman filter* vol 2 (Berlin, Heidelberg: Springer)
[20] Houtekamer P L and Zhang F 2016 *Mon. Weather Rev.* **144** 4489-532
[21] Miyoshi E, Ohno M, Shibuta Y, Yamanaka A and Takaki T 2021 *Mater. Des.* **210** 110089
[22] Yamanaka A, Maeda Y and Sasaki K 2019 *Mater. Des.* **165** 107577
[23] Yamanaka A and Takahashi K 2020 *Mater. Today Commun.* **25** 101331
[24] Imai Y, Sakane S, Ohno M, Yasuda H and Takaki T 2023 *IOP Conf. Ser.: Mater. Sci. Eng.* (in press)
[25] Yamamura A, Sakane S, Ohno M, Yasuda H and Takaki T 2023 *IOP Conf. Ser.: Mater. Sci. Eng.* (in press)
[26] Yamamura A, Sakane S, Ohno M, Yasuda H and Takaki T 2022 *Comput. Mater. Sci.* **215** 111776
[27] Ishii A, Yamanaka A, Miyoshi E, Okada Y and Yamamoto A 2021 *Modell. Simul. Mater. Sci. Eng.* **29** 065012
[28] Ishii A, Yamanaka A, Miyoshi E and Yamamoto A 2022 *Mater. Today Commun.* **30** 103089
[29] Ohno M, Oka Y, Sakane S, Shibuta Y and Takaki T 2020 *Phys. Rev. E* **101** 052121
[30] Eggleston J J, McFadden G B and Voorhees P W 2001 *Phys. D: Nonlinear Phenom.* **150** 91-103
[31] Karrasch C 2016 Solidification Kinetics in Undercooled Pure Iron and Iron-Boron Alloys under Different Fluid Flow Conditions *PhD Thesis* Ruhr-Universität Bochum
[32] Borg R J and Lai D Y F 1970 *J. Appl. Phys.* **41** 5193-200
[33] Cui S and Jung I-H 2017 *Calphad* **56** 108-25

IOP Conf. Series: Materials Science and Engineering 1281 (2023) 012050 doi:10.1088/1757-899X/1281/1/012050

On the occurrence of buoyancy-induced oscillatory growth instability in directional solidification of alloys

Josep Maria Barbera[1,2], **Thomas Isensee**[1,3] **and Damien Tourret**[1]

[1] IMDEA Materials, Getafe, Madrid, Spain
[2] Universidad Politecnica de Madrid, ETSI Industriales, Madrid, Spain
[3] Universidad Politecnica de Madrid, ETSI Caminos, Canales & Puertos, Madrid, Spain

E-mail: damien.tourret@imdea.org

Abstract. Recent solidification experiments identified an oscillatory growth instability during directional solidification of Ni-based superalloy CMSX4 under a given range of cooling rates. From a modeling perspective, the quantitative simulation of dendritic growth under convective conditions remains challenging, due to the multiple length scales involved. Using the dendritic needle network (DNN) model, coupled with an efficient Navier-Stokes solver, we reproduced the buoyancy-induced growth oscillations observed in CMSX4 directional solidification. These previous results have shown that, for a given alloy and temperature gradient, oscillations occur in a narrow range of cooling rates (or pulling velocity, V_p) and that the selected primary dendrite arm spacing (Λ) plays a crucial role in the activation of the flow leading to oscillations. Here, we show that the oscillatory behavior may be generalized to other binary alloys within an appropriate range of (V_p, Λ) by reproducing it for an Al-4at.%Cu alloy. We perform a mapping of oscillatory states as a function of V_p and Λ, and identify the regions of occurrence of different behaviors (e.g., sustained or damped oscillations) and their effect on the oscillation characteristics. Our results suggest a minimum of V_p for the occurrence of oscillations and confirm the correlation between the oscillation type (namely: damped, sustained, or noisy) with the ratio of average fluid velocity $\overline{V}$ over V_p. We describe the different observed growth regimes and highlight similarities and contrasts with our previous results for a CMSX4 alloy.

1. Introduction

Directional solidification (DS) is critical to manufacture a range of high-value components. A prominent example is that of single-crystal Ni-based superalloy turbine blades for aeronautical applications [1]. They are arguably among the most advanced manmade structural metallic components, capable to withstand extreme thermal, mechanical, and chemical conditions all at once. However, various defects may emerge, such as segregated channels and freckles [2, 3, 4, 5], which may lead to a significant amount of manufactured parts being discarded. Therefore, a better fundamental understanding of the conditions leading to stable and homogeneous DS is critical to the production of the next-generation of directionally solidified components.

Recently, experiments using X-ray *in situ* radiography revealed the existence of oscillatory instabilities during DS of Ni-based superalloy CMSX4 [6]. Over the years, a range of oscillatory instabilities during DS had been previously investigated, e.g., related to growth in a narrow channel [7], coordinated "breathing modes" in cellular array growth [8, 9, 10], sidebranching at high primary spacing [11], or banding instabilities in rapid solidification [12, 13]. However, while all of these naturally occur within a purely diffusive thermo-solutal transport regime,

IOP Conf. Series: Materials Science and Engineering 1281 (2023) 012050 doi:10.1088/1757-899X/1281/1/012050

the oscillations observed in CMSX4 stem from gravity-induced buoyancy [6], making them more challenging to investigate. The effect of buoyant convection on dendritic growth and microstructure heterogeneities has been acknowledged for decades [14, 15, 16, 17, 18]. However, due to the various scales to be incorporated together in the problem, quantitative modeling of dendritic growth in presence of fluid flow remains computationally demanding [19, 20].

Using a multiscale dendritic needle network (DNN) approach [21, 22], extended to include fluid flow in the liquid phase [23, 24], we recently managed to reproduce the growth oscillations observed in CMSX4 [24]. Beyond reproducing experimental observations — e.g., the transition from oscillatory to damped oscillations when increasing the cooling rate (or equivalent pulling velocity V_p) — our simulations highlighted the importance of the primary dendritic spacing Λ on the occurrence (at high Λ) or inhibition (at low Λ) of these oscillations, and the fact that sustained oscillations seemed to occur when the average fluid velocity $\overline{V}$ was close to V_p.

Computational simulations of this oscillatory behavior naturally open the way for a deeper and systematic exploration of the conditions of occurrence and characteristics of oscillations as a function of alloy parameters and processing conditions. Extending our previous study [24], here we perform a mapping of this buoyancy-induced oscillatory behavior as a function of primary spacing and pulling velocity. An underlying objective is also to uncover whether this phenomenon is generalizable to other alloys, provided the appropriate combination of temperature gradient G, pulling velocity V_p, and primary spacing Λ. Hence, we focus on a different alloy with relatively well-known phase diagram and thermophysical properties, namely Al-4at.%Cu.

2. Methods

The model, its implementation, and the simulations are similar in nature to those presented elsewhere [24, 22, 23]. Therefore, here we only summarize their main features (see Ref. [24] for further details).

2.1. Model

The dendritic needle network (DNN) model alleviates the burden of explicitly tracking the morphologically complex solid-liquid interface by representing growing crystals as hierarchical networks of thin parabolic-shaped branches [22, 23, 24]. The temperature field, $T(\mathbf{x}, t)$, is usually imposed as a boundary condition throughout the domain, and the evolution of the solute concentration field, $c(\mathbf{x}, t)$, is solved considering either diffusive [22] or convective [23, 24] conditions in the liquid phase, while the concentration along the solid-liquid interface, i.e., along the needle network, is set at the equilibrium concentration at the local temperature.

We consider a dilute binary alloy with constant liquidus slope, $m < 0$, and solute partition coefficient, $0 < k < 1$, and introduce the reduced (dimensionless) concentration field, $U \equiv (c_l^0 - c)/[(1-k)c_l^0]$, where c_l^0 is the liquid equilibrium concentration at the reference (i.e., solidus) temperature, T_0. Directional solidification conditions are represented via the frozen temperature approximation $T = T_0 + G(x - V_p t)$, where G is the strength of the temperature gradient and V_p is the pulling (or isotherm) velocity, both imposed along the x direction. Neglecting curvature and kinetic undercooling contributions, the equilibrium dimensionless concentration, U_i, along the solid-liquid interface is expressed as

$$U_i = (x - V_p t)/l_T, \tag{1}$$

with $l_T = |m|(1-k)c_l^0/G$ the thermal length separating liquidus and solidus temperatures.

The liquid velocity field, $\boldsymbol{v}$, is calculated using Navier-Stokes equations

$$\rho\left[\partial_t \boldsymbol{v} + (\boldsymbol{v} \cdot \nabla)\boldsymbol{v}\right] = \boldsymbol{F} - \nabla p + \eta \nabla^2 \boldsymbol{v}, \tag{2}$$

where ρ is the fluid density, p its pressure, η its viscosity and $\boldsymbol{F}$ corresponds to external forces. The liquid is assumed incompressible with $\nabla \cdot \boldsymbol{v} = 0$ and a null velocity is imposed along the solid-liquid interface, i.e., along the needles. Buoyant forces are included using the Boussinesq approximation, considering only solute-dependent terms

$$\boldsymbol{F} = \rho^l_\infty \boldsymbol{g} \left[1 - \beta_c(c - c_\infty)\right], \tag{3}$$

where $\beta_c \equiv -(\partial\rho/\partial c|_{c=c_\infty})/\rho^l_\infty$ is a solutal expansion coefficient, with ρ^l_∞ the fluid density at the alloy nominal concentration c_∞ and $\boldsymbol{g}$ is the gravity acceleration. The transport of solute in the liquid with fluid velocity $\boldsymbol{v}$ is thus described by the advection-diffusion equation, with diffusion coefficient D,

$$\partial_t U + \nabla \cdot (\boldsymbol{v} U) = D\nabla^2 U. \tag{4}$$

The instantaneous tip radius, $R(t)$, and velocity, $V(t)$, of each needle-like branch is computed using the microscopic solvability condition

$$R^2 V = 2Dd_0/\{[1 - (1-k)U_t]\sigma\}, \tag{5}$$

where $d_0 = \Gamma/\left[|m|(1-k)c_l^0\right]$ is the capillary length at T_0 with Γ the interface Gibbs-Thomson coefficient, $U_t = (x_t - V_p t)/l_T$ is the equilibrium concentration at the tip position, x_t, and σ is the tip selection parameter [25, 26], combined with a solute conservation statement in the vicinity of the parabolic tip

$$RV^2 = 2D^2\mathcal{F}^2/\{\left[1-(1-k)U_t\right]^2 d_0\}, \tag{6}$$

where the flux intensity factor $\mathcal{F} \equiv \int_{\Gamma_0} (\partial_n U)\, dS/\left(4\sqrt{a/d_0}\right)$ measures the incoming flux and can be calculated along any contour Γ_i (instead of directly on the solid-liquid interface along Γ_0) as

$$4\mathcal{F}\sqrt{a/d_0} = \int_{\Gamma_i} (\partial_n U)\, \mathrm{d}S + \frac{V}{D}\int_{\Sigma_i} (\partial_x U)\, \mathrm{d}A, \tag{7}$$

with $\partial_n U$ the outward normal solute gradient, Σ_i the surface enclosed between Γ_0 and Γ_i, and a the distance between the tip and the intersection of the integration contour with the parabolic tip (i.e., the location where Γ_0 and Γ_i meet) [22, 23].

The model is solved similarly as in Refs [23, 24], i.e., using a mostly explicit finite difference scheme on a staggered grid, an upwind discretization scheme for convective terms, a projection method for the resolution of the Navier-Stokes equations [27], and an iterative successive over-relaxation (SOR) method [28, 29] for the incompressibility condition. The code is implemented in C-based CUDA language to leverage acceleration using Graphics Processing Units (GPUs).

2.2. Simulations

In Ref. [24], our simulations aimed at approaching specific experimental conditions of CMSX4 directional solidification [6]. In contrast, here, an underlying objective is to investigate whether the oscillatory behavior may be general to any arbitrary alloy, provided the appropriate set of (G, V_p, Λ) conditions. Therefore, we consider a different alloy, namely a binary Al-4at.%Cu, with relatively well known parameters [24, 30], but so far no report of such oscillatory growth behavior. Specifically, we use alloy parameters as in Refs [24] (Section 3.1 therein) and [30], i.e., $c_\infty = 4.0\,$at.%, $D = 3\times10^{-9}$ m^2/s, $k = 0.14$, $m = 1.6$ K/at.%, $\nu = \eta/\rho = 5.7\times10^{-7}$ m^2/s, $\Gamma = 2.4\times10^{-7}$ Km, $\beta_c = -10^{-2}$/at%, and $\sigma \approx 0.153$ (corresponding to an interfacial excess free energy anisotropy $\epsilon = 0.02$). In an Al-Cu alloy (unlike in CMSX4), buoyant currents and plumes are caused by the rejection of the heavier element (Cu). Therefore, while the growth direction is kept as $x+$ and represented upwards for consistency with Ref. [24], gravity forces also have a $x+$ direction (i.e. pointing upwards in figures below). We fix the temperature gradient at

$G = 10$ K/mm, and explore velocities from $V_p = 40$ to 100 µm/s within a range of primary dendrite arm spacing Λ from about 90 to 350 µm.

The radius of integration of the flux intensity factor is set to $r_i/R_s = 4$ and the truncation radius of needles far behind the tip to $r_{\max}/R_s = 5$, with R_s the theoretical steady state tip radius in the diffusive regime [22, 24]. For each velocity V_p, the grid spacing is chosen between $\Delta x/R_s = 0.7$ and 1.24, ensuring that the diffusion length included a sufficient number of grid points — namely with $D/V_p/\Delta x$ ranging from 6 ($V_p = 100$ µm/s) to 16 ($V_p = 40$ µm/s). Other numerical parameters are similar to those listed in Ref. [24] (Table 1 therein, with $K_{\Delta t} = 0.5$).

With these parameters, the simulations are similar in nature to those presented and discussed in Ref. [24] (Sections 3.1 and 4 therein). Within a domain of height H (x-direction) and width W (y-direction), we set an array of N evenly spaced primary needles at the bottom of the domain growing in the $x+$ direction and with their tips initially aligned in x along the liquidus temperature location. Boundary conditions (BCs) are periodic laterally (in the y direction) for all fields. No-flux ($\partial U/\partial x = 0$) and free-slip ($v_x = 0$) conditions are applied along the top and bottom boundaries (in the x direction). The simulations use a moving frame in the x-direction, such that the most advanced needle tip in x remains at a fixed location — namely at a distance between 30% and 65% of the domain height from the bottom boundary.

The domain size in x, i.e., its height H, is chosen long enough for BCs to have negligible effect on the flow pattern — typically adjusted by trial-and-error, ensuring that the liquid length is always greater than $5D/V_p$ and the solid length greater than $8D/V_p$. The domain size in y is set to either 630 or 1470 inner grid points, for convenience (both numbers are divisible by a broad range of integer needle number N, as required to impose periodic arrays) and performance (accounting for the two extra points used to impose periodic BCs, both 632 and 1472 are divisible by 8, thus allowing nearly optimal GPU block sizes). Each simulation is initialized with between 7 and 30 needle-like primary dendrites, in order to probe different values of primary spacing Λ. The array growth was simulated for a duration of between 90 (high V_p) and 120 (low V_p) seconds, so as to obtain enough oscillation periods to analyze, when relevant. Main parameters of the resulting 44 simulations are summarized in Table 1. Each simulation is performed using a single GPU (Nvidia RTX3090, in most cases) and each required at most 92 h to complete (wall time).

Table 1. Simulation parameters for the mapping of V_p and Λ.

Velocity V_p (µm)	Grid spacing $\Delta x/R_s$	Δx (µm)	Height $H/\Delta x$	Number of needles, N $W/\Delta x = 630$	$W/\Delta x = 1470$	PDAS range Λ (µm)
40	1.24	7.06	638	14, 18, 21, 30	-	(148, 318)
50	1.10	5.56	638	10, 14, 15, 18, 21, 30	-	(117, 350)
60	1.00	4.59	510	9, 10, 14, 15, 18, 21, 30	-	(96, 321)
70	0.94	3.98	510	7, 9, 10, 14, 15, 18, 21	-	(119, 358)
80	0.90	3.54	398, 510	7, 9, 10, 14, 15, 18, 21	-	(106, 319)
90	0.80	2.96	398, 510	7, 9, 10, 14, 15, 18	14	(103, 310)
100	0.70	2.44	398	7, 9, 10, 14, 18	14	(86, 257)

2.3. Post-processing

In order to classify the simulated growth behaviors and extract oscillation characteristics, we analyzed the velocity evolution, $V(t)$, of every needle tip in each simulation. First, we discarded simulations leading to the elimination of one or more dendrites, which is expected to happen when Λ is too low due to intense solutal interaction among dendrites [11, 21, 31]. All remaining simulations thus retain a homogeneous spacing Λ above the lower stability limit Λ_{min}.

Individual needle velocities, $V(t)$, were then fitted to either of these two functions

$$v_1(t) = V_0 - A\cos\left[2\pi(t-t_0)f\right]\exp(-t/\tau), \tag{8}$$

$$v_2(t) = V_0 - A\left\{(1-S)\cos\left[2\pi(t-t_0)f\right] + 2S\left|\cos\left[\pi(t-t_0)f\right]\right| - S\right\}, \tag{9}$$

via the following fitting parameters: mid-range velocity V_0, time origin t_0, oscillation amplitude A, oscillation frequency f, and either a characteristic damping time τ (Eq. (8)) or a "spikiness" factor S (Eq. (9)). The latter, bounded to $0 \leq S \leq 1$, allows fitting signals showing burst-like "spiky" maxima, yet with stable-frequency oscillations. As illustrated in Figure 1, $S = 0$ leads to a regular cosine function and $S = 1$ to a spiky cosine-like function of similar amplitude and frequency. (Note that, when $S > 0$, the mid-range velocity V_0 differs from the average velocity V_p.) We chose to fit to both functions and select the best fit out of the two for convenience, after noticing that damped oscillations typically had smooth (non-spiky) behaviors while sustained oscillations had a range of behaviors from smooth to spiky. For each needle, the fitting was performed using a custom Python script that dynamically scans different fitting ranges and selects the one leading to the best fit, i.e., the highest coefficient of determination r^2, while ensuring that the range included a minimum of 5 oscillation periods (when relevant) or at least 20 seconds. Since raw $V(t)$ data may be prone to numerical oscillations as the needle progresses through the grid [23], in order to facilitate the fitting, the velocities were smoothed applying a moving average to the time derivative of the needle length prior to fitting (both raw and smooth $V(t)$ are shown in later figures). The initial guess for the frequency f, prior to its fitting, was based on a preliminary estimate via fast Fourier transform of the $V(t)$ signal.

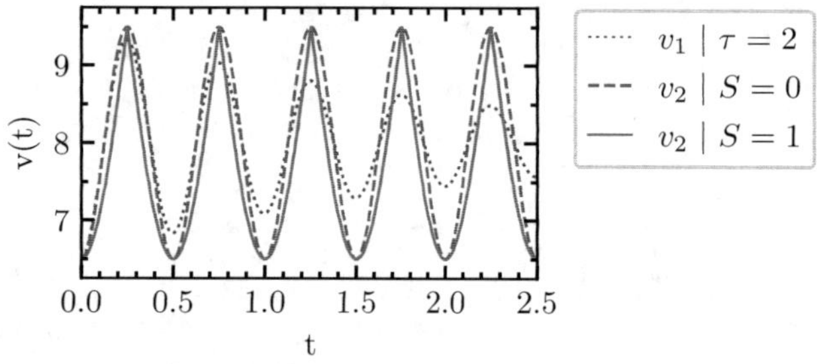

Figure 1. Illustration of functions $v_1(t)$ and $v_2(t)$ for $V_0 = 8.0$, $A = 1.5$, $f = 2.0$, $t_0 = 0.0$ and different values of τ (Eq. (8)) or S (Eq. (9)).

Growth velocities $V(t)$ leading to a poor fit to both functions, namely if $r^2 < 0.8$, where classified as *noisy* (rather than rigorously *non-oscillatory* since their behavior was usually closer to noisy oscillations than to erratic bursts as reported in Ref. [24]). When $r^2 \geq 0.8$, behaviors were classified as *damped* if the fit was better with Eq. (8) than Eq. (9), or *sustained* otherwise. (Damped oscillations were all quite straightforward to identify from visual inspection of $V(t)$.)

3. Results and Discussion

Figure 2 shows the resulting map of growth behaviors as a function of pulling velocity V_p and primary spacing Λ. As a first conclusion, it was indeed possible to obtain sustained oscillatory growth for several simulations over a range of (V_p, Λ). Oscillations were only found above a minimum velocity, here for $V_p \geq 60$ µm/s. Most cases (all but one) leading to sustained oscillations (• symbols) fall within a spacing range $190 \leq \Lambda/\text{µm} \leq 270$, while damped oscillations (▼ symbols) emerge within $120 \leq \Lambda/\text{µm} \leq 190$. While different symbol types denote the classification criteria mentioned in Section 2.3, their color follows a discrete distribution as a function of the ratio of average fluid velocity $\overline{V}$ over pulling velocity V_p, namely: light green

($\overline{V}/V_p \leq 0.05$), medium blue ($0.05 \leq \overline{V}/V_p \leq 2.5$), or dark red ($\overline{V}/V_p \geq 2.5$). Except for a few data points around the sustained/noisy transition, the reasonable match between symbol types and colors shows that we can correlate the occurrence of damped, sustained, or noisy oscillations to the ratio $\overline{V}/V_p$. However, the current threshold values for $\overline{V}/V_p$, here denoted $\xi_1 \approx 0.05$ (damped/sustained) and $\xi_2 \approx 2.5$ (sustained/noisy), differ from those identified in our previous study focused on CMSX4 ($\xi_1 \approx 1.0$, $\xi_2 \approx 2.0$), such that they may depend upon alloy parameters and/or processing conditions (e.g., temperature gradient). Below, we describe and discuss in further details the different growth behaviors observed in our simulations.

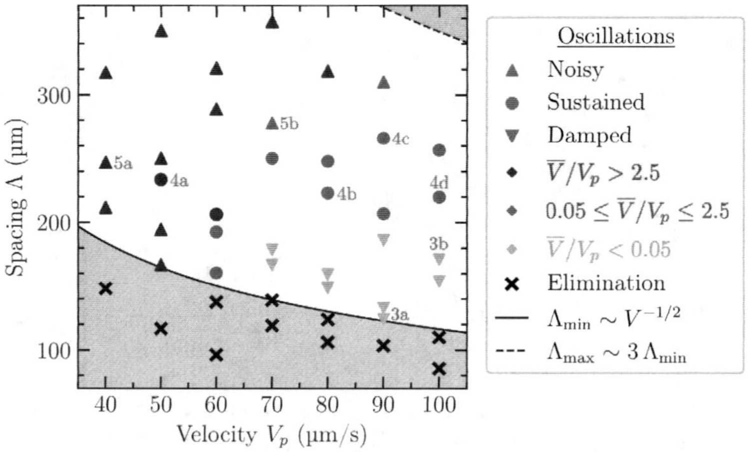

Figure 2. Oscillatory behavior map. Symbols types denote different growth behaviors: (×) elimination events for $\Lambda < \Lambda_{\min}$ (black line) or (▼) damped, (•) sustained, or (▲) noisy oscillations. Symbols are colored according to the ratio $\overline{V}/V_p$. Gray text labels next to symbols mark cases highlighted in the following figures and discussion.

Elimination ($\Lambda < \Lambda_{\min}$). As expected, for each V_p, the lowest spacings lead to the elimination of at least one primary dendrite, marking the lower limit of the stable spacing range, $\Lambda_{\min}$. This minimal stable spacing appears to reasonably match the expected power law $\Lambda_{\min} \sim V^{-1/2}$ (solid black line). The ratio between the lower and upper limits of the stable spacing range $\Lambda_{\max}/\Lambda_{\min}$ typically varies between 2 and 5 [11, 21, 31]. Since the current simulations do not include sidebranching, they cannot predict the upper limit (as done, e.g., in Refs [21, 31]). Hence, once we had an estimate of $\Lambda_{\min}$ for a given V_p, we approximated $\Lambda_{\max} \approx 3\,\Lambda_{\min}$ (black dashed line) and did not perform any simulation above this $\Lambda_{\max}$, because such a situation would in reality likely lead to spacing reduction by tertiary sidebranching [21, 31].

Damped oscillations ($\Lambda \gtrsim \Lambda_{\min}$). Above a certain velocity, here ≈ 75 μm/s, the lowest stable spacings just above $\Lambda_{\min}$ lead to damped oscillations. These are illustrated in Figure 3 for two representative case, namely for (a) $V = 90$ μm/s and $\Lambda = 124$ μm/s and for (b) $V = 100$ μm/s and $\Lambda = 171$ μm. They respectively correspond to the lowest ($\tau = 16.7$ s) and highest ($\tau = 3.13$ s) damping rates obtained among the 44 simulations. While we did not collect sufficient data to extract meaningful scaling laws, we observed that the damping rate typically increases (i.e., τ decreases) when either V_p or Λ increases. In this regime, the behavior of all needles in the array is homogeneous and synchronized (typically oscillating in phase with one another). From Figure 3 on, velocity fields are shown via the flow streamlines, of which the opacity decreases progressively when the fluid velocity is lower than $0.2V_{\max}$, with $V_{\max}$ the maximum fluid velocity over the entire domain. This allows illustrating that, within the damped oscillation regime (Fig. 3), while some convective currents appear around and between the dendrites, the fluid velocity vanishes within a narrow boundary layer ahead of the solidification front. Transport of solute beyond this boundary layer occurs then primarily through diffusion.

Sustained oscillations (intermediate Λ). Within the inspected range, for a sufficient pulling velocity, at intermediate primary spacings Λ, sustained oscillations occur. Typical cases are illustrated in Figure 4 for (a) $V = 50$ μm/s and $\Lambda = 234$ μm, (b) $V = 80$ μm/s and $\Lambda = 223$ μm,

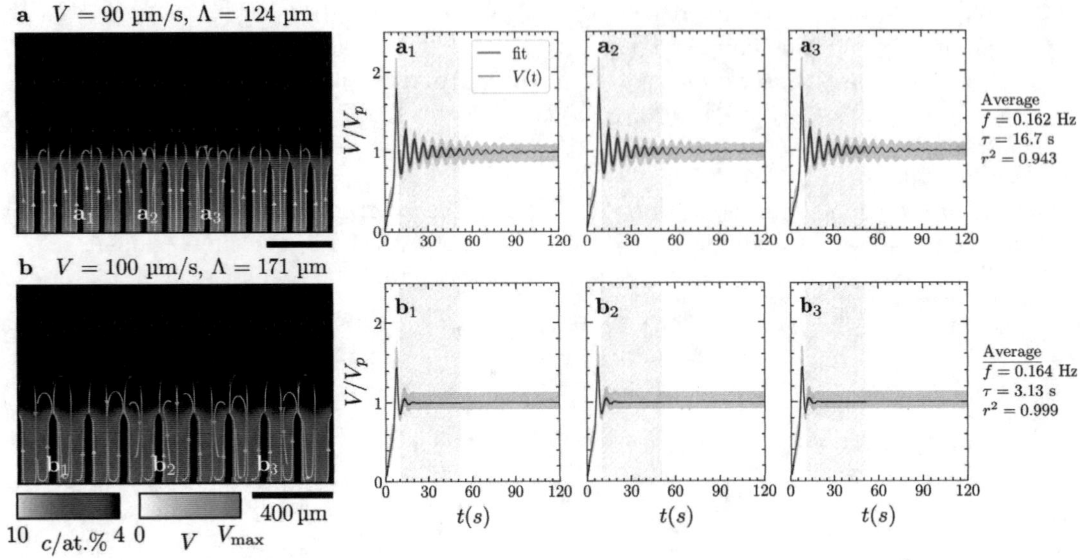

Figure 3. Damped oscillations (a) for $V = 90$ µm/s and $\Lambda = 124$ µm and (b) for $V = 100$ µm/s and $\Lambda = 171$ µm. (left) Concentration field (color map) and fluid flow (streamlines) at $t = 120$ s. Streamlines are progressively transparent toward lowest velocities (from opaque at $V/V_{\text{max}} \geq 0.2$ to 90% transparent at $V = 0$). (right) Tip velocities of selected needles, showing raw (gray) and smoothed (black) $V(t)$ as well as fitted function (thick red) and range (shaded background).

(c) $V = 90$ µm/s and $\Lambda = 266$ µm, and (d) $V = 100$ µm/s and $\Lambda = 220$ µm. Among these, Fig. 4c is a good illustration of a typical sustained oscillatory regime. Therein, all needles exhibit an oscillatory growth, with a small variability of oscillation amplitudes and frequencies throughout the array. As seen on the leftmost panel, prominent convection vortices appear in the liquid, and needle oscillations are desynchronized (not in phase) with one another.

In some cases, illustrated by Fig. 4b and d, the oscillatory growth regime emerges after a usually short (b) but potentially long (d) transient period, momentarily more akin to a damped oscillation regime before a subsequent amplification of the oscillation amplitude toward a steady value. (For this reason, several runs classified as damped were performed for longer durations in order to ascertain that they were indeed not in a transient state.) Such simulations exhibiting a transient regime were observed for the lowest Λ data point classified as sustained at V = 80, 90, and 100 µm/s, i.e., close to the edge of the transition between damped and sustained oscillations (Fig. 2). The corresponding convective patterns are also intermediate, with vanishing velocities toward the top of the domain (like for damped oscillations at lower Λ) and the emergence of small nascent convection rolls (like for sustained oscillations at higher Λ).

Finally, Fig. 4a shows the other end of the sustained oscillation spectrum, at the edge of the transition toward noisy oscillations. This specific case corresponds to the lone outlier data point classified as "sustained" for $V = 50$ µm/s in Fig. 2, due to its $r^2 = 0.847$ higher than the chosen threshold of 0.8. However, in spite of a relatively good fit to $v_2(t)$, Fig. 4a shows that resulting $V(t)$ look relatively noisy (see, e.g., Fig. 4a$_3$). In such cases, strong convection vortices are present within the liquid.

Among the sustained oscillations, measured frequencies range from 0.155 to 0.230 Hz (i.e., periods from 4.35 to 6.45 s), without any conclusive dependence upon Λ or V_p emerging within the investigated range of conditions. Naturally, the oscillation amplitude scales approximately like V_p and tends to increase toward high Λ when the $V(t)$ signals become more spiky.

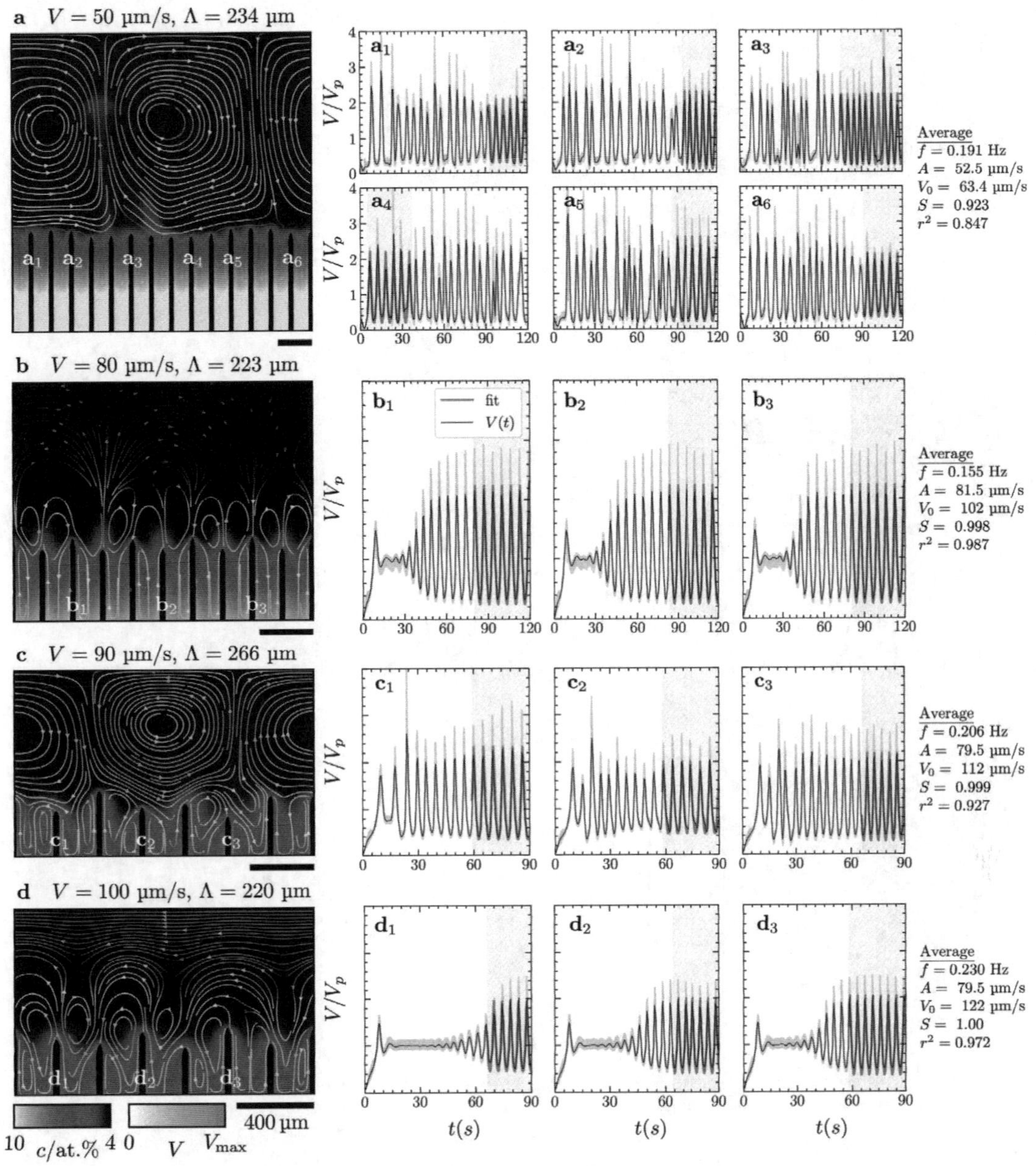

Figure 4. Sustained oscillations for (a) $V = 50$ µm/s and $\Lambda = 234$ µm, (b) $V = 80$ µm/s and $\Lambda = 223$ µm, (c) $V = 90$ µm/s and $\Lambda = 266$ µm, (d) $V = 100$ µm/s and $\Lambda = 220$ µm. (left) Concentration field (color map) and fluid flow (streamlines, from opaque at $V/V_{\max} \geq 0.2$ to 90% transparent at $V = 0$) at $t =$ (a,b) 120 s and (c,d) 90 s. (right) Tip velocities of selected needles, showing raw (gray) and smoothed (black) $V(t)$ as well as fitted function (thick red) and range (shaded background).

Noisy oscillations (high Λ). As shown in Figure 5, at high primary spacings, or across all stable spacings for low V_p, the velocities deviate more prominently from the prototypical behaviors illustrated in Fig. 1. Figure 5a shows the worst encountered fit with $r^2 = 0.590$ and Fig. 5b a case closer to the edge of the transition between the "sustained" and "noisy" regions of Fig. 2. While the $V(t)$ signals are clearly noisy and cannot be assimilated to any clear periodic function, both of them still exhibit an emerging frequency close to that of the sustained oscillations, namely with $0.134 < \mathrm{f/Hz} < 0.255$ (i.e., periods between 3.92 and 7.46 s) across the investigated range of noisy oscillations. This behavior remains quite different from the burst-like growth encountered for CMSX4 simulations at low V_p [24]. All cases of noisy oscillations exhibit prominent convection rolls in the liquid phase.

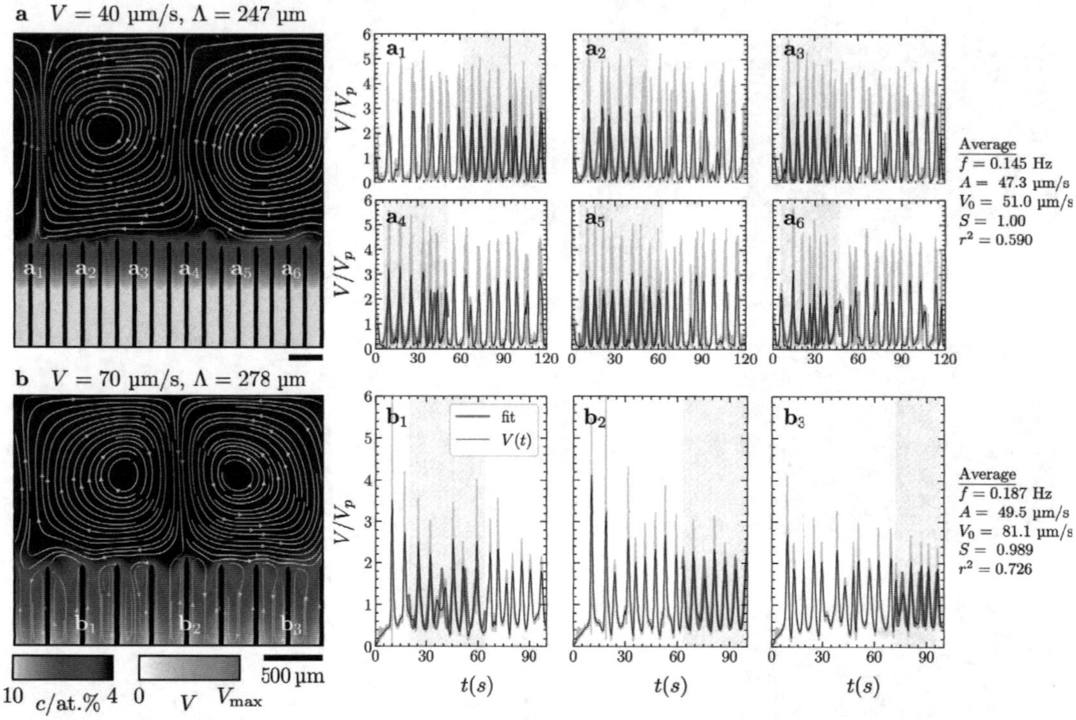

Figure 5. Noisy oscillations for (a) $V = 40$ μm/s and $\Lambda = 247$ μm, and (b) $V = 70$ μm/s and $\Lambda = 278$ μm. (left) Concentration field (color map) and fluid flow (streamlines, from opaque at $V/V_{\max} \geq 0.2$ to 90% transparent at $V = 0$) at $t =$ (a) 120 s and (b) 100 s. (right) Tip velocities of selected needles, showing raw (gray) and smoothed (black) $V(t)$ as well as fitted function (thick red) and range (shaded background).

4. Summary and Perspectives

We investigated the conditions of occurrence and characteristics of buoyancy-driven oscillations of primary dendrite growth velocities in directional solidification using multiscale dendritic needle network (DNN) simulations. While it was previously studied experimentally [6] and computationally [24] for a Ni-based alloy, here we reproduce comparable behaviors using a binary Al-4at.%Cu alloy. These results suggest that this phenomenon may be general to any binary alloy, provided the appropriate (G, V_p, Λ) conditions. We confirmed that the the ratio between average fluid velocity $\overline{V}$ and pulling velocity V_p provides a reasonable indicator of the oscillatory regime (namely: damped, sustained, or noisy). However, the threshold values separating regimes differ from those identified for CMSX4 under different processing conditions [24]. Here, the transition from damped to sustained oscillations occurs together with the onset of appearance of convection vortices in the fluid. This is also in contrast with previous results for CMSX4 [6, 24], exhibiting well-established convection rolls in both regimes. Our results also did not capture the further expected transition from sustained to damped oscillations when increasing V_p (i.e., cooling rate [6, 24]), perhaps because it occurs at higher V_p than those explored here. Moreover, the effect of the null velocity imposed at the solid-liquid interface – in contrast, e.g., to imposing the growth velocity, accounting for the solid-liquid density change, or tracking the solid fraction into a mushy region – remains to be investigated.

Some open questions remain on buoyancy-driven oscillations in directional solidification. Among other things, it remains unclear whether the phenomenon is relevant to bulk samples or is promoted by the 2D configuration (or quasi-2D thin-sample confinement in experiments). Ongoing work, directly following up from the present study, will clarify the dependence of oscillatory behavior upon different alloy parameters (such as diffusion coefficient D and partition coefficient k). Scanning a broader range of alloy parameters and processing conditions will allow us to extract more meaningful scaling laws for the oscillation characteristics, and to establish general rules governing the stability of dendritic fronts in directional solidification processes.

Acknowledgments

This study was supported by the Spanish Ministry of Science and Innovation through the María de Maeztu seal of excellence of IMDEA Materials Institute (CEX2018-000800-M) and a Ramón y Cajal Fellowship (RYC2019-028233-I).

References

[1] Pollock T M and Tin S 2006 *J. Propul. Power* **22** 361–374
[2] Copley S *et al.* 1970 *Metall. Trans.* **1** 2193–2204
[3] Sample A and Hellawell A 1984 *Metall. Trans.* **15** 2163–2173
[4] Hellawell A, Sarazin J and Steube R 1993 *Phil. Trans. Roy. Soc. London. A: Phys. Eng. Sci.* **345** 507–544
[5] Pollock T and Murphy W 1996 *Metall. Mater. Trans. A* **27** 1081–1094
[6] Reinhart G *et al.* 2020 *Acta Mater.* **194** 68–79
[7] Karma A and Pelcé P 1989 *Phys. Rev. A* **39** 4162
[8] Georgelin M and Pocheau A 1997 *Phys. Rev. Lett.* **79** 2698
[9] Bergeon N *et al.* 2013 *Phys. Rev. Lett.* **110** 226102
[10] Tourret D *et al.* 2015 *Phys. Rev. E* **92** 042401
[11] Echebarria B, Karma A and Gurevich S 2010 *Phys. Rev. E* **81** 021608
[12] Kurz W and Trivedi R 1996 *Metall. Mater. Trans. A* **27** 625–634
[13] Ji K *et al.* 2023 *Physical Review Letters* **130** 026203
[14] Mehrabian R, Keane M and Flemings M 1970 *Metall. Mater. Trans. B* **1** 1209–1220
[15] Dupouy M, Camel D and Favier J 1989 *Acta Metall.* **37** 1143–1157
[16] Jamgotchian H *et al.* 2001 *Phys. Rev. Lett.* **87** 166105
[17] Bogno A *et al.* 2011 *Acta Mater.* **59** 4356–4365
[18] Shevchenko N *et al.* 2013 *Metall. Mater. Trans. A* **44** 3797–3808
[19] Sakane S *et al.* 2017 *J. Cryst. Growth* **474** 154–159
[20] Sakane S *et al.* 2020 *Comput. Mater. Sci.* **178** 109639
[21] Tourret D and Karma A 2013 *Acta Mater.* **61** 6474–6491
[22] Tourret D and Karma A 2016 *Acta Mater.* **120** 240–254
[23] Tourret D, Francois M M and Clarke A J 2019 *Comput. Mater. Sci.* **162** 206–227
[24] Isensee T and Tourret D 2022 *Acta Mater.* **234** 118035
[25] Langer J S 1980 *Rev. Mod. Phys.* **52** 1
[26] Kurz W, Fisher D J and Trivedi R 2019 *Int. Mater. Rev.* **64** 311–354
[27] Chorin A J 1968 *Math. Comp.* **22** 745–762
[28] Frankel S P 1950 *Math. Comp.* **4** 65–75
[29] Young D 1954 *Trans. Am. Math. Soc.* **76** 92–111
[30] Steinbach I 2009 *Acta Mater.* **57** 2640–2645
[31] Bellon B *et al.* 2021 *Acta Mater.* **207** 116686

IOP Conf. Series: Materials Science and Engineering 1281 (2023) 012051 doi:10.1088/1757-899X/1281/1/012051

Diffuse modelling of pearlite growth in Calphad-coupled multicomponent multi-phase-field simulations

J Eiken, B Böttger and M Apel

Access e.V., Intzestraße 5, 52072 Aachen, Germany

E-mail: j.eiken@access.rwth-aachen.de

Abstract. During cooling of steels and cast irons, austenite can decompose by a eutectoid transformation into pearlite, a two-phased mixture of ferrite and cementite. Since the internal lamellar structure is commonly too fine to be distinguished on the scale of the austenite grain structure, pearlite is often modelled as an effective, pseudo-single phase. Such a pragmatic treatment would also be desirable to reduce the computational effort of large-scale multi-phase-field simulations, but a fundamental hindrance is that no consistent thermodynamic description exists for effective pearlite in multicomponent databases. Alternatively, we here propose to model pearlite as diffuse mixture of two phases with individual local fractions and concentrations, such that solute partitioning and thermodynamic driving forces can be consistently derived from standard Calphad databases. The essential computational advantage is that only the outer interfaces of the pearlite nodules have to be numerically resolved, which allows for increased grid spacing and time-steps. The impact of the unresolved lamellar structure on the curvature undercooling is modelled analytically based on a characteristic spacing, which may be calibrated either experimentally or by small-scale simulations. The potential and the limitations of the new approach, implemented in the frame of the Micress® software, shall be discussed.

1. Introduction

Spatially and time-resolved phase-field simulations contribute to a better understanding and control of microstructure evolution and associated material properties. The multicomponent multi-phase-field (MMPF) method [1, 2] implemented in the Micress® software [3] addresses microstructure evolution of technical alloys during all stages of material production such as solidification, solid-state transformation, recrystallization and thermal processing. An important feature is the accurate evaluation of thermodynamic phase properties and quasi-equilibrium conditions by coupling to Calphad databases. Calphad-coupled MMPF simulations have become state of the art for steel grades, e.g. [4-16] and recently also for solidification of cast irons [17-19]. A general complexity of iron-based alloys is that the primary solidification structure does not remain stable, but austenite decomposes into ferrite and cementite during further cooling. Below the eutectoid temperature, cooperative growth of both phases results in fine lamellar structures [15, 16], too fine to be resolved on the scale of the initial austenite grain structure. A common practice is to model pearlite as an effective, pseudo-single phase, which however does not account for the local fractions, concentrations and crystallographic orientations of the individual phases, required for consistent physical predictions. To overcome this problem, we here propose an ansatz to model pearlite nodules as a diffuse mixture of two phases with unresolved internal morphology, but still having individual properties. The paper starts with the model description and afterwards test applications to pearlite growth in Fe-C alloys are presented and discussed.

IOP Conf. Series: Materials Science and Engineering 1281 (2023) 012051 doi:10.1088/1757-899X/1281/1/012051

2. The multicomponent multi-phase-field model (MMPF)

The diffuse mixture ansatz is formulated as an extension to the standard MMPF model [1, 2], which is based on multiple phase-field parameters ϕ_α to map the spatial distribution of $\alpha = 1...n$ phases or grains of different crystallographic orientation. The region in which a phase-field equals unity represents the bulk of the corresponding phase or grain, while interfaces are described by a continuous change to zero. Each interface can be assigned with an individual mobility $M_{\alpha\beta}$ and energy $\sigma_{\alpha\beta}$. The temporal evolution of ϕ_α within the interface regions is computed by:

$$\dot{\phi}_\alpha = \Sigma_\beta^{\check{n}} \left[M_{\alpha\beta} \left(b_{\alpha\beta}\, \Delta g_{\alpha\beta} - \sigma_{\alpha\beta}\, K_{\alpha\beta} + \Sigma_\gamma^{\check{n}}\, J_{\alpha\beta\gamma} \right) \right] \qquad \text{with } b_{\alpha\beta} = \frac{\pi}{\eta} (\phi_\alpha + \phi_\beta)\, (\phi_\alpha\, \phi_\beta)^{0.5} \tag{1}$$

where $\check{n}$ denote the number of locally interacting phases or grains. The thermodynamic driving forces $\Delta g_{\alpha\beta}$ are evaluated as function of composition and temperature from Calphad data. Of importance for the understanding of the diffuse model extension is the definition of the interface terms $K_{\alpha\beta,}$ which on the one hand model the interfacial curvature and on the other hand, additionally stabilize a sinusoidal profile with numerical thickness η between adjacent phases (or grains).

$$K_{\alpha\beta} = \frac{1}{\check{n}} \left[\frac{\pi^2}{\eta^2} (\phi_\beta - \phi_\alpha) + (\nabla^2\phi_\beta - \nabla^2\phi_\alpha) \right] \tag{2}$$

The higher order junction torque forces $J_{\alpha\beta\gamma}$ in equation (1) are defined by:

$$J_{\alpha\beta\gamma} = \frac{1}{\check{n}} \left[\left(\frac{\pi^2}{\eta^2} \phi_\gamma + \nabla^2\phi_\gamma \right) \cdot (\sigma_{\beta\gamma} - \sigma_{\alpha\gamma}) \right] \tag{3}$$

In the framework of the Micress® code, the basic multi-phase-field equations (1-3) can be combined with a huge variety of functionalities to model crystallography, solute redistribution, temperature evolution, nucleation, etc. [3].

3. Diffuse mixture ansatz for unresolved two-phase regions

The diffuse mixture ansatz enables modelling the eutectoid decomposition of austenite into pearlite on the scale of the austenitic grain size without resolving the fine-lamellar ferrite-cementite structure. A specific advantage of the present approach is that ferrite and cementite are still described as individual phases with own local fractions, concentrations and crystallographic orientations, such that almost all other features of the multi-phase-field code are kept unaffected. The most significant change is a reformulation of the interface term $K_{\alpha\beta}$ for the unresolved phases in the diffuse regions. This is here illustrated for a simplified system with the three phases: austenite (a), ferrite (f) and cementite (c). The basic idea is to set the internal interface term K_{fc} and the junction torque J_{afc} to zero, thus preventing phase separation between ferrite and cementite, while the outer interface terms with austenite (K_{af} and K_{ac}) are replaced by combined formulations:

$$K_{fc} = 0, \; J_{afc} = 0 \tag{5}$$

$$K_{af} = \frac{1}{\check{n}} \left[\frac{\pi^2}{\eta^2} (\phi_f - \phi_a - w_{af}\, \phi_c) + (\nabla^2\phi_f - \nabla^2\phi_a - w_{af}\, \nabla^2\phi_c) \right] + K_{afc} \quad \text{with } w_{\alpha f} = \phi_a + \phi_f \tag{6}$$

$$K_{ac} = \frac{1}{\check{n}} \left[\frac{\pi^2}{\eta^2} (\phi_c - \phi_a - w_{ac}\phi_f) + (\nabla^2\phi_c - \nabla^2\phi_a - w_{ac}\nabla^2\phi_f) \right] + K_{acf} \quad \text{with } w_{\alpha c} = \phi_a + \phi_c \tag{7}$$

The terms K_{afc} and K_{acf} denote analytical expressions for the curvature of the unresolved lamellae at the interface to austenite. They are evaluated as function of a user-defined spacing λ_{fc}, which may be calibrated by experiment or a small-scale simulation. In the absence of the third phase, equations (6) and (7) reduce to the standard formulation in equation (2). This is important to consistently describe growth of pro-eutectoid ferrite and cementite. The weighting factors w_{af} and w_{af} prevent an instantaneous wetting of the pro-eutectoid grains directly after nucleation of the third phase. Within the diffuse areas, the phase parameters are solely controlled by the thermodynamic driving forces, which can still be evaluated in the standard way based on local composition and temperature.

4. Model environment and sub-models

The diffuse mixture ansatz for two-phase regions (described in section 3) represents an extension to the standard MMPF [1, 2], implemented in the Micress® software [3]. In this framework, it can be coupled via the TQ interface of the ThermoCalc software [20] to Calphad databases and be combined with a large variety of existing solvers and sub-models with respect to e.g. nucleation, solute redistribution and temperature evolution. Since the model extension mainly affects the phase-field solver, most parts of the code can be kept unchanged, while some need small adjustment. For example, thin-interface corrections [21] are still applicable to all interfaces with well-defined profile, but partially not valid within the diffuse two-phase regions, where especially the anti-trapping current has to be defined as zero. It is noteworthy that the general diffuse mixture ansatz is not limited to eutectoid phase transformation, but has been applied before to simulate eutectic solidification, e.g. in Al-Si based alloys [22, 23]. The aim of the present study is to test the extended applicability to eutectoid austenite decomposition in iron-carbon alloys. A specific challenge of the solid-state application is to simulate polycrystalline nucleation and interaction of both pro-eutectoid and diffuse eutectoid phases under consideration of crystallographic orientation relationship and anisotropy. These features are already addressed by dedicated sub-models in Micress®, but require some code adjustment for compatibility with the diffuse mixture ansatz. These changes are implemented in a beta-version and are here tested before becoming part of the official release.

A critical issue for application to multicomponent iron-carbon alloys is to model the incomplete partitioning of slow-diffusing substitutional elements at the phase interfaces. In this context, Micress® offers two alternative sub-models: a) Para-equilibrium, which assumes that only carbon is in local equilibrium, while manganese remains completely undisturbed by the interface motion and has no growth restricting contribution to the driving force, b) 'Negligible partitioning local equilibrium' (NPLE), which also assumes identical manganese concentration on both sides of the moving interface, but full local equilibrium within the interface, where manganese is supposed to form a local build-up due to finite interface diffusion at the atomic length [24]. In Micress®, both sub-models make use of a virtual interface composition c^{Mn*} which is used for the redistribution step instead of the local composition c^{Mn}, thus selecting the tie-line and driving-force of the respective model:

$$\text{Para-equilibrium:} \quad c^{Mn,*} = c^{Mn} + \tfrac{1}{2}\phi_\alpha k_{\alpha\beta}^{Mn}\left(c^{Mn} - c_\beta^{Mn}\right) + \tfrac{1}{2}\phi_\beta k_{\beta\alpha}^{Mn}\left(c^{Mn} - c_\alpha^{Mn}\right) \tag{8}$$

$$\text{NPLE } (\alpha\rightarrow\beta): \quad c^{Mn,*} = \phi_\alpha c^{Mn} + \phi_\beta \left(c_\beta^{Mn} + k_{\beta\alpha}^{Mn}\left(c^{Mn} - c_\alpha^{Mn}\right)\right) \tag{9}$$

Here, c_α^{Mn}, c_β^{Mn} and $k_{\alpha\beta}^{Mn}$ are the uncorrected phase compositions and partition coefficients, respectively. Notably, the NPLE sub-model is asymmetric and thus depends on the moving direction of the interface. This latter can either be determined directly, by evaluation of the interface velocity, or indirectly by the rate of temperature. In this study, the compatibility between the diffuse mixture ansatz and both sub-models shall be tested on the example of a ternary Fe-C-Mn alloy.

5. Simulation scenario and input data

A series of 2D test simulations is performed to validate the applicability of the diffuse mixture ansatz to predict pearlite growth in iron-carbon alloys. All simulations are using identical input parameters, except for changing initial compositions: a) Fe-0.6 C, b) Fe-0.8 C, c) Fe-1.0 C and d) Fe-0.6 C-0.2 Mn (all in weight %). The essential computational advantage of the diffuse mixture ansatz is that only the outer interfaces of the pearlite nodules have to be numerically resolved, which allows for a relative large grid spacing $\Delta x = 0.25$ µm and a diffuse interface width of $\eta = 4\Delta x$. The side-length of the quadratic simulation domain with periodic boundary conditions is 100 µm. All phase-specific thermodynamic data is evaluated from the databases TCFe11 [25] and the diffusion data from MobFe6 [26].

IOP Conf. Series: Materials Science and Engineering 1281 (2023) 012051 doi:10.1088/1757-899X/1281/1/012051

Grain boundary energies are set to $\sigma_{aa} = \sigma_{ff} = \sigma_{cc} = 0.7\ Jm^{-2}$, and phase interface energies to $\sigma_{af} = 0.6\ Jm^{-2}$, $\sigma_{ac} = 0.7\ Jm^{-2}$ and $\sigma_{fc} = 0.5\ Jm^{-2}$, with indices a, f and c corresponding to austenite, ferrite and cementite. Grain boundary mobilities are defined by $M_{aa} = M_{ff} = 10^{-17}\ m^4J^{-1}s^{-1}$ and $M_{cc} = 0$. Phase interface mobilities are set to $M_{af} = 10^{-12}\ m^4J^{-1}s^{-1}$, $M_{ac} = 7\cdot10^{-12}\ m^4J^{-1}s^{-1}$ and $M_{fc} = 10^{-14}\ m^4J^{-1}s^{-1}$, these are automatically corrected in the thin-interface limit [21]. The mobility of austenite/ferrite interfaces M_{af} is defined with a prefactor of $f=5\cdot10^{-4}$ in case of special orientation relationship (OR). The interface properties σ_{ac} and M_{ac} are defined with facetted anisotropy, among other reasons to test the compatibility of this feature. The characteristic spacing of pearlite is assumed to be $\lambda_{fc} = 1\mu m$. It is emphasized that all input values are more or less arbitrarily chosen without any claim of correctness.

The simulations start from an initial polycrystalline austenite grain structure with random orientation at an initial temperature of T = 1020 K, which is homogeneously reduced by a constant cooling rate of 5 K/s. The critical undercooling for nucleation of ferrite is arbitrarily set to $\Delta T_{f\rightarrow aa} = 5$ K on grain boundaries, and to $T_{f\rightarrow ac} = 10$ K for nucleation on cementite. Following a classic hypothesis [27], orientations are modelled in the first case with OR to a parent austenite grain, but in the latter with a different OR to a parent cementite grain. The specific OR are here simply mimicked by arbitrary 2D angles, but shall in future simulations be replaced by realistic 3D formulations. For sake of simplicity and to allow for a wide spectrum of orientations in the relative small simulation domain, cementite nucleation is modelled with identical critical undercooling $\Delta T_{c\rightarrow aa} = 5$ K and $T_{c\rightarrow af} = 10$ K and random orientation.

6. Computation and visualization of simulation results

Based on the input parameter given in section 5, test simulations are performed for binary Fe-C with three different concentrations: a) 0.6 wt%, b) 0.8 wt%, and c) 1.0 wt% and additionally for ternary Fe-0.6 C-0.2Mn under assumption of two different sub-models for manganese partitioning: a) para-equilibrium and b) NPLE [24]. The simulations, performed with the Micress® beta-version, take about two hours computation time on 4 cores of a workstation. In figures 1-4, selected simulation results are visualized to illustrate the evolution of the microstructure at five different stages during pro-eutectoid and eutectoid transformation. A basic feature of the diffuse mixture approach is that two phases numerically coexist in the same position, which requires some adjustment to the graphical visualization. The decision, which phase-specific data to display in the diffuse two-phase regions differs in the individual output fields. The phase-state is depicted in figure 1a-4a by a fine checkerboard pattern with the numbers of the two coexisting phases alternating in each pixel. As the individual colours are be too fine to be distinguished, the diffuse pearlite nodules appear as grey regions. In contrast, the local fractions of the coexisting phases are displayed separately in figures 1b-3b, and figures 1c-3c, respectively. The carbon concentration shown in figures 1d-3d are mixture entities referring to the carbon content of the multiphase mixture. The crystallographic orientation is visualised in the same way as the phase-state by a checkerboard pattern to distinguish changes in both ferrite and cementite orientation. This is important to identify the complex pearlitic polycrystalline substructure, consisting of blocks (areas of identical ferrite orientation), and colonies (areas of identical cementite and ferrite orientation), which significantly control the mechanical properties [28, 29]. Note that all phase-specific data of eutectoid ferrite and cementite remains individually accessible for statistic evaluation or transfer to subsequent models.

7. Discussion of simulation results for binary Fe-C

7.1. Discussion of simulation results for hypo-eutectoid Fe-0.6C

The initial alloy composition in the first simulation (figure 1) is hypo-eutectoid and consequently, ferrite becomes stable first and nucleates at the austenite grain boundaries. In agreement with a classic hypothesis [27], the pro-eutectoid ferrite nucleates with a special orientation relationship (OR) to a parent austenite grain. The interface to the parent is here modelled with low mobility, such that the ferrite only grows into the adjacent grain to which it has no OR. When cementite becomes stable, it also nucleates on the grain austenite boundaries, but close to ferrite. Getting in contact, both phases grow together in a cooperative mode, forming pearlite.

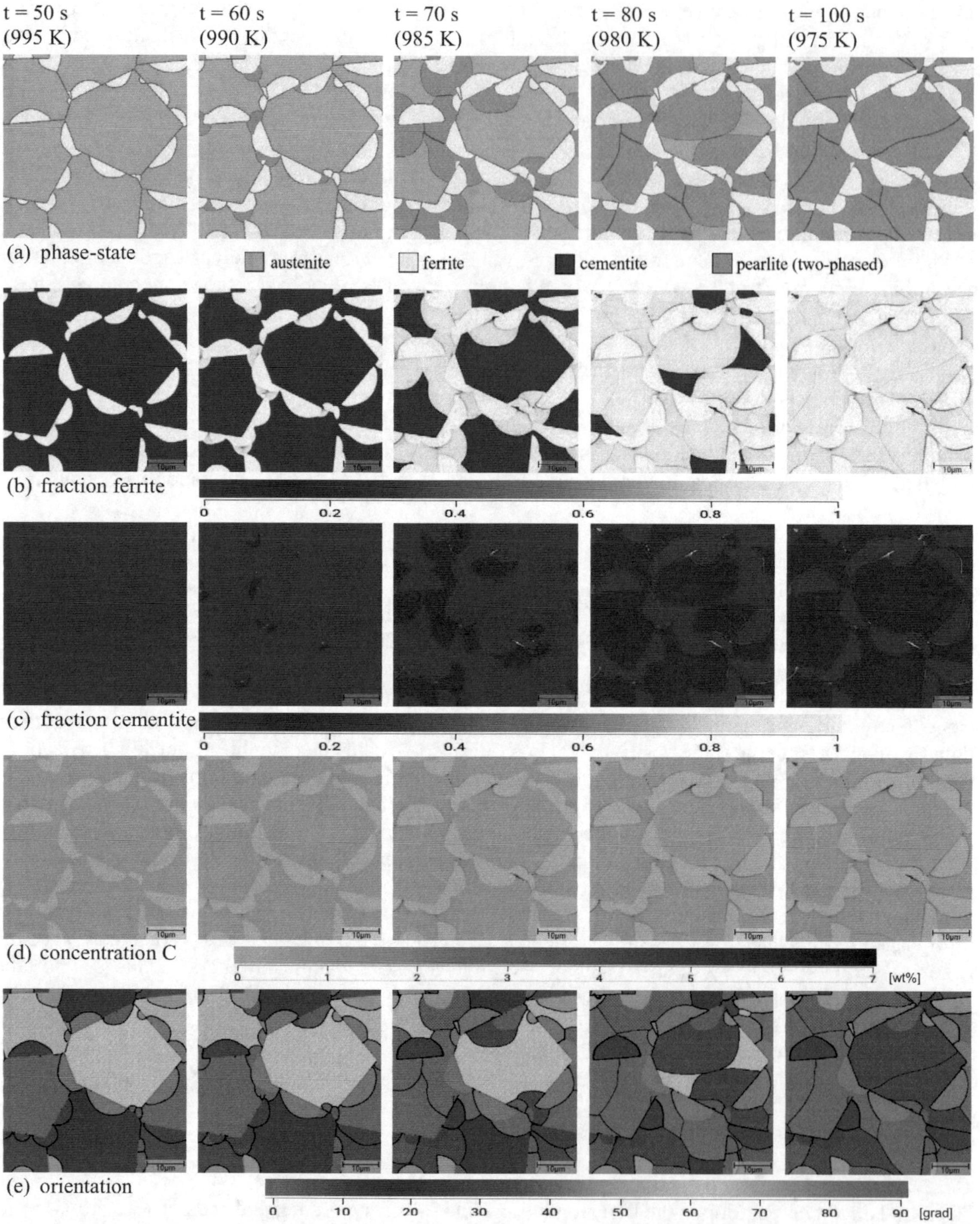

Figure 1. Simulation results for hypo-eutectoid Fe-0.6 C showing the phase-state (a), the fraction of austenite (b) and cementite (c) the carbon concentration (d) and the grain orientation (e). Ferrite forms first as pro-eutectoid phase. After nucleation of cementite, both phases proceed in a cooperative mode forming pearlite, which is here successfully modelled by the diffuse mixture ansatz.

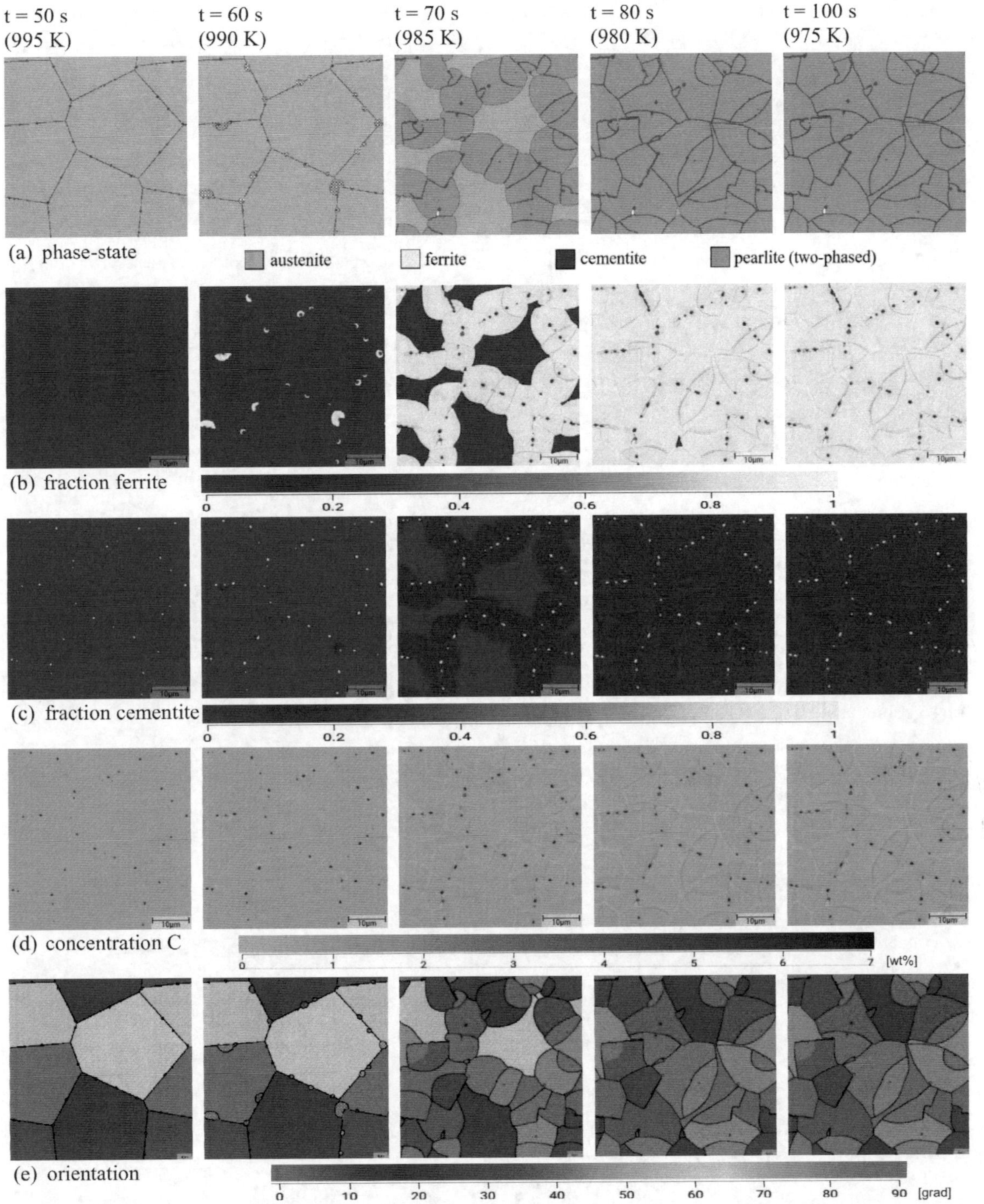

Figure 2. Simulation results for near-eutectoid Fe-0.8 C showing the phase-state (a), the fraction of austenite (b) and cementite (c) the carbon concentration (d) and the grain orientation (e). The initial carbon concentration is just slightly above the eutectoid composition (0.76 wt %), such that nucleation of cementite is directly followed by ferrite and almost the whole austenite decomposes to pearlite.

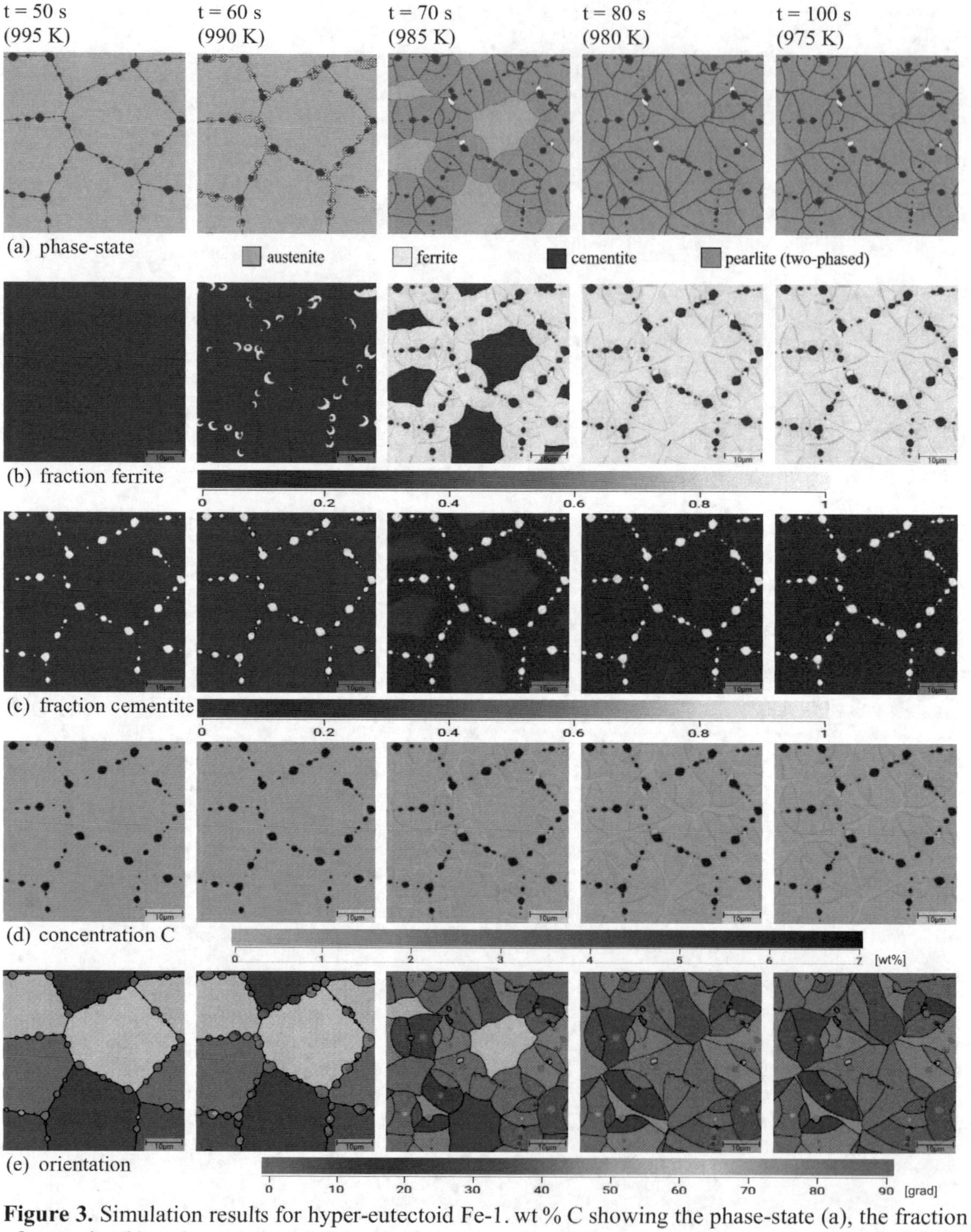

Figure 3. Simulation results for hyper-eutectoid Fe-1. wt % C showing the phase-state (a), the fraction of austenite (b) and cementite (c) the carbon concentration (d) and the grain orientation (e). Cementite forms first as pro-eutectoid phase, covering the grain boundaries. After nucleation of ferrite, both phases grow in a cooperative mode inwards the bulk of the grains.

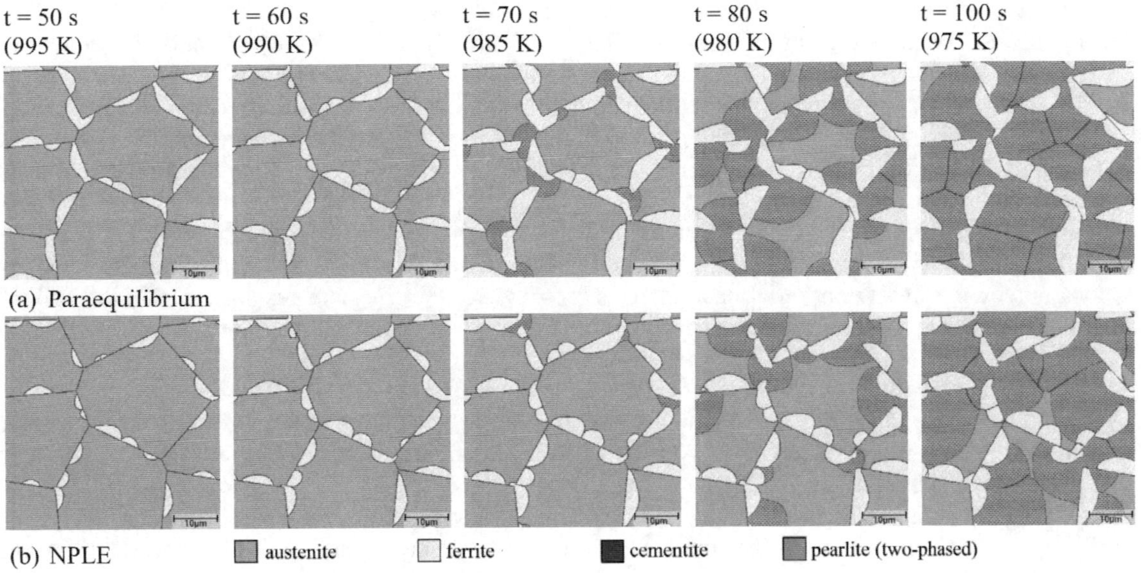

Figure 4. Simulation results for hypo-eutectoid Fe-0.6 C-0.2 Mn in comparison under assumption of para-equilibrium (a) and negligible-partitioning-local-equilibrium (b).

The results demonstrate that nucleation and growth of pro-eutectoid ferrite, as well as its transition to eutectoid ferrite are consistently simulated according to the defined conditions and specified OR. Also, nucleation and growth of eutectoid cementite is numerically stable and no unphysical instantaneous wetting of the pro-eutectoid ferrite occurs. This artefact has been observed in early simulations, but is now avoided by the weighting factor w_{af} in equation (6). Pearlite nodules are successfully described as diffuse mixture of ferrite and cementite with individual fractions, and orientations (figure 1b,c, d), but defined outer interfaces to austenite. Note that a synchronized motion of both eutectoid phases cannot be taken for granted, but requires stable eutectoid conditions and a sufficiently high mobility of the austenite cementite interface. If it is too low, cementite lags behind and becomes overgrown by ferrite.

7.2. Discussion of simulation results for near-eutectoid Fe-0.8C

The initial concentration of carbon in the second test simulation (figure 2) is only slightly above the eutectoid equilibrium concentration c_{eq} = 0.76 wt% [20], such that nucleation of cementite is directly followed by ferrite and almost the whole austenite decomposes into pearlite. Two different morphologies of pearlite nodules are observed: the first crossing the grain boundaries and the second only evolving on one side of the boundary. This behaviour is attributed to the different nucleation modes for ferrite, which can either nucleate as pro-eutectoid ferrite with OR to austenite, or directly as eutectoid ferrite with OR to a parent cementite grain. A specific challenge of polycrystalline transformation is to consistently describe the impingement of pearlite nodules and prevent unphysical penetration of cementite in neighbouring nodules. For this reason, each cementite grain is numerically linked to the first ferrite grain in gets in contact with and its interfacial mobility to all other ferrite grains is strongly reduced. Ferrite grains, in contrast, may interact and coexist with multiple cementite grains. Figure 2e demonstrates that a hierarchically substructure is obtained as known from experimental observations [28]. The areas surrounded by black lines mark zones with the same orientation of both cementite and ferrite, commonly called *pearlite colonies*. Several colonies form larger *pearlite blocks,* i.e. zones with same ferrite orientation [28, 29].

7.3. Discussion of simulation results for hyper eutectoid Fe-1.0C

The initial alloy composition in the third simulation (figure 3) is hyper-eutectoid and consequently, cementite becomes stable first and nucleates as pro-eutectoid phase, covering major parts of the grain boundaries. Pro-eutectoid growth of cementite is modelled in this study with facetted anisotropy, mainly to validate the numerical stability of this feature in context with the diffuse mixture ansatz. The results confirm that nucleation and growth of eutectoid cementite and its transition to eutectoid cementite is well described according to the defined conditions. During pro-eutectoid cementite growth, the surrounding austenite becomes strongly depleted in carbon. Here, ferrite nucleates either directly on the cementite interface or independently on the austenite boundaries. In a few spots with strongly increased carbon content, it grows temporarily uncoupled as pro-eutectoid ferrite. The austenite within the bulk of the grains completely decomposes into pearlite with a fine polycrystalline substructure.

7.4. Discussion of simulation results for hyper eutectoid ternary Fe-0.6C-0.2Mn

In order to test the applicability of the diffuse mixture ansatz to multicomponent iron-carbon alloys, simulations are performed for ternary Fe-0.6 C-0.2 Mn. Manganese is selected as a typical example of a slow-diffusing substitutional component, which does not -or only incompletely- partition at the moving phase interfaces. Figure 4 shows the results of two simulations performed using a) the Para-equilibrium model and b) the NPLE model (section 4). Comparing the results for para-equilibrium (figure 4a) with those of binary Fe-0.6 C (figure 1a), it is found that the eutectoid transition occurs retarded and is shifted to lower temperatures by about 4 K. Under assumption of NPLE, the shift is even further increased to approx. 8 K, which can be explained by the additional contribution of manganese to the driving force [24]. The simulations confirm that the multicomponent extension of the diffuse mixture ansatz is straightforward without any need for changes in the Micress® framework. In the para-equilibrium simulation, it occasionally occurs that thermodynamic calculations temporarily do not properly converge, which still needs further investigation. A critical point in applying the NPLE sub-model is, that it depends on the moving direction of the interface. This can either be determined directly by the interface velocity, or indirectly by the rate of temperature. At present, only the second option is compatible with the diffuse mixture ansatz.

8. Conclusions

A diffuse mixture ansatz has been developed which can be used in the framework of the multi-phase-field method for scale-bridging simulations of pearlite growth on the length scale of the austenite grain structure. The specific feature is that the internal fine-lamellar structure of the pearlite nodules is not resolved, but modelled as diffuse mixture of coexisting ferrite and cementite still having individual properties. Implemented in the Micress® software, the approach can be linked to Calphad databases and existing sub-models with respect to nucleation, solute redistribution and crystallography. The test simulations presented in this paper demonstrate the capability of the ansatz to consistently describe nucleation, growth and polycrystalline interaction of pro-eutectoid grains and pearlite nodules. Further tests are planned to validate the numerical stability and performance in 3D, the applicability to cast irons, as well as the compatibility with the elastic solver. A major motivation for the model development is to simulate pearlitic microstructures of technical cast iron and steel grades in a representative volume element in order to derive mechanical properties.

Acknowledgements

The authors acknowledge funding by the Deutsche Forschungsgemeinschaft (DFG, German Research Foundation) within the project "*Experimentell gestützte Modellierung der Korrelation zwischen metallurgischer Prozessführung, 3D-Gefügeentwicklung und mechanischen Eigenschaften von perlitischem Gusseisen mit Kugelgraphit*" under grant n° 504974025.

References

[1] Eiken J, Böttger B and Steinbach I 2006 *Phys. Rev. E* **73(6)** 066122.
[2] Eiken J 2010 *A Phase-Field Model for Technical Alloy Solidification*, Shaker Verlag ISBN 978-3-8322-9010-8.
[3] MICRESS® documentation *https://docs.micress.rwth-aachen.de.*
[4] Militzer M 2011 *Curr. Opin. Solid State Mater. Sci.* **15** 106–115.
[5] Militzer M, Mecozzi MG, Sietsma J and Van der Zwaag S 2006 *Acta Mater.* ***54***(15) 3961-3972.
[6] Mecozzi MG, Eiken J, Santofimia MJ, Sietsma J 2016 *Comput. Mater. Sci.* **112** 245-256.
[7] Böttger B, Apel M, Eiken J, Schaffnit P and Steinbach I 2008 *Steel Res. Int.* **79**(8) 608-616.
[8] Slater C, Bandi B, Dastur P and Davis C 2022 *Met. Mater. Trans. A* **53** 2286–2299.
[9] Kwok TWJ, Slater C, Xu X, Davis C and Dye D 2022 *Met. Mater. Trans. A* **53**(2) 585-596.
[10] Shen Y, Liu J and Xu H 2022 *Met. Mater. Trans B*, 1-15.
[11] Köhnen P, Ewald S, Schleifenbaum JH, Belyakov A and Haase C 2020 *Addit. Manuf.* **35** 101389.
[12] Mueller JJ, Matlock DK, Speer JG and De Moor E 2019 *Metals* **9**(9) 926.
[13] Ma Y, Zheng R, Gao Z, Krupp U, Luo HW, Song W and Bleck W 2021 *Int. J. Min. Met. Mater.* **28**(5) 847-853.
[14] Papadioti I, Bellas I,Tzini MIT, Christodoulou PI and Aravas N 2020 *Materials*, **13**(2) 458.
[15] Nakajima K, Apel M, Steinbach I 2006 *Acta Mater* **54** 3665–3672
[16] Li S, Yang ZN, Enomoto M and Yang ZG 2019 *Acta Mater.* **177** 198-208.
[17] Eiken J, Subasic E, Lacaze J. 2020 *Materialia* **9** 100538.
[18] Eiken J, Böttger B 2018. *Trans. Indian. Ins. Met.* **71**(11) 2725-2729.
[19] Eiken J 2020 *IOP Conf. Ser.: Mater. Sci. Eng.* **861** 012055.
[20] Thermo-Calc software, *www.thermocalc.com.*
[21] Carré A, Böttger B, Apel M 2013 *J of Crystal Growth* **380** 5-13.
[22] B Zhou, Laschet G . Eiken J, Behnken H and Apel M 2020 *IOP Conf. Ser.: Mater. Sci. Eng.* **861** 012034.
[23] Jakumeit J, Behnken H, Laqua R, Eiken J and Brachmann, J 2020 *IOP Conf. Ser.: Mater. Sci. Eng.* **861** 012035.
[24] Vitek JM, Babu SS. and Kozeschnik E 2003 *Mater. Sci. Technol.* 139-148.
[25] TCFe11, TCS Steels/Fe-Alloys Database version 11.0.
[26] MobFe6, TCS Steels/Fe-Alloys Mobility Database, version 6.0.
[27] Dippenaar J and *Honeycombe WR* 1973 *Proc. Roy. Soc.* **A 333** 455-467.
[28] Takahashi T, Nagumo M, Asano Y 1978 *Trans. Jpn. Inst. Met.* **42** 716-723.
[29] Zhang F, Zhao Y,Tan Y, Ji X, Xiang S, 2019 *Metals* **9**(11) 1133.

IOP Conf. Series: Materials Science and Engineering 1281 (2023) 012052 doi:10.1088/1757-899X/1281/1/012052

Modelling of major phases formation during solidification of Ferro-Silicon-Magnesium

S Gouttebroze[1], A Marthinsen[1], B Kroka[2], A Götz[2], E Ott[2]

[1] SINTEF Industry, Forskningsveien 1, 0373 Oslo, Norway
[2] Elkem Silicon Product Development, Fiskåveien 100, 4621 Kristiansand S, Norway

sylvain.gouttebroze@sintef.no

Abstract. Ductile cast iron, also known as nodular cast iron, is a graphite-rich cast iron with high impact and fatigue resistance, due to its nodular graphite inclusions. Ductile cast iron is produced by incorporating additives (often FeSi alloys) to the iron base metal at different production steps to obtain the desired graphite shape. A crucial step is the addition of Magnesium to promote the spheroidization of the graphite. The most common method is by adding crushed and sized Ferro-Silicon-Magnesium (FSM). The alloy composition, microstructure, and sizing are assumed to affect the key parameters of this reaction, namely, reactivity, recovery, and slag formation. Therefore, the study of the solidification of FSM is important to understand and predict its performance at the foundries. The present work aims at understanding and predicting numerically the formation of the major phases during the solidification process. Two approaches have been used: thermodynamic calculations through Thermo-Calc solver and phase field modelling using MICRESS. The models have been calibrated by comparison with advanced statistical characterization of the microstructure. The results indicate a competitive growth of the major phases and transformation of phases in solid state that can be emulated by the model.

1. Introduction

Ferrosilicon magnesium (FSM) alloys are widely used in cast iron industry as an additive in production of ductile cast iron to promote the formation of spheroidal graphite nodules. The presence of spheroidal graphite results in a material with both good ductility and high strength. [1]

Though FSM alloys are used abundantly in the production of ductile cast iron and their composition is assumed to impact important parameters of the melt treatment (such as reactivity, recovery and slag formation), there is scarce knowledge about how variation in the composition, microstructure and subsequent sizing of the FSM alloys affect their performance in the ductile iron casting process. To bridge this knowledge gap, it is important to understand the solidification process of the FSM alloys in order to tailor their microstructure and resulting properties.

FSM alloys cover a wide range of compositions, the main elements being Fe, Si and Mg. The amount of Si typically varies between 40-50 wt%, whereas Mg is in the range 5-10 wt%. In addition, the FSM alloys often contain small amounts of additional elements, such as Ca, Al, Ti, and Rare Earth Elements (like La and Ce). This large compositional space implies that different phases are present for different alloys, in different amounts and with different morphologies. Processing conditions (in particular cooling rate) can also affect the microstructure. The main phases, which are present in nearly every FSM alloy, include FeSi, $FeSi_2$, Mg_2Si and pure Si. The $FeSi_2$ phase has a high temperature form (α-$FeSi_2$) which has tetragonal crystal structure and a low temperature form (β-$FeSi_2$) which has an orthorhombic crystal structure. The high temperature α-$FeSi_2$, is also rich in Fe vacancies, making the stoichiometry closer to $FeSi_{2.33}$, and later denoted as Fe_3Si_7.

In this work, two different modelling approaches have been used to predict and understand the formation of the main phases in FSM alloys. A python-based thermodynamic solidification model was developed to predict the solidification path and resulting phase fractions by relying on Thermo-Calc software [2], and a phase field software, Micress [3], was used to build a more comprehensive understanding about the solidification process and the effects of phases morphology. Two different FSM alloys have been considered and the model predictions compared with experimental data.

2. Methodology

2.1. Experimental work and microstructure characterization

The FSM alloys were produced at the Elkem Pilot Station using the Casting Simulator setup. A charge mix of 50 kg consisting of ferrosilicon with some alloying elements was molten in a 250 kW induction furnace. The melt was cast via a graphite runner onto a traversing water-cooled copper plate which was divided into three sections allowing to cast variable thicknesses in a single operation. Target casting thicknesses were 15 and 30 mm. Vertical samples were taken through the cast layer thickness while avoiding non-representative areas with oxides or slag inclusions (see Figure 1).

Figure 1. As-cast FSM material (left) and sample embedded in epoxy for SEM characterization (right)

The analysis of the material focused mostly on chemical composition and microstructure. The chemical composition what obtained by XRF analysis. The microstructure characterization was performed using the Advanced Mineral Identification and Characterization System (AMICS) [4], which is a new method that merges scanning electron microscopy (SEM), imaging with X-ray spectroscopy (EDS), and phase identification. It allows automated and very fast mapping of phases covering large areas (up to 30 mm^2 in this study). The core feature of AMICS is the use of image analysis of the backscatter electron images (BSE) to segment regions of the microstructure to limit the number of necessary X-ray measurements to map the individual phase areas. This is combined with fast search/match through a database of X-ray spectra specific to each material. This database of X-ray spectra has been constructed using samples from both the production plants and the pilot casting simulator to cover a larger range of cooling rate and chemical composition. AMICS provides (x, y) coordinates of electron signals, EDS spectra and phase ID's and merges these into regions where size and neighborhoods are quantified.

Using AMICS, FSM microstructures are evaluated and quantified on plane sections through the entire cast thickness with a spatial resolution down to 1µm. Very small grains could be missed or considered part of another phase but the accuracy for the large phases is very good. In addition, the operator needs to carefully evaluate possible overlap in BSE contrast between phases and adjust accordingly the segmentation parameters. AMICS overviews are also used to locate interesting features or phases configurations for detailed EDS analysis of phase compositions down to trace elements level. The resulting data are used to provide input for microstructure model calibration and to correlate process data with microstructure evolution.

IOP Conf. Series: Materials Science and Engineering 1281 (2023) 012052 doi:10.1088/1757-899X/1281/1/012052

2.2. Thermodynamic modelling

Thermodynamic simulations were performed using the Calphad Software Thermo-Calc [2]. An in-house database was used in the simulations, as none of the commercial databases available at SINTEF could properly handle solidification of the FSM alloys, given the very different Fe and Si compositions compared with steel (focus of TCFE commercial database). The in-house database was built from thermodynamic literature data. It includes all the binary systems, and also some ternary systems assessments: Fe-Si-Ca, Fe-Si-Mg, Fe-Si-Al, Mg-Si-Al, Mg-Si-Ca. A python-based modelling framework was built and coupled to Thermo-Calc through the python module TCPython, to facilitate high-throughput calculations, storage, and analysis of output data. As part of the modelling framework, a solidification model was developed to extend the capabilities by including solid-solid reactions, and accounting for the thermal history of the solidified microstructure.

The solid-solid reactions module implemented in the python modelling framework takes as input the temperature history of the solidified microstructure, i.e., a list of times and temperatures as well as an initial state of the system including phases fraction and compositions. For each temperature step, the model runs an equilibrium calculation through the Thermo-Calc solver for the composition of the liquid part of the system as in a Scheil calculation. If no solid-solid phase transformation is activated, the resulting solidification path is exactly equal to the Scheil calculation integrated in Thermo-Calc. Otherwise, between time step *t* and *t+dt*, the user-defined solid-solid reactions are computed, and the solid domain is updated.

The kinetics of solid-solid phase transformations are modelled using the Johnson-Mehl-Avrami-Kolmogorov (JMAK) equation, where the ratio between the fraction of a given phase θ at time *t* and the equilibrium phase fraction at a given temperature *T* is given by:

$$\frac{f_\theta(t)}{f_\theta(eq)} = 1 - \exp\left(-K(T)t^n\right), \text{ with } K(T) = K_0 \exp\left(-\frac{Q_r}{RT}\right) \tag{1}$$

where K is the reaction rate constant, K_0 is the frequency factor, Qr is the activation energy, and n is the Avrami exponent, where $n=4$ under the assumption of nucleation and growth in three dimensions. The cooling curve is discretized in iso-thermal steps and the additivity ule is applied.

2.3. Phase field modelling

To study the effect of phase morphology on the formation of the two major phases in FSM, FeSi and Fe_3Si_7, and to get a more comprehensive picture of the solidification process, a phase field model was developed using the software MICRESS [3]. FeSi and Fe_3Si_7 formation determines the main features of the solidified microstructure as they are often the first to nucleate and represent a large volume fraction (>80%). In contrast to thermodynamic simulations, phase field models allow to study the effect of local phenomena on the solidification process, including local composition gradients, elemental diffusion, and nucleation and growth of phases.

The phase field model in Micress was coupled to the same thermodynamic database used for the thermodynamic simulations with Thermo-Calc. A significant number of input parameters were needed for the phase-field model to describe the different phases and their interactions with neighboring phases and the liquid phase. Extensive parameter testing was therefore done to optimize the set of parameters to fit experimental observations of the FSM morphology. The simulation domains were limited to 2D rectangular domains due to the much higher computational cost of running 3D simulations, with typical size in the range of 200-300 μm and grid size of 1-2 μm.

3. Results

3.1. Experimental measurements and observations

In this study, two FSM alloys were considered and compared. Their compositions mainly differ in the Fe/Si concentration, where Alloy 2 is richer in Si than Alloy 1. The concentrations of the main elements in the two alloys (Fe, Si and Mg) are listed in Table 1.

Table 1: Concentration in wt% of main elements (Fe, Si, Mg) in Alloy 1 and Alloy 2

Alloy name	Fe	Si	Mg	Other
Alloy 1 (LowSi)	47.91	43.49	5.81	2.79
Alloy 2 (HighSi)	44.57	47.16	5.45	2.82

The cross-sections through the thickness of the samples were analyzed with AMICS. The phases spatial distribution for Alloy 1 and Alloy 2 are shown in Figure 2 at three locations: one from the top of the cross section, one from the middle and one from the bottom, close to the copper plate on which the alloys were cast. The bottom of the cast has solidified at a higher cooling rate than the middle and top segments, which results in a finer grain structure. The composition change yields a visually significant variation in the phase fractions and morphology. The lower Si content of Alloy 1 results in a larger fraction of FeSi (yellow) compared to $FeSi_2$ (green) as expected from the liquidus projection in Figure 3 that shows to increasing proximity of the boundary to Fe_3Si_7 domain. The variation between alloys is much larger than through the thickness of the samples. This indicates that this material does not undergo significant macrosegregation at these cooling rates.

Furthermore, the morphology of the FeSi phase, usually dendritic, is somewhat more globular in Alloy 1 than in Alloy 2. The $FeSi_2$ phase on the other hand, forms elongated rod-like grains. In both alloys, the FeSi phase is largely encapsulated by the FeSi2 phase. In some areas, the FeSi phase is seen in the middle of the bulk of $FeSi_2$, whereas in other areas, only a thin layer of $FeSi_2$ encapsules the FeSi. Only the low temperature form of $FeSi_2$ is observed.

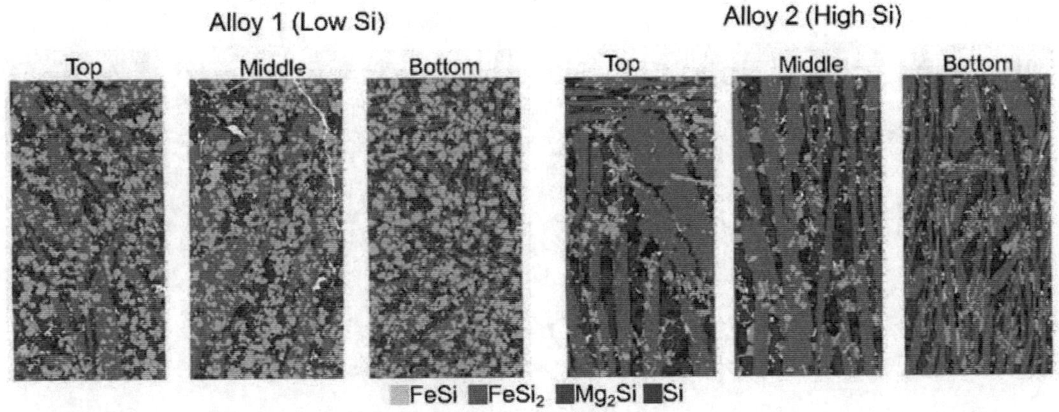

Figure 2 Experimental phases maps generated by AMICS of Alloy 1 (left) and Alloy 2 (right). The domain width is approximately 800 μm.

These features are even clearer in the zoomed images presented in Figure 4. FeSi could form large dendrites (left image) that will disturb the growth of $FeSi_2$ phase. In the central image, all the small remaining FeSi areas are coated with $FeSi_2$, present as large bulky rods and smaller excrescence near FeSi. On the right picture, the $FeSi_2$ rods are growing independently following the complex temperature gradient present at the surface (the radiation cooling combined with the possible presence of oxide layer could leads to non-uniform, high cooling rates).

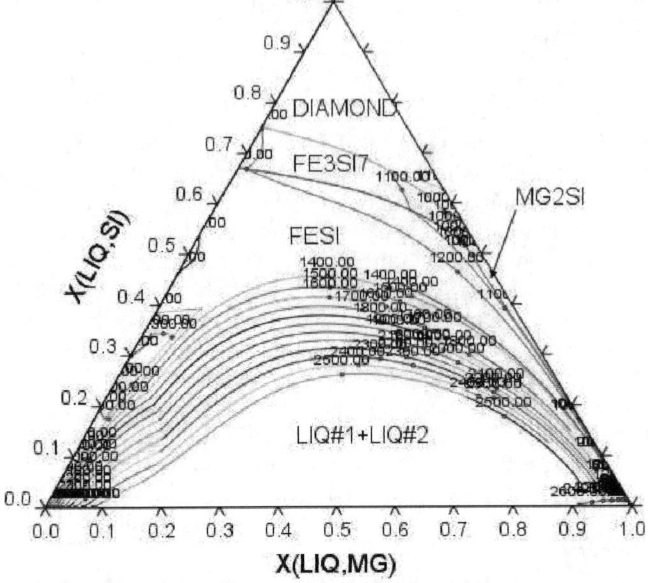

Figure 3. Liquidus projection computed from SINTEF database using Thermo-Calc

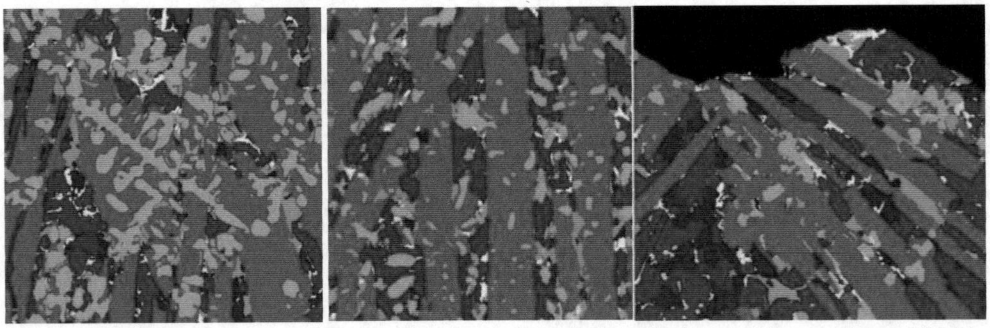
Figure 4. Main features observed in the microstructure taken from other samples (Composition very close to Alloy 1 but produced in the plant)

3.2. Thermodynamic modelling results

Since the experimental samples have solidified under normal casting conditions, it is expected that the resulting phase fractions lie somewhere between the limiting cases of equilibrium and the Scheil approximations. Equilibrium solidification assumes infinite diffusivity of elements in the solid region, whereas Scheil solidification assumes zero diffusion in the solid region. Infinite diffusivity in the liquid region is assumed in both cases. Comparisons of the predicted phase fractions of these two limiting cases with experimental phase fractions quantified with AMICS are shown in Figure 5.

As expected, neither of the two provides an accurate estimate. Under equilibrium conditions, the amount of $FeSi_2$ is overestimated. Furthermore, a small fraction of pure Si phase, which is found experimentally for both samples (Alloy 1 and Alloy 2), is not present in equilibrium solidification simulations. On the other hand, in the Scheil approximation, too much of the FeSi phase is present at the end of the solidification. Additionally, only a very small fraction of the low-temperature β-$FeSi_2$ phase is formed at the end of solidification as no transformation of the high temperature α-$FeSi_2$ phase (present in the thermodynamic database as Fe_3Si_7) could occur (it would require diffusion in the solid phase because of the different stoichiometry).

Simulated evolution of the phase fractions with decreasing temperature under equilibrium and Scheil conditions are shown in Figure 5 (b-c) for Alloy 1 and Figure 5 (e-f) for Alloy 2. The FeSi phase forms first at approximately 1500 K. At ~1400 K, the high-temperature Fe_3Si_7 phase starts to form. Under equilibrium conditions, the formation of Fe_3Si_7 consumes some of the FeSi during formation, in a peritectic reaction with the liquid. At 1250 K, the low temperature $FeSi_2$ phase forms, which fully transforms Fe_3Si_7 into its low temperature form by partly consuming FeSi. Finally, at 1220K the Mg_2Si phase forms and the system reaches full solidification (neglecting the other elements). Within the Scheil approximation on the other hand, no solid-solid transformation can take place. This prevents solid transformation of FeSi as seen during equilibrium solidification, which increases the Si concentration in the liquid, resulting in the formation of a fraction of pure Si phase at the end of the solidification, however, overestimated compared to experimental values.

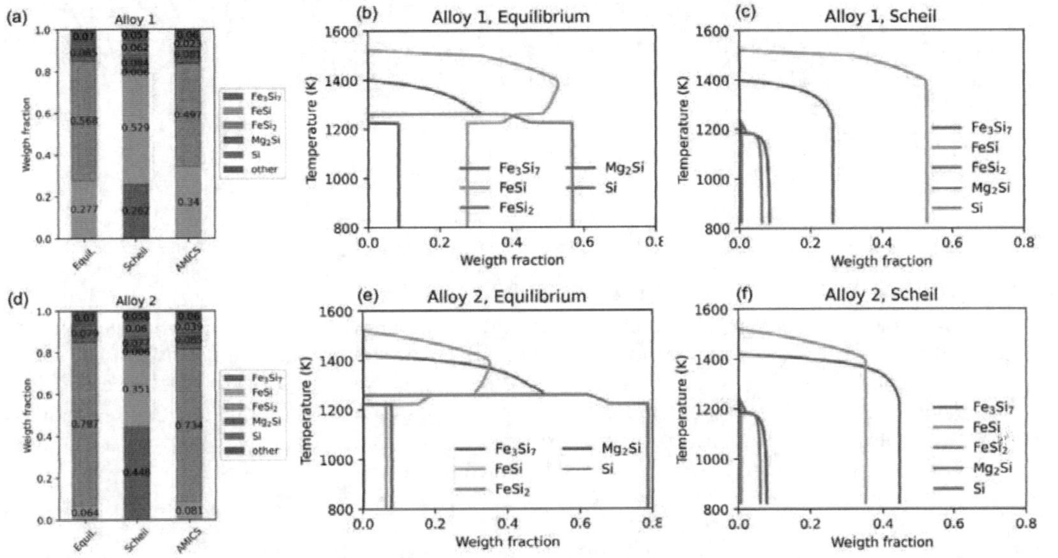

Figure 5 Phase fractions of main phases formed in Alloy 1 and Alloy 2 during solidification. (a/d) Comparison of resulting phase fractions at end of solidification between simulations under Equilibrium and Scheil approximations, and experimental phase fractions from AMICS. (b/e) Evolution of phases in Alloy 1/Alloy 2 with decreasing temperature under equilibrium conditions. (c/f) Evolution of phases in Alloy 1/Alloy 2 with decreasing temperature under Scheil conditions.

To fit the experimental measurements more accurately, a model intermediate to Scheil and Equilibrium is needed. Experimental observations shown in Figure 4 suggest the transformation of FeSi to $FeSi_2$. Indeed, part of the FeSi2 phase is not present as big rods but has the same morphology and seems to replace the FeSi phase. Thus, we need to include solid-solid reactions to obtain an improved model for the FSM alloys. Initially, one solid-solid reaction was implemented yielding $FeSi_2$ as a product, according to:

$$Fe_3Si_7 + FeSi \rightarrow 4\ FeSi_2 \qquad \text{(Reaction 1)}$$

This reaction aligns with the equilibrium solidification simulation at ~1250 K, where Fe_3Si_7 is fully consumed to produce $FeSi_2$. As accurate experimental data for the thermal history of the alloys was not available, a constant cooling rate of 2 K/s was assumed in the model calibration. This assumption does not affect the methodology and trends presented in the following paragraphs, only the calibration of the model parameters. As an initial approach the model was calibrated with Alloy 1, by varying the reaction rate constant *K(T)* of Reaction 1, which enters the JMAK equation (Equation 1). The resulting phase fractions were compared with AMICS data of Alloy 1. The rate constant is dependent on the frequency factor *K0* and the activation energy *Qr*. To simplify calibration and in the absence of multiple cooling rates, *Qr* was fixed and K_0 was varied from 10^{-9} to 10^{-7} as seen in Figure 6(a). A larger value of K_0 corresponds to a higher reaction rate.

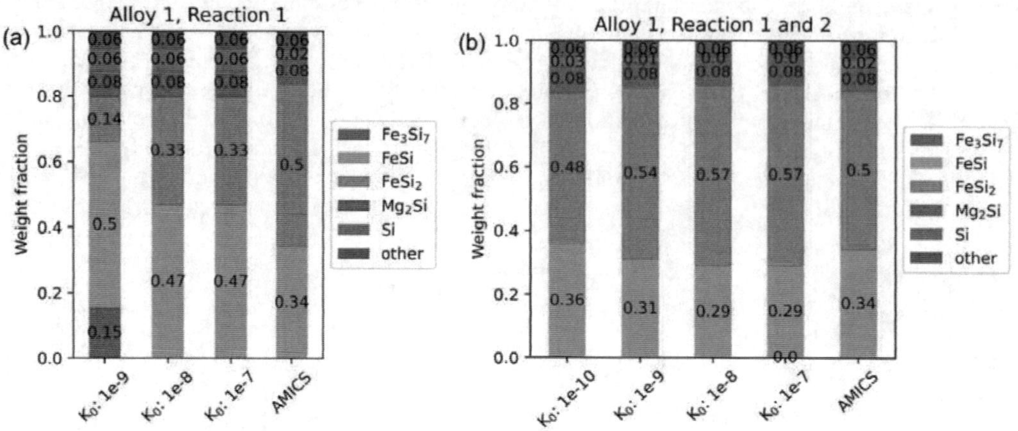

Figure 6 Comparison of calculated phase fractions in Alloy 1, by variation of the frequency factor K0 with experimental phase fractions from AMICS (a) when including Reaction 1, (b) when including both Reaction 1 and Reaction 2.

The experimental phase fractions from AMICS are shown to the right in Figure 6(a). A K_0 value of 10^{-9} results in a too low reaction rate to consume all the high temperature Fe_3Si_7. By increasing the K_0 value to 10^{-8}, this results in full consumption of the Fe_3Si_7 phase. However, continuing to increase K_0 beyond this value has no effect, since all Fe_3Si_7 is consumed. It is evident that this reaction alone does not accurately predict the experimental phase fractions, suggesting that additional reactions should be included. Including only Reaction 1, the amount of Si is overestimated whereas the amount of $FeSi_2$ is underestimated. A second reaction was therefore included in the model, according to:

$$FeSi + Si \rightarrow FeSi_2 \qquad \text{(Reaction 2)}$$

Where FeSi and Si are consumed to yield $FeSi_2$. Figure 4(b) shows the resulting phase fractions when varying K_0 for Reaction 2 between 10^{-10} and 10^{-7}. When including this second reaction, a very good agreement with experimental phase fractions is achieved with $K_0=10^{-10}$, for all main phases.

Even though including both Reaction 1 and Reaction 2 in the model gives a very good correspondence with the experimental data, it is seen from the temperature evolution of phases in the equilibrium simulation, that also a peritectic reaction with FeSi and Si from the liquid to form Fe_3Si_7 at high temperature takes place:

$$3\,FeSi + Si\,(L) \rightarrow Fe_3Si_7 \qquad \text{(Reaction 3)}$$

This would result in formation of more Fe3Si7 at high temperature, than is predicted in the current model, which also would result in more formation of FeSi2 by Reaction 1 and also less pure Si later on in the solidification. Thus, even though including Reaction 2 in our model gives very good agreement with experiment, it may not reflect the actual reactions taking place, simply that Reaction 2 results in the same phase fractions at the end of solidification as Reaction 3 combined with Reaction 1. It is also possible that both Reaction 2 and Reaction 3 take place to some extent. In the absence of additional experimental results like interrupting solidification with quenching, it is not possible to conclude on the prevalence of these reactions.

Nevertheless, the phase fractions at end of solidification are well captured for Alloy 1. The optimized rate constants for Reaction 1 and Reaction 2 for Alloy 1 were applied to Alloy 2, to test the validity or robustness of these parameters across different alloys. Figure 7 shows the resulting simulated phase fractions of Alloy 1 (left) and Alloy 2 (right), compared with AMICS. While the amount of Si is slightly underestimated and FeSi overestimated in the simulation, there is a very good agreement between model and experiment.

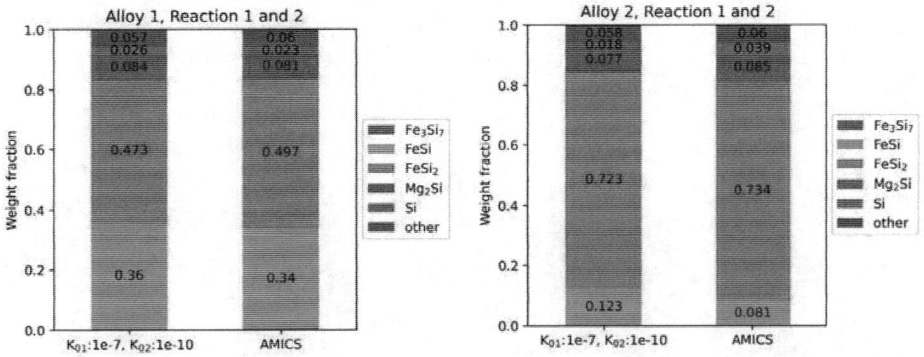

Figure 7 Comparison of modelled phase fractions and experimental phase fractions from AMICS in Alloy 1 (left) and Alloy 2 (right) with calibrated JMAK parameters for Reaction 1 and Reaction 2.

3.3. *Phase field modelling results*

Parts of the solidification process were further investigated by phase field modelling using Micress software [3]. To this end, extensive calibration of numerical parameters was needed as it is often difficult to relate parameters with the physical properties of the phases. The phase field model was first calibrated to reproduce the dendritic morphology of the FeSi phase upon initial nucleation and growth, by tuning the diffusion parameters in the liquid, and the properties of the FeSi/liquid interface: surface energy and kinetic coefficients. Parameters were first tuned to reproduce the FeSi morphology of Alloy 2 and further applied to Alloy 1 to check the validity of the parameter set across different alloys. The simulation domain boundary conditions are periodic in phase and composition, and uniform temperature with an imposed, constant cooling rate or heat flux. The resulting morphologies of the formed FeSi dendrites at 1450K upon parameter calibration for alloys 1 and 2 are shown in Figure 8. The same set of parameters can correctly produce the more dendritic morphology of the FeSi phase observed in experimental samples of Alloy 2 while also yielding the rounder FeSi morphology as seen in experimental samples of Alloy 1. The seed density has been adjusted to be realistic but not really calibrated to replicate the experimental results.

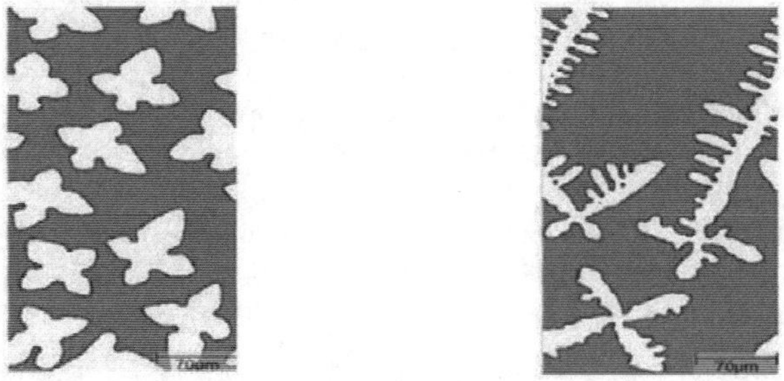

Figure 8 Nucleation and initial growth of FeSi (white) in liquid (red) at 1450K for Alloy 1 (left) and Alloy 2 (right)

Following the calibration of FeSi phase formation, the phase field model was modified to include the formation of Fe_3Si_7. The $FeSi_2$ phase is observed experimentally to have a more elongated, rod-like, phase morphology due to its crystalline structure. It is assumed that the high temperature form is also similar. This was obtained by setting tetragonal anisotropy parameters within the model.

Upon establishing a phase field model with parameters that correctly reproduced the morphology of the FeSi and Fe_3Si_7 phase, the nucleation and growth of Fe_3Si_7 in the presence of FeSi was investigated. Simulations were run from an initial configuration where some FeSi grains were present in the liquid (Figure 9(left)). As seen in Figure 9 (middle) Fe_3Si_7 (white phase) preferentially nucleates in the close vicinity of an FeSi grain (orange phase). This can be reasoned by the change in the local liquid concentration where the FeSi phase first nucleates and grows. Locally, the formation of FeSi results in a surplus of Si in the liquid close to FeSi grains, which creates a driving force for nucleation of Fe_3Si_7, as the Si concentration in this phase is higher than the average in the liquid. This nucleation mode was activated by defining the FeSi/Liquid interface as a nucleation site in addition to the random seed distribution.

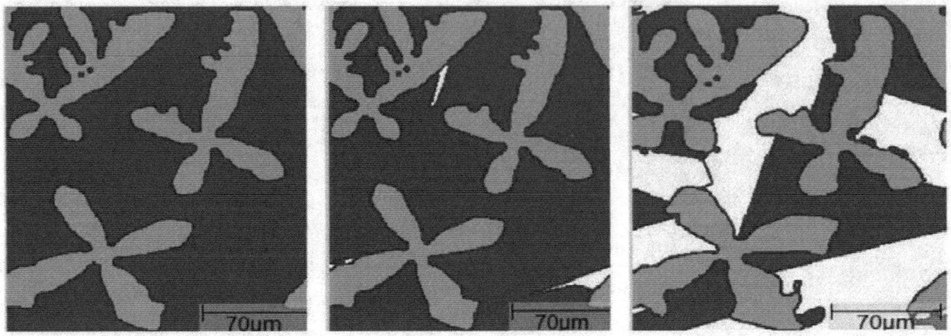

Figure 9 Initial structure of FeSi (orange), nucleation of $FeSi_2$ (white) on top of said structure, growth of $FeSi_2$

As seen from the equilibrium solidification path for the two alloys (Figure 5), the FeSi phase and the Fe_3Si_7 phase both grows simultaneously down to around 1380K. Below this temperature, the amount of Fe_3Si_7 continues to increase whereas FeSi decreases, suggesting some degree of competitive growth between the two phases. The phase interaction between FeSi and Fe_3Si_7 below 1380K was further investigated by the phase field model to explore the effect of morphology on this interaction. Three simulations were set up to study FeSi, Fe_3Si_7 and liquid interactions with three different initial idealized morphologies, as shown in Figure 10(left panel). In the first case (upper left), 85% of the FeSi surface is covered by Fe_3Si_7. The remaining 15% area of FeSi is in contact with the liquid. In the second case (middle left) 45% of the FeSi surface is covered with Fe_3Si_7, whereas 55% is in contact with the liquid. In the latter case (lower left) 100% of the FeSi surface is in contact with the liquid, such that there is no contact between the FeSi and the Fe_3Si_7 phase. In all three simulations, the FeSi phase has initial contact with the liquid. Simulations were initialized with a temperature of 1360 K and a constant cooling rate was applied. No new nucleation was allowed. The evolution of the FeSi and Fe_3Si_7 from the three different initial morphologies is seen in Figure 10 (right panel).

Table 2 lists the resulting phase fractions of FeSi and Fe_3Si_7 with the three levels of coverage of the FeSi phase. The differences indicate that the positioning of the phases relative to each other has a significant influence on the rest of the solidification process. In the two upper configurations, where FeSi is in contact with Fe_3Si_7 and the liquid, the Fe_3Si_7 phase grows by reaction with FeSi and the liquid, resulting in a partial consumption of the FeSi phase as well as formation of an envelope of Fe_3Si_7 around the FeSi phase, thus sealing it away from the liquid. This is also observed in the AMICS maps where every FeSi grain, even those far from the bulk of the $FeSi_2$, is covered by a thin layer of $FeSi_2$. At this point further growth of the FeSi phase is largely hindered. This is because the solid-solid phase boundaries are much less mobile than the solid-liquid phase boundaries. Thus, when FeSi is no longer in direct contact with the liquid, further growth would rely on solid state diffusion which is orders of magnitude smaller than liquid diffusion of elements. In the latter case, the FeSi phase has continued access to the liquid throughout the whole solidification interval considered. This simple case study illustrates the importance of nucleation and growth sequence on the final microstructure.

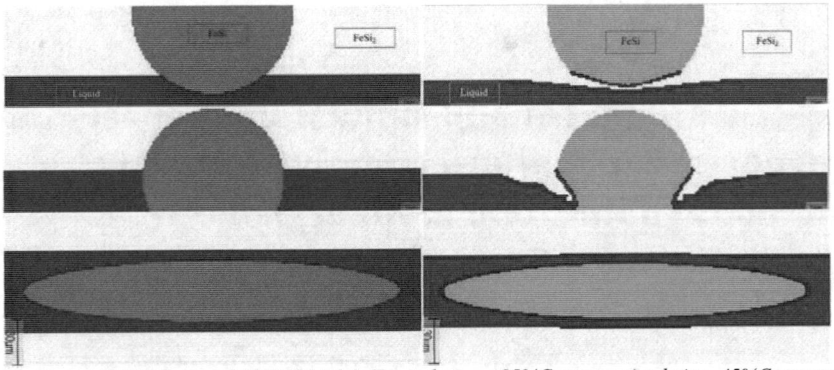

Figure 10. Simulations at 1360K (left) and 1250 K (right). Top to bottom: 85%Coverage simulation, 45%Coverage simulation, 0%Coverage simulation. The phase noted Fe_3Si_7 corresponds to high temperature phase Fe_3Si_7.

Table 2 Resulting phase fractions of FeSi and Fe_3Si_7 with the three levels of coverage of the FeSi phase.

	FeSi mass fraction	Fe_3Si_7 mass fraction
0% Coverage	0.238	0.661
45%Coverage	0.301	0.560
85%Coverage	0.304	0.552

4. Concluding remarks

Thermodynamic modelling and phase field modelling was applied to the solidification of FSM alloys. The thermodynamic database developed at SINTEF covers well the range of composition encountered in commercial alloys. Dedicated experiments were carried out and the microstructure characterized on large areas to get proper statistics of phase fractions using the AMICS. The results showed the necessity to consider solid-solid transformation to accurately predict the phase distribution. The phase field model has enabled us to study the effect of phase morphology and spatial distribution during the solidification process and the consequences on the final phase fractions.

This work is only a first step and additional key elements should be included with the phase they are forming. Inclusion of rate earth elements is particularly important, but it requires an extension of the thermodynamic database and addition of phases not properly documented.

Declaration of competing interest

The authors declare that they have no known competing financial interests or personal relationships that could have appeared to influence the work reported in this article.

Acknowledgments

Part of the work was done by Marc Latournarie and Arthur Wienhold during their internship at SINTEF in 2021 and 2022. This research work is funded through the innovation project FlexSi "Flexible Design and Production of Specialized Ferro-Silicon Materials" with support from the Research Council of Norway and the industrial partners (grant number 309990).

References

[1] Hartung C., White D., Copi K. et al. The Continuing Evolution of MgFeSi Treatments for Ductile and CG Irons. Inter Metalcast 8, 7–15 (2014). https://doi.org/10.1007/BF03355577
[2] ThermoCalc, https://thermocalc.com/
[3] Micress, https://micress.rwth-aachen.de/
[4] AMICS, https://www.bruker.com/en/products-and-solutions/elemental-analyzers/eds-wds-ebsd-SEM-Micro-XRF/software-amics-automated-mineralogy-system.html

IOP Conf. Series: Materials Science and Engineering 1281 (2023) 012053 doi:10.1088/1757-899X/1281/1/012053

Phase-field lattice Boltzmann simulation of three-dimensional settling dendrite with natural convection during nonisothermal solidification of binary alloy

S Sakane[1], T Aoki[2] and T Takaki[1]

[1] Faculty of Mechanical Engineering, Kyoto Institute of Technology, Matsugasaki, Sakyo-ku, Kyoto 606-8585, Japan
[2] Global Scientific Information and Computing Center, Tokyo Institute of Technology, 2-12-1-i7-3, Ookayama, Meguro-ku, Tokyo 152-8550, Japan

E-mail:sakane@kit.ac.jp

Abstract. Understanding the motion and growth behaviors of equiaxed dendrites during solidification is important for predicting macrosegregation. In this study, we develop a phase-field lattice Boltzmann (PF–LB) simulation method for the settling and growth of an equiaxed dendrite during the nonisothermal solidification of a binary alloy. The PF–LB computations are accelerated by employing parallel computation using multiple graphic processing units (GPUs) and the octree block-structured adaptive mesh refinement method, which incorporates multiple mesh and time increment methods. By using the developed method, we can simulate the three-dimensional long-distance settling dendrite while considering the effects of latent heat release and natural convection. From the simulation results, we confirm that the natural convection due to the high solute concentration around a dendrite reduces the settling velocity. In addition, we observe that the temperature increase owing to latent heat release slows dendrite growth, which in turn slightly slows the settling velocity. From these results, we confirm that the effects of latent heat release and natural convection are not negligible in the quantitative evaluation of settling dendrites.

1. Introduction

In casting, the solidification microstructure is generally formed through dendrite growth in an undercooled melt. This is a complex multiphysics phenomenon that involves solid–liquid phase transformation, advection and diffusion of heat and a solute, melt convection, and solid motion [1, 2]. Among these, the settling of equiaxed dendrites, which is caused by the buoyancy forces due to the solid–liquid density difference, leads to negative macrosegregation in the lower part of ingots. Thus, it is essential to understand the motion and growth behavior of equiaxed dendrites with high accuracy [3-5].

The settling of equiaxed dendrites has been studied through in-situ observations of transparent organic materials such as NH_4Cl–H_2O [6-8] and SCN–acetone [9, 10]. The observations are relatively easy because we can use an optical camera. For metallic materials, in-situ X-ray observations on thin samples have been performed [11-14]. However, the settling phenomena in the thin sample are quite different from those in the bulk. Although X-ray tomography can be used to capture the change in the three-dimensional (3D) dendrite morphology of cylindrical samples with time [15-19], it still lacks the spatiotemporal resolution required to investigate the detailed growth and motion behaviors of equiaxed dendrites.

The cellular automata (CA) [20-22] and phase-field (PF) models [23-25] are representative computational methods for studying dendrite growth. CA [26-29] and PF [30-40] models have also been applied to dendrite growth with solid motion and liquid flow. The CA model is capable of dendrite growth simulations for a larger domain than the PF model. However, it lacks accuracy in computing the solid–liquid interface curvature, which is important for dendrite growth. Although the PF model is more accurate than the CA, it is more computationally expensive. Thus, 3D PF simulations have been limited [41-43]. Monas et al. [42, 43] performed 3D PF simulations of dendrites grown with solid rotation under shear flow using parallel computation with the message passing interface (MPI). They used the lattice Boltzmann (LB) method to model melt convection, which has a simple computational algorithm and easy parallelization. Meng et al. [41] performed phase–field lattice Boltzmann (PF–LB) simulations of the motion and growth of several dendrites. The efficiency of the simulations was improved by implementing an adaptive mesh refinement (AMR) method that dynamically assigned fine meshes only around a solid–liquid interface and parallel computation using multiple CPU cores. However, the above-mentioned simulations are limited to a narrow computational domain. We enabled simulations of the long-distance sedimentation of a 3D equiaxed dendrite [44] by employing the PF–LB model [38, 45] and AMR method [45, 46] coupled with parallel computation using multiple graphics processing units (GPUs) [47-50] (parallel-GPU AMR). Comparing the simulation results to those from in-situ observation [9, 10], the simulated settling velocity [44] was found to be approximately twice the experimental settling velocity. Possible reasons for the disagreement are that isothermal conditions were assumed and natural convection was neglected in the simulations [44].

In this study, we investigated the contributions of natural convection and latent heat release to the settling behavior of an equiaxed dendrite. To enable this, the PF–LB method [44] was extended to a new model accounting for latent heat release, the advection and diffusion of heat, and natural convection. Thermal diffusion, which is several orders of magnitude faster than solute diffusion, was efficiently computed by introducing multiple mesh and time increment methods [47] by coupling them with the parallel-GPU AMR computation framework [44]. Their contributions to the dendrite morphology and settling velocity were assessed using equiaxed dendrite settling simulations, both with and without considering the effects of latent heat release and natural convection.

2. Simulation method

We construct a simulation method for the free growth of a settling equiaxed dendrite with natural convection during the nonisothermal solidification of a binary alloy.

2.1. Phase-field lattice Boltzmann model

Dendrite growth is represented using the quantitative PF model for the nonisothermal solidification of a dilute binary alloy [51]. The time-evolution equations for the PF variable, ϕ, dimensionless temperature, θ, and dimensionless solute concentration, u, are as follows:

$$\tau\xi\left(\frac{\partial\phi}{\partial t}+\boldsymbol{U}_s\cdot\nabla\phi\right)=\nabla[W^2\nabla\phi]+\sum_{r=x,y,z}\frac{\partial}{\partial r}\left[W\frac{\partial W}{\partial\phi_r}|\nabla\phi|^2\right]+\phi(1-\phi^2)-\lambda(1-\phi^2)^2(\theta+Mu), \quad (1)$$

$$\left(\frac{\partial\theta}{\partial t}+\boldsymbol{U}\cdot\nabla\theta\right)=D_T\nabla^2\theta+\frac{1}{2}\frac{\partial\phi}{\partial t}, \quad (2)$$

$$\frac{[1+k-(1-k)\phi]}{2}\left(\frac{\partial u}{\partial t}+\boldsymbol{U}\cdot\nabla u\right)=\nabla(D_l q(\phi)\nabla u-\boldsymbol{J}_{AT})+\frac{[1+(1-k)u]}{2}\frac{\partial\phi}{\partial t}-\nabla\cdot\boldsymbol{J}, \quad (3)$$

where $\phi = 1$ and $\phi = -1$ are the solid and liquid phases, respectively. $\xi = Le^{-1}+[1+(1-k)u]M$, where $Le = D_T/D_l$ is the Lewis number, which is the ratio of the thermal and solute diffusivities in the liquid, and $M = d_0/d_s$ is the ratio of the thermal and chemical capillary lengths. τ is the relaxation time, and W is the interface thickness. To stably compute equation (1) with a large PF interface width, we apply nonlinear preconditioning, $\phi = \tanh(\psi/\sqrt{2})$ [52], and solve the time evolution equation for ψ. $\theta = (T - T_0)/(L/c_p)$,

IOP Conf. Series: Materials Science and Engineering 1281 (2023) 012053 doi:10.1088/1757-899X/1281/1/012053

where T is the temperature, T_0 is the reference temperature, L is the latent heat, and c_p is the specific heat. $u = (C_l - C_l^e)/[(1-k)C_l^e]$, where C_l and C_l^e are the solute and equilibrium solute concentrations in the liquid phase, respectively, and k is the partition coefficient. $q(\phi) = [kD_s + D_l + (kD_s - D_l)\phi]/(2D_l)$ is the interpolation function of the solute diffusion coefficient, where D_l and D_s are the solute diffusion coefficients in the liquid and solid phases, respectively. $\boldsymbol{J}_{\mathrm{AT}}$ is the antitrapping current term [51], which is expressed as follows:

$$\boldsymbol{J}_{AT} = -\frac{1}{2\sqrt{2}}\left(1 - \frac{kD_s}{D_l}\right) W_0[1 + (1-k)u]\frac{\partial \phi}{\partial t} \cdot \boldsymbol{n}, \tag{4}$$

where $\boldsymbol{n}$ is the normal unit vector which is computed using $\boldsymbol{n} = \nabla\psi/|\nabla\psi|$. $\nabla \cdot \boldsymbol{J}$ is the noise term for generating the higher-order branching of the dendrite [53]. The second terms on the left-hand side of equations (2) and (3) are the advection terms of the heat and solute, respectively, where $\boldsymbol{U}$ is the flow velocity in the liquid phase. The time evolution of the flow field is computed using the two-relaxation-time LB model [54] as follows:

$$f_i(\boldsymbol{x} + \boldsymbol{c}_i\Delta t, t + \Delta t) = f_i(\boldsymbol{x}, t) - \frac{f_i^+(\boldsymbol{x},t) - f_i^{+(eq)}(\boldsymbol{x},t)}{\tau_{LBM}^+} - \frac{f_i^-(\boldsymbol{x},t) - f_i^{-(eq)}(\boldsymbol{x},t)}{\tau_{LBM}^-} + F_i\Delta t, \tag{5}$$

where $f_i(\boldsymbol{x}, t)$ is the particle distribution function at time t, position $\boldsymbol{x}$, and velocity $\boldsymbol{c}_i$. $f_i^{\pm}(\boldsymbol{x}, t) = (f_i(\boldsymbol{x}, t) \pm f_{\bar{i}}(\boldsymbol{x}, t))/2$, where $\bar{i}$ is the discrete velocity direction number for which $\boldsymbol{c}_{\bar{i}} = -\boldsymbol{c}_i$. $f_i^{\pm\mathrm{eq}}(\boldsymbol{x}, t) = (f_i^{\mathrm{eq}}(\boldsymbol{x}, t) \pm f_{\bar{i}}^{\mathrm{eq}}(\boldsymbol{x}, t))/2$, where $f_i^{\mathrm{eq}}(\boldsymbol{x}, t)$ is the equilibrium particle distribution function. The density, ρ, and flow velocity, $\boldsymbol{U}$, in the liquid phase are computed as follows:

$$\rho(\boldsymbol{x}, t) = \sum_{i=0}^{Q-1} f_i(\boldsymbol{x}, t), \tag{6}$$

$$\boldsymbol{U}(\boldsymbol{x}, t) = \frac{1}{\rho(\boldsymbol{x},t)}\sum_{i=0}^{Q-1} \boldsymbol{c}_i f_i(\boldsymbol{x}, t), \tag{7}$$

where Q is the number of discrete velocity directions. The following second-order discrete external force term is added to equation (5) to represent the buoyancy force caused by the unevenly distributed temperature and solute concentration in the liquid phase:

$$F_i = \rho w_i \left[\frac{3(\boldsymbol{c}_i - \boldsymbol{U})}{c^2} + \frac{9(\boldsymbol{c}_i \cdot \boldsymbol{U})\boldsymbol{c}_i}{c^4}\right] \cdot \boldsymbol{F}, \tag{8}$$

where w_i is the weight function and c is the lattice velocity which is computed using $c = \Delta x/\Delta t$, where Δx is the mesh width and Δt is the time increment. Buoyancy $\boldsymbol{F}$ in equation (8) is expressed as $\boldsymbol{F} = -\rho_0\boldsymbol{g}[\beta_c(C - C_0) + \beta_T(T - T_0)]$, where β_T and β_c are the heat and solute expansion coefficients, respectively, $\boldsymbol{g}$ is gravity, ρ_0 is the reference density, T_0 is the reference temperature, and C_0 is the reference solute concentration. The nonslip boundary condition is imposed at the solid–liquid interface by applying the second-order interpolated bounce-back method [55] at $\phi = 0$. τ_{LBM}^+ and τ_{LBM}^- are the relaxation times of the LB model. $\tau_{\mathrm{LBM}}^+ = 3\nu\Delta t/(\Delta x^2) + 0.5$, where ν is the kinetic viscosity. $\tau_{\mathrm{LBM}}^- = (3\tau_{\mathrm{LBM}}^+ + 3)/(4\tau_{\mathrm{LBM}}^+ - 2)$, and it suppresses the excessive velocity slippage at the solid surface [56]. The second term on the left-hand side of equation (1) is the advection term used to express the motion of the solid. The velocity of motion, $\boldsymbol{U}_s$, at each position of the solid is expressed as follows:

$$\boldsymbol{U}_S = \boldsymbol{U}_T + \boldsymbol{\omega}_S \times (\boldsymbol{x} - \boldsymbol{x}_S), \tag{9}$$

where $\boldsymbol{x}_s$ is the center of gravity and $\boldsymbol{U}_T$ and $\boldsymbol{\omega}_s$ are the translational and rotational velocities of the solid, respectively. The time evolutions of $\boldsymbol{U}_T$ and $\boldsymbol{\omega}_s$ are expressed as follows:

$$M_S\frac{d\boldsymbol{U}_T}{dt} = \boldsymbol{G}_S, \tag{10}$$

$$\boldsymbol{I}_S\frac{d\boldsymbol{\omega}_S}{dt} = \boldsymbol{T}_S, \tag{11}$$

where M_s and $\boldsymbol{I}_s$ are the mass and inertia tensors of the solid, respectively. The force, $\boldsymbol{G}_s$, and moment, $\boldsymbol{T}_s$, acting on the solid are defined as follows:

$$\boldsymbol{G}_s = \int_\Omega \sum_{i=0}^{Q-1} \boldsymbol{G}_i \, dV + \left(1 - \frac{\rho_l}{\rho_s}\right) M_s \boldsymbol{g}, \tag{12}$$

$$\boldsymbol{T}_s = \int_\Omega (\boldsymbol{x} - \boldsymbol{x}_s) \times \sum_{i=0}^{Q-1} \boldsymbol{G}_i \, dV, \tag{13}$$

where ρ_l and ρ_s are the densities of the liquid and solid phases, respectively. $\boldsymbol{G}_i$ is the force acting on the solid by a particle bouncing back at the solid–liquid interface, and it is computed using the Galilean invariant momentum exchange method [57].

Equations (1), (2), (3), (10), and (11) are computed using the finite-difference method. We employed the same schemes as in our previous study [44], except for equation (2). The first-order upwind difference is used for the time derivative. The central second-order difference method is used for the second-order spatial differentiation, and the third-order essentially nonoscillatory method [58] is used to compute the advection term in equation (2). The D3Q19 model is used to compute equation (5).

2.2. AMR with multiple mesh and time increment methods in parallel computation

Parallel computations are implemented using multiple GPUs to accelerate the computations. We reduce the computational cost by employing the octree-type block-structured AMR method [44, 46], which is suitable for GPU computation [59, 60]. In the AMR method, the computational domain is first divided into several cubic blocks, as shown in figure 1(a). The blocks at the locations where fine numerical meshes are required are recursively divided on the basis of the octree structure. The coarsest block is defined as refinement level 0, and the level is increased by one for each octree division of a block. The upper limit of the level is set in advance as $n_{\max}$. Each block is divided into a numerical mesh of N_{PF}^3, and equations (1) and (3) are computed on the mesh points. Because the spatial resolution required for the computation of the temperature field (equation (2)) and flow field (equation (5)) is less than that required for the PF computation (equation (1)), we employ the multiple mesh method [47] in which the block is divided into N_T^3 and N_{LB}^3 meshes for computing equations (2) and (5), respectively.

Equations (1) and (3) are computed for every time increment, Δt_{PF}, on the meshes of all levels. Equation (5) is computed using the time increment $\Delta t_{LB,n} = 2^{n_{max}-n}\Delta t_{\mathrm{PF}}$ on the meshes of level n to maintain lattice velocity c constant on the meshes of all levels [61]. We apply the multiple time increment method because thermal diffusion is 2–3 orders of magnitude larger than solute diffusion and requires a smaller time increment [47]. In this method, only the temperature field (equation (2)) is computed with a time increment (Δt_{TF}) that is smaller than Δt_{PF} on the meshes of all levels.

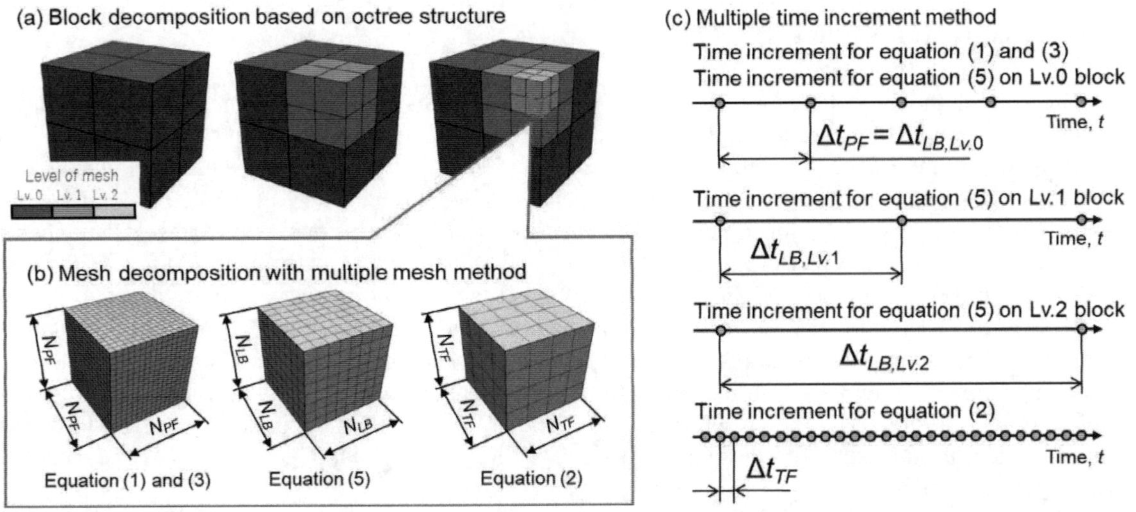

Figure 1. Illustration of the implemented AMR method: (a) Block decomposition of the computational domain based on an octree structure, (b) Mesh decomposition of AMR block using multiple mesh method, (c) Multiple time increment method for the AMR–PF–LB model.

We implement multi-GPU parallel computation to achieve large-scale computation without being limited by the memory size of a single GPU. The computational domain is dynamically divided based on the slice-grid method [44] such that the number of mesh points in each subdomain is equal. One GPU is allocated to each subdomain. The computation code is written using CUDA (NVIDIA), and OpenMPI is used for internode communication.

3. Computational conditions

We simulated the growth of a single equiaxed dendrite settling in an undercooled melt. The selected material is SCN–2.96 wt.% acetone, which is the same as in the experiment by Badillo et al. [9, 10]. The material properties used in the present simulations are listed in table 1.

Table 1. Material properties of SCN–2.96 wt.% acetone and computational conditions.

Property	Symbol	Value	Ref.
Melting point of SCN (K)	T_m	331.23	[62]
Gibbs–Thomson coefficient (K•μm)	Γ	6.525×10^{-2}	[63]
Strength of anisotropy (SCN)	ε_4	0.0055	[64]
Thermal diffusivity (μm²/s)	D_T	1.134×10^5	[63]
Solute diffusivity in the liquid (μm²/s)	D_l	1.27×10^3	[65]
Solute diffusivity in the solid (μm²/s)	D_s	1.27×10^{-1}	
Unit supercooling (K)	L/c_p	23.13	[63]
Liquidus slope (K/mol frac.)	m	−216	[65]
Partition coefficient	k	0.103	[65]
Kinetic viscosity (m²/s)	ν	6.044×10^{-7}	[66]
Thermal expansion factor (1/K)	β_T	8.028×10^{-4}	[67]
Solute expansion factor (1/mol frac.)	β_c	4.476×10^{-1}	[67]
Ratio of solid–liquid density	ρ_s/ρ_l	1.033	[67]
Minimum mesh size for PF (μm)	$\Delta x_{\min}^{PF}$	0.500	
Interface thickness (μm)	W_0	0.625	

The computational domain was set as 2.048 mm × 2.048 mm × 6.144 mm. The zero-Neumann condition was applied for ϕ and u, the nonslip condition for $\boldsymbol{U}$, and the Dirichlet condition ($T = T_0$) for θ. At the beginning of the simulations, a spherical solid seed with a radius of 3 μm was placed at the position (1.024 mm, 1.024 mm, 5.5296 mm) in an undercooled melt with $\Delta T = 0.82$ K. Three simulations were performed: Case A: nonisothermal solidification with natural convection; Case B: isothermal condition (θ = const.) with natural convection ($\beta_T = 0$); and Case C: isothermal condition (θ = const.) without natural convection ($\beta_T = 0, \beta_c = 0$). Case C has the same conditions as those used in our previous study [44], except for the following three parameters, C_0 = 3.0 wt%, ΔT = 0.87 K, and ε_4 = 0.02. In addition, to evaluate the contribution of natural convection in Case A, we simulated the growth of an immobile dendrite under the conditions of Case A and designated it as Case A'. Furthermore, to compare the dendrite morphology without melt convection and solid motion, the simulation was performed by omitting them from Case C and designated as Case C'. In Cases A' and C', the initial seed is placed in the center of the domain.

As AMR criteria, level $n_{\max}$ blocks were assigned only to the interface region ($-0.9996 < \phi < 0.9996$) [44]. The coarsest possible blocks were assigned to the other regions with the constraint that the level difference between adjacent blocks must be less than 1. Each block was divided into meshes of (N_{PF}^3,

$N_{LB}{}^3$, $N_T{}^3$) = (16^3, 8^3, 4^3) [47]. The corresponding minimum mesh sizes on level $n_{\max}$ blocks were $\Delta x_{\min}{}^{\mathrm{PF}}$, $\Delta x_{\min}{}^{\mathrm{LB}}$, and $\Delta x_{\min}{}^{\mathrm{TF}}$, and the relations $\Delta x_{\min}{}^{\mathrm{LB}} = 2\Delta x_{\min}{}^{\mathrm{PF}}$ and $\Delta x_{\min}{}^{\mathrm{TF}} = 4\Delta x_{\min}{}^{\mathrm{PF}}$ were satisfied. $\Delta t_{\mathrm{PF}} = (\Delta x_{\min}{}^{\mathrm{PF}})^2/(7D_l) = 2.812 \times 10^{-5}$ s to satisfy the von Neumann stability condition for the diffusion term in equation (3). $\Delta t_{TF} = \Delta t_{\mathrm{PF}}/6 = 4.687 \times 10^{-6}$ s < $(\Delta x_{\min}{}^{\mathrm{TF}})^2/(6D_T)$ to satisfy the von Neumann stability condition, and Δt_{PF} was an integer multiple of Δt_{TF}.

The TSUBAME3.0 GPU supercomputer at the Tokyo Institute of Technology was used for the simulations. TSUBAME3.0 contains approximately 2000 GPUs (NVIDIA GPU Tesla P100). The simulations were completed in approximately 16 h using 64 GPUs.

4. Results and discussion

Figure 2 shows the time dependent changes in the dendrite morphology and position, distribution of the solute concentration, and temperature on the central cross section in the *x* direction for Cases A and A' (nonisothermal solidification with the natural convection and with/without solid motion, respectively). The results for Case A' are shown only around the dendrite in figure 2(b). Figure 2(c) and (d) show enlarged views around the dendrite at t = 14.06 s, and the flow field is indicated by arrows. In Case A, as the dendrite settles and grows, the solute and latent heat are released from the solid and diffused. While the solute released from the solid remained behind the dendrite, the released latent heat diffused rapidly. Thus, the temperature increase was only around the dendrite and was less than 0.1 K at the interface. As shown in figure 2(c), downward flow occurs around the dendrite by dragging the settling dendrite. On either side of the dendrite, circulating vortex-like flows are observed. On the upper side, away from the dendrite, the upward flow due to buoyancy caused by the solute-rich liquid is observed. In Case A', as shown in figure 2(b) and (d), upward flow due to natural convection occurs around the dendrite. In contrast, the latent heat diffuses spherically around the dendrite because thermal diffusion is faster than the velocity of natural convection. Note that the temperature change due to latent heat release was small, and its buoyancy was negligible—less than 5% of the buoyancy of the solute. In Case A', the maximum velocity of the upward flow is ~70 μm/s, which is smaller than the maximum velocity of 500 μm/s in Case A.

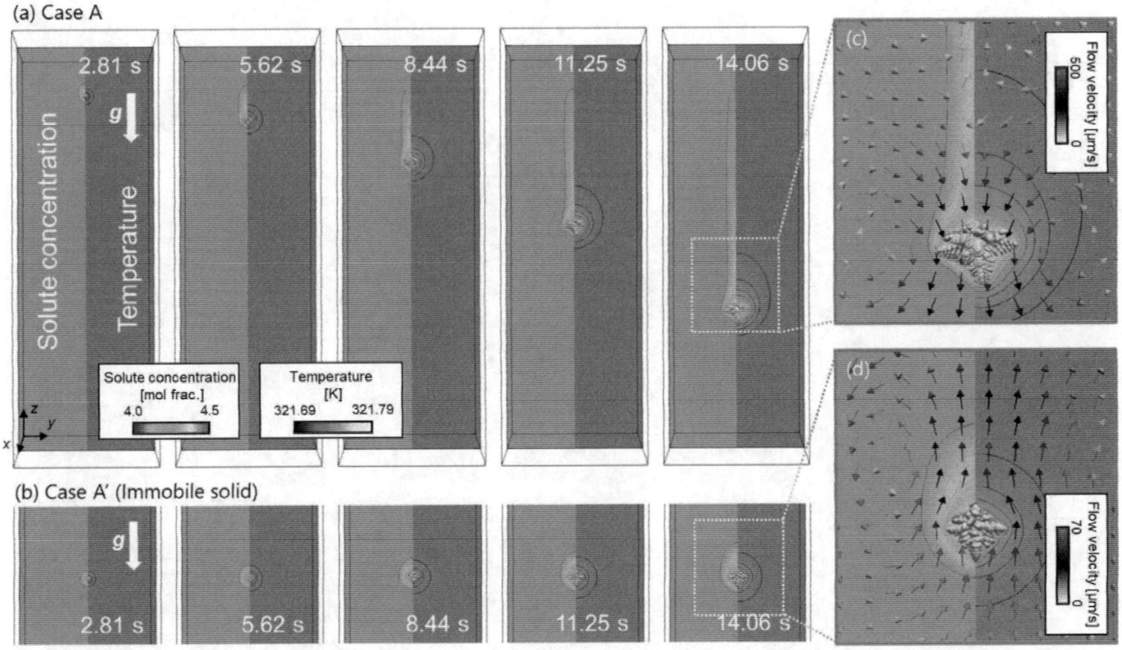

Figure 2. Changes in the morphology and position of the dendrite with time and distributions of temperature and solute concentration on the central cross section in the *x* direction for Cases (a,c) A and (b,d) A'.

IOP Conf. Series: Materials Science and Engineering 1281 (2023) 012053 doi:10.1088/1757-899X/1281/1/012053

Figure 3 shows the change in the dendrite morphologies at t = 2.81, 5.62, 8.44, 11.25, and 14.06 s for Cases A–C. For comparison, the results for Case C' (isothermal solidification without melt convection and solid motion) are also shown. In Case C', the protuberances grow in various directions at t = 2.81 s in the early growth stage and then preferentially grow in the <100> direction, confirming a hyperbranched dendrite structure. This is due to the small strength of anisotropy, $\varepsilon_4 = 0.0055$. In Cases A–C, where solid motion was considered, similar growth behavior was obtained. The primary arms facing downward, which is the direction of settlement, grow preferentially. The primary arms growing perpendicular to the direction of settlement are slightly tilted downward, and the secondary arms grow only in a downward direction. The growth of the arm is retarded on the upper side of the dendrite, which is downstream. These morphologies are similar to those of dendrites growing under forced convection [68]. Cases B and C have nearly identical morphology and size, indicating that only Case A has delayed lateral arm growth. This is because the driving force for interfacial migration is reduced by the temperature increase due to latent heat release from the solid–liquid interface. At t = 14.06 s, the total length of the dendrite in the y direction is approximately 0.42 mm in Case A, which is shorter than the lengths of Cases B and C, 0.46 and 0.48 mm, respectively.

Figure 4 shows the change in the solid volume V_s, the specific interfacial area S_s, which is the solid–liquid interfacial area A_{SL} normalized by V_s, and the settling velocities with time for Cases A–C. As shown in figure 4(a), the volume of the dendrite increases quadratically and monotonically in all cases. Here, V_s of Cases B and C show almost the same time change. In contrast, V_s of Case A is at most 18% smaller than that of Cases B and C. This is because the growth is slowed by the temperature increase due to latent heat release, as shown in figure 3. As shown in figure 4(b), the change in S_s show almost the same behavior, despite the slight difference in dendrite morphology in Cases A–C. In the initial stage of the simulations, when the solid grows as a sphere, S_s drops sharply as V_s increases in proportion to the cube of the radius and A_{SL} to the square of the radius. S_s subsequently turned to an upward trend as the primary arms began to grow, and after t = 5 s, when secondary arms began to grow, S_s remained constant. Figure 4(c) depicts the settling velocity as a function of time. In all cases, the settling velocity

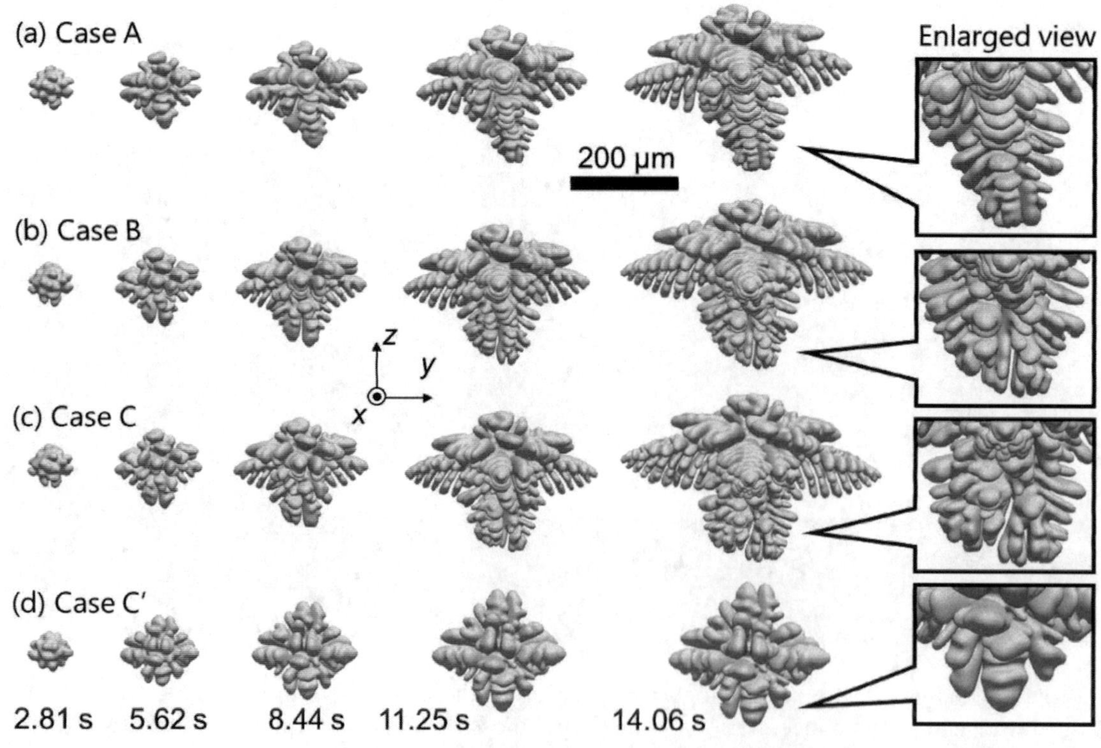

Figure 3. Change in dendrite morphologies with time for Cases (a) A, (b), B, (c) C, and (d) C'.

IOP Conf. Series: Materials Science and Engineering 1281 (2023) 012053 doi:10.1088/1757-899X/1281/1/012053

increases monotonically. In the early stages, the ratio of drag to buoyancy is also larger due to the larger S_s, and the slope of the settling velocity is lower than in the other time intervals. Then, the slope of the settling velocity increases as S_s decreases, and after t = 5 s, when S_s becomes constant, the settling velocity increases linearly. The settling velocity is slower in the order of Cases C, B, and A. In Cases C and B, the morphology (figure 3), V_s (figure 4(a)), and S_s (figure 4(b)) are almost the same; however, the settling velocity in Case B is at most 16% slower. This is because in Case B, where natural convection is considered, the settling velocity is reduced by the buoyancy, which generates the upward flow of natural convection in figure 3(d). Comparing Cases A and B, the smaller V_s of Case A results in a smaller buoyancy acting on the solid, which slows the settling velocity by approximately 7.6%.

In the experiment [9, 10], the observation was made for 80 s while a settling dendrite of 3 mm in total length grew to 5 mm in an 8 cm diameter cylinder. This dendrite size was approximately 10 times the scale of the present simulations. Direct comparison is difficult due to the difference in scale between the experiment and simulation. A linear acceleration of the settling velocity was observed, in both the experiment [9, 10] and the present simulations. Therefore, we compare the slopes of the settling velocity in the experiment and in the simulations. The average slopes of the settling velocity in the 0–14 s interval for Cases A, B, and C are 36.2, 39.2, and 46.6 μm/s^2, respectively. Although these values are larger than the experimental value of 19.8 μm/s^2 [9, 10], the simulation results are close to the experimental results obtained considering natural convection and latent heat release. These results indicate that the effects of latent heat release and natural convection are not negligible in the quantitative evaluation of settling dendrite growth. Additionally, a dendrite with well-developed branch that clearly show <100> anisotropy is observed in the experiment [9, 10], which is quite different from the dendrite morphology shown in figure 3. In fact, the morphology observed in the previous simulation assuming ε_4 = 0.02 [44] is closer to that of the experiment. Nevertheless, as shown in figure 4(c), a comparison of the settling velocity of Case C and that of the previous study [44] shows that ε_4 does not strongly affect the settling velocity. Another possible reason for discrepancies is uncertainty in the material properties used in the simulations. In particular, the buoyancy force acting on a solid is proportional to density differences; therefore, even small errors in density of the solid and liquid phases can strongly affect the settling velocity. Moreover, changes in material properties with temperature and solute concentration over time were neglected in the current simulations. Data assimilation [69] can identify dominant factors that cause discrepancies between simulated and experimental results. Thus, the simulation method developed in this study will be used for data assimilation in the future.

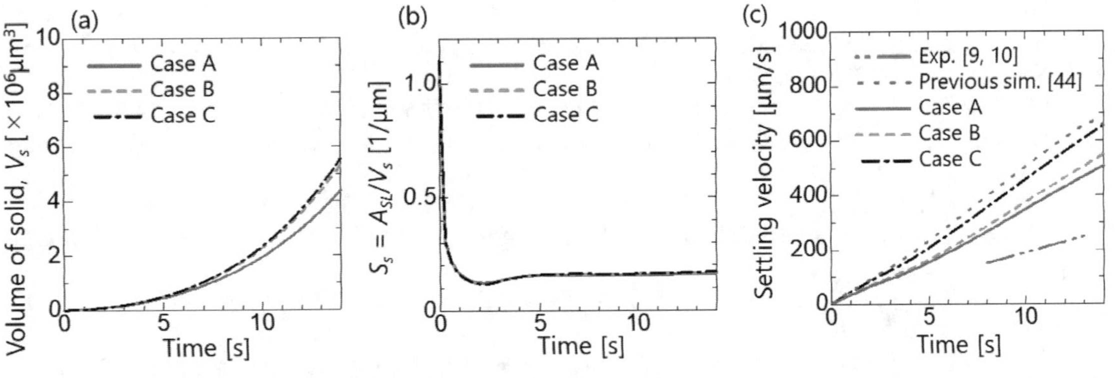

Figure 4. Change in (a) solid volume V_s, (b) specific interfacial area S_s, which is the solid–liquid interfacial area A_{SL} normalized by V_s, and (c) settling velocities with time during the simulation of the settling equiaxed dendrite for Cases A–C.

5. Conclusions

We developed a PF–LB method for simulating the settling and growth of an equiaxed dendrite during the nonisothermal solidification of a binary alloy by considering natural convection. The PF–LB computations were accelerated by employing parallel computation using multiple GPUs and the AMR method, which incorporated multiple mesh and time increment methods. By using the developed method, we performed the 3D simulations with and without considering the latent heat release and natural convection. We confirmed that natural convection due to the high solute concentration around the dendrite reduced the settling velocity. In addition, the release of latent heat slightly changed the dendrite morphology and slightly reduced the settling velocity. The prediction results were closer to the experimental results when natural convection and latent heat release are taken into account; however, quantitative agreement with the experimental results has not yet been obtained. Uncertainty in material properties is thought to be the main cause of discrepancies, and utilizing data assimilation is a promising approach for addressing this issue.

Acknowledgments

This work was supported by KAKENHI Grant-in-Aid for Young Scientists 21K14041, Grant–in–Aid for Scientific Research (A) 20H00217, and Grant-in-Aid for Scientific Research (S) 19H05613 from the Japan Society for the Promotion of Science. This study also used the TSUBAME3.0 computational resources provided by the Tokyo Institute of Technology through the HPCI System Research Project (Project ID: hp220089).

References

[1] Kurz W, Rappaz M and Trivedi R 2020 *Int. Mater. Rev.* **66** 30-76
[2] Kurz W, Fisher D J and Trivedi R 2018 *Int. Mater. Rev.* **64** 311-54
[3] Zhang H, Wu M, Rodrigues C M G, Ludwig A, Kharicha A, Rónaföldi A, Roósz A, Veres Z and Svéda M 2022 *Acta Mater.* **241** 118391
[4] Ge H, Li J, Han X, Xia M and Li J 2016 *J. Mater. Process. Technol.* **227** 308-17
[5] Flemings M C 2000 *ISIJ Int.* **40** 833-41
[6] Appolaire B, Albert V, Combeau H and Lesoult G 1999 *ISIJ Int.* **39** 263-70
[7] Appolaire B, Albert V, Combeau H and Lesoult G 1998 *Acta Mater.* **46** 5851-62
[8] Ramani A and Beckermann C 1997 *Scr. Mater.* **36** 633-8
[9] Badillo A, Ceynar D and Beckermann C 2007 *J. Cryst. Growth* **309** 197-215
[10] Badillo A, Ceynar D and Beckermann C 2007 *J. Cryst. Growth* **309** 216-24
[11] Wang J, Lin X, Fautrelle Y, Nguyen-Thi H and Ren Z 2018 *Metall. Mater. Trans. B* **49** 861-5
[12] Gibbs J W, Tourret D, Gibbs P J, Imhoff S D, Gibbs M J, Walker B A, Fezzaa K and Clarke A J 2016 *JOM* **68** 170-7
[13] Abou-Khalil L, Salloum-Abou-Jaoude G, Reinhart G, Pickmann C, Zimmermann G and Nguyen-Thi H 2016 *Acta Mater.* **110** 44-52
[14] Shevchenko N, Boden S, Gerbeth G and Eckert S 2013 *Metall. Mater. Trans. A* **44** 3797-808
[15] Narumi T, Kawarasaki T, Kato Y, Morishita K and Yasuda H 2020 *Journal of Japan Institute of Light Metals* **70** 339-46
[16] Nakano K, Narumi T, Morishita K and Yasuda H 2020 *Mater. Trans.* **61** 596-604
[17] Yasuda H, Kawarasaki T, Tomiyori Y, Kato Y and Morishita K 2019 *IOP Conf. Ser.: Mater. Sci. Eng.* **529**
[18] Guo E, Phillion A B, Cai B, Shuai S, Kazantsev D, Jing T and Lee P D 2017 *Acta Mater.* **123** 373-82
[19] Cai B, Wang J, Kao A, Pericleous K, Phillion A B, Atwood R C and Lee P D 2016 *Acta Mater.* **117** 160-9
[20] Zhang Y, Zhou J, Yin Y, Shen X and Ji X 2021 *J. Mater. Res. Technol.* **14** 1862-72
[21] Yu Y, Li Y, Lin F and Yan W 2021 *Addit. Manuf.* **47** 102284
[22] Kao A, Krastins I, Alexandrakis M, Shevchenko N, Eckert S and Pericleous K 2018 *Jom* **71** 48-58
[23] Ohno M 2020 *ISIJ Int.* **60** 2745-54
[24] Plapp M 2016 *J. Indian. Inst. Sci.* **96** 179-98
[25] Takaki T 2014 *ISIJ Int.* **54** 437-44

[26] Zhang S, Li C and Li R 2022 *Mater. Today Commun.* **31** 103342
[27] Wang Q, Wang Y, Zhang S, Guo B, Li C and Li R 2021 *Cryst.* **11** 1056
[28] Bai Y, Wang Y, Zhang S, Wang Q and Li R 2020 *Cryst.* **10** 70
[29] Liu L, Pian S, Zhang Z, Bao Y, Li R and Chen H 2018 *Comput. Mater. Sci.* **146** 9-17
[30] Wang X, Mao S, Wang J and Sun D 2022 *Int. J. Therm. Sci.* **178** 107592
[31] Rojas R, Sotomayor V, Takaki T, Hayashi K and Tomiyama A 2022 *Comput. Math. Appl.* **114** 180-7
[32] Liu X, Tang F, Zhao W, Cai J and Wei Y 2022 *Comput. Mater. Sci.* **204** 111182
[33] Sakane S, Takaki T, Ohno M, Shibuta Y and Aoki T 2020 *Comput. Mater. Sci.* **178** 109639
[34] Ren J-k, Chen Y, Cao Y-f, Sun M-y, Xu B and Li D-z 2020 *J. Mater. Sci. Technol.* **58** 171-87
[35] Sakane S, Takaki T, Ohno M and Shibuta Y 2019 *Comput. Mater. Sci.* **164** 39-45
[36] Rátkai L, Pusztai T and Gránásy L 2019 *npj Comput. Mater.* **5** 1-10
[37] Takaki T, Sato R, Rojas R, Ohno M and Shibuta Y 2018 *Comput. Mater. Sci.* **147** 124-31
[38] Rojas R, Takaki T and Ohno M 2015 *J. Comput. Phys.* **298** 29-40
[39] Medvedev D, Varnik F and Steinbach I 2013 *Procedia Computer Science* **18** 2512-20
[40] Do-Quang M and Amberg G 2008 *J. Comput. Phys.* **227** 1772-89
[41] Meng S, Zhang A, Guo Z and Wang Q 2020 *Comput. Mater. Sci.* **184** 109784
[42] Subhedar A, Steinbach I and Varnik F 2015 *Phys. Rev. E* **92** 023303
[43] Monas A, Shchyglo O, Kim S-J, Yim C D, Höche D and Steinbach I 2015 *Jom* **67** 1805-11
[44] Sakane S, Aoki T and Takaki T 2022 *Comput. Mater. Sci.* **211** 111542
[45] Sakane S, Aoki T and Takaki T 2022 *Comput. Mater. Sci.* **211** 111507
[46] Sakane S, Takaki T and Aoki T 2022 *Mater. theory* **6** 1-19
[47] Sakane S, Takaki T, Ohno M, Shibuta Y and Aoki T 2019 *Model. Simul. Mat. Sci. Eng.* **27** 054004
[48] Sakane S, Takaki T, Rojas R, Ohno M, Shibuta Y, Shimokawabe T and Aoki T 2017 *J. Cryst. Growth* **474** 154-9
[49] Sakane S, Takaki T, Ohno M, Shimokawabe T and Aoki T 2015 *IOP Conf. Ser.: Mater. Sci. Eng.* **84** 012063
[50] Shimokawabe T, Aoki T, Takaki T, Endo T, Yamanaka A, Maruyama N, Nukada A and Matsuoka S, 2011 *Proceedings of International Conference for High Performance Computing, Networking, Storage and Analysis* 1-11
[51] Ohno M 2012 *Phys. Rev. E* **86** 051603
[52] Glasner K 2001 *J. Comput. Phys.* **174** 695-711
[53] Echebarria B, Karma A and Gurevich S 2010 *Phys. Rev. E* **81** 021608
[54] Sakane S and Takaki T 2021 *Comput. Mater. Sci.* **186** 110070
[55] Bouzidi M h, Firdaouss M and Lallemand P 2001 *Phys. Fluids* **13** 3452-9
[56] Seta T, Rojas R, Hayashi K and Tomiyama A 2014 *Phys. Rev. E* **89** 023307
[57] Wen B, Zhang C, Tu Y, Wang C and Fang H 2014 *J. Comput. Phys.* **266** 161-70
[58] Shu C-W and Osher S 1988 *J. Comput. Phys.* **77** 439-71
[59] Watanabe S and Aoki T 2021 *Comput. Phys. Commun.* **264** 107871
[60] Schive H-Y, ZuHone J A, Goldbaum N J, Turk M J, Gaspari M and Cheng C-Y 2018 *Mon. Notices Royal Astron. Soc.* **481** 4815-40
[61] Dupuis A and Chopard B 2003 *Phys. Rev. E* **67** 066707
[62] Huang S C and Glicksman M E 1981 *Acta Metallurgica* **29** 701-15
[63] Koss M B, LaCombe J C, Tennenhouse L A, Glicksman M E and Winsa E A 1999 *Metall. Mater. Trans. A* **30** 3177-90
[64] Muschol M, Liu D and Cummins H Z 1992 *Phys. Rev. A* **46** 1038-50
[65] Lipton J, Glicksman M E and Kurz W 1987 *Metall. Mater. Trans. A* **18** 341-5
[66] Tin P and de Groh III H C 2004 *Int. J. Thermophys.* **25** 1143-53
[67] Ceynar D L and Beckermann C 2001 *J. Cryst. Growth* **222** 380-91
[68] Okamoto T, Kishitake K and Bessho I 1975 *J. Cryst. Growth* **29** 131-6
[69] Yamamura A, Sakane S, Ohno M, Yasuda H and Takaki T 2022 *Comput. Mater. Sci.* **215**

IOP Conf. Series: Materials Science and Engineering 1281 (2023) 012054 doi:10.1088/1757-899X/1281/1/012054

Investigation of diffusive grain interactions during equiaxed dendritic solidification

Abdelhalim Chirouf[1,2], Benoît Appolaire[1], Alphonse Finel[2], Yann Le Bouar[2], Miha Založnik[1]

[1] Université de Lorraine, CNRS, IJL, F-54000 Nancy, France

[2] Université Paris-Saclay, ONERA, CNRS, LEM, F-92322 Châtillon, France

E-mail: `abdelhalim.chirouf@univ-lorraine.fr`

Abstract. Grain growth during the solidification of an alloy is accompanied by the enrichment of the surrounding liquid phase in chemical species. The resulting diffusion fields induce interactions between the grains that are strongly dependent on their spatial arrangement. The influence of these diffusive interactions is not limited to the shapes and sizes of grains but also controls the overall kinetics of the transformation. In this study we investigate the role of the spatial arrangement of equiaxed grains in the average growth kinetics of an ensemble of grains. We perform full-field simulations of different grain ensembles using the mesoscopic grain envelope model (GEM). We show that the growth kinetics of randomly arranged ensembles is fundamentally different from that of periodic arrangements. We provide an explanation of this difference through analyses of the behaviour of individual grains in their local neighbourhood, defined by a Voronoi tessellation. Finally, we compare the full-field GEM simulations to state-of-the-art mean-field models and we point out the limitations of the latter in the prediction of dendritic growth.

1. Introduction

Modeling the solidification of metallic alloys can provide valuable insights into the microstructure of castings in various industrial processes, helping to optimize and improve these processes. One of the important phenomena to consider during solidification is the growth of equiaxed dendritic grains. During their growth, the liquid melt surrounding them is enriched in chemical species. Their growth is governed by the concentration field in their vicinity. When these grains are close to each other, their solute diffusion fields overlap, which leads to a modification of their growth kinetics [1, 2]. These interactions depend to a large extent on the spatial arrangement of the grains, in other words, they depend on how the grains are positioned relative to each other, especially in their close neighbourhood. Commonly used volume-averaged mean-field models for equiaxed solidification [3, 4] are based on overly simplistic assumptions on grain interactions. They describe them via ad hoc constitutive relations while ignoring the spatial arrangement aspect. Consequently, this limits their predictive capabilities.

In this paper, the aim is to highlight the influence of the spatial arrangement of equiaxed grains on their growth kinetics during solidification. Our approach consists in performing full-field numerical simulations of an ensemble of about 100 equiaxed grains using the grain envelope model (GEM) [5]. We compare simulation results for different grain arrangements, namely periodic and random arrangements and notice a major difference in terms of the overall growth

IOP Conf. Series: Materials Science and Engineering 1281 (2023) 012054 doi:10.1088/1757-899X/1281/1/012054

kinetics of an ensemble of grains. We then analyse individual grains in the ensemble and their interactions with their neighbourhood. Based on this analysis, we propose an explanation of the difference in behaviour between random and periodic arrangements. Furthermore, we highlight the shortcomings of state-of-the-art mean-field models in predicting growth, arising from the lack of consideration of grain arrangement.

2. Mesoscopic full-field simulations

2.1. Grain envelope model

The mesoscopic grain envelope model (GEM) [5, 6, 7] is a simulation tool for grain growth during solidification, developed with the aim to bridge the gap between microscopic and macroscopic solidification models. The description of dendrites by an envelope requires much coarser mesh in comparison to phase-field models since it does not describe the complexity of the solid-liquid interface. The envelope is a smooth hypothetical surface that connects the active dendrite branches, implying that there are both solid and liquid phases inside the envelope. The growth of the dendrites is determined by the movement of the envelope surface. The envelope velocity is given by a stagnant-film formulation of the Ivantsov solution for the dendrite tip velocity. This analytical solution relates the Péclet number of the tip to the local supersaturation in the liquid at a distance δ (the stagnant-film thickness) from the envelope. The supersaturation in the liquid outside the envelope (extragranular liquid) is obtained from the numerical solution of solute transport. The local supersaturation field in the vicinity of a grain is influenced by the solute enrichment caused by the growth of nearby grains. Therefore, the stagnant-film formulation accounts for the influence of solutal interactions between the grains on the envelope growth velocity. The liquid inside the envelope (intragranular liquid) is assumed to be at thermodynamic equilibrium and the fraction of the solid phase is calculated using a Scheil model.

2.2. Simulation parameters

We performed 3D numerical simulations of equiaxed grain growth during solidification of a binary alloy in isothermal conditions. The crystal structure has a cubic symmetry and dendrite branches growth along the $\langle 100 \rangle$ cubic directions. The physical parameters are the initial supersaturation, $\Omega_0 = (C_\mathrm{l}^* - C_0) / (C_\mathrm{l}^* - C_\mathrm{s}^*)$, and the grain density, N_V (i.e., the number of grains per unit volume). C_l^* and C_s^* are the liquid and solid equilibrium concentrations, respectively, and C_0 is the initial concentration of the liquid. Lengths are expressed in units of $L_\mathrm{diff} = D_\mathrm{l}/V_\mathrm{Iv}$, which is a characteristic diffusion length at the primary dendrite tip, where D_l is the diffusion coefficient in the liquid and V_Iv is the steady-state (Ivantsov) velocity of a primary dendrite tip growing into an infinite melt at a supersaturation Ω_0. The characteristic time is defined as $\tau = D_\mathrm{l}/V_\mathrm{Iv}^2$. The simulations were carried out for an initial supersaturation of $\Omega_0 = 0.05$ and for two grain densities, $N_V = 0.125 L_\mathrm{diff}^{-3}$ and $N_V = 0.35 L_\mathrm{diff}^{-3}$, called low and high density, respectively.

In order to highlight the influence of the spatial arrangement of the grains on the overall growth kinetics, we performed 4 simulations for each grain density: 3 with different periodic grain arrangements and one with a random grain arrangement. In the periodic cases, the grains were arranged in simple cubic (SC), body centred cubic (BCC) and face centred cubic (FCC) lattices with the $\langle 100 \rangle$ directions aligned with the cubic directions of the crystal lattice. The random cases had 125 grains, with both random positions and orientations. The grain arrangements are displayed in Fig. 1.

The stagnant-film thickness of the GEM is $\delta = 0.2 L_\mathrm{diff}$ and we used a regular discretisation grid with a mesh spacing $\Delta x = L_\mathrm{diff}/10$. All grains are initialised at $t = 0$ by spherical envelopes of radius $R_0 = 0.3 L_\mathrm{diff}$ and the simulation domain walls were considered to be symmetry planes.

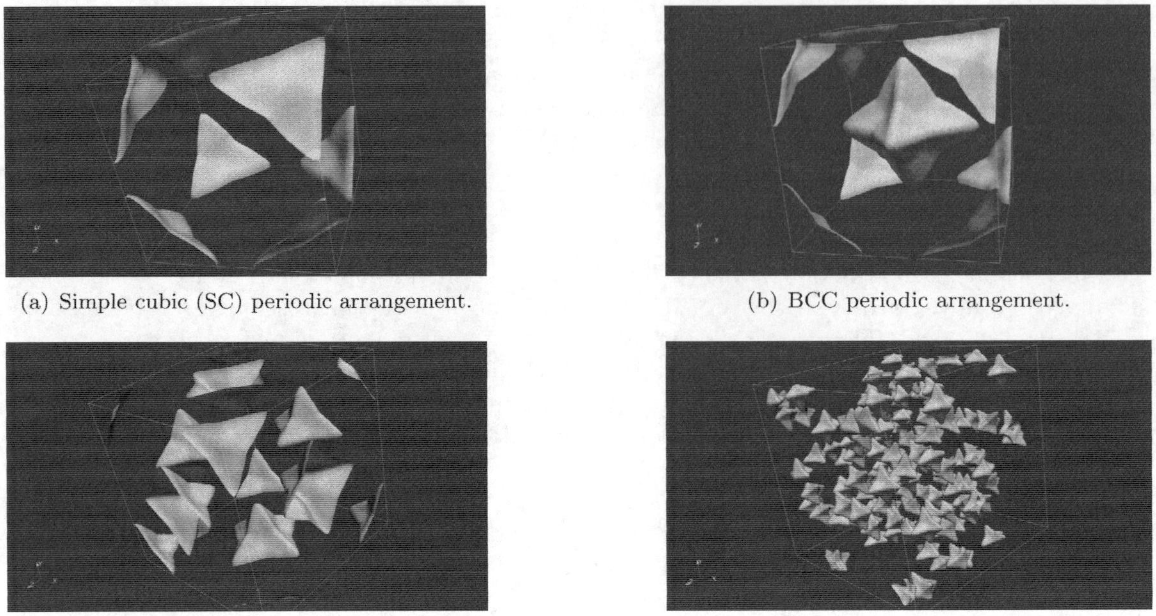

(a) Simple cubic (SC) periodic arrangement. (b) BCC periodic arrangement.

(c) FCC periodic arrangement. (d) Random arrangement.

Figure 1. Spatial arrangements of grains used in the simulations.

3. Simulation results

In this section we compare the evolution of average quantities of the ensembles for periodic and random arrangements. The link between the average quantities can be represented by the global solute mass balance, written in dimensionless form as:

$$g_e \frac{\partial \langle \Omega \rangle}{\partial t} = -\langle \Omega \rangle \frac{\partial g_e}{\partial t} + \frac{1}{V_0} j_S \tag{1}$$

where $\langle \Omega \rangle$ is the average supersaturation in the extragranular liquid, g_e is the extragranular liquid fraction and V_0 is the dimensionless volume of the system. The solute flux j_S diffusion from the grains to the extragranular liquid is given by:

$$j_S = -\int_{S_{\text{env}}} \vec{\nabla}\Omega \cdot d\vec{S} \tag{2}$$

where S_{env} is the surface of the envelopes in the system, and $d\vec{S}$ points outward from the envelopes.

In the GEM, the evolution of the envelope fraction is related to the tip velocity. Then, the evolution of the average supersaturation $\langle \Omega \rangle$ is given by the solute balance Eq. (1) and results from both the envelope growth and the solute flux diffusion from the grains to the extragranular liquid.

To compare the growth of crystalline grains with different initial grain arrangements, we will first examine the evolution of the average size of the grain envelopes. Next, we will investigate the average supersaturation in the extragranular liquid, as it plays a key role in determining the growth kinetics. Lastly, we will look at the average solute flux density across the grain envelopes to understand its impact on the evolution of the solute supersaturation in the extragranular liquid.

3.1. Average grain size

The average grain size is expressed as the radius of a sphere whose volume is equal to the envelope volume. The evolution of this quantity is shown in Fig. 2. For any given grain density, the increase of the equivalent radius is much slower for the random arrangement. When comparing two grain densities, we note faster growth for the lower density. As expected, the final equivalent radius is larger for the lower grain density.

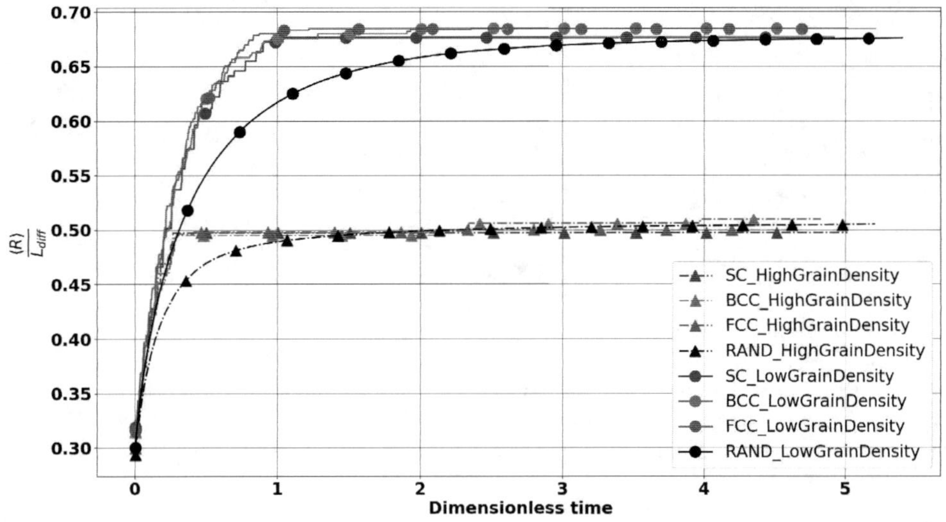

Figure 2. The evolution of the equivalent radius for different spatial arrangements of grains: SC (blue), BCC (red), FCC (green), and random (black). Low and high grain densities are indicated by circular and triangular symbols, respectively. The systems sizes in terms of edge length for the high density are: SC(1.41Ldiff), BCC(1.78Ldiff), FCC(2.23Ldiff), random(7.07Ldiff) for the low grain density the length of every edge is multiplied by $2^{1/2}$

3.2. Average supersaturation

The average supersaturation in the extragranular liquid evolves from the initial supersaturation $\Omega_0 = 0.05$ and decreases gradually to zero. The evolution is shown in Fig. 3. It appears that the lower the grain density the slower the decrease of the average supersaturation, regardless of the arrangement. In addition, for a given grain density, the decrease of the supersaturation is much slower in the random arrangement than in the periodic arrangements. The systems with BCC and FCC arrangements have almost identical evolutions. Only a slight difference is observed for the system with SC arrangement, for which the growth direction of the dendrites is towards the first neighbours of the grain arrangement. We also note that this difference is more pronounced for the low grain density.

3.3. Average solute flux density

We now compute the average solute flux density j_S/S_{env} diffused from the envelopes to the extragranular liquid using Eq. 2. The result is presented in Fig. 4. As expected, we observe a decrease of the flux density. The serrations, related to the discretization of the GEM model, do

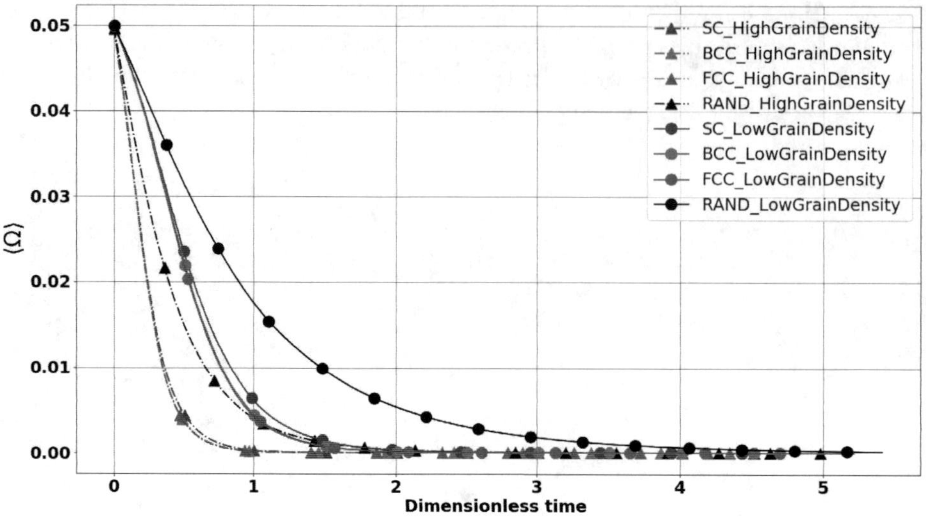

Figure 3. The evolution of the average supersaturation in the extragranular liquid for different spatial arrangements of grains: SC (blue), BCC (red), FCC (green), and random (black). Triangle mark for high grain density and circle for low grain density.

not hinder the comparison between the different cases under study and will not be discussed in this paper.

Comparing this flux density for different grain arrangements and a given grain density, we notice a slightly weaker flux density for the random arrangement. Subsequently this tendency is reversed and the flux density becomes higher for the random arrangement. Moreover, if we compare two grain densities for the same arrangement, we notice a stronger flux density for the lower grain density.

Overall, we notice that ensembles with a random arrangement of grains evolve differently from those with periodic arrangements. This disparity is due to the influence of grain arrangement on diffusive interactions between the grains. In the next section we present an analysis of individual grains in order to explain the origin of these differences.

4. Investigation of individual grains in a random arrangement

In a periodic grain arrangement all grains have exactly the same environment and their growth is therefore identical and equal to the global average. In a random arrangement, on the other hand, each grain has a different environment and grows differently in terms of size and shape. The results in the preceding section show that the overall (average) growth kinetics of the randomly arranged ensemble is not equivalent to that of a single grain with a size equal to the ensemble average. In this section we analyse the growth of individual grains in an attempt to gain understanding on how the contributions of individual grains result in the behaviour of the ensemble.

4.1. The range of interactions during growth

To analyse individual grains, we first need to define the environment that affects their growth. All grains interact via the diffusion field in the liquid, but we expect that the influence of

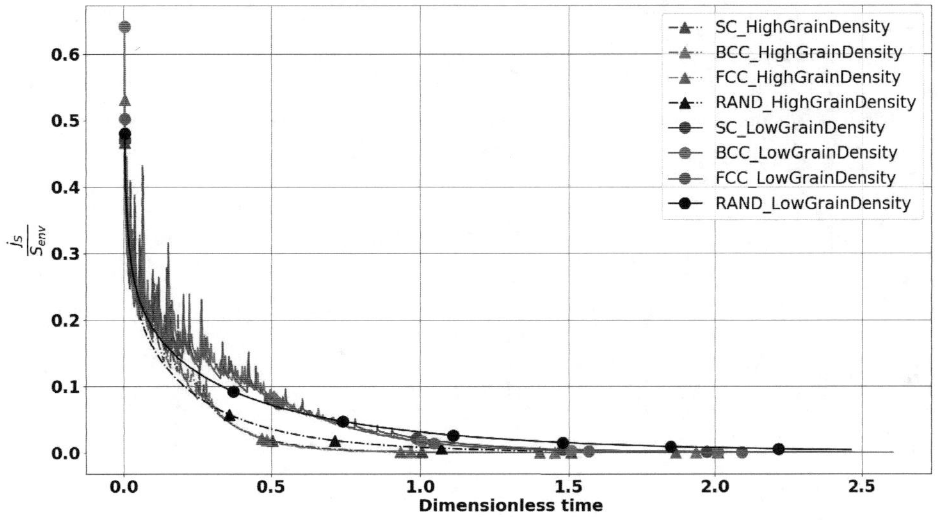

Figure 4. The evolution of the dimensionless global average solute flux density for different spatial arrangements of grains: SC (blue), BCC (red), FCC (green), and random (black). Low and high grain densities are indicated by circular and triangular symbols, respectively.

distant grains is screened off by closer grains. As a consequence, we expect that the strongest interactions are with the first nearest neighbour grains and we will focus on these interactions.

A convenient way to define the nearest neighbours is by a Voronoi tessellation constructed from the initial grain seeds. A Voronoi cell's nearest neighbors are those that share common cell faces. This space decomposition method is an effective way to define the local environment of grains, as the Voronoi cell of a grain represents the region of points that are closer to that grain than to any other grain. Furthermore, faces of a Voronoi cell are halfway between the neighboring grains and therefore the diffusion flux over the cell faces can be expected to be small. In order to test the validity of this assumption, we have calculated the solute flux densities across the Voronoi cell faces for each grain. Figure 5 shows the evolution of these flux densities for four different cells.

By comparing these solute flux densities to those from the grain envelopes to the extragranular liquid within the Voronoi cells, we can conclude that the solute exchange between Voronoi cells is insignificant. This can be attributed to the fact that the solute diffused by each pair of neighboring grains reaches the shared Voronoi cell face at approximately the same time. As the solute fluxes arriving at this face have similar magnitudes and are in opposite directions, they result in a negligible perpendicular component.

Because the solute flux across the Voronoi cell faces is negligible, we can consider each Voronoi cell to be independent from the rest of the system. Hence, We can consider that the interaction of a grain with the ensemble depends on its Voronoi cell volume. For instance, the larger the cell the weaker the interaction since the solute enrichment caused by the grain growth diffuses into a larger volume of liquid. Consequently, the solute concentration in the vicinity of the grain increases more slowly, resulting in a weaker impact on the concentration gradient at the envelope. This is corroborated by the observations we have made in the previous section on the influence of grain density on the evolution of the average supersaturation.

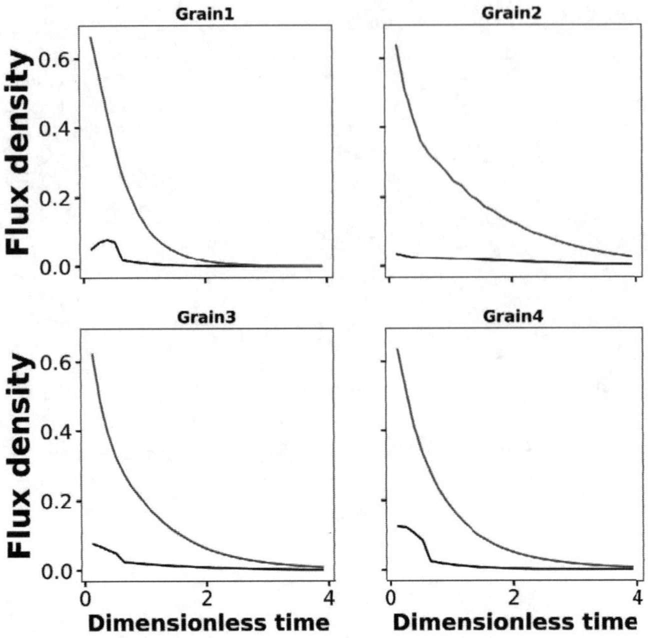

Figure 5. The evolution of the average solute flux density from the grain to the extragranular liquid inside the cell (blue) and of the average solute flux density across cell faces (black). These extractions were made on the low grain density simulation.

4.2. The evolution of individual grains

In order to highlight the disparity in behaviour among randomly dispersed grains, we have extracted the evolutions of the grain envelope size in terms of equivalent radius (Fig. 6) and of the average supersaturation in the extragranular liquid (Fig. 7), for the grains with the largest and the smallest Voronoi cells.

We observe a significant difference between grains within the largest and smallest cells. The grain in the largest cell exhibits faster growth and attains a larger final size. Conversely, the average supersaturation in the smallest cell drops rapidly, leading to an early attainment of equilibrium. Despite the stronger interactions with neighboring grains resulting in slower growth for the grain in the smallest cell, the liquid within the cell reaches equilibrium quickly, ultimately halting growth.

This is due to the fact that the time needed for the solute to diffuse within a cell of diameter L, scales as L^2. Therefore, as there is essentially no flux between the cells, the time needed for a cell of diameter L to reach equilibrium should approximately scale as L^2. This is in agreement with a simple analysis of Fig. 6 and Fig. 7. Indeed, if we note R_1 the final grain size in the smallest cell, we observe that the average final grain size is close to $2R_1$ and the final grain size in the largest cell is close to $3R_1$. These size relations that reflect the size relations between the corresponding cells, should lead to equilibrium times for the average and largest cells which are, respectively, approximately 4 times and 9 times larger than the the one observed for the smallest cell. This is indeed what we observe in Fig. 7.

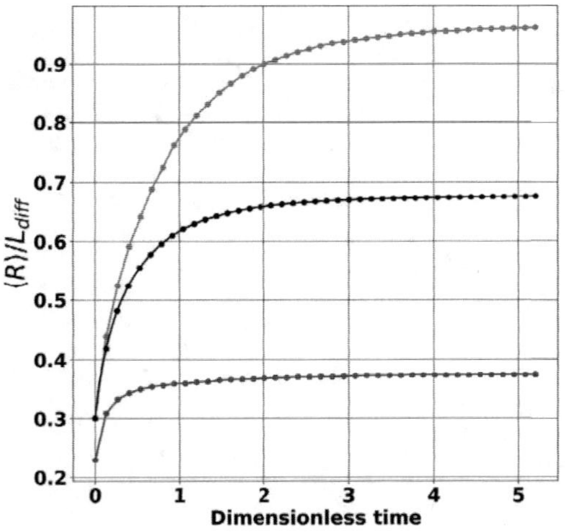

Figure 6. Evolutions of the grain envelope radius of the grain inside the largest Voronoi cell (red), the grain inside the smallest cell (blue) and the overall average (black). These extractions are for the low grain density simulation.

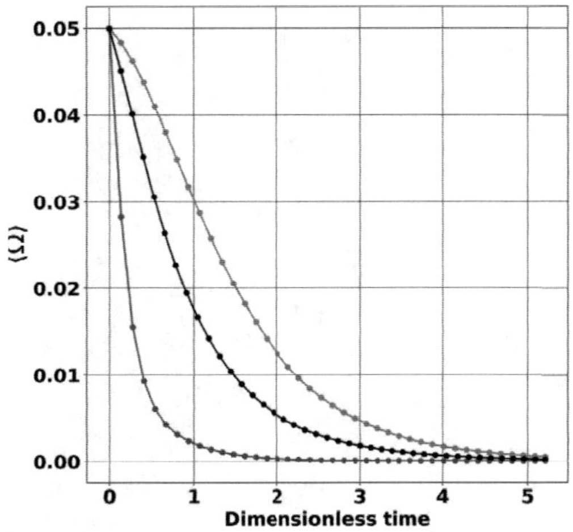

Figure 7. Evolutions of the average supersaturation in the extragranular liquid inside the largest Voronoi cell (red), inside the smallest cell (blue) and the global average in the system (black). These extractions are for the low grain density simulation.

4.3. The origin of the disparity between periodic and random grain arrangements

Figure 8 illustrates the distribution of Voronoi cell sizes in a random arrangement through a histogram. The vertical dashed lines divide the histogram into five classes, each representing one fifth of the total number of cells (125). The volume occupied by cells in each class is shown in red at the top of the figure. The results show that the largest 20% of cells in the distribution occupy around 37% of the total volume of the system, while the smallest 20% of cells only occupy 8% of the total volume. It should be noted that larger cells take longer to reach equilibrium, as shown in Figure 7. This means that a larger fraction of the total volume of the random system is occupied by large cells, which exhibit slow decrease in supersaturation, compared to small cells that reach equilibrium quickly.

Therefore, the overall behavior of the random ensemble is primarily influenced by the larger cells. Additionally, the average grain radius in the random arrangement takes longer to reach its final size compared to the periodic arrangement. This can be attributed to the long time needed by the large cells to reach equilibrium, despite the faster growth rate in large cells due to weaker interactions. Because the large cells are dominant in the random arrangement, this prolongs the driving force for growth of the ensemble.

5. Comparison with mean-field models

Mean-field models are commonly used in simulating phenomena at the process scale due to their low computational cost compared to full-field models. However, these models are based on simplifying assumptions that can compromise the accuracy of predictions. One such assumption is that they represent a population of grains with a single average grain. Additionally, grain interactions are described using a simplified description of solute diffusion around the grains. One of the goals of the present study is to demonstrate the limitations of mean-field models of

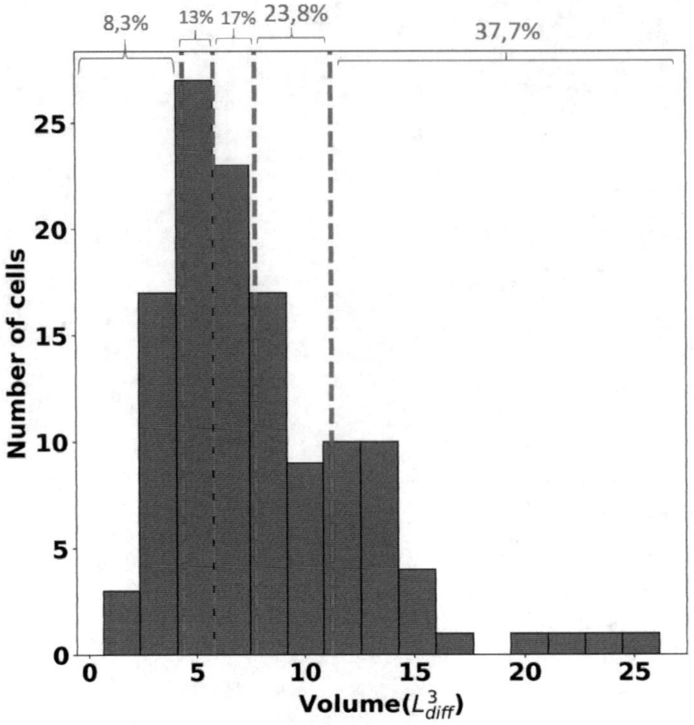

Figure 8.
Distribution of Voronoi cell sizes in the random grain distribution for low grain density. The vertical dashed lines in the graph divide the data into five classes, with each class containing 1/5 of the total number of cells. The volume occupied by the cells in each class is presented at the top of the graph.

equiaxed grain growth that result from the lack of descriptors for grain arrangement.

We compare the evolution of the global grain envelope fraction obtained by the full-field GEM simulations in isothermal conditions with predictions from two mean-field models for equiaxed solidification (Fig. 9). The "Classical model", proposed by Bedel et al. [8], is based on the so-called three-phase model of Wang & Beckermann [4], with an enhancement in terms of the constitutive relation for solute diffusion length in the extragranular liquid. The three phases are the solid, the intragranular liquid (assumed at equilibrium), and the extragranular liquid (considered supersaturated), where the two liquids are separated by the dendrite envelope. The solute exchange between the phases is described by volume-averaged solute balances at the interfaces. The model thus uses similar concepts as the GEM, but describes everything in a volume-averaged sense. Major assumptions in the constitutive relations are that: (i) the envelope has a fixed octahedral shape; (ii) the diffusion flux from the envelope to the extragranular liquid is approximated from an analytical solution of the concentration field around a sphere growing in a closed spherical domain, (iii) the envelope velocity is calculated from an LGK-type model of a freely growing dendrite tip, approximating the far-field concentration by the the average concentration in the extragranular liquid.

The second mean-field model, referred to as the "Upscaled model", was proposed by Torabi Rad et al. [9]. The main concepts are the same as in the Classical model: volume-averaged description, three phases, interface balances. The key difference is that in the Upscaled model the constitutive relations were obtained by upscaling from full-field GEM simulations. The upscaling was done by averaging over a simulated ensemble of equiaxed grains arranged periodically on a BCC lattice [9]. The upscaled constitutive relations describe: (i) the variation of the envelope sphericity during growth, (ii) a diffusion flux fitted to the full-field simulations, (iii) the dependence of the envelope velocity on an effective concentration, different from the extragranular concentration.

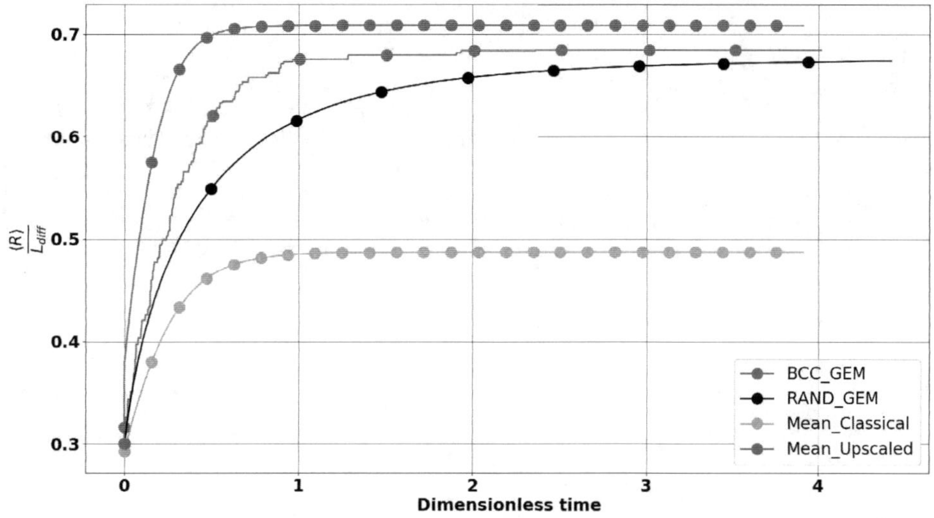

Figure 9. Comparison of evolutions of the equivalent envelope radius for the full-field GEM simulations with random arrangement (black) and BCC (red) arrangements, and for the Classical (orange) and Upscaled (magenta) mean-field models.

The comparisons of the two mean-field with full-field GEM simulations for random and BCC arrangements are shown in Fig. 9. The Classical mean-field model strongly underestimates both the growth rate and the final grain size. This shows that the assumptions contained in this model may be too simplistic to realistically reproduce growth. The Upscaled mean-field model provides a satisfactory prediction for the periodic BCC arrangement, however, it is unable to accurately simulate the growth of the random ensemble. This outcome is expected as the constitutive relations of this model were derived by upscaling simulations for BCC grain arrangements.

6. Conclusions and perspectives

Three dimensional numerical simulations for equiaxed grain growth were performed with a full-field model, for three periodic and for random grain arrangements. A major difference in terms of growth evolution was observed between periodic and random arrangements. Additionally, an investigation on the scale of the individual grains was carried out in order to explain the observations. Furthermore, a description of grain interactions was proposed and used to explain the results. Besides, a comparison with two mean-field models was performed showing the limitations of their predictions. Finally, the description of grains arrangement that we have proposed can pave the way for the enhancement of mean field grain growth models by integrating the grain arrangement and interactions aspect while keeping a reasonable computational cost.

Acknowledgement
The authors acknowledge the support by the French State through the program "Investment in the future" operated by the National Research Agency (ANR) and referenced by ANR-11 LABX-0008-01 (LabEx DAMAS). A part of the required high performance computing resources was provided by the EXPLOR center hosted by the Université de Lorraine.

References
[1] Murphy A, Mathiesen R, Houltz Y, Li J, Lockowandt C, Henriksson K, Zimmermann G, Melville N and Browne D 2016 *Journal of Crystal Growth* **440** 38–46
[2] Bogno A, Nguyen-Thi H, Reinhart G, Billia B and Baruchel J 2013 *Acta Materialia* **61** 1303–1315
[3] Rappaz M and Thevoz P 1987 *Acta Metallurgica* **35** 1487–1497
[4] Wang C and Beckermann C 1993 *Metallurgical and Materials Transactions A* **24** 2787–2802
[5] Steinbach I, Beckermann C, Kauerauf B, Li Q and Guo J 1999 *Acta Materialia* **47** 971–982
[6] Delaleau P, Beckermann C, Mathiesen R H and Arnberg L 2010 *ISIJ International* **50** 1886–1894
[7] Souhar Y, De Felice V F, Beckermann C, Combeau H and Založnik M 2016 *Computational Materials Science* **112** 304–317
[8] Bedel M 2014 *Étude de la formation des structures de solidification et des macroségrégations en coulée semi-continue d'aluminium* Ph.D. thesis université de Lorraine
[9] Torabi Rad M, Založnik M, Combeau H and Beckermann C 2019 *Materialia* **5** 100231

Investigation of silicon sublattice substitution within $(Al,Si)_3Zr$ intermetallics via DFT simulations

Juan-Ricardo Castillo-Sánchez[1,2], Aïmen E. Gheribi[1], Paul Lafaye[1], Georges Salloum-Abou-Jaoude[2], Jean-Philippe Harvey[1*]

[1]Centre for Research in Computational Thermochemistry (CRCT), Department of Chemical Engineering, Polytechnique Montréal, C.P. 6079, Succursale "Downtown", Montréal, Québec H3C 3A7, Canada

[2]Constellium Technology Center, Parc Economique Centr'alp, 725 rue Aristide Bergès, CS10027, Voreppe 38341 CEDEX, France

E-mail: `jean-philippe.harvey@polymtl.ca`

Abstract. Aluminum alloys commonly contain Si as an impurity or alloying element. The energetic behavior of Si within multiple compounds and solutions is incorporated inside thermochemical packages, such as FactSage. This tool allows determining the Si partitioning within complex multiphasic systems. Recent experimental research suggests that Si can be found within Al_3Zr-based intermetallics. Nevertheless, current FactSage databases do not consider the potential substitution of Si within the Al_3Zr-$D0_{23}$ solid solution. In this work, Si substitution within the $(Al,Si)_3Zr$-$D0_{23}$ phase was investigated by means of first-principles calculations. Replacement of Al atoms by Si resulted in a negative enthalpy of mixing, indicating that Si substitution is energetically enabled. The density of states (DOS) for both a Si-diluted $(Al,Si)_3Zr$ and a non-Si-doped (Al_3Zr) simulation cells were analyzed. It is shown that (even in dilution), Si significantly impacts the electronic structure of the Al_3Zr-$D0_{23}$ structure. Specifically, the presence of Si localizes electrons in the p orbital of Al, and increases the DOS of the d_{xy}, d_{xz}, and d_{yz} sub-orbitals of Zr at low energies. Thus, yielding a coupled effect that stabilizes the $D0_{23}$ intermetallic. These findings are a benchmark for the future integration of a Si-based end-member within the Al_3Zr-$D0_{23}$ solid solution of FactSage databases.

Keywords: Al_3Zr, Si-substitution, DFT, enthalpy of mixing, density of states

1. Introduction

The energetic description of peritectic phases such as the Al_3Zr intermetallic is highly important for aluminum alloy design [1]. In complex systems (i.e. with multiple alloying elements), these intermetallics precipitate as ordered multicomponent solid solutions upon solidification. The substitution of compatible elements within Al_3Zr intermetallics can be thermodynamically described using Gibbs free energy models, such as the Compound Energy Formalism (CEF)[2]. These models can be integrated into self-consistent databases of computational thermochemistry packages. To construct these models, mixing parameters are required. As an example, the heat of mixing can be obtained by *in-situ* synthesis [3] for specific stability ranges (as a function of the concentration, temperature, or pressure) of a given phase. Nowadays, the enthalpy of mixing between end-members making up a phase (e.g. using the CEF approach) can be obtained using computational quantum mechanics methods. For instance, Density Functional Theory (DFT) calculations are broadly employed to describe stable and meta-stable phases [4, 5]. Computational-based methods are at the core of the next generation of thermodynamic models [6]. Ultimately, these tools allow a constant updating of thermochemical databases, which is a critical aspect to study new alloys. For casting, thermochemistry is employed for two main purposes: on the one hand, the precipitation of desired primary phases can be easily anticipated by calculating isoplethal cross-sections as a function of temperature and with variations in the alloying elements. On the other side, the formation of secondary phases can be approximated by performing Scheil-Gulliver Cooling calculations [7]. Micro-segregation inside phases can also be predicted with the Scheil-Gulliver technique. Constant improvement on thermochemical databases are therefore required to properly describe the solidification of aluminium alloys.

Recent experimental observations indicate that Si can be found in Al_3Zr-based intermetallics. As an example, Salloum-Abou-Jaoude *et al.* [8] reported the presence of $Al_3(Zr_{1-x-y}, Ti_x, Si_y)$ intermetallics at the core of α-grains of AA6082 Direct Chill (DC)-cast billets. In an independent work, Jo *et al.* [9] indicated that Si can substitute Al sites in the complex $(Al,Si)_3(Zr,Ni,Fe)$ $D0_{23}$-like intermetallic, which was observed in the Al-14Si-3Cu-4.5Ni alloy. These findings, coupled with numerical tests, are required to model Si occupancy within the $D0_{23}$ solution of FactSage [10]. Other thermochemical-software developers can benefit from these results if Si has not been yet included in their $D0_{23}$ solution models, which is the case of the $(Al, Li, Mg)_3(Hf, Sc, Ti, Zr)$-$D0_{23}$ phase model in the Al-based Alloy Database (TCAL8) of Thermo-Calc [11].

In this work, the energetic description of the partially ordered $(Al,Si)_3Zr$-$D0_{23}$ solid solutions is carried out using DFT calculations. Supercells of Al_3Zr, Si_3Zr, and $(Al,Si)_3Zr$ (Si/Al-diluted) of the pseudo-binary solution are generated and equilibrated. Results are later integrated into an enthalpy of the mixing model derived from the partial enthalpies of infinite dilution. Following an equivalent methodology (i.e. with an infinite dilution approach), the lattice parameters of the tetragonal phase are calculated. Finally, the electronic density of states (DOS) of a Si-doped structure is analyzed and compared to a non-Si-doped Al_3Zr-$D0_{23}$ structure. The effect of the dopant (Si) on the electronic distribution of the p orbital of Al, and the d orbital of Zr are discussed.

2. Methodology

The Vienna ab initio Simulation Package (VASP) [12, 13, 14, 15] was used to execute plane wave Density Functional Theory (DFT) calculations to estimate the enthalpy of mixing of the $(Al,Si)_3Zr$-$D0_{23}$ solid solution. Pseudopotentials built with the projector-augmented wave (PAW) approach [16, 17] were utilized for Si, Al and Zr. The exchange-correlation functionals were described through the generalized gradient approximation (GGA) [18, 19]. Convergence in the cell volume and energy was accomplished with a cut-off energy of 520 eV. Sampling of the Brillouin zone was conducted using the Monkhorst-Pack scheme with a $7 \times 7 \times 2$ k-point mesh for the $D0_{23}$ configurations. An accuracy of more than 0.01 meV in the system's energy was achieved by using a Gaussian smearing parameter σ of 0.02 eV. The convergence criterion for self-consistent field (SCF) electronic calculations was $1\text{x}10^{-5}$ and 0.02 eV/Å for ionic relaxation loop. The total energy was estimated in the NPT statistical ensemble, i.e the atomic positions, the equilibrium lattice, cell shape and cell volume were allowed to be relaxed. Each simulation box contained 128 atoms. Al and Si were randomly substituted to generate the pseudo-binary $D0_{23}$-based configurations. According to the methodology reported by Sluiter & Kawazoe [20], the corresponding solution enthalpies ($\Delta h^{\text{Al-rich}}_{soln}$ and $\Delta h^{\text{Si-rich}}_{soln}$) can be computed as:

$$\Delta h^{\text{Al-rich}}_{soln} = \left[\frac{\partial \Delta h_{mix}}{\partial x}\right]_{x=0} \qquad \Delta h^{\text{Si-rich}}_{soln} = -\left[\frac{\partial \Delta h_{mix}}{\partial x}\right]_{x=1} \tag{1}$$

The molar enthalpy of mixing (Δh_{mix}) can be approximated as follows [20]:

$$\Delta h_{mix} \approx \Delta h_{soln}^{\text{Si-rich}} x^2(1-x) + \Delta h_{soln}^{\text{Al-rich}} x(1-x)^2 \tag{2}$$

x stands for the Si substitional ration [Si/(Al+Si)] within the $(Al_{1-x},Si_x)_3Zr$ pseudo-binary solid solution. In an equivalent manner, variations in the lattice parameters (Δa and Δc) were computed using infinite dilution approximations: $\Delta a \approx \Delta a_{soln}^{\text{Si-rich}} x^2(1-x) + \Delta a_{soln}^{\text{Al-rich}} x(1-x)^2$ and $\Delta c \approx c_{soln}^{\text{Si-rich}} x^2(1-x) + \Delta c_{soln}^{\text{Al-rich}} x(1-x)^2$, respectively. Where:

$$\Delta a_{soln}^{\text{Al-rich}} = \left[\frac{\partial a}{\partial x}\right]_{x=0} \qquad \Delta a_{soln}^{\text{Si-rich}} = -\left[\frac{\partial a}{\partial x}\right]_{x=1} \tag{3}$$

$$\Delta c_{soln}^{\text{Al-rich}} = \left[\frac{\partial c}{\partial x}\right]_{x=0} \qquad \Delta c_{soln}^{\text{Si-rich}} = -\left[\frac{\partial c}{\partial x}\right]_{x=1} \tag{4}$$

3. Results

The enthalpy of mixing for the $(Al_{1-x},Si_x)_3Zr$-$D0_{23}$ pseudo-binary solid solution is presented in Figure 1. Acceptance of Si atoms within Al sites lowers the energy of the intermetallic phase. The amount of Si added into an aluminum alloy will therefore impact the precipitation of this phase upon solidification. A fine-tuning accounting for this phenomenon is therefore required whether to promote or to suppress the formation of the $D0_{23}$ phase. This is particularly important for recycling, where Si (along with Cu and Fe) is one of the most critical alloying elements for the recovery and revaluation of aluminum alloys [21]. The enthalpy of mixing of the $(Al_{1-x},Si_x)_3Zr$-$D0_{23}$ solid solution modeled from the DFT calculations of the partial enthalpies (solid line in Figure 1) does not take into account the influence of the chemical ordering of the solid solution. Therefore the model exhibits different predictions than those using the Al:Al:Si:Zr and Al:Si:Si:Zr ordered compounds [22] (black squares in Figure 1). It is particularly interesting to note that the ordered configurations from [22] have a more negative enthalpy of mixing values than those predicted by the partial-enthalpy-based model. Those calculations hence show that the substitution of Al by Si has a stabilizing effect on the $(Al,Si)_3Zr$-$_{23}$ solid solution; and that the substitution is ordered [22]. Other elements play an important role in the stability of the $D0_{23}$ phase, such as Ti. Karpets *et al.* [23] have reported that only a fraction of Zr/(Zr+Ti)=0.1 is required to stabilize the $D0_{23}$ phase in equilibrium with the Al_3Ti-$D0_{22}$-based intermetallic for Al-Ti-Zr alloys. Another example is the case of V, whose substitution has already been studied by Han *et al.* [24]. Note that the substitution of Ti and V (among other elements) is already incorporated within the multi-component $D0_{23}$ solution model inside FactSage databases [10]. Nevertheless, acceptance of Si has not been integrated yet. As such, the enthalpy of mixing calculated in this work can be used to account for this event using the Compound Energy Formalism (CEF)[2]. Likewise, current experimental evidence of Si acceptance within the Al_3Zr phase [8, 9], coupled with additional experimental work, can be used to tune the thermodynamic model of this solid solution.

The evolution of the a and c lattice parameters is presented in Figure 2. The lattice constants for the reference Al_3Zr-$D0_{23}$ (no Si-doped) structure were a=4.012 Å and c=17.374 Å, which are consistent with the values reported the literature: a=4.015 Å and c=17.454 Å from Hu *et al.* [25]; a=4.018 Å and c=17.349 Å from Huang *et al.* [26]; a=4.008 Å and c=17.297 Å from Ghosh *et al.* [27]. On the other side, lattice parameters for the metastable Si_3Zr-$D0_{23}$ compound were a=3.691 Å and c=19.385 Å. The addition of Si within the solid solution decreased the a parameter, whereas the c constant was increased. Variations in both a and c parameters are necessary to quantify the lattice misfit of intermetallic phases with respect to the Al-FCC matrix [28]. For hyper-peritectic compositions, this information is useful to understand the grain refinement mechanisms [29]. In this case, precipitation of the α-Al is believed to be promoted by the phase coherency of the intermetallic and the Al-FCC [29]. Meaning that the

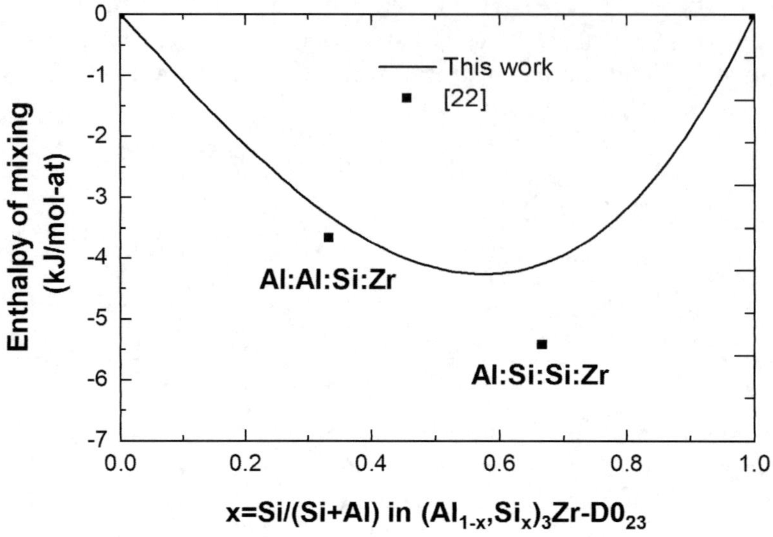

Figure 1. Enthalpy of mixing for the $(Al_{1-x},Si_x)_3Zr$ pseudo-binary solid solution at 0 K (solid line) compared to DFT-GGA data from Castillo-Sánchez *et al.* [22] (squares). For the latter, site occupancy is presented as A:B:C:D where A=4c, B=4c, C=4e and D=4e sites of the $D0_{23}$ structure

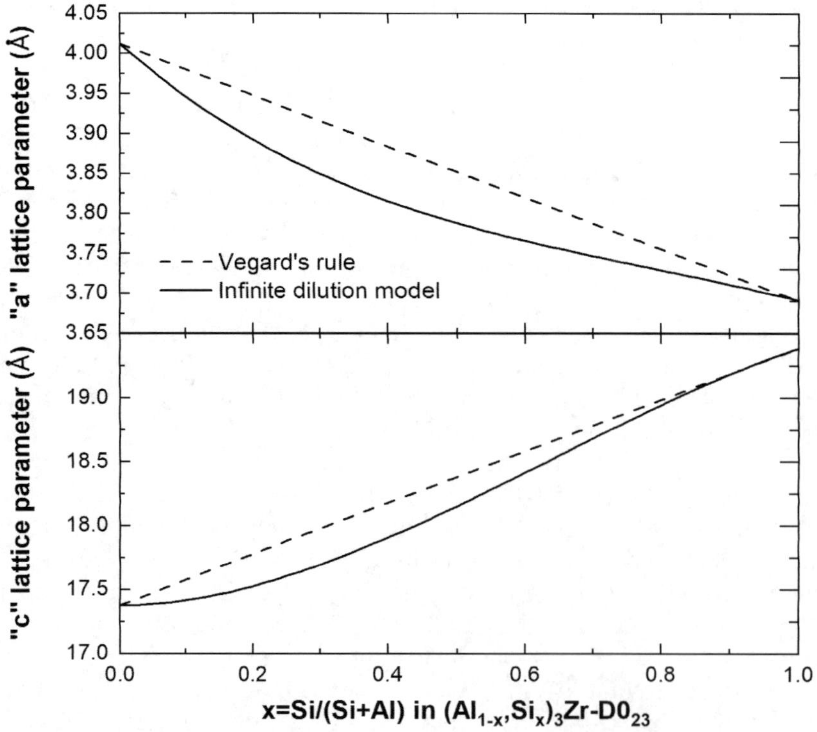

Figure 2. *a* and *b* lattice parameters of the $(Al_{1-x},Si_x)_3Zr$ pseudo-binary solid solution.

IOP Conf. Series: Materials Science and Engineering 1281 (2023) 012055 doi:10.1088/1757-899X/1281/1/012055

intermetallic phases can potentially act as nucleation sites [8]. Here, the complete pseudo-binary solid solution is presented. Nevertheless, experiments are required to correlate the variation of the lattice parameter according to the Si occupancy from the chemical analysis. For this case, crystallography orientation relationships can also be derived based on the lattice parameters of both Al-FCC and intermetallic phases. The lattice parameter of pure Al is reported as 4.046 Å from DFT calculations using the Generalized Gradient Approximation (GGA) [30]. In a rough manner, it can be observed that acceptance of Si increases the mismatch of a(Al-FCC)|a($D0_{23}$) and $4a$(Al-FCC)|c($D0_{23}$).

The stabilization effect of Si on the $D0_{23}$ phase has furthermore been studied from an electronic point of view. In the first place, the substitution of one atom of Si in an Al site of the $D0_{23}$ supercell (128 atoms) resulted in a considerable decrease of the Fermi energy ($E_{fermi}^{\text{Si-doped}}$-$E_{fermi}^{\text{Non-Si-doped}}$=-0.6875 eV). The total and partial density of states (DOS) of a reference Al_3Zr-$D0_{23}$ structure (non-Si-doped) and the Si-doped $D0_{23}$ simulation cell are displayed in Figures 3 and 4, respectively. At first glance, the total DOS is more populated at lower energy due to the presence of Si (Figure 3).

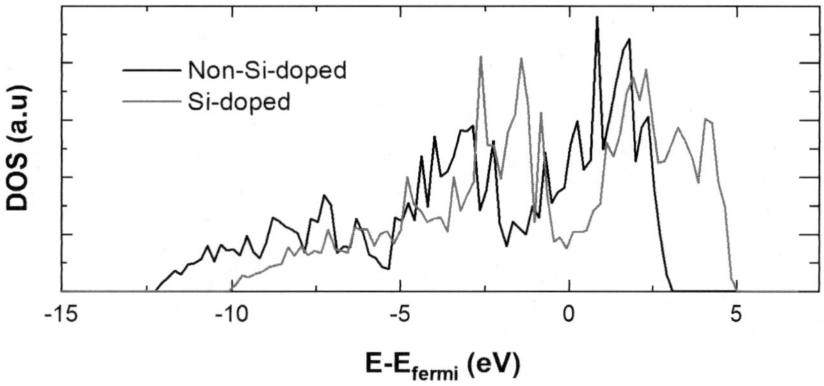

Figure 3. Total density of states (DOS) for a non-Si-doped Al_3Zr-$D0_{23}$ structure (black line), compared to and Si-doped [$(Al,Si)_3Zr$-$D0_{23}$] simulation cell.

Changes on the partial DOS of the p orbital of Al and the d orbital of Zr induced by the Si-sublattice substitution are reported in Figure 4. For the p orbital of Al, the electronic density of states increased at lowers energies when Si is incorporated. Consequently, electrons are localized at lower energy levels, which is characterized by the sharpening of the DOS at low energies and the flattened trend at higher energies. This is reflected in the strong Al-Si interactions promoting the stabilization of the $D0_{23}$ structure (for a Si-diluted scenario). Furthermore, DOS at low energies are also promoted for the d orbital of Zr (Figure 4). The p-suborbitals of Al adjusted in an equivalent manner; however, the d_{xy}, d_{xz}, and d_{yz} suborbital of Zr were the most altered (Figure 5). For such suborbitals, the addition of Si shifted the density of states to lower energy values when compared to the non-Si-doped structure. Thus, Zr also has an important role in the combined stability effect of the $D0_{23}$ phase induced by Si.

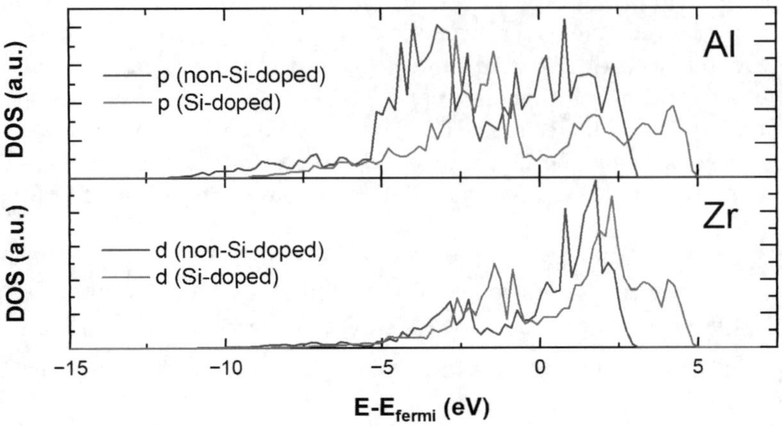

Figure 4. Partial density of states (DOS) for: *p* orbital of aluminium (top) and *d* orbital of Zirconium (bottom).

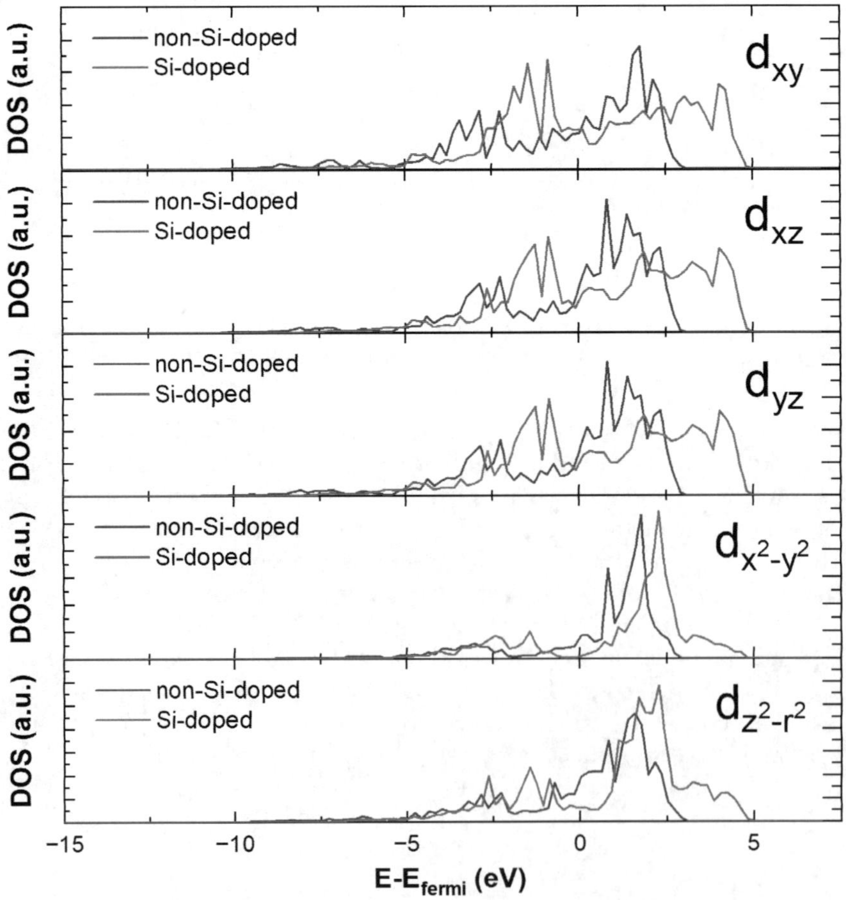

Figure 5. Density of states (DOS) of *d*-suborbitals of Zirconium.

4. Conclusion and perspectives

The enthalpy of mixing of the pseudo-binary $(Al,Si)_3Zr$-$D0_{23}$ solid solution was calculated using Density Functional Theory (DFT) calculations. Substitution of Si within Al sites of the $D0_{23}$-rooted intermetallic displayed a negative heat of mixing, indicating that such substitution is energetically favorable. The a lattice parameter decreases as Si is accepted into the tetragonal phase; the opposite is observed for the c lattice constant. Finally, the electronic density of states (DOS) for the p-orbital of Al and the d-orbital of Zr were evaluated for a Si-diluted scenario (one atom of Al replaced by Si in a 128-atom simulation cell). For both cases, the DOS was increased at low energies by the presence of Si, compared to a non-Si-doped simulation cell. The aforementioned implies that Al and Zr play an important role in the combined stabilization effect of Si on the $D0_{23}$ intermetallic. This is particularly achieved by the localization of electrons of the p-orbital of Al; and a noticeable increase on the DOS at low energies of the d_{xy}, d_{xz}, and d_{yz} sub-orbitals of Zr. These results are the basis for further integration of Si within the thermodynamic modeling of the Al_3Zr-$D0_{23}$-based solid solution inside FactSage databases.

References

[1] Sigli C 2004 *Materials Forum* vol 28 pp 1353–1358
[2] Hillert M 2001 *Journal of Alloys and Compounds* **320** 161–176
[3] Lafaye P, Oishi K, Bourdon M and Harvey J P 2022 *Journal of Alloys and Compounds* **920** 165779
[4] Schönecker S, Li X, Koepernik K, Johansson B, Vitos L and Richter M 2015 *RSC advances* **5** 69680–69689
[5] Jin M, Miao N, Zhao W, Zhou J, Du Q and Sun Z 2018 *Computational Materials Science* **148** 27–37
[6] Harvey J P, Gheribi Ä E, Rincent A, Jofré J and Lafaye P 2020 *Physical Chemistry Chemical Physics* **22** 19999–20013
[7] Pelton A D, Eriksson G and Bale C W 2017 *Metallurgical and Materials Transactions A* **48** 3113–3129
[8] Salloum-Abou-Jaoude G, Eskin D, Barbatti C, Jarry P, Jarrett M and Fan Z 2017 *Light Metals 2017* (Springer) pp 997–1003
[9] Jo M S, Cho Y H, Lee J M, Kim S H, Kang J Y, Jung J G, Kim S B and Jang J i 2020 *Intermetallics* **117** 106667
[10] Bale C W, Chartrand P, Degterov S, Eriksson G, Hack K, Mahfoud R B, Melançon J, Pelton A and Petersen S 2002 *Calphad* **26** 189–228
[11] Andersson J O, Helander T, Höglund L, Shi P and Sundman B 2002 *Calphad* **26** 273–312
[12] Kresse G and Hafner J 1993 *Physical Review B* **47** 558
[13] Kresse G and Hafner J 1994 *Physical Review B* **49** 14251
[14] Kresse G and Furthmüller J 1996 *Computational Materials Science* **6** 15–50
[15] Kresse G and Furthmüller J 1996 *Physical Review B* **54** 11169
[16] Blöchl P E 1994 *Physical Review B* **50** 17953
[17] Kresse G and Joubert D 1999 *Physical Review B* **59** 1758
[18] Perdew J P, Burke K and Ernzerhof M 1996 *Physical Review Letters* **77** 3865
[19] Perdew J P, Burke K and Ernzerhof M 1997 *Phys. Rev. Lett.* **78**(7) 1396–1396
[20] Sluiter M and Kawazoe Y 2002 *EPL (Europhysics Letters)* **57** 526
[21] Soo V K, Peeters J R, Compston P, Doolan M and Duflou J R 2019 *Procedia Manufacturing* **33** 639–646
[22] Castillo-Sánchez J R, Salloum-Abou-Jaoude G, Gheribi A E, Masse J P, Bousser E, L'Espérance G, Lafaye P, Oishi K and Harvey J P 2023 *submitted manuscript*
[23] Karpets M, Milman Y V, Barabash O, Korzhova N, Senkov O, Miracle D, Legkaya T and Voskoboynik I 2003 *Intermetallics* **11** 241–249
[24] Han S, Park S, Huh J S, Lee Z H and Lee H 1997 *Materials Science and Engineering: A* **230** 100–106
[25] Hu H, Zhao M, Wu X, Jia Z, Wang R, Li W and Liu Q 2016 *Journal of Alloys and Compounds* **681** 96–108
[26] Huang H, Wang W, Yuan Q, Rao X, Jing Y, Yi G, Luo L and Liu Y 2019 *Philosophical Magazine* **99** 971–991
[27] Ghosh G, Vaynman S, Asta M and Fine M 2007 *Intermetallics* **15** 44–54
[28] Saha S, Todorova T and Zwanziger J 2015 *Acta Materialia* **89** 109–115
[29] Chen Z and Yan K 2020 *Scientific Reports* **10** 1–8
[30] Fordell T, Salo P and Alatalo M 2002 *Physical Review B* **65** 233408

IOP Conf. Series: Materials Science and Engineering 1281 (2023) 012056 doi:10.1088/1757-899X/1281/1/012056

High-performance GPU computing of phase-field lattice Boltzmann simulations for dendrite growth with natural convection

Tomohiro Takaki[1], Shinji Sakane[1], and Ryosuke Suzuki[2]

[1] Faculty of Mechanical Engineering, Kyoto Institute of Technology, Matsugasaki, Sakyo-ku, Kyoto 606-8585, Japan.
[2] Graduate School of Science and Technology, Kyoto Institute of Technology, Matsugasaki, Sakyo-ku, Kyoto 606-8585, Japan.

Email: takaki@kit.ac.jp

Abstract. The effect of natural convection on dendrite morphology is investigated through three-dimensional large-scale phase-field lattice Boltzmann simulations using a block-structured adaptive mesh refinement scheme with the mother-leaf method in a parallel-GPU environment. The simulations confirmed that downward buoyancy enhances the growth of the primary and secondary arms, and upward buoyancy delays the growth of those arms. In addition, the effect of natural convection on the solidification morphologies gradually decreased as the primary arm tip reached the top of the computational domain and finally stopped. Furthermore, in the longer simulation under purely isothermal diffusive conditions, detachment of the secondary arms owing to curvature-driven fragmentation was observed. A large-scale non-isothermal dendrite growth simulation was also conducted, wherein it was observed that the tip growth rate of the primary arm was delayed, and the secondary arm spacing was larger than that in the isothermal condition.

1. Introduction

In the solidification of metals and alloys in a terrestrial environment, natural convection is caused by buoyancy owing to the density gradient of the liquid, which changes the solidification morphologies. Numerous studies have been conducted to experimentally estimate the primary arm spacing in the directional solidification of alloys on Earth, where thermosolutal natural convection is caused by gravity [1-4]. Directional solidification experiments were conducted under microgravity conditions to remove natural convection, and it was confirmed that the primary arm spacing agreed well with the analytical model [5]. In the case of upward directional solidification on Earth, downward natural convection, which occurs when the solute is heavier than the solvent, expands the primary arm spacing, whereas upward natural convection slightly reduces it [6-8]. When the upward natural convection exceeds a critical value, the upward flow becomes unstable and causes channel segregation, such as freckles [9, 10].

Natural convection affects dendrite fragmentation, which has the potential to cause a columnar-to-equiaxed transition (CET) [11, 12] and refinement of solidified microstructures [13-15]. With the advent of in-situ X-ray observation techniques, dendrite fragmentation of alloys was directly observed in the early 2000s. [16-20]. Currently, dendrite fragmentation is often observed during X-ray characterizations [21, 22]. Possible dendrite-fragmentation mechanisms [23-25] include capillary-driven [26-28], flow-driven [22, 29-32], phase-transformation-driven [25, 33], and deformation-driven [34-37] fragmentation.

Gibbs et al. [32] and Liotti et al. [29] discovered that the frequency of arm fragmentation is highly dependent on the direction of natural convection through upward and downward directional solidification experiments. Abou-Khalil et al. indicated that changes in the dendrite-fragmentation frequency and dendrite-growth rate depend on the gravity level [38]. These results show that local solute enrichment transported by natural convection causes dendrite fragmentation. However, direct observation of the liquid flow in the mushy region is not straightforward, except for specific alloys [39-41]. Recent in-situ X-ray observations have reported that these mechanisms act in combination to cause dendrite fragmentation [42]. Therefore, a computational study to determine the detailed mechanism causing dendrite fragmentation and to clarify the effect of natural convection on dendrite fragmentation is indispensable.

The phase-field (PF) method is an accurate method for investigating the dendrite growth. Neumann-Heyme et al. investigated capillary-driven sidearm pinch-off using PF and compared the PF results with in-situ X-ray observations [26, 27]. Although the PF simulations of dendrite growth with natural convection have been conducted, those have been limited to 2D problems or small 3D domains owing to the enormous computational cost of considering liquid flow [7, 43, 44]. Parallel computing using multiple graphics processing units (GPUs) of the PF-lattice Boltzmann (LB) model [45, 46], coupled with the PF and LB methods enabled us to compute the growth of multiple columnar dendrites with natural convection [8, 47, 48]. Adaptive mesh refinement (AMR) is a powerful numerical scheme for reducing the computational cost of PF simulations [49-53]. Recently, we implemented AMR in parallel computing with multiple GPUs (parallel-GPU AMR) for the PF and PF-LB methods [54-56]. Here, block-structured AMR, dynamic load balancing, and multiple-grid methods were introduced to further reduce the computational cost of PF-LB simulations. The development of the PF-LB parallel-GPU AMR was time-consuming; however, it was an extremely effective tool, specifically in the case of a small solid fraction of dendrite growth in a large computational domain. This condition is encountered when simulating dendrite growth with natural convection.

In our previous study, we performed large-scale simulations of 3D columnar dendrite growth with natural convection by applying PF-LB parallel-GPU AMR [57]. Although the 3D simulations were conducted on thin samples, these were the first 3D PF simulations that considered natural convection. It was concluded that the growth of dendrites with natural convection in 2D and 3D was significantly different. Moreover, the growth of the secondary arms was strongly affected by the flow direction (upward or downward). However, dendrite fragmentation was not observed in these simulations. In this study, simulations similar to those in [57] were conducted for a longer time to investigate the effect of natural convection on dendrite morphology and fragmentation. Moreover, we introduced the mother-leaf method for block-structured AMR to further accelerate the PF-LB parallel-GPU AMR simulation. Additionally, we simulated the columnar dendrite growth under non-isothermal conditions.

2. Simulation

Four simulations of columnar dendrite growth of an Al–3 wt.% Cu alloy were conducted using the domain shown in Figure 1. These correspond to (a) isothermal conditions with downward buoyancy (downward gravity), (b) isothermal purely diffusive conditions, (c) isothermal conditions with upward buoyancy (upward gravity), and (d) non-isothermal purely diffusive conditions and are referred to as Case A, Case B, Case C, and Case D, respectively.

2.1. Phase field-lattice Boltzmann (PF-LB) model

The quantitative PF models of isothermal [58] and non-isothermal [59] solidification in a binary alloy were used for Cases B and D, respectively, where the PF variable ϕ and dimensionless supersaturation u were numerically solved in Case B, and the variables ϕ, u, and dimensionless undercooling θ were numerically solved in Case C. The PF-LB model, where the quantitative PF model of isothermal solidification [58] and the LB model with two relaxation times [60, 61] were coupled and used for Cases A and C, is described in references [57, 61]. In Cases A and C, the liquid flow field $\mathbf{U}$

was solved using the LB method, in addition to ϕ and u. Nonlinear preconditioning was employed in all cases [62].

2.2. Parallel-GPU adaptive mesh refinement (AMR)

Large-scale simulations were accelerated by parallel-GPU AMR [54, 55], in which an octree-based block-structured AMR scheme [63, 64] and parallel computing in multiple GPUs with a dynamic load-balancing scheme were employed. The multigrid scheme [65] was introduced to efficiently conduct a simulation that couples multiple fields: PF, solute concentration, liquid flow, and temperature. Furthermore, we introduced the mother-leaf method for block-structured AMR [66] and achieved approximately 1.5 times speed-up compared with the previous parallel-GPU AMR.

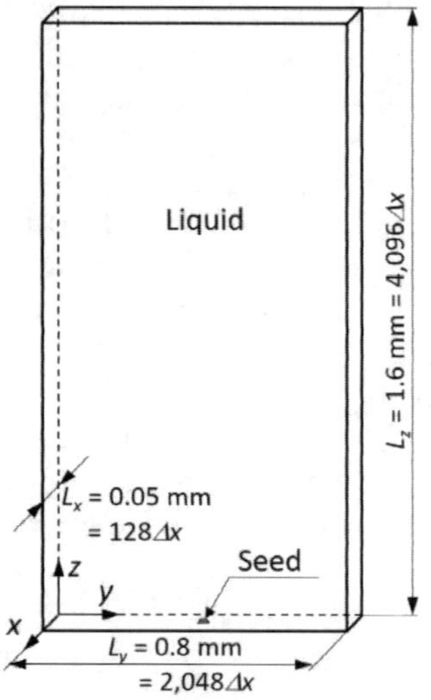

Figure 1. Computational domain and initial conditions.

Table 1. Material parameters and computational conditions

Parameter	Symbol	Value
Solute diffusivity in liquid	D_l	3×10^{-9} m²/s
Solute diffusivity in solid	D_s	3×10^{-13} m²/s
Thermal diffusivity	D_T	4.25×10^{-5} m²/s
Partition coefficient	k	0.14
Anisotropic strength	ε_4	0.02
Gibbs–Thomson constant	Γ	0.24×10^{-6} Km
Liquidus slope	m	−620 K/at. frac.
Melting temperature of pure Al	T_m	933.25 K
Solute expansion coefficient	β_c	−2.25 /at.frac.
Thermal expansion coefficient	β_T	1.245×10^{-4} /K
Minimum grid size	Δx	0.390625 μm
Interface thickness	W_0	Δx
Time increment for ϕ and u	Δt	7.280×10^{-6} s

2.3. Simulation conditions

Figure 1 shows the computational domain and initial conditions. The material parameters of the Al–3 wt% Cu (0.013 at. frac.) and the computational conditions are listed in Table 1. The computational domain size was set to $L_x \times L_y \times L_z = 0.05 \times 0.8 \times 1.6$ mm^3. The number of grids for the domain was $N_{UM} = 128 \times 2{,}048 \times 4{,}096 = 1{,}024^3$ when using a uniform grid size of $\Delta x = 0.390625$ μm. Periodic boundary conditions were set along the x and y directions, and, in the z direction, the non-slip condition for liquid flow, zero Neumann condition for ϕ and u, and Dirichlet condition with $T = 923.06$ K for θ were applied. Initially, the computational domain was filled with a supersaturated liquid with $u_0 = -0.25$ ($\Delta T = 2.2$ K, $T_0 = 923.06$ K), and the solidification started from a small solid seed placed at the center of the bottom surface. In Cases A and C, the gravitational acceleration vectors were set as $\boldsymbol{g} = (0, 0, -g_0)$ and $(0, 0, g_0)$ with $g_0 = 9.81$ m/s^2, respectively. The computational conditions for Cases A–C were the same as those in [57] except for the computational domain size and time.

Based on our previous estimates, the minimum grid sizes for the fields $\mathbf{U}$ and θ in the multigrid method were set as $\Delta x_{\mathrm{LB}} = 2\Delta x$ and $\Delta x_\theta = 4\Delta x$, respectively, where Δx was the grid size used for the fields ϕ and u [65]. The time increments were set as $\Delta t = 7.280 \times 10^{-6}$ s for ϕ and u and $\Delta t_\theta = \Delta t/886$ for θ, which means that the thermal conduction equation was solved 886 times for each step of ϕ and u. To solve the LB equation on the AMR grids, the time increment must be varied with the grid size. Thus, the time increments for the LB equation were set to $\Delta t_{\mathrm{LB}}{}^3 = \Delta t$, $\Delta t_{\mathrm{LB}}{}^2 = 2\Delta t$, $\Delta t_{\mathrm{LB}}{}^1 = 4\Delta t$, and $\Delta t_{\mathrm{LB}}{}^0 = 8\Delta t$, where the superscript indicates the block-level number, as blocks with four levels (levels 0 to 3) were used in the current AMR computing. Each block contained 16^3, 16^3, 8^3, and 4^3 grids to solve for ϕ, u, $\mathbf{U}$, and θ, respectively. The number of grids assigned to each block is different from that of the conventional block structure AMR owing to the introduction of the mother-leaf method [54, 55, 57]. AMR operation, refinement, and coarsening of the blocks were conducted every 104 steps. The simulations were conducted with 3.68×10^6 steps (26.790 s, Case B), 2.22×10^6 steps (16.162 s, Cases A and C), and 0.66×10^6 steps (4.805 s, Case D) in the time increment Δt. Here, the blocks with grids with $-0.998 < \phi < 0.998$ were refined, and the other blocks were coarsened. The D3Q19 model was used to solve the LB equation. The computational domain was decomposed into 4×8 subdomains in the y and z directions, and one GPU was assigned to each subdomain. Parallel GPU computing with 32 GPUs (NVIDIA Tesla P100) was performed using the GPU supercomputer TSUBAME3.0 at the Tokyo Institute of Technology.

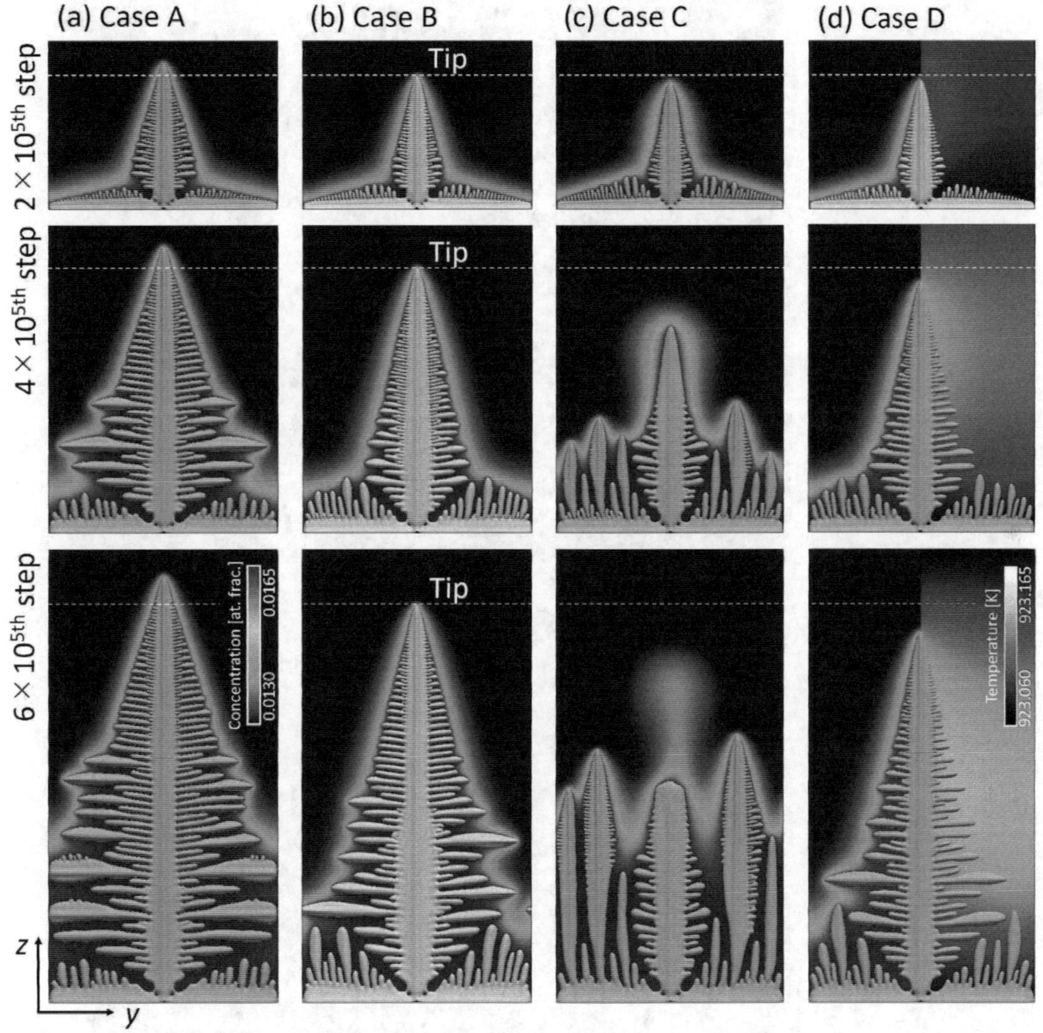

Figure 2. Time evolution of columnar dendrite growth until the $6 \times 10^{5\text{th}}$ step ($t = 4.368$ s) for (a) Case A: isothermal solidification with downward buoyancy, (b) Case B: isothermal purely diffusive condition, (c) Case C: isothermal solidification with upward buoyancy, and (d) Case D: non-isothermal purely diffusive condition. The color map indicates the solute concentration. The right-half region of (d) indicates the temperature distribution. The dotted line indicates the tip position in the z-direction for Case B in each time step.

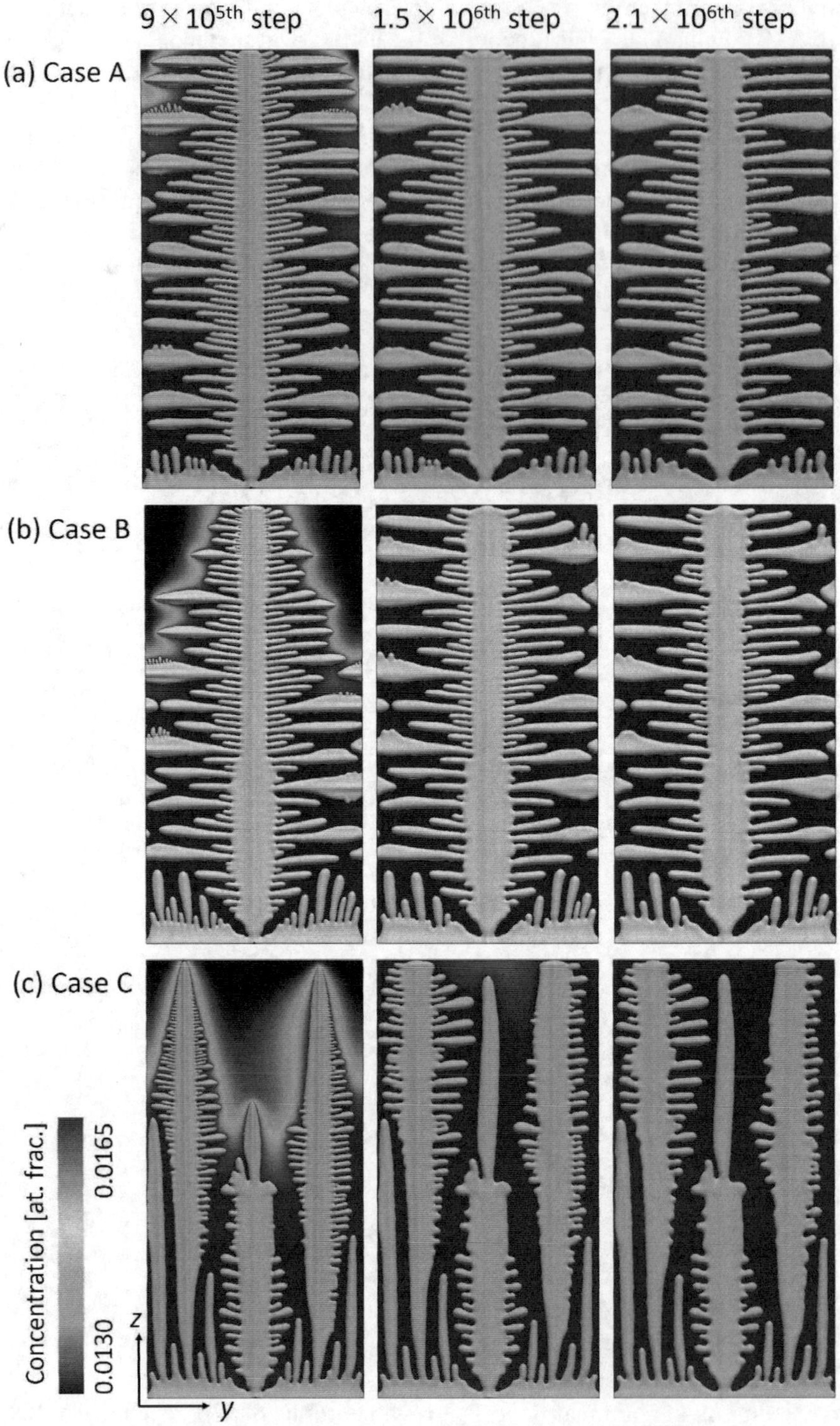

Figure 3. Time evolution of columnar dendrite growth from $9 \times 10^{5\text{th}}$ step (t = 6.552 s) to $2.1 \times 10^{6\text{th}}$ step (t = 15.288 s) for (a) Case A: isothermal solidification with downward buoyancy, (b) Case B: isothermal purely diffusive condition, and (c) Case C: isothermal solidification with upward buoyancy. The color map indicates the solute concentration.

3. Results and discussion

Figure 2 shows the time evolution of the columnar dendrite growth until the 6×10^{5th} step ($t = 4.368$ s). The gray surface indicates the isosurface with $\phi = 0$ and the color indicates the solute concentration. The right half shows the temperature distribution only for Figure 2(d). The horizontal dotted line in each figure indicates the tip position in the *z*-direction for Case B (Figure 2(b)). The tip growth speed was in the following order: Cases A, B, D, and C. In Case A, the downward flow enhanced the growth of the primary and secondary arms, as shown in Figure 2(a). In Case C (Figure 2(c)), the upward flow of the solute-rich liquid prevented the growth of the primary and secondary arms, and finally, the growth of the primary arm stopped. This phenomenon of stopping the growth of the primary arm was not observed in our previous simulation [57]. The reduced tip growth rate was because of the lower sample thickness than in [57] (0.05 mm vs. 0.10 mm, respectively). In Figure 2(d), the latent heat released from the solid–liquid interface slightly retarded the growth of the primary and secondary arms in Case D. The spacing of the secondary arms was wider than that in Case B (Figure 2(b); isothermal condition). Meanwhile, the temperature change in the computational domain was quite small.

Figure 3 shows the later stages of the simulations for Cases A–C. Further computation for Case D was difficult owing to computational resources. At the 9×10^{5th} step for all cases, the primary arms reached the top of the domain, and the liquid concentration tended to be uniform. After the 9×10^{5th} step, retraction and coarsening of the secondary arms were observed [26]. The retraction rate of the thin secondary arms with a large tip curvature was faster, and the shorter and thinner secondary arms disappeared because of retraction. Additionally, at the 2.1×10^{6th} step, necks were observed at the roots of the secondary arms. At this stage, as shown in Figures 3(a) and (c), the liquid flow almost stopped. Therefore, morphological changes due to the liquid flow cannot be expected henceforth.

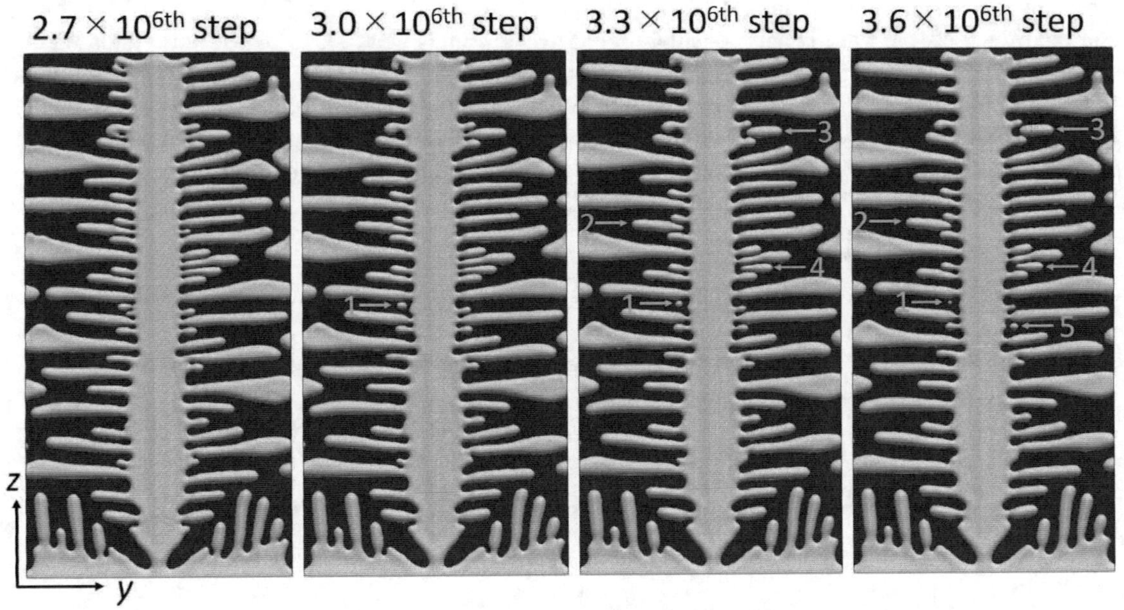

Figure 4. Time evolution of columnar dendrite growth from 2.7×10^{6th} step ($t = 19.656$ s) to 3.6×10^{6th} step ($t = 26.208$ s) for Case B: isothermal purely diffusive condition. The pinched-off arms are indicated by arrows and numbers.

The simulation was continued only for Case B, which is a purely diffusive isothermal condition. The results after the 2.7×10^{6th} step are shown in Figure 4. In these stages, the coarsening was not significant. Instead, constriction of the neck at the root of the secondary arms became more prominent. The first pinching-off was observed at the 3.0×10^{6th} step and is marked by the number 1. The number of pinched arms increased over time, which is the so-called curvature-driven fragmentation or pinching-off [26, 27].

Figure 5 compares the computational performance of Cases A–D. The changes in the number of grids in the AMR simulation, N_{AM}, are shown in Figure 5(a), where $N_{UM} = 128 \times 2{,}048 \times 4{,}096$ when using uniform grids with Δx^3. Until approximately 1×10^{6th} step, N_{AM} rapidly increased linearly with time. This is the growth stage of the primary and secondary arms, as shown in Figures 2 and 3, where the solid–liquid interface area increased. The change in N_{AM} was similar to that of the solid–liquid interface because the fine grids were arranged only around the solid–liquid interface. In fact, at all times, the value of N_{AM} decreased in the following order: Cases A, B, and C; N_{AM} of Case D was similar to that of Case C. From Figure 2, because the growth rate of Case D was slower than that of Case B, and the secondary arm spacing of Case D was wider than Case B, the result in Figure 5 for Case D is convincing. The good agreement between Cases C and D is coincidental because the growth morphology is quite different, as shown in Figure 2. In the range from the 1×10^{6th} step to the 2.5×10^{6th} step, N_{AM} was reduced. This is a coarsening stage, and the solid–liquid interface area gradually decreases, as shown in Figure 3. After approximately 2.5×10^{6th} step, the reduction rate of the N_{AM} became significantly smaller for Case B. In this stage, the morphological changes were quite slow, as shown in Figure 4. Figure 5(b) shows the changes in execution time for every 10,000 steps as a function of N_{AM}/N_{UM}. Moreover, it can be observed that the execution times of Cases A–C vary linearly for $N_{AM}/N_{UM} > 0.1$, which is desirable for parallel-GPU AMR computing. The execution times of Cases A and C when solving the LB equation were approximately three times that of Case B. In our previous study, the difference in execution time between PF and PF-LB simulations was approximately 4.4 [67]. Thus, the current result, which is approximately three times, is significantly smaller than 4.4 and is due to the use of the multi-grid method. The computational cost in Case D was considerably larger (20 times that of Case B) because it also required solving the temperature field. Even though we used coarser grids ($\Delta x_\theta = 4\Delta x$) to solve the temperature field, 886 computations of the heat conduction equation had to be performed for every step of the PF equation. Although we conducted a large-scale simulation of Case D (see Figure 2), solving the temperature field in the dendrite growth PF simulations is challenging even now.

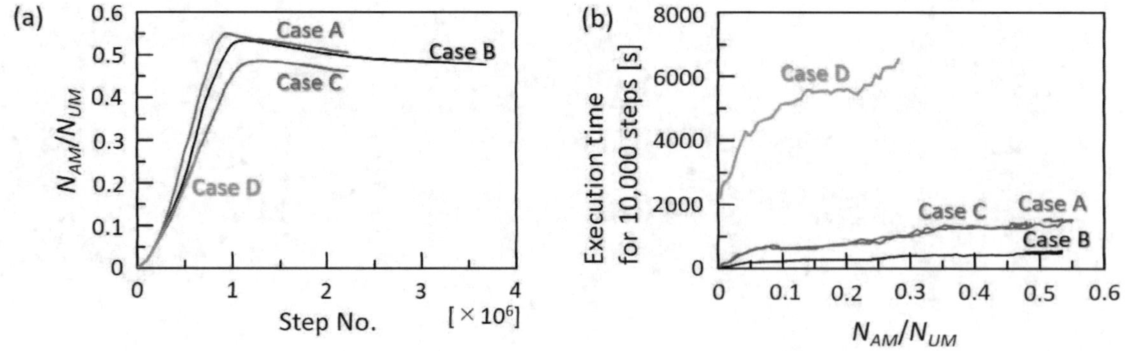

Figure 5. Estimation of the high-performance computing scheme: (a) change of the number of grids in the AMR simulations, N_{AM}, where $N_{UM} = 128 \times 2{,}048 \times 4{,}096$ when using uniform grids with Δx^3, and (b) relations between computational execution time at every 10,000 steps and the number of grids.

4. Conclusions

Columnar dendrite growth with natural convection was numerically investigated using high-performance phase-field lattice-Boltzmann (PF-LB) simulations. PF-LB simulations were conducted using the block-structured AMR scheme with the mother-leaf method in a parallel-GPU environment (parallel GPU AMR). In the simulations with natural convection, the downward buoyancy enhanced the growth of the primary and secondary arms; however, the upward buoyancy delayed the growth of those arms. In this study, the upward buoyancy stopped the growth of the primary arm. After the primary arm tip reached the top of the computational domain, the effect of natural convection on the solidification morphologies gradually decreased and finally stopped. In the longer simulation under an isothermal purely diffusive condition, detachment of the secondary arms caused by curvature-driven fragmentation was observed. Additionally, we conducted a non-isothermal dendrite growth simulation on a 128 × 2,048 × 4,096 grid. The tip growth rate of the primary arm was delayed, and the secondary arm spacing was larger than that in the isothermal condition.

Flow-driven fragmentation was not observed in the present study. To express flow-driven fragmentation, we need a condition in which the interdendritic liquid continuously flows even in the coarsening stage of the secondary arms. On a separate note, consideration of the thermal field in dendrite growth simulations remains a challenging topic.

Acknowledgement

This research was partly supported by a Grant–in–Aid for Scientific Research (A) 23H00160, (A) 20H00217, and (S) 19H05613 and a grant for Early-Career Scientists 21K14041. This study also relied on the computational resources of the HPCI system provided by the Tokyo Institute of Technology through the HPCI System Research Project (Project ID: hp220089).

References

[1] McCartney D G and Hunt J D 1981 *Acta Metall.* **29** 1851-63
[2] Trivedi R and Somboonsuk K 1984 *Mater. Sci. Eng.* **65** 65-74
[3] Dupouy M D, Camel D and Favier J J 1989 *Acta Metall.* **37** 1143-57
[4] Gündüz M and Çadirli E 2002 *Mater. Sci. Eng., A* **327** 167-85
[5] Ghods M, Upadhyay S R, Rajamure R S, Tewari S N, Grugel R N, Poirier D R and Lauer M 2021 *J. Cryst. Growth* **562** 126077
[6] Apel M, Diepers H J and Steinbach I 2006. In: *Modeling of Casting, Welding and Advanced Solidification Processes - XI,* pp 505 - 12
[7] Steinbach I 2009 *Acta Mater.* **57** 2640-5
[8] Takaki T, Sakane S, Ohno M, Shibuta Y and Aoki T 2020 *Comput. Mater. Sci.* **171** 109209
[9] Schneider M C, Gu J P, Beckermann C, Boettinger W J and Kattner U R 1997 *Metall. Mater. Trans. A* **28** 1517-31
[10] Felicelli S D, Poirier D R and Heinrich J C 1998 *Metall. Trans. B* **29** 847-55
[11] Gandin C A 2000 *ISIJ Int.* **40** 971-9
[12] Hunt J D 1984 *Mater. Sci. Eng.* **65** 75-83
[13] Hellawell A, Liu S and Lu S Z 1997 *JOM* **49** 18-20
[14] Campanella T, Charbon C and Rappaz M 2004 *Metall. Trans. A* **35** 3201-10
[15] Flemings M C 2019 *Solid State Phenom.* **285** 3-11
[16] Yasuda H, Ohnaka I, Kawasaki K, Sugiyama A, Ohmichi T, Iwane J and Umetani K 2004 *J. Cryst. Growth* **262** 645-52
[17] Yasuda H, Yamamoto Y, Nakatsuka N, Nagira T, Yoshiya M, Sugiyama A, Ohnaka I, Umetani K and Uesugi K 2008 *Int. J. Cast Met. Res.* **21** 125-8
[18] Mathiesen R H, Arnberg L, Ramsøskar K, Weitkamp T, Rau C and Snigirev A 2002 *Metall. Mater. Trans. B* **33** 613-23
[19] Mi J 2018 *Solidification Processing of Metallic Alloys Under External Fields,* ed D G Eskin and J Mi (Cham: Springer International Publishing) pp 19-74

[20] Horn B K P and Schunck B G 1981 *Artif. Intell.* **17** 185-203
[21] Zhang Z, Wang C, Koe B, Schlepütz C M, Irvine S and Mi J 2021 *Acta Mater.* **209** 116796
[22] Ruvalcaba D, Mathiesen R H, Eskin D G, Arnberg L and Katgerman L 2007 *Acta Mater.* **55** 4287-92
[23] Zhang H, Wu M, Rodrigues C M G, Ludwig A, Kharicha A, Rónaföldi A, Roósz A, Veres Z and Svéda M 2022 *Acta Mater.* **241**
[24] Zimmermann G, Pickmann C, Hamacher M, Schaberger-Zimmermann E, Neumann-Heyme H, Eckert K and Eckert S 2017 *Acta Mater.* **126** 236-50
[25] Yasuda H, Morishita K, Nakatsuka N, Nishimura T, Yoshiya M, Sugiyama A, Uesugi K and Takeuchi A 2019 *Nat. Commun.* **10** 3183
[26] Neumann-Heyme H, Shevchenko N, Lei Z, Eckert K, Keplinger O, Grenzer J, Beckermann C and Eckert S 2018 *Acta Mater.* **146** 176-86
[27] Neumann-Heyme H, Eckert K and Beckermann C 2015 *Phys. Rev. E* **92** 060401
[28] Cool T and Voorhees P W 2017 *Acta Mater.* **127** 359-67
[29] Liotti E, Lui A, Kumar S, Guo Z, Bi C, Connolley T and Grant P S 2016 *Acta Mater.* **121** 384-95
[30] Liotti E, Lui A, Vincent R, Kumar S, Guo Z, Connolley T, Dolbnya I P, Hart M, Arnberg L, Mathiesen R H and Grant P S 2014 *Acta Mater.* **70** 228-39
[31] Shevchenko N, Boden S, Gerbeth G and Eckert S 2013 *Metall. Trans. A* **44** 3797-808
[32] Gibbs J W, Tourret D, Gibbs P J, Imhoff S D, Gibbs M J, Walker B A, Fezzaa K and Clarke A J 2016 *JOM* **68** 170-7
[33] Yasuda H, Morishita K, Yoshiya M and Narumi T 2020 *ISIJ Int.* **60** 2755-64
[34] Pilling J and Hellawell A 1996 *Metall. Mater. Trans. A* **27 A** 229-32
[35] Aveson J W, Reinhart G, Goddard C J L, Nguyen-Thi H, Mangelinck-Noël N, Tandjaoui A, Davenport J R, Warnken N, di Gioacchino F, Lafford T A, D'Souza N, Billia B and Stone H J 2019 *Metall. Trans. A* **50** 5234-41
[36] Ananiev S, Nikrityuk P and Eckert K 2009 *Acta Mater.* **57** 657-65
[37] Takaki T and Kashima H 2011 *J. Cryst. Growth* **337** 97-101
[38] Abou-Khalil L, Salloum-Abou-Jaoude G, Reinhart G, Pickmann C, Zimmermann G and Nguyen-Thi H 2016 *Acta Mater.* **110** 44-52
[39] Shevchenko N, Neumann-Heyme H, Pickmann C, Schaberger-Zimmermann E, Zimmermann G, Eckert K and Eckert S 2017 *IOP Conf. Ser.: Mater. Sci. Eng.* **228** 012005
[40] Shevchenko N, Roshchupkina O, Sokolova O and Eckert S 2015 *J. Cryst. Growth* **417** 1-8
[41] Boden S, Eckert S, Willers B and Gerbeth G 2008 *Metall. Trans. A* **39** 613-23
[42] Abou-Khalil L, da Cruz K S, Reinhart G, Mangelinck-Noël N and Nguyen-Thi H 2022 *Acta Mater.* **241** 118370
[43] Viardin A, Zollinger J, Sturz L, Apel M, Eiken J, Berger R and Hecht U 2020 *Comput. Mater. Sci.* **172** 109358
[44] Zhang A, Meng S, Guo Z, Du J, Wang Q and Xiong S 2019 *Metall. Trans. B* **50** 1514-26
[45] Rojas R, Takaki T and Ohno M 2015 *J. Comput. Phys.* **298** 29-40
[46] Takaki T, Rojas R, Ohno M, Shimokawabe T and Aoki T 2015 *IOP Conf. Ser.: Mater. Sci. Eng.* **84** 012066
[47] Takaki T, Rojas R, Sakane S, Ohno M, Shibuta Y, Shimokawabe T and Aoki T 2017 *J. Cryst. Growth* **474** 146-53
[48] Takaki T, Sakane S, Ohno M and Shibuta Y 2019 *Modell. Simul. Mater. Sci. Eng.* **27** 054001
[49] Provatas N, Goldenfeld N and Dantzig J 1998 *Phys. Rev. Lett.* **80** 3308-11
[50] Tönhardt R and Amberg G 1998 *J. Cryst. Growth* **194** 406-25
[51] Takaki T, Fukuoka T and Tomita Y 2005 *J. Cryst. Growth* **283** 263-78
[52] Chen C C and Lan C W 2010 *J. Cryst. Growth* **312** 1437-42
[53] Zhang A, Jiang B, Guo Z, Du J, Wang Q, Pan F and Xiong S 2021 *Adv. Theory Simul.* **4** 2000251

[54] Sakane S, Takaki T and Aoki T 2022 *Materials Theory* **6** 3
[55] Sakane S, Aoki T and Takaki T 2022 *Comput. Mater. Sci.* **211** 111542
[56] Sakane S, Aoki T and Takaki T 2022 *Comput. Mater. Sci.* **211** 111507
[57] Takaki T, Sakane S and Aoki T 2023 *ISIJ Int.* **63** 83-90
[58] Ohno M and Matsuura K 2009 *Phys. Rev. E* **79** 031603
[59] Ohno M 2012 *Phys. Rev. E* **86** 051603
[60] Seta T, Rojas R, Hayashi K and Tomiyama A 2014 *Phys. Rev. E* **89** 023307
[61] Sakane S and Takaki T 2021 *Comput. Mater. Sci.* **186** 110070
[62] Glasner K 2001 *J. Comput. Phys.* **174** 695-711
[63] Watanabe S and Aoki T 2021 *Comput. Phys. Commun.* **264** 107871
[64] Watanabe S, Aoki T and Takaki T 2020 *Parallel Comput.* **97** 102647
[65] Sakane S, Takaki T, Ohno M, Shibuta Y and Aoki T 2019 *Modell. Simul. Mater. Sci. Eng.* **27** 054004
[66] Hasegawa Y and Aoki T 2016 *IPSJ Transactions on Advanced Computing Systems* **9** 34-45
[67] Sakane S, Takaki T, Rojas R, Ohno M, Shibuta Y, Shimokawabe T and Aoki T 2017 *J. Cryst. Growth* **474** 154-9

IOP Conf. Series: Materials Science and Engineering 1281 (2023) 012057 doi:10.1088/1757-899X/1281/1/012057

Application of a meshless space-time adaptive approach to phase-field modelling of polycrystalline solidification

T Dobravec[1], B Mavrič[1,2] and B Šarler[1,3]

[1] Faculty of Mechanical Engineering, University of Ljubljana, Aškerčeva cesta 6, Ljubljana, SI-1000, Slovenia

[2] Division of Scientific Computing, Department of Information Technology, Uppsala University, P.O. Box 337, Uppsala, SE-751 05, Sweden

[3] Institute of Metals and Technology, Lepi pot 11, Ljubljana, SI-1000, Slovenia

E-mail: tadej.dobravec@fs.uni-lj.si

Abstract. We have developed a 2-D numerical meshless adaptive approach for phase-field modelling of dendritic solidification. The quadtree-based approach decomposes the computational domain into quadtree sub-domains of different sizes. The algorithm generates uniformly-distributed computational nodes in each quadtree sub-domain. We apply the meshless radial basis function generated finite difference method and the forward Euler scheme to discretise governing equations in each computational node. The fixed ratio between the characteristic size and the node spacing of a quadtree sub-domain ensures space adaptivity. The adaptive time-stepping accelerates the calculations further. In the framework of previous research studies, we used the approach to solve quantitative phase-field models for single dendrite growth in pure melts and dilute binary alloys. In the present study, we upgrade the solution procedure for the modelling growth of multiple differently oriented dendrites. Along with the space-time adaptive approach, we apply non-linear preconditioning of the phase-field equation to increase computational efficiency. We investigate a novel numerical approach's accuracy and computational efficiency by simulating the equiaxed dendrite growth from a dilute binary alloy.

1. Introduction

Numerical modelling of dendritic solidification represents an important task for predicting the microstructure evolution of many metals at different casting conditions [1,2]. The microstructure evolution is closely linked to the material properties of the solidified material. The numerical models are, therefore, crucial for the design and production of high-quality castings. In the present study, we employ the phase-field (PF) method for modelling dendritic solidification. The PF method [3, 4] represents a powerful tool for solving many free boundary problems in material science and engineering [5].

In the previous studies, we consider PF models for simulating single dendrite growth in pure melts and dilute binary alloys [6,7]. In this paper, we extend the test case for the single dendrite solidification of dilute binary alloys [8] to polycrystalline solidification. There are two main PF approaches for modelling this situation, which is important for the practice. The first approach is the multi-PF model [9–11], where each of the grains is represented by its own PF. This

IOP Conf. Series: Materials Science and Engineering 1281 (2023) 012057 doi:10.1088/1757-899X/1281/1/012057

approach provides a detailed framework for analysing grain boundary evolution. However, multi-PF simulations experience significant memory usage and long computational times. The second approach is the orientation field method [12–15]. This approach complements the PF model with information on the grain orientation. As a result, the related orientation field method is computationally much less demanding and can be used for PF modelling of many simultaneously growing dendrites [16]. The latter approach, however, experiences a limitation, i.e., the standard governing equation for the orientation field is incorrect in coherent polycrystalline matter [17]. In this paper, we consider early-stage equiaxed solidification of dilute binary alloys. In this case, the orientation field method does not experience the mentioned limitation [17]. Our primary focus is the solidification of multiple differently-oriented dendrites before the impingement. Therefore, we do not have to account for the physics of grain boundaries. For this purpose, the front-tracking (FT) algorithm [18] represents a very suitable and computationally efficient algorithm for tracking the preferential growth direction of each dendrite. Additionally, we employ the non-linear preconditioning of the PF equation to increase computational efficiency [19–21].

Researchers developed many different computational approaches for the PF modelling of dendritic growth in the last 20 years, e.g., adaptive mesh refinement [22], parallel simulations using graphic processing units [23], implicit time-stepping and multi-grid approaches [24], etc. In this paper, we employ the space-time adaptive algorithm to accelerate the meshless PF modelling of dendritic solidification [7]. Meshless methods represent a fast-growing family of numerical methods that do not rely on the division in space in elements like finite element or finite volume methods. The representatives of this family are the element-free Galerkin method [25], the smoothed particle hydrodynamics method [26], the meshless-collocation method [27], etc. In the present study, we employ the local radial basis function collocation method (LRBFCM) [28], also known as the radial basis function generated finite difference (RBF-FD) method [29]. The meshless RBF-FD method is especially suitable for orientation-insensitive PF modelling of dendritic solidification in arbitrary preferential growth directions [6].

2. Governing equations

We consider the isothermal solidification of a dilute supersaturated binary alloy in the 2-D computational domain Ω with the boundary Γ. A simplified case is studied where the solute diffusivity in the liquid phase D_ℓ, the chemical capillary length d_c, and the partition coefficient k_0 are constant. The solute diffusion in the solid phase is neglected. We define supersaturation U as

$$U = \frac{C_\ell - C_\ell^e}{(1-k_0)C_\ell^e}, \tag{1}$$

where C_ℓ and C_ℓ^e stand for the solute concentration in the liquid phase and the equilibrium solute concentration in the liquid phase, defined from the phase diagram, respectively. We use the dimensionless PF model [8], where the spatial and temporal coordinates are measured in units of the PF interface thickness and the PF characteristic attachment time, respectively. The PF interface thickness is defined as

$$W_0 = d_c \frac{1}{\alpha_1} \lambda, \tag{2}$$

where α_1 stands for a constant and λ for the free parameter of the PF model. The PF characteristic attachment time is given as

$$\tau_0 = \frac{d_c^2}{D_\ell} \frac{\alpha_2}{\alpha_1^2} \lambda^3, \tag{3}$$

where α_2 stands for a constant of the PF model. The PF constants are equal to $\alpha_1 = 0.8839$ and $\alpha_2 = 0.6267$ [30]. The selection of free parameter λ has to yield W_0 much smaller than the diffusion length of solidification to ensure a valid PF model. The PF model [8] constrains the PF ϕ in the interval $-1 \leq \phi \leq 1$, where $\phi = 1$ and $\phi = -1$ denote solid and liquid phases, respectively. We use the preconditioned PF variable ψ, defined as [19]

$$\psi = \sqrt{2}\tanh^{-1}(\phi), \tag{4}$$

to increase numerical stability for larger node spacings.

2.1. Equations for ψ and U

The governing equation for ψ reads [20]

$$a^2(\boldsymbol{n})\frac{\partial\psi}{\partial t} = \sqrt{2}\left(\phi - \lambda(1-\phi)U\right) + 2a(\boldsymbol{n})\nabla a(\boldsymbol{n})\cdot\nabla\psi \\ - \sqrt{2}\phi\nabla\psi\cdot\boldsymbol{a}(\boldsymbol{n}) + \nabla\cdot\boldsymbol{a}(\boldsymbol{n}) + a^2(\boldsymbol{n})\left(\nabla^2\psi - \sqrt{2}\phi|\nabla\psi|^2\right), \tag{5}$$

where $a(\boldsymbol{n})$ and $\boldsymbol{a}(\boldsymbol{n})$ stand for the anisotropy function and vector, respectively. They depend on the solid-liquid interface normal

$$\boldsymbol{n} = -\frac{\nabla\psi}{|\nabla\psi|}, \tag{6}$$

and are defined in the following subsection. The governing equation for U reads as

$$\frac{1}{2}\left(1 + k_0 - (1-k_0)\phi\right)\frac{\partial U}{\partial t} = D\frac{1-\phi}{2}\nabla^2 U - D\frac{1-\phi^2}{2\sqrt{2}}\nabla U\cdot\nabla\psi \\ - \frac{1}{2\sqrt{2}}\left((1+(1-k_0)U)\nabla G\cdot\boldsymbol{n} + G(1-k_0)\nabla U\cdot\boldsymbol{n}\right) \\ - \frac{G}{2}\left[1 + (1-k_0)U\left(1 - \frac{\nabla\cdot\boldsymbol{n}}{\sqrt{2}}\right)\right], \tag{7}$$

where $D = D_\ell\tau_0/W_0^2$ is the dimensionless solute diffusivity in the liquid phase, and the scalar G is defined as

$$G = \frac{1-\phi^2}{\sqrt{2}}\frac{\partial\psi}{\partial t}. \tag{8}$$

2.2. Anisotropy of the surface energy

Let's denote the anisotropy function and vector in a dendrite's coordinate system as $\tilde{a}(\boldsymbol{m})$ and $\tilde{\boldsymbol{a}}(\boldsymbol{m})$, where $\boldsymbol{m} = (m_x, m_y)$ represents the normal to the interface in a dendrite's coordinate system. We employ the cubic anisotropy function

$$\tilde{a}(\boldsymbol{m}) = 1 - 3\epsilon_4 + 4\epsilon_4\left(m_x^4 + m_y^4\right), \tag{9}$$

where ϵ_4 stands for the anisotropy strength of the interface energy. Function $\tilde{a}(\boldsymbol{m})$ yields anisotropy vector [20]

$$\tilde{\boldsymbol{a}}(\boldsymbol{m}) = 16\epsilon_4|\nabla\psi|\tilde{a}(\boldsymbol{m})\left[m_x(m_x^4 + m_y^4 - m_x^2), m_y(m_x^4 + m_y^4 - m_y^2)\right]. \tag{10}$$

Function $a(\boldsymbol{n})$ from equation (5) is obtained with the following relation

$$a(\boldsymbol{n}) = \tilde{a}(\boldsymbol{R}^T\boldsymbol{n}), \quad \boldsymbol{R} = \begin{bmatrix}\cos(\theta) & -\sin(\theta)\\ \sin(\theta) & \cos(\theta)\end{bmatrix}, \tag{11}$$

where $\boldsymbol{R}$ is the rotation matrix and θ stands for the preferential growth direction angle. We employ the following relation to obtain $\boldsymbol{a}(\boldsymbol{n})$ from equation (5)

$$\boldsymbol{a}(\boldsymbol{n}) = \boldsymbol{R}\tilde{\boldsymbol{a}}(\boldsymbol{R}^T\boldsymbol{n}). \tag{12}$$

IOP Conf. Series: Materials Science and Engineering 1281 (2023) 012057 doi:10.1088/1757-899X/1281/1/012057

3. Solution procedure

3.1. Solution of governing equations for ψ and U

We solve the governing equations for ψ and U by our previously developed space-time adaptive approach [7]. It is based on dynamic quadtree domain decomposition, as seen in figure 1. Regular or scattered node distribution with fixed node spacing is generated in each quadtree sub-domain; in the present study, we employ regular node distribution. The constant ratio m_Ω between a quadtree sub-domain's side length and node spacing ensures space adaptivity, as seen on the left in figure 1. The free parameters of the space-time adaptive approach are the minimum spacing h, the ratio m_Ω, the maximum number of different node spacings n_h, the maximum number of different time steps $n_{\Delta t}$, and the overlapping parameter $n_{overlap}$ [7]. In this paper, we set $n_{overlap} = 1$ and $n_{\Delta t} = 2$; such configuration yields good accuracy and computational efficiency [7]. In subsection 4.1, we analyse the influence of minimum spacing h on the accuracy.

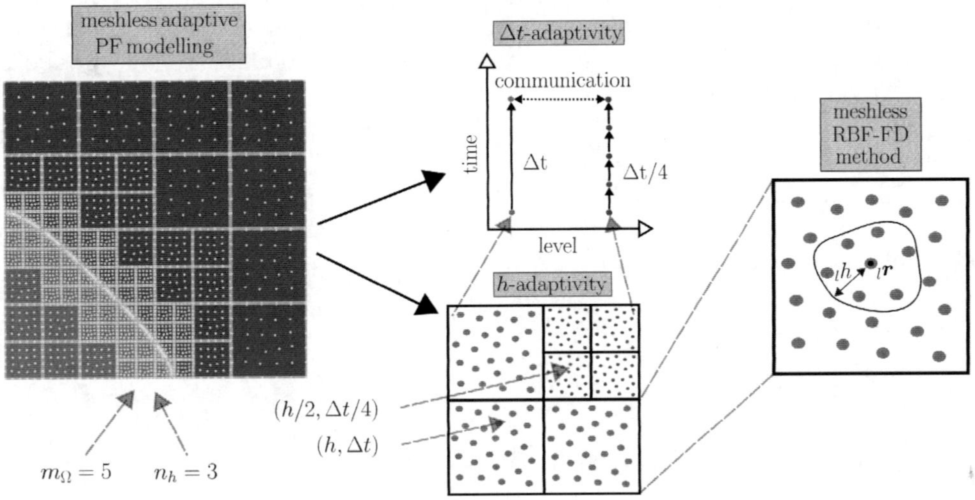

Figure 1. A scheme of space-time adaptive meshless solution procedure. The quadtree-based approach ensures a high density of computational nodes and fine time-stepping at and near the solid-liquid interface. The solution procedure employs the meshless RBF-FD method for spatial discretisation of equations for ψ and U in each quadtree sub-domain.

We apply the forward Euler scheme and the meshless RBF-FD method to discretise the governing equations for ψ and U in the computational nodes from a quadtree sub-domain. We calculate the minimum stable time step in the forward Euler scheme as

$$\Delta t = \alpha_{\Delta t} \frac{1}{4} \frac{h^2}{\max\left(D, 1/(1-\epsilon_4)\right)}, \tag{13}$$

where $\alpha_{\Delta t}$ stands for the time step stability parameter. Value $\alpha_{\Delta t} = 0.2$ yields sufficiently small time steps for solving the considered PF model, i.e., further refinement of $\alpha_{\Delta t}$ has a negligible effect on accuracy. We employ adaptive time-stepping to increase computational efficiency. The stable time step depends on node spacing; hence, larger time steps can be used in quadtree sub-domains with larger node spacing, as seen in figure 1. We employ fifth-degree polyharmonic splines as radial basis functions, second-order augmentation with monomials, and thirteen nodes in a local support domain in the meshless RBF-FD method. We have previously shown that such a configuration is very suitable out of many tested configurations for the PF modelling of dendritic growth [6, 7].

A PF model can be accurately resolved using larger node spacings when employing preconditioning [19–21]. However, when we use space adaptivity, the preconditioning yields stability issues far from the solid-liquid interface, where very large node spacings can not resolve the model. Solving a non-preconditioned PF model by space adaptive algorithm does not experience this problem since the PF is a constant far away from the solid-liquid interface. We tackle this problem by applying the following restriction [16]

$$\psi = \begin{cases} \psi_c & \psi > \psi_c \\ -\psi_c & \psi < -\psi_c \\ \psi & -\psi_c \le \psi \le \psi_c \end{cases}, \tag{14}$$

where ψ_c is the numerical cut-off parameter. The adaptive algorithm ensures the minimum node spacing h where $|\psi| < \psi_c - 1.0$. Regions where $|\psi| > \psi_c - 0.5$ can be de-refined. Care is taken to keep the quadtree balanced in the refinement/de-refinement procedure. Details of the space-time adaptive algorithm can be found in [7]. In subsection 4.1, we analyse the influence of parameter ψ_c on the accuracy.

3.2. Front-tracking of the grain orientation

We use the FT algorithm [18] to track the preferential growth direction angle of each dendrite in Ω. Along with $\psi(\boldsymbol{r})$ and $U(\boldsymbol{r})$, we also define position-dependent orientation field $\theta(\boldsymbol{r})$, which is used for calculating $a(\boldsymbol{n})$ and $\boldsymbol{a}(\boldsymbol{n})$ in equations (11) and (12). When we consider single dendrite growth or multiple dendrite growth with the same preferential growth direction θ_0, $\theta(\boldsymbol{r}) = \theta_0$ throughout simulation.

Suppose a particle with a radius r_0, centre $\boldsymbol{r}_0$, and orientation angle $\theta_0 \in (0, \pi/2]$ nucleates in Ω. The initial condition for ψ reads as

$$\psi(\boldsymbol{r}, t = 0) = r_0 - |\boldsymbol{r} - \boldsymbol{r}_0|. \tag{15}$$

We define the initial condition for θ as

$$\theta(\boldsymbol{r}, t = 0) = \begin{cases} \theta_0 & \big| r_0 - |\boldsymbol{r} - \boldsymbol{r}_0| \big| \le \psi_{FT} \\ 0 & \big| r_0 - |\boldsymbol{r} - \boldsymbol{r}_0| \big| > \psi_{FT} \end{cases}, \tag{16}$$

where ψ_{FT} is the FT thickness of the boundary layer around the solid-liquid interface. Value $\theta = 0$ represents the bulk of the solid or liquid phase, while values $\theta > 0$ represent the FT boundary layer surrounding the solid-liquid interface.

As ψ evolves, the FT algorithm applies in a computational node $\boldsymbol{r}_i \in \Omega$ to track the orientation of a growing dendrite. Suppose $\{\boldsymbol{r}_j\}_{\boldsymbol{r}_i}$ is a set of nodes nearest to $\boldsymbol{r}_i$. The FT algorithm sets the orientation field as

$$\theta(\boldsymbol{r}_i, t) = \begin{cases} 0 & |\psi(\boldsymbol{r}_i, t)| > \psi_{FT} \\ \theta(\boldsymbol{r}_i, t) & \theta(\boldsymbol{r}_i, t) > 0 \\ \theta(\boldsymbol{r}_j, t) & \theta(\boldsymbol{r}_i, t) = 0 \end{cases}, \tag{17}$$

where $\theta(\boldsymbol{r}_j, t) > 0$ and $|\psi(\boldsymbol{r}_j, t)| \le \psi_{FT}$ for $\boldsymbol{r}_j \in \{\boldsymbol{r}_j\}_{\boldsymbol{r}_i}$. The algorithm performs a loop over a set of nearest nodes $\{\boldsymbol{r}_j\}_{\boldsymbol{r}_i}$ when checking the last condition from equation (17). When conditions $\theta(\boldsymbol{r}_j, t) > 0$ and $|\psi(\boldsymbol{r}_j, t)| \le \psi_{FT}$ are fulfilled for $\boldsymbol{r}_j \in \{\boldsymbol{r}_j\}_{\boldsymbol{r}_i}$, the algorithm sets $\theta(\boldsymbol{r}_i, t) = \theta(\boldsymbol{r}_j, t)$, exits the loop over $\boldsymbol{r}_j \in \{\boldsymbol{r}_j\}_{\boldsymbol{r}_i}$, and moves to the next computational node $\boldsymbol{r}_i \in \Omega$. In subsection 4.1, we analyse the influence of parameter ψ_{FT} to the accuracy.

4. Results

We verify the present numerical model by solving the test case for a single dendrite growth [8]. The influence of the free parameters on the accuracy of modelling growth in arbitrary growth directions is investigated. Next, we extend the test case for a single dendrite growth [8] to the solidification of multiple randomly-oriented dendrites.

4.1. Verification of the numerical model

The test case [8] considers the growth of a single dendrite from a supersaturated dilute binary alloy in a square computational domain $\Omega = [-L/2, L/2] \times [-L/2, L/2]$, where L stands for the size of the computational domain. We set the initial condition for ψ according to equation (15). The initial condition for U reads as $U(\boldsymbol{r}, t = 0) = -\Delta$, where Δ stands for the initial supersaturation. Zero flux Neumann boundary conditions are applied for ψ and U

$$\nabla\psi\Big|_{\Gamma} \cdot \boldsymbol{n}_\Gamma = 0, \quad \nabla U\Big|_{\Gamma} \cdot \boldsymbol{n}_\Gamma = 0, \tag{18}$$

where $\boldsymbol{n}_\Gamma$ is the outward-facing normal to Γ. Table 1 contains the simulation parameters used in verification of our numerical model.

Table 1. Simulation parameters for verifying the numerical model.

Computational domain	
Size of domain (L)	537.6
Physical problem	
Strength of anisotropy (ϵ_4)	0.02
Initial supersaturation (Δ)	0.55
Center of nucleus ($\boldsymbol{r}_0$)	$(0, 0)$
Radius of nucleus (r_0)	$22\alpha_1\alpha_2/2$
Partition coefficient (k_0)	0.15
PF model	
Constant (α_1)	0.8839
Constant (α_2)	0.6267
Coupling parameter (λ)	$2/\alpha_2$
Dimensionless solute diffusivity (D)	2

We analyse first the influence of the minimum node spacing h on the accuracy. In this analysis, the orientation field is constant in the whole Ω, i.e., we turn off the FT algorithm and set $\theta(\boldsymbol{r}) = \theta_0$. The reason for the turned-off FT algorithm is to analyse only the influence of spacing h on the accuracy. We test the performance for the minimum node spacings $h = 0.4$, $h = 0.8$, $h = 1.2$, and $h = 1.4$. The following sets of free parameters are used ($n_h = 6, m_\Omega = 21$), ($n_h = 5, m_\Omega = 21$), ($n_h = 5, m_\Omega = 14$), and ($n_h = 5, m_\Omega = 12$) for $h = 0.4$, $h = 0.8$, $h = 1.2$, and $h = 1.4$, respectively. We set the numerical cut-off parameter to $\psi_c = 12$. Figure 2 shows the dendrite growth velocity as a function of time at $\theta_0 = 0$ for $h = 0.4$. The figure also shows Karma's reference solution [8], obtained using the finite-difference method with the node spacing $h = 0.4$ and the forward Euler scheme with the time step $\Delta t = 0.008$. We can see that our velocity agrees well with the reference solution. Figure 3 shows the steady-state dendrite growth velocity as a function angle θ_0 for four node spacings h. The steady-state velocity for $h = 0.4$ and $h = 0.8$ is approximately 1.9 % above the reference solution and almost independent of θ_0, while $h = 1.2$ and $h = 1.4$ yield θ_0-dependent results. There are three main reasons for minor discrepancy with the reference solution at $h = 0.4$ and $h = 0.8$. First, contrary to [8], we employ preconditioned PF model. Second, we use different numerical methods to solve the PF model. And third, our algorithm for calculating growth velocity is probably not the same as in [8]. The steady-state velocity for $h = 1.2$ is closer to the reference solution. The difference between the maximum and the minimum velocity at $h = 1.2$ is approximately 1.6 %. Nevertheless, we believe that $h = 1.2$ ensures high enough accuracy and is used for calculations in the rest of the paper. The value $h = 1.4$ is unsuitable for calculating this test case.

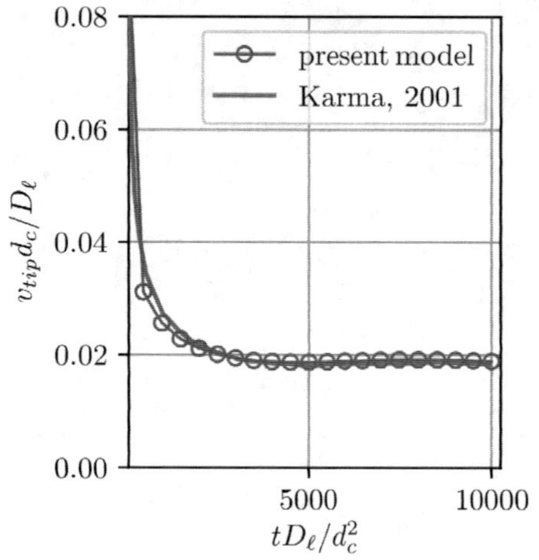

Figure 2. Dendrite growth velocity as a function of time.

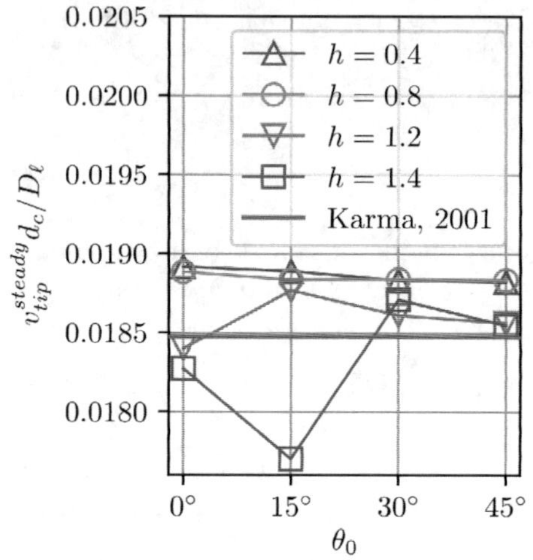

Figure 3. Steady-state dendrite growth velocity as a function of preferential growth direction angle θ_0 at four values of the node spacing h.

We next analyse the influence of the numerical cut-off parameter ψ_c on the accuracy. The FT algorithm is still turned off in this analysis. The reason for the turned-off FT algorithm is to analyse only the influence of parameter ψ_c on the accuracy. Figure 4 shows the steady-state dendrite growth velocity as a function angle θ_0 for five parameters ψ_c. Too small values of ψ_c significantly decrease the accuracy. The accuracy is no longer increased for $\psi_c \geq 12$. Therefore, we employ $\psi_c = 12$ for calculations in the rest of the paper. Finally, we turn on the FT algorithm and analyse the influence of the FT interface thickness ψ_{FT} to the accuracy. We set the initial condition for θ according to equation (16). Figure 5 shows the steady-state dendrite growth velocity as a function angle θ_0 for four parameters ψ_{FT}. As for parameter ψ_c, too small values of ψ_{FT} significantly decrease the accuracy. The accuracy is no longer increased for $\psi_{FT} \geq 5$ Therefore, we employ $\psi_{FT} = 5$ for calculations in the rest of the paper. To conclude, parameters $h = 1.2$, $\psi_c = 12$, and $\psi_{FT} = 5$ represent a good compromise between the accuracy and computational efficiency for the PF modelling of dendritic growth in arbitrary growth directions. We use this selection of parameters to simulate polycrystalline solidification in the following subsection.

4.2. Polycrystalline solidification

In this subsection, we extend the test case from the previous subsection to the growth of multiple randomly-oriented dendrites. We use the simulation parameters from table 1 with two differences. First, the size of Ω is set to $L = 1075.2$. And second, we have multiple centres

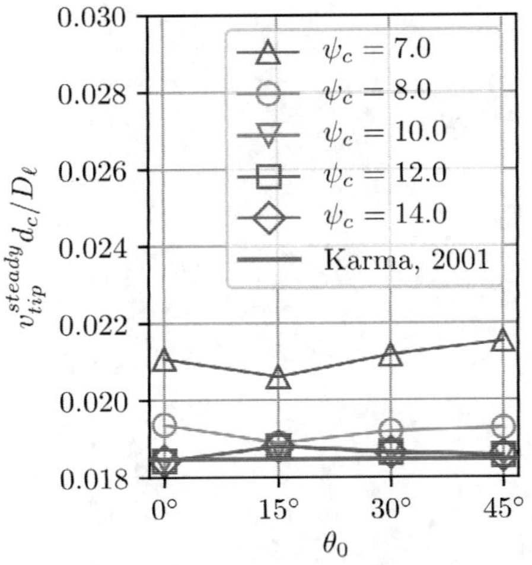

Figure 4. Steady-state dendrite growth velocity as a function of preferential growth direction angle θ_0 at five values of the cut-off parameter ψ_c.

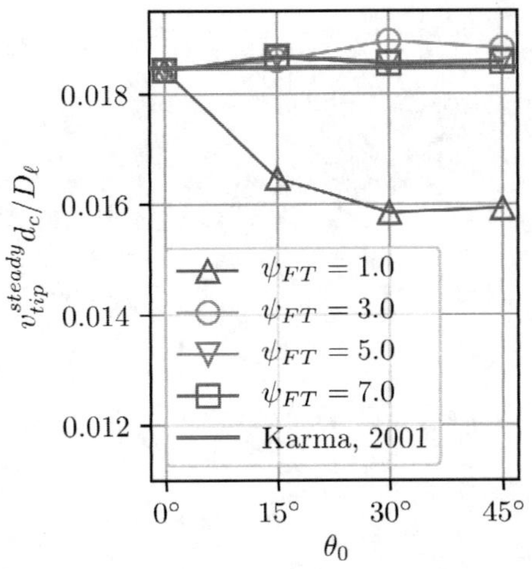

Figure 5. Steady-state dendrite growth velocity as a function of preferential growth direction angle θ_0 at four values of the orientation parameter ψ_{FT}.

of nuclei since we consider polycrystalline solidification. We use the same boundary condition for ψ and U as in the previous subsection. The initial condition for U is also the same. However, the initial conditions for ψ and θ have to be modified to describe the nucleation of multiple dendrites.

Nuclei are quasi-randomly positioned in Ω. For instance, if the number of nuclei is 4, we first divide a square Ω into 4 square sub-domains $\tilde{\Omega}$ of equal size. Suppose $\tilde{\boldsymbol{r}}_\Omega$ is the centre of $\tilde{\Omega}$. The quasi-random position of the nucleus $\boldsymbol{r}_0 \in \tilde{\Omega}$ is set as

$$\boldsymbol{r}_0 = \tilde{\boldsymbol{r}}_\Omega + 0.2\tilde{L}(\mathcal{R}_x, \mathcal{R}_y), \tag{19}$$

where $\tilde{L}$ is the side length of $\tilde{\Omega}$ while $\mathcal{R}_x \in [-1, 1]$ and $\mathcal{R}_y \in [-1, 1]$ represent random numbers. The random orientation angle is similarly set as $\theta_0 = \pi/2\mathcal{R}_\theta$, where $\mathcal{R}_\theta \in (0, 1]$ is a random number. After we determine the sets of positions $\{\boldsymbol{r}_0\}$ and corresponding orientations $\{\theta_0\}$, we use the following equation for the initial condition for ψ

$$\psi(\boldsymbol{r}, t = 0) = \max_{\boldsymbol{r}'_0 \in \{\boldsymbol{r}_0\}} (r_0 - |\boldsymbol{r} - \boldsymbol{r}'_0|). \tag{20}$$

The initial condition for θ is set as $\theta(\boldsymbol{r}, t = 0) = \theta'_0$ if $|r_0 - |\boldsymbol{r} - \boldsymbol{r}'_0|| \leq \psi_{FT}$ where $\boldsymbol{r}'_0$ and θ'_0 belong to the same nucleus. This condition is checked for all $\boldsymbol{r}'_0 \in \{\boldsymbol{r}_0\}$ and $\theta'_0 \in \{\theta_0\}$. The initial condition for θ is set as $\theta(\boldsymbol{r}, t = 0) = 0$ if $|r_0 - |\boldsymbol{r} - \boldsymbol{r}'_0|| > \psi_{FT}$ for all $\boldsymbol{r}'_0 \in \{\boldsymbol{r}_0\}$.

Figure 6 shows rescaled concentration field C/C_ℓ^e in Ω in the case of four (top row), nine (middle row), and sixteen (bottom row) randomly oriented and quasi-randomly-positioned growing dendrites. In the first stage of solidification, dendrites freely grow into the melt. Their growth is constrained after they approach neighbouring dendrites. In the final stage, the dendrites become coarser and coarser. Naturally, the average dendrite's size and primary trunk thickness decrease as the number of nuclei increases.

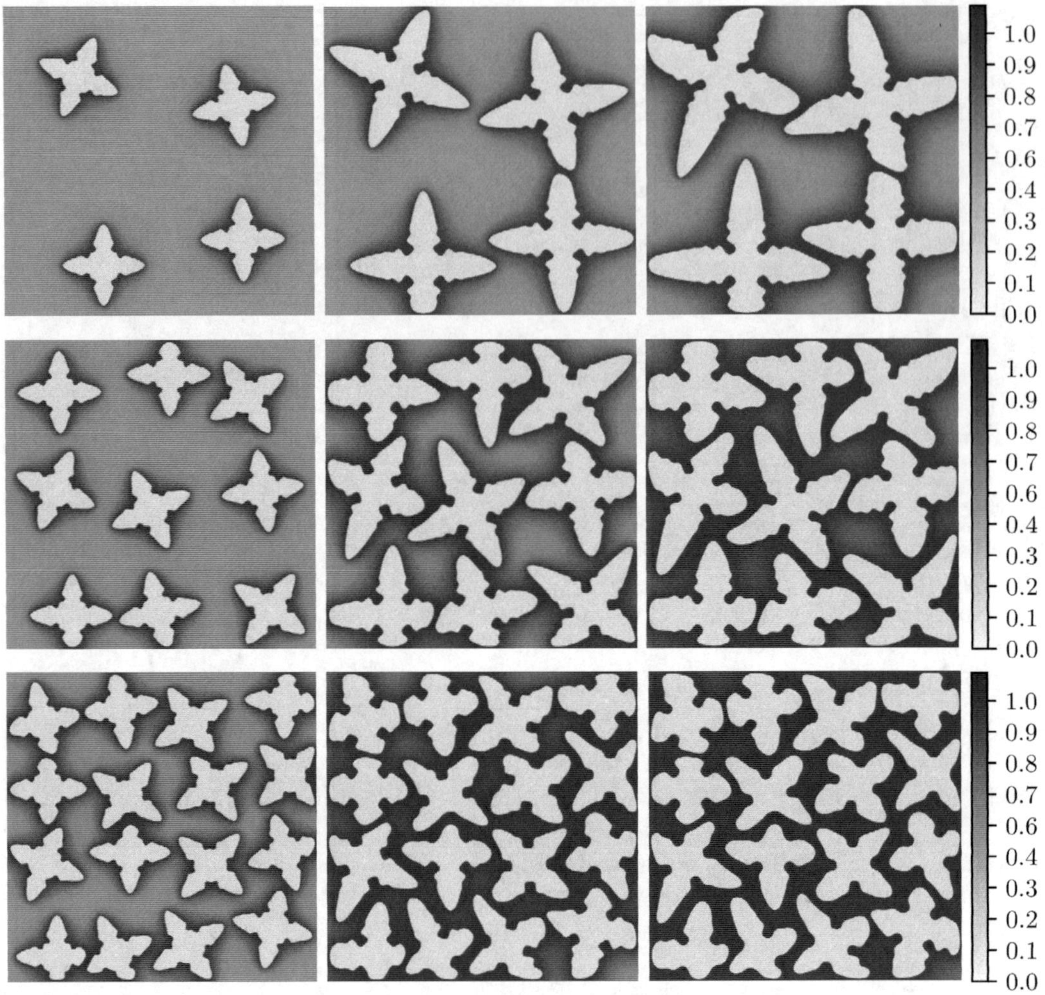

Figure 6. Rescaled concentration field C/C_ℓ^e in computational domain $\Omega = [-537.6, 537.6] \times [-537.6, 537.6]$ in the case of four (top row), nine (middle row), and sixteen (bottom row) randomly oriented and quasi-randomly-positioned growing dendrites. The field is plotted at $t = 1000$ (left column), $t = 2000$ (middle column), and $t = 3000$ (right column).

5. Conclusions

This paper demonstrates the use of meshless-based space-time adaptive numerical approach for the PF modelling of polycrystalline early-stage isothermal solidification of supersaturated dilute binary alloys. The approach, developed initially for modelling single dendrite growth, is extended to model the growth of multiple differently-oriented dendrites. We analyse the influence of the minimum node spacing h, the numerical cut-off parameter ψ_c, and the FT interface thickness ψ_{FT} to the accuracy in the case of single dendrite growth at different preferential growth directions. We show that the selection of parameters $h = 1.2$, $\psi_c = 12$, and $\psi_{FT} = 5$ represents the best compromise between accuracy and computational efficiency. We then simulate polycrystalline solidification in the case of four, nine, and sixteen randomly-oriented nuclei and obtain expected results.

The present paper shows that our numerical approach is suitable for the PF modelling of dendritic solidification. The approach can be easily extended to model the solidification of pure melts and multi-component alloys. In the case of dilute binary and multi-component alloys, the model can be easily extended to simulate non-isothermal solidification in the frozen-temperature approximation, i.e., when the temperature is considered as an input parameter. In the future, the numerical approach will be extended to 3-D using an octree instead of a quadtree algorithm.

Acknowledgments

This work was supported by the Slovenian Research Agency as post-doctoral basic project Z2-4479, programme group P2-0162, and applied project L2-3173, also supported by Štore Steel company.

References

[1] Kurz W, Fisher D J and Trivedi R 2019 *Int. Mater. Rev.* **64** 311–354
[2] Kurz W, Rappaz M and Trivedi R 2021 *Int. Mater. Rev.* **66** 30–76
[3] Chen L Q 2002 *Annu. Rev. Mater. Res.* **32** 113–140
[4] Boettinger W J, Warren J A, Beckermann C and Karma A 2002 *Annu. Rev. Mater. Res.* **32** 163–194
[5] Steinbach I 2009 *Model. Simul. Mat. Sci. Eng.* **17** 073001
[6] Dobravec T, Mavrič B and Šarler B 2020 *Comput. Mater. Sci* **172** 109166
[7] Dobravec T, Mavrič B and Šarler B 2022 *Comput. Math. Appl.* **126** 77–99
[8] Karma A 2001 *Phys. Rev. Lett.* **87** 115701
[9] Steinbach I, Pezzolla F, Nestler B, Seeßelberg M, Prieler R, Schmitz G J and Rezende J L L 1996 *Phys. D: Nonlinear Phenom.* **94** 135–147
[10] Moelans N, Blanpain B and Wollants P 2008 *Phys. Rev. B* **78** 024113
[11] Ofori-Opoku N and Provatas N 2010 *Acta Mater.* **58** 2155–2164
[12] Kobayashi R, Warren J A and Craig Carter W 2000 *Phys. D: Nonlinear Phenom.* **140** 141–150
[13] Warren J A, Kobayashi R, Lobkovsky A E and Craig Carter W 2003 *Acta Mater.* **51** 6035–6058
[14] Granasy L, Pusztai T and Warren J A 2004 *J. Phys. Condens. Matter* **16**
[15] Pusztai T, Bortel G and Gránásy L 2005 *EPL* **71** 131
[16] Gong T Z, Chen Y, Cao Y F, Kang X H and Li D Z 2018 *Comput. Mater. Sci* **147** 338–352
[17] Plapp M 2011 *Philos. Mag.* **91** 25–44
[18] Chen Y, Qi X B, Li D Z, Kang X H and Xiao N M 2015 *Comput. Mater. Sci* **104** 155–161
[19] Glasner K 2001 *J. Comput. Phys* **174** 695–711
[20] Boukellal A K, Rouby M and Debierre J M 2021 *Comput. Mater. Sci* **186** 110051
[21] Tourret D and Karma A 2015 *Acta Mater.* **82** 64–83
[22] Provatas N, Goldenfeld N and Dantzig J 1998 *Phys. Rev. Lett.* **80** 3308–3311
[23] Takaki T, Shimokawabe T, Ohno M, Yamanaka A and Aoki T 2013 *J. Cryst. Growth* **382** 21–25
[24] Bollada P C, Goodyer C E, Jimack P K, Mullis A M and Yang F W 2015 *J. Comput. Phys.* **287** 130–150
[25] Belytschko T, Lu Y Y and Gu L 1994 *J. Numer. Methods Eng.* **37** 229–256
[26] Liu G R and Liu M B 2003 *Smoothed Particle Hydrodynamics: A Meshfree Particle Method* (Singapore: World Scientific)
[27] Kansa E J 1990 *Comput. Math. with Appl.* **19** 127–145
[28] Šarler B and Vertnik R 2006 *Comput. Math. with Appl.* **51** 1269–1282
[29] Flyer N, Fornberg B, Bayona V and Barnett G A 2016 *J. Comput. Phys.* **321** 21–38
[30] Karma A and Rappel W J 1998 *Phys. Rev. E* **57** 4323

Phase field model of microstructure formation during cooling slope slurry generation of novel Al-15Mg_2Si-4.5Si composite

Prosenjit Das*[a]

[a]Advanced Manufacturing and Materials Processing Laboratory, Department of Materials Engineering, Indian Institute of Science Bangalore, Bengaluru-560012, India.

*Email: prosenjitdas@iisc.ac.in

Abstract. The present study discusses the development of a two-dimensional phase field model of semi solid microstructure evolution of the novel Al-15Mg_2Si-4.5Si composite during cooling slope rheoprocessing. The superheated melt of the said composite undergoes partial solidification during its shear flow over the slope free surface, prior to get collected within the isothermal slurry holding furnace. Seed undercooling based nucleation model is adopted in the present PF model to simulate slurry microstructure of the composite. Experimentally determined cooling rate values are considered to simulate slurry microstructure for different process conditions. Micrographs of melt samples collected during experimentation establishes the developed PF model as an accurate microstructure prediction tool for the composite slurry. Apart from morphological features such as size and degree of sphericity values of the evolving primary Mg_2Si and α-Al grains, the model predictions includes melt temperature as well as solid content at any given instant of slurry processing.

1. Introduction

Phase-field (PF) models have proven their importance over the years to accurately capture the microstructure formation mechanism of different alloys and composites. There are several variants of phase field models proposed till date, such as two phase models [1-5], and subsequent multiphase models [6-9]. The multiphase models allow for consideration of interactions between the neighbouring grains/ phases. Here, two dimensional multiphase filed model is developed to investigate the physics associated with the microstructure formation of novel Al-15Mg_2Si-4.5Si composite (Chemical constituents of individual composite compositions are listed below in Table 1), during cooling slope semi solid slurry generation process. Brief overview of different slurry preparation techniques may be seen from our earlier publications [10-13]. Cooling slope technique holds an edge over other techniques owing to its low capital investment and ease of customisation [10-13]. Although the literature available on cooling slope technique mostly deals with the aluminium alloys, however, being in-situ, the present composite is an ideal candidate for cooling slope rheoprocessing.

The rheocast (using cooling slope) Al-15Mg_2Si-4.5Si composite offers higher strength to weight ratio and higher stiffness, compared to cast Al-Si alloys [10-13] and other composites of Al-Mg-Si series [14-16]. This is due to the presence of Mg of approx. 10 wt%, and due to the globularisation of microstructural features attained by rheoprocessing. In-situ formed Mg_2Si particles facilitates higher strength and improved wear resistance of the developed composite. Furthermore, other alloying elements of the composite i.e., extra Si facilitates enhanced ductility to the composite. It should be noted here that the brittle nature of the hard Mg_2Si particles are tackled in the composite by its engulfment by the primary Al grains, which facilitates simultaneous improvement in strength, ductility, fracture toughness and crack arrest capabilities of the composite. There is no literature available on mesoscopic simulation of microstructure formation of the above mentioned composite, at the best of authors knowledge on the topic. Earlier work on phase filed modelling of slurry microstructure formation, in case of cooling slope processing, were focused on Al-Cu [17] and Al-Si alloys [18-20]. The present study is taken up to bridge the above gap. Thermodynamic and mobility description is provided to the present PF model via thermodynamic database TCAL1 and Mobal2.

Table 1: Chemical composition of Al-15Mg_2Si-4.5Si composite.

Composite specifications	Chemical constituents		
	Al	Mg	Si
Al-15Mg_2Si-4.5Si	0.78	0.10	0.12

2.Phase field model

2.1 Governing equations

The multiphase filed theory speaks of tracking spatial positioning of grains of different phases having independent thermodynamic properties or of different crystallographic orientations, with the help of phase-field variable $\varphi_\alpha(x,t)$. The said variable assumes different values at different locations within the computational domain (Fig. 1), for example, it takes the value of 0 when encounters liquid phase and takes the value of 1 when encounters solid phase, whereas it assumes values in between 0 and 1 based on solid and liquid content at a particular location within the interface thickness η. Here, the assumed value of interface thickness is notably higher compared to atomic length scale of solid-liquid interface thickness.

Temporal evolution of the PF variable $\varphi_\alpha(x,t)$, within the simulation domain, is derived from the free energy functional F. Whereas F is estimated here by integrating the density functional f over the computational domain Ω, and the corresponding mathematical expression is given below [19, 20].

$$F\left(\{\phi_\alpha\},\{\overrightarrow{c_\alpha}\}\right) = \int_\Omega f\left(\{\phi_\alpha\},\{\overrightarrow{c_\alpha}\}\right) \tag{1}$$

$$f = \int_\Omega f^{\text{int}}(\{\phi_\alpha\}) + f^{chem}\left(\{\phi_\alpha\},\{\overrightarrow{c_\alpha}\}\right) \tag{2}$$

The above equation (eq^n 2) shows that, there are two components of the density functional i.e., interface component (f^{int}) and chemical component (f^{chem}). Both chemical as well as interfacial free energy functional depends on phase field variable $\varphi_\alpha(x,t)$, wherein brackets $\{\ \}$ shown in eq^n 1 and 2 points toward all the phases available within the computational domain and not any particular phase α [19, 20]. Unlike interfacial free energy, chemical free energy also depends on global phase composition field $\{\overrightarrow{c_\alpha}\}$. Further details on model description are depicted elsewhere [19, 20]. In this study, thermodynamic database TCAL1 is employed to solve the phase field as well as diffusion equations, adopting quasi-equilibrium phase concentration values. Further, isotropic heat extraction is assumed throughout the simulation domain, wherein temperature profile is estimated via latent heat averaging technique. Further details on data extrapolation and temperature evolution technique are reported elsewhere [19, 20].

2.2 Assumptions

Following assumptions forms the backbone of the present simulation study.

- Here, microstructure evolution simulation of Al-15Mg_2Si-4.5Si composite terminates at semi solid stage, hence, eutectic nucleation from bulk liquid is not considered.
- Considering rapid cooling experienced by the melt during its travel down the slope, convection and cross-diffusion effects are neglected here, in line with Vusanovic et al. [21-22].
- The developed 2D PF model deals with the anisotropic interfacial properties, which are described here via a vector which is normal to the solid-liquid as well as solid-solid interface $n = (n_x, n_y, n_z)$. The anisotropic functions considered here in case of Face centered cubic (FCC) α-Al [17-20] and primary Mg_2Si [23-24] grains are depicted below.

Static: $a_{\sigma} = 1 - \delta_{\sigma} \cos 4\phi$ (3)

Kinematic: $a_{\mu} = 1 + \delta_{\mu} \cos 4\phi$ (4)

where, φ, $\delta_{\sigma*}$, and δ_{μ}denotes the Euler angle of grain orientation, static and kinematic anisotropy coefficients (non dimensional), respectively. Further details on the adopted anisotropic model for the α-Al and primary Mg_2Si grains may be seen from elsewhere [19, 20]. Table 2 reports the values considered for the anisotropic model parameters, such as surface energy, anisotropy coefficients etc.

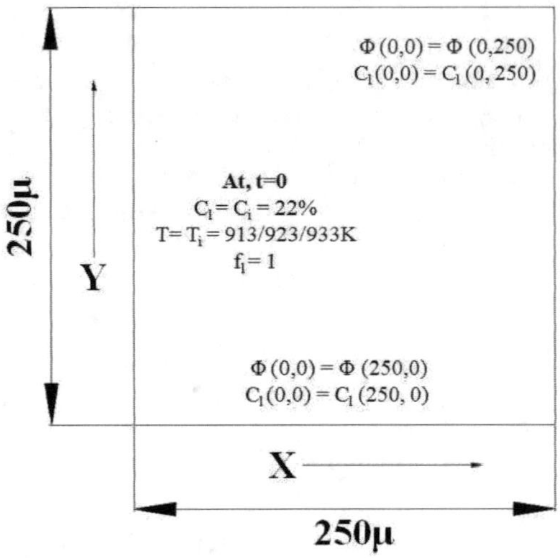

Figure 1. Computational domain for slurry microstructure evolution of the Al-15Mg_2Si-4.5Si composite.

Table 2: Input parameters considered for the present simulation study.

Composite specifications	Primary Mg_2Si anisotropy coefficients		α-Al anisotropy coefficients		Liquid/solid surface energy (J/cm^2)		Primary Mg_2Si formation start temp. (K)	α-Al formation start temp. (K)
	Static	Kinematic	Static	Kinematic	Primary Mg_2Si	α-Al		
650°C	0.29	0.1	0.3	0.3	1.6E-08	1E-07	906	863

2.3 Nucleation model

The present 2D PF model deals with the seed undercooling based nucleation model [25] to simulate the evolution of α-Al and primary Mg_2Si grains from the liquid melt, during cooling slope rheoprocessing of the Al-15Mg_2Si-4.5Si composite. As stated before, MICRESS software is used here to perform the phase field simulation studies, wherein nucleation of primary grains is governed by a specified critical undercooling criterion (seed undercooling). Here, nucleation of α-Al and primary Mg_2Si grains from the parent melt becomes possible when the local undercooling at a particular site within the melt surpasses the specified critical undercooling values of the above mentioned phases. It is to be noted here that, separate critical undercooling values are specified as nucleation barrier for α-Al and primary Mg_2Si phases. Except local undercooling, there are other parameters, considered in the present PF model, which govern the nucleation process of primary solid grains, such as shield distance, shield time as well as the frequency of nucleation check. It is to be noted here that the composition and temperature of the melt in the vicinity governs its local undercooling. Detailed description of the seed undercooling model may be seen from here [26].

Once formed, the nucleated primary grains of Mg_2Si and α-Al undergoes ripening driven growth, wherein the growth rate for individual process conditions depend on the applied cooling rate. Here, experimentally estimated cooling rate (via K type thermocouples) value for the melt pouring temperature of 650°C is depicted below in Table 3.

Table 3: Experimentally estimated slurry cooling rate value for the melt pouring temperature of 650°C P.T.

Processing condition	Temperature (K) (Experimental)		Temperature drop ΔT upto the slope exit (K)	Cooling rate (Temperature drop/avg. cooling time of 12 seconds) (K/s)
	Melt pouring temperature into the slope	Slope exit temperature		
650°C	923	864	59	4.92

3. Experimental procedure
A resistance heating furnace was used to melt the commercially available Al-11.43Si alloy ingots at 730°C. Then the melt was held for 15-20 minutes, for uniform temperature distribution and homogenization of the prepared melt. After this, small pieces of pure magnesium ingot (99.91% purity) are plunged into the melt. Argon gas atmosphere was maintained before charging the pure magnesium to avoid burning loss of Mg. Subsequently, degassing was performed by both the degasser tablet and supplying argon gas for a period of 5 minutes. After melt preparation of the composite, slag removal was performed at 660°C. Thereafter, the composite melt was poured into the slope at 650°C. The melt experiences shear flow over the slope free surface and exits cooling slope as a semi soli slurry, to get stored in an isothermal slurry holding bath. The cooling slope rheoprocessing facility used in this study is shown below in Fig. 2, along with the microstructure formation mechanism [27, 28].

4. Solution technique
Microstructure evolution within the Al-15Mg_2Si-4.5Si composite during semi solid processing is studied here by employing two dimensional phase field models, using the platform of commercial phase field solver MICRESS. The computational domain of 250μm×250μm, having 390,625 square cells, is considered here to perform the present simulations. Whereas material dependent input data/parameters considered here are depicted in Table 2 and Table 4. Here, chemical and thermophysical properties of the constituent primary Mg_2Si and α-Al phases such as local composition, free energy and temperature values are obtained from the Thermo Scientific provided TCAL1 thermodynamic database. The thermodynamic coupling between the phase filed solver and that of the database file is made via Micress TQ interface. Whereas keeping in view of computational efficiency, the developed phase field model is simplified by assuming constant values of grain boundary mobility ($M_{\alpha\alpha}$) and interface mobility ($M_{\alpha\beta}$) coefficients. The values are determined by performing trial numerical experiments, in an iterative manner. Similar approach was followed earlier by Seki et al. [29]. The values determined for mobility coefficients of primary Mg_2Si and α-Al phases, as well as interface/boundary mobility coefficient values are depicted in Table 4. The agreement between numerical predictions and experimental measurements of grain size values confirms reasonable accuracy of the adopted iterative technique (shown later in Fig. 4c). Detailed description on the adopted solution technique may be seen from our earlier publications [19-20].

Table 4: Input parameters considered at the solid/liquid and solid/solid interface.

Melt pouring temperature	Seed count within the computational domain of size 250μm×250 μm		Solid-Liquid interface thickness in terms of cell count	Kinetic coefficient at the solidification interface ($M_{\alpha\beta}$) (cm^4/Js)		Kinetic coefficient between the neighbouring solid grains ($M_{\alpha\alpha}$) (cm^4/Js)	
	Primary Mg_2Si	α-Al		Primary Mg_2Si	α-Al	α-Al and α-Al	α-Al and Mg_2Si
650°C	15	95	4	0.92E-03	2.84E-05	1.41E-06	0.42E-04

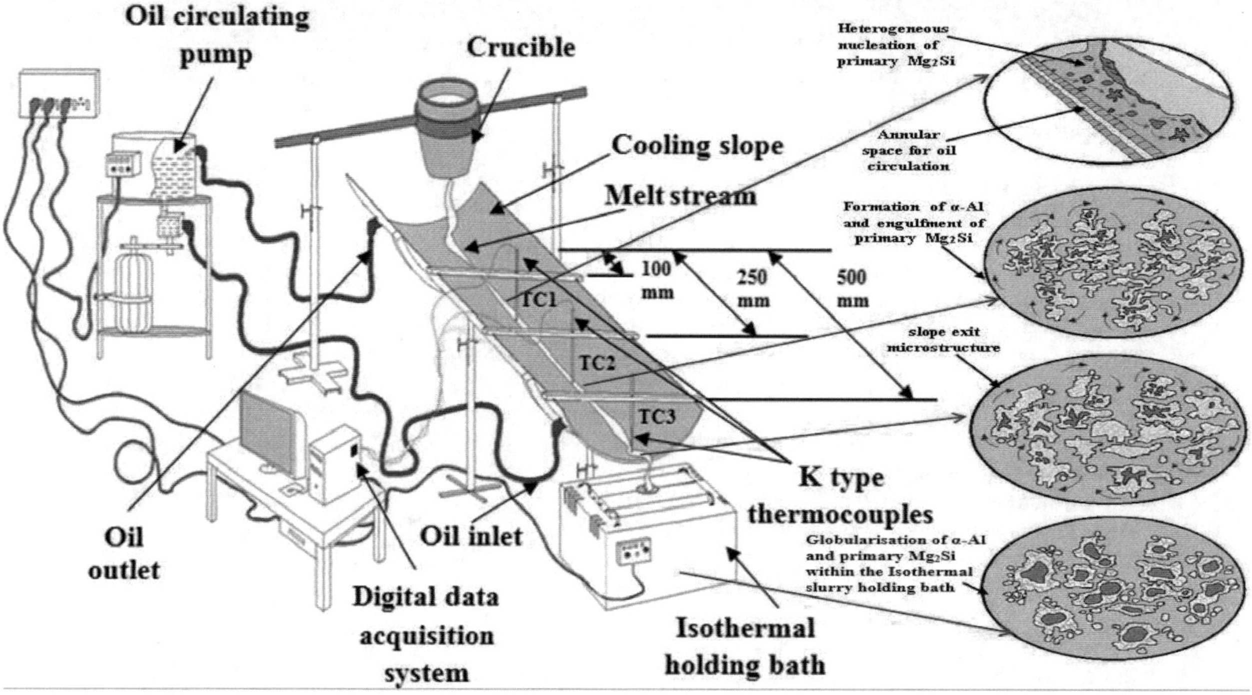

Figure 2. Sketch showing functional parts of the cooling slope slurry preparation set up as well as mechanism of slurry microstructure formation in case of Al-15Mg_2Si-4.5Si composite.

5. Results & discussion

5.1 Microstructure formation

The 2D phase field simulations performed here unlocks the physics behind the microstructure formation as well as predicts the microstructure of the composite slurry during cooling slope processing. Values of grain boundary mobility ($\boldsymbol{M_{\alpha\alpha}}$) and interface mobility ($\boldsymbol{M_{\alpha\beta}}$) coefficients of the solidifying primary grains (Mg_2Si and α-Al) are chosen carefully to achieve the above purpose. Apart from prediction of microstructural morphology (Fig. 3), the PF model provides quantitative information on grain dia. and shape factor of solidifying primary grains at different junctures of the adopted process technology, for a chosen melt pouring temperature of 650°C (923K) (Fig. 4). The simulation begins from the impingement of the composite melt over the slope free surface at 650°C. The present PF model operates with seed undercooling nucleation model, wherein no predefined seed size (radii) distribution is provided to the solver. Seeds of primary Mg_2Si nucleates within the bulk melt (shown in deep yellow), whereas α-Al grains nucleate within the interfacial region between the bulk melt (shown in red) and the evolving primary Mg_2Si particles (shown in light yellow). In this study, local undercooling and solute concentration within the melt determines growth or dissolution of the nucleated embryos of α-Al and primary Mg_2Si. Figure 3 also shows comparative experimental micrographs, along with their simulated counterparts, for different processing stages, wherein micrograph of the slope exit slurry shows highest homogeneity of microstructural features. Quantitative findings of the slurry morphology are depicted later within the manuscript.

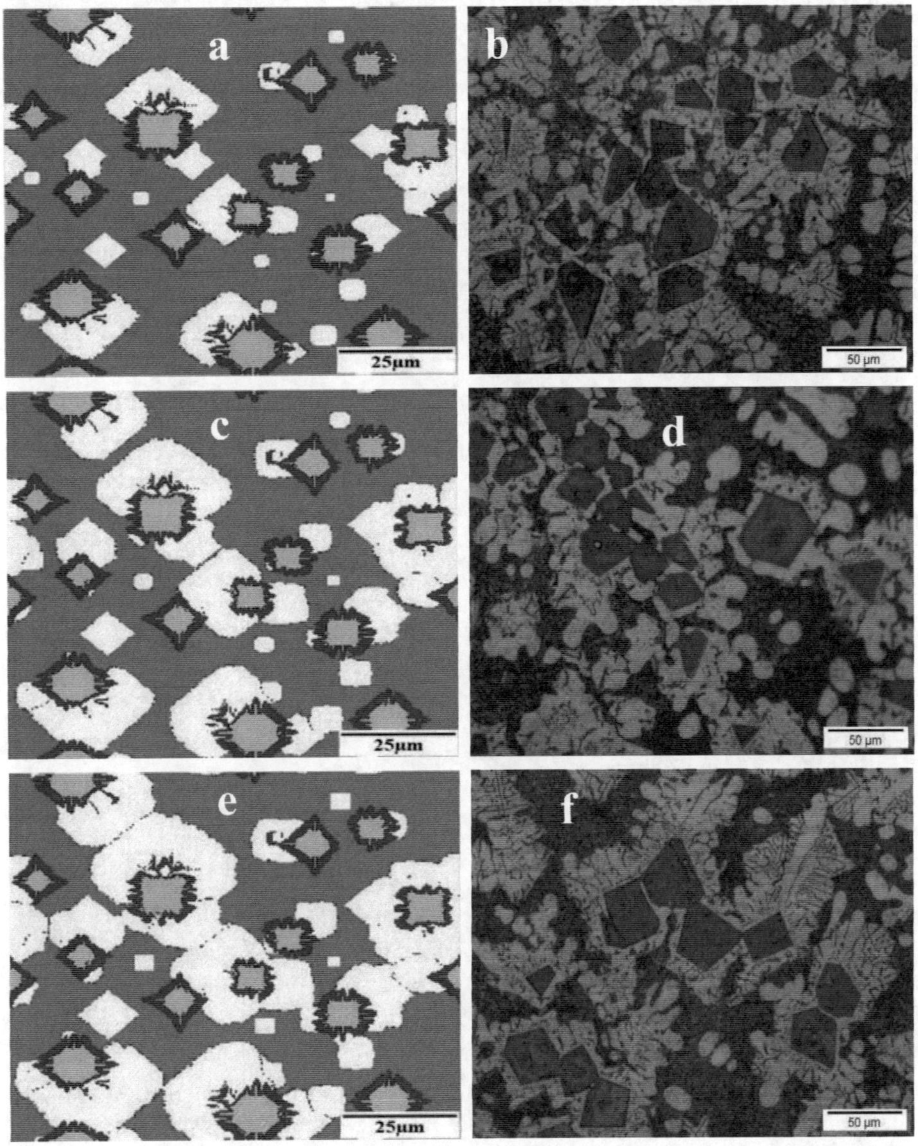

Figure 3. Simulated and experimental micrographs of the composite slurry corresponding to the melt pouring temperature of 650°C: (a and b) numerical and experimental micrograph of start sample, (c and d) numerical and experimental micrograph of middle sample, and (e and f) numerical and experimental micrograph of exit sample.

5.2 Solid content and morphology of primary solid

Increasing solid content within the composite melt, with the increasing simulation time, is evident in Fig. 4. Progressive nuclei formation and grain growth of primary Mg_2Si and α-Al grains causes gradual increase in slurry solid content with the progressing simulation time. Whereas numerical findings of temporal evolution of avg. grain dia. and shape factor of the primary grains (Mg_2Si and α-Al) are plotted in Fig. 4b, c, and d, for the entire simulation regime of slurry generation. Figure 4 also shows experimental validation in terms of grain size comparison with the experimental measurement of grain size values. It is to be noted here that the experimental measurements are obtained from the optical micrographs of melt samples, collected from chosen locations over the slope. Here, diffusion driven growth of primary grains causes the increase in grain size values with the progressing simulation time.

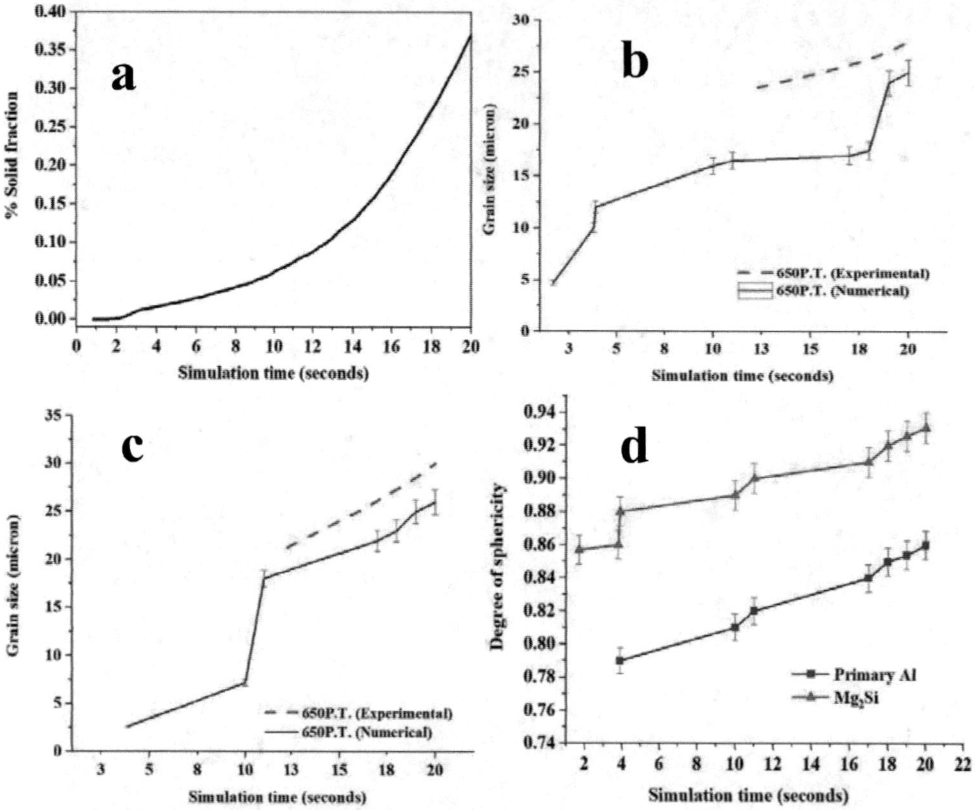

Figure 4. (a) Solid fraction vs simulation time curve for 650°C P.T., comparative numerical and experimental grain size evolution curves of (b) primary Mg_2Si and (c) α-Al grains, and (d) comparative α-Al and primary Mg_2Si sphericity (simulation findings).

5.3 Temporal evolution of grain size distribution curves

Normalised distribution of grain dia. evolution of primary grains (Mg_2Si and α-Al) is shown below in Fig. 5, during melt travel along the slope and subsequent slurry formation. It is observed that the particle count of primary Mg_2Si increases till 11 seconds of simulation time and then decreases. The above phenomenon is attributed to the formation of individual primary Mg_2Si nuclei's in the initial stages of slurry generation, whereas afterwards coalescence, impingement of neighbouring Mg_2Si grains facilitates the decrease in particle count. Here, bimodal distribution of primary Mg_2Si grain dia. causes the peak shift observed in Fig. 5a. However, increasing uniformity in grain size distribution is evident with the increasing simulation time. Here, variability of grain size distribution of primary Mg_2Si grains is primarily caused by the difference in growth rate of neighbouring grains due to the varying solute concentration and change in the local thermo physical parameters. Figure 5b shows The temporal evolution of normalised grain dia. distribution shown in Fig. 5b corresponds to α-Al grains. It has been observed that the peak height of particle count decreases with the simulation progress, which in turn triggers peak broadening. This is due to the decrease in no. density of independent globules of α-Al with the progressive solid formation during slurry generation, which is attributed to the impingement, coalescence and coagulation of neighbouring α-Al grains/globules. This is further confirmed from the progressive shift of the particle count peak towards the right side of the median.

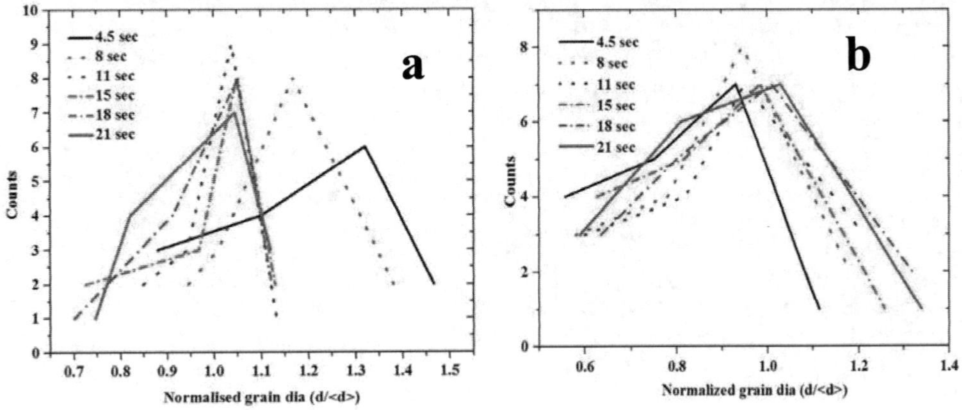

Figure 5. Temporal evolution of normalised grain size distribution: (a) primary Mg_2Si grains and (b) α-Al grains, at 650°C P.T.

5.4 Experimental validation
To validate the numerical findings, melt samples were scooped from the top, middle and exit section of the cooling slope. The scooped and quenched slurry samples were studied under optical microscope to measure the grain size and sphericity values of Mg_2Si and α-Al phases. Table 5 compares the simulated as well as experimental findings of slurry morphology at slope exit state. The developed 2D PF model is found to be capable of near accurate prediction of microstructural as well as thermophysical properties of the slurry, which firmly establishes its usefulness as a process control tool to develop engineering components.

6. Conclusions
The concluding remarks originated from the study are listed below.
i. The developed PF model is found to be capable of predicting the thermophysical (solid content, temperature) as well as morphological (grain dia., shape factor, and size distribution of primary Mg_2Si and α-Al grains) properties of the cooling slope processed slurry of the novel Al-15Mg_2Si-4.5Si composite.
ii. Reasonable accuracy of model predictions with that of experimental measurements offers promise to use the developed PF model as a process control tool to develop microstructurally engineered, complex contoured cast components.
iii. The present PF model can be used in conjunction with the thermodynamic and mobility databases to predict the semi solid slurry morphology for a variety composite composition bearing to the Al-Mg_2Si-Si composite family.

Acknowledgements
The author would like to thank SERB, DST, Government of India for their financial support to this work vide grant no. TTR/2021/000113. Author also like to acknowledge former Project Assistant Mr. Subrata Sarkar for his support with the data analysis. Finally, the author would like to express his gratitude to Director, IISc for his encouragements.

Table 5: Comparative numerical and experimental findings of thermophysical and morphological properties of the composite slurry.

Thermophysical and Morphological properties		Numerical result	Experimental result
Temperature		583.5°C	591°C
Solid fraction		0.37097	0.394
Primary Mg_2Si	Diameter	25 μm	28 μm
	Sphericity	0.93	0.797
α-Al	Diameter	26 μm	25.375 μm
	Sphericity	0.86	0.693

References

[1] Wheeler AA, Boettinger WJ, McFadden GB. "Phase-field model for isothermal phase transitions in binary alloys" Phys. Rev. A, vol. 45, pp. 7424-7439, 1992.

[2] Karma, A., Lee, Y.H., Plapp, M. "Three-dimensional dendrite-tip morphology at low undercooling" Phys. Rev. E, vol. 61, 3996-4006, 2000.

[3] Kobayashi, R. "Modeling and numerical simulations of dendritic crystal growth", Physica D: Nonlinear Phenomena, vol. 63, pp. 410-423, 1993.

[4] Kim, S.G., Kim, W.T., Suzuki, T. "Phase-field model for binary alloys" Phys. Rev. E, vol. 60, 7186-7197, 1999.

[5] Bottger B, Eiken J, Apel M. "Multi-ternary extrapolation scheme for efficient coupling of thermodynamic data to a multi-phase-field model", Computational Materials Science, vol. 108, pp. 283-292, 2015.

[6] Steinbach, I., Pezzolla, F., Nestler, B., Seeßelberg, M., Prieler, R., Schmitz, G.J., Rezende, J.L.L "A phase field concept for multiphase systems" Physica D, vol. 94, pp. 135-147, 1996.

[7] Steinbach, I., Pezzolla, F. "A Generalized Field Method for Multiphase Transformations Using Interface Fields" Physica D, vol. 134, pp. 358-393, 1999.

[8] Nestler, B., Wheeler, A.A. "A multi-phase-field model of eutectic and peritectic alloys: numerical simulation of growth structures" Physica D, vol. 138, pp. 114, 2000.

[9] Tiaden J., Nestler B., Diepers H.J., Steinbach I. "The multiphase-field model with an integrated concept for modelling solute diffusion", Physica D, vol. 115, pp. 73-86, 1998.

[10] Das, P., Samanta, S.K., S. Bera, Dutta, P. "Microstructure Evolution and Rheological Behavior of Cooling Slope Processed Al-Si-Cu-Fe Alloy Slurry." Metall. Trans. A., vol. 47, pp. 2243-2256, 2016.

[11] Das, P., Samanta, S.K., Das, R., Dutta, P. "Optimization of degree of sphericity of primary phase during cooling slope casting of A356 Al alloy: Taguchi method and regression analysis." Measurement, vol. 55, pp. 605–615, 2014.

[12] Das, P., Kumar, M., Samanta, S.K., Dutta, P., Ghosh, D. "Semi-solid processing of A380 Al alloy using Cooling Slope." Materials and Manufacturing Processes, vol. 29, pp. 422–428, 2014.

[13] Das, P., Samanta, S.K., Mondal, B., Dutta, P. "Multiphase model of semi solid slurry generation and isothermal holding during cooling slope rheoprocessing of A356 Al alloy" Metall. Trans. B., vol. 49, pp. 1925-1944, 2018.

[14] Zhang, J., Fan, Z., Wang, Y. Q., Zhou, B. L. "Effect of cooling rate on the microstructure of hypereutectic Al-Mg_2Si alloys." Journal of Materials Science Letters, vol. 19, pp. 1825 – 1828, 2000.

[15] Qingxiu, J., Caixia, Z., Xiaodong, H. “Study on in-situ Mg_2Si/Al-Si composites with different compositions.” China Foundry, vol. 6, pp. 133-136, 2009.
[16] Srinivas, V., Singh, V. “Development of In Situ As Cast Al–Mg_2Si Particulate Composite: Microstructure Refinement and Modification Studies.” Trans Indian Inst Met, vol. 65, pp. 759–764, 2012.
[17] Das, P., Samanta, S.K., Kumar, P., Dutta, P. "Phase Field Simulation of Equiaxed Microstructure Formation during Semi-solid Processing of A380 Al Alloy" ISIJ International, vol. 54 (7), pp. 1601–1610, 2014.
[18] Bunck, M., Warnken, N., Polaczek, A. B. "Microstructure evolution of rheo-cast A356 aluminium alloy in consideration of different cooling conditions by means of the cooling channel process" Journal of Materials Processing Technology, vol. 210, pp. 624-630, 2010.
[19] Das, P., Dutta, P. "Phase field modelling of microstructure evolution and ripening driven grain growth during Cooling Slope processing of A356 Al alloy" Computational Materials Science, vol. 125, pp. 8-19, 2016.
[20] Das, P., Dutta, P. "Three-dimensional phase field simulation of spheroidal grain formation during semi solid processing of Al-7Si-0.3Mg alloy" Computational Materials Science, vol. 184, 109856, 2020.
[21] Vusanovic I., Voronjec D., Krane M. J. M. "Microsegregation phenomena in Al-Cu-Mg alloy with considering of diffusion phenomena in primary phase" Facta Universitatis Series: Mechanical Engineering, vol.1(8), pp. 965-980, 2001.
[22] Vusanovic I., Sarler B., Krane M. J. M. "Microsegregation during the solidification of an Al–Mg–Si alloy in the presence of back diffusion and macrosegregation" Materials Science and Engineering A, vol. 413–414, pp. 217–222, 2005.
[23] Ghandvar, H., Jabbar, K. A., Idris, M. H., Ahmad, N., Jahare, M. H., Koloor, S. S. R., Petru, M. "Influence of barium addition on the formation of primary Mg_2Si crystals from Al-Mg-Si melts", Journal of Materials Research and Technology, vol. 11, pp. 448-465, 2021.
[24] Chen, L., Wang, H.Y., Li, Y.J., Zha, M., Jiang, Q.C. "Morphology and size control of octahedral and cubic primary Mg_2Si in an Mg–Si system by regulating Sr contents", Cryst Eng Comm, vol. 16, pp. 448–454, 2014.
[25] Wei, M., Tang, Y., Zhang, L., Sun, W., Du, Y. "Phase-Field Simulation of Microstructure Evolution in Industrial A2214 Alloy During Solidification" Metallurgical and Materials Transactions A, vol. 46, pp. 3182-3191, 2015.
[26] www.micress.rwth-aachen.de
[27] Kumar, S., Das, P., Tiwari, S., Mondal, M.K., Bera, S., Roy, H., Samanta, S. K. "Study of Microstructure Evolution during Semi-Solid Processing of an in-Situ Al Alloy Composite" Materials and Manufacturing Processes, vol. 29, pp. 422–428, 2014.
[28] Das, P., Islam, S.T., Samanta, S.K., Das, S. "Microscale Deformation behaviour of Rheocast Al-7Si-0.3Mg alloy " Proceedings of the Institution of Mechanical Engineers, Part L: Journal of Materials: Design and Applications, vol. 230 (6), pp. 1041–1061, 2016.
[29] Seki, A., Sawada, M., Moriguchi, K., Shirai, Y. "Simulations of Grain Growth in Titanium and Stainless Steels" Nippon Steel & Sumitomo Metal Technical Report No. 106, pp. 41-46, 2014.

IOP Conf. Series: Materials Science and Engineering 1281 (2023) 012059 doi:10.1088/1757-899X/1281/1/012059

Numerical study on the formation of spurious grains and freckles during the directional solidification of superalloys

H Zhang[1], M Wu[1,*], X Liu[2,3], D Ma[2,3], M Song[2], A Ludwig[1], A Kharicha[1]

[1] Chair for Simulation and Modelling of Metallurgy Processes, Department of Metallurgy, University of Leoben, A-8700 Leoben, Austria
[2] State Key Laboratory of Powder Metallurgy, Central South University, Changsha 410083, China
[3] Wedge Central South Research Institute, Shenzhen 518045, China

E-mail: menghuai.wu@unileoben.ac.at

Abstract. Freckles, a significant issue encountered during the directional solidification of superalloys, are recognised by a trail of equiaxed grains parallel to the direction of gravity accompanied by local eutectic enrichment. In the current study, a mixed-columnar-equiaxed multiphase volume-average solidification model was employed to study the formation of freckles in superalloy casting. Fragments produced via flow-driven and capillary-driven fragmentation mechanisms are considered as the source of spurious grains. The transport and the growth/remelting of the fragments are considered. According to the simulation results, some segregation channels develop at the corners of the casting. Flow-driven fragments are produced in/around the segregation channels, whereas capillary-driven fragments are produced at a certain depth of the mushy zone across the entire section of the casting. The fragmentation rate caused by the flow-driven mechanism is several orders of magnitude larger than that caused by the capillary-driven mechanism, i.e. the flow-driven fragmentation mechanism is dominant for the currently investigated sample. After the solidification process, four freckles formed at the casting corners on the shadowed side, whereas it was freckle-free on the bright side.

1. Introduction

Single-crystal superalloy turbine blades fabricated using directional solidification (DS) are key components of aeroengines and gas turbines. Such components are susceptible to flow-induced freckle defects, which are recognised by a trail of spurious/stray grains parallel to the direction of gravity, accompanied by the local enrichment of eutectics [1]. The freckling mechanism is well-understood experimentally. The thermo-solutal convection due to the liquid density inversion in the mushy zone is responsible for the onset of freckles [2,3]. The segregation of the alloying elements makes the interdendritic liquid less dense than the bulk liquid. In the event of an upward DS, the light interdendritic liquid tends to flow upward to release its gravitational potential, and thereby, some plumes form at the solidification front. After merging and competing with each other, only a few plumes survive and further develop into stable segregation channels. The solute-enriched liquid is continuously sucked into the channel. The roots of higher-order sidearms can be remelted by flow-induced solute enrichment or capillary effects [4,5], that is, dendrite fragmentation occurs.

Near the open mouth of the segregation channel, small fragments are transported out of the mushy zone by upwelling (upward flow of solute-enriched liquid enclosed in the plume) and finally remelted in the top hot liquid [6]. Deep in the mushy zone, the fragments sink and continue to grow into spurious grains [2], and the residual solute-enriched liquid solidifies as eutectics. The formation of the freckle is a consequence of a long and complex process involving the aforementioned solidification physics. The quantitative prediction of such a freckling process through an experimental study is still not feasible. Alternatively, numerical modelling has gradually made it possible to visualise the casting process and accurately predict the formation of freckles. According to state-of-the-art simulations [7–10], the freckles were indirectly predicted through the calculated segregation channels. The significant role of the fragments, which are the main source of the spurious grains, was ignored or roughly treated [11].

In this study, a three-phase mixed columnar-equiaxed solidification model [12] was used to simulate the DS of superalloy casting. The model was recently extended to consider flow- and capillary-driven fragmentation mechanisms [5]. The remelting and solidification of the as-formed solid phases (i.e. columnar dendrites and equiaxed grains) were also considered [13]. The main goal of this study was to numerically investigate the formation of spurious grains produced by different fragmentation mechanisms, and hence the onset of freckles.

2. Model configuration and calculation method

A cluster of Ni-based superalloy specimens was directionally solidified in a Bridgman-type vacuum furnace. The shell mould produced by the normal investment casting procedure was placed on a copper-chill base. A typical spiral grain selector, including the starter block, was used to realise the single-crystal solidification of the casting. The shell mould was preheated first, and then the hot melt was poured into the mould. The solidification of the casting was triggered by withdrawing the sample downwards from the heated zone to the cold zone at a constant velocity of 3 mm/min. The composition of the superalloy is listed in Table I. After the casting process, the superalloy casting was knocked out from the shell mould and mechanically separated from the casting cluster, followed by metallographic analysis.

Table I. Main composition of the considered superalloy.

Elements	Ni	Ta	Cr	Al	Co	W	Re	Ti	Mo
Content (wt.%)	-	8.07	3.39	5.69	5.97	6.52	4.89	0.15	0.41

A full-scale geometry of the Bridgman furnace and casting system was constructed for the current simulation. As shown in Figure 1(a), only a sector of 30° of the furnace containing one casting was considered. The dimensions of the casting are shown in Figures 1(b) and (c). The DS process was simulated in two steps. The first step is to calculate the global thermal field in the Bridgman furnace, including the casting system, and the second step is to calculate the flow and solidification in the casting. The former was performed using the commercial software ProCAST, whereas the latter was performed using our previously developed multiphase volume-average-based (VA) solidification model. The thermal calculation was based on the full-scale geometry (Figure 1(a)), and the flow-solidification calculation was restricted to the casting body (Figure 1(b)).

For the thermal calculation in ProCAST, the top hot and bottom cold chambers were set to fixed temperatures of 1773 K and 353 K, respectively. The radiation was considered between the inducting wall and shell mould, and the withdrawing process was realised by raising the furnace. The calculation started from a mould fully filled with the superalloy melt, i.e. the filling process was ignored. The conservation of enthalpy was solved for the shell mould and casting, but only latent heat owing to solidification of the casting was treated with an equivalent specific heat method [10].

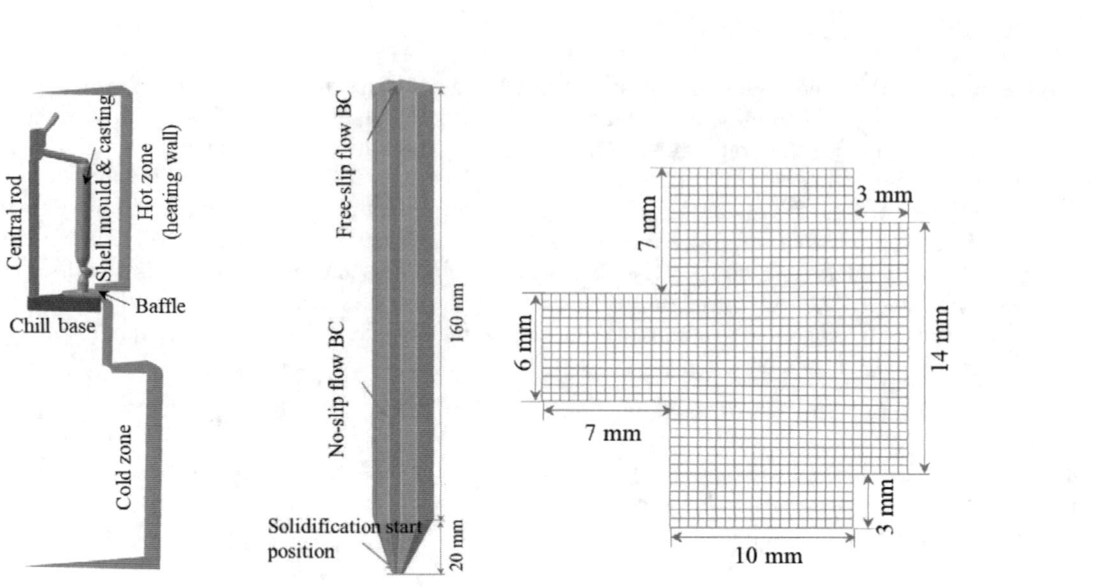

Figure 1. (a) Layout of the Bridgman furnace and the casting system; (b) shape of the sample casting; (c) dimension of the casting.

Flow and solidification were solved using the VA multiphase solidification model. Three hydrodynamic phases were involved: the liquid melt, equiaxed grains, and columnar dendrites with volume fractions of up to one ($f_\ell + f_e + f_c = 1$). The morphology of the columnar dendrites was approximated by growing cylinders. The equiaxed dendrites were simplified as spheres. Thermodynamic equilibrium was assumed at the solid–liquid interface. Diffusion-governed growth kinetics was considered to calculate the solidification/remelting rate. The concentration difference between the thermodynamic equilibrium concentration of the liquid at the interface (c_ℓ^*) and the volume-averaged liquid concentration (c_ℓ) served as the driving force for solidification and remelting. The columnar structure was initiated only from the bottom of the casting, and its tip front was traced using the Lipton–Glicksman–Kurz (LGK) model [14]. In the current study, equiaxed grains were assumed to originate from dendrite fragmentation, and two fragmentation mechanisms, i.e. capillary- and flow-driven fragmentation mechanisms, were considered. Remelting and destruction of the equiaxed grains were also considered. The conservation of the number density of the equiaxed grains was solved considering the transport of equiaxed grains:

$$\frac{\partial}{\partial t} n_e + \nabla \cdot (\vec{u}_e n_e) = N_{\text{frag}}^{\text{capillary}} + N_{\text{frag}}^{\text{flow}} + N_{\text{des}}, \quad (1)$$

where n_e and $\vec{u}_e$ are the number density and velocity of the equiaxed grains; $N_{\text{frag}}^{\text{capillary}}$ and $N_{\text{frag}}^{\text{flow}}$ are the fragmentation rates due to capillary- and flow-driven mechanisms; and N_{des} is the destruction rate of the grains through remelting. The formulations for these terms and the meanings of the relevant symbols are summarised in Table II.

The production rate of fragments via the capillary-driven mechanism ($N_{\text{frag}}^{\text{capillary}}$) was calculated based on equation (2) [15,16], in which a is an alloy-dependent constant, and S_V is the interfacial area density. A general expression for S_V was shown in equation (3) [17], which accounted for the effects of dendrite growth, coarsening, and interface coalescence. The initial size of the formed fragments ($d_{e,\text{frag}}^0$) was assumed to be inversely proportional to S_V, equation (4) [18]. Since a spherical morphology was assumed for the equiaxed grains, the corresponding mass transfer rate ($M_{\text{frag}}^{\text{capillary}}$) from the columnar to the equiaxed due to the capillary-driven fragmentation can be calculated by equation (5) [16].

Table II. Formulations for $N_{\text{frag}}^{\text{capillary}}$, $N_{\text{frag}}^{\text{flow}}$, and N_{des}

Terms	Equations	Symbols	Refs.
$N_{\text{frag}}^{\text{capillary}}$	**Capillary-driven fragmentation:** $N_{\text{frag}}^{\text{capillary}} = \frac{\mathrm{d}\left(a \cdot S_V^{\,3}\right)}{\mathrm{d}t}$ (2) $S_V = f_c(1-f_c)^{\tilde{r}}\left(\left(S_{S0}^{\,-1}\right)^3 + K_0 t_s\right)^{-1/3}$ (3) $d_{\text{e,frag}}^0 = f_c \frac{1.6}{S_V}$ (4) $M_{\text{frag}}^{\text{capillary}} = N_{\text{frag}}^{\text{capillary}}\left(\rho_e \frac{\pi}{6}\left(d_{\text{e,frag}}^0\right)^3\right)$ (5)	S_V: interfacial area density a, $\tilde{r}$, S_{S0}, K_0: alloy-dependent constants f_c: volume fraction of columnar phase $d_{\text{e,frag}}^0$: initial size of the fragment $M_{\text{frag}}^{\text{capillary}}$: mass transfer rate from columnar to equiaxed due to capillary-driven fragmentation ρ_e : equiaxed density	[15,16,18]
	$a = 1.0\times10^{-5}$, $\tilde{r} = 0.4$, $S_{S0}^{\,-1} = 246.0$ µm, $K_0 = 23.5$ µm³/s, $\rho_e = 7646.0$ kg/m³		
$N_{\text{frag}}^{\text{flow}}$	**Flow-driven fragmentation:** $M_{\text{frag}}^{\text{flow}} = -\gamma \cdot (\vec{u}_\ell - \vec{u}_c) \cdot \nabla c_\ell \cdot \rho_e \cdot f_c$ (6) $N_{\text{frag}}^{\text{flow}} = \frac{M_{\text{frag}}^{\text{flow}}}{\frac{\pi}{6}\left(d_{\text{e,frag}}^0\right)^3 \rho_e}$ (7)	$M_{\text{frag}}^{\text{flow}}$: mass transfer rate from columnar to equiaxed due to flow-driven fragmentation γ : fragmentation coefficient $\vec{u}_\ell$: liquid velocity $\vec{u}_c$: columnar velocity ∇c_ℓ : liquid concentration gradient	[5,19]
	$\gamma = 2.0\times10^{-5}$		
N_{des}	**Grain destruction:** $\frac{\mathrm{d}n_e}{\mathrm{d}x} = \frac{n_e}{\sqrt{2\pi}\sigma_d x} e^{-\frac{1}{2}\left(\frac{\ln(x)-\ln(\hat{d}_e)}{\sigma_d}\right)^2}$ (8) $N_{\text{des}} = v_{\text{Re}} \cdot \left.\frac{d(n_e)}{d(x)}\right\rvert_{x=d_{\text{e,critical}}}$ (9) $v_{\text{Re}} = \frac{D_\ell}{l_\ell} \cdot \frac{(c_\ell^* - c_\ell)}{(1-k)c_\ell^*}$ (10)	x: grain diameter of various size classes v_{Re}: grain remelting speed σ_d : geometric standard deviation $\hat{d}_e$: geometric mean of the grain diameter D_ℓ : liquid diffusion coefficient l_ℓ: liquid diffusion length c_ℓ^* : equilibrium liquid concentration c_ℓ : volume-averaged liquid concentration k: solute partition coefficient	[5,13]
	$\sigma_d = 0.873$, $d_{\text{e,critical}} = 50$ µm, $D_\ell = 3.6\times10^{-9}$ m²/s, $k = 0.569$		

Based on the fact that a flow in the growth direction of primary columnar dendrites in the mushy zone promotes re-melting, the mass transfer rate ($M_{\text{frag}}^{\text{flow}}$) from the columnar to the equiaxed due to the flow-induced fragmentation was assumed to be proportional to the value of $-(\vec{u}_\ell - \vec{u}_c)\cdot\nabla c_\ell$, equation (6) [19]. The fragmentation coefficient γ is to bridge all other unknown contributors to $M_{\text{frag}}^{\text{flow}}$. Because of the assumed spherical morphology for the equiaxed grains, the production rate of flow-driven fragments can be calculated via equation (7). Refer to equation (4) for the initial size of the fragments.

The equiaxed grains, and the newly produced fragments, shall be remelted and destroyed if they are exposed to a hot liquid. Because the grain remelting was assumed to be governed by diffusion, equation (10) was valid to calculate the remelting speed v_{Re}. In each calculation cell, the equiaxed grains were assumed to follow a lognormal size distribution, equation (8) [13]. When the grain diameter was melted to smaller than a critical value $d_{e,critical}$, the grain would be eliminated from the casting domain, i.e. grain destruction occurred. The destruction rate of the equiaxed grains was calculated via equation (9). More fundamentals of the VA multiphase solidification model have been detailed in [12].

The cooling history of the temperature distribution on the casting surface, as calculated by ProCAST, was used as the Dirichlet thermal boundary condition for the flow-solidification calculation using VA solidification model. The thermal coupling method has been introduced in [10]. As shown in Figure 1(b), a no-slip flow boundary condition was applied to the lateral wall of the sample, and a free-slip flow boundary condition was applied to the top surface. The alloy was assumed to be incompressible, with constant density and viscosity.

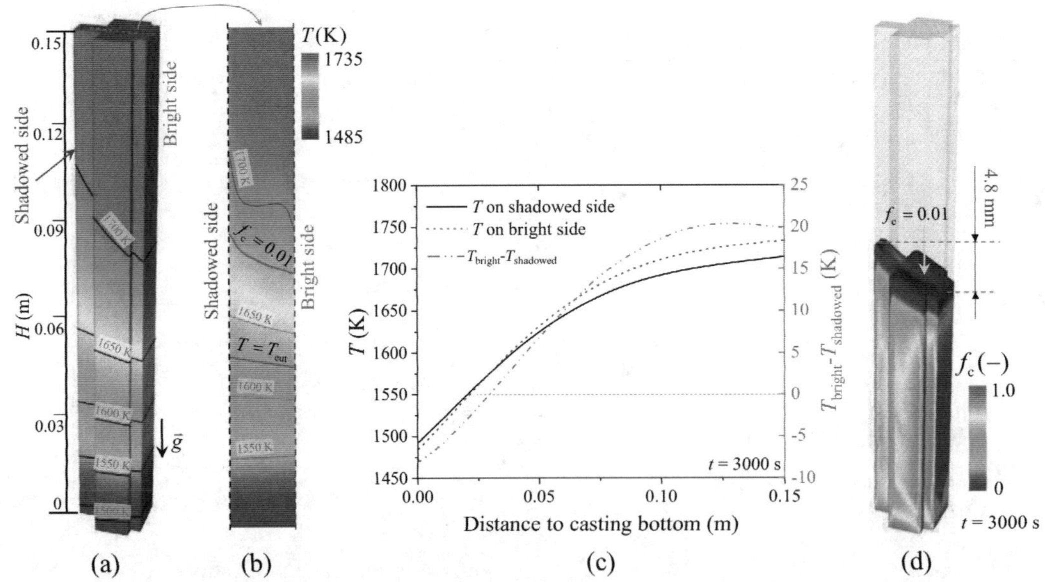

Figure 2. Analysis of the calculated temperature (T) in the casting. (a)-(b) T field on the casting surface and casting vertical section. (c) T profiles along the two vertical dashed lines in (b), and the temperature difference between the two lines. (d) Calculated volume fraction of the columnar phase f_c, and the shape of the solidification front ($f_c = 0.01$).

3. Interpretation of simulation results

3.1. Temperature field

The calculated temperature on the casting surface at $t = 3000$ s is shown in Figure 2(a). The bottom of the sample is 250 K cooler than the top. Owing to the shadow effect in the furnace, the shadowed side facing the central rod is cooler than the bright side facing the inducting wall. The temperature on the central vertical section of the casting is presented in Figure 2(b), where the solidification front (f_c = 0.01) and eutectic isotherm ($T = T_{eut}$) are marked. The isotherms (e.g. T = 1700 K) are bent by the liquid convection in the top liquid region. The temperature profiles along the two dashed lines in Figure 2(b) are plotted in Figure 2(c), and the temperature difference between the two lines ($T_{bright} - T_{shadowed}$) is also overlaid. From the top to the bottom of the sample, the temperature difference changes from positive to negative gradually. In the top part of the sample, the temperature on the bright side is 20 K higher than that on the shadowed side. Owing to this temperature difference, the solidification front is significantly inclined. According to the current simulation results shown in Figure 2(d), the solidification front on the shadowed side is 4.8 mm higher than that on the bright side.

3.2. Liquid flow pattern

Typical solidification results at t = 2500 s are shown in Figure 3. Figure 3(a) presents the contour of the columnar fraction (f_c) and the shape of the solidification front. The liquid flow of the top hot liquid is overlaid on the vertical section. Figure 3(b) depicts the liquid concentration (c_ℓ) on the casting surface, on which the isolines for f_c=0.01 and T=T_{eut} are drawn. The liquid density (ρ_ℓ) is shown in Figure 3(c). The liquid flow in the plumes and freckles is depicted in Figure 3(d). It should be addressed that ρ_ℓ is a post-processed result according to local T and c_ℓ, i.e. $\rho_\ell = \rho_\ell^{\text{ref}} \cdot \left(1 + \beta_T \cdot \left(T - T^{\text{ref}}\right) + \beta_c \cdot \left(c_\ell - c^{\text{ref}}\right)\right)$, in which ρ_ℓ^{ref}, T^{ref}, and c^{ref} are the reference density, temperature, and liquid concentration; β_T and β_c are the thermal and solutal expansion coefficients.

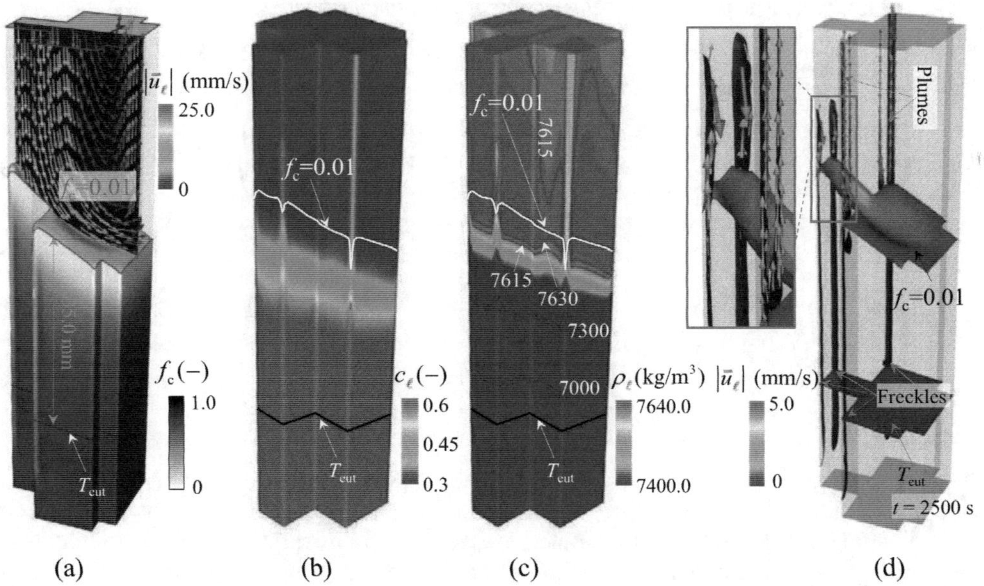

Figure 3. Typical solidification results at t = 2500 s. (a) Contour of the volume fraction of columnar phase f_c. The liquid convection of the top hot liquid is shown by the colourful vectors on the vertical section. (b) Contour of the liquid concentration c_ℓ. (c) Contour of the liquid density ρ_ℓ. (d) Iso-surfaces in black are to show the distribution of the plumes and freckles. The fluid flow in the plumes and freckles is shown by the vectors.

Because of the horizontal temperature gradient, the melt flows downwards on the shadowed side and then flows upwards on the bright side, Figure 3(a). Under the current cooling condition, a 35 mm thick mushy zone developed. In the mushy zone, the solute was rejected from the solid to the liquid owing to the alloy solidification. As shown in Figure 3(b), c_ℓ increases with the depth of the mushy zone. Because $\rho_\ell = \rho_\ell^{\mathrm{ref}} \cdot \left(1+\beta_{\mathrm{T}} \cdot \left(T-T^{\mathrm{ref}}\right)+\beta_{\mathrm{c}} \cdot \left(c_\ell - c^{\mathrm{ref}}\right)\right)$, as depicted in Figure 3(c), above the solidification front, ρ_ℓ decreases with the height of the casting owing to the increased temperature; below the solidification front, ρ_ℓ decreases with the depth of the mush owing to the solute enrichment; and the maximal $\rho_\ell\left(7630\ \mathrm{kg/m^3}\right)$ is found near the solidification front. The inversion of ρ_ℓ makes the liquid hydrodynamically unstable. The solute-enriched liquid in the mushy zone tends to flow upward owing to its low density, which leads to the onset of the plumes. As depicted in Figure 3(d), some plumes develop on the shadowed side, and the light solute-enriched liquid flows upward along the plumes. The vertical transport of the solute-enriched liquid raises c_ℓ in and near the plumes (Figure 3(b)), and thereby drops ρ_ℓ locally (Figure 3(c)).

Note that plumes only formed on the shadowed (left) side of the casting. The onset of plumes is highly related to the shape of the solidification front. At the peak of the inclined solidification front, the solute-enriched interdendritic liquid exhibited the highest gravitational potential energy to flow upwards, making it the preferred position for the formation of the plumes. This has been theorised in our previous work [10]. As shown in Figure 3(d), four plumes located as the casting corners further developed into stable freckles below the solidification front. On the bright (right) side, it is free of plumes and freckles.

3.3. Formation of spurious grains via different fragmentation mechanisms

Two fragmentation mechanisms, i.e. flow- and capillary-driven mechanisms, have been considered in the current study, and the calculated $N_{\mathrm{frag}}^{\mathrm{flow}}$ and $N_{\mathrm{frag}}^{\mathrm{capillary}}$ are depicted in Figures 4(a) and (b), respectively. The flow-driven fragments are produced near the solidification front and around the segregation channels, but the capillary-driven fragments are generated at a distance below the solidification front. Additionally, the capillary-driven fragmentation occurs throughout the whole sample cross-section except the freckled areas. The production efficiency of the two fragmentation mechanisms is also quite different. According to the current simulation results, $N_{\mathrm{frag,max}}^{\mathrm{flow}} = 7.5\times10^8\ (\mathrm{m^{-3}\cdot s^{-1}})$, whereas $N_{\mathrm{frag,max}}^{\mathrm{capillary}} = 1073\ \mathrm{m^{-3}\cdot s^{-1}}$. Different from the results of $N_{\mathrm{frag}}^{\mathrm{flow}}$, a negative value of $N_{\mathrm{frag}}^{\mathrm{capillary}}$ is predicted in the lower part of the mushy zone (Figure 4(b)), which indicates that some fragments vanished because of dendrite coarsening (coalescence or remelting) [5].

4. Discussions

The freckles are recognised by a trail of equiaxed (spurious) grains parallel to the direction of gravity, and the dendrite fragmentation is believed to be the main source for the equiaxed grains [2,4]. According to state-of-the-art simulations [7–10], due to the lack of a reliable fragmentation model, the freckles were indirectly predicted through the calculated segregation channels by ignoring the formation of equiaxed grains. In this study, the formation of equiaxed grains via two different fragmentation mechanisms [5,15,19] are numerically investigated. The calculation of flow and solidification in the superalloy casting was coupled with radiation in the Bridgman furnace.

As shown in the inset of Figure 3(d), the liquid in the freckles beneath the solidification front flows upward. Despite the low flow strength, some fragments created near the open mouths of the freckles were transported out of the mushy zone into the top hot liquid. The remelting and grain destruction occurred subsequently. Remelting of the equiaxed grains also happened in the freckled areas owing to the increase in liquid concentration. The rest of the formed fragments sedimented in the freckles and continued to solidify. These fragments blocked the horizontal growth of columnar dendrites toward the freckles. After the solidification process, the fragments grew into coarse spurious grains, and the residual liquid solidified as eutectics.

Comparing Figures 4(a) and (b), in addition to the flow-driven fragments in/around the segregation channels, a small number of capillary-driven fragments were generated deep in the mushy zone. Since columnar dendrites were fully developed there, the capillary-driven fragments could hardly be transported. Their growth was also suppressed by the as-developed columnar dendrites. Based on the current numerical study, in the main body of the casting, the casting was dominated by a columnar structure ($f_c \approx 0.9$) with a small amount of eutectics ($f_{eut} \approx 0.1$). In the freckled areas, the sum of the volume fractions of the equiaxed grains and eutectics is up to 0.75, whereas $f_c < 0.25$. Such simulation results are consistent with our experimental observations. Further analysis of the simulation results and a quantitative comparison between the simulation and the experiment will be presented in the following journal paper.

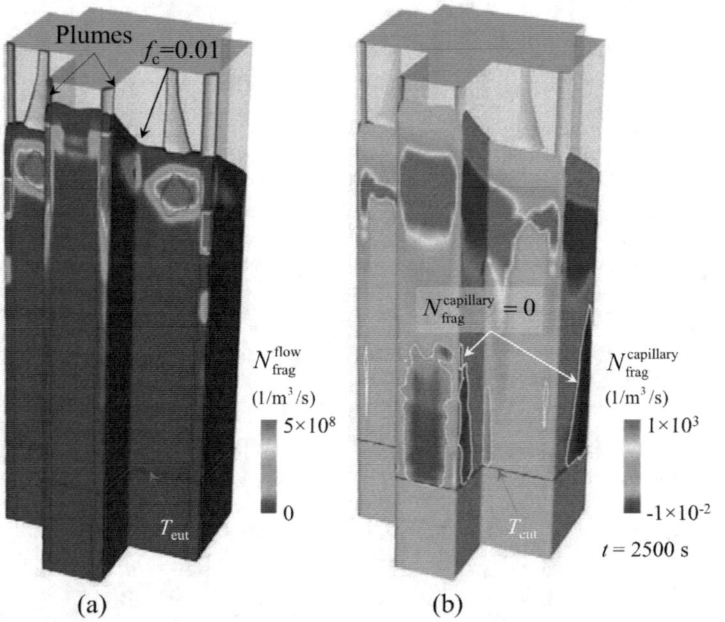

Figure 4. Analysis of the formation of spurious grains via dendrite fragmentation. The yellow shapes on the top indicate the as-developed plumes. (a) Contour of the flow-driven fragmentation rate N_{frag}^{flow}. (b) Contour of the capillary-driven fragmentation rate $N_{frag}^{capillary}$. The isoline for $N_{frag}^{capillary} = 0$ is overlaid. $N_{frag}^{capillary}$ is negative inside the white line and positive outside the white line.

5. Conclusions

The formation of spurious grains and freckles during directional solidification of superalloys was numerically studied using a mixed-columnar-equiaxed volume-average solidification model. Two different fragmentation mechanisms, i.e. capillary-driven and flow-driven fragmentation mechanisms, were considered to produce spurious grains. The following conclusions were drawn for the currently studied sample:

1) Flow-driven fragmentation primarily occurred in/around the segregation channels, whereas capillary-driven fragmentation occurred throughout the entire cross-section of the casting, except in the segregation channel areas.

2) The fragmentation rate caused by the flow-driven mechanism was several orders of magnitude larger than that of the capillary-driven mechanism. The flow-driven fragmentation mechanism contributed mostly to the formation of spurious grains in the freckles.

3) Owing to the shadow effect of the Bridgman furnace, the shadowed side of the casting could be 20 K cooler than the bright side. Thus, the solidification front was significantly inclined. Plumes only formed on the shadowed side, and four plumes developed further into freckles at the casting corners. On the bright side, it was free of plumes and freckles.

6. Acknowledgements

This work was financially supported by the Austrian Science Fund in the framework of the FWF-NKFIN joint project (FWF, I4278-N36).

References

[1] Giamei A F and Kear B H 1970 On the nature of freckles in nickel base superalloys *Metall. Trans.* **1** 2185–92
[2] Hellawell A, Sarazin J R and Steube R S 1993 Channel convection in partly solidified systems *Philos. Trans. - R. Soc. London, A* **345** 507–44
[3] Auburtin P, Wang T, Cockcroft S L and Mitchell A 2000 Freckle formation and freckle criterion in superalloy castings *Metall. Mater. Trans. B* **31** 801–11
[4] Copley S M, Giamei A F, Johnson S M and Hornbecker M F 1970 The origin of freckles in unidirectionally solidified castings *Metall. Trans.* **1** 2193–204
[5] Zhang H, Wu M, Rodrigues C M G, Ludwig A, Kharicha A, Rónaföldi A, Roósz A, Veres Z and Svéda M 2022 Dendrite fragmentation mechanism under forced convection condition by rotating magnetic field during unidirectional solidification of AlSi7 alloy *Acta Mater.* 118391
[6] Hellawell A, Liu S and Lu S Z 1997 Dendrite fragmentation and the effects of fluid flow in castings *Jom* **49** 18–20
[7] Schneider M C, Gu J P, Beckermann C, Boettinger W J and Kattner U R 1997 Modeling of micro- and macrosegregation and freckle formation in single-crystal nickel-base superalloy directional solidification *Metall. Mater. Trans. A* **28** 1517–31
[8] Felicelli S D, Heinrich J C and Poirier D R 1998 Three-dimensional simulations of freckles in binary alloys *J. Cryst. Growth* **191** 879–88
[9] Wu M, Stefan-Kharicha M, Kharicha A and Ludwig A 2021 Flow-solidification interaction: A numerical study on solidification of NH4Cl – 70 wt.%H2O solution in a water-cooled mould with a large sample thickness *Int. J. Heat Mass Transf.* **164** 120566
[10] Zhang H, Liu X, Ma D, Song M, Ludwig A, Kharicha A and Wu M 2023 Digital twin for directional solidification of a single-crystal turbine blade *Acta Mater.* **244** 118579
[11] Gu J P, Beckermann C and Giamei A F 1997 Motion and remelting of dendrite fragments during directional solidification of a nickel-base superalloy *Metall. Mater. Trans. A Phys. Metall. Mater. Sci.* **28** 1533–42
[12] Wu M and Ludwig A 2006 A three-phase model for mixed columnar-equiaxed solidification *Metall. Mater. Trans. A* **37** 1613–31
[13] Zhang H, Wu M, Schumacher P, Rodrigues C M G, Ludwig A and Kharicha A 2021 Modelling melting and grain destruction phenomena during globular equiaxed solidification *Appl. Math. Model.* **97** 821–38
[14] Lipton J, Glicksman M E and Kurz W 1987 Equiaxed dendrite growth in alloys at small supercooling *Metall. Trans. A* **18** 341–5
[15] Cool T and Voorhees P W 2017 The evolution of dendrites during coarsening: Fragmentation and morphology *Acta Mater.* **127** 359–67
[16] Rodrigues C M G, Wu M, Zhang H, Ludwig A and Kharicha A 2021 Bridging Capillary-Driven

Fragmentation and Grain Transport with Mixed Columnar-Equiaxed Solidification *Metall. Mater. Trans. A* **52** 4609–22
[17] Neumann-Heyme H, Eckert K and Beckermann C 2017 General evolution equation for the specific interface area of dendrites during alloy solidification *Acta Mater.* **140** 87–96
[18] Li Q and Beckermann C 1998 Scaling behavior of three-dimensional dendrites *Phys. Rev. E* **57** 3176–88
[19] Zheng Y, Wu M, Kharicha A and Ludwig A 2018 Incorporation of fragmentation into a volume average solidification model *Model. Simul. Mater. Sci. Eng.* **26** 015004

IOP Conf. Series: Materials Science and Engineering 1281 (2023) 012060 doi:10.1088/1757-899X/1281/1/012060

Phase field simulation of segregation and precipitates formation during steel solidification

B Santillana[1,*], V Paruchuri[2], and U Prahl[3]

[1] Tata Steel R&D, The Netherlands

[2] RWTH Aachen University, Germany

[3] Freiberg University, Germany

[*]corresponding author: begona.santillana@tatasteeleurope.com

Abstract: Multi-phase field modelling using MICRESS® has been used to perform simulations of the nucleation and evolution microstructure in a 2D domain as function of fundamental thermo-physical parameters, time, cooling rate and chemical composition during solidification. In order to have a better understanding of primary precipitates evolution during solidification in a case hardening steel and the precipites chosen were niobium-carbides (NbC) and manganese sulphides (MnS), simulations were done including the nucleation and growth of precipitates and coupled through thermodynamic databases. The results show a good comparison with the experimental microstructure measured with SEM-EDX measurements. The simulated time dependent segregation profiles of Nb and Mn in the interdendritic regions, and the precipitates composition, have a good similarity with the measurements. Even when the size and morphology of the NbC and MnS particles has a difference in the both simulated and experimental situations, this can be explained by the low resolution used in the simulation. This publication shows the modelling results and the comparison with the experimental measurements.

1. Introduction

Mechanical properties of metals during solidification depend strongly on the as-cast microstructure [1]. During the last stages of solidification in the brittle temperature range (BTR) mechanical state of alloys is extremely low and when the solidification process is completed a sudden and marked change in strength and ductility of metal can be observed by many researches[1-4]. It means that as long as liquid phase is present, metal will fail in a brittle manner. In alloys, presence of this liquid in between the dendrite arms is an effect of segregated solute elements which have low solidification temperature than the initial melt. So, when considering the mechanical properties of any casting, solidification-structure and microsegregation are of great importance.

Phase field simulations have the advantage of having spatial resolution, which means it is possible to not only predict the size of the precipitates, but also the location at which they are expected to appear. Both are not possible with a 1D simulation. Therefore, without the use of phase field it will not be possible to predict their size or even the location where the precipitates are expected to be formed, i.e. interdendritic or at grain boundaries. In this study the phase field model is set up to simulate two well know precipitates (MnS and NbC) and validate the results with sample analysis.

1.1 Introduction to multiphase field method

A phase field model is a mathematical model for solving interfacial problems, mainly applied to solidification dynamics. Such model takes into account all relevant boundary conditions at the liquid-solid interface. In the present work the solidification microstructure and in particular the segregation and precipitation of NbC and MnS are both simulated using the MICRESS® software.

Technical alloys are comprise multiple of components, multiple phases, and multiple grains. Their description in a numerical model requires that several phases have to be considered in the phase field equations therefore this approach is the so-called multi-phase field [2]. The method used for the multiphase field modelling means that beforehand equilibrium calculations have to be performed to not only accurately predict the phases to be modelled, but mostly to reduce the temperature range at which the transformations are expected (in this case the precipitation) in order to reduce the computational time of the simulation.

2. Materials and methods

Phase field is a known method used to simulate microstructure evolution and phase transformations during solidification and other materials processing steps. It takes into account all the boundary conditions at the liquid-solid interface, including anisotropy. In general, a multi-phase field method not only considers liquid and solid phases, but also intermetallic and any other phases that may be present. Previous research [6-9] has shown that morphologies computed with the phase-field method display most of the features observed experimentally.

In the present work an effort is made to quantify the solidification microstructure of case hardening steel (Table-1) computed using MICRESS® and experimentally simulate the casting conditions and microstructure from the Mould Cracking Simulator (MCS) trials. The experimental procedure and features of the MCS trials are explained in a previous publication by Santillana et al. [3].

2.1 Precipitation of Nb(C,N) and MnS

In steels, the classification of other non-metallic phases or intermetallics is different than in any other metallic alloy. The so-called non-metallic particles or inclusions are considered as the results of chemical reactions during processing, such as deoxidation, alloying or process-related like slag entrapment. Within this classification, non-metallic particles that can appear in steels are in fact the result of alloy composition, like sulphides, carbides and nitrides precipitates. Manganese sulphides (MnS) are still classified as non-metallic inclusions because in in some cases Mn is used to deoxidise and de-sulphurise the steel. But all carbides and nitrides are considered precipitates since they are expected to form in micro-alloyed steels or HSLA (High Strength Low Alloyed) steels. These HSLA steels have the addition of B, Nb, V, Ti, or Al to form the carbo-nitrides to enable the grain refinement during further processing and thus improve toughness, plasticity, wear- and corrosion-resistant properties [4]. The carbo-nitrides are formed at the final stages of solidification and their precipitation at grain boundaries can be detrimental for cracking at the second ductility trough [4].

Precipitates can form at different rates, stages and locations during steel processing, i.e. in the mushy-zone between dendrites due to rapid diffusion in segregated liquid, at solid state inside the grains due to slow diffusion, or at the grain boundaries due to faster diffusion [5, 6]. These different mechanisms cause the precipitate particles to show a variety of compositions, morphologies and size distributions [7]. During continuous casting of micro-alloyed steels, precipitates could form in the interdendritic segregated liquid because the micro-alloyed elements such as Nb, Ti and V are not only strongly segregating but mostly due to their affinity to form nitrides and carbides [5]. On the other hand, MnS can also form at the interdendritic region for very similar reasons: both Mn and S segregate and have a strong affinity with each other, but the MnS particles are not considered as precipitates but non-metallic inclusions due to their size and composition that differ from classical precipitates that at mostly nanometres size.

Table 1. Chemical composition of the steel considered in this study.

[wt.%]	C	Si	Mn	Cr	Mo	Ni	S	Nb	N
Case hardening steel	0.280	0.180	1.40	1.40	0.565	0.030	0.0048	0.043	0.0230

In the present study, precipitates of Nb(C,N) along with non-metallic inclusions of MnS are observed in the metallographic samples of post mortem samples from previously solidified and mechanical tested (MCS samples) case hardening steel with Nb=0.043wt% (full composition is listed in table 1). Since, the location and morphology of the precipitates coincide with the theory that they may be formed at the interdendritic segregated liquid, it revels the fact that these are formed during the last stages of solidification.

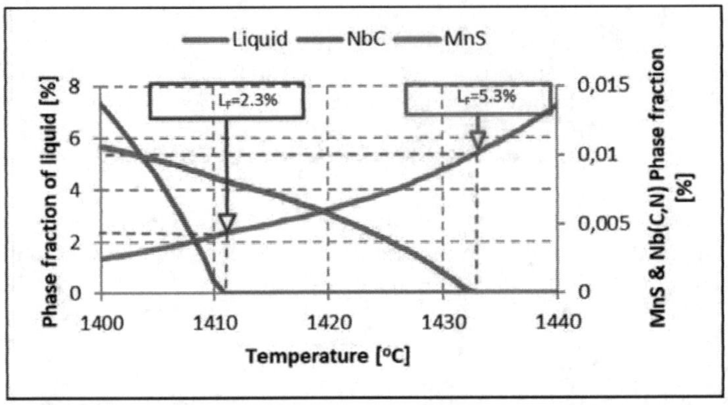

Figure 1. Scheil simulations showing the phase fraction of precipitates.

A non-equilibrium Scheil calculation shows that the MnS and NbC seem to form when there is still approx. 2-5 % liquid phase present as shown in figure 1. Therefore, primary precipitation of Nb(C,N) and MnS could influence the hot tearing susceptibility during last stages of solidification.

At the micrographic analysis, precipitates of Nb(C,N) are observed at the crack initiation points and in the segregated regions between the secondary dendrite arms, as shown in figure 2, together with the EDX analysis on several points. To increase the conductivity of the electrons during SEM the samples were carbon coated. So, the composition of these particles can only be quantitatively analysed. Likewise, it is observed that the Nb(C,N) have a small amount of Molybdenum due to presence of minute solubility in that phase.

Due to irrelevance of some of the elements like nickel on the formation of precipitates and their influence to the cracking susceptibility, they were not included in the EDX analysis.

Formation of these precipitates in the segregated interdendritic region during solidification can have a positive effect on the hot tearing and mechanical properties of the partially solidified shell. The carbides and nitrides precipitate in a chain-like form at the dendrites boundaries and as stated in many hot tearing theories [6, 8, 9] when these precipitates nucleate in the segregated liquid between the dendrites and grow by substituting the liquid phase, a drastic decrease in the liquid fraction and increase in solid fraction can result in improving the coalescence between the dendrite arms and also strength of the material. Previously, this theory has also been proposed and observed by the author of the study [8] in the case of TiN precipitation.

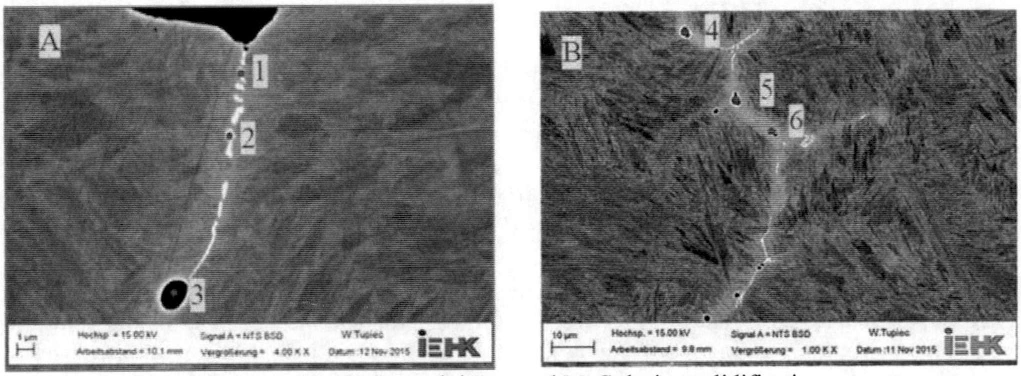

Figure 2. Nb(C,N) precipitates and MnS during solidification.

According to the observations made in this study, it appears that the cracks are generated and passing through these chain-like precipitates which appears to contradict the above theory of strengthening. This is due to the fact that, although when solid Nb(C,N) precipitates out of liquid and improves the connectivity and strength material, due to the low grain boundary binding force between the Nb(C,N) and dendrite interface, they still can act as the possible crack nucleation sites, but at higher stress concentration comparted to the interdendritic regions with liquid. Thus, formation of Nb(C,N) during the last stages of solidification could still significantly improve the strength of the as-cast structure when compared to the presence of segregated liquid between the dendrites.

Table 2. EDX of points from figure 2.

Point	C [wt%]	N[wt%]	Si[wt%]	P[wt%]	S[wt%]	Cr[wt%]	Mn[wt%]	Fe[wt%]	Nb[wt%]	Mo[wt%]
-1	21.42	0.62	0.17	0.23	0	2.45	1.14	46.28	21.82	5.87
-2	24.55	0.57	0.23	0	0	2.17	0.66	27.7	34.77	7.99
-3	14.46	0	0.38	0.41	19.49	1.92	36.4	24.21	0.22	0
-4	6.52	0.04	0.2	0.19	15.05	1.94	29.59	42.87	0	0
-5	4.94	0	0	0	28.09	0.5	53.78	3.94	0	0
-6	3.21	0.15	0.25	0.13	2.43	3.12	7.51	80.69	0	1.92

2.2 MICRESS® phenomenology

The multiphase-field theory is a computational approach which describes the evolution of phase-fields $\phi_\propto(\vec{x},t), \alpha = 1, \ldots, v.$ in time and space [10]. These fields describe the distribution of either different phases or of grains with different orientations [11]. If the respective phase or grain locally exists, *then* $\phi_\propto = 1$ else $\phi_\propto = 0$. At the interfaces, the phase field variables will change continuously from 1 to 0, thus the interfaces will have a thickness g, which can be chosen to be large compared to the atomic interface thickness, but still small compared to the microstructure length scale [2]. The phase field equation in its simplest form, is described as follows:

$$\dot{\phi}(\vec{x},t) = \mu\left[\sigma\left(\nabla^2\phi - \frac{(1-\phi)(1-2\phi)\phi}{\eta^2}\right) + \frac{1}{\eta}\Delta G\phi(1-\phi)\right] \quad (1)$$

For example, for the phase field parameters when only two phases are present such as in a solid-liquid system, the order parameter ϕ can be distinguished as, the fraction of solid $\phi_S(\vec{X})$ and liquid fraction $\phi_l(\vec{X})$. The sum of the phase field parameters of all phases is equal to one [12], therefore:

$$\phi_l(\vec{X}) = 1 - \phi_s(\vec{X}) \quad (2)$$

The MICRESS software solves the coupled system of partial differential equations for all phase fields and for all external fields affecting the phase transition [16]. By directly coupling it to a thermodynamic database via Thermo-Calc, where the thermodynamic driving force ΔG can be calculated using a quasi-equilibrium approach, the software becomes then independent of the thermodynamic data and is not restricted by the number of components and phases involved.

Technical alloys comprise multiple grains, multiple phases, and multiple components. Their description in numerical model requires several phases considered and thus extra phase field equations are added and it is the so-called multi-phase field [16].

2.3 Boundary conditions

Heat flux has been defined as boundary condition at the bottom of the simulation domain and calculated from the experimental conditions of MCS trials [3, 13]. According to the design of the MCS, different layers of thermal resistances exist between the semi-solid shell forming and the cooling water inside the block which are taken in to account to calculate the heat flux. The schematic representation of MCS and boundary conditions of the domain are shown in figure 3. The cylindrical shell is formed around the mould, thermally shrinks in the radial direction, making a tight contact with the mould. Therefore, the thermal resistance due to the air gap between the solid flux and mould is not taken in to account.

$$Heat\ Flux\ \dot{q} = \frac{T_s - T_w}{\frac{S_{Sol\ shell}}{\lambda_{Sol\ shell}} + \frac{S_{S-flux}}{\lambda_{S-flux}} + \frac{S_{l-flux}}{\lambda_{S-flux}} + \frac{S_{Mold}}{\lambda_{Mold}} + \frac{S_{Block}}{\lambda_{Block}} + \frac{1}{\propto_{Water}}} \quad [\frac{W}{m^2}] \tag{3}$$

Where:

T_S=Temperature of liquid steel [K]

T_W=Temperature of cooling water [K]

$\lambda_{Sol\ shell}$=Thermal conductivity of steel formed [W/(m*K)]

$S_{Sol\ shell}$= thickness of the solid shell [m]

λ_{s-flux}= conductivity of the solid flux [W/(m*K)]

S_{s-flux}= thickness of liquid flux[m]

S_{Mold}= thickness of copper mold [m]

λ_{Mold}=conductivity of the copper mold[W/(m*K)]

S_{Block}=thickness of the internal cooling block [m]

λ_{Block}= conductivity of the internal cooling block [W/(m*K)]

$\propto_{Water}$=convective heat transfer coefficient of water [W/(m^2*K)]

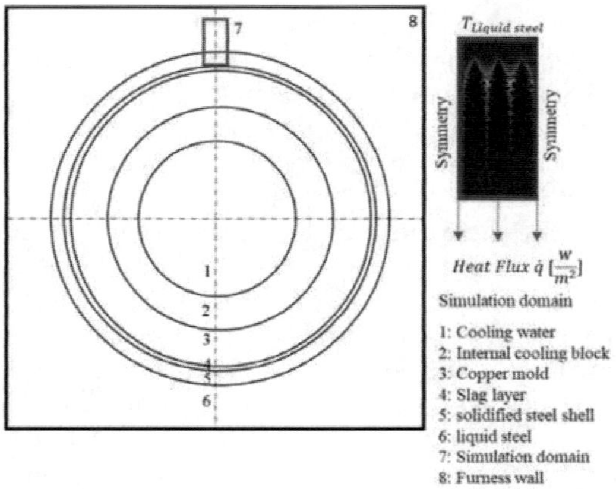

Figure 3. Schematic representation of MCS Top view[10].

It is assumed during the calculation, that the steel shell and solid flux increase/grows over the time. The growth rate of the solid shell is taken form the experimental observations and growth rate of solid-flux is assumed to be 0.1565 mm/√Sec.

2.4 Model input parameters

From the compositional point of view, the steel grade does not show any unexpected elements in the composition, except that the Mn concentration is somewhat higher than other standard steels that have been simulated until now like a regular carbon steel grade. All the numerical input parameters used in the current model are listed in the table 3 below.

Table 3. simulation input parameters.

Parameters		values	Parameters		values
Chemical composition		C, Mn, Nb, S (Table 1)	Interfaces Mobility $(\mu)\left[\frac{cm^4}{JS}\right]$	$\mu_{L/\delta} = \mu_{L/\gamma}$	$5.75 * 10^{-3}$
Initial temperature or casting temperature of MCS trials [K]		1800		$\mu_{l/\gamma}$	$2 * 10^{-5}$
Simulations duration [S]		2		$\mu_{L/NbC} = \mu_{L/MnS}$	$1 * 10^{-5}$
Number of cells [x*y*z]		1500*1*3000		$\mu_{\gamma/NbC} = \mu_{\gamma/MnS} = \mu_{NbC/MnS}$	$1 * 10^{-12}$
Cell dimension,Δx [μm]		1	Interface thickness, η		3 cells (1 μm)
Interfacial energy $(\sigma)\left[\frac{J}{Cm^2}\right]$	$\sigma_{L/\delta} = \sigma_{L/\gamma}$	$2.5 \cdot 10^{-5}$	Boundary conditions for phase field & Boundary conditions for phase field in [W E B T] borders.		[ssii] i (insulation); s (symmetric)
	$\sigma_{L/\gamma}$	$1 * 10^{-4}$	shield distance [μm]		25
	$\sigma_{L/NbC} = \sigma_{L/MnS}$	$5.04 * 10^{-5}$			
	$\sigma_{\gamma/NbC} = \sigma_{\gamma/MnS} = \sigma_{NbC/MnS}$	$1 * 10^{-6}$			

2.5 Interface mobility calibration

The Phase-field model could show artifacts with wrong kinetics if the thickness of the diffuse interface is not small enough compared to the diffusion length towards the melt. The practical consequences of this numerical problem is the need for high resolution to get the correct diffusion-limited growth kinetics. However, running a simulation with high resolution will tremendously increase the simulation time. So, to reduce the simulation time, a lower resolution can be used with corrected interface kinetics by calibration of the interface mobility using the high resolution reference simulation.

In the present study, a similar approach followed by Böttger et al. [14] is used to calibrate the interface mobility. High resolution (Δx=0.05 μm) simulations are repeated with different interface mobilities starting from 0.005 and increasing, until a constant height of the dendrite tip reached. This indicates that diffusion-limited growth was reached and that the grid resolution was sufficiently high [14]. Height of the dendrite tips of all the simulations domains are compared at a constant simulation time of 0.1s as shown in Figure 4. Here the obtained height of the dendrite tip is taken as a reference for low resolution simulations with the same domain size and Δx=0.333 μm. Low resolution simulations were also simulated until 0.1s. By this systematic variation of mobility for low resolution domain, a calibrated mobility value was obtained which has the similar height of the dendrite tip but also the shape of the dendrite as defined by the high-resolution reference, see figure 4. This mobility value (μ=0.00575 $\left[\frac{cm^4}{Js}\right]$) and cell dimension Δx=0.333 μm have been further taken into account for final simulations with larger grid size.

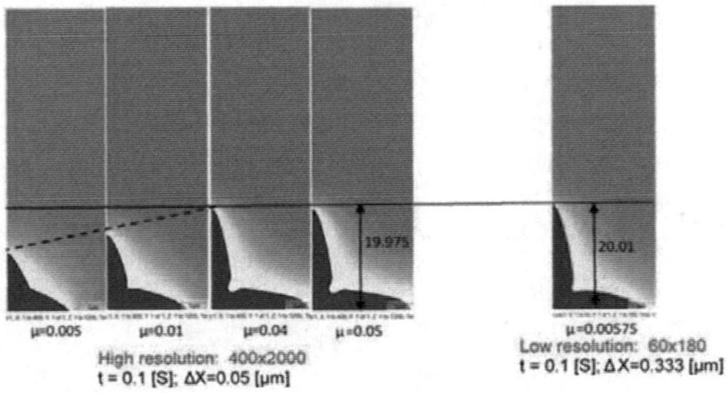

Figure 4. Mobility calibration using a high resolution reference simulations.

2.6 Nucleation model

The MICRESS software uses two nucleation models, the seed density model and seed undercooling model. Seed density describes nucleation from the melt, triggered by the seeding particles which may be added initially or which may exist as impurities [6]. In contrast to seed-density model, in seed-undercooling model, no explicit potential nucleation sites are predefined and the number of grains nucleated during the simulation is not limited [6].

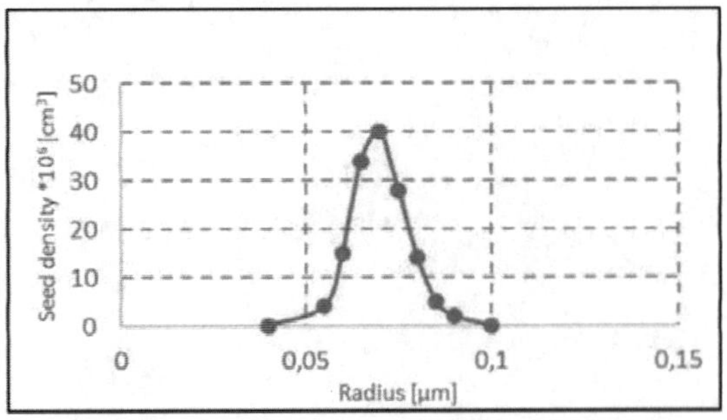

Figure 5. Seed density used in the simulations.

A seed density model was used for the region between 0-30 μm from the surface along the length of the domain. The seed density distribution is taken from **[14]** as shown in figure 5. For the remaining region of the domain a seed undercooling model was used with an assumed undercooling value of 30K.

Table 4. Seed undercooling values for each phase.

Phase	Nucleation model	Range
δ-Ferrite	Seed density	0-30 [μm] (From bottom)
δ-Ferrite	Seed Undercooling	30 [K]
γ-Austenite	Seed Undercooling	2 [K]
MnS	Seed Undercooling	3 [K]
NbC	Seed Undercooling	3 [K]

For austenite, MnS and NbC seed undercooling model was also used with undercooling values as shown in table 4. The temperature dependent thermal conductivity λ for each phase, as calculated using JMatPro, was given as an input to the simulation.

3. Simulation results

For the case hardening steel only 4 elements (C, Mn, Nb, S) were considered. Figure 6 shows the evolution of the dendritic microstructure with the simulation time. Due to the seed density model used near the surface, after the melt is thermally undercooled, it activates the initial seeds with radius specified in figure 5. Due to large heat flux at the surface the nucleated equiaxed seed take the shape of columnar dendrites. When the liquid at the solidification front is thermally or constitutionally undercooled the new seeds are activated. At 0.40 seconds of the simulation, figure 6C) nucleation of new grains near the solidification front is observed, this is because of the specified seed undercooling model for rest of the domain without seed density.

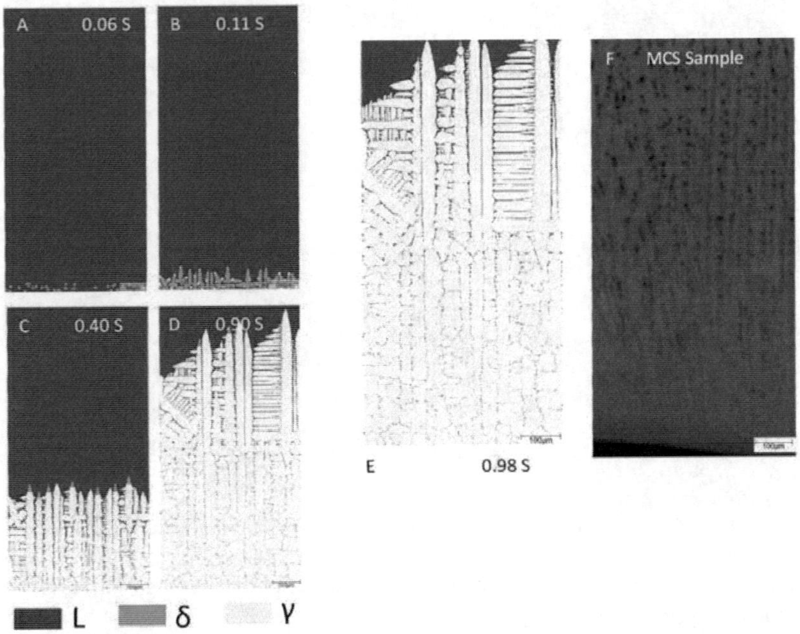

Figure 6. Microstructure simulation of the case hardening steel.

The simulated microstructure is compared with the experimental microstructure in figure 6E and 6F. The results show that the morphology of dendrites in the simulated microstructure, matches with experimental microstructure. Dendrite arm spacing was also measured on the simulated micrograph and compared with the experimental micrograph. The average primary dendrite arm spacing value of 30-35 μm was observed for the simulated microstructure while the value of 30-47 μm has been measured for the experimental micrograph (figure 6F). The dendrite arm spacing is a function of composition and the cooling rate, therefore, the difference between measured and simulated can be attributed to the heat extraction rate and compositional difference since only four elements are considered in the simulation.

3.1 Microsegregation and formation of NbC

According to the phase field simulations, all elements show a positive segregation, i.e. they are enriched at the interdendritic regions during solidification[6]. However, after the complete solidification, elements like C, N show a smooth distribution because of their high diffusivity in solid, unless they have already formed carbides or nitrides[6]. The other slow diffusion elements like Nb, Mn that segregate in the interdendritic regions are shown in figure 7 and figure 8.

Transition of niobium rich interdendritic liquid to niobium carbides is observed by measuring the time dependent segregation profiles of niobium during solidification, as shown in figure 7. The broad segregation profile at time 0.55s is translated to a narrow profile, enriched in niobium and finally forming Nb(C) by the end of solidification.

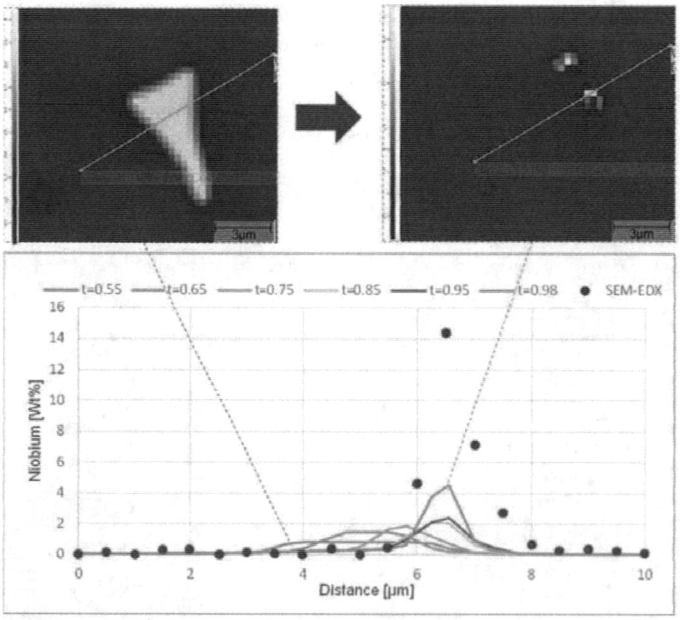

Figure 7. Comparison of microsegregation of Nb between MICRESS and SEM-EDX.

3.2 Microsegregation and formation of MnS

Further, concentration of Manganese in the interdendritic region is also measured and compared in both cases (SEM-EDX and virtual EDX) as shown in the figure 8. The result showed a good comparison between both methods. Since the cooling rate has an influence on the segregation, both measurements are made at same height from the surface.

IOP Conf. Series: Materials Science and Engineering 1281 (2023) 012060 doi:10.1088/1757-899X/1281/1/012060

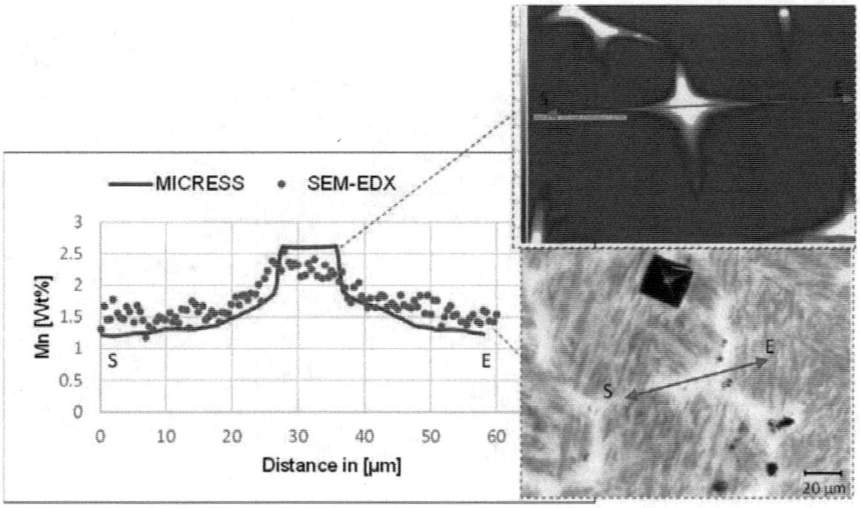

Figure 8. Comparison of microsegregation of Mn between MICRESS and SEM-EDX.

4. Discussion

The simulated morphology of NbC and MnS has been compared with the morphologies observed in the MCS experimental samples. It was found that, the size and morphology of the particles have a large difference in both cases. For instance, considering the NbC, small and cubic particles have been observed on the simulated micrographs, while the relatively large and chain-like morphologies are observed on the experimental micrographs. A similar difference has been observed for MnS particles as shown in figure 9.The reason for the morphological difference is because MICRESS has limitation in predicting the morphology of particles on nano-scale. Moreover, to reduce the simulation time, the cell dimension used were larger than the expected particle sizes of NbC and MnS.

The reason for small size particles is that the simulation is a 2-D structure and the real solidification is a 3-D phenomenon. In 3-D structure, the interdendritic regions are bigger than in the 2-D structure. Therefore, the particles formed in a 3-D interdendritic regions have large area to grow and also increasing the diffusivity of elements into the nucleated particles from all the sides and angles. Moreover, too high interface energy (precipitates/austenite) and too low interface mobility (precipitates/austenite) can also be the reasons for small size particles. In spite of all the differences, MICRESS model is able to simulate the formation of NbC and MnS at temperatures comparable with the Scheil models.

The concentration of niobium in the NbC measured by EDX in the experimental samples shows a higher value when compared to "virtual EDX measurement" made on the simulated micrograph as shown in figure 7. This difference is because the precipitates in simulated micrograph are still in growing state at high temperatures and thus their composition is not the final one like in the precipitates in the experimental samples at room temperature that are fully grown. To understand the evolution of NbC over time and temperature, a simulation with a larger temperature range has to be performed but in this study, the simulation is run only till end of the solidification.

IOP Conf. Series: Materials Science and Engineering 1281 (2023) 012060 doi:10.1088/1757-899X/1281/1/012060

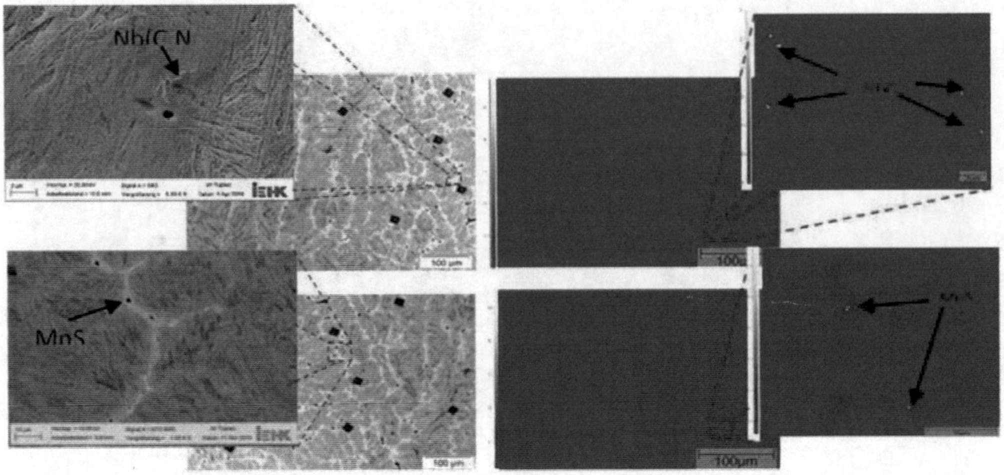

Figure 9. Formation of Niobium carbides and Manganese sulphides: SEM-LOM vs MICRESS.

5. Conclusions

In conclusion, the multi-phase field simulations using MICRESS give a good understanding about nucleation and solidification processes and how the thermo-physical parameters influence the morphology.

The simulated microstructure shows a good similarity with the experimental microstructure although there are minor differences in the dendrite arm spacing due to the input parameters in the model like heat extraction rate and composition of the melt.

Measurement of the time dependent segregation profiles of niobium, showed the transition of niobium rich interdendritic liquid to NbC during solidification. It was observed that the size and morphology of the particles has a large difference in both, simulated and experimental situations, which is predominantly due to the low resolution used in the simulation and due to the 3-D structure and not showing the fully grown and developed chemistry of such precipitates, since the simulations were only performed till the end of solidification.

6. References

[1] Kurz W, Solidification Processes and Properties of Castings, *5th Int. Conf. . Vacuum Met. and ESR, Leybold-Heraeus, Hanau,* 1976.

[2] Mikio S, Makoto Y, Toshihiko E, In-Situ Measurementof Fracture Strength ot Solidifying to Predict Upper Limit of Casting Speedin Continuous Oscillating Mold Steel Shells Caster with Oscillating Mold, *ISIJ International,* Vols. 37, No:4, pp. 375-382, 1997.

[3] Bernard C, Hiebler H, Mechanical properties and crack susceptibility of the steel during solidification, *Steel research,* vol. 70, pp. 349-355, 1999.

[4] Bernard C, Hiebler H, Schuetzenhoefer W, Physical and methematical modelling of the initial shell strength in continuous casting, *Metal,* vol. 15, 17 May 2001.

[5] MICRESS User guide, Volume 0: Micress Phenomelogy., vol. 2.

[6] Böttger B, Apel M, Santillana B and Eskin D, Relationship Between Solidification Microstructure and Hot Cracking Susceptibility for Continuous Casting of Low Carbon and High-Strength Low-Alloyed Steels A Phase Field Study, *Metallurgical and materials transactions A,* vol. 44A, p. 3765, 2013.

IOP Conf. Series: Materials Science and Engineering 1281 (2023) 012060 doi:10.1088/1757-899X/1281/1/012060

[7] Bottger B, Eiken J and Steinbach I, Phase field simulation of equiaxed solidification in technical alloys, *Acta Materialia 54 ()* , p. 2697–2704, 2006.

[8] Böttger B, Apel M, Santillana B and Eskin D, Phase-field modelling of microstructure formation during the solidification of continuously cast low carbon and HSLA steels, *Materials Science and Engineering 33* , 2012.

[9] Bottger B, Eiken J and Steinbach I, Schaffnit P, Phase-Field Simulation of Solidification and Solid-State Transformations in Multicomponent Steels., S*teel research int 79, ,* vol. No.8 , p. 608, 2008.

[10] Paruchuri V, Kripak V, Prahl U, Santillana B, Assessment of Shell Strength During Solidification in the Mold Cracking Simulator (MCS) Test, *Metallurgical and Materials Transactions,* no. Issue 1/2019, pp. 142-150, 30 10 2018.

[11] Abbel G, Santillana B, Non-metallic Particles Benefit or Burden for AHSS?, *Advanced High Strength Steel* pp 171–180, ISBN 978-981-10-7891-0

[12] Crowther D, The Effects of Microalloying Elements on Cracking During Continuous Casting, *The Use of Vanadium in Steel - Proceedings of the Vanitec Symposium,* pp. 99-131, 2007.

[13] Espinosa Ruiz A, Prevention of transversal corner cracking in continuous casting of micro-alloyed steels, PhD Thesis, University of Cataluna, 2014.

[14] Mintz B, Vue S, Jonas J, Hot ductility of steels and its relationship to the problem of transverse cracking during continuous casting, *Int. Mater. Rev,* vol. 36, no. 5, p. 187–219, 1991.

[15] Santillana B, Thermo-mechanical properties and cracking during solidification of thin slab cast steel, PhD Thesis, Delft University of Technology, 2013.

[16] Xu K, Thomas B, Prediction of Grain Size ,Precipitation and Crack Susceptibility in Continuous Casting, in *AISTech Steelmaking Conference Proceeding*, 2009.

[17] B Bottger B, Eiken J, Apel M, Phase-field simulation of microstructure formation in technical castings –A self-consistent homoenthalpic approach to the micro–macro problem, *Journal of Computational Physics- 228* , p. 6784–6795, 2009 .

[18] http://www.thermocalc.com..

[19] Böttger B, Schmitz G, Santillana B, Multi-Phase-Field Modeling of Solidification in Technical Steel Grades, *Transactions of the Indian Institute of Metals*: Vol 65, Issue 6 (2012), Page 613-615

IOP Conf. Series: Materials Science and Engineering 1281 (2023) 012061 doi:10.1088/1757-899X/1281/1/012061

Modelling of spatial distribution of segregations in large steel ingots by analysis of centimetric segregation maps

L Gutman[1,2], J R Kennedy[2], A.K Boukellal, F Roch[1], A Marceaux dit Clément[1], M Založnik[2] and J Zollinger[2,*]

[1] Framatome DTI, F-92400 Courbevoie, France
[2] Université de Lorraine, CNRS, IJL, F-54000 Nancy, France

* corresponding author: julien.zollinger@univ-lorraine.fr

Abstract. Mesosegregation appears during the solidification of low-alloyed steels at the scale of a few grains. It causes chemically segregated bands in forgings manufactured from steel ingots. As this may affect the mechanical properties of the produced parts, it is crucial to understand the formation of such segregation patterns. While both microsegregation and macrosegregation are well understood and can be readily characterized, little is known about the formation of mesosegregation. In this paper we present a data analysis method to identify the characteristic scales and patterns of mesosegregation. Segregation was mapped on centimetric samples, using micro X-Ray fluorescence (μXRF). The fine sampling grid used reveals segregation patterns at different scales. To identify and distinguish the mesoscale segregation patterns from smaller and larger present patterns, a numerical data analysis technique based on spatial filtering was used. It can identify characteristic scales of mesosegregation patterns on 2D serial cut samples, leading to 3D reconstruction. This approach combined with simulation studies, will ultimately pave the way to a comprehensive understanding of the formation of mesosegregation.

1. Introduction

Segregation is a very common phenomenon that occurs during alloy solidification. It consists in a local departure from the nominal concentration of the alloying elements in the liquid. It is positive when the local liquid is enriched, or negative when it is depleted. During steel solidification, the common alloying elements (for instance C, Mn or Mo) are rejected in the liquid phase [1]. Consequently, chemical heterogeneity appears in the solid phase at different scales. At the dendrite scale (from μm to a few hundreds of μm), *microsegregation* is caused by the solute rejection into the liquid during the dendrite growth. It is generally affected by the partition coefficient k at the solid-liquid interface and the solute diffusion in the two phases [2]. If the partition coefficient k is smaller than 1 and diffusion is faster in the liquid, as with C, Mn and Mo in Fe, microsegregation creates depleted solid and enriched liquid. Such segregated phases move at the ingot scale (m) through convection, grain movement, solidification shrinkage and mushy zone deformation, leading to *macrosegregation* [3]. Heat treatment processes can reduce microsegregation through diffusion at the dendrite scale. Macrosegregation, however, cannot be reduced, as it is at a much larger scale. As a result, macrosegregation persists in the semi-products and even the final parts, thus affecting final mechanical properties [4].

Solid state transformations may be influenced by segregation, resulting in non-uniform microstructures associated with different local enrichments of the solid. Final mechanical properties are affected by microstructure, which can be related to certain segregation patterns. Hence segregation can affect the final mechanical properties. The shape, size and distribution of segregations are affected by the forging steps. Their intensity or number remain unchanged however [6]. During forging and hot rolling certain segregation patterns are transformed (in shape and size) to banded patterns which can lead to anisotropic mechanical properties. In banded patterns, impact toughness is highly anisotropic: the measured toughness depends on the direction of sampling and its depth in the forged piece [7,8]. Hardness of the steel is linked to it carbon content; hence the hardness varies in banded segregations patterns, leading to hardness anisotropy in forged parts [9]. Residual stress in welds are also affected by banding since their distribution depends on phase transformation kinetics, hence chemical segregation [10]. Banding is caused by segregations at an intermediate scale between microsegregation (dendrite scale) and macrosegregation (ingot scale) [11]. Such intermediate scale segregations, probably at the grain-scale, are called *mesosegregations*. To control such banding phenomena, a thorough understanding of mesosegregation is key.

Understanding mesosegregation starts with comprehensive characterization of mesosegregated material. Different characterisation techniques have been developed to track chemical heterogeneities at different scales. Investigation of microsegregation is commonly carried out with an Electron Probe Micro Analyser (EPMA) which enables the quantification of the chemical composition with an interaction volume of roughly 1 μm^3 [9]. In large steel ingots, macrosegregation can be measured with Optical Emission Spectrometry (OES) or Infrared Combustion Analysis (ICA) for Carbon and Sulphur investigation. Macrosegregation can also be detected by X-Ray Fluorescence (XRF) spectroscopy [12]. Manganese (Mn) and Molybdenum (Mo) macrosegregation patterns can easily be distinguished, with a 5 mm analysis step and minimal sample preparation [13]. However, these techniques cannot be adapted to intermediate-scale segregation in the case of large ingots due to their spatial resolution being to large to resolve centimetric details. Solidification times in large ingots, such as those used in the nuclear industry, are very long (on the order of magnitude of ~10 ~100hours). As a result, investigated samples from these ingots present secondary dendrite arm spacing of 800 µm. To investigate the occurrence of mesosegregations in such ingots, chemical composition should be mapped at the centimetric, grain, scale. At this scale, none of the aforementioned techniques are suitable. On the fine scale, EPMA analysis is impractical: it provides more detail than necessary, is thus too time consuming, and it is non-trivial to analyze samples of the needed scale within conventional equipment. Conversely, the OES and ICA provide less details than needed due to their coarse, manual sampling methodology.

In this paper we analyzed the spatial distribution of segregation in a volume of 115 x 95 x 145 mm^3 based on a new sampling and characterisation methodology recently developed [14]. This new method, based on micro-X-Ray Fluorescence (µXRF), is promising for the chemical analysis of mesosegregation in steel ingots and enables the observation of channel segregates present within the sample volume. First, thirty chemical segregation maps were obtained using µXRF on serial cut samples. These maps were filtered to improve the detection of channel segregates. Then the filtered segregation maps were treated to detect intermediate scale segregation and channel segregates. Finally, these filtered and treated maps were used to reconstruct, in 3D, channel segregates and intermediate scales segregations present in the sample volume.

2. Material and methods

2.1. Samples

The low-alloy steel 16MND5 (0.16 %wt. C, 1.45 %wt. Mn, 0.725 %wt. Ni and 0.5 %wt. Mo) was investigated. Samples were taken from the equiaxed region of a 116 t ingot.

As the aim of the study is to observe centimetric scale segregation volumetrically, centimetric scale sample plates were serial cut with a 5 mm step using wire electron discharge machining. The studied surfaces of the plates, the x and y directions, are perpendicular to the top of the ingot, while the z direction aligns with the height of the ingot and gravity. In total 30 plates were analyzed each with surface of 115 x 95 mm, x and y, and a distance of 5 mm, along the z direction, between each. In total a volume of 115 x 95 x 145 mm was analyzed. Samples are numbered successively from the bottom of the stack, N°1 being the lowest, closest to the bottom of the ingot, and N°30 being the highest, closest to the top of the ingot.

Several selected samples were etched with 3% Nital to reveal and characterize their solidification structure. On Nital etched surfaces, several structures overlap: both the solidified structure and solid-state transformed structure can be observed. With repeated etching procedure, the solidification structure is highlighted, however primary austenitic grain boundaries are also visible as long bright lines crossing dendritic arms. The Nital etching procedure does not require a metallographic surface finish, samples were ground flat with MD discs (SiC abrasive in a resin matrix) to P1200 without diamond polishing. Figure 1 shows an example of the structures observed after such Nital etching.

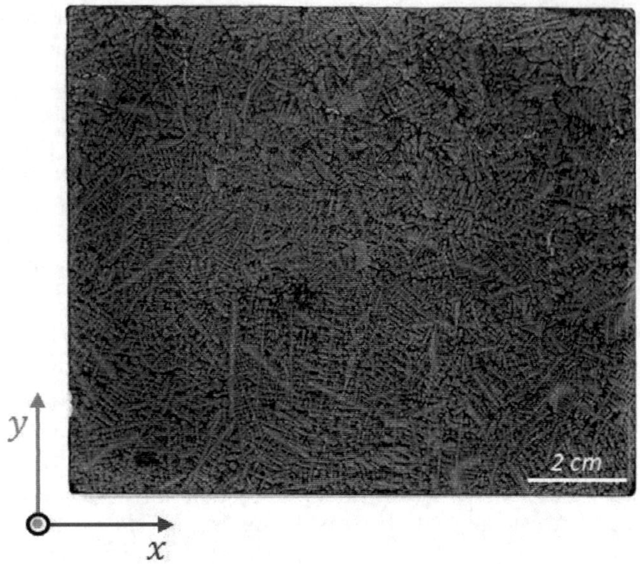

Figure 1. Optical macrograph of the lowest sample of the stack. Solid depleted in solute appears in light grey, while enriched regions appear in dark. The thin bright lines are primary austenitic grain boundaries.

2.2. Micro-X-Ray Fluorescence (µXRF)

A Bruker M4 Tornado device equipped with a Rh X-Ray source (30 W power) was used for Micro-X-Ray Fluorescence (µXRF) analysis. On both sides of the primary X-Ray source, there is an Energy Dispersive Spectroscopy detector. The samples were placed on a stage which moves dynamically below the fixed X-Ray source and detectors in a vacuum chamber (20 mbar vacuum). The spot size focused on the sample is 20 µm.

Optimization of the primary X-Ray sources conditions leads to a minimisation of artefacts and an improvement of the output signal. Several tests were performed to define the best conditions for the analysis of low alloyed centimetric-scale steel samples, with the objective to improve both Mn and Mo signals [13]. The best outcoming signal was obtained with an acceleration voltage of 50 keV, a 200 µA current and a 100 µm thick Al filter. The use of the filter allows analysis with a 20 % dead time (an indicator of minimized artefacts) without decreasing the current, and thus without weakening the output signal.

Once the X-Ray source conditions were selected, the acquisition time and analysis step size were defined. This choice is influenced by the size of the sample, the size of the structures desired to analyzed and the experimental time available. The sampling effect on centimetric scale samples has been investigated previously and the provided guidelines for sampling [13] were followed in the present work. A balance must be found between accuracy, obtained with an increase of analysis time and a decrease of analysis step, and reasonable experimental time. The analysis step set for all the samples investigated here was 100 µm, with an analysis time of 45 ms on each point. These settings enable the analysis of a single sample per night.

Comparison with EPMA measurement and between the different samples shows that the Mo-maps are more reliable than the Mn-maps. This can be explained by the proximity between the Mn peak and the iron (Fe) peak in the obtained XRF spectra. In low-alloyed steels, the Fe concentration is much higher than the concentration of any other element, and as such the Fe peaks are much larger than others, elements with X-ray peaks close to Fe, such as Mn may then be affected by the Fe peak. One the other

IOP Conf. Series: Materials Science and Engineering 1281 (2023) 012061 doi:10.1088/1757-899X/1281/1/012061

hand, the Mn peaks are isolated are not influenced by other elements within the alloy. Hence the present study will be based on data solely from the Mn peaks.

2.3. Analysis tools

The segregation ratio allows the comparison of segregation intensities, patterns or occurrence on different samples from different steel grades. For each chemical segregation map, the segregation ratio, S_i, for a specific element i is calculated for each point of the map with the formula:

$$S_i = \frac{C_i - \bar{C}_i}{\bar{C}_i} \quad (1)$$

where C_i is the local composition for the element i and $\bar{C}_i$ the average composition on the map.

The use of a local segregation ratio, processed with the average composition from a single sample plate focuses the analysis on segregation at the sample scale. However, this methodology assumes that segregation between the sample plates is negligible. In the present study, this sample-based local segregation ratio is used, and thus, it is assumed that there is no macro or global segregation between the sample plates. Experimental data can be filtered to reduce small scale spatial variations. Different filters were tested to treat segregation maps obtained from μXRF measurements. Filters assign a new value to each point of the map. With a Gaussian filter, a weighted average is calculated using a Gaussian weighting function that depends on the distance between from the point, thus the close neighbourhood of the point has a greater influence on the weighted average than more distant areas.

3. Results and discussion

3.1. Chemical segregation maps

Thirty successive chemical segregation maps were collected with μXRF, giving access to 3D information on segregations. In Figure 2, two example maps are shown: the lowest in the ingot and another which was cut 5 mm above the first one. On both maps, the small acquisition step (100 μm) allows the observation of microsegregation with evident enrichment between secondary dendritic arms. In the bottom left quadrant of both maps, there are two strongly enriched regions. These regions exist on numerous successive plates and correspond to channel segregates which have been cut transversely. An example channel segregate transverse section is indicated in red on the maps in Figure 2. In the top right quadrant of Figure 2(a), enrichment is observed at a larger scale than microsegregation which is present between dendrite arms, but does not appear on the second map and is smaller than the channel segregation sections, as such this is another kind of segregation.

The present study aims at the reconstruction in 3D of the intermediate and large scale segregations. The first step is to eliminate the effect of microsegregation and highlight mesosegregations and channel segregates on each of the segregation maps.

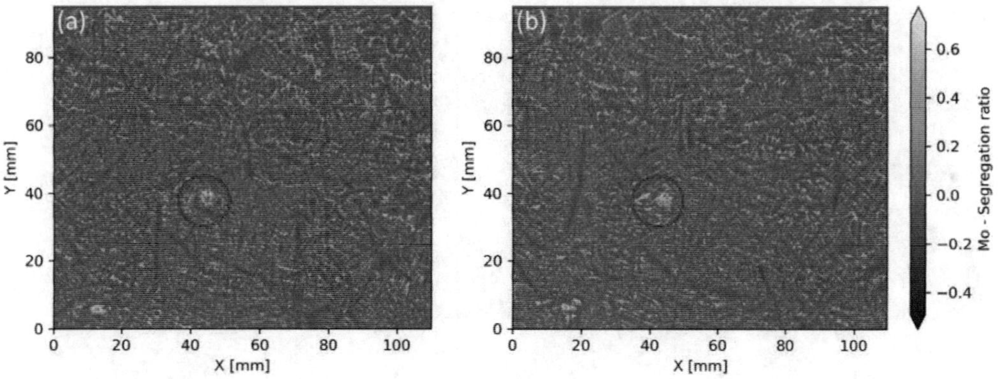

Figure 2. Mo segregation ratio maps of sample (a) N°1 and (b) N°2, the lowest and second lowest samples of the stack.

3.2. Filtration of raw chemical segregation maps

Filters were used to eliminate microsegregation patterns from the segregations maps to study larger segregations patterns. As the filter influences the final map obtained, it is important to choose it appropriately, thus several filters were studied. In Figure 3, a map obtained from μ-XRF measurements and three filtered maps are presented. The filter used is a truncated Gaussian filter. The filtered segregation ratio S_i^f is given by:

$$S_i^f(x_j, y_j) = \frac{1}{0.911} \int_{y_j-a/2}^{y_j+a/2} \int_{x_j-a/2}^{x_j+a/2} \frac{1}{2\pi\sigma^2} \exp\left(-\frac{(x-x_j)^2+(y-y_j)^2}{2\sigma^2}\right) S_i(x,y)\, dx\, dy \quad (2)$$

The truncation of the Gaussian kernel is by a square with a side of $a = 4\sigma$, the coefficient 0.911 is obtained after integration of the truncated Gaussian kernel.Three different standard deviations of the kernel were tested: 13, 25, or 50 pixels, as one pixel is 100 μm the standard deviations tested are 0.13, 0.25 and 0.50 cm.

With a 0.13 cm standard deviation Gaussian kernel (Figure 3(b)), the segregation between dendritic arms is removed, however the depleted dendritic trunks are still visible. The second filter tested, with a 0.25 cm standard deviation Gaussian kernel, leads to a microsegregation free map on which channel segregates are clearly identifiable (Figure 3(c)). Finally, with the last filter tested the channel segregates shown in Figure 3(d) can no longer all be identified and the enrichment and depletion intensities were decreased to level below which useful information can be gathered.

The treatment of the successive segregation maps was then done using the selected 0.25 cm standard deviation Gaussian kernel to eliminate microsegregation from the maps while retaining useful information on the prioritized analysis of the channel segregates. The following 3D reconstruction is based on binary images obtained by the application of a suitable threshold to these filtered maps.

Figure 3. Mo segregation map from sample N°1 with different Gaussian filters applied. The truncation of the Gaussian kernel is 4σ and 3 standard deviation values are presented ((a) unfiltered map, (b) $\sigma = 0.13$ cm, (c) $\sigma = 0.25$ cm, (b) $\sigma = 0.50$ cm).

3.3. 3D reconstruction and observation of segregations

The filtered maps were binarized with a segregation ratio threshold of 0.029, i.e., any pixel with a segregation ratio superior to 0.029 was considered to be a part of the segregations to observe in 3D and kept in the map, whereas any value below was eliminated. This segregation ratio was fixed so the volume of segregations reconstructed in 3D corresponds to 5% of the total volume analysed. A higher threshold would reduce the reconstructed segregation patterns and it would not reveal continuous channel segregates. On the other hand, a lower threshold increases the number and size of 3D segregation patterns which hides the channel segregates in the images. In Figure 4, the segregation ratio map of the lowest sample is presented after filtration (Figure 4(a)) and after binarization using this threshold (Figure 4(b)). On the segregation map (Figure 4(a)), the red lines show the 0.029 segregation ratio threshold. After binarization, the segregation ratio map of sample N°1 turns to the image displayed in Figure 4(b). Each of the segregation maps from the sample plates were filtered and made binary, afterwards the free, open source software 3DSlicer [15] was used to reconstruct the segregation volume.

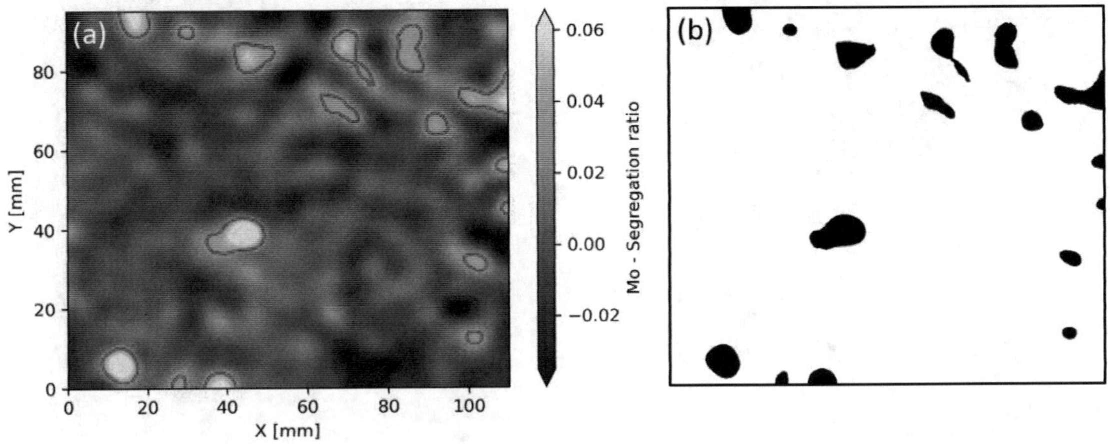

Figure 4. Binarization process for reconstruction, (a) the Gaussian-filtered chemical segregation map, with the segregation ratio threshold of 0.029 highlighted in red (b) the map obtained after binarization with this threshold.

The 3D reconstruction presented in Figure 5 was made by interpolation between each of the 30 binarized maps. The orientation of the reconstruction follows the previously defined co-ordinates also shown in Figure 1, i.e., the x-direction is the radius of the ingot and the z-direction is the height.

The 3D reconstruction shows different segregation patterns at different scales. Several channel segregates exist within the volume analysed, some go through the entire volume from bottom to top, appearing in each sample plate while others merge together or appear from within the volume. Other smaller segregations can be observed within the volume, appearing only on limited number of sample plates, without the classical shape of the observed channel segregates. These segregations are larger than microsegregation since they exist at a larger scale than the secondary dendritic arm spacing, however they are smaller than the channel segregates and cannot be linked to macrosegregation at the ingot scale. These smaller segregations observed in the volume are intermediate scale segregation. The characterisation of these segregates, such as their size, distribution, occurrence, or shape using such tools as geostatistical analysis remains a key task to better understand the occurrence of mesosegregation.

4. Conclusion

The observation of segregation in a 3D volume is key to understand their formation. The objective of this work was to analyze different segregation patterns in 3 dimensions at a centimetric scale. The analysis of serial cut samples with μXRF reveals segregations patterns at different scales. Image treatment techniques such as Gaussian filters were used to eliminate microsegregation patterns and prioritize investigations on intermediate scale and channel shape segregation. The maps of these remaining segregations were then binarized with a segregation ratio threshold of 0.029 to clearly identify channel segregates on successive samples and enable their reconstruction in 3D. The volume representation of segregations shows the co-existence of several channel segregates with intermediate scale segregations, which cannot be identified as micro- or macrosegregation. These mesoscale segregations require a comprehensive study to quantify their size, shape, distribution and occurrence. This study can be based on the present experimental results, analyzed with geostatistics.

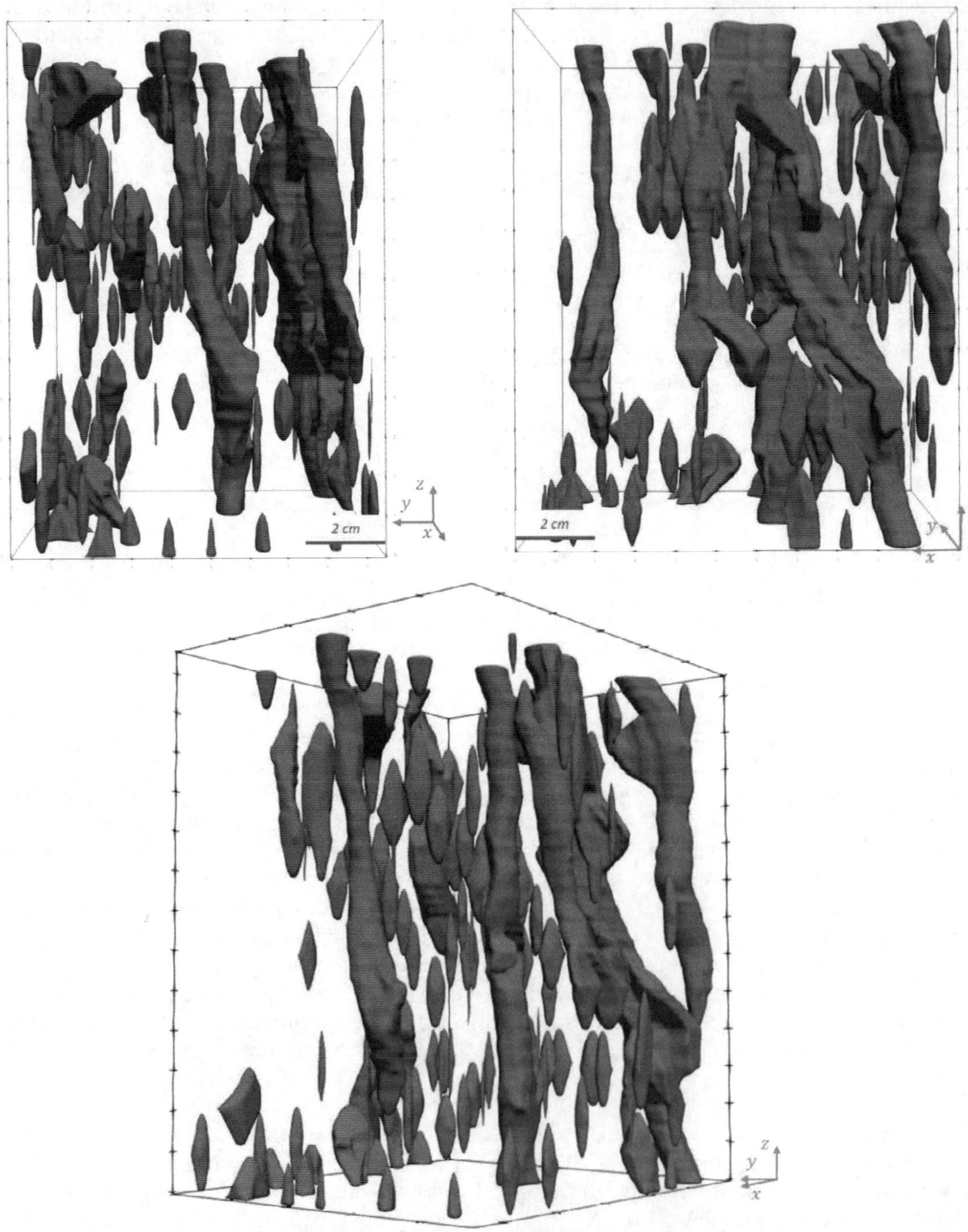

Figure 5. 3D reconstruction of sample volume showing channel and intermediate scale segregation.

IOP Conf. Series: Materials Science and Engineering 1281 (2023) 012061 doi:10.1088/1757-899X/1281/1/012061

Acknowledgments
This research work is supported by ANRT France under a CIFRE Ph.D. fellowship (Grant Number 2020/1234), through the Industrial chair Solidification at Institut Jean Lamour, supported by ArcelorMittal, ArcelorMittal Industeel, CEA, EDF and Framatome, and though μXRF usage provided by SCMEM at the Universite de Lorraine. A part of the required high performance computing resources was provided by the EXPLOR center hosted by the University of Lorraine.

References
[1] Lesoult G 1986 *Techniques de l'ingénieur. Matériaux métalliques.* **M58** 1
[2] Dantzig J A and Rappaz M 2009 *Solidification.* (Lausanne: EPFL Press) p 394
[3] Pickering E J 2013 *ISIJ international.* **53** 935–49
[4] Ludwig A, Wu M and Kharicha A 2015 *Metall. Mater. Trans. A* **46** 4854–67
[5] Thornthon P A and Colangelo V J 1976 *Metall. Mater. Trans. B.* **7** 425–33
[6] Pickering E J and Bhadeshia H K D H 2014 *J. Press. Vessel Technol. Trans. ASME.* **136**
[7] Salimi A, Zadeh H M, Toroghinejad M R, Asefi D and Ansaripour A 1996 *Materiali in tehnologije.* **47** 385
[8] Hong S, Song, J, Kim MC, Choi KJ and Lee BS 2015 *Met. Mater. Int.* **22** 196–203
[9] Yan, G *et al.* 2017 *Metallurgical and Materials Transactions A.* **48**, 3470-81.
[10] Vasileiou A N *et al.* 2021 *Materials and Design.* **209** 109924
[11] Yamashita T, Torizuka S and Nagai K 2003 *ISIJ international.* **43** 1833–41
[12] Pickering E J *et al.* 2015 *Metallurgical and Materials Transactions B.* **46** 1860–74
[13] Pickering E J and Holland M 2014 *Ironmaking & Steelmaking.* **41** 493–9
[14] Gutman L *et al.* 2023 *IOP Conf. Ser. : Mater. Sci. Eng.* **1274** 012049
[15] Fedorov A *et al.* 2012 *3D Slicer as an Image Computing Platform for the Quantitative Imaging Network. Magnetic Resonance Imaging.* 1323-41. PMID: 22770690. PMCID: PMC3466397.

IOP Conf. Series: Materials Science and Engineering 1281 (2023) 012062 doi:10.1088/1757-899X/1281/1/012062

Solidification path for rapid solidification – Application to multicomponent alloys for L-PBF

Paul Martin[1,2], Gildas Guillemot[1], Michel Bellet[1], François Pichot[2], Nicolas Leriche[2], Yaasin Mayi [2] and Charles-André Gandin[1]

[1]Mines Paris – PSL University, CEMEF UMR CNRS 7635, 06904 Sophia Antipolis, France
[2]Safran Additive Manufacturing Campus, rue de Touban – BP 90053, 33185 Le Haillan, France

E-mail: paul.martin@minesparis.psl.eu

Abstract. Laser Powder Bed Fusion (L-PBF) is seen as a process of interest by aeronautical industry to develop new engine components. Nevertheless, the reliability and durability of parts produced by L-PBF depend on the possibility to suppress the occurrence of defects. Among them, hot cracking represents a key issue. These cracks are due to the liquid film remaining between grains at the end of the solidification stage combined with stresses and strains endured by the mushy domain. A microsegregation model providing relevant prediction of the solidification path during L-PBF is thus required for coupling with a thermomechanical analysis. As an answer to the industrial need, a new model is proposed and applied in cooling conditions encountered in L-PBF. It includes the initial solidification conditions and follows the phases, and their composition in the interdendritic liquid region to predict the brittle temperature range. Both dendrite tip growth model and kinetic phase diagram due to non-equilibrium interface phenomena are considered. Cross-diffusion of solute species in the liquid phase is accounted for, as well as thermodynamic coupling with CALPHAD. The model will be applied to IN718, a nickel-based superalloy widely used in the aeronautic industry.

1. Introduction

Additive manufacturing, also known as "3D printing", has become an increasingly popular technology in the aeronautic industry. A significant challenge is to produce parts that can withstand the extreme conditions they will be subjected to in an aircraft, such as high pressure, heat, vibrations and exposure to a corrosive atmosphere. One solution could be to consider the use of highly durable materials, such as nickel-based superalloys, in combination with an additive manufacturing process. On the one hand, superalloys have demonstrated their ability to withstand both mechanical and chemical degradation at temperatures approaching their melting points, making them suitable for use in these applications [1, 2]. On another hand, additive manufacturing allows for the production of customized and complex designs, which may not be possible with traditional manufacturing methods. It is also more cost effective than traditional manufacturing processes of complex parts as only the necessary amount of material is used [3]. In the present activity, we focus our interest in the development of aeronautic parts based on the Laser Powder Bed Fusion (L-PBF) process.

However, applications of L-PBF to nickel-based superalloys sometimes lead to the occurrence of defects such as cracks or porosities which make the parts unusable in industrial sectors such as aeronautics regarding the high quality expectation for manufactured parts. To address this issue, it is necessary to develop a methodology that can prevent the appearance of such defects.

IOP Conf. Series: Materials Science and Engineering 1281 (2023) 012062 doi:10.1088/1757-899X/1281/1/012062

Consequently, numerical simulation may assess the safety and flawless of components before their manufacturing.

To pursue this goal, we focus our activity on the modeling of the solidification path, which could allow to identify the different thermodynamic phases forming to better understand why defects such as cracks may appear.

In this article, we only present the development of a truncated Gulliver-Scheil-type microsegregation model which considers the curvature, chemical and kinetics undercooling that emerge from the rapid development of dendritic structures. The present multicomponent - multiphase model is fully coupled with thermodynamic computations based on the CALPHAD approach.

2. Modeling

2.1. Solidification path

The Gulliver-Scheil equation can be derived by considering the conservation of solute in a closed elementary volume and assuming that the diffusion in the solid is negligible [4, 5]. If we consider the deposition of a single solid phase, the solute balance can be expressed as follows considering an elementary transformation:

$$\mathrm{d}f^{\mathrm{s}} \cdot \left({}^{\mathrm{e}}x_{\mathrm{i}}^{\ell/\mathrm{s}} - {}^{\mathrm{e}}x_{\mathrm{i}}^{\mathrm{s}/\ell}\right) = \mathrm{d}^{\mathrm{e}}x_{\mathrm{i}}^{\ell/\mathrm{s}} \cdot (1 - f^{\mathrm{s}}) \quad \forall\, \mathrm{i} \in \{2 : \mathrm{N}\} \tag{1}$$

where f^{s} is the solid fraction, ${}^{\mathrm{e}}x_{\mathrm{i}}^{\mathrm{s}/\ell}$ and ${}^{\mathrm{e}}x_{\mathrm{i}}^{\ell/\mathrm{s}}$ are respectively the interfacial solid and liquid compositions taken on an equilibrium phase diagram and the index i characterizes each solute specie, $\mathrm{i} \in \{2 : \mathrm{N}\}$, with $\mathrm{i} = 1$ being defined as the solvent ($1 - \sum_{\mathrm{i}=2}^{\mathrm{N}} x_{\mathrm{i}}^{\phi} = x_1^{\phi}$, with x_{i}^{ϕ} corresponding to the composition of i in any phase ϕ). This simple microsegragation path is relatively well adapted to rapid solidification because it does not consider diffusion in the solid phases which could be a relatively correct hypothesis in the present rapid solidification conditions. However, as the cooling rate increases, the hypothesis of complete diffusion in the liquid inherent to Eq. 1, becomes less valid. During solidification, the solute elements will pile up or deplete in front of the solid-liquid interface due to a difference in solubility between the solid and liquid phases. This phenomenon also lead to a delay of solidification as dendrite tips start at a temperature lower than the liquidus temperature of the alloy. The step-by-step integration of Eq. 1 from the liquidus temperature neglects undercooling occurring in solidification.
In rapid cooling, predicting this starting point is of prime importance to determine the composition of the first solid phase, which will affect the solidification path and then the final properties of the material. This also requires to account for non equilibrium phenomenon that can happen at the solid-liquid interface, hence modification of the phase diagram. Consequently, in addition to the Gulliver-Scheil path, we have developed a model to predict the dendritic growth kinetics and undercooling.

2.2. Dendrite tip kinetics

Most industrial alloys used in the aeronautical industry solidify in conditions promoting a dendritic regime. One of the most popular theories characterizing the dendritic growth was developed for binary alloys by Kurz, Giovanola and Trivedi [6] and is widely used in literature, because it includes a description of all contributions leading to the dendrite tip undercooling , i.e. solutal, curvature and kinetics, the latter referring to the departure from equilibrium by using the theory of Aziz [7]. However, the need for a more comprehensive theory to address the complex compositions of industrial alloys has been recognized. In 2001, Hunziker [8] presented a model for the growth of dendrite structures in alloys with any number of components, which accounted for the diffusion interactions between species (cross diffusion). More recently, Guillemot et al. [9] extended the Hunziker model to make it fully coupled with the CALPHAD methodology and suitable for industrial multicomponent alloys. The model was demonstrated for IN718, including both solutal and curvature undercooling. However, dendritic growth was assumed to occur under equilibrium conditions, i.e. the interface compositions were deduced from the equilibrium phase diagram. During rapid solidification, as in the L-PBF process, thermodynamic equilibrium at the solid-liquid interface is not relevant. Thermodynamic concepts designed for rapid solidification are required, as presented hereafter.

2.3. Kinetic phase diagram

Various theories have been proposed to describe the interface kinetics in non-equilibrium conditions. They could be distinguished by the theoretical approach used by the authors to describe associated phenomena. Nearly all of these theories agree on the phenomena that should be predicted :

- "Solute trapping" occurs when atoms do not have enough time to rearrange at the interface between the solid and liquid phases during rapid solidification, causing the chemical potential of each species to be unequal between the phases ($\mu_i^{s/\ell} \neq \mu_i^{\ell/s}$). The atoms are "trapped" in the solidifying liquid, resulting in solid and liquid phases having compositions closer to each other in comparison with equilibrium conditions. As a result, the solidus and liquidus surfaces on the phase diagram tend to each other when increasing the velocity, leading to a segregation coefficient, i.e. ratio $x_i^{s/\ell}/x_i^{\ell/s}$, approaching unity. Solute trapping depends on the diffusion speed of chemical species across the solid-liquid interface. Indeed, the velocity of the solid-liquid interface is limited by the speed at which the atoms can diffuse through it. The diffusive speed of each solute species i is equal to $v_i^D = D_i^I/\delta$ with D_i^I and δ being respectively the diffusion coefficient in the interface and the thickness of the interface. In other words, the higher the diffusion speed, the more the segregation coefficient will tend towards its equilibrium value due to an easier diffusion of the solute species. However, the lower the diffusion speed, the harder it will be for the chemical species to diffuse and the more the segregation coefficient will tend towards unity. For the sake of simplicity, we assume that all species i have the same diffusive speed, v^D.
- "Attachment kinetics" refers to the transition time required for atoms to evolve from a disordered structure in the liquid phase to an ordered structure in the solid phase. This process involves the attachment of atoms to the growing solid phase and the rearrangement of the atoms into an ordered structure. Attachment kinetics can be affected by various factors such as temperature, cooling rate, and the presence of defects onto the interface. It depends, in the same way as before, on a physical limit which is the maximum speed for crystallization, called v_0. This speed corresponds to the maximum speed of displacement of the interface allowing the liquid metal to crystallize. Beyond this limit, the atomic structure created becomes amorphous.
- "Solute drag" characterizes the fact that the displacement of an interface (solid-solid or liquid-solid) is affected by a dissipative phenomenon which reduces its speed. The first theory on this subject was developed in 1957 by Lücke and Detert [10] for problems of recrystallization (solid-solid transformation). It was not until the mid 1970's that Hillert and Sundman [11] established a unified treatment characterizing the solute drag for both solid-solid and liquid-solid transformations.

Several theories in the field of non-equilibrium thermodynamics aim to model the migration of interfaces during rapid solidification conditions also considering the phenomena described above. Among them, the Aziz and Kaplan model [12], also called Continuous Growth Model (CGM), could predict solute trapping with/without the effect of solute drag. It was developed for binary alloys and improved by Aziz and Boettinger [13] to include a parameter that adjusts the impact of solute drag. The CGM model is based on irreversible thermodynamics, and

IOP Conf. Series: Materials Science and Engineering 1281 (2023) 012062 doi:10.1088/1757-899X/1281/1/012062

uses a phenomenological approach to describe how the state of a system changes through the redistribution of solute-solvent atoms at the solid-liquid interface. Ludwig [14] extended the model of Aziz and Kaplan by adapting it to multicomponent alloys. However, the effect of solute drag was not considered in his approach. Wang et al. [15] developed a model to study the rapid solidification of multicomponent alloys based on thermodynamic extremal principle and extended irreversible thermodynamics. However, although the model takes into account the effect of the solute drag, it is not possible to adjust its impact. We propose to use the recent approach proposed by Hareland et al. [16] whose main principles are summarized hereafter. To account for the influence of solute drag, an interfacial composition is introduced, denoted x_i^{eff}. This composition is determined using a parameter λ that controls the extent to which solute drag impacts the transformation.

$$x_{\mathrm{i}}^{eff} = \lambda x_{\mathrm{i}}^{\ell/\mathrm{s}} + (1-\lambda) x_{\mathrm{i}}^{\mathrm{s}/\ell} \quad \forall\, \mathrm{i} \in \{2:\mathrm{N}\} \tag{2}$$

where $x_{\mathrm{i}}^{\mathrm{s}/\ell}$ and $x_{\mathrm{i}}^{\ell/\mathrm{s}}$ are respectively the interfacial composition of the solid and liquid phases. x_{i}^{eff} represents the solute concentration at the interface between liquid and solid phases. There is only a single value of λ in the current formulation in order to maintain $\sum_{\mathrm{i}=1}^{\mathrm{N}} x_{\mathrm{i}}^{eff} = 1$. During rapid growth of the solid phase, the interface will attempt to match the composition of the liquid parent phase. However, if the solute concentration at the interface differs from that of the solid phase (i.e. $x_{\mathrm{i}}^{eff} \neq x_{\mathrm{i}}^{\mathrm{s}/\ell}$), a certain amount of solute must be rejected into the liquid phase in order to maintain the desired composition of the solid phase. This process, known as solute drag, can slow down the growth of the interface. The value of x_{i}^{eff} is dependent on the degree of solute drag, ranging from complete solute drag ($\lambda = 1$) where the solute concentration at the interface is equal to that of the liquid phase (with $x_{\mathrm{i}}^{eff} = x_{\mathrm{i}}^{\ell/\mathrm{s}}$ the quantity of solute that must be desorbed in the liquid phase is $x_{\mathrm{i}}^{\ell/\mathrm{s}} - x_{\mathrm{i}}^{\mathrm{s}/\ell}$), to no solute drag ($\lambda = 0$) where the concentration at the interface is equal to that of the solid phase ($x_{\mathrm{i}}^{eff} = x_{\mathrm{i}}^{\mathrm{s}/\ell}$ no solute need to be desorbed). It must be emphasized that the solute drag parameter λ will have an influence on solute drag, solute trapping and attachment kinetics. Therefore, varying lambda means modifying the influence of these three physical phenomena at the same time. Then, based on theories that define the movement of an interface in a pure metal, it is possible to express the velocity of the interface in a multicomponent alloy, v, by rearranging the Velocity Response Function (VRF) [16] :

$$v = \frac{v_0}{RT} \sum_{\mathrm{i}=1}^{\mathrm{N}} x_{\mathrm{i}}^{eff} [\![\mu_{\mathrm{i}}]\!] \tag{3}$$

where v_0 is the maximum speed of crystallization, R in the gas constant, T is the system temperature and $[\![\mu_{\mathrm{i}}]\!]$ is a difference in chemical potential between two phases, $[\![\mu_{\mathrm{i}}]\!] = \mu_{\mathrm{i}}^{\ell/\mathrm{s}} - \mu_{\mathrm{i}}^{\mathrm{s}/\ell}$. A final set of equations must be established to make the connection between the solute balance at a moving interface (in our case, the solid-liquid interface) and its equivalent described using thermodynamic laws. By rearranging the Composition Response Function (CRF) [16] :

$$v\left(x_{\mathrm{i}}^{eff} - x_{\mathrm{i}}^{\mathrm{s}/\ell}\right) = \sum_{\mathrm{j}=2}^{\mathrm{N}} \left[-x_{\mathrm{i}}^{eff} x_{\mathrm{j}}^{eff} \frac{v^{\mathrm{D}}}{RT} \left([\![\mu_1]\!] - [\![\mu_{\mathrm{j}}]\!]\right)(1-\delta_{\mathrm{ij}})\right] + \left(x_{\mathrm{i}}^{eff}\left(1 - x_{\mathrm{i}}^{eff}\right)\frac{v^{\mathrm{D}}}{RT}\right)\left([\![\mu_1]\!] - [\![\mu_{\mathrm{i}}]\!]\right) \quad \forall\, \mathrm{i} \in \{2:\mathrm{N}\} \tag{4}$$

where v^{D} is the diffusive speed at the interface and δ_{ij} is the kronecker symbol.

IOP Conf. Series: Materials Science and Engineering 1281 (2023) 012062 doi:10.1088/1757-899X/1281/1/012062

With this set of equations, it is theoretically possible to draw a phase diagram for any multi-component alloys, at least for the first primary solid phase growing in the liquid and given a solidification speed and pressure. Therefore, the equations can be used to determine non-equilibrium tie lines, which can be incorporated in the analysis of various physical phenomena, such as chemical diffusion and solidification path, that were typically studied under the assumption of thermodynamic equilibrium at the interface between the solid and liquid phases. Proceeding with the analysis, it is possible to evaluate another quantity that is widely used in the studies of phase transformations, i.e. the T_0 line. This line is associated with the notion of solute trapping as it characterises, for a given alloy, the temperature at which the liquid is "frozen" or "trapped" in its nominal composition, without any segregation. In other words, in this situation $x_{\mathrm{i}}^{\mathrm{s}/\ell} = x_{\mathrm{i}}^{\ell/\mathrm{s}}$ and $\mu_{\mathrm{i}}^{\mathrm{s}/\ell} \neq \mu_{\mathrm{i}}^{\ell/\mathrm{s}}$. To calculate the T_0 line, Eq. 3 will evolve as follow :

$$T_0 = \frac{v_0}{Rv} \sum_{\mathrm{i}=1}^{\mathrm{N}} x_{\mathrm{i}}^{\mathrm{s}/\ell} [\![\mu_{\mathrm{i}}]\!] \tag{5}$$

2.4. Coupling strategy

We aim at combining theories associated with the Gulliver-Scheil solidification path, dendrite growth velocity and non-equilibrium transformation to create a simple and straightforward tool. However, it is necessary to outline the strategy that will be used to calculate a truncated Gulliver-Scheil solidification path [17]. The starting point for the solidification path, which depends on the dendrite tip undercooling, is calculated using the dendrite tip kinetics. For a given solidification speed, the tip temperature is stored and transferred to the Gulliver-Scheil solidification path calculation. At this previously recorded temperature (tip temperature), an equilibrium calculation is performed on the equilibrium phase diagram (blue tie line in Fig. 1). An initial phase fraction is determined through a Lever-Rule approach, and the rest of the solidification process proceeds by incrementally increasing the initial solid fraction according to the equilibrium diagram using Eq. 1. Fig. 1 highlights the main flaw in the adopted strategy: continuity of compositions is not ensured between the operating point of the dendrite and the beginning of the solidification path (jump between the kinetic diagram in red and the equilibrium diagram in black). However, it is assumed that the growth velocity of the solid-liquid interface below the dendrite tip on the side of the parabola and for secondary branching return to an equilibrium diagram. Another initialization strategy consists of initializing the solid fraction from the kinetic phase diagram to be as close as possible to the working point of the dendrite (green tie line in Fig. 1). This strategy is characterized by the purple tie line in Fig. 1.

3. Results and Discussion

The goal of this section is to put into practice the concepts previously presented: 1) to studying the growth of dendritic structures in relation to the solidification rate and 2) examining the outcomes of the truncated Gulliver-Scheil simulation, hence determining the impact of the solidification rate on the solidification path. To pursue this objective the IN718 nickel based superalloy will be used. The nominal composition as well as the parameters for the simulations are listed in Table 1 and Table 2, respectively. It should be noted that the thermal conductivity, as well as the temperature gradient, were assumed to be identical in the solid and liquid phases, thus eliminating the need for explicit specification.

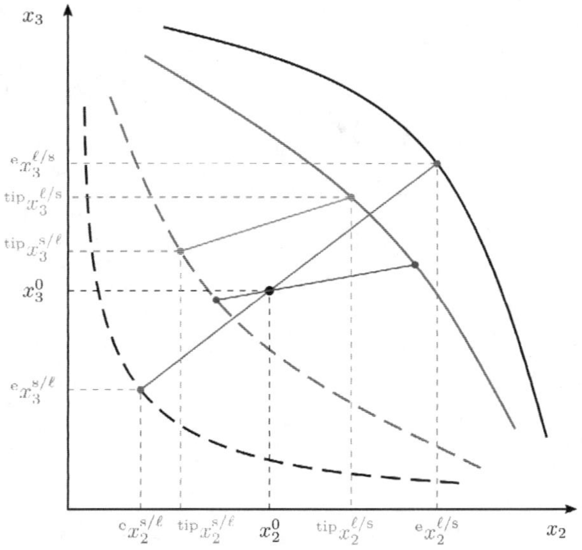

Figure 1. Schematic isothermal–isobaric cross-sectional view of a ternary phase diagram that illustrates the strategy employed to initialize a truncated Gulliver-Scheil solidification path. The dashed curves correspond to the solidus boundaries, while the solid curves denote the liquidus boundaries. The black curves in the diagram depict the thermodynamic equilibrium phase diagram, and the red curves represent the kinetic phase diagram. At high velocity the operating point of the dendrite tip is defined by the green tie line on the red kinetic diagram. The initial solid fraction and the solidification path are computed using the blue tie line on the equilibrium phase diagram. Another strategy to determine the initial solid fraction is shown by the purple line, which involves using a tie line on the kinetic phase diagram that passes through the nominal composition of the alloy.

Table 1. Chemical composition of the IN718 [18]

Chemical elements	Cr	Fe	Nb	Mo	Ti	Al	C	Ni
Nominal composition [wt.%]	18.2	18.9	5.1	3.1	0.9	0.29	0.025	Balance

Table 2. Material and thermodynamic data used for the simulations.

Property	Symbol	Value	Unit	Ref
Thermodynamic data : TCNI10, MOBNI5				[19, 20]
Diffusion matrix at ${}^{e}T_L$	$\mathbf{D}^{l}$	-	$[\mathrm{m}^2\,\mathrm{s}^{-1}]$	[9]
Interfacial energy	γ	0.37	$[\mathrm{J\,m}^{-2}]$	[9]
Selection parameter	σ	$1/(4\pi^2) \simeq 0.02533$	$[-]$	[6]
Temperature gradient in solid and liquid	G	$1 \cdot 10^7$	$[\mathrm{K\,m}^{-1}]$	[6]
Drag parameter	λ	0.5	[-]	
Maximum crystallisation speed	v_0	850	$[\mathrm{m\,s}^{-1}]$	
Diffusion speed	v^D	0.8	$[\mathrm{m\,s}^{-1}]$	

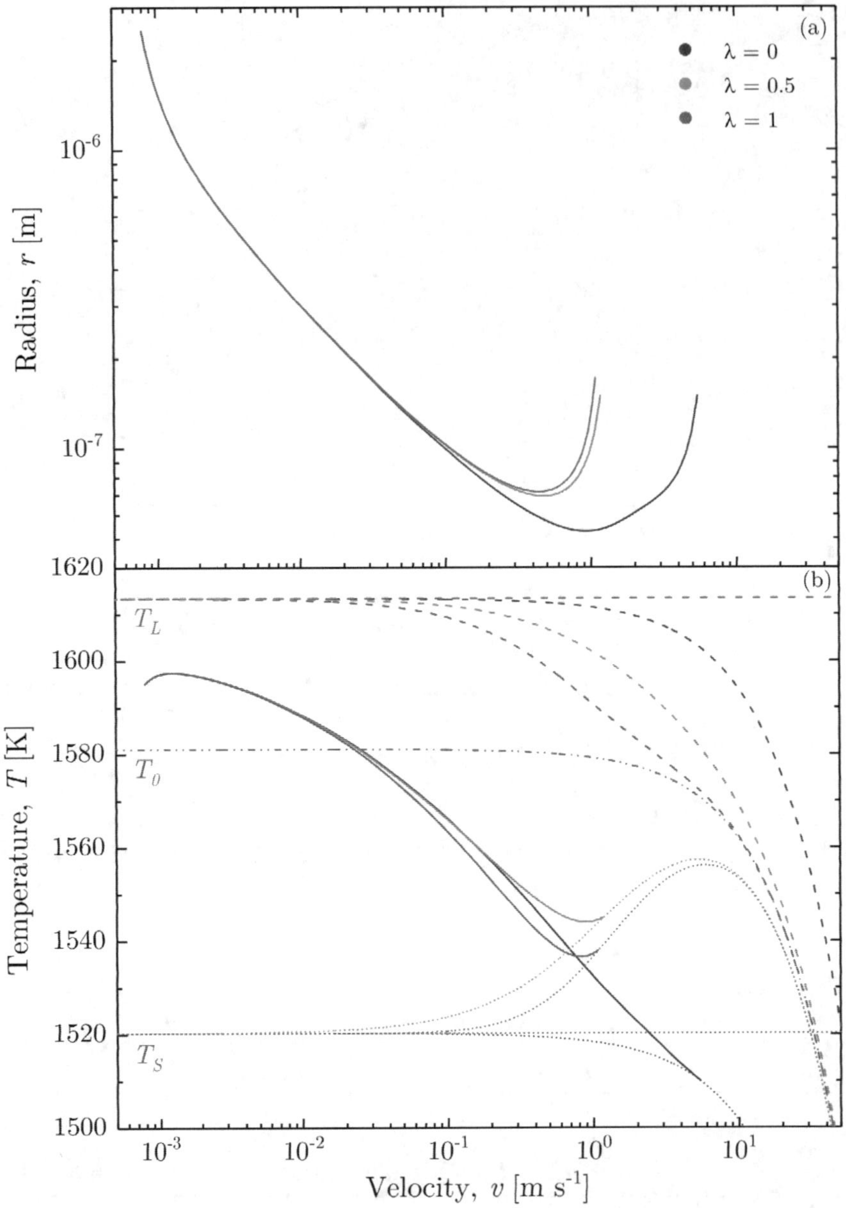

Figure 2. Evolution of the dendrite tip radius (a) and temperature (b) with respect to the velocity of the solid-liquid interface. The figure also highlights the influence of the solute drag parameter on the working point of the dendrite tip. The dashed and dotted curves represent respectively the evolution of the solidus and liquidus of the alloy as a function of the velocity of the solid-liquid interface. The dashed-dotted curve represents the evolution of the T_0 curve as a function of the velocity of the solid-liquid interface.

Fig. 2 illustrates the development of a dendritic structure as a function of the velocity of the solid-liquid interface. The dendrite tip radius is shown in Fig. 2 (a). A typical trend is observed, indicating that the dendrite tip radius first decreases when increasing the velocity of the solid-liquid interface. At higher velocity, the opposite trend is seen, corresponding to a dendrite to planar front growth. Outside the computed curves, the solid-liquid interface is planar. The solid curves in Fig. 2 (a) correspond to the solid curves in Fig. 2 (b) where the temperature of the solid-liquid interface is drawn. It can be observed that the solute drag parameter, λ, affects the range of dendrite growth, shifting the transition to planar front at lower speeds when λ tend to unity.

Fig. 2 (b) reveals the evolution of the liquidus (dashes curves) and solidus (dotted curves) of IN718 with respect to the velocity of the solid-liquid interface and the parameter λ. It is observed that at low velocity the kinetic solidus and liquidus (colored curves in red, green or blue, depending on the value of lambda) are almost perfectly superimposed with their equilibrium values (constant values in gray). However, as the velocity increases, the kinetic solidus and liquidus tend to converge towards the T_0 curve (represented by the dashed-dotted curve), which is a direct consequence of the solute trapping phenomenon. Nevertheless, a state of complete solute trapping occurs at much higher velocities than the limit of stability for dendrite growth. As depicted in Fig. 2 (b), as the velocity increases, the dendrite tip temperature decreases, leading to a greater degree of undercooling. The dendrite tip temperature decreases in a non-monotonic manner, up to reaching the temperature of the kinetic solidus, beyond which the structure shifts to a planar front. It is noteworthy that without kinetic effects, this phenomenon would not be observable, as the dendrite curve would terminate at the equilibrium solidus temperature rather than at the kinetic solidus. Fig. 2 (b) also illustrates the variation of the different quantities

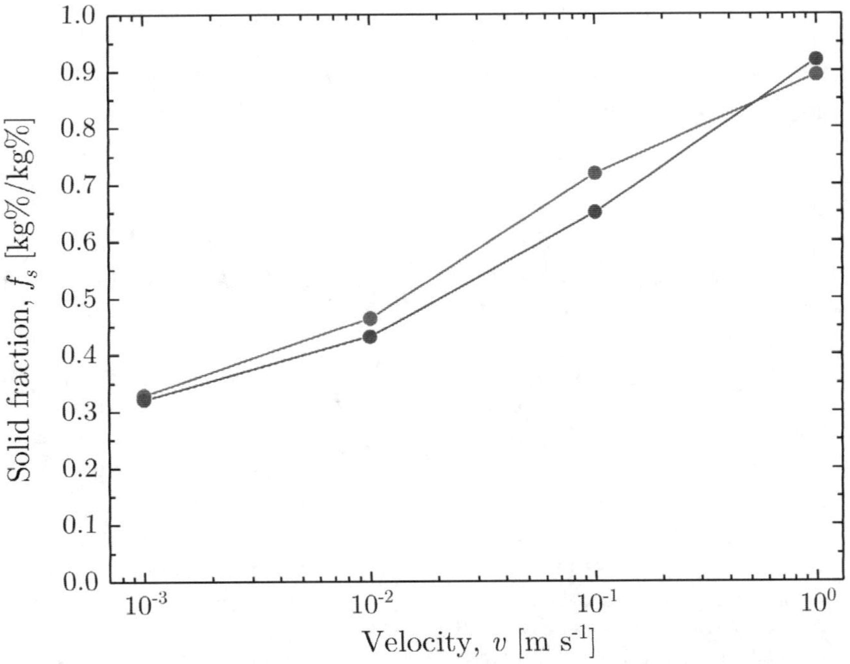

Figure 3. Evolution of the initial solid phase fraction in relation to the solidification speed (with $\lambda = 0.5$). The blue and purple curves correspond respectively to the evolution of the solid fraction, for our model, initialized on the equilibrium phase diagram (blue tie line on Fig. 1) and on the kinetic phase diagram (purple tie line on Fig. 1).

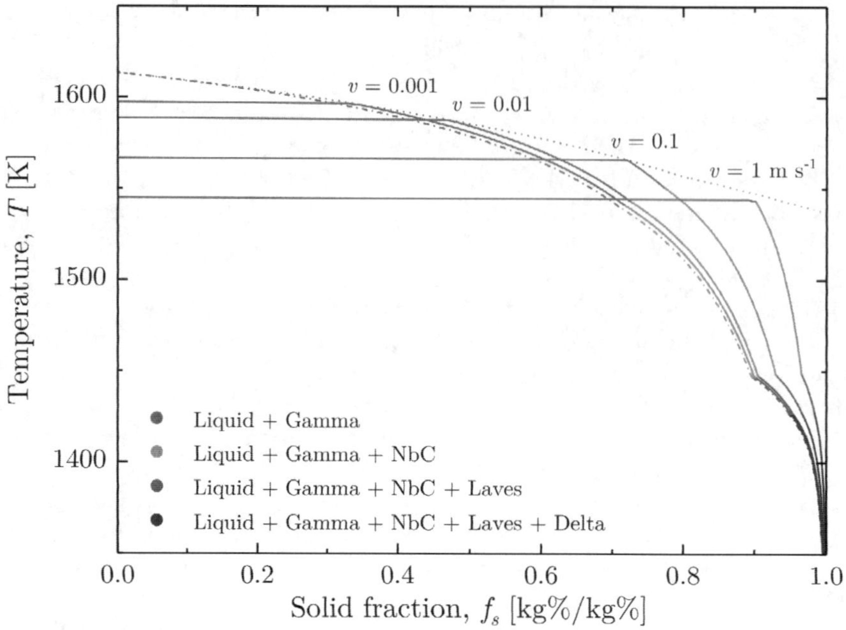

Figure 4. The dotted and dashed-dotted curves correspond to the evolution of a Lever-Rule and Gulliver-Scheil solidification path respectively. The solid curves correspond to our model for different solidification rate. The calculations were performed using $\lambda = 0.5$.

in relation to the solute drag parameter, λ. When $\lambda = 0$, the solidus and liquidus curves do not converge even at very high velocities, indicating that a full solute trapping effect is not achieve. However, for $\lambda \neq 0$, the solidus and liquidus curves converge relatively easily. The evolution of the solidus and dendrite tip curves are non-monotonic with λ, in contrast to the evolution of the liquidus curve.

As previously explained, two strategies can be implemented to determine the first solid fraction formed. In the following section, the calculations will be performed with $\lambda = 0.5$. Fig. 3 compares the evolution of the initial solid fraction initialized, on the one hand, using the equilibrium phase diagram and, on the other hand, calculated on the kinetic phase diagram. The results indicate that the variation between the two methodologies is relatively small for this alloy.

The first methodology using the equilibrium solid fraction is used in Fig. 4 to compute the solidification path of IN718 at different solidification speeds. The calculations were performed by selecting the solid phases FCC_L12 (Gamma), FCC_L12#3 (NbC), C14_LAVES (Laves) and NI3TA_D0A (Delta) [21]. The results demonstrate that as the solidification speed increases, the temperature of the dendrite tip decreases as shown in Fig. 2 (b), resulting in a more significant effect of undercooling on the onset of solidification. Furthermore, the quantity of initial solid formed also increases as the solidification speed increases. The combined effect of these two phenomena results in a greater difference between the Gulliver-Scheil (dashed-dotted curve) and truncated Gulliver-Scheil (plain curves) solidification path. The graph illustrates that at the highest speeds ($v = 1$ m s^{-1}), the initial stage of solidification can be three-phase, with the Liquid + Gamma + NbC phase becoming the first equilibrium band. The established strategy does not take into account the potential competition among the primary solid phases. This choice is reflected in the graph, where the quantity of the first solid formed at $v = 1$ m s^{-1} does not match perfectly to the lever rule curve, unlike the other solidification paths at lower speeds. If the formation of the NbC phase had been prevented, the fraction of Gamma initially

IOP Conf. Series: Materials Science and Engineering 1281 (2023) 012062 doi:10.1088/1757-899X/1281/1/012062

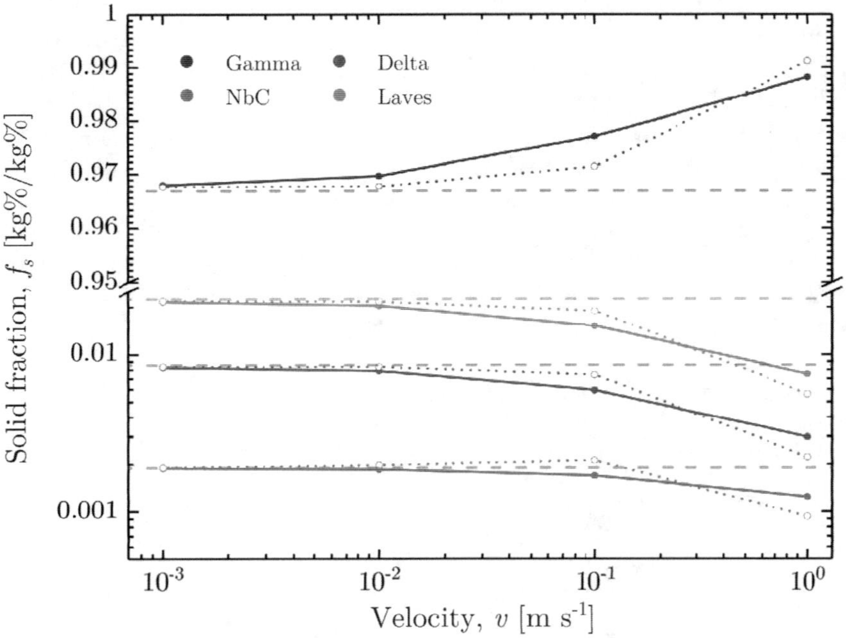

Figure 5. Evolution of the final solid phase fraction in relation to the solidification rate. The solid and dotted curves correspond to the evolution of the solid fraction, for our model, initialized on the equilibrium phase diagram and on the kinetic phase diagram respectively. The dashed curves correspond to the final solid phase fraction of a Gulliver-Scheil solidification path. The calculations were performed using $\lambda = 0.5$.

computed would have been in agreement with the metastable Lever-Rule equilibrium.

The graph in Fig. 5 shows the relationship between the velocity of the solid-liquid interface and the fraction of each phase at the end of solidification. Despite the tendency for the primary Gamma phase fraction to increase, the variation of the secondary phases (NbC, Delta, and Laves) remain limited. This stability may be due to the hypothesis of the solidification model used, which does not take into account time and length scales. An alternative model that considers interface velocity and primary arm spacing could potentially yield different results in terms of the amount and types of phases formed [22]. Additionally, an alternative strategy that predicts the onset of solidification using not only temperature but also the solid fraction of the kinetic phases diagram is proposed, as represented by the dashed curves in Fig. 5. The results and conclusions are the same as before, even with a different initialization, there is no significant evolution of the solid fraction quantity with respect to the velocity.

4. Conclusions

Using the theory developed by Hareland et al. [16] and including it into the dendritic kinetics model developed by Guillemot et al. [9], it is possible to study the dendritic growth of industrial nickel-based superalloy while accounting for non-equilibrium rapid solidification phenomena. The results showed that the kinetic effects can significantly impact the dendrite tip kinetic. The model retrieves the occurrence of solute trapping thanks to the description of kinetic solidus and liquidus temperatures. By coupling the dendrite tip kinetic model with the Gulliver-Scheil model, it is possible to create a truncated solidification path. The findings indicate that, as solidification speed increases, the level of undercooling and the amount of the primary Gamma phase increase. However, the final fraction of solid phases appears to be relatively insensitive to both the velocity and the initial solid fraction. The approach presented for the calculations of the dendrite tip kinetics is innovative since it considers the effects of interface kinetics for undiluted multicomponent alloys while being fully coupled with the CALPHAD method. The microsegregation path, thanks to its simplicity, can be implemented quickly and easily for industrial use. To complete and improve the approach presented in this study, it would be necessary to extend the microsegregation model to characterize the dimensions of the studied domain (i.e. interdendritic spacing) and to take into account the non-stationary character of the interface displacement (by computing the evolution of the diffusion profile as a function of the geometry and the velocity of the interface as proposed by Guillemot and Gandin [23]).

References

[1] Reed R C 2006 *The Superalloys: Fundamentals and Applications* (Cambridge: Cambridge University Press) ISBN 978-0-521-07011-9
[2] Donachie M J and Donachie S J 2002 *Superalloys: a technical guide* 2nd ed (Materials Park, OH: ASM International) ISBN 978-1-61503-064-4
[3] Yadroitsev I, Yadroitsava I, Plessis A D and MacDonald E 2021 *Fundamentals of Laser Powder Bed Fusion of Metals* (Elsevier) ISBN 978-0-12-824091-5
[4] Boettinger W J 2016 *Journal of Phase Equilibria and Diffusion* **37** 4–18
[5] Dantzig J A and Rappaz M 2016 *Solidification* 2nd ed Engineering sciences (Lausanne: EPFL Press) ISBN 978-2-940222-97-1
[6] Kurz W, Giovanola B and Trivedi R 1986 *Acta Metallurgica* **34** 823–830
[7] Aziz M J **53** 1158–1168 publisher: American Institute of Physics
[8] Hunziker O 2001 *Acta Materialia* **49** 4191–4203
[9] Guillemot G, Senninger O, Hareland C A, Voorhees P W and Gandin C A 2022 *Calphad* **77** 102429
[10] Lücke K and Detert K 1957 *Acta Metallurgica* **5** 628–637
[11] Hillert M and Sundman B 1976 *Acta Metallurgica* **24** 731–743
[12] Aziz M J and Kaplan T 1988 *Acta Metallurgica* **36** 2335–2347
[13] Aziz M J and Boettinger W J 1994 *Acta Metallurgica et Materialia* **42** 527–537
[14] Ludwig A 1998 *Physica D: Nonlinear Phenomena* **124** 271–284
[15] Wang K, Wang H, Liu F and Zhai H 2013 *Acta Materialia* **61** 1359–1372
[16] Hareland C A, Guillemot G, Gandin C A and Voorhees P W 2022 *Acta Materialia* **241** 118407
[17] Sarreal J A and ABbaschian G J 1986 *Metallurgical Transactions A* **17** 2063–2073
[18] Wang Z, Guan K, Gao M, Li X, Chen X and Zeng X 2012 *Journal of Alloys and Compounds* **513** 518–523
[19] 2021 Thermocalc database TCNI10 \Ni-Alloys version v10.0, Thermo-Calc Software AB, Sweden URL `https://thermocalc.com`
[20] 2021 Thermocalc database MOBNI5 \Ni-Alloys Mobility version v5.1, Thermo-Calc Software AB, Sweden URL `https://thermocalc.com`
[21] Pautrat A 2013 *Experimental quantitative studies that the solidification of the superalloys 718 in investment casting* Phd thesis ParisTech
[22] Tong X and Beckermann C 1998 *Journal of Crystal Growth* **97** 289–302
[23] Guillemot G and Gandin C A 2015 *Acta Materialia* **97** 419–434

IOP Conf. Series: Materials Science and Engineering 1281 (2023) 012063 doi:10.1088/1757-899X/1281/1/012063

Analysis of the melting and solidification process of aluminum in a mirror furnace using Fiber-Bragg-Grating and numerical models

Constantin Bauer[1*], Maximilian Erber[1], Georg Fuchs[1], Tobias Brügge[1], Christoph Hartmann[1] and Wolfram Volk[1]

[1]Institute of Metal Forming and Casting, Technical University Munich, Walther-Meissner-Str. 4, D-85748 Garching, Germany

[*] Corresponding author.

E-mail: c.bauer@tum.de

Abstract. In the search of an adequate real time strain measurement method in aluminum casting, the use of Fiber-Bragg-Grating (*FBG*) is being investigated with great interest. In order to do so, the behaviour of glass fiber sensors in a liquid aluminium alloy at temperatures up to 750°C is experimentally analysed in a laboratory environment. For better process understanding a simulation of the fiber alloy composite is conducted. FBG is an optical measurement method, which uses engraved Bragg reflectors in a 125 μm in diameter thick glass fiber. This reflector transmits most of the wavelengths but only reflects one specific wavelength. This specific wavelength can be measured and changes due to the axial strain on the grating by the fluid alloy reaction and by the changes in temperature. Using a so-called mirror furnace, several experiments with the fiber alloy composite are evaluated. These measurements are also the basis for the further understanding of hot tearing. The data gathered during the measurement campaign - both numerical and experimental - is used to parameterize a simulation. As a result, the understanding of the fiber alloy composite behaviour is expanded and a digital twin is modeled with MATLAB's partial differential equation toolbox.

1. Introduction

Understanding the solidification process helps to determine properties of the casting such as its mechanical behaviour [1]. A main influencing factor on solidification processes is the heat flow in the alloy [2]. Analysing the solidification process of aluminum cast parts poses some mayor challenges. Due to the material's complex behaviour and the temperature range at which aluminum alloys liquefy, appropriate measurement devices need to be applied. Only then reliable data can be collected to describe the process accurately. A common way of collecting basic information about the melt is to use thermal measurement devices, which allow the temperature of the melt to be determined at specific positions inside or outside a casting. Other parameters such as the strains in the casting demand different measurement methods. One example is the use of neutron diffraction, which allows data collection during solidification [3]. However, such measurements can only be carried out and conducted within a laboratory environment. An alternative way of measuring both temperature and strain at the same time is the use of optical methods [4]. FBGs is such a method to perform experiments in the liquid phase of aluminum alloys [5][6]. In this measurements, sensors are introduced into the melt and acquire signals during the cooling phase. The sensors also remain in the casting to act as monitoring devices later on [7][8].

1.1. Objective

The introduction leads to the objective of this paper, which is to prove that the process of melting and remelting in several cycles can also be successfully measured with the use of FBGs. Such measurements are useful because with cyclic measurements of the same geometry a larger set of data is gathered of the solidification and melting process of aluminum. Therefore, in a first step the understanding of this process is simulated to gain a better understanding of the temperature values at the fiber and the values of the axial strain resulting from the temperature load. In the second step the experiment of placing a fiber inside an aluminum specimen and melting it onto the the fiber is conducted. The combination of the experimental and the numerical results lead to an improved insight of the composite behaviour of the FBG and the aluminum. In order to compete this task the following two subsections present the basis of Bragg grating and the numerical approach chosen.

1.2. Fiber-Bragg-Grating

With the aid of an interrogator a spectrum of a multi wavelength is generated and put into the optical fiber. At the Bragg grating, a specific wavelength λ_B gets reflected. The reflected wavelength is recorded using an optical detector and calculated according to equation 1 from twice the product of the effective refractive index n_{eff} of the grating plane spacing Λ. [9]

$$\lambda_B = 2n_{eff}\Lambda \tag{1}$$

A change of the wavelength $\lambda_{B,0}$ by a value $\Delta\lambda_B$ allows a conclusion about external and internal influences, such as a change in the elongation of the fiber. This elongation change of the fiber in axial direction as a combination of mechanical and thermal strain and can be calculated according to equation 2. The proportionality constant is denoted by κ [4]. This describes the interface influences between the glass fiber and the alloy. This constant is multiplied by the strain in the axial direction along the fiber and is equal to the quotient of the sum of the spectral drifts and the rest-bearing wavelength [10]. Every other spectrum transmits through the fiber's end and gets lost. From the resulting drift of the spectrum the external parameters influencing can be retraced [11].

$$\kappa\epsilon_z = \frac{\Delta\lambda_B + \Delta\lambda_{B,th}}{\lambda_{B,0}} \tag{2}$$

1.3. Numerical model

Multidomain geometries can be described analytically as well as numerically [12]. In this case MATLAB's partial differential equation toolbox is used to analyse the thermal stress of a multidomain geometry consisting of a glass fiber, an aluminum body, a gap of argon and a graphite crucible [13]. Therefore, a two step workflow is necessary. First the temperature distribution due to the process is computed. Second the strain due to the thermal expansion is calculated. The approach builds on the calculation methods for heat transport by heat conduction. If there is a local temperature gradient in the direction of the spatial coordinates x, the heat flux density in pure heat conduction depends only on the so-called heat conductivity λ of the material. The solution is developed using Fourier's Law [14]. Where the parameter T describes the temperature, λ describes the coefficient of conduction and x the direction, see

IOP Conf. Series: Materials Science and Engineering 1281 (2023) 012063 doi:10.1088/1757-899X/1281/1/012063

equation 3.

$$\dot{q} = -\lambda \frac{\partial T}{\partial x} \tag{3}$$

In the case of the fiber-aluminum simulation the model is reduced to conduction only, since the influencing of conduction and radiation are negligible in comparison [15]. This is described in equation 4 and 5 [16]. Whereas T_{solid} describes the temperature of the material and T_{fluid} describes the temperature of the surrounding medium which is cooling the solid material by convection.

$$\dot{q}_{convection} = \alpha(T_{solid} - T_{fluid}) = 0 \tag{4}$$

For equation 5, the Stefan–Boltzmann constant is represented by σ and ϵ is the factor of emissivity.

$$\dot{q}_{radiation} = \epsilon\sigma T^4 = 0 \tag{5}$$

In the case of transient heat conduction, time is taken into account by applying equation 6 [17].

$$\frac{\partial T}{\partial t} = a\nabla^2 T + \frac{\dot{q}}{\rho c} \tag{6}$$

Results from non-uniform temperature fields can be applied to solve for the thermoelastic problem induced by thermal stresses [18]. The material properties are assumed to be isotropic and the problem is defined as static solid, even though the results are calculated for different temperature distributions. For the 3D linear elastic case the approach used by the toolbox is stated in equation 7 and 8, where ϵ is the strain, u is the displacement and f is the body force of the model [19] [20]. Equation 8 is valid for linearized, small-displacement and strain-displacement assumptions.

$$-\nabla(c \otimes \nabla u) = f \tag{7}$$

$$\epsilon = \frac{1}{2}(\nabla u + \nabla u^T) \tag{8}$$

2. Axissymmetric analysis of a multidomain model

A numerical approach is chosen to model the process of a cylindrical specimen being melted onto the glas fiber. The result is a compound structure with the FBG sensor along the axisymmetric center of the casting. The four different materials are pictured in figure 1.

While heating up there is also an air gap filled with argon gas between the fiber and the alloy. This air gab will be closed as soon as the aluminum becomes liquefied. At this stage the simulation simply changes the material properties from argon to aluminum in the cell region C1. Figure 1 (a) shows the described setup of the four cell regions. Although, cell region C4 has 62.5 μm in radius, this is the area of interest. In this region the FBG measurement takes place in the real experiment. In figure 1 (b) the mesh of the simulation is shown. To reduce the numerical effort, the mesh is refined in the cell region of the fiber and the size increases in radial direction of the cylinder. For the numerical calculation the system is modeled using MATLAB's Partial Differential Equation Toolbox. This toolbox allows to solve the heat transfer and structural problem by the use of the finite element method.

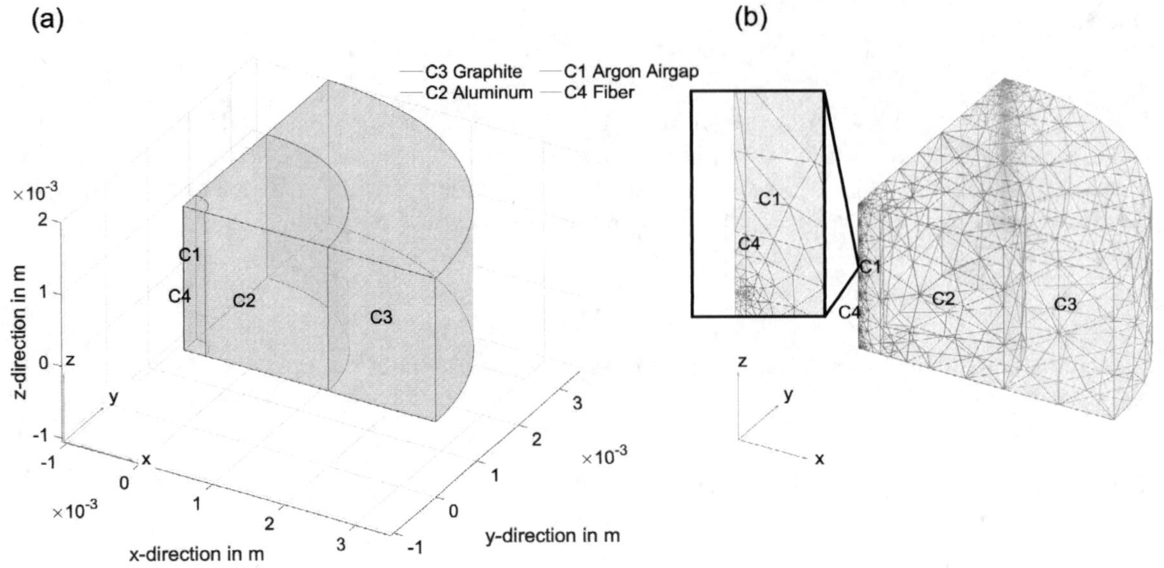

Figure 1. Showing the multi domain problem using symmetric relationships to reduce the numerical effort; (a), Cell domain plot; (b) Mesh of the four different materials C1 to C4. [21]

2.1. Transient thermal analysis with nonuniform heatflux

Modeling the thermal behaviour of the experiment, several approximations are considered to describe the process. The values of thermal conductivity, mass density and specific heat coefficient change with temperature, in order to reflect the temperature dependency of the material properties. The simulation takes this into account for each material by representing the thermal properties as function of temperature. The initial properties at room temperature and at the state of 750°C are listed in table 1. Other values between these temperatures are also implemented but not shown because of the limited space of the table. Every other value is then linearly interpolated. Several boundary conditions are applied to the mesh from figure 1.

Table 1. Material properties at room temperature left column and 750 °C right column used to numerically approximate the thermal behaviour of the mirror furnace experiment. [22] [23] [24]

	Thermal conductivity (W/mK)	Mass density (kg/m^{-3})	Specific heat coeff. (J/kg K)	Thermal conductivity (W/mK)	Mass density (kg/m^{-3})	Specific heat coeff. (J/kg K)
Aluminum	151.20	2.6e03	875.6	69	2.4e03	11490
Graphite	114.47	1800	793.4	64	1800	1758
Fiber	1.18	2538	805.6	14	2479	1654
Argon	0.018	1.784	520.0	0.018	1.784	520

A function to model the heat source over time is applied on the outer surface of cell region C3. This function is an approximation of the measured temperature curve of the conducted mirror furnace experiment and is represented by equations 9 and 10. The constants of these equations are a result from the temperature measured in the mirror furnace.

$$Q = 21 + 36.45 t_{heat,i} \quad t_{heat,i} < 20 \tag{9}$$

IOP Conf. Series: Materials Science and Engineering 1281 (2023) 012063 doi:10.1088/1757-899X/1281/1/012063

$$Q = 21 + 36.45t_{heat,end} - 1.92(t_{cool,i} - t_{heat,end}) \quad t_{cool,i} \geq 20 \tag{10}$$

The remaining outer surfaces are governed by the boundary condition of symmetry. As an initial value the system's temperature is set to 21° C. The resulting temperature distribution of the specimen at selected time steps is shown in figure 2. All materials are at room temperature at the initial state. Eventually the increasing temperature at the outer surface causes the materials to be heated. Due to the thermal resistance, the heating of the inner layers in the sample is delayed compared to the outer layers. For the analysis of the fiber temperature, its mean temperature over time is separately shown in figure 2. With a maximum temperature peak at around 20 s, the boundary condition of the model changes from heating to cooling. The process of cooling has a smaller gradient than the heating. This is because in the mirror furnace the specimen mainly cools down because of convectional processes, whereas the process of heating is mainly driven by direct radiation.

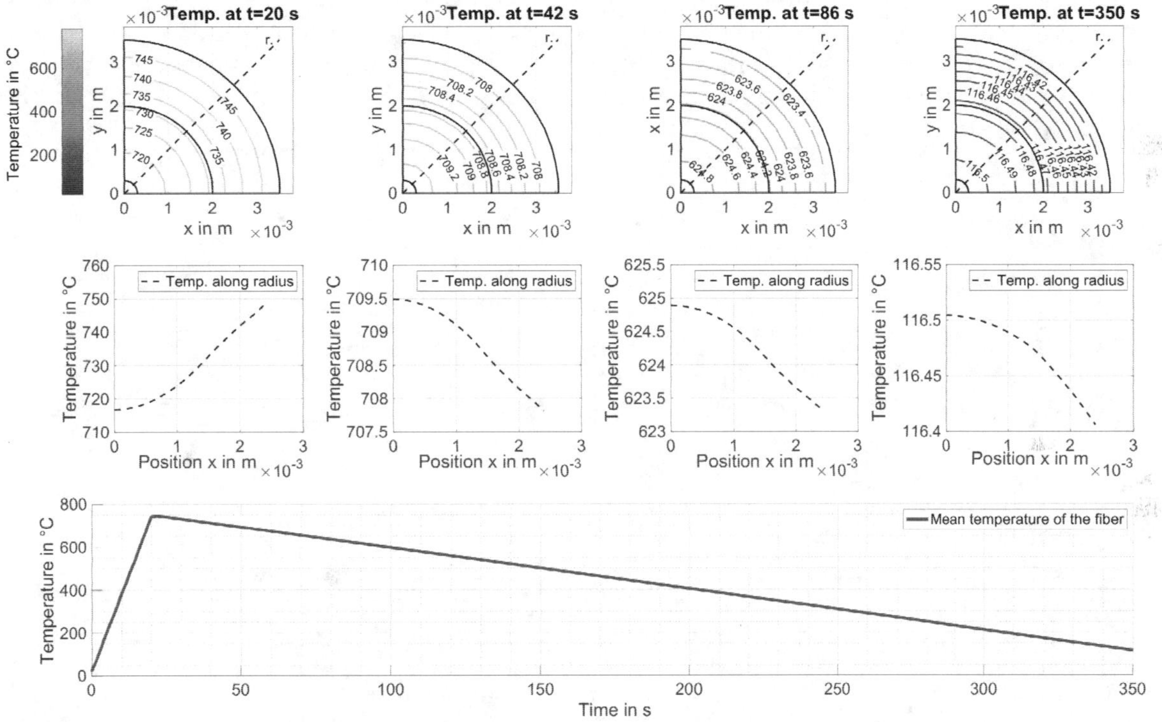

Figure 2. Temperature graphs from the partial differential equation simulation showing the temperature distribution of the four different cell regions at specific time steps, the temperature along the radius and the mean temperature of the fiber over time. [21]

2.2. Structural analysis

The same model as before is used to perform a quasistatic structural analysis to compute the thermal stresses at several states of time. To accomplish this, the results of the selected time steps are copied to the structural model and structural properties are applied. The characteristic values for material properties are listed in table 2.

The model is constrained to prevent rigid body motion, i.e. The surfaces of the crucible and every bottom surface are fixed. The structural model's reference temperature is assumed to be

Table 2. Material properties used to analyse the structural model.[25] [26]

	Youngs modulus (MPa)	Poissons ratio (-)	Thermal expansion coeeficient (1/°C)
Aluminum alloy	66080	0.3	2.15e-5
Graphite	500	0.33	1.00e-6
Fiber	95800	0.33	1.1125e-5

534°C. This is the temperature at which the alloy is around the solidus temperature [25] [26]. At this state, the stress in every direction is close to zero in the aluminum alloy because it is not solid anymore. The authors choose this point because they had specific values for this state. This causes an error but the use of the results give information to get a rough understanding of the structural behaviour of the model. Above this temperature the thermal expansion rapidly changes. In order to satisfy the symmetry boundary conditions, the remaining surfaces are constrained accordingly. The applied structural body load is the temperature field for the specific time step. Figure 3 and figure 4 show the results starting at the state after 20 s to the final state after 350 s. Figure 3 shows the axial strain along the z-direction. This directional

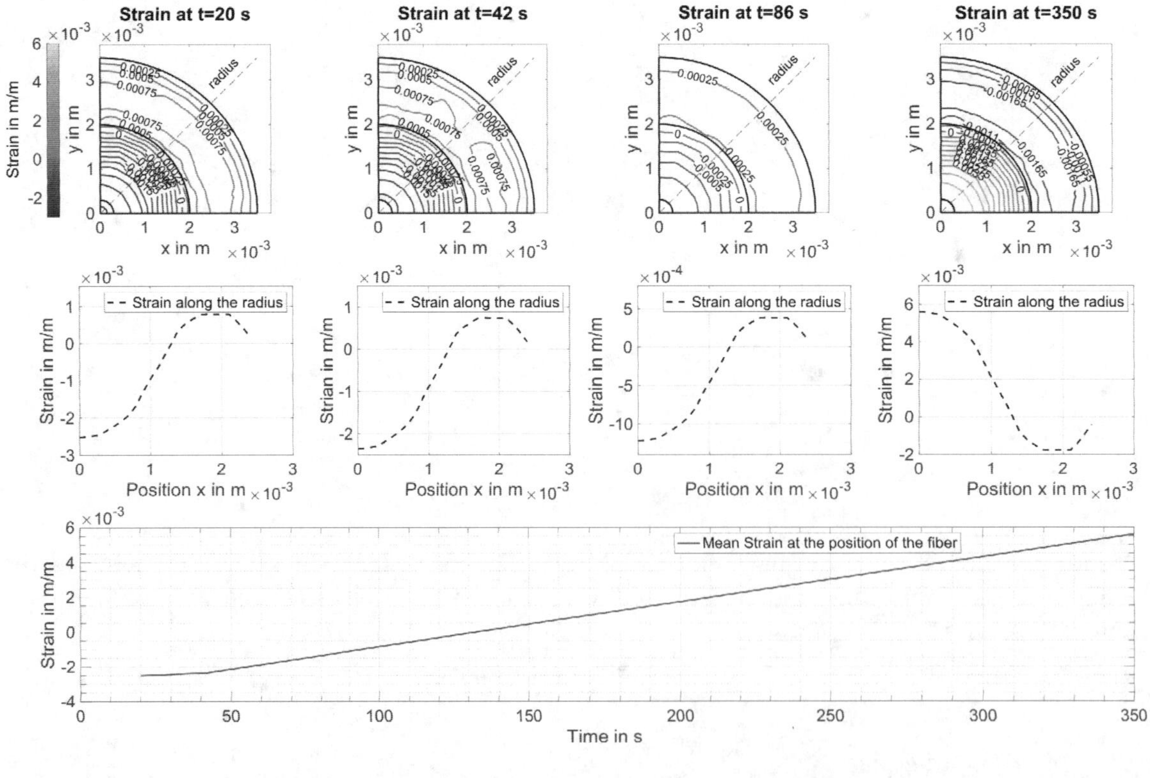

Figure 3. Strain graphs showing the distribution of axial strain at the center spatial location of the specimen, the axial strain along the radius and the overall strain occurring at the position of the fiber over time. [18]

strain is the main influencing value in the FBG measurement together with the temperature value of the fiber. In the final state at 350 s, the strain at the position of the fiber is positive.

Looking at figure 6 the wavelength shift of the fiber for specimen 4 is also positive, which is an expected result. For the remaining specimen the results have to be discussed in the next section. Looking at the Von Mises Stress, see figure 4, the overall result of the simulation can be summarized by stating that the stress builds up mostly driven by the process of solidification and the different structural properties of the materials. Especially towards the center regions, higher Von Mises Stress build up compared to the outside cells. Looking at the last time step of 350 s very high values for the Von Mises Stress are calculated. These values need be looked at carefully. MATLAB's partial differential toolbox does not work for temperature dependent material properties in the defined case. This causes the error to increase for higher temperature differences between the considered material properties temperature and the real temperatures. Another error leading to relatively high values at 350 s is the fact that the structural static-solid definition works best for small temperature differences. At time 350 s the temperature difference between the reference temperature and the temperature field from the heat transfer results are higher compared to the other states. Nevertheless, the results help to qualitatively understand, how the specimen reacts to the different temperature results. For example, the resulting positive strain at 350 s occur also in the FBG measurement of specimen number four, see figure 6, where the drift of the spectrum indicates this positive strain.

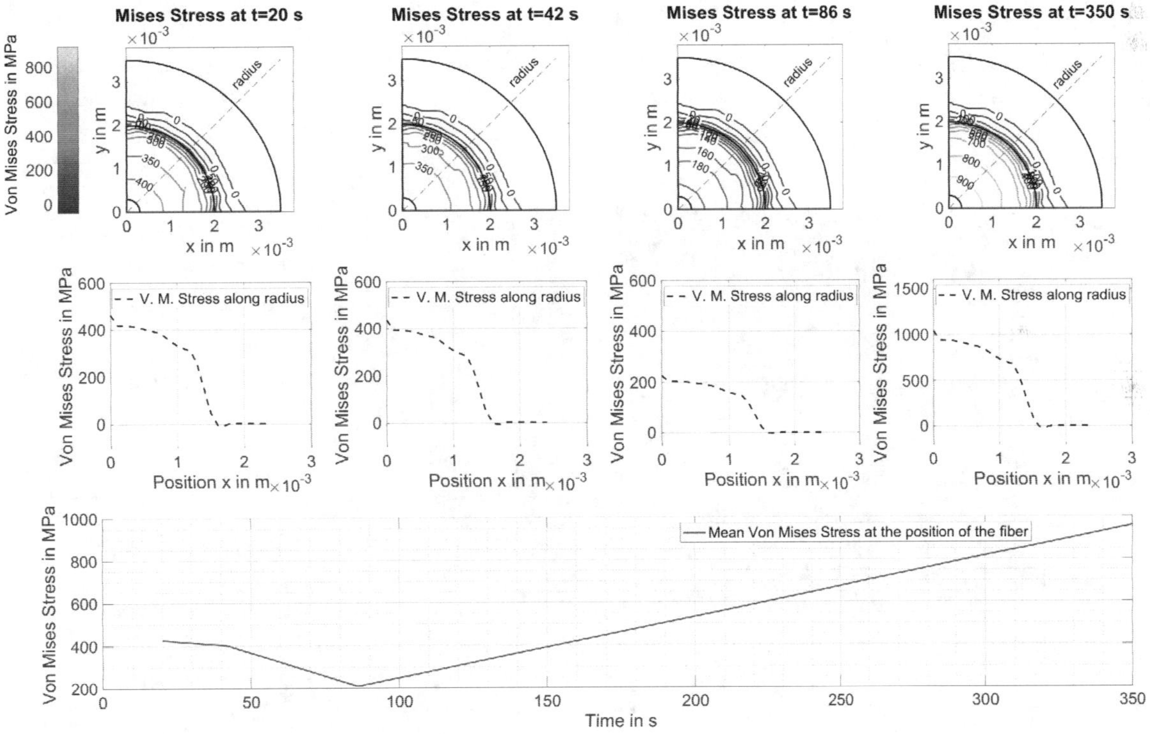

Figure 4. Von Mises graphs showing the distribution of Von Mises Stress at the center spatial location of the specimen, the Von Mises Stress along the radius and the overall Von Mises Stress occurring at the position of the fiber over time. [18]

3. Mirror furnace experiments

The alloy of interest is a typical aluminum cast alloy consisting of the following elements listed in table 3. From this material, a cylindrical specimen was milled including a hole with a diameter

Table 3. List of the percentage amount of elements inside the used aluminum alloy.

Al in %	*Si in %*	*Fe in %*	*Cu in %*	*Mn in %*	*Mg in %*	*others in %*
90.87	8.44	0.131	0.07	0.005	0.23	0.26

of 0.6 mm. Figure 5 shows the experimental setup. The figure 5 (a) shows the CAD drawing of the mirror furnace with the four halogen lamps placed in a mirror setup. These lamps heat up the specimen at the center of the furnace chamber. A photo of the heating process is shown in figure 5 (b). In order to be able to put the fiber through the aluminum alloy, the setup in figure 5 (c) is used. The different parts and the crucible are screwed together. In this way, the fiber is placed inside the aluminum with the FBG at the center of the cylinder. The experiment is divided into two steps. The first step is the process of melting. In this step the alloy liquefies, the melt closes the gap between the fiber and the alloy to create a compound structure. The second step is the solidifying of the melt. The data collected during these steps is shown in figure 6. The left ordinate is labeled with the wavelength in *nm*. The associated plots show the results measured in the three experiments. The right ordinate is labeled with the temperature in °C and corresponds to the orange curve. The abscissa of the graph is the time in seconds. The hysteresis in top right corner shows the change of the wavelength over the temperature.

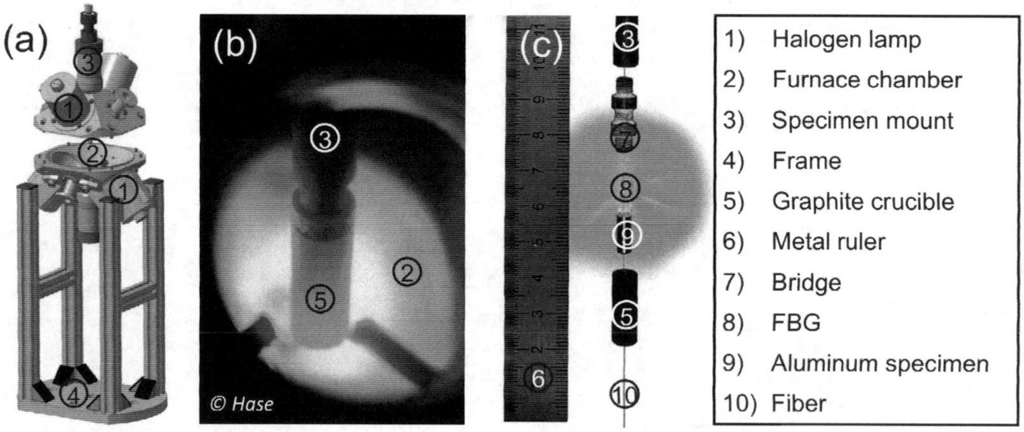

Figure 5. Experimental setup including the mirror furnace; (a), CAD drawing of the mirror furnace with its four halogen lamps; (b), Graphite crucible with specimen inside; (c), Installation of the FBG into the aluminum specimen

Looking at the legend, it can be seen that the graph shows three different specimen. These specimen are indicated with the numbers four, five and seven. In total seven fibers got prepared to conduct the experiments. During the first three experiments the authors realised a problem with the experimental setup. This caused the results of the measurement to be incomparable. The process was changed afterwards this led to an improvement which caused the specimen to be comparable. The following list gives an overview of the specimen. The loss of one of the fibers also shows the difficult preparation of the experiment.

- Original process: Specimen number 1-3
- Improved process: Specimen number 4-7
- Broken fiber: Specimen number 6

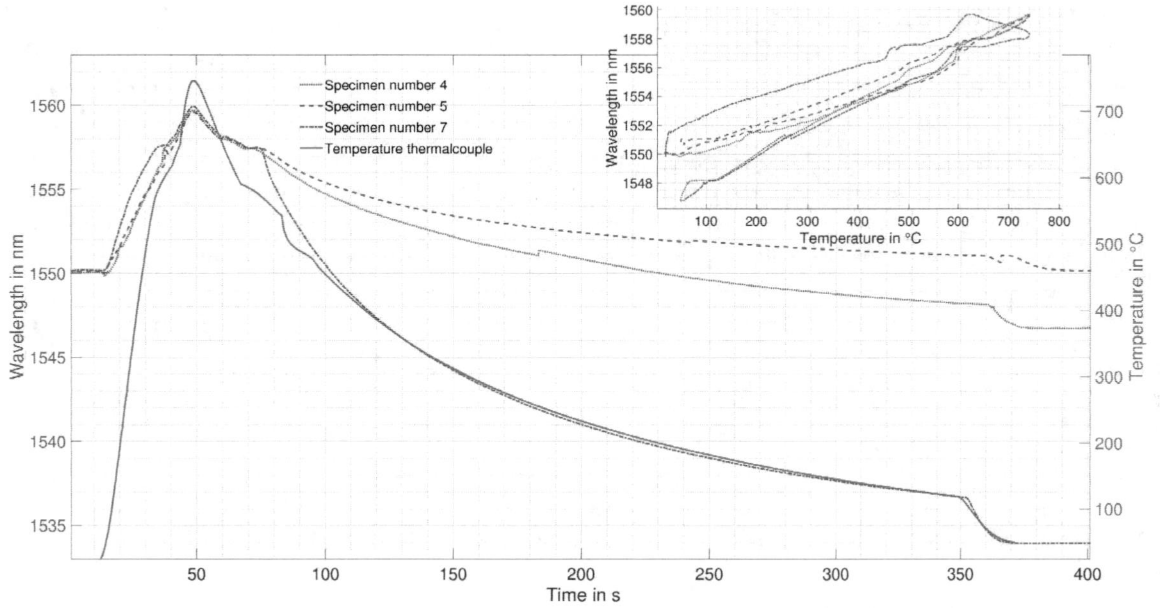

Figure 6. Showing the wavelength change of the FBG during the melting and solidification process over time of the cylindrical aluminum body onto the silica fiber.

3.1. Melting

As can be seen in figure 7, the process occurs during the time interval of 22 s to 49 s and between the temperature interval of about 21°C to 740°C. The dotted curves representing the FBG measurements show a similar behaviour in this specific interval. At the beginning the fiber is free to move inside the specimen. Therefore, strain can not be transferred from the alloy to the fiber. The resulting elongation along the axial direction is a result of the temperature increase of the fiber. Especially the phase change of the alloy, resulting in an endothermic reaction at the time 35 seconds, is indicated by the slight drop of the gradient in the measured wavelength. At this state, there is also no strain influencing the FBG, despite the alloy now being in contact with the fiber, since the alloy is still liquid at this point. The reason of the wavelength shift is a result of the increasing temperature, which is conducted from the heat source to the fiber. The reference measurement uses a thermalcouple and shows that the phase change occurs at 620°C. The curves of the FBG measurement and the thermalcouple show a similar behaviour. This indicates that the FBG measurement is appropriate for measuring the temperature, if calibrated at this specific time. At 746°C the mirror furnace stops heating and the melting process ends.

3.2. Solidification

Through the graphite crucible, heat is conducted to the outer surface. As soon as the halogen lamps are switched off, the system continuously cools down. This happens between the time interval of 49 s to 350 s and between the temperature interval of about 740°C and 120°C. Beyond this point the mirror furnace uses a gas lance to quickly cool down the specimen using a stream of argon. While cooling down, strain builds up between the fiber and the alloy resulting in a shift of the wavelength. Looking at specimen number seven this shift is stronger than the shift measured while heating. This difference is comes from the residual stress remaining in the specimen.

3.3. Remelting

As the solidification ends, the specimen remains in the mirror furnace together with the fiber. At this state, the fiber and the alloy build a compound structure. As such, they are heated up again to perform cyclic measurements. Figure 7 shows the three curves of the specimen number four, five and seven. The double arrows indicate the residual stress remaining in the specimen

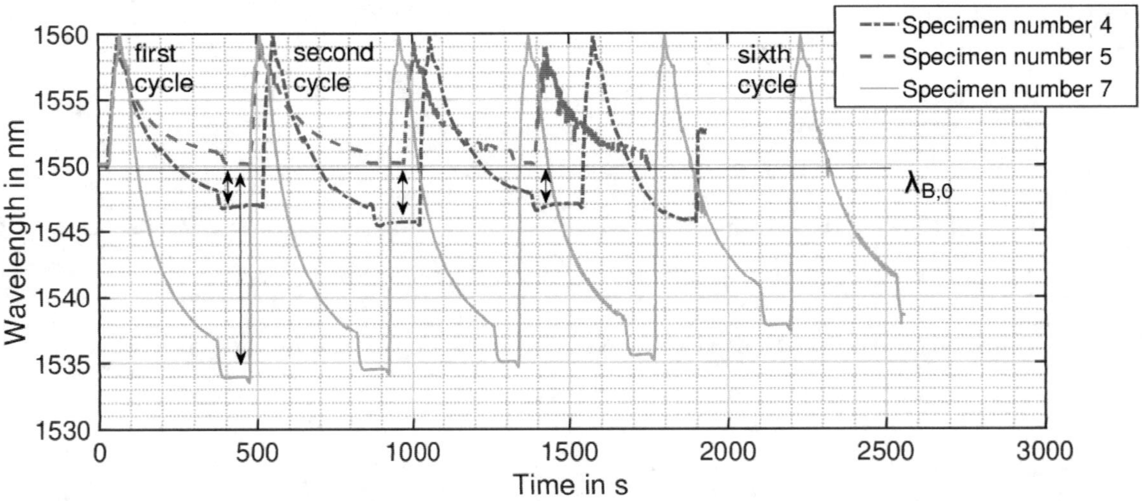

Figure 7. Showing the wavelength change of the three specimen over time for several cycles.

number five and seven after each solidification. Over each cycle the residual stress built up by specimen number seven decreases. Looking at specimen number four an opposite behaviour in the remaining stress is observed. The reason for this result can possibly be the fact that the FBG was not placed accurately at the same position as specimen number seven. This error is caused by the challenging installation process shown in figure 5 (c). The remelting of the specimen eventually causes all three FBGs to break at some stage. For specimen number five this happens after the fourth cycle. Looking at the last cycle the breaking of the fiber is already indicated by strong noise in the measurement. Specimen number seven on the other hand breaks after six cycles. This behaviour is promising, because it shows that with FBGs several remelting cycles can be conducted. The maximum peaks which occur at the highest temperature of the specimen have a very similar wavelength which is around 1560 *nm*. This indicates that the mirror furnace accurately heats the specimen to the same temperature in every cycle. This shows that the oven with the halogen lamps is an appropriate controlled environment. This leads to the assumption that at hottest stage all specimen were in the same initial situation. Otherwise the maximum peak wavelength would differ.

4. Conclusion and Future Scope

Research in the past always focuses on the measurement of the solidification using FBGs. The new insight gained by the experiments is that FBG measurements are in fact capable of collecting data of the melting, solidification and remelting process of the fiber alloy compound. However, remelting increases the chance of breaking the fragile fiber. The maximum number of achieved remelting processes was six. Thereby, the mirror furnace proves to provide a controlled environment in order to regulate the outside heatflux and the cooling rate. Thus the experiments can be repeated accurately. In combination with the transient multi domain heat transfer simulation and the structural finite element model, the behaviour of the interaction between the fiber and the alloy is described. This model still needs improvement to give accurate results. Since this model was implemented in MATLAB it can be used as a base for further enhancements. The remelting of the fiber alloy compound was not investigated in the past, the novel results could be substantiated in further experiments. One interesting field for experimental analysis using FBG is the development of hot tears occurring during solidification. Other experiments, which will be conducted with FBGs rely on the approved fact that this work shows that the cyclic remelting using FBGs inside aluminium castings is possible.

Acknowledgments

A special thanks to the German Research Foundation (*DFG*) for supporting this research and for providing measurement equipment, such as the interrogator under grant 456662088.

References

[1] Campbell J. In: Complete casting handbook: metal casting processes, metallurgy, techniques and design. Butterworth-Heinemann; 2015. p. 945.

[2] Flemings MC. Solidification Processing. Materials Science and Engineering Series. McGraw-Hill; 1974.

[3] Saal PG. Quantitative Phasenanalyse von ausferritischem Gusseisen mithilfe der Neutronendiffraktometrie [Dissertation]. München: Technische Universität München; 2017.

[4] Heilmeier F, Koos R, Hornberger P, Hiller J, Weraneck K, Jakobi M, et al. Calibration of cast-in fibre Bragg gratings for internal strain measurements in cast aluminium by using neutron diffraction. Measurement. 2020;163:107939. Available from: https://www.sciencedirect.com/science/article/pii/S0263224120304772.

[5] Corre Y, Laffont G, Pocheau C, Cotillard R, Gaspar J, Roussel N, et al. Integration of fiber Bragg grating temperature sensors in plasma facing components of the WEST tokamak. Review of Scientific Instruments. 2018;89(6):063508.

[6] Heilmeier FM. Ermittlung schwindungsbedingter Gussteilspannungen mit Hilfe eingegossener, faseroptischer Dehnungssensoren [Dissertation]. München: Technische Universität München; 2020.

[7] Bian Q, Bauer C, Stadler A, Buchfellner F, Jakobi M, Volk W, et al. Monitoring strain evolution and distribution during the casting process of AlSi9Cu3 alloy with optical fiber sensors. Journal of Alloys and Compounds. 2023;935:168146.

[8] Heilmeier F, Koos R, Singer M, Bauer C, Hornberger P, Hiller J, et al. Evaluation of Strain Transition Properties between Cast-In Fibre Bragg Gratings and Cast Aluminium during Uniaxial Straining. Sensors. 2020;20(21). Available from: https://www.mdpi.com/1424-8220/20/21/6276.

[9] Gdoutos EE. In: Optical Fiber Strain Sensors. Cham: Springer International Publishing; 2022. p. 219-33. Available from: https://doi.org/10.1007/978-3-030-89466-$5_1$0.

[10] Gafsi R, El-Sherif MA. Analysis of induced-birefringence effects on fiber Bragg gratings. Optical fiber technology. 2000;6(3):299-323.

[11] Lindner M, Stadler A, Hamann G, Fischer B, Jakobi M, Heilmeier F, et al. Fiber Bragg Sensors Embedded in Cast Aluminum Parts: Axial Strain and Temperature Response. Sensors. 2021 Mar;21(5):1680.

[12] Singh S, Jain PK, Rizwan-uddin Ru. Analytical solution for three-dimensional, unsteady heat conduction in a multilayer sphere. Journal of Heat Transfer. 2016;138(10).

[13] Burstein L. PDE Toolbox Primer for Engineering Applications with MATLAB Basics. Milton: Taylor & Francis Group; 2022.

[14] Baehr H, Stephan K. Heat and Mass Transfer, Springer. Berlin, Heidelberg, New York. 2006.

[15] Böckh Pv. Wärmeübertragung. 6th ed. Wetzel T, editor. Berlin ; Heidelberg: Springer Vieweg; 2015.

[16] Bronstein IN MGMH Semendjajew KA. Handbook of Mathematics, Springer. 2007, 5th edition.

[17] Gnielinski V. Berechnung mittlerer Wärme-und Stoffübergangskoeffizienten an laminar und turbulent überströmten Einzelkörpern mit Hilfe einer einheitlichen Gleichung. Forschung im Ingenieurwesen A. 1975;41(5):145-53.

[18] Adamowicz A. AXISYMMETRIC FE MODEL TO ANALYSIS OF THERMAL STRESSES IN A BRAKE DISK. Journal of Theoretical and Applied Mechanics. 2015;53(2). Available from: http://ptmts.org.pl/jtam/index.php/jtam/article/view/1965.

[19] Lurie AI. Theory of elasticity. Springer Science & Business Media; 2010.

[20] Palaniappan D. A general solution of equations of equilibrium in linear elasticity. Applied Mathematical Modelling. 2011;35(12):5494-9. Available from: https://www.sciencedirect.com/science/article/pii/S0307904X11000552.

[21] MATLAB PDT. version 9.12.0 (R2022a Update 4). Natick, Massachusetts United States: The MathWorks Inc.; 2022.

[22] Lemmon EW, Jacobsen RT, Penoncello SG, Friend DG. Thermodynamic Properties of Air and Mixtures of Nitrogen, Argon, and Oxygen From 60 to 2000 K at Pressures to 2000 MPa. Journal of Physical and Chemical Reference Data. 2000;29(3):331-85. Available from: https://doi.org/10.1063/1.1285884.

[23] V VDI. VDI-Wärmeatlas. VDI-Wärmeatlas. Springer Berlin Heidelberg; 2013.

[24] Wagner W. New vapour pressure measurements for argon and nitrogen and a new method for establishing rational vapour pressure equations. Cryogenics. 1973;13(8):470-82. Available from: https://www.sciencedirect.com/science/article/pii/0011227573900039.

[25] Bäckerud L, Król E, Tamminen J. Solidification Characteristics of Aluminium Alloys Vol. 1: Skanaluminium. Universitetsforlaget AS, Oslo, Norway. 1986.

[26] Richter F. Die physikalischen Eigenschaften von metallischen Werkstoffen. Metall (Berlin, West). 1991;45(6):582-9.

Reconstruction of dendritic growth by fast tomography and phase field filtering

H Yasuda[1], A Nishiguchi[1], H Xue[1], M Nonomura[1], K Nakano[1], R Katsube[1], T Narumi[1], T Takaki[2]

[1] Department of Materials Science and Engineering, Kyoto University, Sakyo, Kyoto 6068501, Japan
[2] Faculty of Mechanical Engineering, Kyoto Institute of Technology, Sakyo, Kyoto 6068585, Japan

yasuda.hideyuki.6s@kyoto-u.ac.jp

Abstract. Three dimensional models of dendritic structures during solidification are valuable for building physical models, validating simulated results, estimating some properties such as permeability in the mushy, simulating semisolid deformation and so on. Thus, it is of interest to observe microstructure evolution in situ. Time-resolved tomography combined with X-ray diffraction has allowed us to observe the evolution of dendritic structures and to measure crystallographic orientation in situ. Reconstruction still proves to be difficult for some alloy systems because of the tradeoff between time and spatial resolution. This paper demonstrates the reconstruction of dendritic structures for three different alloy systems (Al-10mass%Cu alloy with a diameter of 4 mm, CrMnFeCoNi alloy with 1 mm, and Zn-4mass%Al alloy with 0.7 mm). The observations were performed in a synchrotron radiation facility SPring-8. A filter using a phase field model was introduced to reconstruct the three-dimensional images. Parameters used in the filtering were consistently determined based on the raw reconstruction images. Evaluation of solid-liquid interface area and curvature was significantly improved by the filter. For the Al-Cu alloy, a three-dimensional model containing approximately 300 million voxels was obtained. For the CrMnFeCoNi alloys, the preferred growth direction <100> was confirmed by tomography and X-ray diffraction. For the Zn-Al alloy, the observed 14 growth directions were not simply defined by the crystallographic orientations, although the directions were consistent with the hexagonal symmetry. This study verifies that time resolved tomography, X-ray diffraction and the filter using a phase field model provide three dimensional models for light metal alloys with rather large diameters and 3d transition-metal alloys with rather large X-ray absorption coefficients. The models are expected to be used for further studies.

1. Introduction

Synchrotron radiation X-rays with high brilliance and coherency have allowed us to observe solidification and related phenomena for various alloy systems over the last two decades. So far, observable alloy systems have been extended from low-melting-temperature alloys (Sn, Zn, Al, Mg alloys) [1-8] to high-melting-temperature alloys (copper, Ni-based superalloy, cast iron, carbon steel, stainless steel)[9-12]. Even a reactive Ti alloy has been achieved [13]. Time-resolved and in situ observations have been recognized as valuable techniques for investigating solidification at high temperatures. Radiography also revealed the deformation of the semisolid [14, 15]. An advantage of time-resolved in situ observation is understanding how solidification/transformation proceeds without a preconceived mold. For example, it was pointed out that a massive-like transformation rather than the peritectic reaction/solidification was dominantly selected in the peritectic steels [10, 16, 17]. In addition, the massive-like transformation induces dendrite fragmentation and consequently influenced the austenite grain structure [18].

Because the three-dimensional observations provide significant information on the microstructure evolution in the three-dimensional space, time-resolved in situ tomography (referred to as 4D-TC) has been developed [19-34]. For example, the late stage of solidification of Al-4mass%Cu alloys (1 mm in diameter, cooling rate of 0.1 K/s) was observed using a pink beam (peak energy of 40 keV) to increase the transmission X-ray intensity [19]. Melting and solidification for Al–Mg–Si-Y_2O_3 alloys were also observed using a pink beam (peak energy of 17.6 keV) [21]. 4D-CT using a monochromatized X-ray (30 keV) was performed for Al-15mass%Cu alloys and the observed microstructure was compared with the phase field simulation [20]. The combination of in situ observation and phase field simulation opens new avenues for understanding solidification phenomena. The time-interlaced model-based iterative reconstruction method, which improves time resolution, has been applied to observe dendrite morphology during the early stage of solidification in Al-24mass%Cu alloys (1 mm in diameter) [22]. 4D-CT observations found transgranular liquation cracking of solid grains in the semisolid state [23]. 4D-CT combined with three-dimensional X-ray diffraction for determining crystallographic orientations of solid grains (3D-XRD) has been used to analyze the translational and rotational motion of solid grains during semisolid deformation [32]. Adding a high attenuation element such as Hf enhances the difference in transmission X-ray intensity between the solid and the liquid phases (referred to as contrast), and solidification in the Hf-containing alloys was observed in situ [24, 27]. Thus, 4D-CT has also been recognized as an observation technique to investigate solidification phenomena.

Particular attention has been paid to image processing in 4D-CT analysis [29, 31]. Our previous study reported a preliminary result for image processing of dendrites in Fe-C alloys and proposed a filtering technique using a phase field model [29]. We have focused on developing image processing based on the proposed method because it is rather difficult to reconstruct the three-dimensional structure without artifacts.

Our motivation for the development of image processing is briefly explained here. Sufficient intensity transmitted through a specimen (transmission intensity) is required for obtaining projection images. Simultaneously, the difference in the transmission intensities between the solid and liquid phases (contrast) is critical for reconstructing the three-dimensional structure from the projection images.

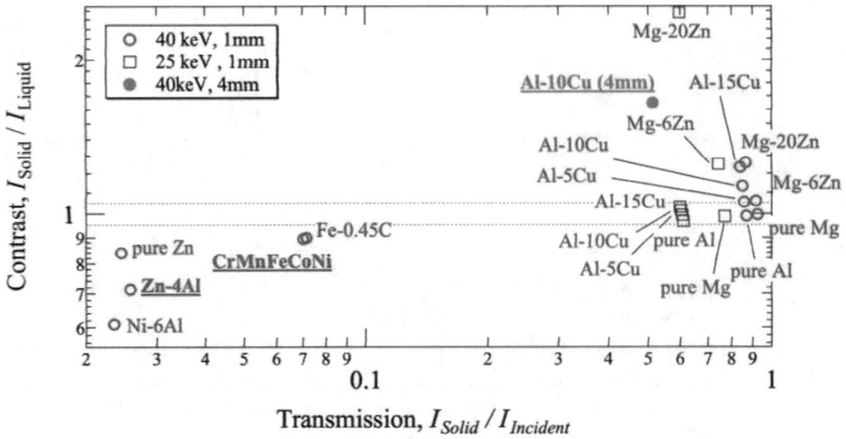

Figure 1. Contrast between solid and liquid phases and intensity of transmission. The alloy systems indicated by underlining are demonstrated in this paper.

Thus, it is essential to optimize the X-ray energy and specimen diameter. The X-ray intensity transmitted through a specimen with a thickness, t, is given by

$$I = I_{\text{Incident}} \cdot \exp[-\bar{\mu}t]. \quad (1)$$

Here, $\bar{\mu}$ is linear absorption coefficient depending on X-ray energy, density, and alloy composition. The linear absorption coefficient, $\bar{\mu}$, for an alloy system is given by

$$\bar{\mu} = \rho \sum_i w_i \left(\frac{\mu}{\rho}\right)_i. \quad (2)$$

Here, ρ is the density of the specimen and w_i and $(\mu/\rho)_i$ are the mass fraction and mass absorption coefficients of the i-th constituent element, respectively. As mentioned before, the feasibility of 4D-CT is considered in terms of transmission intensity and contrast. Figure 1 shows the ratio (contrast), $I_{\text{Solid}}/I_{\text{Liquid}}$, as a function of the intensity transmitted through a solid phase (1mm in thickness). The contrast is simply expressed by

$$\frac{I_{\text{Solid}}}{I_{\text{Liquid}}} = \exp[-(\mu_{\text{Solid}} - \mu_{\text{Liquid}})t]. \quad (3)$$

The values were calculated using a database of X-ray mass attenuation coefficients [35-37]. It should be noted that the contrast in the projection images of 4D-CT is much smaller than the contrast given by Eq. (3) because X-rays penetrate solid grains and liquids in the transmission direction. In the case of 25 keV, the transmission intensity through Al and Mg-based alloys is sufficiently high, although the contrast depends on the alloy system. Because of the larger density of the solid phase and the partition of copper into the liquid phase in Al-Cu alloys, the contrast at 25 keV is low. Thus, performing 4D-CT for the Al-Cu alloys at 25 keV is unsuitable. Although some exceptions exist, such as the Al-Cu alloys, lower X-ray energy is generally suitable for light metal alloys. If the contrast is sufficiently large, the reconstruction is not critical.

The transmission for transition metals of period 3 in the periodic table is less than 0.01 at 25 keV. Thus, higher X-ray energies must be selected for Fe, Ni, Cu, and Zn alloys with a diameter of 1 mm. Note that a diameter of approximately 1 mm is generally suitable for observing dendritic growth. Even at 40 keV, the transmission, $I_{\text{Solid}}/I_{\text{Incident}}$, is less than 0.1. Although the transmission increases with increasing X-ray energy, the contrast generally decreases. In addition, the available energy range can be limited by X-ray optics. Thus, a balance between transmission and contrast is necessary within possible conditions. From our experience, the transmission and the contrast are not sufficiently large. However, it is still possible to obtain the projection images of 4D-CT at an X-ray energy of approximately 40 keV.

In a specimen with a large diameter, the interactions between the growing dendrites can influence the microstructure evolution. Observation of solidification in Al-Cu alloys with a diameter of 4 mm, as indicated by the green point in Figure 1, is expected to observe the influence. A critical issue arising in the observation is to avoid the degradation of reconstructed images and/or the decrease in time resolution because many dendrite arms exist in the cross-section. Therefore, developing a novel reconstruction procedure is valuable for observing solidification structures in transition metal alloys (typically 1 mm in diameter) and in light metal alloys (more than several mm in diameter).

In this paper, an image-filtering technique using a phase field model is demonstrated. In addition, this paper presents three-dimensional observations for an Al-10mass%Cu (4 mm in diameter), equimolar CrMnFeCoNi high entropy alloy (1 mm in diameter), and a Zn-4mass%Al (0.7 mm in diameter).

2. Experimental procedures

2.1. Time-resolved tomography

4D-CT experiments were performed at beamlines BL20XU, BL47XU and BL20B2 of SPring-8 (Synchrotron radiation facility, Hyogo, Japan). An X-ray of 37.7 keV was obtained by a Si (111) double-crystal monochromator at BL20XU and BL47XU. Typical beam sizes at BL20XU and BL47XU are approximately 3 mm (horizontal) × 1 mm (vertical) and 1 mm × 0.6 mm, respectively. An X-ray of 40 keV, obtained by a multilayer reflection spectrometer at BL20B2, was used for the Al-10mass%Cu alloys (4 mm in diameter). The beam size was approximately 5 mm × 5 mm.

Figure 2 shows a schematic illustration of 4D-CT combined with X-ray diffraction (referred to as XRD) [38-40]. The sample was heated by a graphite heater in a vacuum chamber (0.1-10Pa). Transmission images were observed by a beam monitor (sample-detector distance: 0.3-0.5 m). XRD was simultaneously observed by another beam monitor (pixel: 0.1 mm × 0.1 mm, frame rate: 30 fps) to determine crystallographic orientation. Observation conditions for projection images of 4D-CT are listed in Table. 1.

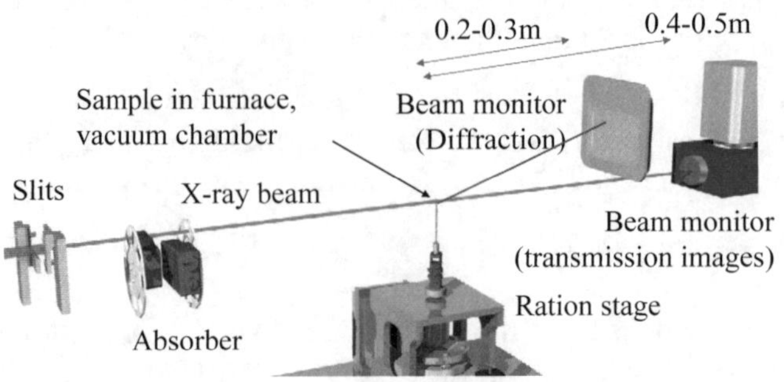

Figure 2. Setup of 4D-CT.

Table 1. Observation conditions of 4D-CT.

Alloy	Diameter	Cooling	X-ray energy	Voxel	Rotation	Frame rate
Al-10mass%Cu	4 mm	20 K/min	40 keV	$(5.5\ \mu m)^3$	0.125 rps	120 fps
CrMnFeCoNi	1 mm	5K/min	37.7 keV	$(5.5\ \mu m)^3$	0.25 rps	100 fps
Zn–4mass%Al	0.7 mm	20 K/min	37.7, 40 keV	$(5.5\ \mu m)^3$	0.5 rps	400 fps

2.2. Reconstruction of three-dimensional images

As a conventional reconstruction method, convolution back-projection (CBP) image reconstruction was performed using projection images over 180° rotation. As listed in Table 1, 480, 200, and 200 frames were used for the Al-Cu alloys, CrMnFeCoNi alloys, and Zn-Al alloys, respectively. A preliminary trial of a filter using a phase field model (referred to as phase field filtering) was reported in a previous study [39]. As mentioned in the previous study [39], the solid-liquid interface in the dendritic structure could be traced by using conventional smoothing filters. However, the remaining noise degraded the evaluation of the solid-liquid interface area and curvature significantly.

The phase field model includes the curvature effect [41-44]. A governing equation of the phase field parameter, ϕ, is generally given by

$$\partial\phi/\partial t = -M[6\phi(1-\phi)\Delta f + 2\phi(1-\phi)(1-2\phi)W - \varepsilon^2\nabla^2\phi]. \quad (4)$$

Here, M is the interface mobility, and Δf is the difference in Gibbs energy between solid and liquid. W and ε are parameters determined by the solid-liquid interface energy (σ) and the thickness of the interface (δ). In this study, $W = 3\sigma/\sqrt{2}\delta$ and $\varepsilon^2 = 3\delta\sigma/2.2$ were used. A radius r was introduced to evaluate the Gibbs energy difference:

$$\Delta f = -2\sigma/r. \quad (5)$$

Although the introduction of the radius is not intrinsic, the radius is easily estimated by considering the size of dendrite and the curvature effect. For image processing, the governing equation given by eq. (4) was transformed into the following equation:

$$\Delta\phi = -(M\Delta t)^*\left[-\frac{4\times 2.2}{\delta_e r_e}\phi(1-\phi) + \frac{2.2\sqrt{2}}{{\delta_e}^2}\phi(1-\phi)(1-2\phi) - {\nabla_e}^2\phi\right]. \quad (6)$$

δ_e and r_e are the interface thickness and the radius normalized by the edge length of voxels (Δx). Δt is time step. Because the interface energy is included only in the term $(M\Delta t)^* = 3\delta_e\sigma(M\Delta t)(\Delta x)^3/2.2$, the interface energy does not explicitly influence image processing. As the filtering is not time evolution,

$(M\Delta t)^*$ is a parameter which determines convergence speed. Thus, the phase field filtering is essentially performed by the two parameters, radius r_e and thickness δ_e. In principle, it is better to adopt a larger δ_e, because a thicker interface with respect to the voxel size well-represents the solid-liquid interface in the phase field model. In the reconstruction of 4D-CT, the number of voxels can be as large as 10^{10} if a fine mesh and a thick interface are used. In this study, a thickness δ_e of 1.5-4 was adopted to reduce the calculation cost. Details in the phase field filtering are demonstrated in the next section.

3. Results and discussion

3.1. Al-10mass%Cu alloy

A part of the reconstructed images was extracted to determine the parameters used in the phase field filtering. In the present case, the center region containing 250 × 250 × 250 voxels (15 million voxels) was extracted to determine the parameters. Figure 3(a) shows a raw reconstructed image of the Al-10mass%Cu alloy with a diameter of 4 mm. Dendrite arms (dark) in the liquid phase (light) are recognizable in the image. However, it is difficult to trace the solid-liquid interface precisely. The distribution of brightness, which is proportional to the X-ray linear absorption coefficient, is shown in Figure 3(b). The peak position of brightness for the liquid phase was determined approximately by analyzing the reconstructed image immediately before solidification. The peak position for the solid phase was determined in the same way. The positions were used as a reference to fit the distribution to double Gaussian functions. Figure 3(b) also shows the fitted curves (double Gaussian functions). The solid fraction was estimated to be 0.51 by integrating the two Gaussian functions. This value was used as the restraint condition in the phase field filtering.

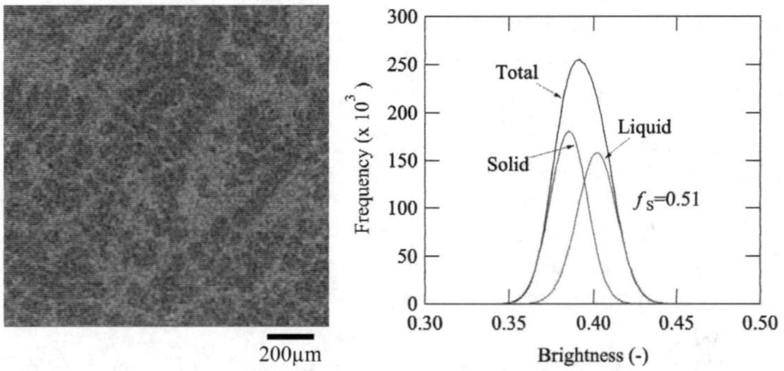

Figure 3. (a) Raw reconstructed image and (b) distribution of brightness (linear absorption coefficient).

In the first step, the calculations of the phase field parameter were performed using Eq. (6) for various radiuses, r_e. Figure 4 shows the change in the solid fraction during the iteration calculation. In the case of $r_e = -256\ (1/r_e = 3.9 \times 10^{-3})$, the solid fraction slightly increases and then nearly converges to 0.5. Thus, the radius which satisfies the restraint condition ($f_S = 0.51$) was determined as the parameter.

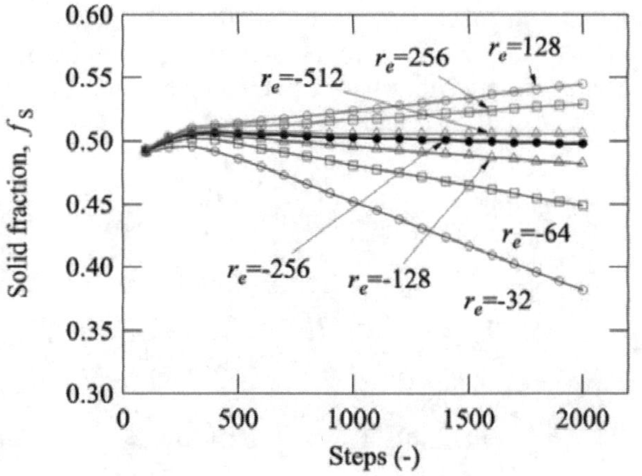

Figure 4. Change in solid fraction as a function of iteration number in phase field filtering.

In the next step, an appropriate iteration number is determined. Figure 5 shows slice images for $r_e = -256$ as a function of the iteration number. The black and the white correspond to the liquid ($\phi = 0$) and the solid ($\phi = 1$),

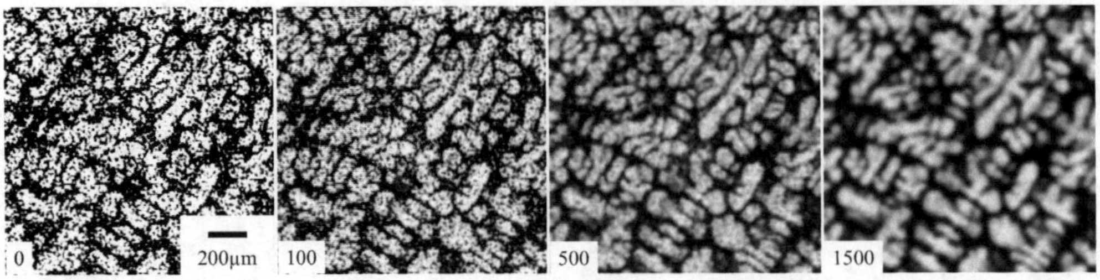

Figure 5. Slice images during the phase field filtering as a function of iteration number.

respectively. As the iteration number increases, the dark spots in the dendrite arm and the light spots in the liquid phase, which are artifacts induced in the reconstruction, are removed, and the solid-liquid interface tends to be smooth. The interface is blurred at an iteration number of 1500 because the phase field parameter continuously changes from 0 to 1 across the interface. The distribution of curvature, defined by $K = \kappa_1 + \kappa_2$ (κ_1 and κ_2 are the principal curvatures), is useful as a criterion for determining the iteration number. Figure 6 shows the distribution of curvature as a function of iteration number. Sharp peaks, which is generated by the voxel-scale noise in the orthogonal coordinate, can distinctly be observed at an iteration number of 2. The 100-iteration calculation almost removed the peaks, but curvature ranges from $10^6 m^{-1}$ to $10^6 m^{-1}$. A curvature of $0.5 \times 10^6 m^{-1}$(curvature radius of $4 \times 10^{-6} m$) is too large because the edge length of the voxel ($5.5 \times 10^{-6} m$) cannot depict curves with such a large curvature. As the iteration number increases, the distribution becomes sharper without the artifact peaks but slightly differs from the Gaussian distribution. The mean curvature and the half maximum full width at 1500 are $1.4 \times 10^4 m^{-1}$and $1.2 \times 10^5 m^{-1}$, respectively. In the present case, the iteration number of 1500 was selected in the phase field filtering, because the change in mean curvature in the iteration becomes small and the distribution is sufficiently small for analysis.

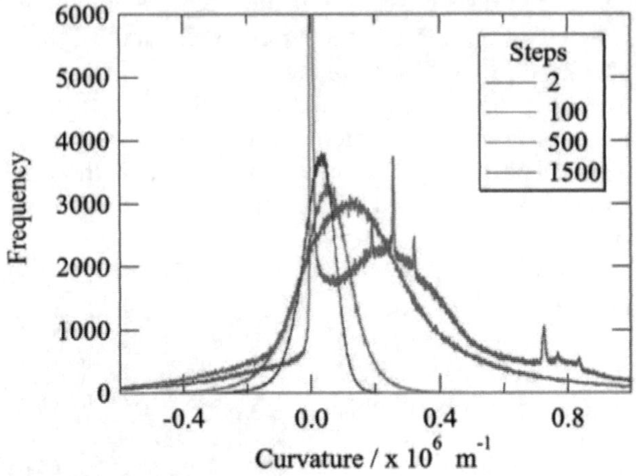

Figure 6. Distribution of interface curvature. Steps in the figure mean iteration number in the phase field filtering.

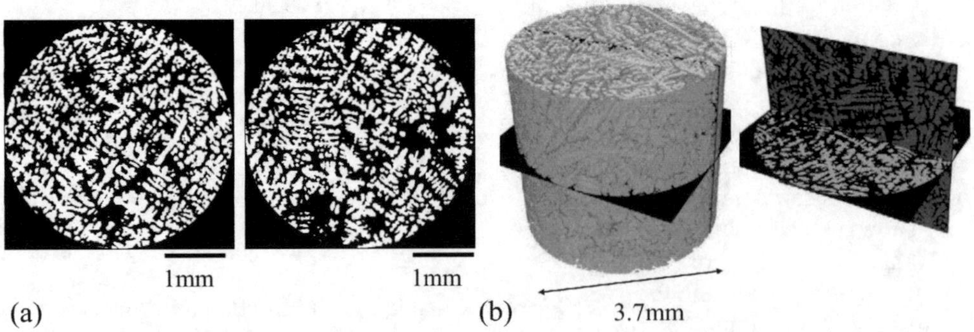

Figure 7. (a) Slice images and (b) Three-dimensional image after binarization.

In the final step, phase field filtering was performed for the entire specimen containing approximately 300 million voxels. Subsequently, a binarization procedure (solid phase: $\phi \geq 0.5$ and liquid phase: $\phi < 0.5$) was performed to build a three-dimensional model of solidifying microstructure. Figure 7 shows the cross-sections and the three-dimensional image after the binarization. The three-dimensional model, of which diameter is approximately 50 times larger than the secondary dendrite arms spacing, was reconstructed, and the solid-liquid interface area and the curvature were evaluated. The present study demonstrates that a standard routine consisting of a conventional reconstruction procedure (e.g., CBP) and phase field filtering allows us to build a three-dimensional model containing more than 100 million voxels. The models are expected to be used to examine melt flow in the mushy region and deformation of semisolids.

3.2. Equimolar CrMnFeCoNi alloy

4D-CT combined with XRD was performed to observe the dendrites and to determine the crystallographic orientation [39]. Figures 8(a) and (b) show three-dimensional images of growing dendrites in the equimolar CrMnFeCoNi alloy (Cantor alloy [45]) with a face-centered cubic (FCC) structure. The dendrite arms are perpendicular to each other, and the growing directions of the arms, indicated by ***p***, ***q***, ***r***, ***s***, ***t***, and ***u***, coincide with the <100> directions, as shown in Figure 8(c). A preferred growth direction of <100> has commonly been observed for typical FCC alloys [46].

According to a previous study [39], if the five constituent elements are randomly distributed in the solid and liquid phases, a random configuration—defined as a type of entropy effect—can reduce the interfacial anisotropy arising from atomic bonding. If this assumption is valid, we may attribute the preferred <100> growth direction of the CrMnFeCoNi alloy to the FCC structure itself.

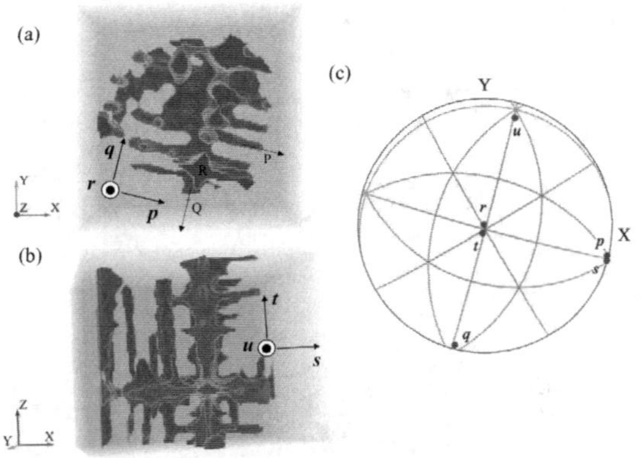

Figure 8. (a)(b) 3D images and (c) Crystallographic orientation of the dendrite

Figure 9 shows the specific interfacial area, S_V, which is defined as the interfacial area normalized by the total volume and the curvature as a function of the solid fraction at a cooling rate of 5 and 20 K/min. The maximum values of the specific interface area are in the order of 10^4 m^{-1}. It has been reported that the specific interfacial area is 2.8×10^4 m^{-1} in the Al-24 mass%Cu alloy solidified at 0.033 K/s [47] and 2.8×10^4 m^{-1} in the Mg-15 mass%Sn alloy solidified at 0.2 K/s [25]. In addition, the dendrite arm spacing is in the same order as those of conventional alloys [39]. Thus, no notable change has been observed for the CrMnFeCoNi alloy except for intense microsegregation.

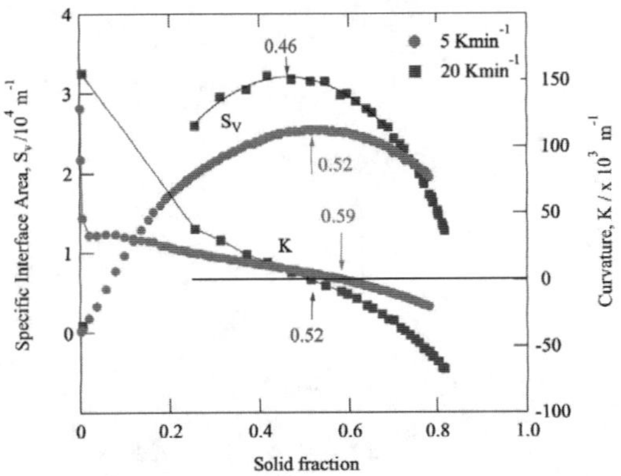

Figure 9. Specific interface area and curvature in the CrMnFeCoNi alloy. The data at 5 K/min was taken from ref [34].

3.3. Zn-4mass%Al alloy

It is of interest to determine the preferred growth directions of dendrites for investigating the anisotropy of the interface energy and for understanding the microstructure evolution as a result of the competition between the dendrites. The growth directions of a hexagonal closed-packed (HPC) alloys are rather complex compared to those of a body centered cubic (BCC) and FCC alloys. It was reported that the growth directions were mainly $< 10\bar{1}0 >$, and rarely $< 11\bar{2}0 >$ and $< 0001 >$ in Zn-0.2mass%Al alloys used for hot-dip galvanized steel [48]. The growth directions were determined by SEM observation and pole figures obtained by XRD. Recently, it has been pointed out that the large range of the inclination angle remains unexplored due to the difficulty [49]. Hence, a third family of $< 1\bar{2}11 >$ was proposed in addition to $< 10\bar{1}0 >$ and $< 0001 >$. This study also analyzed the preferred growth directions by SEM/EBSD. Namely, the directions were determined from the solidified structure. As an alternative method, 4D-CT combined with XRD is expected to provide new insights into the dendritic growth of hexagonal alloys.

Figure 10 shows equiaxed solidification in the Zn-4mass%Al alloy. The Zn-rich HCP phase is preferably nucleated on the surface and inside the specimen. Because the number of grains is too large, the crystallographic orientation for each grain could not be determined. The large X-ray absorption of Zn at 40 keV degrades the reconstruction images. However, the phase field filter improved the reconstructed images, and the dendrites can be recognized in the three-dimensional images.

Figure 11 shows an equiaxed grain that freely grew in the three-dimensional space of the specimen. The sixfold symmetry as shown in Figure 11(b) suggests that the directions correspond to $< 10\bar{1}0 >$ or $< 1\bar{1}20 >$. As shown in Figure 11(c), the sixfold arm tends to split toward the c-axis. As a result, 14 growth directions were identified in the Zn-4mass%Al alloy. The angle between the growth directions of the split arms is smaller than 30 degrees. The angle between the $[10\bar{1}1]$ and $[10\bar{1}\bar{1}]$ planes is 50 degrees, and the angle between the $[1\bar{2}11]$ and $[1\bar{2}1\bar{1}]$ planes is 30 degrees. Thus, the well-defined crystallographic directions cannot explain the split directions. The twofold symmetry of the a-c plane is relatively low, while the sixfold symmetry of the basal plane is high. The stiffness of the solid-liquid interface, $\tilde{\sigma}$ is defined as $\tilde{\sigma} = \sigma + \partial^2\sigma/\partial\theta^2$, the anisotropy of the stiffness in the a-c plane can be relatively small. It is worth discussing if the small anisotropy in the plane induces the split and consequently results in complicated growth directions. 4D-CT combined with XRD is required for further discussion.

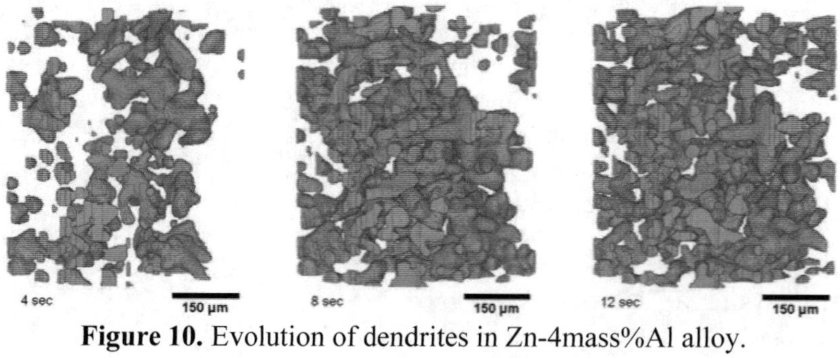

Figure 10. Evolution of dendrites in Zn-4mass%Al alloy.

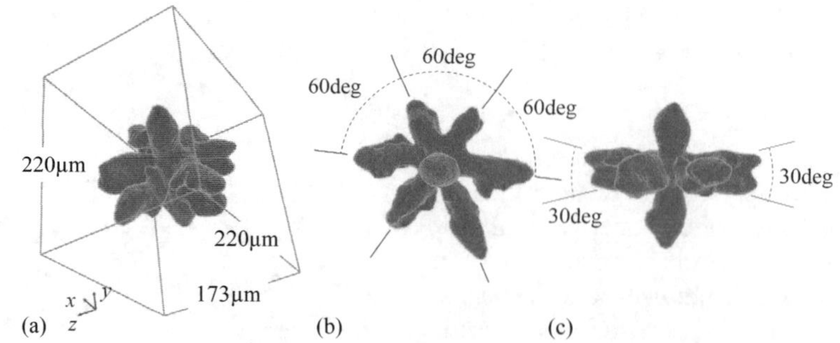

Figure 11. Equiaxed grain in Zn-4mass%Al alloy.

4. Summary

The reconstruction of a dendritic structure during solidification was performed using 4D-CT and phase field filtering. The parameters used in the phase field filtering process were consistently determined using raw reconstructed images. The filtering significantly improved the quality of the reconstructed images and consequently allowed us to evaluate the solid-liquid interface area and curvature. It should be noted that the filtering procedures did not depend on the person because the parameters were determined according to the rules presented in this paper. For an Al-10mass%Cu alloy with a diameter of 4 mm, three-dimensional models containing approximately 300 million voxels were obtained. The three-dimensional model will be used for modeling and simulation in the future. 4D-CT combined with XRD allowed us to observe dendrites in CrMnFeCoNi alloys. The observations showed that the preferred growth direction of dendrites is <100>. The specific interfacial area and the curvature are in the same order as those of Al-Cu and Mg-Sn alloys. Dendritic growth was observed even for the Zn-4mass%Al, characterized by a large X-ray absorption. The equiaxed dendrite grows in 14 directions. The directions are consistent with a hexagonal structure. However, the directions are not simply defined by the crystallographic orientations. As demonstrated in the three alloy systems, 4D-CT, XRD and phase filed filter will provide the three-dimensional model of dendritic structures during solidification. We expect that the models are used for understanding solidification phenomena.

Acknowledgment

In situ observations using synchrotron radiation X-rays were performed as general projects (Nos. 2020A1564, 2020A1567, 2021A1485, 2021A1502, 2021B1561, 2021B1537, and 2022A1612) at BL20XU, 20B2 and 47XU of SPring-8 (JASRI), Japan. This work was supported by JSPS KAKENHI Grant Numbers 17H06155, 20H00217, 22H04963, 20K15071, and 22K14522. The author, HY, would like to thank Prof. A Sugiyama, Osaka Sangyo University, for advice in observations and fruitful desiccations. We thank Liwen Bianji (Edanz) (www.liwenbianji.cn) for editing the English text of a draft of this manuscript.

References

[1] Mathiesen R H, Arnberg L, Mo F, Weitkamp T, Snigirev A 1999 *Phys Rev Lett* **83** 5062-5
[2] Yasuda H, Ohnaka I, Kawasaki K, Sugiyama A, Ohmichi T, Iwane J, Umetani K 2004 *J Cryst Growth* **262** 645-52
[3] Li B, Brody H, Kazimirov A 2004 *Phys Rev E* **70** 062602-
[4] Schenk T, Thi H N, Gastaldi J, Reinhart G, Cristiglio V, Mangelinck-Noël N, Klein H, Härtwig J, Grushko B, Billia B, Baruchel J 2005 *J Cryst Growth* **275** 201-8
[5] Mangelinck-Noël N, Nguyen-Thi H, Reinhart G, Schenk T, Cristiglio V, Dupouy M D, Gastaldi J, Billia B, Härtwig J, Baruchel J 2005 *J Phys D* **38** A28-A32
[6] Li B, Brody H D, Kazimirov A 2006 *Metall Mater Trans A* **37** 1039-44
[7] Li B, Brody H D, Black D R, Burdette H E, Rau C 2006 *J Phys D* **39** 4450-6
[8] Peng L, Zeng G, Su T C, Yasuda H, Nogita K, Gourlay C M 2019 *JOM* **71** 2235-44
[9] Yasuda H, Yamamoto Y, Nakatsuka N, Nagira T, Yoshiya M, Sugiyama A, Ohnaka I, Umetani K, Uesugi K 2008 *Int J Cast Met Res* **21** 125-8
[10] Yasuda H, Nagira T, Yoshiya M, Nakatsuka N, Sugiyama A, Uesugi K, Umetani K 2011 *ISIJ Int* **51** 402-8
[11] Yamane K, Yasuda H, Sugiyama A, Nagira T, Yoshiya M, Morishita K, Uesugi K, Takeuchi A, Suzuki Y 2015 *Metall Mater Trans A* **46** 4937-46
[12] Reinhart G, Grange D, Abou-Khalil L, Mangelinck-Noël N, Niane N T, Maguin V, Guillemot G, Gandin C-A, Nguyen-Thi H 2020 *Acta Mater* **194** 68-79
[13] Nishimura T, Ishida H, Yasuda H 2021 *ISIJ Int* **61** 1-7
[14] Gourlay C M, Dahle A K, Nagira T, Nakatsuka N, Nogita K, Uesugi K, Yasuda H 2011 *Acta Mater* **59** 4933-43

[15] Nagira T, Gourlay C M, Sugiyama A, Uesugi M, Kanzawa Y, Yoshiya M, Uesugi K, Umetani K, Yasuda H 2011 *Scr Mater* **64** 1129-32
[16] Nishimura T, Matsubayashi R, Morishita K, Yoshiya M, Nagira T, Yasuda H 2019 *ISIJ Int* **59** 459-65
[17] Yasuda H, Morishita K, Yoshiya M, Narumi T 2020 *ISIJ Int* **60** 2755-64
[18] Yasuda H, Morishita K, Nakatsuka N, Nishimura T, Yoshiya M, Sugiyama A, Uesugi K, Takeuchi A 2019 *Nat Comm* **10** 3183-
[19] Ludwig O, Dimichiel M, Salvo L, Suéry M, Falus P 2005 *Metall Mater Trans A* **36** 1515-23
[20] Aagesen L K, Fife J L, Lauridsen E M, Voorhees P W 2011 *Scr Mater* **64** 394-7
[21] Daudin R, Terzi S, Lhuissier P, Salvo L, Boller E 2015 *Materials & Design* **87** 313-7
[22] Gibbs J W, Mohan K A, Gulsoy E B, Shahani A J, Xiao X, Bouman C A, De Graef M, Voorhees P W 2015 *Sci Rep* **5** 11824
[23] Karagadde S, Lee P D, Cai B, Fife J L, Azeem M A, Kareh K M, Puncreobutr C, Tsivoulas D, Connolley T, Atwood R C 2015 *Nat Comm* **6** 8300-
[24] Cai B, Wang J, Kao A, Pericleous K, Phillion A B, Atwood R C, Lee P D 2016 *Acta Mater* **117** 160-9
[25] Shuai S, Guo E, Phillion A B, Callaghan M D, Jing T, Lee P D 2016 *Acta Mater* **118** 260-9
[26] Sillekens W H, Casari D, Mirihanage W U, Terzi S, Mathiesen R H, Salvo L, Daudin R, Lhuissier P, Guo E, Lee P D 2016 *JOM* **68** 3042-50
[27] Azeem M A, Lee P D, Phillion A B, Karagadde S, Rockett P, Atwood R C, Courtois L, Rahman K M, Dye D 2017 *Acta Mater* **128** 241-8
[28] Kageyama K, Adziman F, Alabort E, Sui T, Korsunsky A M, Reed R C 2018 *Metall Mater Trans A* **49** 4274-89
[29] Yasuda H, Kawarasaki T, Tomiyori Y, Kato Y, Morishita K 2019 *IOP Conf Ser Mater Sci Eng* **529** 012023-
[30] Mirihanage W U, Falch K V, Casari D, Mcfadden S, Browne D J, Snigireva I, Snigirev A, Li Y J, Mathiesen R H 2019 *Materialia* **5** 100215
[31] Elder K L M, Stan T, Sun Y, Xiao X, Voorhees P W 2020 *Scr Mater* **182** 81-5
[32] Narumi T, Nakata T, Kobayashi R, Yasuda H 2021 *ISIJ Int* **61** 1567-78
[33] Wang Y, Shuai S, Huang C, Jing T, Chen C, Hu T, Wang J, Ren Z 2021 *Acta Metall Sinica (English Letters)*
[34] Tolnai D, Gavras S, Wilde F, Hammel J U, Bruns S 2021 *Adv Eng Mater*
[35] Hubbell J H 1982 *Int J App Rad Iso* **33** 1269-90
[36] Seltzer S M 1993 *Rad Res* **136**
[37] X-Ray Mass Attenuation Coefficients [Internet]. Available from: https://www.nist.gov/pml/x-ray-mass-attenuation-coefficients.
[38] Narumi T, Nakata T, Yasuda H 2020 *IOP Conf Ser Mater Sci Eng* **861** 012065-
[39] Nakano K, Narumi T, Morishita K, Yasuda H 2020 *Mater Trans* **61** 596-604
[40] Chatcharit K, Sugiyama A, Morishita K, Narumi T, Kajiwara K, Yasuda H 2020 *Int J Metalcast* **14** 794-801
[41] Kobayashi R 1993 *Physica D* **63** 410-23
[42] Wang S-L, Sekerka R F 1996 *J Comput Phys* **127** 110-7
[43] Karma A, Rappel W-J 1996 *Phys Rev E* **53** R3017-R20
[44] Mcfadden G B, Wheeler A A, Braun R J, Coriell S R, Sekerka R F 1993 *Phys Rev E* **48** 2016-24
[45] Cantor B, Chang I T H, Knight P, Vincent A J B 2004 *Mater Sci Eng A* **375-377** 213-8
[46] Dantzig J A, Rappaz M. Solidification 2nd edition: EPFL Press; 2016.
[47] Gibbs J W, Mohan K A, Gulsoy E B, Shahani A J, Xiao X, Bouman C A, De Graef M, Voorhees P W 2015 *Sci Rep* **5** 1-9
[48] Strutzenberger J, Faderl J 1998 *Metall Mater Trans A* **29** 631-46
[49] Kim S G, Huh J-Y, Chung G-J, Hwang H-S, Kim S-H 2019 *Metall Mater Trans A* **50** 3186-200

Tracking the evolution of hot tears in aluminium alloys using high-speed X-ray imaging

Insung Han[1], Shikang Feng[1], Andrew Lui[1], Fabian Wilde[2], Patrick S Grant[1] and Enzo Liotti[1]*

[1]Department of Materials, University of Oxford, Oxford OX1 3PH, United Kingdom
[2]Institute of Materials Research, Helmholtz-Zentrum Hereon, Max-Planck-Straße 1, 21502 Geesthacht, Germany

E-mail: enzo.liotti@materials.ox.ac.uk

Abstract. Hot tears are detrimental defects forming during the final stage of solidification when the remaining liquid loses the capacity to compensate for liquid to solid volume shrinkage. Although a mature semi-quantitative description of hot tearing has been developed, little is known about the dynamic evolution of hot tears as experimental studies have been conducted mostly post-solidification or in semi-static *in-situ* conditions. Here, we present a methodology to investigate the evolution of hot tears with high spatial and temporal resolution using synchrotron-based X-ray radiography. We develop a novel hot tear detection and tracking algorithm for quantification of hot tear density, area fraction and merging from the analysis of radiographic sequences of the solidification of thin metal samples. The methodology is demonstrated for an Al-5wt%Cu alloy and examples of the results and new insights that can be achieved are described.

1. Introduction

Hot tearing is a shrinkage-induced defect often observed during the last stage of solidification and caused by the pressure drop due to a lack of liquid feeding of solidification shrinkage in the mushy zone [1]. Hot tears significantly degrade the mechanical properties of casting alloys, as they act as stress concentrators that encourage the formation and propagation of cracks [2]. So far, due to the technical difficulties in observing the propagation of hot tears in real time within opaque metal alloys, most of the experimental work on hot tearing has been conducted *ex-situ* [3–5], with relatively few *in-situ* studies [6–9]. Early *in-situ* approaches were based on transparent organic alloys, such as succinonitrile-acetone, to replicate dendritic structures equivalent to those found in metallic alloys [6]. However, thanks to the advancement of synchrotron technologies in the last 15 years, *in-situ* X-ray imaging techniques have been increasingly employed for the study of solidification phenomena in metallic materials, including hot tearing. Puncreobutr et al. [9] investigated the effects of intermetallic phases on liquid permeability of the interdendritic channel, which influenced the pressure drop evolution in the liquid within the mush, using time-resolved tomography. Sistaninia et al. [8] combined a 3D granular modeling and *in-situ* X-ray tomography to study crack propagation during tensile deformation in the semi-solid microstructure as a model of the mushy zone at high solid fraction. In this paper we present a novel methodology for capturing the onset of hot tears formation and their subsequent growth and merging in an Al-5wt%Cu alloy using fast (up to 100 Hz) synchrotron-based X-ray radiography and advanced computer vision techniques.

2. Experimental methods

2.1. In situ solidification experiment

Radiographic videos of the solidification of grain refined Al-5wt%Cu samples were collected at the P05 beamline at the Petra III synchrotron (Germany) using our bespoke Bridgman furnace [10]. The furnace comprises two small ceramic plate heaters, which are independently controlled, enclosed in a chamber with Ar atmosphere. The arrangement of the rig in the beamline and the configuration of the heating plates are shown in figure 1. The experiments were carried out in near isothermal conditions by setting the same temperature for both heaters during cooling. In each sequence, thin alloy samples were melted and homogenised at 690 °C and then cooled at a constant cooling rate to a temperature of 510 °C, acquiring images at a variable rate to maintain a constant temperature step of 0.04 °C/frame. Cooling rates ranged from 0.5 °C/s to 4 °C/s and the corresponding acquisition rates from 12.5 Hz to 100 Hz respectively. A 19 keV monochromatic beam illuminated the alloy samples and the transmitted signal collected using a 200 µm LuAG:Ce screen combined with either a Ximea CB500MG or a KIT CMOS cameras. The field of view ranged between 4.51 mm × 3.38 mm and 6.55 mm × 4.92 mm with spatial resolutions of 1.1 µm/pixel and 1.3 µm/pixel. A large number of videos sequences were collected and we present illustrative examples of some of the key features to demonstrate the technique and analysis method.

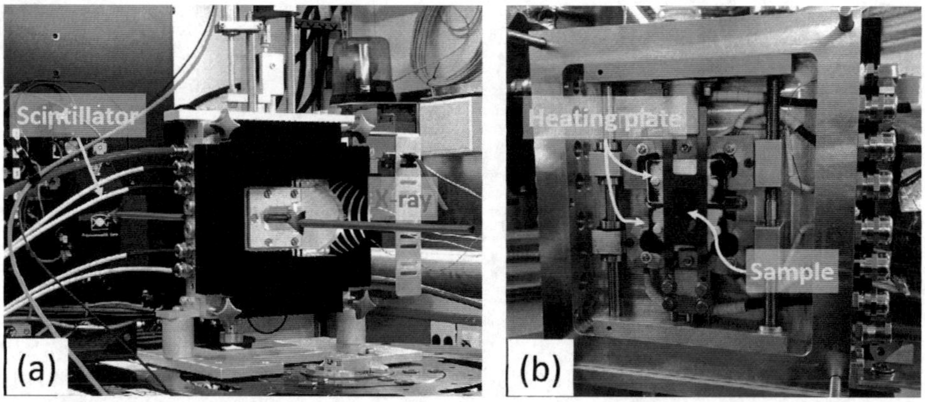

Figure 1. (a) Bridgman furnace setup at P05 beamline (Petra III). (b) The inside of the chamber showing the arrangement of the heating plates and sample.

2.2. Sample preparation

Alloys with composition Al-5wt%Cu were prepared using high purity Al (99.999%) and Al-80Cu master alloys. 0.1 wt% of Al-5Ti-1B and 1 wt% of pure Pb (99.999%) were also added for grain refinement and to form a fine emulsion of micron-scale Pb droplets to be used as flow tracers in the radiographic videos. Bulk samples were melted at 750 °C and homogenized for 10 minutes in an induction furnace under an Ar atmosphere. The molten metal was then poured into a water-cooled Cu mold. The small solid ingot was sectioned, ground and polished into thin foils with dimensions of 10 mm (width) × 20 mm (height) × 0.2 mm (thickness) for the synchrotron radiography experiments.

3. Hot tear detection and tracking

3.1. Algorithm

An algorithm was written in MATLAB language to detect hot tears as they appeared in the radiographic images, track their growth and characterise their merging behaviour. The algorithm is composed of two separated steps: in the first step the frames are analysed to detect all the hot tearing regions; in the second step the segmented images are further analysed to track the temporal evolution of each individual hot tear and any hot tears interactions such as merging.

3.1.1. Hot tear detection

Hot tears appeared as bright regions in the radiographs due to their low X-ray absorption compared with the surrounding liquid or solid metal. Hot tear appearance is demonstrated in Figs. 2(a-b) showing two sequential frames of a region of interest (ROI) in which a hot tear formed. In the detection algorithm, hot tears were detected by segmenting a difference image, which was obtained by subtracting the previous frame from the current frame. Image subtraction was needed to enhance the contrast between hot tears and the liquid in the interdendritic channels. The subtracted image was further processed to minimize the noise signal and then the obtained image was segmented using an intensity-based thresholding approach. Figure 2(c) shows the results of the segmentation procedure applied to the ROI in figure 2(a-b).

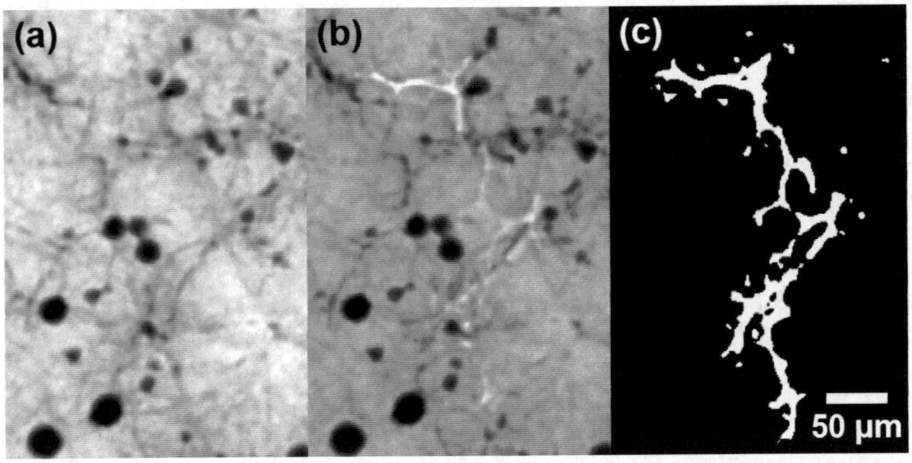

Figure 2. (a, b) Consecutive radiography images, at time t, prior to hot tearing and at $t + 0.02$ s, after hot tearing, respectively. (c) A hot tear identified by subtracting image (a) from image (b), followed by appropriate thresholding.

3.1.2. Unique identifiers for hot tear tracking

The segmented images were further processed by applying a dilation morphological operation with a 10 pixel × 10 pixel structuring element to all the foreground objects. This operation was needed to correctly identify each hot tear in the segmented images. Without the dilation routine, the hot tears often appeared fragmented into smaller pieces only because of limitations in spatial resolution and blurring. An example of this operation is shown in figure 3(a). The white pixels represent the hot tearing region detected by segmentation. By applying dilatation, only two unconnected regions remain (in blue and red) and these were assigned a unique identifier (UI). For each UI, detailed information, such as hot tear size, location and formation temperature, was recorded every frame from the first detection. As hot tears were spatially fixed, frame to frame UI matching was achieved simply by testing if each new detection included an existing hot tear. When there was overlap between dilated hot tears with different UIs, a merging event was assumed, a new UI was assigned to the merged tear, and the prior hot tears no longer tracked. Instead, the newly merged hot tear was tracked. An example is shown in figure 3(b) where the dilated hot tear (magenta color) now contained the information associated with the pre-formed hot tears in figure 3(a).

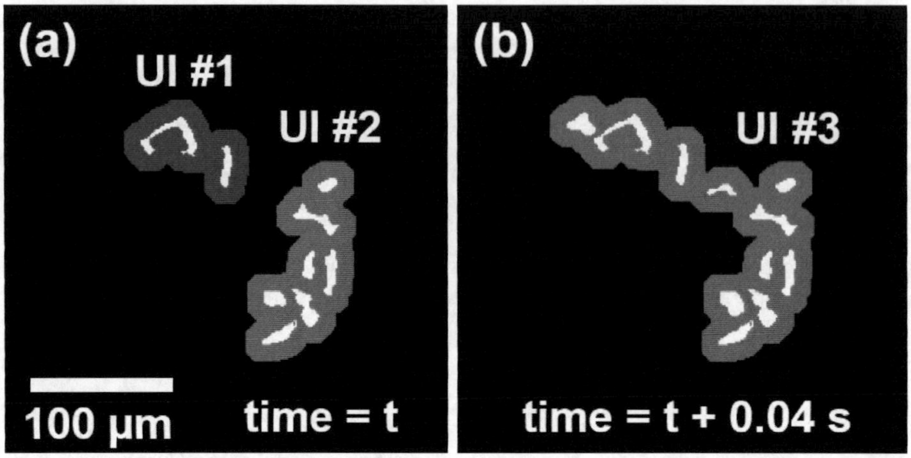

Figure 3. (a) Segmented hot tears (white pixels) with their own UIs (blue and red pixels). (b) A merged hot tear from the pre-formed hot tears in (a), with a newly assigned UI (magenta pixels).

3.2. Application: hot tear formation and growth analysis

Figures 4(a)-(d) are a typical image sequence from the late stage of the solidification of Al-5wt%Cu cooled at $2\,°C\,s^{-1}$ showing the formation of hot tears. The instantaneous temperature in each image was determined by knowing the temperature and frame at which primary α-Al dendrites in Al-5wt%Cu were first detected (liquidus temperature = 647.2 °C) and then recalculating the temperature of each subsequent frame according to the cooling rate and time elapsed since the first dendrite was observed. Using the Al-Cu phase diagram, the solid fraction was estimated as 0.88 and the average Al grain size was ≈200 µm. The black Pb droplets in figure 4, formed by monotectic reaction in the early stage of solidification and were located and entrained within the interdendritic channels. The Pb droplets were also tracked to help characterise the liquid flow in the last stage of solidification, as described in [11]. Hot tears, highlighted in cyan for clarity, first appeared at a temperature of 552.5 °C, and increased in number thereafter. Although hot tears formed across the entire FOV, they were, in this case, clustered in the bottom left region of figure 4(d).

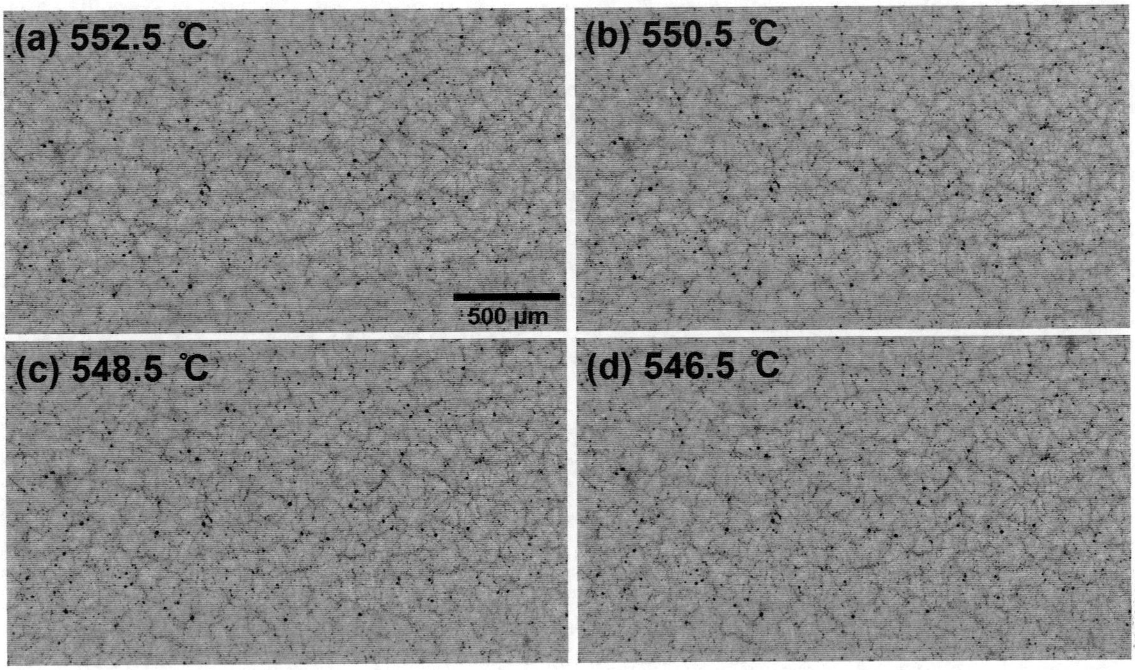

Figure 4. Sequence of radiographs of the solidification of Al-5wt%Cu cooled at 2 °C s^{-1} showing the microstructure at increasing solid fraction in the last stages of solidification. Hot tears are highlighted in cyan (and are dilated for visualization purposes). Pb droplets (black) flowing in the interdendritic regions are also visible.

The hot tear density was determined by counting the number of active hot tears (i.e. with separeate UIs) per unit area in each frame and is plotted in figure 5 (blue curve). The hot tear density increased steadily up to 10 hot tear/mm^2, and then plateaued. The occasional slight reductions in hot tear density in figure 5 below 549 °C were attributed to the merging of hot tears. The overall hot tear area fraction was calculated by summing the white pixels within the field of view in each frame and is the orange curve in figure 5. The curve follows the trend of the hot tear density, but delayed in time indicating that in the early stage of hot tear formation, growth was primarily due to nucleation of new hot tears, whereas the later stage was primarily due to growth. Both curves had an approximate Avrami *S-shape*, typical of classical nucleation and growth phase transformations.

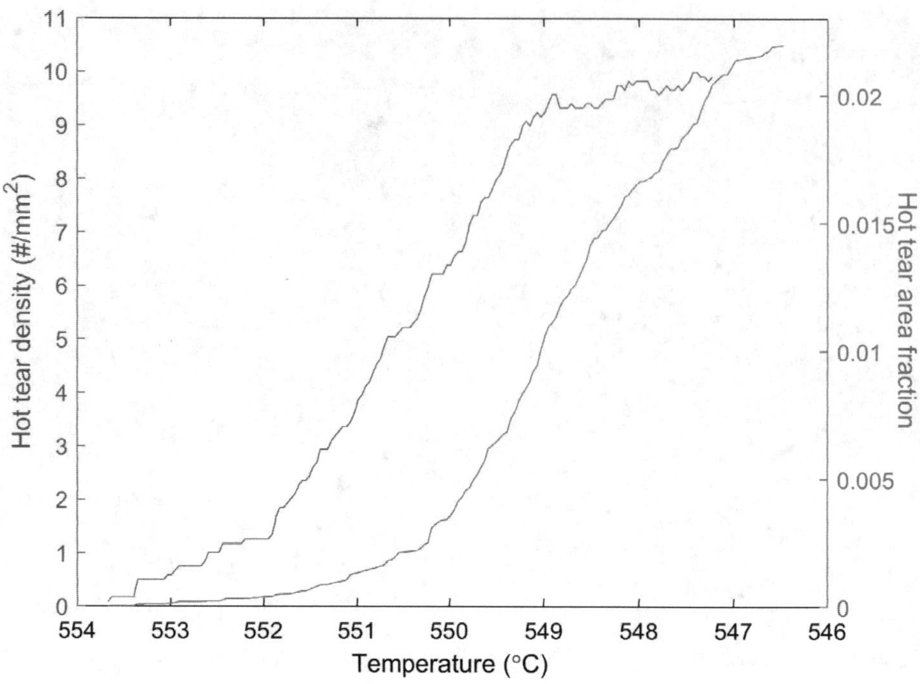

Figure 5. Hot tear density (blue) and area fraction (orange) as a function of temperature during the solidification of Al-5wt%Cu at a cooling rate of $2\,°C\,s^{-1}$

The hot tear detection algorithm was also used to investigate the sequence in which hot tears merged. For example, figure 6 shows a single, mm-scale hot tear that was formed by merging of thirty-six distinct and separate hot tears (each color indicates an originally independent hot tear). Although more data is required, one insight that the approach suggests – and which is unavailable from post-solidification analysis – is that large hot tears normally assumed to be a single tear may instead be due to the merging of a large number of smaller, initially separate tears.

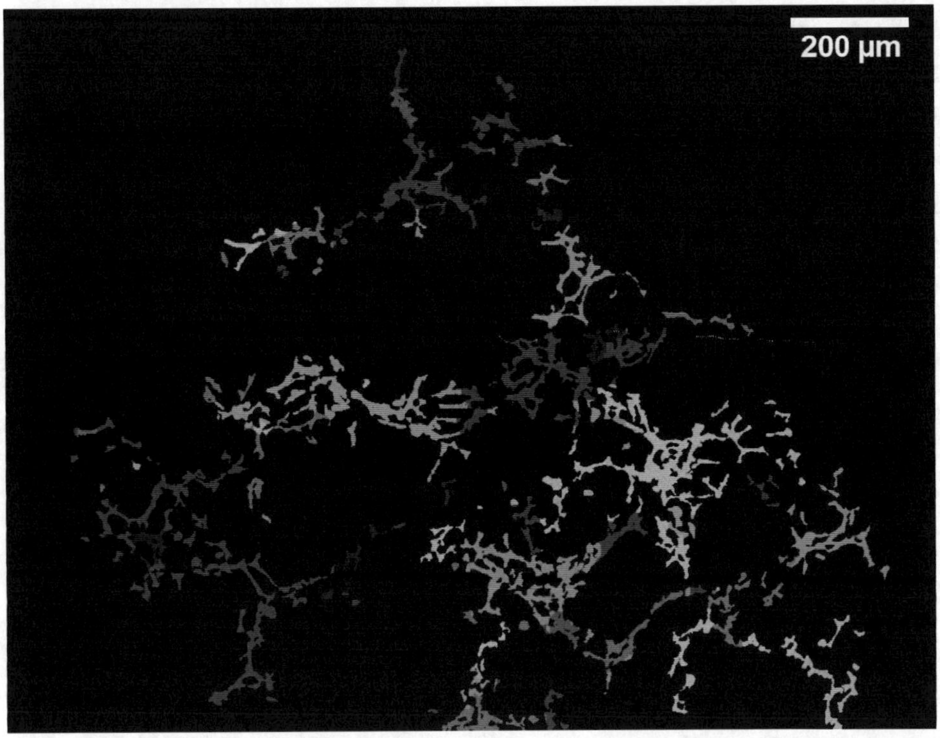

Figure 6. A large single inter-connected hot tear formed by the progressive merging of thirty-six initially separate (and differently coloured) tears, each with a different initial UI.

4. Conclusions

A novel algorithm was developed to extract quantified information of the evolution of hot tears from X-ray radiography video sequences of the solidification of metal alloys. The capabilities have been demonstrated using video sequences of a solidifying Al-5wt%Cu alloy. The algorithm allowed for the measurement of the formation and growth of hot tears for an ensemble of hot tears, each tracked in detail, and including analysis of interactions between existing hot tears such as merging. Further work is underway to extend the analysis to other fast X-ray radiography video sequences of other alloys with different solidification sequences, to give more robust insights into phenomena controlling hot tearing. We are also considering the development of algorithms for 3D analysis of high speed synchrotron tomographic sequences to overcome the constrains arising with thin samples. Continuing efforts that combine X-ray imaging experiments and automated computational analyses will likely help researchers decipher further the complexities of defect formation and solidification phenomena.

Acknowledgments

The work was funded by EPSRC (UK) under grant number EP/N007638/1 (Future Liquid Metal Engineering Hub). The synchrotron work was enabled by beamtime I-20210153 EC at the Petra III (Desy) beamline (Germany).

IOP Conf. Series: Materials Science and Engineering 1281 (2023) 012065 doi:10.1088/1757-899X/1281/1/012065

References

[1] Campbell J 2015 *Complete casting handbook: metal casting processes, metallurgy, techniques and design* (Butterworth-Heinemann)

[2] Eskin D, Katgerman L *et al.* 2004 *Progress in Materials Science* **49** 629–711

[3] Zhang J and Singer R 2002 *Acta Materialia* **50** 1869–1879

[4] Easton M, Wang H, Grandfield J, St John D and Sweet E 2004 *Materials Forum* vol 28 pp 224–229

[5] Cao G P and Kou S D 2006 *Metallurgical and Materials Transactions A* **37** 3647–3663

[6] Farup I, Drezet J M and Rappaz M 2001 *Acta Materialia* **49** 1261–1269

[7] Davidson C, Viano D, Lu L and StJohn D 2006 *International Journal of Cast Metals Research* **19** 59–65

[8] Sistaninia M, Terzi S, Phillion A, Drezet J M and Rappaz M 2013 *Acta Materialia* **61** 3831–3841

[9] Puncreobutr C, Phillion A, Fife J and Lee P 2014 *Acta Materialia* **64** 316–325

[10] Feng S, Han I, Lui A, Vincent R, Ring G, Grant P S and Liotti E 2022 *Metals* **12** 395

[11] Liotti E, Lui A, Connolley T and Grant P S 2022 *Acta Materialia* **240** 118298

IOP Conf. Series: Materials Science and Engineering 1281 (2023) 012066 doi:10.1088/1757-899X/1281/1/012066

Revealing and simulating the effect of hidden damage on local and full-field deformation behaviour of cast aluminium

Jakob Olofsson*[,1], Toni Bogdanoff[1] and Murat Tiryakioğlu[2]

[1] Department of Materials and Manufacturing, School of Engineering, Jönköping University, Box 1026, SE-551 11 Jönköping, Sweden

[2] School of Engineering and Technology, Jacksonville University, Jacksonville, Florida, FL 32211, USA

E-mail: jakob.olofsson@ju.se

Abstract. Pores have been the main focus of quality assurance in castings. Latest research has shown that in aluminium castings pores can form only if there is existing entrainment damage, i.e., pores are merely the visible parts of the entrainment damage, and usually invisible damage is much more extensive. However, its effect on deformation behaviour has not been previously established or observed in-situ. This work applies 2D Digital Image Correlation (DIC) to an in-situ full-field stress-strain analysis of tensile samples with a non-conventional heterogeneous stress distribution. The observations reveal that the effect of hidden damage extends far beyond its impact on fracture behaviour and is responsible for initiating local strain concentrations during deformation. By extracting local stress-strain data, FE simulations have been performed to mimic the effect of local hidden damage on the heterogeneous stress-strain field. SEM and FIB-SEM analysis has been applied to investigate the cause for the strain concentrations. The combined results show that hidden damage in the form of oxide films is not only responsible for premature fracture, but also affects the deformation behaviour of tensile samples by introducing dispersed strain concentrations.

1. Introduction

The quality and performance of aluminium castings are commonly discussed based on different aspects of porosity. This is encouraged by extensive research on the effect of pores, which in different ways aims to relate the characteristics, as size and morphology, of pores to the mechanical performance of the material. Different approaches have been applied to assess these characteristics, either in solid material using various x-ray techniques [1] or in post-mortem analyses of fracture surfaces [2]. Extensive research has also been targeted at understanding the growth of pores during solidification using modelling and simulation [3].

Recent research has shown that pores can only form if there is existing entrainment damage, i.e., pores bypass nucleation and grow from existing, folded-over oxide films (bifilms) [4]. Pores can thus be interpreted as merely a visible part of the damage caused by entrainment. The total damage includes the effect of the invisible damage, i.e., the oxide films that are not revealed in x-ray techniques and not always easily identified, even in fracture surface analyses. The traditionally strong focus on porosity might be misleading and suggest that reducing porosity will always lead to high-quality material. However, evaluations of the material quality in terms of mechanical performance have shown that even in the absence of visible pores, cast aluminium typically reaches only about 30% of its ductility potential

IOP Conf. Series: Materials Science and Engineering 1281 (2023) 012066 doi:10.1088/1757-899X/1281/1/012066

[5]. This has been attributed to the presence of invisible damage, and it has been noted that reduced structural quality not only impairs elongation to failure but also reduces the initial plastic hardening behaviour [6–9].

Here we aim to reveal and visualise the effect of invisible damage on the full-field straining behaviour of an aluminium alloy die casting. We apply Digital Image Correlation (DIC) to observe the straining behaviour on the surface of the specimen, a technique which has previously been applied to reveal the effect of local variations in solidification conditions on the strain distribution in cast aluminium using standard tensile bars [10]. In the current study, we instead use a test specimen geometry which is traditionally intended for bending fatigue testing. This geometry induces a heterogeneous stress-strain field over the geometry when subjected to tensile load. From a mechanical point-of-view, the samples are thus expected to fail in the middle of the specimen, where the highest stress concentration and the highest level of strain is expected. The DIC results reveal that local strain concentrations occur in the samples that cannot be explained by the sample geometry or visible pores. Using Scanning Electron Microscopy (SEM) in combination with Focused Ion Beam (FIB) and Scanning Transmission Electron Microscopy (STEM) techniques, the origin of these strain concentrations is identified as oxide films. This provides new insights into the effect and significance of invisible damage on the performance and failure of cast aluminium.

2. Methods

2.1. Material

A standard industrial High Pressure Die Casting (HPDC) alloy was used. The chemical composition of the material, measured with optical spectroscopy, is shown in table 1. The material follows the EN AC 46 000 standard alloy specification.

The samples were cast at a foundry in a standard HPDC machine using a metal mould with the specimen geometry (200 mm length, 6 mm thickness, 35 mm wide, and two 75 mm radii to create a thinnest width of 20 mm in the specimen centre). The foundry's standard production method was used. In total more than 300 samples were produced to ensure conditions similar to serial production in terms of steady-state mould temperature etc. All samples were inspected using an industrial X-ray with a resolution of 0.1 mm. 10 samples with no visible indication of porosity were investigated for the current study. To ensure parallel surfaces, achieve conditions closer to plane stress conditions, and reduce the effect of cast surface-related defects, the samples were machined on the front and back sides to a thickness of 3 mm. Visual inspection was performed on the surfaces to make sure that no cracks or defects could be identified on the surface of the specimen prior to testing.

Table 1. Chemical composition of the investigated alloy .

Element	Si	Fe	Cu	Mn	Mg	Cr	Zn	Al
Content (wt.%)	10.2	1.05	3.14	0.26	0.16	0.04	0.63	Bal.

2.2. Tensile testing

The specimens were tested at room temperature in a Zwick/Roell Z100 tensile testing machine. Displacement was controlled by a constant crosshead speed of 0.35 mm/min, and force was logged as a function of crosshead displacement.

2.3. Digital Image Correlation setup

To enable DIC measurements, a pattern was manually sprayed on the surface of the specimen, starting with a white bottom layer followed by a black random speckle pattern. The samples were tested within an hour of colour application to avoid decohesion between the metal and the paint.

A Stereo-DIC setup was used to observe the deformation of the speckle pattern. The use of a Stereo-DIC setup ensures that also out-of-plane motions are identified in the measurements. Two AVT Stingray F201B 8-bit cameras were used, with Tamron 25 mm lenses. The cameras were mounted on a bar, with one camera perpendicular to the observed specimen surface and the second on the same height with an angle of about 20°. Stereo calibration was done before each test. Images with a resolution of 1624 × 1232 pixels were captured at a frame rate of 1 Hz. The software MatchID™ Grabber was used for capturing the images, and the MatchID™ Stereo software was used for the correlation. The noise level was measured to about 0.85%. For displacement calculations, a subset size of 17 pixels and a step size of 5 pixels was used. The spatial resolution was 17 pixels, or about 0.7 mm. For strain calculations, a quadratic polynomial (Q4) smoothing method was used with a strain window size of 15 pixels, a Virtual Strain Gauge Size of 87 pixels (about 3.5 mm) and a logarithmic Euler-Almansi strain tensor.

2.4. *Microscopy*
The microstructural characterisation was carried out by optical microscopy and a TESCAN Lyra 3 Scanning Electron Microscopy (SEM) equipped with Energy Dispersive Spectroscopy (EDS) and Scanning Transmission Electron Microscopy (STEM). For the analysis of the cracked area, FIB-SEM slicing was used, providing more information about the bulk material than was obtained using two-dimensional investigations of the sample surface. Locations at the end of the open crack were selected to produce lamellas for STEM investigation. Rough milling of ~1.7 μA at 30 kV made a trench around the area of interest until a lamella of around 1.2 μm could be lifted out. Fine polishing of the lamella was then done down to 0.2 μm using a FIB current of ~30 nA at 30 keV. STEM EDS was performed on the lamella to identify the secondary phases and their constituents.

2.5. *Simulation setup*
To determine the stress distribution throughout the samples at different levels of crosshead displacement, a FEM simulation of the tensile test was made. A 3D geometry was created in Abaqus CAE using second-order hexagonal elements with reduced integration (C3D20R) and an approximate element size of 1 mm. Simulations were first performed with a homogeneous material definition, with plastic material behaviour defined with a piecewise linear plastic material definition. FEM simulations were also performed to mimic the effect of heterogeneous material behaviour observed in the DIC. The same FEM model as for the homogeneous simulation was used, in which elements were then manually selected corresponding to the locations of the observed strain concentrations.

3. Results and discussion
The DIC analyses reveal that several of the investigated samples exhibit local strain concentrations in the material when subjected to load. Three different categories of results can be identified, which will be discussed in more detail in the following:

- Single strain concentration
- Multiple strain concentrations
- Samples with no clear local strain concentrations

3.1. *Single strain concentration*
Figure 1 shows the effective strain field evaluated in sample A at a load of 11.9 kN (figure 1(a)) and at 15.4 kN (figure 1(b)). As can be seen already at the lower load level, the strain field is not symmetric, as is to be expected from the sample geometry, but the left edge shows a different straining behaviour than the right. As the load increases, figure 1(b), a clear strain concentration is seen at a point located about 5 mm in from the left edge and about 5 mm above the thinnest cross-section of the sample. This strain concentration is not explained by the geometry of the sample. At the next timestep, figure 2(a), the fracture has occurred and is clearly seen to have propagated at the height of this strain concentration rather than through the middle of the specimen. The SEM investigation of the fracture surface reveals

the presence of an oxide film in this specific location, see figure 2(b)-(c), which is draped over the dendritic microstructure.

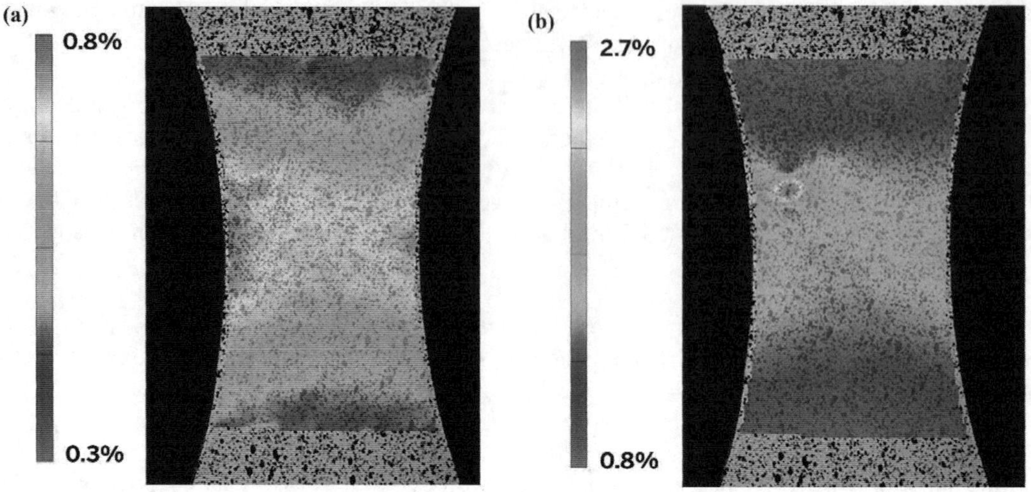

Figure 1. Full-field distributions of von Mises equivalent strain in sample A at a load of (a) 11.9 kN and (b) 15.4 kN.

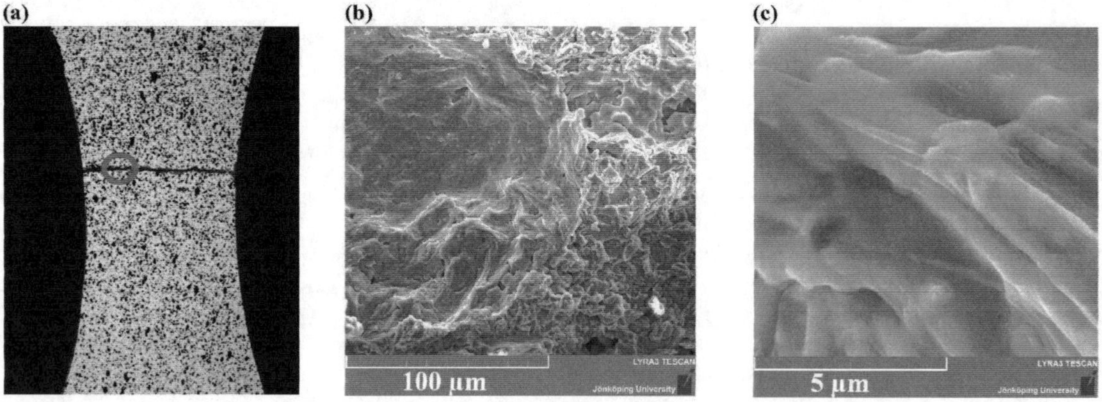

Figure 2. (a) Sample A after fracture. Red ring indicates the location of the strain concentration seen in figure 1(b). (b) SEM image of the oxide film found in the strain concentration area. (c) Increased magnification of the oxide draped over the microstructure.

By inserting virtual extensometers in the DIC results, the straining behaviour around the oxide can now be compared to the hardening of material without notable oxides. To get the simultaneous hardening behaviour of material exhibiting a comparable stress state during the deformation, an area placed symmetrically, around the vertical symmetry line of the specimen, from the strain concentration is used as a comparison. Extracting the local strains at the different loads during the tensile test, the force-strain curve in figure 3 is obtained. It is seen that the strain concentration already at low loads displays a higher strain level, and as the load level increases, the difference between the two points also increases. The strain level before failure is around 2.7% in the strain concentration, compared to approximately 1.4% in the symmetric position. The material around the oxide thus exhibits a strain level twice as high as expected at this load level. The strain at the edge at the thinnest width of the specimen is, at the same time, around 1.5%. The strain concentration caused by the oxide is thus much more dominant than the geometric stress concentration.

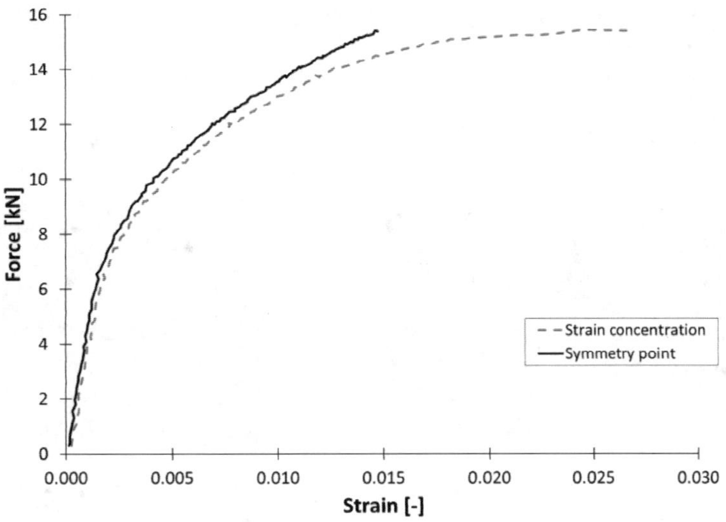

Figure 3. Comparison between the straining behaviour in the strain concentration compared to material in the symmetry point.

Further investigation of the DIC results shows that the strain concentration is also associated with an unsymmetric distribution of the lateral strain, figure 4(a), with an increased lateral strain to the left and right sides of the strain concentration in figure 1(b). Calculating the local variation in the Poisson's ratio, see figure 4(b), also shows that there is a strong local decrease in the fraction lateral strain compared to the axial strain in the strain concentration. The oxide film has thus been causing a structural discontinuity in the material where there is no bond between the material above and below the oxide film. When the load was applied, the oxide opened so that the material separated and acted as a crack initiation which caused the failure.

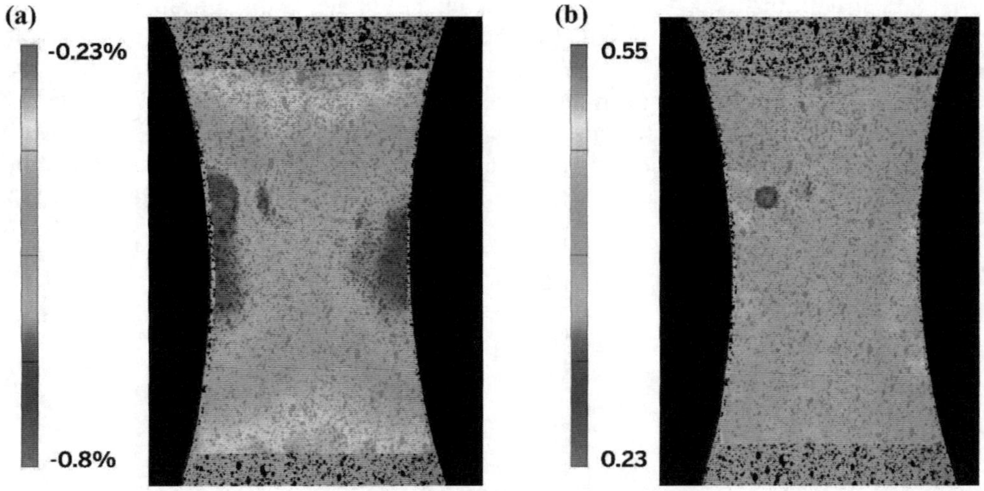

Figure 4. (a) Lateral strain and (b) local Poisson's ratio at the last time step before fracture.

IOP Conf. Series: Materials Science and Engineering 1281 (2023) 012066 doi:10.1088/1757-899X/1281/1/012066

3.2. Multiple strain concentrations

Some samples display several strain concentrations in the strain field, where not all are associated with the failure progression. Figure 5 shows the strain distribution in sample B at a load of 15.9 kN, just before fracture. Compared to figure 1(b), the strains are, in general, more evenly dispersed. Strain concentrations are observed both across the middle section but also in a location higher up, close to the vertical centreline. The SEM analysis shows two oxide films (one of them associated with a small internal pore below the detection limit of the x-ray) are present in the fracture surface and have interacted to cause and propagate the failure.

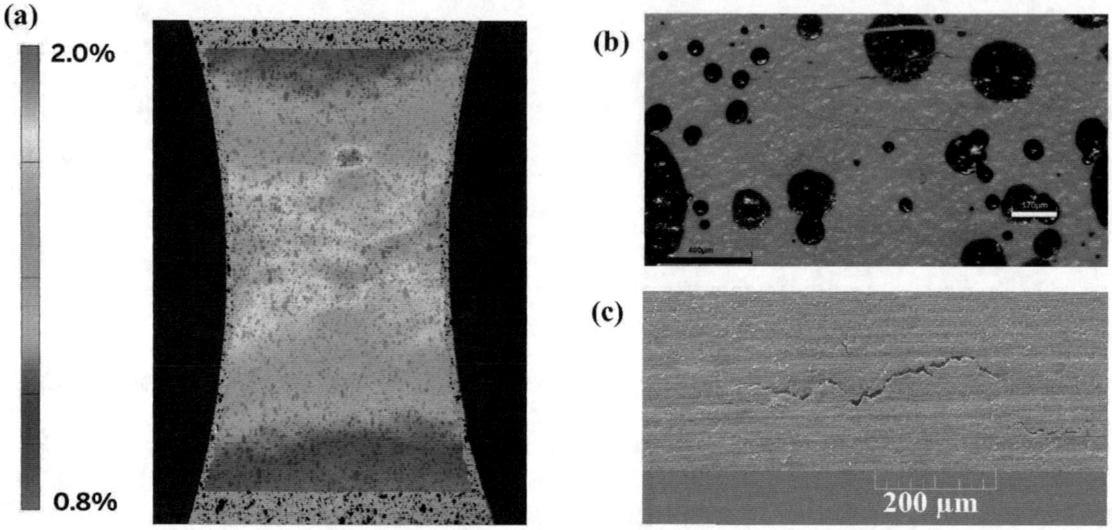

Figure 5. (a) DIC strain field in Sample B at a load of 15.9 kN, just before fracture. (b) Close-up on the upper strain concentration after failure. (c) Visible crack after paint removal.

We now focus on the upper strain concentration, which has not been part of the failure. A closer investigation of the surface reveals a small crack in the paint has formed in this area, figure 5(b). The DIC images confirm the crack was not present on the surface before the testing. The analysis also shows that the remaining deformation in this area after unloading is higher in this area than in the surroundings. After removing the painted pattern, the crack in the bulk material is clearly seen, see figure 5(c). To further investigate the cause for this crack, FIB milling is applied to extract a thin lamella in front of the visible crack figure 6(a)-(b). The 0.2 µm thin lamella is then examined using STEM EDS analysis. As seen in figure 6(c), the analysis clearly shows a layer of oxygen on the inside of the crack, verifying that both surfaces are indeed part of a folded oxide film. This confirms that the strain concentration observed in the DIC analysis is indeed caused by an oxide film which has opened up when subjected to load.

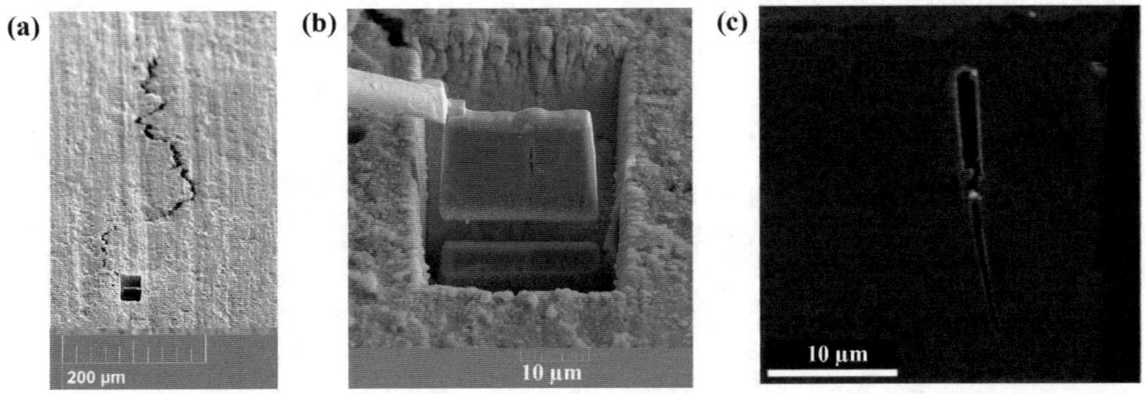

Figure 6. (a) Location of the lamella in front of the crack. (b) Extraction of lamella. (c) STEM-EDS map of oxygen distribution in the lamella, revealing the oxide film on the inside of the crack.

3.3. *Samples with no clear strain concentrations*

Some of the samples fail without showing any clear strain concentrations on the observed surface. However, fracture surface analysis reveals that also in these samples the fracture is associated with the presence of oxide films and invisible damage. These have been located deeper below the surface or on the opposite side of the specimen. The samples also exhibit the same levels of maximum force before failure. Even though they were not visible on the surface under investigation, it is thus logical to suggest that the same type of local strain concentrations as observed in the other samples has also occurred in these samples. This can be motivated by the stochastic distribution of these defects within the material.

3.4. *FEM simulations*

As a first attempt to simulate the local effects of oxides film in FEM analyses, a simulation of the test of the specimen is set up using a standard isotropic elasto-plastic material model. The bulk material is assigned to a standard elasto-plastic description for the material obtained from regular tensile testing. To mimic the strain concentration area in sample A, a defect area is created by selecting four neighboring elements forming a cube (2x2x1 elements, i.e., approx. 2x2x1 mm) in the outer layer of the three layer thick 3D mesh. These elements are assigned to elasto-plastic material data scaled according to the straining behaviour of the strain concentration, shown in figure 3. Note that the use of a standard material model implies the use of a constant Poisson's ratio, and does not resemble the effect seen in the DIC results that the Poisson's ratio changes with increasing load. Similarly, a FEM simulation to mimic the observed multi-defect behaviour in sample B is set up with three such defect areas: two areas in the middle section and one above.

The results of the simulated strain fields are shown in figure 7. It should be noted that the figures of DIC results and FEM simulations are not directly quantitatively comparable since they have different colour scales, filtering, spatial resolution and mesh size and display results at slightly different deformation levels. The simulated strain distributions in figure 7(a) and (b) are also not directly comparable to each other as they, similarly to figure 1(b) and figure 5(b), represent different levels of crosshead displacements and loads. Nevertheless, the simulations help to understand and verify the observations in the DIC analysis.

For the case of sample A, it is seen both in the FEM simulations, figure 8(a), and DIC results, figure 8(b), that as the defect area exhibits an increased straining compared to its surrounding material, an area above and below the defect area display a lower strain. This leads to a diamond-like strain field with shear bands forming at approximately 45° from the defect area. As multiple defects are introduced in the case of sample B, figure 8(c)-(d), the effects of the different defect areas interact and meet each other. This behaviour much resembles the local deformations and strain localisation and strain bands that have previously been found associated with graphite nodules in ductile iron [11] as well as in other cast iron morphologies [12]. Here the oxides are found to similarly act as inclusions that induces localised strains.

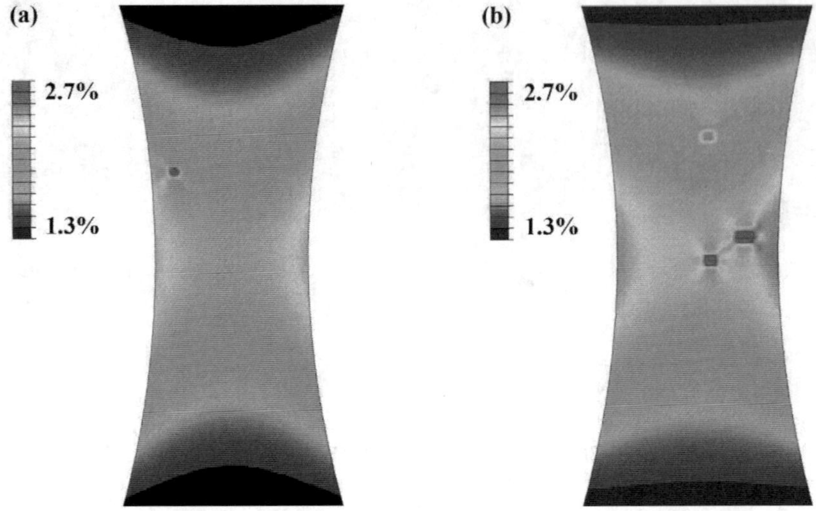

Figure 7. FEM simulations to mimic the straining behaviour seen in (a) sample A with one strain concentration and (b) sample B with multiple strain concentrations.

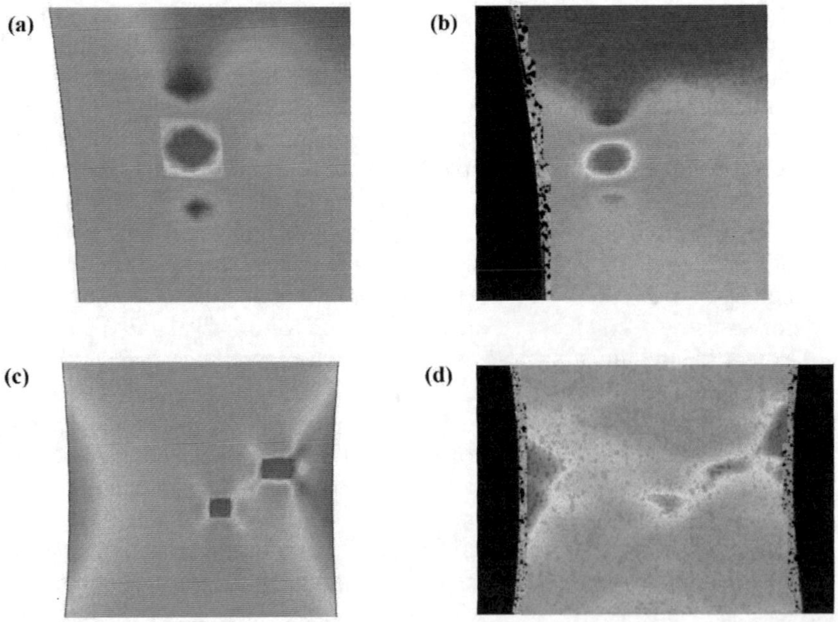

Figure 8. Comparison between details in the strain fields in FEM simulations and DIC observations. (a) FEM results and (b) DIC observations in sample A. (c) FEM and (d) DIC results for sample B.

These findings demonstrate that invisible damage is an important aspect of the material performance and quality of aluminium. The results also highlight the importance, but also complexity, of modelling and simulating the entrainment and presence of oxide films in castings. Future work in the field aims to investigate approaches to extract data for material models to simulate the mechanical behaviour and damage caused by oxides opening up, e.g., the Virtual Fields Method (VFM) [13].

4. Conclusions

- DIC observations reveal that the presence of invisible entrainment damage initiates local strain concentrations in the material during the deformation of cast aluminium.
- It is verified that the strain concentrations are caused by oxide films that open up during deformation.
- Invisible damage can initiate final fracture but can also cause local strain concentrations and crack opening in lesser loaded areas.
- Multiple strain concentrations can cause interacting local strain bands.
- X-ray inspection does not reveal the hidden damage and can give a false impression about the material quality in the casting.

Acknowledgement
The authors thank the Swedish funding agency VINNOVA for financial support of the project EVIDENT, Johan Börjesson for help with the FIB-SEM analyses and Jacob Steggo for support with the DIC measurements.

References

[1] Lashkari O, Yao L, Cockcroft S and Maijer D 2009 X-ray microtomographic characterization of porosity in aluminum alloy A356 *Metall Mater Trans A Phys Metall Mater Sci* **40** 991–9
[2] Ma Z, Samuel A M, Samuel F H, Doty H W and Valtierra S 2008 A study of tensile properties in Al–Si–Cu and Al–Si–Mg alloys: Effect of β-iron intermetallics and porosity *Materials Science and Engineering: A* **490** 36–51
[3] Zhang A, Guo Z, Wang Q and Xiong S 2020 Multiphase-field modelling of hydrogen pore evolution during alloy solidification *IOP Conference Series: Materials Science and Engineering* vol 861
[4] Yousefian P and Tiryakioğlu M 2018 Pore Formation During Solidification of Aluminum: Reconciliation of Experimental Observations, Modeling Assumptions, and Classical Nucleation Theory *Metall Mater Trans A Phys Metall Mater Sci* **49** 563–75
[5] Tiryakioglu M, Campbell J and Alexopoulos N 2009 On the Ductility of Cast Al-7 Pct Si-Mg Alloys *Metallurgical and Materials Transactions A* **40** 1000–7
[6] Tiryakioglu M, Campbell J and Alexopoulos N 2009 Quality Indices for Aluminum Alloy Castings: A Critical Review *Metallurgical and Materials Transactions B* **40** 802–11
[7] Tiryakioglu M, Campbell J and Staley J T 2003 The influence of structural integrity on the tensile deformation of cast Al-7wt.%Si-0.6wt.%Mg alloys *Scr Mater* **49** 873–8
[8] Tiryakioglu M, Campbell J and Staley J T 2004 Evaluating structural integrity of cast Al-7% Si-Mg alloys via work hardening characteristics: 1. Concept of target properties *Materials Science and Engineering A* **368** 205–11
[9] Tiryakioglu M, Staley Jr J T and Campbell J 2008 The effect of structural integrity on the tensile deformation characteristics of A206-T71 alloy castings *Materials Science and Engineering A* **487** 383–7
[10] Olofsson J, Svensson I L, Lava P and Debruyne D 2014 Characterisation and investigation of local variations in mechanical behaviour in cast aluminium using gradient solidification, Digital Image Correlation and finite element simulation *Mater Des* **56** 755–62
[11] Xu C L, Andriollo T, Zhang Y B, Hernando J C, Hattel J and Tiedje N 2020 Micromechanical

impact of solidification regions in ductile iron revealed via a 3D strain partitioning analysis method *Scr Mater* **178** 463–7

[12] Olofsson J, Salomonsson K, Dahle A K and Mathiesen R H 2019 Three-dimensional study of nodule clustering and heterogeneous strain localization for tailored material properties in ductile iron *IOP Conference Series: Materials Science and Engineering* vol 529

[13] Devivier C, Pierron F and Wisnom M R 2013 Impact damage detection in composite plates using deflectometry and the Virtual Fields Method *Compos Part A Appl Sci Manuf* **48** 201–18

Integrated fatigue life predictions of aluminium castings using simulated local microstructure and defects

Jakob Olofsson

Department of Materials and Manufacturing, School of Engineering, Jönköping University, Box 1026, SE-551 11 Jönköping, Sweden

E-mail: jakob.olofsson@ju.se

Abstract. In this work, an integrated simulation approach previously developed for static FE analyses is extended to microstructure- and defect-based fatigue life assessments of castings. The approach, *the closed chain of simulations for cast components*, combines casting process simulation with microstructure modelling and local material characterisation to generate heterogeneous material data for FE analysis and fatigue life assessment. The method is demonstrated on a High-Pressure Die Cast aluminium component. Areas with a high risk of defects are identified based on the simulated solidification conditions, and heterogeneous material data for the fatigue life analysis is generated. Fatigue testing has been performed with different levels of porosities to quantify the effect of defects on the element-specific Wöhler curves. Pore characteristics are assessed using 2D X-ray, fracture surface analysis and Kitagawa diagram. The results highlight the importance of taking the risk of defect formation into consideration when designing industrial aluminium castings subjected to fatigue loads.

1. Introduction

As the electrification and lightweight trends in the automotive area push towards increased use of aluminium castings, the importance of considering the casting process already during the early design stages become essential. As many of these castings are designed for endurance and fatigue loads, there is an increasing interest in simulation-based tools to understand better and predict the fatigue performance of castings in use. Preferably, it should be possible to estimate the fatigue life early in the product realisation process and use fatigue performance as a design criterion in the virtual product development process.

The fatigue life of aluminium castings is strongly related to the solidification conditions and the structural quality of the material[1]. There is thus a strong focus in the literature on understanding the intrinsic and extrinsic effects in cast aluminium alloys [2], and extensive work has been made in the modelling of defects [3] and to understand and predict the impact of porosities on mechanical properties and fatigue life [4–7]. However, at an early design stage, where only a preliminary geometry and type of casting process are known, an attempt to precisely predict the distribution and morphology of porosities and many other defects is an impossible task. The current contribution instead aims to provide a simplistic but physically based approach to help designers identify areas prone to pore formation and evaluate if pores in these areas might be critical from a fatigue perspective when the part is subjected to load.

Different simulation-based approaches have been presented that aim to consider the impact of defects on the mechanical performance of aluminium casting [8–12]. In this work, *the closed chain of*

IOP Conf. Series: Materials Science and Engineering 1281 (2023) 012067 doi:10.1088/1757-899X/1281/1/012067

simulations for cast components previously presented by the author and coworkers [11,12] forms the basis. Here casting process simulation with microstructure modelling is used, and material characterisation models are applied to predict local microstructure-based mechanical behaviour [12]. The results are incorporated as local material data into Finite Element Method (FEM) simulations of the casting in use. The approach was initially developed to handle local static elasto-plastic material behaviour based on local variations in microstructure [11] These local variations were shown to have an important effect on the stress-strain distribution in ductile iron castings [13]. This effect is also observed in aluminium samples [14], but in HPDC castings the effect of residual stresses and defects were found to have a larger impact on the performance of the casting [12]. The local porosity levels predicted by the casting process simulation were then applied to a local quality factor, reducing the achievable elongation to failure. By introducing a local damage indicator, defect areas prone to fracture could be revealed [15]. Other additions to the approach include thermomechanical properties, thermophysical properties, and the strength-differential effect on tensile and compressive behaviour [16,17].

In this work, the first steps in extending *the closed chain of simulations for cast components* towards integrated simulations with local defect-based fatigue performance are presented. An experimental series to investigate the effect of visible damage and pore characteristics of an HPDC aluminium alloy is first introduced. The results are evaluated using Wöhler and Kitagawa diagrams. Since the approach aims to enable a methodology for early design phases, the use of so-called defect indicators is discussed and introduced to identify areas where the risk of pore formation is considered high. The approach is applied to a demonstrator, and future steps in the development are outlined.

2. Methods

2.1. Material

Table 1 shows the chemical composition of the investigated alloy as measured with optical spectroscopy. The alloy is a standard HPDC alloy that meets the EN AC 46 000 alloy specification. The material was produced at an HPDC foundry and cast using a conventional HPDC machine directly into the final geometry, as shown in figure 1(a). The production parameters, in particular the intensification pressure and injection speed, were varied to obtain different amounts and characteristics of pores. More than 300 samples were produced and inspected using an industrial 2D X-ray with a resolution of 0.1 mm. Based on the X-ray images, the samples were qualitatively categorised into four different categories:

- I: No visible pores
- II: Small and few pores
- III: Few bigger pores or clusters of small pores
- IV: Large pores concentrated in clusters.

Examples of samples in the different categories are shown in figure 2. From each of the four categories, 25 samples were selected: 5 for static tensile testing and 20 for fatigue testing.

Table 1. Chemical composition of the investigated alloy .

Element	Si	Fe	Cu	Mn	Mg	Cr	Zn	Al
Content (wt.%)	10.2	1.05	3.14	0.26	0.16	0.04	0.63	Bal.

2.2. Tensile testing

From the cast geometry, tensile tests with a parallel gauge length of 90 mm and thickness of 5.2 mm were machined, see figure 1(b). Tensile testing was performed at room temperature in a Zwick/Roell Z100 machine using a constant crosshead speed of 0.35 mm/min.

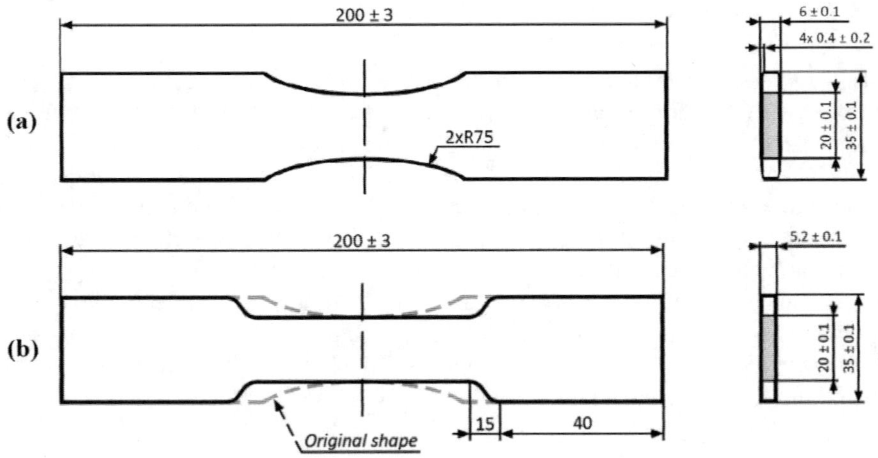

Figure 1. The geometries of (a) as-cast bar for fatigue testing and (b) tensile bar after machining.

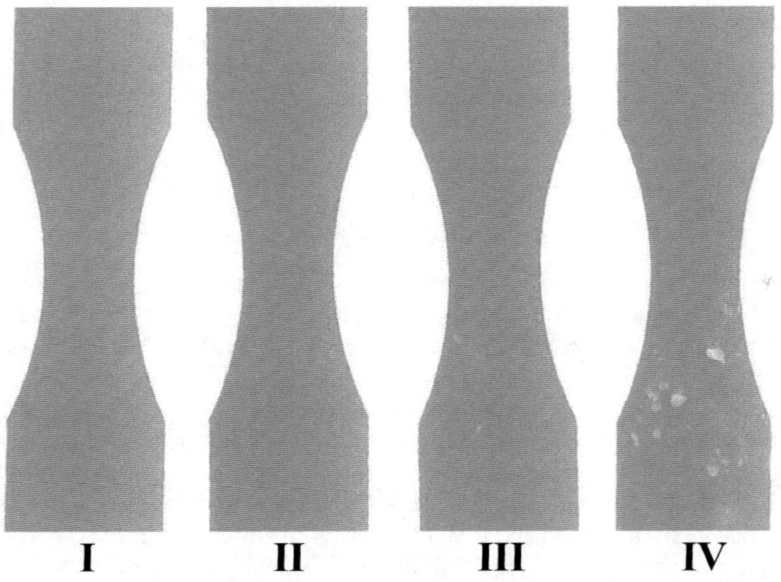

Figure 2. Examples of 2D X-ray images of samples from the four material categories I-IV.

2.3. Fatigue testing

Bending fatigue testing was performed at room temperature using a fully reversed load ($R = -1$). Samples surviving more than 5 million cycles were considered runouts. Runouts were later on fractured using tensile testing to enable characterisation of the crack initiating defects.

Wöhler curves were generated from the fatigue results using a Basquin approach. A bilinear Basquin relation was fitted with a least squares method on the data to an equation on the form

IOP Conf. Series: Materials Science and Engineering 1281 (2023) 012067 doi:10.1088/1757-899X/1281/1/012067

$$\sigma_A(N) = \begin{cases} -A\log(N) + B, & N < N_c \\ \sigma_{fl}, & N \geq N_c \end{cases} \tag{1}$$

Here $\sigma_A(N)$ is the stress amplitude at N number of cycles, and N_c the number of cycles where the curve is assumed linear with the fatigue limit σ_{fl}. A and B are empirical constants determined by least square fitting. These constants were evaluated separately for the four different categories.

A Kitagawa diagram was used to evaluate the effect of the size on the fatigue limit. The Kitagawa diagram is based on linear elastic fracture mechanics (LEFM) theory and can be described as a bilinear diagram relating the stress amplitude to defect size, A. Two lines separate a region where no fatigue crack propagation is expected from one where crack propagation will lead to fatigue failure. The first section corresponds to the fatigue strength obtained with minor or no defects. Above a critical defect size A_{cr}, a second line is identified where the fatigue strength will decrease proportionally to the square root of the defect size and can be evaluated using the LEFM theory. The Kitagawa diagram can then be expressed using the following equation:

$$\sigma_A(A) = \begin{cases} \sigma_{fl}, & A < A_{cr} \\ \dfrac{\Delta K_{th}}{\alpha\sqrt{\pi\sqrt{A}}}, & A \geq A_{cr} \end{cases} \tag{2}$$

The multiplier factor for calculation of stress intensity factor α is from LEFM given as 0.5 for internal cracks and 0.65 for surface cracks. The value of the threshold stress intensity factor ΔK_{th} is then determined from the experimental data points plotted in the Kitagawa diagram. Kitagawa diagrams in different variants have previously been applied to evaluate the effect of pores in aluminium castings, e.g., by Linder *et al.*[18] and Garb *et al.* [19].

3. Results and discussion

3.1. Tensile testing

The tensile testing results clearly show that the strain hardening behaviour remains similar for all the categories while the elongation decrease with increased levels of visible porosity. Figure 3 shows the obtained stress-strain curves for one sample from each quality category. The elongation for category I, without any visible pores, reaches an average of about 1.9%. The presence of small visible pores, category II, reduces the average elongation by about 36% to 1.2%. In the case of category IV, the elongation has dropped almost 70% to an average of 0.6%. However, the spread in the results is also highest for category I, with a standard deviation of 0.4%, decreasing to 0.2% in category IV. This indicates that even in the absence of visible damage, there are also other structural features and effects, e.g. oxide films, that play an important role in determining the properties. This was confirmed by fractography in optical microscopy, where the fractures in category I were found to initiate at large oxide films or surface defects, while the failures in categories II-IV originated from visible pores. This topic is further investigated in an accompanying work at this conference [20].

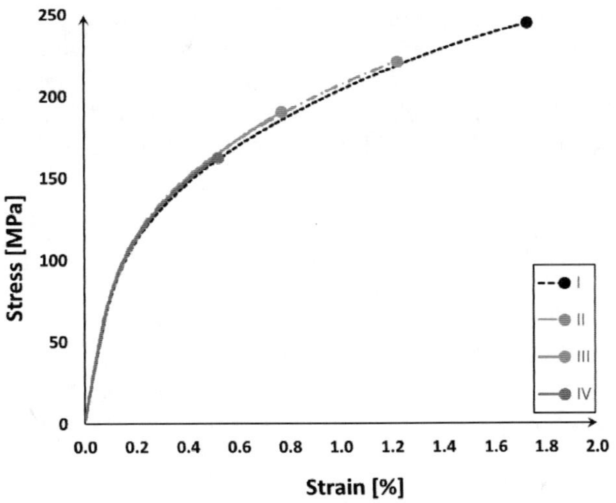

Figure 3. Tensile results for the different quality categories.

3.2. Fatigue testing

The resulting Wöhler curves are shown in figure 4. Note that the obtained fatigue stress levels are not presented in the current work for proprietary reasons, but the results are presented qualitatively. The results show that, similar to the tensile properties, the fatigue life decreases with increased porosity levels. The Wöhler curves are shifted downwards for each category, but the slope of the initial part is very similar. Comparing the fatigue limits, the limit decreases by about 25% from category I to II due to the presence of small pores. This decrease continues as the level of porosities increases, and category IV only exhibit 50% of the fatigue limit of category I.

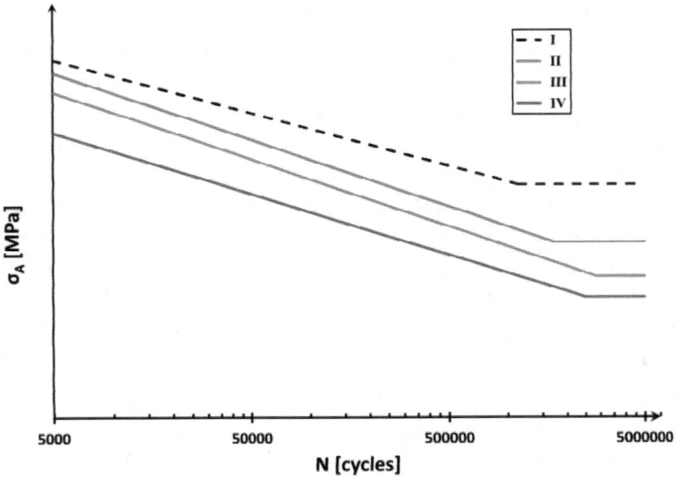

Figure 4. Wöhler curves obtained for the four different categories.

To further evaluate the effect of different pore sizes on the fatigue limit, fractographic analysis was performed where the cross-section area of the defects initiating failure was determined. The coordinates of the position of the defect within the samples were also measured, and the corresponding local stress in that exact location was extracted from a FEM simulation of the test. From this data, a Kitagawa diagram was derived, see figure 5. The first part of the diagram shows the fatigue limit of category I, while the second line is obtained from curve fitting towards the experimental points. The fitting has been made to be conservative, i.e., the runouts should be on or below the curve while failures are on or above the line. The intersection between these two lines is identified as the critical defect size A_{cr}. In this study, the critical defect size was found to be just below the detection limit of the industrial x-ray used. Thus, all pores identified in the x-ray can be considered as having a detrimental effect on the fatigue performance of the material.

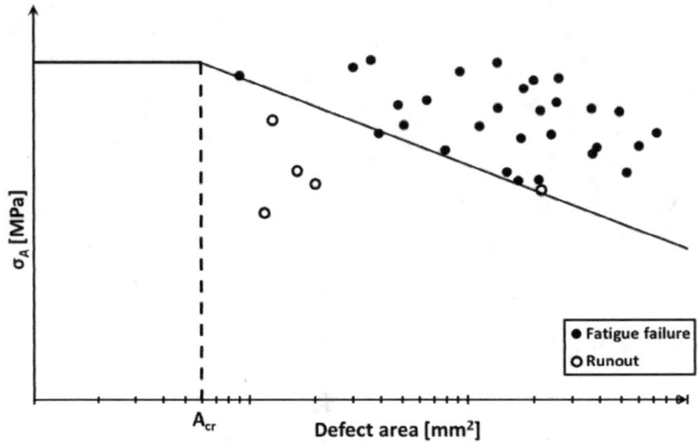

Figure 5. Kitagawa diagram for the fatigue testing results.

3.3. *Simulation methodology*

3.3.1. *Defect indicator*

Since the Wöhler curve and the Kitagawa diagram provide quantitative relationships between porosity level and material performance, the experimental results can now be incorporated into the closed chain of simulations for cast components. The material properties are here treated on a mesoscale level, where each element of the FEM mesh of the casting is assigned to its own local material data. As long as a quantitative defect indicator of the porosity level or defect size in the element is provided, the Wöhler curves, respectively the Kitagawa diagram, can be used to generate an element-specific local Wöhler curve to be used in the fatigue life evaluation. Ideally, this indicator would provide information about the exact location and morphology corresponding to the actual pore in a real casting. In the scope of the current work, where the aim is to rapidly provide a reasonable but approximated view of the fatigue life distribution for comparison of different preliminary designs, it is not feasible to perform a complete process simulation. In this context, we instead focus on only using the given geometry and a simulation of solidification. This simplistic approach will not be able to model or consider all the complex phenomena associated with defect formation in an actual casting process. Still, it can give a designer a good view of how the solidification of the geometry will occur and highlight areas of the geometry that needs to be addressed either in geometrical design or in the process layout. As a first step, the simulation result *Hot Spot FSTime* is here used as an indicator of critical areas with a high risk of pore formation. This result indicates isolated volumes based on their macroscopic feeding capability during solidification. It is thus an indicator of regions that, if not adequately considered in the casting process setup, are likely to contain porosity after solidification. This indicator does not provide any information about the volume fraction or size of the porosity. Instead, user-defined ranges are here used to numerically relate different values of the indicator to different material categories.

3.3.2. *Implementation*

Solidification is simulated using MAGMAOSFT™ using the MAGMAnonferrous module™ to model microstructure formation. The results of Secondary Dendrite Arm Spacing (SDAS) and Hot Spot

FSTime are mapped to the FEM mesh of the casting using MAGMAlink™. The SDAS results are applied to generate microstructure-based local material definitions to the FEM input file using an in-house script [11]. For simulations with residual stress, these are included as initial conditions for the load case. The FEM simulation of the casting subjected to static load is performed in Abaqus.

Local fatigue data in the form of element-specific Wöhler curves is then generated. Each element of the FEM mesh is first assigned to a material category based on its value of the defect indicator and the user-defined limits corresponding to categories II-IV. Elements that are not in a Hot Spot are assigned to category I. Elements connected in the same hot spot region are identified and assigned to a material quality according to the maximum value of the defect indicator in that region. The fatigue data can generally be provided in different formats, e.g. for specialised software such as FEMFAT™, which can consider several load cases, apply mean stress correction etc. For ease of simplicity, the fatigue life calculation in this demonstration is performed as a post-processing step in Abaqus™. The fatigue life is estimated by extracting the predicted value of von Mises stress and calculating the number of cycles expected at this stress level using the Basquin relation in equation (1), with parameters defined for each element. The flow of data is schematically illustrated in figure 6.

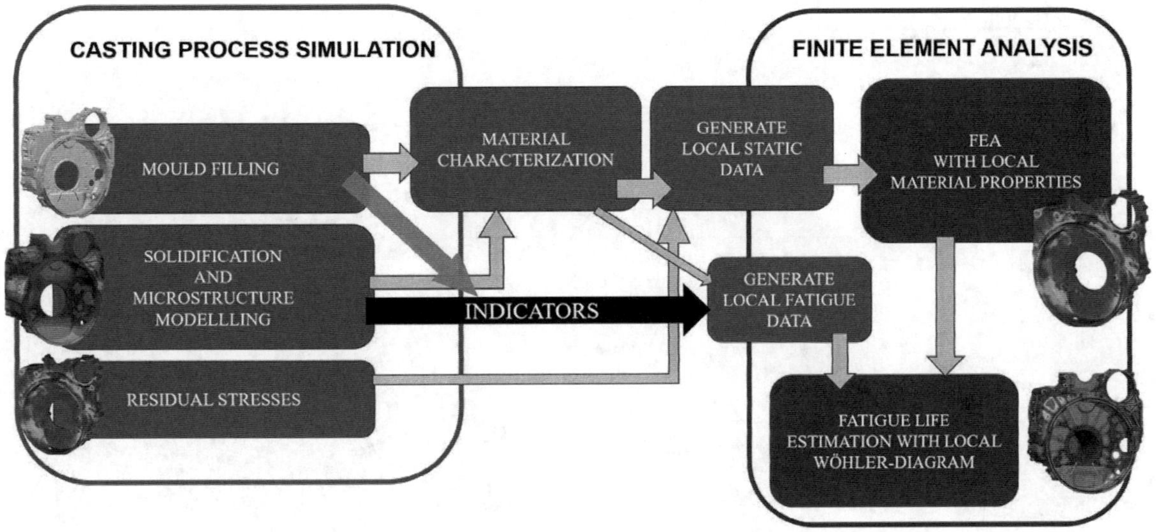

Figure 6. The general flow of numerical information in the closed chain of simulations for fatigue predictions. The black arrow indicates the step presented in the current work, and the grey arrow is a work in progress.

3.3.3. Application

The application of the methodology is demonstrated on a geometry previously used in other demonstrations of the closed chain of simulations [11,13]. The casting process simulation reveals several hot spots distributed throughout the geometry, figure 7(a). These regions are, as expected, found in thicker areas of the geometry or geometrical transitions. Though the entire part is simulated, the results from a cross-section through one of the hot spots, see figure 7(b), are further described in the following. To investigate the expected fatigue life if large pores occur in this hot spot, material categorisation is applied so that elements in this region are assigned to material data corresponding to category III, figure 8(a). The simulated distribution of von Mises stress at a given loadcase is shown in figure 8(b). If the risk of porosities is not considered, this stress level falls below the fatigue limit of category I. Thus, the entire region's fatigue life is estimated to be above 5 million cycles, figure 8(c). However, when the risk of defects is considered and the material is categorised as category III, the estimated

IOP Conf. Series: Materials Science and Engineering 1281 (2023) 012067 doi:10.1088/1757-899X/1281/1/012067

fatigue life in this region locally decreases to about 500 000 cycles, figure 8(d). Taking the risk of local defects into consideration thus provides a significantly lower estimation of the fatigue life.

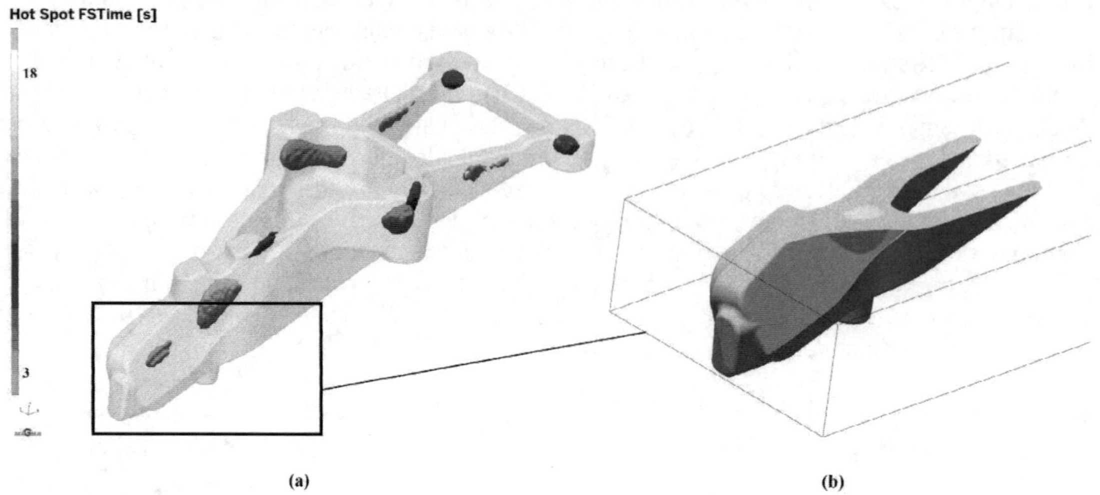

Figure 7. Results of Hot Spot FSTime over (a) the whole part and (b) the investigated cross-section.

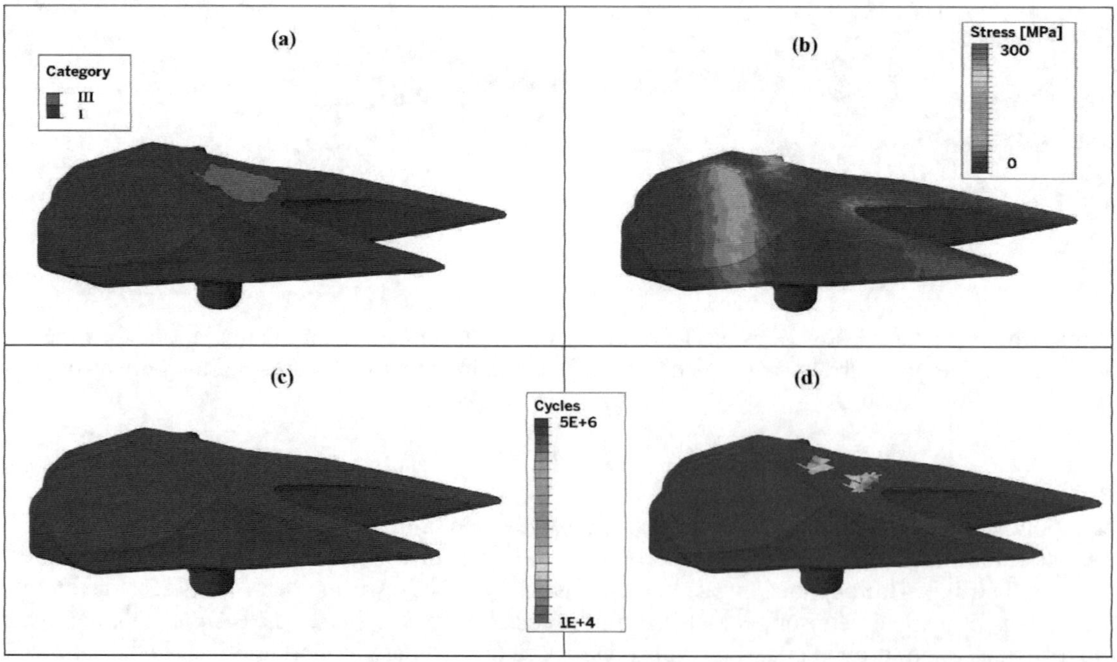

Figure 8. Output from the FEM simulation includes (a) material categorisation, (b) distribution of von Mises effective stress, (c) number of cycles to failure assuming constant fatigue life, and (d) number of cycles to fatigue failure with local consideration of porosities.

3.4. *Future outlook*

The approach enables an integrated transfer of material data from casting process simulations to FEM analyses and fatigue life predictions. This work has provided a step towards integrated simulations for fatigue life predictions but also highlights areas for future work. For instance, the current classification of materials based on the defect indicator is an acceptable approach at an early design stage. It enables designers to understand better the impact of different levels of porosities and material quality on the performance of the casting. A quantitative relation between predicted pore size and fatigue life would be preferred for parts in later design stages and in production. Castings also exhibit several other types of defects, e.g., oxide films, pre-solidified particles and inclusions. Future work in this field is aimed at understanding the role of invisible defects as oxide films on static and dynamic material behaviour, e.g., as presented in a related work in the current proceedings [20]. As for hot spots, melt quality and local flow-related aspects can also be considered local defect indicators. As indicated in the information flow in figure 6 (by a grey arrow), this is currently a topic for work in progress.

4. Conclusions

- The *closed chain of simulations* is extended to incorporate microstructure- and defect-based material properties into the industrial product realisation process
- The impact of different levels of porosity on the static behaviour and bending fatigue performance of a HPDC alloy is investigated and quantified.
- The experimental data is incorporated to enable defect-based fatigue properties in numerical fatigue life predictions.
- The numerical flow of data is outlined, and the application of the methodology is demonstrated.

Acknowledgement
The author would like to express his gratitude to Alexander Ahrnens and Patrik Engström for their master thesis work performed in this study. The Swedish Knowledge Foundation (KK-stiftelsen) is greatly acknowledged for the financial support of the research project ODISSEE2.

References

[1] Özdes H and Tiryakioglu M 2017 The Effect of Structural Quality on Fatigue Life in 319 Aluminum Alloy Castings *J Mater Eng Perform* **26** 736–43
[2] Tiryakioğlu M 2020 Intrinsic and extrinsic effects of microstructure on properties in cast al alloys *Materials* **13**
[3] Jolly M and Katgerman L 2022 Modelling of defects in aluminium cast products *Prog Mater Sci* **123**
[4] Gao Y X, Yi J Z, Lee P D and Lindley T C 2004 The effect of porosity on the fatigue life of cast aluminium-silicon alloys *Fatigue Fract Eng Mater Struct* **27** 559–70
[5] Weidt M, Hardin R A, Garb C, Rosc J, Brunner R and Beckermann C 2018 Prediction of porosity characteristics of aluminium castings based on X-ray CT measurements *International Journal of Cast Metals Research* **31** 289–307
[6] Nourian-Avval A and Fatemi A 2020 Fatigue life prediction of cast aluminum alloy based on porosity characteristics *Theoretical and Applied Fracture Mechanics* **109** 102774
[7] Olofsson J, Bjurenstedt A and Seifeddine S 2019 *On the effects of defects and imperfections on tensile toughness of a secondary aluminium alloy*
[8] Olofsson J 2012 Simulation of mechanical behaviour of cast aluminium components *International Journal of Cast Metals Research* **25** 319–27
[9] Dørum C, Laukli H I and Hopperstad O S 2009 Through-process numerical simulations of the structural behaviour of Al-Si die-castings *Comput Mater Sci* **46** 100–11
[10] Ridgeway C D, Gu C, Ripplinger K, Detwiler D, Ji M, Soghrati S and Luo A A 2020 Prediction of location specific mechanical properties of aluminum casting using a new CA-FEA (cellular

automaton-finite element analysis) approach *Mater Des* **194** 108929

[11] Olofsson J and Svensson I L 2012 Incorporating predicted local mechanical behaviour of cast components into finite element simulations *Mater Des* **34** 494–500

[12] Olofsson J and Svensson I L 2014 Closed chain simulations of a cast aluminium component - incorporating casting process simulation and local material characterization into stress-strain simulations *ISIJ International* **54**

[13] Olofsson J and Svensson I L 2012 Casting and stress-strain simulations of a cast ductile iron component using microstructure based mechanical behavior *Modeling of Casting, Welding and Advanced Solidification Processes XIII*

[14] Olofsson J, Svensson I L, Lava P and Debruyne D 2014 Characterisation and investigation of local variations in mechanical behaviour in cast aluminium using gradient solidification, Digital Image Correlation and finite element simulation *Mater Des* **56** 755–62

[15] Olofsson J 2018 Local microstructure-based material performance and damage in design and finite element simulations of cast components *J Comput Des Eng*

[16] Jansson J, Olofsson J and Salomonsson K 2019 On the use of heterogeneous thermomechanical and thermophysical material properties in finite element analyses of cast components *IOP Conference Series: Materials Science and Engineering* vol 529

[17] Jansson J, Olofsson J and Salomonsson K 2020 Simulation-driven product development of cast components with allowance for process-induced material behaviour *J Comput Des Eng* **7** 78–85

[18] Linder J, Arvidsson A and Kron J 2006 The influence of porosity on the fatigue strength of high-pressure die cast aluminium *Fatigue Fract Eng Mater Struct* **29** 357–63

[19] Garb C, Leitner M, Stauder B, Schnubel D and Grün F 2018 Application of modified Kitagawa-Takahashi diagram for fatigue strength assessment of cast Al-Si-Cu alloys *Int J Fatigue* **111** 256–68

[20] Olofsson J, Bogdanoff T and Tiryakioğlu M 2023 Revealing and simulating the effect of hidden damage on local and full-field deformation behaviour of cast aluminium *Modeling of Casting, Welding and Advanced Solidification Processes XVI*

IOP Conf. Series: Materials Science and Engineering 1281 (2023) 012068 doi:10.1088/1757-899X/1281/1/012068

Hot tear prediction in large sized high alloyed turbine steel parts – experimental based calibration of mechanical data and model validation

J Klinkhammer[1], J Thorborg[1], M Bernhard[2], J Winkler[2], C Bernhard[2], R Hanus[3] and G Tischler[3]

[1]MAGMA GmbH, Kackertstraße 16-18, 52072 Aachen, Germany
[2]Montanuniversität Leoben, Austria
[3] voestalpine Giesserei Linz GmbH, Linz, Austria

E-mail: J.Klinkhammer@magmasoft.de

Abstract. The main defects in heavy steel castings are related to hot tear formation during solidification. Depending on the steel grade, design, and local solidification conditions, it is possible to predict regions with higher risk of hot tear formation during the casting process. However, steels containing Boron show more complex crack and defect patterns compared to common steel casting alloys. The mechanisms behind the Boron induced hot tearing is investigated in this work to understand the influence of Boron enrichment during solidification and the influence on hot tearing. The experimental work includes the investigation of phase diagrams and the corresponding fractions of the solid and liquid phases depending on temperature using thermal analysis e.g. DSC and HT-LSCM. The hot tearing sensitivity and mechanical properties during solidification are obtained in the Submerged Split Chill Test, SSCT. In addition IMC-B 3-point bending tests are performed to determine high-temperature material properties in the solid state. The work is part of a research project where the final goal is to improve the hot tear predictions based on experimental work and carry out a benchmark simulation of a real sized casting and use it to show the agreement between the numerical results and extensive non-destructive testing from industrial observations.

1. Introduction – Industrial Motivation

Manufacturing of heavy steel castings used for gas and steam turbines involves several process steps, including Non-Destructive Testing (NDT) methods, defect excavation and repair welding, see Figure 1. The casting is removed from the mold after a typical solidification and cooling time of 2-6 weeks depending on the wall thickness, complexity, and material. Risers are removed by thermal cutting and subsequently the requested microstructure and mechanical properties are ensured by the quality heat treatment process. After rough machining the parts are inspected with various NDT methods and also as part of the final quality assurance after the stress relief.

Depending on the steel grade, the casting design and the constraints from the molding material, different levels and types of defects can be found. Typical heavy steel casting defects are:

IOP Conf. Series: Materials Science and Engineering 1281 (2023) 012068 doi:10.1088/1757-899X/1281/1/012068

- Cavities and surface sinks due to shrinkage where feeding has not been sufficient.
- Surface defects with accumulated non-metallic inclusions, which typically occur on the upper horizontal surfaces of the mold and cores.
- Cold cracks due to thermal stresses during solid state cooling.
- Hot tearing can lead to cracks with different levels of severity. They are considered to be very critical when they are located just below and perpendicular to the surface.

The NDT methods used for detecting defects are visual inspection, surface testing like Magnetic Particle Inspection and Liquid Penetrant Testing, and volume testing like Ultrasonic Testing and Radiographic Testing.

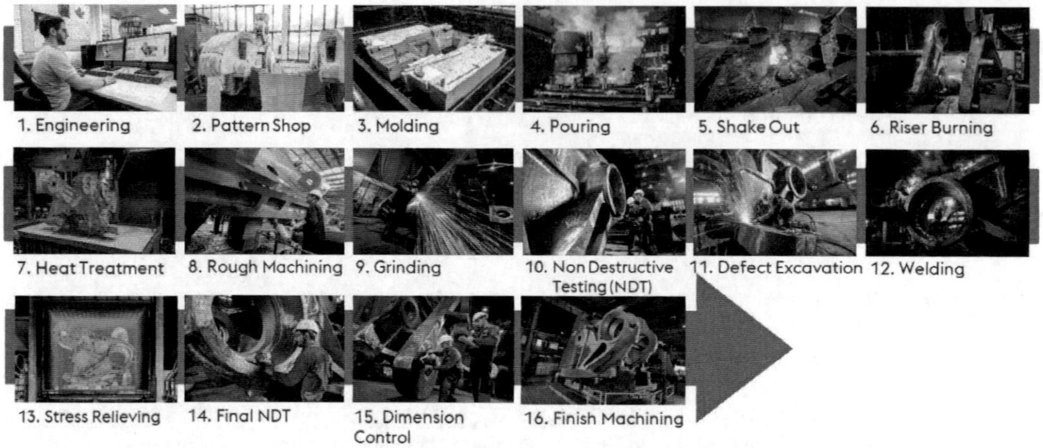

Figure 1. Typical production schedule for heavy steel castings

The demand for high temperature, high performance creep resistant steel grades used to further increase the efficiency of power plants, has gradually pushed the requirements to the manufacturing process where new complex alloyed steel grades are more sensitive to hot tear defects. The development in steel grades is summarized in Figure 2.

The casting materials most commonly used for turbine components of sub- and super critical power plants are G20Mo5, G17CrMo5-5, G17CrMo9-10, G17CrMoV5-10 steel grades (alloyed with Mo, CrMo and CrMoV) [1]. These low-alloy, creep-resistant materials are used in the temperature range up to 550°C. Today, especially in the field of medium pressure turbines, temperatures of up to 625°C are applied in ultra super critical (USC) power plants. Higher Cr contents and the use of elements such as V, Nb, W, Co, N, B significantly increase creep resistance. These grades are generally known as 9–10% Cr steels.

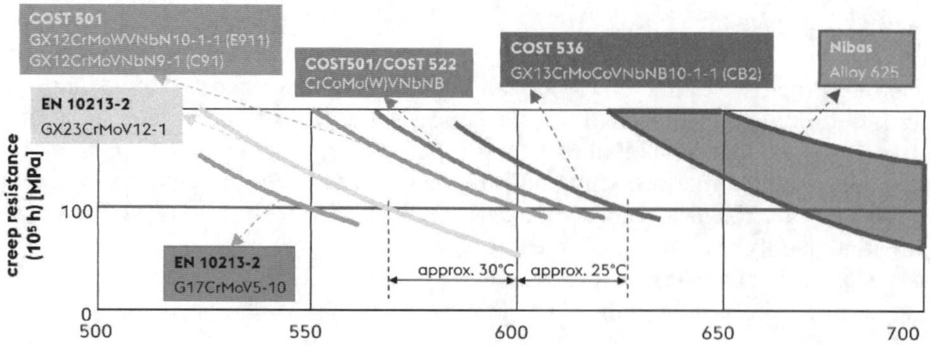

Figure 2. Limits of use for high-temperature turbine materials

The work presented in this paper is based on the on-going activities inside the research project Casting QPS, which is part of the Austrian funded project ICMPPE mentioned in the acknowledgment. The project is split into two branches focusing on various aspects of simulation and process control of heavy steel casting and continuous casting, respectively. In the considered heavy steel casting part of the project; in-situ measurements are used to calibrate thermomechanical simulation models to improve prediction of hot tearing for new high performance steel grades. The research project is motivated by the continuous development of steel grades, which requires a deeper understanding and control of the increased risk of hot tearing. The current state of technology for commercial production of heavy steel casting components is at the moment the CB2 steel grade with service load temperature up to 630 °C as shown in Figure 2.

2. Calibration of simulation model

2.1. Unified Creep Model

Thermomechanical modelling of the casting process is a challenge, where the main concern is to sufficiently model the response of the material at different temperature levels, on different time scales and often with different strain rates, which is governed by different deformation mechanisms. To have a reasonable compromise between the stability in the results and consistency between simulation results and experimental creep and tensile tests, a unified creep formulation is used as the fundamental constitutive law, [2], for the matrix material. The model is based on Norton's power law and includes by that, strain rate sensitivity and the possibility to describe creep at elevated temperature. Isotropic strain hardening is described by a classical power law, see (1) below.

$$\dot{\bar{\varepsilon}} = A\, exp\left(\frac{-Q}{RT}\right)\left(\frac{\sigma_y}{\sigma_{ref}}\right)^{\frac{1}{m}} \quad and \quad \sigma_{ref} = \sigma_{0,ref}\left(1 + \frac{E\, h}{n\, \sigma_{0,ref}}\right)^{n} \tag{1}$$

The Arrhenius expression scales the response according to the temperature dependency and the model is therefore applicable in a wide temperature range. The reference stress, σ_{ref}, describes the isotropic strain hardening governed by the hardening parameter h, the two temperature dependent properties, initial reference stress, $\sigma_{0,ref}$, and the hardening exponent, n.

2.2. Ductile damage model at high volume fractions of solid material

At high volume fractions of solid the material behaviour is described by a ductile damage model, i.e. a porous plasticity model, derived for a viscoplastic metal matrix containing spherical voids. Although the matrix material is plastically incompressible the compound material is plastically compressible due to the presence of voids. Instead of working with the (macroscopic) stress tensor σ_{ij} and its deviator s_{ij} respectively, it is convenient to work with the mean stress $\sigma_m = \sigma_{kk}/3$ and the equivalent stress $\sigma_e = \sqrt{3s_{ij}s_{ij}/2}$. Then the yield function Φ and the associated plastic potential $\Pi \equiv \Phi$ can be written in the general form $\Pi \equiv \Phi = \hat{\Phi}(\sigma_m, \sigma_e, \sigma_y, f, m)$ where m is the rate sensitivity and σ_y the rate-dependent (dynamic) yield stress of the matrix material. The void volume fraction f serves here as a scalar damage parameter. The (macroscopic) inelastic strain-rate tensor at yield is given by the flow rule $\dot{\varepsilon}_{ij}^{in} = \dot{\gamma}\frac{\partial \Pi}{\partial \sigma_{ij}}$ where $\dot{\gamma}$ is the strain-rate multiplier. Equivalently the volumetric and deviatoric plastic strain-rates defined by $\dot{\varepsilon}_p = \dot{\gamma}\frac{\partial \Pi}{\partial \sigma_m}$ and $\dot{\varepsilon}_q = \dot{\gamma}\frac{\partial \Pi}{\partial \sigma_e}$ will be used. It is assumed that the equivalent plastic strain varies according to the plastic work expression (plastic dissipation equation)

$$(1 - f)\sigma_y \dot{\bar{\varepsilon}} = \sigma_{ij}\dot{\varepsilon}_{ij}^{in} = \sigma_m \dot{\varepsilon}_p + \sigma_e \dot{\varepsilon}_q$$

and the evolution of the void volume fraction or damage f is given in the simplest case by

$$\dot{f} = (1-f)\dot{\varepsilon}_p.$$

We use the yield condition in the quadratic (elliptic) form

$$\Phi = \frac{A_2}{A_1}\sigma_m^2 + \sigma_e^2 - \frac{A_3}{A_1}\sigma_y^2 = 0$$

as proposed by [3]; the coefficients therein are functions of f and r and we choose the expressions $A_1 = 1 + \frac{2}{3}f$, $A_2 = \frac{9}{4}\left(\frac{rf^r(1-f)}{1+f^r}\right)^{\frac{2}{1+r}}$, $A_3 = (1-f)^{\frac{2}{1+r}}$ as given in [4]. In the limit of an undamaged material ($f = 0$) we retain the behavior of the plastically incompressible matrix material as in the unified creep case, (1).

2.3. Ductile damage model for solidification – interpretation of porosity

Application of a ductile damage model describing a viscoplastic metal matrix containing spherical voids in the context of solidification and hot tearing as proposed by [5] requires some considerations regarding the different phases and types of porosities involved.

We use the following simplified approach. Given the volume fraction of solid as function of temperature, $f_s = \hat{f}_s(T)$, the semi-solid region ,$f_s \in \left(f_s^{coalescence}, f_s^{coherence}\right)$, is split by the no-feeding value $f_s^{nf} = \hat{f}_s(T_{nf})$ in two parts:

Open pore space, $\left(f_s \leq f_s^{nf}\right)$. (Pore space saturated with liquid).

- The melt pressure at the solid-liquid interfaces is neglected and everything except the volume fraction of the solid is considered as porosity: $f = f_l = 1 - f_s$.
- Because of possible liquid-feeding neither the creation of shrinkage porosity nor damage porosity from the melt is possible, i.e. $f = f_l$, $f_p^{shrinkage} = f_p^{damage} = 0$.
- Due to the reduction of porosity/liquid the material is hardening during solidification. Nevertheless the liquid can be seen as a softening agent in this phase, [6].

Closed pore space, $\left(f_s > f_s^{nf}\right)$. Isolated pockets surrounded by solid material. (Pockets partially filled with liquid and voids).

- Dilation, $\left(\sigma_m > 0,\ \dot{\varepsilon}_p > 0\right)$:
  The volume fraction, that is not solid, is entirely used as porosity, i.e. $f = 1 - f_s = f_l + f_p^{damage} + f_p^{shrinkage} + f_p^{gas} + \ldots$, and therefore contributes to the growth of damage porosity. At present only liquid and damage porosity are considered, shrinkage and gas porosities are ignored.
- Compression, $\left(\sigma_m < 0,\ \dot{\varepsilon}_p < 0\right)$:
  The embraced volume fraction of liquid is treated as incompressible and only the remaining damage porosity can be reduced or completely removed in compression: $f = 1 - f_p^{damage}$.

3. In-situ characterization of hot tearing and thermo-mechanical behaviour of steel

3.1. IMC-B test – In-situ Materials Characterization

The IMC-B-test provides the bending of a sample (200x60x20 mm) after solidification and controlled cooling at temperatures between 700 and 1150 °C, Figure 3. In former research work, mainly defect formation in the continuous casting of steel has been addressed so far, [7]. Possible experimental procedures include bending (pre-defined deformation rate and maximum deflection) and creep testing by applying a constant load and measuring the deflection versus time at a constant temperature. The results are here used to calibrate the applied constitutive model, equation (1).

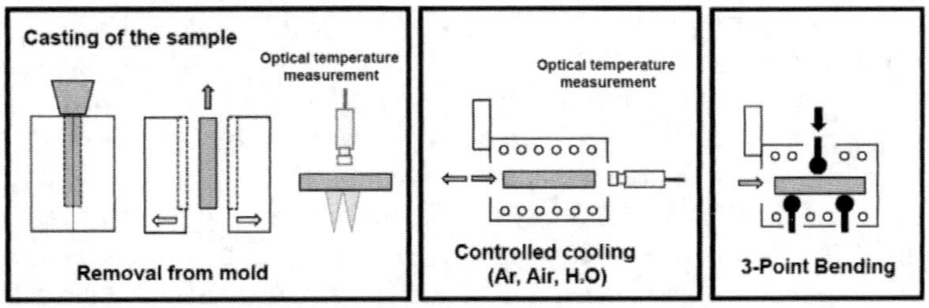

Figure 3. Schematic illustration of the IMC-B experimental procedure.

Within the present work, creep tests and bending tests on steel with 0.17 %C, 0.4 %Si and 1.5 %Mn and the alloy G17CrMoV5-10-2 were performed at temperatures of 700, 800, 900, 1000 and 1100 °C. Figure 4 shows selected bending force curves versus deflection. The influence of the moderate additions of Cr, Mo and V is remarkable: The bending forces become twice as high in the temperature range of 800 to 1000 °C. The measurement serves to calibrate the material law as discussed later.

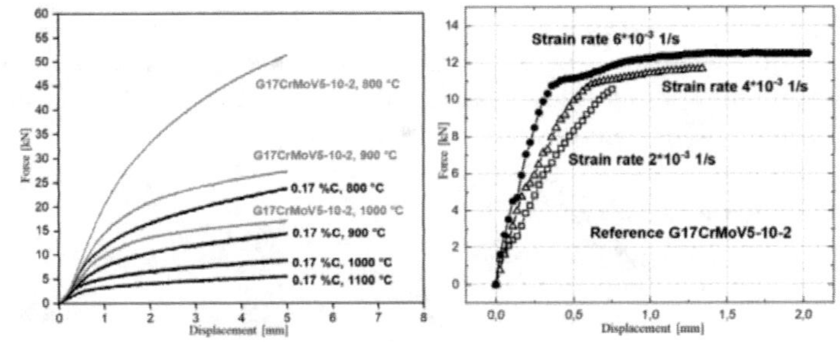

Figure 4. Force displacement curves. Left: The two considered materials at various temperatures. Right: One material for 3 selected strain rates.

3.2. IMC-B test – Data calibration

The heated 3-point bending system is modelled like the definition of the experimental setup. The thermal boundary conditions control the different temperature levels and the bending bar is supported on each end while a piston is moved from the upper side by applying a displacement boundary condition as a function of time.

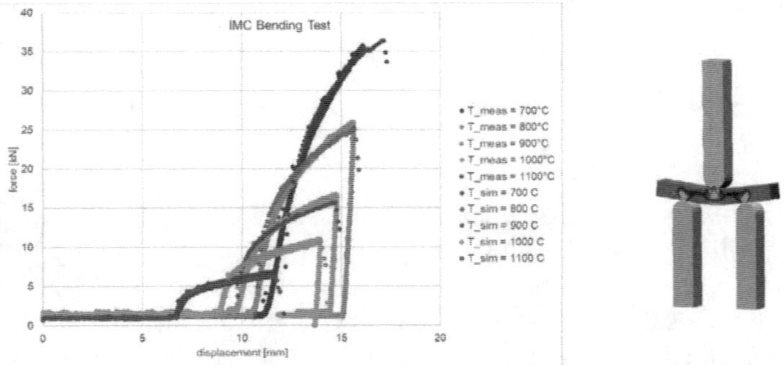

Figure 5. Left: Calibration of IMC bending results, measurements and simulation results. Right: Simulation setup

An in-house dataset was used as a starting point to describe the thermophysical data. Further calibration of equation (1) was done by adjusting the initial reference stress and the strain hardening exponent to obtain agreement between the simulated and measured curves.

3.3. SSCT test – short description

The experimental equipment was originally developed by Ackermann et al. [8] for *in-situ* testing of Al-alloys. In the past decades, the Submerged-Split-Chill-Tensile (SSCT) test has been successfully applied for laboratory characterization of hot tearing close to continuous casting conditions [9] by adopting the testing procedure to steel alloys. A similar apparatus is also used by Suzuki et al. for steel [10]. The cylindrical specimen consists of a fixed upper part and a moveable lower part. The lower part is usually coated by Y-stabilized ZrO_2 material. The coating thickness (here 0.40 mm) defines the heat transfer and, as a consequence, the cooling rate in the test. The experimental procedure is illustrated in Figure 6. At stage I, the steel composition and superheat of the melt are specifically adjusted in an induction furnace with a capacity of 25 kg. The steel bath's temperature is controlled by individual thermocouple measurements. During stage II, the specimen is submerged into the liquid steel. The initial solidification starts and the shell growth proceeds for a pre-defined period of time. At stage III, the setup is changed to distance-controlled mode and the actual tensile test starts by moving the lower part downwards with a defined speed v. In this testing period, the solidification process still continues. When the tensile test is completed, the specimen is lifted out of the melt (stage IV).
Thermocouples of type k record the temperature 2 mm below the specimen surface during the whole experiment. The measurements are used as thermal boundary conditions in the heat transfer model. The force between the upper and lower part is measured by a load cell, the actual position of the lower part is determined by an inductive position sensor. If only a low number of hot tears are formed, or, the minor defect pattern is analysed within the metallographic work, the measured force-elongation curves can be used for calibrating thermo-mechanical models.

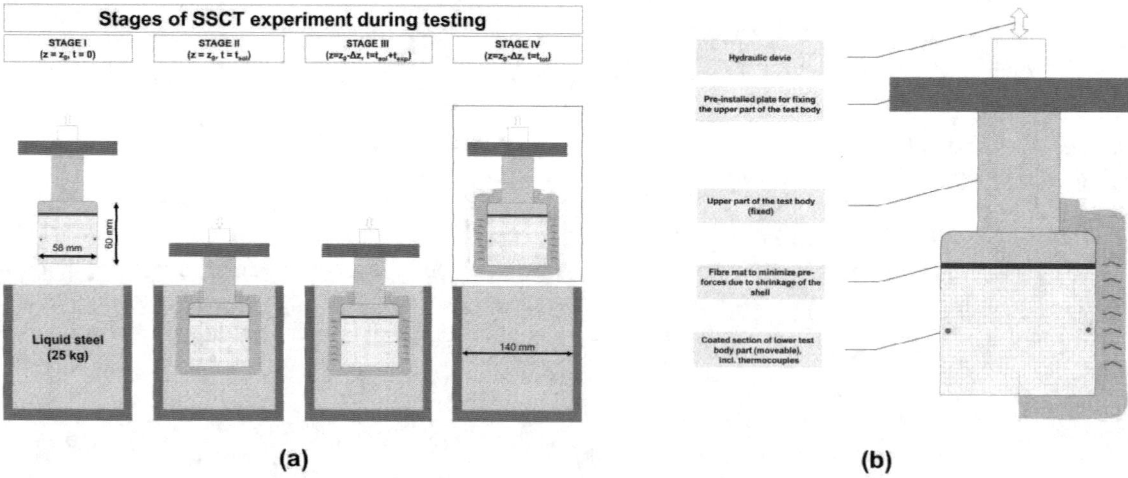

Figure 6: Left: Experimental procedure of the SSCT test. Right: Enlarged view of the cylindrical specimen with upper and lower part

3.4. SSCT test – simulation setup and data calibration

The complete system of liquid steel, outer wall, upper tool and the moving piston is defined in the simulation model to be able to calculate the temperature distribution in the system and the mechanical loading and response of the solidifying shell. The details of the geometry is shown in Figure 7.
The thermal and mechanical simulation steps are fully coupled by using the mechanical solution to determine the thermal interface conditions between the liquid steel and the piston. In this way the

IOP Conf. Series: Materials Science and Engineering 1281 (2023) 012068 doi:10.1088/1757-899X/1281/1/012068

nominal[1] heat transfer coefficient is made gap dependent; as shown below a constant nominal heat transfer coefficient of 1500 $Wm^{-1}K^{-1}$ turned out to be sufficient to reproduce the temperature readings from thermo-couples located 2 mm below the surface inside the tool for the cases where the tool was coated. Similar results could also be obtained in a more complicated way by calibrating a temperature dependent heat transfer coefficient with optimization. Standard convection and radiation equivalent boundary conditions are used to the surrounding.

The mechanical loading of the system is done with the specified strain rate from the experiments and applied via a displacement boundary condition on top of a small block above the piston, see Figure 7 for the indications of the "Z-displacement" curve. The block is initially lifted while the solidifying shell is forming, i.e. before the load is applied to the piston. This is done to compensate for the thermal expansion of the tool and the piston, while being submerged into the liquid steel. The initial lift is calibrated to just have the interface between the block and the piston closed when the loading is started.

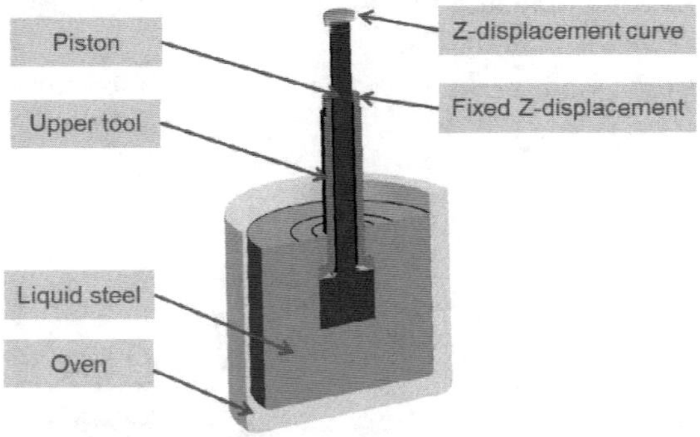

Figure 7: Simulation model of the SSCT test. The dimensions can be seen on the drawing in Figure 6.

The temperature result 2mm into the tool is compared to 5 measurements in Figure 8. The result is inside the spread of the measurements and the overall temperature distribution is used for the coupled thermomechanical analysis. The shape and size of the solidified shell is shown next to the curves and corresponds well to the size of the measured shell thickness.

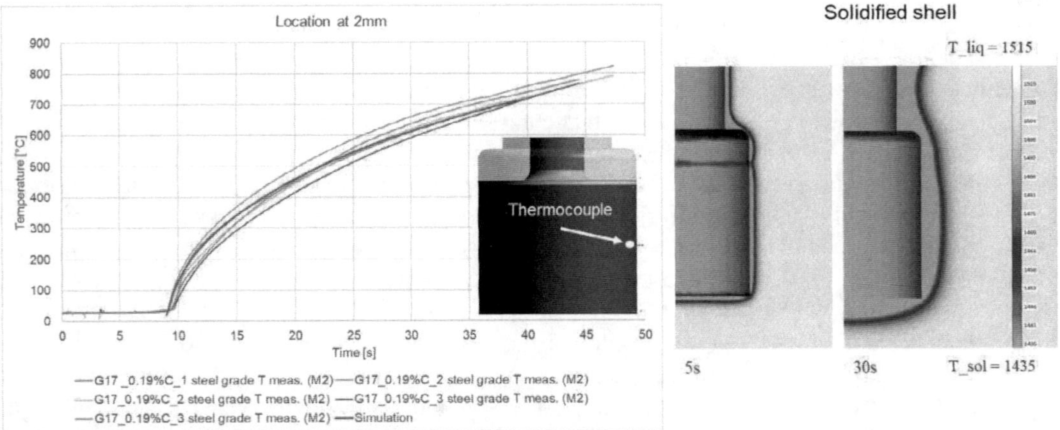

Figure 8: Left: Temperature results in the tool, 2mm from the interface – comparing to measurements. Right: Solidified shell thickness at the end of the experiment.

[1] *Nominal* heat transfer coefficient means the heat transfer coefficient that is assigned in the program.

Multiple experiments have been done with different holding times and different strain rates. The elongation curve and measured force response curve are shown in Figure 9 for one experiment.

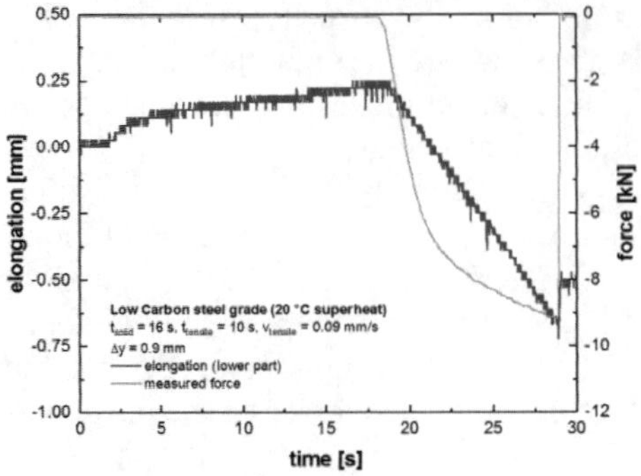

Figure 9. Elongation and force curve as recorded during the test.

Two experiments and the corresponding simulation results, i.e. the stress level inside the piston, are shown in Figure 10. Similar agreement has been found for several different load cases.

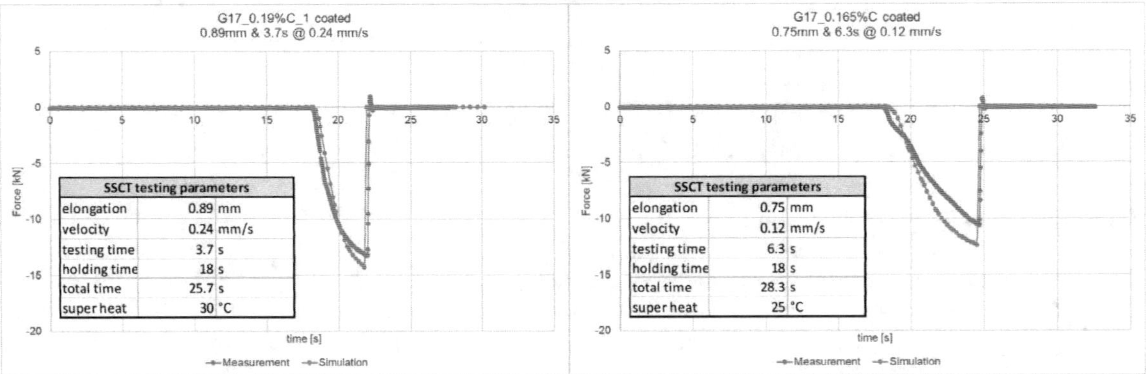

Figure 10. Force response for two different experiments compared to the corresponding simulation results.

4. Outlook – first measurement of the influence of Boron

The SSCT tests on the G17CrMoV5-10-2 indicate a relatively low sensitivity for forming hot tears and hot tear segregation. Severe defect formation requires more than 1.5 mm elongation in the SSCT test, and the detected defects are mostly hot tear segregation. Defect formation in real castings becomes nevertheless possible if a coarse solidification microstructure and an unfavourable geometry of the casting results in strain localizations. The consistency between measured force elongation curves for the SSCT and IMC-B-test and the simulations is excellent, as well as the qualitative indication of the formation of hot tears in the SSCT experiment.

An example of an element with a more complex influence on defect formation is Boron, commonly used to increase the creep resistance of large steel castings at elevated temperatures. Boron is a strongly segregating element. Even additions of only a few ppm of Boron significantly lower the solidus temperature. Thermodynamic investigations and in-situ observations by high-temperature confocal microscopy also indicate a remarkable further effect of increasing Boron content, the so-called "retrograde-melting" [11]. For steel with 0.1 %C, 150 ppm of Boron result in the formation of a liquid phase after solidification at temperatures below 1200 °C. Boron may then not only result in an increasing hot tear sensitivity within the mushy zone due to a widening of the critical temperature range but also cause the formation of defects at temperatures between 1150 and 1200 °C.

The addition of Boron to 0.17 %C steel with 0.4 %Si and 0.7 %Mn (equivalent to the G17CrMoV5-10-2 without Cr. Mo and V) significantly influences the response of the solidifying shell during the SSCT test, Figure 11. The addition of 35 ppm Boron to the same steel results in a thinner shell due to the lower solidus temperature. In consequence, the measured force is also lower. The number of detected defects increases, but the morphology of the defects is still that of a classical hot tear. Figure 11 shows the position and length of the detected defects in the shell and an example of the typical morphology of a segregated hot tear. The dotted lines indicate the solidus isotherms for a solid fraction of 0.96 and 1.0. The mechanism of the formation of these defects is still in accordance with the common understanding.

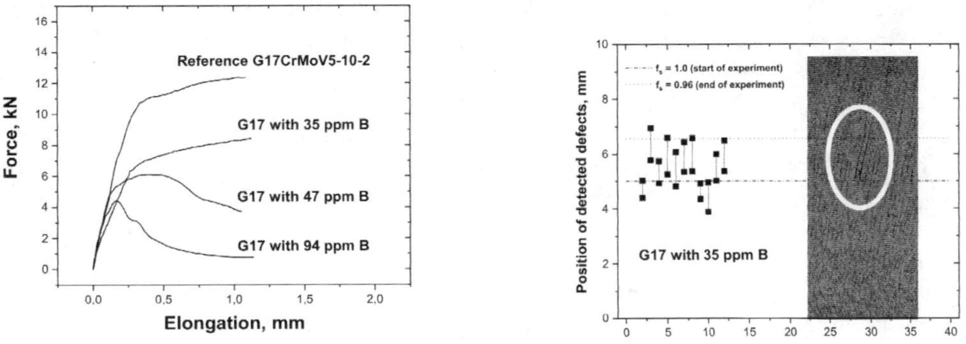

Figure 11. Left: Force-elongation curves for G17 with increasing Boron content. Right: Position of detected defects with an example of the typical morphology.

For 47 ppm of Boron, the measured force achieves a maximum at 0.4 – 0.5 mm elongation and decreases afterwards, Figure 11. This behaviour is no longer the consequence of hot tearing but must be related to massive shell damage during the experiment. For 94 ppm of Boron, this behaviour becomes even more pronounced, and the elongation of only 0.2 mm is already high enough to exceed the limit for massive shell damage. Figure 12 indicates the position of the detected defects and gives examples of the morphological changes: The cracks appear much closer to the surface, indicating at least a partial formation after solidification. The scattering of the position is mainly due to the local unevenness of shell thickness, the nucleation and growth of the cracks during and after solidification.

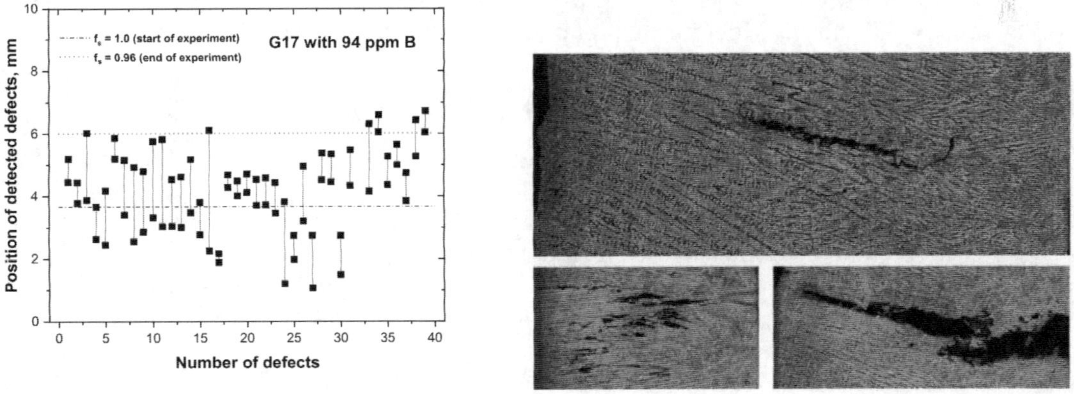

Figure 12. Left: Position of detected defects with isotherms for solid fractions 0.96 and 1.0. Right: Micrographs with examples of the typical morphology of defects in SSCT samples for steel with 94 ppm Boron.

Boron influences the defect formation phenomena significantly, but so far, conventional damage models are unable to consider these effects, last but not least, due to the lack of thermodynamic data.

The detected shell damage and softening behaviour in the force response curve is showing the fundamental need for the presented damage model in section 2.2, which cannot be sufficiently covered by the unified creep model above the coalescence point.

Summary

Hot tearing in heavy steel castings is an ongoing challenge in the foundries. In-situ and NDT methods present an opportunity to gain insight into the behaviour of newly developed steels, the calibration of material models, and the improvement of their predictive capabilities. The presented work is focused on extending the modelling capabilities to take large cracks and softening phenomena into account, which are observed for high performance turbine steels. In-situ testing has been used to measure the force response of the SSCT and IMC-B tests. The measurements have been used to calibrate the thermophysical data for the implemented unified creep model. Recent measurements have shown the crack sensitivity to Boron, where shell damage and softening are caused by low melting phases and different grain sizes, which influence the behaviour in the mushy zone. A previously presented porous damage formulation is implemented with the objective to model the observed behaviour with the intention to improve the prediction of hot tearing in the heavy steel cast parts as used in power plants.

Acknowledgment

The authors gratefully acknowledge the financial support under the scope of the COMET program within the K2 Center "Integrated Computational Material, Process and Product Engineering (ICMPPE)" (Project no. 859480). This program is supported by the Austrian Federal Ministries for Transport, Innovation, and Technology (BMVIT) and for Digital and Economic Affairs (BMDW), represented by the Austrian research funding association (FFG), and the federal states of Styria, Upper Austria, and Tyrol.

References

[1] Hanus R, Process-Development and Experience for Materials C91, C911, CB2 and Alloy 625 for Heavy Walled Cast Components, 45. Vortragsveranstaltung, Forschungsvereinigung Warmfeste Stähle und Hochtemperaturwerkstoffe (FVWHT), Düsseldorf , Germany, 2022.
[2] Thorborg J, Hattel J H, Bellini A. "Thermomechanical conditions in heat treated aluminium cast parts." In Gandin C, editor, Modelling of Casting, Welding and Advanced Solidification Processes: MCWASP XI. Vol. 11. p. 193-200. 2006.
[3] Cocks A C F. "Inelastic deformation of porous materials." Journal of the Mechanics and Physics of Solids, 37, 6, 693-715, 1989.
[4] Michel J C, Suquet P. "The constitutive law of nonlinear viscous and porous materials." Journal of the Mechanics and Physics of Solids, 40, 783-812, 1992.
[5] Monroe C A. "A Modeling and Experimental Study of Deformation and Hot Tearing in Steel." Ph.D. thesis, University of Iowa, 2008.
[6] Martin C L, Favier D, Suery M. "Viscoplastic Behaviour of Porous Metallic Materials Saturated with Liquid. Part I: Constitutive Equations." Int. Journal of Plasticity, 13, 3, 215-235, 1997.
[7] Krobath R, Bernhard C. "Experimental Quantification of Critical Parameters for Prediction of Surface Crack Formation in Continuous Casting." Steel research int., 91: 2000234. https://doi.org/10.1002/srin.202000234.
[8] Ackermann P, Kurz W, Heinemann W. "In Situ Tensile Testing of Solidifying Aluminium and Al-Mg Shells." Materials Science and Engineering, 75, 79-86, 1985.
[9] Pierer R and Bernhard C. "The nature of internal defects in continuously cast steel and their impact on final product quality." Proceedings of AIST, Pittsburgh, USA, 2010, 193-203.
[10] Suzuki M, Suzuki M, Yu C, Emi T. "In-Situ Measurement of Fracture Strength of Solidifying Steel Shells to Predict Upper Limit of Casting Speed in Continuous Caster with Oscillating Mold." ISIJ International, 37, 4, 375-382, 1997.
[11] Yin H, Blazek K, Lanzi O. " "In-situ" Observation of Remelting Phenomenon after Solidification of Fe–B Alloy and B-bearing Commercial Steels." ISIJ International, 49, 10, 1561-1567, 2009.

IOP Conf. Series: Materials Science and Engineering 1281 (2023) 012069 doi:10.1088/1757-899X/1281/1/012069

Modelling hot distortion of inorganic bonded sand cores and application on complex 3D printed automotive cores

J Thorborg[1], J Wendling[2], J Klinkhammer[1] and M Heitzer[1]

[1]MAGMA GmbH, Kackertstraße 16-18, 52072 Aachen, Germany
[2]BMW AG, Landshut, Germany

E-mail: J.Thorborg@magmasoft.de

Abstract. Inorganic bonded sand cores are getting increasing attention in industry due to their environmental advantages, and they are now widely used for series production in automotive applications. The advantages of these binder systems are nevertheless associated with a greater sensitivity of the core quality in relation to the manufacturing and storage process before casting, which is also having a major impact on the core strength. This paper presents the current work in integrating the modelling of the binder decomposition and evaporation of binder water with the mechanical performance of partially hardened and dried sand cores. This allows a local description of mechanical properties, considering different behaviour in compression and tension which depends on temperature and different levels of moisture content through the core. Recent work has shown the importance of including creep effects in the numerical modelling of 3D Printed (3DP) inorganic sand cores. Measurements of the mechanical behaviour of 3DP cores compared to shot cores have shown some fundamental differences in strength due to different density and binder content. This has been investigated in detail in collaboration with a major automotive manufacturer and the application of the model extension was done on a complex water jacket core.

1. Introduction

Casting is one of the most flexible manufacturing processes to produce parts of different shapes and sizes. The process is applicable for a wide range of alloys and the obtained material quality and dimensional tolerances can be controlled by modifying either the part design or the casting process. The flexibility in design often leads to complex mold and core layouts, which requires high precision manufacturing of the shaping materials and stable performance during the casting process.

State of the art simulation of the casting process is today used to analyze and optimize both the casting design and the casting process by combining the thermal simulation of the casting process with an integrated prediction of stress and distortion. It is possible to analyze how the thermal gradients, cooling time and constraints from the mold and cores affect the risk of stress related defects and unwanted distortion. Application on industrial examples show the significant advantages of being able to systematically analyze different designs and casting layouts prior to series production, to avoid quality problems in production and fulfill dimensional tolerance requirements.

Traditionally, the focus has been on the cast part itself, the alloy and variations in the process conditions. However, in recent years it has been increasingly important to pay more systematical attention to the thermomechanical performance of the shaping materials to control the tight tolerance requirements and the stability during the casting process.

IOP Conf. Series: Materials Science and Engineering 1281 (2023) 012069 doi:10.1088/1757-899X/1281/1/012069

Many foundries are switching to inorganic bonded cores or looking into the possibility to make the switch to benefit from the more environmentally friendly binder system, [1]. The advantages of these binder systems are associated with a greater sensitivity of the core quality in relation to the manufacturing and storage before the casting process. The drying process as well as the moisture distribution and the storage conditions of the core have a major impact on core strength and core gas-related casting defects. In addition to the traditional core shooting process, 3DP sand cores have gained an increasing interest in the later years. The design freedom of 3DP sand cores makes the technology very attractive for highly complex geometries where other methods are not applicable or would require assembly of several sub components. 3DP sand cores are based on a powder bed technology where a re-coater distribute layers of sand followed by a printing step, based on the binder-jetting process to form the final geometry of the core layer-by-layer. Compared to the traditional manufacturing of sand cores the method is known to have directional dependencies in the mechanical properties and the density is typically reported to be lower compared to methods where the sand is compacted during the core shooting process. The final conditions of the sand core are also dependent on microwave drying and hardening.

The objective of this paper is to have a closer look on the differences in mechanical performance and mechanical properties of shot and 3DP inorganic sand cores. The work was motivated by an industrial case where a 3DP water jacket core showed significant deformation in thin-walled sections, which required pre-shaping of the core to stay inside the specified tolerances. Simulation of hot distortion of the complex core was intensively used to analyze the influence of thermomechanical data, variations in the casting process and reviewing the differences in the behavior of the shot and 3DP cores. Mechanical tests were performed to provide mechanical data to the implemented constitutive models, according to the procedures presented in [2].

2. Mechanical model

During core manufacturing loose sand is bonded as a porous material with high strength in compression and weak and brittle behavior in tension, where tensile strength depends on the binder type. The fundamental behavior furthermore depends on sand type, grain size distribution and grain morphology. The sand types used in the foundry have different levels of thermal expansion with possible contributions from phase transformations at various temperature levels. The sand changes state during manufacturing from a loose granular material to a bonded core. Due to binder degradation it changes partially back to a granular state during casting, which is also wanted in order to be able to remove the sand material from the casting.

Due to the different behavior in compression and tension it is not sufficient to use a standard J_2 plasticity formulation, which is only formulated in the deviatoric stress and strain space. Instead a soil plasticity model has been considered to model the mechanical behavior of the bonded sand, based on the original Drucker Prager Cap model. The model has a pressure dependent yield surface, with a high strength for high confining pressure and a lower strength for tensile conditions, [3]. The pressure (mean stress) is related to the volumetric plastic strain, which describes the inelastic response of the porous material. The classical Drucker Prager Cap model is fundamentally defined by two regions which define the yield surface of the model. One region describes how the material dilates during plastic deformation and the other region defines how the material is compressed during plastic deformation. The first region is described by the failure envelope, equation (1), which is defined by the line with the slope M and the pressure offset, Pt. The second region is described by the cap, equation (2), which limits the deformation in compression. The cap is typically defined by an ellipse. The classical model is illustrated in Figure 1 below. To improve the convergence behavior of the implemented model, smaller modifications were done to smoothen the interface between the cap and the failure line and the failure line and the pressure axis. Smoothing the two interfaces reduces the unwanted influence of the vertices on the yield surface.

$$\Phi^D(\sigma) \equiv \sqrt{J_2\big(s(\sigma)\big)} + \sqrt{2/3}\,M[p(\sigma) - p_t] \quad (1)$$

$$\Phi^C(\sigma, a) \equiv \frac{1}{\beta^2}[p(\sigma) - p_t + a]^2 + \left[\frac{q(\sigma)}{M}\right]^2 - a^2 \quad (2)$$

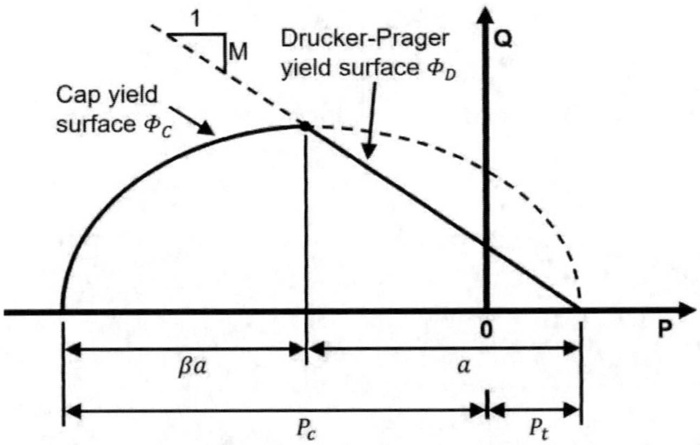

Figure 1. Sketch of the classical Drucker Prager Cap model. The yield surface is defined by the failure envelope and the cap. The slope M defines the failure envelope and the ellipse defines the cap.

Time dependent deformation was observed for the 3DP sand cores at high temperature which corresponds well with the large deformations measured on the real water jacket core. A rheological model was used to describe the time dependent behaviour in the constitutive model. Based on the various creep tests it was possible to extract a viscosity type deformation behaviour of the material. It is assumed that softening at high temperature is associated to additives and to the glass transition temperature of the sand and binder mixture used for the 3DP cores.

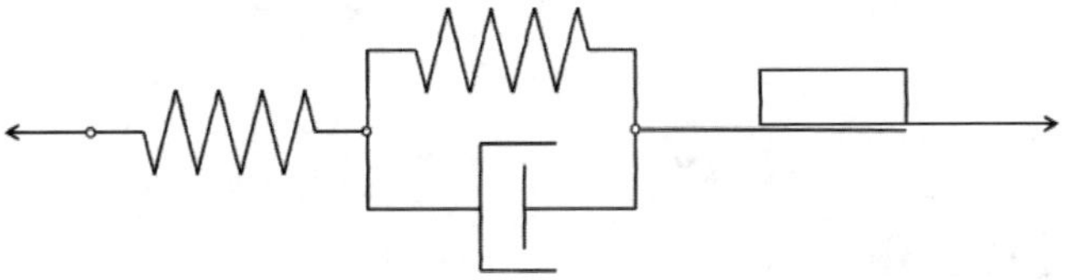

Figure 2. General rheological body to describe the creep behaviour in the considered 3DP core

The strain tensor in the dashpot can be generalized in two different ways, either in the same way as the elastic strain tensor, which includes both a deviatoric and a volumetric contribution or the deviatoric part only

$$\dot{\varepsilon}_{ij}^{vi} = \frac{1}{\eta}\{(1+v)\sigma_{ij} - v\delta_{ij}\sigma_{kk}\} \quad \text{or} \quad \dot{e}_{ij}^{vi} = \frac{1}{\eta}s_{ij} = \frac{1}{\eta}\left\{\sigma_{ij} - \frac{1}{3}\delta_{ij}\sigma_{kk}\right\} \quad (3)$$

where, σ_{ij}, s_{ij}, and η are the stress in the Maxwell element, and the viscosity of the dashpot, respectively.

Overall, the strain contribution from the rheological model is added to the plastic strain component from the soil plasticity model to comprise a combined response, governed by both yielding and creep. The creep response from the rheological model is made pressure sensitive with a pressure dependent viscosity. The strain decomposition equation can easily be extended to include the plastic strain contribution from the soil plasticity model, which gives a corresponding change to the stress update equation. The simple derivation of adding the Maxwell element to the constitutive behaviour is presented in [2].

IOP Conf. Series: Materials Science and Engineering 1281 (2023) 012069 doi:10.1088/1757-899X/1281/1/012069

3. Mechanical tests – shot and 3DP sand cores

Different mechanical tests are needed to provide sufficient input data for the Drucker Prager Cap model. The tests have to provide strength information for different stress states in terms of pressure and tension. For the considered application the failure line is the main focus and the corresponding mechanical data are determined from three conventional mechanical tests, which have different stress states and by that different locations on the pressure axis, [4]. The 3 point bending test provides input to the tensile region, while two cylinder tests provide information to the compressible region by performing compression tests and indirect tensile tests. The loading and displacement response are used to calculate the corresponding pressure, P and effective stress, Q, which is used to extract the properties for the failure line. The pressure cap is determined from triaxial measurements, but the details are not address here.

Figure 3. Left: 3 point bending test, middle: Compression test, right: Indirect tensile test

For the 3DP sand cores the mechanical behaviour depends on the print direction. The 3 basic mechanical tests, the evaluation of thermal expansion, Young's modulus and Poisson's ratio were all done on samples printed in different directions. The 3 tests and the different print directions are sketched in Figure 4.

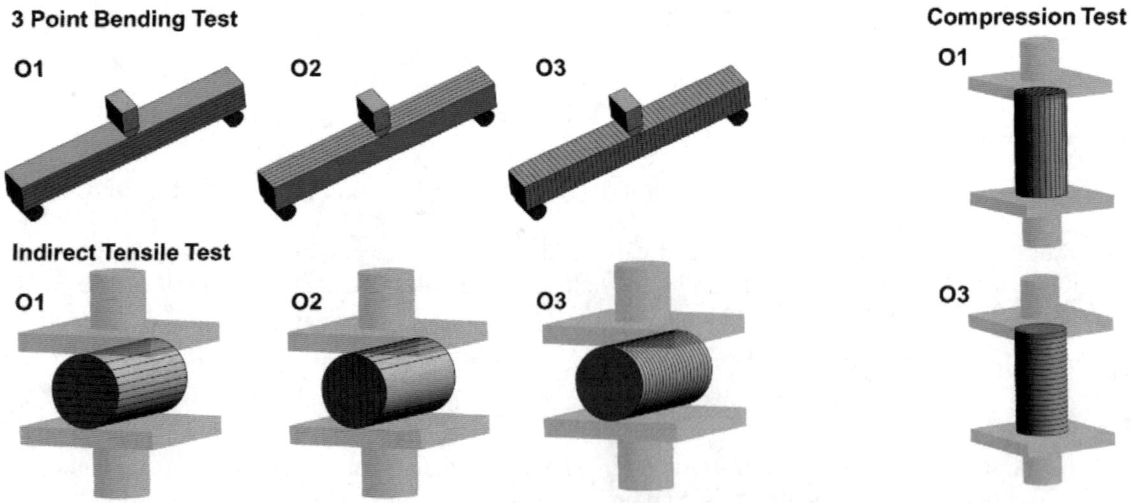

Figure 4. Top left: 3 Point bending test in 3 directions, Lower left: Indirect tensile test in 3 directions, right: Compression test in 2 directions. The lines indicate the 'layers' due to the print direction.

The information of the performed tests and the results of the tests used for extracting the mechanical data are not published yet, but currently being prepared for publication by the co-author J. Wendling. The outcome of the experiments however, are summarized in the subsequent sections.

3.1. Binder and sand types

Two different silica sands were used for the shot cores and the 3DP cores, GS14 and H32, respectively. The AFS fineness number calculated from the size distribution (American Foundry Society), the mean grain size, and the bulk density are listed in Table 1.

A BFD binder system from Reinsicht GmbH was used for the printing process with the fluid additive TREXEA added together with a powder additive by the re-coater. The cores were printed on a ExOne Exerial 3D printer. The Inotec binder system from ASK was used for the shot cores and they were manufactured on a Laempe L1 core shooting machine.

Table 1. Details of the two considered sand types

	GS14	H32
AFS (fineness number)	99	45
Mean Grainsize in μm	130	320
Bulk Density in t/m^3	1.32	1.40

3.2. Basic mechanical tests and data

To extract data for the Drucker Prager Cap model the core material was generally tested at different temperature levels from room temperature to 500 °C. The measurements are collected in the 3 graphs in Figure 5. Each curve represents the average of 5 tests for the indicated direction and for the shot cores. The curves are shortly commented below the figure. The values are used to collect P-Q information for the Drucker Prager Cap model used for the simulation.

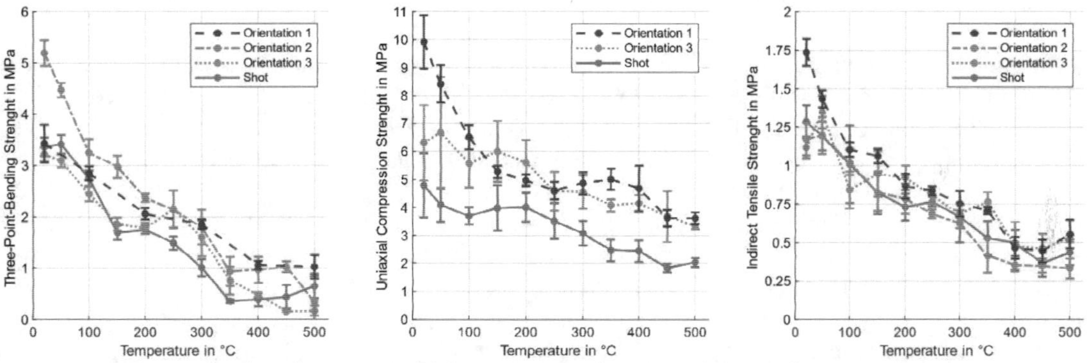

Figure 5. Left: 3 point bending test, middle: Compression test, right: Indirect tensile test

3.2.1. 3 Point Bending test. The Bending strength of the shot cores at RT is $(3.36 \pm 0.18)\ MPa$, which was also used as reference for the calibration of binder content of the 3D-printing process. The strength of O1 at room temperature is $(3.42 \pm 0.37)\ MPa$, which correspond well to the strength of the shot cores. The strength is assumed to be partially governed by the 'quality' of the lower printed layer. For O2 all layers are loaded equally and the strength is $(5.19 \pm 0.24)\ MPa$. The strength of O3 is $(3.22\ \pm 0.13)\ MPa$ which is similar to O1 although the tensile stresses are induced perpendicular to the weak connection between layers. For increasing temperature the strength generally drops and especially O2 and O3 drops below the strength of O1 and the shot cores.

3.2.2. Indirect tensile test. O1 is the strongest direction at RT which is 35% higher than the strength of the shot cores. The two other directions, O2 and O3 are 9% and 13% weaker than the shot cores, respectively. O2 and O3 have some oscillations in the strength level at lower temperatures, but all measurements generally decreases for increasing temperature. O2 is decreasing to the lowest level compared to the other measurements, which corresponds well to the tensile stress response being perpendicular to the printed layers.

3.2.3. Compression test. The compression strength O1 and O2 of the 3DP cores are significantly higher than the compression strength of the shot cores at RT, 107 % and 32 % for O1 and O2, respectively. The standard deviation for both orientations are quite high, which can be explained by an unusually big variation in the fracture pattern of the tests. For increasing temperature the strength in O1 and O3 gets closer and stays well above the strength of the shot cores. The level generally drops, but the compression strength of the 3DP cores at 500 C is almost at the level of the compression strength of the shot cores at RT.

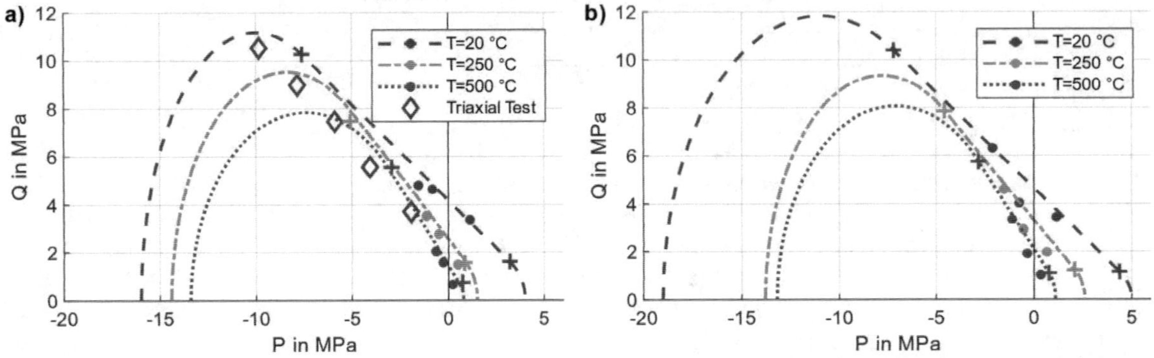

Figure 6. Drucker Prager Cap yield surfaces for different temperature levels. Left: shot cores. Right: 3DP cores

Data from the three mechanical tests were used to provide P-Q data for the Drucker Prager Cap model according to the equations for the different load cases, [4]. The combined data are presented in Figure 6, for both the shot cores and 3DP cores at various temperature levels. Regarding 3DP cores, a conservative approach was chosen, that means the weakest orientations for all tests were selected. The 3 point bending strength is represented by O1, the indirect tensile strength by O2 and the compression strength by O3. The cap region was estimated from some earlier triaxial measurements performed on organic bonded cores. However, the cap region is not considered important for the current application and it is therefore not further addressed in this paper.

3.3. Thermal expansion, density and Young's modulus

The two orientations O1 and O3 corresponds to the orientations for the two compression tests sketched in Figure 4.

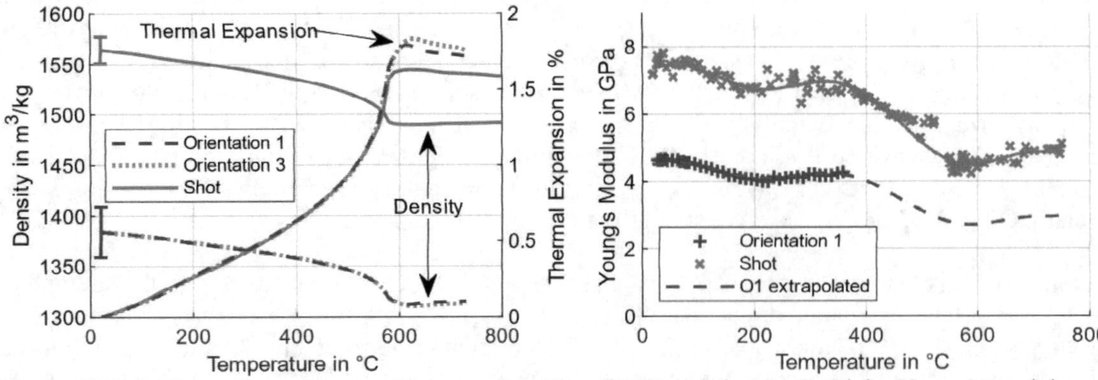

Figure 7. Left: Thermal expansion coefficient and density of 3DP and shot cores. Right: Young's modulus of 3DP and shot cores

IOP Conf. Series: Materials Science and Engineering 1281 (2023) 012069 doi:10.1088/1757-899X/1281/1/012069

The thermal expansion coefficients α for 3DP and shot cores show the expected quartz phase transition at $573\ °C$. Until $550\ °C$ α is the same for all specimens and above this temperature level α increases slightly more for O1 and O3, which is likely to be caused by the different binder systems. The density is generally quite different for the shot cores compared to the 3DP cores. The density at room temperature is $(1384 \pm 25) kg/m^3$ and $(1566\ \pm 13)\ kg/m^3$ for 3DP and shot cores respectively. This leads to a $11.6\ \%$ lower density for 3DP cores. Due to the nearly identical bulk density of GS14 and H32 this difference is attributed to the lower compression level for the 3D-printing process.
Young's modulus is measured with the acoustic method for the shot cores and the O1 3DP cores. At room temperature the stiffness of the shot cores and 3DP cores are 7.42 and $4.64\ GPa$, respectively. The lower stiffness is assumed to be related to the lower density. Unfortunately, it was only possible to perform measurements on the 3DP cores until $350\ °C$. Above this temperature level sand grains from the rough printed surface contaminated the equipment. It was decided to shift the measurements of the shot cores to the level of 3DP cores for the temperature above $350\ °C$ due to the similarities seen in the curvatures for the lower temperature levels.

4. Application – Water jacket core

The extensive focus on the mechanical performance of 3DP inorganic cores was originally motivated by measured high level deformations in a water jacket exhaust core. With 3D printing it was possible to make a combined core which is today in series production. The geometry of the core is shown in Figure 8, showing the deviation between the optical measurement of the sand core and the CT scan of the casting. The geometry of the 3DP core is scaled according to the thermal strain in the casting process to have a comparable reference/target. The evaluation of these measurements shows a large dimensional deviation in the thin-walled area of the ventilation strip, which is almost 4.5mm in the middle region, VP4.

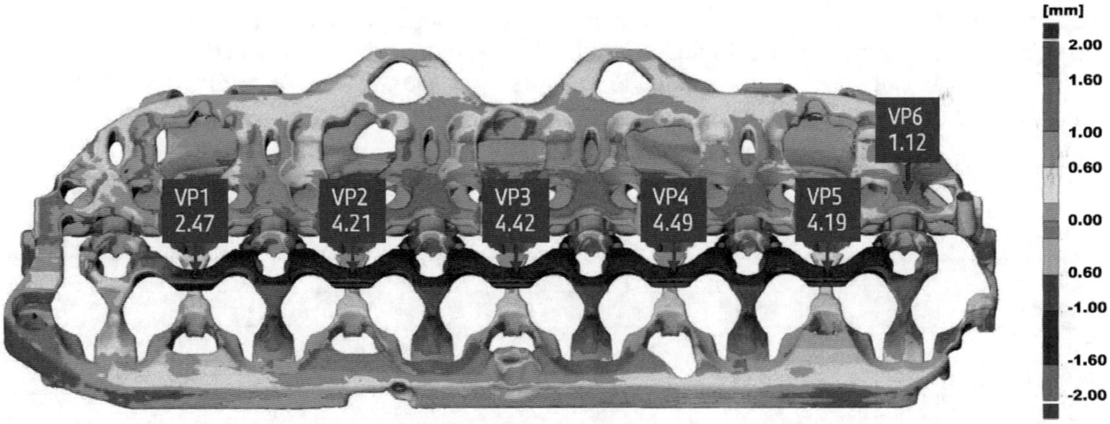

Figure 8. Deviation between the optical measurement of the 3DP sand core and the CT scan of the casting. The considered evaluation area of the thin-walled cooling channel around the outlet valves and the ventilation strip is highlighted with the blue box.

The first simulations with the Drucker Prager Cap model using the measured mechanical data did not show this high level of deformation and several simulations were performed to show the sensitivity to the thermomechanical data and the process conditions. The outcome of those simulations showed a deformation level in the range of 1.5 to 2.3 mm, which corresponds well to the input data and earlier experience from similar simulations and measurements for shot inorganic cores. Figure 9 shows some of the evaluated variations and the corresponding differences in the predicted deformations in the thin-walled area around the outlet valves and the ventilation strip.

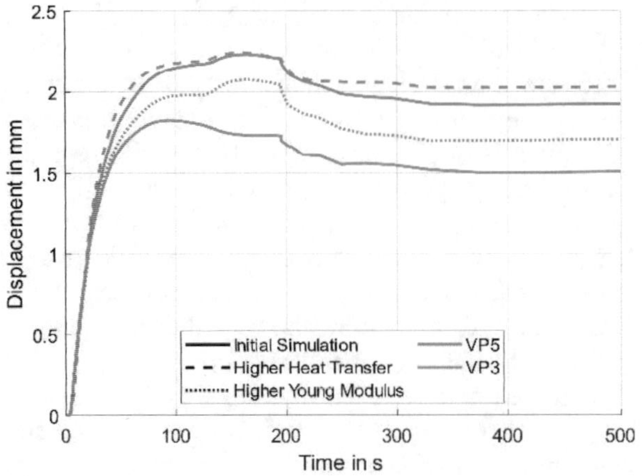

Figure 9. Sensitivity to variations in Young's modulus and the heat transfer coefficient. The initial simulation, the higher heat transfer coefficient and the higher Young's modulus simulations are shown for VP5. The initial simulation is only shown for VP3. The line type is shown in black in the legend.

Based on the experience from working with organic cores it was decided to evaluate the creep sensitivity of the 3DP cores and by that include time dependent deformation and buoyancy effects in the simulations. The influence of buoyancy can be rather pronounced for similar organically bonded cores, but it was originally not considered in this work since creep effects have not been detected for shot inorganic cores. Nevertheless, the foundry suspected buoyancy to play an important role and generally suspected the 3DP cores to be weaker at high temperature compared to the shot cores.

Creep measurements confirmed this suspicion and significant time dependent deformation was measured at high temperature, see Table 2. Details of these measurements and details of the performed simulations will be published in the planned PhD thesis by the co-author J. Wendling and based on this work a planned paper as a follow-up.

Table 2. Creep measurements for shot and 3DP inorganic cores; X: Failure within the 15 min measuring time; + … +++: grade of deflection; 0: No deflection. Details and creep curves are planned to be published later

Load / Temperature	3D-printing			Core Shooting		
	56 g	**28 g**	**0 g**	**56 g**	**28 g**	**0 g**
450 °C	X	++		0	0	
500 °C	X	+++		0	0	
525 °C	X	X				
550 °C	X	X	0	0		
600 °C		X	0		0	
650 °C			+			
700 °C			+			0

The early work and current status of applying the new creep measurements are shown in Figure 10. Several simulations were performed using different interface coherency temperature (T_{coh}) levels, where the solidifying aluminum is assumed to constrain the hot distortion of the sand core, by enabling the contact algorithm locally when the temperature drops below T_{coh}. The very high displacement level indicates a weak core in thin areas and the simulations are therefore also very sensitive to the combination of creep and the T_{coh} level. Some of the remaining differences might be due to local variations in quality or additional physical phenomena that influence the deformation level and/or the temperature level where the creep and buoyancy forces are reduced due to the solidifying aluminum. Further calibration of the thermal simulation to match the production condition could also shift the creep level and the T_{coh} evaluation.

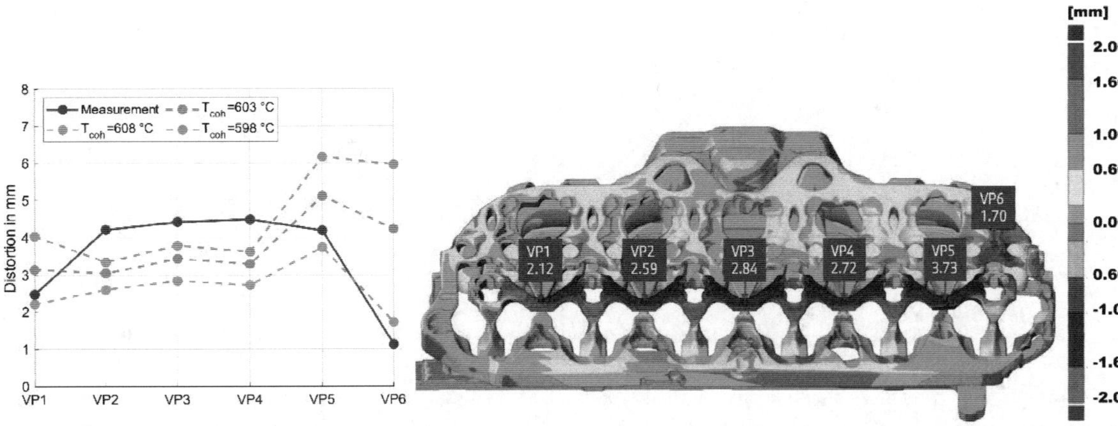

Figure 10. Left: Sensitivity to variations in the interface coherency temperature. Right: Simulation compared to CT scan ($T_{coh} = 608\,°C$).

5. Summary

Casting wall thickness and tolerance requirements become challenging due to light weight casting designs and therefore core geometries become more complex, thin walled and fragile due to functional requirements and hence more sensitive to distortion. 3D printing of sand cores makes it possible to combine complex cores and adds even more flexibility in the design of cores used in series production. Knowing the level of core shrinkage factors and core distortion during the casting process are of prime concern for tool makers as well as the core shop to secure an optimized core box layout or printing setup where pre-shaping of the core design is taken into account to compensate for deformations during the casting process. This paper focused on a 3DP inorganic water jacket core where significant deformations were measured in the thin-wall ventilation strip. Simulation were used to analyse the hot distortion of the core having a special focus on the time dependent deformation and buoyancy forces during the early state of solidification. Creep measurements were performed to evaluate and describe the time dependent behaviour. The early results show the improved predictions and the sensitivity to the interface coherency temperature. Future work will focus on local phenomena to improve the correlation including the influence of anisotropic behaviour for 3DP cores.

References

[1] Polzin H, Inorganic binders: For mould and core production in the foundry, first edition, Fachverlag Schiele Schön, Berlin, 2014.

[2] Thorborg J, Kumar S, Wagner I, Sturm J C. “The Virtual Core - Modelling and Optimization of Core Manufacturing and Application.” IOP Conference Series: Materials Science and Engineering 861, no. 1 (May 1, 2020): 012004. https://doi.org/10.1088/1757-899X/861/1/012004.

[3] Desai C S, Siriwardane H J. “Constitutive Laws for Engineering Materials. With Emphasis on Geologic Material.” Prentice-Hall. 1984.

[4] Brewin P R et al., Modelling of Powder Die Compaction. 2008.

General thermomechanical model of FSW based on a characteristic temperature for deformation and heat transfer

Sofia Salazar[1,2], **Diego Rojas**[1], **Patricio F Mendez**[2] **and Karem E Tello**[3]

[1]Universidad de Chile, Avenida Beauchef 850, Santiago, Chile, 8370459
[2]University of Alberta, 9107 116th St., Edmonton, AB, Canada T6G 2V4
[3]Universidad Técnica Federico Santa María, Dept. Ciencia de Materiales, Avenida España 1680, Valparaíso, Chile

E-mail: sofiaisa@ualberta.ca

Abstract. This paper presents a coupled model of heat transfer and plastic deformation in friction stir welding (FSW), accounting for the temperature profile in the substrate near the pin. This approach is analogous to the boundary layer analysis in fluid mechanics and is based on the methodology of scaling and calibration based on published data. A model focusing on common conditions in FSW, such as relatively slow translation and high rotation velocities, a thin shear layer and the influence of the shoulder on the maximum temperature was reformulated. This paper extends previous work by considering the heat flow into the pin and an improved criterion for determining the temperature at the edge of the shear layer. The results are a set of updated closed-form expressions for the maximum temperature, the thickness of the shear layer, the shear stress around the pin, torque and thermal effect of the shoulder, applicable to all metals. The predictions from this model are verified against a comprehensive database of published experiments. Applications of this model also include the accelerated determination of procedure variables and the generalization of maps of process limits.

1. Introduction

A previous coupled model of heat transfer and plastic deformation proposed by the authors [1] describes the heat generation and plastic deformation around the pin using fundamental continuum mechanics equations approximated via scaling analysis. That original model incorporates the heat losses by conduction into the pin by considering a heat efficiency lesser than 100% despite assuming no temperature gradient in the substrate next to the pin. This inconsistency was motivated by simplicity in the mathematical treatment, and it enabled the accurate treatment of a larger scope model of FSW, matching published measurements and simulations [2], especially for aluminium. Alloys with low thermal conductivity such as titanium or stainless steels are known to have much lower efficiencies than aluminium [3-6], introducing a potentially significant effect not fully considered in the original treatment.

This paper addresses the inconsistency between temperature profile and efficiency considerations by accounting properly for the temperature profile in the substrate. This work also uses an improved treatment of the constitutive relationship between strain rate, stress, and temperature. As a result, an in-depth analysis within the shear layer is obtained with insight about the location of the maximum temperature in the shear layer, and limits to process efficiency. Consistently with the previous work, the analysis performed here aims at developing closed-form expressions based only on process parameters known a priori. All relevant physics are considered, and the full thermomechanical coupling is preserved. The solution to the mathematical problem is obtained through asymptotic scaling and calibration based on published data.

1.1. *Asymptotic modelling of coupled heat and and plastic deformation*

The model treated in [1] made four main approximations: i) The travelling pin can be considered a steady state, slow moving heat source. ii) There is small advance per revolution, ii) the shear layer is thin, and iv) the shoulder has a secondary role on the maximum temperature. All these approximations are studied in [1] , and are expressed as:

$$\mathrm{Pe}_a = \frac{Va}{2\alpha_\delta} \ll 1 \tag{1}$$

$$Va \ll \omega a \delta \tag{2}$$

$$\delta \ll a \tag{3}$$

$$T_p - T_\infty \ll T_{max} - T_\infty \tag{4}$$

Where a is the pin radius, V and ω are the welding speed and angular velocity of the pin respectively, α_δ is the thermal diffusivity of the base plate at the critical temperature T_δ, δ is the shear layer thickness, T_p is the preheat temperature due to the shoulder, T_∞ is the temperature of the plate far from the pin and the shoulder, and T_{max} is the temperature of the base material at the interface with the pin; all these temperatures will be explained in detail below. The first three conditions imply that heat transfer and plastic deformation around the tool can be approximated as a one dimensional problem, with the coordinate originating at the pin-substrate interface and perpendicular to it. The fourth condition establishes that the coupling between the thermal effect of the shoulder and that of the pin is weak. In addition to the mathematically quantified approximations described above, an additional important approximation is the consideration of a threadless cylindrical pin, which yields general insight into the fundamental mechanisms of coupled heat and plastic deformation shared by all FSW processes.

2. Heat transfer in the shear layer

Figure 1 illustrates the temperature profile in the shear layer when heat losses into the pin are accounted for. The temperature increases gradually from temperature T_S at the interface ($x = 0$) into the shear layer where the maximum temperature (T_{max}) is achieved at a distance x_{max} from the interface. Then, it decreases monotonically from T_{max} to T_δ. Temperature T_δ is used as the characteristic value that captures the transition from the shear layer into the "undeformed" substrate at $x = \delta$; above T_δ the model considers there is significant plastic deformation, while below T_δ the plastic deformation is negligible. This transition is not sharp, but occurs within a relatively narrow range. Temperature T_δ fulfills a similar role to T_0 used in [1] , and their relationship will be analyzed in detail later in this paper. T_P was studied in detail Dr.Mendez et al. paper [1].

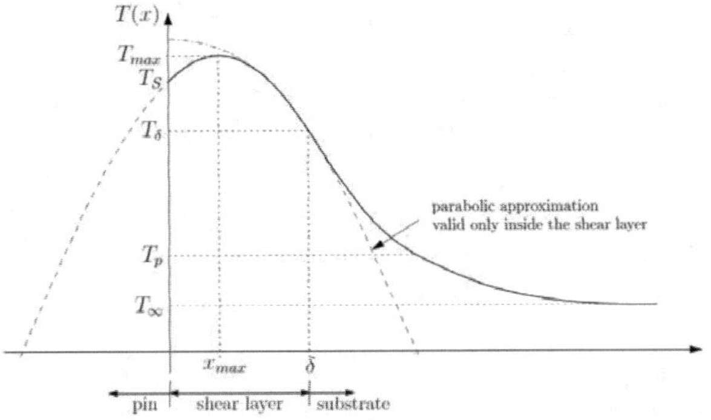

Figure 1. Temperature profile in the shear layer and surrounding material.

Following the analysis detailed in [1] for the realistic conditions described by equations (1) to (4), heat transfer in the shear layer is dominated by advective heat transfer in the circumferential direction due

IOP Conf. Series: Materials Science and Engineering 1281 (2023) 012070 doi:10.1088/1757-899X/1281/1/012070

to material movement in the shear layer and heat transfer by conduction in the radial direction [7]. The dominant modes of heat transfer are corroborated by analysing the Peclet number. The fast rotational flow results in a relatively flat temperature profile in the circumferential direction, and the relatively thin shear layer behaves in a way analogous to a thermal boundary layer.

The governing equation and boundary conditions for the pseudo-steady-state, low-Peclet, one-dimensional formulation of the problem are:

$$\frac{d^2T}{dx^2} + \frac{q''(x)}{k(T)} = 0 \tag{5}$$

$$T|_{x=\delta} = T_\delta \tag{6}$$

$$\left[k(T)\frac{dT}{dx}\right]_{x=0} = q''_{out} \tag{7}$$

Where $T(x)$ is the temperature profile in the shear layer, $q'''(x)$ is the volumetric heat generation due to plastic deformation, $k(T)$ is the thermal conductivity of the base metal, and q''_{out} is the heat flow that goes out from the shear layer into the pin. The lower maximum temperature and the positive gradient $(dT/dx|_{x=0} > 0)$ at the pin/shear layer interface (see figure 1) consider the effect of heat losses by conduction into the tool.

The dependent and independent variables in equation (5) will be scaled for further treatment by an estimation of their characteristic values [8–11]. The normalized governing equation 5 can then be expressed by:

$$\left(\frac{d^2T}{dx^2}\right)_c \left(\frac{d^2T}{dx^2}\right)^* + \frac{q'''_c}{k_\delta}\left(\frac{q'''}{k}\right)^* = 0 \tag{8}$$

where $\left(d^2T/dx^2\right)_c$ and q'''_c are unknown estimates of the characteristic values for the maximum second derivative of temperature, and the maximum volumetric heat generation respectively. The magnitude k_δ is the thermal conductivity at T_δ, which can be obtained from tabulated data. The quantities with an asterisk are dimensionless functions which extreme values approximately -1 o 1.

Heat losses that decrease process efficiency follow three paths: through the pin, through the shoulder and to the backing plate (This one is considered in equation 42). Contact of the shoulder with the base plate is discussed in detail in [1]. The analysis presented here is a conservative analysis in which shoulder heat losses are lumped with those of the pin, maximizing the temperature gradient in the material in contact with the pin. It will be shown that even in this case, the effect of positive gradient in the calculations is small. The process efficiency (η) can be expressed in terms of the heat flow that is dissipated out of the shear substrate and into the pin (q''_{out}) and heat flow into the substrate from the shear region (q''_{in}):

$$\eta = \frac{q''_{in}}{q''_{in} + q''_{out}} \tag{9}$$

where q''_{out} is introduced in equation 7, and q''_{in} is:

$$q''_{in} = \left[-k(T)\frac{dT}{dx}\right]_{x=\delta} \tag{10}$$

Equation 9 is valid when the shear layer is thin, such that the heat flow in the axial direction is small, and the surfaces through which heat is transferred out to the pin and into the substrates are nearly the same size. In the above equations, x=delta, means the thickness of the share layer (see figure 1). At this position, $T(x) = T_\delta$, and $k(T) = k_\delta$. The shear layer's temperature profile can be scaled as follows:

$$T(x) = T_\delta + (T_{max} - T_\delta)\Theta^*(x^*) \tag{11}$$

where $x = \delta x^*$. From the schematic evolution in figure 1, the temperature evolution can be approximated as the following parabolic variation:

$$\Theta^*(x^*) \approx \left[1 - \left(\frac{x^* - x^*_{max}}{1 - x^*_{max}}\right)^2\right] \tag{12}$$

where $x^*_{max} = x_{max}/\delta$ and T^*_{max} are the location and value of the maximum temperature which need to be calculated. This dimensionless temperature has a maximum value of 1 at $x^* = x^*_{max}$ and a minimum value of 0 at $x^* = 1$ when $x^*_{max} \geq 0.5$. This approximation is used to calculate $[dT/dx]_{x^*=0}$ for q''_{out} in equation 7 and $[dT/dx]_{x^*=\delta}$ for q''_{in} in equation 10:

$$\frac{dT}{dx} = \frac{dT}{d\Theta^*}\frac{d\Theta^*}{dx^*}\frac{dx^*}{dx} = (T_{max} - T_\delta)\frac{d\Theta^*}{dx^*}\frac{1}{\delta} \tag{13}$$

where

$$\frac{d\Theta^*}{dx^*} = \frac{-2(x^* - x^*_{max})}{(1 - x^*_{max})^2} \quad (14)$$

therefore:

$$q''_{out} = k_\delta \frac{T_{max} - T_\delta}{\delta} \frac{2x^*_{max}}{(1 - x^*_{max})^2} \quad (15)$$

$$q''_{in} = k_\delta \frac{T_{max} - T_\delta}{\delta} \frac{2}{(1 - x^*_{max})} \quad (16)$$

Replacing equation 15 and 16 in equation 9, an unexpectedly simple relationship between the total efficiency of the process and the normalized location of the maximum temperature is obtained equation 17, which establishes one of the two unknowns in the temperature profile (x^*_{max}) in the shear layer as a function of process efficiency, which is often known or it is measurable. To determine the other unknown parameter in the temperature distribution in the shear layer (T^*_{max}), the parabolic approximation is used to estimate $\left(d^2T/dx^2\right)_c$ in equation 8:

$$\eta = 1 - x^*_{max} \quad (17)$$

$$\frac{d^2T}{dx^2} = \frac{d}{dx^*}\left(\frac{dT}{d\Theta^*}\frac{d\Theta^*}{dx^*}\frac{dx^*}{dx}\right)\frac{dx^*}{dx} \quad (18)$$

$$= \frac{T_{max} - T_\delta}{\delta^2}\frac{-2}{(1 - x^*_{max})^2} \quad (19)$$

This expression is independent of the location within the shear layer, which is consistent with the constant value obtained from the second derivative of a parabola. Considering that $x^*_{max} = 1 - \eta$, $\left(d^2T/dx^2\right)_c$ can be estimated as:

$$\left(\frac{d^2T}{dx^2}\right)_c = 2\frac{T_{max} - T_\delta}{\delta^2\eta^2} \quad (20)$$

The negative sign in equation 19 was dropped because by definition characteristic values are positive [11]. The normalized equation of heat conduction in the shear layer is then expressed by:

$$\frac{2(T_{max} - T_\delta)}{\delta^2\eta^2}\left(\frac{d^2T}{dx^2}\right)^* + \frac{q'''_c}{k_\delta}\left(\frac{q'''}{k}\right)^* = 0 \quad (21)$$

Following [8–11], equation 21 yields the following algebraic equation:

$$-\frac{2(\widehat{\Delta T_{max}})}{\widehat{\delta}^2\eta^2} + \frac{\widehat{q'''_c}}{k_\delta} = 0 \quad (22)$$

where $\Delta T_{max} = T_{max} - T_\delta$, and the values with $\widehat{}$ represent estimations based on algebraic equations considering only dominant factors. Similarly to [1] , this expression involves three unknowns $\widehat{\Delta T_{max}}$, $\widehat{\delta}$, and $\widehat{q'''_c}$. In addition, the estimation of the temperature at the pin/shear layer interface, $\widehat{T_s}$, can be calculated evaluating equation 11 at $x = 0$:

$$\widehat{T_s} = \left(\frac{2\eta - 1}{\eta^2}\right)\widehat{\Delta T_{max}} + T_\delta \quad (23)$$

3. Heat generation in the shear layer

The revised temperature profile in the shear layer requires a study of the estimation of heat generated by plastic deformation in the shear layer. Following [1], the volumetric heat generated by plastic deformation is expressed by:

$$q'''(x) = \eta_s \tau(x)\dot{\gamma}(x) \quad (24)$$

where η_s is the efficiency that accounts for the fraction of the mechanical energy converted into heat excluding the small amount of mechanical energy stored in the form of lattice defects, $\tau(x)$ is the shear stress experienced by an element of volume at coordinate x, and $\dot{\gamma}$ is the rate of plastic deformation.

The normalized counterpart of equation 24 is:

$$q'''_c q'''^*(x^*) = \eta_s \tau_c \tau^*(x^*)\dot{\gamma}_c\dot{\gamma}^*(x^*) \quad (25)$$

where $\widehat{q'''_c}$, $\widehat{\tau_c}$, and $\widehat{\dot{\gamma}}_c$ are unknown estimates considering a parabolic temperature variation within the shear layer. The associated algebraic equation resulting from scaling is:

$$\widehat{q'''_c} = \eta_s \widehat{\tau_c}\widehat{\dot{\gamma}}_c \quad (26)$$

In this equation, there is no obvious way of estimating $\widehat{\dot\gamma}_c$; but the kinematic relationships from continuum mechanics indicate that the shear rate can also be expressed as $\dot\gamma = -(dv/dx)$, where $v(x)$ is the transverse velocity of the metal experiencing plastic shear. This way, the following relationships can be stated:

$$\dot\gamma_c = -\left(\frac{dv}{dx}\right)_c \tag{27}$$

$$\dot\gamma^* = -\left(\frac{dv}{dx}\right)^* \tag{28}$$

These relationships are useful because the rotational velocity of the pin can be used to calculate the unknown characteristic value, considering that at the position $x = \delta$, the velocity of plastic deformation can be neglected, this means that $v(\delta) \approx 0$.

$$v(\delta) - v(0) = \int_0^\delta \frac{dv}{dx} dx = 0 - \omega a \tag{29}$$

$$\omega a = \widehat{\delta}\left(\frac{dv}{dx}\right)_c \int_0^1 \dot\gamma^* dx^* \tag{30}$$

The calculation of the integral can be greatly simplified because at the high temperatures typical of FSW, the temperature and shear rate in equation 36 have a near-linear relationship [1] resulting in

$$\dot\gamma^* \approx \frac{T - T_\delta}{T_{max} - T_\delta} = \Theta^*(x^*) \tag{31}$$

Replacing equation 12 and 31 into (30) the characteristic value $(dv/dx)_c$ can be calculated as follows:

$$\widehat{\delta}\left(\frac{dv}{dx}\right)_c \int_0^1 \left[1 - \left(\frac{x^* - x^*_{max}}{1 - x^*_{max}}\right)^2\right] dx^* = \omega a \tag{32}$$

$$\left(\frac{dv}{dx}\right)_c = \frac{\omega a}{\widehat{\delta}\zeta} \tag{33}$$

where ζ is a function of the total efficiency of the process represented in figure 2, and given by:

$$\zeta = 1 - \frac{1}{\eta^2}\left[\frac{1}{12} + \left(\eta - \frac{1}{2}\right)^2\right] \tag{34}$$

Equation 26 can then be rewritten as:

$$\widehat{q_c'''} = \eta_s \widehat{\tau}_c \frac{\omega a}{\widehat{\delta}\zeta} \tag{35}$$

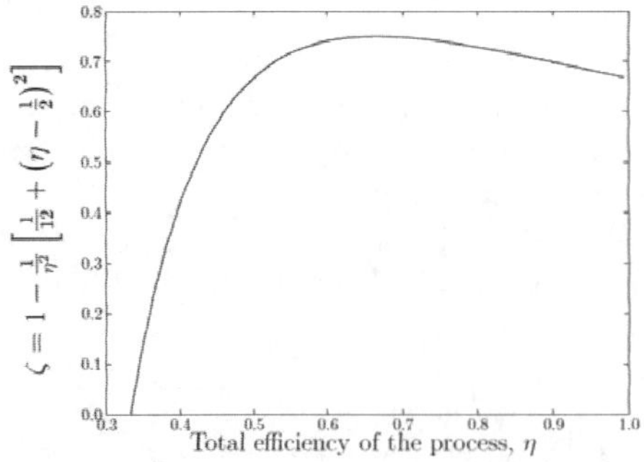

Figure 2. Function ζ to calculate the effect of heat losses on the temperature profile. Its maximum is 0.75, for $\eta = 2/3$. For $\eta = 1$, $\zeta = 0.666$. Typical values of ζ are 0.666 for aluminium, where ($\eta = 0.50$) [12], 0.74 for plain carbon steel ($\eta = 0.75$) [13] and 0.666 for stainless steel ($\eta = 0.5$) [3].

Equation 22 and (35) involve four unknowns $\widehat{T_{max}}$, $\widehat{\delta}$, $\widehat{q_c'''}$, and $\widehat{\tau}_c$. Following the analysis in previous work, the two missing equations are obtained from analysis of the constitutive behavior in the shear layer and heat conduction outside of it.

4. Constitutive behaviour in the shear layer

The mechanical response of the substrate is model as a Zenner-Hollomon constitutive model which relates the effects of temperature, stress, and strain rate: [14]

$$\dot{\gamma} = A\left(\frac{\tau}{\tau_R}\right)^n \exp\left(-\frac{Q}{RT}\right) \tag{36}$$

where Q is the activation energy and together with,A and n are the parameters of the constitutive model of Zener-Hollomon, R is the gas constant, and τ_R is an arbitrary reference stress. Reference [1] justifies the selection of this constitutive behaviour and explains its relationship with the more sophisticated Sellars-McTegart hyperbolic sin model [15], but it will be briefly explain later. For typical FSW materials, the difference in models is only significant for Mg alloys [16].

Figure 3 illustrates schematically the behavior of the substrate material; the vertical axis indicates the ratio of shear strain rate to that corresponding to the nominal solidus temperature. For ratios below ε, it will be considered that the plastic deformation is negligible ("no shearing"). The limit between the shear layer and the rest of the substrate will then be considered this shear rate, and its corresponding temperature T_δ.

Similarly as it was proposed in [1], the final expressions will be of best practical use when the constitutive law of equation 36 is linearized. The proposed linearization is illustrated in figure 3 and can be expressed as

$$\exp\left(-\frac{Q}{RT}\right) \approx \begin{cases} 0 & \text{if } T \leq T_\delta \\ \dfrac{T - T_\delta}{T_1 - T_\delta} \exp\left(-\dfrac{Q}{RT_1}\right) & \text{if } T > T_\delta \end{cases} \tag{37}$$

where T_1 is the temperature at which the linearization is tangent to the actual curve of the constitutive law and it its typically close to the solidus temperature of the base material (T_m). The relationship between T_1 and T_δ is

$$T_1 - T_\delta = \frac{RT_1^2}{Q} \tag{38}$$

In [1], it was considered that $T_\delta = T_0 = T_m(1 - RT_m/Q)$, such that $T_1 = T_m$. The choice of T_0 as a threshold resulted in temperature estimates that were consistently higher than measurements, suggesting that at T_0, the shear rate might not be negligible as assumed. The value of T_δ is obtained using the constitutive law of equation 36; in this case, for a given choice of ε, the estimation of T_δ is independent of the shear stress:

$$T_\delta = T_m\left(1 - \frac{RT_m}{Q}\ln\varepsilon\right)^{-1} \tag{39}$$

Replacing the estimates obtained above into equation 36, and using only the characteristic values, we obtain the following algebraic equation:

$$\frac{\omega a}{\widehat{\delta\zeta}} \approx AB'\left(\frac{\widehat{\tau_c}}{\tau_R}\right)^n \frac{\widehat{\Delta T_{max}}}{\Delta T_1} \tag{40}$$

where $\Delta T_1 = T_1 - T_\delta$ and $B' = \exp(-Q/RT_1)$. The values of T_δ and T_1 depend only on the materials properties (which are known) and on the choice of threshold shear rate (captured by ε). The value of ε is unknown and in this work will be determined empirically by best match to available data, with the caveat that it has an important physical meaning that should be explored in more detail in future research. Equation 40 does not introduce any new unknown characteristic value.

5. Heat conduction outside the shear layer

In this region, heat is transmitted by conduction into the substrate. For the slow-moving heat sources considered, Rosenthal's thin-plate solution [17,18] results in circular isotherms matching the pin geometry. The scaled equation of heat conduction outside the shear layer is given by [1]:

$$\Delta T_\delta = T_\delta - T_\infty = \eta\eta_s \frac{\omega a^2 \widehat{\tau_c}}{k_\delta} K_0(\xi) \tag{41}$$

with

$$\xi = r\sqrt{\left(\frac{V}{2\alpha_\delta}\right)^2 + \frac{h + h'}{k_\delta d}} \tag{42}$$

where h and h' are the coefficients for convection heat losses on the top and bottom surfaces of the plate, d is the plate thickness, x' and r correspond to a distance in the direction of travel and a radial distance from the line heat source respectively. The symbol K_0 corresponds to the modified Bessel function of a second kind and order 0. Equation 41 does not add any new unknown characteristic value.

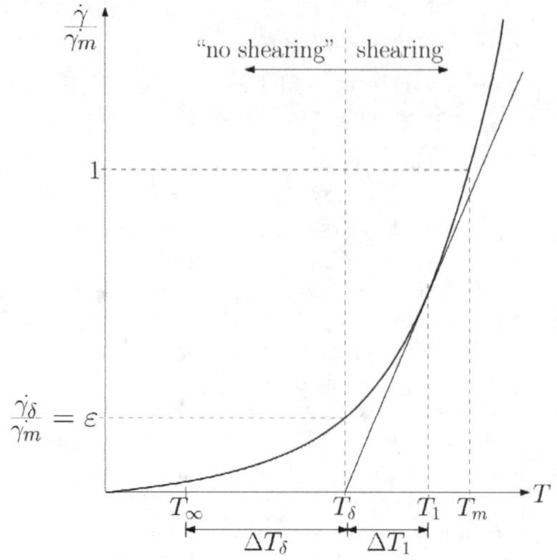

Figure 3: Constitutive behavior of the substrate material. Strain rates below $\dot{\gamma}_\delta/\dot{\gamma}_m = \varepsilon$ are considered negligible. The corresponding temperature T_δ divides the regions in which plastic shear is important and negligible.

6. Scaling expressions for the coupled thermomechanical system

Equation 22, (35), (40), and (41) constitute a complete system of four equations with four unknowns. The solution to this set of equations is:

$$\widehat{\delta} = a\left[2\frac{\Delta T_1}{AB'}\left(\frac{a^2\tau_R\eta_s}{k_\delta}\right)^n\left(\frac{K_0\omega}{\Delta T_\delta}\right)^{n+1}\eta^{n-1}\right]^{\frac{1}{2}} \tag{43}$$

$$\widehat{\Delta T_{max}} = \Delta T_1\left[\frac{1}{2AB'\Delta T_1\zeta^2}\left(\frac{K_0}{\Delta T_\delta}\right)^{n-1}\left(\frac{a^2\tau_R\eta_s}{k_\delta}\right)^n(\omega\eta)^{n+1}\right]^{\frac{1}{2}} \tag{44}$$

$$\widehat{q_c'''} = \left[\frac{1}{2}\frac{AB'}{\Delta T_1\zeta^2}\left(\frac{1}{\tau_R\eta_s}\right)^n\left(\frac{1}{\omega\eta}\right)^{n+1}\left(\frac{k_\delta}{a^2}\right)^{n+2}\left(\frac{\Delta T_\delta}{K_0}\right)^{n+3}\right]^{\frac{1}{2}} \tag{45}$$

$$\widehat{\tau_c} = \frac{k_\delta\Delta T_\delta}{\eta\eta_s\omega a^2K_0} \tag{46}$$

These equations are very similar to those derived in [1], with T_δ replacing T_0, T_1 replacing T_m, and presence of the function of the function of total efficiency ζ.

7. Methodology

To test the validity of the equations, the estimations are compared with results of published simulation or experimental tests. Every relevant parameter value reported in each test will be considered to generate the estimations.

Following the publication of Mendez [1], this work considers an updated version of the database. A total of 280 experiments and simulations are considered. Therefore, the materials in the database are: seventeen aluminium alloys belonging to the seriesAA1XXX, AA2XXX, AA5XXX, AA6XXX and AA7XXX, four carbon steels (Interstitial Free Steel, from series AISI1XXX), one stainless steel (304SS), one titanium alloy (Ti-6Al-4V) and one magnesium alloy (AZ31).

For maximum temperature data, measurements are usually taken using thermocouples embedded in the work piece and sometimes in the pin. To standardize the measurements between this two positions, a conclusion drawn in [19] is applied, where the temperature difference between the work piece and the pin is 296.8°K on average. Thus, for the data taken from publications [20], [21], [22], where the temperature was measured at the pin, 296.8°K is subtracted to the measured value.

7.1. *Optimization for estimations*

As it was explained before, for equation 39 the value of ϵ is obtained empirically from the data available. This value is obtained through the optimization of the objective function in equation 47. In this equation,

IOP Conf. Series: Materials Science and Engineering 1281 (2023) 012070 doi:10.1088/1757-899X/1281/1/012070

ϵ is not shown explicitly, however it is within T_δ as equation 39 shows. Then, the value of ϵ is obtained by minimizing equation 47 varying ϵ. So, the value of ϵ is the one that gives the minimum of the objective function in equation 47. It is important to mention that only experimental data was considered in this optimization. This was done to obtain a value based on real experiments and not on calculations or simulations.

$$\sum_i \left[\ln T_{max} - \ln\left(\Delta\widehat{T_{max}} + T_\delta\right)\right]_i^2 \tag{47}$$

7.2. *Ratio of measurement to estimation*

To evaluate how the estimations of maximum temperature compare to measurements and numerical results, the ratio between both these values is used.

$$\Theta = \frac{T_{max} - T_\infty}{\widehat{T_{max}} - T_\infty} \tag{48}$$

The ratio in equation 48 is plotted against Pe, $V/\omega\widehat{\delta}$, $\widehat{\delta}/a$ and $(T_p - T_\infty)/(T_{max} - T_\infty)$ in four different graphs, each one representing the simplifications stated at the beginning. In each of these graphs, only a subset of the data available is used. The subset in each graph consists of the data that fulfills all the simplifications except the one that is being used. So, for example, for the Θ vs Pe graph, $V/\omega\widehat{\delta}$, $\widehat{\delta}/a$ and $(T_p - T_\infty)/(T_{max} - T_\infty)$ must be lower than one.

7.3. *Correction Factors*

To calculate the correction factors, which leads to the trend line values, the following function is used:

$$f\left(\frac{\widehat{\delta}}{a}\right) = C_1\left(1 + C_2\frac{\widehat{\delta}}{a}\right)^{C_3} \Leftrightarrow 1.077\left(1 + 0.3319\frac{\widehat{\delta}}{a}\right)^{-1.0099} \tag{49}$$

As it was mentioned before, when the points tend to a constant value (therefore simplifications become negligible), only C_1 is considered and both C_2 and C_3 are equal to 0. Now, if the points deviate from a constant value (therefore the simplifications affect the results), all constants from equation 49 are considered. In addition, from this equation and by observing the figures 4, 5, and 7 it can be verify that the trend-lines have a value of 1.077 in the vertical axis.

8. Comparison with experimental and simulated results

This section shows the ratio theta, which compares the temperature from the experiments, divided by the mathematical model results. For both values, the temperature in the farthest point of the plate is subtracted (48). No calibrations where applied to figures 4, 5, 6, and 7, so the data without any kind of calibrations can be studied with their respective references to material properties. However the correction factors for this study will be mentioned later. It can be seen that all the graphs have a tendency to be align around the value of one in the vertical axis, this is the ideal of the graphs because it is expected that the value obtained by the model is similar to the value of the real experiments.

All the values presented here, present a measurement placed in the base, for which some of them where calculated according to [19], therfore the uncertainty stemming from the response time of the thermocouple and its placement in a deforming material, we also have to consider that the mathematical conversion could not be as exact as measuring the temperature directly from the base. Because the errors related to the experiments are inherent to the measurement or simulations are expected to be small, they will only be considered in the correction factors at the end of the discussion section.

Figure 4 shows the normalized maximum temperature, with the ratio theta in the base material as a function of the Peclet number. All the values presented here are lower than one in the abscissa axis, therefore it can be established that FSW is in reality a slow moving heat source, regardless the kind of material. The same statement can be maintained for figure 7, where the estimated preheating effect of the shoulder is illustrated and only two exceptions are presented. In this figure the preheat temperature given by the shoulder is calculated using a value of $f = 0.1$ [1]. Figure 5 represent the normalized maximum temperature in function of the hypothesis 2, where most of the experiments are align around one in the vertical axis, but the fact that so many experiments are beyond one in the horizontal axis, demonstrate that FSW might not always satisfy the same difference between the travel and rotational speed for all the materials. Finally we can establish the same for figure 6, where once again the shear layer thickness will vary depending on the reached temperature, and this also depend on the material properties, so for this two figures the hypothesis is not always true.

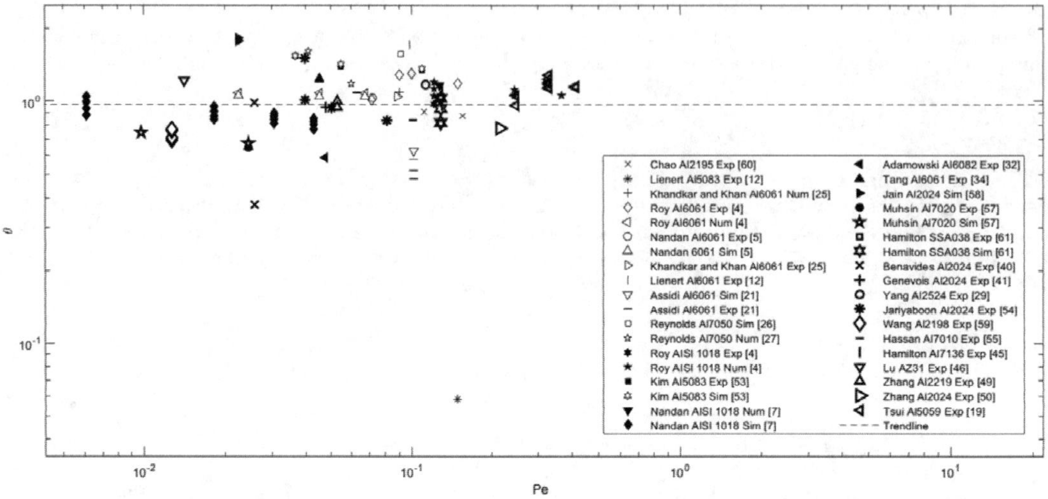

Figure 4. Ratio of maximum temperature in the base plate against hypothesis 1.

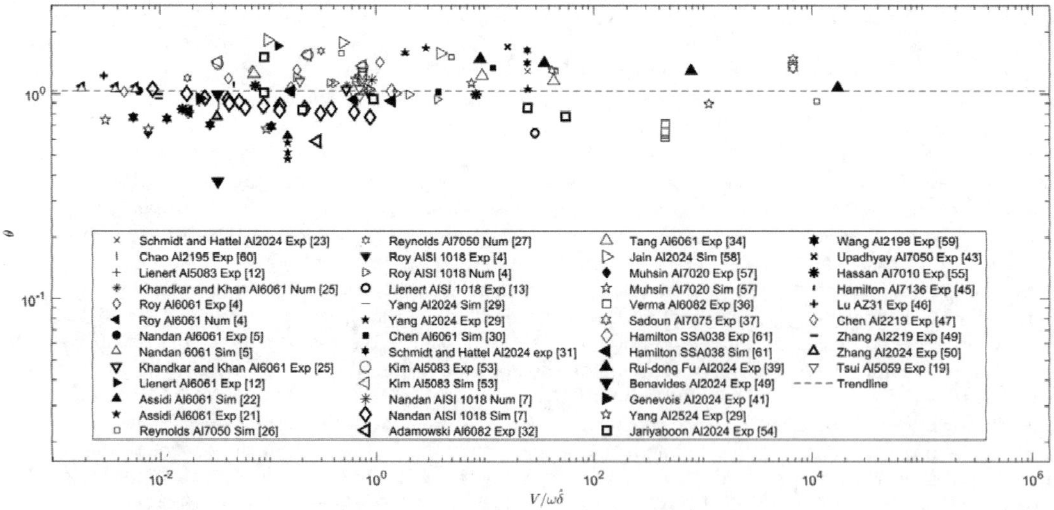

Figure 5. Ratio of maximum temperature in the base plate against hypothesis 2.

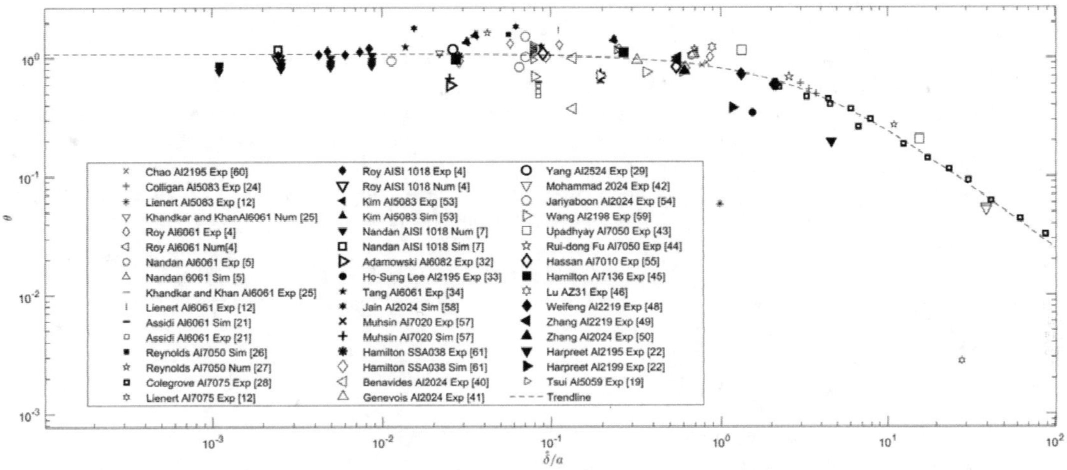

Figure 6. Ratio of maximum temperature in the base plate against hypothesis 3.

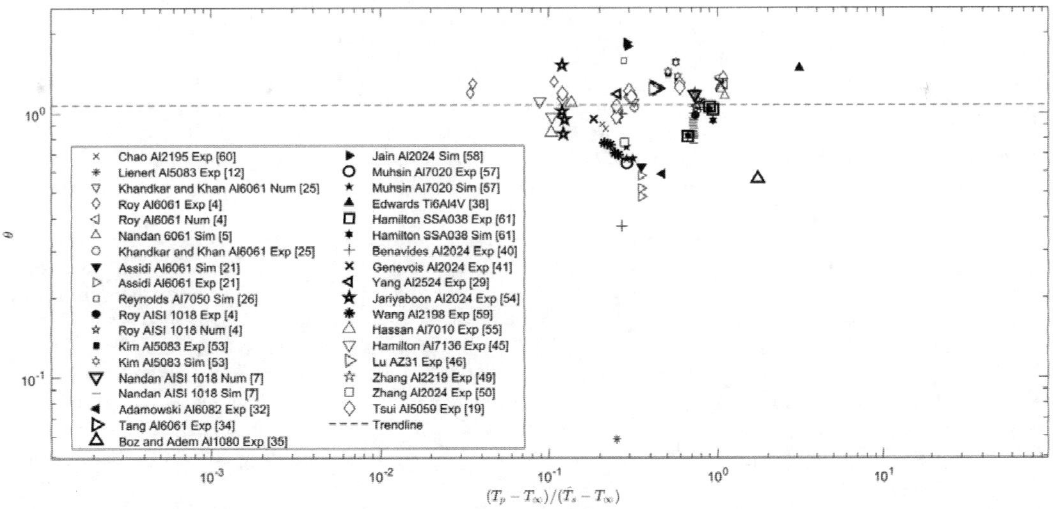

Figure 7. Ratio of maximum temperature in the base plate against hypothesis 4.

IOP Conf. Series: Materials Science and Engineering 1281 (2023) 012070 doi:10.1088/1757-899X/1281/1/012070

9. Discussion

Figures 4, 5, 6, and 7 shows the behaviour of the collected data in relation with the each of the assumptions made. In the first figure 4, it can be seen that all the assumption where lower than one, so as it was explained in the last section no matter the alloy or the conditions, FSW can always be considered a steady state, slow moving heat source. Besides this, there is a mayor concentration of the experiments around one, as the line of tendency indicates, so the model is predicting successfully the real results obtained by the experiments. In this figure, two kind of behaviour can be study: the first one can be seen with the experiment of Nandan AISI 1018 Sim [7], Kim Al5083Sim [51], Roy AISI1018 Num [4], Jariyaboon Al2024 Exp [52], Hasan Al7010 [53], Chao Al2195 Exp [54] and Muhsin Al7020 Sim [55], where theta decreases as the Pe increases, and the other case where theta decreases with a constant Pe, this happened with the experiments Jain Al2024 Sim [56], Nandan AISI 1018 Sim [7], Tsui Al5059 Exp [19], Hamilton SSA038 Sim [45], Assidi Al6061 Exp [21], and Wang Al2198 Exp [57]. The experiment of Nandan AISI 1018 SIM [7] will be used to explain this differences because it present both behaviours. The experiment of Nandan AISI 1018 Sim [7] present four rotational speed, and for each rotational speed there are four variations in the travel speed. Given the equation of Pe presented in 1 it can be seen that the four groups in the horizontal axis correspond to the variation in the travel speed, while the variation in the vertical axis correspond to the variation in the rotational speed, this was verified by testing different values. So, as the travel speed increases, the model estimates a higher temperature compared to the temperatures collected from the literature. On the other hand, as the rotation speed increases the maximum temperature estimated by equation 44 is smaller than the temperatures collected resulting in higher theta values.

In figure 5 the data collected from the experiment Nandan AISI 1018 Sim [7] present a decreasing behaviour as v/$w\omega$ increases, but this time is not separated in four groups because there is a mathematical relation between the travel and rotational speed in the horizontal axis, given by the equation 2 but as in the previous figure if the rotational speed decrease the value of theta also decrease and the value of v/$w\omega$ increases. In the same way if the travel speed increase the value of v/$w\omega$ also increases and theta decreases. This can be explained because the travel speed is used to calculate the Bessel function. It can be establish that there is an equilibrium between the rotational and travel speed for the model to be more closer to the experimental results, and that the model still needs to consider some other phenomena to be completed. From the figure 6, it can also be seen that the model assumptions might need some arrangement because the behaviour of the trending line has a negative slope as δ/a increases. Once again the experiment Nandan AISI 1018 Sim [7] help to visualize the effect of the different speed in the behaviour of the model because the theta decrease as the horizontal axis increases due to the increase of the rotational speed. This is related to the fact that in both graph the assumptions are not accomplished (in both cases the experiments present values higher than one in the horizontal axis) meaning that for the FSW process, not all the alloys satisfied equally the second and third hypothesis explained by the equations 2 and 3. To verify this, the experiments of Sadoun Al7075 Exp [37] has a constant rotational speed of 500 RPM and a travel speed of 0.000833[m/s], by dividing both velocities, the ratio of 600,240.09 is obtained, meanwhile the experiment Khandkar and Khan Al6061 Exp [25] has a ratio of 162,500, being 3.7 times lower.

Lastly, in figure 7, the importance of the preheat temperature of the shoulder is studied. Here as it is expected if the preheat temperature increase the experiments tend to locate in higher values of the x axis, and the steels tends to locate also in the right side of the graph because their preheat temperature are also higher(around the 1020K) v/s the 400K of the aluminium materials. But there two aluminium in the right side of this figure which is, once again Nandan AISI 1018 Sim [7] and the Roy Al6061 Num [4]. This two experiments both present a preheat temperature around the 670°K, for all the aluminium the maximum temperature is very similar, and the other parameters such as pin and shoulder radius are very similar, the real difference is in the thickness of the plate, which is twice for Nandan AISI 1018 Sim [7] and the Roy Al6061 Num [4]. So by changing the plate thickness it can be seen the impact on the preheat temperature, this make sense considering that the heat generated by the shoulder is dissipated over a larger area than the pin, but if this area is lower then the preheat temperature is higher and the torque made by the shoulder should decrease.

The simplification of a slow-moving heat source represented by the low Peclet number was valid for all cases. Regarding the slow advance per revolution simplification, for most cases the simplification was valid, although positive results were obtained when it was not the case, but even though the simplifications are not always satisfied in all the figures but figure6 the model still predicts the maximum temperature reached. As expected, if the shear layer is too thick, in particular bigger than the pin radius, the model does not satisfied completely the assumptions. Finally, almost all points tend to show that the heat from the shoulder does not affect significantly the peak temperature, this happens because, even if the heat loss is only considered for the dissipated in the pin, and not in the shoulder, the equation of the total efficiency (34) does consider it.

Regarding to the assumptions taken to build the model the most relevant ones, in terms to be considered to improve the model are; 1) The cylindrical pin tool: is important to extend the model where the threaded or Trivex pin are used, because of their impact in the friction and stir in the material. This would increase the maximum temperature reached in the interface and therefore the thickness of the shear layer, and 2) In this work it is assumed that the plastic shear stress experienced by the pin is the same as the experienced by the shoulder, however the shear stress at different points under the shoulder can involve large variations, due to the push angle and plunge force, therefore a more detailed study of the shoulder is required. Even though the physical phenomena that are known to exist but where neglected because their effect to the process are considered small, they do have some influence on scatter behaviour of the graphs, and also the scatter due to measurement or numerical errors will still remain. The model under predicts the estimation of maximum temperature by approximately 7% with an associated error of 9.835, as figures 4 to 7 have shown. For the correction function, the conclusion reached in Dr.Karem Tello's thesis [2] is used, and the result was shown in equation49.

10. Conclusions

Heat transfer and plastic deformation around the pin in FSW has been model using scaling analysis. For this model, new considerations were taken into account as the temperature profile of the work-piece near the pin accounts for heat loss through the tool. The model generated results in closed-form expressions for the thickness of the shear layer (equation 43), the maximum temperature (equation 44), the volumetric heat generation (equation 45) and the shear stress in the metal (equation 46).The expressions obtained depend only on process parameters, which are known a priori. This model involves an empirical parameter ε, that intends to capture a shear rate that can be considered as a threshold of deformation. This parameter has been found to depend on the alloy family.Future iterations of the model will revisit this criterion to find greater generality.

This work aims to formulate a model that can help further development in FSW, particularly reducing time and resources upon testing by providing an accurate parameter selection for an specific desired joint.

11. Acknowledgements

This work was partially supported by the Canadian Government through the scholarship "Emerging Leaders in the Americas Program".

References

[1] P. F. Mendez, K. E. Tello, and T. J. Lienert, *Acta Mater*, vol. 58, no. 18, pp. 6012–6026, 2010.
[2] K. E. Tello, Master of Science Thesis, Colorado School of Mines, Golden, CO, May 2009.
[3] X. K. Zhu and Y. J. Chao, *J. Mater. Process. Technol.*, vol. 146, no. 2, pp. 263–272, 2004. 0924-0136.
[4] G. G. Roy, R. Nandan, and T. DebRoy, *Sci. Technol. Weld. Joining*, vol. 11, pp. 606–608, 2006.
[5] R. Nandan, G. G. Roy, and T. Debroy, *Metall. Mater. Trans. A*, vol. 37A, no. 4, pp. 1247–1259, 2006.
[6] T. Lienert, J. W. L. Stellwag, and L. R. Lehman, Unpublished manuscript, 2008.
[7] R. Nandan, G. G. Roy, T. J. Lienert, and T. Debroy, *Acta Materialia*, vol. 55, pp. 883–895, 2007.
[8] J. A. Dantzig and C. L. Tucker III, *Modeling in Materials Processing*. Cambridge University Press, 2001.
[9] W. B. Krantz, . John Wiley & Sons, INC., 2007.
[10] P. F. Mendez, in *Sohn International Symposium Advanced Processing of Metals and Materials* (F. Komgoli and R. G. Reddy, eds.), vol. 7, pp. 393–404, TMS, 2006.
[11] P. F. Mendez, *J. Appl. Mech.-Trans. ASME*, vol. 77, NOV 2010.
[12] T. J. Lienert, W. E. Stellwag, and H. Shao, Summary Report SR0019, EWI, 2000.
[13] T. J. Lienert, J. W. L. Stell, B. B. Grimmett, and R. W. Warke, *Welding Journal*, pp. 1–9, January 2003.
[14] C. Zener and J. Hollomon, *J. Appl. Phys.*, vol. 15, January 1944.
[15] C. M. Sellars and W. J. McTegart *Acta metall. mater.*, vol. 14, p. 9, 1966.
[16] K. Tello, A. Gerlich, and P. Mendez, *Science and Technology of Welding and Joining*, vol. 15, no. 3, pp. 260–266, 2010.
[17] D. Rosenthal, *Trans. ASME* pp. 849–866, 1946.
[18] Ø. Grong, *Metallurgical Modelling of Welding-First Edition*. London: Institute of Materials, 1994.
[19] J. Tsui, 2016.
[20] T. Lienert, W. Stellwag, and H. Shao, *EWI*, 2000.
[21] M. Assidi, L. Fourment, S. Guerdoux, and T. Nelson, *Int. J. Mach. Tools Manuf*, vol. 50, no. 2, pp. 143–155, 2010.
[22] R. S. M. Harpreet Sidhar, *ScienceDirect*, 2016.
[23] H. Schmidt and J. Hattel, *Modell. Simul. Mater. Sci. Eng.*, vol. 13, no. 1, pp. 77–93, 2005.
[24] K. J. Colligan, *Friction Stir Welding Processing IV* (R. S. Mishra, M. W. Mahoney, T. J. Lienert, and K. V. Jata, eds.), vol. 4, pp. 39–54, TMS, August 2007.
[25] M. Z. H. Khandkar, J. A. Khan, and A. P. Reynolds, *Sci. Technol. Weld. Joining*, vol. 8, no. 3, pp. 165–174, 2003.
[26] A. P. Reynolds, Z. Khandkar, T. Long, W. Tang, and J. Khan, *Mater. Sci. Forum*, vol. 426-432, pp. 2959–2964, 2003.
[27] A. P. Reynolds, W. Tang, Z. Khandkar, J. A. Khan, and K. Lindner, *Sci. Technol. Weld. Joining*, vol. 10, no. 2, pp. 190-199, 2005.
[28] P. A. Colegrove and H. R. Shercliff, *Sci. Technol. Weld. Joining*, vol. 8, no. 5, pp. 360–368, 2003.
[29] B. Yang, J. Yan, M. A. Sutton, and A. P. Reynolds, *Mater. Sci. Eng., A*, vol. 364, pp. 55–65, 2004.
[30] C. M. Chen and R. Kovacevic, *Int. J. Mach. Tools Manuf*, vol. 43, pp. 1319–1326, 2003.
[31] H. Schmidt and J. Hattel, *Sci. Technol. Weld. Joining*, vol. 10, no. 2, pp. 176–186, 2005.
[32] J. Adamowski and M. Szkodo, *Journal of Achievements of Materials and Manufacturing Engineering*, vol. 20, 2007.
[33] H.-S. Lee, J.-H. Yoon, J.-T. Yoo, and K. No, *Journal of Materials Processing and Manufacturing Science*, vol. 149, pp. 62–66, 2016.

[34] J. M. W. Tang, X. Guo and L. Murr, *Journal of Materials Processing and Manufacturing Science* , vol. 7, 1998
[35] M. Boz and A. Kurt, *Materials and Design* , vol. 25, pp. 343–347, 2004.
[36] S. Verma, Meenu, and J. P. Misra, *Materials and Design* , vol. 4, pp. 1350–1356, 2017.
[37] A. Sadoun, A. Wagih, A. Fathy, and A. Essa, *Science and Technology of Welding and Joining* , vol. 15, 2019.
[38] P. Edwards and M. Ramulu, *Science and Technology of Welding and Joining* , vol. 15, pp. 468–472, 2013.
[39] R. dong Fu, J. feng Zhang, Y. jun Li, J. Kang, H. jie Liu, and F. cheng Zhang, *Materials Science and Engineering* , vol. 559, pp. 319–324, 2013.
[40] L. M. D. B. S. Benavides, Y. Li and J. McClure, *Scripta Materialia* , vol. 41, pp. 809–815, 1999.
[41] C. Genevois, D. Fabregue, and W. J. P. Alexis Deschamps, *Materials Science and Engineering* vol. 441, pp. 39–48, 2006.
[42] Mohammadtaheri, M. Haddad-sabzevar, M. Mazinani, M. Motlagh, and E. Bahrami, *Metallurgical and Materials,* vol. 44, pp. 738–743, 2013.
[43] P. Upadhyay and A. Reynolds, *Materials Science and Engineering* , pp. 1537–1543, 2010.
[44] R. dong Fu, Z. qiang Sun, R. cheng Sun, Y. Li, and L. L. Hui-jie Liu, *Materials and Design*, vol. 32, pp. 4825–4831, 2011.
[45] Hamilton, C., Dymek, S., Kalemba, I., and M. Blicharski, *Sci. Technol. Weld. Joining* , vol. 13, no. 8, pp. 714–720, 2008.
[46] S. Lu, D. L. Yang, S. Y. Xiao, and S. J. Chen, *Proceedings of the 1st Int. Joint Symposium on Joining and Welding,* pp. 67–72, 2013.
[47] Y. Chen, J. Feng, and H. Liu, *Materials Characterization* , vol. 60, pp. 476–481, 2009.
[48] W. Xu, J. Liu, G. Luan, and C. Dong, *Materials and Design* , vol. 30, pp. 1886–1893, 2009.
[49] H. Zhang, H. Liu, , and L. Yu, *Journal of Materials Engineering and Performance volume*, vol. 21, p. 1182–1187, 2011.
[50] H. Zhang, H. Liu, , and L. Yu, *Acta Materialia*, vol. 73, pp. 227–239, 2014.
[51] D. Kim, H. Badarinarayan, J. H. Kim, C. Kim, K. Okamoto, R. H. Wagoner, and K. Chung, *Eur J Mech A-Solid,* vol. 29, pp. 204–215, Mar. 2010.
[52] M. Jariyaboon, A. Davenport, R. Ambat, B. Connolly, S. Williams, and D. Price, pp. 877–909, 2007.
[53] K. A. A. Hassan, P. B. Prangnell, A. F. Norman, D. A. Price, and S. W. Williams, *Science and Technology of Welding and Joining*, vol. 8, no. 4, 2013.
[54] Y. J. Chao, X. Qi, and W. Tang, *Journal of Manufacturing Science and Engineering*, vol. 125, pp. 138–145, 2003.
[55] M. J. J., M. H. Tolephih, and M. A. M., *Science and Technology of Welding and Joining*, 2012.
[56] R. Jain, S. K. Pal, and S. B. Singh, *Science and Technology of Welding and Joining*, vol. 23, pp. 278–286, 2016.
[57] F. Wang, W. Li, J. Shen, S. Hub, and J. dos Santos, *Science and Technology of Welding and Joining* , vol. 86, 2015.

AUTHOR INDEX